THE UNIVERSITY OF ARIZONA SPACE SCIENCE SERIES

Richard P. Binzel, General Editor

Enceladus and the Icy Moons of Saturn
P. M. Schenk, R. N. Clark, C. J. A. Howett, A. J. Verbiscer, and J. H. Waite, editors, 2018, 475 pages

Asteroids IV
P. Michel, F. E. DeMeo, and W. F. Bottke, editors, 2015, 895 pages

Protostars and Planets VI
Henrik Beuther, Ralf S. Klessen, Cornelis P. Dullemond, and Thomas Henning, editors, 2014, 914 pages

Comparative Climatology of Terrestrial Planets
Stephen J. Mackwell, Amy A. Simon-Miller, Jerald W. Harder, and Mark A. Bullock, editors, 2013, 610 pages

Exoplanets
S. Seager, editor, 2010, 526 pages

Europa
Robert T. Pappalardo, William B. McKinnon, and Krishan K. Khurana, editors, 2009, 727 pages

The Solar System Beyond Neptune
M. Antonietta Barucci, Hermann Boehnhardt, Dale P. Cruikshank, and Alessandro Morbidelli, editors, 2008, 592 pages

Protostars and Planets V
Bo Reipurth, David Jewitt, and Klaus Keil, editors, 2007, 951 pages

Meteorites and the Early Solar System II
D. S. Lauretta and H. Y. McSween, editors, 2006, 943 pages

Comets II
M. C. Festou, H. U. Keller, and H. A. Weaver, editors, 2004, 745 pages

Asteroids III
William F. Bottke Jr., Alberto Cellino, Paolo Paolicchi, and Richard P. Binzel, editors, 2002, 785 pages

Tom Gehrels, General Editor

Origin of the Earth and Moon
R. M. Canup and K. Righter, editors, 2000, 555 pages

Protostars and Planets IV
Vincent Mannings, Alan P. Boss, and Sara S. Russell, editors, 2000, 1422 pages

Pluto and Charon
S. Alan Stern and David J. Tholen, editors, 1997, 728 pages

Venus II—Geology, Geophysics, Atmosphere, and Solar Wind Environment
S. W. Bougher, D. M. Hunten, and R. J. Phillips, editors, 1997, 1376 pages

Cosmic Winds and the Heliosphere
J. R. Jokipii, C. P. Sonett, and M. S. Giampapa, editors, 1997, 1013 pages

Neptune and Triton
Dale P. Cruikshank, editor, 1995, 1249 pages

Hazards Due to Comets and Asteroids
Tom Gehrels, editor, 1994, 1300 pages

Resources of Near-Earth Space
John S. Lewis, Mildred S. Matthews, and Mary L. Guerrieri, editors, 1993, 977 pages

Protostars and Planets III
Eugene H. Levy and Jonathan I. Lunine, editors, 1993, 1596 pages

Mars
Hugh H. Kieffer, Bruce M. Jakosky, Conway W. Snyder, and Mildred S. Matthews, editors, 1992, 1498 pages

Solar Interior and Atmosphere
A. N. Cox, W. C. Livingston, and M. S. Matthews, editors, 1991, 1416 pages

The Sun in Time
C. P. Sonett, M. S. Giampapa, and M. S. Matthews, editors, 1991, 990 pages

Uranus
Jay T. Bergstralh, Ellis D. Miner, and Mildred S. Matthews, editors, 1991, 1076 pages

Asteroids II
Richard P. Binzel, Tom Gehrels, and Mildred S. Matthews, editors, 1989, 1258 pages

Origin and Evolution of Planetary and Satellite Atmospheres
S. K. Atreya, J. B. Pollack, and Mildred S. Matthews, editors, 1989, 1269 pages

Mercury
Faith Vilas, Clark R. Chapman,
and Mildred S. Matthews, editors, 1988, 794 pages

Meteorites and the Early Solar System
John F. Kerridge and Mildred S. Matthews, editors, 1988, 1269 pages

The Galaxy and the Solar System
Roman Smoluchowski, John N. Bahcall,
and Mildred S. Matthews, editors, 1986, 483 pages

Satellites
Joseph A. Burns and Mildred S. Matthews, editors, 1986, 1021 pages

Protostars and Planets II
David C. Black and Mildred S. Matthews, editors, 1985, 1293 pages

Planetary Rings
Richard Greenberg and André Brahic, editors, 1984, 784 pages

Saturn
Tom Gehrels and Mildred S. Matthews, editors, 1984, 968 pages

Venus
D. M. Hunten, L. Colin, T. M. Donahue,
and V. I. Moroz, editors, 1983, 1143 pages

Satellites of Jupiter
David Morrison, editor, 1982, 972 pages

Comets
Laurel L. Wilkening, editor, 1982, 766 pages

Asteroids
Tom Gehrels, editor, 1979, 1181 pages

Protostars and Planets
Tom Gehrels, editor, 1978, 756 pages

Planetary Satellites
Joseph A. Burns, editor, 1977, 598 pages

Jupiter
Tom Gehrels, editor, 1976, 1254 pages

Planets, Stars and Nebulae, Studied with Photopolarimetry
Tom Gehrels, editor, 1974, 1133 pages

Enceladus and the Icy Moons of Saturn

Enceladus and the Icy Moons of Saturn

Edited by

Paul M. Schenk
Roger N. Clark
Carly J. A. Howett
Anne J. Verbiscer
J. Hunter Waite

With the assistance of

Renée Dotson

With 84 collaborating authors

THE UNIVERSITY OF ARIZONA PRESS
Tucson

in collaboration with

LUNAR AND PLANETARY INSTITUTE
Houston

About the front cover:

Artist's conception of south polar eruptions along a fracture on Enceladus as seen from nearby ice fields. The foreground was painted on site, from nature, in Kilauea caldera, Hawaii Volcanoes National Park, using lava fractures and textures as a proxy for ice formations along fracture zones on Enceladus (with colors changed appropriately). The fractured slabs also resemble fractured ice masses visited by the artist on Arctic Ocean coastal zones in Barrow, Alaska. Subtle colors in the icy "gravel" suggest possible organic compounds erupted from the fracture. Near the south pole, Saturn's rings would be nearly parallel to the horizon. The satellite Mimas can be seen in front of the rings. As this book goes to press, the original Kilauea landscape is being altered by explosive eruptions from adjacent Halemaumau pit crater. Painting copyright 2006 by William K. Hartmann.

About the back cover:

Perspective views of Saturn's icy moons reveal a surprising variety of geologic landforms and processes. Clockwise from upper left: (1) *Mimas* — Cratered terrains, including the 125-km Herschel crater at left, and an irregular trough at right. (2) *Rhea* — Young parallel graben-style fractures in the cratered plains (largest graben is ~20 km across). (3) *Hyperion* — Rolling cratered terrains, showing dark and bright materials (scene width is ~50 km). (4) *Phoebe* — Rolling cratered terrain; largest craters are ~5 km across. (5) *Iapetus* — Bright and dark materials along the great equatorial ridge (largest bright patch is ~15 km across). (6) *Dione* — Smooth leading hemisphere plains, including relaxed craters in foreground, fractures and possible volcanic calderas at upper right (large crater ~35 km across). All these views are derived from global color mosaics at resolutions of 100–400-m/pixel scales, combined with global and local topographic maps derived from Cassini stereo images. Credit: P. Schenk, Lunar and Planetary Institute.

The University of Arizona Press
in collaboration with the Lunar and Planetary Institute

∞ This book is printed on acid-free, archival-quality paper.
Manufactured in the United States of America

23 22 21 20 19 18 6 5 4 3 2 1

Library of Congress Cataloging-in-Publication Data
Names: Schenk, Paul M., editor. | Clark, Roger N. (Roger Nelson), editor. | Howett, Carly J. A., editor | Verbiscer, Anne J., editor | Waite, J. H. (John H.), editor. | Dotson, Renée.
Title: Enceladus and the icy moons of Saturn / edited by Paul M. Schenk, Roger N. Clark, Carly J. A. Howett, Anne J. Verbiscer, J. Hunter Waite ; with the assistance of Renée Dotson.
Other titles: Space science series.
Description: Tucson : The University of Arizona Press in collaboration with Lunar and Planetary Institute, Houston, 2018. | Series: The University of Arizona space science series | Includes bibliographical references and index.
Identifiers: LCCN 2017056126 | ISBN 9780816537075 (cloth : alk. paper)
Subjects: LCSH: Enceladus (Satellite) | Saturn (Planet)—Satellites.
Classification: LCC QB405 .E53 2018 | DDC 523.9/86—dc23 LC record available at https://lccn.loc.gov/2017056126

Contents

PART 3: SATURN'S ICY MOONS

PART 4: ASTROBIOLOGY AND EXPLORATION OF ENCELADUS

List of Contributing Authors

Acknowledgment of Reviewers

The editors gratefully acknowledge the following individuals, as well as several anonymous reviewers, for their time and effort in reviewing chapters in this volume:

Michael Bland
Tim Cassidy
Julie C. Castillo-Rogez
Gaël Choblet
Charles Cockell
Joshua Colwell
Anton I. Ermakov
Larry W. Esposito
Gianrico Filacchione
Brett Gladman
Jay Goguen
Will Grundy
Kevin P. Hand
Candice Hansen
Bruce Hapke
Mihaly Horanyi
Hsiang-Wen Hsu
Andrew P. Ingersoll
Robert E. Johnson
JJ Kavelaars
Jack Lissauer
Michael Manga
Emily S. Martin
Alfred McEwen
Melissa A. McGrath
Jay Melosh
Francis Nimmo
Robert Pappalardo
Alyssa Rhoden
James H. Roberts
Julien Salmon
Brent Sherwood
Everett Shock
Linda J. Spilker
Peter C. Thomas

Foreword

Enceladus has been an enigmatic moon from the start. Early telescopic observations following its 1789 discovery by William Herschel revealed a bright world that would sometimes increase in brightness on its trailing side. Two centuries later, Voyager flybys in the early 1980s revealed Enceladus' bright surface to be smooth plains composed primarily of water ice and surprisingly devoid of craters. Such a crater-free surface implies a young and active surface that must be renewing itself, but how? While the Voyager flyby glimpses did not yield answers for the surface renewal or the long-noticed brightness variations, these first images ushered in hypotheses for intrinsic activity and meteoritic bombardment as possible explanations.

Decoding the mysteries of Enceladus was one of the many high-priority objectives for the Cassini mission to Saturn, whose arrival in 2004 marked the achievement of the first spacecraft ever to orbit this ringed jewel of our solar system. Being in orbit meant that the Cassini spacecraft would have many flyby opportunities for close examination of Saturn's satellites, and the earliest encounters did not disappoint: Enceladus was quickly revealed as an intrinsically active world with geyser-like jets emanating from its previously unseen south pole. With such a richness of activity and abundance of water, likely sourced from a subsurface ocean, Cassini's mission plan evolved to enable a total of 23 close Enceladus flybys over the course of the entire mission, including seven flybys through the jets. Outcomes from the resulting investigations, and more, fill the pages of this volume as a testament to the human spirit of exploration and the ingenuity to achieve it.

Eleven chapters in the first two sections of the book cover a broad range of topics that encompass Cassini's revelations during the first three Enceladus flybys followed by the maturing understanding from the successive encounters. The sum of the knowledge gained forms the current thinking for the origin of Enceladus and allows a detailed characterization of its surface geochemistry and likely subsurface ocean. Measurements of these chemical and bulk properties feed the state-of-the-art models for the interior and the complex tidal-thermal evolution that it experiences, thereby building scenarios for a long-lived ocean. Both endogenic and exogenic processes drive the geologic history on Enceladus, including the orientation of the south polar terrain. Three chapters focus on Enceladus' plume, including its composition, origin, dynamics, and plumbing. Other chapter topics include the E ring's connection to Enceladus and Enceladus' influence on Saturn's magnetosphere.

Particularly exciting, and largely unexpected, has been how Cassini's discoveries about Enceladus have fundamentally altered our concepts of where life might be found in our own solar system and beyond. With organic molecules, and tidal heating producing a global subsurface liquid water ocean and a possible hydrothermal vent system on the seafloor, Enceladus' ocean could harbor the ingredients for life. As such a discovery was totally unexpected, the Cassini orbiter did not carry the instruments needed for life detection. Cassini's most enduring triumph may be its trailblazing discovery that Enceladus may be the most promising place in the solar system to search for extant life beyond Earth. Foundational knowledge set forth in this volume lays the cornerstone upon which future life-seeking exploration missions may be built.

Linda J. Spilker, Cassini Project Scientist
NASA Jet Propulsion Laboratory/California Institute of Technology
Pasadena, California
January 2018

Preface

Ever since Giovanni Cassini discovered Iapetus (in 1671) and correctly surmised that it had a dark and a bright side, the family of icy moons surrounding Saturn has been a source of puzzlement. The fast passages of the two Voyager spacecraft through the Saturn system in 1980 and 1981 were revolutionary and revealed Saturn's icy moons as distinct worlds with discrete and somewhat familiar geologic features and histories, but they did not fully prepare us for what the Cassini orbiter and Huygens probe would reveal decades later. From new rings and moons to Titan's lakes, Cassini astounded us with discoveries throughout its 13-year mission as Saturn's only artificial satellite. Even on Saturn's mid-sized "regular" icy moons, the main subject of this volume, Cassini found thermal bands formed by surface damage from high-energy electrons along the equator; deposition of circum-satellite ring debris on two of these moons; extensive geologic resurfacing and fracturing on Rhea, Dione, and Tethys; and of course the ongoing activity and subsurface ocean on Enceladus, just to name a few. Many of these features were subtly or marginally visible in the Voyager images but unrecognizable in form or origin. Cassini brought all of these and more into sharp focus and in doing so revealed processes that were unsuspected before its arrival in 2004. To paraphrase the olde lament, "Saturn, we hardly knew ye."

In July 2016, scientists gathered in Boulder, Colorado, toward the end of the Cassini mission to discuss the scientific results related to the icy satellites of Saturn at the "Enceladus and the Icy Moons of Saturn" conference. One of the outcomes of that meeting was this volume, which encapsulates the current state of the art in our understanding of the chemical, physical, and geologic processes occurring on and within the mid-sized icy satellites of Saturn. Most of the authors who have contributed to this volume have been active in Saturn system science for decades and represent a cross-section of the Saturn science community.

Understandably, much of Cassini's investigations focused on Enceladus and its ongoing eruptive plume activity. These dramatic eruptions of water vapor and other volatiles provide a direct window into the geophysics and geochemistry of Enceladus' subsurface ocean. Like Io before Voyager, there were hints that Enceladus was different. The chapter by Dougherty et al. reviews the history of Enceladus discovery from early telescopic days through the first years of the Cassini mission. The complexity of the satellite and ring systems allows several competing hypotheses for the formation of Enceladus and the other mid-sized satellites. Chapters by McKinnon et al. and by Castillo-Rogez et al. explore the compositional constraints on these origin scenarios.

The discoveries at Enceladus (and perhaps Dione) dramatically confirmed the existence and importance of ice-covered ocean worlds in our solar system, thereby joining Europa and perhaps more worlds to come as tantalizing destinations for future exploration. The astounding discovery of eruptive plumes at Enceladus in the first year of the mission radically altered the course of the Cassini mission. Questions that Cassini tried to address during its extended mission included whether or not these plumes were related to a subsurface ocean and whether such an ocean might be global and potentially capable of supporting prebiotic or organic chemistry. Geophysical evidence for an ocean is reviewed in the chapter by Hemingway et al. The geophysics of Enceladus, including the heat source(s) for maintaining an ocean, are discussed by Nimmo et al. Cassini's passage through the plumes at altitudes less than 100 kilometers have provided the first direct (albeit processed) sampling of a global ocean on any solar system object other than Earth. Glein et al. address the geochemistry of the ocean and McKay et al. explore the astrobiological potential implied by this chemistry.

The eruptive plumes of Enceladus represent a complex geochemical and geophysical system. Two chapters deal directly with plume phenomenology, from the generation and diurnal modulation of plume activity between the ocean and the surface in the chapter by Spencer et al., to the dynamics of plume materials as they leave the surface and are vented into the near-space environment in the chapter by Goldstein et al. A third chapter by Kempf et al. addresses the ultimate fate of this ejected material, some of which returns to paint the surface of Enceladus itself, and some of which is distributed throughout the inner satellite system via the E ring. Postberg et al. review the composition of Enceladus' plume and surface, while Patterson et al. present Enceladus' complex geology.

Enceladus does not operate in a (complete) vacuum, being in a complex gravitational and magnetospheric environment as part of Saturn's family of icy moons. Cassini devoted considerable time to mapping and investigating these mid-sized icy satellites. All these moons are embedded with the saturnian magnetic and radiation fields and are bathed in particles, both charged and uncharged. Some of these particles, energetic electrons in particular, even visibly damage satellite surfaces, as discussed in the chapter by Howett et al. The effects of these materials on the optical and physical properties of the moons are also discussed in the chapters by Hendrix et al. and Verbiscer et al. Enceladus itself is the source of much of this material, ejecting large volumes of water and other minor species into the E ring, as discussed in the chapter by Smith et al. The other moons also affect their local environment, ejecting low volumes of material via sputtering and altering the magnetic field of Saturn in their vicinity as described in the chapter by Teolis et al.

Saturn's other moons, from Mimas to Iapetus, also proved to be geologically complex, with levels of geologic activity ranging from heavily cratered (Iapetus) to geologically complex (Dione). Geologic processes range from recently formed faulting to volcanic resurfacing and redeposition of circum-satellite ring debris. The geologic features and histories of these moons are described in the chapter by Schenk et al. Saturn's smaller irregularly shaped moons are not forgotten. The surprisingly complex, irregularly shaped inner moons, including the ring shepherd moons, are described in the chapter by Thomas et al., while the distant, outer irregular moons are described by Denk et al.

Sometimes overlooked in the rush of new discoveries is how Cassini fundamentally rewrote the textbook not only on satellite geology and interactions with the space environment, but also on their basic physical parameters, summarized in tables in the Castillo-Rogez et al., Thomas et al., and Denk et al. chapters. Among the more surprising updates is that Tethys has a mean bulk density of 0.984 grams per cubic centimeter, essentially that of uncompressed water ice. This density sets Tethys apart and indicates an interior composed of almost all water ice or ice with negligible amounts of rock in a porous state, in contrast to neighboring Dione, which is composed of as much as 50% non-ice materials. Meanwhile, active Enceladus has the highest density and highest "rock" content among the inner moons. How these proximal moons could have such different internal compositions will have to be answered if their origins are to be understood.

The great leaps forward made by Cassini were enabled by time, proximity, and hard work. Cassini spent more than 13 years in Saturn orbit. This longevity gave Cassini the time and maneuverability to get much closer to its icy targets than any previous spacecraft in order to complete mapping and implement investigations of new discoveries. What's more, the resolving power and scope of its 12 instruments, which spanned much more of the electromagnetic spectrum, enabled Cassini to map these moons with much higher fidelity than the Voyager spacecraft could. With several *in situ* instruments, direct sampling of ejected materials from Enceladus as well as the other moons was also possible (as described in several chapters, including those by Goldstein et al. and Spencer et al.). A lesson from Cassini is that even though Flagship-class missions require greater investment, in the long run they are potentially more cost effective in terms of fully exploring a complex planetary system such as Saturn's in comprehensive detail. Lessons learned at Enceladus and the other icy moons are already being applied to Europa as plans are in motion to return to that apparently active world.

It is worth noting that the five mid-sized uranian satellites bear more than a passing resemblance to Saturn's mid-sized icy satellites, as described in chapters throughout this volume. Tectonically and volcanically disrupted Miranda and Ariel are as compelling geologic targets as Dione and Enceladus, while Umbriel, Titania, and Oberon were only partially viewed by Voyager 2. Neptune's icy Triton, a near twin of complex Pluto, is very likely a geologically active ocean world and thus a priority exploration target. In light of the results of the Cassini mission, it seems likely that a return to the Uranus and Neptune planetary systems on a similar scale will yield almost as great a treasure of discovery, not only in terms of the verdant planets that Voyager 2 saw briefly, but also their ring systems and families of diverse icy satellites.

The nature of exploration is that our current state of knowledge will have a limited shelf life, as new models and interpretations come to the fore, and new analytical techniques are developed. With this in mind, the authors endeavored to set forth in each chapter not just our current state of knowledge of the features and processes occurring in the Saturn system, but also the outstanding unresolved issues, many of which should be valuable avenues of research in the future. While some will yield to improvements in modeling techniques or new ideas, others will require new flight missions to the Saturn system to acquire additional data. Cassini has set the stage for more focused future exploration of Enceladus, Titan, and even the lesser moons, as discussed in the chapter by Lunine et al.

The editors would like to thank General Editor Richard Binzel for keeping this book on track through hurricanes and the editing and review process, Lunar and Planetary Institute (LPI) Directors Stephen Mackwell and Louise Prockter for facilitating the production of this book in the Space Science Series, and especially Renée Dotson for her tireless efforts in the compilation and production of the book during a difficult time. Other invaluable LPI editorial assistance was provided by Elizabeth Cunningham, Linda Chappell, Sandra Cherry, and Linda Garcia. We thank chapter reviewers for their invaluable and timely assistance in making sure the content was fresh and useful and complete. Special thanks to William Hartmann for his original painting that graces the book's cover. We also wish to thank the hundreds of engineers, scientists, and administrators at the Jet Propulsion Laboratory, NASA, the European Space Agency, and the instrument teams around the world for making the Cassini mission such a great success, the results of which are (partially) compiled in this volume.

Paul Schenk, Roger Clark, Carly Howett,
Anne Verbiscer, and Hunter Waite
July 2018

Part 1:

Enceladus Geophysics, Geology, and Geochemistry

Dougherty M. K., Buratti B. J., Seidelmann P. K., and Spencer J. R. (2018) Enceladus as an active world: History and discovery. In *Enceladus and the Icy Moons of Saturn* (P. M. Schenk et al., eds.), pp. 3–16. Univ. of Arizona, Tucson, DOI: 10.2458/azu_uapress_9780816537075-ch001.

Enceladus as an Active World: History and Discovery

M. K. Dougherty
Imperial College London

B. J. Buratti
Jet Propulsion Laboratory

P. K. Seidelmann
University of Virginia

J. R. Spencer
Southwest Research Institute

Enceladus was discovered by William Herschel in 1789, although it wasn't named until more than half a century later. During the twentieth century, there were some unusual telescope observations of Enceladus showing an increase in brightness on its trailing side, but our understanding of this moon improved little until the Voyager spacecraft flybys in the early 1980s. An extensive diffuse ring, now known as the E ring, was discovered in 1966 and found in 1980 to peak in density near Enceladus. The Cassini observations made on two distant flybys in 2005 by the magnetometer instrument are described; these observations led to a lowering of the altitude of the third flyby and the confirmation and discovery by multiple Cassini instruments of the water vapor plume emanating from cracks at the south polar surface.

1. INTRODUCTION

This chapter discusses Saturn's moon Enceladus and observations taken since the late eighteenth century to date. These observations culminated in the discovery in 2005 of a water vapor-, dust-, and organic-filled plume of material emanating from cracks on its surface at the south pole. Enceladus, the sixth largest moon of Saturn, was discovered by William Herschel in 1789 (*Herschel,* 1790). Our understanding of the moon improved little after this initial discovery until the Voyager spacecraft flybys in the early 1980s. During the Voyager 2 flyby in August 1981, the first detailed observations of the surface were made, revealing relatively few craters and numerous smooth areas, as well as some extensive linear cracks. The smooth, reflective surface of Enceladus, which is dominated by water ice, pointed to some type of resurfacing taking place (*Verbiscer et al.,* 2007). At about the same time, the extensive diffuse E ring, discovered in 1966 (*Feibelman,* 1967), which extends between the orbits of Mimas and Titan, was found in groundbased observations during the 1980 ring plane crossing to peak in density near the orbit of Enceladus (*Baum et al.,* 1981). These discoveries, coupled with the expected short lifetime of E-ring particles (*Jurac et al.,* 2002), led to the suspicion that Enceladus might be the source of this E ring, perhaps even as a result of geyser-like activity (*Haff et al.,* 1983; *Pang et al.,* 1984a). When Cassini arrived at the Saturn system in mid-2004, the team wondered what unusual findings they might observe at Enceladus.

In early 2005, the Cassini spacecraft made two distant targeted flybys of Enceladus (under a month apart), at 1265 km and 500 km from the surface. Observations from the magnetometer instrument (MAG) showed draping of the Saturn magnetic field lines upstream of Enceladus, as well as an increase in ion cyclotron wave activity in the vicinity of the moon, driven by water group ions (*Dougherty et al.,* 2006). Based on these observations, the MAG team made the case to the Cassini Project that there was potentially an atmosphere at Enceladus made up of water group ions that was holding off the Saturn field lines from the Enceladus surface, and requested the third flyby, due to take place four months later, be lowered from 1000 km to fly much closer to the Enceladus surface to investigate this possibility. The Cassini Project approved this change and the third Cassini

Enceladus flyby took place in July 2005 at an altitude of 173 km. On this flyby, multiple Cassini instruments obtained definitive evidence for active ejection of water vapor and ice particles from the south pole of Enceladus.

2. EARLY ENCELADUS OBSERVATIONS

2.1. Initial Discovery of Enceladus

William Herschel, a German-born British astronomer, was the founder of using sidereal astronomy to systematically observe the night sky. He began observing Saturn as early as 1774 using reflecting telescopes that he built himself. He initially focused on the planet and the changing view of its rings over time (and with season), as well as the thinness of the rings. When viewing Saturn in the plane of the rings, he repeatedly saw numerous satellites, five of which had already been discovered by other astronomers. An observing complication regarding Saturn is that small moons are difficult to observe due to the glare from both the planet and the rings and therefore, particularly with small telescopes, are only possible to see during Saturn's equinox. This is as a result of the reduced glare from the rings due to Earth's perspective of the planet. It was on August 28, 1789, that Herschel, using a new telescope (the largest reflecting telescope of the time, with a 1.26-m primary mirror and a 12-m focal length), discovered a new, sixth satellite of Saturn. During the following month, on September 17, 1789, he discovered the seventh (*Herschel,* 1790).

The sixth satellite, Enceladus, was in fact only named some 60 years later by Herschel's son, John Herschel (*Lassell,* 1848), and was named after a Greek mythological giant. Each of the first seven discovered satellites were named by John Herschel [Tethys, Dione, Rhea, Titan, Iapetus, Enceladus, and Mimas (the seventh satellite and second to be discovered by his father)], and he chose these names since Saturn (known in Greek mythology as Cronos) was the leader of the Titans.

2.2. Earliest Hints of Activity on Enceladus

The earliest hints that Enceladus might be unusual were produced by telescopic observations of the moon, particularly during times when the maximum excursion in southern latitude was attained. This maximum is about 32°, sufficiently south so that terrestrial observers obtain a glimpse of what we now know is the most active region of Enceladus. Percival Lowell and E. C. Slipher, observing on the 24-inch (0.6-m) telescope at Lowell Observatory in 1912–1913, reported a 0.3-magnitude increase in the brightness of Enceladus at western elongation (the trailing side) when the subobserver latitude was –32° (*Slipher and Slipher,*1914). Also, observing from Lowell Observatory in 1972–1973, *Franz and Millis* (1975) measured a 0.3-magnitude increase in the brightness of the lightcurve of Enceladus at western elongation (the trailing side) when the subobserver latitude was also –32°. It is unlikely that groundbased observers ever saw the plume of Enceladus, as it is so forward-scattering (unless, of course, Enceladus exhibited sporadic large outbursts).

3. SATURN'S DIFFUSE E RING

3.1. Early Observations of Saturn's E Ring

Well prior to any spacecraft flybys of the Saturn system, there were several reports of observations of an outer faint Saturn ring in the early twentieth century. The first known observation occurred in September 1907, when G. Fournier, while observing from the Jarry-Desloges Observatory on Mont Revard in Savoy, reported a new dark ring composed of a luminous zone containing swiftly moving luminous particles outside Saturn's main rings. In this report, he did not refer to a faint or dusty ring, and it was not observed on subsequent nights (*Jarry-Desloges Observatory,* 1907). The following year, in October 1908, E. Schaer from the Geneva Observatory announced the existence of a dusky brown ring surrounding the bright rings of Saturn (*Schaer,* 1908), although the work did not refer to the observations by Fournier in the previous year. In the same month (October 10–15), at the Royal Greenwich Observatory, Bowyer and Lewis observed Saturn's rings to be dusky on the outer edge; they also reported a trace of a faint, fuzzy ring at both the north and south ring edges over several nights. At the same time, Eddington (October 15, 1908) also reported the appearance of a narrow dusky ring surrounding the bright ring on the north edge (*Greenwich Observatory,* 1908). A few months later, in January 1909, Schaer reported that the new dusky ring was easier to observe (*Schaer,* 1909). The use of the terms "dusty" or "dusky" was not based on particle sizes, as there was no knowledge concerning the source of the images. In 1908 and 1909, E. Barnard, who was known for observing faint objects, used the 40-inch (1-m) telescope at Yerkes Observatory to observe Saturn. He searched carefully for the dusky ring, but he could see "nothing abnormal anywhere" (*Barnard,* 1909). All these early observations and the problems involved were discussed by R. M. Baum in 1954 (*Baum,* 1954; *Alexander,* 1962).

In 1909 the dusky ring observations (as well as lack of them) were described (*Antoniadi,* 1909a,b), with the conclusion being that it was very probable that extraplanar particles existed and that their discovery should be credited to Fournier. This work also recalled observations of extraplanar particles by Wray as early as 1861 as well as a telegram by Lowell in November 1907 that mentioned such particles (*Alexander,* 1962). At this stage, the resulting consensus was that an outer ring did not exist, although as described below, observers continued to report viewing such a ring. A decade later, on February 28, 1919, Rev. Ellison used a 10-inch (25-cm) refractor at Armagh Observatory to observe an eclipse of Iapetus by the rings of Saturn, and, due to bad weather at the time, was the only observer able to view the eclipse. What he found was that the satellite gradually increased in brightness both before and after the predicted time of emergence from the A ring, indicating that the ring system extended further than generally thought, and supporting the existence of the suspected outer dusky ring (*Alexander,* 1962).

On April 1, 1952, R. M. Baum observed dusky nebulous matter in an additional ring beyond the A ring, with further observations of the dusky fringe also seen on April 11 and 16 of the same year. The following year, this exterior ring was again observed, but appeared fainter than in the previous year (*Baum,* 1954). Further evidence of this outer dark ring was obtained from Mt. Wilson observations (*Cave and Cragg,* 1952).

The first definitive evidence for what we now know as the E ring was obtained between October 27, 1966, and January 16, 1967, when fifty 5–10- and 30-minute photographs of Saturn were taken with the 30-inch (76-cm) refractor at Allegheny Observatory, when the ring plane was edge-on to Earth. A very thin line was observed extending more than twice the known ring diameter distance with an approximate equivalent brightness of magnitude ~15/sq arcsec or fainter. An interpretation of these observations was that the thin line was a tenuous outer ring, only visible when the ring plane was nearly edge-on to our line of sight (*Feibelman,* 1967). Kuiper inspected photographs from 61-inch (1.5-m) telescopes from the same period and confirmed several exposures showed a faint extension reaching at least 42 inches (1 m) from the center, about to the orbit of Tethys (*Kuiper,* 1974). Feibelman named this ring the D ring, since at that time only the A, B, and C rings were known. In 1969, *Guerin* (1970) discovered a previously unseen ring and it was confirmed and named the D ring. In planning for the Pioneer 11 encounter with Saturn, the outer dusky ring was referred to as the E ring (*Smith,* 1978).

3.2. Observations from the 1980 Ring Plane Crossing

For the 1980 passage of Earth through the plane of Saturn's rings, charge-coupled devices (CCDs) were just becoming available, and the Hubble Space Telescope (HST) Wide Field/Planetary Camera (WF/PC) Team had a traveling camera system available for team members to both learn how to use CCDs as well as make observations. W.A. Baum, Currie, Pascu, and Seidelmann used the CCD camera on the 61-inch (1.5-m) telescope at the U.S. Naval Observatory Flagstaff Station to observe Saturn. They used an instrument that they constructed to mask off unwanted light from the planet, rings, and satellites at the focal plane as well as suppress diffracted light by a coronagraphic system. These observations revealed the outer ring to have a relatively sharp maximum near the Enceladus orbit and a faint outer tail reaching out more than 8 Saturn radii (R_S) that was more spatially diffuse than the maximum region. The distribution suggested the ring to be associated with Enceladus and potentially with material ejected from Enceladus. The spread of material above and below the ring plane was greater in its tenuous outskirts than in its denser inner region, suggesting that the E ring may be at an early stage in its evolution (*Baum et al.,* 1981) (see Fig. 1). This result was sufficient to prompt an illustration of Io-like water plumes on Enceladus in a popular article (*Gore,* 1981) (see Fig. 2). The unusual blue color of the E ring (*Terrile and Tokunaga,* 1980; *Larson et al.,* 1982), in striking contrast to the reddish color of Saturn's other rings, indicated that the E-ring particles were dominantly micrometer-sized. *Baum et al.* (1984) concluded that the evidence for a physical association with Enceladus was overwhelming and that the E ring is either young or frequently replenished. Due to the incomplete collisional flattening process following a conjectured single ejection event on Enceladus some 100 years ago, erosional processes would not have permitted the particles to survive to the present day (*Durisen et al.,* 1982; *Haff,* 1982; *Morfill et al.,* 1982). If the E ring is being frequently, or continuously, replenished, then it appeared that interactions with charged particles and fields play a role in shaping the spatial distribution of the E ring.

Reitsema et al. (1980) also used the HST WF/PC CCD camera to observe Saturn, and observed the E ring and found a peak in the distribution, but did not place it at the Enceladus orbit. *Lamy and Mauron* (1980) found a maximum in the E ring near the orbit of Enceladus, but described it as

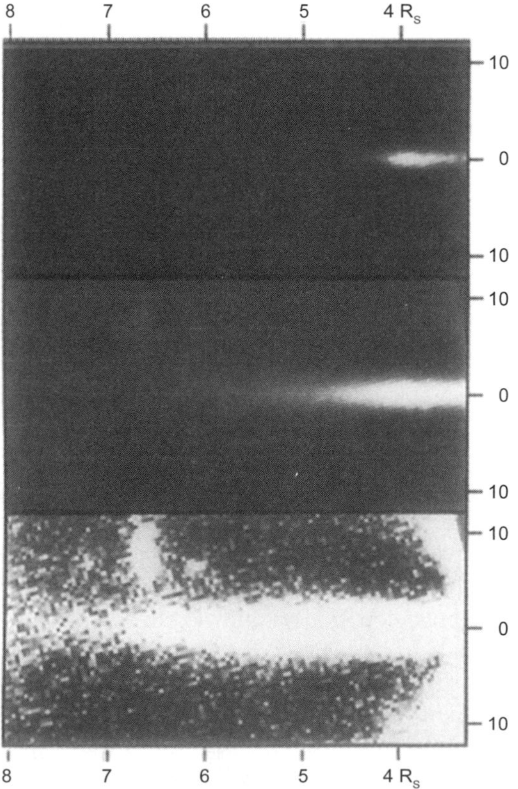

Fig. 1. Taken from *Baum et al.* (1981), with three versions of part of a sample CCD frame (240-set exposure, 1006 UT on March 14, 1980) showing the west side of the E ring. The pronounced peak brightness at about 3.8 R_S is easily seen in the top version, while the outskirts become apparent at large R in the contrast-stretched bottom version.

Fig. 2. See Plate 1 for color version. Io-like water plumes on Enceladus drawn in a popular article published in 1981 (*Gore,* 1981). Credit: Lloyd K. Townsend Jr./National Geographic Creative.

a broad feature and not a sharp peak. *Brahic et al.* (1980) found two broad maxima in the E ring, but neither was at the Enceladus orbit, and *Dollfus and Brunier* (1980a,b) reported no peak. *Terrile and Tokanaga* (1980) made infrared observations of the E ring and found a brightness profile that fell off rapidly beyond Enceladus, similar to that of *Baum et al.* (1981). *Larson et al.* (1982) observed the E ring photographically in 1981 and found a brightest region 8000 ± 4000 km outside the orbit of Enceladus. In order to better understand the early E-ring observations, the 1966 discovery plates of Feibelman were digitized, computer-enhanced, and examined further (*Feibelman and Klinglesmith,* 1980). Photographs of Saturn taken in the red spectral region were reexamined (*Feibelman,* 1984) and the E ring observed.

3.3. Observations from Spacecraft Flybys

The Pioneer 11 flyby of Saturn in September 1979 did not yield direct evidence for the E ring, although there were at least four impacts recorded by the micrometeoroid sensors near the time of ring passage, inferring that the E ring may be 1800 km thick near 2.9 R_S with an optical thickness greater than 10^{-8} (*Humes et al.,* 1980), and the charged-particle detectors observed large-scale signatures (*McDonald et al.,* 1980; *Hood,* 1983; *Burns et al.,* 1984). Prior to the press conference following the Pioneer 11 flyby of Saturn in 1979, at which it was announced that Saturn's F ring had been discovered, a decision was made to skip the letter E and propose F as the name for the newly discovered ring due to the certainty that the E ring would later be detected. This decision was validated some months later when Voyager 1 provided the first E-ring images (L. Esposito, personal communication). Voyager observations of Enceladus itself will be described in detail in section 4, but in regard to the E ring, a Voyager 1 image at a very large phase angle showed a region of maximum intensity as a broad diffuse band between 210,000 and 300,000 km from the center of Saturn (*Smith et al.,* 1981), with radial and vertical structures of the E ring being consistent with *Baum et al.* (1984). Data were also acquired by the Voyager planetary radio-astronomy (PRA) experiment when it passed through the E ring, with dust grain impacts and plasma waves being detected. The power spectral signal produced by the dust was dominant in a region of ~12,000 km vertical extent with a maximum ~5000 km southward of the equator at 6.1 R_S from Saturn. The PRA signal had a full vertical width at half-maximum of ~8000 km, which is 2.3 times less than the optical model (*Meyer-Vernet et al.,* 1996). In addition to the confirmation of the existence of the E ring, librational satellites in the orbits of Tethys and Dione were discovered from CCD observations in 1980 (*Seidelmann et al.,* 1984). A review of planetary probe observations of the E ring can be found in *Pang et al.* (1984a).

Cheng et al. (1982); *Morfill et al.* (1983) and *Haff et al.* (1983), determined from ice sputtering rates, inferred from the Voyager and Pioneer plasma measurements, that the lifetime against sputtering of micron-sized E-ring particles was only a few hundred years or less, requiring active replenishment, perhaps by geyser activity. *Pang et al.* (1984a) used the E-ring color to infer an unusually narrow particle size distribution peaked near about 2 μm diameter, very different from the power-law size distributions typical of collision-generated rings. Pang et al. also used the E-ring phase function to infer dominantly spherical particle shapes, and suggested that the sizes and shapes of the particles indicated a possible origin from freezing of liquid, saying "we conclude that the E-ring particles are formed from a supply of liquid and/or gas within Enceladus." The conclusion of a peculiarly narrow size distribution, perhaps generated by geyser activity, was reinforced by a more comprehensive analysis of then-available photometry by *Showalter et al.* (1991) in preparation for the Cassini orbital mission, although they did not confirm *Pang et al.*'s (1984a) conclusion that the particles were dominantly spherical. *Showalter et al.* (1991) concluded that the E-ring density peak was at the Enceladus orbit, with a radial offset of <3000 km. The vertical thickness of the ring increases with distance from Saturn, from 6000 km at its inner bound to 40,000 km at its outer bound. At its density peak there was a decrease in thickness by 30%. There appeared to be a narrow distribution of slightly nonspherical particles of radius 1.0 ± 0.3 μm, which indicated that the ring does not originate from collisional or disruptive processes. There was also evidence for time-variable phenomena in the E-ring: *Roddier et al.* (1998) observed an apparent short-lived arc of material near Enceladus' orbit, with a sharp boundary 76° ahead of Enceladus, unlike anything subsequently seen by Cassini.

An important breakthrough came when *Shemansky et al.* (1993), using HST, discovered neutral OH emission in the vicinity of the E ring, presumably derived from H_2O and therefore implying an unusual source of molecular gas as well as dust. They inferred that neutral OH was one of the dominant species in the magnetosphere, "implying a source rate for H_2O twenty times current theoretical estimates." The variation in OH density with distance from Saturn (see Fig. 3) required a sharp peak in the H_2O source rate near Enceladus' orbit (*Richardson et al.,* 1998; *Jurac et al.,* 2001), proposing that impacts alone might be able to explain the E-ring particle size distribution, concentration at Enceladus, and associated OH.

3.4. Hypotheses on the Origin and Lifetime of Saturn's E Ring

Once the existence of the E ring was confirmed from groundbased CCD observations, thoughts turned to the origin and lifetime of such a ring, with numerous theories and ideas being postulated. *Yoder* (1979) noted a possible tidal heating of the interior of Enceladus due to a forced eccentricity from a 2:1 resonance with Dione. *Hill and Mendis* (1982) suggested that interplanetary micrometeoroids, captured by Saturn, and the secondary particles that they blast off the surface of Enceladus, make up the E ring. *Cheng et al.* (1982) and *Morfill et al.* (1983) found that the E-ring particles should have a life expectancy less than 100 years due to interactions with ambient plasma. *McKinnon* (1983) proposed that a giant meteoroid had recently impacted Enceladus, which created fluids that had solidified into the E-ring particles, although in order to create the mass of the E ring, the meteoroid mass would have had to be so large as to make such an event very rare. Voyager 2 images of Enceladus had revealed a surface younger than those of the other Saturn satellites and regions devoid of craters (*Plescia and Boyce,* 1983). Cook and Terrile (unpublished manuscript, 1984) suggested that meteoroid impacts could puncture the Enceladus crust to expose liquid water, permitting outgassing which in turn would create the E ring.

The purity of the ice spheres in the E ring seemed to be consistent with young or recycled water, and this was supported by the discovery of tectonic activity on Enceladus (*Kargel,* 1984). Methane or ammonia clathrate hydrate have lower melting points than water ice, so methane or ammonia gas jets could launch ice particles from Enceladus (*Wilson and Head,* 1984). Previous orbital resonances could have provided tidal heating rates that melted the Enceladus interior (*Lissauer et al.,* 1984), although the orbital evolution of the ring particles ejected from Enceladus and their interactions with Saturn plasma remained an outstanding problem (*Seidelmann et al.,* 1984; *Haff et al.,* 1983; *Morfill et al.,* 1983; *Pang et al.,* 1984b). *Hamilton and Burns* (1994) also suggested a nongeyser origin and *Jurac et al.* (2001) had difficulty explaining the magnitude of the required source, suggesting "orbital collisions between icy fragments, possibly the remains of a disrupted satellite near Enceladus," as a means of replenishing the E ring.

3.5. Observations from the 1995 Ring Crossing

In 1995, Earth again passed through the Saturn ring plane, providing an opportunity to observe the E ring once more. *Roddier et al.* (1998) observed in the H and J band on August 12 with the Canada-France-Hawaii Telescope. They detected particles orbiting Saturn close to the Enceladus distance and scattered in an arc with a maximum near the Enceladus L4 point. This appeared to be a transient phenomenon, as it was not observed two days earlier and

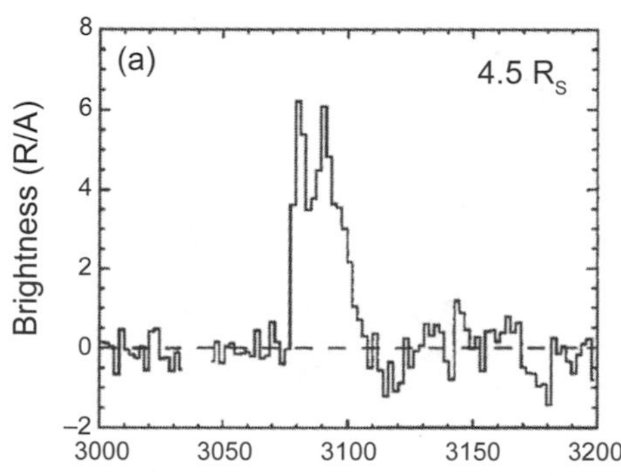

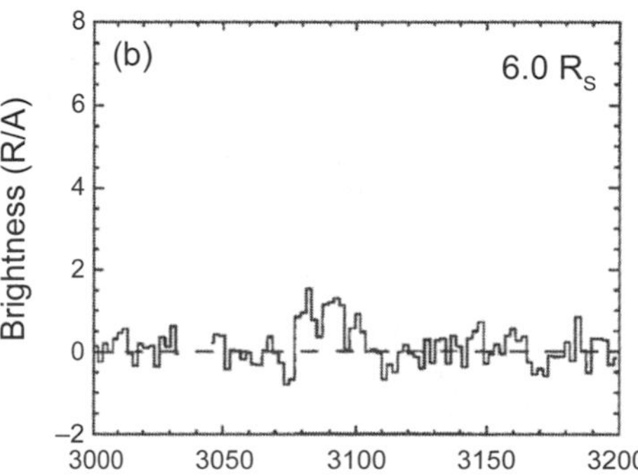

Fig. 3. Ultraviolet OH emission **(a)** near Enceladus' orbit (3.95 R_S) and **(b)** decreasing rapidly at greater distances, as seen by the Hubble Space Telescope. The x-axes show wavelength (in angstroms). From *Richardson et al.* (1998).

decreased during the observations. One interpretation was that the material originated from Enceladus and became part of the E ring. *de Pater et al.* (1996) reported observations made with the Keck telescope at 2.26-μm wavelength, which confirmed the blue color of the E ring and its extension further into the infrared than previously thought. The K band was only 40% of that reported by *Baum et al.* (1981, 1984) at visual wavelengths, although the radial and vertical profiles were consistent. Observations with the HST WFPC2 in August and November 1995 concluded that the broad, blue E ring flares to a maximum thickness of about 15,000 km at 7.5 R_S and has a spatially uniform particle size distribution (*Nicholson et al.,*1996).

4. VOYAGER-ERA OBSERVATIONS OF ENCELADUS

4.1. Voyager Remote Sensing Observations

The first direct indication of possible activity on Enceladus resulted from spacecraft observations provided by Voyagers 1 and 2. The highest-resolution image captured during a distant flyby by Voyager 1 shows an intriguing high-albedo feature [see Fig. 4, an enhanced image of observations first described in *Buratti* (1988)]. This feature is about 15% higher in albedo than Enceladus' already bright surface, and Voyager scientists speculated that it might be a volcanic feature. No one ever published that specific opinion, but instead more conservative statements such as "the high albedo suggests [the features] are the result of a geologic process that produces fresh ice" (*Buratti,* 1988). Although north of the now well-known tiger stripes, this feature could more conservatively be an illuminated trough at the edge of the tiger stripe active network. It coincides with the intersection of the Khorasan and Bishangath Fossae near a longitude of 240° and latitude of 50°S (see Fig. 4).

The measurement of an accurate size by the two Voyager spacecraft enabled the determination of the albedo of Enceladus: a very surprising number near unity (*Buratti and Veverka,* 1984). Almost immediately, there was speculation that an albedo this high could only mean ongoing activity on its surface, possibly due to volcanos or geysers. *Herkenhoff and Stevenson* (1984) made a connection between possible volcanism on Enceladus and the E-ring of Saturn.

Later images by Voyager 2 revealed a wide range of geologic units on Enceladus, ranging from heavily cratered terrains that date from the tail end of the late heavy bombardment (~3.9 b.y. ago) to crater-free plains that could be no older than 800 m.y. (*Smith et al.,* 1982) (see Fig. 5). The unusual fact about these terrains is that they were fairly uniform in albedo, with normal reflectances near unity, as if some process had coated the surface of Enceladus (*Buratti,* 1988). This apparent coating by the E ring of the other main inner moons of Saturn as well added to the growing evidence for likely ongoing activity. Voyager and groundbased rotational lightcurves of the moons exterior to Enceladus were all brighter on the leading hemisphere, as if particles from the E ring were being swept up by the moons (*Buratti*

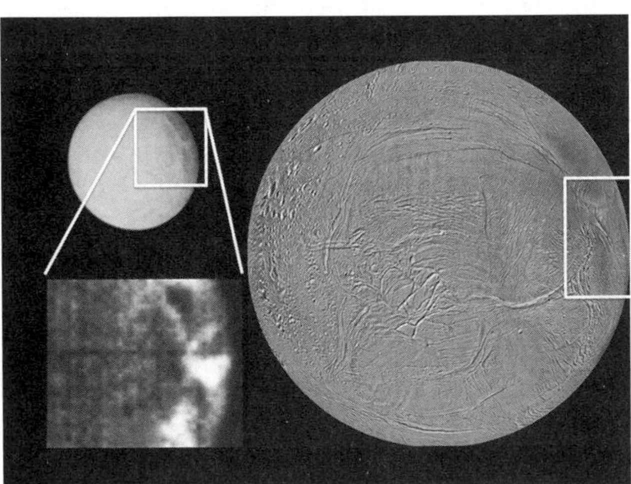

Fig. 4. See Plate 2 for color version. *Left:* An enhanced image of Enceladus from frame FDS 3493157 obtained by Voyager 1 of the southern hemisphere of Enceladus (whose limb is visible on the righthand side of the image) showing high-albedo markings on its trailing side at southern latitudes around 50°. *Right:* The Cassini map of Enceladus shows that this region is near the edge of the tiger stripe region. In each image the south pole is just off to the right of each image. Credit: NASA/JPL/Caltech; Cassini map is PIA18435.

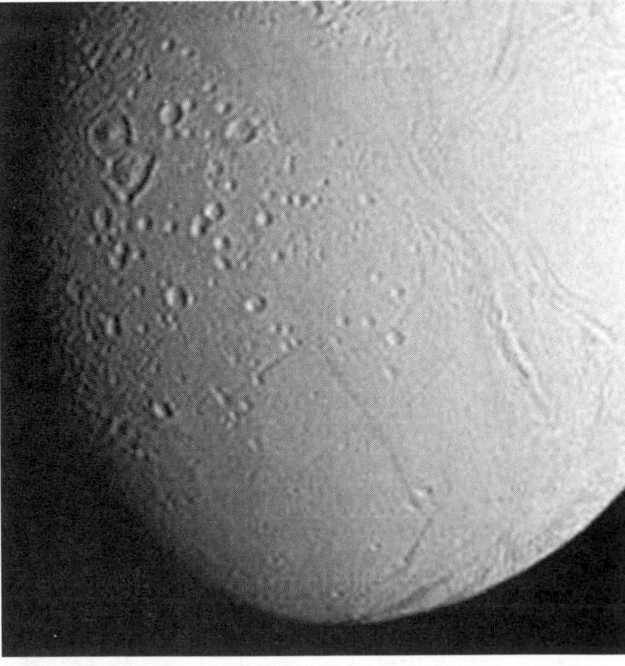

Fig. 5. The highest-resolution Voyager 2 image of Enceladus, showing both heavily cratered old terrains and smooth, crater-free plains. All terrains have high albedos, near unity. Credit: NASA/JPL/Caltech; PIA01950.

and Veverka, 1984; *Noland et al.,* 1974; *Franz and Millis,* 1975; *Koutchmy and Lamy,* 1975). Moreover, the amplitude of the lightcurves increased as the distance from Enceladus increased, suggesting that the outer ones had greater regions of native, uncoated terrain.

Meanwhile, work continued on the understanding of Enceladus itself, in particular its youthful appearance revealed by the remarkable Voyager images. Relaxed craters indicated a high heat flow (*Passey,* 1983), and the large variations in crater density revealed a long geological history extending to recent times (*Kargel and Pozio,* 1996). *Squyres et al.* (1983) examined possible tidal heating of Enceladus due to its orbital eccentricity maintained by the 2:1 mean-motion resonance with Dione, and concluded that while this was the likely cause of the activity, maintaining subsurface liquid water was not easy, and might require NH_3 to lower the melting temperature and clathrates to reduce the ice shell conductivity. *Ross and Schubert* (1989) determined that the Dione resonance could produce very high tidal dissipation, but that steady-state heating rates were limited by the likely rate of dissipation within Saturn itself.

4.2. Further Remote Sensing Observations of Enceladus Prior to Cassini

Further analysis of Voyager images suggested the particles from the E ring affected the optical properties of the five main moons of Saturn. Several references showed that the leading sides of the moons exterior to Enceladus were brighter and less red (in the visible) (*Buratti et al.,* 1990), as revealed in Fig. 6, and have since been mapped out in great detail with Cassini data by *Schenk et al.* (2011). Both observables suggested that the moons were being coated predominantly on the leading side of E-ring particles. Voyager results for Mimas were less clear, and Enceladus had a very low amplitude to its lightcurve, consistent with its uniform albedo (*Buratti and Veverka,* 1984; *Verbischer and Veverka,* 1994; *Verbiscer et al.,* 2005). Using HST observations, *Verbiscer et al.* (2007) showed that the high geometric albedos of the other inner moons, which decreased as a function of distance from Enceladus, could be explained by the presence of E-ring particles impacting their surfaces. Observations of the lightcurves of the moons during the Saturn ring plane crossing in 1995 showed that Enceladus and Mimas were brighter on their trailing sides, unlike the moons exterior to Enceladus (*Buratti et al.,* 1998), which is consistent with dynamical modeling of the E ring (*Hamilton and Burns,* 1994).

Decades prior to the discovery of cryovolcanism on Enceladus, observers noted evidence for it and were making predictions of its existence. However, there were alternate theories to the placement of bright icy particles on the surface of Enceladus and the surrounding moons. For example, the *Hamilton and Burns* (1994) dynamical model had self-sustaining meteoritic bombardment maintaining the E ring. In addition to its high albedo, other photometric properties of Enceladus were unusual. In an application of photometric modeling to bright icy moons, *Buratti* (1985) found a very high single-scattering albedo and a more isotropically scattering surface than other icy moons. A full photometric analysis of Enceladus found a single-scattering albedo approaching unity, and an extraordinarily low mean slope angle of rough features of only 6° ± 1° (*Verbiscer and Veverka,* 1994). The latter result suggested infilling of cavities by volcanic deposits, or some type of recent resurfacing event.

HST observations from 0.34 to 1.04 μm of the opposition surge of Enceladus and fits to Hapke's photometric model (*Hapke,* 1981, 1984, 1986, 2002) found that the surface of Enceladus is moderately fluffy (*Verbiscer et al.,* 2005). Since Hapke's photometric model did not fully account for partial illumination of shadows, the surface may be even more loosely compacted than the models suggest. Enceladus also possesses a sharply peaked coherent backscatter solar phase curve, which is expected for a high-albedo body. Verbiscer et al. also showed that the trailing side of Enceladus is about 6% brighter than its leading side, in good agreement with the observations (*Franz and Millis,* 1975; *Verbiscer and Veverka,* 1994; *Buratti et al.,* 1998; *Verbiscer et al.,* 2007). The geometric albedo at visible wavelengths (0.549 μm) is an astounding 1.41 ± 0.03. Spectral observations between 0.8 and 2.5 μm of the leading and trailing sides of Enceladus suggested that magnetospheric bombardment of the trailing side led to particles that were smaller or annealed in a complex fashion (*Verbiscer et al.,* 2006). Water ice absorption bands were also muted on the trailing hemisphere. The same study also identified ammonia hydrate as a possible component of the surface of Enceladus.

5. CASSINI-ERA OBSERVATIONS

5.1. Discovery of an Ionized Cloud at Enceladus from Cassini Magnetometer Observations

The first targeted Cassini flyby of Enceladus occurred on February 17 2005 [Day of Year (DOY) 048] and was a relatively distant flyby with a closest approach altitude of 1265 km above Enceladus' orbital plane [~4.7 Enceladus radii (R_E)]. MAG observations revealed a clear perturbation near Enceladus, one interpretation of which was that the nearly co-rotating saturnian plasma and frozen-in magnetic field were being deflected around Enceladus, which was acting as an obstacle to the flow, as well as being slowed down. Furthermore, there was a clear increase in ion cyclotron wave activity during this flyby at the gyrofrequency of water group ions, with the implication being that Enceladus was adding water group ions to the flowing magnetospheric plasma. The lefthand elliptically polarized ion cyclotron waves travel at small angles to the magnetic field and are generated by pickup ions created by the ionization of water group molecules being generated in the vicinity of Enceladus. The MAG team's initial instinct was that there was potentially a diffuse atmosphere around Enceladus being generated by processes as yet unknown. Less than a month later, a second, already planned flyby of the moon took place

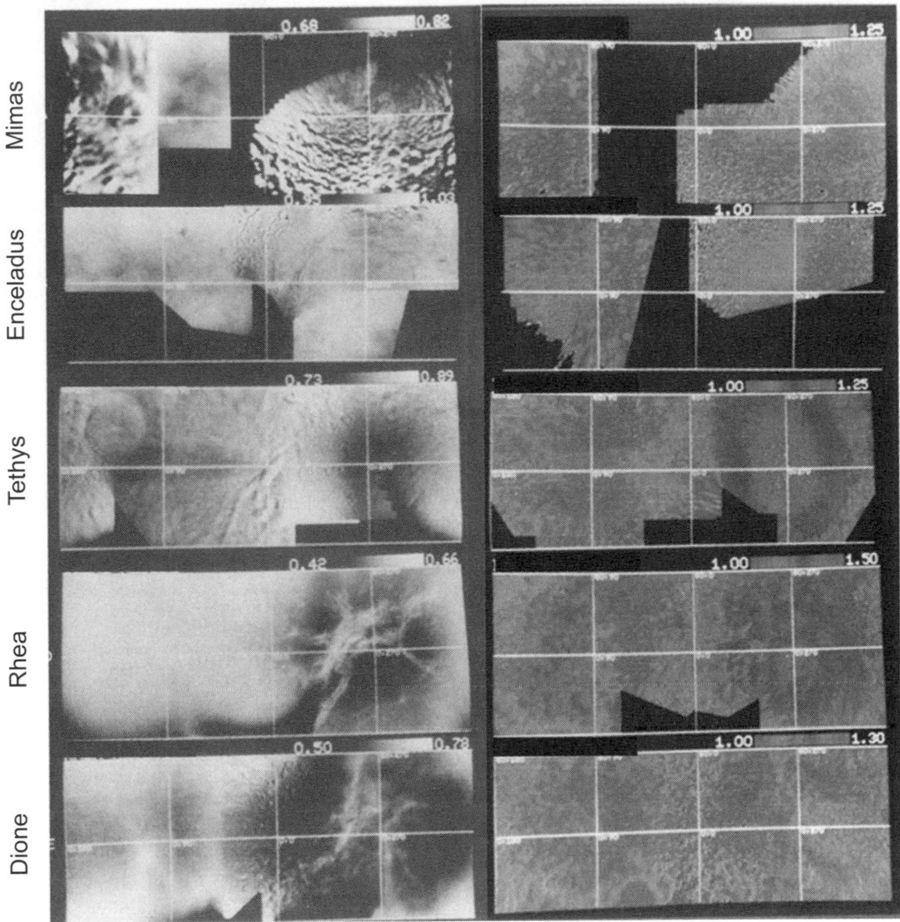

Fig. 6. See Plate 3 for color version. Voyager maps of the main inner moons of Saturn. The maps on the left depict the normal reflectance in the Voyager clear filter (0.47 μm), corrected for all the effects of viewing geometry (current values are higher because of a better knowledge of their opposition surges). The right set of maps are color ratios of the Voyager orange filter (0.59 μm; in the case of Tethys, the green filter at 0.56 μm) to the Voyager violet filter (0.41 μm). Adapted from *Buratti et al.* (1990).

on March 9, 2005 (DOY 068) at a closest approach altitude of 500 km and below the moon at a distance of ~1.5 R_E. Very similar perturbations of the magnetic field (both the draping signature as well as increased ion cyclotron wave activity) were observed during this second flyby, confirming the team's sense of the presence of an atmospheric interaction, which, since the Enceladus' gravitational field is not strong, would require a strong source to be maintained between the two flybys. Moreover, there also seemed to be an additional signature around the closest approach of the second flyby (south of the moon), almost as if the magnetic field was being pulled in toward the moon, as if Enceladus was acting as an amplifier of the Saturn magnetic field.

The observations can be seen in Fig. 7, taken from *Dougherty et al.* (2006). The data is shown in the Enceladus Interaction Coordinate System (ENIS) where X is along the direction of saturnian co-rotational flow, Y is positive toward Saturn, and Z is along Saturn's spin axis. The direction of the spacecraft motion during the flybys is shown with an arrow, revealing that for the first flyby (DOY 048) the spacecraft was moving away from Saturn and for the second (DOY 068) was moving inbound toward Saturn, with both trajectories in the same direction as the co-rotational flow. Overlain on the trajectories are the residuals of the magnetic field (after the magnetic field due to Saturn has been subtracted from the data). Both of these datasets reveal similar signatures in the Y component of the magnetic field, with a positive deflection on the Saturn side of the moon and a negative deflection on the anti-Saturn side. Such signatures would be expected to arise when the saturnian co-rotating magnetospheric plasma is being slowed and deflected around the moon, which is behaving as if it is a larger obstacle (than its physical size) to the plasma flow, such as an atmosphere where ionization leads to mass loading near the moon. On the assumption that an atmospheric interaction was arising, Alfven wings are expected above the conducting obstacle (Enceladus), which are generated by currents driven through the atmosphere as seen in Fig. 2 of *Dougherty et al.* (2006).

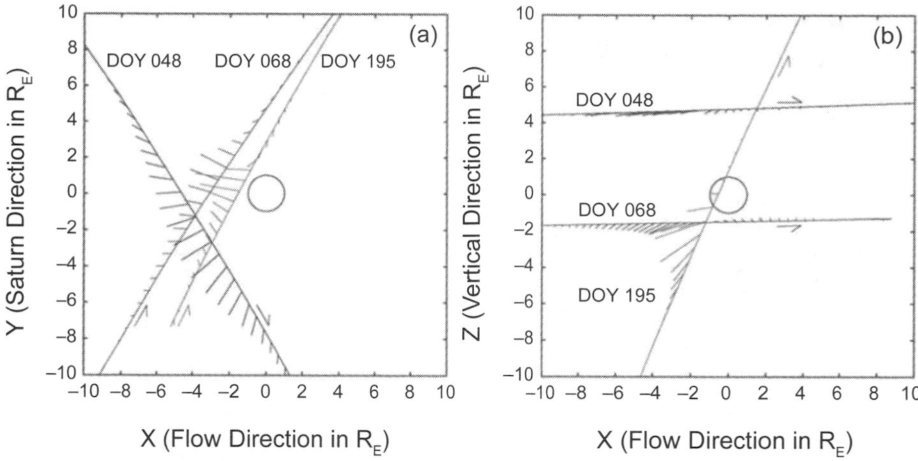

Fig. 7. The three flyby trajectories in **(a)** the (X,Y) plane and **(b)** the (X,Z) plane of the ENIS coordinate system. Overlain on the trajectories are the magnetic field residuals, scaled such that 4.5 nT = 2 R_E (from *Dougherty et al.,* 2006).

These Alfvenic disturbances couple the Enceladus plasma to the ionosphere of Saturn and enhance the magnetic field upstream of the obstacle and drape it around the obstacle. *Dougherty et al.* (2006) discusses the details of the perturbed magnetic field signature and highlights the inconsistency of the observations between the first and second flybys, where the second flyby data is not as easily explained by the "simple" symmetric atmospheric interaction scenario, but rather pointed to a more complex interaction where the major perturbation appeared to be south of the moon.

Based on these two sets of flyby observations, the MAG team proposed to the Cassini Project that Enceladus was surrounded by a diffuse extended atmosphere, as shown in the schematic in Fig. 8. Presentations were made to both the Project and to the Satellite Orbital Science Team (SOST), the group responsible for the planning of all icy satellite observations during the mission. These presentations described the results and the interpretation of the team as well, requesting that the closest-approach altitude be lowered for the third Enceladus flyby [planned for July 14, 2005 (DOY 195)], enabling the spacecraft to be much closer to the surface and thereby able to conduct a more detailed study of this unexpected interaction.

During the subsequent discussions with the Cassini Project to change the trajectory to lower the flyby altitude (*Buratti,* 2017) and to allow an Ultraviolet Imaging Spectrograph (UVIS) star occultation that passed above the south polar limb (UVIS had already had an occultation that detected no atmosphere at equatorial latitudes), previous imaging results from the Imaging Science Subsystem (ISS) that showed possible light scattering off the limb played an important role in the considerations, even though at that stage there was no agreement within the ISS team of their significance. Several early images of Enceladus (January and February 2005) had shown what could be interpreted as an atmosphere, plume, or jet, but the ISS team was also investigating explanations such as scattered light or ghost images. One such image, obtained on January 16, 2005, is shown in Fig. 9, and has been enhanced to show the plume-like structure. After the firm discovery of active geologic processes at the south pole of Enceladus in July 2005 by other Cassini instruments, images such as the one in Fig. 9 were unequivocally interpreted as jets coming from active regions on the moon.

A complication was that the four years' worth of observations for the nominal Cassini mission (mid-2004 to mid-2008) had already been planned in detail with instrument pointing designs, data rates, and spacecraft attitude already agreed, and a change in flyby altitude would have a ripple effect on some of the subsequent observations. The potential

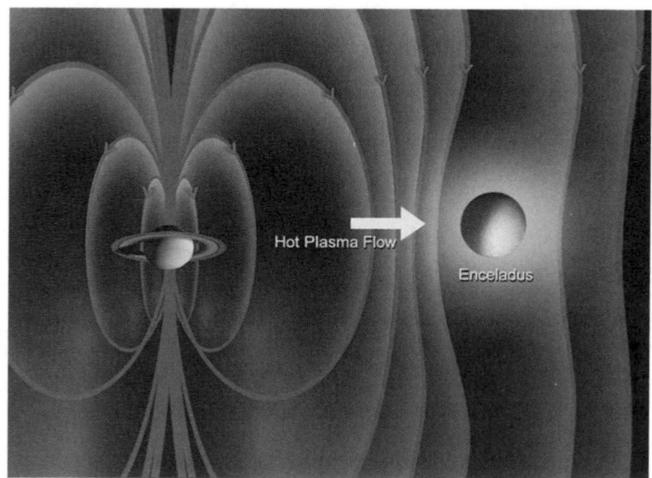

Fig. 8. See Plate 4 for color version. A schematic (where Saturn and Enceladus are not to scale) showing the co-rotating Saturn magnetic field and plasma being draped ahead of Enceladus by a diffuse extended atmosphere.

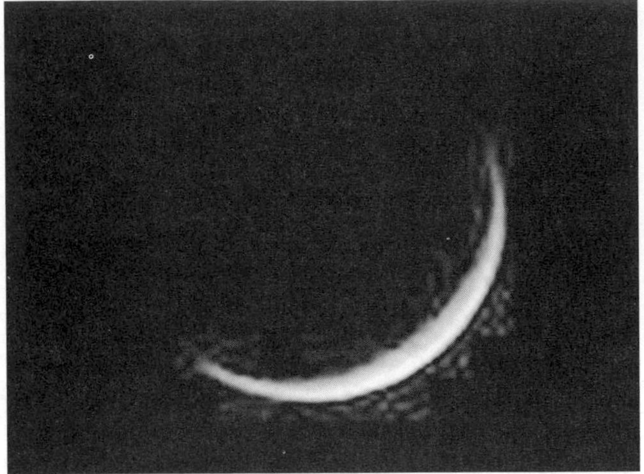

Fig. 9. An enhanced ISS image obtained of Enceladus in January 2005, with a plume or atmospheric-like structure off the illuminated limb.

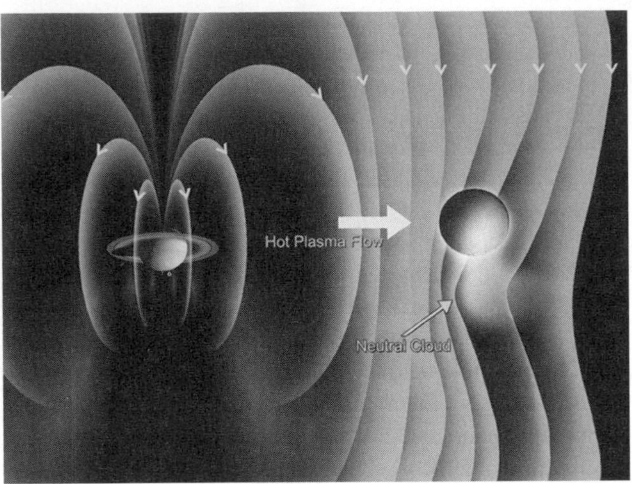

Fig. 10. See Plate 5 for color version. A schematic (where Saturn and Enceladus are not to scale) showing the corotating Saturn magnetic field and plasma being perturbed by the polar plume of water vapor generated at the south pole of Enceladus.

existence of a dynamic atmosphere at Enceladus was agreed to be important enough that such changes should be made, and hence the flyby altitude for the third flyby four months later was substantially decreased from 1000 km to only 173 km above the surface.

5.2. Multi-Instrument Discovery of Plume Activity in July 2005

The magnetic field perturbations arising during the third lowered Enceladus flyby as well as the spacecraft trajectory are included in Fig. 7 and shown by the notation "DOY 195," where the closest approach was 0.7 R_E below the southern pole of Enceladus and the trajectory cut from south to north. These close observations revealed that the largest magnetic field perturbation was in the X component and the perturbation vectors were very similar to those from DOY 068, even though for DOY 195 the spacecraft was moving very rapidly in the Z direction. The BX perturbation is resulting from a slowdown of the plasma flow below the spacecraft, which in turn was below Enceladus, the implication being that the outgassing source on the moon was at high southern latitudes. The MAG team updated the overview schematic to reveal instead a south polar plume of water vapor emanating from the polar region of the Enceladus surface, and indeed consistent with the additional unknown "amplifier" signal observed on the second flyby [see Fig. 10, taken from *Dougherty et al.* (2006)].

Further details of the analysis and interpretation of the magnetic field observations from the first three Enceladus flybys by the Cassini spacecraft can be found in *Dougherty et al.* (2006) and *Khurana et.al.* (2007), and a focus on the ion cyclotron wave analysis in *Leisner et al.* (2011).

The discovery of an ionized gas cloud via the MAG data was due to a south polar water vapor plume emanating from Enceladus, as simultaneously revealed by many Cassini instruments during the third, close flyby. UVIS observed stellar occultations on two of the three 2005 flybys, one which had a nondetection of an atmospheric signature, whereas the close flyby confirmed the existence, composition, and regionally confined nature of a water vapor plume in the south polar region of Enceladus. This work also confirmed that the plume provides an adequate amount of water to resupply losses from Saturn's E ring and to be the dominant source of the neutral OH and atomic oxygen that fill the saturnian system (*Hansen et al.,* 2006).

ISS images identified a geologically active province at the south pole of Enceladus that was circumscribed by a chain of folded ridges and troughs at 55°S latitude. The terrain southward of this boundary was distinguished by its albedo and color contrasts, elevated temperatures, extreme geologic youth, and narrow tectonic rifts (dubbed "tiger stripes") that exhibit coarse-grained ice (*Porco et al.,* 2006). The work also describes how jets of fine icy particles that supply Saturn's E ring emanate from this province, carried aloft by water vapor probably venting from subsurface reservoirs of liquid water, and that the shape of Enceladus suggests a possible intense heating epoch in the past by capture into a 1:4 secondary spin/orbit resonance. The ice particle plumes were also directly imaged by Cassini at high phase angles in visible light (*Porco et al.,* 2006), including some observations taken before the July 2005 flyby but not understood at the time.

Cassini's Composite Infrared Spectrometer (CIRS) detected the hottest temperatures on Enceladus coinciding with the tiger stripes. Between 3 to 7 GW of thermal emission from the south polar troughs at temperatures up to 145 K or higher were observed, making Enceladus only the third known solid planetary body (after Earth and Jupiter's volcanic moon, Io) that is sufficiently geologically active for its internal heat to be detected by remote sensing. If the plume is generated by the sublimation of water ice and if the sublimation source is visible to CIRS, then sublimation temperatures of at least 180 K are required (*Spencer et al.,* 2006), and this value was later revised upward (e.g., see the chapter in this volume by Spencer et al.).

Observations using Cassini's Visual and Infrared Mapping Spectrometer instrument (VIMS) revealed that Enceladus' surface is composed mostly of nearly pure water ice except near its south pole, where there are light organics, CO_2, and amorphous and crystalline water ice, particularly in the tiger stripes region. An upper limit of 5 precipitable nanometers was derived for CO in the atmospheric column above Enceladus, and 2% for NH_3 in global surface deposits. Upper limits of 140 K were derived for the temperatures in the tiger stripes (*Brown et al.,* 2006).

Before and during the closest approach, a substantial atmospheric plume and coma were observed, detectable in the Ion and Neutral Mass Spectrometer (INMS) data out to a distance of over 4000 km from Enceladus. INMS data indicated that the atmospheric plume and coma are dominated by water, with significant amounts of carbon dioxide, an unidentified species with a mass-to-charge ratio of 28 Da (either carbon monoxide or molecular nitrogen), and methane. Trace quantities of acetylene and propane also appeared to be present and ammonia was present at a level <0.5% (*Waite et al.,* 2006).

The High Rate Detector of the Cosmic Dust Analyzer (CDA) registered micrometer-sized dust particles enveloping the moon. The dust impact rate peaked about 1 minute before the closest approach, consistent with a locally enhanced dust production in the south polar region of Enceladus (*Spahn et al.,* 2006).

The synergistic multiple instrument discovery of the south polar plume emanating from cracks at the south polar surface revealed interior heat was leaking out from the moon as well as dust and organic material. The morphology of the plume was determined, and the Enceladus water source explained previous observations of hydroxyl and oxygen atoms in the Saturn system (*Shemansky et al.,* 1993; *Jurac et al.,* 2002), as well as the discovery of an atomic O torus by Cassini UVIS on Saturn approach (*Esposito et al.,* 2005). This discovery led to a focus on Enceladus in the Cassini extended missions and resulted in major changes to the Cassini mission plans, including in the original prime mission (July 2004–July 2008). Some examples of such changes include the addition of extensive distant monitoring of the plume from ISS in the prime mission; the formation of a Cassini Project plume working group to evaluate the risks to the spacecraft from deeper passage through the plume in subsequent orbits, starting with the last prime mission encounter in Rev (orbit) 61; and optimization of the Rev 61 encounter on December 3, 2008 (which was during Saturn eclipse), for plume sampling and south polar thermal emission studies. Following the prime mission the continued study of Enceladus, with many additional targeted encounters of the south pole, became a major driver for the tour design for the first mission extension (the Cassini Equinox Mission, July 2008–October 2010) as well as the second mission extension (the Cassini Solstice Mission, October 2012–September 2017). Table 1 shows the details of all the targeted Enceladus encounters from the Cassini mission, including some of the science highlights (*Spencer et al.,* 2009), and Fig. 11 shows the geometries of these flybys relative to Enceladus.

5.3. The Importance of Groundbased and Spacebased Observations

Missions represent a snapshot in time: groundbased campaigns greatly enlarge the temporal excursion of any celestial phenomenon observed by a spacecraft. Observations of activity on Europa may play out in a manner similar to that of Enceladus.There have been at least three observations of possible plume or heat activity on Europa, even though there have been no detections by spacecraft (there is, however, an abundance of geologic evidence for past and ongoing activity from the Voyagers and from the Galileo spacecraft). There was a tentative brief thermal event identified from 1981 groundbased observations (*Tittemore and Sinton,* 1989), observing at NASA's Infrared Telescope Facility (IRTF) on Mauna Kea, Hawaii. More than a generation later, possible vapor plumes over 100 km high were imaged with the HST (*Roth et al.,* 2014), and evidence for off-limb continuum absorption as Europa transited Jupiter has recently been presented (*Sparks et al.,* 2016, 2017). Europa's potential activity may be more sporadic and less energetic than that of Enceladus. It is essential to continue groundbased monitoring studies of Enceladus to understand the temporal evolution of its plume and jets and its heat balance through time. It is important to devise new methods of observing the heat flux and evolution of its plume activity following the Cassini end of mission in September 2017.

6. SUMMARY

The focus on Enceladus and its E ring began in the eighteenth century, and as time has passed, we have learned a great deal about this tiny moon and its young surface as well as confirmed that it is the source of Saturn's diffuse E ring. Well prior to the discovery of cryovolcanism on Enceladus, scientists were describing evidence for it and predicting that such activity was taking place. The discovery by the Cassini MAG measurements of a water vapor plume emanating from the south pole of the moon was confirmed by numerous other Cassini instrument measurements and resulted in a focus on Enceladus throughout Cassini's extended missions. The confirmation of a liquid water ocean beneath its surface, organic material and dust within the plume, and modulation of the plume and potentially thermal activity by Saturn's tides are all discussed in further detail in other chapters in this volume. Interest in Enceladus as a potential habitat is very high, as well as the desire for new future planetary missions to confirm and explore this potential habitability.

TABLE 1. Details of all the targeted Enceladus encounters from the Cassini mission.

			Closest Approach								
Orbit	Name	Mission Phase	Date	Time	Speed (km s^{-1})	Altitude (km)	Sub-S/C Long	Sub-S/C Lat	−1 Hour Sub-S/C Lat	+1 Hour Sub-S/C Lat	Closest Approach Science Highlights
3	E0	Prime	17 Feb 2005	03:30	6.7	1260	239°W	51°N	1°N	3N	First high-resolution imaging, magnetic perturbation observed
4	E1	Prime	09 Mar 2005	09:08	6.6	497	304°W	30°S	1°S	0°S	High-resolution imaging, magnetic perturbation observed
11	E2	Prime	14 Jul 2005	19:55	8.2	166	327°W	22°S	48°S	47°N	First high-resolution south polar remote sensing; discovery of activity by multiple instruments
61	E3	Prime	12 Mar 2008	19:06	14.4	47	135°W	21°S	70°N	70°S	*In situ* plume sampling, high-resolution thermal mapping of south pole
80	E4	Equinox	11 Aug 2008	21:06	17.7	46	98°W	28°S	62°N	63°S	Very-high-resolution south polar remote sensing
88	E5	Equinox	09 Oct 2008	19:06	17.8	20	97°W	31°S	62°N	63°S	Closest flyby; Plume sampling including identification of salt-rich low-speed grains
91	E6	Equinox	31 Oct 2008	17:14	17.7	191	97°W	27°S	62°N	63°S	Very high-resolution south polar remote sensing
120	E7	Equinox	02 Nov 2009	07:41	7.7	99	159°W	88°S	1°S	0°N	Low-altitude, low-speed *in situ* plume sampling
121	E8	Equinox	21 Nov 2009	02:09	7.7	1602	112°W	82°S	3°S	3°S	High-resolution remote sensing of Baghdad Sulcus
130	E9	Equinox	28 Apr 2010	00:10	6.5	99	147°W	89°S	1°S	0°S	First gravity pass: south pole
131	E10	Equinox	18 May 2010	06:04	6.5	434	304°W	34°S	0°S	1°S	UVIS solar occultation of plume: Non-detection of N_2
136	E11	Solstice	13 Aug 2010	22:31	6.9	2550	29°W	78°S	17°S	5°N	High-resolution remote sensing of Damascus Sulcus in darkness
141	E12	Solstice	30 Nov 2010	11:53	6.3	47	53°W	62°N	0°N	0°N	Second gravity pass, northern hemisphere
142	E13	Solstice	21 Dec 2010	01:08	6.2	47	231°W	62°N	0°N	0°N	Northern hemisphere close pass: *in situ* observations
154	E14	Solstice	01 Oct 2011	13:52	7.4	99	204°W	89°S	0°S	1°S	Low-altitude, low-speed *in situ* plume sampling
155	E15	Solstice	19 Oct 2011	09:22	7.5	1231	114°W	14°S	0°S	1°S	UVIS double plume occultation
156	E16	Solstice	06 Nov 2011	04:58	7.4	495	295°W	30°S	0°S	1°S	RADAR SAR observations of southern mid-latitudes
163	E17	Solstice	27 Mar 2012	18:30	7.5	74	138°W	86°S	0°N	1°S	Low-altitude, low-speed *in situ* plume sampling
164	E18	Solstice	14 Apr 2012	14:01	7.5	74	290°W	87°S	0°N	1°S	Low-altitude, low-speed *in situ* plume sampling: CIRS and VIMS very-high-resolution observations of Baghdad Suclus emission
165	E19	Solstice	02 May 2012	09:31	7.5	73	291°W	72°S	0°N	1°S	Third gravity pass: southern hemisphere
223	E20	Solstice	14 Oct 2015	10:41	8.5	1838	105°W	79°N	3°N	3°N	Remote sensing of north polar region
224	E21	Solstice	28 Oct 2015	15:22	8.5	48	66°W	84°S	0°N	1°S	Lowest-altitude sampling through the plume: H_2 discovered by INMS
228	E22	Solstice	19 Dec 2015	17:49	9.5	4999	353°W	88°S	6°S	10°S	CIRS, VIMS observations of thermal emission from winter pole

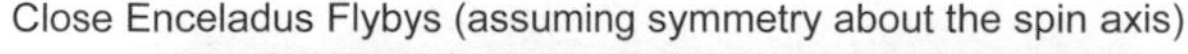

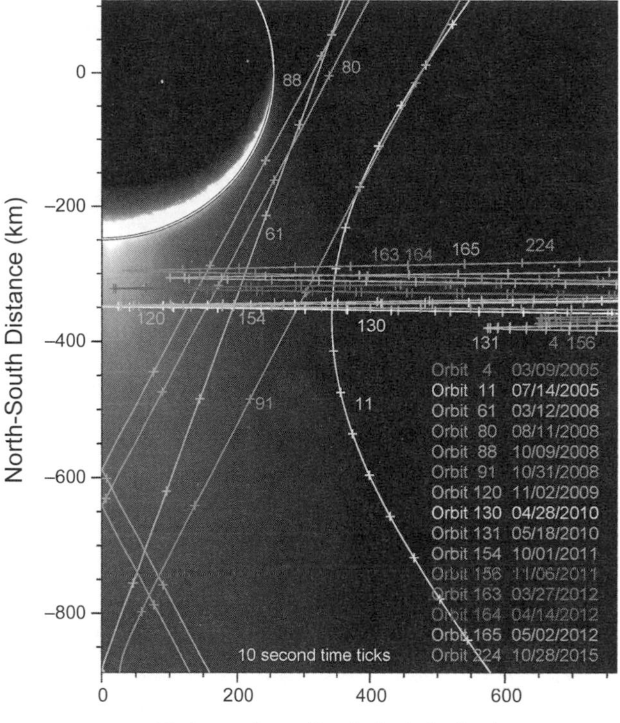

Fig. 11. See Plate 6 for color version. Schematic revealing the geometries of the targeted flybys relative to Enceladus, where the different colors show the different flybys and the date of their closest approach.

REFERENCES

Alexander A. F. O'D. (1962) *The Planet Saturn.* MacMillan, New York.

Antoniadi E.-M. (1909a) Corpuscules en dehors du plan de l'anneau de Saturne. *Bull. Soc. Astron. de France, 23,* 448–450.

Antoniadi E.-M. (1909b) Corpuscules en dehors du plans des anneaux des Saturne. *Bull. Soc. Astron. France, 23,* 506.

Barnard E. E. (1909) Saturn, recent observations of the rings and their bearing upon some of the phenomena of the disappearance of the rings in 1907. *Mon. Not. R. Astron. Soc., 69,* 621.

Baum R. M. (1954) On observations of the reported dusky ring outside the bright rings of the planet Saturn. *J. British Astron. Assoc., 64,* 192–196.

Baum W. A., Kreidl T., Westphal J. A., Danielson G. E., Seidelmann P. K., Pascu D., and Currie D. G. (1981) Saturn's E ring I. CCD observations of March 1980. *Icarus, 47,* 84–96.

Baum W. A., Kreidl T., and Wasserman L. H. (1984) Saturn's E ring. In *Planetary Rings* (A. Brahic, ed.), pp. 103–108. IAU Colloquium 75, Centre National d'Etudes Spatiales, Toulouse, France.

Brahic A., Lecaceux J., and Sicardy B. (1980) Observations of Saturn's edgewise rings in France. *Bull. Am. Astron. Soc., 12,* 700.

Brown R. H. and 24 co-authors (2006) Composition and physical properties of Enceladus' surface. *Science, 311,* 1425–1428.

Buratti B. J. (1985) Application of a radiative transfer model to bright icy satellites. *Icarus, 61,* 208–217.

Buratti B. J. (1988) Enceladus: Implications of its unusual photometric properties. *Icarus, 75,* 113–126.

Buratti B. J. (2017) *Worlds Fantastic, Worlds Familiar: A Guided Tour of the Solar System.* Cambridge Univ., New York.

Buratti B. J. and Veverka J. (1984) Voyager photometry of Rhea, Dione, Tethys, Enceladus and Mimas. *Icarus, 58,* 254–264.

Buratti B. J., Mosher J. A., and Johnson T. V. (1990) Albedo and color maps of the saturnian satellites. *Icarus, 87,* 339–357.

Buratti B. J., et al. (1998) Near-Infrared photometry of the saturnian satellites during ring plane crossing. *Icarus, 136,* 223–231.

Burns J. A., Showalter M. R., and Morfill G. (1984) The ethereal rings of Jupiter and Saturn. In *Planetary Rings* (R. Greenberg and A. Brahic, eds.), pp. 200–272. Univ. of Arizona, Tucson.

Cave T. R. and Cragg T. (1952) *Mitteilungen fur Planetenbrobachter, 6,* 16–17 (also published in *Strolling Astronomer, 6,* 160).

Cheng A. F., Lanzerotti L. J., and Pirronello V. (1982) Charged particle sputtering of ice surfaces in Saturn's magnetosphere. *J. Geophys. Res., 87,* 4567–4570.

de Pater I., Showalter M. R., Lissauer J. J., and Graham J. R. (1996) Keck infrared observations of Saturn's E and G rings during Earth's 1995 ring plane crossings. *Icarus, 121,* 195–198.

Dollfus A. and Brunier S. (1980a) Le coronographe focal et l'observation d'un nouvel anneau de Saturne. *C. R. Acad. Sci. Paris Ser. B, 290,* 261–263.

Dollfus A. and Brunier S. (1980b) Searches for Saturn outer ring and inner satellites with a focal coronagraph. *Bull. Am. Astron. Soc., 12,* 728.

Dougherty M. K., Khurana K. K., Neubauer F. M., Russell C. T., Saur J., Leisner J. S., and Burton M. E. (2006) Identification of a dynamic atmosphere at Enceladus with the Cassini Magnetometer. *Science, 311,* 1406–1409.

Durisen R. H., Cramer M. L., Mulliken T. L., and Cuzzi N. (1982) The evolution of Saturn's rings due to particle erosion mechanisms. In *Saturn,* p. 60. Conference abstracts, Univ. of Arizona, Tucson.

Esposito L. W. et al. (2005) Ultraviolet imaging spectroscopy shows an active saturnian system. *Science, 307,* 1251.

Feibelman W. A. (1967) Concerning the D ring of Saturn. *Nature, 214,* 793–794.

Feibelman W. A. (1984) Saturn's E-ring in the red spectral region. In *Planetary Rings* (A. Brahic, ed.). IAU Colloquium 75, Centre National d'Etudes Spatiales, Toulouse, France.

Feibelman W. A. and Klinglesmith D. A. (1980) Saturn's E ring revisited. *Science, 209,* 277–279.

Franz O. G. and Millis R. L. (1975) Photometry of Dione, Tethys, and Enceladus on the UBV system. *Icarus, 24,* 433–442.

Guerin P. (1970) The new ring of Saturn. *Sky and Telescope, 40,* 88.

Gore R. (1981) Voyager 1 at Saturn: Riddles of the rings. *National Geographic, 160,* 3–31.

Greenwich Observatory (1908) Note on the appearance of Saturn's rings. *Mon. Not. R. Astron. Soc., 69,* 39–41.

Haff P. K. (1982) Vapor transport and regolith differentiation on the icy Saturn satellites. In *Saturn,* p. 77. Conference abstracts, Univ. of Arizona, Tucson.

Haff P. K., Siscoe G. L., and Eviatar A. (1983) Ring and plasma — The enigmae of Enceladus. *Icarus, 56,* 426–438.

Hamilton D. P. and Burns J. A. (1994) Origin of Saturn's E ring: Self-sustained, naturally. *Science, 264,* 550–553.

Hansen C. J., Esposito L., Stewart A. I. F., Colwell J., Hendrix A., Pryor W., Shemansky D., and West R. (2006) Enceladus' water vapor plume. *Science, 311,* 1422–1425.

Hapke B. (1981) Bidirectional reflectance spectroscopy. 1. Theory. *J. Geophys. Res., 86,* 3039–3054.

Hapke B. (1984) Bidirectional reflectance spectroscopy. 3. Correction for macroscopic roughness. *Icarus, 59,* 41–59.

Hapke B. (1986) Bidirectional reflectance spectroscopy. 4. The extinction coefficient and the opposition effect. *Icarus, 67,* 264–280.

Hapke B. (2002) Bidirectional reflectance spectroscopy, 5: The coherent backscatter opposition effect and anisotropic scattering. *Icarus, 157,* 523–534.

Herkenhoff K. and Stevenson D. (1984) Formation of Saturn's E-ring by evaporation of liquid from the surface of Enceladus. In *Lunar and Planetary Science XV,* pp. 361–362. Lunar and Planetary Institute, Houston.

Herschel W. (1790) Account of the discovery of a sixth and seventh satellite of the planet Saturn; with remarks on the construction of its ring, its atmosphere, its rotation on an axis and its spheroidical figure. *Phil. Trans. R. Soc. London, 80,* 1–20.

Hill J. R. and Mendis D. A. (1982) The origin of the E-ring of Saturn (abstract). *Eos Trans. AGU, 63,* 1019.

Hood I. L. (1983) Radial diffusion in Saturn's radiation belts: A modeling analysis assuming satellites and ring E absorption. *J. Geophys. Res., 88,* 808–818.

Humes D. H., O'Neal R. L., Kinard W. H., and Alvarez J. M. (1980) Impact of Saturn ring particles on Pioneer 11. *Science, 207,* 443–444.

Jarry-Desloges Observatory (1907) *Observations des Surfaces Planetaires, 1,* 106–107.

Jurac S., Johnson R. E., and Richardson J. D. (2001) Saturn's E ring and production of the neutral torus. *Icarus, 149,* 384–396.

Jurac S., McGrath M. A., Johnson R. E., Richardson J. D., Vasyliunas V. M., and Eviatar A. (2002) Saturn: Search for a missing water source. *Geophys. Res. Lett., 29,* DOI: 10.1029/2002GL015855.

Kargel J. S. (1984) A crater chain on Enceladus: Evidence for explosive volcanism. In *Lunar and Planetary Science XV,* pp. 427–428. Lunar and Planetary Institute, Houston.

Kargel J. and Pozio S. (1996) The volcanic and tectonic history of Enceladus. *Icarus, 119,* 385–404.

Khurana K. K., Dougherty M. K., Russell C. T., and Leisner J. S. (2007) Mass loading of Saturn's magnetosphere near Enceladus. *J. Geophys. Res., 112,* A08203, DOI: 10.1029/2006JA0121220.

Koutchmy S. and Lamy P. L. (1975) Study of the inner satellites of Saturn by photographic photometry. *Icarus, 25(3),* 459–465.

Kuiper G. P. (1974) On the origin of the solar system I. *Cel. Mech., 9,* 321–348.

Lamy P. L. and Mauron N. (1980) The new satellite Dione B and outer ring of Saturn. *Bull. Am. Astron. Soc., 12,* 728–729.

Larson S. M., Fountain W., Smith B. A., and Reitsema H. J. (1982) Observations of the Saturn ring and a new satellite. *Icarus, 47,* 288–290.

Lassell W. (1848) Names. *Mon. Not. R. Astron. Soc., 8,* 42–43.

Leisner J. S., Russell C. T. Wei H. Y., and Dougherty M. K. (2011) Probing Saturn's ion cyclotron waves on high-inclination orbits: Lessons for wave generation. *J. Geophys. Res., 116,* A09235. DOI: 10.1029/2011JA016555.

Lissauer J. J., Peale S., and Cuzzi N. (1984) Ring torque on Janus and melting on Enceladus. *Icarus, 58,* 150–168.

McDonald F. B., Schardt A. W., and Trainor J. H. (1980) If you've seen one magnetosphere, you haven't seen them all: Energetic particles observations in the Saturn's magnetosphere. *J. Geophys. Res., 85,* 5813–5830.

McKinnon W. B. (1983) Origin of the E ring: Condensation of impact vapor or boiling of impact melt? In *Lunar and Planetary Science XIV,* pp. 478–488. Lunar and Planetary Institute, Houston.

Meyer-Vernet N., Lecacheux A., and Pedersen B. M. (1996) Constraints on Saturn's E ring from Voyager 1 radio astronomy instrument. *Icarus, 123,* 113–128.

Morfill G., Grun E., and Johnson T. V. (1982) Interaction of Saturn's rings with plasma. In *Saturn,* p. 59. Conference abstracts, Univ. of Arizona, Tucson.

Morfill G., Grun E., and Johnson T. V. (1983) Saturn's E, G, and F rings: Modulated by the plasma sheet? *J. Geophys. Res., 88,* 5573–5579.

Nicholson P. D., Showalter M. R., Dones L., French R. G., Larson S. M., et al. (1996) Observations of Saturn's ring-plane crossing in August and November 1995. *Science, 272,* 509–516.

Noland M. J. et al. (1974) Six color photometry of Iapetus, Titan, Rhea, Dione and Tethys. *Icarus, 23,* 334–354.

Pang K. D., Voge C. C., Rhoads J. W., and Ajello J. M. (1984a) The E ring of Saturn and satellite Enceladus. *J. Geophys. Res., 89,* 9459–9470.

Pang K. D., Voge C. C., and Rhoads J. W. (1984b) Macrostructure and microphysics of Saturn's E-ring. In *Planetary Rings* (A. Brahic, ed.). IAU Colloquium 75, Centre National d'Etudes Spatiales, Toulouse, France.

Passey Q. (1983) Viscosity of the lithosphere of Enceladus. *Icarus, 53,* 105–120.

Plescia J. B. and Boyce J. M. (1983) Crater numbers and geological histories of Iapetus, Enceladus, Tethys, and Hyperion. *Nature, 301,* 666–670.

Porco C. C. and 24 co-authors (2006) Cassini observes the active south pole of Enceladus. *Science, 311,* 1393–1401.

Reitsema H. J., Smith B. A., and Larson S. M. (1980) The E ring of Saturn. *Bull. Am. Astron. Soc., 12,* 701.

Richardson J. D., Eviatar A., McGrath M. A., and Vasyliunas V. M. (1998) OH in Saturn's magnetosphere: Observations and implications. *J. Geophys. Res., 103,* 20245–20256.

Roddier C., Roddier F., Graves J. E., and Northcott M. (1998) Discovery of an arc of particles near Enceladus' orbit: A possible key to the origin of the E ring. *Icarus, 136,* 50–59.

Ross M. N. and Schubert G. (1989) Viscoelastic models of tidal heating in Enceladus. *Icarus, 78,* 90–101.

Roth L. et al. (2014) Transient water vapor at Europa's south pole. *Science, 343,* 171–174.

Schaer E. (1908) A new Saturn ring. *J. British Astron. Assoc., 19,* 55.

Schaer E. (1909) Observations de Saturne et de ses Anneaux. *Astron. Nach., 181,* 177–180.

Schenk P. M., Hamilton D. P., Johnson R. E., McKinnon W. B., Paranicas C., Schmidt J., and Showalter M. R. (2011) Plasma, plumes and rings: Saturn system dynamics as recorded in global color patterns on its mid-sized icy satellites. *Icarus, 211,* 740–757, DOI: 10.1016/j.icarus.2010.08.016.

Seidelmann P. K., Harrington R. S., and Szebehely V. (1984) Dynamics of Saturn's E ring. *Icarus, 58,* 169–177.

Shemansky D. E., Matheson P., Hall D. T., Hu H.-Y., and Tripp T. M. (1993) Detection of the hydroxyl radical in the Saturn magnetosphere. *Nature, 363,* 329–331.

Showalter M. R., Cuzzi J. N., and Larson S. M. (1991) Structure and particle properties of Saturn's E ring. *Icarus, 94,* 451–473.

Slipher E. C. and Slipher V. (1914) Mimas and Enceladus turn always the same face to Saturn. *Lowell Observatory Bulletin No. 62. Vol. II,* 70–72.

Smith B. A.(1978) The D and E rings of Saturn. In *The Saturn System,* pp. 105–111. JPL Publication N79-16758 07-91.

Smith B. A., Soderblom L., Beebe R., Boyce G., Briggs G., et al. (1981) Encounter with Saturn: Voyager 1 imaging science results. *Science, 212,* 182–184.

Smith B. et al. (1982) A new look at the Saturn system — The Voyager 2 images. *Science, 215,* 504–537.

Spahn F. and 15 colleagues (2006) Cassini dust measurements at Enceladus and implications for the origin of the E ring. *Science, 311,* 1416–1418.

Sparks W. B., Hand K. P., McGrath M. A., Bergeron E., Cracraft M., and Deustua S. E. (2016) Probing for evidence of plumes on Europa with HST/STIS. *Astrophy. J., 829,* DOI: 10.3847/0004-637X/829/2/121.

Sparks W. B. et al. (2017) Active cryovolcanism on Europa? *Astrophys. J. Lett., 839,* L18.

Spencer J. R., Pearl J. C., Segura M., Flasar F. M., Mamoutkine A., Romani P., Buratti B. J., Hendrix A. R., Spilker L. J., and Lopes R. M. C. (2006) Cassini encounters Enceladus: Background and the discovery of a south polar hot spot. *Science, 311,* 1401–1405.

Spencer J. R., Barr A. C., Esposito L. W., Helfenstein P., Ingersoll A. P., Jaumann R., McKay C. P., Nimmo F., and Waite H. J. (2009) Enceladus: An active cryovolcanic satellite. In *Saturn from Cassini-Huygens* (M. K. Dougherty et al., eds.), pp. 683–724. Springer, Berlin.

Squyres S. W., Reynolds R. T., Cassen P. M., and Peale S. J. (1983) The evolution of Enceladus. *Icarus, 53,* 319–331.

Terile R. J. and Tokanaga A. (1980) Infrared photometry of Saturn's E ring. *Bull. Am. Astron. Soc., 12,* 701.

Tittemore W. C. and Sinton W. (1989) Near-infrared photometry of the Galilean satellites. *Icarus, 77,* 82–97.

Verbiscer A. J. and Veverka J. (1994) A photometric study of Enceladus. *Icarus, 77,* 155–164.

Verbiscer A. J., French R. G., and McGhee C. A. (2005) The opposition surge of Enceladus: HST observations 338–1022 nm. *Icarus, 173,* 66–83.

Verbiscer A. J. et al. (2006) Near infrared spectra of the leading and trailing hemispheres of Enceladus. *Icarus, 182,* 211–223.

Verbiscer A. J. et al. (2007) Enceladus: Cosmic graffiti artist caught in the act. *Science, 315,* 815–817.

Waite J. H. and 13 colleagues (2006) Cassini Ion and Neutral Mass Spectrometer: Enceladus plume composition and structure. *Science, 311,* 1419–1422.

Wilson L. and Head W. (1984) Aspects of water eruption on icy satellites.In *Lunar and Planetary Science XV,* pp. 924–925. Lunar and Planetary Institute, Houston.

Yoder C. F. (1979) How tidal heating in Io drives the Galilean orbital resonance locks. *Nature, 279,* 767–770.

McKinnon W. B., Lunine J. I., Mousis O., Waite J. H., and Zolotov M. Y. (2018) The mysterious origin of Enceladus: A compositional perspective. In *Enceladus and the Icy Moons of Saturn* (P. M. Schenk et al., eds.), pp. 17–38. Univ. of Arizona, Tucson, DOI: 10.2458/azu_uapress_9780816537075-ch002.

The Mysterious Origin of Enceladus: A Compositional Perspective

William B. McKinnon
Washington University in St. Louis

Jonathan I. Lunine
Cornell University

Olivier Mousis
Aix Marseille Université, CNRS

J. Hunter Waite
Southwest Research Institute

Mikhail Yu. Zolotov
Arizona State University

Enceladus must have accreted in a disspative, prograde rotating, coplanar disk about Saturn, along with other mid-sized satellites, but beyond this there is little agreement. It may have accreted over an extended period of time (>0.1–1 m.y.) in a circumsaturnian nebula formed during the end stage of the protosolar nebula. Alternatively, it may have formed later from an accumulation of solid ring material. It may even be a reaccreted moon, formed when an earlier mid-sized satellite system went unstable. Whatever its true origin, chemical and isotopic data from Enceladus, as well as Titan, imply that the solids from which Enceladus formed were essentially protosolar in nature, trapped in satellitesimals that plausibly had a common origin with comets and perhaps some classes of carbonaceous asteroids. These solids would have been transported into the inner mid-sized satellite feeding zone without significant subsequent chemical or isotopic interaction with early saturnian subnebular gas. If Enceladus is primordial, vaporization of highly volatile ices may have accompanied infall from the protosolar nebula into the saturnian subnebula, but NH_3 ice survived and was accreted along with carbon-bearing ices (CO_2 and CH_4), which sets an upper limit on subnebular temperatures. Carbonaceous matter accreted as well, macromolecular CHONSP (carbon, hydrogen, oxygen, nitrogen, sulfur, and phosphorus) being an important additional source for these biogenic elements, but the relative amounts of the various components are uncertain. Because the Enceladus we see today may be a surviving remnant of a larger, precursor satellite (or Centaur) torn apart by saturnian tides or impacts, or possibly birthed by a long evolving and much more massive ring system, the moon may be substantially younger than 4.5 b.y. old. We discuss possible tests of these ring origin and catastrophic scenarios.

1. INTRODUCTION

Enceladus is a remarkable world. From Cassini mass spectrometry of plume gas (*Waite et al.,* 2009, 2017; *Glein et al.,* 2015), ice and silica grain analyses (*Postberg et al.,* 2008, 2009, 2011; *Hsu et al.,* 2015), and gravity (*Iess et al.,* 2014; *McKinnon,* 2015; *Beuthe et al.,* 2016) and libration (*Thomas et al.,* 2016) measurements, the existence of an internal ocean in contact with a rock core that may be undergoing hydrothermal activity is strongly indicated (*Choblet et al.,* 2017). Active hydrothermal areas provide a habitable environment on Earth and have been implicated as sites where early life emerged (e.g., *Sullivan and Baross,* 2007). In this chapter we put these compositional analyses and structural inferences into the context of origin scenarios and aspects of internal evolution models (including those of Titan), and by doing so constrain these models where practical. The principal question we address is when and how Enceladus formed around Saturn. While we cannot claim to answer this question, we hope this review will stimulate further work and analyses that may lead to its resolution.

This chapter is organized as follows. In section 2 we survey proposed origin scenarios for Saturn's mid-sized satellites: traditional, revisionist, and even catastrophist. In section 3 we summarize the likely compositions of Enceladus' building blocks. Chemical and isotopic observations of Enceladus and Titan are presented in section 4 and constraints and implications discussed in section 5. Finally, in section 6 we suggest future measurements and modeling that may lead to the clarification of the issues and answers to the questions discussed.

2. FORMATION MODELS FOR REGULAR SATELLITES OF SATURN

The satellites of Saturn offer the most extensive, complex, and variegated system of regular satellites in the solar system. Especially notable among these bodies are Titan and, of course, Enceladus. Details of their masses, densities, orbital parameters, etc., can be found in the chapter in this volume by Castillo-Rogez et al.

2.1. Overview of Primordial Formation Models for Saturn's Circumplanetary Disk

The coplanar, prograde, saturnian regular icy satellites, including Enceladus, almost certainly formed in a dynamically cold, dissipative disk *of some kind*. Traditionally, this disk has been viewed as a circumplanetary accretion disk (hereafter CPD) that surrounded Saturn at the end of its formation (e.g., *Coradini et al.,* 1995; *Peale and Canup,* 2015). But this inference also applies to revisionist formation schemes that posit birth from an massive ring system, the key dynamical idea — energy dissipation in an angular-momentum-conserving rotating disk or cloud — remaining unchanged. We begin, however, with the traditional, primordial view. At the beginning of the formation of Saturn (and Jupiter, to which most studies have been aimed), the total radius of the CPD would have extended up to the planet's Hill radius, $a(t)[M_S(t)/3M_{Sun}]^{1/3}$, where a(t) and $M_S(t)$ are Saturn's semimajor axis and mass at time t, a(t) not necessarily the same as today. At the end of the growth of Saturn, however, the planet's envelope would have shrunk and its total mass would have been sufficient to have opened at least a partial gap in the protosolar nebula. Accretion of gas then (nominally) proceeded through streamers, which coalesced and formed a prograde-rotating CPD surrounding Saturn (Fig. 1, top) (*Lubow et al.,* 1999; *D'Angelo et al.,* 2003b; *Bate et al.,* 2003; *Klahr and Kley,* 2006; *Machida et al.,* 2008; *Ayliffe and Bate,* 2009; *Tanigawa et al.,* 2012).

Time-dependent models suggest that the CPDs around Jupiter and perhaps Saturn evolved similarly and followed two distinct phases (*Alibert et al.,* 2005; *Alibert and Mousis,* 2007). The first phase corresponds to the period when the protosolar nebula has not dissipated. The CPD is fed by gas and gas-coupled material originating from the protosolar nebula (*Canup and Ward,* 2002, 2006; *Ward and Canup,* 2010). Moreover, during this phase temperature and pressure conditions in the CPD could have been high enough to induce gas-gas and gas-solid chemical reactions in the satellites' formation region. The thermodynamic conditions encountered in the CPD at this time may match those required to induce gas phase conversion of CO to CH_4 and N_2 to NH_3 prior to the condensation of the satellites' "building blocks," or satellitesimals [see Figs. 2 and 3 in *Prinn and Fegley* (1989)]. The presence of Fe-Ni grains coupled to the gas could speed up these chemical reactions (*Sekine et al.,* 2005). Nevertheless, it is unclear whether satellitesimals would have the time to condense from the hot gas present in the chemically active zone of the CPD before said subnebular gas viscously accreted onto Saturn (*Mousis et al.,* 2006).

As the protosolar nebula dissipated and the accretion rate onto Saturn declined, the CPD nominally entered into the second phase of its evolution. It progressively attenuated with time as it both accreted to the forming giant planet and expanded radially due to angular momentum conservation. During this period, the CPD became cool enough so as to not alter the composition of embedded solids that formed in the protosolar nebula (*Mousis et al.,* 2002).

The detailed formation conditions of Enceladus (or, possibly, an earlier precursor satellite or satellites) in Saturn's CPD are very poorly known. The satellite's building blocks might have condensed *in situ* during the first phase of the CPD's evolution, implying that they possibly ultimately incorporated species such as NH_3 and CH_4 that resulted from thermochemistry in the subnebula (*Prinn and Fegley,* 1981, 1989). Alternative models suggest that the satellite-forming planetesimals were initially formed in the protosolar nebula and were (depending on size) partly devolatilized during gas-drag-induced migration within the CPD toward Saturn (*Mousis and Gautier,* 2004; *Alibert et al.,* 2005; *Mousis et al.,* 2009c; *Ronnet et al.,* 2017) or after direct accretion onto the disk (Fig. 1, bottom) (e.g., *Canup and Ward,* 2006; *Ayliffe and Bate,* 2009). Either of these effects could explain the increase in rock mass fraction, at Jupiter, among the Galilean satellites with the decreasing distance from the planet. In this view, the ice-to-rock ratios of Ganymede and Callisto reflect the values acquired by bodies formed in the protosolar nebula, whereas those of Io and Europa result from the processing of their building blocks (i.e., vaporization and removal of ices) in the warm inner zone of the CPD (*Mousis and Gautier,* 2004; *Canup and Ward,* 2002, 2006).

The problem of course is that the above scenario is not obviously applicable to the Saturn satellite system. Titan is the sole major satellite of Galilean satellite scale, and the inner mid-sized moons (which include Enceladus) do not show a simple monotonic density (or ice/rock) trend: the densities of Mimas, Enceladus, Tethys, Dione, and Rhea are 1.15, 1.61, 0.985, 1.48, and 1.24 g cm^{-3}, respectively (*Thomas,* 2010). Moreover, substantial water ice is found across the Saturn system, from the rings to Iapetus. Possibly, the conditions that led to the formation of a CPD were sufficiently different at Saturn from those at Jupiter, such as described next.

Figure 2a shows how a gap opening depends on the mass of the embedded giant planet, from *Fung and Chiang* (2016). The difference between a 1 M_J (1 Jupiter mass) and 0.5 M_J planet is striking, more than an order of magnitude difference in surface density in the gap well. Saturn, at 0.3 M_J, is even less massive, so the gap opening should be even less

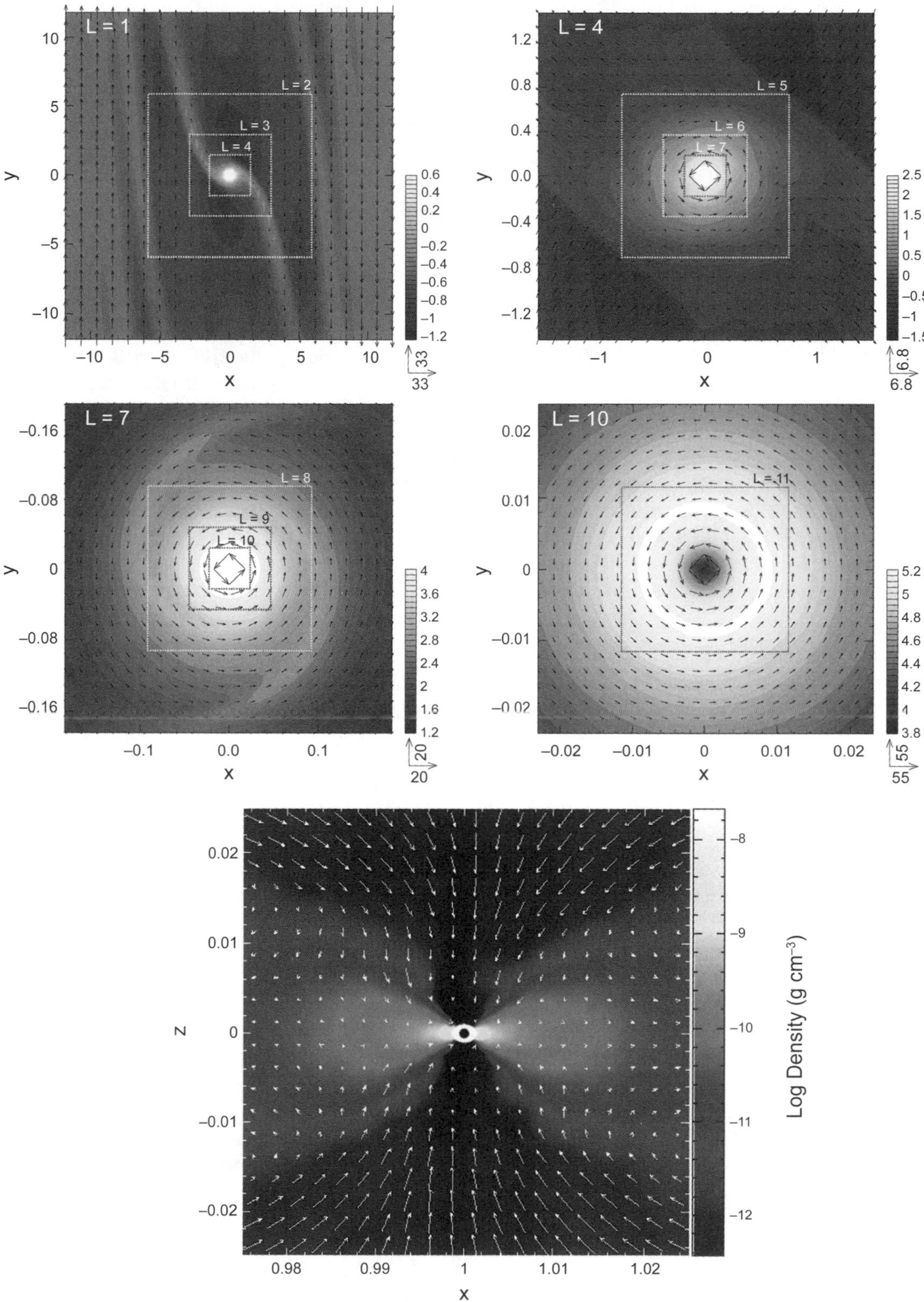

Fig. 1. See Plate 7 for color version. *Top:* Velocity (arrows) and log density (color) (both non-dimensional) of the flow around an accreting "Jupiter" in the protosolar nebular midplane, and at nested spatial scales (each factor of L is a power of 2); x and y are distances from the planet in units of the planet's Hill radius. From three-dimensional hydrodynamic calculations of *Tanigawa et al.* (2012). *Bottom:* Similar calculation for an accreting "Jupiter" from *Ayliffe and Bate* (2009), but for a vertical slice, showing rain of accreting gas and entrained solids onto surface of a circumplanetary accretion disk (CPD); x is normalized distance from the Sun, z is vertical to the mid-plane.

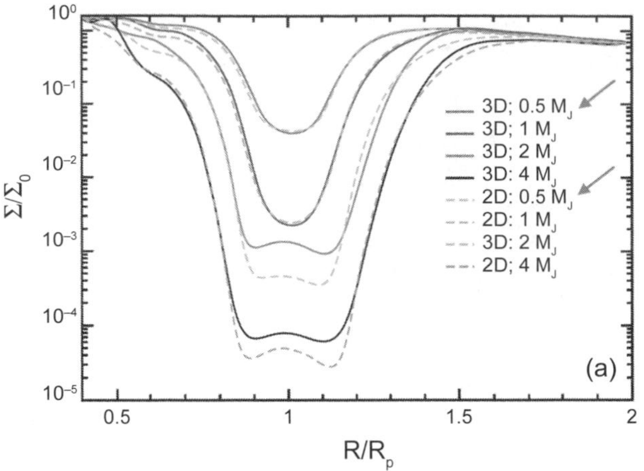

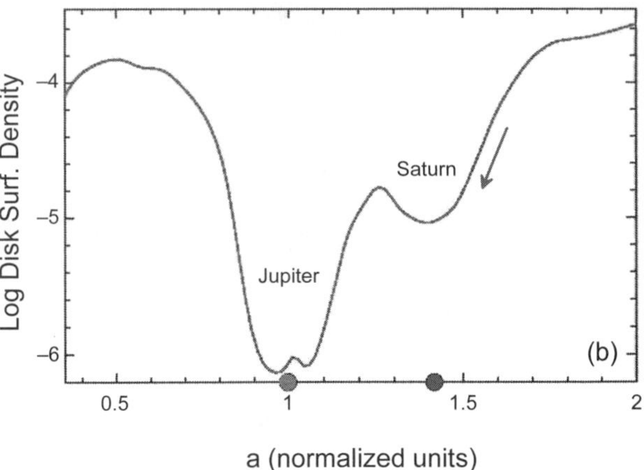

Fig. 2. See Plate 8 for color version. **(a)** Protosolar nebular surface density profiles, as a function of normalized radius from the Sun, averaged azimuthally except for a small area around a giant planet and averaged in time after steady-state conditions are achieved (after 1000 orbits); modified from *Fung and Chiang* (2016). **(b)** Nebular surface density with an embedded Jupiter and Saturn near but outside their 3:2 mean-motion resonance (A. Morbidelli, personal communication, 2016, and see text). Local density peak near Jupiter indicates formation of a circumplanetary accretion disk whereas the lack of a similar peak near Saturn signifies a more distended, circumplanetary envelope.

deep and less well defined. It is tempting to imagine that the pressure-temperature conditions in Saturn's CPD (or better, circumplanetary envelope) would have been much closer to ambient conditions in the nearby protosolar nebula, which itself could have been quite cold. This may have allowed incoming icy solids to remain so throughout the circumsaturnian disk or envelope (with some highly volatile exceptions; see below). On the other hand, the lack of a well-defined gap (*Sasaki et al.,* 2010) likely meant different mass accretion rates onto the jovian and saturnian CPDs (higher for Saturn) or at least different time histories (*Heller and Pudritz,* 2015), so Saturn's CPD could in contrast have initially been hotter than Jupiter's, as in the radiative, three-dimensional hydrodynamic envelope simulations of *Szulágyi et al.* (2016). Such a hot envelope would ultimately have to cool toward the end of its lifetime, but no specific numerical studies have been directed at the unique circumstances of Saturn's CPD (to our knowledge), so its evolutionary pressure and temperature track is unknown.

The calculations illustrated in Fig. 2a are for an isolated giant planet embedded in a nebular disk appropriate to a solar-mass star. This may be directly relevant to Saturn's situation in that Saturn's core could have formed at some exterior distance from Jupiter and later drifted inward toward Jupiter via a combination of density-wave-torque-driven type I and type II migration (*D'Angelo et al.,* 2003a, 2005; *Crida and Morbidelli,* 2007; *Crida,* 2009; cf. *Raymond and Izidoro,* 2017). Saturn may also have formed much closer to Jupiter, possibly in mean-motion resonance as in the Grand Tack model (*Walsh et al.,* 2011). In this case Jupiter exerts a strong influence on the properties of the Saturn gap. Figure 2b is based on the calculations of *Morbidelli and Crida* (2007) and illustrates not only the shallowness of the gap in such a configuration but its asymmetry. Accretion of gas and entrained solids by viscous evolution onto Saturn is plausibly biased toward (or preferred from) the exterior region of the protosolar nebula illustrated. If there were a compositional gradient across the protosolar nebula in the Jupiter-Saturn region (discussed in section 3), this could show up in the compositions of Enceladus and Titan vs. Ganymede and Callisto.

2.2. Saturn Satellite Accretion Models

The complexity of the Saturn system, with its major satellite, Titan, plethora of mid-sized satellites, and extensive icy rings, has prompted a broad variety of satellite origin scenarios, far more numerous and more elaborate than those proposed for the Galilean satellites of Jupiter. Here we summarize and comment on these specific models: primordial, ring, and catastrophic.

2.2.1. Minimum mass subnebula (MMSN). Although largely of historical interest now, such models postulated a quasi-steady-state circumplanetary nebula in which satellite ice + rock are augmented with H + He to bring the total to solar composition, such as in the classic model for the Jupiter system by *Lunine and Stevenson* (1982). This subnebula then cooled, leading to condensation of solids and ultimately to the accretion of satellites (*Pollack and Consolmagno,* 1984). It was never clearly demonstrated how this essentially static (inviscid) subnebula was supposed to be created, but that was not really the point, which was, rather, to examine for the first time and in analogy with then current protosolar nebula models a plausible satellite formation scenario. Possibly the gravest problem MMSN models faced was that accreting satellitesimals would be continuously lost to the central planet by gas drag, as pressure support in the subnebula guaranteed sub-Keplerian rotation so that all satellitesimals faced stiff "headwinds." The relative timescales for satellite accretion vs. loss to gas drag were not favorable (*Stevenson et al.,* 1985).

2.2.2. Solids enhanced minimum mass (SEMM) subnebula. One potential solution to the gas-drag crisis is to substantially decrease the ratio of subnebular gas to solids. This decreases the potency of gas drag and opens up the possibility that the largest satellites (e.g., Titan) can open gaps in the subnebula, which means their orbital migration will be type II and thus controlled by the overall viscous evolution of the CPD (*Mosqueira and Estrada,* 2003a,b). Naturally, as long as the subnebula is inviscid (or possesses a very low Shakura-Sunayev viscosity parameter $\alpha \lesssim 10^{-6}$), then satellites such as Titan can outlast the gas phase. And there is no shortage of potential extra solid mass for the CPD. Once Jupiter or Saturn reached the critical, high-mass phase in their evolution of continuous protosolar nebular gas accretion, any planetesimals in nearby solar orbit would have been strongly scattered into crossing orbits (*Raymond and Izidoro,* 2017; *Ronnet et al.,* 2018). Such planetesimals passing through Saturn's CPD or envelope could have been ablated or even captured directly if small enough (*Mosqueira et al.,* 2010; *Ronnet et al.,* 2018). In addition, the protosolar nebula itself may have been enhanced in solids (and Z-elements generally), due to photoevaporative loss of H + He at the exposed nebular surface (*Guillot and Hueso,* 2006).

The SEMM model as described also presumes a quasi-static (low α) circumsaturnian gas disk that is ultimately lost by photoevaporation (along with the entire protosolar nebula), as opposed to viscous spreading (*Mosqueira and Estrada,* 2003a,b). But it is hard to justify such a low-viscosity subnebula given the continuing although declining infall of new gas across the extent of the subnebula (as in Fig. 1, bottom), which would have acted as a source of turbulence. We emphasize that although the shallower gap at Saturn favors the formation of a slowly rotating circumplanetary envelope, the ultimate formation of a true CPD at Saturn would still be favored as long as the gas temperature became low enough (*Szulágyi et al.,* 2016). The prevalence of ice in the Saturn system from the rings outward is clear evidence that by the end stages of the protosolar nebula, the required low temperatures were realized (that is, if the rings are ancient; alternatives are discussed below). Moreover, vertical shear instabilities were likely operative in the saturnian CPD; all that is really required is a radial temperature gradient (*Nelson et al.,* 2013; *Stoll and Kley,* 2014; *Umurhan et al.,* 2016), which plausibly provided a measure of turbulent diffusion ($\alpha \sim 10^{-4}$ or greater). Hence, an inviscid SEMM subnebula appears unlikely for Saturn, but further inquiries into sources of turbulence, potential "dead zones," etc., remain warranted for both the jovian and saturnian subnebulae.

2.2.3. Gas-starved accretion disk. Detailed consideration of inflow of gas and entrained particles across the gap opened up by a giant planet naturally leads to a model in which a CPD is continually fed from the protosolar nebula, although at a diminishing rate through time, forming an accretion disk; this accretion disk evolves viscously, spreading both outward and inward (*Canup and Ward,* 2002, 2006; *Ward and Canup,* 2010; *Sasaki et al.,* 2010; *Ogihara and Ida,* 2012). At its inner edge the disk is accreted by the planet (e.g., Saturn), but throughout its evolution entrained solids accrete into satellitesimals — and satellites — that decouple from the surrounding gas. These larger solid bodies migrate radially by gas drag or type I or even type II migration, but more slowly than in the MMSN model, as the gas density at any given time is much less than in the MMSN model and can even be less than in the SSEM model. Satellites continue to accrete even as large amounts of nebular gas are processed and ultimately lost to the central planet. In this sense the subnebula at any moment is "gas-starved" and the ratio of total, accumulated solids to gas is much higher than the solar ratio.

Sources of solids include those described in the SSEM model above [e.g., planetesimals (*Raymond and Izidoro,* 2017; *Ronnet et al.,* 2018)], but critically, at this end stage of the protosolar nebula, there should also be plenty of residual, unaccreted small-scale solids, i.e., grains and "pebbles" (e.g., *Johansen et al.,* 2014, 2015; *Ronnet et al.,* 2017). Their motion is coupled to the gas, and the entrained solids will continue to flow onto or into the saturnian CPD as long as sufficient protosolar nebular gas exists. We note that considerable complexity may be obtained for heliocentric planetesimal deposition into the CPD, depending on planetesimal size [see *Tanigawa et al.* (2014) for a two-dimensional simulation and *Ronnet et al.* (2018) for a three-dimensional one].

Numerical (N-body) simulations show that the Canup-Ward model can result both in Galilean-satellite-like systems and in Saturn-like systems dominated by a single, large external satellite (a Titan) along with numerous interior and exterior mid-sized moons (Fig. 3), depending on model parameters such as α and the gas/solid mass ratio of infalling material (*Canup and Ward,* 2006). A characteristic of these models is that most infalling material is first intercepted at relatively large distances from the planet, so moons are born and then migrate inward in the CPD as they grow in mass: Most are lost, and only the final suite of satellites is left standing as the protosolar nebula dissipates. Inner mid-sized moons such as depicted in Fig. 3 are thus predicted to be both rock and ice rich (as Titan is). This is a problem for the gas-starved model in that Saturn's mid-sized moons are much icier than Titan. In particular, the rings and innermost moons (Mimas, Enceladus, and Tethys) are together extremely icy, the next two out (Dione and Rhea) are less icy, and Titan is the most rock-rich of all. This is the true broad-scale compositional gradient in the Saturn satellite system, opposite to that at Jupiter, and must be explained. We recognize Enceladus is quite rocky compared with Mimas and Tethys, but Tethys is extremely icy and ~6× as massive as Enceladus [Table 6 in *Hussmann et al.* (2010)], and thus defines the broad-scale trend. We also note, for completeness, that the SEMM model, with its concentration of mass in the inner nebula, implies several large and potentially very rock-rich moons in these orbits, i.e., a Jupiter-like system, not Saturn-like.

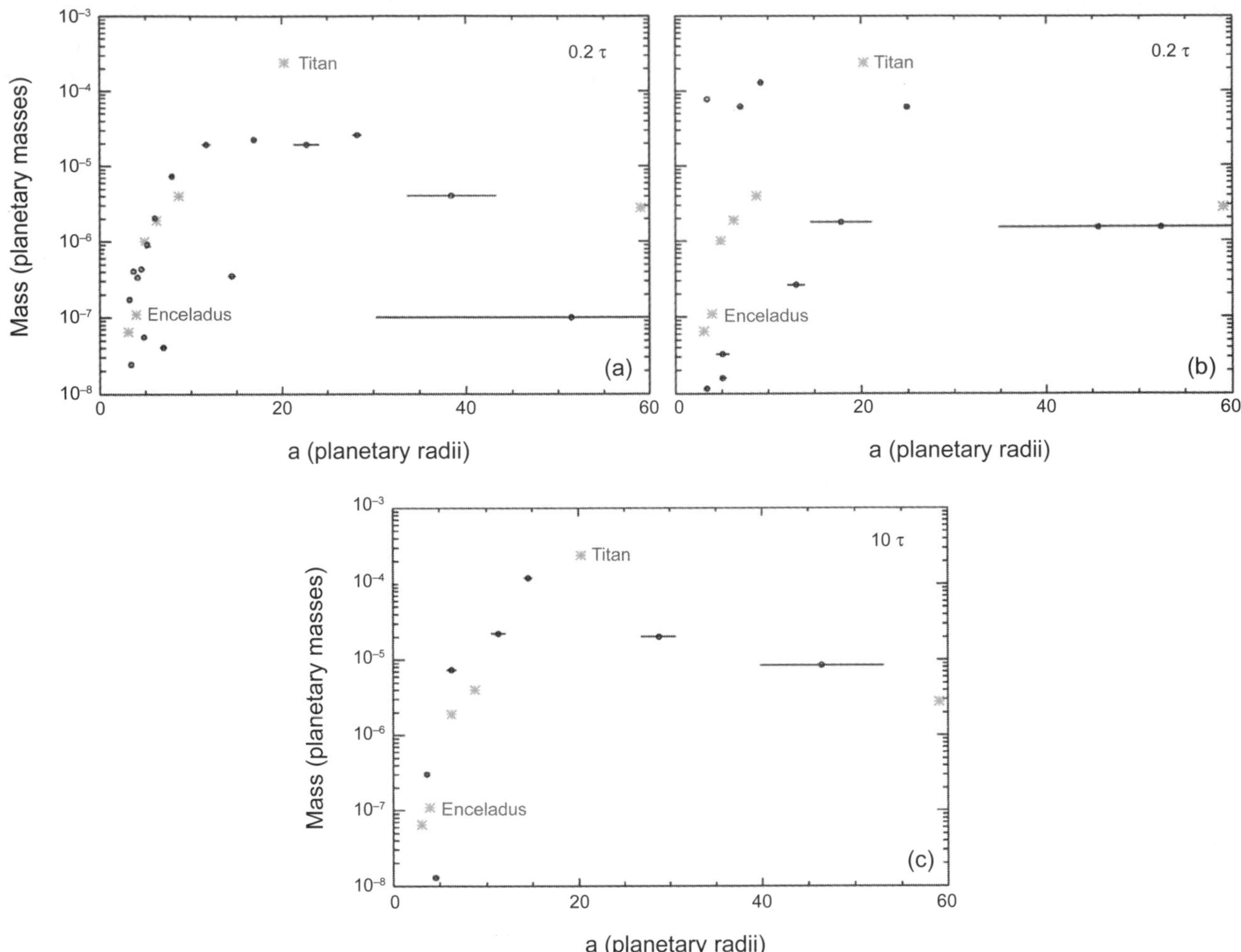

Fig. 3. Three time steps from a satellite accretion simulation that produced a Saturn-like system of satellites ($\alpha = 6 \times 10^{-3}$, gas/solid mass ratio = 100). The inflow to the gas-starved CPD is assumed to decrease exponentially with a time constant τ. Black circles show the simulated satellites (horizontal lines are proportional to orbital eccentricities) and Saturn's satellites are shown as green stars. **(a)** Multiple satellites form and begin to migrate inward as solid material flows into the disk. **(b)** The system resembles Jupiter's Galilean system, with four similarly sized large satellites, the inner three ultimately lost to collision with Saturn. **(c)** The final system has a single large Titan-like satellite orbiting at 15 planetary radii and several inner mid-sized moons. Modified from *Canup* (2010).

2.2.4. Birth from a massive ring. *Canup* (2010) took the gas-starved, satellite accretion model to the next level. Given that Titan-sized satellites are lost to Saturn during the simulations, single-particle hydrodynamic (SPH) calculations showed that as a Titan-like satellite migrates inward to the Roche limit, tides may strip away the icy mantle, that is, *if* the satellite had previously differentiated into a rock core and icy mantle (Fig. 4a). Being dense, the rocky core resists tidal disruption and remains more or less intact but is ultimately lost to a hot, bloated proto-Saturn due to tides (the rocky core being well inside synchronous altitude at this time), whereas the massive icy ring formed from the disrupted mantle spreads both outward and inward. As the ring evolves, its mass decreases and icy moons are spawned from its outer edge by the physical process described in *Charnoz et al.* (2010). The total ejected mass was estimated by *Canup* (2010) to be consistent with the sum of Saturn's ice-rich mid-sized moons.

Charnoz et al. (2011) then carried this idea to its logical conclusion, that *all* the inner mid-sized moons (i.e., out to and including Rhea) were spawned in this manner (see also the chapter in this volume by Castillo-Rogez et al.). The progressive increase in mid-sized satellite mass with distance from Saturn (*Thomas,* 2010) is a characteristic signature of such a ring origin (*Crida and Charnoz,* 2012) (Fig. 4b), because the proto-ring progressively decreases in mass as each satellite forms and as the ring spreads onto the planet. However, such a mid-sized satellite mass trend with orbital radius can also be seen in the models of *Canup and Ward* (2006) and in Fig. 3. The stochastic variation in rock content among the moons, specifically Enceladus' rockiness, which is not directly predicted by this ring-origin scenario, was

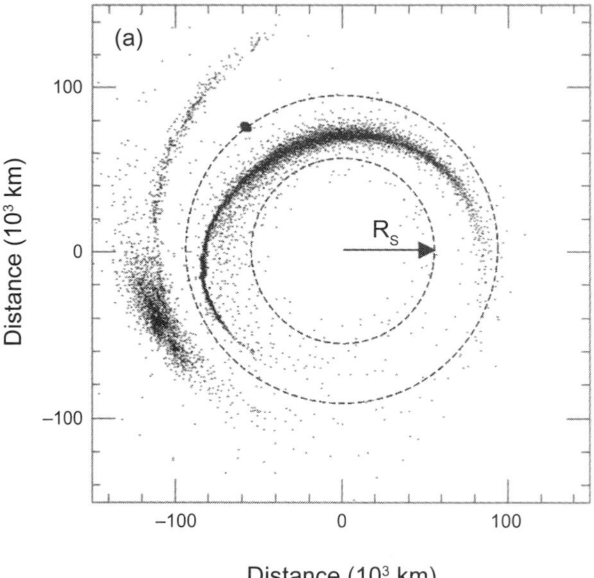

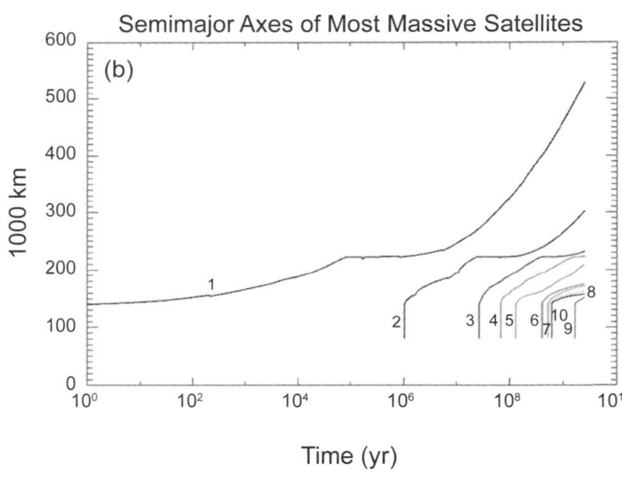

Fig. 4. (a) Snapshot of an SPH simulation illustrating the tidal stripping of icy fragments from a differentiated Titan-like body just inside proto-Saturn's Roche limit (outer dashed line). Mutual collisions between the fragments should create a massive icy ring, although it is only the ring exterior to the tidally decaying rocky core that survives to be the ultimate mass reservoir for mid-sized satellite formation. R_s is Saturn's radius at the end of gas accretion, much larger than today's 30,150 km. From *Canup* (2010). **(b)** Example orbital evolution of satellites accreted at or near Saturn's Roche limit from a ring whose initial mass is four times that of Rhea. Satellites are numbered in terms of decreasing mass (Enceladus would correspond to 4), and evolve rapidly away due to a low and frequency independent Saturn tidal Q = 1680; resonant interactions between the satellites are ignored for simplicity. T = 0 is arbitrary. Adapted from *Charnoz et al.* (2011).

attributed by *Charnoz et al.* (2011) to early tidal ejection of large rock-rich fragments from the massive proto-ring. These fragments were hypothesized by them to be a consequence of the incomplete differentiation of the Titan-like precursor; partial tidal disassembly of a Titan-like core is an alternative.

Subsequently, *Salmon and Canup* (2017) showed that sufficient rock could have been added to any originally all-ice innermost mid-sized satellites (Mimas through Tethys) by a heliocentric heavy bombardment, but they ascribed Enceladus' relative rockiness to stochastic variation of this input; the almost pure ice composition of Saturn's rings is explained by them as due to the low heliocentric capture efficiency of the rings (their low surface density).

The detailed N-body simulations in *Salmon and Canup* (2017), which include resonant interactions between the innermost satellites and rings *and* between the satellites themselves (and Dione and Rhea), confirm that a ring origin for Enceladus (and Mimas and Tethys) is dynamically plausible. The important consequence of a ring origin for Enceladus is that the moon's age (its proximate formation age) *is less than the age of Saturn. Charnoz et al.* (2011) state that a billion years could have elapsed between the formation of Rhea and bodies like Mimas (e.g., Fig. 4b). If we place the formation of the original proto-ring as coincident with Saturn [as in *Canup* (2010)], Enceladus might have fully formed only *after* the time of the putative late heavy bombardment (about 3.8 G.y. ago).

If Saturn's rings are not a product of the planet's accretion, but stem from the tidal stripping during a close pass by a large but differentiated body (possibly from the Kuiper belt, i.e., a Centaur), as originally proposed by *Dones* (1991), then Enceladus and other mid-sized icy moons could have formed at almost any time in Saturn's history. Tidal stripping of such a heliocentric body is an intrinsically low probability event, of course (discussed further below), and would only make sense probabilistically during the sort of heavy bombardment that would have accompanied giant planet migration. While giant planet migration has been linked to, and proposed as an explanation for, the late heavy bombardment of the terrestrial planets (*Gomes et al.,* 2005), such migration may have occurred much earlier in solar system history, within 100 m.y. of solar nebula dispersal (*Nesvorný and Morbidelli,* 2012; *Nesvorný and Vokrouhlický,* 2016). SPH modeling by *Hyodo et al.* (2017) of such tidal encounters shows that tidal stripping under favorable circumstances could result in mass capture efficiencies between 0.01 and 0.1. Losses of captured debris to collisions with Titan and Saturn were considered by them, but solar tides and Kozai-Lidov oscillations (*Benner and McKinnon,* 1995) were apparently not, which could dramatically increase collisional loses to Saturn. Even without such loss considerations, *Hyodo et al.* (2017) find that to ultimately yield the inner mid-sized satellites of Saturn requires the fortuitous encounter with a body much more massive than Pluto.

2.2.5. Catastrophic reaccretion. An even more revisionist view takes its cue from a putative young age for the rings. It has long been known from groundbased radio and radar studies that Saturn's rings are nearly pure water ice (e.g., *Cuzzi et al.,* 2010). Water ice is essentially transparent at radar wavelengths, and Cassini 2.2-cm radiometry has detected only a fraction of percent of non-icy material within the rings [see *Cuzzi* (2018) for a summary]. Given measurements of the incoming non-icy dust flux from the Cassini Dust Analyzer and the total mass of ring from Doppler radio tracking (L. Iess, personal communication, 2017), the "exposure age" of the rings has been estimated to be on the order of 100 to 200 m.y., much shorter than the age of the solar system (*Cuzzi,* 2018). This conclusion should be viewed with caution, however. The non-icy dust flux into the rings could be enhanced at present for some reason. Moreover, whatever the mass flux, the resulting ring impacts are extremely high speed [>30 km s^{-1} (*Zahnle et al.,* 2003)]. Such impacts should vaporize both the impacting particles and substantial ring ice besides. Ionization during the impact or after would then lead to magnetospheric acceleration and loss from the Saturn system.

It is nonetheless useful to consider the implications of such a young ring for the origin of Enceladus and the other inner mid-sized moons. The *recent* tidal disruption of a large, passing comet or transneptunian body to form the rings is not only extremely unlikely probabistically, but it is physically difficult to realize given Saturn's low density and its narrow Roche zone [see supplementary discussion in *Canup* (2010)]. A more physically likely possibility is the impact disruption of an icy Mimas-like satellite within Saturn's Roche zone (*Harris,* 1984), which is improbable but not physically impossible (*Canup,* 2010). Such a catastrophic ring origin would not necessarily impact the above interpretations of Enceladus' origin, because it would only involve the ring and its associated small ring-moons. However, a catastrophic and recent (~100 m.y. ago) disruption and reaccretion of all the inner mid-sized moons has been proposed by *Ćuk et al.* (2016). Their scenario relies on strong tidal evolution of all the moons due to a low tidal Q for Saturn, and could in principle explain Saturn's rings as the outcome of a collision between two icy moons (cf. *Asphaug and Reufer,* 2013). Whether this is really the case is debatable (one might expect evidence for a recent bombardment spike on Titan, for example). The Q relevant to Enceladus' diurnal tidal period may have monotonically declined over geological time (*Fuller et al.,* 2016), effectively keeping Enceladus and the other mid-sized moons safe from catastrophic collision with a neighboring satellite. See the chapter in this volume by Nimmo et al. for a more detailed discussion of tidal evolution scenarios.

2.2.6. Gas-free planetesimal capture. For completeness we note the possibly of post-nebula, and therefore gas-free, accretion of heliocentric planetesimal debris (*Estrada and Mosqueira,* 2006). This Safronov-style model, based on heliocentric planetesimals colliding within Saturn's Hill sphere and some fraction of the debris being permanently bound, suffers from the usual problems associated with such capture physics (*Peale and Canup,* 2015). Foremost is the nearly zero total angular momentum of the expected ensemble as prograde and retrograde captures average out. Nor is there any obvious or simple explanation for the compositional gradient/variations in the Saturn satellite and ring system.

2.2.7. Summary. The variety of models and scenarios proposed to explain the origin of Saturn's satellites and rings is broad if daunting. The best-developed models, and ones that are most broadly applicable to the other giant planet satellite systems, are the gas-starved accretion disk model and its extension to satellite origin at the outer edge of a spreading, massive ring.

3. BUILDING BLOCKS: SOLID COMPOSITIONS AT SATURN

3.1. General Inferences for Enceladus' Composition

The above primordial accretion scenarios can be organized according to their gas/dust ratios and, correspondingly, to their implied degrees of thermochemical processing (Fig. 5, left). The gas/rock ratio in the *Lunine and Stevenson* (1982) MMSN model is ~100 (with the total gas/dust ratio smaller due to ice condensation). Thermochemical equilibration in the cooling subnebula predicts rock mineralogies akin to those in *Prinn and Fegley* (1981). For SEMM models (*Mosqueira and Estrada,* 2003a,b) with their enhanced dust content or any models with markedly lower gas surface densities (e.g., *Ward and Canup,* 2010), the likelihood of more limited thermochemical equilibration due to kinetic barriers or sequestering in satellitesimals goes up. In the Canup-Ward gas-starved accretion disk model, even though the inflowing solar material may or may not have a solar gas/dust ratio, at any given time the bulk gas/(rock + ice) ratio in the CPD should be much less than 100 (*Canup and Ward,* 2002, 2006). And naturally, capture of planetesimals after nebular dissipation (section 2.2.6) does not involve any gas, and the solids that do manage to be captured may be essentially unprocessed protosolar nebular and interstellar solids (although we cannot rule an intermediate stage of aqueous alteration in heliocentric planetesimal parent bodies; see section 3.2).

The Canup-Ward model for the formation of giant planet satellites (nominally including Enceladus) is a leading model at present in the literature (and is certainly so for massive satellites such as Io, Europa, Ganymede, Callisto, and Titan). It is detailed and has been calculated from a number of perspectives (e.g., *Canup and Ward,* 2002, 2006; *Sasaki et al.,* 2010; *Ogihara and Ida,* 2012). For Saturn, plausible inflow and CPD parameters can be found that result in relatively cool temperatures and water-ice stability (an obvious prerequisite). How cool the saturnian circumplanetary disk got and whether even more volatile ices (e.g., NH_3 and CH_4) were stable cannot be decided from a theoretical standpoint alone. Data from Enceladus and Titan (discussed in section 4) provide critical guidance. But the reader is advised that much is unsettled and much is insecure. Many aspects of

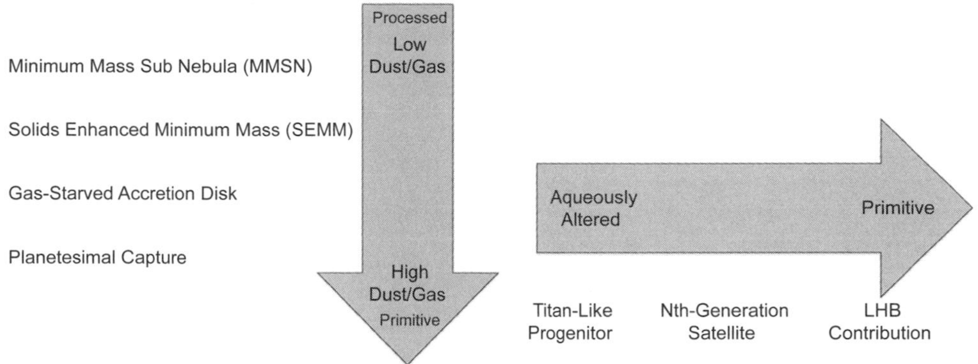

Fig. 5. *Left:* Schematic of proposed primordial, nebular origin scenarios for the satellites of Saturn in general, organized by implied gas/dust mass ratio. The minimum mass sub nebula (MMSN), as the minimum mass protosolar nebula models it mimics, could have a dust/gas ratio as low as ~1/65 (if all but H_2, He, and Ne condense). Planetesimal capture subsequent to dispersal of the protosolar nebula naturally is all "dust." The implication for Enceladus' composition is a corresponding variation in thermochemical processing: condensation of solids from a cooling CPD in the first case to accretion of unprocessed (primitive) planetesimal fragments in the latter. *Right:* Enceladus itself may not be primordial satellite, but birthed from a massive ring derived from a differentiated Titan-like body (itself primordial); it may also be a moon reborn from one or more catastrophic collisions with other mid-sized satellites or heliocentric impactors. The latter could have contributed to Encleadus' rock component during heavy bombardment. The degree of processing of rock, organics, and icy volatiles in these progenitor bodies varies widely.

the Canup-Ward formalism remain to be examined in greater numerical detail (inflow rates, temperatures, and planetesimal deposition having already been mentioned), and the growing literature on extrasolar giant planet formation (and their potential exomoons) should prove helpful.

A ring origin for Enceladus and at least Mimas and Tethys (and the small ring-moons of course) has received the most attention in recent years (*Canup,* 2010; *Charnoz et al.,* 2010, 2011; *Crida and Charnoz,* 2012; *Salmon and Canup,* 2017). The most plausible, or at least self-consistent, version links the origin of the inner mid-sized satellites to the creation of a primordial massive ring (*Canup,* 2010). In this case the thermochemical considerations above for the saturnian CPD apply to the solids that accreted to form the Titan-like world that was subsequently tidally disrupted. But this intermediate step requires an episode of rock-from-ice differentiation [driven by accretional and tidal heating (*Canup,* 2010)], which carries with it the presumption of water-rock interaction, volatile ice separation, etc. (Fig. 5, right). Even if the massive ring is due to tidal disruption of a passing heliocentric body (*Dones et al.,* 1991; *Hyodo et al.,* 2017), that body would need to be differentiated as well, implying a similar if not protracted history of internal activity. Of course, in that case the composition and chemistry might better reflect that of the Kuiper belt (bodies accreted beyond the orbit of Neptune), and would ostensibly be distinct from that of Titan. Similarly, capture addition of heliocentric solids into the inner satellite region during early heavy bombardment (*Salmon and Canup,* 2017) would imply the addition of similar "exotic" materials, albeit not necessarily aqueously processed ones.

One issue stands out in particular: the rock-rich nature of Enceladus vs. the iciness of the moons immediately interior (Mimas) and exterior (Tethys). If all these moons were born from a massive ring (*Canup,* 2010; *Charnoz et al.,* 2011; *Salmon and Canup,* 2017), a critical question is, when? Enceladus could be younger than Saturn by at least 1 b.y. in the *Charnoz et al.* (2011) analysis. And any ring origin for at least the inner trio of mid-sized moons must account for their necessarily extensive orbital evolution to their present positions, which includes crossing of important mean-motion resonances (*Peale and Canup,* 2015; *Salmon and Canup,* 2017). As alluded to above, it is not yet clear that this is a physically feasible scenario (see the discussion of Saturn's tidal Q in the chapter in this volume by Nimmo et al.). *Salmon and Canup* (2017) have shown that the tidal evolution of the inner triplet, if eccentricity damping by tides is included, can be regular and stable (i.e., no intersatellite collisions) over ~10^9 yr and lead to mid-sized satellite systems that broadly resemble Saturn's. But many numerical outcomes leave the mid-sized satellites in stable, low-order mean-motion resonances that they are not presently in. Full numerical simulations that posit all the mid-sized satellites out to Rhea are ringspawned, driven by strong tidal interaction with Saturn, have not been attempted. Dynamical instabilities during such evolution could lead to mid-sized satellite collisions and resorting of ice/rock ratios — if the satellites were differentiated prior, which would also imply a "prehistory" of water-rock interaction for Enceladus (Fig. 5, right).

We end this subsection with a final caveat. Much effort, as described above, has gone into explaining the ice-rich nature of Saturn's rings and inner mid-sized satellites, with Enceladus standing out as a rock-rich "sore thumb." But there is another mid-sized moon, Iapetus, in a distant orbit beyond Titan that is also quite icy (density = 1.09 g cm^{-3}). Whether the satellites of Saturn were built from the outside in (*Canup and Ward,* 2006), the inside out (*Crida and Charnoz,*

2012), or a combination of the two (*Salmon and Canup,* 2017), Iapetus is an anomaly (*Mosqueira et al.,* 2010), and a reminder that we do not have the full picture.

3.2. "Rock"

3.2.1. Meteorite analogs. Carbonaceous chondrites (CCs) represent the most primitive class of chondritic materials, so it is often assumed that they are most representative of the rocky materials that were incorporated into the jovian and saturnian satellites. CI carbonaceous chondrites (CCs), which contain the full complement of elements at solar abundance levels (except H, C, O, N, and the noble gases) (*Lodders,* 2003), is the class often associated with outer planet satellite formation (e.g., *Shock and McKinnon,* 1993; *Zolotov,* 2012). CI chondrites are also the most aqueously altered chondrites. It is not clear, however, if icy moons accreted CC-type material or their anhydrous precursors (or perhaps both). *Kargel et al.* (2000), for example, suggested that dehydration of CC building blocks that ostensibly accreted to form Europa supplied the water and ice that we see in its outer layers today. Alternatively or additionally, Europa could have accreted anhydrous CC-like materials together with water ice (*McKinnon and Zolensky,* 2003; *Zolotov and Kargel,* 2009).

Compared with Jupiter and main-belt asteroids, Saturn and its satellites could have accreted much more anhydrous material that was not affected by aqueous processes in heliocentric planetesimals: (1) A later formation of Saturn suggests accretion of planetesimals with lesser amounts of heat-producing ^{26}Al; (2) Saturn-forming planetesimals are also plausibly more (ice + organic)/rock rich, which would tend to disfavor their aqueous alteration (less bulk heating); less hydration is observed in D and P asteroids, which become abundant in the peripheral parts of the main asteroid belt (*DeMeo and Carry,* 2014); and (3) anhydrous interplanetary dust particles (IDPs) could be cometary solids (*Bradley,* 2014). Returned cometary dust (Stardust samples) and *in situ* studies of comets (e.g., *Davidsson et al.,* 2016) do not indicate aqueous alteration within cometary bodies. Apparently, late and slow formation of comets (*Davidsson et al.,* 2016) was responsible for their little or no thermal processing after accretion. We note that the late formation of comets also accounts for the incorporation of earlier processed chondritic materials, including some aqueously formed phases (e.g., Fe-rich olivine, magnetite), through mixing within the solar nebula (*Brownlee,* 2014). Thus, *these* aqueously processed grains could have contributed to the Enceladus' bulk composition or that of a progenitor satellite.

Although previous researchers have assumed that CCs are likely representative of rocky formation materials at Jupiter and Saturn, we have no definitive measurements of materials from these satellites to confirm this assumption. This lack of direct information about the starting materials of the icy satellites in general, and Enceladus in particular, leads us below to identify key isotopic measurements for understanding the origin of Enceladus and other icy satellites (section 6).

3.2.2. The jovian gap. The last few years have seen something of a cosmochemical revolution, one that has profound implications for the formation of Jupiter's and Saturn's satellites. Whole-rock Δ^{17}O (which defines the degree of mass independent fractionation in oxygen isotopes with respect to terrestrial values) and nucleosynthetic isotopic variations for chromium, titanium, nickel, and molybdenum in meteorites define two isotopically distinct populations: CCs and some achondrites, pallasites, and irons in one population and all other chondrites and differentiated meteorites (including irons) and Mars and Earth in the other (non-CCs) (*Warren,* 2011; *Kruijer et al.,* 2017; *Scott et al.,* 2018; *Yin et al.,* 2018). These isotopic distinctions are not simply the result of temporal evolution in the solar nebula, but define two spatially distinct arenas of early, iron meteorite parent body accretion and melting followed by distinct suites of chondrite parent body accretion, CC and non-CC. From the isotopic dating of these various events, *Kruijer et al.* (2017) offer that the logical cause of the partitioning of the solar nebula into two distinct reservoirs, and all within 1 m.y. of the condensation of the first calcium-aluminum-rich inclusions (CAIs), was the formation of proto-Jupiter, blocking radial transport of nebular solids, and culminating in the opening of its tidal gap (as in Figs. 1 and 2).

In this picture, CC parent bodies originally accreted beyond Jupiter. As Jupiter's "core" grew, its gravity increasingly perturbed the orbits of any remaining planetesimals inside and outside its orbit and the same would have occurred for proto-Saturn (*Raymond and Izidoro,* 2017; *Ronnet et al.,* 2018). Planetesimals would have been scattered on increasingly wide orbits as both proto-giant planets grew and the protosolar nebula density declined (which reduces gas drag damping); if the two giant planets migrated, as in the Grand Tack scenario (*Walsh et al.,* 2011), scattering would have markedly increased as well (*Raymond and Izidoro,* 2017). Carbonaceous chondrite planetesimals should have been scattered from Jupiter's accretion zone into Saturn's CPD and from Saturn's into Jupiter's CPD (*Ronnet et al.,* 2018). This would have been in addition to the local inflows of nebular gas and solids (grains and pebbles) to their respective satellite feeding zones (*Peale and Canup,* 2015). One interesting implication is that the grains, pebbles, and planetesimals that did so need not *all* have been primitive CCs, but some could have been remnants of aqueously altered and differentiated CC-group parent bodies.

The jovian gap hypothesis of meteorite parent body genesis does not explicitly address the early environment at Saturn. A detailed elaboration by *Desch et al.* (2018), focused on the distribution of CAIs in meteorites, argues that CI chondrites are sufficiently enhanced in water and organics yet lack CAIs and chondrules and so likely accreted much farther from the Sun. They posit that CI parent bodies may have accreted beyond proto-Saturn's gap (i.e., Fig. 2b, right). *Scott et al.* (2018) argue that the very low abundance of CAIs and chondrules in CI chondrites may be more apparent than real, and mostly due to brecciation and alteration. Regardless, that CI-like materials may be characteristic of Saturn- and Saturn-satellite-forming solids is not unreasonable. What isn't known is whether any of the planetesimals that formed near Saturn (and contributed to the formation of that planet's core) accreted early enough so as to undergo strong thermal evolution, as apparently happened near Jupiter (*Kruijer et al.,* 2017).

3.3. Ices (Frozen Volatiles)

Our best representatives of the icy volatiles likely to be accreted into mid-sized moons such as Enceladus come from determinations of cometary compositions. This is an imperfect match, of course, as comets have not been thought to represent the Saturn formation region per se but either formed (in the classical view) mainly nearer to Uranus and Neptune before being ejected, becoming Oort cloud comets (e.g., *Fernández and Ip,* 1981; *Dones et al.,* 2004), or they formed even farther out, in the exterior planetesimal disk that ultimately sourced the Kuiper belt proper and scattered disk beyond, becoming potential Jupiter-family comets and Centaurs (e.g., *Duncan and Levison,* 1997; *Levison et al.,* 2008). Furthermore, the scattered disk, under the gravitational control of Neptune, supplies comets to the Oort cloud as well (*Fernández et al.,* 2004), and many original Oort cloud comets may have come from other stars in the Sun's birth cluster (*Levison et al.,* 2010) (although whether that translates into a significant difference in bulk chemistry is unclear).

Our pre-Rosetta understanding of cometary bulk volatiles is summarized in Fig. 6 (from *Mumma and Charnley,* 2011). The data mainly represent Oort cloud comets, and so may better characterize comets from the giant planet region [at least in classical models of the Oort cloud genesis (see *A'Hearn et al.* (2012) for an alternative view]. All the volatile abundance observations are for sublimated gases in cometary comae. We do not know at present if the compositions of the comae are similar or significantly fractionated from that of surface ices or the bulk cometary nucleus from which they emanate. Here, we make the conventional and justifiable assumption that the comae abundances are representative of accreted abundances of volatiles. Volatiles may be sublimated at rates proportional to their surface abundances if surface temperatures at perihelion are sufficiently high (everything evaporates).

Several aspects in Fig. 6 are notable: (1) Water ice is the dominant volatile, as befits oxygen's status as the third most abundant element in the Sun; (2) cometary volatiles are a chemically unequilibrated mélange, as opposed to a mixture of relatively simple compounds representing a consistent oxidation state and sulfur fugacity; (3) oxidized carbon as CO_2 *and* CO are both generally relatively abundant, at the several percent level compared with water, and sometimes much more so, reaching 20–30% (*A'Hearn et al.,* 2012); (4) reduced carbon and nitrogen, as CH_4 and NH_3, respectively, are important at the percent level; (5) methanol, CH_3OH, is variable but also present at the percent level; and (6) H_2S is the dominant but not sole sulfur-bearing ice, at the 0.1–1% level.

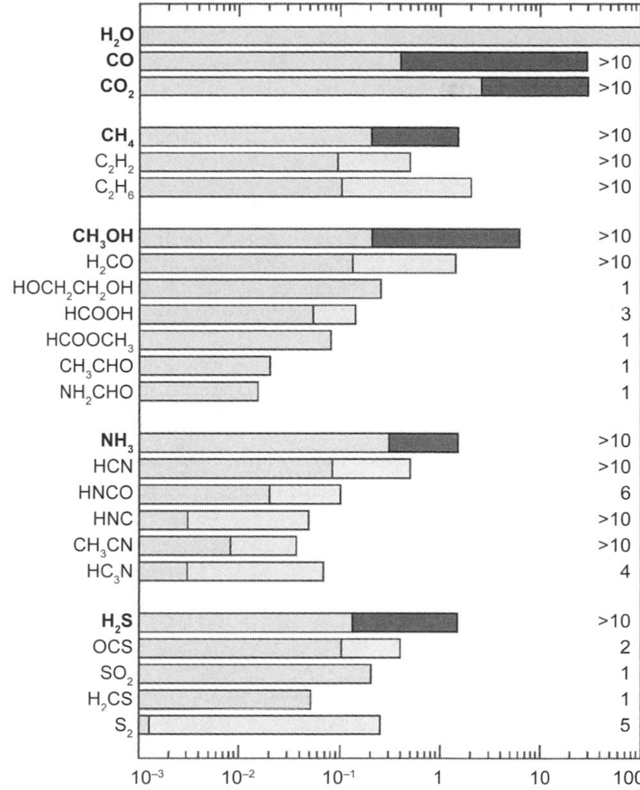

Fig. 6. Volatile abundances relative to water in comets, modified from *Mumma and Charnley* (2011). The range of measured values is shown by the rightmost portion of each bar, with darker highlighting for particularly abundant molecules. The number of comets for which data is available is given on the right. For CO, abundances refer to total CO [both nuclear and distributed (coma) sources]. The CO upper limit is updated from *A'Hearn et al.* (2012), who list numerous AKARI satellite detections (*Ootsubo et al.,* 2012).

Most of the information in Fig. 6 comes from groundbased studies, augmented by *in situ* mass spectrometer measurements at Comet 1P/Halley (by Giotto and Vega) and close-in infrared emission spectra of Comet 103P/Hartley 2 [by EPOXI (*A'Hearn et al.,* 2011)]. It predates the Rosetta mission to Comet 67P/Churyumov-Gerasimenko (hereafter 67P), but results from the latter mission are generally consistent for the major volatiles. The important differences appear to be that 67P is particularly CO_2 rich and is depleted in NH_3 compared with other comets (*Le Roy et al.,* 2015; *Bockelée-Morvan et al.,* 2016). At 67P a suite of S-bearing volatiles were detected representing the full range of oxidation states with $H_2S > S_x + SO_y$ (*Calmonte et al.,* 2016). This by itself is not shocking (see Fig. 6), but *Calmonte et al.* (2016) argue that the total (volatiles plus dust) S/O ratio in 67P is solar. If 67P can be taken as representative, then during Enceladus' accretion it could have drawn on solids hosting a full solar complement of S — an important biogenic element. The question of Enceladus' sulfur abundance is returned to in section 4.1.

Rosetta's ability to station-keep near 67P allowed detection and accurate measurements of many minor volatile species, obviously a major advance. Notable was determination of $N_2/CO = 5.70$ ($\pm$ 0.66) $\times 10^{-3}$ (2σ) in the coma by the ROSINA mass spectrometer (*Rubin et al.,* 2015). Nitrogen as N_2 has been notoriously hard to detect in comets astronomically, and this measurement confirms that it is very underabundant given that N_2 and CO ices have very similar volatilities. Whether 67P never had much condensed or trapped N_2 or lost it due to very modest thermal processing (*Rubin et al.,* 2015), N_2 ice is unlikely to have been a dominant form of nitrogen contributed to Enceladus. Phosphorus (another important biogenic element) was also detected in the coma gas, a first for a comet, although the parent molecule (possibly PH_3) could not be determined (*Altwegg et al.,* 2016).

3.4. Organic Matter

CI and CM CCs, proxies for the rocky accretionary material of Enceladus (as discussed above), contain both soluble organic material such as carboxylic and amino acids, nitrogen heterocycles, etc., and a dominant component of insoluble macromolecular organic compounds (>70% of the organic carbon) (e.g., *Sephton,* 2005; *Pizzarello et al.,* 2006; *Fegley and Schaefer,* 2009). Soluble organic compounds can also be released through aqueous and thermal activity from the insoluble component [in CC parent bodies if they were sufficiently heated (section 3.2.2), and later in an active satellite] (*Shock and Schulte,* 1990; *Sephton and Gilmour,* 2000; *Yabuta et al.,* 2007; *Alexander et al.,* 2017).

The icy component, if similar in composition to comets (as just discussed), would have contained a variety of hydrocarbons, nitriles, and amines (*Bocklée-Morvan et al.,* 2004; *Mumma and Charnley,* 2011; *Altwegg et al.,* 2016). And comets are notably rich in relatively non-volatile organic matter, often referred to as CHON or CHONSP to denote its elemental composition (in order of decreasing elemental abundance) (*Jessberger et al.,* 1988; *Fomenkova,* 1999; *Fray et al.,* 2016). Rosetta data for 67P show that at least some high molecular weight organic material is similar to chondritic insoluble organic matter (IOM) but less altered (*Fray et al.,* 2016). Chondritic IOM contains ammonia as well, at the 0.1–1% level by mass (*Pizzarello and Williams,* 2012). Similar macromolecular carbonaceous materials should have accreted into the saturnian CPD and thus, by one path or another, into Enceladus.

For Enceladus, due to radiogenic and tidal heating, the organic content of its core, ocean, and ice shell could be altered (and possibly augmented) by the thermal and aqueous processing of accreted organic material. Synthesis of hydrocarbons such as CH_4 from inorganic species (i.e., methanogenesis) may be thermodynamically favorable in Enceladus' ocean and core (*Waite et al.,* 2017). We note, however, that there is scant evidence for such organic synthesis in CCs, as opposed to transformation of existing organics (*Alexander et al.,* 2007), and that there is a lack of geochemical evidence for the active synthesis of CH_4 at hydrothermal vents on Earth as well (*McDermott et al.,* 2015). These last two observations are consistent with the kinetic inhibition of CH_4 synthesis over the timescales relevant to CC parent bodies and seafloor hydrothermal systems respectively (*McCollom,* 2016). But the interpretation of at least moderate temperature (>90°C) water-rock reactions in Enceladus' core (*Hsu et al.,* 2015) means that there may be conditions where abundances of organic compounds are under thermodynamic or kinetic control (e.g., *Shock et al.,* 2013).

4. OBSERVATIONS IN THE SATURN SYSTEM

4.1. Deuterium/Hydrogen and Plume Volatiles at Enceladus

The volatile composition at Enceladus is derived largely from the plume composition measured by the Cassini Ion Neutral Mass Spectrometer (INMS). The predominant source region for the plume gases and particles appears to be the moon's interior ocean (see the chapters in this volume by Glein et al. and Goldstein et al.). Table 1 summarizes our present knowledge of the volatile composition. The major plume species are reported in *Waite et al.* (2017) and listed in Table 1 as unambiguously detected. The primary molecule is water (H_2O) with an abundance exceeding 96% mixing ratio by volume (or number), which is consistent with an ocean source. Furthermore, based on what we infer from the plume composition, the major nitrogen-bearing volatile in the ocean is ammonia (NH_3), whereas the major carbon-bearing volatile is carbon dioxide (CO_2) followed closely by methane (CH_4).

Minor species of moderate to high ambiguity in the plume are reported in *Magee and Waite* (2017) and also listed in Table 1. The ambiguity is largely due to the limited mass resolution of the INMS instrument, which due to the complex gas mixture introduces ambiguity of compound identification through the mass deconvolution process (Magee et al., in preparation, 2018). The occurrences and concentrations of compounds with volume mixing ratios of 100 to 2000 ppm (C2 hydrocarbons, HCN, CH_2O, CO, N_2, NO) are thus uncertain. In particular, there is a clear mass-28 signal for all the Cassini encounters in which the INMS instrument operated in its nominal, closed source configuration (e.g., *Waite et al.,* 2009). The final INMS measurements, during the E21 pass of Enceladus, were made in a novel open-source mode in an effort to detect native H_2 uncontaminated from hydrogen produced by high-speed collisions of H_2O molecules and grains with the titanium antechamber used in the closed-source mode (*Waite et al.,* 2017). During this encounter, no mass-28 signal was seen to the noise limit, which rules out native CO, N_2, or C_2H_2 abundances (or their sum) greater than 5×10^{-4} (2σ). It also implies that earlier detections at mass 28 were most likely of fragmentation products of larger but unidentified molecules.

Oxidized sulfur is not readily identified by INMS for technical reasons, but S may have been identified as a volatile as H_2S [at the 20-ppm level (*Waite et al.,* 2009)].

TABLE 1. Summary of Enceladus' volatile plume compositions as measured by Cassini INMS.

Major Species No Ambiguity >0.1%	Minor Species I Moderate Ambiguity <0.2% and >100 ppm	Minor Species II High Ambiguity <100 ppm				
		Hydrocarbons	N-bearing	O-bearing	NO-bearing	Others
H_2O (96–99%)	CO	C_3H_4	CH_5N	O_2	C_2H_7NO	Ar
	C_2H_2	C_3H_6	C_2H_3N	CH_3OH	$C_2H_5NO_2$	H_2S
	C_2H_4	C_3H_8	C_2H_7N	C_2H_2O	$C_3H_7NO_2$	PH_3
CO_2 (0.3–0.8%)	C_2H_6	C_4H_8	$C_2H_6N_2$	C_2H_4O		C_3H_5Cl
	N_2	C_4H_{10}	C_4H_9N	C_2H_6O		
CH_4 (0.1–0.3%)	HCN	C_5H_{10}	$C_4H_8N_2$	C_3H_6O		
	CH_2O	C_5H_{12}	$C_6H_{12}N_4$	C_3H_8O		
	NO	C_8H_{18}		$C_2H_4O_2$		
NH_3 (0.4–1.3%)				$C_2H_6O_2$		
				$C_4H_{10}O$		
H_2 (0.4–1.4%)				$C_4H_6O_2$		

Formally, this identification is highly ambiguous (Table 1). Phosphorus may also have been seen in the form of phosphine (PH_3), but this identification is also highly ambiguous. We note that H_2S is not expected to be abundant in an alkaline ocean (*Zolotov,* 2007), which Enceladus' ocean appears to be (*Glein et al.,* 2015).

Although the practical lower limit cutoff for sensitivity with INMS in the plume of Enceladus is ~1 ppm, there are signs for even heavier molecules at volume mixing ratios below 100 ppm. These heavier molecules include hydrocarbons up to C8, as well as a host of nitrogen- and oxygen-bearing organic compounds. Species with masses above 50 amu are largely semivolatiles imbedded in ice grains that appear to be sourced from the ocean spray. Such condensed organic compounds have been detected in E-ring particles (*Postberg et al.,* 2008). For details on the ice grain composition, see the chapter in this volume by Postberg et al.

The high signal-to-noise of the E5 flyby combined with the high relative velocity of the spacecraft with respect to Enceladus and the impingement of ice grains on the titanium walls of the closed source antechamber allowed a one-time measurement of the D/H in water (*Waite et al.,* 2009) (Fig. 7a). This was enabled by the reaction of the raw Ti with the oxygen in the water liberating H_2, which was measured to determine the D/H ratio. The observed value was $2.9^{+1.5}_{-0.7} \times 10^{-4}$, which is similar to the value found at Comet 1P/Halley and many other comets (Fig. 7b, discussed further below). All other isotope values were not of adequate precision to provide meaningful constraints. Argon-40 was inferred from mass spectra on the very high signal-to-noise E5 flyby (*Waite et al.,* 2009), but was not seen on any subsequent flyby (*Waite et al.,* 2017). The isotopic results are shown in Table 2, reproduced from *Waite et al.* (2009).

4.2. Stable Isotope Measurements at Titan: Implications for Enceladus

*4.2.1. **Nitrogen/primordial argon.*** Titan is the largest satellite by far in the saturnian system and, like Enceladus, a part of the regular satellite system. If by assumption Enceladus formed in the same or a related sequence of events by which Titan was formed, then the very extensive chemical clues present in Titan's atmosphere might be used to assess the conditions under which Enceladus formed (e.g., *Atreya et al.,* 2009). This is important because, while Enceladus' plume has provided a wealth of data, the tenuous nature of Enceladus' plume combined with the limits to the sensitivity of Cassini's mass spectrometers made determination of certain species, such as ^{36}Ar, and isotopic ratios such as those of nitrogen, impossible. These isotopic species are key to understanding both Enceladus' and Titan's source materials. Mass spectrometers ready for flight today (e.g., *Reh et al.,* 2016) can make these measurements in Enceladus' plume, but until they fly, use of the Titan data — with the proviso that common formation is an assumption and with due attention to possible fractionation and other evolutionary effects — is instructive.

Titan's ratio of primordial argon (^{36}Ar) to total nitrogen is $(2.1 \pm 0.8) \times 10^{-7}$ (*Niemann et al.,* 2010). According to the classic test proposed by T. Owen (*Owen,* 1982), this very low ratio precludes the origin of nitrogen in Titan's atmosphere as being from primordial molecular nitrogen trapped in the icy planetesimals from which Titan formed (otherwise, ^{36}Ar would have been trapped as well). It points strongly, instead, to the original molecular carrier of nitrogen into Titan being ammonia. The presence of accreted NH_3 is consistent with the observation of ammonia in Enceladus' plume (*Waite et al.,* 2009, 2017) at a level of 0.4–1.3% relative to water. We do not know the plume N_2 abundance, other than that it is $<5 \times 10^{-4}$, as noted above. *If* the NH_3 abundance in Enceladus is primordial and there is negligible N_2, then the Enceladus ammonia-to-water value is consistent with the upper limit of 3% ammonia-to-water in Titan based on orbital evolution and other considerations (*Tobie et al.,* 2005). Moreover, 0.4–1.3% ammonia in Enceladus' plume (Table 1) translates into ~30–90× as much N_2 than is required to supply Titan's present-day atmosphere. Thus, there is complete self-consistency in the idea that the primordial

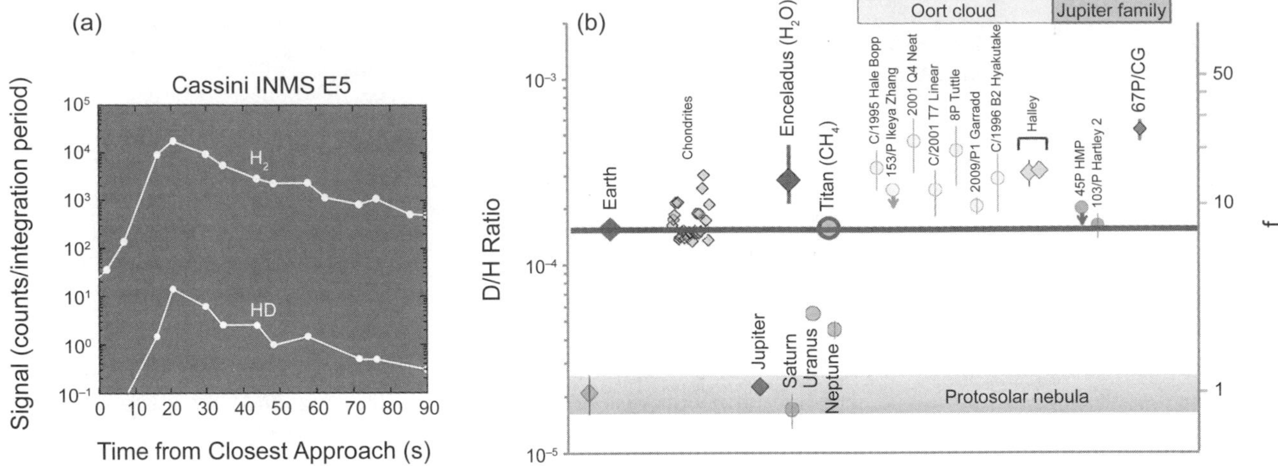

Fig. 7. See Plate 9 for color version. **(a)** The bulk H_2 signal seen during the E5 flyby, shown here, is representative of plume H_2O. During the time of maximum plume material influx into the INMS, signal-representing HD rose above noise levels for approximately 60 seconds, allowing a direct comparison of H_2 (black) to HD (red) to obtain the D/H ratio in water shown in Table 2. **(b)** D/H ratios in different objects of the solar system. Diamonds represent data obtained by means of *in situ* mass spectrometry measurements, and circles refer to data obtained with astronomical methods. The protosolar nebula (PSN) D/H value (lower left) is estimated to be 2.1 (± 0.5) × 10^{-5} based on measurements of D/H and $^3He/^4He$ in the atmosphere of Jupiter (*Mahaffy et al.,* 1998) and $^3He/^4He$ in the solar wind and meteorites (*Geiss and Gloeckler,* 1998). The fractionation factor f is defined as $[D/H]_{object}/[D/H]_{PSN}$. Modified from *Altwegg et al.* (2015).

TABLE 2. Stable H and O isotope determination from water vapor in Enceladus' plume, compared with representatives of major reservoirs of volatiles and Earth (*Waite et al.,* 2009).

Object Name	D/H	$^{18}O/^{16}O$	$^{12}C/^{13}C$
Enceladus	$2.9^{+1.5}_{-0.7} \times 10^{-4}$	$2.1^{+4.0}_{-0.2} \times 10^{-3}$	84 ± 13
Comet 1P/Halley	$3.1^{+0.4}_{-0.5} \times 10^{-4}$	$1.93 \pm 0.12 \times 10^{-3}$	90 ± 10
Protostar	$2.1 \pm 0.4 \times 10^{-5}$		
Earth	1.56×10^{-4}	2×10^{-3}	89
Oort cloud comets	$2–6 \times 10^{-4}$		
Kuiper belt comets	1.6×10^{-4}		

molecular carrier of nitrogen for both Titan and Enceladus was ammonia. This argument holds true even if there is substantial ammonia/water fractionation due to copious H_2O plume vapor freezing on the "tiger stripe" fracture walls during venting, increasing Enceladus' apparent NH_3/H_2O abundance (*Glein et al.,* 2015; *Waite et al.,* 2017).

An alternative primordial source for Enceladus' NH_3 is condensed organic compounds. The presence of organic matter in satellite-forming materials is consistent with the detection of organic compounds in the atmosphere and surface materials on Titan (*Strobel et al.,* 2008; *Lorentz et al., 2008*; *Soderblom et al.,* 2009) and in E-ring particles emitted from Enceladus (*Postberg et al.,* 2008). Organic matter is abundant in CCs and especially in comets, both of which have certain affinities to Enceladus, as previously discussed. Specifically, NH_3 is present in CCs at the 100-ppm level (*Pizzarello and Williams,* 2012).

4.2.2. Nitrogen isotopes. The $^{14}N/^{15}N$ value in N_2 in Titan's atmosphere is 167.7 ± 1.7 (*Niemann et al.,* 2010), more than a factor of 2 heavier (more ^{15}N-rich) than solar, jovian, or saturnian atmospheric values [≈440, ~435, and >500, respectively (*Füri and Marty,* 2015)], but consistent with the value for ammonia seen in several Jupiter-family and Oort cloud comets [Table 1 of *Mandt et al.* (2014) and references therein]. *Mandt et al.* (2014) argued that the Titan value is essentially its primordial one, given that atmospheric processes are inefficient at altering it, and they interpreted the difference from protosolar values to mean that Titan's molecular nitrogen came from primordial ammonia ice rather than molecular nitrogen, consistent with the arguments in the previous subsection. Furthermore, they argued that the ammonia ice must have been formed in the protosolar nebula rather than any saturnian subnebula that was chemically and isotopically connected with the proto-saturnian atmosphere. This inference is consistent with the discussion in section 2 that argued for limited if any thermochemical equilibration between saturnian satellitesimals and Saturn's CPD.

The conclusion of *Mandt et al.* (2014) assumes fractionation processes cannot evolve Titan's NH_3 or N_2 from what might have been lighter, solar $^{14}N/^{15}N$ values to the value observed today. Preferential escape of ^{14}N over geologic time can drive Titan's $^{14}N/^{15}N$ ratio down, but the process is inefficient (*Mandt et al.,* 2014). Thus, that Titan's nitrogen is intrinsically isotopically heavy appears firm. An additional source of heavy nitrogen for Titan's atmosphere could be input from ^{15}N-rich organic matter (released during hydrothermal evolution of Titan's core). Insoluble organic matter in CCs is enriched in ^{15}N, although generally not as much as cometary NH_3 (*Alexander et al.,* 2007; *Pizzarello and Williams,* 2012) [and see Fig. 2 in *Füri and Marty* (2015)]. This does beg the question of how nitrogen in protosolar ammonia ice, represented by comets, or that in CCs, becomes fractionated relative to bulk solar system material, but the answer seems to be very-low-temperature formation in interstellar cloud environments followed by limited or no exchange with other reservoirs (*Füri and Marty,* 2015; *Shinnaka and Kawakita,* 2016; cf. *Alexander et al.,* 2017).

More recently, $^{14}N/^{15}N$ has been determined for NH_2 (NH_3 being the parent molecule) in a suite of 18 comets, yielding a more statistically precise average value of 136 ± 6 (*Shinnaka et al.,* 2016), close to and statistically consistent with the ensemble average for cometary $C^{14}N/C^{15}N$ of 148 ± 6 (*Mumma and Charnley,* 2011). Both are substantially heavier than the solar or jovian/saturnian values but also somewhat heavier than Titan's ratio of 168 ± 2. Although the isotopic difference between Titan and cometary NH_3 does not appear great, in comparison with solar values (thus smacking of hair splitting), it is of interest because it is now clear that cometary NH_3 is actually isotopically heavier than Titan's atmospheric N_2, not lighter. So rather than focusing on mechanisms to isotopically fractionate Titan's N_2 to heavier values, explanations are being explored that do the opposite. One is that the carbonaceous component in Titan has contributed isotopically lighter nitrogen (*Miller et al.,* 2017), *in this case meaning lighter than cometary*. This appears quite plausible [again see Fig. 2 in *Füri and Marty* (2015), specifically the values for CCs]. Alternatively, *Krasnopolsky* (2016) argues that photolysis- and radiolysis-driven formation of HCN in Titan's atmosphere causes ^{15}N-rich HCN to form and precipitate, driving Titan's atmospheric N_2 to be somewhat more ^{14}N-enriched over geologic time. In addition, it may be that there was, after all, modest isotopic exchange between ammonia ice accreted to Saturn's CPD and isotopically light, subnebular N_2 gas. Regardless, all these explanations support or are at least consistent with the idea that the dominant contributor to Titan's N_2 was ammonia ice.

4.2.3. Deuterium/hydrogen. The deuterium-to-hydrogen ratio (D/H) in methane (from CH_3D/CH_4) in Titan's atmosphere supports the conclusion from the nitrogen isotopes that Titan's volatiles were derived from heavy-isotope-enriched solar material that subsequently experienced little isotopic equilibration in the saturnian CPD. Titan's D/H value is $(1.58 \pm 0.16) \times 10^{-4}$, 5.5–9.5× the protosolar estimate (*Abbas et al.,* 2010) and the same as terrestrial seawater within measurement uncertainty [Vienna Standard Mean Ocean Water (VSMOW), 1.56×10^{-4}] (Fig. 7b); cf. *Nixon et al.* (2012). Early work ascribed this value to enrichment associated with photochemistry (*Pinto et al.,* 1986; *Lunine et al.,* 1999), but more recent studies have challenged the ability of either photochemistry or atmospheric escape to explain more than about half the observed enrichment in D relative to protosolar values (*Cordier et al.,* 2008).

Serpentinization reactions with water in Titan's interior cannot account for the deuterium enrichment either, unless the D/H in Titan's interior water is isotopically distinct (much lighter) than that measured by Cassini emanating from Enceladus (*Mousis et al.,* 2009a; *Glein,* 2015). *Mousis et al.* (2009a) instead invoke methane directly captured from the solar system's molecular birth cloud, where CH_4 would be deuterium-rich, by the icy grains that became the building blocks of Titan's solid body and atmosphere. While this scenario may not be unique, it is consistent with the view that derives from the noble gas and nitrogen isotopic data on Titan. It would be of great interest to measure the D/H of Titan's water ice, both for comparison with Enceladus, and because H_2O represents the largest H reservoir on both bodies, and so is more fundamental and less susceptible to fractionation effects.

5. CONSTRAINTS ON FORMATION MODELS FROM OBSERVATIONS

Enceladus' volatile and isotopic compositions are constraints on its formation and evolution, provided such measurements can be made accurately and precisely. In terms of the origin scenarios outlined in section 2, some are more easily discriminated than others. For example, if Enceladus formed as a primordial body, its chemical and isotopic signatures would not necessarily be very different than if it were a second-generation moon formed from the catastrophic breakup and reassembly of a more or less similar progenitor body — the potential energy of reaccretion is low for a body the size of Enceladus (e.g., *Hussmann et al.,* 2010). If, however, the progenitor was a Titan-like satellite disrupted by tides (*Canup,* 2010), then we might expect some evolutionary geochemical effects but not necessarily isotopic effects. For example, D/H in water ice would hardly change, but CH_3D/CH_4 might evolve due to aqueous alteration of organics in the Titan-like body's core. That said, present-day tidal heating in Enceladus' core (e.g., *Choblet et al.,* 2017) would likely drive the most important evolutionary geochemical effects overall, and regardless of origin.

Clearer chemical and isotopic signals would accrue to some of the less likely origin scenarios. We have already commented on how thermochemical equilibration of accreting solids in a dense saturnian CPD is not favored, either theoretically or from Titan data (section 4.2). Presumably, if Enceladus and the other mid-sized icy moons were actually remnants of a tidally stripped body from the Kuiper belt (*Hyodo et al.,* 2017), we should be able to tell, or at least note, the important, if unexplained, differences between Enceladus and Titan. But this depends on knowing well the isotopic and chemical compositions of Kuiper belt objects

(Jupiter-family comets, Centuars, Pluto, Triton, etc.), which is clearly a work in progress.

In this section we first address what we can infer from Enceladus' D/H, which for the present is the only useful isotopic measurement we have for this moon. We then assess the degree to which Enceladus' volatiles match those of comets, and whether Enceladus and Titan are similar or dissimilar samples of the saturnian CPD. We close the section with a summary and further implications.

5.1. Interpretation of Enceladus' Deuterium/Hydrogen Ratio

Cassini measurements, summarized above, allow us to place a reasonably firm constraint on the CPD phase during which Enceladus or a progenitor body would have formed. The D/H ratio in H_2O measured in Enceladus' plume by the INMS instrument onboard the Cassini spacecraft is close to the value inferred for several Oort cloud comets [$\sim 2.9 \times 10^{-4}$ (*Waite et al.,* 2009; *Mumma and Charnley,* 2011)] and to that of Jupiter-family comet 67P (Fig. 7). This measurement precludes the possibility that Enceladus' building blocks condensed in an initially warm and dense saturnian subnebula, according to the scenario proposed by *Prinn and Fegley* (1981). In the latter, the D/H ratio acquired by the water ice accreted by Enceladus should be close to the protosolar value [$\approx 2.1 \times 10^{-5}$ (*Geiss and Gloeckler,* 1998)], because the high temperature and pressure conditions of the CPD would have favored a fast reequilibration of deuterium between hydrogen, the dominant deuterium reservoir in the subnebula, and H_2O, a minor deuterium reservoir compared to H_2 (*Kavelaars et al.,* 2011). Also, the conversion of nebular CO into CH_4 would have produced water possessing the same D/H ratio as protosolar hydrogen, driving the total D/H in water ice in that direction.

Instead, the D/H ratio measured in H_2O in Enceladus suggests that its building blocks have condensed in the protosolar nebula at the same formation location as Oort cloud comets (*Waite et al.,* 2009). This conclusion also applies to Titan, because its atmospheric molecular nitrogen shares a similar $^{14}N/^{15}N$ ratio as cometary ammonia, as described in section 4.2.2. This measurement suggests that molecular nitrogen was initially present in the form of primordial ammonia in the interior of Titan before being outgassed to the atmosphere, and also implies that the satellite's building blocks likely originated (formed) in the protosolar nebula (*Rousselot et al.,* 2014).

We note that Comet 103P/Hartley 2's terrestrial D/H not withstanding (*Hartogh et al.,* 2011), the enriched D/H values of most comets including those determined *in situ* by mass spectrometry (for Halley and 67P) appear secure (Fig. 7). The consistent enrichment in cometary D/H is likely interstellar in origin (*Cleeves et al.,* 2014). We also note that Enceladus' highly deuterated water ice cannot be the result of fractionation during hydrothermal alteration [the effect is too small (*Saccocia et al.,* 2009)], a point returned to in section 5.3.

In detail, given that D/H indicates Enceladus' H_2O ice never significantly exchanged deuterium with H_2 gas in the saturnian CPD, we can conceive two possible scenarios: (1) There was no sublimation of infalling water ice, or (2) there was sublimation of water ice but accompanied by only a limited isotopic exchange with H_2 gas. Scenario (1) does not preclude a partial devolatilization of Enceladus' building blocks, meaning that some volatiles may remain primordial, but not others. Scenario (2) means that the full volatile phase was vaporized. The chemistry in the disk would necessarily have been very limited, but some volatiles may have not recondensed because of a moderately warm formation temperature for Enceladus.

Can we tell the difference between these two scenarios, and in particular, can we set a more definitive limit on the upper temperature of Enceladus' accretion zone? One possibility is to look at volatility trends for Enceladus and Titan. The strong evidence cited that Titan's N_2 atmosphere ultimately derives or mainly derives from NH_3 ice as opposed to N_2 trapped in water ice could be taken to imply that satellitesimals in the saturnian CPD were heated modestly (to no more than ~50 K), heated enough to drive off their N_2 and species of similar volatility (*Mousis et al.,* 2019b,c). However, the evidence that 67P is quite depleted in N_2 (*Rubin et al.,* 2015) (section 3.3), as well as the comet's very low ^{36}Ar abundance (*Balsiger et al.,* 2015), leads to a simpler hypothesis: that the infalling icy grains, pebbles, and planetesimals that ultimately built Enceladus and Titan were initially depleted in N_2 and ^{36}Ar compared with solar values. This depletion in N_2 and ^{36}Ar is likely due to the low trapping efficiency of these particular supervolatiles in icy planetesimals and possibly to early but mild ^{26}Al radiogenic heating of the same (*Mousis et al.,* 2012, 2017). To take this argument further we need to consider the composition of Enceladus' plume volatiles (next section).

Clark et al. (2017) have recently argued from a significant wavelength recalibration of Cassini VIMS spectra and laboratory work that the D-O stretch in water ice can be accurately measured remotely, and in particular that this implies a D/H for the water ice in Saturn's rings (and by implication Enceladus) that is closer to terrestrial (VSMOW) rather than cometary. It is perhaps too early to tell whether this is really the case, but the possibility that in the future we may be able to determine remotely the D/H in water ice is most encouraging.

5.2. Enceladus' Plume: Consistency with Cometary Chemistry?

The secure detection of CO_2, CH_4, and NH_3 in Enceladus' plume vapor (*Waite et al.,* 2017) and the broader identification of families of organic compounds earlier (*Waite et al.,* 2006, 2009) prompted an analogy with cometary composition (*Waite et al.,* 2006). CO_2, CH_4, and NH_3 are among the most prominent of cometary volatiles, and both the relative abundance of NH_3 to H_2O and that $CO_2 > CH_4$ are consistent with cometary ices. The absolute abundance of

CO_2 with respect to water is not consistent, however, as it is rather low on Enceladus compared with comets. Nor is CO, another important cometary volatile, seen in the INMS data (section 4.1).

In this matter, it is important to consider that the volatiles erupting from the south pole of Enceladus have almost certainly fractionated with respect to their oceanic sources. The heat flow seen by the Cassini CIRS instrument is clearly concentrated along the tiger stripe fissures (see the chapter by Spencer et al. in this volume), and is plausibly derived from the latent heat of condensation of H_2O vapor along the fracture surfaces (*Ingersoll and Pankine,* 2010; *Kite and Rubin,* 2016). To account for Enceladus' heat flow, *Ingersoll and Pankine* (2010) calculate that 10× as much H_2O vapor must condense as vented in the plume gas, the point being that water vapor condenses in fractures and the volatile/H_2O(g) ratio changes (increases) along the eruption path toward the surface. More detailed aqueous geochemical models of CO_2 speciation in Enceladus' ocean are presented in *Glein et al.* (2015) and *Waite et al.* (2017); they find even lower abundances (concentrations) of CO_2, and other volatiles as well assuming the various volatile/CO_2 ratios remain unchanged. Put simply, the volatile abundances in Table 1 should be divided by at least 10 to get more realistic oceanic abundances. In this sense the ocean at Enceladus is less primordial-soup-like and more ocean-like (i.e., dilute). In detail, of course, the volatilities of individual species matter. For example, NH_3 may be somewhat less fractionated than, say, CH_4, because NH_3 may also plate out on the fracture walls, near the coldest portions closest to Enceladus' surface.

Fractionation during ascent and eruption does not in and of itself rationalize Enceladus' plume CO_2 abundance as a cometary volatile. Reducing it to, e.g., 300–800 ppm in fact goes the other way. The low CO_2/H_2O ratio in the ocean inferred through the fractionation assumption does not, however, indicate unabundant bulk "CO_2" in the ocean, because inorganic C is mainly represented by carbonate and bicarbonate ions (*Glein et al.,* 2015; *Zolotov,* 2007). Abundant dissolved inorganic carbon (CO_2 + HCO_3^- + CO_3^{-2}) in Enceladus' ocean would be consistent with cometary ice source building blocks. Sodium and potassium carbonate and bicarbonate plume and E-ring particles were identified in Cassini Dust Analyzer (CDA) mass spectra. Because these salts must derive directly from the ocean (*Postberg et al.,* 2009, 2011), the inference is that Enceladus' ocean *is* relatively abundant in dissolved inorganic carbon (see the chapter by Glein et al. in this volume).

Assuming a fractionation factor of 10 reduces the plume abundance of CH_4 in Table 1 to 100–300 ppm in the ocean. Such an abundance is easier to reconcile with the limited CH_4 solubility at Enceladus ocean pressures [see Table S7 in *Waite et al.* (2009)]. *Waite et al.* (2017) estimate that the CH_4 concentration in the ocean may actually be less than 1 ppm, definitely soluble under the appropriate pressure conditions (at least 0.5 MPa under 5 km of south polar terrain ice). The methane partial pressures implied in all these cases are insufficient for sequestration by clathrate formation (*Bouquet et al.,* 2015), thus the measured methane abundance in the plume can be taken to be directly indicative of, or linked to, the oceanic abundance.

The apparent absence (i.e., non-detection) of CO in the plume could have several explanations: (1) CO was lost by mild heating of satellitesimals in the saturnian CPD (*Mousis et al.,* 2009b,c). This cannot be ruled out by any existing compositional or isotopic data. (2) CO is even less soluble in water than CH_4 and in the absence of resupply Enceladus could simply have lost all its primordial CO to space over geologic time through fractures in the icy shell. This would imply, however, that the CH_4 seen at Enceladus (which would otherwise have been lost as well) would need to be resupplied by Enceladus' core, such as through methanogenesis ($CO_2 + H_2 \rightarrow CH_4 + H_2O$) accompanying serpentinization (*Waite et al.,* 2017). This would also serve to draw down the dissolved CO_2. (3) CO could have been consumed by hydrolysis (conversion to formate, $HCOO^-$, and carbonate/bicarbonate ions), Fischer-Tropsch type (CO + $H_2 \rightarrow$ hydrocarbons + H_2O), or other reactions (e.g., *Shock and McKinnon,* 1993; cf. *Zolotov,* 2007).

Ultimately, a definitive comparison of oceanic and cometary carbon abundances must be viewed with caution, because initial C species were likely transformed and redistributed during Enceladus' differentiation and thermal evolution. But the overall conclusion from the CO_2, CH_4, and CO (and NH_3) abundances is one of consistency with a cometary source. Moreover, if comets contributed substantially to the composition of Enceladus, we would also expect a major fraction of low-density, high-molecular-weight organic matter such as seen in cometary dust (*Jessberger et al.,* 1988; *Fomenkova,* 1999; *Fray et al.,* 2016). Cassini INMS results (*Waite et al.,* 2006, 2009) and the organic grains identified in CDA mass spectra (*Postberg et al.,* 2008) are both consistent with this prediction, even if the specific organic molecules have not been definitively identified. Finally, the native H detected in the plume (Table 1) is very unlikely to be primordial (*Waite et al.,* 2017), and so does not directly enter into the discussion here.

5.3. Summary and Implications for Enceladus

What then, is the lesson of all this for Enceladus? A self-consistent argument can be made that the material from which the saturnian satellites formed was essentially protosolar in nature, trapped in planetesimals that might have had a common origin with some classes of comets, and transported into the satellite feeding zone without significant subsequent chemical or isotopic interaction with the gas of the saturnian subnebula — assuming that gas contained a significant signature of the bulk saturnian envelope itself. The argument rests on utilizing evidence from both Titan and Enceladus, and so it cannot easily be used to assess whether these two bodies formed coevally from the same reservoir of material. But it is compelling that several separate types of chemical and isotopic data can be used to paint a self-consistent picture. It also emphasizes the potential complexity of the saturnian subnebula itself, as a structure that did not isotopically and chemically homogenize or equilibrate the material that passed through it or condensed from it, but rather would have carried a significant signature of the external protosolar nebula environment, at least in the icy grains that formed Titan and Enceladus.

The protosolar signature pertains to the water ice, icy volatiles, and organic materials measured by Cassini at Enceladus. From these chemical and isotopic clues alone, it is not obvious that we can presently rule out a later, tidal capture and reaccretion origin for Enceladus and the inner mid-sized satellites (section 2.5.5), or Enceladus being contaminated by carbonaceous rock from the Kuiper belt (*Salmon and Canup,* 2017), given the presumed chemical link (if not equivalence) between comets and bodies in the Kuiper belt. Future observations and measurements will be required to confirm or refute these catastrophic and contamination scenarios.

Regarding Enceladus' rock component, we again refer to the D/H measurements for CCs shown in Fig. 7. These are either consistent with or somewhat elevated in D with respect to VSMOW, which has been used for some time to argue for an asteroidal source for Earth's water (e.g., *Alexander et al.,* 2012). In section 3.2 we emphasized the gathering isotopic evidence that CCs formed beyond the position of proto-Jupiter in the protosolar nebula, with the implication being that planetesimals of similar composition must have contributed to the formation of Saturn's satellites, and may have dominated if CI chondrites formed near the position of Saturn [as advocated by *Desch et al.* (2018)]. If so, we might expect the D/H in CI chondrites to reflect the high D/H values in cometary water and Enceladus plume vapor as measured by Cassini (*Waite et al.,* 2009). This does not appear to be the case, however, which represents a puzzle.

During aqueous alteration in CC parent bodies, as well as in Enceladus' core today (*Hsu et al.,* 2015; *Waite et al.,* 2017), there will be isotopic fractionation between pore water and the OH in hydrated silicates. Experiments have shown that the silicate minerals become isotopically lighter. In terms of δD, which refers to the relative change in D/H in parts per mil, serpentine- and talc-forming reactions result in a shift of about –30‰ in the minerals (*Saccocia et al.,* 2009). The corresponding shift in δD in the water will be positive (more deuterated) but dependent on the water/rock ratio (W/R). Because Enceladus' bulk W/R is ~40/60 by mass (e.g., *Barr and McKinnon,* 2007), the isotopic shift would be smaller (δD ~ few per mil) and hardly noticeable on Fig. 7 (the difference between VSMOW and the Enceladus value is a rather substantial 860_{-450}^{+960}‰ in δD). Thus there is truly an inconsistency between Enceladus' measured D/H (*Waite et al.,* 2009) and the presumption that CI chondrites are specifically representative of rock in the Saturn system [although we note that Tagish Lake, an ungrouped CM-like CC (*Brown et al.,* 2000; *Zolensky et al.,* 2000) has a bulk δD ~ 500‰ (*Alexander et al.,* 2017)]. If the INMS measurement is accepted as correct, then possible explanations are (1) CC-type parent bodies were not able to form near Saturn (meaning planetesimals that did form remained as unequilibrated mixtures of anhydrous grains and ices) or (2) CC-type meteorites with cometary D/H did form near Saturn but are not yet recognized in our collections.

6. TOWARD UNDERSTANDING ENCELADUS' ORIGIN

We have highlighted the different scenarios for the origin of Enceladus: (1) Primordial formation in Saturn's subnebula along with Titan and other satellites. This includes several subscenarios (sections 2.2.1–2.2.3), of which only the gas-starved accretion disk model is considered viable from a theoretical and observational point of view. (2) Late formation from accumulation of ring material (section 2.2.4–2.2.5), including formation of all the moons out to Rhea; formation of Mimas, Enceladus, and Tethys only; and even formation via tidal stripping of a heliocentric body during heavy bombardment. (3) Formation 100–200 m.y. ago after an ancient satellite system went unstable (section 2.2.5). As discussed at the close of section 2, there are many open issues regarding the formation of satellites around giant planets. Much remains to be learned from a theoretical and modeling standpoint, guided both by the details of the jovian and saturnian satellite systems and by observations of extrasolar planets and protoplanetary nebulae.

For Enceladus, what are crucially needed are additional chemical and isotopic data. Advanced mass spectrometers flown through the Enceladus plume in the future should strive for unambiguous determinations of the species therein and higher sensitivity. For example, the D/H for Enceladus water should be determined much more precisely to clearly link it to or discriminate from cometary D/H values. Measuring the CH_3D/CH_4 ratio could in principle determine whether it had a cometary heritage or was a product of organic synthesis reactions within the body. The $^{14}N/^{15}N$ ratio should also be measured for Enceladus. If Enceladus' NH_3 is indeed primordial, then $^{14}N/^{15}N$ in NH_3 should be consistent with the average value measured in comets (from NH_2) and in Titan's N_2 (since it comes mostly from primordial NH_3). Potential differences between the values for Titan and Enceladus could prove instructive; for example, atmospheric fractionation processes (section 4.2.2) would not operate at Enceladus. Greater mass resolution and sensitivity should determine the abundances (or meaningful upper limits) for a host of minor species such as N_2, CO, ^{36}Ar, ^{40}Ar, CH_3OH, H_2CO, and various other aliphatic and aromatic organics (cf. Table 1). Measurements of volatile organics in particular could determine the degree of unaltered cometary provenance vs. aqueous and thermal alteration. All these measurements could of course have important geological, geophysical, and astrobiological consequences as well (see other chapters in this book), not just implications for origins.

Note added in proof: R. N. Clark (personal communication, 2018) estimates from Cassini VIMS spectra a D/H for Enceladus' surface ice that is, within the uncertainties, consistent with the INMS measurement (Table 2). It is lower than the mean Enceladus value in Fig. 7 but elevated with respect to VSMOW.

Acknowledgments. We thank J. Castillo-Rogez and an anonymous reviewer for their comments and suggestions, which led to a substantially improved manuscript. W.B.M. thanks NASA's Cassini Data Analysis Program for past support, the Cassini INMS team for stimulating collaboration, and C. Glein for input to section 3.3. J.L. is grateful to the Cassini Project for support. O.M. acknowledges support from CNES. Additional support from NASA's Europa Clipper mission through the MASPEX Europa instrument team has also considerably sharpened our understanding of the issues behind compositional and isotopic measurements and interpretations.

REFERENCES

Abbas M. M., Kandadi H., and LeClair A. (2010) D/H ratio of Titan from observations of the Cassini/Composite Infrared Spectrometer. *Astrophys. J., 708,* 342–353.

A'Hearn M. A., Belton M. J. S., and Delamere W. A. (2011) EPOXI at Comet Hartley 2. *Science, 332,* 1396–1400.

A'Hearn M. F., Feaga L. M., Keller H. U., et al. (2012) Cometary volatiles and origin of comets. *Astrophys. J., 758,* 29.

Alexander C. M. O.'D., Fogel M., Yabuta H., and Cody G. D. (2007) The origin and evolution of chondrites recorded in the elemental and isotopic compositions of their macromolecular organic matter. *Geochim. Cosmochim. Acta, 71,* 4380–4403.

Alexander C. M. O'D., Bowden R., Fogel M. L., Howard K. T., Herd C. D. K., and Nittler L. R. (2012) The provenances of asteroids, and their contributions to the volatile inventories of the terrestrial planets. *Science, 337,* 721–723.

Alexander C. M. O'D., Cody G. D., De Gregorio B. T., Nittler L. R., and Stroud R. M. (2017) The nature, origin and modification of insoluble organic matter in chondrites, the major source of Earth's C and N. *Chem. Erde, 77,* 227–256.

Alibert Y. and Mousis O. (2007) Formation of Titan in Saturn's subnebula: Constraints from Huygens probe measurements. *Astron. Astrophys., 465,* 1051–1060.

Alibert Y., Mousis O., and Benz W. (2005) Modeling the jovian subnebula. I. Thermodynamic conditions and migration of proto-satellites. *Astron. Astrophys., 439,* 1205–1213.

Altwegg K., Balsiger H., Bar-Nun A., et al. (2015) 67P/Churyumov-Gerasimenko, a Jupiter family comet with a high D/H ratio. *Science, 347,* DOI: 10.1126/science.1261952.

Altwegg K., Balsiger H., Bar-Nun A., et al. (2016) Prebiotic chemicals — amino acid and phosphorus — in the coma of Comet 67P/Churyumov-Gerasimenko. *Sci. Adv., 2,* e1600285.

Asphaug E. and Reufer A. (2013) Late origin of the Saturn system. *Icarus, 223,* 544–565.

Atreya S. K., Lorenz R. D., and Waite J. H. (2009) Volatile origin and cycles: Nitrogen and methane. In *Titan from Cassini-Huygens* (R. H. Brown et al., eds.), pp. 77–99. Springer, New York.

Ayliffe B. A. and Bate M. R. (2009) Circumplanetary disc properties obtained from radiation hydrodynamical simulations of gas accretion by protoplanets. *Mon. Not. R. Astron. Soc., 397,* 657–665.

Balsiger H., Altwegg K., Bar-Nun A., et al. (2015) Detection of argon in the coma of Comet 67P/Churyumov-Gerasimenko. *Sci. Adv., 1,* e1500377.

Barr A. C. and McKinnon W. B. (2007) Convection in Enceladus' ice shell: Conditions for initiation. *Geophys. Res. Lett., 34,* L09202.

Bate M. R., Lubow S. H., Ogilvie G. I., and Miller K. A. (2003) Three-dimensional calculations of high- and low-mass planets embedded in protoplanetary disks. *Mon. Not. R. Astron. Soc., 341,* 213–229.

Benner L. A. M. and McKinnon W. B. (1995). Orbital evolution of captured satellites: The effect of solar gravity on Triton's post-capture orbit. *Icarus, 114,* 1–20.

Beuthe M., Rivoldini A., and Trinh A. (2016) Enceladus' and Dione's floating ice shells supported by minimum stress isostasy. *Geophys. Res. Lett., 43,* 10,088–10,096.

Bockelée-Morvan D., Crovisier J., Mumma M. J., and Weaver H. A. (2004) The composition of cometary volatiles. In *Comets II* (M. C. Festou et al., eds.), pp. 391–423. Univ. of Arizona, Tucson.

Bockelée-Morvan D., Crovisier J., Erard S., et al. (2016) Evolution of CO_2, CH_4, and OCS abundances relative to H_2O in the coma of comet 67P around perihelion from Rosetta/VIRTIS-H observations. *Mon. Not. R. Astron. Soc., 462,* S170–S183.

Bouquet A., Mousis O., Waite J. H., and Picaud S. (2015) Possible evidence for a methane source in Enceladus' ocean. *Geophys. Res. Lett., 42,* 1334–1339.

Bradley J. P. (2014) Early solar nebula grains — Interplanetary dust particles. In *Treatise on Geochemistry, 2nd edition, Vol. 1: Meteorites and Cosmochemical Processes* (A. M. Davis, ed.), pp. 287–308. Elsevier, Amsterdam.

Brown P. G., Hildebrand A. R., Zolensky M. E., et al. (2000) The fall, recovery, orbit, and composition of the Tagish Lake meteorite: A new type of carbonaceous chondrite. *Science, 290,* 320–325.

Brownlee D. (2014) The Stardust mission: Analyzing samples from the edge of the solar system. *Annu. Rev. Earth Planet. Sci., 42,* 179–205.

Calmonte U., Altwegg K., Balsiger H., et al. (2016) Sulphur-bearing species in the coma of Comet 67P/Churyumov-Gerasimenko. *Mon. Not. R. Astron. Soc., 462,* S253–S273.

Canup R. M. (2010) Origin of Saturn's rings and inner moons by mass removal from a lost Titan-sized satellite. *Nature, 468,* 943–946.

Canup R. M. and Ward W. R. (2002) Formation of the Galilean satellites: Conditions of accretion. *Astron. J., 124,* 3404–3423.

Canup R. M. and Ward W. R. (2006) A common mass scaling for satellite systems of gaseous planets. *Nature, 441,* 834–839.

Charnoz S., Salmon J., and Crida A. (2010) Origin of Saturn's small moons and F ring: Recent accretion at the ring's outer edge. *Nature, 465,* 752–754.

Charnoz S., Crida A., Castillo-Rogez J. C., Lainey V., Dones L., Karatekin Ö., Tobie G., Mathis S., Le Poncin-Lafitte C., and Salmon J. (2011) Accretion of Saturn's mid-sized moons during the viscous spreading of young massive rings: Solving the paradox of silicate-poor rings versus silicate-rich moons. *Icarus, 216,* 535–550.

Choblet G., Tobie G., Sotin C., Běhounková M., Čadek O., Postberg F., and Souček O. (2017) Powering prolonged hydrothermal activity inside Enceladus. *Nature Astron., 1,* 841–847.

Clark R. N., Brown R. H., Swayze G. A., and Cruikshank D. P. (2017) Detection of deuterium in icy surfaces and the D/H ratio of icy objects. *AAS/Division for Planetary Sciences Meeting Abstracts, 49,* #210.01.

Cleeves L. I., Bergin E. A., Alexander C. M. O'D., Du F., Graninger D., Öberg K. I., and Harries T. J. (2014) The ancient heritage of water ice in the solar system. *Science, 345,* 1590–1593.

Coradini A., Federico C., Forni O., and Magni G. (1995) Origin and thermal evolution of icy satellites. *Surv. Geophys., 16,* 533–591.

Cordier D., Mousis O., Lunine J. I., Moudens A., and Vuitton V. (2008) Photochemical enrichment of deuterium in Titan's atmosphere: New insights from Cassini-Huygens. *Astrophys. J. Lett., 689,* L61–L64.

Crida A. (2009) Minimum mass solar nebulae and planetary migration. *Astrophys. J., 698,* 606–614.

Crida A. and Charnoz S. (2012) Formation of regular satellites from ancient massive rings in the solar system. *Science, 338,* 1196–1199.

Crida A. and Morbidelli A. (2007) Cavity opening by a giant planet in a protoplanetary disc and effects on planetary migration. *Mon. Not. R. Astron. Soc., 377,* 1324–1336.

Ćuk M., Dones L., and Nesvorný D. (2016) Dynamical evidence for a late formation of Saturn's moons. *Astrophys. J., 820,* 97.

Cuzzi J. N. (2018) Saturn's rings after Cassini. In *Lunar and Planetary Science XLIX,* Abstract #1632. Lunar and Planetary Institute, Houston.

Cuzzi J. N. et al. (2010) An evolving view of Saturn's dynamic rings. *Science, 327,* 1470–1475.

D'Angelo G., Bate M. R., and Lubow S. H. (2005) The dependence of protoplanet migration rates on co-orbital torques. *Mon. Not. R. Astron. Soc., 358,* 316–332.

D'Angelo G., Kley W., and Henning T. (2003a) Orbital migration and mass accretion of protoplanets in three-dimensional lobal computations with nested grids. *Astrophys. J., 586,* 540–561.

D'Angelo G., Henning T., and Kley W. (2003b) Thermohydrodynamics of circumstellar disks with high-mass planets. *Astrophys. J., 599,* 548–576.

Davidsson B. J. R., Sierks H., Güttler C., et al. (2016) The primordial nucleus of Comet 67P/Churyumov-Gerasimenko. *Astron. Astrophys., 592,* A63.

DeMeo F. E. and Carry B. (2014) Solar system evolution from compositional mapping of the asteroid belt. *Nature, 505,* 629–634.

Desch S. J., Kalyaan A., and Alexander C. M. O'D. (2018) The effect of Jupiter's formation on the distribution of refractory elements and inclusions in meteorites. *ArXiV e-prints,* arXiv:1710.03809v2.

Dones L. (1991) A recent cometary origin for Saturn's rings? *Icarus, 92,* 194–203.

Dones L., Weissman P. R., Levison H. F., and Duncan M. J. (2004) Oort cloud formation and dynamics. In *Comets II* (M. C. Festou et al., eds.), pp. 153–174. Univ. of Arizona, Tucson.

Duncan M. J. and Levison H. F. (1997) A disk of scattered icy objects and the origin of Jupiter-family comets. *Science, 276,* 1670–1672.

Estrada P. R. and Mosqueira I. (2006) A gas-poor planetesimal capture model for the formation of giant planet satellite systems. *Icarus, 181,* 486–509.

Fegley B. and Schaefer L. (2009) Cosmochemistry of the biogenic elements C, H, N, O, and S. In *Astrobiology: Emergence, Search and Detection of Life* (V. A. Basiuk, ed.), pp. 23–49. American Scientific, Valencia, California.

Fernández J. A. and Ip W.-H. (1981) Dynamical evolution of a cometary swarm in the outer planetary region. *Icarus, 47,* 470–479.

Fernández J. A., Gallardo T., and Brunini A. (2004) The scattered disk population as a source of Oort cloud comets: Evaluation of its current and past role in populating the Oort cloud. *Icarus, 171,* 372–381.

Fomenkova M. N. (1999) On the organic refractory component of cometary dust. *Space Sci. Rev., 90,* 109–114.

Fray N., Bardyn A., Cottin H., et al. (2016) High-molecular-weight organic matter in the particles of Comet 67P/Churyumov-Gerasimenko. *Nature, 538,* 72–74.

Fuller J., Luan J., and Quataert E. (2016) Resonance locking as the source of rapid tidal migration in the Jupiter and Saturn moon systems. *Mon. Not. R. Astron. Soc., 458,* 3867–3879.

Fung J. and Chiang E. (2016) Gap opening in 3D: Single-planet gaps. *Astrophys. J., 832,* 105.

Füri E. and Marty B. (2015) Nitrogen isotope variations in the solar system. *Nature Geosci., 8,* 515–522.

Geiss J. and Gloeckler G. (1998) Abundances of deuterium and helium-3 in the protosolar cloud. *Space Sci. Rev., 84,* 239–250.

Glein C. R. (2015) Noble gases, nitrogen, and methane from the deep interior to the atmosphere of Titan. *Icarus, 250,* 570–586.

Glein C. R., Baross J. A., and Waite J. H. (2015) The pH of Enceladus' ocean. *Geochim. Cosmochim. Acta, 162,* 202–219.

Gomes R., Levison H. F., Tsiganis K., and Morbidelli A. (2005) Origin of the cataclysmic late heavy bombardment period of the terrestrial planets. *Nature, 435,* 466–469.

Guillot T. and Hueso R. (2006) The composition of Jupiter: Sign of a (relatively) late formation in a chemically evolved protosolar disc. *Mon. Not. R. Astron. Soc., 367,* L47–L51.

Harris A. W. (1984) The origin and evolution of planetary rings. In *Planetary Rings* (R. Greenberg and A. Brahic, eds.), pp. 641–659. Univ. of Arizona, Tucson.

Hartogh P., Lis D. C., Bocklelée-Morvan D., et al. (2011) Ocean-like water in the Jupiter-family Comet 103P/Hartley 2. *Nature, 478,* 218–220.

Heller R. and Pudritz R. (2015) Water ice lines and the formation of giant moons around super-jovian planets. *Astrophys. J., 806,* 181.

Hsu H.-W., Postberg F., Sekine Y., Shibya T., Kempf S., Horányi M., Juhász A., Altobelli N., Suzuki K., Masaki Y., Kuwatani T., Tachiban S., Sirono S.-I., Moragas-Klostermeyer G., and Srama R. (2015) Ongoing hydrothermal activities within Enceladus. *Nature, 519,* 207–210.

Hussmann H., Choblet G., Lainey V., Matson D. L., Sotin C., Tobie G., and Van Hoolst T. (2010) Implications of rotation, orbital states, energy sources, and heat transport for internal processes in icy satellites. *Space Sci. Rev., 153,* 317–348.

Hyodo R., Charnoz S., Ohtsuki K., and Genda H. (2017) Ring formation around giant planets by tidal disruption of a single passing large Kuiper belt object. *Icarus, 282,* 195–213.

Iess L., Stevenson D. J., Parisi M., Hemingway D., Jacobson R. A., Lunine J. I., Nimmo F., Armstrong J. W., Asmar S. W., Ducci M., and Tortora P. (2014) The gravity field and interior structure of Enceladus. *Science, 344,* 78–80, DOI: 10.1126/science.1250551.

Ingersoll A. P. and Pankine A. A. (2010) Subsurface heat transfer on Enceladus: Conditions under which melting occurs. *Icarus, 206,* 594–607.

Jessberger E. K., Christoforidis A., and Kissel J. (1988) Aspects of the major element composition of Halley's dust. *Nature, 332,* 691–695.

Johansen A., Blum J., Tanaka H., Ormel C., Bizzarro M., and Rickman H. (2014) The multifaceted planetesimal formation process. In *Protostars and Planets VI* (H. Beuther et al., eds.), pp. 547–570. Univ. of Arizona, Tucson.

Johansen A., Mac Low M.-M., Lacerda P., and Bizzaro M. (2015) Growth of asteroids, planetary embryos, and Kuiper belt objects by chondrule accretion. *Sci. Adv., 1,* e1500109 .

Kargel J. S., Kaye J. Z., Head J. W., et al. (2000) Europa's crust and ocean: Origin, composition, and the prospects for life. *Icarus, 148,* 226–265, DOI: 10.1006/icar.2000.6471.

Kavelaars J. J., Mousis O., Petit J.-M., and Weaver H. A. (2011) On the formation location of Uranus and Neptune as constrained by dynamical and chemical models of comets. *Astrophys. J. Lett., 734,* L30.

Kite E. S. and Rubin A. M. (2016) Sustained eruptions on Enceladus explained by turbulent dissipation in tiger stripes. *Proc. Natl. Acad. Sci. USA, 113,* 3972–3975.

Klahr H. and Kley W. (2006) 3D-radiation hydro simulations of disk-planet interactions I. Numerical algorithm and test cases. *Astron. Astrophys., 445,* 747–758.

Krasnopolsky V. (2016) Isotopic ratio of nitrogen on Titan: Photochemical interpretation. *Planet. Space Sci., 134,* 61–63.

Kruijer T. S., Burkhardt C., Budde G., and Kleine T. (2017) Age of Jupiter inferred from the distinct genetics and formation times of meteorites. *Proc. Natl. Acad. Sci. USA, 114,* 6712–6716.

Le Roy L., Altwegg K., Balsiger H., et al. (2015) Inventory of the volatiles on Comet 67P/Churyumov-Gerasimenko from Rosetta/ ROSINA. *Astron. Astrophys., 583,* A1.

Levison H. F., Morbidelli A., VanLaerhoven C., Gomes R., and Tsiganis K. (2008) Origin of the structure of the Kuiper belt during a dynamical instability in the orbits of Uranus and Neptune. *Icarus, 196,* 258–273.

Levison H. F., Duncan M. J., Brasser R., and Kaufman D. E. (2010) Capture of the Sun's Oort cloud from stars in its birth cluster. *Science, 329,* 187–190.

Lodders K. (2003) Solar system abundances and condensation temperatures of the elements. *Astrophys. J., 591,* 1220–1247.

Lorenz R. D., Mitchell K. L., Kirk R. L., et al. (2008) Titan's inventory of organic surface materials. *Geophys. Res. Lett., 35,* L02206.

Lubow S. H., Seibert M., and Artymowicz P. (1999) Disk accretion onto high-mass planets. *Astrophys. J., 526,* 1001–1012.

Lunine J. I. and Stevenson D. J. (1982) Formation of the Galilean satellites in a gaseous nebula. *Icarus, 52,* 14–39.

Lunine J. I., Yung Y. L., and Lorenz R. D. (1999) On the volatile inventory of Titan from isotopic abundances in nitrogen and methane. *Planet. Space Sci., 47,* 1291–1303.

Machida M. N., Kokubo E., Inutsuka S.-I., and Matsumoto T. (2008) Angular momentum accretion onto a gas giant planet. *Astrophys. J., 685,* 1220–1236.

Magee B. A. and Waite J. H. (2017) Neutral gas composition of Enceladus' plume — Model parameter insights from Cassini-INMS. In *Lunar and Planetary Science XLVIII,* Abstract #2974. Lunar and Planetary Institute, Houston.

Mahaffy P. R., Donahue T. M., Atreya S. K., Owen T., and Niemann H. B. (1998) Galileo Probe measurements of D/H and $^3He/^4He$ in Jupiter's atmosphere. *Space Sci. Rev., 84,* 251–263.

Mandt K. E., Mousis O., Lunine J., and Gautier D. (2014) Protosolar ammonia as the unique source of Titan's nitrogen. *Astrophys. J. Lett., 788,* L24, DOI: 10.1088/2041-8205/788/2/L24.

McCollom T. M. (2016) Abiotic methane formation during experimental serpentinization of olivine. *Proc. Natl. Acad. Sci. USA, 113,* 13965–13970.

McDermott J. M., Seewald J. S., German C. R., and Sylva S. P. (2015) Pathways for abiotic organic synthesis at submarine hydrothermal fields. *Proc. Natl. Acad. Sci. USA, 112,* 7668–7672.

McKinnon W. B. (2015) Effect of Enceladus's rapid synchronous spin on interpretation of Cassini gravity. *Geophys. Res. Lett., 42,* 2137–2143, DOI: 10.1002/2015GL063384.

McKinnon W. B. and Zolensky M. E. (2003) Sulfate content of Europa's ocean and shell: Evolutionary considerations and some geological and astrobiological implications. *Astrobiology, 3,* 879–897.

Miller K. E., Glein C. R., and Waite J. H. (2017) A new source for Titan's N_2 atmosphere: Outgassing from accreted organic-rich dust in Titan's interior. In *Lunar and Planetary Science XLVIII,* Abstract #2072. Lunar and Planetary Institute, Houston.

Morbidelli A. and Crida A. (2007) The dynamics of Jupiter and Saturn in the gaseous protoplanetary disk. *Icarus, 191,* 158–171.

Mosqueira I. and Estrada P. R. (2003a) Formation of the regular satellites of giant planets in an extended gaseous nebula I: Subnebula model and accretion of satellites. *Icarus, 163,* 198–231.

Mosqueira I. and Estrada P. R. (2003b) Formation of the regular satellites of giant planets in an extended gaseous nebula II: Satellite migration and survival. *Icarus, 163,* 232–255.

Mosqueira I., Estrada P. R., and Charnoz S. (2010) Deciphering the origin of the regular satellites of gaseous giants — Iapetus: The Rosetta ice-moon. *Icarus, 207,* 448–460.

Mousis O. and Gautier D. (2004) Constraints on the presence of volatiles in Ganymede and Callisto from an evolutionary turbulent model of the jovian subnebula. *Planet. Space Sci., 52,* 361–370.

Mousis O., Gautier D., and Bockelée-Morvan D. (2002) An evolutionary turbulent model of Saturn's subnebula: Implications for the origin of the atmosphere of Titan. *Icarus, 156,* 162–175.

Mousis O., Alibert Y., Sekine Y., Sugita S., and Matsui T. (2006) The role of Fischer-Tropsch catalysis in jovian subnebular chemistry. *Astron. Astrophys., 459,* 965–968.

Mousis O., Lunine J. I., Pasek M., Cordier D., Waite J. H., Mandt K. E., Lewis W. S., and Nguyen M.-J. (2009a) A primordial origin for the atmospheric methane of Saturn's moon Titan. *Icarus, 204,* 749–751.

Mousis O., Lunine J. I., Thomas C., et al. (2009b) Clathration of volatiles in the solar nebula and implications for the origin of Titan's atmosphere. *Astrophys. J., 691,* 1780–1786.

Mousis O., Lunine J. I., Waite J. H., Magee B., Lewis W. S., Mandt K. E., Marquer D., and Cordier D. (2009c) Formation conditions of Enceladus and origin of its methane reservoir. *Astrophys. J. Lett., 701,* L39–L42.

Mousis O., Guilbert-Lepoutre A., Lunine J. I., Cochran A. L., Waite J. H., Petit J.-M., and Rousselot P. (2012) The dual origin of the nitrogen deficiency in comets: Selective volatile trapping in the nebula and postaccretion radiogenic heating. *Astrophys. J., 757,* 146.

Mousis O., Drouard A., Vernazza P., et al. (2017) Impact of radiogenic heating on the formation conditions of Comet 67P/Churyumov-Gerasimenko. *Astrophys. J. Lett., 839,* L4.

Mumma M. J. and Charnley S. B. (2011) The chemical composition of comets — Emerging taxonomies and natal heritage. *Annu. Rev. Astron Astrophys., 49,* 471–524.

Nelson R. P., Gressel O., and Umurhan O. M. (2013). Linear and non-linear evolution of the vertical shear instability in accretion discs. *Mon. Not. R. Astron. Soc., 435,* 2610–2632.

Nesvorný D. and Morbidelli A. (2012) Statistical study of the early solar system's instability with four, five, and six giant planets. *Astron. J., 144,* 117.

Nesvorný D. and Vokrouhlický D. (2016) Neptune's orbital migration was grainy, not smooth. *Astrophys. J., 825,* 14.

Niemann H. B., Atreya S. K., Demick J. E., Gautier D., Haberman J. A., Harpold D. N., Kasprzak W. T., Lunine J. I., Owen T. C., and Raulin F. (2010) Composition of Titan's lower atmosphere and simple surface volatiles as measured by the Cassini-Huygens probe gas chromatograph mass spectrometer experiment. *J. Geophys. Res., 115,* E12006, DOI: 10.1029/2010JE003659.

Nixon C. A., Temelso B., and Vinatier S., et al. (2012) Isotopic ratios in Titan's methane: Measurements and modeling. *Astrophys. J., 749,* 159.

Ogihara M. and Ida S. (2012) N-body simulations of satellite formation around giant planets: Origin of orbital configuration of the Galilean moons. *Astrophys. J., 753,* 60–77.

Ootsubo T., Kawakita H., Hamada S., et al. (2012) AKARI near-infrared spectroscopic survey for CO_2 in 18 comets. *Astrophys. J., 752,* 15.

Owen T. C. (1982) The composition and origin of Titan's atmosphere. *Planet. Space Sci., 30,* 833–838.

Peale S. J. and Canup R. M. (2015) The origin of the natural satellites. In *Treatise on Geophysics, 2nd edition, Vol. 10: Planets and Moons* (G. Schubert, ed.), pp. 559–604. Elsevier, Boston.

Pinto J. P., Lunine J. I., Kim S.-J., and Yung Y. L. (1986) D to H ratio and the origin and evolution of Titan's atmosphere. *Nature, 319,* 388–390.

Pizzarello S. and Williams L. B. (2012) Ammonia in the early solar system: An account from carbonaceous chondrites. *Astrophys. J., 749,* 161.

Pizzarello S., Cooper G. W., and Flynn G. J. (2006) The nature and distribution of the organic material in carbonaceous chondrites and interplanetary dust particles. In *Meteorites and the Early Solar System II* (D. S. Lauretta and H. Y. McSween, eds.), pp. 625–651. Univ. of Arizona, Tucson.

Pollack J. B. and Consolmagno G. (1984) Origin and evolution of the Saturn system. In *Saturn* (T. Gehrels and M. S. Matthews, eds.), pp. 811–866. Univ. of Arizona, Tucson.

Postberg F., Kempf S., Hillier J. K., Srama R., Green S. F., McBride N., and Grün E. (2008) The E ring in the vicinity of Enceladus: II. Probing the moon's interior — the composition of E-ring particles. *Icarus, 193,* 438–454.

Postberg F., Kempf S., Schmidt J., Brilliantov N., Beinsen A., Abel B., Buck U., and Srama R. (2009) Sodium salts in E-ring ice grains from an ocean below the surface of Enceladus. *Nature, 459,* 1098–1101.

Postberg F., Schmidt J., Hillier J., Kempf S., and Srama R. (2011) A salt-water reservoir as the source of a compositionally stratified plume on Enceladus. *Nature, 474,* 620–622.

Prinn R. G. and Fegley B. (1981) Kinetic inhibition of CO and N_2 reduction in circumplanetary nebulae: Implications for satellite composition. *Astrophys. J., 249,* 308–317.

Prinn R. G. and Fegley B. Jr. (1989) Solar nebula chemistry: Origin of planetary, satellite, and cometary volatiles. In *Origin and Evolution of Planetary and Satellite Atmospheres* (S. K. Atreya et al., eds.), pp. 78–136. Univ. of Arizona, Tucson.

Raymond S. N. and Izidoro A. (2017) Origin of water in the inner solar system: Planetesimals scattered inward during Jupiter and Saturn's rapid gas accretion. *Icarus, 297,* 134–148.

Reh K., Lunine J. I., Cable M. L., Spilker L., Waite J. H., Postberg F., and Clarke K. (2016) Enceladus Life Finder: The search for life in a habitable Moon. *IEEE Aerospace Conference,* Big Sky, Montana.

Ronnet T., Mousis O., and Vernazza P. (2017) Pebble accretion at the origin of water in Europa. *Astrophys. J., 845,* 92.

Ronnet T., Mousis O., Vernazza P., Lunine J. I., and Crida A. (2018) Saturn's formation and early evolution at the origin of Jupiter's massive moons. *Astron. J., 155,* 224.

Rousselot P., Olivier P., Emmanuël J., et al. (2014) Toward a unique nitrogen isotopic ratio in cometary ices. *Astrophys. J. Lett., 780,* L17.

Rubin M., Altwegg K., Balsiger H., et al. (2015) Molecular nitrogen in Comet 67P/Churyumov-Gerasimenko indicates a low formation temperature. *Science, 384,* 232–235.

Saccocia P. J, Seewald J. S., and Shanks W. C. (2009) Oxygen and hydrogen isotope fractionation in serpentine — water and talc — water systems from 250 to 450°C, 50 MPa. *Geochim. Cosmochim. Acta, 73,* 6789–6804.

Salmon J. and Canup R. M. (2017) Accretion of Saturn's inner mid-sized moons from a massive primordial ice ring. *Astrophys. J., 836,* 109.

Sasaki T., Stewart G. R., and Ida S. (2010) Origin of the different architectures of the jovian and saturnian satellite systems. *Astrophys. J., 714,* 1052–1064.

Scott E. R. D., Krot A. N., and Sanders I. S. (2018) Isotopic dichotomy among meteorites and its bearing on the protoplanetary disk. *Astrophys. J., 854,* 164.

Sekine Y., Sugita S., Shido T., Yamamoto T., Iwasawa Y., Kadono T., and Matsu T. (2005) The role of Fischer-Tropsch catalysis in the origin of methane-rich Titan. *Icarus, 178,* 154–164.

Sephton M. (2005) Organic matter in carbonaceous meteorites: Past, present, and future research. *Phil. Trans. R. Soc. A, 363,* 2729–2742, DOI: 10.1098/rsta.2005.1670.

Sephton M. and Gilmour I. (2000) Macromolecular organic materials in carbonaceous chondrites: A review of their sources and their role in the origin of life on the early Earth. In *Impacts and the Early Earth* (I. Gilmour and C. Koeberl, eds.), pp. 27–49. Lecture Notes in Earth Sciences, Vol. 91, Springer, New York.

Shinnaka Y. and Kawakita H. (2016) Nitrogen isotopic ratio of cometary ammonia from high-resolution optical spectroscopic observations of C/2014 Q2 (Lovejoy). *Astron. J., 152,* 145.

Shinnaka Y., Kawakita H., Emmanuël J., Decock A., Hutsemékers D., Manfroid J., and Arai A. (2016) Nitrogen isotopic ratios of NH_2 in comets: Implication for ^{15}N-fractionation in cometary ammonia. *Mon. Not. R. Astron. Soc., 462,* S195–S209.

Shock E. L. and McKinnon W. B. (1993) Hydrothermal processing of cometary volatiles — Applications to Triton. *Icarus, 106,* 464–477.

Shock E. L. and Schulte M. D. (1990) Amino acid synthesis in carbonaceous meteorites by aqueous alteration of polycyclic aromatic hydrocarbons. *Nature, 343,* 728–731.

Shock E. L., Canovas P., Yang Z., Boyer G., Johnson K., Robinson K., Fecteau K., Windman T., and Cox A. (2013) Thermodynamics of organic transformations in hydrothermal fluids. *Rev. Mineral. Geochem., 76,* 311–350.

Soderblom L. A., Barnes J. W., Brown R. H., et al. (2009) Composition of Titan's surface. In *Titan from Cassini-Huygens* (R. H. Brown et al., eds.), pp. 141–175. Springer, Dordrecht.

Stevenson D. J., Harris A. W., and Lunine J. I. (1985) Origins of satellites. In *Satellites* (J. A. Burns and M. S. Matthews, eds.), pp. 39–88. Univ. of Arizona, Tucson.

Stoll M. H. R. and Kley W. (2014) Vertical shear instability in accretion disc models with radiation transport. *Astron. Astrophys., 572,* A77.

Strobel D. F., Atreya S. K., Bézard B., et al. (2008) Atmospheric structure and composition. In *Titan from Cassini-Huygens* (R. H. Brown et al., eds.), pp. 235–257. Springer, Dordrecht.

Sullivan W. T. and Baross J. A., eds. (2007) *Planets and Life: The Emerging Science of Astrobiology.* Cambridge Univ., Cambridge. 604 pp.

Szulágyi J., Masset F., Legal E., Crida A., Morbidelli A., and Guillot T. (2016) Circumplanetary disc or circumplanetary envelope? *Mon. Not. R. Astron. Soc., 460,* 2853–2861.

Tanigawa T., Ohtsuki K., and Machida M. N. (2012) Distribution of accreting gas and angular momentum onto circumplanetary disks. *Astrophys. J., 747,* 47–63.

Tanigawa T., Maruta A., and Machida M. (2014) Accretion of solid materials onto circumplanetary disks from protoplanetary disks. *Astrophys. J., 784,* 109.

Thomas P. C. (2010) Sizes, shapes, and derived properties of the saturnian satellites after the Cassini nominal mission. *Icarus, 208,* 395–401,

Thomas P. C., Tajeddine R., Tiscareno M. S., Burns J. A., Joseph J., Loredo T. J., Helfenstein P., and Porco C. (2016) Enceladus's measured physical libration requires a global subsurface ocean. *Icarus, 264,* 37–47.

Tobie G., Grasset O., Lunine J. I., Mocquet A., and Sotin C. (2005) Titan's internal structure inferred from a coupled thermal-orbital model. *Icarus, 175,* 496–502, DOI: 10.1016/j.icarus.2004.12.007.

Umurhan O. M., Nelson R. P., and Gressel O. (2016) Linear analysis of the vertical shear instability: Outstanding issues and improved solutions. *Astron. Astrophys., 586,* A33.

Waite J. H., Combi M. R., Ip W.-H., et al. (2006) Cassini Ion and Neutral Mass Spectrometer: Enceladus plume composition and structure. *Science, 311,* 1419–1422.

Waite J. H., Lewis W. S., Magee B. A., Lunine J. I., McKinnon W. B., Glein C. R., Mousis O., Young D. T., Brockwell T., Westlake J., Nguyen M. J., Teolis B. D., Niemann H. B., McNutt R. L., Perry M., and Ip W. H. (2009) Liquid water on Enceladus from observations of ammonia and ^{40}Ar in the plume. *Nature, 460,* 487–490. DOI: 10.1038/nature08153.

Waite J. H., Glein C. R., Perryman R., Teolis B. D., Magee B. A., Miller G., Grimes J., Perry M. E., Miller K. E., Bouquet A., Lunine J. I., Brockwell T., and Bolton S. J. (2017) Cassini finds molecular hydrogen in the Enceladus plume: Evidence for hydrothermal processes. *Science, 356,* 155–159.

Walsh K. J., Morbidelli A., Raymond S. N., O'Brien D. P., and Mandell A. M. (2011) A low mass for Mars from Jupiter's early gas-driven migration. *Nature, 475,* 206–209.

Ward W. R. and Canup R. M. (2010) Circumplanetary disk formation. *Astron. J., 140,* 1168–1193.

Warren P. (2011) Stable-isotopic anomalies and the accretionary assemblage of the Earth and Mars: A subordinate role for carbonaceous chondrites. *Earth Planet. Sci. Lett., 311,* 93–100.

Yabuta H., Williams L. B., Cody G. D., Alexander C. M. O'D., and Pizzarello S. (2007) The insoluble carbonaceous material of CM chondrites: A possible source of discrete organic compounds under hydrothermal conditions. *Meteoritics & Planet. Sci., 42,* 37–48.

Yin Q.-Z., Sanborn M. E., Goodrich C. A., Zolensky M., Fioretti A. M., Shaddad M., Kohl I. E., and Young E. D. (2018) Nebula scale mixing between non-carbonaceous and carbonaceous chondrite reservoirs: Testing the Grand Tack model with Almahata Sitta stones. In *Lunar and Planetary Science XLIX,* Abstract #2083. Lunar and Planetary Institute, Houston.

Zahnle K., Schenk P., Levison H., and Dones L. (2003) Cratering rates in the outer solar system. *Icarus, 163,* 263–289.

Zolensky M. E., Nakamura K., Gounelle M., Mikouchi T., Kasama T., Tachikawa O., and Tonui E. (2002) Mineralogy of Tagish Lake: An ungrouped type 2 carbonaceous chondrite. *Meteoritics & Planet. Sci., 37,* 737–761.

Zolotov M. Y. (2007) An oceanic composition on early and today's Enceladus. *Geophys. Res. Lett., 34,* L23203, DOI: 10.1029/2007/GL031234.

Zolotov M. Yu. (2012) Aqueous fluid composition in CI chondritic materials: Chemical equilibrium assessments in closed systems. *Icarus, 220,* 713–729.

Zolotov M. Yu. and Kargel J. S. (2009) On the chemical composition of Europa's icy shell, ocean, and underlying rocks. In *Europa* (R. T. Pappalardo et al., eds.), pp. 431–457. Univ. of Arizona, Tucson.

Glein C. R., Postberg F., and Vance S. D. (2018) The geochemistry of Enceladus: Composition and controls. In *Enceladus and the Icy Moons of Saturn* (P. M. Schenk et al., eds.), pp. 39–56. Univ. of Arizona, Tucson, DOI: 10.2458/azu_uapress_9780816537075-ch003.

The Geochemistry of Enceladus: Composition and Controls

C. R. Glein
Southwest Research Institute

F. Postberg
University of Heidelberg

S. D. Vance
NASA Jet Propulsion Laboratory/California Institute of Technology

Enceladus is the first world beyond Earth for which we have diagnostic observational data concerning the composition of an extant ocean. The abundances of chemical species in the plume of Enceladus, which is sourced by the ocean, provide valuable constraints on the temperature, major ion composition, pH, and reduction-oxidation (redox) chemistry of its ocean. The pressure profile is determined by the internal structure of the moon, which also constrains temperatures in Enceladus' rocky core. The data indicate that the ocean is a relatively alkaline and reduced solution of dissolved sodium, chloride, and bicarbonate/carbonate ions just below the freezing point of pure water. Observations of silica nanoparticles and molecular hydrogen from Enceladus can be explained in terms of hydrothermal sources. We present a model in which the dissolution of quartz in hot water yields high concentrations of silica, and reduced rocks tidally heated in the presence of water supply abundant hydrogen. Future studies should seek to address many of the geochemical questions raised by the Cassini era of Enceladus exploration.

1. ENCELADUS AS A GATEWAY TO EXTRATERRESTRIAL OCEANOGRAPHY

Numerous lines of evidence made possible by the Cassini-Huygens mission (*Porco et al.,* 2006, 2014; *Spencer et al.,* 2006; *Collins and Goodman,* 2007; *Nimmo et al.,* 2007; *Zolotov,* 2007; *Glein et al.,* 2008; *Schmidt et al.,* 2008; *Tobie et al.,* 2008; *Postberg et al.,* 2009, 2011; *Waite et al.,* 2009, 2017; *Glein and Shock,* 2010; *Ingersoll and Pankine,* 2010; *Patthoff and Kattenhorn,* 2011; *Běhounková et al.,* 2012; *Matson et al.,* 2012; *Iess et al.,* 2014; *Bouquet et al.,* 2015; *Hsu et al.,* 2015; *McKinnon,* 2015; *Travis and Schubert,* 2015; *Beuthe et al.,* 2016; Čadek et al., 2016; *Ingersoll and Nakajima,* 2016; *Kite and Rubin,* 2016; *Nakajima and Ingersoll,* 2016; *Thomas et al.,* 2016; *Choblet et al.,* 2017; see also chapters by Hemingway et al. and Spencer et al. in this volume) support the paradigm that Saturn's moon, Enceladus, has a global subsurface ocean of liquid water, which erupts into space forming a south polar plume of gases and ice grains. From a geochemical perspective, what is most remarkable about Enceladus is that its plume provides access to the chemical composition of its ocean. The relative ease of obtaining constraints on the geochemistry of the subsurface of Enceladus is unprecedented in the field of planetary science, where detailed information on composition usually necessitates landers.

Why might one be interested in the geochemistry of Enceladus? First, it is of broad interest because the composition of the environment imposes boundary conditions on the possible origin, evolution, and persistence of life (*McKay et al.,* 2008). Second, the accessibility of its ocean enables Enceladus to serve as a model for the geochemistry of oceans inside other icy worlds, such as Ganymede (*Vance et al.,* 2014), Callisto, Mimas, Dione, Titan (*Glein,* 2015), Triton (*Shock and McKinnon,* 1993), and Pluto (*Neveu et al.,* 2015). Third, knowledge of the geochemistry of Enceladus allows us to frame the subject of aqueous geochemistry in a more universal context. This will help us to better understand the processes that led to similarities and differences in the observed compositions of water-bearing bodies across the solar system, including Earth (*Lowenstein et al.,* 2014; *German and Seyfried,* 2014), Mars (*Ehlmann et al.,* 2013; *Niles et al.,* 2013), carbonaceous chondrite parent bodies (*Brearley,* 2006), Ceres (*Zolotov,* 2017), and Europa (*Zolotov and Kargel,* 2009; *Vance et al.,* 2016). Fourth, this is new science; the thrill of exploration motivates the burgeoning field of extraterrestrial chemical oceanography.

In this chapter, we examine key aspects of the geochemistry of Enceladus. The chapter contains a mixture of review material and more refined interpretations. We begin with an overview of the relevant observational data (see section 2). Then, we organize our discussion of enceladean geochemistry into two parts: marine geochemistry (see section 3) and hydrothermal geochemistry (see section 4). Figure 1 shows the locations of these types of geochemistry with respect to the vertical structure of Enceladus. In reality, the boundaries may not be sharp. Our goals in these discussions are to (1) introduce some of the basic concepts of geochemistry that are useful for interpreting chemical data from Enceladus, and (2) present contemporary models for the compositions of the ocean and suspected hydrothermal fluids. We conclude this chapter with a set of critical questions to take us from Cassini to the next era of exploration (see section 5).

2. CONSTRAINTS FROM PLUME COMPOSITION

Because the plume of Enceladus is sourced from a liquid water ocean (see section 1), the plume's composition can provide insights into the geochemistry of Enceladus' ocean and deeper hydrothermal fluids. The Cosmic Dust Analyzer (CDA) and the Ion and Neutral Mass Spectrometer (INMS) were the primary instruments onboard the Cassini spacecraft that measured the composition of the plume (see the chapter by Postberg et al. in this volume). *In situ* measurements were performed when the spacecraft flew through the plume, or sampled Saturn's E ring, which is maintained by the delivery of plume materials (*Haff et al.,* 1983; see the chapter by Kempf et al. in this volume).

The CDA instrument discovered that the plume contains three compositionally distinct populations of ice grains (*Postberg et al.,* 2008, 2009, 2011, 2018). Type I is nearly pure water ice, Type II contains organic compounds, and Type III is rich in salts. The largest mass line in CDA spectra of Type III grains is from Na^+; smaller mass lines from K^+, and sodium clusters with OH^-, Cl^-, and/or CO_3^{-2}, were observed (*Postberg et al.,* 2009). The salt-rich ice grains are thought to form by flash freezing of ocean water. The large abundance of Na in these grains is indicative of extensive water-rock interaction (*Zolotov,* 2007), consistent with the

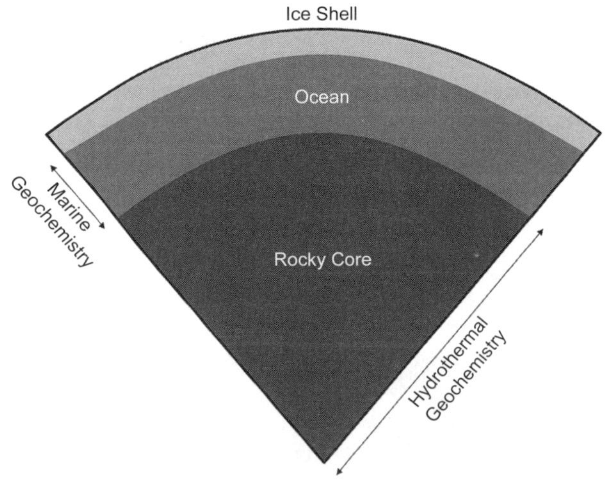

Fig. 1. The two general categories of geochemical processes shown in a schematic cross section of Enceladus.

presence of a rocky core with a low density (*Iess et al.,* 2014; *McKinnon,* 2015; see the chapter by Hemingway et al. in this volume), resulting from silicate hydration.

Postberg et al. (2009) used a laser to generate gaseous ions from solutions containing NaCl and $NaHCO_3$ or Na_2CO_3. Laser ionization in the laboratory was meant to serve as a proxy for impact ionization in the CDA instrument. It was found that CDA spectra of salt-rich ice grains can be reproduced from solutions containing 0.05–0.2 mol NaCl per kg H_2O and 2–5 times more chloride than bicarbonate or carbonate (throughout this chapter, molality (mol/kg H_2O) is the preferred unit for the concentrations of aqueous species). There is currently ambiguity as to whether HCO_3^- or CO_3^{-2} is predominant in the ocean source of the plume, because solutions containing either of these species are able to reproduce a mass peak attributed to $(Na_2CO_3)Na^+$ (*Postberg et al.,* 2009). The salt composition derived by *Postberg et al.* (2009) is summarized in Table 1.

The CDA detected silicon-rich, nanometer-scale particles escaping from the saturnian system, during the approach of the spacecraft to Saturn (*Kempf et al.,* 2005). These particles are thought to originate as inclusions in Type II or III E-ring ice grains, and then be released from the grains by plasma-sputtering erosion (*Hsu et al.,* 2011). Therefore, their source is inferred to be the plume of Enceladus. The Si-rich stream particles are poor in metal cations, which implies that they are composed of nearly pure silica (SiO_2). It is difficult to determine the concentration of silica in ocean-derived ice grains, as silica was not measured in the plume or E ring. The current best estimate is a SiO_2 concentration of ~150–3900 ppm by mass, based on the dynamical modeling of *Hsu et al.* (2015).

The INMS made measurements of gas molecules in the plume, using both closed- and open-source modes of operation. The former mode provides greater sensitivity, but some species undergo chemical reactions inside the instrument before they are able to reach the detector. In open-source mode, collisions and thus chemical reactions of plume materials are minimized. The initial set of closed-source data revealed that the most abundant plume constituent is water vapor (*Waite et al.,* 2006), consistent with observations by the Ultraviolet Imaging Spectrograph (UVIS) (*Hansen et al.,* 2006). Subsequent INMS measurements showed that the detected composition depends on the flyby velocity, indicative of reactions of certain plume materials (e.g., organic compounds) induced by high-energy impacts of ice grains on the walls of the instrument (*Waite et al.,* 2009). Hydrogen

TABLE 1. Major composition of salt-rich ice grains in Enceladus' plume.

Constituent	Concentration (mol/kg H_2O)
NaCl	0.05–0.2
$NaHCO_3 + Na_2CO_3$	0.01–0.1
KCl	$(0.5–2) \times 10^{-3}$

Data from *Postberg et al.* (2009, 2011).

gas in the closed source was inferred to be produced by interactions between water molecules and titanium metal in the INMS antechamber (*Waite et al.,* 2009, 2017).

The final set of measurements, made in open-source mode, provide the most reliable discrimination between the true composition vs. the impact-modified composition (*Waite et al.,* 2017). These data indicate the presence of H_2O, H_2, and CO_2. The identification of CO_2 in the plume gas is further supported by observations of CO_2 in the tiger stripes region on the surface of Enceladus by the Visible and Infrared Mapping Spectrometer (VIMS) (*Brown et al.,* 2006). No mass-28 species were detected in open-source mode, consistent with the lack of detection of CO (*Hansen et al.,* 2008) and N_2 (*Hansen et al.,* 2011) by UVIS. However, a mass-28 species was detected using the closed source (*Waite et al.,* 2006, 2009).

The discrepancy between the open- and closed-source data with regards to the detection of mass 28 (*Waite et al.,* 2017) implies that a closed-source detection may be considered tentative if an open-source measurement was not made. Two major species fall into this category: NH_3 and CH_4. Observations of NH_3 on Enceladus' surface are presently equivocal (*Brown et al.,* 2006; *Hendrix et al.,* 2010), but the discovery of N^+ in Saturn's inner magnetosphere by the Cassini Plasma Spectrometer (CAPS) suggests that plume NH_3 may be the parent species (*Smith et al.,* 2008; see the chapter by Postberg et al. in this volume). This supports the identification of NH_3 in the plume. For CH_4, we are not aware of any additional observational evidence for its presence at Enceladus. The gas composition derived by *Waite et al.* (2017) is summarized in Table 2.

3. MARINE GEOCHEMISTRY

3.1. Pressure and Temperature in the Ocean

Pressures in the ocean of Enceladus can be estimated from models of the internal structure of the satellite. These models are constrained by Cassini measurements of gravity, topography, and libration (*Iess et al.,* 2014; *Thomas et al.,* 2016; see the chapter by Hemingway et al. in this volume). Here, we consider the model of *McKinnon* (2015), which has a rocky core of density 2450 kg m^{-3} and radius 190 km. We do not distinguish between the slightly different densities of water ice (~920 kg m^{-3}) and ocean water (~1030 kg m^{-3}), but instead adopt a uniform density of ρ_h = 1000 kg m^{-3} for the whole hydrosphere overlying the rocky core. For such a two-layer model in which the interior can be approximated as being in a state of hydrostatic equilibrium, the pressure (P) at radius r between the core radius and the mean surface radius (R_E = 252 km) can be computed analytically using (*Turcotte and Schubert,* 2002)

$$P = \frac{4}{3}\pi\rho_h G R_c^3 (\rho_c - \rho_h)\left(r^{-1} - R_E^{-1}\right) + \frac{2}{3}\pi G \rho_h^2 \left(R_E^2 - r^2\right) \quad (1)$$

where G refers to the gravitational constant (6.674 × 10^{-11} m^3 kg^{-1} s^{-2}), R_c the radius of the core, and ρ_c its density. Figure 2 shows the pressure profile in the hydrosphere. This model yields pressures in the ocean ranging between ~2 and ~74 bar (1 bar = 0.1 MPa). Pressures are relatively low even under a deep ocean, because of Enceladus' weak gravity (the surface acceleration due to gravity g = 0.113 m s^{-2}).

Because the ocean of Enceladus is covered by water ice, its temperature ought to be affected by phase equilibrium between the ocean and ice. However, the freezing point of the ocean is depressed because the water is not pure, but contains dissolved salts that decrease the fugacity of H_2O. The fugacity can be thought of as a thermodynamically corrected partial pressure (*Anderson,* 2005). Freezing-point depression depends in general on the total concentration of solute particles but not their identities. The depression of the freezing point (ΔT_f) of the Enceladus ocean can be approximated by

$$\Delta T_f = -K_f\left(2m_{NaCl} + 2m_{NaHCO_3} + 3m_{Na_2CO_3}\right) \quad (2)$$

TABLE 2. Major composition of gases in Enceladus' plume.

Constituent	Molar Percentage
H_2O	96–99
H_2	0.4–1.4
CO_2	0.3–0.8
NH_3	0.4–1.3
CH_4	0.1–0.3

Data from *Waite et al.* (2017).

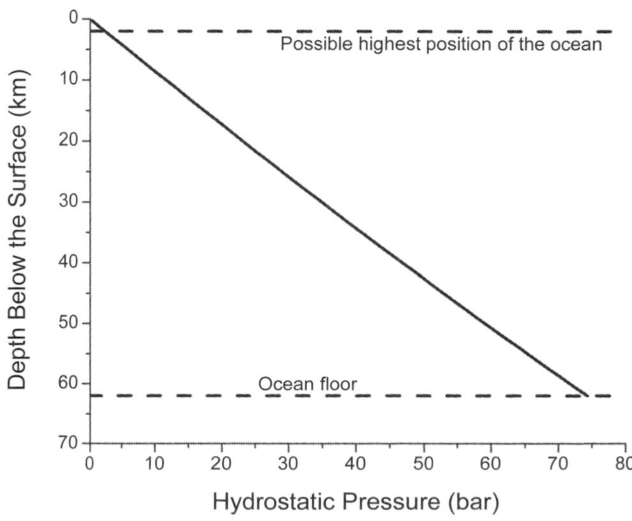

Fig. 2. Pressures in the ice shell and ocean of Enceladus based on the internal structure model of *McKinnon* (2015). The thickness of the ice shell varies with latitude and longitude, but it could be as thin as ~2 km in the south polar region (*Beuthe et al.,* 2016; Čadek et al., 2016; see the chapter by Hemingway et al. in this volume). The depth of the ocean floor follows from the core radius of ~190 km.

where K_f stands for the cryoscopic constant of water (1.86°C kg mol^{-1}), and m_i the molality of the ith salt. The coefficients (called van't Hoff factors) account for the number of ions that are released per formula unit of the corresponding salt.

Figure 3 depicts results from equation (2) for endmember salt compositions from Table 1. This plot indicates that the temperature of the ocean can be expected to be only slightly below the freezing point of pure water. The minimum temperature is about −1°C. It should be noted that ion pairing (e.g., NaCl,aq) is ignored in the present treatment, but this effect would lessen the freezing-point depression. This reinforces the conclusion that the ocean should not be much cooler than ~0°C.

An alternative frigid ocean maintained at the eutectic point of the NH_3-H_2O system (*Squyres et al.,* 1983) is implausible on Enceladus, because the plume is not rich in NH_3 (*Porco et al.,* 2006). Assuming that NH_3 is indeed present (see section 2), the NH_3/H_2O ratio in the plume gas ranges between ~0.004 and ~0.014 (Table 2). For the eutectic composition of ~35 mol% NH_3 and ~65 mol% H_2O at ~175 K, the saturation vapor pressures are predicted to be ~1×10^{-4} bar NH_3 and ~2×10^{-8} bar H_2O (*Tillner-Roth and Friend,* 1998). This corresponds to an NH_3/H_2O ratio of ~5000 in the gas phase, which is inconsistent with the data from INMS by many orders of magnitude. An analogous argument applies to methanol and other volatile antifreezes (*Waite et al.,* 2009).

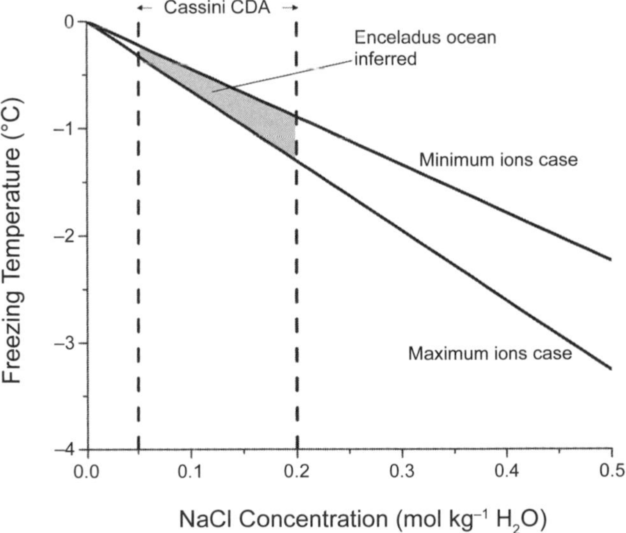

Fig. 3. Temperature of the Enceladus ocean (gray region) based on the freezing point depression of salt solutions (see section 3.1). The minimum ions case has a $NaHCO_3$/NaCl molal ratio of 0.2 (and no Na_2CO_3), while the molal ratio of Na_2CO_3/NaCl is 0.5 (with no $NaHCO_3$) in the maximum ions case (*Postberg et al.,* 2009). The dashed lines indicate the range of NaCl concentration in salt-rich plume particles (Table 1), which would be similar to that in the ocean if the water is flash frozen as a result of boiling (*Ingersoll and Nakajima,* 2016; *Nakajima and Ingersoll,* 2016).

3.2. Composition of the Ocean

3.2.1. ***Major ions.*** As a first step toward understanding the geochemistry of Enceladus' ocean, it can help to compare what we think we know about its composition to various natural waters on Earth. We compiled literature sets of representative geochemical data (Table 3) for eight terrestrial sites that may serve as physical (ice-covered) or chemical (carbonate-rich, basic pH, or reducing conditions) analogs of Enceladus' ocean. These sites are not meant to be exhaustive of all possible analogs. Rather, the intent is to introduce a group that can help us get our bearings in terms of discussing the composition of the ocean from an empirical point of view.

Our representation of seawater is the reference composition given by *Millero* (2013). Lost City is a low-temperature hydrothermal system that is located near the Mid-Atlantic Ridge (*Kelley et al.,* 2001, 2005). The fluid composition is derived from the circulation of seawater through ultramafic (magnesium- and iron-rich) rocks. This leads to hydration and oxidation (called "serpentinization") of the rocks. The Cedars is a site in northern California where ultramafic rocks are being serpentinized on a continent (*Morrill et al.,* 2013). Ronda Peridotite is another continental ultramafic site that is located in southern Spain (*Bucher et al.,* 2015). The data in Table 3 for Ronda Peridotite refer to brooks whose water has participated in extensive weathering.

Ikka Fjord is a small fjord in southwestern Greenland (*Buchardt et al.,* 1997). Towers composed of ikaite ($CaCO_3 \cdot 6H_2O$) precipitate at the bottom of the fjord, where carbonate-rich spring water seeps into the fjord and mixes with calcium-bearing seawater. Lake Magadi is a soda lake in the East African Rift Valley in Kenya. It is a brine that is mostly covered by trona ($Na_2CO_3 \cdot NaHCO_3 \cdot 2H_2O$). Lake Vida is a perennially ice-covered brine in the McMurdo Dry Valley of East Antarctica (*Murray et al.,* 2012). Its water chemistry may be controlled by reactions with igneous rocks in the surroundings or lake sediments, as well as freezing of water that concentrates soluble salts. Lake Untersee is another East Antarctic lake (*Wand et al.,* 1997), but it is a freshwater lake because its ice cover (~3 m) is much thinner than that over Lake Vida (~30 m).

We specify a nominal empirical composition for the Enceladus ocean to facilitate comparisons between its major ion chemistry and those of the possible analogs. We adopt a recommended pH range of 9 to 11 (see section 3.2.2). The rest of the nominal composition is as follows: (1) 0.1 molal Cl^-, (2) $HCO_3^- + CO_3^{-2}$ concentration of 0.03 molal, (3) Na/K molal ratio of 100, and (4) H_2/CO_2 molal ratio of 1.6. These values are consistent with the data in Tables 1 and 2 (see also section 3.2.3). Speciation calculations are performed at 0°C using the thermo.com.V8.R6+ database in The Geochemist's Workbench (*Bethke,* 2008). The present model does not include Ca^{+2}, Mg^{+2}, or SO_4^{-2} as these species have not been detected in the plume or E ring. However, this does not mean that these species are absent from the ocean. We also do not consider NH_3 because it is presently unclear

TABLE 3. Geochemical properties of some potential Earth analogs of the Enceladus ocean, compared to nominal models of the latter (see section 3.2.1).

Quantity	Seawater	Lost City	The Cedars	Ronda Peridotite	Ikka Fjord	Lake Magadi	Lake Vida	Lake Untersee	Enceladus pH 9	Enceladus pH 11
T (°C)	25	90	17.2	20	4	35	−13.4	0.4	~0	~0
log a_{H_2}	−44.3	−2.0	−4.7	−45.2	−48.0	−42.8	−4.9	−48.8	−3.9	−6.8
pH	8.1	10.6	11.5	8.5	10.5	10.5	6.2	10.6	9	11
Na^+	486	511	0.94	0.17	175	6230	2090	1.98	130	154
K^+	10.6	10.8	0.01	0.005	1.66	53.5	89.2	0.08	1.30	1.54
Ca^{+2}	10.7	28.3	1.28	0.02	0.20	~0	32.4	1.04	ND	ND
Mg^{+2}	54.7	~0	0.008	3.23	~0	~0	716	0.04	ND	ND
Cl^-	566	559	0.93	0.19	22.5	2460	3510	0.87	100	100
OH^-	0.002	12.3	2.58	0.002	8e-5	0.69	5e-6	0.05	0.002	0.16
HCO_3^-	1.78	2e-4	1e-4	5.88	26.4	86.4	38.4	0.04	28.3	4.22
CO_3^{-2}	0.25	0.01	0.006	0.23	61.4	1845	0.22	0.09	1.62	25.8
SO_4^{-2}	29.3	3.31	0.001	0.04	2.74	23.6	62.9	1.43	ND	ND
I_S (molal)	0.72	0.61	0.005	0.01	0.24	8.1	4.5	0.007	0.13	0.18
Reference	*	†	‡	§	¶	**	††	‡‡	§§	§§

* *Millero* (2013).
† *Seyfried et al.* (2015).
‡ *Morrill et al.* (2013).
§ *Bucher et al.* (2015).
¶ *Buchardt et al.* (2001).
** *Jones et al.* (1977).
†† *Murray et al.* (2012).
‡‡ *Wand et al.* (1997).
§§ This work.

Oxidation state is expressed in terms of the activity of H_2 (a_{H_2}). Values for the aqueous species are stoichiometric concentrations (mmoles per kilogram of water) that include the free ion and ion pairs. The ionic strength is given as the stoichiometric ionic strength (I_S). ND means not detected.

how the plume abundance of NH_3 (Table 2) translates to an ocean concentration. Because NH_3 is relatively non-volatile, it probably freezes out during transport and thus does not exhibit conservative behavior between the ocean and plume.

The concentrations of major ions in Table 3 reveal similarities and differences between the natural waters on Earth and Enceladus' ocean. Seawater is similar to Enceladus' ocean in terms of being dominated by NaCl. However, seawater is poorer in carbonate species, and may be richer in Mg^{+2} and SO_4^{-2} than the Enceladus ocean if the lack of detection of these species by CDA argues against the ocean containing more than ~10 mmolal $MgSO_4$. The serpentinization systems at Lost City and the Cedars both have relatively high concentrations of dissolved calcium hydroxide. This drives a decrease in the concentrations of carbonate species in these systems, unlike the case of Enceladus. Ronda Peridotite also appears to differ from the Enceladus ocean, as the former is dominated by Mg^{+2} and HCO_3^-. In contrast, both Ikka Fjord and Lake Magadi seem similar to Enceladus' ocean in terms of being rich in Na^+, Cl^-, and carbonate species. Lake Vida may have a larger enrichment in Mg^{+2} than the ocean of Enceladus, if it is assumed that this species has a concentration below ~10 mmolal in the latter. Finally, the large relative contributions from Ca^{+2} and SO_4^{-2} to the ion chemistry of Lake Untersee may represent a key difference between this environment and Enceladus' ocean.

We also observe that some of the potential analogs are similar to the models of the Enceladus ocean with respect to ionic strength (an indicator of total salt content), while others are dramatically different (Table 3). Both seawater and Lost City fluid (derived from seawater) are modestly higher in ionic strength than modeled ocean water on Enceladus, whereas rainwater-derived fluids at the Cedars and Ronda Peridotite are markedly lower in ionic strength. Ikka Fjord presents an interesting case that appears to have a similar ionic strength as the ocean of Enceladus. It is apparent that waters from both Lake Magadi and Lake Vida are much higher in ionic strength than the Enceladus ocean (Table 3). Conversely, the low ionic strength of Lake Untersee makes this body of glacial meltwater much "fresher" than the ocean of Enceladus.

Based on the preceding comparisons, Ikka Fjord and seawater may be the closest analogs of the Enceladus ocean out of the eight cases considered, with respect to the composition of major ions. We caution that these analogs are not perfect matches to Enceladus (nor should we expect them to be given the different geological contexts). But between the two of them, they adequately mirror the dominant Na-Cl-HCO_3/CO_3 chemistry of the Enceladus ocean.

3.2.2. ***pH.*** The pH of an aqueous solution is given by the following equation

$$pH = -\log\left(a_{H^+}\right) \tag{3}$$

where a_{H^+} stands for the activity of the hydrogen ion [the standard state for aqueous species, which defines an activity of unity, is a hypothetical one molal solution referenced to infinite dilution at any temperature and pressure (*Anderson,* 2005)]. A solution with a low pH is acidic because the activity of H^+ is high. Self-ionization of water relates the acidity and basicity of the solution via

$$H_2O, aq \leftrightarrow H^+, aq + OH^-, aq \tag{4}$$

which has an equilibrium constant

$$K_w = \frac{a_{H^+} a_{OH^-}}{a_{H_2O}} = 10^{-14.94} \text{ at } 0°C \text{ and } 1 \text{ bar} \tag{5}$$

The activities (a) of H^+ and OH^- are inversely related. The activity of H_2O is usually close to unity (the pure standard state) in natural waters, with the exception of brines. From equation (5), it can be deduced that a basic solution with a high activity of OH^- must have a low activity of H^+ and thus a high pH. A solution is said to be neutral if the activities of H^+ and OH^- are equal. The neutral pH at 0°C is ~7.5.

The pH is generally the most important compositional variable in aquatic systems. It has a powerful influence on the geochemical behavior of almost all the elements, except for the noble gases, the alkali metals, and the halogens. The pH governs the speciation of many systems, as species interconvert by releasing or taking up H^+. The carbonate system provides a classic example that is relevant to the geochemistry of Enceladus. The chief equilibria are

$$CO_2, aq + H_2O, aq \leftrightarrow H^+, aq + HCO_3^-, aq \tag{6}$$

and

$$HCO_3^-, aq \leftrightarrow H^+, aq + CO_3^{-2}, aq \tag{7}$$

As another example relevant to Enceladus, the pH regulates the availability of metals derived from minerals that have equilibrated with the solution. For the case of talc dissolution, we have

$$\begin{aligned} &Mg_3Si_4O_{10}(OH)_2, \text{talc} + 6\ H^+, aq \leftrightarrow \\ &3\ Mg^{+2}, aq + 4\ SiO_2, aq + 4\ H_2O, aq \end{aligned} \tag{8}$$

By applying Le Chatelier's principle, it can be deduced that an increase in the activity of H^+ would shift the equilibrium to the right. Therefore, there should tend to be more Mg^{+2} in solution at lower pH.

There have been several attempts to estimate the pH of Enceladus' ocean. *Zolotov* (2007) performed calculations of chemical equilibrium between liquid water and rock of CI carbonaceous chondrite bulk composition. The latter may be representative of rocks in Enceladus' core if it has not undergone igneous differentiation by partial melting (*Médard and Kiefer,* 2017), which is unlikely for a body as small as Enceladus [vigorous hydrothermal circulation may also make it difficult to reach magmatic temperatures (*Travis and Schubert,* 2015; *Choblet et al.,* 2017)]. *Zolotov*'s (2007) calculations indicate that the rocks would be hydrated to primarily Mg-phyllosilicates during alteration, and a Na-Cl-CO_3-HCO_3 solution with a pH of ~10.9 (at 0°C for a water/rock mass ratio of 1) would be produced. The general character of the predicted solution is consistent with subsequent measurements by Cassini CDA (Table 1). Of course, we do not know whether water-rock equilibrium is reached, which is a crucial assumption.

Postberg et al. (2009) attempted to determine the pH of the ocean from CDA spectra of salt-rich ice grains from the plume. They fired a laser at laboratory solutions of known pH in experiments simulating impact ionization of these grains. The relative abundances of salt clusters [$Na(NaOH)_n^+$ and $Na(NaCl)_n^+$, where n signifies a positive integer] from the analog experiments were correlated to the pattern of clusters in the Enceladus spectra. The concept can be illustrated by considering the following reaction that occurs in the high-energy zone of a laser or ice grain impact

$$Na(NaCl)^+, g + OH^-, g \rightarrow Na(NaOH)^+, g + Cl^-, g \tag{9}$$

At higher pH, OH^- is more abundant, so a larger peak area ratio of $Na(NaOH)^+$ to $Na(NaCl)^+$ is observed. The pronounced peaks from $Na(NaOH)_{1-3}^+$ in the Enceladus spectra therefore imply a relatively basic pH. *Postberg et al.* (2009) inferred a pH of 8.5–9. A potential concern is the issue of whether laser ionization in the laboratory is a quantitative analog for impact ionization in the CDA instrument. The energy density of the laser can be tuned to match that of ice grain impacts, but it is presently unclear if there should be a 1:1 correspondence, or if the laser-calibrated pH could be offset from the pH of flash frozen ocean droplets that impact the instrument.

Marion et al. (2012) performed speciation calculations in the carbonate system (see equations (6) and (7)) to estimate the pH. In their model, they adopted input conditions of 0.2 molal chloride and an alkalinity ($\approx m_{HCO_3^-} + 2m_{CO_3^{-2}}$) of 0.05 eq kg^{-1} H_2O. These values are consistent with the CDA data (Table 1). Their interpretation of the INMS data (Table 2) was that the ocean is rich in dissolved gases. This led to the assumption that gas (clathrate) hydrates might be controlling the fugacity of CO_2. Hence, a high fugacity of CO_2 (0.349 bar) was adopted as an input parameter. Marion et al. computed the speciation of the system using the FREZCHEM code. The modeled pH ranged between ~5.7 and ~6.8. However, this estimate is probably too low because the adopted fugacity of CO_2 is too high (see below). The weakly acidic pH in their model is caused by the formation of carbonic acid at relatively high fugacities of CO_2. Rainwater on Earth represents a classic example of this phenomenon (*Drever,* 1997).

Hsu et al. (2015) suggested that the presence of nanosilica in Enceladus' ocean constrains the pH of the ocean. In their interpretation (see section 4.2.1), a cooled hydrothermal fluid that is supersaturated in amorphous silica can produce nanometer-sized particles only if the ocean has a moderately alkaline pH. Above a pH of ~10.5, amorphous silica becomes too soluble to maintain a stable colloidal phase, as implied by the reaction below

$$SiO_2, \text{amorphous} + OH^-, aq \rightarrow HSiO_3^-, aq \quad (10)$$

These considerations led *Hsu et al.* (2015) to propose a pH of ~8.5–10.5. A possible concern with this approach is that it depends on nanosilica being present in the ocean. *Hsu et al.* (2015) provided a number of supporting arguments that the nanosilica measured by CDA originates from E-ring ice grains, but nanosilica was not directly observed in either the plume or the E ring.

Glein et al. (2015) improved upon the carbonate speciation approach of *Marion et al.* (2012) in their effort to infer the pH. *Glein et al.* (2015) attempted to constrain the activity of CO_2 in the ocean not directly from the plume abundance of CO_2 (Table 2), but instead by trying to account for the evolution of the CO_2/H_2O ratio in the gas as it migrates from the ocean to space. A large amount of H_2O condenses during migration, as the tiger stripes (~200 K) are much colder than the ocean (~273 K). If it is assumed that CO_2 does not condense or condenses to a lesser extent owing to its greater volatility, then the CO_2/H_2O ratio in the plume should be greater than the ratio in the ocean source region. *Glein et al.* (2015) adopted this no condensation of CO_2 endmember, and used the sublimation curve of water ice to estimate the CO_2/H_2O ratio at the ocean. This allowed them to make estimates of the activity of CO_2, which were combined with the salt data (Table 1) to calculate the carbonate speciation of the ocean. The model returned a pH range of ~10.8–13.5. A caveat is that the assumption of no CO_2 condensation is at best an approximation, and could lead to a pH overestimate of perhaps ~1 unit. This is a concern because VIMS data demonstrate the condensation of CO_2 (*Brown et al.,* 2006), although the fraction that condenses is still being worked out (*Matson et al.,* 2018).

Table 4 summarizes the previously published estimates for the pH of the ocean of Enceladus. If the inconsistent values from *Marion et al.* (2012) are discarded, then we are left with a range from pH 8.5 to 13.5. This is the total uncertainty. The range is not small, but it is not a trivial problem to determine the pH of an ice-covered ocean using measurements made in space. Looking at the ranges in the remaining studies, we suggest a "best-fit" pH of ~9–11. This can be regarded as a compromise between the studies, and it is proposed as a working model for the Enceladus ocean.

3.2.3. Apparent oxidation state. In general, low-temperature natural waters do not have a unique oxidation state. Individual redox couples comprising reduced and oxidized forms of an element can be quantified by an oxidation state parameter (see below), but different redox couples will have different parameter values if the entire chemical system is not at redox equilibrium. The latter is almost always the case in low-temperature environments where the rates of electron transfer reactions are too sluggish to permit the attainment of equilibrium. Nevertheless, it is useful to define oxidation state parameters, as they give us a scale that can be used to make general comparisons between different geochemical environments in terms of whether reduction or oxidation is thermodynamically favored. This is essential for rationalizing the behavior of elements that can exist in multiple formal oxidation states (e.g., Fe, S, C, N).

Because of the great abundance of water in Enceladus' ocean, it makes sense to choose a redox parameter based on hydrogen or oxygen. Here, we choose the activity of H_2 (a_{H_2}) because H_2 was measured in Enceladus' plume, and a model has been developed to derive the H_2 molality (~activity) in the ocean from the plume measurement (*Waite et al.,* 2017). A higher H_2 activity means that there is a stronger thermodynamic drive for H_2 to reduce other species. The choice of H_2 activity to define a redox scale is simply one of convenience. Values of other commonly used redox parameters can be calculated using thermodynamic relations. For example, the O_2 fugacity (f_{O_2}) (*Frost,* 1991) is related to the H_2 activity via the disproportionation of water

$$2\ H_2O, aq \leftrightarrow 2\ H_2, aq + O_2, g \quad (11)$$

which has an equilibrium constant

$$K_{11} = \frac{a_{H_2}^2 f_{O_2}}{a_{H_2O}^2} = 10^{-98.3} \text{ at } 0°C \text{ and } 1 \text{ bar} \quad (12)$$

The minuscule value of this equilibrium constant implies that measurable amounts of H_2 and O_2 cannot coexist at equilibrium at 0°C. Therefore, if a measured H_2 or O_2 parameter is used to derive the other parameter, the derived parameter will be unphysically small, and is not "real" but an abstraction (*Anderson,* 2005).

Another widely used redox variable is the reduction potential (Eh) referenced to the standard hydrogen electrode (SHE). On this electrochemical scale, SHE (comprised of unit fugacity of H_2 gas and unit activity of H^+) is defined as the zero point. A half-cell reaction that allows one to relate the activity of H_2 to the corresponding Eh can be written as

$$2\ H^+, aq + 2\ e^-, aq \leftrightarrow H_2, aq \quad (13)$$

From the stoichiometry of this reaction, the Eh can be calculated using an appropriate form of the Nernst equation as shown below

$$Eh = E^\circ - \frac{2.3026RT}{nF}\left(\log a_{H_2} + 2pH\right) \quad (14)$$

TABLE 4. Estimates for the pH of Enceladus' ocean.

Reference	pH Value
Zolotov (2007)	10.9
Postberg et al. (2009)	8.5–9
Marion et al. (2012)	5.7–6.8
Hsu et al. (2015)	8.5–10.5
Glein et al. (2015)	10.8–13.5
Recommendation (this work)	~9–11

where E° designates the standard reduction potential (–0.0983 V at 0°C and 1 bar), R the gas constant (8.3145 J mol^{-1} K^{-1}), T the absolute temperature, n the number of electrons transferred, and F Faraday's constant (96,485 C mol^{-1}).

It is not easy to determine the concentration of H_2 in the ocean of Enceladus from the mixing ratio of H_2 in the plume. One approach is to assume that the H_2/H_2O ratio is the same in the ocean as in the plume. However, this is problematic because water vapor condensation during transport should increase the gas-phase ratio of H_2/H_2O. Modeling can be performed to try to account for this effect, but that requires an assumption to be made with regard to the quench temperature for solid-vapor equilibrium of water in the tiger stripes (*Glein et al.,* 2015). A more robust approach is to find a different volatile ratio (i.e., one with a much less condensable reference species than H_2O) that would be minimally fractionated between the ocean and plume. An issue, however, is that deriving absolute concentrations in the ocean (e.g., the molality of H_2) from ratios of volatile gases in the plume requires a constraint on the absolute concentration of the reference species.

Waite et al. (2017) considered this state of affairs, and proposed an approach centered on CO_2 as the reference species. Carbon dioxide is much less susceptible to condensation than water because of its much greater volatility. Thus, it may be reasonable to assume that there is minimal condensation of CO_2 during transport, although some condensation must occur (*Brown et al.,* 2006). Another reason why CO_2 is a useful reference species is that its absolute concentration in the Enceladus ocean can be estimated using existing data from Enceladus. This can be done by evaluating the speciation of the carbonate system (see equations (6) and (7)) for specified values of $HCO_3^- + CO_3^{-2}$ concentration and pH. Constraints on these parameters can be found in Tables 1 and 4, respectively. To ascertain the range in H_2 activity (~molality) consistent with the present data, we consider high and low H_2 endmembers. For the high H_2 endmember, we determine the speciation for a solution containing 0.2 molal Cl^-, 0.1 molal $HCO_3^- + CO_3^{-2}$, and an aqueous H_2/CO_2 ratio of 4.7 (Table 2). The adopted composition for the low H_2 endmember is 0.05 molal Cl^-, 0.01 molal $HCO_3^- + CO_3^{-2}$, and H_2/CO_2 = 0.5 (Table 2). By solving the carbonate speciation problem [using The Geochemist's Workbench (*Bethke,* 2008)] over the recommended pH range of 9–11 at 0°C, the molality of CO_2 can be obtained. The activity of H_2 can then be estimated using the scaling relation

$$a_{H_2} \approx m_{H_2} = \left(\frac{H_2}{CO_2}\right)_{\text{plume}} \times m_{CO_2} \tag{15}$$

Figure 4 shows derived values for the activity of H_2 in the ocean of Enceladus. Strictly speaking, these should be called apparent activities because they are model outputs based on plume data. The H_2 activity from this model is between ~2 × 10^{-8} and ~9 × 10^{-4}. The implied concentration range is ~20 nmolal to ~0.9 mmolal. The ranges in f_{O_2} and Eh are ~$10^{-92.2}$ to ~$10^{-83.0}$ bar [i.e., ~4.4 log f_{O_2} units less oxidized than the fayalite-magnetite-quartz (FMQ) buffer to ~4.8 log f_{O_2} units more oxidized than FMQ; see section 4.2.2], and –0.504 to –0.488 V, respectively. The trend of decreasing activity of H_2 with increasing pH (Fig. 4) reflects lower concentrations of aqueous CO_2 at higher pH. This relationship between pH and H_2 activity is a consequence of how the model is set up.

Three possible complications have been identified. First, freezing out ocean-derived CO_2 would decrease the apparent H_2 activity, while subliming CO_2 from the ice shell would increase the apparent H_2 activity. Second, the present lack of explanation for the observations of H_2 spikes by INMS (*Waite et al.,* 2017) could imply that they are generated by an instrumental effect. If an unknown instrumental effect can produce the rest of the H_2 signal (interpreted to be from native H_2) under different circumstances from the spikes, then the apparent H_2 activity would represent an upper limit [however, see *Waite et al.* (2017) for a detailed analysis of the instrument background at mass 2]. Third, the apparent H_2 activity is for the plume source region at the top of the ocean. Because H_2 is escaping from this region, the deeper ocean can be expected to have a higher activity of H_2. In this case, the derived values in Fig. 4 would be lower limits for the bulk ocean.

With constraints on the pH and activity of H_2 in Enceladus' ocean, we can place this environment into a broader context by comparing its conditions of pH-log a_{H_2} to the possible Earth analogs from section 3.2.1. To add the ana-

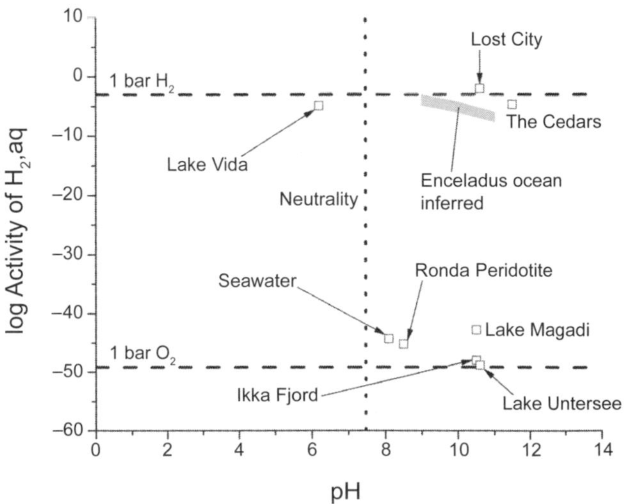

Fig. 4. Redox-pH conditions in Enceladus' ocean vs. potential terrestrial analogs (Table 3). The pH range for Enceladus is from Table 4, and the range in H_2 activity is determined by the range in H_2 mixing ratio in the plume according to the geochemical model described in section 3.2.3. Neutral pH at 0 °C is shown. The 1 bar H_2 line represents the solubility of 1 bar of H_2 from Henry's law, while the 1 bar O_2 line indicates the H_2 activity that would be in equilibrium with 1 bar of O_2.

logs to Fig. 4, we calculated the H_2 activity from measurements of the concentration of H_2 in fluids from Lost City (*Seyfried et al.,* 2015), the Cedars (*Morrill et al.,* 2013), and Lake Vida (*Murray et al.,* 2012). For the well-oxygenated other sites, we assumed that O_2 is the dominant controller of their redox chemistry, and we derived thermodynamic values of the H_2 activity consistent with equations (11) and (12) for the temperature of the site of interest (Table 3). In pH-log a_{H_2} space, Lost City and the Cedars appear to be most similar to the Enceladus ocean (Fig. 4). This suggests that reactions between water and ultramafic rocks could be influencing these properties of the ocean (*Glein et al.,* 2015) (see section 4.2.2). Lake Vida overlaps the Enceladus ocean in H_2 activity, but the former is significantly more acidic than the latter. Conversely, we find that the other sites are fairly consistent with the inferred pH of Enceladus' ocean, but the availability of abundant atmospheric O_2 on Earth makes them much more oxidized compared with the Enceladus ocean.

Overall, the composition of the Enceladus ocean resembles a hybrid of Ikka Fjord, seawater (see section 3.2.1), and Lost City/the Cedars. None of these sites capture all the essential features of Enceladus' ocean, but they each exhibit at least one key similarity to the inferred geochemical properties of the latter. Cassini discovered an ocean of sea salt and soda serpentinizing the underlying rocks.

4. HYDROTHERMAL GEOCHEMISTRY

4.1. Pressure and Temperature in the Rocky Core

We continue using the interior model of *McKinnon* (2015) to estimate pressures in the rocky core of Enceladus. Pressures inside the core can be calculated using the following formula (*Turcotte and Schubert,* 2002)

$$P = \frac{2}{3}\pi G\rho_c^2\left(R_c^2 - r^2\right) + \frac{2}{3}\pi G\rho_h^2\left(R_E^2 - R_c^2\right) + \frac{4}{3}\pi\rho_h G R_c^3\left(\rho_c - \rho_h\right)\left(R_c^{-1} - R_E^{-1}\right) \quad (16)$$

which is applicable to $0 \le r \le R_c$ (see section 3.1 for definitions of the parameters). Figure 5 shows the pressure profile below the ocean floor. According to this model, the pressure increases from an ocean floor value of ~74 bar to a maximum of ~377 bar at the center of the moon. The highest pressure in Enceladus is comparable to the average seafloor pressure on Earth (~400 bar).

It is more difficult to determine temperatures in the core (see the chapter by Castillo-Rogez et al. in this volume). Here, we attempt to set some conservative upper limits. In sections 4.2.1 and 4.2.2, we consider the temperature implications of models for hydrothermal SiO_2 and H_2, respectively. A first constraint is imposed by the liquid-vapor saturation curve of water. For a core with ~20–30% water-filled porosity (*Choblet et al.,* 2017; *Waite et al.,* 2017; *Vance et al.,* 2018), the boiling point represents the maximum temperature for pressures less than the critical pressure of H_2O (~221 bar). Figure 5 indicates that boiling would control the maximum temperature of hydrothermal fluids down to a depth of ~54 km beneath the ocean floor. Once a fluid is heated to the appropriate boiling temperature, any additional input of heat would not increase its temperature but would go into vaporizing water. At the ocean floor, the boiling temperature is ~290°C. This is the maximum temperature of aqueous fluids issuing into the ocean from hydrothermal vents. A large portion of the rocky core is subject to the boiling limit. Owing to the spherical geometry of the core, ~63% of its volume experiences pressures less than the critical pressure of H_2O. In this region (~74–221 bar; Fig. 5), the relationship between the pressure and maximum temperature (T_{max}) can be parameterized as (*Wagner and Pruß,* 2002)

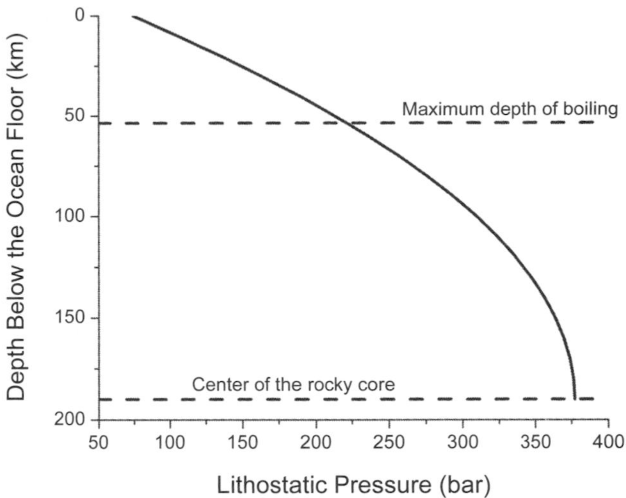

Fig. 5. Pressures in the rocky core of Enceladus based on the internal structure model of *McKinnon* (2015). The maximum depth of boiling is where the pressure equals the critical pressure of water (~221 bar).

$$\log P(\text{bar}) = 5.504 - \frac{2047}{T_{max}(\text{K})} \quad (17)$$

The boiling limit does not apply deep in the core (Fig. 5), so a different approach must be taken to constrain temperatures there. Because the density of Enceladus' core is rather low [~2450 kg m^{-3} (*Iess et al.,* 2014; *McKinnon,* 2015)], the core should be rich in hydrated silicates [although see *Vance et al.* (2018) for an anhydrous case]. This imposes a constraint on the temperature given that hydrated minerals undergo dehydration at sufficiently high temperatures. Dehydrating the rocks would make the core too dense. The argument can be quantified by performing a simple geochemical analysis. Here, we approximate the dehydration process by tracking the release of water from antigorite serpentine. Antigorite ($Mg_{48}Si_{34}O_{85}(OH)_{62}$) is the polymorph of serpentine that is most stable during metamorphism. Serpentine minerals are likely to dominate the budget of mineral-bound water (i.e., OH) in the deep interior (*Zolotov,* 2007). To show that this is likely to be the case, one must go a step beyond the consensus of a heavily hydrated core, and consider specific hydroxylated silicates that could be present. *Waite et al.* (2017) developed a normative model for the mineralogy of possible rocks in the Mg-Si-Fe-S-O-H system on Enceladus. In their reduced hydrous rock, ~68% of the mineral-bound water is in Mg-serpentine, ~31% is in Fe(II)-serpentine, and ~1% is in talc/saponite. Their oxidized hydrous rock has ~69% of the mineral-bound water in Mg-serpentine and ~31% in talc/saponite. Thus, the mass balance supports a focus on the dehydration of antigorite to elucidate an approximate upper temperature limit.

The classic dehydration sequence of antigorite (*Tracy and Frost,* 1991) can be represented by the following reactions

$$\begin{aligned} &Mg_{48}Si_{34}O_{85}(OH)_{62}, \text{antigorite} \rightarrow \\ &4\ Mg_3Si_4O_{10}(OH)_2, \text{talc} + \\ &18\ Mg_2SiO_4, \text{forsterite} + 27\ H_2O, \text{aq} \end{aligned} \quad (18)$$

$$\begin{aligned} &4\ Mg_3Si_4O_{10}(OH)_2, \text{talc} + \\ &1.778\ Mg_2SiO_4, \text{forsterite} \rightarrow \\ &2.222\ Mg_7Si_8O_{22}(OH)_2, \text{anthophyllite} + \\ &1.778\ H_2O, \text{aq} \end{aligned} \quad (19)$$

and

$$\begin{aligned} &2.222\ Mg_7Si_8O_{22}(OH)_2, \text{anthophyllite} + \\ &2.222\ Mg_2SiO_4, \text{forsterite} \rightarrow \\ &20\ MgSiO_3, \text{enstatite} + 2.222\ H_2O, \text{aq} \end{aligned} \quad (20)$$

The curves in Fig. 6 show where these reactions would occur in pressure-temperature space. By summing equations (18) through (20), it can be found that the complete dehydration of 1 mole of antigorite releases 31 moles of water. Because equation (18) releases ~87% of the total H_2O, we deduce that this reaction effectively determines the content of bound water in rocks, and therefore their densities. To maintain a relatively low core density, temperatures in the core should be lower than the appropriate dehydration temperature of antigorite. The dehydration temperature increases with pressure (Fig. 6). For a maximum pressure of ~377 bar, the corresponding temperature is ~445°C. This can be regarded as an apparent (see below) upper limit for the present core.

Core temperatures could have been higher than ~445°C in the past if conduction was the dominant mode of heat transfer (*Schubert et al.,* 2007). In this case, the rocks would have needed to experience subsequent hydration to end up with a consistent density. Alternatively, advection of hydrothermal fluids could have always been the dominant heat transfer mechanism (*Choblet et al.,* 2017). The greater efficiency of advection provides a way of explaining the apparent persistence of hydrated minerals, by modulating increases in temperature. Nevertheless, it remains a possibility for a portion of the present core to be hotter than the antigorite limit if anhydrous rock exists at the center (*Malamud and Prialnik,* 2016).

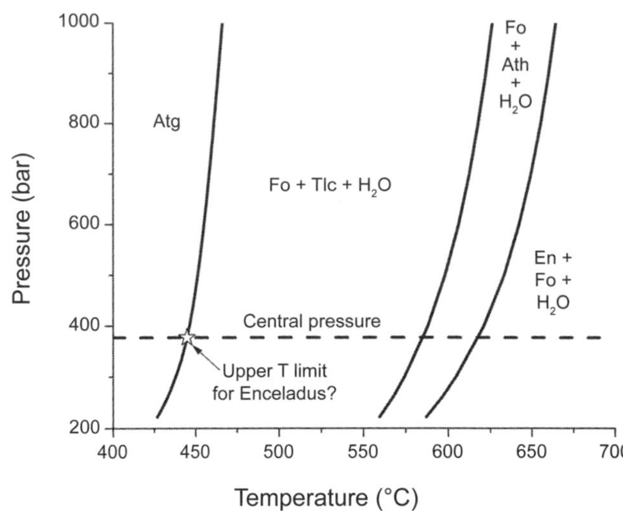

Fig. 6. Pseudosection phase diagram showing the stable phases for the initial composition of antigorite as a function of temperature and pressure. Central pressure refers to the pressure at the center of Enceladus' rocky core. A core rich in hydrated rock may not exceed the dehydration temperature of antigorite (see section 4.1). Mineral abbreviations: Atg, antigorite; Ath, anthophyllite; En, enstatite; Fo, forsterite; Tlc, talc. Thermodynamic data used to construct this diagram were taken from the SUPCRT database (*Helgeson et al.,* 1978).

4.2. Proposed Hydrothermal Species

4.2.1. SiO_2. *Hsu et al.* (2015) identified silica as the only significant constituent in saturnian stream particles, and inferred that these particles are being erupted inside ice grains from the Enceladus plume. They proposed that silica nanoparticles form at the ocean floor of Enceladus, where Si-enriched hydrothermal fluids mix with the cold ocean. The solubility of amorphous silica increases with temperature, so cooling promotes supersaturation, which is necessary to form colloidal silica.

Would enceladean hydrothermal fluids contain a sufficient concentration of dissolved silicon? Hydrothermal experiments with a mixture of olivine and orthopyroxene were performed to address this question (*Sekine et al.,* 2015). These minerals are thought to have been the most abundant silicates in accreted rocks. However, the experiments did not include Na- or Ca-rich silicates, which can have a strong influence on the speciation of fluids (*Zolotov and Postberg,* 2014). *Sekine et al.* (2015) found that the Si concentration in their experimental hydrothermal fluid was similar to that predicted for a fluid in equilibrium with serpentine and talc. Indeed, X-ray diffraction analysis revealed that serpentine and saponite (a trioctahedral phyllosilicate compositionally similar to the Al-free endmember talc) are the major alteration minerals. To constrain the formation temperature of the silica nanoparticles, *Sekine et al.* (2015) performed speciation calculations under the assumption that the activity of SiO_2 in hydrothermal fluids on Enceladus would be similar to that determined by a serpentine-talc buffer, such as

$$\begin{aligned}&0.5\ Mg_3Si_4O_{10}(OH)_2, \text{talc} + 0.5\ H_2O, aq \leftrightarrow \\ &0.5\ Mg_3Si_2O_5(OH)_4, \text{chrysotile} + SiO_2, aq\end{aligned} \tag{21}$$

In alkaline systems, $HSiO_3^-$ and $NaHSiO_3$ can be more abundant than aqueous SiO_2. There is a pH effect as indicated by the following equilibria

$$SiO_2, aq + H_2O, aq \leftrightarrow H^+, aq + HSiO_3^-, aq \tag{22}$$

and

$$Na^+, aq + HSiO_3^-, aq \leftrightarrow NaHSiO_3, aq \tag{23}$$

which imply that the concentrations of both $HSiO_3^-$ and $NaHSiO_3$ should increase at higher pH. The minimum temperature required to precipitate amorphous silica as a result of cooling depends on the pH of the hydrothermal fluid and that of ocean water. *Sekine et al.* (2015) used their speciation results to map out these relationships. They suggested that the minimum temperature would be ~150°–200°C if the compositions of the hydrothermal fluid and ocean water are controlled by the same water-rock equilibria. If the two fluids are treated as pH-decoupled systems, then the model of *Sekine et al.* (2015) indicates that the minimum temperature could be decreased to a value as low as ~50°C if the hydrothermal fluid pH is ~2 units higher than the ocean pH.

To test if rocks containing serpentine and talc can generate sufficiently Si-rich fluids that would precipitate amorphous silica upon cooling, the pH of the fluids must be modeled self-consistently. One way to do this is to assume that the rocks would also contain calcite ($CaCO_3$) and tremolite ($Ca_2Mg_5Si_8O_{22}(OH)_2$). Downwelling of ocean water into the rocky core would bring carbonate species (Table 1) in contact with rocks deeper in the core, facilitating the formation of calcite. For a deep rock endmember (see section 4.2.2), we envision that some of the calcium is still in silicate minerals, thus tremolite is chosen as a model mineral to simulate the pH effect. The assemblage chrysotile-talc-calcite-tremolite sets the activity of CO_2 according to

$$\begin{aligned}&3\ Ca_2Mg_5Si_8O_{22}(OH)_2, \text{tremolite} + \\ &2\ Mg_3Si_2O_5(OH)_4, \text{chrysotile} + 6\ CO_2, aq \leftrightarrow \\ &6\ CaCO_3, \text{calcite} + 7\ Mg_3Si_4O_{10}(OH)_2, \text{talc}\end{aligned} \tag{24}$$

To complete the model, it is assumed that the hydrothermal fluid inherits the chlorinity and carbonate alkalinity of the downwelling ocean water. We adopt a nominal chlorinity of 100 mmolal and a total carbonate concentration of 30 mmolal (see section 3.2.1), which translates to a carbonate alkalinity of 31 and 56 meq kg^{-1} H_2O for pH 9 and 11, respectively. With constraints on the activity of CO_2 and carbonate alkalinity, the pH can be evaluated from the speciation of the hydrothermal fluid. Here, the speciation is computed using the GEOCHEQ code (*Zolotov,* 2012).

Hydrothermal fluids with compositions described by this geochemical model would be relatively poor in dissolved silicon (Fig. 7). The chrysotile-talc buffer (see equation (21)) does not yield high activities of SiO_2. The pH of the hydrothermal fluid is predicted to decrease with temperature from ~10.5 (~4 units above neutral) at 50°C to ~8 (~2 units above neutral) at 350°C. There is more silica in the modeled fluid at the higher alkalinity because the pH is ~0.1–0.2 units higher. The modeled fluid is generally deficient in dissolved silicon for the case of a pH 9 ocean, and the model is definitely incapable of achieving amorphous silica saturation for the case

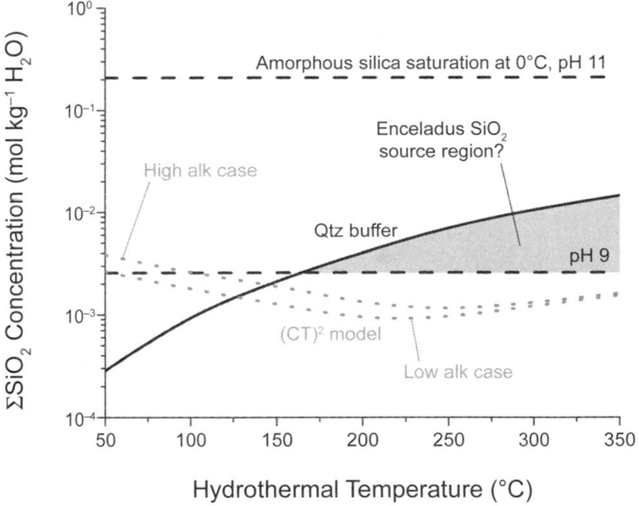

Fig. 7. Total concentration of silica species ($\Sigma SiO_2 \approx SiO_2,aq + HSiO_3^- + NaHSiO_3,aq$) in hydrothermal fluids on Enceladus from equilibrium with quartz (Qtz; solid curve), or with chrysotile-talc-calcite-tremolite [$(CT)^2$; dotted gray curves]. The dashed horizontal lines show the saturation state of amorphous silica for the indicated ocean pH at a constant temperature of 0°C. In this representation (see Fig. 9), hydrothermal fluids leach Si from silicates in the core, and then cooling of the fluids in the ocean causes amorphous silica to precipitate if the hydrothermal equilibrium curve lies above the appropriate saturation line. The low and high alk curves for the $(CT)^2$ model should be compared to the values for pH 9 or 11, respectively, while the quartz curve can be compared to either of the saturation values (see section 4.2.1). The gray region provides a lower limit on temperatures of amorphous silica-forming hydrothermal fluids from the rocky core. Amorphous silica rather than quartz is the relevant precipitating phase of SiO_2 for fluids that undergo rapid cooling (as in siliceous sinter-depositing hot springs at Yellowstone National Park), because the disordered structure is kinetically/mechanistically easier to form.

of a pH 11 ocean (Fig. 7). The latter observation supports the conclusion of *Hsu et al.* (2015) that the pH of the ocean should not be too high to allow nanophase silica to form.

The $(CT)^2$ model can provide enough silica if the ocean pH is lower than ~9 and if the hydrothermal temperature is below ~50°C; the enhancement in dissolved silicon follows from the high pH of such low-temperature hydrothermal fluids. This underscores the possible role of a gradient between high pH fluids in the core and lower pH ocean water driving the formation of amorphous silica (*Sekine et al.,* 2015). Similarly, NaOH-rich core fluids if present would dramatically increase the solubility of silicates, especially at low temperatures (*Zolotov,* 2012). At temperatures above ~50°C, the concentration of dissolved silicon predicted by the pH 9 ocean model is generally a factor of ~2–3 lower than required (Fig. 7). We interpret this to mean that achieving consistency may not be impossible, but hydrothermal alteration of chondritic rock inside Enceladus' core would not provide a robust mechanism for making amorphous silica. This conclusion is further supported by the more detailed geochemical modeling of *Zolotov and Postberg* (2014), who showed that fluids in equilibrium with altered chondritic rock could be even more deficient in dissolved silicon (by up to about 1 order of magnitude).

We wish to clarify that it is a misconception to assume that fluids compositionally similar to those at Lost City would precipitate amorphous silica at Enceladus. There is not enough dissolved silicon in Lost City fluids, by ~1–2 orders of magnitude (*Seyfried et al.,* 2015). Amorphous silica at Enceladus can be used to argue for the existence of hydrothermal systems (see below), but not systems that are geochemically analogous to Lost City (i.e., serpentinization systems).

There is motivation to identify more robust mechanisms for producing amorphous silica inside Enceladus. We introduce a possibility that involves hydrothermal processing of quartz-bearing rocks. This is usually how amorphous silica is produced by hydrothermal processes on Earth (*Fournier and Rowe,* 1966; *Von Damm et al.,* 1985). Quartz-bearing rocks could be present as a weathering crust on the core of Enceladus. Such rocks can be formed via the process of carbonation (*Klein and Garrido,* 2011), where carbonate derived from CO_2 serves as a sink of divalent cations such as magnesium, leaving behind Si-enriched phases. As an example of such chemistry, *Streit et al.* (2012) reported a remarkable occurrence of quartz that apparently formed by carbonation of serpentinized peridotite at ambient temperatures on Earth. Carbonation of rocks on Enceladus would be promoted by the accretion of a large amount of CO_2, as observed at numerous comets (*Ootsubo et al.,* 2012). The relatively high abundance of (bi)carbonate salts in many plume particles (Table 1) provides evidence for the reaction of CO_2 with rocks. Enceladus' ocean floor has presumably experienced the most intense chemical weathering, because that is where the volatiles-to-rock ratio should be largest.

The solubility curve of quartz in Fig. 7 corresponds to the following equilibrium

$$SiO_2, \text{quartz} \leftrightarrow SiO_2, \text{aq} \tag{25}$$

The curve does not account for variable pH, but GEOCHEQ calculations showed that neutral SiO_2 would be the dominant Si species (>80% of dissolved Si) in heated ocean water that had equilibrated with quartz, magnesite ($MgCO_3$), and talc (a possible carbonation assemblage in the weathering crust). Hydrothermal fluids in equilibrium with quartz would be rich in dissolved silicon (Fig. 7). Quartz-buffered fluids at temperatures above ~165°C would contain sufficient silica to precipitate amorphous silica into a pH 9 ocean. This is a lower limit on temperatures in the source region of the silica nanoparticles. The pH of the ocean would also be constrained; if we assume a hydrothermal temperature of ~350°C, then amorphous silica precipitation from a quartz-buffered fluid is possible only if the ocean has a pH lower than ~10. The range in the concentration of silica from the quartz model (~3–15 mmolal from ~165° to 350°C; Fig. 7) is consistent with the observationally based estimate (~2.5–65 mmolal) of *Hsu et al.* (2015).

4.2.2. H_2. The INMS instrument detected H_2 in the plume (Table 2), and it was concluded that the H_2 is native to Enceladus (*Waite et al.,* 2017). In addition to water-rock processes such as serpentinization (*Vance et al.,* 2007), there are several other candidate sources of H_2 that must be considered. One possibility is that the H_2 is primordial, and was acquired by gravitational capture from the saturnian subnebula, or by trapping in cold amorphous ices. However, these sources can be ruled out based on the lack of detection of ^4He ($^4He/H_2O < 6 \times 10^{-5}$) and ^{36}Ar ($^{36}Ar/H_2O < 4 \times 10^{-6}$), respectively, in the plume gas (*Waite et al.,* 2017). Another possible source of H_2 is thermal cracking of NH_3 to N_2 and H_2 (*Matson et al.,* 2007). But, this mechanism is unattractive owing to the non-detection of N_2 ($N_2/H_2O < 5 \times 10^{-4}$). Alternatively, the H_2 could be radiolytic, and might have been produced from H_2O by radiation chemistry at the surface or in the interior (*Bouquet et al.,* 2017). Both of these processes should be ongoing, but kinetic calculations suggest that neither of them can make enough H_2 to account for the observations (*Waite et al.,* 2017). Conversely, mass balance calculations indicate that huge amounts of H_2 can be generated by reactions between water and reduced minerals, or by pyrolysis of organic matter (*Waite et al.,* 2017). The H_2-generating potential of both of these sources is sufficient to sustain the present level of outgassing over the history of the solar system. *Waite et al.* (2017) therefore suggested that the H_2 is likely produced by hydrothermal processing of rocks containing both ferrous iron-bearing silicates and organic materials. They further argued that this process is occurring today, as the H_2/CH_4 ratio may be much lower than the observed value (~1–14) if the H_2 had been stored in impermeable rocks and released recently.

The following questions remain: (1) How can mineral vs. organic sources of H_2 be discriminated? (2) Would the formation of carbonate minerals affect the yield of H_2? (3) How does the H_2 abundance in the plume relate to hydrothermal conditions in the rocky core? We develop a model that can be used to predict the mixing ratio of H_2 in the plume gas based on geochemical and geophysical properties of a globally averaged hydrothermal system. The rate at which H_2 is delivered via hydrothermal vents to Enceladus' ocean (δ_{H_2}) is given by the product of the molal concentration of H_2 in hydrothermal fluids (m_{H_2}) times the mass flow rate of hydrothermal fluids (Q_{H_2O})

$$\delta_{H_2} = m_{H_2} Q_{H_2O} \tag{26}$$

We consider three simplified source materials for making H_2: (1) rocks buried deep in the core ("deep rocks"), (2) rocks near the ocean floor ("shallow rocks"), and (3) organic materials ("organics"). For deep rocks, we assume that their equilibrium oxidation state is similar to that of rocks on Jupiter's moon Io. Io's rocks might be representative of relatively reduced silicate assemblages in the outer solar system. We adopt an oxidation state that is 2 log f_{O_2} units below the fayalite-magnetite-quartz buffer (*Zolotov and Fegley,* 2000). This is equivalent (see equation (12)) to an H_2 activity that is 1 log a_{H_2} unit above FMQ. The FMQ buffer can be represented by

$$\begin{aligned} &1.5\ Fe_2SiO_4, \text{fayalite} + H_2O, \text{aq} \leftrightarrow \\ &Fe_3O_4, \text{magnetite} + 1.5\ SiO_2, \text{quartz} + H_2, \text{aq} \end{aligned} \tag{27}$$

Using thermodynamic data from *Helgeson et al.* (1978) and *Shock et al.* (1989) and assuming that aqueous H_2 behaves ideally, we derive the following relationship for this model of deep rocks on Enceladus

$$\begin{aligned} &\log m_{H_2}(\text{deep rocks}) = \\ &-37.87 + \frac{644.4}{T(K)} + 5.56 \times \ln(T(K)) \end{aligned} \tag{28}$$

which is applicable to temperatures from 273 to 623 K (0°–350°C).

For shallow rocks in the core, we consider the possibility of an extensively carbonated ocean floor (see section 4.2.1), and therefore assume that Fe-bearing carbonate (e.g., siderite ($FeCO_3$)) is the key mineral carrier of iron (*Ueda et al.,* 2016). The oxidation of Fe(II) in siderite to Fe(III) in magnetite can generate H_2 if this process is coupled to the reduction of H_2O, which could occur as follows

$$\begin{aligned} &3\ FeCO_3, \text{siderite} + Mg_3Si_4O_{10}(OH)_2, \text{talc} \leftrightarrow \\ &Fe_3O_4, \text{magnetite} + 3\ MgCO_3, \text{magnesite} + \\ &4\ SiO_2, \text{quartz} + H_2, \text{aq} \end{aligned} \tag{29}$$

The equilibrium concentration of H_2 for this reaction is given by (*Helgeson et al.,* 1978; *Shock et al.,* 1989)

$$\begin{aligned} &\log m_{H_2}(\text{shallow rocks}) = \\ &-54.48 - \frac{364.3}{T(K)} + 8.02 \times \ln(T(K)) \end{aligned} \tag{30}$$

from 273 to 623 K (0°–350°C).

The concentration of H_2 in hydrothermal fluids cooking organic matter in Enceladus' core can be represented by

$$m_{H_2}(\text{organics}) = \frac{(O/R)}{(W/R)} Y_{H_2} \tag{31}$$

where O/R and W/R correspond to the organic/rock and water/rock mass ratios, respectively; and Y_{H_2} designates the yield of H_2 in mol per kilogram of organic matter. Here, we adopt W/R = 1 for hydrothermally active regions in Enceladus' core, and O/R = 0.4, a value intermediate between CI chondrites (*Alexander et al.,* 2007) and dust particles from Comet 67P (*Bardyn et al.,* 2017). The parameterization of *Waite et al.* (2017), based on the Murchison meteorite (*Okumura and Mimura,* 2011), is used to estimate the pyrolytic yield of H_2. Equation (31) then becomes

$$\log m_{H_2}(\text{organics}) = 3.55 - \frac{4023}{T(K)} \tag{32}$$

The H_2 yield was fit to experimental data from 350° to 800°C, but this equation can be extrapolated to lower temperatures, as the functional form is consistent with thermodynamic (van't Hoff equation) or kinetic (Arrhenius equation) control of the H_2 yield in the experiments of *Okumura and Mimura* (2011).

The mass flow rate of hydrothermal fluids (see equation (26)) can be estimated using an approach similar to that of *Lowell and DuBose* (2005). The basis of the approach is energy conservation for heat that is transported by hydrothermal fluids into the Enceladus ocean. We found that heat capacities (C_p) for liquid water at 300 bar can be represented by (*Wagner and Pruß,* 2002)

$$C_p = \alpha + \beta \times T^9 \tag{33}$$

with α = 4130 J kg^{-1} K^{-1} and β = 1.53 × 10^{-22} J kg^{-1} K^{-10} for 273–623 K (0°–350°C). For this functional form of the heat capacity, the rate of heat transfer (H) is related to the mass flow rate by

$$H = Q_{H_2O} \left[\alpha(T_{hydro} - T_{ocean}) + 0.1 \times \beta(T_{hydro}^{10} - T_{ocean}^{10}) \right] \tag{34}$$

where T_{hydro} refers to the hydrothermal fluid temperature, and T_{ocean} the temperature of Enceladus' ocean. We can obtain Q_{H_2O} as a function of T_{hydro} by specifying a value for H. Below, we consider a range of 1–10 GW based on *Choblet et al.*'s (2017) model of tidal dissipation and fluid flow in Enceladus' core.

If the concentration of H_2 in Enceladus' ocean has reached a steady state between hydrothermal input and output from the plume, then the latter rate is also given by equation (26). The molar ratio of H_2/H_2O that would be expected in the plume gas can be computed from the rate of H_2 output from the preceding hydrothermal model, and the emission rate of water vapor from the plume. A value of 200 kg s^{-1} (11,000 mol s^{-1}) is adopted for the latter quantity (*Hansen et al.,* 2011).

Figure 8 compares H_2/H_2O ratios predicted to be in the plume gas from the three sources vs. the observed range. The models of organic pyrolysis, and hydrothermal processing of shallow rocks, do not provide enough H_2, as their fluids are too dilute in H_2 (<1 mmolal). Making these models consistent with the H_2 observations apparently requires over ~50 GW of heat to be transferred by advection of these fluids. In contrast, the deep rock model can explain the measurement of H_2 if the hydrothermal temperature is higher than ~170°C (Fig. 8). Higher temperatures would be needed if H is lower, and vice versa. However, a value of H less than ~1 GW may not yield sufficient H_2. Therefore, the observations of H_2 in the plume appear to provide support to the idea that tidal heating is taking place in Enceladus' core, and a large amount of heat is being transferred by fluid circulation (*Choblet et al.,* 2017). It should be noted, however, that hydrothermal activity would not have to be as vigorous if the rock is more reduced than the Io model. The H_2 measurement also imposes a limit on the oxidation state of deep rocks. They should not be more oxidized than the FMQ buffer; otherwise, the fluid would not be sufficiently rich in H_2 (>10 mmolal) to account for the measurement. These calculations suggest that H_2 in the plume (*Waite et al.,* 2017) can be taken as evidence for chemical interactions between hot fluids and reduced minerals relatively deep in the core of Enceladus.

4.2.3. Self-consistency? There is a potential inconsistency between the explanations for hydrothermal SiO_2 and H_2. The quartz buffer in section 4.2.1 is part of the shallow rock model in section 4.2.2. Hydrothermal processing of this rock can be the source of the silica (Fig. 7) detected by CDA (*Hsu et al.,* 2015), but not the molecular hydrogen (Fig. 8) detected by INMS (*Waite et al.,* 2017). On the other hand, the $(CT)^2$ model (section 4.2.1) is an unfavorable source of SiO_2 (Fig. 7), but the corresponding deep rock model in section 4.2.2 is the favored source of H_2 (Fig. 8). Theory shows that a low SiO_2 activity is one of the chief factors that enables a high H_2 activity to be created during serpentinization (*Frost and Beard,* 2007). This general expectation is borne out by an inverse correlation between the concentrations of H_2 and Si in ultramafic-hosted hydrothermal fluids (*Seyfried et al.,* 2011). In our modeling, one rock needs to be invoked to explain the SiO_2 observation, while the other rock is required to explain the H_2 observation; neither rock can explain both observations. What does this mean?

There could be multiple source rocks of hydrothermally derived species. Figure 9 shows a scenario illustrating how this might occur in a self-consistent manner. Hydrothermal fluids may react with reduced rocks if the fluids descend sufficiently. This would lead to the production of H_2, which dissolves in the fluids under pressure. As they rise back to the ocean floor, the fluids may intersect and react with quartz-bearing rocks, which would enrich the fluids in dissolved silicon. The H_2 produced at depth may survive passage through the shallow rock layer if iron there is sequestered in carbonate minerals (see section 4.2.2). Finally, fluids rich in both H_2 and SiO_2 may vent into the ocean of Enceladus, which could explain the observations by Cassini. The complexity of this hypothesis may accord with what might be expected of a geologically active world. As an example of this type of multi-rock model, *Seyfried et al.* (2015) found that Lost City fluids contain more silica than the very low concentrations expected for peridotite alteration. These researchers therefore suggested that Lost City

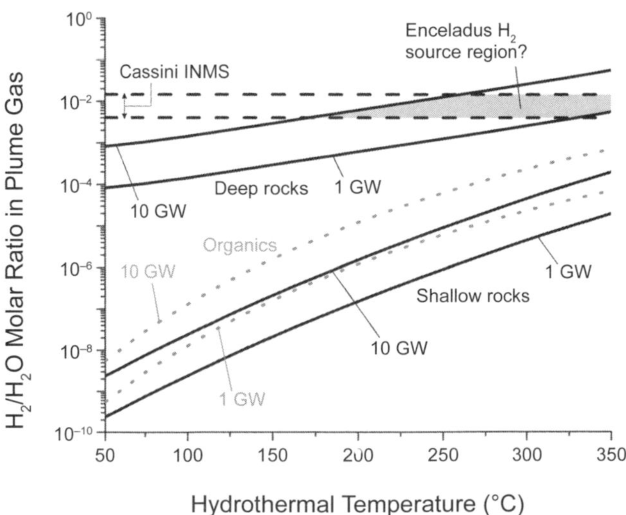

Fig. 8. Predicted abundance of H_2 in the Enceladus plume from hydrothermal models for H_2 production (see section 4.2.2) from deep rocks (top set of solid curves), organics (dotted gray curves), and shallow rocks (bottom set of solid curves). The dashed lines indicate the observational range for H_2 (Table 2). The gray region provides a lower limit on temperatures of H_2-producing hydrothermal systems in the rocky core.

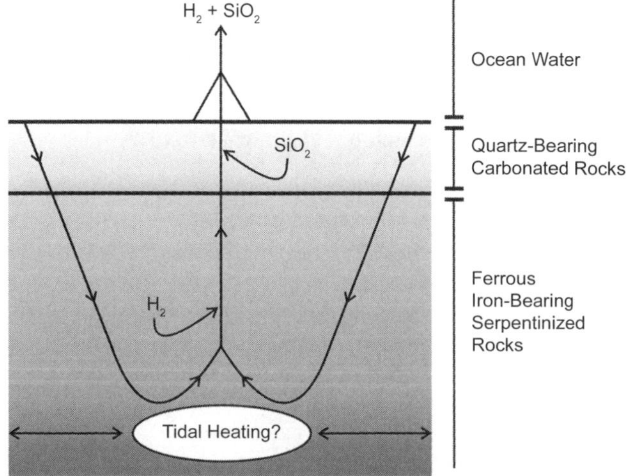

Fig. 9. Conceptual model for the formation of high H_2-high SiO_2 hydrothermal fluids by convective fluid flow through two different types of rocks under the ocean floor of Enceladus.

peridotite was intruded by more silicic rocks (e.g., gabbro), as often observed in peridotites dredged from the seafloor.

5. SOME GEOCHEMICAL QUESTIONS FOR THE FUTURE

Cassini gave humanity our first "taste" of enceladean geochemistry. There is still much to learn. In our view, finding ways to address the following questions should be considered a high priority for future studies.

How accurately can the composition of the ocean be inferred from that of the plume? Approaches have been developed to estimate the concentrations of ions (*Postberg et al.,* 2009) and gases (*Waite et al.,* 2017) in Enceladus' ocean using mass spectrometry data from salt-rich ice grains and the plume gas, respectively. However, the exclusion of a portion of salts during the freezing of ocean water, or vapor condensation onto grains in the tiger stripes, will inevitably fractionate them with respect to the composition of the ocean. Likewise, the gas composition could be fractionated between the ocean and plume by adsorption on ice, or clathrate formation (*Bouquet et al.,* 2015) or decomposition (*Kieffer et al.,* 2006; *Fortes,* 2007). We do not know at present how large these effects might be, nor their dependence on the physical properties of individual species. This necessitates a deeper understanding of how dynamical processes leading to plume formation influence the composition of the plume (see the chapter by Goldstein et al. in this volume). One path forward is to build a plume in the lab to investigate the plume's dynamical effects.

What are the concentrations of minor species in the ocean? A first step is to set upper limits on currently undetected ions using plume or E-ring data from Cassini CDA. Such constraints would be geochemically useful. Important species that have yet to be quantified include Ca^{+2}, Mg^{+2}, and SO_4^{-2}. Both Ca^{+2} and Mg^{+2} would provide information on mineral controls (e.g., carbonates) of ocean composition, while SO_4^{-2} could serve as a tracer of oxidant production or delivery to the ocean (*Ray et al.,* 2017). A direct measurement of the concentration of silica in ice grains from the plume (*Hsu et al.,* 2015) should be a high priority. This could be used to evaluate the saturation indices of different silicate minerals in contact with Enceladus' ocean. A wet chemistry laboratory on a south polar lander could allow silica and other aqueous species to be measured.

What is the detailed redox chemistry of the ocean? Enceladus appears to have a relatively reduced ocean as H_2, NH_3, and CH_4 are abundant in the plume gas (Table 2). However, CO_2 is also present and the CO_2-CH_4 couple is apparently out of equilibrium with the H_2O-H_2 couple (*Waite et al.,* 2017). Because there is geophysical activity to drive fluid mixing, other redox couples can be expected to be maintained in disequilibrium states in a cold ocean. This is vital for establishing the availability of chemical energy sources that could support life. How can we address this fundamental question? First, we should corroborate that the ocean is indeed reduced. CH_4, NH_3, and H_2 could be remeasured at slower flyby speeds (<5 km s^{-1}) to verify that they are not produced from ice-grain impacts on spacecraft instrumentation (*Waite et al.,* 2009). A complementary suite of reduced species are recommended to be measured in salt-rich plume particles, which could include Fe^{+2} and formate ($HCOO^-$). Second, our knowledge of redox disequilibria in the Enceladus ocean would be revolutionized if a comprehensive survey of gaseous and ionic H, O, C, N, and S species in the plume is performed.

Are there gradients in the composition of the ocean or the rocky core? Degassing could make the top of the ocean compositionally distinct (e.g., higher pH, lower H_2) from lower layers. Heterogeneous heating could create complex patterns of fluid flow that result in a redistribution of chemical elements. The development of coupled geophysical-geochemical models of reactive transport inside Enceladus would help the scientific community to better understand the consequences of such processes. For example, a rigorous assessment of the hypothesis outlined in section 4.2.3 requires quantitative results for the compositional evolution of both fluids and rocks as fluid parcels interact with rocks along reaction paths underneath the ocean floor (*McCollom and Shock,* 1998). Detailed geophysical modeling of both hydrothermal (*Choblet et al.,* 2017) and ocean circulation is needed to constrain rates of mixing and thus residence times of fluids in different geochemical environments on Enceladus. Modeling studies of these types offer opportunities for testing, refining, and differentiating between competing hypotheses. They therefore represent an important bridge between Cassini and plans to design future missions to Enceladus (see the chapter by Lunine et al. in this volume). A notable example of a useful probe of material interfaces in the subsurface and their associated compositional gradients would be a landed seismometer (*Stähler et al.,* 2018).

What geochemical evolutionary pathway did Enceladus take? The bulk composition of rocks accreted by Enceladus is commonly assumed to have been similar to that of CI chondrites, and Enceladus might have accreted volatiles in proportions similar to those in some comets (see the chapter by McKinnon et al. in this volume). It is still an open question as to how an initial mixture of chondritic rock and cometary ices (*Zolotov,* 2012) relates to the current composition of Enceladus as summarized in this chapter. To obtain insights into the formation conditions and accreted composition of Enceladus, additional cosmochemical data from Enceladus are needed such as a more precise value for the D/H ratio in H_2O (*Waite et al.,* 2009), a value for the $^{15}N/^{14}N$ ratio in NH_3, the abundances of primordial noble gases, and triple oxygen-isotope ($^{18}O/^{16}O$, $^{17}O/^{16}O$) measurements in samples of water ice from the plume. Considerable progress toward understanding the history of Enceladus could be made by elucidating the specific reactions and processes that would link plausible models of Enceladus' initial composition to its present geochemical state. Geochemical mass transfer modeling should be performed to learn what can happen if initial mixtures of frozen volatiles and ultramafic rocks were subjected to heating, water-rock separation, fluid circulation, and outgassing from the plume. Can the ocean floor really become heavily carbonated? How would Fe(II) and Fe(III) partition among minerals, and what are the implications for H_2 production during serpentinization?

How representative are thermodynamic models of water-rock interaction for conditions relevant to Enceladus? We are advocating for experimental work that could be done to assess if chemical equilibrium is an appropriate assumption for aqueous solutions and altered rocks inside Enceladus. *Neveu et al.* (2017) took a step in this direction by comparing chemical equilibrium predictions to literature results from hydrothermal experiments with olivine or basalt. They found general agreement in the solid and fluid compositions, but the error in the predicted pH for the basalt experiments was ~1 unit. This may reflect inaccuracies in the thermodynamic properties of clay minerals in existing databases. Further comparisons between models and experiments may suggest strategies for improving the models. As a next step, it would help if such comparisons were made for volatile-rich chondritic compositions (e.g., a melted lab-made comet), particularly at temperatures closer to 0°C.

What new observations can be made to test for hydrothermal activity? The observations of SiO_2 and H_2 made by the Cassini spacecraft represent the first recognized indications of hydrothermal geochemistry (*Hsu et al.,* 2015; *Waite et al.,* 2017). Additional complementary data are desired, and their promise beckons us to go back to Enceladus. We should search for other chemical species (e.g., H_2S, CO) that may exhibit anomalous enrichments in the plume, consistent with the presence of fluids that are hot and reduced. A defining feature of hydrothermal activity is an elevated temperature relative to the ambient environment. Specific constraints on temperatures of core fluids on Enceladus can be obtained using tools of geothermometry, including isotopic ratios in simple volatiles [e.g., H_2-H_2O (*Horibe and Craig,* 1995)], ratios of certain organic compounds (e.g., ethylene/ethane; *Seewald,* 1994), and "clumping" of rare isotopes [e.g., $^{13}CH_3D$ (*Stolper et al.,* 2014)]. Measurements of radiogenic (mainly rock-derived) noble gases (primarily ^{40}Ar and ^{4}He) are also recommended as indicators of the extent and vigor of hydrothermal transport from the core to the hydrosphere.

What is the organic geochemistry of Enceladus? Data from both INMS and CDA show that the plume contains organic materials (*Waite et al.,* 2009; *Postberg et al.,* 2018). This is consistent with the association of an organic spectral signature with the tiger stripes (*Brown et al.,* 2006). The organic molecules detected by CDA are massive, rich in unsaturated carbon atoms in the form of unfused benzene rings, contain O- or N-bearing groups, and are probably phase-separated as solids from ocean water (*Postberg et al.,* 2018). At high spacecraft flyby velocities (>14 km s^{-1}), these molecules at least partially decompose to the smaller unsaturated molecules observed by INMS (*Waite et al.,* 2009). What is the origin of Enceladus' organic matter? Possibilities include accreted materials analogous to insoluble organic matter in chondrites (*Alexander et al.,* 2007); hydrothermal synthesis from small molecule precursors such as CO_2, HCN, formaldehyde, or methanol; and biological carbon fixation (see the chapter by McKay et al. in this volume). We emphasize that composition is the key to origin. Clues are needed in the form of measurements of the elemental and isotopic compositions of the particulate organic matter, as well as its functional group chemistry and structure. Pyrolysis-gas chromatography-mass spectrometry of south polar surface samples would be useful in this respect. We are also in need of detailed compositional characterizations of hydrocarbon gases (*Waite et al.,* 2009), and forms of dissolved organic carbon (e.g., fatty and amino acids) that may be present in the plume. Only with a more complete dataset will we be able to arrive at an integrated understanding of the geochemistry of Enceladus.

Acknowledgments. C.R.G. is grateful to J. Baross, M. Cable, J. Castillo-Rogez, F. Klein, J. Lunine, W. McKinnon, K. Miller, C. Porco, K. Rogers, E. Shock, N. Sleep, and H. Waite for numerous discussions on the geochemistry of Enceladus. C.R.G. wishes to express his appreciation to M. Zolotov for giving him a copy of the GEOCHEQ code. C.R.G. would also like to give thanks to the organizers of the 2nd Annual Ocean Worlds meeting at Woods Hole, where some of the new ideas in this chapter were first presented. J. Dao and R. Menchaca deserve recognition for their superb assistance in the preparation of Figs. 1 and 9. The work of C.R.G. was supported by the Cassini project (NAS703001TONMO711123). F.P. acknowledges funding from the German Research Foundation (DFG projects PO 1015/2-1, /3-1, /4-1), and the European Research Council (ERC Grant 724908). S.D.V.'s contribution was supported by the Icy Worlds node of NASA's Astrobiology Institute (13-13NAI7_2-0024). His part of the research was carried out at the Jet Propulsion Laboratory, California Institute of Technology, under a contract with the National Aeronautics and Space Administration.

REFERENCES

Alexander C. M. O'D. et al. (2007) The origin and evolution of chondrites recorded in the elemental and isotopic compositions of their macromolecular organic matter. *Geochim. Cosmochim. Acta, 71,* 4380–4403.

Anderson G. M. (2005) *Thermodynamics of Natural Systems, 2nd edition.* Cambridge Univ., Cambridge. 648 pp.

Bardyn A. et al. (2017) Carbon-rich dust in Comet 67P/Churyumov-Gerasimenko measured by COSIMA/Rosetta. *Mon. Not. R. Astron. Soc., 469,* S712–S722.

Běhounková M. et al. (2012) Tidally-induced melting events as the origin of south-pole activity on Enceladus. *Icarus, 219,* 655–664.

Bethke C. M. (2008) *Geochemical and Biogeochemical Reaction Modeling, 2nd edition.* Cambridge Univ., New York. 543 pp.

Beuthe M. et al. (2016) Enceladus's and Dione's floating ice shells supported by minimum stress isostasy. *Geophys. Res. Lett., 43,* 10,088–10,096.

Bouquet A. et al. (2015) Possible evidence for a methane source in Enceladus' ocean. *Geophys. Res. Lett., 42,* 1334–1339.

Bouquet A. et al. (2017) Alternative energy: Production of H_2 by radiolysis of water in the rocky cores of icy bodies. *Astrophys. J. Lett., 840,* L8, DOI: 10.3847/2041-8213/aa6d56.

Brearley A. J. (2006) The action of water. In *Meteorites and the Early Solar System II* (D. S. Lauretta and H. Y. McSween, eds.), pp. 587–624. Univ. of Arizona, Tucson.

Brown R. H. et al. (2006) Composition and physical properties of Enceladus' surface. *Science, 311,* 1425–1428.

Buchardt B. et al. (1997) Submarine columns of ikaite tufa. *Nature, 390,* 129–130.

Buchardt B. et al. (2001) Ikaite tufa towers in Ikka Fjord, southwest Greenland: Their formation by mixing of seawater and alkaline spring water. *J. Sed. Res., 71,* 176–189.

Bucher K. et al. (2015) Weathering crusts on peridotite. *Contrib. Mineral. Petrol., 169,* 52, DOI: 10.1007/s00410-015-1146-3.

Čadek O. et al. (2016) Enceladus's internal ocean and ice shell constrained from Cassini gravity, shape, and libration data. *Geophys. Res. Lett., 43,* 5653–5660.

Choblet G. et al. (2017) Powering prolonged hydrothermal activity inside Enceladus. *Nature Astron., 1,* 841–847.

Collins G. C. and Goodman J. C. (2007) Enceladus' south polar sea. *Icarus, 189,* 72–82.

Drever J. I. (1997) *The Geochemistry of Natural Waters: Surface and Groundwater Environments, 3rd edition.* Prentice Hall, Upper Saddle River. 436 pp.

Ehlmann B. L. et al. (2013) Geochemical consequences of widespread clay mineral formation in Mars' ancient crust. *Space Sci. Rev., 174,* 329–364.

Fortes A. D. (2007) Metasomatic clathrate xenoliths as a possible source for the south polar plumes of Enceladus. *Icarus, 191,* 743–748.

Fournier R. O. and Rowe J. J. (1966) Estimation of underground temperatures from the silica content of water from hot springs and wet-steam wells. *Am. J. Sci., 264,* 685–697.

Frost B. R. (1991) Introduction to oxygen fugacity and its petrologic importance. *Rev. Mineral. Geochem., 25,* 1–9.

Frost B. R. and Beard J. S. (2007) On silica activity and serpentinization. *J. Petrol., 48,* 1351–1368.

German C. R. and Seyfried W. E. (2014) Hydrothermal processes. In *Treatise on Geochemistry, 2nd edition, Vol. 8: The Oceans and Marine Geochemistry* (M. J. Mottl and H. Elderfield, eds.), pp. 191–233. Elsevier, Amsterdam.

Glein C. R. (2015) Noble gases, nitrogen, and methane from the deep interior to the atmosphere of Titan. *Icarus, 250,* 570–586.

Glein C. R. and Shock E. L. (2010) Sodium chloride as a geophysical probe of a subsurface ocean on Enceladus. *Geophys. Res. Lett., 37,* L09204, DOI: 10.1029/2010GL042446.

Glein C. R. et al. (2008) The oxidation state of hydrothermal systems on early Enceladus. *Icarus, 197,* 157–163.

Glein C. R. et al. (2015) The pH of Enceladus' ocean. *Geochim. Cosmochim. Acta, 162,* 202–219.

Haff P. K. et al. (1983) Ring and plasma: The enigmae of Enceladus. *Icarus, 56,* 426–438.

Hansen C. J. et al. (2006) Enceladus' water vapor plume. *Science, 311,* 1422–1425.

Hansen C. J. et al. (2008) Water vapour jets inside the plume of gas leaving Enceladus. *Nature, 456,* 477–479.

Hansen C. J. et al. (2011) The composition and structure of the Enceladus plume. *Geophys. Res. Lett., 38,* L11202, DOI: 10.1029/2011GL047415.

Helgeson H. C. et al. (1978) Summary and critique of the thermodynamic properties of rock-forming minerals. *Am. J. Sci., 278-A,* 1–229.

Hendrix A. R. et al. (2010) The ultraviolet reflectance of Enceladus: Implications for surface composition. *Icarus, 206,* 608–617.

Horibe Y. and Craig H. (1995) D/H fractionation in the system methane-hydrogen-water. *Geochim. Cosmochim. Acta, 59,* 5209–5217.

Hsu H-W. et al. (2011) Stream particles as the probe of the dust-plasma magnetosphere interaction at Saturn. *J. Geophys. Res.–Space Phys., 116,* A09215, DOI: 10.1029/2011JA016488.

Hsu H-W. et al. (2015) Ongoing hydrothermal activities within Enceladus. *Nature, 519,* 207–210.

Iess L. et al. (2014) The gravity field and interior structure of Enceladus. *Science, 344,* 78–80.

Ingersoll A. P. and Nakajima M. (2016) Controlled boiling on Enceladus. 2. Model of the liquid-filled cracks. *Icarus, 272,* 319–326.

Ingersoll A. P. and Pankine A. A. (2010) Subsurface heat transfer on Enceladus: Conditions under which melting occurs. *Icarus, 206,* 594–607.

Jones B. F. et al. (1977) Hydrochemistry of the Lake Magadi basin, Kenya. *Geochim. Cosmochim. Acta, 41,* 53–72.

Kelley D. S. et al. (2001) An off-axis hydrothermal vent field near the Mid-Atlantic Ridge at 30°N. *Nature, 412,* 145–149.

Kelley D. S. et al. (2005) A serpentinite-hosted ecosystem: The Lost City hydrothermal field. *Science, 307,* 1428–1434.

Kempf S. et al. (2005) Composition of saturnian stream particles. *Science, 307,* 1274–1276.

Kieffer S. W. et al. (2006) A clathrate reservoir hypothesis for Enceladus' south polar plume. *Science, 314,* 1764–1766.

Kite E. S. and Rubin A. M. (2016) Sustained eruptions on Enceladus explained by turbulent dissipation in tiger stripes. *Proc. Natl. Acad. Sci., 113,* 3972–3975.

Klein F. and Garrido C. J. (2011) Thermodynamic constraints on mineral carbonation of serpentinized peridotite. *Lithos, 126,* 147–160.

Lowell R. P. and DuBose M. (2005) Hydrothermal systems on Europa. *Geophys. Res. Lett., 32,* L05202, DOI: 10.1029/2005GL022375.

Lowenstein T. K. et al. (2014) The geologic history of seawater. In *Treatise on Geochemistry, 2nd edition, Vol. 8: Oceans and Marine Geochemistry* (M. J. Mottl and H. Elderfield, eds.), pp. 569–622. Elsevier, Amsterdam.

Malamud U. and Prialnik D. (2016) A 1-D evolutionary model for icy satellites, applied to Enceladus. *Icarus, 268,* 1–11.

Marion G. M. et al. (2012) Modeling ammonia-ammonium aqueous chemistries in the solar system's icy bodies. *Icarus, 220,* 932–946.

Matson D. L. et al. (2007) Enceladus' plume: Compositional evidence for a hot interior. *Icarus, 187,* 569–573.

Matson D. L. et al. (2012) Enceladus: A hypothesis for bringing both heat and chemicals to the surface. *Icarus, 221,* 53–62.

Matson D. L. et al. (2018) Enceladus' near-surface CO_2 gas pockets and surface frost deposits. *Icarus, 302,* 18–26.

McCollom T. M. and Shock E. L. (1998) Fluid-rock interactions in the lower oceanic crust: Thermodynamic models of hydrothermal alteration. *J. Geophys. Res.–Solid Earth, 103,* 547–575.

McKay C. P. et al. (2008) The possible origin and persistence of life on Enceladus and detection of biomarkers in the plume. *Astrobiology, 8,* 909–919.

McKinnon W. B. (2015) Effect of Enceladus's rapid synchronous spin on interpretation of Cassini gravity. *Geophys. Res. Lett., 42,* 2137–2143.

Médard E. and Kiefer W. S. (2017) Differentiation of water-rich planetary bodies: Dehydration, magmatism and water storage. In *Lunar and Planetary Science XLVIII,* Abstract #2749. Lunar and Planetary Institute, Houston.

Millero F. J. (2013) *Chemical Oceanography, 4th edition.* CRC, Boca Raton. 591 pp.

Morrill P. L. et al. (2013) Geochemistry and geobiology of a present-day serpentinization site in California: The Cedars. *Geochim. Cosmochim. Acta, 109,* 222–240.

Murray A. E. et al. (2012) Microbial life at –13°C in the brine of an ice-sealed Antarctic lake. *Proc. Natl. Acad. Sci., 109,* 20,626–20,631.

Nakajima M. and Ingersoll A. P. (2016) Controlled boiling on Enceladus. 1. Model of the vapor-driven jets. *Icarus, 272,* 309–318.

Neveu M. et al. (2015) Prerequisites for explosive cryovolcanism on dwarf planet-class Kuiper belt objects. *Icarus, 246,* 48–64.

Neveu M. et al. (2017) Aqueous geochemistry in icy world interiors: Equilibrium fluid, rock, and gas compositions, and fate of antifreezes and radionuclides. *Geochim. Cosmochim. Acta, 212,* 324–371.

Niles P. B. et al. (2013) Geochemistry of carbonates on Mars: Implications for climate history and nature of aqueous environments. *Space Sci. Rev., 174,* 301–328.

Nimmo F. et al. (2007) Shear heating as the origin of the plumes and heat flux on Enceladus. *Nature, 447,* 289–291.

Okumura F. and Mimura K. (2011) Gradual and stepwise pyrolyses of insoluble organic matter from the Murchison meteorite revealing chemical structure and isotopic distribution. *Geochim. Cosmochim. Acta, 75,* 7063–7080.

Ootsubo T. et al. (2012) AKARI near-infrared spectroscopic survey for CO_2 in 18 comets. *Astrophys. J., 752,* 15, DOI: 10.1088/0004-637X/752/1/15.

Patthoff D. A. and Kattenhorn S. A. (2011) A fracture history on Enceladus provides evidence for a global ocean. *Geophys. Res. Lett., 38,* L18201, DOI: 10.1029/2011GL048387.

Porco C. C. et al. (2006) Cassini observes the active south pole of Enceladus. *Science, 311,* 1393–1401.

Porco C. et al. (2014) How the geysers, tidal stresses, and thermal emission across the south polar terrain of Enceladus are related. *Astron. J., 148,* 45, DOI: 10.1088/0004-6256/148/3/45.

Postberg F. et al. (2008) The E-ring in the vicinity of Enceladus: II. Probing the moon's interior — The composition of E-ring particles. *Icarus, 193,* 438–454.

Postberg F. et al. (2009) Sodium salts in E-ring ice grains from an ocean below the surface of Enceladus. *Nature, 459,* 1098–1101.

Postberg F. et al. (2011) A salt-water reservoir as the source of a compositionally stratified plume on Enceladus. *Nature, 474,* 620–622.

Postberg F. et al. (2018) Macromolecular organic compounds from the depths of Enceladus. *Nature, 558,* 564–568.

Ray C. et al. (2017) Oxidation in Enceladus' ocean. Abstract P51F-08 presented at 2017 Fall Meeting, AGU, San Francisco, California, 11–15 December.

Schmidt J. et al. (2008) Slow dust in Enceladus' plume from condensation and wall collisions in tiger stripe fractures. *Nature, 451,* 685–688.

Schubert G. et al. (2007) Enceladus: Present internal structure and differentiation by early and long-term radiogenic heating. *Icarus, 188,* 345–355.

Seewald J. S. (1994) Evidence for metastable equilibrium between hydrocarbons under hydrothermal conditions. *Nature, 370,* 285–287.

Sekine Y. et al. (2015) High-temperature water-rock interactions and hydrothermal environments in the chondrite-like core of Enceladus. *Nature Commun., 6,* 8604, DOI: 10.1038/ncomms9604.

Seyfried W. E. et al. (2011) Vent fluid chemistry of the Rainbow hydrothermal system (36°N, MAR): Phase equilibria and *in situ* pH controls on subseafloor alteration processes. *Geochim. Cosmochim. Acta, 75,* 1574–1593.

Seyfried W. E. et al. (2015) The Lost City hydrothermal system: Constraints imposed by vent fluid chemistry and reaction path models on subseafloor heat and mass transfer processes. *Geochim. Cosmochim. Acta, 163,* 59–79.

Shock E. L. and McKinnon W. B. (1993) Hydrothermal processing of cometary volatiles — Applications to Triton. *Icarus, 106,* 464–477.

Shock E. L. et al. (1989) Calculation of the thermodynamic and transport properties of aqueous species at high pressures and temperatures: Standard partial molal properties of inorganic neutral species. *Geochim. Cosmochim. Acta, 53,* 2157–2183.

Smith H. T. et al. (2008) Enceladus: A potential source of ammonia products and molecular nitrogen for Saturn's magnetosphere. *J. Geophys. Res.–Space Phys., 113,* A11206, DOI: 10.1029/2008JA013352.

Spencer J. R. et al. (2006) Cassini encounters Enceladus: Background and the discovery of a south polar hot spot. *Science, 311,* 1401–1405.

Squyres S. W. et al. (1983) The evolution of Enceladus. *Icarus, 53,* 319–331.

Stähler S. C. et al. (2018) Seismic wave propagation in icy ocean worlds. *J. Geophys. Res.–Planets, 123,* 206–232.

Stolper D. A. et al. (2014) Formation temperatures of thermogenic and biogenic methane. *Science, 344,* 1500–1503.

Streit E. et al. (2012) Coexisting serpentine and quartz from carbonate-bearing serpentinized peridotite in the Samail Ophiolite, Oman. *Contrib. Mineral. Petrol., 164,* 821–837.

Thomas P. C. et al. (2016) Enceladus's measured physical libration requires a global subsurface ocean. *Icarus, 264,* 37–47.

Tillner-Roth R. and Friend D. G. (1998) A Helmholtz free energy formulation of the thermodynamic properties of the mixture {water + ammonia}. *J. Phys. Chem. Ref. Data, 27,* 63–96.

Tobie G. et al. (2008) Solid tidal friction above a liquid water reservoir as the origin of the south pole hotspot on Enceladus. *Icarus, 196,* 642–652.

Tracy R. J. and Frost B. R. (1991) Phase equilibria and thermobarometry of calcareous, ultramafic and mafic rocks, and iron formations. *Rev. Mineral. Geochem., 26,* 207–289.

Travis B. J. and Schubert G. (2015) Keeping Enceladus warm. *Icarus, 250,* 32–42.

Turcotte D. L. and Schubert G. (2002) *Geodynamics, 2nd edition.* Cambridge Univ., New York. 472 pp.

Ueda H. et al. (2016) Reactions between komatiite and CO_2-rich seawater at 250° and 350°C, 500 bars: Implications for hydrogen generation in the Hadean seafloor hydrothermal system. *Prog. Earth Planet. Sci., 3,* 35, DOI: 10.1186/s40645-016-0111-8.

Vance S. et al. (2007) Hydrothermal systems in small ocean planets. *Astrobiology, 7,* 987–1005.

Vance S. et al. (2014) Ganymede's internal structure including thermodynamics of magnesium sulfate oceans in contact with ice. *Planet. Space Sci., 96,* 62–70.

Vance S. D. et al. (2016) Geophysical controls of chemical disequilibria in Europa. *Geophys. Res. Lett., 43,* 4871–4879.

Vance S. D. et al. (2018) Geophysical investigations of habitability in ice-covered ocean worlds. *J. Geophys. Res.–Planets, 123,* 180–205.

Von Damm K. L. et al. (1985) Chemistry of submarine hydrothermal solutions at 21°N, East Pacific Rise. *Geochim. Cosmochim. Acta, 49,* 2197–2220.

Wagner W. and Pruß A. (2002) The IAPWS formulation 1995 for the thermodynamic properties of ordinary water substance for general and scientific use. *J. Phys. Chem. Ref. Data, 31,* 387–535.

Waite J. H. et al. (2006) Cassini Ion and Neutral Mass Spectrometer: Enceladus plume composition and structure. *Science, 311,* 1419–1422.

Waite J. H. et al. (2009) Liquid water on Enceladus from observations of ammonia and ^{40}Ar in the plume. *Nature, 460,* 487–490.

Waite J. H. et al. (2017) Cassini finds molecular hydrogen in the Enceladus plume: Evidence for hydrothermal processes. *Science, 356,* 155–159.

Wand U. et al. (1997) Evidence for physical and chemical stratification in Lake Untersee (central Dronning Maud Land, East Antarctica). *Antarct. Sci., 9,* 43–45.

Zolotov M. Y. (2007) An oceanic composition on early and today's Enceladus. *Geophys. Res. Lett., 34,* L23203, DOI: 10.1029/2007GL031234.

Zolotov M. Y. (2012) Aqueous fluid composition in CI chondritic materials: Chemical equilibrium assessments in closed systems. *Icarus, 220,* 713–729.

Zolotov M. Y. (2017) Aqueous origins of bright salt deposits on Ceres. *Icarus, 296,* 289–304.

Zolotov M. Y. and Fegley B. (2000) Eruption conditions of Pele volcano on Io inferred from chemistry of its volcanic plume. *Geophys. Res. Lett., 27,* 2789–2792.

Zolotov M. Y. and Kargel J. S. (2009) On the composition of Europa's icy shell, ocean, and underlying rocks. In *Europa* (R. T. Pappalardo et al., eds.), pp. 431–457. Univ. of Arizona, Tucson.

Zolotov M. Y. and Postberg F. (2014) Can nano-phase silica originate from chondritic fluids? The application to Enceladus' SiO_2 particles. In *Lunar and Planetary Science XLV,* Abstract #2496. Lunar and Planetary Institute, Houston.

Hemingway D., Iess L., Tajeddine R., and Tobie G. (2018) The interior of Enceladus. In *Enceladus and the Icy Moons of Saturn* (P. M. Schenk et al., eds.), pp. 57–77. Univ. of Arizona, Tucson, DOI: 10.2458/azu_uapress_9780816537075-ch004.

The Interior of Enceladus

Douglas Hemingway
University of California, Berkeley

Luciano Iess
Università La Sapienza, Rome

Radwan Tajeddine
Cornell University

Gabriel Tobie
Université de Nantes

Knowledge of internal structure places constraints on the origin, evolution, and present behavior of Enceladus. Although the interior cannot be observed directly, much can be deduced from the magnitude of tidal and rotational asymmetries in the shape and gravity field and from how these asymmetries affect the rotational dynamics. The data and models suggest a differentiated interior consisting of a ~2400 kg m^{-3} rocky core, a 20–50-km mean thickness global subsurface liquid water ocean, and a 20–30-km mean thickness icy shell. The presence of an internal liquid water reservoir had been suggested by a series of observations early in the Cassini mission and was ultimately confirmed by two crucial and independent analyses related to (1) determination of Enceladus' low-order gravity field based on Doppler tracking of the Cassini spacecraft during dedicated flybys and (2) measurement of forced physical librations based on control point tracking through seven years of Cassini images. Although Cassini-based observations have yielded a good, basic picture of the interior of Enceladus, many unresolved issues remain — in particular, relating to the structure and stability of the icy shell and related implications for the energy budget and thermal evolution of Enceladus — and will need to be addressed through further modeling and, eventually, additional observations.

1. INTRODUCTION

1.1. Motivation

The internal structure of a planetary body provides essential clues about its origin and evolution. Based on its internal density structure, what is the likely composition of the materials from which the body formed (see the chapter in this volume by McKinnon et al.)? Did the interior become sufficiently warm to enable differentiation, with heavier components sinking toward the center? How quickly or slowly did the body form? What does the structure of the near surface and deeper interior tell us about the nature of the topography and the distribution of geologic provinces across the surface?

The interior of Enceladus is of particular interest because of the complex tectonic structures that cover much of its unusually youthful surface (see chapter in this volume by Patterson et al.) and, most strikingly, because of the ongoing eruptions of water vapor and ice grains from its south polar region (see chapters in this volume by Postberg et al., Goldstein et al., and Spencer et al.). Are these eruptions sourced from a potentially habitable subsurface liquid water reservoir? If so, how deep is the reservoir beneath the icy surface? How thick is the icy crust? What is the structure of the crust and how does it support the surface topography? How is the liquid water, which is not buoyant with respect to the ice, transported to the surface? How does the interior respond to tidal stresses? What modulates the eruptive activity at the south pole? How large is the internal sea or ocean, and how does it keep from freezing solid? What is the composition and nature of the core, and what is its capacity for contributing to the heating of the interior?

Although the internal structure cannot be observed directly, a variety of techniques are available for building a picture of the interior that can help to address many of these questions. Seismic imaging has proven to be the definitive technique for constraining the structure of Earth's interior, and seismic stations placed on the Moon during the Apollo program have likewise helped to improve our understanding of the Moon's interior. However, although seismic stations may soon be deployed on Mars and other worlds, the study of planetary interiors aside from Earth and the Moon has so far had to rely on other methods. In particular, as we will discuss in detail in this chapter, studies of shape, gravitational fields, and rotational dynamics can place constraints on planetary interiors, and have enabled significant advancements in our emerging understanding of Enceladus.

1.2. Key Observations and Basic Picture

Prior to the Cassini spacecraft's arrival in the Saturn system in 2004, our knowledge of Enceladus was limited to what could be inferred from telescopic observations and the data collected during the Saturn system flybys of Pioneer 11 in 1979 and the two Voyager spacecraft in 1980 and 1981. However, even these early observations led to considerable interest in Enceladus due to its unusually high albedo (*Buratti and Veverka,* 1984) and apparently youthful icy surface (*Smith et al.,* 1982; *Clark et al.,* 1983; *Squyres et al.,* 1983), as well as its relationship to Saturn's E ring (*Baum et al.,* 1981; *Haff et al.,* 1983; *Pang et al.,* 1984), all of which pointed to the incredible possibility of recent, or even ongoing, geologic activity on this tiny moon. Given the apparent history of geologic activity, and incorporating mass estimates based on astrometric observations (*Kozai,* 1976; *Jacobson,* 2004), basic interior models were developed for a differentiated Enceladus (*Zharkov et al.,* 1985), but were limited by the large uncertainties regarding its size and shape.

Beginning in 2005, the Cassini mission, which included numerous flybys of Enceladus, led to a series of remarkable observations that reinforced Enceladus' status as one of the most compelling exploration targets in the solar system, and provided clues about the nature of its interior. In particular, several of Cassini's instruments detected water vapor and organic molecules in the vicinity of Enceladus' south pole (*Brown et al.,* 2006; *Dougherty et al.,* 2006; *Hansen et al.,* 2006; *Spahn et al.,* 2006; *Spencer et al.,* 2006; *Waite et al.,* 2006), and spectacular back-lit images revealed the source of this material to be a series of ongoing eruptions concentrated along four major fissures spanning the highly tectonized south polar terrain (SPT) (*Porco et al.,* 2006). Subsequent observations identified high heat flow emanating from the south polar region and especially at the sites of the most prominent eruptions (*Spitale and Porco,* 2007; *Howett et al.,* 2011). The ongoing activity led to the suggestion that a subsurface sea might be feeding the eruptions (*Collins and Goodman,* 2007) and producing a topographic depression at the south pole, helping to explain the anomalous shape (*Thomas et al.,* 2007). The finding that the erupted ice grains contained salts (*Postberg et al.,* 2009, 2011) further supported the notion of an internal liquid water reservoir. On the other hand, the limited presumed available tidal heating energy (*Meyer and Wisdom,* 2007), together with the rapid heat loss, suggested that an internal sea should have frozen solid in just tens of millions of years (*Roberts and Nimmo,* 2008).

A major advancement in our knowledge of the interior came with Doppler tracking of the Cassini spacecraft during close encounters with Enceladus (*Rappaport et al.,* 2007), and especially with the determination of Enceladus' quadrupole gravity field and hemispherical asymmetry, once enough flyby results had been accumulated (*Iess et al.,* 2014). In combination with improved models of the topography (*Thomas et al.,* 2007; *Nimmo et al.,* 2011), the gravity data provided the first opportunity to effectively probe the interior, yielding an estimate of the degree of differentiation, and identifying mass anomalies suggestive of a (possibly global) subsurface liquid ocean. Subsequent measurement of the forced physical librations (*Thomas et al.,* 2016) independently confirmed that the icy crust must be fully decoupled from the deep interior, indicating that the internal ocean must indeed be global. Although detailed interpretation of these observations is complicated and remains an active area of research (e.g., *Čadek et al.,* 2016, 2017; *van Hoolst et al.,* 2016; *Beuthe et al.,* 2016; *Hemingway and Mittal,* 2017), the general conclusion is that Enceladus is differentiated, with an icy shell covering a global subsurface liquid water ocean, overlying a low-density rocky core (Fig. 1).

In this chapter, we examine the relevant theory and discuss the constraints that can be placed on the internal structure of Enceladus based mainly on the gravity and libration observations (sections 2 and 3, respectively). We also discuss the implications for the thermal state of Enceladus and, in particular, the structure and dynamics of its icy shell (section 4). Finally, we discuss the remaining open

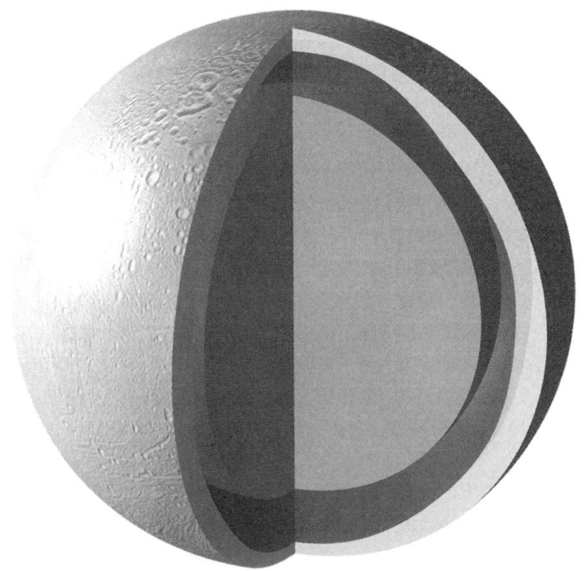

Fig. 1. Approximate interior structure of Enceladus (to scale). Its 252-km radius comprises a ~190-km-radius core with a density of ~2400 kg m^{-3}, surrounded by layers of liquid and solid H_2O making up the remaining ~60 km. The ice shell and ocean layer thicknesses vary laterally, with the thinnest part of the ice shell (and thickest part of the ocean) being centered on the south pole (see Table 3).

questions and how future observations and modeling might help to address them (section 5).

2. IMPLICATIONS FROM GRAVITY

Gravity is a powerful tool for studying planetary interiors because a body's gravitational field is ultimately a function of its internal mass distribution. While the problem of inverting gravity for interior structure suffers from inherent non-uniqueness, with a few reasonable assumptions, it is nevertheless possible to draw useful conclusions. For example, it is often reasonable to assume that large planetary bodies have interiors that are weak on long timescales, meaning that they tend to relax to near spherical symmetry, and that their internal densities do not generally increase with radial position — a situation that would be gravitationally unstable.

A straightforward approach to interior modeling is therefore to treat the body as a series of concentric spherical shells of different densities, with the requirement that the densities increase monotonically inward. The mean density, $\bar{\rho}$, of such a body is given by

$$\bar{\rho} = \sum_i \Delta\rho_i \left(\frac{R_i}{R}\right)^3 \quad (1)$$

where R is the body's full radius, R_i is the outer radius of the ith layer, and $\Delta\rho_i$ is the density contrast between layer i and the layer above it. This formulation implicitly assumes uniform density within each layer, which is appropriate only when internal pressures are small enough not to cause significant compression. For larger bodies, where compression may be important, equations of state are required to model the radial dependence of density on temperature and pressure. We proceed here assuming that compression is not important in the interior of Enceladus.

The body's normalized mean moment of inertia (or moment of inertia factor) is given by

$$\frac{I}{MR^2} = \frac{2}{5}\sum_i \frac{\Delta\rho_i}{\bar{\rho}} \left(\frac{R_i}{R}\right)^5 \quad (2)$$

Although there are generally more than two unknowns, necessitating additional assumptions, these two equations provide fundamental constraints on the internal structure. The mean density can be determined directly from the mass and radius, and immediately restricts the range of possible bulk compositions. But the moment of inertia, which reflects density stratification, must be determined via other means.

The three principal moments of inertia can be defined as

$$\begin{aligned} A &= \int \rho\left(y^2 + z^2\right) dV \\ B &= \int \rho\left(x^2 + z^2\right) dV \\ C &= \int \rho\left(x^2 + y^2\right) dV \end{aligned} \quad (3)$$

where ρ is the density of a volume element dV, at position x, y, z within the body. Here, the principal frame is a coordinate frame centered on the body with the z-axis coincident with the spin pole, the x-axis pointing toward the prime meridian, and the y-axis completing the righthanded system (for synchronous satellites in circular orbits, the x-axis points along the long axis, toward the parent body). Differences in these principal moments of inertia are important because they can, in some cases, be related to observable rotational dynamical effects such as spin pole precession, obliquity, nutation, and librations (e.g., *van Hoolst,* 2015) — the subject of section 3 of this chapter; or to asymmetries in the gravitational field — the subject of the rest of this section. In the limit of perfect spherical symmetry, a body's gravitational field would be equivalent to that of a point mass located at its center and, apart from the mean density, little could be concluded about the interior. Real planetary bodies, however, exhibit asymmetries that effectively carry information about their internal structures.

The case of Enceladus is interesting and instructive for several reasons. Because of the short period of its orbit around Saturn (1.37 days), the asymmetries in its shape and gravity (primarily due to tidal and rotational deformation) are large, permitting unprecedented relative precision in their measurement. This precision presents both challenges (it forces us to abandon the usual simplifying assumption of hydrostatic equilibrium) and opportunities (it allows us to estimate the thickness of the icy crust via analysis of the non-hydrostatic gravity and topography), forcing the development of new methods for interior modeling from shape and gravity (*Hemingway et al.,* 2013b; *Iess et al.,* 2014; *McKinnon,* 2015; *Čadek et al.,* 2016; *Beuthe et al.,* 2016). Finally, the special circumstance of the substantial north-south polar asymmetry associated with the SPT provides an opportunity for independent analysis of long-wavelength gravity and topography.

In section 2.1, we discuss how rotational and tidal effects lead to characteristic asymmetries in a body's shape and, consequently, its gravitational field, and how the magnitude of those asymmetries is related to the internal mass distribution. To do this, we introduce several concepts, including gravitational potential and the ways it is affected by tidal and rotational forces; hydrostatic equilibrium figure theory; and Love numbers and their connection to the principal moments of inertia and, ultimately, the density stratification within the body. In section 2.2, we discuss how the gravitational field of Enceladus has been measured based on radio tracking of the Cassini spacecraft during three close flybys. Finally, in section 2.3, we discuss the interpretation of those measurements, including some of the subtleties and challenges associated with interior modeling.

2.1. Theory

2.1.1. Hydrostatic equilibrium. For planetary bodies that are sufficiently large, the combination of high internal pressures and low internal viscosities (especially true when there is appreciable internal heating) results in relaxation to a figure that approaches the expectation for a strengthless fluid body, in which the inward acceleration due to gravity is everywhere balanced by the gradient in fluid pressure — a condition referred to as hydrostatic equilibrium and described by the equation $dp = -\rho g dr$, where p, ρ, g, and r are the pressure, local density, local gravity, and radial position, respectively.

Under the influence of self-gravitation alone, the hydrostatic equilibrium figure would be a sphere. Because of their rotation, however, planetary bodies also experience centrifugal flattening. In addition, satellites that are in synchronous rotation with their parent bodies (i.e., tidally locked) also experience permanent elongation along the static tidal axis. This applies to most of the large natural satellites in the solar system because of the short timescale associated with tidal locking (e.g., *Gladman et al.,* 1996; *Murray and Dermott,* 1999). The tidal and rotational deformation results in an asymmetry in the equilibrium figure — and consequently the gravitational field — of synchronous satellites like Enceladus.

2.1.2. Gravitational potential. A body's Newtonian gravitational potential, U, resolved at an arbitrary position **r**, is a function of its internal mass distribution, and is given by

$$U(\mathbf{r}) = -G\int_V \frac{\rho(\mathbf{r}')}{|\mathbf{r}-\mathbf{r}'|}dV' \tag{4}$$

where $\rho(\mathbf{r}')$ is the density at position $\mathbf{r}'$, where G is the universal gravitational constant, and where the integral is performed over the body's entire volume, V. The potential represents the work per unit mass done by the gravitational field and gives rise to the gravitational acceleration $\mathbf{g} = -\nabla U$.

Everywhere outside the body, the gravitational potential satisfies Laplace's equation, $\nabla^2 U(r) = 0$, and can be expressed as a linear combination of spherical harmonic functions as

$$U(r,\theta,\phi) = -\frac{GM}{r}\sum_{l=0}^{\infty}\sum_{m=-l}^{l}\left(\frac{R_{ref}}{r}\right)^l C_{lm}Y_{lm}(\theta,\phi) \tag{5}$$

where M is the total mass of the body, r is the radius at which the potential is to be expressed, θ is colatitude, ϕ is longitude, C_{lm} are the degree-l and order-m spherical harmonic expansion coefficients representing the dimensionless gravitational potential at the reference radius R_{ref}, and where Y_{lm} are the spherical harmonic functions — the natural set of orthogonal basis functions on a sphere (e.g., *Wieczorek,* 2015).

2.1.3. Tidal/rotational disturbing potential. The tidal and rotational forces that cause planetary bodies to deviate from spherical symmetry can be described in terms of disturbances in the potential field. For instance, the centrifugal acceleration at a point on the surface of a spherical body with radius R and spin rate ω is $\omega^2\mathbf{x}$, where $\mathbf{x} = R\sin\theta\,\hat{\mathbf{x}}$ is a vector pointed outward from, and perpendicular to, the axis of rotation and reaching the surface at colatitude θ. This acceleration can be expressed as the negative gradient of a centrifugal disturbing potential, $-\nabla V^{cf}$, where

$$V^{cf} = -\tfrac{1}{2}\omega^2R^2\sin^2\theta \tag{6}$$

For convenient comparison with equation (5), we can rewrite the centrifugal disturbing potential in terms of the degree-2 Legendre polynomial, $P_2(\cos\theta) = \frac{1}{2}(3\cos^2\theta - 1)$, as $V^{cf}(\theta,\phi) = \frac{1}{3}\omega^2R^2(P_2(\cos\theta) - 1)$. Since the last term is a constant, its gradient is zero, meaning that it gives rise to no deforming forces. As such, many authors do not include it, and we will likewise disregard it from here on. Finally, we can write the centrifugal disturbing potential (excluding the constant term) in terms of the degree-2 zonal spherical harmonic function, $Y_{20}(\theta,\phi)$, as

$$V^{cf}(\theta,\phi) = \tfrac{1}{3}\omega^2R^2Y_{20}(\theta,\phi) \tag{7}$$

In addition to the effects of rotation, satellites experience a tidal disturbing potential due to the spatially varying difference between their gravitational acceleration toward the parent body and the outward centrifugal acceleration associated with their orbital motion. To second order, it can be shown (e.g., *Murray and Dermott,* 1999, p. 133) that the net tidal disturbing potential is

$$V^{tid}(\psi) = -\frac{GmR^2}{a^3}P_2(\cos\psi) \tag{8}$$

where m is the mass of the parent body, a is the distance between the centers of the two bodies, and ψ is the angle from the axis connecting the centers of the two bodies to an arbitrary point on the satellite's surface. If the parent body is much more massive than the satellite ($m \gg M$), and the satellite's orbit is synchronous with its mean rotation rate, then $\omega^2 \approx Gm/a^3$. And if the satellite's spin axis is nearly normal to its orbital plane (i.e., its obliquity is close to zero), then we can rewrite the angle ψ in terms of colatitude and longitude as $\cos\psi \approx \cos\phi\sin\theta$. Both of these conditions are satisfied in the present case because Saturn is more than a million times as massive as Enceladus, whose obliquity is <0.001 (*Chen and Nimmo,* 2011; *Baland et al.,* 2016). Incorporating these approximations, it can be shown that

$$V^{tid}(\theta,\phi) = \omega^2R^2\left(\tfrac{1}{2}Y_{20}(\theta,\phi) - \tfrac{1}{4}Y_{22}(\theta,\phi)\right) \tag{9}$$

Adding equations (7) and (9) yields the combined effect of the tidal and rotational disturbing potentials

$$V(\theta,\phi) = \omega^2R^2\left(\tfrac{5}{6}Y_{20}(\theta,\phi) - \tfrac{1}{4}Y_{22}(\theta,\phi)\right) \tag{10}$$

2.1.4. Equilibrium figure theory. To the extent that the body can be treated as a hydrostatic fluid, its equilibrium figure will conform to an equipotential. The effect of the disturbing potential, V(θ,ϕ), is that the surface of the initially spherically symmetric body will no longer be equipotential. To first order, and before any deformation has taken place, the resulting change in elevation of the equipotential surface is $-V(\theta,\phi)/g$ (since g is the gradient of the gravitational potential). In response to the disturbing potential, the body's figure will relax toward the new equipotential surface. However, in so doing, the body's mass distribution is altered, effectively causing a small additional disturbance in the gravitational potential. Accounting for this self-gravitation effect, the hydrostatic equilibrium figure can be written

$$H(\theta,\phi) = R - \frac{V(\theta,\phi)}{g} h_f \tag{11}$$

where h_f is the fluid Love number for the figure, a scalar quantity describing the magnitude of the body's long timescale (zero frequency) deformation in response to the disturbing potential. An infinitely rigid body would have $h_f = 0$, whereas a perfectly homogeneous fluid body has $h_f = \frac{5}{2}$ (the value is larger than unity because of the aforementioned self-gravitation); fluid bodies with some internal density stratification have $0 < h_f < \frac{5}{2}$. The resulting asymmetries in the gravitational potential are given by

$$U(\theta,\phi) = V(\theta,\phi) k_f \tag{12}$$

which defines the potential fluid Love number k_f. For perfectly fluid bodies, $h_f = 1 + k_f$. The fluid Love numbers are of great importance for interior modeling because they can be related to the body's internal density distribution, as discussed in the next section. Since the disturbing potential in equation (10) is a degree-2 spherical harmonic function, we are here concerned with the degree-2 fluid Love numbers, h_{2f} and k_{2f}.

Note that the fluid Love numbers should not be confused with the tidal Love numbers, which characterize the body's viscoelastic response in both shape (h_{2t}) and gravitational potential (k_{2t}) due to eccentricity tides — variations in tidal potential associated with the body's changing proximity to its parent body. Ignoring these short-timescale variations in shape and gravity, a synchronous satellite in hydrostatic equilibrium thus has a mean figure described by

$$H(\theta,\phi) = R - Rqh_{2f}\left(\tfrac{5}{6}Y_{20}(\theta,\phi) - \tfrac{1}{4}Y_{22}(\theta,\phi)\right) \tag{13}$$

and a gravitational field whose asymmetries are described by

$$U(\theta,\phi) = \frac{GM}{R} qk_{2f}\left(\tfrac{5}{6}Y_{20}(\theta,\phi) - \tfrac{1}{4}Y_{22}(\theta,\phi)\right) \tag{14}$$

where we have introduced the commonly used parameter

$$q = \frac{\omega^2 R}{g} = \frac{mR^3}{Ma^3} \tag{15}$$

representing the relationship between the inward gravitational acceleration and the outward centrifugal acceleration at the body's equator.

It is clear from equation (14) that, for a synchronous satellite in hydrostatic equilibrium, the non-central part of the gravitational field described by equation (5), and resolved at the reference radius, has just two non-zero coefficients

$$\begin{aligned} J_2 &= -C_{20} = \tfrac{5}{6} qk_{2f} \\ C_{22} &= \tfrac{1}{4} qk_{2f} \end{aligned} \tag{16}$$

where we have followed the convention of expressing the zonal gravity coefficient according to $J_l = -C_{l0}$. For hydrostatic synchronous satellites, the ratio of the gravity coefficients is thus $J_2/C_{22} = 10/3$.

Likewise, if the body's shape is expressed in spherical harmonics as

$$H(\theta,\phi) = \sum_{l=0}^{\infty} \sum_{m=-l}^{l} H_{lm} Y_{lm}(\theta,\phi) \tag{17}$$

where H_{lm} are the degree-l and order-m spherical harmonic expansion coefficients representing the shape (with $H_{00} = R$), then the hydrostatic equilibrium figure is asymmetric in precisely the same way, with the coefficients

$$\begin{aligned} H_{20} &= -\tfrac{5}{6} qRh_{2f} \\ H_{22} &= \tfrac{1}{4} qRh_{2f} \end{aligned} \tag{18}$$

which, again, have the characteristic ratio $H_{20}/H_{22} = -10/3$. It is also common to describe the figure using the semiaxes ($a > b > c$) of an (approximately) equivalent triaxial ellipsoid, by evaluating equation (17) at the pole and at two points on the equator (at longitudes 0 and $\pi/2$), yielding

$$\begin{aligned} a &= R\left(1 + \tfrac{7}{6} qh_{2f}\right) \\ b &= R\left(1 - \tfrac{1}{3} qh_{2f}\right) \\ c &= R\left(1 - \tfrac{5}{6} qh_{2f}\right) \end{aligned} \tag{19}$$

This leads to another commonly referenced characteristic ratio for synchronous satellites in hydrostatic equilibrium

$$\frac{a-c}{b-c} = 4 \tag{20}$$

The forgoing describes the condition of hydrostatic equilibrium in terms of both the expected equilibrium figure and the resulting asymmetries in the gravitational field. However, whereas the first-order approximation used to obtain equation (11) is adequate in most cases, and sufficient for a basic interpretation of the gravity data (e.g., *Iess et al.,* 2014), higher-order approximations become important for more precise modeling of fast rotating bodies like Enceladus (*Tricarico,* 2014; *McKinnon,* 2015), as we will discuss in section 2.3.

2.1.5. Moments of inertia. In the previous section, we showed that, in the case of a body in perfect hydrostatic equilibrium, measuring either of the two degree-2 gravity coefficients, J_2 or C_{22}, is sufficient to recover the fluid Love number k_{2f}. This is very useful because k_{2f} can be related to the body's moment of inertia using the Radau-Darwin equation (e.g., *Darwin,* 1899; *Murray and Dermott,* 1999)

$$\frac{C}{MR^2} = \frac{2}{3}\left(1 - \frac{2}{5}\sqrt{\frac{4-k_{2f}}{1+k_{2f}}}\right) \tag{21}$$

where C is the polar moment of inertia (i.e., about the spin pole, or c-axis).

The asymmetric nature of the tidally and rotationally deformed body means that its principal moments of inertia are not equal (see equation (3)). For a hydrostatic synchronous satellite, $A < B < C$; in the case of a non-synchronously rotating body, with no static tidal bulge, $A = B < C$.

It can be shown (e.g., *Hubbard,* 1984, p. 79) that the moment of inertia asymmetries are related to the degree-2 gravity coefficients according to

$$\begin{aligned} J_2 &= \frac{C - \frac{1}{2}(A+B)}{MR^2} \\ C_{22} &= \frac{B - A}{4MR^2} \end{aligned} \tag{22}$$

from which it follows that the mean moment of inertia is

$$\frac{I}{MR^2} = \frac{C}{MR^2} - \frac{2}{3}J_2 \tag{23}$$

Having obtained the mean moment of inertia factor, we are finally in a position to constrain the internal density structure described by equations (1) and (2).

2.1.6. Nonhydrostatic effects. An additional complication arises from the fact that solid planetary bodies, even large ones like Earth, have exteriors that are cold and rigid enough to support some non-hydrostatic topography, even over long timescales. In general, the measured shape and gravity are therefore a reflection of a mostly hydrostatic body, superimposed with some relatively smaller non-hydrostatic topography (and corresponding non-hydrostatic gravity). The challenge is that it is not necessarily obvious how to separate the hydrostatic and non-hydrostatic parts of these signals.

In previous work [e.g., Europa (*Anderson et al.,* 1998)], it was not possible to obtain independent constraints on J_2 and C_{22} or to measure H_{20} and H_{22} with sufficient precision to confirm whether or not they conform to the hydrostatic expectation. Instead, the condition of hydrostatic equilibrium was assumed, allowing direct calculation of k_{2f} from C_{22} via equation (16), and yielding an estimate of the moment of inertia via equation (21). The Cassini mission, however, has enabled more precise and nearly independent measurements of each of these four quantities for the saturnian satellites Titan, Enceladus, Dione, and Rhea, and has revealed that the condition of hydrostatic equilibrium is generally not satisfied. For example, Titan's figure exhibits considerable excess flattening, with $-H_{20}/H_{22} = 4.9 \pm 0.1$ (*Zebker et al.,* 2012; *Hemingway et al.,* 2013a) in spite of its nearly hydrostatic gravity, for which $J_2/C_{22} = 3.32 \pm 0.02$ (*Iess et al.,* 2010, 2012). Similarly, Enceladus has considerable excess flattening, with $-H_{20}/H_{22} = 4.2 \pm 0.2$ (*Nimmo et al.,* 2011; *Thomas et al.,* 2016; *Tajeddine et al.,* 2017), while its gravity field is only modestly non-hydrostatic, with $J_2/C_{22} = 3.51 \pm 0.05$ (*Iess et al.,* 2014). For Dione, both the figure and gravity are considerably non-hydrostatic, with $-H_{20}/H_{22} = 5.2 \pm 0.6$ and $J_2/C_{22} = 4.00 \pm 0.06$ (*Thomas et al.,* 2007; *Nimmo et al.,* 2011; *Hemingway et al.,* 2016). While Rhea's shape is not as well determined, with $-H_{20}/H_{22} = 3.4 \pm 1.0$, its gravity field is substantially non-hydrostatic, with $J_2/C_{22} = 3.91 \pm 0.10$ (*Tortora et al.,* 2016).

This deviation from hydrostatic equilibrium complicates the interpretation, as we discuss below, but the combination of both shape and gravity observations allows some of the ambiguities to be resolved, and additionally provides information about how such unrelaxed topography is supported in the relatively stiff exterior. For example, the large non-hydrostatic topography and small non-hydrostatic gravity of Titan and Enceladus are suggestive of isostatic compensation (see section 2.3).

2.2. Observations

2.2.1. Gravity determination. Cassini carried out gravity measurements of Enceladus during three flybys on April 28, 2010; November 30, 2010; and May 2, 2012 (labeled as E9, E12, and E19). Flyby altitude and latitude were selected to enhance the signature of a hemispherical asymmetry of the gravity field, characterized mainly by the harmonic coefficient J_3. E9 and E19 occurred over the south polar region (latitude 89°S and 72°S), at closest approach altitudes of 100 and 70 km, respectively. E12 was over the north polar region (latitude 62°N), at an altitude of just 48 km. The small closest approach distances were required by the need to maximize the spacecraft accelerations due to Enceladus' gravity field.

The estimation of the mass and gravity field was obtained solely from measurements of the spacecraft range rate (Doppler). Range rate is approximately equal to twice the line-of-sight projection of the spacecraft velocity with respect to a ground antenna. This observable quantity is a standard product of the radio tracking system and is crucial for the accurate navigation of the spacecraft. Although other radiometric data are used for orbit reconstruction (such as range), range rate is by far the most valuable for geodesy applications. The range rate of Cassini is measured at a ground antenna of NASA's Deep Space Network (DSN) after establishing a coherent, two-way, microwave link at X-band (7.2–8.4 GHz). In this configuration, the ground antenna transmits a radio signal (a monochromatic carrier, possibly modulated for telecommands and ranging) generated by a highly stable frequency reference. The signal is then received onboard by means of a transponder, then coherently retransmitted back to Earth. The frequency of the beat tone between the incoming and the outgoing signal provides the Doppler shift and the (two-way) range rate.

The Cassini X-band radio system delivers measurement accuracies up to 0.01 mm s^{-1} at 60-s integration times under favorable conditions (although 2–4 times higher noise is more frequently encountered). On timescales relevant to Cassini gravity investigations, the noise is dominated by path delay variations due to interplanetary plasma and tropospheric water vapor. During the three gravity flybys of Enceladus, the (two-way) range rate noise was 0.017 mm s^{-1} (E9), 0.027 mm s^{-1} (E12), and 0.036 mm s^{-1} (E19) at 60 s integration time.

The estimation of Enceladus's gravity field with Cassini poses several challenges, all traceable to three factors. First, the small number of flybys strongly limits the sampling of the gravity field and does not permit breaking the correlations between some estimated parameters. Second, the spacecraft acceleration due to the moon's gravity is quite small (surface gravity is only ~0.11 m s^{-2}), and so is the change in the spacecraft range rate measured by the ground antenna. Third, the gravitational interaction time of the spacecraft with the small moon is short.

The velocity variation induced by the monopole gravity, and the zonal harmonics of degree 2 and 3 can be computed from elementary arguments. Neglecting factors on the order of unity, we simply multiply the acceleration due to each harmonic by the interaction time r/V (r being the Cassini-Enceladus distance and V the nearly constant relative velocity) and obtain

$$
\begin{aligned}
dV^{(0)} &\approx \frac{GM}{rV} \\
dV^{(2)} &\approx \frac{GM}{rV}\left(\frac{R}{r}\right)^2 J_2 \\
dV^{(3)} &\approx \frac{GM}{rV}\left(\frac{R}{r}\right)^3 J_3
\end{aligned}
\tag{24}
$$

Here, GM is the gravitational parameter of Enceladus and R is the reference radius (252.1 km). Note that, for close flybys ($r - R \ll R$), the velocity variation scales with the product ρR^2, where ρ is the mean density. Using the values reported by *Iess et al.* (2014), the velocity changes due to GM, J_2, and J_3 are, respectively, about 3–4 m s^{-1}, 1.1–1.4 cm s^{-1}, and 0.2–0.3 mm s^{-1}. Although all these values are well above the noise threshold of the Cassini radio system, the various contributions are not fully separable with only three flybys.

Gravity and orbits of the spacecraft and Enceladus were recovered by fitting the range rate data using orbit determination software developed at JPL for deep space navigation. These tools (ODP and the recently developed MONTE) generate computed observables using a fully relativistic model of the spacecraft dynamics and radio signal propagation. The residuals (observed range rate minus computed range rate) are minimized through an iterative process where a set of free parameters is adjusted by linearizing the observation equations. The output of the process is a set of estimated parameters and their covariances. The estimated parameters included the full satellite degree-2 harmonic coefficients, J_3, the position and velocity of Cassini at each flyby, and corrections to the mass and the orbital elements of Enceladus. In the nominal solution, the tesseral and sectorial components of the degree-3 field were assumed to be zero. This set of parameters was the minimum set able to fit the data. As a test of the stability of the model, an alternate solution, which also estimated the remaining degree-3 terms, was obtained. The result is statistically identical to the nominal result, although the uncertainties increase. All the additional degree-3 terms are compatible with zero within 2σ or less, and thereby add no new information (*Iess et al.,* 2014, section S2.1). The non-gravitational accelerations due to the anisotropic thermal emission from Cassini's radioisotope thermoelectric generators and the solar radiation pressure were accounted for using previous estimates obtained during the saturnian tour. The associated uncertainties were considered in the construction of the covariance matrix.

A good orbital fit would not be possible without also including the small but non-negligible aerodynamic drag experienced by Cassini when it flew through the plumes in the southern polar flybys E9 and E19. The dynamical effect was modeled as a nearly impulsive, vectorial acceleration at closest approach, to be estimated together with the other parameters. An equivalent method relies on models of the neutral particle density and estimates the aerodynamic coefficient of the spacecraft. Both methods lead to statistically identical solutions for the gravity field. The estimated, aerodynamic ΔV is almost parallel to the spacecraft velocity V (as expected for a drag force), with a magnitude of 0.25 mm s^{-1} for E9 and 0.26 mm s^{-1} for E19. The ΔV experienced by Cassini is comparable to the effect of the J_3 harmonic.

As expected, the gravity field is dominated by the J_2 and C_{22} harmonics, associated with the rotational and tidal deformation of Enceladus (Table 1). The estimate of J_2 and C_{22} is quite precise (about 1%, 1σ). $C_{2,1}$, $C_{2,-1}$, and $C_{2,-2}$ are consistent with a null value at the 2σ level. $C_{2,2}$ is about one order of magnitude larger than J_3, which is estimated to a relative accuracy of about 20%. The positive sign of J_3 implies a negative mass anomaly at the south pole (consistent with the observed topography), but its small value indicates the presence of a compensating positive mass anomaly at depth (see section 2.3).

2.2.2. Shape determination. The shape of Enceladus has been determined based on analysis of limb profiles collected over the course of the Cassini mission (*Thomas et al.,* 2007;

TABLE 1. Gravity model (*Iess et al.,* 2014) (unnormalized, dimensionless potential coefficients, for R_{ref} = 252.1 km).

Parameter	Value ±1σ
GM	7.210443 ± 0.00003 km^3 s^{-2}
$J_2 = -C_{2,0}$	5526.1 ± 35.5 × 10^{-6}
$C_{2,1}$	9.4 ± 11.8 × 10^{-6}
$C_{2,-1}$	40.5 ± 22.8 × 10^{-6}
$C_{2,2}$	1575.7 ± 15.9 × 10^{-6}
$C_{2,-2}$	23.0 ± 7.5 × 10^{-6}
$J_3 = -C_{3,0}$	−118.2 ± 23.5 × 10^{-6}
$J_2/C_{2,2}$	3.51 ± 0.05

Thomas, 2010; *Nimmo et al.,* 2011; *Tajeddine et al.,* 2017). The best fitting triaxial ellipsoidal figure exhibits substantial polar flattening, with a–c ≈ 8.5 km and a significant departure from hydrostatic equilibrium, with (a–c)/(b–c) ≈ 2.8 (recall that the hydrostatic ratio is 4). Similarly, in terms of spherical harmonic coefficients, the ratio of the degree-2 zonal to sectorial terms is ~4.2, much higher than the ~3.3 expected if Enceladus were in hydrostatic equilibrium (Table 2). The limb profiles provide sufficient spatial coverage to constrain global spherical harmonic models up to degree and order 8 (*Nimmo et al.,* 2011); the recent addition of stereogrammetric measurements has extended the latest topography models to degree and order 16 (*Tajeddine et al.,* 2017). One of the most significant features of the shape is the topographic depression associated with the SPT, having an elevation that is ~1.1 km lower than at the north pole, an effect that is expressed mainly in the H_{30} term (Table 2).

TABLE 2. Shape model (*Nimmo et al.,* 2011) (unnormalized).

Parameter	Value ±1σ
R	252.1 km
$H_{2,0}$	−3846 ± 179 m
$H_{2,1}$	0 ± 52 m
$H_{2,-1}$	−65 ± 52 m
$H_{2,2}$	917 ± 19 m
$H_{2,-2}$	−39 ± 19 m
$H_{3,0}$	384 ± 5 m
$-H_{2,0}/H_{2,2}$	4.20 ± 0.22

2.3. Modeling and Interpretation

2.3.1. Basic interpretation. Since the gravity field of Enceladus was first measured (*Iess et al.,* 2014), its interpretation has been an active area of research (*Iess et al.,* 2014; *McKinnon,* 2015; *Čadek et al.,* 2016, 2017; *van Hoolst et al.,* 2016; *Beuthe et al.,* 2016; *Hemingway and Mittal,* 2017). In part because of Enceladus' small size and rapid rotation rate, interpretation of the data is unusually sensitive to modeling details, as we discuss below. Nevertheless, there are a few basic observations that are uncontroversial.

The first is that while the shape is substantially nonhydrostatic, with $H_{20}/H_{22} = -4.20 \pm 0.22$ (1σ), the gravity field is only modestly so, with $J_2/C_{22} = 3.51 \pm 0.05$ (1σ). This situation is immediately suggestive of compensation. Without compensation, the large non-hydrostatic topography should give rise to a correspondingly large non-hydrostatic gravity signal. Likewise, the observed J_3 gravity anomaly, $(-118 \pm 23) \times 10^{-6}$, is substantially smaller than would be expected if the ~1-km north-south polar elevation difference were completely uncompensated ($\approx -375 \times 10^{-6}$).

A compensation mechanism is therefore required. Although there are several possibilities, including systematic lateral variations in density (i.e., Pratt isostasy) or nonhydrostatic topography on the surface of the rocky core (see section 2.3.5), the most straightforward mechanism, which naturally yields a small non-hydrostatic gravity signal in spite of the large non-hydrostatic topography, is Airy-type isostatic compensation. That is, Enceladus' long wavelength topography appears to be supported, at least in part, by the displacement of a relatively low viscosity, higher density material beneath the crust — e.g., a subsurface liquid ocean.

2.3.2. Admittance analysis. The presence of nonhydrostatic degree-2 gravity and topography prevents direct calculation of the moment of inertia via equations (16) and (21). Instead, it is necessary to first separate the observed degree-2 gravity and topography signals into their hydrostatic and non-hydrostatic parts (*Iess et al.,* 2014). Assuming these components can be separated linearly (a reasonable assumption as long as the non-hydrostatic parts are small compared to the hydrostatic parts), we can write

$$\begin{aligned} J_2^{obs} &= J_2^{hyd} + J_2^{nh} \\ C_{22}^{obs} &= C_{22}^{hyd} + C_{22}^{nh} \\ H_{20}^{obs} &= H_{20}^{hyd} + H_{20}^{nh} \\ H_{22}^{obs} &= H_{22}^{hyd} + H_{22}^{nh} \end{aligned} \tag{25}$$

where the superscripts refer to the observations (obs), the hydrostatic expectation (hyd) given by equations (16) and (18), and the non-hydrostatic components (nh). For a given assumed moment of inertia, the only unknowns in equation (25) are the non-hydrostatic components. The relationship between these non-hydrostatic gravity and topography signals can be quantified by the ratios

$$\begin{aligned} Z_{20} &= -J_2^{nh} / H_{20}^{nh} \\ Z_{22} &= C_{22}^{nh} / H_{22}^{nh} \end{aligned} \tag{26}$$

This is effectively a component-wise version of the spectral admittance (e.g., *Wieczorek,* 2015), a quantity related to the degree and depth of isostatic compensation: Greater and/or shallower compensation results in a more muted gravity signal and thus a smaller admittance. The advantage of this approach is that, provided the ice shell's mechanical properties are not significantly variable laterally, we should expect that $Z_{20} = Z_{22}$. Hence, we obtain self-consistent results when we find a moment of inertia that, via equations (21), (16), (18), (25), and (26), yields $Z_{20} = Z_{22}$.

It is worth emphasizing, however, that, for a body like Enceladus, the expectation that $Z_{20} = Z_{22}$ is an assumption that has not been explicitly tested in the literature to date. Future work to better justify this assumption, or to investigate the effect of relaxing it to some degree, may be valuable.

Considering a wide range of possible mean moments of inertia, *Iess et al.* (2014) produced the equivalent of Fig. 2a. This result follows from the first-order equilibrium figure theory described by equation (11) — an approximation that becomes increasingly poor for fast-rotating bodies like Enceladus (*Tricarico,* 2014). To account for this effect, *McKinnon* (2015) repeated the analysis of *Iess et al.* (2014) using a fourth-order theory of figures approach (*Tricarico,* 2014), yielding instead the equivalent of Fig. 2b. Whereas the first-order results indicate a preferred admittance of $Z_2 \approx 11.8 \pm 1.6$ mGal km^{-1} and a mean moment of inertia factor of ~0.333 ± 0.002, the fourth-order results suggest instead $Z_2 \approx 15.7 \pm 2.0$ mGal km^{-1} and ~0.331 ± 0.002. While the difference may appear subtle, the implications for compensation depth (and hence ice shell thickness) are significant.

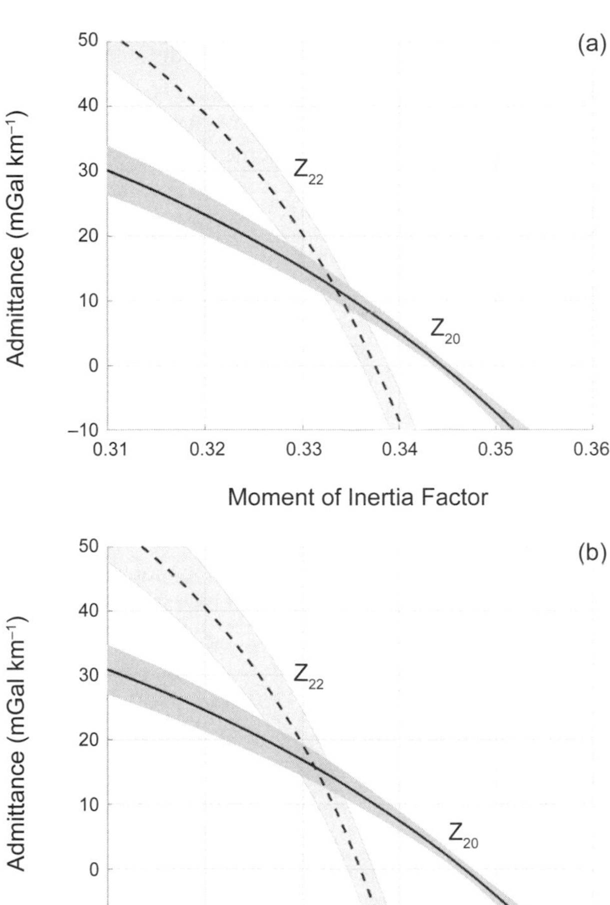

Fig. 2. Comparison of degree-2 zonal (Z_{20}, solid line) and sectorial (Z_{22}, dashed line) admittances across a range of possible mean moments of inertia. Shaded bands indicate 1σ uncertainties propagated from both the shape and gravity models. The point of intersection represents the admittance and moment of inertia combination that yields self-consistent results. **(a)** Replication of the *Iess et al.* (2014) results, following from first-order equilibrium figure theory (except using gravity coefficients from Table 1). **(b)** Replication of *McKinnon* (2015) results, employing fourth-order methods of *Tricarico* (2014). Unlike in *Iess et al.* (2014) and *McKinnon* (2015), admittances are here given in the more commonly used units of mGal km^{-1} (where 1 mGal = 10^{-5} m s^{-2}), obtained by converting the dimensionless gravitational potential coefficients in equation (26) to acceleration by multiplying them by (l+1)GM/R^2.

In addition to the degree-2 signals, the zonal terms of the degree-3 gravity and topography are available and can be used to obtain the zonal part of the degree-3 admittance, $Z_{30} \approx 14.0 \pm 2.8$ mGal km^{-1}. The degree-3 observations provide an independent admittance estimate that also has the advantage of not being complicated by tidal and rotational deformation (the theoretical hydrostatic equilibrium figure has no degree-3 components), and is therefore independent of moment of inertia.

To put these admittance values in context, assuming a density of $\rho_c = 925$ kg m^{-3} for the icy crust, the degree-2 and -3 admittances expected in the case of zero compensation are $Z_2 \approx 47$ mGal km^{-1} and $Z_3 \approx 44$ mGal km^{-1}. The fact that the observed admittances are much smaller is another indication that the topography is significantly compensated. Although the degree-2 admittance values depend on the assumed mean moment of inertia, even without requiring that $Z_{20} = Z_{22}$, the small values of both Z_{20} and Z_{22} suggest compensation independently. Although Z_{22} becomes compatible with uncompensated topography for moments of inertia below ~0.32, this seems unlikely, as it would require Enceladus to be one of the most strongly differentiated solid bodies in the solar system. This observation also provides a clue about the spatial extent of the subsurface liquid layer (assuming Airy type compensation): Whereas a south polar regional sea could account for the smallness of the zonal admittance terms (Z_{20} and Z_{30}), the small value of Z_{22} requires a more extensive, perhaps global, subsurface liquid layer.

2.3.3. Compensation model. Further interpretation of the admittance values requires a compensation model. While most authors to date have favored Airy isostasy to explain the compensation of Enceladus' long wavelength topography (*Iess et al.,* 2014; *McKinnon,* 2015; *Čadek et al.,* 2016; *Beuthe et al.,* 2016; *Hemingway and Mittal,* 2017), there are multiple ways to conceive of Airy compensation, leading to subtle but consequential differences between the models and their implications. Moreover, the introduction of elastic stresses to the models widens the parameter space considerably, and can have a significant effect on the results — an issue we discuss further below.

The non-hydrostatic topography gives rise to non-hydrostatic gravitational acceleration according to (*Jeffreys,* 1976; *Burša and Peč,* 1993)

$$g_{lm}(R) = \frac{l+1}{2l+1} 4\pi G \sum_i \Delta\rho_i H_{ilm} \left(\frac{R_i}{R}\right)^{l+2} \tag{27}$$

where R is the body's mean radius, G is the gravitational constant, H_{ilm} describes the degree-l and order-m non-hydrostatic topography at the top of the ith layer, whose mean outer radius is R_i, and where $\Delta\rho_i$ is the density contrast between the ith layer and the layer above it.

Assuming Airy compensation, and ignoring the role of elastic support for the moment, the relief at the base of the ice shell mirrors the non-hydrostatic surface relief with the amplitude scaled by $\rho_c/\Delta\rho$, where ρ_c is the density of the crust (or ice shell) and $\Delta\rho$ is the density contrast between the ice shell and the underlying ocean. In this case, the non-hydrostatic topography at the top (t) and bottom (b) of the crust are related by

$$H_{blm} = -H_{tlm} \frac{\rho_c}{\Delta\rho} \gamma \tag{28}$$

Here, the dimensionless factor γ is a placeholder allowing for various conceptions of Airy isostasy and different types of elastic support, as discussed below.

In general, the model admittance (gravity/topography ratio) can thus be written

$$Z(l) = \frac{l+1}{2l+1} 4\pi G \rho_c \left(1 - \gamma \left(\frac{R_b}{R_t} \right)^{l+2} \right) \tag{29}$$

where R_t and R_b are the radii corresponding to the top and bottom of the ice shell, respectively (the mean shell thickness being $d = R_t – R_b$).

Equations (27) and (29) implicitly treat the non-hydrostatic topography as a surface density anomaly — an approximation that is good when the topography is sufficiently small. In the present case, however, the small density contrast between the ice shell and the ocean leads to substantial relief at the ice/ocean interface. Taking this finite-amplitude into account (e.g., *Martinec,* 1994; *Wieczorek and Phillips,* 1998), one obtains a slightly smaller admittance. Although we neglect this effect in our discussion here, within the context of any particular compensation model, the effect can be significant, and so some authors do take it into account (e.g., *Čadek et al.,* 2016; *Hemingway and Mittal,* 2017).

In the limit of zero elastic support, and when isostasy is defined as requiring equal masses in columns of equal solid angle, then $\gamma = (R_t/R_b)^2$ (e.g., *Wieczorek,* 2015); the requirement of equal pressures at equal depths instead leads to $\gamma = (g_t/g_b)$, where g_t and g_b are the mean gravitational acceleration at the top and bottom of the ice shell, respectively (*Hemingway and Matsuyama,* 2017); when the total weight of the surface topography is required to equal the buoyancy force of the basal relief, then $\gamma = (g_t/g_b)(R_t/R_b)^2$ (e.g., *Čadek et al.,* 2016). Unfortunately, these different definitions of isostasy lead to significantly different estimates of the compensation depth.

Assuming a crustal density of ρ_c = 925 kg m^{-3}, the observed degree-3 admittance suggests a compensation depth of between 18 km and 30 km, depending on how one defines isostasy [the "equal masses" model leads to the largest compensation depth estimate; the "equal pressures" model leads to the smallest (*Hemingway and Matsuyama,* 2017)], in agreement with the result following from the degree-2 admittance obtained using first-order equilibrium figure theory (*Iess et al.,* 2014). The degree-2 admittance obtained using fourth-order equilibrium figure theory (*Tricarico,* 2014; *McKinnon,* 2015) instead suggests a compensation depth of between 24 km and 47 km, again, depending on how isostasy is defined. Accounting for uncertainties in the observed shape and gravity, the range of possible compensation depths expands even further: from ~14 km at the low end of the estimated degree-3 admittance to ~53 km at the high end of the estimated degree-2 admittance.

When elastic support is included, γ decreases in the case of surface loading (e.g., due to impacts or sedimentation/erosion at the surface) and increases in the case of basal loading (e.g., freezing/melting/relaxation at the base of the crust), and can have a significant effect on the resulting compensation depth estimate (*Hemingway and Mittal,* 2017). For instance, adding even an effective 200-m elastic layer can decrease the estimated compensation depth by more than a factor of 2 (*Čadek et al.,* 2016), assuming the shell thickness variations are generated at the surface; the compensation depth estimate increases if the shell thickness variations are instead generated at the base of the shell (*Beuthe et al.,* 2016; *Hemingway and Mittal,* 2017).

Further details of elastic/viscoelastic support models are beyond the scope of this chapter, but it should be noted that this remains an active area of research (*Čadek et al.,* 2016; *Soucek et al.,* 2016; *Hemingway and Mittal,* 2017). In particular, it is not yet clear how elastic support of the long-wavelength topography is affected by factors such as lateral variations in the shell's elastic properties (*Beuthe,* 2008; *Čadek et al.,* 2016, 2017), or the way bending and membrane stresses are transmitted across the SPT given the presence of the major fracture systems (i.e., the Tiger Stripes) in that region (e.g., *Soucek et al.,* 2016).

Although the range of possible mean ice shell thicknesses (i.e., mean compensation depths) can be narrowed by arguing for one or another compensation model, researchers have yet to converge on the best approach, making it difficult to make definitive statements about the ice shell thickness. However, assuming some version of Airy compensation, a few things can be stated with confidence. First, the thinnest part of the ice shell must be located beneath the large topographic depression at the south pole. Since the shell thickness is necessarily greater than zero there, this provides an effective lower bound on the mean shell thickness. Depending on the definition of isostasy, and the assumed ice shell and ocean densities, one finds that the mean shell thickness must be at least ~18 km in order to ensure nonzero shell thickness at the south pole. An upper bound on the shell thickness is harder to establish. However, as long as the degree-3 observations are taken into account, the mean shell thickness is unlikely to exceed ~44 km. Shell thicknesses greater than 44 km would lead to degree-3 admittances more than 2σ larger than the observed value, even when using the isostasy model that produces the smallest admittances [i.e., the "equal masses" model, with $\gamma = (R_t/R_b)^2$]. Again, however, the possibility of substantial elastic support of basal topography can allow for larger mean shell thicknesses.

2.3.4. Internal structure. The simplest approach to modeling the internal density structure is to start by assuming a two-layer body. If the density of the outer H_2O layer (ocean plus ice shell) is prescribed, then with the known bulk density (1609 kg m^{-3}) and the moment of inertia obtained from the admittance analysis, we can use equations (1) and (2) to solve for the core radius and density (Fig. 3). A mean core radius of ~192 km leaves ~60 km for the mean thickness of the H_2O layer. If the mean compensation depth (i.e., ice shell thickness) is ~30 km, this leaves a mean ocean thickness of ~30 km, neglecting the effect of the small difference in densities between the ice shell and the ocean (*Iess et al.,* 2014; *McKinnon,* 2015). More sophisticated three-layer models have also been constructed (*Čadek et al.,* 2016; *Hemingway and Mittal,* 2017), yielding broadly similar results (Table 3).

A few basic conclusions emerge immediately. First, the low density of the (presumably silicate) core suggests hydrothermal alteration (e.g., serpentinization) and/or substantial porosity, which may be easy to maintain given the modest internal pressures (Fig. 4). In particular, an unconsolidated rubble pile core, with water-filled pores, is possible (*Roberts,* 2015). The combined thickness of the H_2O layers is reasonably well constrained, but the models permit some tradeoff between the thicknesses of the ice shell and the ocean (e.g., *Hemingway and Mittal,* 2017). Although these tradeoffs have little effect on the density and therefore the internal pressure and gravity profiles (Fig. 4), they do make it difficult to determine the shell thickness with precision. This is regrettable since the thickness of the ice shell has a number of important implications. We discuss the ice shell's possible structure, dynamics, and related implications in section 4.

2.3.5. Additional considerations. In the models discussed so far, the core shape has been assumed to conform to an equipotential, as expected if the core shape was established when the interior was warm. The small internal gravity and pressure, however, means that the surface of the core could, in principle, support considerable non-hydrostatic topography over long timescales. Indeed, an irregular core shape had previously been proposed to account for the non-hydrostatic shape of Enceladus' surface (*Thomas et al.,* 2007; *McKinnon,* 2013) and to explain the rotation state of the similarly sized Mimas (*Tajeddine et al.,* 2014). However, while the presence of non-hydrostatic core topography would contribute to the observed gravity, there is no reason to expect it to do so in precisely such a way as to offset the gravity anomalies associated with the surface topography. That is, there is no reason to expect the non-hydrostatic core topography to effectively mirror the non-hydrostatic surface topography, as would be required for it to be substantially responsible for the compensating effect observed in the gravity data. In fact, it is just as likely, if not more so, that non-hydrostatic core topography would have the opposite effect, as would be expected in the case of an overly oblate or tidally elongated core. While the core shape may not be precisely hydrostatic, the hydrostatic assumption may be justified on the grounds that it is the most parsimonious. Although it has not been done in the literature to date, some unknown non-hydrostatic core topography could be added as a free parameter, and would increase uncertainties in the other estimated parameters.

Whereas all the models discussed so far assume some version of Airy-type isostasy, in which the surface topography is supported in part by lateral variations in the ice shell's thickness, it should be noted that the topography could also be supported in part by lateral variations in density [i.e., Pratt-type isostasy (e.g., *Besserer et al.,* 2013; *Tajeddine et al.,* 2017)]. This mechanism is unlikely to be primarily responsible for the observed compensation, however, because it would require parts of the heavily cratered plains to exhibit a higher crustal density, which is the opposite of what one would expect (e.g., *McKinnon,* 2015; *Hemingway and Mittal,* 2017) and because it would require the density variations to extend through an unrealistically large depth of ~40 km (*Hemingway and Mittal,* 2017). Nevertheless, the potential compensating effect of lateral variations in density cannot be ruled out entirely.

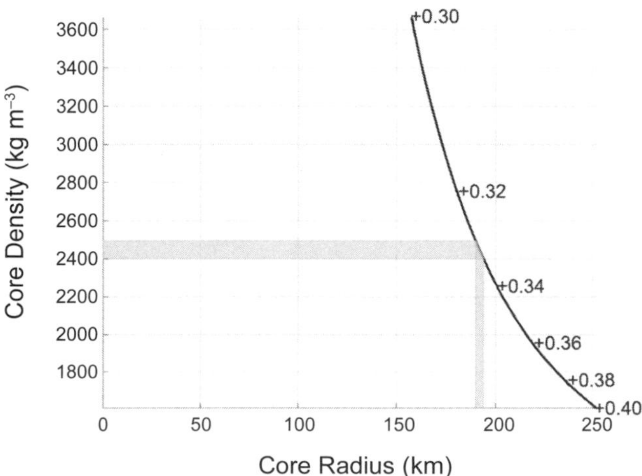

Fig. 3. Core radius and density over a range of possible mean moments of inertia (I/MR^2 = 0.30–0.40), assuming a two-layer body with outer layer density fixed at ρ = 950 kg m^{-3}. The estimated mean moment of inertia (0.331 ± 0.002) corresponds to a core radius of ~192 km and a core density of ~2450 kg m^{-3}.

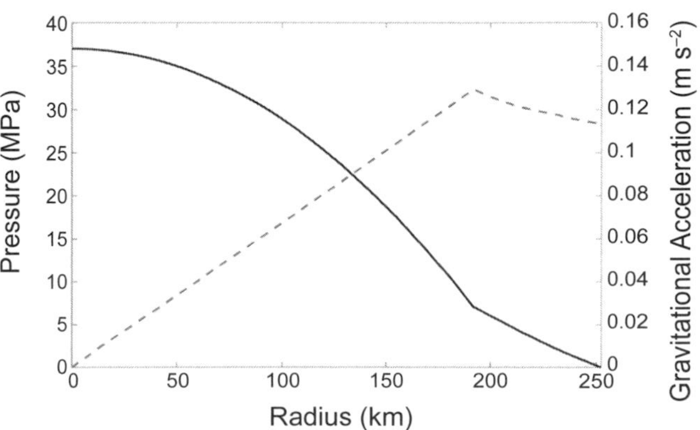

Fig. 4. Interior pressure (solid line) and gravity (dashed line) profiles for a three-layer model of Enceladus with 22-km mean thickness ice shell (900 kg m^{-3}), 38-km mean thickness ocean (1030 kg m^{-3}), and 192-km radius core (~2450 kg m^{-3}).

TABLE 3. Comparison of published interior models.

Model	Core	Ocean	Ice Shell	Notes
Iess et al. (2014)	~190 km (~2400 kg m^{-3})	10–30 km (1000 kg m^{-3}*)	30–40 km (920 kg m^{-3}*)	*Based on gravity and topography:* Hydrostatic terms from first-order equilibrium figure theory; elastic support deemed insignificant; both regional and global oceans discussed, but left as an open question.
McKinnon (2015)	190–195 km (~2450 kg m^{-3})	~10 km (1007 kg m^{-3}*)	~50 km (925 kg m^{-3}*)	*Based on gravity and topography:* Hydrostatic terms from fourth-order equilibrium figure theory (*Tricarico,* 2014); elastic support deemed insignificant; ice shell thickness determination is based on degree-2 data only (degree-3 data deemed to have only regional significance).
Thomas et al. (2016)	~200 km (~2300 kg m^{-3})	26–31 km (1000 kg m^{-3}*)	21–26 km (~850 kg m^{-3})	*Based on physical libration amplitude:* Three-layer model with layer interfaces assumed to be hydrostatic and defined by first-order equilibrium figure theory; ice shell thickness constrained by libration amplitude. Core and ice shell densities adjusted to fit the moment of inertia reported by *Iess et al.* [2014].
Čadek et al. (2016)	180–185 km (~2450 kg m^{-3})	~50 km (~1030 kg m^{-3})	18–22 km (925 kg m^{-3}*)	*Based on gravity, topography, and librations:* Elastic support (assuming top loading) introduced in order to bring gravity-based shell thickness estimate into agreement with *Thomas et al.* (2016); Airy compensation model modified to account for radial variation in gravity; finite-amplitude correction included (*Martinec,* 1994).
Van Hoolst et al. (2016)	170–205 km (2158–2829 kg m^{-3})	21–67 km (950–1100 kg m^{-3})	14–26 km (900–1000 kg m^{-3})	*Based on physical libration amplitude:* Similar to *Thomas et al.* (2016) but additionally accounting for uncertainties on libration amplitude arising from rigidity and viscoelastic behavior of the ice shell, a wider range of ice and ocean densities, and the possibility of core topography; relaxes assumption of Airy compensation.
Beuthe et al. (2016)	186–196 km (2350–2480 kg m^{-3})	34–42 km (1020 kg m^{-3}*)	19–27 km (925 kg m^{-3}*)	*Based on gravity and topography:* Three-layer model with hydrostatic terms from second-order equilibrium figure theory (*Zharkov,* 2004); isostatic compensation achieved by minimizing deviatoric stresses (e.g., *Dahlen,* 1982).
Hemingway and Mittal (2017)	188–205 km (2200–2450 kg m^{-3})	12–36 km (1000–1100 kg m^{-3})	22–41 km (850–950 kg m^{-3})	*Based on gravity and topography:* Three-layer model with equilibrium figures determined numerically (*Tricarico,* 2014); Airy compensation model modified to ensure equal pressures at depth (*Hemingway and Matsuyama,* 2017); using newer shape model (*Tajeddine et al.,* 2017); finite-amplitude correction included (*Wieczorek and Phillips,* 1998).

*Values that were prescribed rather than derived.

Thicknesses refer to mean layer thickness; lateral thickness variations are required by all models.

3. IMPLICATIONS FROM LIBRATIONS

In the previous section, we saw how tidal and rotational forces deform synchronous satellites like Enceladus into approximately triaxial ellipsoidal figures. We discussed how the resulting asymmetries in the mass distribution lead to differences among the principal moments of inertia, how these moments of inertia are related to observable asymmetries in the gravitational field, and how the moments of inertia are related to the satellite's internal structure. Here we look at how a satellite's rotational dynamics depend on the principal moments of inertia, focusing specifically on how the amplitude of forced physical librations can help to constrain the interior structure.

Most satellites in the solar system are locked in a 1:1 spin:orbit resonance, where the satellite's mean rotation rate is equal to its orbital mean motion. In this configuration, and assuming a circular orbit, the long axis of the satellite (i.e., its static tidal bulge) always points toward the parent body (the central planet). The eccentricity of the satellite's orbit, however, causes it to move faster at pericenter and slower at apocenter, even as the rotation rate remains (nearly) constant, leading to misalignments between the satellite's long axis and the line connecting the two bodies. To a first approximation, the long axis of the satellite points at the empty focus of the satellite's elliptical orbit (*Murray and Dermott,* 1999, p. 44). In a frame fixed on the planet, the satellite thus appears to rock back and forth about its spin axis as it orbits the planet — these apparent oscillations, called "optical librations," explain in part why more than 50% of the nearside of the Moon is visible from Earth.

Because of the periodic misalignments between the satellite's long axis and the satellite-planet line, the planet exerts gravitational torques on the satellite's static tidal bulge, creating additional oscillations called "physical longitudinal librations." Whereas the amplitude of the optical libration depends only on the orbital eccentricity, the physical libration amplitude additionally depends on the satellite's moments of inertia, thus making its measurement a valuable tool for constraining the internal structure [e.g., for the Moon (*Dickey et al.,* 1994), Mercury (*Margot et al.,* 2007), Phobos (*Willner et al.,* 2010; *Nadezhdina and Zubarev,* 2014), Janus and Epimetheus (*Tiscareno et al.,* 2009), Mimas (*Tajeddine et al.,* 2014), and Enceladus (*Thomas et al.,* 2016)].

3.1. Theory

The rotation of a satellite orbiting a central planet with mass m is described by the three Euler equations of motion (*Danby,* 1988; *Murray and Dermott,* 1999)

$$\begin{aligned} A\dot{\omega}_x - (B-C)\omega_y\omega_z &= 3Gm(C-B)YZ/r^5 \\ B\dot{\omega}_y - (C-A)\omega_z\omega_x &= 3Gm(A-C)ZX/r^5 \\ C\dot{\omega}_z - (A-B)\omega_x\omega_y &= 3Gm(B-A)XY/r^5 \end{aligned} \tag{30}$$

where A, B, and C are the satellite's principal moments of inertia; ω_x, ω_y, and ω_z are the projections of the spin vector onto the satellite's principal axes, x, y, and z; X, Y, and Z represent the coordinates of the planet in the satellite's principal frame; and r is the distance between the satellite and the planet.

When the satellite's obliquity is negligible (i.e., when its spin pole is normal to its orbital plane), then $\omega_x = \omega_y = 0$, and the problem is reduced to two dimensions (Fig. 5). The small obliquity expected for Enceladus (*Chen and Nimmo,* 2011; *Baland et al.,* 2016) makes this a reasonable approximation for purposes of our discussion.

If we define ψ as the angle between the satellite's long axis and the line connecting the two bodies, then we have $X/r = \cos\psi$ and $Y/r = \sin\psi$, which reduces the third line in equation (30) to

$$\ddot{\theta} = -\frac{3}{2}\frac{(B-A)}{C}\frac{Gm}{r^3}\sin 2\psi \tag{31}$$

where we now use $\ddot{\theta}$ to represent the variation in the satellite's angular velocity. Here, θ is the orientation of the satellite's long axis with respect to a fixed inertial reference frame

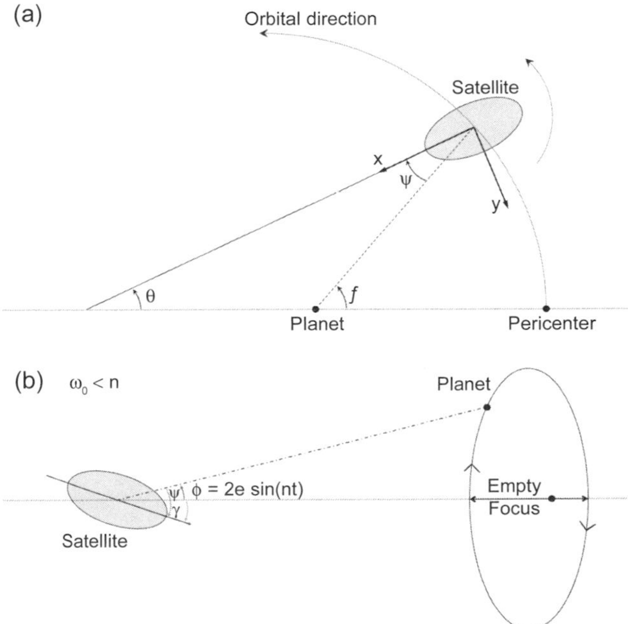

Fig. 5. The geometry of a satellite subject to librations illustrated **(a)** in a frame fixed on the planet; and **(b)** in a frame fixed on the satellite and oriented with the x-axis pointing toward the empty focus of the satellite's orbit. In **(a)**, f represents the true anomaly for the satellite, while θ represents the orientation of its long axis with respect to a fixed inertial frame. In **(b)**, ϕ represents the angle between the satellite-planet line and the line connecting the satellite with the empty focus of its orbit [this is the optical libration angle, not shown in **(a)**]; γ is the physical libration. After *Tiscareno et al.* (2009).

(where $\theta = 0$ corresponds to pericenter). This variation in angular velocity goes as (B–A)/C because the longitudinal torques the planet exerts on the satellite are proportional to the satellite's moment of inertia difference, B–A, and because the polar moment of inertia, C, represents the resistance to those torques.

We also define ϕ to be the angle between the line connecting the two bodies and the line connecting the satellite with the empty focus of its orbit (Fig. 5b) — this is the optical libration angle. In the absence of longitudinal torques, the satellite's long axis would remain pointed at the empty focus, so that ψ would be equal to ϕ. In this case, the satellite's orientation would be given simply by the mean anomaly $\theta(t) = nt$. The presence of longitudinal torques, however, causes small additional variations in the satellite's orientation so that $\theta(t) = nt + \gamma(t)$, where γ is the forced physical libration. Assuming for now that n is constant (although a discussion on perturbations from other satellites is provided below), we have $\ddot{\theta} = \ddot{\gamma}$.

The satellite's true anomaly (Fig. 5a), expanded as a Fourier series and limited to first order in eccentricity (*Murray and Dermott,* 1999), is given by

$$f = nt + 2e \sin nt + O\left(e^2\right) \tag{32}$$

It is clear from Fig. 5a that $f - \theta = \psi$, so that $\psi = 2e \sin nt - \gamma$. Using the small angle approximation for ψ, equation (31) can be rewritten

$$\ddot{\theta} = \frac{3(B-A)}{C}\frac{Gm}{r^3}(2e \sin nt - \gamma) \tag{33}$$

With the approximation that $Gm \approx r^3n^2$, and given that $\ddot{\theta} = \ddot{\gamma}$, this yields

$$\ddot{\gamma} = 2e\omega_0^2 \sin nt - \gamma\omega_0^2 \tag{34}$$

where ω_0 is the natural frequency, defined as

$$\omega_0 = n\sqrt{\frac{3(B-A)}{C}} \tag{35}$$

Finally, substituting $\gamma = \gamma_0 \sin nt$ into equation (34), the solution becomes

$$\gamma = \frac{2e}{1-(n/\omega_0)^2} \sin nt \tag{36}$$

The amplitude of the forced physical libration thus depends on the natural frequency ω_0, which is a function of the moments of inertia via equation (35), and hence the satellite's interior structure. Note that the phase of the physical libration is a function of n and ω_0: If $n < \omega_0$, then the libration would be in phase with the torque, while if $n > \omega_0$, the libration and the torque would be 180° out of phase (*Murray and Dermott,* 1999, p. 216), which is the case for Enceladus and the case illustrated in Fig. 5b.

So far, we have considered only the two-body problem, where the satellite's motion is not perturbed by other satellites. The more general expression of the physical libration is (*Rambaux et al.,* 2010)

$$\gamma = A_d \sin(\omega_d t + \alpha_d) e^{-\lambda t} + \sum_i \frac{H_i}{1-(\omega_i/\omega_0)^2} \sin(\omega_i t + \alpha_i) \tag{37}$$

The first term in equation (37) represents the free libration of the satellite, usually induced by episodic events such as impacts. The free libration oscillates at $\omega_d = \sqrt{\omega_0^2 - \lambda^2}$, also called the free libration frequency, very close to the satellite's natural frequency ω_0 [λ is a damping constant (*Rambaux et al.,* 2010)]. The amplitude A_d and the phase α_d are integration constants that are functions of the initial conditions related to the perturbing event. Since dissipation in the satellite will damp the free librations on a timescale of tens of years (*Rambaux et al.,* 2010; *Noyelles et al.,* 2011), we can assume that the observed librations of Enceladus are forced.

The second term in equation (37) represents the forced libration, including gravitational perturbations from other satellites. Such perturbations introduce additional librations, where H_i, ω_i, and α_i are the magnitude, frequency, and phase of the ith perturbation, respectively. Equation (36) is a special case of equation (37), corresponding to the orbital frequency, n, and an amplitude of $H_i = 2e$. Using the JPL Horizons Ephemeris (*http://ssd.jpl.nasa.gov/horizons.cgi*), *Rambaux et al.* (2010) identified three major libration frequencies for Enceladus, with periods of 1.37 days, 3.89 years, and 11.05 years, and respective amplitudes of –0.028°, 0.189°, and 0.259° (based on a solid-body interior model). The first signal, at the orbital period of Enceladus, is due to the mean anomaly perturbation. The comparison to the semi-analytical model of Enceladus' orbit (*Vienne and Duriez,* 1995) suggests that the remaining two signals are due to the Enceladus-Dione resonance: The first is related to Dione's precession of pericenter, and the second is related to the orbital libration argument of the Dione-Enceladus resonance.

Equation (37) indicates that the forced libration amplitudes, at all frequencies, depend on the natural frequency ω_0, and thus on the satellite's internal structure. However, because the amplitude depends on $1-(\omega_i/\omega_0)^2$, the effect is only significant for the libration at the orbital frequency, for which $(\omega_i/\omega_0)^2$ is non-negligible (Table 4). For this reason, in attempting to constrain the interior structure, we focus only on the libration at the satellite's orbital period.

3.2. Measuring Physical Librations

With the exceptions of Mercury (*Margot et al.,* 2007) and the Moon (*Dickey et al.,* 1994), all measurements of physical libration amplitudes have been carried out by photogrammetry using spacecraft imagery (*Tiscareno et al.,* 2009; *Willner et al.,* 2010; *Nadezhdina and Zubarev,* 2014; *Tajeddine et al.,* 2014; *Thomas et al.,* 2016). This technique uses images to

TABLE 4. Dominant signals involved in the forced librations of Enceladus (*Rambaux et al.,* 2010).

Period (days)	Amplitude (deg)	$(\omega_i/\omega_0)^2$
1.37	–0.028	16.3
1418.93	0.188	1.5×10^{-5}
4035.64	0.259	1.9×10^{-6}

Because $(\omega_i/\omega_0)^2$ is negligibly small for the longer-period librations, only the libration at the orbital period is significantly determined by the internal structure, via equation (35).

build three-dimensional coordinates of recognizable surface features, called control points, by observing them from various angles. With sufficient coverage, a global control point network can be built and used for mapping and for generating digital terrain models (DTMs). Like any observation-based technique, control points are subject to reconstruction errors. While some are random, others are systematic, such as errors in spacecraft position, camera pointing, and the body's rotation model (including the effect of any physical libration).

During its time in the Saturn system, the Cassini spacecraft collected tens of thousands of images of Enceladus. Observations were made from various phase angles and distances, in order to study the plumes, surface geology, and orbital and rotational dynamics. The amplitude of the physical libration of Enceladus is on the order of hundreds of meters of surface displacement at the equator. The spacecraft must be close enough to obtain images of sufficiently high resolution to permit libration detection. However, the spacecraft must not be too close, because when only a small part of the surface is in view, all the predicted positions of control points projected onto that image have nearly the same systematic errors, preventing the fitting software from distinguishing between libration-related offsets and camera-pointing errors. Provided that a sufficient latitudinal range is in view, however, the ambiguity is removed because, unlike camera pointing errors, the libration-related offsets vary with latitude.

To measure the physical librations of Enceladus, *Thomas et al.* (2016) used a network of 488 control points (mostly craters) across the surface, tracking them through a series of 340 Cassini Imaging Science Subsystem (ISS) images, totaling 5873 measurements. Starting from the approximately known positions of each control point in the satellite's reference frame, the control points were first converted to J2000, a fixed frame based on the initial conditions of Earth's rotational state on January 1, 2000, at 12:00 Terrestrial Time. This requires knowledge of the satellite's rotational parameters (i.e., orientation of the spin axis, rotation rate, librations, etc.), accounting for the three major libration frequencies (Table 4). As discussed above, the amplitudes of the long-period librations are well known since their dependence on Enceladus' interior is negligible. The amplitude of the signal at the orbital period of Enceladus (including the physical libration), however, is a parameter that needs to be fitted. Next, the coordinates are converted to the reference frame of the Cassini ISS Narrow Angle Camera (NAC), requiring instantaneous information of the spacecraft's position and the camera's pointing angle. This information is available at NASA's Navigation and Ancillary Information Facility (NAIF) SPICE website (*http://naif.jpl.nasa.gov/naif*). The twist angle (orientation about the axis along the line of sight) of the Cassini NAC is well determined, and the position of the spacecraft is known to within 1–10 km. The camera's right ascension and declination, however, are not well determined and must be fitted.

Next, the three-dimensional coordinates of the control points were adjusted in order to minimize the χ^2 residuals between the observed and predicted positions of the control points projected onto the camera image plane. To avoid degeneracies associated with going from three dimensions to two, each control point must have been observed from at least two sufficiently different viewing geometries. The more a control point is observed in different images, the lower the uncertainties in its position. Similarly, increasing the number of control points in an image reduces the uncertainties on camera-pointing angle. Finally, the reconstructed control point network still depends on the assumed rotational model for the satellite. When the model does not describe the satellite's rotation accurately, additional errors will result in the reconstructed positions of control points. Therefore, the libration amplitude at the orbital period must be varied in order to minimize the reconstruction errors.

Following this procedure, *Thomas et al.* (2016) obtained a best-fit physical libration amplitude of –0.120° (the minus sign indicates that the libration is out of phase by 180°) with a 2σ uncertainty of 0.014°. This translates to 528 ± 60 m in surface displacement at the equator. The smallness of the uncertainty is a function of the fact that the control points are based on ellipses fitted to craters, permitting subpixel accuracy in their positions and, more importantly, the large number of data points (since the uncertainty goes as $N^{-1/2}$).

3.3. Interior Modeling and Interpretation

Having measured the libration amplitude, we can now compute the satellite's natural frequency via equation (34) and hence determine its dynamical triaxiality, (B–A)/C, via equation (35). The value corresponding to the observed libration amplitude of 0.120° is (B–A)/C ≈ 0.0607. Although fixing this quantity does not yield a unique interior model, various simple trial models can be tested to determine the approximate internal structure.

Assuming a multi-layered body consisting of nested triaxial ellipsoids, the principal moments of inertia, defined in equation (3), become

$$\begin{aligned} A &= \frac{4\pi}{15}\sum_i \Delta\rho_i\left(b_i^2 + c_i^2\right)a_i b_i c_i \\ B &= \frac{4\pi}{15}\sum_i \Delta\rho_i\left(a_i^2 + c_i^2\right)a_i b_i c_i \\ C &= \frac{4\pi}{15}\sum_i \Delta\rho_i\left(a_i^2 + b_i^2\right)a_i b_i c_i \end{aligned} \tag{38}$$

where a_i, b_i, and c_i are the semiaxes of the ith ellipsoid, and $\Delta\rho_i$ is the density contrast between layer i and the layer above it. In the simplest case of a homogeneous interior, the dynamical triaxiality is simply

$$\frac{B-A}{C}=\frac{a^2-b^2}{a^2+b^2} \tag{39}$$

Based on their shape model, in which a = 256.2 km, and b = 251.4 km, *Thomas et al.* (2016) found that the libration amplitude expected for a homogenous interior was ~0.032° — more than 10σ smaller than the observed value. In the case of a two-layer body, the triaxiality is given by

$$\frac{B-A}{C}=\frac{V\rho_s\left(a_s^2-b_s^2\right)+V_c\left(\rho_c-\rho_s\right)\left(a_c^2-b_c^2\right)}{V\rho_s\left(a_s^2+b_s^2\right)+V_c\left(\rho_c-\rho_s\right)\left(a_c^2+b_c^2\right)} \tag{40}$$

where the subscripts c and s indicate the rocky core and the icy shell, respectively, V is the volume, and ρ is the density. Unlike the case of a homogeneous interior, models with two or more layers require additional assumptions about the shape and densities of each layer. The known bulk density (1609 kg m^{-3} for Enceladus) and the inferred moment of inertia factor constrain, via equations (1) and (2), the layer thicknesses such that the core radius can be computed once the layer densities are specified. Computing the shapes of internal layers is less straightforward but can be done in a number of ways. *Thomas et al.* (2016) and *van Hoolst et al.* (2016), for instance, integrate the Clairaut equation (*Clairaut,* 1743; *Danby,* 1988) to first order in terms of the polar and equatorial flattening to obtain expressions that depend only on the densities and mean radii of the layers, which can then be related to the semiaxes of the triaxial ellipsoidal shape (for details, see *Tajeddine et al.,* 2014). Similarly, *Tricarico* (2014) describes methods for computing higher-order nested ellipsoidal figures recursively or numerically in terms of polar and equatorial eccentricities, again, starting from the specified densities and mean radii of each layer.

Making the assumption that the core shape conforms to the hydrostatic expectation (i.e., its surface is equipotential), and considering a range of ice shell densities between 700 kg m^{-3} and 930 kg m^{-3}, *Thomas et al.* (2016) determined that the libration amplitude for the two-layer model would be between 0.032° and 0.034°, still much smaller than the observed value. Thomas et al. further showed that a regional subsurface sea centered on the south pole has little effect on the libration amplitude because of its symmetry about the axis of rotation. Similar results were obtained by *van Hoolst et al.* (2016), confirming the incompatibility of the libration observations with interior models involving a core that is physically coupled to the icy mantle.

As another check, one can use the above dynamical triaxiality in combination with the observed sectorial gravity harmonic, C_{22} (Table 1), and equation (22) to compute the polar moment of inertia. The result is $C \approx 0.104M\ R^2$. This unrealistically small polar moment of inertia is not compatible with the internal density structure inferred from the gravity observations [roughly 0.22M R^2 (*Iess et al.,* 2014; *McKinnon,* 2015)] and is another indication that Enceladus is not behaving like an entirely solid body.

The large observed libration amplitude evidently requires a more radically different interior model. *Thomas et al.* (2016) thus argued for a model with a global subsurface ocean that completely decouples the rocky core from the icy shell. In this case, the icy shell and the rocky core experience and respond to gravitational torques nearly independently. In the limit of a spherical core, for example, only the shell itself experiences external gravitational torques and, because the polar moment of inertia for the shell alone, C_s, is smaller than that of the body as a whole, C, the libration amplitude should be larger.

Thomas et al. (2016) and *van Hoolst et al.* (2016) applied a more generalized analysis, which includes the effects of torques between the shell and the core, as well as the effects of the ocean pressure on both the core and the shell. For a body with a subsurface ocean decoupling the shell from the core, the libration amplitude of the shell is given by (e.g., *van Hoolst et al.,* 2009; *Rambaux et al.,* 2011; *Richard et al.,* 2014)

$$\gamma_s=\frac{2e\left[K_s\left(K_c+2K_{int}-n^2C_c\right)+2K_{int}K_c\right]}{C_cC_s\left(n^2-\omega_1^2\right)\left(n^2-\omega_2^2\right)} \tag{41}$$

where K_s, K_c, and K_{int} are the planet-shell, planet-core, and core-shell torques, respectively. The first two torques are functions of the moments of inertia of the shell and the core, as well as the ocean pressure on each layer; K_{int} depends on the densities and the dimensions of each layer; and ω_1 and ω_2 are the system's natural frequencies (for details, see *Richard et al.,* 2014).

This expression shows that the libration amplitude once again depends on the densities assumed for the internal layers as well as their shapes. Assuming hydrostatic figures for each layer and assuming the ice shell and ocean densities to be 850 kg m^{-3} and 1000 kg m^{-3}, respectively, *Thomas et al.* (2016) found the best agreement with the observed librations for interior models with mean ice shell thicknesses in the range 21–26 km and mean ocean thicknesses in the range 26–31 km. The densities of the ice shell and the core were adjusted (within the constraints noted above) to 850 kg m^{-3} and 2300 kg m^{-3}, respectively, in order to be consistent with the moment of inertia inferred from the gravity observations (*Iess et al.,* 2014). Using similar methods, but considering different ranges of core, ice shell, and ocean densities (2158–2829 kg m^{-3}, 900–1000 kg m^{-3}, and 950–1100 kg m^{-3}, respectively), and allowing for non-hydrostatic figures for the core and the ocean/ice interface, *van Hoolst et al.* (2016) obtained mean ice shell and ocean thicknesses in the ranges 14–21 km and 24–67 km, respectively.

Although results vary slightly with the assumptions that go into such models, it is clear that the observed physical libration amplitude places powerful constraints on the internal structure. Most importantly, these libration studies demonstrate that the ice shell must be fully decoupled from the deeper interior, requiring the subsurface ocean to be global. Additionally, the libration studies provide a means of estimating the ice shell's thickness in a manner that is independent from the gravity-based analysis discussed in section 2, increasing confidence in both approaches.

4. ICE SHELL STRUCTURE AND DYNAMICS

4.1. Structure

The structure of the ice shell is of particular interest because it bears on the thermal state of Enceladus, the means for supporting its topography, the mechanics of the major fissures in the south polar region, and the nature of the ongoing eruptions. Determining the precise thickness of the ice shell, however, is not necessarily straightforward. For example, analysis of Enceladus' gravity field (section 2) has yielded mean shell thickness estimates ranging from 20–40 km (*Iess et al.,* 2014) at the low end, to 40–60 km (*McKinnon*, 2015) at the high end (although the latter estimate is based only on the degree-2 gravity). The large amplitude of the diurnal forced physical librations (section 3) (*Thomas et al.,* 2016; *van Hoolst et al.,* 2016), on the other hand, suggests a mean shell thickness of 15–25 km. The apparent discrepancy is not as problematic as it may seem, however, as it can be resolved in a number of ways involving details of the elastic/isostatic compensation model (*Čadek et al.,* 2016; *Beuthe et al.,* 2016; *Hemingway and Mittal,* 2017), as discussed in section 2.3.3.

What is common among the various models is that they take the long-wavelength topography to be related to lateral variations in the thickness of the icy shell, which is, to some degree, supported isostatically (i.e., it is "floating" on a subsurface ocean). Hence, the large topographic basin at the south pole implies a regional thinning of the ice shell, as predicted by *Collins and Goodman* (2007) based on the observed shape (*Thomas et al.,* 2007) and the localized geologic activity. The gravity observations (*Iess et al.,* 2014) confirmed this prediction by showing that the small magnitude of the corresponding south polar gravity anomaly implied substantial compensation (i.e., thinning at the base of the ice shell). Given the ~2-km topographic depression at the south pole (measured with respect to an equilibrium figure) (*Nimmo et al.,* 2011), and making reasonable assumptions about the ice shell and ocean densities, the total crustal thinning (relative to the mean shell thickness) at the south pole must be roughly 16–18 km, assuming complete Airy-type isostatic compensation (see equation (28) in section 2.3.3) (*Čadek et al.,* 2016; *Hemingway and Mittal,* 2017). Since the shell thickness is evidently greater than zero at the south pole, this may be regarded as an approximate lower bound on the mean shell thickness. However, the precise amplitude of lateral shell thickness variations also depends on the assumed compensation model (i.e., pressure vs. force balance, inclusion of partial elastic support, lateral variations in crustal density, etc.) (*Čadek et al.,* 2016; *Hemingway and Matsuyama,* 2017; *Hemingway and Mittal,* 2017; *Tajeddine et al.,* 2017).

The various models also agree that the thickest part of the shell is around the equator, and especially near the sub- and anti-saturnian points, where the thickness may be 30–40 km, or even more, depending on the preferred compensation model (*Čadek et al.,* 2016; *Hemingway and Mittal,* 2017). The overall pattern (Fig. 6) may be the result, at least in part, of tidal heating. Tidal dissipation is strongest at the poles and weaker in the equatorial regions (*Ojakangas and Stevenson,* 1989). The equilibrium shell thickness based on tidal dissipation and insolation results in a pattern with power at degrees 2 and 4 (*Ojakangas and Stevenson,* 1989; *Hemingway and Mittal,* 2017). Higher-order structure in the shell thickness variations may be a result of heterogeneities in the shell and/or mode coupling associated with non-Newtonian flow of the ice as it relaxes (*Nimmo,* 2004).

4.2. Thermal and Mechanical Stability

The structure of the ice shell raises questions about the stability of the current configuration. Lateral shell thickness variations induce stresses that will tend to relax away those thickness variations over time, with the ice flowing from the thicker regions (i.e., the equator) to the thinner regions (i.e., the poles). The observed structure is thus stable only if the relaxation rate is counterbalanced by spatial variations in ice melting (associated with tidal dissipation) and ocean crystallization rates (e.g., *Collins and Goodman,* 2007; *Kamata and Nimmo,* 2017). The rate at which the deflected ice/ocean interface relaxes toward an equipotential surface is controlled by the viscosity structure of the ice shell. Following the approach of *Lefevre et al.* (2014), developed for Titan, *Čadek et al.* (2016) determined that the relaxation velocity remains smaller than a few centimeters per year only if the ice shell is conductive with a bottom viscosity of at least 10^{14}–10^{15} Pa s. Counterbalancing such a relaxation rate requires heat flux variations between the equatorial and polar regions of at least 100–200 mW m^{-2}, which is comparable with the estimated heat flux in the SPT.

It is also possible that the current configuration is transient, with the observed surface topography being the partially relaxed remnants of a previous state with larger thickness variations (*Čadek et al.,* 2016). *Čadek et al.* (2017) have argued, using a viscoelastic relaxation model, that the observed surface topography may be presently relaxing toward a new equilibrium if Enceladus experiences changes in heat production and ice melting/ocean crystallization on timescales comparable to the relaxation timescale (which may range between 1 and 100 m.y. for lithospheric viscosities between 10^{22} and 10^{24} Pa s). Indeed, the intense

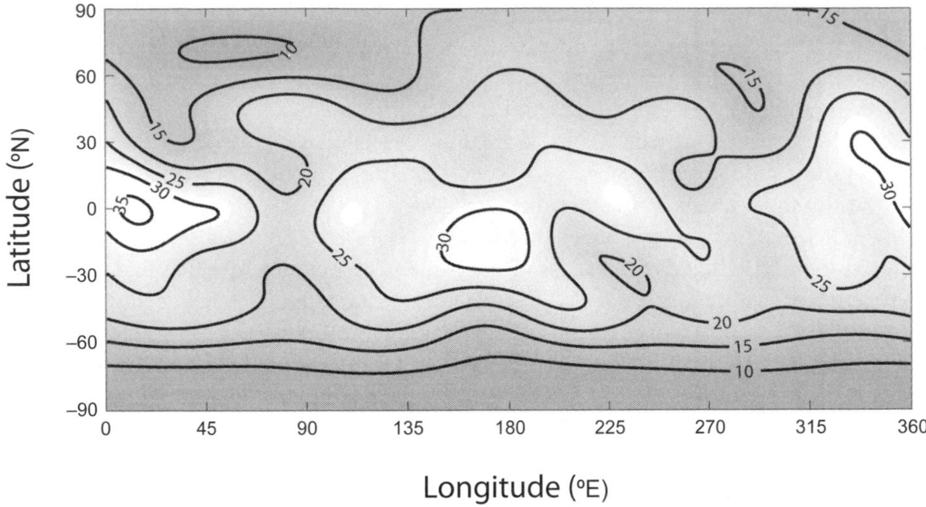

Fig. 6. Lateral variations in ice shell thickness assuming complete Airy compensation of all known topography [up to spherical harmonic degree 8 (*Nimmo et al.,* 2011)]. Contours indicate shell thickness in kilometers. In this example, the ice shell thickness ranges from ~6 km at the south pole to ~36 km at the sub- and anti-saturnian points along the equator, with a mean thickness of 22 km. While the mean shell thickness is somewhat model-dependent, the amplitude of lateral variation does not vary significantly among the models (e.g., *Čadek et al.,* 2016; *Hemingway and Mittal,* 2017). The addition of partial elastic support, however, would alter the amplitude of shell thickness variations, especially at the shortest wavelengths.

surface activity in the south polar region is an indication of a dynamical world where the ice shell structure may be evolving on even shorter timescales.

4.3. Implications

The shell thickness implies a small Rayleigh number, making convection unlikely and suggesting a thermally conductive ice shell. Heat flux, F, can be related to the temperature structure and thickness of a conductive ice shell according to

$$F = \frac{c}{d} \ln\left(\frac{T_s}{T_b}\right) \tag{42}$$

where d is the shell thickness, T_s and T_b are the surface and basal temperatures, respectively, and where c is an empirically derived constant, taken to be 567 W/m (e.g., *Klinger,* 1980; *Nimmo,* 2004). For an ice shell thickness of about 30–40 km in the equatorial region, the diffusive heat flux typically ranges between 15 and 20 mWm^{-2}. As the diffusive heat loss is inversely proportional to the ice shell thickness, cooling is even faster away from the equator, where the shell is thinner. For an average ice shell thickness between 20 and 25 km, the total diffusive heat loss outside the SPT is about 20 GW. In addition to the passive heat loss, one must also consider the heat loss associated with the intense activity at the south pole, which is estimated to be up to 15 GW (*Howett et al.,* 2011). For ice shell thicknesses ranging between 3 and 5 km in the SPT, about 5 to 10 GW is lost just by thermal diffusion through the ice shell, with the rest of the power being associated with the eruption activity itself. In total, then, some ~35 GW must be generated to maintain the present-day state of Enceladus.

Such a large power is consistent with the latest estimates of the dissipation function in Saturn (*Lainey et al.,* 2012; *Lainey,* 2016), which indicates that a strong dissipation (20 GW or even more) may be sustained during relatively long periods of time before the orbital eccentricity would be damped. However, the mechanism for generating such power within Enceladus is less clear (see the chapter in this volume by Nimmo et al.). Such heat production in the ice shell would require a very low bottom viscosity [<5 × 10^{13} Pa s (*Roberts and Nimmo,* 2008)], which would result in rapid relaxation of the basal topography, incompatible with the observed thickness variations. On the other hand, for a bottom viscosity ranging between 10^{14} and 10^{15} Pa s, heat flow due to tidal dissipation in the conductive ice shell would barely exceed a few milliwatts per square meter (*Roberts and Nimmo,* 2008; *Čadek et al.,* 2016), too weak to counterbalance the diffusive heat loss and to prevent the ocean from freezing (the timescale to crystallize 10 km of ice in the thickest part of the ice shell is on the order of 10 m.y.). The present-day state may result from enhanced tidal dissipation due to higher eccentricities in the recent past, as suggested by *Běhounková et al.* (2012). This would imply that the ocean is presently crystallizing with a maximum crystallization rate in the equatorial region where tidal heating is at a minimum.

An alternative solution would be strong dissipation in the deeper interior that counterbalances the diffusive heat loss through the ice shell. Two candidate processes would be dissipation of resonant tidal waves in the ocean (*Tyler,* 2011; *Matsuyama,* 2014; *Hay and Matsuyama,* 2017) or tidal friction in the unconsolidated core (*Roberts,* 2015). The former process requires a thin ocean, contrary to the ocean thickness inferred by recent studies (*Čadek et al.,* 2016; *van Hoolst et al.,* 2016; *Beuthe et al.,* 2016; *Hemingway and Mittal,* 2017). For the latter process, an unconsolidated rocky core filled with water ice, the power produced is too low (*Roberts,* 2015). Dissipation in a water-filled porous core might be more efficient, but more work is needed to demonstrate whether this would be sufficient to counterbalance the diffusive heat loss. Interestingly, strong dissipation in the core may also provide the energy source to power the hydrothermal activity, as suggested by the detection of nanosilica emitted from Enceladus (*Hsu et al.,* 2015), and to sustain long-term circulation of hot water in the core, as modeled by *Travis and Schubert* (2015).

An interesting consequence of the reduced ice shell thickness in the SPT is a strong increase of tidal deformation and associated tidal friction. Compared to models with constant thickness of 25 km, the reduction in the SPT to less than 5 km in thickness results in an amplification of the tidal stresses by a factor of 4 (*Soucek et al.,* 2017). This is a consequence of the small size of Enceladus, which makes the amplitude of tidal deformation much more sensitive to the ice shell thickness than in larger moons like Europa and Titan, where tidal deformation depends only slightly on the ice shell thickness (*Tobie et al.,* 2005). *Soucek et al.* (2016) have also demonstrated that the presence of faults further enhances the tidal deflections by at least a factor of 2. The resulting stress patterns are much more complex than those predicted from standard tidal deformation models based on a thin-shell approximation and neglecting the presence of faults (e.g., *Hurford et al.,* 2007; *Nimmo et al.,* 2007). Enhanced tidal stresses in the SPT, together with lithospheric stresses resulting from ice shell melting and subsequent relaxation, may help explain the tectonic patterns observed in the SPT (e.g., *Patthoff and Kattenhorn,* 2011; *Yin et al.,* 2016). However, future modeling efforts are required to better understand the complex interplay between the ice shell evolution and the tectonic patterns.

5. SUMMARY AND OPEN QUESTIONS

The Cassini mission has helped to answer many questions about the interior of Enceladus. In particular, as we discussed in this chapter, Cassini-derived measurements of the low-order gravity field and rotational dynamics have led to a good, albeit basic, understanding of the internal structure (Fig. 1). We now know, for instance, that Enceladus has a large, low-density core that is in contact with a global subsurface liquid water ocean containing some 10^7 km^3 of water (section 2.3.4) — comparable to the volume of Earth's Arctic Ocean. The icy shell that covers the ocean may be vanishingly thin at the south pole, but is at least a few tens of kilometers thick in the equatorial regions (Fig. 6).

Nevertheless, the thickness, structure, and dynamics of the ice shell are not yet sufficiently well constrained to give us a clear understanding of the thermal history and evolution of the interior. Is the current configuration stable, or are we seeing Enceladus in the midst of a transition? If the present configuration is stable, how is the long-wavelength topography maintained in spite of the relatively high temperatures at the base of the ice shell? How and where is the internal heat generated, and can we match the predicted input power with the estimated heat loss? In the near term, additional modeling efforts may help to better address these questions. In the longer term, when spacecraft next visit Enceladus, and especially with dedicated missions that can orbit or even land on Enceladus, new observations — which may include higher-order gravity measurements, improved heat flow measurements, ice penetrating radar, and even seismology — are sure to be the most powerful drivers in advancing our understanding of the interior of Enceladus.

Acknowledgments. D.H. was supported by the Miller Institute for Basic Research in Science at the University of California, Berkeley. R.T. was supported by the Cassini mission. L.I. acknowledges support from the Italian Space Agency. G.T. was supported by the European Research Council under the European Community's Seventh Framework Programme, FP7/20077-2013, grant agreement 259285. We thank F. Nimmo and A. Ermakov for constructive reviews and M. Manga for helpful comments.

REFERENCES

Anderson J. D., Schubert G., Jacobson R. A., Lau E. L., Moore W. B., and Sjogren W. L. (1998) Europa's differentiated internal structure: Inferences from four Galileo encounters. *Science, 281(5385),* 2019–2022, DOI: 10.1126/science.281.5385.2019.

Baland R. M., Yseboodt M., and Van Hoolst T. (2016) The obliquity of Enceladus. *Icarus, 268,* 12–31, DOI: 10.1016/j.icarus.2015.11.039.

Baum W. A., Kreidl T., Westphal J. A., Danielson G. E., Seidelmann P. K., Pascu D., and Currie D. G. (1981) Saturn's E ring: I. CCD observations of March 1980. *Icarus, 47(1),* 84–96, DOI: 10.1016/0019-1035(81)90093-2.

Běhounková M., Tobie G., Choblet G., and Čadek O. (2012) Tidally-induced melting events as the origin of south-pole activity on Enceladus. *Icarus, 219(2),* 655–664, DOI: 10.1016/j.icarus.2012.03.024.

Besserer J., Nimmo F., Roberts J. H., and Pappalardo R. T. (2013) Convection-driven compaction as a possible origin of Enceladus's long wavelength topography. *J. Geophys. Res.–Planets, 118(5),* 908–915, DOI: 10.1002/jgre.20079.

Beuthe M. (2008) Thin elastic shells with variable thickness for lithospheric flexure of one-plate planets. *Geophys. J. Intl., 172(2),* 817–841, DOI: 10.1111/j.1365-246X.2007.03671.x.

Beuthe M., Rivoldini A., and Trinh A. (2016) Enceladus' and Dione's floating ice shells supported by minimum stress isostasy. *Geophys. Res. Lett., 43,* DOI: 10.1002/2016GL070650.

Brown R. H. et al. (2006) Composition and physical properties of Enceladus' surface. *Science, 311(5766),* 1425–1428, DOI: 10.1126/science.1121031.

Buratti B. and Veverka J. (1984) Voyager photometry of Rhea, Dione, Tethys, Enceladus and Mimas. *Icarus, 58,* 254–264, DOI: 10.1016/0019-1035(84)90042-3.

Burša M. and Peč K. (1993) *Gravity Field and Dynamics of the Earth.* Springer, New York. 333 pp.

Čadek O. et al. (2016) Enceladus's internal ocean and ice shell constrained from Cassini gravity, shape and libration data. *Geophys. Res. Lett., 43,* DOI: 10.1002/2016GL068634.

Čadek O., Běhounková M., Tobie G., and Choblet G. (2017) Viscoelastic relaxation of Enceladus's ice shell. *Icarus, 291,* 31–35, DOI: 10.1016/j.icarus.2017.03.011.

Chen E. M. A. and F. Nimmo (2011) Obliquity tides do not significantly heat Enceladus. *Icarus, 214(2),* 779–781, DOI: 10.1016/j.icarus.2011.06.007.

Clairaut A. C. (1743) *Theorie de la figure de la Terre, Tirée des principes de l'Hydrostatique,* Paris.

Clark R. N., Brown R. H., Nelson M. L., and Hayashi J. N. (1983) Surface composition of Enceladus. *Bull. Am. Astron. Soc., 15,* 853.

Collins G. C. and Goodman J. C. (2007) Enceladus' south polar sea. *Icarus, 189(1),* 72–82, DOI: 10.1016/j.icarus.2007.01.010.

Dahlen F. A. (1982) Isostatic geoid anomalies on a sphere. *J. Geophys. Res., 87(B5),* 3943–3947, DOI: 10.1029/JB087iB05p03943.

Danby J. M. A. (1988) *Fundamentals of Celestial Mechanics.* Willman-Bell, Richmond. 348 pp.

Darwin G. H. (1899) The theory of the figure of the Earth carried to the second order of small quantities. *Mon. Not. R. Astron. Soc., 60,* 82–124.

Dickey J. O. et al. (1994) Lunar Laser Ranging: A continuing legacy of the Apollo program. *Science, 265(5171),* 482–490.

Dougherty M. K., Khurana K. K., Neubauer F. M., Russell C. T., Saur J., Leisner J. S., and Burton M. E. (2006) Identification of a dynamic atmosphere at Enceladus with the Cassini magnetometer. *Science, 311(5766),* 1406–1409, DOI: 10.1126/science.1120985.

Gladman B., Quinn D. D., Nicholson P., and Rand R. (1996) Synchronous locking of tidally evolving satellites. *Icarus, 122,* 166–192, DOI: 10.1006/icar.1996.0117.

Haff P. K., Eviatar A., and Siscoe G. L. (1983) Ring and plasma: The enigmae of Enceladus. *Icarus, 56(3),* 426–438, DOI: 10.1016/0019-1035(83)90164-1.

Hansen C. J., Esposito L., Stewart A. I. F., Colwell J., Hendrix A., Pryor W., Shemansky D., and West R. (2006) Enceladus' water vapor plume. *Science, 311,* 1422–1425, DOI: 10.1126/science.1121254.

Hay H. C. F. C. and Matsuyama I. (2017) Numerically modelling tidal dissipation with bottom drag in the oceans of Titan and Enceladus. *Icarus, 281,* 342–356, DOI: 10.1016/j.icarus.2016.09.022.

Hemingway D. J. and Matsuyama I. (2017) Isostatic equilibrium in spherical coordinates and implications for crustal thickness on the Moon, Mars, Enceladus, and elsewhere. *Geophys. Res. Lett., 44,* 7695–7705, DOI: 10.1002/2017GL073334.

Hemingway D. J. and Mittal T. (2017) What explains the structure of Enceladus's ice shell and can it be in equilibrium? Abstract P43B-2877 presented at 2017 Fall Meeting, AGU, New Orleans, Louisiana, 11–15 December.

Hemingway D., Nimmo F., Zebker H., and Iess L. (2013a) A rigid and weathered ice shell on Titan. *Nature, 500(7464),* 550–552, DOI: 10.1038/nature12400.

Hemingway D., Nimmo F., and Iess L. (2013b) Enceladus' internal structure inferred from analysis of Cassini-derived gravity and topography. Abstract P53E-03 presented at 2013 Fall Meeting, AGU, San Francisco, California, 9–13 December.

Hemingway D. J., Zannoni M., Tortora P., Nimmo F., and Asmar S. W. (2016) Dione's internal structure inferred from Cassini gravity and topography. In *Lunar and Planetary Science XLVII,* Abstract #1314. Lunar and Planetary Institute, Houston.

Howett C. J. A., Spencer J. R., Pearl J., and Segura M. (2011) High heat flow from Enceladus' south polar region measured using 10–600 cm^{-1} Cassini/CIRS data. *J. Geophys. Res.–Planets, 116(3),* 1–15, DOI: 10.1029/2010JE003718.

Hsu H.-W. et al. (2015) Ongoing hydrothermal activities within Enceladus. *Nature, 519(7542),* 207–210, DOI: 10.1038/nature14262.

Hubbard W. B. (1984) *Planetary Interiors.* Van Nostrand Reinhold, New York. 334 pp.

Hurford T. A., Helfenstein P., Hoppa G. V., Greenberg R., and Bills B. G. (2007) Eruptions arising from tidally controlled periodic openings of rifts on Enceladus. *Nature, 447(7142),* 292–294, DOI: 10.1038/nature05821.

Iess L., Rappaport N. J., Jacobson R. A., Racioppa P., Stevenson D. J., Tortora P., Armstrong J. W., and Asmar S. W. (2010) Gravity field, shape, and moment of inertia of Titan. *Science, 327(5971),* 1367–1369, DOI: 10.1126/science.1182583.

Iess L., Jacobson, Ducci M., Stevenson D. J., Lunine J. I., Armstrong J. W., Asmar S. W., Racioppa P., Rappaport N. J., and Tortora R. A. P. (2012) The tides of Titan. *Science, 337,* 457–459, DOI: 10.1126/science.1219631.

Iess L. et al. (2014) The gravity field and interior structure of Enceladus. *Science, 344(6179),* 78–80, DOI: 10.1126/science.1250551.

Jacobson R. A. (2004) The orbits of the major saturnian satellites and the gravity field of Saturn from spacecraft and Earth-based observations. *Astron. J., 128,* 492–501, DOI: 10.1086/421738.

Jeffreys H. (1976) *The Earth: Its Origin, History, and Physical Constitution,* 6th edition. Cambridge Univ., New York. 612 pp.

Kamata S. and Nimmo F. (2017) Interior thermal state of Enceladus inferred from the viscoelastic state of the ice shell. *Icarus, 284,* 387–393, DOI: 10.1029/2011JE003835.

Klinger J. (1980) Influence of a phase transition of ice on the heat and mass balance of comets. *Science, 209(4453),* 271–272.

Kozai Y. (1976) Masses of satellites and oblateness parameters of Saturn. *Publ. Astron. Soc. Japan, 28,* 675–691.

Lainey V. (2016) Quantification of tidal parameters from solar system data. *Cel. Mech. Dyn. Astron., 126(1),* 1–12, DOI: 10.1007/s10569-016-9695-y.

Lainey V. et al. (2012) Strong tidal dissipation in Saturn and constraints on Enceladus' thermal state from astrometry. *Astrophys. J., 752(14),* DOI: 10.1088/0004-637X/752/1/14.

Lefevre A., Tobie G., Choblet G., and Cadek O. (2014) Structure and dynamics of Titan's outer icy shell constrained from Cassini data. *Icarus, 237,* 16–28, DOI: 10.1016/j.icarus.2014.04.006.

Margot J. L., Peale S. J., Jurgens R. F., Slade M. A, and Holin I. V. (2007) Large longitude libration of Mercury reveals a molten core. *Science, 316(5825),* 710–714, DOI: 10.1126/science.1140514.

Martinec Z. (1994) The density contrast at the Mohorovičić discontinuity. *Geophys. J. Intl., 117(2),* 539–544, DOI: 10.1111/j.1365-246X.1994.tb03950.x.

Matsuyama I. (2014) Tidal dissipation in the oceans of icy satellites. *Icarus, 242,* 11–18, DOI: 10.1016/j.icarus.2014.07.005.

McKinnon W. B. (2013) The shape of Enceladus as explained by an irregular core: Implications for gravity, libration, and survival of its subsurface ocean. *J. Geophys. Res.–Planets, 118,* 1775–1788, DOI: 10.1002/jgre.20122.

McKinnon W. B. (2015) Effect of Enceladus's rapid synchronous spin on interpretation of Cassini gravity. *Geophys. Res. Lett., 42,* DOI: 10.1002/2015GL063384.

Meyer J. and Wisdom J. (2007) Tidal heating in Enceladus. *Icarus, 188(2),* 535–539, DOI: 10.1016/j.icarus.2007.03.001.

Murray C. D. and Dermott S. F. (1999) *Solar System Dynamics.* Cambridge Univ., Cambridge.

Nadezhdina I. E. and Zubarev A. E. (2014) Formation of a reference coordinate network as a basis for studying the physical parameters of phobos. *Solar System Res., 48(4),* 269–278, DOI: 10.1134/S003809461404008X.

Nimmo F. (2004) Non-Newtonian topographic relaxation on Europa. *Icarus, 168(1),* 205–208, DOI: 10.1016/j.icarus.2003.11.022.

Nimmo F., Spencer J. R., Pappalardo R. T., and Mullen M. E. (2007) Shear heating as the origin of the plumes and heat flux on Enceladus. *Nature, 447(7142),* 289–291, DOI: 10.1038/nature05783.

Nimmo F., Bills B. G., and Thomas P. C. (2011) Geophysical implications of the long-wavelength topography of the Saturnian satellites. *J. Geophys. Res., 116,* E11001, DOI: 10.1029/2011JE003835.

Noyelles B., Karatekin O., and Rambaux N. (2011) The rotation of Mimas. *Astron. Astrophys., 536,* 1–13, DOI: 10.1051/0004-6361/201117558.

Ojakangas G. W. and Stevenson D. J. (1989) Thermal state of an ice shell on Europa. *Icarus, 81(2),* 220–241, DOI: 10.1016/0019-1035(89)90053-5.

Pang K. D., Voge C. C., Rhoads J. W., and Ajello J. M. (1984) The E ring of Saturn and satellite Enceladus. *J. Geophys. Res., 89(B11),* 9459, DOI: 10.1029/JB089iB11p09459.

Patthoff D. A. and Kattenhorn S. A. (2011) A fracture history on Enceladus provides evidence for a global ocean. *Geophys. Res. Lett., 38(18),* 1–6, DOI: 10.1029/2011GL048387.

Porco C. C. et al. (2006) Cassini observes the active south pole of Enceladus. *Science, 311(5766),* 1393–1401, DOI: 10.1126/science.1123013.

Postberg F., Kempf S., Schmidt J., Brilliantov N., Beinsen A., Abel B., Buck U., and Srama R. (2009) Sodium salts in E-ring ice grains from an ocean below the surface of Enceladus. *Nature, 459,* 1098–1101, DOI: 10.1038/nature08046.

Postberg F., Schmidt J., Hillier J., Kempf S., and Srama R. (2011) A salt-water reservoir as the source of a compositionally stratified plume on Enceladus. *Nature, 474(7353),* 620–622, DOI: 10.1038/nature10175.

Rambaux N., Castillo-Rogez J. C., Williams J. G., and Karatekin Ö. (2010) Librational response of Enceladus. *Geophys. Res. Lett., 37(4),* 1–6, DOI: 10.1029/2009GL041465.

Rambaux N., Van Hoolst T., and Karatekin Ö. (2011) Librational response of Europa, Ganymede, and Callisto with an ocean for a non-Keplerian orbit. *Astron. Astrophys., 527,* A118, DOI: 10.1051/0004-6361/201015304.

Rappaport N. J., Iess L., Tortora P., Anabtawi A., Asmar S. W., Somenzi L., and Zingoni F. (2007) Mass and interior of Enceladus from Cassini data analysis. *Icarus, 190(1),* 175–178, DOI: 10.1016/j.icarus.2007.03.025.

Richard A., Rambaux N., and Charnay B. (2014) Librational response of a deformed 3-layer Titan perturbed by non-Keplerian orbit and atmospheric couplings. *Planet. Space Sci., 93–94,* 22–34, DOI: 10.1016/j.pss.2014.02.006.

Roberts J. H. (2015) The fluffy core of Enceladus. *Icarus, 258,* 54–66, DOI: 10.1016/j.icarus.2015.05.033.

Roberts J. H. and Nimmo F. (2008) Tidal heating and the long-term stability of a subsurface ocean on Enceladus. *Icarus, 194(2),* 675–689, DOI: 10.1016/j.icarus.2007.11.010.

Smith B. A. et al. (1982) A new look at the Saturn system: The Voyager 2 images. *Science, 215(4532),* 504–537.

Soucek O., Hron J., Behounkova M., and Cadek O. (2016) Effect of the tiger stripes on the deformation of Saturn's moon Enceladus. *Geophys. Res. Lett., 43,* 7417–7423, DOI: 10.1002/2016GL069415.

Soucek O., Behounkova M., Cadek O., Tobie G., and Choblet G. (2017) Tidal deformation of Enceladus' ice shell with variable thickness and Maxwell rheology. In *19th EGU General Assembly,* EGU2017-16357.

Spahn F. et al. (2006) Cassini dust measurements at Enceladus and implications for the origin of the E ring. *Science, 311(5766),* 1416–1418, DOI: 10.1126/science.1121375.

Spencer J. R., Pearl J. C., Segura M., Flasar F. M., Mamoutkine A., Romani P., Buratti B. J., Hendrix A. R., Spilker L. J., and Lopes R. M. C. (2006) Cassini encounters Enceladus: Background and the discovery of a south polar hot spot. *Science, 311(5766),* 1401–1405, DOI: 10.1126/science.1121661.

Spitale J. N. and Porco C. C. (2007) Association of the jets of Enceladus with the warmest regions on its south-polar fractures. *Nature, 449(7163),* 695–697, DOI: 10.1038/nature06217.

Squyres S. W., Reynolds R. T., Cassen P. M., and Peale S. J. (1983) The evolution of Enceladus. *Icarus, 53(2),* 319–331, DOI: 10.1016/0019-1035(83)90152-5.

Tajeddine R., Rambaux N., Lainey V., Charnoz S., Richard A., Rivoldini A., and Noyelles B. (2014) Constraints on Mimas' interior from Cassini ISS libration measurements. *Science, 346(6207),* 322–324.

Tajeddine R., Soderlund K. M., Thomas P. C., Helfenstein P., Hedman M. M., Burns J. A., and Schenk P. M. (2017) True polar wander of Enceladus from topographic data. *Icarus, 295,* 46–60, DOI: 10.1016/j.icarus.2017.04.019.

Thomas P. C. (2010) Sizes, shapes, and derived properties of the saturnian satellites after the Cassini nominal mission. *Icarus, 208(1),* 395–401, DOI: 10.1016/j.icarus.2010.01.025.

Thomas P., Burns J., Helfenstein P., Squyres S., Veverka J., Porco C., Turtle E., McEwen A., Denk T., and Giese B. (2007) Shapes of the saturnian icy satellites and their significance. *Icarus, 190(2),* 573–584, DOI: 10.1016/j.icarus.2007.03.012.

Thomas P. C., Tajeddine R., Tiscareno M. S., Burns J. A., Joseph J., Loredo T. J., Helfenstein P., and Porco C. (2016) Enceladus's measured physical libration requires a global subsurface ocean. *Icarus, 264,* 37–47, DOI: 10.1016/j.icarus.2015.08.037.

Tiscareno M. S., Thomas P. C., and Burns J. A. (2009) The rotation of Janus and Epimetheus. *Icarus, 204(1),* 254–261, DOI: 10.1016/j.icarus.2009.06.023.

Tobie G., Mocquet A., and Sotin C. (2005) Tidal dissipation within large icy satellites: Applications to Europa and Titan. *Icarus, 177(2),* 534–549, DOI: 10.1016/j.icarus.2005.04.006.

Tortora P., Zannoni M., Hemingway D., Nimmo F., Jacobson R. A., Iess L., and Parisi M. (2016) Rhea gravity field and interior modeling from Cassini data analysis. *Icarus, 264,* 264–273, DOI: 10.1016/j.icarus.2015.09.022.

Travis B. J. and Schubert G. (2015) Keeping Enceladus warm. *Icarus, 250,* 32–42, DOI:10.1016/j.icarus.2014.11.017.

Tricarico P. (2014) Multi-layer hydrostatic equilibrium of planets and synchronous moons: Theory and application to Ceres and to solar system moons. *Astrophys. J., 782(2),* 99, DOI: 10.1088/0004-637X/782/2/99.

Tyler R. (2011) Tidal dynamical considerations constrain the state of an ocean on Enceladus. *Icarus, 211,* 770–779, DOI: 10.1016/j.icarus.2010.10.007.

van Hoolst T. (2015) Rotation of the terrestrial planets. In *Treatise on Geophysics, 2nd edition, Vol. 10: Physics of Terrestrial Planets and Moons* (G. Schubert, ed.), pp. 121–151. Elsevier, Amsterdam.

van Hoolst T., Rambaux N., Karatekin Ö., and Baland R. M. (2009) The effect of gravitational and pressure torques on Titan's length-of-day variations. *Icarus, 200(1),* 256–264, DOI: 10.1016/j.icarus.2008.11.009.

van Hoolst T., Baland R.-M., and Trinh A. (2016) The diurnal libration and interior structure of Enceladus. *Icarus, 277,* 311–318, DOI: 10.1016/j.icarus.2016.05.025.

Vienne A. and Duriez L. (1995) Ephemerides of the major saturnian satellites. *Astron. Astrophys., 297,* 588–605.

Waite J. H. et al. (2006) Cassini Ion and Neutral Mass Spectrometer: Enceladus plume composition and structure. *Science, 311(5766),* 1419–1422, DOI: 10.1126/science.1121290.

Wieczorek M. A. (2015) Gravity and topography of the terrestrial planets. In *Treatise on Geophysics, 2nd edition, Vol. 10: Physics of Terrestrial Planets and Moons* (G. Schubert, ed.), pp. 153–193. Elsevier, Amsterdam.

Wieczorek M. A. and Phillips R. J. (1998) Potential anomalies on a sphere: Applications to the thickness of the lunar crust. *J. Geophys. Res., 103(E1),* 1715–1724.

Willner K., Oberst J., Hussmann H., Giese B., Hoffmann H., Matz K. D., Roatsch T., and Duxbury T. (2010) Phobos control point network, rotation, and shape. *Earth Planet. Sci. Lett., 294(3–4),* 541–546, DOI: 10.1016/j.epsl.2009.07.033.

Yin A., Zuza A. V., and Pappalardo R. T. (2016) Mechanics of evenly spaced strike-slip faults and its implications for the formation of tiger-stripe fractures on Saturn's moon Enceladus. *Icarus, 266,* 204–216, DOI: 10.1016/j.icarus.2015.10.027.

Zebker H. A., Iess L., Wall S. D., Lorenz R. D., Lunine J. I., and Stiles B. W. (2012) Titan's figure fatter, flatter than its gravity field. Abstract P23F-01 presented at 2012 Fall Meeting, AGU, San Francisco, California, 3–7 December.

Zharkov V. N. (2004) A theory of the equilibrium figure and gravitational field of the Galilean satellite Io: The second approximation. *Astron. Lett., 30(7),* 496–507, DOI: 10.1134/1.1774402.

Zharkov V. N., Leontjev V. V., and Kozenko A. V. (1985) Models, figures, and gravitational moments of the Galilean satellites of Jupiter and icy satellites of Saturn. *Icarus, 61,* 92–100, DOI: 10.1016/0019-1035(85)90157-5.

Nimmo F., Barr A. C., Běhounková M., and McKinnon W. B. (2018) The thermal and orbital evolution of Enceladus: Observational constraints and models. In *Enceladus and the Icy Moons of Saturn* (P. M. Schenk et al., eds.), pp. 79–94. Univ. of Arizona, Tucson, DOI: 10.2458/azu_uapress_9780816537075-ch005.

The Thermal and Orbital Evolution of Enceladus: Observational Constraints and Models

Francis Nimmo
University of California, Santa Cruz

Amy C. Barr
Planetary Science Institute

Marie Běhounková
Charles University

William B. McKinnon
Washington University

Enceladus possesses a global subsurface ocean beneath an ice shell a few tens of kilometers thick, and is observed to be losing heat at a rate of ~10 GW from its south polar region. Two major puzzles are the source of the observed heat, and how the ocean could have been maintained. Tidal dissipation in Enceladus is ultimately controlled by the rate of dissipation within Saturn, parameterized by the factor Q_p. A Q_p of about 2000 is indicated by astrometric measurements and generates an equilibrium heating rate at Enceladus sufficient to explain the observed heat and maintain an ocean indefinitely if the ice shell is conductive. If constant, this Q_p would indicate an age for Enceladus much less than that of the solar system. An alternative, however, termed the "resonance-locking" scenario, is that the effective Q_p is time-variable, such that the heating rate is almost constant over geological time. This scenario can explain the long-term survival of the ocean and the present-day heat flux without requiring Enceladus to have formed recently.

1. INTRODUCTION

Enceladus, despite its limited size, is one of the most surprising bodies in the solar system, with a south polar terrain (SPT) containing localized active tectonics, remarkable geysers connected to four prominent fracture sets (the "tiger stripes"), and extreme heat flow. Due to these unique characteristics and a high astrobiological potential, Enceladus has become one of the primary targets of the Cassini mission. However, despite a decade of Cassini observations, Enceladus's thermal state, its long-term evolution, and even its internal structure still remain rather puzzling.

One major puzzle is the large heat flow measured at the SPT. The source of this heat must ultimately be tidal dissipation, arising from the periodic deformation Enceladus experiences in its elliptical path around Saturn. But how this process works in detail, and what it is telling us about the long-term thermal and orbital evolution of Enceladus, is not yet clear.

As discussed below (and in more detail in the chapter in this volume by Hemingway et al.), there is strong evidence that Enceladus possesses a global ocean beneath its ice shell. The second major puzzle is how Enceladus, a small and thus rapidly cooling body, could maintain such an ocean. Tidal heating is a good way of maintaining such an ocean, but again, the details of how this actually works are not well understood.

A particularly difficult aspect of satellite evolution is that the thermal and orbital evolution of satellites are often intimately coupled. This is because the amount of tidal heating depends on the satellite's mechanical properties, many of which are temperature-dependent. There is thus feedback between tidal heating and temperature evolution (e.g., *Hussmann and Spohn,* 2004; *Ojakangas and Stevenson,* 1986). As discussed later, this feedback can lead to complex, nonmonotonic behavior.

The goal of this chapter is to discuss, and attempt to provide an answer to, the two puzzles outlined above. In the second section, we review the available observational constraints on interior structure, heat production and loss, and surface geology. The third section deals with estimates of the present-day tidal heat production in the ice shell, the ocean, and the core. In the fourth section, the tidal evolution of Enceladus' orbit, and its implications for the thermal history, are discussed. In sections 5 and 6 we discuss how to answer the puzzles of the heat output and the ocean survival, respectively. We conclude with a summary and suggestions for future work. Other chapters in this volume relevant to this one include those on the interior of Enceladus (Hemingway et al.), its geological history (Patterson et al.), and the nature of the plume source (Spencer et al.) and ocean (Glein et al.). A recent general review may be found in *Spencer and Nimmo* (2013).

2. OBSERVATIONAL CONSTRAINTS

Most observational constraints, with the exception of the surface geology, are relevant to the present-day state of Enceladus. One of the challenges in understanding the long-term evolution of Enceladus is therefore how to best use these present-day constraints to infer past behavior.

2.1. Internal Structure

Models of Enceladus's interior structure rely mainly on shape (*Thomas et al.,* 2007), long-wavelength topography (*Nimmo et al.,* 2011), and gravity field (*Iess et al.,* 2014) measurements, and also on the detection of librations — periodic changes in the body's rotation rate (*Thomas et al.,* 2016). These constraints are discussed in detail in the chapter by Hemingway et al., so only a brief summary is given here.

Based on its bulk density and observed icy surface, Enceladus is thought to consist of a silicate core overlain by an H_2O layer. Both the global gravity and regional gravity variations at the SPT are smaller than those expected from the surface topography (*Iess et al.,* 2014), suggesting that the topography is compensated. One way of achieving such compensation is to appeal to a subsurface ocean and a reduced ice shell thickness at the SPT (*Collins and Goodman,* 2007). Joint analyses of gravity and topography can separate out the effects of shell thickness variations from those arising from the deeper internal structure (*Iess et al.,* 2014; *McKinnon,* 2015; *Beuthe et al.,* 2016). These models suggest the presence of a silicate core that is relatively large ($\approx$190 km radius) and of low density ($\approx$2500 kg m^{-3}). The mean ice shell thickness is probably in the range 25–60 km, with libration results (see below) supporting the thinner values, and the ocean is at least 10 km thick everywhere. Although these models commonly assume the presence of a global ocean, they cannot, strictly speaking, be used to prove that such an ocean actually exists.

Even prior to the gravity measurements, subsurface liquid water was widely assumed to exist on Enceladus. This is because of the detection of sodium and potassium salts in ice grains ejected by geysers at the SPT (*Postberg et al.,* 2009, 2011). Such salts could most naturally be explained by the presence of liquid water that had interacted with silicates. Furthermore, silica-rich nanoparticles may be indicative of hydrothermal reactions between water and silicates (*Hsu et al.,* 2015); hydrothermal activity is also suggested by the detection of hydrogen in the plume of Enceladus (*Waite et al.,* 2017). However, these arguments do not distinguish between a regional sea and a global ocean.

In contrast, the detection of a large amplitude physical libration (*Thomas et al.,* 2016) is strong evidence for the existence of a global ocean, because only a fully decoupled ice shell could satisfy the observed amplitude. The amplitude of libration moreover suggests that the ice shell is at the thin end (21–26 km) of the results derived from the gravity/topography data alone (*Thomas et al.,* 2016). Several ways of reconciling the gravity/topography results with the librations have been proposed, including elastic or other nonhydrostatic effects (*Čadek et al.,* 2016; *van Hoolst et al.,* 2016), a delayed relaxation to equilibrium (*Čadek et al.,* 2017), or a modified definition of isostasy (*Beuthe et al.,* 2016). [Note that (*Čadek et al.,* 2016) incorrectly mix top-loading and bottom-loading in their calculations; see *Beuthe et al.* (2016).] In any event, the libration results are strong evidence for an ice shell only about 20–25 km thick on average, and perhaps as small as a few kilometers beneath the south pole. Figure 1 summarizes the constraints, showing the low-density silicate core, the ice shell and ocean, and the reduced shell thickness beneath the SPT.

One important consequence of the thinned shell at the SPT is that the ice will tend to flow laterally and fill in the hole (*Collins and Goodman,* 2007). The rate of inflow depends mainly on the shell thickness and temperature structure. The fact that inflow is not complete is consistent with the shell being locally thin, and might also indicate a relatively cold ocean and conductive ice shell. Alternatively, inflow could be balanced by melting, which would require there to be a heat source either within the ice shell or below its base (*Kamata and Nimmo,* 2017).

Another geophysical constraint on Enceladus' internal structure arises from the fact that the Enceladus plume is modulated on a tidal timescale (*Hedman et al.,* 2013; *Nimmo et al.,* 2014). Assuming that the activity is controlled by the normal tidal stress along the faults (*Hurford et al.,* 2007),

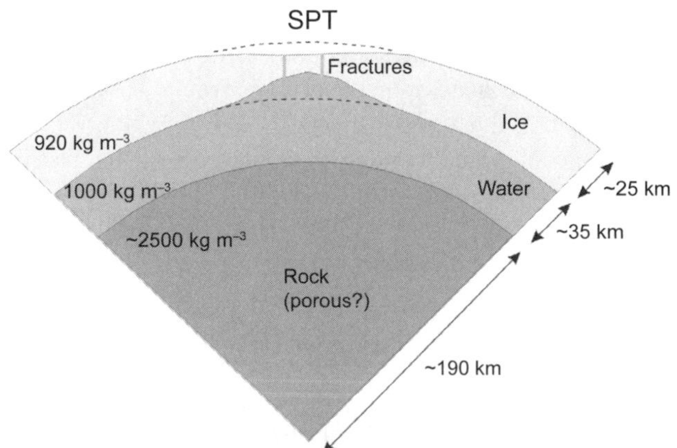

Fig. 1. Sketch of likely internal structure of Enceladus derived from gravity, topography, and libration observations (see text). The south polar terrain (SPT) has a reduced shell thickness.

the exact timing is sensitive to interior properties, most notably the viscosity of the ice, the ice shell thickness, and the extent of the internal ocean. A low viscosity (lower than 5×10^{13} Pa s) at least beneath the SPT and an ice shell thickness larger than 60 km are required to explain the timing for a global ocean (*Běhounková et al.,* 2015). We note, however, that this constraint is weaker than those above, because this explanation for the timing is not unique (and it contradicts the other evidence for shell thickness). The timing could instead be due to the delayed ascent of water through fractures (e.g., *Kite and Rubin,* 2016), or perhaps a delayed response of the ocean beneath.

2.2. Heat Flow

Observations of Enceladus by the Cassini Composite Infrared Spectrometer (CIRS) show that the SPT is a region of locally high heat flow (*Spencer et al.,* 2006, 2013; *Howett et al.,* 2011). Initial CIRS observations estimated the total power output to be 5.8 ± 1.9 GW (*Spencer et al.,* 2006), but subsequent higher-resolution observations incorporating data from a broader range of infrared wavelengths suggest that the power output is a factor of ~3 higher, 15.9 ± 3.1 GW (*Howett et al.,* 2011). The most recent analysis, using CIRS observations that can spatially resolve the tiger stripes, indicate an endogenic power of 4.2 GW (*Spencer et al.,* 2013), with an additional 0.5 GW being liberated as latent heat in the plume and possible thermal emission coming from between the tiger stripes. Divided by the ~70,000 km^2 surface area of the SPT (*Porco et al.,* 2006), the 4.7 GW corresponds to a heat flux of $F_{obs} \approx 70$ mW m^{-2}, comparable to terrestrial heat fluxes. [*Le Gall et al.* (2017) report even higher heat flows, in excess of 1 W m^{-2}, from analysis of Cassini microwave radiometry of the SPT (although outside the tiger stripe region proper).]

What could be the source of such high heat flows? If all the radiogenic heating in Enceladus were, for some reason, concentrated in the SPT, could F_{obs} be provided by radiogenic heating alone? The mean density of Enceladus, $\bar{\rho} = 1607$ kg m^{-3} (*Porco et al.,* 2006), indicates that the satellite has a significant amount of rock. Based on the core radius and density inferred from gravity/ topography (see section 2.1), the mass fraction of rock in the interior of Enceladus is $m_r \approx 0.7$. With a present-day chondritic heating rate $H = 4.5 \times 10^{-12}$ W kg^{-1} (*Spohn and Schubert,* 2003) and R_E (the radius of Enceladus) set to 252 km, the total power from radiogenic heat alone is $P_{tot} = [(4/3)\pi R_E^3 \bar{\rho}]$ $m_r H = 0.34$ GW, at least an order of magnitude lower than the total power measured by CIRS.

If the ice shell at the SPT is only a few kilometers thick, as suggested in *Čadek et al.* (2016), then the local heat flux there (F_{obs}) could be explained purely by conductive heat transfer across the thin shell. A perhaps more likely situation is that the shell is somewhat thicker, and the observed heat flux is a combination of regional conductive heat plus local heat sources (e.g., warm water being advected from depth) along the major fractures.

If the mean shell thickness is $\bar{d}$, then neglecting a small correction for curvature the global conductive heat loss is

$$\dot{E}_{cond} \approx 4\pi R_E^2 \frac{k\Delta T}{\bar{d}} = 21 \text{ GW}\left(\frac{25 \text{ km}}{\bar{d}}\right) \quad (1)$$

where k is the mean ice shell conductivity (3.5 W m^{-1} K^{-1}), ΔT is the temperature contrast across the ice shell (about 190 K), and parameter values assumed for this and subsequent equations are given in Table 1. Note that if the ice shell is convecting, or some heat is being advected upward along fractures, the total heat loss will be higher.

Such a high rate of heat loss suggests that refreezing of the ocean should be rapid. For instance, if a 20-GW cooling rate is not balanced by any heat production, a 30-km-thick ocean would completely freeze in about 10 m.y. Numerical calculations (*Roberts and Nimmo,* 2008) yield the same result. Thus, either we are seeing Enceladus at a special time (the ocean is a recent phenomenon), or there is some heat source capable of slowing or halting the refreezing. We return to this issue in section 6.

Whether the shell is convecting or conductive is unclear. Many models have assumed that convection is operating (e.g., *Nimmo and Pappalardo,* 2006; *Barr and McKinnon,* 2007; *Mitri and Showman,* 2008; *Roberts and Nimmo,* 2008; *Stegman et al.,* 2009; *Han et al.,* 2012). However, if the mean shell thickness is really ≈25 km, then a conductive state is more likely (*Barr and McKinnon,* 2007) and is also required to permit incomplete lateral flow (see section 2.1).

2.3. Geological Constraints

The surface geology of Enceladus records a complex history (*Nahm and Kattenhorn,* 2015), and provides potentially important constraints on how Enceladus has evolved (see the chapter in this volume by Patterson et al.). The relative sequence of events can be established without too much difficulty using a combination of cross-cutting relationships and crater counting. The big problem is in deriving absolute ages, because these depend on inferred impactor fluxes. Even in the conventional picture of (primarily heliocentric) impactors

TABLE 1. List of parameter values assumed.

Quantity	Symbol	Value	Units	Equation
Thermal conductivity	k	3.5	W m^{-1} K^{-1}	(1)
Temperature drop	ΔT	190	K	(1)
Mean motion	n	5.31×10^{-5}	s^{-1}	(2)
Eccentricity	e	0.0047	—	(2)
Saturn Love number	k_{2P}	0.34	—	(5)
Enceladus radius	R_E	252	km	(1)
Saturn mass	M_P	5.68×10^{26}	kg	(2)
Semimajor axis	a	238,000	km	(2)
Saturn radius	R_P	60,300	km	(5)
Saturn spin frequency	Ω_P	1.65×10^{-4}	s^{-1}	(10)

accumulating since solar system formation, uncertainties in age can exceed a factor of 3 (*Zahnle et al.,* 2003). If the satellites formed more recently from planetocentric debris (*Ćuk et al.,* 2016), the uncertainties are even larger, although the apparent detection of longitudinal crater density variations at Rhea and Iapetus (*Hirata,* 2016) suggests that at least for these satellites, planetocentric debris was not dominant.

The surface geology of Enceladus may be divided into four regions that exhibit a strong symmetry around the present-day tidal and rotational axes (*Crow-Willard and Pappalardo,* 2015). The northern cratered plains are the oldest and do not show evidence of tectonic deformation, although some craters are relaxed, indicating high heat flows (see below). The trailing and leading hemisphere areas are more deformed, with the former being the older of the two, while the SPT is most heavily deformed and youngest. The cratered plains are perhaps about 4 G.y. old and the leading hemisphere 1–3 G.y. old, although there are large uncertainties (*Kirchoff and Schenk,* 2009).

Schenk and McKinnon (2009) identified several deep basins scattered over Enceladus that have no apparent expression in surface geology. The origin of these basins is unclear, but one possibility is that they arise from compaction of surface pores driven by localized enhanced heat fluxes (*Besserer et al.,* 2013).

Geology and topography both thus suggest that Enceladus has experienced several episodes of localized activity, with the present-day locus (the SPT) being the most recent. In some cases, estimates of the heat flux associated with these episodes can be derived. In an area of the trailing hemisphere, *Giese et al.* (2008) identified an apparently flexural feature and used the derived elastic thickness to infer a heat flux of at least 45 mW m^{-2}. *Bland et al.* (2007, 2012, 2015) investigated the heat fluxes required to cause crater relaxation or characteristic deformation wavelengths and deduced paleo-heat fluxes of a few hundred milliwatts per square meter. Despite the ranges and uncertainties in these estimates, the heat fluxes derived are similar to the present-day SPT flux measurements, and suggest that different regions of Enceladus (including the cratered plains) have experienced high heat fluxes in the past. Unfortunately, it is not yet clear whether there were long intervals of quiescence between these heating episodes, or whether the heating was continuous but simply changed location over time.

3. TIDAL HEATING AT THE PRESENT DAY

3.1. Tidal Heating: The Basics

Most of the regular satellites orbiting giant planets are in a "synchronous state," in which the satellite's spin period is equal to its orbital period. Because the giant planet is so massive, its gravity deforms the satellite, raising a tidal bulge that points along the line connecting the center of the satellite to the center of the planet.

If the planet is orbited by multiple satellites, pairs or groups of satellites can enter a resonance, in which the gravitational tugs on pairs of satellites occur at regular intervals. The most common resonance occurring in multiple satellite systems in the outer planets is a mean-motion resonance (MMR), in which the orbital periods of satellites are integer multiples of each other (see section 4.1.1). Enceladus is presently in a 2:1 MMR resonance with Dione, meaning that Enceladus orbits Saturn twice each time Dione orbits once. In this configuration, the gravitational perturbations from Dione keep the eccentricity of Enceladus at a forced value e = 0.0047.

The small eccentricity of Enceladus' orbit means that its distance to Saturn varies by a small amount during the course of its P = 1.37-day orbit around Saturn. So the height of the tidal bulge raised on Enceladus varies by a small amount as well. The cyclical raising and lowering of the tidal bulge results in tidal deformation on a 1.37-day period. This motion is resisted by Enceladus' own internal friction. It is this friction that results in tidal heat.

The rate of energy dissipation in Enceladus can be related to the properties of its orbit and interior (*Peale and Cassen,* 1978)

$$\dot{E} = \frac{21}{2}\frac{k_2}{Q}\frac{R_e^5 G M_p^2 n}{a^6} e^2 \tag{2}$$

where G is the gravitational constant, M_P is the mass of Saturn, n = (2π/P) is the mean motion, and a is the semimajor axis of Enceladus. The ratio k_2/Q in equation (2) describes how the interior of Enceladus responds to the applied tidal potential, and depends sensitively on Enceladus' interior structure. The value k_2 is the Love number describing how the tidal potential of Enceladus responds to the applied tidal potential, where the subscript 2 indicates the harmonic degree on which the tidal forcing occurs, and Q is the tidal quality factor.

The value of the ratio k_2/Q, is uncertain, and depends on the details of how the mechanical deformation of Enceladus' interior is converted into heat. But one logical question to ask is, what is the value of k_2/Q that could explain the observed heat loss, and how does this compare to the measured values of k_2/Q for other solar system bodies? Evaluating equation (2) for Enceladus yields (*Spencer and Nimmo,* 2013)

$$\dot{E} = 15\text{GW}\left(\frac{(k_2/Q)}{0.01}\right)\left(\frac{e}{0.0047}\right)^2 \tag{3}$$

implying, for the present value of the eccentricity of Enceladus and taking the global heat flow to be 3.9–19 GW, that $0.0026 < k_2/Q < 0.0127$. Note that this calculation assumes all the measured heat is being produced at the present day. An independent constraint arises from the present-day characteristics of the Enceladus-Dione resonance, which gives $34\,(k_{2_p}/Q_p) < k_2/Q < 41\,(k_{2_p}/Q_p)$ (*Zhang and Nimmo,* 2009), where the subscript p indicates the primary (Saturn). For comparison, fits to lunar laser ranging data indicate that

k_2/Q for Earth's Moon ~6.4×10^{-4} (*Williams and Boggs,* 2015), a factor of 4 to 20 lower than implied for Enceladus, whereas a similar astrometric analysis for Io gave $k_2/Q = 0.015 \pm 0.003$ (*Lainey et al.,* 2009). These comparisons suggest that Enceladus has a warm interior that can readily deform in response to the applied tidal potential, but still retains enough internal friction to result in strong dissipation.

3.2. Where is the Heat Dissipated?

The details of how deformation gives rise to heat in a solid planet, or a body like Enceladus that probably has an internal ocean, are not well understood. Enceladus has probably fully differentiated over the course of its history (*Schubert et al.,* 2007), and so its interior likely harbors a central rock core and a shell of rock-free ice. Any ocean in Enceladus would sit perched between the base of the ice shell and the top of the rock core (see Fig. 1).

It is not known whether the tidal heat occurs mostly in the solid portions of Enceladus (e.g., the ice shell or rock core), the ocean, or potentially at the interfaces between these layers (e.g., frictional dissipation from the ocean sloshing against the surface of the rock core). It is also not known at what spatial scale the dissipation occurs. The most common type of tidal heating model assumes that dissipation occurs at the microphysical level: The ice/rock is assumed to have a viscoelastic rheology (*Findley et al.,* 1989), where the deformation of the material is thought to be accommodated by one of the usual microphysical mechanisms [e.g., volume diffusion, easy slip, dislocation creep; see, e.g., *McCarthy and Castillo-Rogez* (2013) for discussion]. Another possibility is that tidal dissipation is localized along the tiger stripes, either as frictional dissipation along faults (*Nimmo et al.,* 2007) or from turbulent dissipation in water-filled cracks (*Kite and Rubin,* 2016). Thus, the tidal heat could be dissipated in the ice shell of Enceladus, in the ocean, in the rock core, or in a combination of or all of these locations. Each of these alternatives has been explored, with varying degrees of success.

If an ocean is not present, the ice shell is then coupled to the core, which is nominally less deformable. As a result, k_2/Q and the amount of tidal dissipation are expected to decrease significantly (*Nimmo et al.,* 2007). It is therefore often argued that the ocean must be a long-lived feature, because if the ocean freezes completely, tidal heating is reduced and the ice shell will never remelt. However, as discussed in section 4.1.3, this argument is not correct in detail, because tidal heating naturally tends toward an equilibrium value, independent of the satellite k_2/Q.

The remainder of this section will describe the numerical and analytical models that have been constructed to explain the magnitude and distribution of heat coming from Enceladus at present. In section 3.3, we describe tidal heating in the ice shell of Enceladus; in section 3.4 we summarize the results of calculations of dissipation in the ocean; and we conclude in section 3.5 by discussing the possibility of tidal heating in the rock core. The time-evolution of heat generation and transport will be discussed in section 4.

3.3. Ice Shell

3.3.1. One-dimensional models. Many models of tidal heating in Enceladus assume that the satellite is radially symmetric, meaning that its physical properties vary as a function of radius, but not location on the satellite. Each layer of the satellite is assumed to have a viscoelastic rheology; the simplest viscoelastic model is that of Maxwell, in which each layer is characterized by a viscosity and rigidity. For a satellite with a viscoelastic rheology, k_2 is a complex number, and Q is related to the imaginary part of k_2 (*Zschau,* 1978; *Segatz et al.,* 1988).

Values of k_2 and Q can be obtained by representing the interior of the satellite as a series of layers, each with its own viscosity and rigidity (*Ross and Schubert,* 1989; *Roberts and Nimmo,* 2008; *Barr,* 2008). Because the viscosity is a strong function of temperature, different assumed temperature profiles (e.g., conductive vs. convective) can lead to quite different dissipation results. Values of k_2 are calculated using the correspondence principle (*Sabadini et al.,* 1982; *Tobie et al.,* 2005; *Roberts and Nimmo,* 2008; *Wahr et al.,* 2009). Calculating k_2/Q for a convecting ice shell, and substituting into equation (2), yields a tidal heating rate about a factor of 10 lower than observed by CIRS (*Barr,* 2008).

An alternative viscoelastic formulation, the Andrade model, may be more appropriate (*McCarthy and Castillo-Rogez,* 2013). Figure 2 shows global tidal heat production generated with such a model (*Běhounková et al.,* 2015). In Fig. 2a the shell is assumed to have a uniform viscosity, appropriate for a shell that is convecting. For sufficiently low viscosities, the total power output can equal or exceed that detected at the SPT (shaded region). Heat production increases with decreasing shell thickness because the amplitude of tidal deformation increases with a thinner shell. For a conductive shell, more likely given the thin shell deduced from libration measurements, the total heat production is lower (Fig. 2b) and is exceeded by the conductive heat loss (dotted line).

A significant problem with all one-dimensional models is that, while they predict spatial variations in tidal heat production, they do not explain why only the south pole is currently a hot-spot. All such models result in a symmetrical distribution of heat production around the equator.

3.3.2. Two-dimensional and three-dimensional models. Two types of more sophisticated models have been proposed to explain the distribution and magnitude of tidal heat coming from the south pole of Enceladus. One possibility is that tidal heating is dissipated on the macroscale, as frictional heating on faults (*Nimmo et al.,* 2007). In this model, heating is generated in the cold, near-surface ice on Enceladus by cyclical strike-slip motion along faults in the centers of the tiger stripes. This model provides a natural explanation for the localization of endogenic heating along the tiger stripes observed by CIRS (*Nimmo et al.,* 2007). However, it is difficult to produce enough heat by this mechanism to explain the 5.8 ± 1.9 GW of power implied by the *Spencer et al.* (2006) analysis (*Ingersoll and Pankine,* 2010), let alone enough heat to explain the higher power output implied

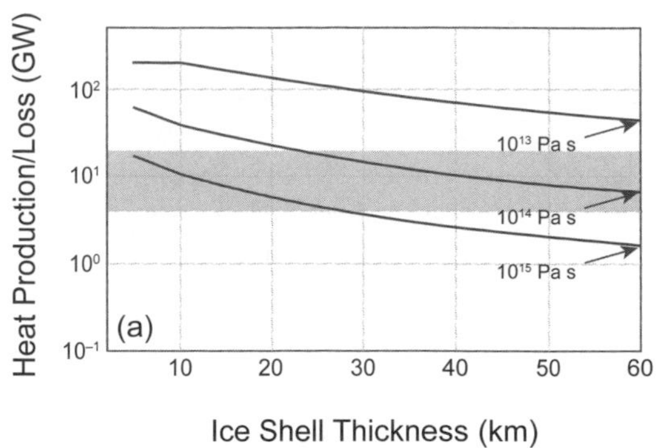

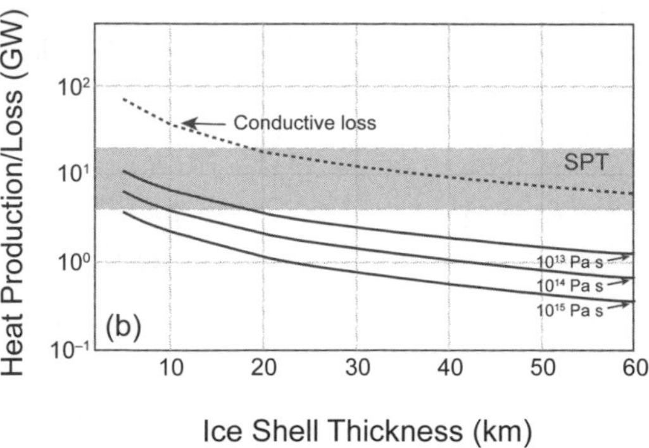

Fig. 2. Present-day tidal heating as a function of ice shell thickness and a viscosity using an Andrade rheology (*Běhounková et al.,* 2015). **(a)** Heat production for a constant viscosity (solid line), **(b)** tidal heating (solid lines) and heat loss (dashed line) for a conductive temperature profile and temperature-dependent viscosity (with maximum viscosity 10^{20} Pa s). Gray area is the present-day heat loss through the SPT (*Howett et al.,* 2011, 2013).

by later observations. Furthermore, this model appealed to direct sublimation of water ice to explain the plume, which is inconsistent with the salts detected there (*Postberg et al.,* 2009, 2011).

A variation on this theme also appeals to dissipation within the tiger stripes, but in the form of turbulent dissipation in ocean water within the tiger stripes (*Kite and Rubin,* 2016). However, this model appears unable to explain the higher power outputs implied by some observations, and even for 5 GW requires a substantially thicker ice shell beneath the SPT than is now considered likely.

A second set of models build on the idea of calculating k_2/Q assuming a Maxwell rheology for the ice and rock, but relaxing the assumption that the properties of Enceladus' interior are radially symmetric. The basic idea is that both the tidal forcing and the temperature structure of the ice shell of Enceladus should vary as a function of latitude and longitude. If the ice shell is convecting, one expects lateral variations in the temperature structure of the shell: Warm convective upwellings might experience more intense tidal dissipation (*Wang and Stevenson,* 2000; *Sotin et al.,* 2002; *Tobie et al.,* 2003; *Mitri and Showman,* 2008; *Běhounková et al.,* 2010, 2012). If the SPT were underlain by a single convective upwelling but the north pole was not (e.g., *Han et al.,* 2012; *Rozel et al.,* 2014), this might explain why all of Enceladus' tidal heat is coming out of one location. Alternatively, the shape of the core might influence where upwellings and dissipation were focused (*Showman et al.,* 2013). As a last resort, one can always appeal to an impact to break the initial symmetry (e.g., *Roberts,* 2016); the impact site would then reorient to whichever pole was closer (*Nimmo and Pappalardo,* 2006).

The most sophisticated model to date (*Souček et al.,* 2016) incorporates aspects of both approaches. These authors introduce weak discontinuities (faults) within the SPT and calculate both the local dissipation and the background tidal response of the shell. Tidal deformation within the SPT is significantly enhanced by the presence of these weak zones, suggesting an additional way of localizing heating. Whether this kind of model is also consistent with the observed modulation of the plume remains to be seen.

3.4. Ocean

In principle, tidal heating in the ocean of Enceladus could be a significant source of heat. In particular, if Enceladus has a nonzero obliquity (tilt), obliquity-driven ocean tides could contribute to the overall heat budget (*Tyler,* 2009). Unfortunately, however, dissipation drives satellites toward a so-called Cassini state, which for Enceladus results in an obliquity far too small to generate significant heat (*Chen and Nimmo,* 2011). Additionally, the resonances responsible for driving ocean dissipation only operate when oceans are thin, typically on the order of 1 km or less (*Tyler,* 2011; *Hay and Matsuyama,* 2017). The results discussed above suggest that the ocean on Enceladus is ≥10 km deep. Finally, the presence of an elastic lid will reduce the ocean dissipation even further (*Beuthe,* 2016). Thus ocean currents are unlikely to be a significant source of tidal heat in present-day Enceladus. It is possible that ocean dissipation could become more important as the shell thickened, thereby preventing total ocean freezing, but even in this case the small obliquity limits the total amount of heat produced.

Another possible heating mechanism is Joule heating, in which electric currents passing through water pockets of low salinity buried kilometers beneath the surface (*Hand et al.,* 2011). Estimates of the magnitude of this heat source imply that the heat source would be small (something like 0.001% to 0.25% of the observed heat flow) (*Hand et al.,* 2011).

3.5. Rock Core

At face value, one would expect the rock core of Enceladus to be too stiff to experience much tidal heating (*Ross and Schubert,* 1989; *Roberts and Nimmo,* 2008). In a Maxwell viscoelastic solid, dissipation is maximized when the period of the cyclical forcing is comparable to the Maxwell time, $\tau_M = \eta/\mu$, where η is the viscosity of the material, and μ is its rigidity. For rock, with $\eta \sim 10^{20}$ Pa s, and μ = 100 GPa (*Barr,* 2008), τ = 31.7 years, far longer than the 1.37-day orbital period of Enceladus. (By comparison, for ice at its melting point, with $\eta \sim 10^{14}$ Pa s and $\mu \sim 3$ GPa, τ = 9 hours.) Thus, it is reasonable to expect that much, or all, of the tidal dissipation would be concentrated in the ice shell of Enceladus.

Evidence reviewed in section 2.1 suggests that the silicate core has a low density, most likely due to porosity. The presence of such (probably water-filled) pores would alter the material properties of the core and potentially render it more conducive to tidal dissipation (*Roberts,* 2015). Tidal flushing of water through the pores could contribute an additional heat source, although the only study on this topic to date (*Vance et al.,* 2007) concludes that the contribution is minimal. Alternatively, a core that started warm might be sufficiently dissipative to maintain its temperature over the long term. We note that a warm core is also suggested by the existence of silica nanoparticles (*Hsu et al.,* 2015) and hydrogen gas (*Waite et al.,* 2017) in the plume. Thus, although little work has been done to date, dissipation in the core remains a possibility.

4. TIDAL AND ORBITAL EVOLUTION

4.1. Introduction and General Picture

At a basic level, it is clear that the energy budget of Enceladus must be dominated by tidal heating. Thus, in order to understand its long-term thermal evolution, its long-term orbital evolution must also be understood. Orbital dynamics is a complicated subject. Below, we try and summarize in relatively simple terms the key processes that are operating. More in-depth explanations can be found in, e.g., *Murray and Dermott* (1999) and *Peale* (1999).

An orbiting satellite will raise tides on the primary. The resulting tidal bulge will not in general point directly toward the satellite (Fig. 3a, inset); there will be a lag, with the magnitude of the lag depending on the rate of energy dissipation in the primary. The lag will also cause torques on the satellite; for satellites beyond the co-rotation point, the torques will cause outward motion. The Moon is moving away from Earth because of tides raised by the Moon on Earth.

The lag angle ($\delta/2$) may be derived using $\delta \approx 1/Q_p$, where Q_p is the so-called dissipation factor of the primary (*Murray and Dermott,* 1999). A small Q_p implies a large lag angle and high dissipation rates. In general, Q_p is expected to be frequency-dependent, with peaks in dissipation at particular frequencies (*Ogilvie and Lin,* 2004; *Wu,* 2005). Since a satellite's orbital frequency depends on its semimajor axis, Q_p can vary from satellite to satellite, and can vary in time for a single satellite.

The tidal torque exerted by the primary is a strong function of semimajor axis. It is therefore often the case that the inner satellites migrate outward faster than more distant satellites. For example, Fig. 3a shows a hypothetical scenario in which Q_p is set to a constant value of 16,000 (dashed lines). The orbits are generally converging, except that Tethys moves outward faster than Enceladus, because Tethys is much more massive and thus raises larger tides on Saturn. This simple calculation suggests that Q_p must exceed 16,000 unless the age of Mimas is less than that of the solar system (but see section 4.2).

4.1.1. Mean-motion resonances. An important consequence of the relative outward motion of the satellites is that they may have encountered commensurate locations, in which the inner orbital frequency is a simple integer ratio

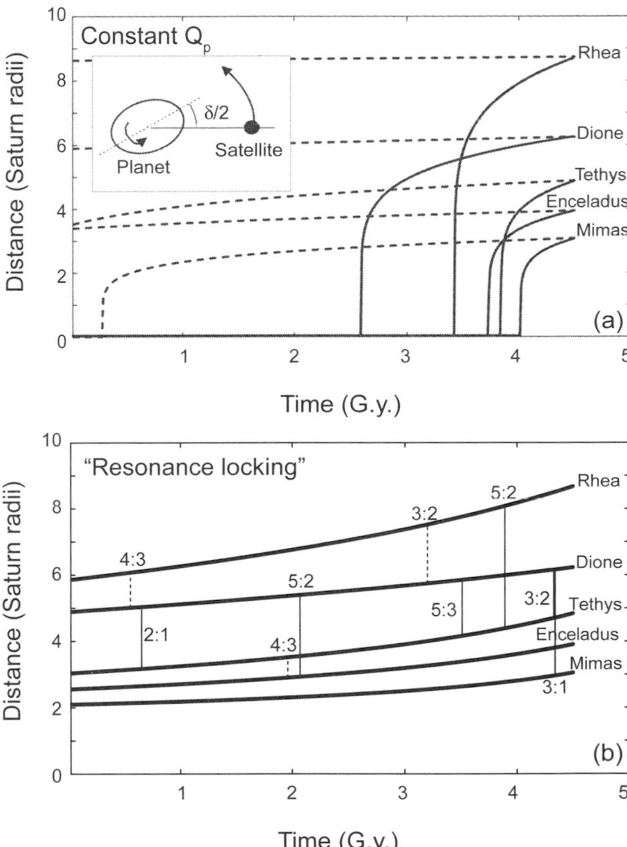

Fig. 3. (a) Outward satellite migration assuming constant Q_p. Dashed lines show evolution when Q_p = 16,000 everywhere. Solid lines show evolution when Q_p is set to a different (constant) value for each satellite, based on astrometric observations of present-day migration rates (section 4.3). Inner- to outer-body values for Q_p are 3000, 1920, 2600, 2540, and 325 respectively and are calculated using equation (9) and the observed migration rate. Inset shows geometry of tidal lag angle $\delta/2$ (see text). **(b)** Outward evolution based on "resonance locking" scenario (section 4.3) (see also *Fuller et al.,* 2016). Present-day migration rate is based on the same present-day Q_p as in **(a)** and evolution of semimajor axis is calculated using equation (10) for each satellite with t_α set to 4, 8, 12, 51, and 49 G.y., respectively. Example mean-motion resonances are marked with dashed lines for diverging orbits and solid lines for converging orbits.

of the outer orbital frequency. For instance, Enceladus and Dione are currently in a 2:1 eccentricity-type MMR. Such resonances are extremely important, because they provide a way of exciting orbital eccentricities and thus a potential long-term heat source (see section 4.1.2). In simple terms, in an MMR the satellites encounter each other at the same point in succeeding orbits; the mutual orbital perturbations thus add coherently, leading to large eccentricities.

In detail, several different resonances exist around a particular commensurate location. For instance, the Enceladus-Dione 2:1 eccentricity commensurability is actually a triplet of resonances: e_D, e_D-e_E, and e_E (the present resonance) (*Meyer and Wisdom,* 2008a; *Zhang and Nimmo,* 2009). For converging, circular orbits passing through a commensurate location, capture into at least one of these resonances is assured as long as the satellites are migrating sufficiently slowly. Whether or not a particular resonance eventually breaks depends on the details of the resonance and the extent to which the satellites' eccentricity is excited (higher eccentricities favor resonance breaking). Thus, for example, a hypothetical ancient Mimas-Enceladus 3:2 e_E resonance cannot have occurred, because there is no apparent way to break the resonance (*Meyer and Wisdom,* 2008a).

A commensurability of j:j–k has order k and is defined by $(j–k)n_i = jn_o$ where j is an integer and n_i and n_o are the mean motions of the inner and outer satellite. Thus, 3:2 and 2:1 are both first-order resonances. Capture into higher-order resonances has a lower probability than capture into first-order resonances (*Dermott et al.,* 1988). On the other hand, higher-order resonances can lead to higher equilibrium heating rates (see section 4.1.2).

In contrast to converging orbits, satellites on diverging orbits will pass through commensurate locations without getting trapped into resonance. The eccentricities may be transiently excited (*Dermott et al.,* 1988), but the effect on the total energy budget will be very small.

4.1.2. Equilibrium eccentricity and heating. For an isolated satellite, the effect of outward motion due to torques raised by the primary also results in eccentricity growth. But a satellite in an eccentric orbit experiences diurnal tides that dissipate energy internally (equation (2)) and cause the orbit to circularize. Because the circularization timescale is usually much shorter than the outward migration timescale, isolated satellites are expected to have near-zero eccentricity.

However, the situation changes if two satellites are in an eccentricity-type MMR. In an MMR eccentricity growth is much more rapid, and can become comparable to the rate of circularization. One possible result of these competing processes is that the eccentricity reaches an equilibrium value, in which eccentricity excitation due to the MMR is balanced by the eccentricity damping in the satellite, and the total tidal heat production rate is constant.

If the satellites are in this equilibrium situation, a very important result arises: The total rate of energy dissipation in the satellites is then independent of the satellite structure. This is extremely useful, because in general the rate of satellite heating requires its k_2/Q, and thus its internal structure, to be known (section 3.1). The partitioning of heat between the two satellites does require their k_2/Q to be specified, and so too does calculation of the actual equilibrium eccentricity, but for thermal evolution calculations, these are relatively minor disadvantages.

The equilibrium tidal heating rate can be determined by consideration of conservation of mass and angular momentum (*Meyer and Wisdom,* 2007; *Fuller et al.,* 2016). The heating rate is given by

$$\dot{E}_{eq} = \frac{n_i T_i}{\sqrt{(1-e_i^2)}} + \frac{n_o T_o}{\sqrt{(1-e_o^2)}} - \frac{T_i + T_o}{L_i + L_o} GM\left(\frac{m_i}{a_i} + \frac{m_o}{a_o}\right) \quad (4)$$

Here n, e, m, a, and L refer to the mean motion, eccentricity, mass, semimajor axis, and orbital angular momentum of a satellite, with subscripts referring to the inner i and outer o body. The angular momentum is given by $m\sqrt{GM_p a(1-e^2)}$ with M_p the mass of the primary. T represents the torque from the primary acting on the satellite, where

$$T = \frac{3}{2}\frac{Gm^2 R_p^5 k_{2p}}{a^6 Q_p} \quad (5)$$

with R_p and k_{2p} the radius and tidal Love number of the primary, respectively. Inspection of equations (4) and (5) show that the satellite k_2/Q is not required and that the major unknown governing the equilibrium heat production rate is Q_p. Physically, what is happening is that the energy dissipated in the satellites is ultimately coming from the rotation kinetic energy of the primary (a large reservoir); the rate at which this energy can be extracted is controlled by the tidal torque, which depends on Q_p.

For Enceladus, the equilibrium heat production can thus be readily calculated (*Meyer and Wisdom,* 2007). If Q_p is constant, the equilibrium heat production rate is

$$\dot{E}_{eq} = 1.1\ \text{GW}\left(\frac{18{,}000}{Q_p}\right) \quad (6)$$

This is an extremely important result, because estimates of the current heat output of Enceladus cluster around 10 GW (see section 2.2). As discussed in more detail below, there are at least three possible resolutions to this apparent problem: (1) Enceladus is currently producing heat in excess of the equilibrium rate; (2) Enceladus is producing heat at the equilibrium rate, but releases it episodically; (3) Q_p is much smaller than 18,000 at the present day.

4.1.3. Periodic behavior. In its current resonance, the eccentricity evolution of Enceladus can be described by an equation of the form (*Ojakangas and Stevenson,* 1986; *Meyer and Wisdom,* 2008b)

$$\frac{de}{dt} = ae^2\left(1 - be^2\right) \quad (7)$$

where a and b are constants, with b depending on the k_2/Q of the satellite. The first term in brackets is the eccentricity excitation associated with the MMR, while the second is the eccentricity damping associated with dissipation in the satellite. The equilibrium eccentricity $e_{eq} = 1/\sqrt{b}$ and is thus dependent on the satellite structure. The satellite structure, however, is itself evolving at a rate that depends on the extent to which temperature changes affect k_2/Q. There are thus two timescales of interest: the timescale for the eccentricity to equilibrate, and the timescale for k_2/Q to change significantly.

Neglecting coefficients on the order of unity, the eccentricity equilibration timescale is (*Meyer and Wisdom,* 2008b)

$$\tau_e \approx \frac{2}{9}\frac{1}{n_i}\frac{1}{e_{eq}}\frac{m_o}{m_i}\frac{a_i}{a_o}\left(\frac{a_i}{R_p}\right)^5\frac{Q_p}{k_{2_p}} \approx 9\ \text{Myr}\left(\frac{Q_p}{18{,}000}\right) \qquad (8)$$

where R_i is the radius of Enceladus, e_{eq} is the equilibrium eccentricity (taken to be the present-day value of 0.0047), and we have taken $k_{2_p} = 0.34$ (*Gavrilov and Zharkov,* 1977). This timescale is fast compared to the thermal diffusion timescale of a body the size of Enceladus; it becomes shorter if Saturn is more dissipative (lower Q_p).

If the timescales of eccentricity equilibration and thermal adjustment are similar, an interesting result occurs: The orbital evolution can experience periodic behavior (*Ojakangas and Stevenson,* 1986; *Hussmann and Spohn,* 2004; *Meyer and Wisdom,* 2008b). The total heat production and the eccentricities of both satellites oscillate around their equilibrium values. Thus, Enceladus might currently be in a high-heat-production part of the cycle. The long-term average heat production, however, is still fixed at the equilibrium value (equation (6)). We will return to these issues in more detail below.

Another important consequence of the long-term average heat production being independent of the satellite k_2/Q concerns the fate of the ocean (section 3.2). If the ocean freezes completely, then k_2/Q of the satellite will be reduced, and so will the instantaneous heat production. However, as equation (7) shows, a reduction in k_2/Q will drive the eccentricity to higher values and thus increase the heat flux back to the equilibrium value. If the equilibrium heating is not enough to prevent ocean freezing, it will presumably not cause remelting. Nonetheless, this example illustrates the importance of considering orbital and thermal evolution together.

4.2. Astrometry

The instantaneous outward evolution of an isolated satellite depends on the torque exerted by the primary (equation (5)) and is given by

$$\frac{1}{a}\frac{da}{dt} = 3\frac{k_{2_p}}{Q_p}\frac{m}{M_p}\left(\frac{R_p}{a}\right)^5 n = \frac{1}{t_{tide}} \qquad (9)$$

Accordingly, if da/dt can be measured, then Q_p (at that particular frequency) can be inferred.

The situation is slightly more complicated if two satellites are in a MMR. For the commensurability to be maintained $(1/a_i)da_i/dt = (1/a_o)da_o/dt$. If the torque on the inner satellite dominates, the outward evolution timescale t_{tide} will be reduced by a factor of $[1 + (m_o\sqrt{a_o}/m_i\sqrt{a_i})]$ because the torque is now changing the angular momentum of both satellites. For Enceladus-Dione this factor is large ($\approx$13) because Dione is so much more massive.

Recently, a combination of long-baseline Earth-based observations and Cassini spacecraft observations has allowed measurements of da/dt to be made for various saturnian satellites (*Lainey et al.,* 2017). These measurements can then be converted to Q_p at different frequencies using equation (9). For Enceladus, two different approaches yield k_{2_p}/Q_p = 18.1 ± 3.1 × 10^{-5} and 27.1 ± 13.5 × 10^{-5}. For the canonical Saturn k_{2_p} of 0.34, the implied Q_p values are 1880 ± 280 and 1260(+1240,–420). These results take the Enceladus-Dione MMR into account (R. Jacobson, personal communication) and confirm earlier astrometric suggestions that $Q_p \approx 2000$ for Saturn (*Lainey et al.,* 2012). Values for Q_p derived from Tethys and Dione observations are similar to those from Enceladus. For Rhea, $Q_p \approx 300$ with an uncertainty of about 10%.

These estimates of present-day Q_p are subject to large uncertainties. Nonetheless, it seems very likely that, in contrast to Jupiter, the present-day Q_p of Saturn is significantly smaller than the canonical, time-independent value of 18,000 (section 4.1).

4.3. "Resonance Locking"

The low Q_p values derived from astrometry (see section 4.2) have two immediate consequences. First, the equilibrium heat production rate in Enceladus (equation (6)) could easily be 10 GW or more. This issue is discussed in more detail in section 5.3. Second, if Q_p is constant, then both Enceladus and Mimas must be much younger than the age of the solar system (Fig. 3a). Recent work, however, shows that a low present-day Q_p does not necessarily require young satellites (*Fuller et al.,* 2016). This is an important result, which we discuss in more detail in the rest of this section.

Because the major saturnian satellites are outside the corotation point, from a reference frame rotating with Saturn the satellites move in a retrograde fashion, with the outer satellites having a more negative synodic frequency. In this reference frame, outward satellite migration thus results in an increase in the absolute (synodic) frequency of the tides raised on Saturn.

As discussed above, the dissipation spectrum of Saturn is likely composed of discrete peaks. A migrating satellite may thus sweep through one of these peaks as its orbit evolves. Alternatively, a satellite may encounter a peak because the location of the peaks changes as Saturn's internal structure evolves. *Fuller et al.* (2016) consider this latter scenario, in the specific case of a peak evolving to higher frequencies (Fig. 4).

As the peak approaches the satellite frequency, dissipation in Saturn increases and the satellite experiences a growing

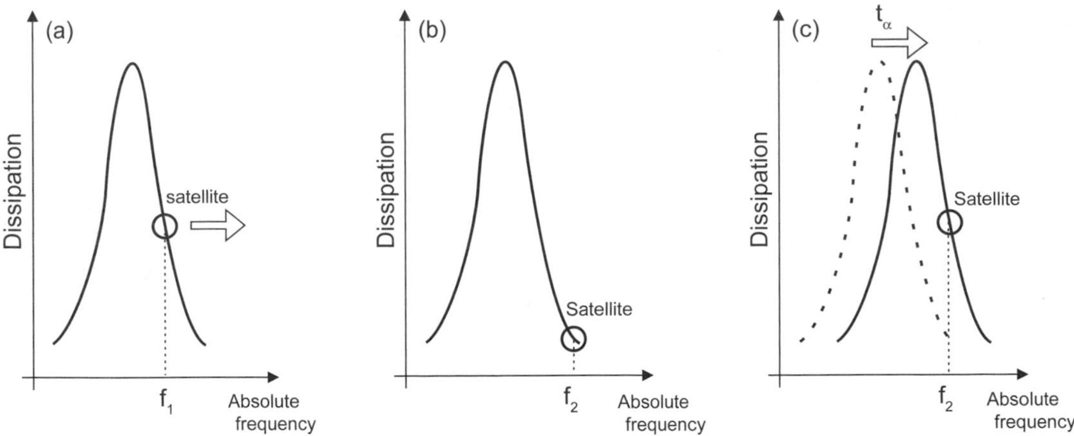

Fig. 4. Cartoon showing how the resonance locking scenario operates. **(a)** Satellite is close to a dissipation peak at absolute synodic frequency f_1 and migrates outward, increasing its frequency. **(b)** At frequency f_2 the satellite experiences less dissipation and stops migrating. **(c)** The peak evolves to higher frequency as Saturn evolves, so dissipation in the satellite increases again. The outward migration rate of the satellite is thus controlled by the timescale t_α for Saturn to evolve.

outward torque (equation (5)), moving the satellite outward and increasing its absolute synodic frequency. This increase, however, moves the satellite away from the dissipation peak, so that its outward motion slows down until the peak catches up again. The rate of outward evolution of the satellite is thus governed by the rate at which the frequency of the saturnian dissipation peak changes. So the satellite orbit evolution timescale (equation (9)) is set by the timescale t_α with which Saturn's interior structure evolves. Note that t_α can in principle be different at different frequencies, although the physical justification for such an assumption seems weak. *Fuller et al.* (2016) term their scenario "resonance locking" because it involves the satellite orbital frequency being pinned to a particular resonance within Saturn.

Fuller et al. (2016) give the resulting orbit evolution timescale t_{tide} as

$$\frac{1}{t_{tide}} = \frac{1}{a}\frac{da}{dt} = \frac{2}{3}\left[\frac{\Omega_p}{n}\left(\frac{1}{t_\alpha} - \frac{1}{t_p}\right) - \frac{1}{t_\alpha}\right] \tag{10}$$

where Ω_p is the rotation angular frequency of Saturn, n is the mean motion of the satellite, and t_p is the timescale over which the rotation rate of Saturn evolves. Because this latter timescale is expected to be long, we will set $t_p = \infty$ in the following.

Equation (10) can be used to derive a simple expression for the change in satellite mean motion

$$n(t) = \Omega_p + \left(n_{now} - \Omega_p\right)\exp\left(\frac{t - t_{now}}{t_\alpha}\right) \tag{11}$$

where the mean motion is set to n_{now} at time t_{now}. Of course, equation (9) can still be used to derive the effective Q_p governing a particular rate of outward evolution. But in this approach, Q_p is not really the fundamental quantity; it is a derived quantity, which depends ultimately on t_α, the timescale of Saturn's evolution.

The behavior described by equation (11) is very different from the conventional picture of satellite orbital migration (Fig. 3a). Instead of evolving outward at a rapidly decreasing rate, the pace of orbital migration is almost constant. Figure 3b shows outward satellite migration calculated using equation (11) with the present-day values of t_α derived from astrometry (*Lainey et al.,* 2017). Crucially, the present-day, relatively rapid, outward migration can still be reconciled with the satellites being as old as the solar system. Another way of interpreting this result is that the *effective* Q_p that a given satellite experiences has decreased substantially over the course of satellite orbital evolution. For instance, Fig. 5 shows that in the case of Enceladus the present-day value of Q_p is 1920 (and t_{tide} = 5.7 G.y.), while at 2 Ga, Q_p was 18,700 (with n = 7.8×10^{-5} s^{-1} and t_{tide} = 10.7 G.y.) (equations (10) and (11)).

A second important consequence of Fig. 3b is that the specific resonances that satellites encounter are quite different from those of conventional, constant Q_p scenarios. For Enceladus in particular, the Mimas:Enceladus 3:2 resonance is never encountered, which is convenient because this particular resonance appears impossible to break (*Meyer and Wisdom,* 2008a). Prior to the present day 2:1 resonance with Dione, the only major resonance Enceladus encountered is a 4:3 with Tethys, but since the two bodies are on diverging orbits, this resonance is unlikely to have played a major role.

A note of caution is required here. First, the astrometric measurements used to derive t_α for the various satellites are subject to large uncertainties. Even small variations in t_α can lead to substantially different orbital histories. For instance, if t_α is set to 50 G.y. for all satellites, then the orbits are all

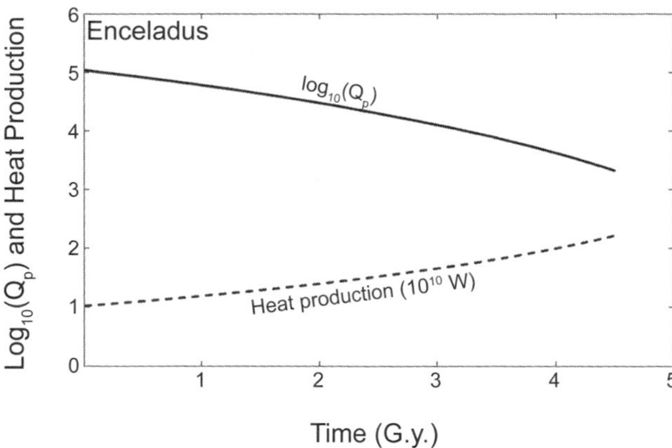

Fig. 5. Variation of equilibrium heat production rate for Enceladus and effective Q_p as a function of time, based on the resonance locking scenario shown in Fig. 3b. Q_p is calculated using equations (9) and (10) and heat production using equation (4) with $T_o = 0$.

diverging, prohibiting establishment of resonances. Second, the evolution shown in Fig. 3b is highly simplified because it ignores the effect of MMRs on orbital evolution. Thus, for instance, if the Enceladus:Dione 2:1 resonance were established early, their evolution could have been quite different. Third, the resonance-locking theory is heuristic, in that it uses a very simple description for how the internal structure evolves. Especially early in Saturn's history, evolution was probably quite rapid; thus, the early stages of orbital evolution and the initial locations of the satellites (Fig. 3b) are highly uncertain. Finally, the behavior invoked by *Fuller et al.* (2016) requires that there exists a dissipation peak close to the frequency being excited by the satellite, and that the peak increases in frequency with time. Detailed modeling will be needed to explore how reasonable these assumptions are for Saturn.

5. EXPLAINING THE PRESENT-DAY HEAT FLUX

As summarized in section 2, the present-day measured heat output of Enceladus is an order of magnitude larger than the equilibrium tidal heating value for a conventional Saturn Q_p of 18,000 (equation (6)). There are at least three possible explanations for this discrepancy: (1) Tidal heating on Enceladus is periodic, with the present-day heating rate exceeding the long-term average. (2) The rate of heat loss on Enceladus is episodic, with the present-day loss rate exceeding the long-term average. (3) The effective Q_p of Saturn is an order of magnitude smaller than the conventional value. Below we discuss each of these possibilities in turn.

In this context, the geological observations (section 2.3) are of great importance. In particular, evidence of ancient (several gigayears ago) heating episodes in different regions of Enceladus can potentially distinguish between different scenarios.

5.1. Is Enceladus Experiencing Periodic Tidal Heating?

Section 4.1.3 noted the possibility that Enceladus might be experiencing time-variable heating. If so, we could happen to be seeing Enceladus at a special time when the tidal heating rate greatly exceeds the equilibrium value. Since the heat flow is about 10 times the conventional equilibrium value, the probability of this eventuality is about 10%.

Meyer and Wisdom (2008b) investigated this possibility and concluded that Enceladus was unlikely to be experiencing periodic behavior, essentially because the eccentricity equilibration timescale (equation (8)) was significantly shorter than the thermal adjustment timescale. More recently, *Shoji et al.* (2014) used a slightly different approach and found that, for some parameter combinations, periodic heating could occur (Fig. 6a). In principle, such periodicity could potentially explain the ancient, high-heat flux terrains as well as the present-day heat loss. However, the deviation around the long-term average heat production value is rather small, and at least for a constant Q_p of 18,000 even the peak heat fluxes are still insufficient to explain the observed heat loss. Furthermore, the interval between heating episodes appears short compared to the geologically-inferred intervals. Although more work could be done, on its own periodicity seems unlikely to be able to explain the present-day heat loss.

5.2. Is Enceladus Experiencing Episodic Heat Loss?

An alternative possibility is that Enceladus builds up heat in its interior at the equilibrium rate of 1.1 GW, and then releases it episodically. The same duty cycle argument applies as in the preceding section: If correct, we are seeing Enceladus in a state it occupies only about 10% of the time.

O'Neill and Nimmo (2010) used numerical convection models to show how this process could work. For a sufficiently low effective yield strength, heat build-up in the interior eventually gives rises to stresses that cause large-scale overturn of the near-surface ice and transiently high heat fluxes (Fig. 6b). These episodes recur every 0.1–1 G.y. and tend to be localized rather than global. An advantage of this model is that it provides a potential explanation for the inferred ancient episodes of relaxation and tectonic deformation, and the localized present-day heat loss. One drawback of this analysis is that it is both two-dimensional and Cartesian, and it is unclear whether the same results would arise in a three-dimensional spherical setting. Additionally, more recent work suggests a thinner present-day ice shell (section 2), which makes convection correspondingly more difficult to achieve. And if convection is really operating, rapid lateral flow of ice would make the maintenance of a thin shell over the SPT hard to explain.

This explanation seems somewhat more promising than invoking periodic heat production, but it still suffers from the problem that it requires us to be seeing Enceladus at a special time. More work could undoubtedly be done in this

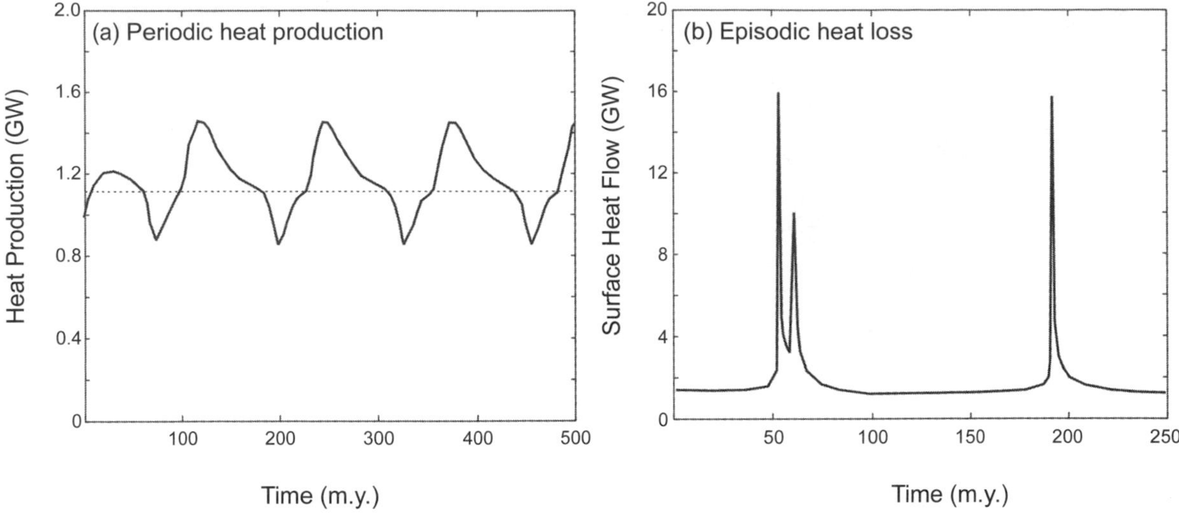

Fig. 6. **(a)** Periodic heat production, modified from *Shoji et al.* (2014). The dashed line gives the equilibrium heat production rate (1.1 GW). (b) Episodic heat loss, modified from *O'Neill and Nimmo* (2010). Heat flux was converted to heat flow by assuming the area of the south polar terrain (10% of the total area).

area; for instance, it would be desirable to self-consistently couple tidal heating into such models, and to explicitly deal with the onset or turn-off of convection as the shell thickens or thins.

5.3. Is the Effective Q of Saturn Small?

If the astrometry results (section 4.2) are correct, then the Q of Saturn is small, about 2000 at Enceladus frequencies. This in turn implies that the equilibrium heat production rate is on the order of 10 GW, roughly consistent with the observed heat flux. The conventional objection to this scenario is that such a low Q_p implies young satellites. But as discussed in section 4.3, if the resonance-locking mechanism is correct, a low present-day Q_p does not require young satellites (Fig. 3b).

As an additional benefit, application of the resonance-locking mechanism means that the equilibrium tidal heat production varies only slowly with satellite semimajor axis — which is not the case with a constant-Q assumption. For example, if the present-day t_{tide} of Enceladus is 5.7 G.y., then neglecting the torque on Dione ($T_o = 0$), equation (4) gives a present-day equilibrium heat production rate of 22 GW. If the bodies were in resonance 2 G.y. ago, t_{tide} = 10.7 G.y. and the corresponding equilibrium heat production rate was 15 GW (Fig. 5). Note that these dissipation rates do not cause any kind of energy crisis for Saturn: 20 GW output over 4 G.y. only represents 2×10^{-7} of Saturn's rotational energy.

We again caution readers that the heat production numbers are subject to considerable uncertainty. First, the astrometrically-derived values (e.g., t_{tide}) themselves are quite uncertain. Second, accounting for the effects of the MMR introduces additional uncertainties. *Meyer and Wisdom* (2007) showed that including the torque on the outer body (equation (4)) can reduce the total heat production by a factor of around 2. Third, the distribution of heat between the inner and outer body cannot be derived without knowing the values of the satellite k_2/Q.

Nonetheless, and despite the uncertainties, in our view the resonance-locking scenario solves many problems. The present-day observed heat loss does not require special pleading, but is simply due to the high equilibrium tidal production at Enceladus. This result by itself does not explain why the heat loss is concentrated in the SPT, or how heat is partitioned between different transport mechanisms. Perhaps more importantly, as we will see below, a high heat flux also helps maintain a liquid ocean for the long term.

6. LONG-TERM THERMAL EVOLUTION

The long-term thermal evolution of Enceladus depends mainly on the competition between heat production and heat loss. The present-day radioactive heat contribution is about 0.3 GW (section 2). Primordial stored heat (e.g., from early ^{26}Al decay) is unlikely to play a role, because the conduction timescale of the silicate core is on the order of 100 m.y., much smaller than the age of the solar system (*Roberts and Nimmo,* 2008). The long-term average tidal heat production is 1.1 GW for a conventional Saturn Q, but could be up to 22 GW in the resonance-locking case (section 5.3).

Global conductive heat losses are given by equation (1). This equation illustrates the difficulty of maintaining a global ocean. For instance, to maintain water at 270 K at the base of a 65-km-thick ice shell (the likely maximum), a power of 8 GW is required. This requirement is only reduced to 5 GW even if the water is at 200 K due to the presence of antifreezes such as salts or ammonia. Surface porosity could help by reducing the near-surface thermal conductivity, but this process is self-limiting (*Besserer et al.,* 2013) and probably makes only a small difference. The real situation is probably even worse, because any heat generated in the shallow ice shell will be conducted out rapidly and will contribute little to the global energy balance. Convective heat transfer, if it is occurring, will also increase the required heat production. These simple calculations make it clear that, for an ocean to survive for more than a few tens of millions of years, the heat production rate must be much greater than the conventional equilibrium value of 1.1 GW.

Figure 7 provides an illustrative demonstration of these results. It models the evolution of a conductive, differentiated Enceladus using the methodology of *Nimmo and Spencer* (2015). The body is initially cold (150 K uniform temperature) and heated only by radioactive decay in the core, insufficient to cause melting. At 1 G.y. the internal heat production in the core is increased by 20 GW to simulate an episode of tidal heating (in reality, tidal heating will likely also occur in the ice shell, but this is more complicated to model). The heat produced exceeds the shell's ability to transport heat conductively (equation (1)) and thus the shell begins to melt. Melting continues until the shell reaches an equilibrium thickness of about 20 km, at which point the heat produced and the conductive heat loss are balanced. If some fraction of the tidal heat were being advected to the near-surface (e.g., through fluid-filled fractures), the equilibrium shell thickness would be larger. In any event, these conditions result in an internal structure similar to that deduced for present-day Enceladus (section 2). Tidal heat production is (arbitrarily) stopped at 2 G.y., at which time the ice shell starts to refreeze, with the freezing process complete in around 100 m.y. A similar calculation using 1.1 GW tidal heating shows that, as expected, melting never occurs, even in the presence of antifreeze.

Of course, this simulation is highly simplified. But it shows that the present-day structure of Enceladus is compatible with an equilibrium situation, where a thin conductive ice shell is being maintained by a high rate of tidal heat production. When this equilibrium was established is unknown: It depends on when Enceladus and Dione entered the 2:1 resonance, and when the orbital frequency became commensurate with some resonant frequency inside Saturn. While the geological evidence suggests that tidal heating has been important over various epochs in Enceladus' history, it does not reveal whether this heating was episodic in nature, or was continuous but changed its location with time.

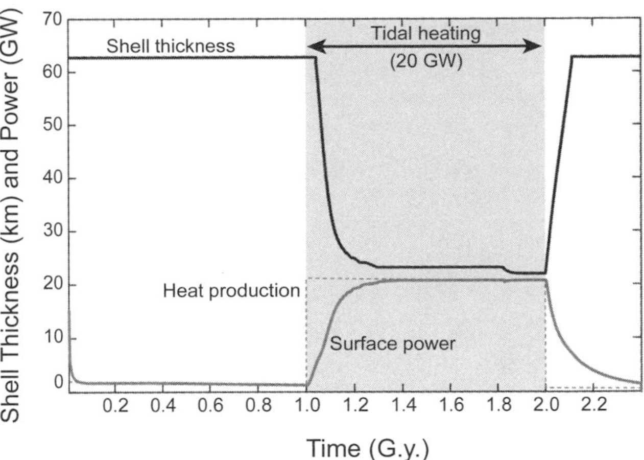

Fig. 7. Thermal evolution of a conductive Enceladus, using the method of *Nimmo and Spencer* (2015). Heat production in the shell is initially radiogenic, and is then supplemented by a tidal contribution of 20 GW over the interval from 1 to 2 G.y. Surface heat flow is plotted by the solid line at bottom, shell thickness by the solid line at top. Ice shell melting commences shortly after tidal heating begins and refreezing is complete within about 0.1 G.y. after the cessation of tidal heating. Model geometry is as in Fig. 1, with thermal conductivities set to 3.5 and 3.0 W m^{-1} K^{-1} for ice and rock, respectively. Initial temperature is 150 K throughout (no ocean initially).

7. DISCUSSION AND CONCLUSIONS

We began by posing two related questions: What is the explanation for the large heat flow observed at Enceladus? And how is the subsurface ocean maintained? At present, the most satisfactory explanation appears to be that the observed heat flow is all being produced at the present day by tidal dissipation, and that the high heating rate is a consequence of the low Q of Saturn. An equilibrium tidal heating rate of 10–20 GW matches the observed heat flow, is sufficient to maintain a global ocean over billions of years, and requires a Q_p consistent with astrometry. Because of the apparently small shell thickness (≈25 km) it appears likely that the shell is conductive, not convective. This helps the ocean to persist, and is probably required to avoid rapid removal of shell topography by lateral flow. If the resonance-locking mechanism of *Fuller et al.* (2016) is correct, a low Q_p does not necessarily require young satellites and also implies a roughly constant rate of equilibrium heat production (Fig. 5).

The resonance-locking model provides a natural way of explaining Enceladus' thin ice shell and high heat flow without requiring either the satellite itself or the subsurface ocean to be recently formed. It thus presents a strongly uniformitarian picture, in contrast to catastrophist ideas for recent formation of the inner saturnian satellites (*Ćuk et al.*, 2016; *Asphaug and Reufer,* 2013). Which of these two pictures is closer to reality will become clear in time, but at least for Enceladus we favor uniformitarianism.

It is currently unclear where the bulk of the dissipation is located. A conductive shell will dissipate mostly toward its base (where the ice is warmest). Dissipation in the deep ocean seems unlikely, but significant heating in the low-density silicate core may also be occurring. It is also not yet clear whether the heat measured at the surface is being transferred mainly by upward advection of warm material at the tiger stripes, conductive heat transfer between the tiger stripes, or a combination of the two.

Several other questions remain to explore. An important question is whether there is any independent evidence for when and how Enceladus (and the other satellites) formed (see the chapter in this volume by McKinnon et al.). The detection of apex-antapex crater asymmetries by *Hirata* (2016) suggests that at least some satellites experienced mostly heliocentric impacts, and did not reform after a catastrophic disruption event (which would produce planetocentric debris); more work on this subject is needed.

The resonance-locking mechanism itself obviously needs to be explored more deeply, but so too do its consequences for the satellites. Equilibrium tidal heating only determines the total rate of heat production, not its location. In order to understand how the total amount of heating is distributed between Enceladus and Dione, coupled thermal-orbital models will need to be developed (cf. section 5.1). Such models should also be able to investigate whether periodic tidal heating is a possibility, as well as the effects of ocean freezing on tidal and thermal evolution. Forward models, based on plausible initial satellite positions, will be required to try and pin down when the current Enceladus:Dione MMR was established. And the ease or difficulty of passing through previously encountered resonances provides additional constraints on the system. Of course, many of these questions are equally applicable to other satellites, such as Tethys, where ancient heating episodes are inferred to have taken place (*Giese et al.,* 2007; Schenk et al., this volume).

Although tidal heating is generally higher at the poles than the equator, an unresolved issue is why current heat loss is concentrated only at one pole. As mentioned in section 3.3.2, there are several ways in which heat production could become focused in one region, while an alternative (section 5.2) is that in a convecting system the heat loss can be spatially variable. Incorporating the geological evidence — which suggests a moving locus of high heat flow — will be important here. If the geology turns out to support several discrete heating episodes, then this makes investigation of potential orbital-thermal feedbacks more pressing.

No further data on Enceladus will be returned by Cassini, and it is likely that the wait for the next spacecraft to visit Saturn will be a long one. Further analysis of the geological and geophysical evidence will undoubtedly be carried out, and Earth-based measurements will continue to play a role. The most important advances, however, are likely to be theoretical in nature. Problems such as dissipation in gas giants and the coupled thermal-orbital evolution of satellites will require sustained effort to tackle, and will keep the next generation of theorists busy for some time.

Acknowledgments. This work was partially supported by NASA grant NNX13AG02G to F.N., who acknowledges helpful discussions with J. Fuller and R. Jacobson. M.B. acknowledges support from the Czech Science Foundation project no. 14-04145S. We thank two anonymous reviewers for helpful comments.

REFERENCES

Asphaug E. and Reufer A. (2013) Late origin of the Saturn system. *Icarus, 223,* 544–565.

Barr A. C. (2008) Mobile lid convection beneath Enceladus' south polar terrain. *J. Geophys. Res., 113,* E07009, DOI: 10.1029/2008JE003114.

Barr A. C. and McKinnon W. B. (2007) Convection in Enceladus' ice shell: Conditions for initiation. *Geophys. Res. Lett., 34,* L09202, DOI: 10.1029/2006GL028799.

Běhounková M., Tobie G., Choblet G., and Čadek O. (2010) Coupling mantle convection and tidal dissipation: Applications to Enceladus and Earth-like planets. *J. Geophys. Res., 115,* E09011, DOI: 10.1029/2009JE003564.

Běhounková M., Tobie G., Choblet G., and Čadek O. (2012) Tidally-induced melting events as the origin of south pole activity on Enceladus. *Icarus, 219(2),* 655–664.

Běhounková M., Tobie G., Čadek O., Choblet G., Porco C. P., and Nimmo F. (2015) Timing of water plume eruptions on Enceladus explained by interior viscosity structure. *Nature Geosci., 8,* 601–604, DOI: 10.1038/ngeo2475.

Besserer J., Nimmo F., Roberts J. H., and Pappalardo R. T. (2013) Convection-driven compaction as a possible origin of Enceladus's long wavelength topography. *J. Geophys. Res., 118,* 908–915, DOI: 10.1002/jgre.20079.

Beuthe M. (2016) Crustal control of dissipative ocean tides in Enceladus and other icy moons. *Icarus, 280,* 278–299.

Beuthe M., Rivoldini A., and Trinh A. (2016) Enceladus' and Dione's floating ice shells supported by minimum stress isostasy. *Geophys. Res. Lett., 43,* 10088–10096.

Bland M. T., Beyer R. A., and Showman A. P. (2007) Unstable extension of Enceladus's lithosphere. *Icarus, 192,* 92–105.

Bland M. T., Singer K. N., McKinnon W. B., and Schenk P. M. (2012) Enceladus' extreme heat flux as revealed by its relaxed craters. *Geophys. Res. Lett., 39,* L17204, DOI: 10.1029/2012GL052736.

Bland M. T., McKinnon W. B., and Schenk P. M. (2015) Constraining the heat flux between Enceladus's tiger stripes: Numerical modeling of funiscular plains formation. *Icarus, 39,* 232–245.

Čadek O., Tobie G., van Hoolst T., Masse M., Choblet G., Lefevre A., Mitri G., Baland R.-M., Běhounková M., Bourgeois O., and Trinh A. (2016) Enceladus's internal ocean and ice shell constrained from Cassini gravity, shape, and libration data. *Geophys. Res. Lett. 46,* 5653–5660.

Čadek O., Běhounková M., Tobie G., and Choblet G. (2017) Viscoelastic relaxation of Enceladus's ice shell. *Icarus, 291,* 31–35.

Chen E. and Nimmo F. (2011) Obliquity tides do not significantly heat Enceladus. *Icarus, 214(2),* 779–781.

Collins G. C. and Goodman J. C. (2007) Enceladus' south polar sea. *Icarus, 189,* 72–82.

Crow-Willard E. N. and Pappalardo R. T. (2015) Structural mapping of Enceladus and implications for formation of tectonized regions. *J. Geophys. Res., 120,* 928–950, DOI: 10.1002/2015JE004818.

Ćuk M., Dones L., and Nesvorný D. (2016) Dynamical evidence for a late formation of Saturn's moons. *Astrophys. J., 820,* 97.

Dermott S. F., Malhotra R., and Murray C. D. (1988) Dynamics of the uranian and saturnian satellite systems — A chaotic route to melting Miranda? *Icarus, 76,* 295–334.

Findley W. N., Lai J. S., and Onaran K. (1989) *Creep and Relaxation of Nonlinear Viscoelastic Materials.* Courier Corp., 971 pp.

Fuller J., Luan J., and Quataert E. (2016) Resonance locking as the source of rapid tidal migration in the Jupiter and Saturn moon systems. *Mon. Not. R. Astron. Soc., 458,* 3867–3879.

Gavrilov S. V. and Zharkov V. N. (1977) Love numbers of the giant planets. *Icarus, 32,* 443–449.

Giese B., Wagner R., Neukum G., Helfenstein P., and Thomas P. C. (2007) Tethys: Lithospheric thickness and heat flux from flexurally supported topography at Ithaca Chasma. *Geophys. Res. Lett., 34,* L21203, DOI: 10.1029/2007GL031467.

Giese B., Wagner R., Hussmann H., Neukum G., Perry J., Helfenstein P., and Thomas P. C. (2008) Enceladus: An estimate of heat flux and lithospheric thickness from flexurally supported topography. *Geophys. Res. Lett., 35,* L24204, DOI: 10.1029/2008GL036149.

Han L., Tobie G., and Showman A. P. (2012) The impact of a weak south pole on thermal convection in Enceladus ice shell. *Icarus, 218(1),* 320–330.

Hand K., Khurana K., and Chyba C. (2011) Joule heating of the south polar terrain on Enceladus. *J. Geophys. Res., 116,* E04010, DOI: 10.1029/2010JE003776.

Hay H. C. F. C. and Matsuyama I. (2017) Numerically modelling tidal dissipation with bottom drag in the oceans of Titan and Enceladus. *Icarus, 281,* 342–356.

Hedman M. M., Gosmeyer C. M., Nicholson P. D., Sotin C., Brown

R. H., Clark R. N., Baines K. H., Buratti B. J., and Showalter M. R. (2013) An observed correlation between plume activity and tidal stresses on Enceladus. *Nature, 500,* 182–184.

Hirata N. (2016) Differential impact cratering of Saturn's satellites by heliocentric impactors. *J. Geophys. Res., 121,* 111–117, DOI: 10.1002/2015JE004940.

Howett C. J. A., Spencer J. R., Pearl J., and Segura M. (2011) High heat flow from Enceladus' south polar region measured using 10–600 cm^{-1} Cassini/CIRS data. *J. Geophys. Res., 116,* E03003, DOI: 10.1029/2010JE003718.

Howett C., Spencer J. R., Spencer D., Verbiscer A., Hurford T., and Segura M. (2013) Enceladus' enigmatic heat flow. Abstract P53E-06 presented at 2013 Fall Meeting, AGU, San Francisco, California, 9–13 December.

Hsu H.-W., Postberg F., Sekine Y., Shibuya T., Kempf S., Hoŕanyi M., Juhász A., Altobelli N., Suzuki K., Masaki Y., Kuwatani T., Tachibana S., Sirono S.-I., Moragas-Klostermeyer G., and Srama R. (2015) Ongoing hydrothermal activities within Enceladus. *Nature, 519,* 207–210.

Hurford T. A., Helfenstein P., Hoppa G. V., Greenberg R., and Bills B. G. (2007) Eruptions arising from tidally controlled periodic openings of rifts on Enceladus. *Nature, 447,* 292–294.

Hussmann H. and Spohn T. (2004) Thermal-orbital evolution of Io and Europa. *Icarus, 171,* 391–410, DOI: 10.1016/j.icarus.2004.05.020.

Iess L., Stevenson D. J., Parisi M., Hemingway D., Jacobson R. A., Lunine J. I., Nimmo F., Armstrong J. W., Asmar S. W., Ducci M., and Tortora P. (2014) The gravity field and interior structure of Enceladus. *Science, 344,* 78–80.

Ingersoll A. P. and Pankine A. A. (2010) Subsurface heat transfer on Enceladus: Conditions under which melting occurs. *Icarus, 206(2),* 594–607.

Kamata S. and Nimmo F. (2017) Interior thermal state of Enceladus inferred from the viscoelastic state of the ice shell. *Icarus, 284,* 387–393.

Kirchoff M. R. and Schenk P. (2009) Crater modification and geologic activity in Enceladus' heavily cratered plains: Evidence from the impact crater distribution. *Icarus, 202,* 656–668, DOI: 10.1016/j.icarus.2009.03.034.

Kite E. S. and Rubin A. M. (2016) Sustained eruptions on Enceladus explained by turbulent dissipation in tiger stripes. *Proc. Natl. Acad. Sci., 113,* 3972– 3975.

Lainey V., Arlot J. E., and van Hoolst T. (2009) Strong tidal dissipation in Io and Jupiter from astrometric observations. *Nature, 459,* 957–959.

Lainey V., Karatekin Ö., Desmars J., Charnoz S., Arlot J.-E., Emelyanov N., Le Poncin-Lafitte C., Mathis S., Remus F., Tobie G., and Zahn J.-P. (2012) Strong tidal dissipation in Saturn and constraints on Enceladus' thermal state from astrometry. *Astrophys. J., 752,* 14.

Lainey V., Jacobson R. A., Tajeddine R., Cooper N. J., Murray C., Robert V., Tobie G., Guillot T., Mathis S., Remus F., Desmars J., Arlot J.-E., De Cuyper J.-P., Dehant V., Pascu D., Thuillot W., Le Poncin-Lafitte C., and Zahn J.-P. (2017) New constraints on Saturn's interior from Cassini astrometric data. *Icarus, 281,* 286–296.

Le Gall A., Leyrat C., Janssen M. A., Choblet G., Tobie G., Bourgeois O., Lucas A., Sotin C., Howett C., Kirk R., Lorenz R. D., West R. D., Stolzenbach A., Masse M., Hayes A. H., Bonnefoy L., Veyssiere G., and Paganelli F. (2017) Thermally anomalous features in the subsurface of Enceladus's south polar terrain. *Nature Astron., 1,* 0063.

McCarthy C. and Castillo-Rogez J. C. (2013) Planetary ices attenuation properties. In *The Science of Solar System Ices* (M. S. Gudipati and J. Castillo-Rogez, eds.), pp. 183–225. Springer, New York.

McKinnon W. B. (2015) Effect of Enceladus's rapid synchronous spin on interpretation of Cassini gravity. *Geophys. Res. Lett. 42,* 2137–2143.

Meyer J. and Wisdom J. (2007) Tidal heating in Enceladus. *Icarus, 188,* 535–539.

Meyer J. and Wisdom J. (2008a) Tidal evolution of Mimas, Enceladus, and Dione. *Icarus, 193,* 213–223.

Meyer J. and Wisdom J. (2008b) Episodic volcanism on Enceladus: Application of the Ojakangas-Stevenson model. *Icarus, 198,* 178–180.

Mitri G. and Showman A. P. (2008) A model for the temperature-dependence of tidal dissipation in convective plumes on icy satellites: Implications for Europa and Enceladus. *Icarus, 195(2),* 758–764.

Murray C. D. and Dermott S. F. (1999) *Solar System Dynamics.* Cambridge Univ., New York.

Nahm A. L. and Kattenhorn S. A. (2015) A unified nomenclature for tectonic structures on the surface of Enceladus. *Icarus, 258,* 67–81, DOI: 10.1016/j.icarus.2015.06.009.

Nimmo F. and Pappalardo R. T. (2006) Diapir-induced reorientation of Saturn's moon Enceladus. *Nature, 441,* 614–616.

Nimmo F. and Spencer J. R. (2015) Powering Triton's recent geological activity by obliquity tides: Implications for Pluto geology. *Icarus, 246,* 2–10.

Nimmo F., Spencer J. R., Pappalardo R. T., and Mullen M. E. (2007) Shear heating as the origin of plumes and heat flux on Enceladus. *Nature, 447,* 289–291.

Nimmo F., Bills B. G., and Thomas P. C. (2011) Geophysical implications of the long-wavelength topography of the saturnian satellites. *J. Geophys. Res., 116,* E11001.

Nimmo F., Porco C., and Mitchell C. (2014) Tidally modulated eruptions on Enceladus: Cassini ISS observations and models. *Astron. J., 148,* 46.

Ogilvie G. I. and Lin D. N. C. (2004) Tidal dissipation in rotating giant planets. *Astrophys. J., 610,* 477–509, DOI: 10.1086/421454.

Ojakangas G. W. and Stevenson D. J. (1986) Episodic volcanism of tidally heated satellites with application to Io. *Icarus, 66,* 341–358.

O'Neill C. and Nimmo F. (2010) The role of episodic overturn in generating the surface geology and heat flow on Enceladus. *Nature Geosci., 3,* 88–91.

Peale S. (1999) Origin and evolution of the natural satellites. *Annu. Rev. Astron. Astrophys., 37(1),* 533– 602.

Peale S. J. and Cassen P. (1978) Contribution of tidal dissipation to lunar thermal history. *Icarus, 36,* 245–269.

Porco C. C., Helfenstein P., Thomas P. C., Ingersoll A. P., Wisdom J., West R., Neukum G., Denk T., Wagner R., Roatsch T., Kieffer S., Turtle E., McEwen A., Johnson T. V., Rathbun J., Veverka J., Wilson D., Perry J., Spitale J., Brahic A., Burns J. A., DelGenio A. D., Dones L., Murray C. D., and Squyres S. (2006) Cassini observes the active south pole of Enceladus. *Science, 311,* 1393–1401, DOI: 10.1126/science.1123013.

Postberg F., Kempf S., Schmidt J., Brilliantov N., Beinsen A., Abel B., Buck U., and Srama R. (2009) Sodium salts in E-ring ice grains from an ocean below the surface of Enceladus. *Nature, 459,* 1098–1101, DOI: 10.1038/nature08046.

Postberg F., Schmidt J., Hillier J., Kempf S., and Srama R. (2011) A salt-water reservoir as the source of a compositionally stratified plume on Enceladus. *Nature, 474,* 620–622, DOI: 10.1038/nature10175.

Roberts J. H. (2015) The fluffy core of Enceladus. *Icarus, 258,* 54–66.

Roberts J. H. (2016) Last refuge of the scoundrel: Effects of a giant impact on the south polar region of Enceladus. Abstract P32A-08 presented at 2016 Fall Meeting, AGU, San Francisco, California, 11–15 December.

Roberts J. H. and Nimmo F. (2008) Tidal heating and the long-term stability of a subsurface ocean on Enceladus. *Icarus, 194,* 675–689.

Ross M. and Schubert G. (1989) Viscoelastic models of tidal heating in Enceladus. *Icarus, 78,* 90–101.

Rozel A., Besserer J., Golabek G. J., Kaplan M., and Tackley P. J. (2014) Self-consistent generation of single-plume state for Enceladus using non-Newtonian rheology. *J. Geophys. Res., 119(3),* 416–439, DOI: 10.1002/2013JE004473.

Sabadini R., Yuen D. A., and Boschi E. (1982) Polar wandering and the forced responses of a rotating, multilayered, viscoelastic planet. *J. Geophys. Res., 87,* 2885–2903.

Schenk P. M. and McKinnon W. B. (2009) One-hundred-km-scale basins on Enceladus: Evidence for an active ice shell. *Geophys. Res. Lett., 36,* L16202, DOI: 10.1029/2009GL039916.

Schubert G., Anderson J. D., Travis B. J., and Palguta J. (2007) Enceladus: Present internal structure and differentiation by early and long-term radiogenic heating. *Icarus, 188,* 345–355.

Segatz M., Spohn T., Ross M. N., and Schubert G. (1988) Tidal dissipation, surface heat flow, and figure of viscoelastic models of Io. *Icarus, 75,* 187–206.

Shoji D., Hussmann H., Sohl F., and Kurita K. (2014) Non-steady state tidal heating of Enceladus. *Icarus, 235,* 75–85.

Showman A. P., Han L., and Hubbard W. B. (2013) The effect of an asymmetric core on convection in Enceladus' ice shell: Implications for south polar tectonics and heat flux. *Geophys. Res. Lett., 40,* 5610–5614.

Sotin C., Head J. W. III, and Tobie G. (2002) Europa: Tidal heating of upwelling thermal plumes and the origin of lenticulae and chaos melting. *Geophys. Res. Lett., 29,* DOI: 10.1029/2001GL013844.

Souček O., Hron J., Běhounková M., and Čadek O. (2016) Effect of the tiger stripes on the deformation of Saturn's moon Enceladus. *Geophys. Res. Lett., 43(14),* 7417–7423, DOI: 10.1002/2016GL069415.

Spencer J. S. and Nimmo F. (2013) Enceladus: An active ice world in the Saturn system. *Annu. Rev. Earth Planet. Sci., 41,* 693–717.

Spencer J. R., Pearl J. C., Segura M., Flasar F. M., Mamoutkine A., Romani P., Buratti B. J., Hendrix A. R., Spilker L. J., and Lopes R. M. C. (2006) Cassini encounters Enceladus: Background and the discovery of a south polar hot spot. *Science, 311,* 1401–1405.

Spencer J. R., Verbiscer A., and Hurford T. A. (2013) Enceladus heat flow from high spatial resolution thermal emission observations. *EPSC Abstracts, Vol. 8,* EPSC2013-8401.

Spohn T. and Schubert G. (2003) Oceans in the icy Galilean satellites of Jupiter? *Icarus, 161,* 456–467.

Stegman D. R., Freeman J., and May D. A. (2009) Origin of ice diapirism, true polar wander, subsurface ocean, and tiger stripes of Enceladus driven by compositional convection. *Icarus, 202,* 669–680.

Thomas P. C., Burns J. A., Helfenstein P., Squyres S., Veverka J., Porco C., Turtle E. P., McEwen A., Denk T., Giese B., Roatsch T., Johnson T. V., and Jacobson R. A. (2007) Shapes of the saturnian icy satellites and their significance. *Icarus, 190,* 573–584.

Thomas P. C., Tajeddine R., Tiscareno M. S., Burns J. A., Joseph J., Loredo T. J., Helfenstein P., and Porco C. (2016) Enceladus's measured physical libration requires a global subsurface ocean. *Icarus, 264,* 37–47.

Tobie G., Choblet G., and Sotin C. (2003) Tidally heated convection: Constraints on Europa's ice shell thickness. *J. Geophys. Res., 108,* 5124, DOI: 10.1029/2003JE002099.

Tobie G., Mocquet A., and Sotin C. (2005) Tidal dissipation within large icy satellites: Applications to Europa and Titan. *Icarus, 177,* 534–549.

Tyler R. (2009) Ocean tides heat Enceladus. *Geophys. Res. Lett., 36(15),* DOI: 10.1029/2009GL038300.

Tyler R. (2011) Tidal dynamical considerations constrain the state of an ocean on Enceladus. *Icarus, 211(1),* 770–779.

Vance S., Harnmeijer J., Kimura J., Hussmann H., de- Martin B., and Brown J. M. (2007) Hydrothermal systems in small ocean planets. *Astrobiology, 7,* 987–1005.

van Hoolst T., Baland R.-M., and Trinh A. (2016) The diurnal libration and interior structure of Enceladus. *Icarus, 277,* 311–318.

Wahr J., Pappalardo R. T., Barr A. C., Crawford Z., Gleeson D., Stempel M. M., Mullen M. E., and Collins G. C. (2009) Modeling stresses on satellites due to nonsynchronous rotation and orbital eccentricity using gravitational potential theory. *Icarus, 200(1),* 188–206.

Waite J. H., Glein C. R., Perryman R. S., Teolis B. D., Magee B. A., Miller G., Grimes J., Perry M. E., Miller K. E., Bouquet A., Lunine J. I., Brockwell T., and Bolton S. (2017) Cassini finds molecular hydrogen in the Enceladus plume: Evidence for hydrothermal processes. *Science, 356,* 155–159.

Wang H. and Stevenson D. J. (2000) Convection and internal melting of Europa's ice shell. *Lunar and Planetary Science XXX,* Abstract #1293. Lunar and Planetary Institute, Houston.

Williams J. G. and Boggs D. (2015) Tides on the Moon: Theory and determination of dissipation. *J. Geophys. Res., 120(4),* 689–724, DOI: 10.1002/2014JE004755.

Wu Y. (2005) Origin of tidal dissipation in Jupiter. II. The value of Q. *Astrophys. J., 635,* 688–710, DOI: 10.1086/497355

Zahnle K., Schenk P., Levison H., and Dones L. (2003) Cratering rates in the outer solar system. *Icarus, 163,* 263–289, DOI: 10.1016/S0019-1035(03)00048-4.

Zhang K. and Nimmo F. (2009) Recent orbital evolution and the internal structures of Enceladus and Dione. *Icarus, 204,* 597–609.

Zschau J. (1978) Tidal friction in the solid Earth: Loading tides versus body tides. In *Tidal Friction and the Earth's Rotation* (P. Brosche and J. Sündermann, eds.), pp. 62–94. Springer-Verlag, Berlin.

Patterson G. W., Kattenhorn S. A., Helfenstein P., Collins G. C., and Pappalardo R. T. (2018) The geology of Enceladus. In *Enceladus and the Icy Moons of Saturn* (P. M. Schenk et al., eds.), pp. 95–125. Univ. of Arizona, Tucson, DOI: 10.2458/azu_uapress_9780816537075-ch006.

The Geology of Enceladus

G. Wesley Patterson
Johns Hopkins University Applied Physics Laboratory

Simon A. Kattenhorn
University of Alaska Anchorage

Paul Helfenstein
Cornell University

Geoffrey C. Collins
Wheaton College

Robert T. Pappalardo
Jet Propulsion Laboratory

The intent of this chapter is to summarize the current state of knowledge regarding the geology of Enceladus. This is accomplished through descriptions of the data sources used to analyze its icy surface, the distribution and morphology of craters, the current classification of the varied tectonic features observed on the satellite, the regional- and local-scale geological relationships that constrain Enceladus' surface evolution, and the possible endogenic and exogenic driving mechanisms for explaining that history. Key questions addressed in this chapter include: (1) How did the south polar terrain (SPT) form and how has it evolved with time? (2) How are the tectonized terrains of the leading and trailing hemispheres related to the SPT? (3) What do the heavily cratered surfaces of Enceladus tell us about its early history?

1. INTRODUCTION

The unusual geology of the saturnian moon Enceladus has drawn the interest of planetary scientists since the first images of the satellite were returned by the Voyager spacecraft during their encounters with Saturn (*Smith et al.,* 1982). Those images revealed a surface with evidence of tectonic activity and episodic partial resurfacing (*Squyres et al.,* 1983), suggesting a geologic history that is remarkably complex for a moon with a mean radius of ~250 km. The earliest geologic map of Enceladus used the Voyager 2 spacecraft's highest resolution coverage (~1 km/pixel) of the satellite, which included the Saturn facing hemisphere and the north polar region (*Smith et al.,* 1982). That coverage was sufficient to show that Enceladus' surface geology is characterized by distinct provinces that show abrupt changes in crater densities and which are confined by major tectonic contacts. The geological provinces identified were cratered terrain, cratered plains, ridged plains, and smooth plains. Several of these provinces were further subdivided based on differences in crater density and tectonic characteristics (*Smith et al.,* 1982; *Squyres et al.,* 1983; *Passey,* 1983; *Plescia and Boyce,* 1983).

The Cassini mission to Saturn provided a wealth of additional information regarding the diverse geology of Enceladus. Most notable was the detection of an active plume containing water vapor, dust, and other materials erupting as jets from fractures near its south pole, informally referred to as "tiger stripes" (*Porco et al.,* 2006; *Dougherty et al.,* 2006; *Hansen et al.,* 2006; *Spencer et al.,* 2006). The tiger stripes, along with the terrain that surround them, are bound by a circumpolar chain of tectonic structures that define a geologic province referred to as the south polar terrain (SPT) (*Porco et al.,* 2006). Analyses of this region have revolutionized our understanding of the evolution of icy satellite surfaces (e.g., *Spencer et al.,* 2006, 2009; *Nimmo and Pappalardo,* 2006; *Collins and Goodman,* 2007; *Barr,* 2008; *Patthoff and Kattenhorn,* 2011). However, Enceladus' south polar terrain tells only the most recent part of the story of this unique icy body.

An initial global-scale geologic map of Enceladus (*Spencer et al.,* 2009) using image data available at the time (*Roatsch et al.,* 2008) suggested the satellite could be divided into four geologic provinces: cratered terrain, western (leading) hemisphere fractured plains, eastern (trailing) hemisphere fractured plains, and SPT. The resolution of image data used in that effort was uneven, with higher-resolution coverage available for the southern and equatorial trailing hemispheres. As a result, the SPT was a particular focus of *Spencer et al.* (2009), with numerous geologic subunits and distinct structures described for the region (Fig. 1; Table 1). Another global-scale geologic map of Enceladus (Fig. 2), using better-resolution data of the surface (*Roatsch et al.,* 2013), was produced by *Crow-Willard and Pappalardo* (2015). That map also divided the surface into four geologic provinces that are broadly consistent with those defined by *Spencer et al.* (2009): cratered terrain, trailing hemisphere terrain, leading hemisphere terrain, and the SPT. Each province was further subdivided into geologic units based on

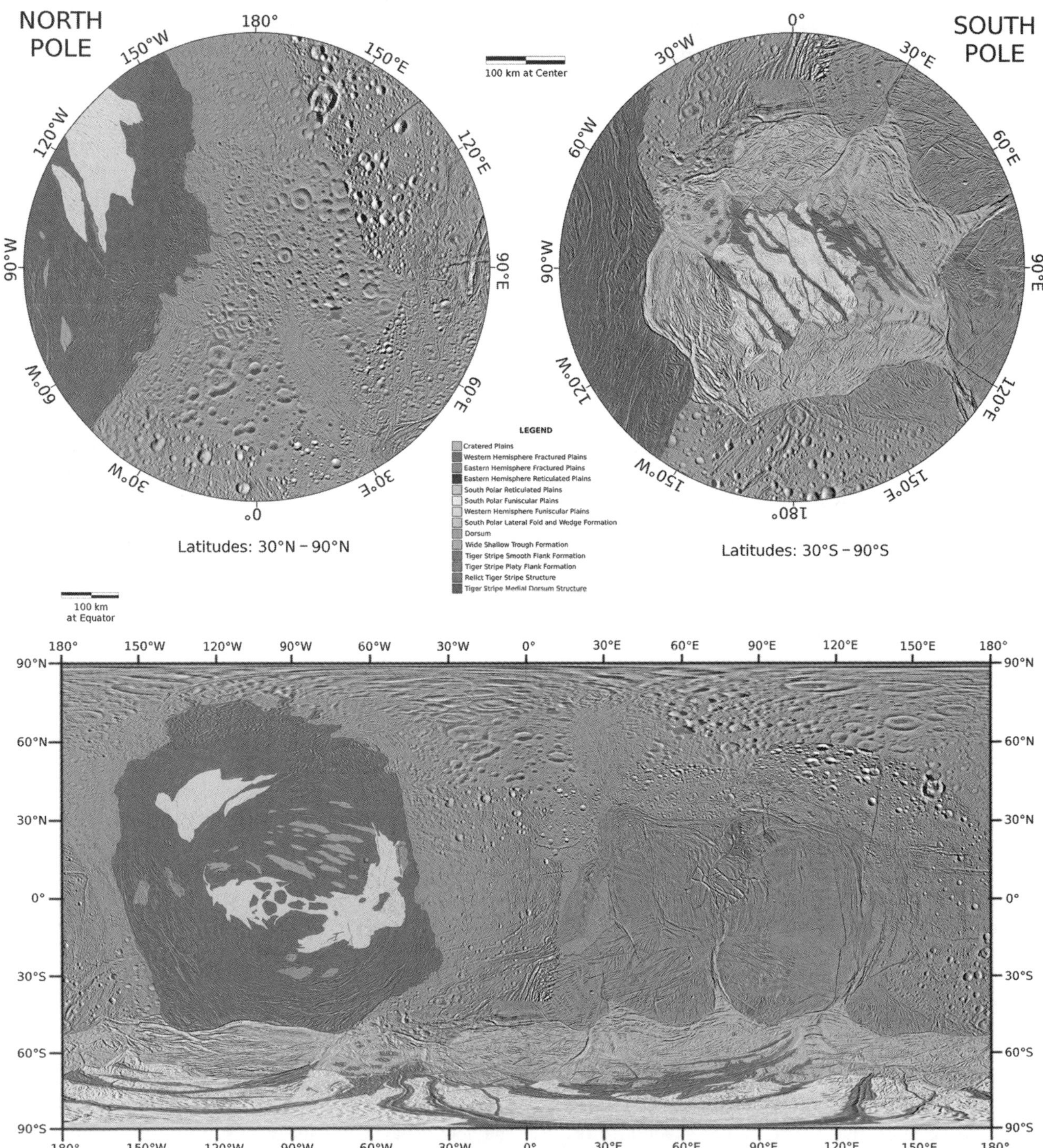

Fig. 1. See Plate 10 for color version. Polar stereographic (above) and simple cylindrical (below) image mosaic (*Becker et al.,* 2016) with superposed terrain unit map from *Spencer et al.* (2009); updated according to *Helfenstein et al.* (2010a,b).

TABLE 1. Comparison of mapped geologic units on Enceladus from *Spencer et al.* (2009) and *Helfenstein et al.* (2010a,b) with mapped units from *Crow-Willard and Pappalardo* (2015).

Spencer et al. (2009), *Helfenstein et al.* (2010a,b)			*Crow-Willard and Pappalardo* (2015)		
Province	Sub-Units		Province	Sub-Units	
	Name	Abbrev.		Name	Abbrev.
Cratered terrain	Cratered plains	CR_p	Cratered terrain	Cratered plains	cp_1
				Subdued cratered plains	cp_2
Eastern hemisphere terrain (EHT)	Eastern hemisphere fractured plains	EH_{frp}	Trailing hemisphere terrain (THT)	Striated plains	sp
	Dorsa	EH_{dsm}		Northern lineated	nl
	Relict tiger stripe structure	TS_{rct}		Trailing hemisphere curvilinear	cl_1
	Eastern hemisphere reticulated plains	EH_{rtp}		Ridged	r
				Transitional	t
Western hemisphere terrain (WHT)	Western hemisphere fractured plains	WH_{frp}*	Leading hemisphere terrain (LHT)	Leading hemisphere smooth	ls
	Western hemisphere funiscular plains	WH_{fun}*		Central leading hemisphere	clh
	Wide shallow trough formation	WH_{wt}†		Leading hemisphere curvilinear	cl_2
	Dorsa	WH_{dsm}			
South polar terrain (SPT)	South polar funiscular plains	SP_{fun}	South polar terrain (SPT)	Central south polar	csp
	Tiger stripe smooth flank formation	TS_{sfk}		Southern curvilinear	cl_3
	Tiger stripe platy flank formation	TS_{pfk}			
	Relict tiger stripe structure	TS_{rct}			
	Tiger stripe medial dorsum structure	TS_{mds}			
	Dorsa	SP_{dsm}			
	South polar reticulated plains	SP_{rtp}			
	South polar lateral fold-and-wedge formation	SP_{fwf}			
	Dorsa	SP_{dsm}			
	Relict tiger stripe structure	TS_{rct}			

**Helfenstein et al.* (2010a,b).

†Unit as described in *Crow-Willard and Pappalardo* (2015).

factors including morphology, the orientations of structures, cross-cutting relationships, etc. (Table 1).

In this chapter, we review the geologic history of Enceladus' observable surface by using available image data to examine the morphology, distribution, and density of observed impact craters on the surface; the distribution, orientations, and cross-cutting relationships of tectonic features across the surface; the local-scale characteristics of key morphological features; and the complex geological relationships among the SPT and other geologic terrains on Enceladus.

2. DATA

Four spacecraft have visited Saturn. The first, Pioneer 11, was capable of obtaining digital imaging data with its imaging photopolarimeter (IPP) instrument, but it did not acquire any surface-resolved images of Enceladus during its flyby of Saturn in September 1979. The Voyager 1 and Voyager 2 flybys of Saturn in November 1980 and August 1981, respectively, imaged the surface of Enceladus' northern and equatorial trailing hemisphere at scales of kilometers per pixel (*Smith et al.,* 1981, 1982). The Cassini mission operated in Saturn orbit from July 2004 through September 2017, acquiring image data at resolutions as high as tens of centimeters per pixel. These data, and products derived from them (e.g., global control point networks, controlled mosaics, stereophotogrammetric data, photoclinometric data, etc.), have provided the primary source of information used to characterize the geology of the satellite and decipher the formation and evolution of surface features.

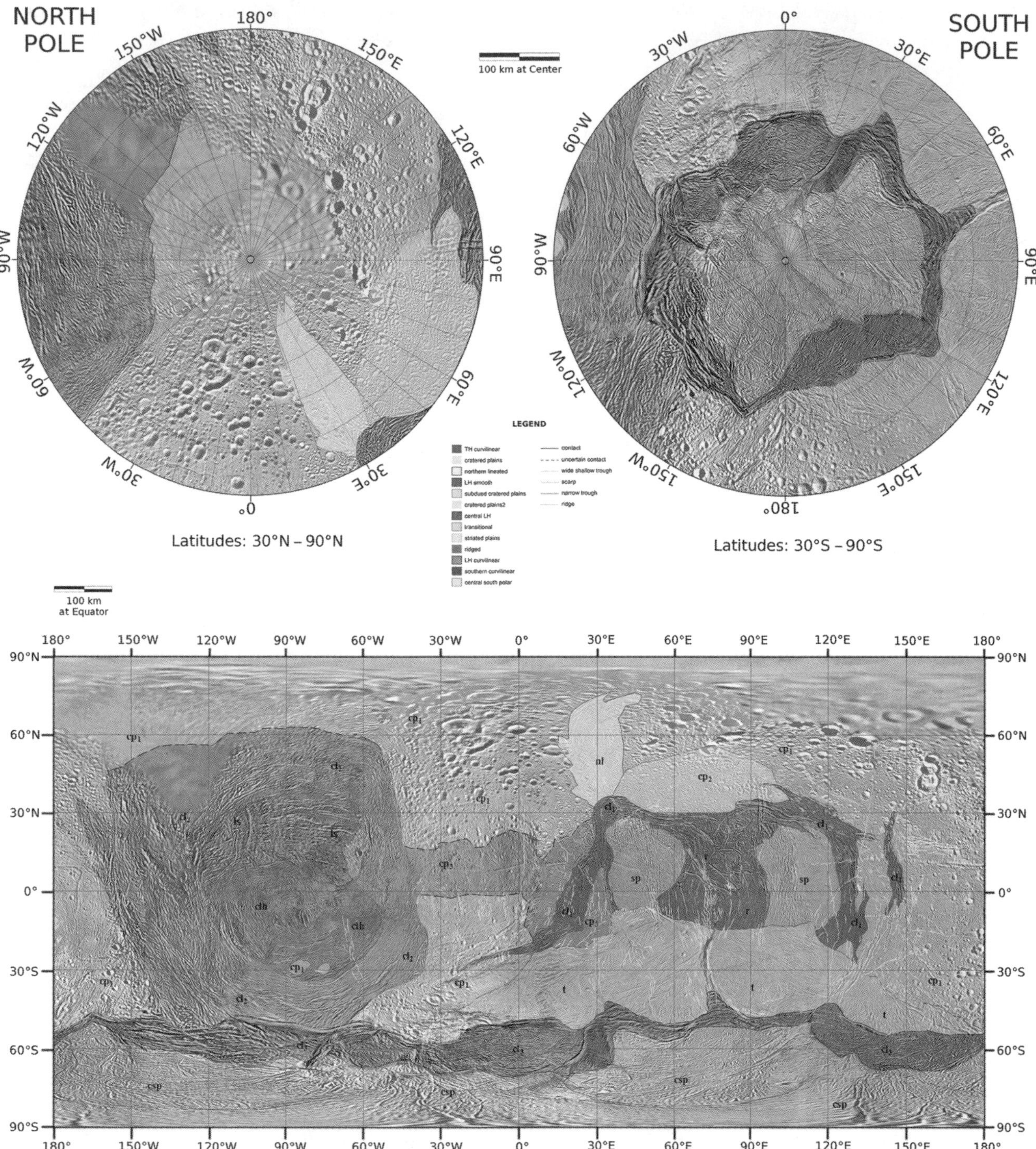

Fig. 2. See Plate 11 for color version. Polar stereographic (above) and simple cylindrical (below) image mosaic (*Becker et al.,* 2016) with superposed terrain unit map from *Crow-Willard and Pappalardo* (2015).

2.1. Image Data

The Voyager 1 and 2 spacecraft were each equipped with wide and narrow angle vidicon cameras (*Benesh and Jepsen,* 1978) that had broadband spectral sensitivity ranging from 280 nm (UV) to 640 nm (orange). Multispectral imaging was obtained using up to eight filters. Voyager 1 returned 22 images of Enceladus. They provided an equatorial view of the trailing hemisphere but the disk of the satellite was less than 100 pixels across in the best of the images. Voyager 2 returned 65 images of Enceladus. The images were used to create a controlled photomosaic of approximately one-third of the surface at 1.1 km/pixel resolution, centered on the northern portion of Enceladus' trailing hemisphere.

The Cassini spacecraft was equipped with Narrow Angle (NA) and Wide Angle (WA) CCD cameras with a broadband spectral sensitivity ranging from UV (200 nm) to the near-IR (1100 nm), collectively referred to as the Imaging Science Subsystem (ISS) (*Porco et al.,* 2004). The NA camera had a 2000-mm focal-length and a field-of-view of 6 mrad, while the WA camera had a 200-mm focal-length and 60-mrad field of view. The ISS WA camera was bore-sighted with the NA camera to provide context for scenes in which the NA and WA cameras are simultaneously shuttered (referred to as BOTSIM images). Each camera had a 1024 × 1024 CCD and dual 12-position filter wheels that operated in tandem within each camera to provide broadband multispectral imaging capability. Additional details regarding the ISS instrument and its performance can be found in *Porco et al.* (2004) and *West et al.* (2010).

The Cassini ISS returned 1981 images with resolutions better than ~1 km/pixel (Appendix A). A control network and global-scale image mosaic of Enceladus, based on subsets of those images, was produced and updated over the course of the Cassini mission (*Roatsch et al.,* 2006, 2008, 2013; *Becker et al.,* 2016; Bland *et al.,* in preparation, 2018). The most recent control network produced for Enceladus (*Becker et al.,* 2016) incorporates position information gathered from 586 ISS images with resolutions between 50 and 500 m/pixel and phase angles <120°. This network has been used to create a controlled global image mosaic at a resolution of 100 m/pixel (Fig. 3) (Bland *et al.,* in preparation, 2018). Cassini image data of Enceladus with spatial resolutions better than 100 m/pixel were obtained on 18 close flybys of the satellite (Table 2; section 5).

2.2. Topographic Data

Cassini-derived topographic data for Enceladus include limb profiles, geodetic shape models, and digital terrain models (DTMs) at local to global scales (*Thomas et al.,* 2007, 2016; *Thomas,* 2010; *Nimmo et al.,* 2011; *Giese et al.,* 2008, 2017, 2010a, 2011; *Schenk and McKinnon,* 2009; *Tajeddine et al.,* 2017). Limb profiles (*Thomas et al.,* 2007; *Thomas,* 2010; *Nimmo et al.,* 2011) provided the first measurements of Enceladus' global figure and long-wavelength topography accurate enough to demonstrate that Enceladus is not in hydrostatic equilibrium. Geodetic shape models based on control point information (*Giese et al.,* 2011; *Thomas et al.,* 2016) constrained the libration of Enceladus and provided evidence for the presence of a global subsurface ocean. Local-scale DTMs of ridge topography (*Giese et al.,* 2008, 2017) yielded estimates of the effective lithospheric thickness of tectonically modified trailing-hemisphere terrains ($0.3 \leq T_e < 0.4$ km) at a time when the inferred thermal flux was comparable to the average modern thermal flux in the SPT (200–270 mW m^{-2}). Regional DTMs derived using stereogrammetry alone (*Giese et al.,* 2010a), and wider-ranging digital terrain maps that were derived from a combination of stereogrammetric methods and photoclinometry (*Schenk and McKinnon,* 2009) revealed the presence numerous ancient basins (section 4.2). These data have also been used to understand the implications of observed crater relaxation (*Bland et al.,* 2012) (section 3.2).

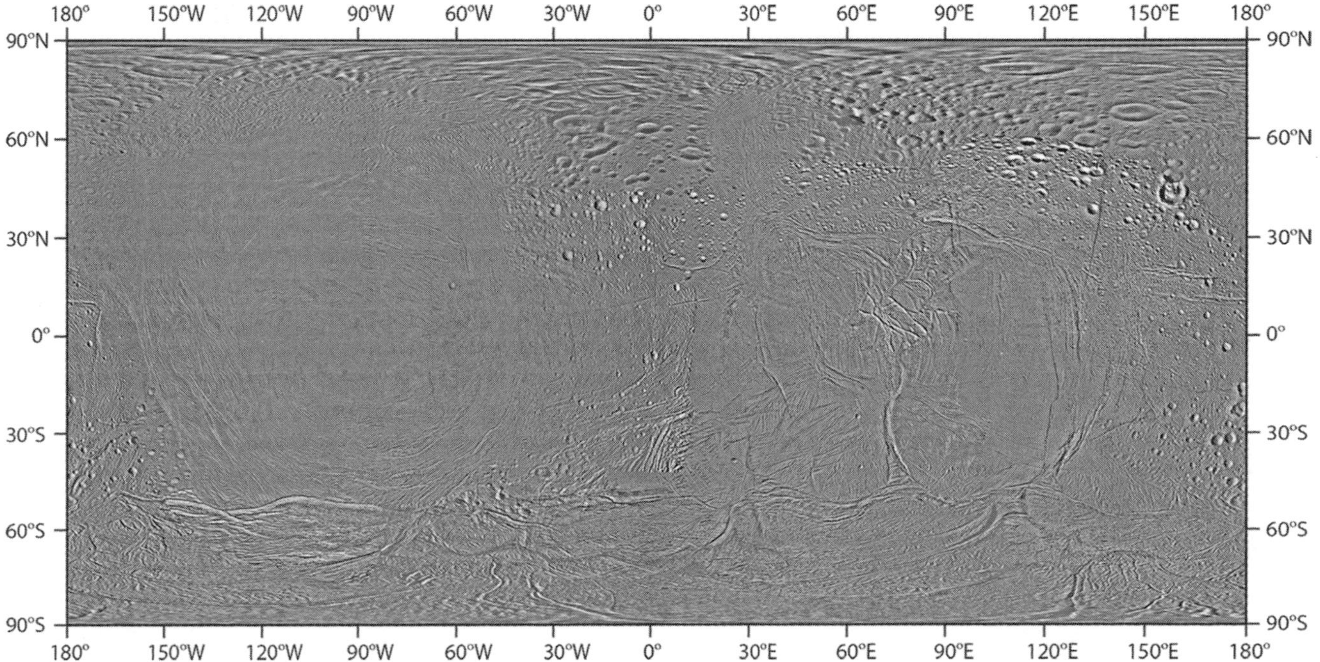

Fig. 3. Simple cylindrical controlled global image mosaic with spatial resolution of 100 m/pixel (*Becker et al.,* 2016).

TABLE 2. High-definition ISS flyby imaging (best resolution finer than 100 m/pixel).

Observation Name	Date	No. of Images	Best Res. m/px	Primary Target*	Tracking Maneuver†
ISS_003EN_LIMTOP004_PRIME	02/17/05	3	63	EH	RWA Track
ISS_003EN_ENCDUST001_CDA	02/17/05	1	89	NH:WH	ORS Drag
ISS_004EN_REGEO002_PRIME	03/09/05	27	24	SPRT	RWA Track
ISS_011EN_N9COL001_PRIME	07/14/05	9	88	SH	RWA Track
ISS_011EN_MORPH002_PRIME	07/14/05	9	67	SPRT	RWA Track
ISS_011EN_ICYEXO001_UVIS	07/14/05	2	9	SPT	ORS Drag
ISS_061EN_FP3HOTSPT001_CIRS	03/12/08	1	97	SPT	RWA Track
ISS_080EN_ENCELCA001_PRIME	08/11/08	33	8	SPT	Skeetshoot
ISS_091EN_ENCELCA001_PRIME	10/31/08	12	9	SPT	Skeetshoot
ISS_120EN_FP3SPMAP001_CIRS	11/02/09	3	64	SPRT	RWA Track
ISS_121EN_PLMHR001_PRIME	11/21/09	10	41	SPT	RWA Track
ISS_121EN_FP3HIRES001_CIRS	11/21/09	27	12	SPT	RCS Track
ISS_136EN_HIRES001_CIRS	08/13/10	25	15	SPT	RWA Track
ISS_141EN_GRAVITY002_RSS	11/30/10	4	0.7	NH:SS	ORS Drag
ISS_164EN_ENCEL18001_INMS	04/14/12	2	0.6	SPT	ORS Drag
ISS_223EN_ENCEL001_PIE	10/14/15	18	33	ASH:NP	RWA Track
ISS_228EN_ENCELOUTB001_CIRS	12/19/15	11	55	TH:SS	RWA Track
ISS_224EN_ENCEL21001_INMS	10/28/15	1	16	57°S, 324°W	ORS Drag

*SPT = south polar terrain; SPRT = south polar reticulated terrain; NH:SS = northern hemisphere:sub-Saturn; NH:WH = northern hemisphere:western hemisphere; SH = southern hemisphere; ASH:NP = anti-Saturn hemisphere:north pole; TH:SS = trailing hemisphere:Samark and Sulci.

†A description of tracking maneuvers used for the acquisition of high-resolution ISS images found in Appendix A.2.

3. CRATERS

3.1. Crater Distribution

The distribution of craters on Enceladus varies significantly with respect to location and provides the basis for our understanding of the satellite's geological history (section 6). As first described in *Porco* et al. (2006), the south polar terrain has the lowest density of impact craters, the equatorial tectonized terrains have a slightly higher density of craters, and a belt of terrain extending from the anti-Saturn point across the north pole to the sub-Saturn point has the highest density of craters. The highly cratered terrain has been referred to as "ancient terrain." Subsequent mapping and crater counting (*Kirchoff and Schenk,* 2009) support the initial observations of *Porco et al.* (2006). Model ages for the ancient terrain range from 1 to 2 Ga (*Porco et al.,* 2006; *Kirchoff and Schenk,* 2009). *Kirchoff and Schenk* (2009) demonstrated that within the ancient terrain, there are fewer craters near the equator than at the mid-latitudes and near the north pole. Craters in the mid-latitude ancient terrain are at or near saturation equilibrium at diameters between 2 and 6 km, with steep decreases in the expected density at both smaller and larger diameters (*Kirchoff and Schenk,* 2009). Due to the limited imaging coverage available at the time, Kirchoff and Schenk's work was necessarily restricted to less than global geographic extent. Recent work by *Kinczyk et al.* (2017) using the latest available controlled global image mosaic (*Becker et al.,* 2016) supports the crater density observations of *Kirchoff and Schenk* (2009). Figure 4 shows a contour map of the density of craters greater than 3 km in diameter on Enceladus, derived from the analysis of *Kinczyk et al.* (2017). The ancient terrain stands out from the rest of Enceladus as having a significant density of craters, and an equatorial belt of lower crater density within the ancient terrain is clearly observed. Additional discussion on the cratering history of Enceladus can be found in section 6 and the chapter by Kirchoff et al. in this volume and references therein.

3.2. Crater Topography

Most impact craters on Enceladus are unusually shallow, and many of the largest craters exhibit a "central mound," which appears as a broad and fractured upwarp of crater floor material, as opposed to the central peaks of complex craters seen on other icy satellites. The upwarped crater floors were recognized in Voyager images and *Passey* (1983) interpreted them to be indicative of viscous relaxation of the initial bowl-shaped crater topography. *Bland et al.* (2012) surveyed 151 craters using Cassini images and stereo topography, finding that all the craters larger than 12 km in diameter have undergone almost complete topographic relaxation and a significant proportion of smaller craters are also topographically relaxed, even down to sizes of 2 km in diameter. While burial by plume material may explain a portion of the topographic shallowing, it would require that the plume sources have moved over time, and it does not explain the pervasive central mounds observed in the larger craters. By modeling the formation of the shallow crater topography by viscous relaxation, *Bland et al.* (2012) were

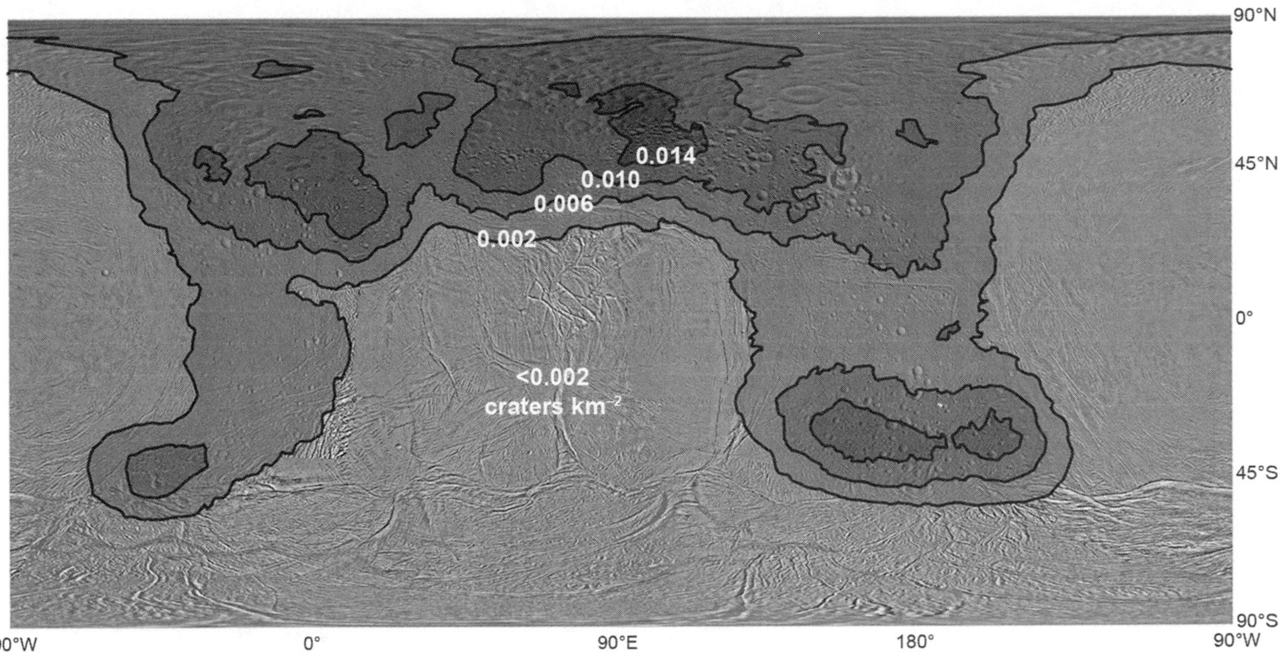

Fig. 4. Simple cylindrical controlled global image mosaic of Enceladus (*Becker et al.,* 2016) overlain with contours of spatial density, in craters per square kilomters, for impact crater diameters >3 km (*Kinczyk et al.,* 2017). Ancient terrain is fully enclosed within the first contour, and the higher-density contour lines show the relative lack of craters in the equatorial zones of the ancient terrain. The map projection uses a center longitude of 90°E to emphasize the equatorial belt of lower crater density.

able to explain the population of craters only by invoking extremely high heat fluxes (>150 mW m^{-2}) in the past.

3.3. Crater-Fracture Interactions

Over much of Enceladus outside the SPT, craters show distinct influences on regional patterns of relatively younger fractures (typically troughs or pit chains; see section 4.1.2). These fracture sets are commonly influenced by craters during fracture growth in such a way that the fractures deviate from their regular trends and converge radially toward the crater (*Barnash et al.,* 2006) (Fig. 5). The fractures then cross through the crater center and exit radially out to some distance (up to tens of kilometers, or about 5 crater diameters) before returning to the original fracture trend (*Martin and Kattenhorn,* 2011, 2012; *Martin,* 2014). These interactions occur with both simple and complex craters, most of which are in various stages of relaxation (*Bland et al.,* 2012). Several mechanisms have been proposed to explain these interactions between craters and later fractures, including topographic influences on local stress states by the crater rim (*Miller et al.,* 2007) or crater depression (*Bray et al.,* 2007), or a stress perturbation associated with a diapiric upwelling of ice beneath the original impact site (*Martin and Kattenhorn,* 2011). Craters smaller than ~5 km in diameter typically do not cause fracture sets to deviate from their regular trends, whereas all observed craters larger than ~7 km induce fracture interactions (*Martin and Kattenhorn,* 2012), indicating that the responsible mechanism for this interaction likely has a crater-size dependence.

4. TECTONICS

4.1. Surface Structures

The surface of Enceladus is pervasively dissected by structural features that, in large swaths of Enceladus, have resulted in complete tectonic resurfacing and the removal of older terrain features such as craters to form new, morphologically distinct terrains (*Crow-Willard and Pappalardo,* 2015). This tectonic dissection appears to have been achieved through a variety of deformation mechanisms, evidenced by a range of morphologically disparate deformation features on the surface. The classification of tectonic features on Earth involves an integrated analysis of the geometries (e.g., morphology, size, location, orientation), kinematics (sense of motion evidence), and mechanical context (the relationship of the structure to driving stresses responsible for development) of the features, resulting in distinct feature types related to the three dominant tectonic environments: extensional, contractional, and translational. However, on Enceladus and other planetary bodies, kinematic evidence and formation mechanics are rarely obtainable through remote observation such as in spacecraft images. Therefore, the classification of tectonic features is necessarily morphology based, with inferences about kinematics and driving

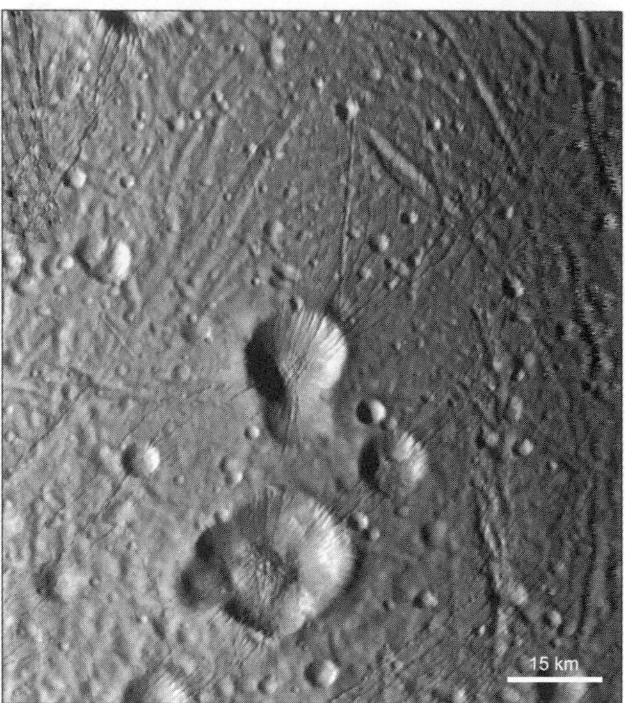

Fig. 5. Crater in the southern hemisphere anti-saturnian cratered terrain of Enceladus that shows induced fracture reorientation. Image N1500060254, from *Martin* (2014).

mechanisms being based on perceived similarities of features to analogs on Earth (e.g., *Kattenhorn and Hurford,* 2009). *Nahm and Kattenhorn* (2015) presented a morphology-based classification scheme for tectonic features on Enceladus, based on five distinct categories of features: ridges, troughs, scarps, chasmata, and bands (Fig. 6). We refer the reader to that paper (and similar terminology usage in *Crow-Willard and Pappalardo,* 2015) for detailed descriptions and interpretations. We summarize this classification scheme here.

4.1.1. Ridges. Ridges are approximately linear structures that stand up to hundreds of meters above the surrounding terrain and can attain lengths of tens of kilometers to more than 100 km. Sets of multiple, arcuate, parallel ridges comprise the interiors of triangular wedges that extend northward away from the cusps in the tectonic boundary surrounding the SPT, and are referred to as *corrugated ridge belts* (Fig. 6a). Corrugated ridge belts have been interpreted as either compressional ridges (*Porco et al.,* 2006; *Schenk and McKinnon,* 2009) or tilt-block (imbricate) normal faults (*Nahm and Kattenhorn,* 2015). *Double ridges* exhibit a central trough hundreds of meters deep, flanked by ridges up to ~100 m high, and typify the eruptive cracks (tiger stripes) in the SPT (Fig. 6b). Double ridges in the SPT generally have been interpreted as resulting from topographic buildups along the margins of opening fractures that may have been subsequently sheared by tidal forcing (*Gioia et al.,* 2007; *Hurford et al.,* 2007, 2009; *Nimmo et al.,* 2007; *Smith-Konter and Pappalardo,* 2008; *Spencer et al.,* 2009; *Patthoff and Kattenhorn,* 2011). In the regions between the double ridges of the SPT, the terrain is defined by closely spaced *funiscular ridges* (Fig. 6b), which are typically oriented parallel to the adjacent tiger stripe double ridges, but are relatively topographically lower. These regularly-spaced ridges are hypothesized to have formed by compression that resulted in regular-wavelength buckling (e.g., *Spencer et al.,* 2009; *Barr and Preuss,* 2010; *Bland et al.,* 2015). *Crow-Willard and Pappalardo* (2015) refer to *ropy ridges* outside the SPT, which are similar in morphology to funiscular ridges but with less consistency in planform orientations to those between the tiger stripes. *Single ridges* with a variety of orientations or bifurcating patterns have been described within tectonized terrains (Fig. 6c) (cf. "dorsa" ridges of *Crow-Willard and Pappalardo,* 2015). Discontinuous single ridges also occur within the ancient cratered terrain (Fig. 6e). Single ridges have been hypothesized to have a cryovolcanic origin (*Spencer et al.,* 2009) or to be the result of deformation associated with thrust faulting (*Pappalardo et al.,* 2010; *Patthoff et al.,* 2015). Tectonized terrains may also exhibit relatively short, almond-shaped *lenticular ridges*, several hundreds of meters high but <35 km long. Based on topography, sinuosity, and lack of lobate features, lenticular ridges have been hypothesized to form as a result of compressional stresses (*Patthoff et al.,* 2015). The tectonized terrain of the trailing hemisphere shows broad regions of parallel ridges and troughs (Fig. 2, Fig. 6c), referred to as *striated plains* by *Crow-Willard and Pappalardo* (2015). *Bland et al.* (2007) describe these periodic, low-wavelength ridged plains as being the result of regional extension accompanied by high heat flows that resulted in localized tectonic resurfacing.

4.1.2. Troughs. Troughs are linear to curvilinear, crack-like depressions that lack bounding ridges and are the most abundant type of tectonic feature on Enceladus (*Nahm and Kattenhorn,* 2015) (Fig. 6d). They appear to exhibit only opening motions, in that the features they crosscut exhibit no evidence of lateral offsets. Troughs have been interpreted as opening cracks related to regional tension in the ice (*Nahm and Kattenhorn,* 2015), possibly caused by tidal forcing (e.g., diurnal, nonsynchronous rotation, libration). Included within this category are *pit chains* (*Martin et al.,* 2017) (Figs. 6e and 7a), which are linear arrangements of individual or partially merged shallow, rimless pits up to 700 m in diameter and spaced less than 500 m apart. Pit chains can extend for tens to hundreds of kilometers, and usually occur in parallel sets spaced <2 km apart. Pits appear to represent draining of loose, surface regolith into an underlying chasm (e.g., *Wyrick et al.,* 2004; *Ferrill et al.,* 2004) and may represent genetic precursors to continuous troughs. Pit chains form the youngest tectonic features outside the SPT in that they cross-cut tectonic boundaries between the terrains, including the polygonal boundary circumscribing the SPT, and have recently begun to dissect the cratered terrain. Pit chains imply localized extension-related opening, which could be associated with dilational normal faulting (*Martin and Kattenhorn,* 2013). Discontinuous single troughs a few kilometers wide occur in the cratered terrain (Fig. 6e). Their morphology is characterized by subdued, shallow walls, and

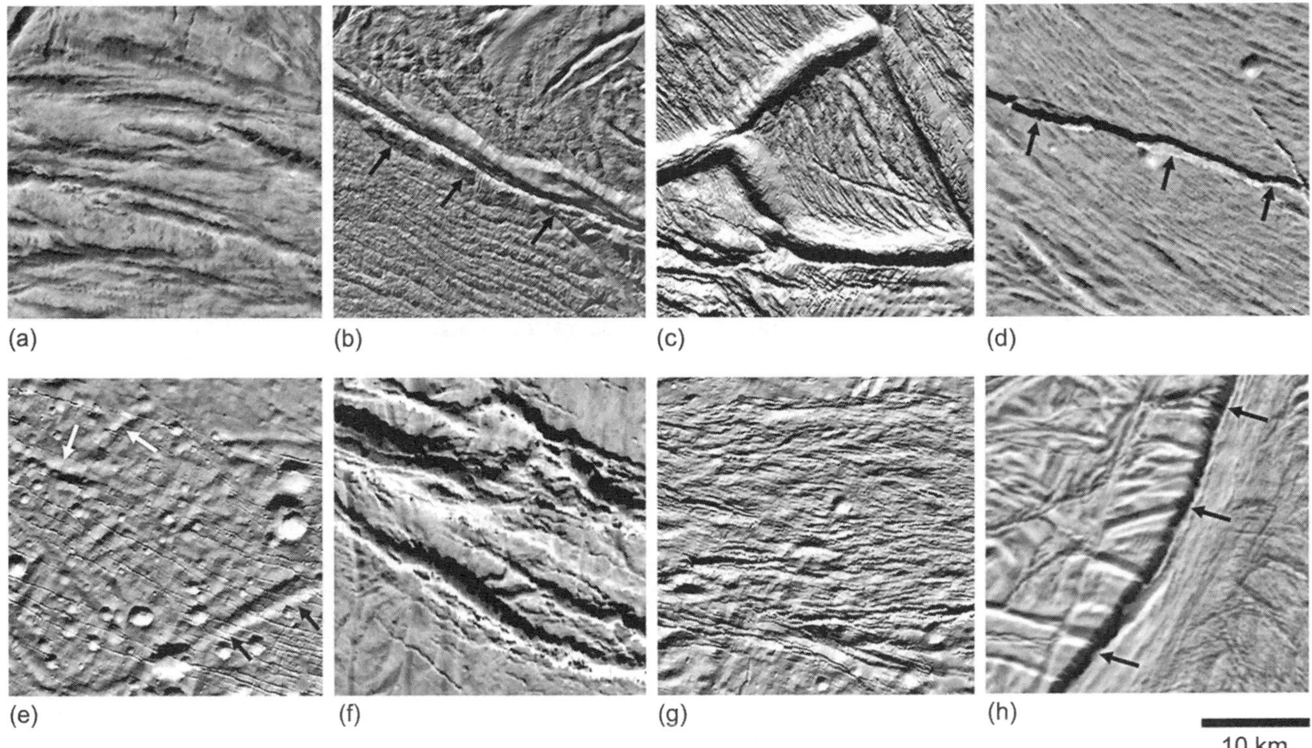

Fig. 6. Representative examples of tectonic features on Enceladus. All subfigures are 30 km across and have been rotated so that the light is coming from the top. **(a)** Corrugated ridge belt at the northern edge of the south polar terrain, 45°S, 120°E. **(b)** Double ridge (black arrows) in the south polar terrain; below this are funiscular ridges, 80°S, 51°E. **(c)** Single ridges surrounded by low-relief striated plains (ridges and troughs), 6°N, 68°E. **(d)** Isolated trough (black arrows) cutting across band, 5°S, 122°E. **(e)** Pit chains are the narrow features trending horizontally across this image (see also Fig. 7a), cutting across ancient troughs (black arrows) and ridges (white arrows), 2°S, 157°E. **(f)** Chasmata located to the north of the south polar terrain boundary, 34°S, 128°E. **(g)** Lineated band displaying closely-spaced scarps, 5°N, 145°E. **(h)** Scarp (black arrows), 59°S, 22°E.

their cratered surfaces indicate they are much older than the other tectonic features discussed here. Other trough-like features referred to as *en echelon* troughs in *Nahm and Kattenhorn* (2015) are discussed in section 4.1.6.

4.1.3. Chasmata. Chasmata are elongate, linear depressions that strike hundreds of kilometers northward from the polygonal boundary of the SPT (Fig. 6f). These tectonic valleys are up to ~3.6 km deep with relatively flat floors and segmented or scalloped edges. They have steep sides reminiscent of faulted topography, implying that chasmata are bounded by scarps (cf. *Crow-Willard and Pappalardo,* 2015) and are analogous to rift valleys (*Helfenstein et al.,* 2006). Chasmata transition along-strike into troughs at their distal ends, pointing to a common origin for the two types of features.

4.1.4. Bands. Although common on other icy bodies such as Europa (*Kattenhorn and Hurford,* 2009; *Prockter and Patterson,* 2009) and Ganymede (*Patel et al.,* 1999; *Pappalardo et al.,* 2004), tectonic bands are relatively rare on Enceladus. Bands that are present have subparallel boundaries and comprise a tabular zone of intense deformation up to 25 km wide (Fig. 6g). *Nahm and Kattenhorn* (2015) propose that the lineated internal texture of bands on Enceladus may be defined by either closely spaced scarps (which they refer to as *lineated bands*) or parallel ridges and troughs (*ridged bands*), but there are insufficient examples to determine consistency in these distinctions. Although smaller in scale, lineated bands on Enceladus more closely resemble those on Ganymede (*Pappalardo et al.,* 2004) than those on

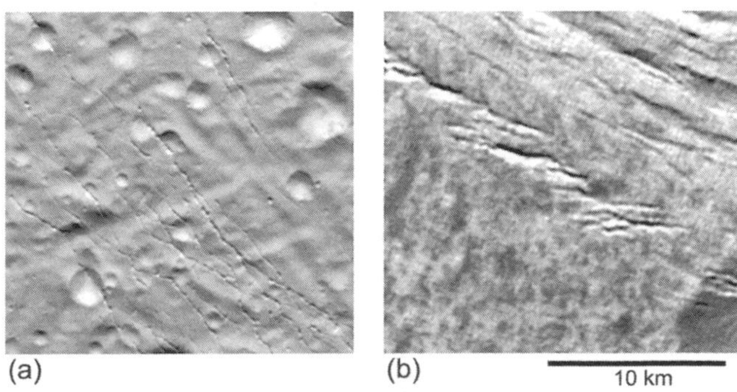

Fig. 7. Representative examples of local-scale tectonic features. Subfigures are 20 km across and have been rotated so that the light is coming from the top. **(a)** Pit chains (8°S, 163°E); **(b)** en echelon troughs (13°S, 154°W).

Europa (*Greeley et al.,* 2000; *Prockter et al.,* 2002) and so may represent localized zones of extensional deformation, predominantly by normal faults. Ridged bands may represent zones of tabular contraction, similar to fold-and-thrust belts on Earth (*Kargel and Pozio,* 1996); however, ridges do not necessarily imply contraction (see section 4.1.1), so an extensional origin cannot be excluded.

4.1.5. Scarps. Scarps are linear or arcuate surface structures across which an abrupt elevation change is apparent in topographic data or spacecraft images based on associated shading. Scarps on Enceladus (Fig. 6h) show a strong resemblance to tectonic scarps on Earth (e.g., *Muirhead et al.,* 2016), Mars (e.g., *Kronberg et al.,* 2007; *Polit et al.,* 2009), Earth's Moon (*Nahm and Schultz,* 2013), and Europa (*Nimmo and Schenk,* 2006) that are related to normal faulting and hence extension, perhaps implying a similar genesis on Enceladus. Isolated scarps are rarely observed on Enceladus (*Nahm and Kattenhorn,* 2015), possibly because of the current resolution limitations of DTMs; however, the boundaries of more abundant chasmata (section 4.1.3) are likely defined by scarps.

4.1.6. Strike-slip faults. Enceladus contains lineaments along which lateral motions have been interpreted to have occurred, including both left-lateral and right-lateral senses of motion (*Martin,* 2016). Some of these faults occur along distinct boundaries between tectonized and cratered terrains, whereas others are embedded within the cratered terrains. Kinematic evidence is in some instances provided by *en echelon* fractures (Fig. 7b) [referred to as *en echelon troughs* in *Nahm and Kattenhorn* (2015)], which form distinctive patterns of fault surface rupture dependent on the slip sense along the strike-slip fault. Strike-slip motions have also been suggested to be important in the geological evolution of the south-polar terrain (*Yin and Pappalardo,* 2015) and potentially influencing the locations of plume eruptions along the tiger stripe cracks (*Hurford et al.,* 2007, 2012; *Smith-Konter and Pappalardo,* 2008; *Olgin et al.,* 2011).

4.2. Topographic Basins

Global topographic mapping of the surface of Enceladus from stereophotogrammetric methods (*Schenk and McKinnon,* 2009; *Giese et al.,* 2010a) and recent broad-scale spherical harmonic fitting to control point data and limb contours (*Tajeddine et al.,* 2017) have identified more than a dozen large (100-km-scale) noncircular topographic basins with depths of hundreds of meters to over a kilometer. *Schenk and McKinnon* (2009) observed that the basins are not correlated with any distinguishable geologic terrain boundary, implying that the basins may have arisen via a deep internal mechanism, such as localized convecting hot spot, perhaps during an early epoch when comparatively high internal heat fluxes were more geographically widespread than today (*Bland et al.,* 2007, 2012; *Giese et al.,* 2010a). *Besserer et al.* (2013) showed that the observed surface topography could be explained by compaction of a shallow ice layer with a plausible internal porosity ~25–30% (cf. *Eluszkiewicz,* 1990; *Durham et al.,* 2005), likely driven by underlying convective hot spots. They concluded that the basins may preserve signs of ancient activity and heating, making them paleobasins that identify the locations of ancient upwellings. Analysis by *Tajeddine et al.* (2017) suggests that the spatial distribution of the basins is evidence for true polar wander of Enceladus' ice shell.

5. LOCAL-SCALE FEATURES

Over the course of the Cassini mission, 198 images with resolutions better than 100 m/pixel were acquired on 18 close flybys of the satellite (Table 2). These local-scale data have provided useful insights regarding how larger-scale terrain features have evolved through time under the influence of degradational processes such as mass wasting, burial by particulate ice ejected from Enceladus geysers and accreted from the E ring, thermal erosion, and space-weathering processes. On Enceladus, representative features observed at local scales include topographic relief and texture, exposed arrangements of ice blocks, ice pinnacles, pits and small craters, networks of fine cracks, cracks along ridge crests, and parallel arrangements of peculiar narrow ridges ("shark fins") that occur along the medial troughs of tiger stripe fractures. Much of Cassini's highest-resolution coverage focused on the south polar terrain (SPT) but images were acquired at spatial resolutions better than 25 m/pixel of other regions of the surface (e.g., Ali Baba crater).

5.1. The South Polar Terrain

Local-scale images of SPT features associated with Baghdad Sulcus (Fig. 8) have been valuable for characterizing possible surface expressions of geyser eruptions (*Goguen et al.,* 2013; *Helfenstein and Porco,* 2015). Morphological evidence for the presence of discrete geysers (*Helfenstein and Porco,* 2015) include radial striations that converge toward a point along a tiger stripe fracture (Fig. 8a), quasi-circular rimmed pits that interrupt the tiger stripe rifts or fractures that cross-cut them (Figs. 8b,c), and local albedo streaks that appear to extend from tiger stripe fractures (Fig. 8d). At least five active geysers identified in a survey by *Porco et al.* (2014) correlate with the locations of local-scale features within a 15-km radius of the feature in Fig. 8a (*Helfenstein and Porco,* 2015).

Other morphological features associated with Baghdad Sulcus that provide insight into local geological processes include pinnacles, flower structures, and ice blocks. Pinnacles have been observed along some narrow ridge crests (Fig. 8e) with a periodicity that may be controlled by the spacing of narrow, gossamer thin cracks that permeate the surface. Scouring, ablation, thermal erosion, and mass-wasting are all factors that may have led to the formation and structural enhancement of these features. It has been suggested the pinnacles in Fig. 8e may be confined to the site of a former geyser eruption along a medial fracture that separates the ridges (*Helfenstein and Porco,* 2015). Parallel

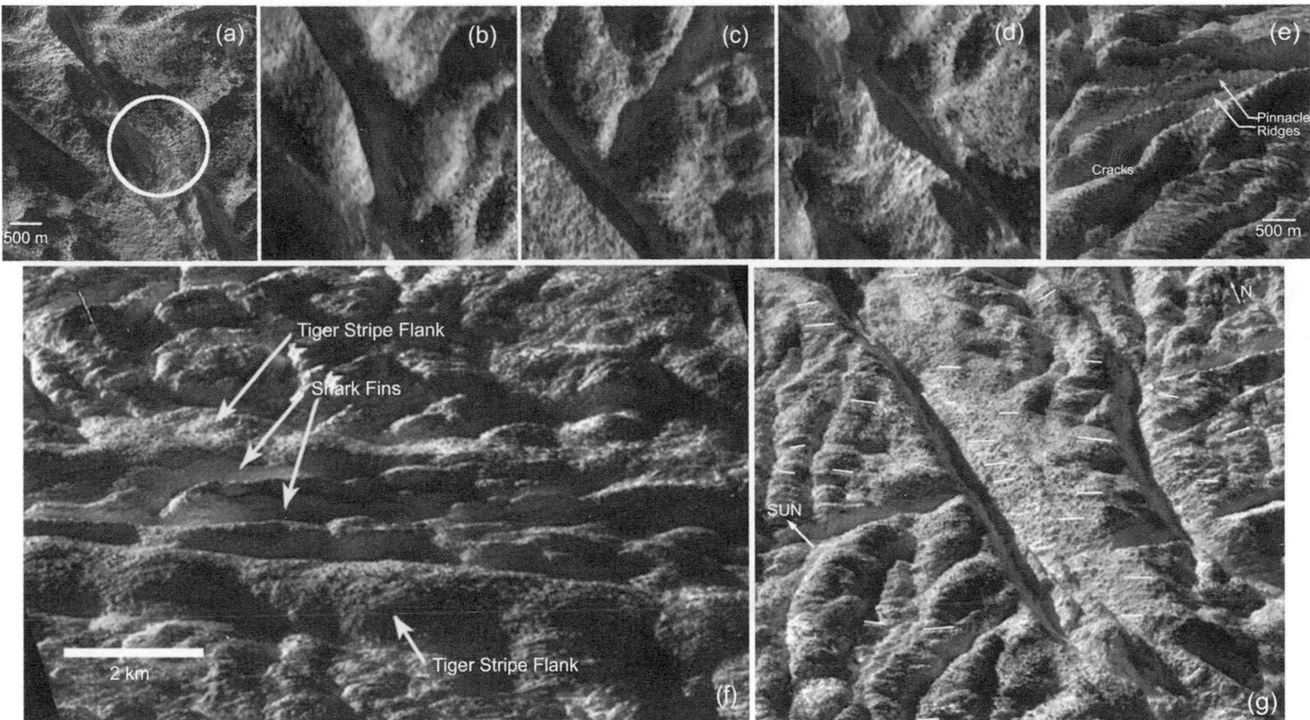

Fig. 8. Tectonically disrupted active section of Baghdad Sulcus. **(a)** Image is centered on the summit of a precipice along an undulating tiger-stripe flank. Circle highlights radial pattern of striations and ice blocks centered on the precipice peak near the location of *Porco et al.*'s (2014) Geyser #27. From image N163746854 (8.3 m/pixel). **(b)** A circular pit (~1.5 km diameter) with a torus-shaped rim occurs near Geyser #29 at the juncture where Baghdad splits into two parallel cracks. **(c)** An ~1.3-km-diameter circular pit near Geyser #26 with a roughly torus-shaped rim is transected by a fracture that is orthogonal to and terminates against Baghdad Sulcus. **(d)** Asymmetric bright albedo pattern near Geyser #28 that projects from the medial fracture across the tiger stripe left flank to the lower left of the image. **(e)** Icy pinnacle ridges branching off of Baghdad Sulcus near 76.6°S, 30.9°W, image N1637462964 (14 m/pixel). **(f)** Obliquely viewed branch of Baghdad Sulcus at 83.6°S, 43.0°W that shows details of prominent "shark fin" ridges. **(g)** Cassini image N1637462964 (77°S, 35°W; north is approximately up) showing a section of Baghdad Sulcus (the large fracture cutting through the center of the image) viewed at high resolution (15 m/pixel). Note: **(a)–(f)** are adapted from *Helfenstein and Porco* (2015).

sets of narrow, segmented ridges that flank the tiger stripe valley floor and form flower structures (informally referred to as "shark fins" — Fig. 8f) have been interpreted to form under transpression and suggest that some discrete geysers may initiate at the intersection points of tiger stripe fissures and cross-cutting fractures (*Helfenstein,* personal communication). Ice blocks ≤10 m in diameter are ubiquitously distributed over active portions of the region (Fig. 8g) and along ridge crests on older structures elsewhere (Figs. 8a–c). The paucity of impact craters in the SPT and block size-distribution statistics (*Landry et al.,* 2014) indicate that they were not likely distributed as impact ejecta (*Martens et al.,* 2015). While their relatively uniform distribution along tiger stripe flanks might suggest that they were emplaced by curtain-like fissure eruptions, modeling demonstrates that the jets erupting from geysers on Enceladus are too tenuous to loft 10-m-diameter ice blocks (*Martens et al.,* 2015). Instead, their origin may be tied to their relationship to small-scale fracturing of the surface (*Martens et al.,* 2015; *Giese et al.,* 2010b; *Landry et al.,* 2014). In this scenario the ice blocks are not isolated structures, but rather represent the exposed protrusions of solid ice through a blanket of fine particulate fallout and detritus that accumulates around their bases. Additional discussion on jets and geysers can be found in section 6.1.2 and the chapters by Spencer et al. and Goldstein et al. in this volume.

5.2. Ali Baba Crater

Image data at scales of hundreds of meters to submeter per pixel were acquired for the crater Ali Baba (56.84°N, 17.51°W; 34 km diameter) on two separate Cassini flybys of Enceladus (Fig. 9). The imaged footprint of these data falls on the fractured flank of the crater's central dome (Fig. 9a). A meter-scale image acquired as a BOTSIM pair shows a series of roughly parallel arcuate troughs and interspersed mesas (~250 m wide) along the up-domed interior of the crater and trending around the dome's circumference (Fig. 9b). Possible exposures of solid ice appear to be present in the highest-resolution image (Fig. 9c) along fracture walls and an adjacent relatively hummocky background terrain with few resolvable ice blocks present in the image

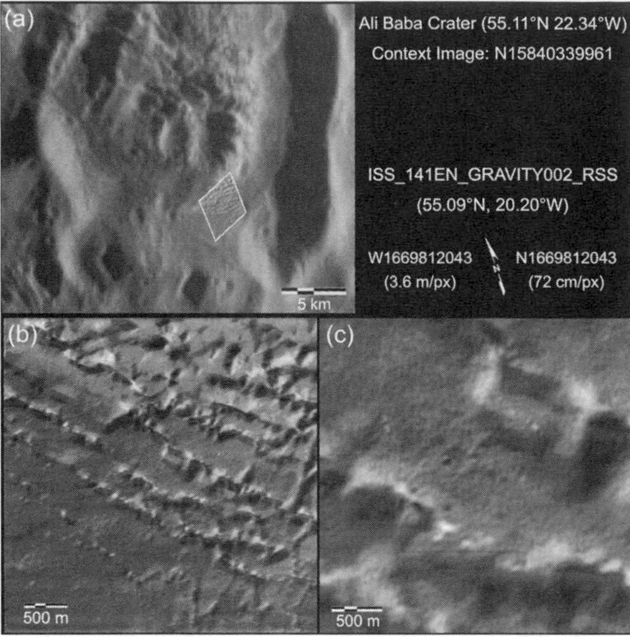

Fig. 9. (a) Context image N15840339961 of Ali Baba crater with placement of high-resolution BOTSIM footprint shown along slope of the crater's central dome. **(b)** WAC BOTSIM image W1669812043, acquired in Saturnshine. The image has been spatially filtered to remove electronic noise and contrast-enhanced. **(c)** NAC BOTSIM image N1669812043, boresighted to WAC image at left. Images adapted from *Helfenstein et al.* (2013).

is apparent. The trough and mesa morphology seen in Ali Baba crater has not been previously observed on internal crater structures of other icy satellites. The regular spacing of the troughs suggests that the flanking dome material may have distinct physical properties to depths of a few hundred meters (*Lachenbruch,* 1962, p. 69).

6. GEOLOGICAL HISTORY

Early image data returned from Cassini (*Porco et al.,* 2006) revealed four distinct terrains on Enceladus that could be divided into three different epochs in the surface history of the satellite. Further exploration and analysis of Enceladus has upheld this basic division of surface history, and in this section we describe the geological history of Enceladus through the terrains that exemplify these epochs. A contemporary epoch encompasses the currently active SPT, its recent development, and its effects on surrounding terrains (section 6.1). An intermediate epoch involves the development of terrains found on the leading and trailing hemispheres of Enceladus (section 6.2). The cratered terrain occupying the north polar region of the satellite and equatorial regions not resurfaced by the formation of the SPT, leading hemisphere terrain (LHT), and trailing hemisphere terrain (THT) tell the story of an ancient epoch in the surface history of the satellite (section 6.3).

6.1. Contemporary Epoch

The SPT is pervasively fractured, geologically young (section 3.1; see the chapter by Kirchoff et al. in this volume), low-lying, and bound poleward of ~55°S (Fig. 10) by a crudely polygonal circumpolar system of south-facing scarps that stand ~1 km higher than the interior SPT basin and are intermittently broken by Y-shaped structural discontinuities (*Porco et al.,* 2006; *Helfenstein,* 2014). An outer annulus of highly fractured terrain that includes corrugated ridge belts (section 4.1.1) encloses a central region within the SPT (Fig. 10). The central region includes the tiger stripe double ridges Damascus Sulcus, Baghdad Sulcus, Cairo Sulcus, and Alexandria Sulcus (Fig. 10; section 4.1.1) and a quasi-parallel system of funiscular ridges (section 4.1.1). It is also associated with anomalously high heat flows (*Spencer et al.,* 2006; see also the chapter by Spencer et al. in this volume) and the only location on Enceladus with confirmed current geologic activity (*Porco et al.,* 2006; *Spitale and Porco,* 2007; *Hansen et al.,* 2008; *Spencer et al.,* 2009).

Models of SPT formation have generally focused on concentrated subsurface tidal heating, possibly associated with ice shell convection (*Barr and McKinnon,* 2007; *Grott et al.,* 2007; *Barr,* 2008; *Roberts and Nimmo,* 2008; *Tobie et al.,* 2008; *Stegman et al.,* 2009; *Zhang and Nimmo,* 2009; *O'Neill and Nimmo,* 2010; *Běhounková et al.,* 2012, 2013; *Han et al.,* 2012; *Rozel et al.,* 2014; *Showman et al.,* 2013). Such heating would result in localized lithospheric thinning and deformation (*Bland et al.,* 2007; *Spencer et al.,* 2009; *Barr and Preuss,* 2010; *Patthoff and Kattenhorn,* 2011; *Yin and Pappalardo,* 2015). The SPT may have formed *in situ* (*Běhounková et al.,* 2012, 2013), given that eccentricity-induced tidal heating is elevated at the poles. Alternatively, subsurface concentration of warm ice and/or melting above a south polar sea might have induced polar wander, moving a mass anomaly from lower latitudes to the satellite's pole (*Nimmo and Pappalardo,* 2006; *Collins and Goodman,* 2007; *Roberts and Nimmo,* 2008).

6.1.1. SPT thermal anomaly. Direct evidence of the anomalously high south polar heating and its relation to tiger-stripe fractures derives from observations obtained from Cassini's Composite Infrared Spectrometer (CIRS) and Visual and Infrared Mapping Spectrometer (VIMS). See the chapter by Spencer et al. in this volume for a detailed discussion. The total heat flow averaged over the entire south pole includes contributions from the active tiger stripes and possibly a contribution from heat conduction through the ice between the tiger stripes. *Howett et al.* (2011) obtained a net south polar heat flow of 15.8 ± 3.1 GW. A more recent estimate of heat flow from the tiger stripes (*Spencer et al.,* 2013) plus the latent heat content of the water vapor plumes (*Ingersoll and Pankine,* 2010) is ~4.7 GW. The significantly larger net south polar heat flow suggests a broad, background contribution that is difficult to accurately model because the thickness of the south polar lithosphere needed for heat conduction calculation is not well-determined — it falls somewhere in the range

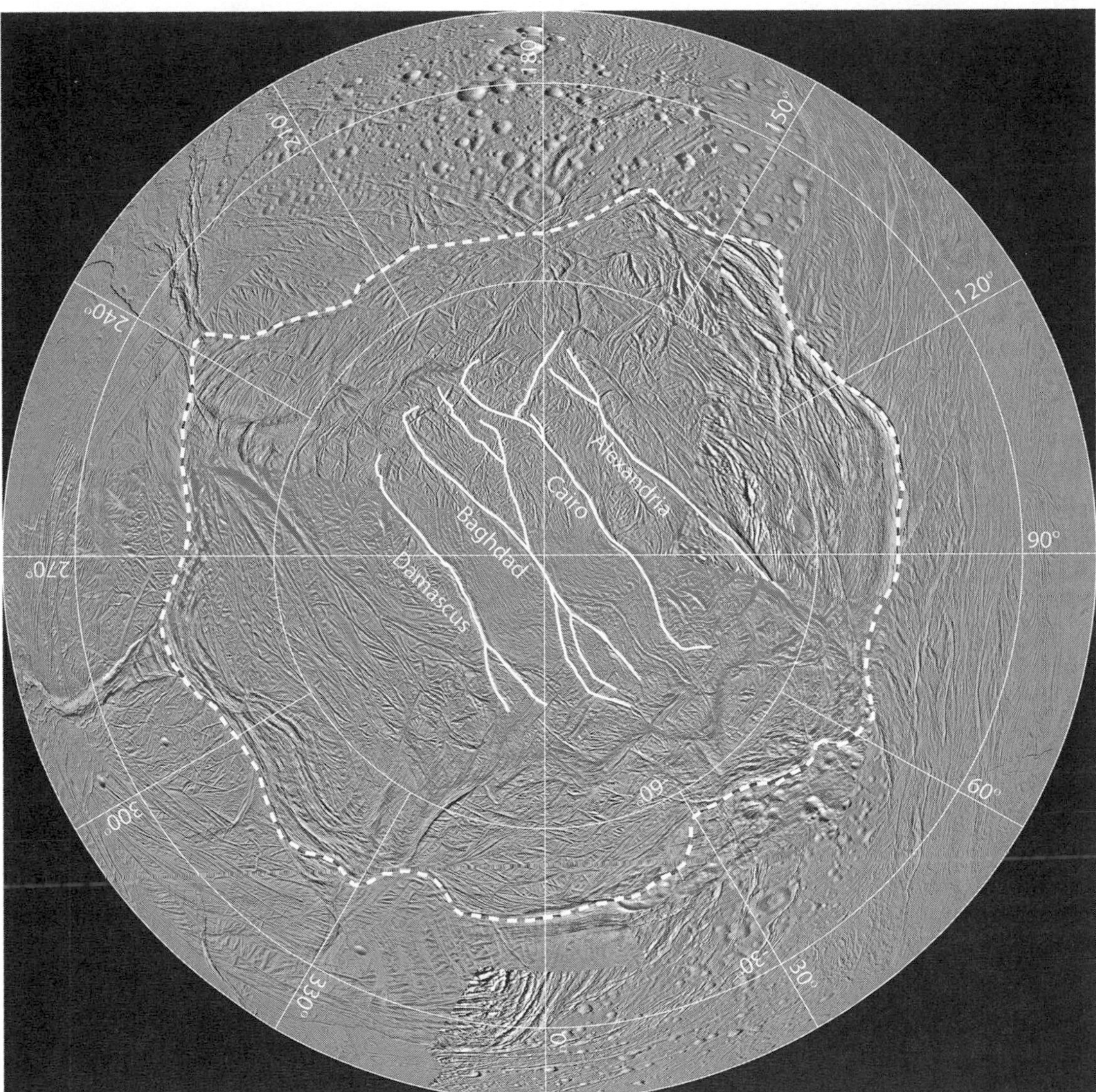

Fig. 10. Orthographic controlled global image mosaic (*Becker et al.,* 2016) for the southern hemisphere of Enceladus (positive west longitude). Prominent fractures (tiger stripes) that source the south polar plume are labeled and indicated with a solid line. The approximate boundary of the SPT is indicated with a dashed line.

of 10 km to 45 km (see the chapter by Hemingway et al. in this volume). The highest temperatures occur along the tiger stripes, but the thermal emissions used to detect them are not uniform. A best-fit isothermal temperature 197 ± 20 K was obtained by VIMS observation of an active spot along Baghdad Sulcus (*Goguen et al.,* 2013). This value is in good agreement with the highest temperature obtained by CIRS (176.7 ± 1.3 K) at the brightest hot spot observed on Damascus Sulcus (*Spencer et al.,* 2011).

6.1.2. Eruptive fracture (tiger stripes) in the south polar terrain. Several researchers have addressed potential driving mechanisms behind the creation of the eruptive fractures in the SPT (e.g., *Kieffer et al.,* 2006; *Gioia et al.,* 2007; *Hurford et al.,* 2007, 2009; *Halvey and Stewart,* 2008; *Smith-Konter and Pappalardo,* 2008) and their potential eruptive mechanisms (see the chapter by Spencer et al. in this volume and references therein). The tiger stripe fractures must extend deep enough into the ice shell to tap into a body of fluid that feeds the eruptive jets and provide a permeable pathway for fluid transport. This is most likely achieved through the development of tensile fractures; therefore, the SPT fractures are commonly interpreted as having formed in tension, with later reactivation by lateral shearing that potentially creates a mechanism for jet eruptions (*Nimmo et al.,* 2007; *Hurford et al.,* 2007, 2009, 2012; *Smith-Konter and Pappalardo,* 2008).

Matson et al. (2012) proposed an exsolving fluid circulation model for the transport of warm water from a 20-km-deep ocean to an icy water table just below a shallow ice cap (also described in the chapter by Spencer et al. in this volume). In this model, fractures in the subsurface ice bounding the ocean act as natural conduits for the water to rise and circulate. This mechanism would provide a means for bringing water and dissolved chemicals to the surface and may also provide a means of bringing localized concentrations of heat under the tiger stripes. More recently, *Matson et al.* (2018) proposed that the exsolving CO_2 could fill natural irregularities and hollows under the ice cap, forming "gas chambers." Penetration of fractures from the surface to such chambers would result in gas and aerosol venting as water fills the evacuating gas chamber. As water continues to rise through the fracture, it would likely boil and freeze into a solid plug. This hypothesis would account for the placement of concentrated CO_2 frost and entrained chemicals near and in between the tiger stripes (Combe et al., in preparation, 2017).

Local-scale image data of active tiger stripe fractures may provide our best available constraints on the eruptive styles that are expressed in the active SPT region (Fig. 8; section 5.1). While there is ample evidence from Cassini imaging and thermal data that tiger stripe fissures are the dominant conduits for water eruptions at the south polar region, the relative importance of discrete geyser eruptions (*Spitale and Porco,* 2007; *Porco et al.,* 2014; *Helfenstein and Porco,* 2015) in comparison to contiguous vent curtains that trace the active fissures (*Spitale et al.,* 2015) has been a subject of recent debate (see also the chapter by Spencer et al. in this volume). From a physical standpoint, the two styles do not appear to be mutually exclusive and it is possible that both may occur or even locally overlap. In addition, it has not been ruled out that at some small-scale, curtain eruptions may be non-uniform, exhibiting some granularity indistinguishable from a series of closely spaced discrete eruptions and/or the presence of cross-cutting fractures (cf. *Helfenstein and Porco,* 2015).

6.1.3. Models for the formation of south polar terrain tectonic features. Numerous models have been proposed to explain the eruption of jets from fractures in the SPT (section 6.1.2; Spencer et al. in this volume), but relatively few models have focused on the types and distribution of tectonic features across the SPT and their significance in the context of Enceladus' tectonic history (e.g., *Patthoff and Kattenhorn,* 2011; *Walker et al.,* 2012; *Yin and Pappalardo,* 2015). The model by *Walker et al.* (2012) draws analogy with terrestrial rift basins to explain the broad-scale tectonic characteristics of the SPT. The models for the formation of the SPT by *Patthoff and Kattenhorn* (2011) and *Yin and Pappalardo* (2015) look in more detail at the formation and evolution of tectonic structures within the SPT under the controlling influence of nonsynchronous tidal stresses or gravitational collapse, respectively, as discussed below.

6.1.3.1. Nonsynchronous rotation model for south polar terrain tectonics: A model by *Patthoff and Kattenhorn* (2011) describes the fracture patterns in the SPT as consisting of a minimum of four distinct fracture sets, implying at least four stages of fracturing at different times and in different orientations (Fig. 11). Each fracture set is inferred to contain hundreds of fractures with almost identical (to within 5°) orientations, implying that the fractures define systematic sets (i.e., fractures of identical type, orientation, age, and formation mechanism). The fracture sets suggest that the SPT records a prolonged and ordered geologic history under regionally consistent but temporally changing states of stress. The most prominent of these fracture sets contains the currently active tiger stripes. Although there are only four named tiger stripes (Fig. 10), there are hundreds of additional fractures in this set (fracture set 1 in Fig. 11), based on near-identical orientations, with measurable lengths as small as tens of meters. The more prominent fractures crosscut all other fractures with contrasting orientations, indicating that the fracture set containing the tiger stripes formed most recently, consistent with their ongoing geologic activity (i.e., eruptive jets). Successively older fracture sets, identified based on crosscutting relationships and identifiable mechanical interaction patterns (e.g., *Cruikshank and Aydin,* 1995), are all oriented in a progressively more clockwise (CW) orientation to the sets that are inferred to have formed later. Some of the fractures in the older sets are geomorphically similar to the currently active tiger stripes in that they have muted double ridge (section 4.1.1) topographic signatures and have a similar regular spacing and cumulative fracture length, raising the possibility that they are ancient, now inactive tiger stripes. Several of these in the intermediate-aged sets are warmer than other fractures in those sets (*Howett et al.,* 2011), and may even contain ongoing eruptive activity (*Spitale and Porco,* 2007; *Porco et al.,* 2014; *Helfenstein and Porco,* 2015), implying reactivation of portions of older fracture sets is possible for the current state of stress in the SPT.

Patthoff and Kattenhorn (2011) assert that the development of systematic sets of fractures is consistent with an extensional interpretation, similar to tectonic joint sets on Earth. The fractures in each set do not differ in orientation by more than 5° from the mean orientation, implying growth in a spatially consistent stress field on the timescale of fracture growth. In this model, stresses in the SPT are likely related to the deformation of the ice shell that results from tidal behavior and orbital evolution. On the diurnal timescale, the tidally responding ice shell experiences a consistent clockwise rotation in principal stress orientations in the SPT, rotating 180° per orbit (*Hurford et al.,* 2009). If these stress rotations occurred on the timescale of fracture growth, resultant fractures should be curved and cuspate, similar to cycloidal fractures on Europa (*Hoppa et al.,* 1999; *Marshall and Kattenhorn,* 2005). Instead, the fracture sets in the SPT are remarkably linear and consistent within each set, implying a stable stress orientation at the timescale of fracture growth, but with unique stress orientations at different points in time relative to the current cartographic coordinate system. *Patthoff and Kattenhorn* (2011) interpret the fracture patterns as being the result of nonsynchronous rotation (NSR) of the ice shell relative to the solid core of Enceladus. In this model, the Saturn-facing point on Enceladus migrates longitudinally westward through time, in response to eastward migration of the ice shell. The stress field induced by the reorientation of the ice shell through the Saturn-locked tidal bulges is likely sufficient to induce fracturing of the ice shell with consistent regional fracture

Fig. 11. Structural map (positive west longitude) indicating orientations of fracture sets in the SPT (modified after *Patthoff and Kattenhorn,* 2011). Four distinct sets are identifiable, with an apparent age progression from younger to older (Set 1 to Set 4), implying a counterclockwise rotation in relative orientations through time. The named tiger stripes are labeled and belong to the youngest fracture set, Set 1.

orientations given that the timescale of fracture growth is likely less than the tens of thousands to millions of years nonsynchronous rotational period. In this NSR framework, when looking directly down at the south pole in the SPT, the principal stresses related to NSR would appear to rotate counterclockwise through time (reflecting a CW rotation of the ice shell through a Saturn-locked stress field). Hence, progressively older fracture sets would be oriented more CW to younger sets, as is observed. A similar process of NSR has been used to explain global lineament orientations on Europa (*Helfenstein and Parmentier,* 1985; *Geissler et al.,* 1998; *Kattenhorn,* 2002). *Patthoff and Kattenhorn* (2011) used this geologic evidence of NSR to infer the existence of a global ocean on Enceladus, above which the ice shell would be able to freely rotate. The presence of this global ocean was later seemingly verified by spacecraft measurements of Enceladus' physical libration (*Thomas et al.,* 2016). A potential problem with this model is the occurrence of Enceladus' other two pervasively tectonized regions along the present leading and trailing axis of the satellite (*Crow-Willard and Pappalardo,* 2015), which must be a coincidental occurrence if NSR is proceeding today.

6.1.3.2. Gravitational collapse model for south polar terrain tectonics: An alternative tectonic model for the SPT

(*Yin and Pappalardo,* 2015) invokes tectonic gravitational collapse to explain SPT tectonics. *Yin and Pappalardo* (2015) mapped the spatial distribution of major structures, their mutual cross-cutting relationships, and features inferred to be indicative of the kinematic development of the mapped structures (Fig. 12). The structures mapped include closely spaced tension cracks and extensional fracture zones, normal faults and fault-bounded grabens, thrusts, folds, and discrete strike-slip faults and distributed ductile shear zones. From these data, *Yin and Pappalardo* (2015) infer that the SPT is bounded by right-slip and left-slip shear zones along the anti-saturnian and sub-saturnian margins, respectively, and by an extensional fault zone and a contractional fold zone along the leading-edge and trailing-edge margins, respectively. The kinematics involved indicate that the SPT has translated from the leading-edge margin toward the trailing-edge margin, a transport direction parallel to the regional topographic slope immediately outside the SPT (i.e., spread gravitationally).

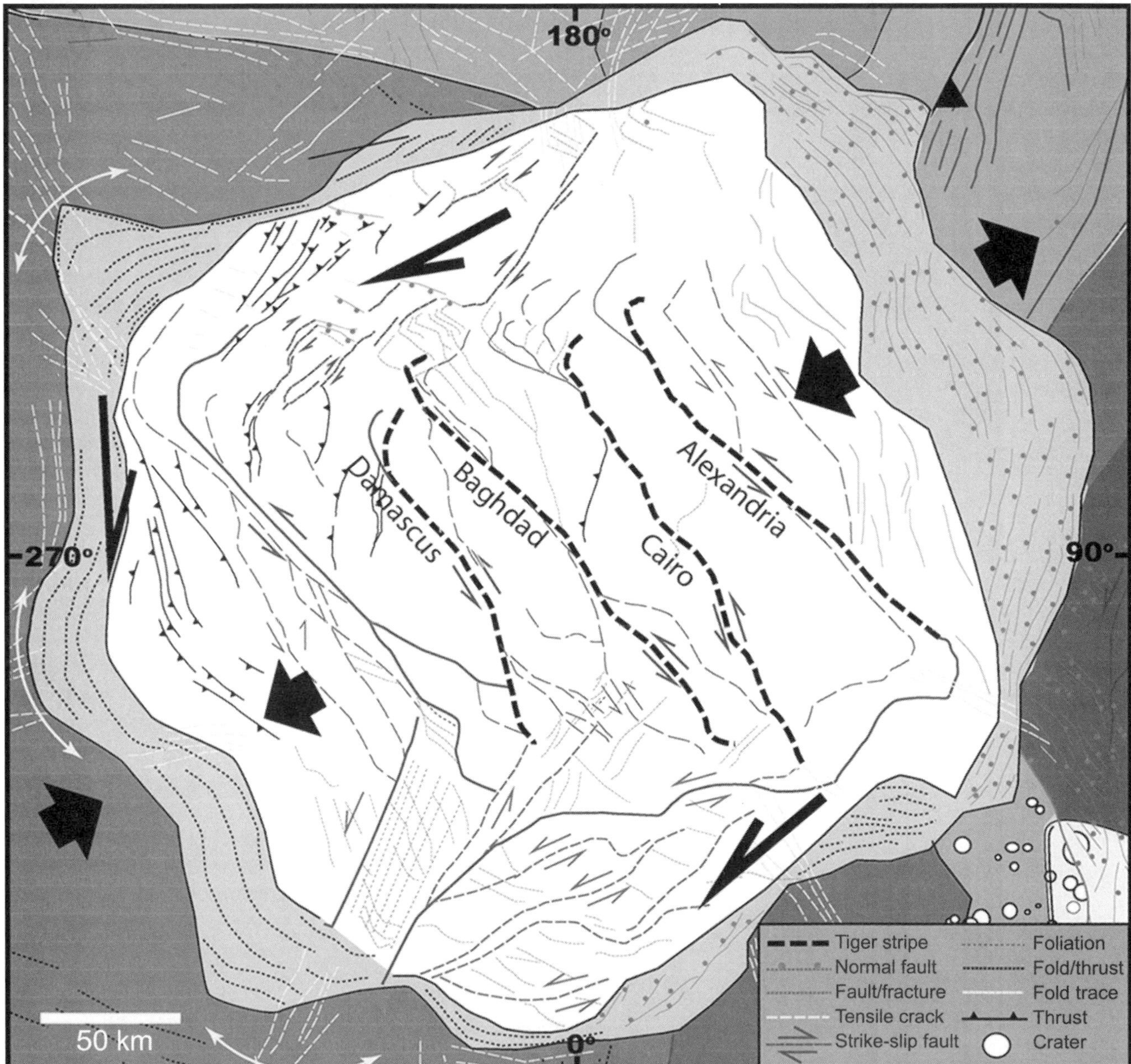

Fig. 12. Structural map (positive west longitude) of the south polar terrain (SPT) in south polar projection, after *Yin and Pappalardo* (2015). The leading-edge margin (upper right) is inferred to be a zone of extension and normal faulting, while the trailing-edge margin (lower left) is inferred as a zone of shortening and folding. The anti-saturnian margin (upper left) and the sub-saturnian margin (lower right) are inferred to be regions of right- and left-lateral shearing, respectively. In this model, the SPT is explained as a block that has translated via gravitational spreading from a topographic high along the leading-edge margin toward the trailing-edge margin, with shear along the intervening margins. This in turn is interpreted to result in a clockwise rotation of the SPT interior relative to the cratered terrains, inducing left-lateral slip along the tiger stripe fractures (bold dashed lines).

During gravitational spreading, the strain state in the ice shell would have been extensional in the upslope section of the SPT interior (toward the leading-edge margin) and contractional in the downslope section of the SPT interior (toward the trailing-edge margin). The anti- and sub-saturnian margins are inferred to have developed coevally with the formation of C-shaped extensional fractures, whereas the trailing-edge margin developed coevally with the formation of Y-shaped fractures (*Yin and Pappalardo,* 2015).

In map view, the SPT deformation is characterized by distributed right-slip shear parallel to the transport direction of the SPT block. This regional shear is inferred to have been induced by non-uniform extension and contraction along the leading- and trailing-edge margins of the SPT. The tiger stripe fractures are inferred to be left-slip structures displaying right-step restraining bends and terminating at hook-shaped fold zones or horsetail fault systems. Regional right-slip shear across the SPT is inferred as responsible for initiating several tiger stripe fractures as left-slip conjugate Riedel shears, along with closely spaced folds in the regions between the tiger stripes. Subsequent left-slip bookshelf-style faulting along the tiger stripe fractures accommodated continued regional right-slip shear and induced clockwise rotation across the SPT.

Locations of the erupting jets along the tiger stripe fractures appear to be independent of the detailed structural settings defined in the *Yin and Pappalardo* (2015) model (e.g., transtensional or transpressional), implying that contemporary jet locations in this tectonic model are unlikely to be controlled by long-term deformation processes expressed by left-slip motion along the tiger stripes. Instead, gravitational spreading has controlled the overall tectonic pattern and evolution of the SPT, while tidal stress assists the diurnal opening and closing of the tiger stripe fractures, modulating the temporal variation of plume fluxes (e.g., *Nimmo et al.,* 2014). The tidal stress may have also modified the fracture patterns across the SPT while reactivating and lengthening the preexisting weaknesses. Segments of the tiger stripe fractures (e.g., the anti-saturnian segment of Cairo Sulcus) are inferred to have locally reactivated or were cut by extensional fractures, suggesting that a young stress regime replaced the older stress regime responsible for initiating and maintaining left-slip motion along the tiger stripes.

In the gravitational collapse model, the inferred flow-like tectonics across the SPT might be explained by the occurrence of a transient thermal event centered near the present south pole (Fig. 13). The thermal event increased the ice shell temperature, which in turn lowered the ice shell viscosity and allowed the release of gravitational potential energy via lateral viscous flow within the ice shell with variable thickness. The transient heating event from below could have been induced by a rising warm ice plume, tidal heating, convection of warm ice, and/or coupling and feedbacks among these processes. Owing to the initially high heat flux of the inferred transient heating event, the ice shell of the SPT was first dominated by early shallow detachment faulting and closely spaced folding, later followed by deep detachment faulting and widely spaced fracturing as a result of decreasing heat flux and consequent increasing ice-shell thickness. The basal sliding surface of the detachment fault in this model most likely follows the brittle ductile transition zone, similar to detachment faults on Earth.

6.2. Intermediate Epoch

Distinct geologic terrains have been recognized on the leading and trailing hemispheres of Enceladus that share characteristics with the SPT (*Spencer et al.,* 2009; *Crow-Willard and Pappalardo*, 2015). Namely, the boundary of each of these two large terrains is defined by a circumferential belt that partially or fully encloses structurally deformed materials with relative crater ages that are less than that of surrounding ancient cratered terrains but more than that of the SPT (section 3.1) (*Kirchoff and Schenk,* 2009; *Crow-Willard and Pappalado,* 2015; *Kinczyk et al.,* 2017; Kirchoff et al., this volume). Previous work mapping the geology of Enceladus has referred to these terrains as the western and eastern hemisphere fractured plains (*Spencer et al.,* 2009) (Table 2), or as the LHT and THT (*Crow-Willard and Pappalardo,* 2015) (Table 2; Fig. 2). We use the latter terminology herein. An assessment by *Crow-Willard and Pappalardo* (2015) and *Kinczyk et al.* (2017) of the crater densities within mapped units of the THT and LHT suggests that the former is generally older than the latter. Analyses of these terrains, their relationship to each other, and their relationship to the SPT have provided insight into the thermal and temporal evolution of the satellite (*Bland et al.,* 2007; *Giese et al.,* 2008), the potential for reorientation of its spin-pole axis, and the potential for variability in the rheological and mechanical properties of its icy shell (*Bland et al.,* 2007; *Giese et al.,* 2008; *Barr,* 2008; *O'Neill and Nimmo,* 2010; *Crow-Willard and Pappalardo,* 2015).

6.2.1. Trailing hemisphere terrain. The THT is centered at ~0°, 285°W (*Crow-Willard and Pappalardo,* 2015) and is bound on its north, east, and west edges by a circumferential belt characterized by long, curvilinear, ridges (section 4.1.1) and troughs (section 4.1.2) (i.e., the Hamah, Harran, and southern Samarkand Sulci; see Figs. 14 and 16), as well as shorter arcuate ridges and troughs, with subparallel to sigmoidal or *en echelon* orientations. The central region within the THT encompasses the named features Diyar Planitia, Sarandib Planitia, Cufa Dorsa (Fig. 6c), and Ebony Dorsum. Diyar Planitia includes a distinct set of north-south-trending troughs up to 4.5 km in width and 20 km in length (*Crow-Willard and Pappalardo,* 2015). Diyar and Sarandib Planitia both include many small (0.6- to 2-km wide) ridges and troughs with approximate north-south trends that, at larger scales, can produce sigmoidal patterns. Located between Diyar and Sarandib Planitia, Ebony Dorsum and the Cufa Dorsa are a set of prominent rounded and convex ridges with an average width of 3 km (*Crow-Willard and Pappalardo,* 2015). The ridges have a predominant northwest-southeast trend but also branch in a variety of orientations and intersect each other.

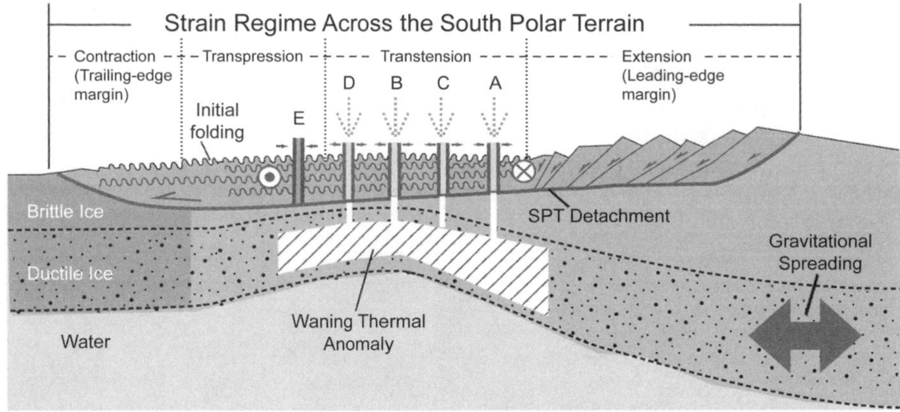

Fig. 13. Gravitational spreading model for the evolution of the south polar terrain (SPT), after *Yin and Pappalardo* (2015). Rising warm ice caused crustal thinning, viscosity reduction, lateral flow of the ice shell from the leading edge toward the trailing edge, and closely spaced folding above a shallow decollement. Initial folding was replaced by left-slip faulting to accommodate distributed simple shear across the SPT, creating the four active tiger stripe fractures (A–D) and one additional prominent fracture zone (E). The state of strain is extensional and transtensional in the upslope SPT, and contractional and transtensional in the downslope SPT.

Morphologies indicative of contraction, extension, and shear are all inferred within the circumferential belt that partially encompasses the THT (*Spencer et al.,* 2009; *Crow-Willard and Pappalardo,* 2015). Analysis of stereo-derived topography along the Harran Sulci indicate the presence of uplift consistent with flexure, implying an effective elastic thickness of as little as 0.3 km at the time of formation (*Giese et al.,* 2008). Portions of Sarandib Planitia and the Harran Sulci have been modeled as resulting from east-west oriented, unstable extensions of the lithosphere (*Bland et al.,* 2007). *Crow-Willard and Pappalardo* (2015) indicate that sigmoidal patterns within Sarandib Planitia also suggest a component of shear and superimposed low, large ridges suggest folding, during or after extension. The formation of dorsa within the THT have been interpreted as linear cryovolcanic extrusions from fractures (*Spencer et al.,* 2009) or as thrust blocks resulting from contractional strain (*Pappalardo et al.,* 2010; *Patthoff et al.,* 2015).

Two additional geologic terrains associated with the THT are defined by *Crow-Willard and Pappalardo* (2015) as a southern transitional unit and a northern lineated unit (Fig. 2; Table 2). The southern transitional unit is characterized by sets of 1–3-km-wide, curvilinear troughs that are subparallel to slightly oblique and crosscut at 20° to 30° angles. Material between the troughs is described as smooth to lineated with several tectonic domains of varying fabric present. The northern lineated unit is characterized by roughly north-south-trending troughs that disrupt ancient cratered terrains north of the Samarkand Sulci. Superposed craters at diameters <1 km are more commonly observed within this unit than for the central regions of the THT.

The southern transitional unit appears to record multiple generations of tectonic deformation, based on the orientations and cross-cutting relationships of troughs at the eastern and western boundaries of the unit as it grades into surrounding ancient cratered terrains (*Crow-Willard and Pappalardo,* 2015). The unit shares characteristics with the SPT and crosscuts all other features within the THT, implying it is transitional in age with respect to the two terrains (*Crow-Willard and Pappalardo,* 2015). However, recent work on the density of craters in these regions (*Kinczyk et al.,* 2017) suggests the transitional unit is distinctly older than the SPT and the LHT (section 6.2.2).

Initial interpretation of the northern lineated unit based on kilometer-scale Voyager images suggested the possibility of cryovolcanism (*Squyres et al.,* 1983). However, Cassini image data of the region clearly show subparallel troughs and dissected impact craters indicative of extensional tectonic deformation. *Crow-Willard and Pappalardo* (2015) interpret this unit as analogous to the kind of tectonic resurfacing proposed for some grooved terrains on Ganymede (cf. *Pappalardo et al.,* 2004).

6.2.2. Leading hemisphere terrain. The LHT is centered close to Enceladus' leading point (*Crow-Willard and Pappalardo,* 2015) and is bound by a subcircular circumferential belt characterized by curvilinear ropy ridges (section 4.1.1) and troughs (section 4.1.2) of variable width (Fig. 15). Structures within the belt can be grouped into domains, consisting of sets of features with similar orientations. *Crow-Willard and Pappalardo* (2015) divide the central region of the LHT into two geologic units: a smooth and a central unit (Fig. 2). They describe the smooth unit as consisting predominantly of subparallel, approximately east-west-trending ridges 10–15 km wide and ~25–80 km long among a background of smaller subparallel ridges and troughs. The smaller ridges and troughs appear analogous to features observed in the THT but with less organization and continuity. A few north-south-trending ridges 8 km wide and 80 km long are also present. They describe the central unit as characterized by ropy and interlaced ridges and troughs

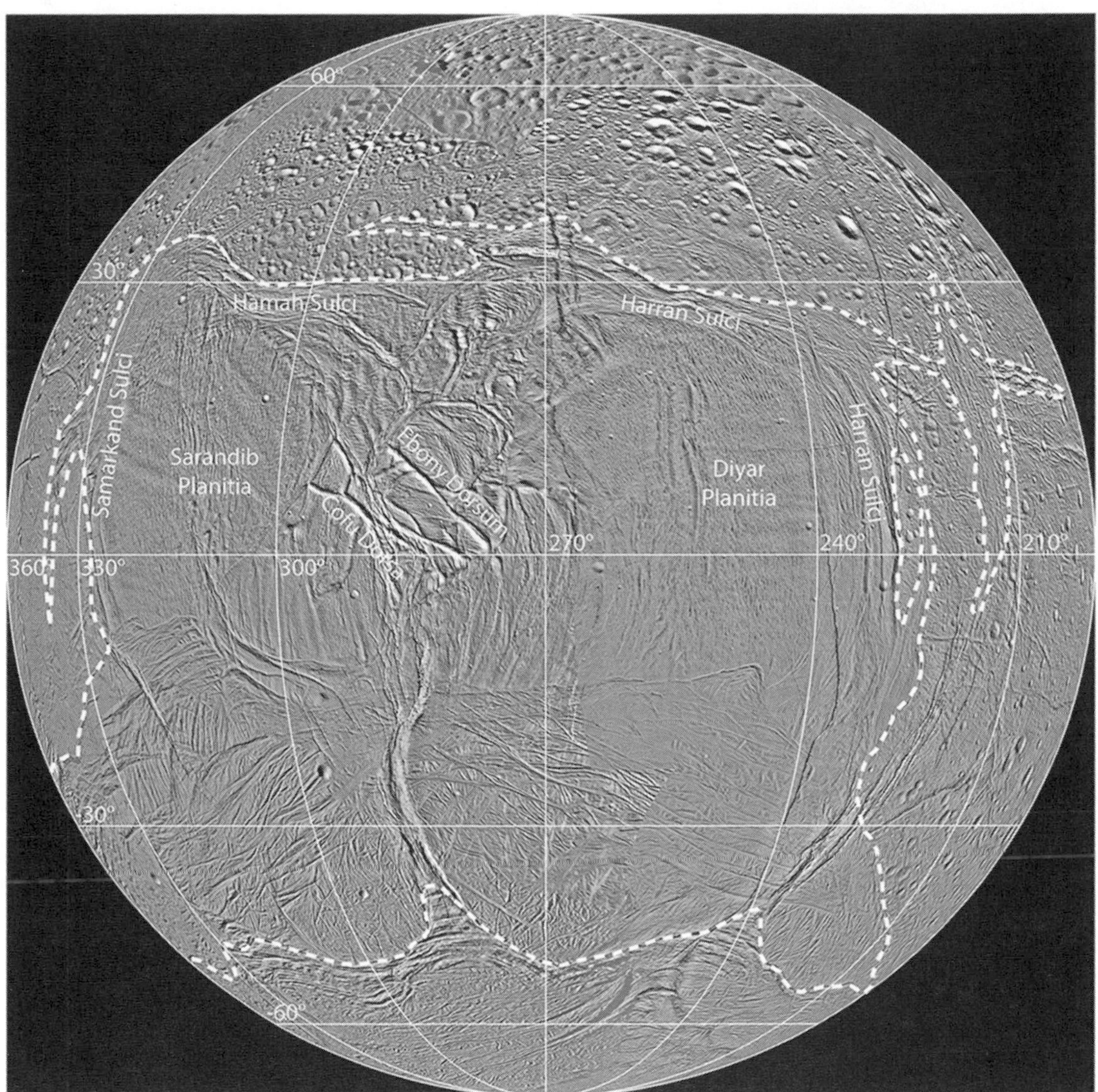

Fig. 14. Orthographic controlled global image mosaic (*Becker et al.,* 2016) for the trailing hemisphere of Enceladus (positive west longitude). Prominent tectonic features that define the region are labeled. The approximate boundary of the THT is indicated with a dashed line.

without a preferred orientation and draw analogy to funicular ridges (section 4.1.1) of the SPT. Structural domains consisting of collections of subparallel structures oriented oblique to their boundaries appear throughout the unit and at least two "islands" of ancient cratered terrain are present within the unit (Fig. 15).

The geology of the LHT suggests a history of intense tectonic deformation (*Helfenstein et al.,* 2010a,b). Morphologies indicative of contraction, extension, and shear all appear present within the circumferential belt bounding the LHT, similar to what is observed for the THT and SPT. The smooth unit described by *Crow-Willard and Pappalardo* (2015) also contains morphological evidence of contraction, extension, and shear. The east-west trend of prominent features suggests a regional stress oriented north-south, and the scale of deformation suggests a relatively thin effective elastic lithosphere. The ropy ridges of the central unit suggest contractional folding (*Helfenstein et al.,* 2010a,b; *Crow-Willard and Pappalardo,* 2015), and the inferred topographic wavelength of the folds is consistent with essentially ductile near-surface behavior.

6.2.3. Models for the formation of trailing hemisphere terrain and leading hemisphere terrain tectonic features. The THT and LHT share broad similarities with both each other and with the SPT. Their areal extents are similar within a factor of 2.5 (*Crow-Willard and Pappalardo,* 2015). All are bound by circumferential, curvilinear tectonized belts that enclose (or partially do, in the case of the THT) central regions with multiple distinct terrains. The belts each contain subparallel ridges and troughs with complex deformation histories that may include extensional, contractional, and shear-related structures (sections 6.2.1 and 6.2.2) (*Crow-Willard and Pappalardo,* 2015). However, the THT and LHT do not share the prominent Y-shaped structural

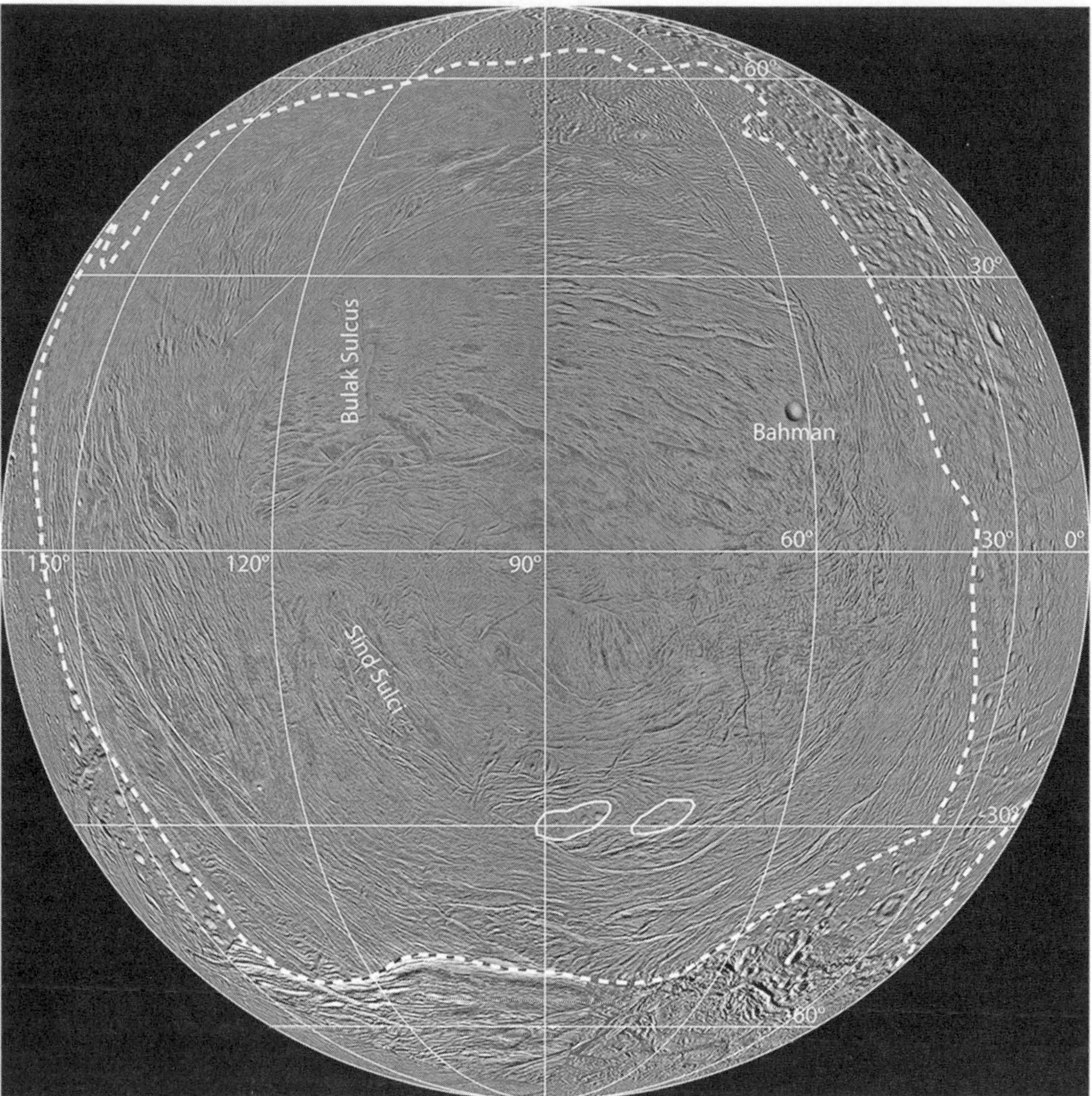

Fig. 15. Orthographic controlled global image mosaic (*Becker et al.,* 2016) for the leading hemisphere of Enceladus (positive west longitude). Prominent tectonic features and a young, fresh crater that define the region are labeled. The approximate boundary of the LHT is indicated with a dashed line and the boundaries of two "islands" of cratered plains are indicated with a solid line.

discontinuities observed in association with the SPT and, despite suggestions to the contrary based on initial structural mapping efforts using lower-resolution data (i.e., *Spencer et al.,* 2009; *Helfenstein et al.,* 2010a), clear evidence for tiger-stripe-like structures within the THT and LHT is not present.

The circumferential belts and distinct interior terrains of the THT, LHT, and SPT appear to share similarities with coronae on Miranda (*Pappalardo et al.,* 1997). The formation of those features have been attributed to vertical loading on a curved plate (e.g., *Janes and Melosh,* 1988; *Pappalardo et al.,* 1997; *Hammond and Barr,* 2014). *Janes and Melosh* (1988) demonstrated that loads associated with a rising or sinking mass anomaly would produce surface deformation in concentric patterns. A rising mass anomaly would yield central, disorganized extension bound by a concentric zone of extension that is, in turn, bound by an outer concentric strike-slip zone. A sinking mass anomaly would yield central, disorganized contraction bound by a concentric zone of contraction that is also, in turn, bound by an outer concentric strike-slip zone. As with the coronae on Miranda, the locations of the THT, LHT, and SPT on Enceladus are aligned with satellite principal axes of inertia. This suggests the presence of long-lived mass anomalies associated with the terrains, and suggests that formation mechanisms involving upwelling, downwelling, or a combination of the two are plausible. To that end, *Crow-Willard and Pappalardo* (2015) have suggested that the geologic histories of the THT and LHT can be broadly characterized with a two-stage scenario that involves an initial upwarping of the region followed by later downwarping. The upwarping stage of formation produces extensional features observed within the central regions of the THT (section 6.2.1) and LHT (section 6.2.2) and the downwarping stage of formation produces contractional features observed within the same regions. Such a scenario would imply reactivation of normal faults as contractional structures within the terrains.

6.3. Ancient Epoch

The cratered plains of Enceladus are the most ancient surfaces observed on the satellite and preserve the only record of geological events on the satellite before ~1 Ga. The cratered plains are not a pristine record of early events; they have also been modified recently by activity that may be related to the resurfacing of terrains at the LHT, THT, and SPT. The cratered plains occupy the interstices of Enceladus geography left over after those three terrains formed, similar to the geography of cratered terrain on Miranda (*Smith et al.,* 1986). This results in a narrow (~300 km wide) strip of cratered terrain stretching north-south, from 50°S in the sub-Saturn hemisphere, through the sub-Saturn point, over the north pole (Fig. 16), through the anti-Saturn point, to 50°S in the anti-Saturn hemisphere. With this geographical distribution, it is reasonable to assume that the cratered plains occupied much more (or perhaps all) of the surface of Enceladus before resurfacing of the LHT, THT, and SPT occurred.

The density of craters on the cratered plains indicates that the surface could be primordial, or at least older than 1 b.y. (*Kirchoff and Schenk,* 2009). Crater density is the highest at high northern and southern latitudes, while an equatorial band within the cratered plains from approximately 25°N to 25°S exhibits a decrease in the density of craters (*Kirchoff and Schenk,* 2009; *Kinczyk et al.,* 2017) (see also Fig. 4). Craters <2 km diameter are underrepresented, which could be due to preferential burial of small craters. *Kirchoff and Schenk* (2009) note that the equatorial band has morphologically "softened" craters, supporting the burial hypothesis, and they note that infall of E-ring material concentrated in the equatorial plane could be driving this burial process. Direct burial by plume material could certainly play a role, but there is no systematic decrease in small craters with increasing southern latitude, as one would expect if burial by plume particles was responsible (*Kempf et al.,* 2010). Indeed, current inferred rates of plume burial near the equator are not high enough to erase craters, even given billions of years (*Bland et al.,* 2012), so other processes must be at work. Viscous relaxation could subdue crater topography, and many large craters in the cratered plains appear to be anomalously shallow (section 3.2). The topography of many large craters in the cratered plains exhibits central floors updomed above the crater rims, which is a classic prediction of viscous relaxation of initial bowl-shaped crater topography. The amount of topographic relaxation observed cannot be explained without high heat flows, comparable to heat flows inferred for the younger, tectonically active terrains (*Bland et al.,* 2012). Craters are particularly flattened in limited areas, such as the northern trailing hemisphere (e.g., the craters Aladdin and Ali Baba in Fig. 16). Taken as a whole, the morphological characteristics of craters within the cratered plains indicate that plume fallout and subsurface heat flow have probably varied dramatically through time and space, so the present activity level is not necessarily indicative of what has happened in the past (i.e., the cratered plains integrate the variable heat flow and particle deposition history of the surface).

Additional evidence for spatially variable heat flow comes from the topography of the cratered plains and surrounding terrain. Several basins 100–200 km across and ~1 km deep are evident in both stereo-derived DEMs and limb profiles (*Schenk and McKinnon,* 2009; *Nimmo et al.,* 2011; *Tajeddine et al.,* 2017). These basins (section 4.2) are uncorrelated with any geological surface features or terrain boundaries. They could represent dynamic topography of downwelling cold ice, or isostatic topography of thin areas in the ice shell (*Schenk and McKinnon,* 2009), or they could be regions where enhanced heat flow from upwelling convective plumes in the ice shell has viscously compacted the porous upper ice shell (*Besserer et al.,* 2013). All these mechanisms posit spatial variations in heat flow, but there is no straightforward relationship between the locations of relaxed craters and the locations of these broad topographic basins, as one might expect if localized regions of heat flow were responsible for both of them.

Although the cratered plains are not as thoroughly tectonized as the rest of the terrains on Enceladus, they do exhibit a variety of tectonic features. *Nahm and Kattenhorn* (2015) proposed a classification scheme for tectonic features on Enceladus, as discussed in section 4; within this scheme the structures in the cratered plains would be classified as chasmata, troughs, pit chains, and ridges. Pit chains (Fig. 7a) and narrow chasmata are the most recent tectonic features, based on the observation that they crosscut other tectonic features and preexisting craters.

Pit chains are predominantly found in the cratered terrain south of 40°N (*Martin et al.,* 2017) and appear to be associated with recent tectonic activity that has occurred outside the cratered plains. *Martin and Kattenhorn* (2014) describe sets of pit chains in antipodal subequatorial locations on the sub-Saturn and anti-Saturn hemisphere of the cratered plains. In each location, at least seven sets of systematic fractures are inferred to have formed in a sequential manner, with each set forming with a different orientation to the other sets. These changing orientations imply that the regional stress field in these broad areas changes through time, but with a regionally consistent pattern that implies a global state of stress in order to create antipodal hemispheric symmetry. Such a pattern of stress through time is consistent with NSR of the ice shell through a tidally locked stress field, resulting in a consistent rotation of the principal stresses through time at the locations of pit chain development. In order to account for the range of orientations of observed systematic sets, *Martin and Kattenhorn* (2014) infer up to 255° of rotation of the ice shell relative to the current Saturn direction [i.e., more than the 153° of rotation required to explain the young fracture sets in the SPT; section 6.1.3.1 (*Patthoff and Kattenhorn,* 2011)].

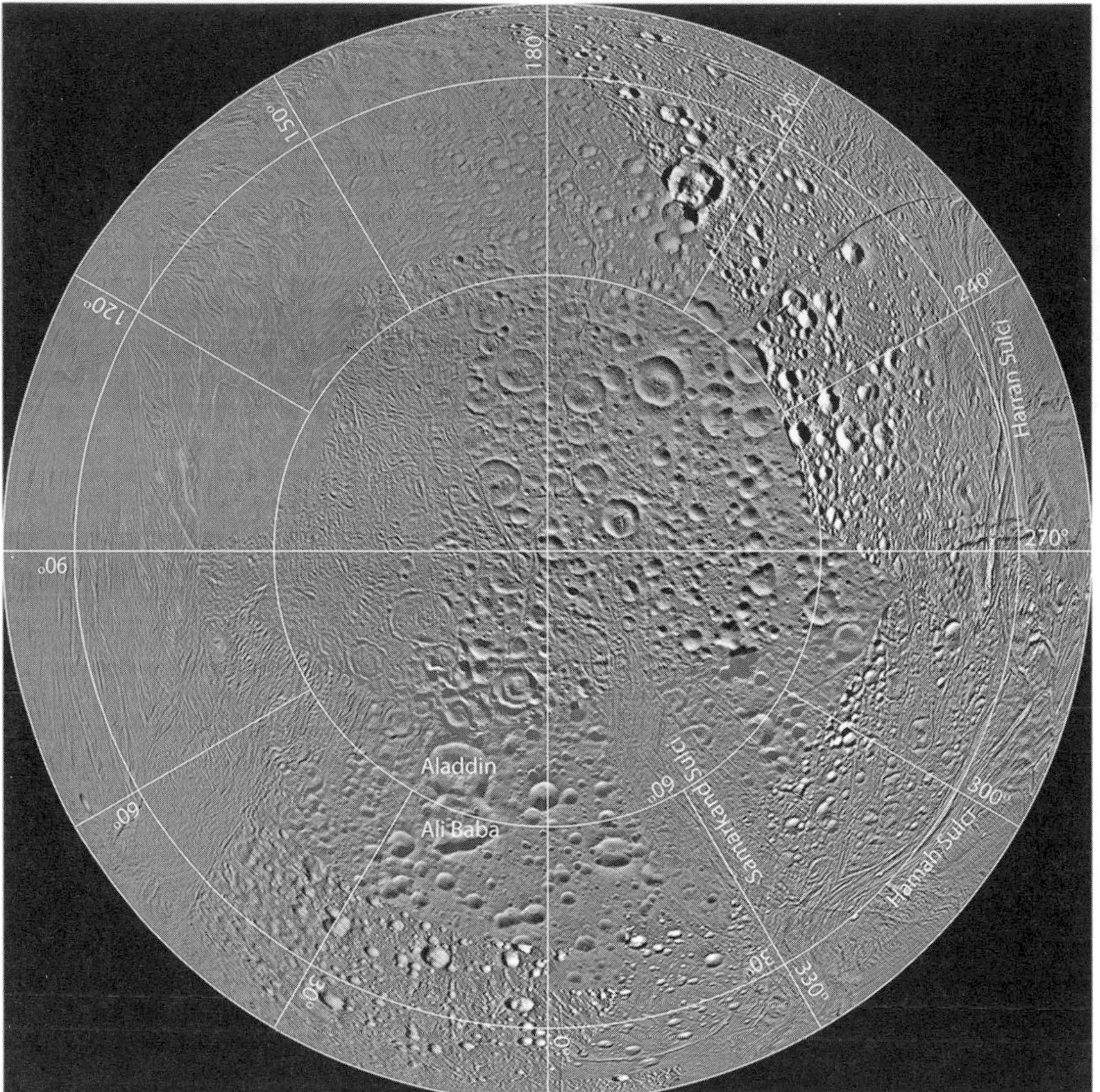

Fig. 16. Orthographic controlled global image mosaic (*Becker et al.,* 2016) for the northern hemisphere of Enceladus (positive west longitude). Prominent tectonic features of the LHT are labeled, for reference, as are the locations of the relaxed craters Aladdin and Ali Baba.

The cratered plains also exhibit several recent tectonic features interpreted to be created through strike-slip motion (*Martin* 2016). Several sets of *en echelon* fractures (section 4.1.6; Fig. 7b) cross the cratered plains or occur at its margins; most notably, a long set of left-lateral *en echelon* fractures bounds the eastern edge of the anti-saturnian cratered plains. A slightly older generation of wider chasmata (section 4.1.3; Fig. 6f) appear to be connected to the margins of the THT and LHT. The THT (*Crow-Willard and Pappalardo,* 2015) in particular has several branching networks of faults that extend out to form rift zones within the cratered plains.

Subdued ridges and troughs (Fig. 17) are the oldest tectonic features in the cratered plains, crosscut by all other tectonic features and overlain by many craters. These features are commonly 2 to 3 km wide, and can typically only be traced for a few tens of kilometers before they are interrupted by younger features. One might interpret the troughs as ancient graben and the ridges as ancient thrust faults, to be consistent with interpretations of more recent similar features elsewhere on Enceladus, but the formation details of the subdued ridges and troughs are obscure. No preexisting features can be mapped out that are cut by these features, so tectonic strain cannot be confidently assessed. Subdued ridges and troughs are most concentrated in the equatorial (±30° latitude) region of the cratered plains. The close relationship between the concentration of ancient tectonic features and the equatorial band of low crater density (section 2.1; Fig. 4) brings up an intriguing possibility: Could the equatorial band of cratered plains have experienced an ancient resurfacing event? No sharp boundaries have been found within the cratered plains to match the sharp boundaries identified elsewhere on Enceladus where resurfacing has been recently active. If tectonic resurfacing helped to reset the crater age in the equatorial cratered plains, either it behaved differently than in recent resurfacing episodes, or the margins of the terrain have become too heavily cratered and mantled by regolith to be recognizable today.

The early geological history preserved in Enceladus' cratered plains can be summarized as follows. (1) Craters accumulated on an ancient surface. (2) Tectonic activity created isolated ridges and troughs, possibly erasing some of the ancient craters in the equatorial regions. (3) Infall of plume and/or E-ring material blanketed the cratered plains and softened the appearance of the early craters, ridges, and troughs; note that this process may have continued through all subsequent stages (perhaps not always at the same rate) and still occurs today. (4) Localized heat pulses relaxed impact craters and possibly created large, shallow basins by compaction and/or melting. The sequence of this step relative to the other steps is uncertain. (5) Tectonic activity in the leading and trailing hemispheres resurfaced some areas presumably covered by cratered plains. Localized rifting associated with resurfacing in the trailing hemisphere created wide chasmata and sets of parallel normal faults. Resurfacing in the leading hemisphere created small rifts and strike-slip faults branching into the cratered plains. (6) Tectonic activity surrounding the south pole resurfaced more of the cratered plains area, and created strike-slip faults, narrow chasmata, and pit chains in the southern and equatorial cratered plains.

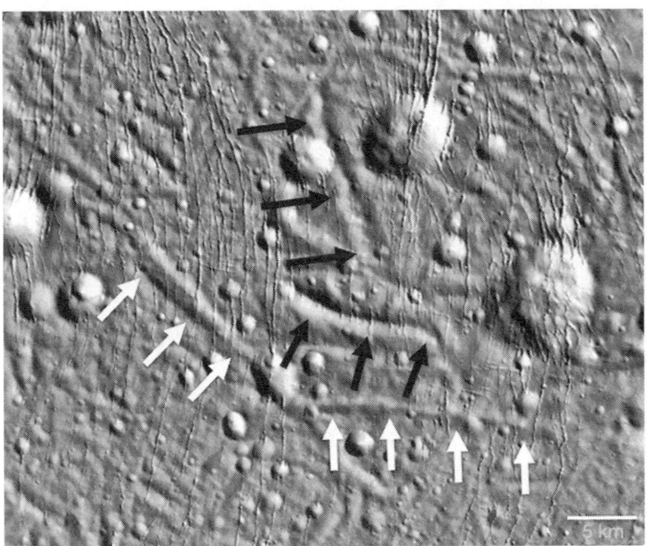

Fig. 17. Examples of subdued troughs (white arrows) and subdued ridges (black arrows) in the ancient terrain of Enceladus, near 0°N, 160°E.

7. SUMMARY

The Cassini mission operated in Saturn orbit from July 2004 through September 2017, acquiring thousands of images of the surface of Enceladus. These data, and products derived from them, have provided the primary source of information used to characterize the diverse geology of this unique icy satellite and to decipher the formation and evolution of its surface features (section 2). The morphology and distribution of observed craters on Enceladus, along with their interaction with structural features, have provided a window into differences in surface age, thermal history, and stress environment across the satellite's surface (section 3; Kirchoff et al., this volume). The range of morphologically disparate deformation features on the surface (section 4) have provided important insight into the formation of distinct terrains observed on the surface. Local-scale image data with pixel scales as high as tens of centimeters (Table 2) have provided useful information regarding how larger-scale terrain features have evolved through time under the influence of degradational processes (section 5). Taken together, this information provides the basis for developing a geologic history of the satellites observable surface (section 6). The observable geologic history of Enceladus provides a record of the internal processes and stressing patterns through time that have shaped the ice shell. The disparity between tectonized terrains on Enceladus (*Spencer et al.,* 2009; *Crow-Willard and Pappalardo,* 2015) and the older cratered terrains, combined with the dynamic geological activity of the tectonically distinct south polar terrain, implies a complex history and perhaps spatially variable endogenic processes in the ice shell. This variability makes the identification of driving mechanisms for tectonic resurfacing of the ice shell difficult to pin down, as these mechanisms are likely to differ. However, based on available data for Enceladus regarding regional- and local-scale geological relationships involving the distribution and morphology of both craters and tectonic features observed on the satellite, we can discuss plausible driving mechanisms for the formation and evolution of the SPT (section 6.1), the tectonized terrains of the leading and trailing hemispheres (section 6.2), and the ancient cratered plains (section 6.3).

The geologic history of the SPT marks the most recent stage of regional geologic activity on Enceladus. The dearth of craters in the SPT indicates a young surface age; eruptive jets of water-ice emanating from fractures indicate a geologically active area. The SPT forms a topographic basin surrounded by a tectonic belt that creates a boundary with cratered or tectonized terrains further north. The low elevations suggest a thinned ice shell above an interior thermal anomaly that results in the highest heat flows on the satellite (~4.7 GW). The SPT may have formed in place or may have been relocated to the south pole as a migrated mass anomaly. The SPT is pervasively internally fractured, including four dominant ridges (the "tiger stripes") that form the locus of the eruptive jets and imply that these cracks extend deep enough into the ice shell to tap into a liquid source. Fractures in the SPT occur in a range of orientations that have been variably interpreted to result from (1) the long-term effects of NSR stresses that formed multiple successive fracture sets during ongoing ice shell rotation, or (2) the effects of gravitational collapse along a shallow detachment fault above the thermal anomaly that resulted in internal dissection of the SPT along multiple sets of shear fractures. Tidal stresses related to diurnal forcing potentially reactivate the tiger stripe fractures daily in both tension and shear, and may be the dominant control on the timing and locations of jet eruptions. Tectonic activity within the SPT likely affected adjacent areas as well, presumably including accounting for the deformation of the polygonal tectonic belt that encloses the SPT, the spoke-like chasmata that extend radially northward from the SPT, and the tectonic dissection of relatively older terrains around the SPT.

The geologic histories of the THT and LHT record an intermediate stage of regional geologic activity. Crater densities within the terrains indicate that they predate the SPT and that the THT is older than the LHT. Similar to the SPT, the THT and LHT are bound by circumferential belts that partially or fully enclose central regions that harbor a range of tectonic structures. The THT includes the distinct Cufa Dorsa and Ebony Dorsum that split the Sarandib Planitia and Diyar Planitia. Its southern margin is unbound by circumferential tectonic structures and instead grades into the surrounding cratered plains. It also shares structural characteristics with SPT and may record multiple generations of tectonic deformation. The LHT more closely resembles a relict SPT, with a complete circumferential belt and a central region that includes ridge and trough morphologies analogous to features within the central SPT. The broad similarities in tectonic structures and areal extents of the three tectonized terrains on Enceladus harbor a passing resemblance to the coronae of Miranda, a uranian satellite of similar size and icy lithospheric composition to Enceladus. As with Miranda's coronae, the tectonized terrains of Enceladus located along principal inertia axes. This suggests that formation mechanisms that include the presence of long-lived mass anomalies are likely. However, connecting the formation of the THT, LHT, and SPT is complicated by a lack of evidence for structure analogous to the distinctive tiger stripes and Y-shaped structural discontinuities of the SPT.

The cratered plains of Enceladus that surround the THT, LHT, and SPT record the earliest stages of geologic activity on the satellite. Crater densities indicate that the cratered plains are greater than 1 b.y. in age, and may be primordial. The presence of topographic basins and topographically relaxed craters within the cratered plains suggest subsurface heat flows that have varied significantly in space and time. Although the cratered plains are not as thoroughly tectonized as the THT, LHT, and SPT, they do exhibit a variety of tectonic features. Subdued ridges and troughs present in equatorial regions of the cratered plains, coupled with anomalously low small-crater densities, suggest the possibility of an ancient resurfacing event. On the other hand, pit chains present within the cratered plains that have orientations indicating they are controlled NSR stresses suggest much more recent activity.

The combination of evidence from geologic structures in both the SPT and the cratered terrains north of the south polar dichotomy boundary suggests that at least some of the tectonic history of Enceladus is related to global tidal deformation of the ice shell. The possibility that significant tidal deformation plays a key role in Enceladus' tectonics has become increasingly plausible given recent evidence for the likely presence of a global ocean (*Thomas et al.,* 2016; *Iess et al.,* 2014). On the diurnal timescale, tidal distortion may be responsible for the cyclic nature of jet eruptions from the SPT tiger stripes as these fracture respond to daily stresses through a combination of shearing and crack-orthogonal motions that provide pathways for fluids to escape into space from the underlying ocean (*Hurford et al.,* 2007, 2009; *Nimmo et al.,* 2007, 2014). On a longer timescale of perhaps $\sim 10^5$–10^6 yr, a consistent nonsynchronous rotation of the ice shell above the global ocean could result in global stress fields that created regionally extensive systematic fracture sets in both the SPT and cratered terrains further north (*Patthoff and Kattenhorn,* 2011; *Martin and Kattenhorn,* 2014).

The wealth of information on Enceladus obtained by the Cassini mission to Saturn has truly revolutionized our understanding of this unique and diminutive icy satellite. However, many fundamental questions remain that will likely require new data to address: How old is the oldest terrain on Enceladus? Has Enceladus been active for a significant portion of its geologic history? What is the heat flow outside the SPT and how has it varied over time? What is the fundamental cause of the deformation that created the tectonized terrains? How are cratered terrains preserved given the very high heat flux implied by the tectonized terrains? What can the long wavelength topography of Enceladus tell us about its geophysical evolution? What can local-scale features tell us about spatial and temporal variability in the geologic evolution of Enceladus?

Acknowledgments. G.W.P., P.H., and S.A.K. gratefully acknowledge support from NASA's Cassini Data Analysis Program. The portion of this work performed by R.T.P. was carried out at the Jet Propulsion Laboratory, California Institute of Technology, under a contract with the National Aeronautics and Space Administration.

REFERENCES

Barnash A. N., Rathbun J. A., Turtle E. P., and Squyres S. W. (2006) Interactions between impact craters and tectonic fractures on Enceladus. *AAS/Division for Planetary Sciences Meeting Abstracts, 38,* #24.06.

Barr A. C. (2008) Mobile lid convection beneath Enceladus' south polar terrain. *J. Geophys. Res., 113,* E07009, DOI: 10.1029/2008JE003114.

Barr A. C. and McKinnon W. B. (2007) Convection in Enceladus' ice shell: Conditions for initiation. *Geophys. Res. Lett., 34,* L09202, DOI: 10.1029/2006GL028799.

Barr A. C. and Preuss L. J. (2010) On the origin of south polar folds on Enceladus. *Icarus, 208,* 499–503.

Becker T. L., Bland M. T., Edmundsen K. L., Soderblom L. A., Takir D., Patterson G. W., Collins G. C., Pappalardo R. T., Roatsch T., and Schenk P. M. (2016) Completed global control network and basemap of Enceladus. In *Lunar and Planetary Science XLVII,* Abstract #2342. Lunar and Planetary Institute, Houston.

Běhounková M., Tobie G., Choblet G., and Cadek O. (2012) Tidally-induced melting events as the origin of south-pole activity on Encealdus. *Icarus, 219,* 655–664.

Běhounková M., Tobie G., Choblet G., and Cadek O. (2013) Impact of tidal heating on the onset of convection in Enceladus's ice shell. *Icarus, 226,* 898–904.

Benesh M. and Jepsen P. (1978) *Voyager Imaging Science Subsystem Calibration Report.* JPL Rep. No. 618-802, 308 pp. Jet Propulsion Laboratory, Pasadena.

Besserer J., Nimmo F., Roberts J. H., and Pappalardo R. T. (2013) Convection-driven compaction as a possible origin of Enceladus's long wavelength topography. *J. Geophys. Res.–Planets, 118,* DOI: 10.1002/jgre.20079.

Bland M. T., Beyer R. A., and Showman A. P. (2007) Unstable extension of Enceladus' lithosphere. *Icarus, 192,* 92–105, DOI: 10.1016/j.icarus.2007.06.011.

Bland M. T., Singer K. N., McKinnon W. B., and Schenk P. M. (2012) Enceladus' extreme heat flux as revealed by its relaxed craters. *Geophys. Res. Lett., 39,* L17204, DOI: 10.1029/2012GL052736.

Bland M. T., McKinnon W. B., and Schenk P. M. (2015) Constraining the heat flux between Enceladus' tiger stripes: Numerical modeling of funiscular plains formation. *Icarus, 260,* 232–245.

Bray V. J., Smith D. E., Turtle E. P., Perry J. E., Rathbun J. A., Barnash A. N., Helfenstein P., and Porco C. C. (2007) Impact crater morphology variations on Enceladus. In *Lunar and Planetary Science XXXVIII,* Abstract #1873. Lunar and Planetary Institute, Houston.

Collins G. C. and Goodman J. C. (2007) Enceladus' south polar sea. *Icarus, 189,* 72–82.

Crow-Willard E. N. and Pappalardo R. T. (2015) Structural mapping of Enceladus and implications for formation of tectonized regions. *J. Geophys. Res.–Planets, 120,* 928–950, DOI: 10.1002/2015JE004818.

Cruikshank K. M. and Aydin A. (1995) Unweaving the joints in Entrada Sandstone, Arches National Park, Utah, U.S.A. *J. Structural Geol., 17(3),* 409–421, DOI: 10.1016/0191-8141(94)00061-4.

Dougherty M. K., Khurana K. K., Neubauer F. M., Russell C. T., Saur J., Leisner J. S., and Burton M. E. (2006) Identification of a dynamic atmosphere at Enceladus with the Cassini magnetometer. *Science, 311,* 1406–1409.

Durham W. B., McKinnon W. B., and Stern L. A. (2005) Cold compaction of water ice. *Geophys. Res. Lett., 32,* L18202, DOI: 10.1029/2005GL023484.

Eluszkiewicz J. (1990) Compaction and internal structure of Mimas. *Icarus, 84,* 215–225, DOI: 10.1016/0019-1035(90)90167-8.

Ferrill D. A., Morris A. P., Wyrick D. Y., Sims D. W., and Franklin N. M. (2004) Dilational fault slip and pit chain formation on Mars. *Geol. Soc. Am. Today, 14(10),* 4–12. DOI: 10.1130/1052-5173(2004)014<4:DFSAPC>2.0.CO;2.

Geissler P. E., Greenberg R., Hoppa G., Helfenstein P., McEwen A., Pappalardo R., Tufts R., Ockert-Bell M., Sullivan R., Greeley R., Belton M. J. S., Denk T., Clark B., Burns J., Veverka J., and the Galileo imaging Team (1998) Evidence for non-synchronous rotation of Europa. *Nature, 391,* 368–370, DOI: 10.1038/34869.

Giese B., Wagner R., Hussmann H., Neukum G., Perry J., Helfenstein P., and Thomas P. C. (2008) Enceladus: An estimate of heat flux and lithospheric thickness from flexurally supported topography. *Geophys. Res. Lett., 35,* L24204.

Giese B. and the Cassini Imaging Team (2010a) The topography of Enceladus. *European Planet. Sci. Congr. 2010,* 675.

Giese B., Helfenstein P., Thomas P. C., Ingersoll A. P., Perry J., Wagner R., Neukum G., and Porco C. C. (2010b) The morphology of an active zone near Enceladus' south pole and implications. *European Geophys. Union General Assembly Conf. Abstracts, 12,* 11085.

Giese B., Hussmann H., Roatsch T., Helfenstein P., Thomas P. C., and Neukum G. (2011) Enceladus: Evidence for librations forced by Dione. *EPSC-DPS Joint Meeting 2011,* 976.

Giese B., Helfenstein P., Hauber E., Hussmann H., and Wagner R. (2017) An exceptionally high standing ridge on Enceladus. *European Planet. Sci. Congr. 2017,* 28.

Gioia G., Chakraborty P., Marshak S., and Kieffer S. W. (2007) Unified model of tectonics and heat transport in a frigid Enceladus. *Proc. Natl. Acad. Sci., 104,* 13578–13581.

Goguen J. D., Buratti B. J., Brown R. H., Clark R. N., Nicholson P. D., Hedman M. M., Howell R. R., Sotin C., Cruikshank D. P., Baines K. H., Lawrence K. J., Spencer J. R., and Blackburn D. G. (2013) The temperature and width of an active fissure on Enceladus measured with Cassini VIMS during the 14 April 2012 south pole flyover. *Icarus, 226,* 1128–1137.

Greeley R., Figueredo P. H., Williams D. A., Chuang F. C., Klemaszewski J. E., Kadel S. D., Prockter L. M., Pappalardo R. T., Head J. W., Collins G. C., Spaun N. A., Sullivan R. J., Moore J. M., Senske D. A., Tufts B. R., Johnson T. V., Belton M. J. S., and Tanaka K. L. (2000) Geologic mapping of Europa. *J. Geophys. Res., 105,* 22559–22578.

Grott M., Sohl F., and Hussmann H. (2007) Degree-one convection and the origin of Enceladus' dichotomy. *Icarus, 191,* 203–210.

Halvey I. and Stewart S. T. (2008) Is Enceladus' plume tidally controlled? *Geophys. Res. Lett., 35,* L12203, DOI: 10.1029/2008GL034349.

Hammond N. P. and Barr A. C. (2014) Global resurfacing of Uranus's moon Miranda by convection. *Geology, 42(11),* 931–934.

Han L., Tobie G., and Showman A. P. (2012) The impact of a weak south pole on thermal convection in Enceladus' ice shell. *Icarus, 218,* 320–330, DOI: 10.1016/j.icarus.2011.12.006.

Hansen C. J., Esposito L., Stewart A. I. F., Colwell J., Hendrix A., Pryor W., Shemansky D., and West R. (2006) Enceladus' water vapor plume. *Science, 311,* 1422–1425.

Hansen C. J., Esposito L. W., Stewart A. I. F., Meinke B., Wallis B., Colwell J. E., Hendrix A. R., Larsen K., Pryor W., and Tian F. (2008) Water vapor jets inside the plume gas leaving Enceladus. *Nature, 456,* 477–479.

Helfenstein P. (2014) Y-shaped discontinuity. In *Encyclopedia of Planetary Landforms* (H. Hargitai et al., eds.), Springer, New York, DOI 10.1007/978-1-4614-9213-9.

Helfenstein P. and Parmentier E. M. (1985) Patterns of fracture and tidal stresses due to nonsynchronous rotation: Implications for fracturing on Europa. *Icarus, 61,* 175–184, DOI: 10.1016/0019-1035(85)90099-5.

Helfenstein P. and Porco C. C. (2015) Enceladus geysers: Relation to geological features. *Astron. J., 150,* 96, DOI:10.1088/0004-6256/150/3/96.

Helfenstein P., Thomas P. C., Veverka J., Rathbun J., Perry J., Turtle E., Denk T., Neukum G., Roatsch T., Wagner R., Giese B., Squyres S., Burns J., McEwen A., Porco C., Johnson T. V., and the ISS Science Team (2006) Patterns of fracture and tectonic convergence near the south pole of Enceladus. In *Lunar and Planetary Science XXXVII,* Abstract #2182. Lunar and Planetary Institute, Houston.

Helfenstein P., Veverka J., Thomas P. C., Perry J., Denk T., Neukum G., Giese B., Roatsch T., Turtle E. P., and Porco C. C. (2010a) The leading side of Enceladus: New views from Cassini ISS. *Bull. Am. Astron. Soc., 42,* 976, 16.01.

Helfenstein P., Giese B., Perry J. E., Roatsch T., Veverka J., Thomas P. C., Denk T., Neukum G., and Porco C. (2010b) The leading side of Enceladus: Clues to early volcanism and tectonism from Cassini ISS. Abstract #P23C-04 presented at 2010 Fall Meeting, AGU, San Francisco, California, 13–17 December.

Helfenstein P., Thomas P. C., and Veverka J. (2013) Enceladus close up: New details recovered from Cassini ISS boresight-drag images. *AAS/Division for Planetary Sciences, 45,* 416.05.

Hoppa G. V., Tufts B. R., Greenberg R., and Geissler P. E. (1999) Formation of cycloidal features on Europa. *Science, 285,* 1899–1902, DOI: 10.1126/science.285.5435.1899.

Howett C. J. A., Spencer J. R., Pearl J., and Segura M. (2011) High heat flow from Enceladus' south polar region measured using 10–600 cm^{-1} Cassini/CIRS data. *J. Geophys. Res., 116,* E03003, DOI: 10.1029/2010JE003718.

Hurford T. A., Helfenstein P., Hoppa G., Greenberg R., and Bills B. (2007) Eruptions arising from tidally controlled periodic openings of rifts on Enceladus. *Nature, 447,* 292–294, DOI: 10.1038/nature05821.

Hurford T. A., Bills B. G., Helfenstein P., Greenberg R., Hoppa G. V., and Hamilton D. P. (2009) Geological implications of a physical libration on Enceladus. *Icarus, 203,* 541–552, DOI: 10.1016/j.icarus.2009.04.025.

Hurford T. A., Helfenstein P., and Spitale J. (2012) Tidal control of jet eruptions on Enceladus as observed by Cassini ISS between 2005 and 2007. *Icarus, 220,* 896–903.

Iess L., Stevenson D. J., Parisi M., Hemingway D., Jacobson R. A., Lunine J. I., Nimmo F., Armstrong J. W., Asmar S. W., Ducci M., and Tortora P. (2014) The gravity field and interior structure of Enceladus. *Science, 344,* 78–80, DOI: 10.1126/science.1250551.

Ingersoll A. P. and Pankine A. A. (2010) Subsurface heat transfer on Enceladus: Conditions under which melting occurs. *Icarus, 206,* 594–607.

Janes D. M. and Melosh H. J. (1988) Sinker tectonics: An approach to the surface of Miranda. *J. Geophys. Res., 93,* 3127–3143.

Kargel J. S. and Pozio S. (1996) The volcanic and tectonic history of Enceladus. *Icarus, 119,* 385–404.

Kattenhorn S. A. (2002) Nonsynchronous rotation evidence and fracture history in the Bright Plains region, Europa. *Icarus, 157,* 490–506, DOI: 10.1006/icar.2002.6825.

Kattenhorn S. A. and Hurford T. (2009) Tectonics of Europa. In *Europa* (R. T. Pappalardo et al., eds.), pp. 199–223. Univ. of Arizona, Tucson.

Kempf S., Beckmann U., and Schmidt J. (2010) How the Enceladus dust plume feeds Saturn's E ring. *Icarus, 206,* 446–457.

Kieffer S. W., Lu X., Bethke C. M., Spencer J. R., Marshak S., and Navrotsky A. (2006) A clathrate reservoir hypothesis for Enceladus' south polar plume. *Science, 314,* 1764–1766, DOI: 10.1126/science.1133519.

Kinczyk M. J., Patterson G. W., Perkins R. P., Collins G. C., Borrelli M., Becker T. L., Bland M. T., and Pappalardo R. T. (2017) Evaluation of impact crater distributions for geological terrains on Enceladus. In *Lunar and Planetary Science XLVIII,* Abstract #2926. Lunar and Planetary Institute, Houston.

Kirchoff M. R. and Schenk P. (2009) Crater modification and geologic activity in Enceladus' heavily cratered plains: Evidence from the impact crater distribution. *Icarus, 206,* 656–668.

Kronberg P., Hauber E., Grott M., Werner S. C., Schafer T., Gwinner K., Giese B., Masson P., and Neukum G. (2007) Acheron Fossae, Mars: Tectonic rifting, volcanism, and implications for lithospheric thickness. *J. Geophys. Res., 112,* E04005, DOI: 10.1029/2006JE002780.

Lachenbruch A. H. (1962) *Mechanics of Thermal Contraction Cracks and Ice-Wedge Polygons in Permafrost.* GSA Special Paper 70, Geological Society of America, New York.

Landry B., Munsill L., Collins G., and Mitchell K. (2014) Observations about boulders on the south polar terrain of Enceladus. In *Lunar and Planetary Science XLV,* Abstract #2317. Lunar and Planetary Institute, Houston.

Marshall S. T. and Kattenhorn S. A. (2005) A revised model for cycloid growth mechanics on Europa: Evidence from surface morphologies and geometries. *Icarus, 177,* 341–366, DOI: 10.1016/j.icarus.2005.02.022.

Martens H. R., Ingersoll A. P., Ewald S. P., Helfenstein P., and Giese B. (2015) Spatial distribution of ice blocks on Enceladus and implications for their origin and emplacement. *Icarus, 245,* 162–176.

Martin E. S. (2014) The fractured ice shell of Saturn's moon Enceladus: Insights into the global stress history and interior structure. Ph.D. thesis, University of Idaho, Moscow. 109 pp.

Martin E. S. (2016) The distribution and characterization of strike-slip faults on Enceladus. *Geophys. Res. Lett., 43,* 2456–2464, DOI: 10.1002/2016GL067805.

Martin E. S. and Kattenhorn S. A. (2011) Crater-fracture interactions on Enceladus: The control of crater size on perturbations of fracture growth. In *Lunar and Planetary Science XLII,* Abstract #2666. Lunar and Planetary Institute, Houston.

Martin E. S. and Kattenhorn S. A. (2012) Crater induced fracture reorientation on Enceladus. In *Lunar and Planetary Science XLIII,* Abstract #2883. Lunar and Planetary Institute, Houston.

Martin E. S. and Kattenhorn S. A. (2013) Probing regolith depths on Enceladus by exploring a pit chain proxy. In *Lunar and Planetary Science XLIV,* Abstract #2047. Lunar and Planetary Institute, Houston.

Martin E. S. and Kattenhorn S. A. (2014) A history of pit chain formation within Enceladus's cratered terrains suggests a nonsynchronous rotation stress field. In *Lunar and Planetary Science XLV,* Abstract #1083. Lunar and Planetary Institute, Houston.

Martin E. S., Kattenhorn S. A., Collins G. C., Michaud R. L., Pappalardo R. T., and Wyrick D. (2017) Pit chains on Enceladus signal the recent tectonic dissection of the ancient cratered terrains. *Icarus, 294,* 209–217.

Matson D., Castillo-Rogez J. C., Davies A. G., and Johnson T. V. (2012) Enceladus: A hypothesis for bringing both heat and chemicals to the surface. *Icarus, 221,* 53–62.

Matson D., Davies A. G., Johnson T. V., Combe J.-P., McCord T. B., Radebaugh J., and Singh S. (2018) Enceladus's near-surface CO_2 gas pockets and surface frost deposits. *Icarus, 302,* 18–26.

Miller D. J., Barnash A. N., Bray V. J., Turtle E. P., Helfenstein P., Squyres S. W., and Rathbun J. A. (2007) Interactions between craters and tectonic fractures on Enceladus and Dione. In *Workshop on Ices, Oceans, and Fire: Satellites of the Outer Solar System,* pp. 95–96. LPI Contribution No. 1357, Lunar and Planetary Institute, Houston.

Muirhead J. D., Kattenhorn S. A., Lee H., Mana S., Turrin B. D., Fischer T. P., Kianji G., Dindi E., and Stamps D. S. (2016) Evolution of upper crustal faulting assisted by magmatic volatile release during early-stage continental rift development in the East African Rift. *Geosphere, 12(6),* DOI: 10.1130/GES01375.1.

Nahm A. L. and Kattenhorn S. A. (2015) A unified nomenclature for tectonic structures on the surface of Enceladus. *Icarus, 258,* 67–81, DOI: 10.1016/j.icarus.2015.06.009.

Nahm A. L. and Schultz R. A. (2013) Rupes Recta and the geological history of the Mare Nubium region of the Moon: Insights from forward mechanical modelling of the 'Straight Wall.' In *Volcanism and Tectonism Across the Inner Solar System* (T. Platz et al., eds.), Geological Society of London Spec. Publ. 401, DOI: 10.1144/SP401.4.

Nimmo F. and Pappalardo R. T. (2006) Diapir-induced reorientation of Saturn's moon Enceladus. *Nature, 441,* 614–616.

Nimmo F. and Schenk P. (2006) Normal faulting on Europa: Implications for ice shell properties. *J. Structural Geol., 28,* 2194–2203, DOI: 10.1016/j.jsg.2005.08.009.

Nimmo F., Spencer J. R., Pappalardo R. T., and Mullen M. E. (2007) Shear heating as the origin of the plumes and heat flux on Enceladus. *Nature, 447,* 289–291.

Nimmo F., Bills B. G., and Thomas P. C. (2011) Geophysical implications of the long-wavelength topography of the saturnian satellites. *J. Geophys. Res., 116,* E11001, DOI: 10.1029/2011JE003835

Nimmo F., Porco C., and Mitchell C. (2014) Tidally modulated eruptions on Enceladus: Cassini ISS observations and models. *Astron. J., 148(46),* DOI: 10.1088/0004-6256/148/3/46.

Olgin J. G., Smith-Konter B. R., and Pappalardo R. T. (2011) Limits of Enceladus's ice shell thickness from tidally driven tiger stripe shear failure. *Geophys. Res. Lett., 38,* L02201. DOI: 10.1029/2010GL044950.

O'Neill C. and Nimmo F. (2010) The role of episodic overturn in generating the surface geology and heat flow on Enceladus. *Nature Geosci., 3,* 88–91.

Pappalardo R. T., Reynolds S. J., and Greeley R. (1997) Extensional tilt blocks on Miranda: Evidence for an upwelling origin of Arden Corona. *J. Geophys. Res., 102,* 13369–13379.

Pappalardo R. T., Collins G. C., Head J. W., Helfenstein P., McCord T. B., Moore J. M., Prockter L. M., Schenk P. M., and Spencer J. R. (2004) Geology of Ganymede. In *Jupiter: The Planet, Satellites and Magnetosphere* (F. Bagenal et al., eds.), pp. 363–396. Cambridge Univ., Cambridge.

Pappalardo R. T., Crow-Willard E. N., and Golombek M. (2010) Thrust faulting as the origin of dorsa in the trailing hemisphere on Enceladus. *Bull. Am. Astron. Soc., 42,* 976, #16.02.

Passey Q. (1983) Viscosity of the lithosphere of Enceladus. *Icarus, 53,* 105–120.

Patel J. G., Pappalardo R. T., Head J. W., Collins G. C., Hiesinger H., and Sun J. (1999) Topographic wavelengths of Ganymede groove lanes from Fourier analysis of Galileo images. *J. Geophys. Res., 104,* 24057–24074.

Patthoff D. A. and Kattenhorn S. A. (2011) A fracture history on Enceladus provides evidence for a global ocean. *Geophys. Res. Lett., 38,* L18201, DOI: 10.1029/2011GL048387.

Patthoff D. A., Pappalardo R. T., Chilton H. T., Thomas P., and Schenk P. (2015) Diverse origins of Enceladus's ridge terrains including evidence for contraction. In *Lunar and Planetary Science XLVI,* Abstract #2870. Lunar and Planetary Institute, Houston.

Plescia J. B. and Boyce J. M. (1983) Crater numbers and geological histories of Iapetus, Enceladus, Tethys and Hyperion. *Nature, 301,* 666–670.

Polit A. T., Schultz R. A., and Soliva R. (2009) Geometry, displacement-length scaling, and extensional strain of normal faults on Mars with inferences on mechanical stratigraphy of the martian crust. *J. Structural Geol., 31,* 662–673, DOI: 10.1016/j.jsg.2009.03.016.

Porco C. C, West R. A., Squyres S., McEwen A., Thomas P., Murray C. D., DelGenio A., Ingersoll A. P., Johnson T. V., Neukum G., Veverka J., Dones L., Brahic A., Burns J. A., Haemmerle V., Knowles B., Dawson D., Roatsch T., Beurle K., and Owen W. (2004) Cassini imaging science: Instrument characteristics and anticipated scientific investigations at Saturn. *Space Sci. Rev., 115,* 363–497.

Porco C. C., Helfenstein P., Thomas P. C., Ingersoll A. P., Wisdom J., West R., Neukum G., Denk T., Wagner R., Roatsch T., Kieffer S., Turtle E., McEwen A., Johnson T. V., Rathbun J., Veverka J., Wilson D., Perry J., Spitale J., Brahic A., Burns J. A., DelGenio A. D., Dones L., Murray C. D., and Squyres S. (2006) Cassini observes the active south pole of Enceladus. *Science, 311,* 1393–1401, DOI: 10.1126/science.1123013.

Porco C. C., DiNino D., and Nimmo F. (2014) How the geysers, tidal stresses, and thermal emission across the south polar terrain of Enceladus are related. *Astron. J., 148,* 45.

Prockter L. M. and Patterson G. W. (2009) Morphology and evolution of Europa's ridges and bands. In *Europa* (R. T. Pappalardo et al., eds.), pp. 237–258. Univ. of Arizona, Tucson.

Prockter L. M., Head J. W. III, Pappalardo R. T., Sullivan R. J., Clifton A. E., Giese B., Wagner R., and Neukum G. (2002) Morphology of europan bands at high resolution: A mid-ocean ridge-type rift mechanism. *J. Geophys. Res., 107,* DOI: 10.1029/2000JE001458.

Roatsch T., Wählisch M., Scholten F., Hoffmeister A., Matz K.-D., Denk T., Neukum G., Thomas P., Helfenstein P., and Porco C. (2006) Mapping of the icy saturnian satellites: First results from Cassini-ISS. *Planet. Space Sci., 54,* 1137–1145.

Roatsch T., Wählisch M., Giese B., Hoffmeister A., Matz K.-D., Scholten F., Kuhn A., Wagner R., Neukum G., Helfenstein P., and Porco C. (2008) High-resolution Enceladus atlas derived from Cassini-ISS images. *Planet. Space Sci., 56,* 109–116.

Roatsch T., Kersten E., Hoffmeister A., Wählisch M., Matz K.-D., and Porco C. (2013) Recent improvements of the saturnian satellites atlases: Mimas, Enceladus, and Dione. *Planet. Space Sci., 77,* 118–125.

Roberts J. H. and Nimmo F. (2008) Tidal heating and long-term stability of a subsurface ocean on Enceladus. *Icarus, 194,* 675–689.

Rozel A., Besserer J., Golabek G. J., Kaplan M., and Tackley P. J. (2014) Self-consistent generation of single-plume state for Enceladus using non-Newtonian rheology. *J. Geophys. Res., 119(3),* 416–439, DOI: 10.1002/2013JE004473.

Schenk P. M. and McKinnon W. B. (2009) One-hundred-km-scale basins on Enceladus: Evidence for an active ice shell. *Geophys. Res. Lett., 36,* L16202, DOI: 10.1029/2009GL039916.

Showman A. P., Han L., and Hubbard W. B. (2013) The effect of an asymmetric core on convection in Enceladus' ice shell: Implications for south polar tectonics and heat flux. *Geophys. Res. Lett., 40,* 5610–5614.

Smith B. A., Soderblom L., Beebe R., Boyce J., Briggs G., Bunker A., Collins S. A., Hansen C. J., Johnson T. V., Mitchell J. L., Terrile R. J., Carr M., Cook A. F., Cuzzi J., Pollack J. B., Danielson E., Ingersoll A., Davies M. E., Hunt G. E., Masusrky H., Shoemaker E., Morrison D., Owen T., Sagan C., Veverka J., Strom R., and Suomi V. E. (1981) Encounter with Saturn: Voyager 1 imaging science results. *Science, 212,* 163–191.

Smith B. A., Soderblom L., Batson R., Bridges P., Inge J., Masursky H., Shoemaker E., Beebe R., Boyce J., Briggs G., Bunker A., Collins S. A., Hansen C. J., Johnson T. V., Mitchell J. L., Terrile R. J., Cook A. F., Cuzzi J., Pollack J. B., Danielson E., Ingersoll A. P., Davies M. E., Hunt G. E., Morrison D., Owen T., Sagan C., Veverka J., Strom R., and Suomi V. E. (1982) A new look at the Saturn system: The Voyager 2 images. *Science, 215,* 504–537.

Smith B. A., Soderblum L., Beebe R., Bliss D., Boyce J. M., Brahic A., Briggs G. A., Brown R. H., Collins S. A., Cook A. F., Croft S. K., Cuzzi J. N., Danielson G. E., Davies M. E., Dowling T. E., Godfrey D., Hansen C. J., Harris C., Hunt G. E., Ingersoll A., Johnson T. V., Krauss R. J., Masursky H., Morrison D., Owen T., Plescia J. B., Pollack J. B., Porco C. C., Rages K., Sagan C., Shoemaker E. M., Sromovsky L. A., Stoker C., Strom R. G., Suomi V. E., Synnott S. P., Terrile R. J., Thomas P., Thompson W. R., and Veverka J. (1986) Voyager 2 in the uranian system: Imaging science results. *Science, 233,* 43–64.

Smith-Konter B. R. and Pappalardo R. T. (2008) Tidally driven stress accumulation and shear failure of Enceladus's tiger stripes. *Icarus, 198,* 435–451.

Spencer J. R., Pearl J. C., Segura M., Flasar F. M., Mamoutkine A., Romani P., Buratti B. J., Hendrix A. R., Spilker L. J., and Lopes R. M. C. (2006) Cassini encounters Enceladus: Background and the discovery of a south polar hot spot. *Science, 311,* 1401–1405.

Spencer J. R., Barr A. C., Esposito L. W., Helfenstein P., Ingersoll A. P., Jaumann R., McKay C. P., Nimmo F., and Waite J. H. (2009) Enceladus: An active cryovolcanic satellite. In *Saturn from Cassini-Huygens* (M. Dougherty et al., eds.), pp. 683–724. Springer, Berlin, DOI: 10.1007/978-1-4020-9217-6-21.

Spencer J. R., Howett C. J. A., Verbiscer A. J., Hurford T. A., Segura M. E., and Pearl J. C. (2011) Observations of thermal emission from the south pole of Enceladus in August 2010. *EPSC-DPS Joint Meeting 2011,* 1630.

Spencer J. R., Howett C. J. A., Verbiscer A., Hurford T. A., Segura M., and Spencer D. C. (2013) Enceladus heat flow from high spatial resolution thermal emission observations. *European Planet. Sci. Congr., 2013,* EPSC2013-840-1.

Spitale J. N. and Porco C. C. (2007) Association of the jets of Enceladus with the warmest regions on its south-polar fractures. *Nature, 449,* 695–697.

Spitale J. N., Hurford T. A., Rhoden A. R., Berkson E. E., and Platts S. S. (2015) Curtain eruptions from Enceladus' south-polar terrain. *Nature, 521,* 57–60.

Squyres S., Reynolds R. T., Cassen P. M., and Peale S. J. (1983) The evolution of Enceladus. *Icarus, 53,* 319–331.

Stegman D. R., Freeman J., and May D. A. (2009) Origin of ice diapirism, true polar wander, subsurface ocean, and tiger stripes of Enceladus driven by compositional convection. *Icarus, 202,* 669–680.

Tajeddine R.., Soderlund K. M., Thomas P. C., Helfenstein P., Hedman M. M., Burns J. A., and Schenk P. M. (2017) True polar wander of Enceladus from topographic data. *Icarus, 295,* 46–60.

Thomas P. C. (2010) Sizes, shapes, and derived properties of the saturnian satellites after the Cassini nominal mission. *Icarus, 208,* 395–401.

Thomas P. C., Burns J. A., Helfenstein P., Squyres S., Veverka J., Porco C., Turtle E. P., McEwen A., Denk T., Giese B., Roatsch T., Johnson T. V., and Jacobson R. A. (2007) Shapes of the saturnian icy satellites and their significance. *Icarus, 190,* 573–584.

Thomas P. C., Tajeddine R., Tiscareno M. S., Burns J. A., Joseph J., Loredo T. J., Helfenstein P., and Porco C. (2016) Enceladus's physical libration requires a global subsurface ocean. *Icarus, 264,* 37–47, DOI: 10.1016/j.icarus.2015.08.037.

Tobie G., Čadek O., and Sotin C. (2008) Solid tidal friction above a liquid water reservoir as the origin of the south pole hotspot on Enceladus. *Icarus, 196,* 642–652.

Walker C. C., Bassis J. N., and Liemohn M. W. (2012) On the application of simple rift basin models to the south polar region of Enceladus. *J. Geophys. Res., 117,* E07003, DOI: 10.1029/2012JE004084.

West R., Knowles B., Birath E., Charnoz S., Di Nino D., Hedman M., Helfenstein P., McEwen A., Perry J., Porco C., Salmon J., Throop H., and Wilson D. (2010) In-flight calibration of the Cassini imaging-science subsystem cameras. *Planet. Space Sci., 58(11),* 1475–1488.

Wyrick D., Ferrill D. A., Morris A. P., Colton S. L., and Sims D. W. (2004) Distribution, morphology, and origins of martian pit crater chains. *J. Geophys. Res., 109,* E06005. DOI: 10.1029/2004JE002240.

Yin A. and Pappalardo R.T. (2015) Gravitational spreading, bookshelf faulting, and tectonic evolution of the south polar terrain of Saturn's moon Enceladus. *Icarus, 260,* 409–439.

Zhang K. and Nimmo F. (2009) Recent orbital evolution and the internal structures of Enceladus and Dione. *Icarus, 204,* 597–609.

APPENDIX A

A.1. Cassini Imaging Science Subsystem Enceladus Observations with Spatial Resolutions <1 km/pixel

Table A1 lists Cassini ISS observation sequences for which the diameter of the whole disk of Enceladus is at least as large as one-half of an ISS NAC field-of-view (i.e., 512 pixels or greater or with a spatial resolutions better than ~1 km/pixel). The observation name, data, and number of images obtained at spatial resolutions better than 1 km/pixel are provided. The best observations were generally acquired during Cassini flybys in which the viewing and illumination geometry dramatically changed from the inbound leg of the observation, through closest approach (C/A), and then through the outbound leg. Thus, for each observation, Table A1 provides three entries each (START, C/A, END) for the spacecraft range to Enceladus and subspacecraft latitude, longitude coordinates. Excluded from the table are the special "Skeetshoot" observations that are listed in Table 2 and described below. Also listed is the latitude, longitude coordinate of the subsolar point, which does not significantly change over the relatively short duration of each observation. Finally, a column noting the ISS filter types that were used is given. In the table, CLR is for CLEAR (CL1:CL2) filter, COL for color filter, and POL for polarization filter. The spatial resolution is not listed because it is image-dependent and varies with the aim-point on the satellite disk. An approximate measure is the spatial resolution at the center of disk, which can be calculated as $(5.86 \times 10^{-6}) \times$ (RANGE–RADIUS), where the coefficient is the angular width of an ISS NAC pixel (in radians), and RANGE and RADIUS are, respectively, the distance from the spacecraft to the center of Enceladus and the assumed radius of Enceladus. Although Enceladus was the listed target of most planned observations, the spacecraft geometry was occasionally favorable for making observations of Enceladus when it was not the primary target of a Cassini flyby (such as ISS_028TE_TETHYSORS001_CIRS).

As of this writing all the image sequences listed are archived in raw VICAR format and they are available from the rings node of NASA's Planetary Data System (*https://pds-rings.seti.org/saturn/*). The collection of desmeared images will be archived in the Cornell University eCommons data archive (*https://ecommons.cornell.edu/*) by December 2018.

A.2. Definition of Cassini Imaging Science Subsystem Tracking Maneuvers

The quality of the highest-definition images acquired by the Cassini ISS (Table 2) strongly depends on whether or how the spacecraft was able to track Enceladus' surface during the camera exposure. Unlike the Voyager spacecraft, which each had an articulating camera, Cassini's ISS instrument was fixed to the body of the spacecraft on an optical remote sensing (ORS) platform. Consequently, obtaining images required pointing the entire Cassini spacecraft at the target. Doing so for an appropriate exposure time was accomplished in one of four ways. The nominal means of tracking an imaging target was to utilize the spacecraft's Reaction Wheel Assembly (RWA track; see Table 2). However, during particularly close flybys, the relative angular motion of Enceladus relative to the spacecraft may have exceeded the RWA's rate and acceleration limits. In marginal cases, Reaction Control System (RCS track; see Table 2) thrusters could be used for tracking. However, at extremely close encounters, neither method may have been sufficient to track the surface of Enceladus. In such situations, there were two possible options: (1) When another instrument team was controlling pointing for a different instrument (for example, if an occultation of a star by Enceladus was being observed), the cameras were pointed to a fixed point in space and the images were exposed while Enceladus quickly passed in front of them (ORS drag; Table 2). (2) In rare cases, a maneuver called a "skeetshoot" (Table 2) was used in which the spacecraft began turning as fast as possible ahead of the predicted path of Enceladus across the celestial sphere. The maneuver was timed so that when Enceladus passed in front of the camera, for a very brief period of time, their relative angular rates were almost perfectly matched and relatively sharp images could be obtained in situations where they would otherwise be hopelessly smeared by image motion. Even so, spatial reconstruction processing was sometimes applied to the images to correct for motion blur after the images were received on Earth.

TABLE A1. Cassini ISS observation sequences for which the diameter of the whole disk of Enceladus is at least as large as one-half of an ISS NAC field of view (i.e., 512 pixels or greater or with a spatial resolution better than ~1 km/pixel).

Observation	Date YYYY-DDD	No. of images	Range to Enceladus (km)			Subspacecraft Point						Subsolar Point		Filter Notes
						Start		C/A		End				
			Start	C/A	End	Lat	Lon	Lat	Lon	Lat	Lon	Lat	Lon	
ISS_003EN_LIMTOP004_PRIME	2005-048	14	89746	10856	10856	1	315	5	320	5	320	–23	331	CLR
ISS_004EN_N4COLR003_PRIME	2005-068	4	244264	143218	143218	0	157	0	157	0	157	–22	201	COL
ISS_004EN_N4COLR004_PRIME	2005-068	4	93910	93041	93041	–1	173	–1	173	–1	173	–22	217	COL
ISS_004EN_NGNPOL001_PRIME	2005-068	24	27672	43220	43220	–1	187	–1	194	–1	194	–22	234	COL/POL
ISS_004EN_REGEO002_PRIME	2005-068	66	31785	4321	4321	–1	199	–5	220	–5	220	–22	247	COL
ISS_004EN_OBSERV002_PRIME	2005-068	24	165395	1677	167990	0	130	0	133	0	133	–22	4	CLR/COL/POL
ISS_011EN_N4COLR003_PRIME	2005-195	4	158909	157880	157880	–39	153	–39	154	–39	154	–21	202	CLR/COL
ISS_011EN_N3CPOL003_PRIME	2005-195	10	149725	145387	145387	–40	156	–40	157	–40	157	–21	205	CLR/COL/POL
ISS_011EN_N4COLR004_PRIME	2005-195	5	112092	110865	110865	–43	167	–43	167	–43	167	–21	216	CLR/COL
ISS_011EN_N3CPOL004_PRIME	2005-195	10	106565	103223	103223	–43	169	–44	170	–44	170	–21	219	CLR/COL/POL
ISS_011EN_NGNPOL001_PRIME	2005-195	16	69161	61331	61331	–46	182	–47	185	–47	185	–21	232	CLR/COL/POL
ISS_011EN_N3COL001_PRIME	2005-195	16	59027	51128	51128	–47	186	–47	189	–47	189	–21	236	CLR/COL
ISS_011EN_REGEO002_PRIME	2005-195	19	39096	30954	30954	–48	194	–48	198	–48	198	–21	243	CLR/COL
ISS_011EN_MORPH001_PRIME	2005-195	3	28891	27179	27179	–48	198	–48	199	–48	199	–21	246	CLR
ISS_011EN_N9COL001_PRIME	2005-195	39	24780	14945	14945	–48	200	–49	205	–49	205	–21	249	CLR/COL
ISS_011EN_MORPH002_PRIME	2005-195	9	13539	11083	11083	–49	206	–49	207	–49	207	–21	252	CLR/COL
ISS_011EN_NCPOL001_PRIME	2005-195	12	41924	41924	45208	47	42	47	42	47	44	–21	273	CLR/COL/POL
ISS_018EN_HIPHAS001_PRIME	2005-331	14	144189	144189	149804	1	131	1	131	1	139	–20	315	CLR
ISS_018EN_PLUMES001_PRIME	2005-331	2	966123	166123	166491	1	157	1	157	1	157	–20	333	CLR
ISS_019EN_FP3HOTSPT020_CIRS	2005-358	4	108485	108290	108290	0	116	0	116	0	116	–19	218	CLR/COL
ISS_020EN_FP3MAP001_CIRS	2006-017	19	161670	649384	149384	–1	201	0	225	0	225	–19	234	CLR/COL/POL
ISS_020EN_GEOLOG004_PRIME	2006-017	49	152563	156194	156194	0	240	0	244	0	244	–19	266	CLR/COL/POL
ISS_028EN_ENCELORS001_CIRS	2006-252	9	64731	64731	66754	84	181	84	181	83	181	–16	252	CLR/COL/POL
ISS_028TE_TETHYSORS001_CIRS	2006-252	2	141090	141090	141493	32	286	32	286	32	286	–16	311	POL
ISS_050EN_COLORF001_PRIME	2007-273	9	101773	101773	101853	13	109	13	109	13	109	–10	18	CLR/COL/POL
ISS_050EN_PHOTOM002_PRIME	2007-273	8	107518	107518	108058	15	109	15	109	15	109	–10	38	CLR/COL/POL
ISS_061EN_FP34MAP001_CIRS	2008-072	4	106369	104590	104590	69	115	69	115	69	115	–8	326	CLR/COL
ISS_061EN_ENCELADUS002_VIMS	2008-072	11	92800	74632	74632	69	117	69	121	69	121	–8	330	CLR/COL/POL
ISS_061EN_ICYMAP002_UVIS	2008-072	4	66444	63818	63818	70	122	70	122	70	122	–8	335	CLR/COL
ISS_061EN_REGMAP002_PRIME	2008-072	3	34058	29628	29628	69	128	69	129	69	129	–8	342	CLR
ISS_061EN_FP3HOTSPT001_CIRS	2008-072	5	16392	16392	53032	–71	319	–71	319	–70	327	–8	285	CLR
ISS_061EN_FP1SECLX001_CIRS	2008-072	4	130592	130592	132797	–69	346	–69	346	–69	346	–8	18	CLR/COL
ISS_061EN_ICYLON006_UVIS	2008-072	14	143969	143969	153593	–69	349	–69	349	–69	352	–8	22	CLR/COL/POL
ISS_074EN_ENCELADUS001_CIRS	2008-182	7	167967	162317	162317	50	26	49	26	49	26	–6	312	CLR/COL/POL
ISS_080EN_ENCEL001_VIMS	2008-224	18	113800	100245	100245	62	79	62	81	62	81	–6	322	CLR/COL/POL
ISS_080EN_ICYLIMB001_UVIS	2008-224	5	54106	50620	50620	62	89	62	89	62	89	–6	332	CLR/COL

TABLE A1. (continued)

Observation	Date YYYY-DDD	No. of images	Range to Enceladus (km) Start	C/A	End	Subspacecraft Point Start Lat	Start Lon	C/A Lat	C/A Lon	End Lat	End Lon	Subsolar Point Lat	Lon	Filter Notes
ISS_080EN_ENCELCA001_PRIME	2008-224	54	41203	8024	29998	62	91	−64	279	−63	283	−6	343	CLR/COL/POL
ISS_080EN_FP1ECLSCN001_CIRS	2008-224	38	48957	48957	132311	−62	286	−62	286	−62	301	−6	308	CLR
ISS_088EN_ENCEL001_CIRS	2008-283	13	152114	143148	143148	62	72	62	73	62	73	−5	313	CLR/COL/POL
ISS_088EN_ICYLIMB001_UVIS	2008-283	4	72160	69105	69105	62	85	62	86	62	86	−5	327	CLR/COL
ISS_088EN_ENCELCA001_PRIME	2008-283	44	23894	23894	45389	−63	281	−63	281	−62	285	−5	345	CLR/COL/POL
ISS_088EN_SECLNX001_CIRS	2008-283	35	46849	46849	89290	−62	285	−62	285	−62	293	−5	350	CLR/COL
ISS_088SC_DFPWBIAS283_ENGR	2008-283	21	94033	94033	113244	−62	294	−62	294	−62	297	−5	357	CLR
ISS_088EN_SECLNX002_CIRS	2008-283	20	117260	117260	166304	−62	298	−62	298	−62	307	−5	21	CLR
ISS_091EN_FP3STARE001_CIRS	2008-305	16	103442	92803	92803	62	80	62	81	62	81	−4	321	CLR/COL/POL
ISS_091EN_ENCEL002_VIMS	2008-305	5	69495	65951	65951	62	85	62	86	62	86	−4	326	CLR/COL
ISS_091EN_ENCELCA001_PRIME	2008-305	42	9151	9151	31594	−65	279	−65	279	−63	283	−4	341	CLR/COL/POL
ISS_091EN_ICYMAP002_UVIS	2008-305	3	41684	41684	45828	−63	284	−63	284	−63	285	−4	345	CLR
ISS_091EN_SECLNX001_CIRS	2008-305	29	54130	54130	159651	−63	287	−63	287	−62	306	−4	173	CLR
ISS_095EN_GEOLOG001_PRIME	2008-337	23	127537	125098	125098	44	220	38	225	38	225	−4	343	CLR/COL/POL
ISS_121EN_ENCEL001_VIMS	2009-324	16	329423	62520	62520	−1	157	−2	177	−2	177	2	15	CLR/COL/POL
ISS_121EN_PLMHR001_PRIME	2009-325	15	17037	5790	5790	−6	194	−18	197	−18	197	2	51	CLR
ISS_121EN_FP3HIRES001_CIRS	2009-325	47	3173	2218	70417	−35	196	−56	192	−1	45	2	65	CLR/COL
ISS_121SC_DFPWBIAS325_ENGR	2009-325	2	57650	77650	78108	−1	47	−1	47	−1	47	2	85	CLR/COL
ISS_121EN_FP3DAYMAP002_CIRS	2009-325	10	67647	97647	103526	−1	52	−1	52	−1	53	2	93	CLR/COL/POL
ISS_121EN_ICYSTARE002_UVIS	2009-325	12	433038	133038	151936	0	59	0	59	0	64	2	106	CLR/COL/POL
ISS_131EN_FP1DRKMAP001_CIRS	2010-138	6	19661	46711	46711	0	193	0	196	0	196	4	40	CLR
ISS_131EN_PLMHR001_PRIME	2010-138	6	47012	13174	13174	−1	211	−1	213	−1	213	4	56	CLR
ISS_131EN_ENCEL001_PRIME	2010-138	36	15369	15369	22961	−2	42	−2	42	−1	46	4	72	CLR/COL
ISS_131EN_ENCEL001_VIMS	2010-138	2	26846	26846	27396	−1	48	−1	48	−1	48	4	75	CLR/COL
ISS_131EN_ENCEL002_PRIME	2010-138	16	49231	39231	43189	−1	53	−1	53	−1	55	4	82	CLR/POL
ISS_131EN_ENCEL002_VIMS	2010-138	7	24331	44331	67968	−1	55	−1	55	−1	64	4	85	CLR/COL
ISS_131EN_FP3DAYMAP001_CIRS	2010-138	16	21887	71887	90245	−1	65	−1	65	−1	71	4	100	CLR/POL
ISS_134EN_GLOBAL001_CIRS	2010-185	40	167422	91649	91649	−37	131	−47	125	−47	125	5	91	CLR/COL/POL
ISS_136EN_DRKPLUME001_CIRS	2010-225	1	97963	97963	97963	−11	173	−11	173	−11	173	6	18	CLR
ISS_136EN_PLMHRHP002_PRIME	2010-225	33	62764	27936	27936	−13	186	−17	202	−17	202	6	40	CLR/COL
ISS_136EN_HIRES001_CIRS	2010-225	64	90976	2890	45826	−19	205	−87	198	8	52	6	67	CLR/COL
ISS_136EN_ENCEL001_VIMS	2010-226	4	41580	93580	94002	9	68	9	68	9	68	6	99	CLR/POL
ISS_141EN_PLMHPHR001_PIE	2010-334	1	96625	96625	96625	0	287	0	287	0	287	7	87	CLR
ISS_141EN_DAYMAP001_CIRS	2010-334	9	45431	45431	58510	0	170	0	170	0	176	7	160	CLR
ISS_141EN_ENCELADUS001_VIMS	2010-334	16	60286	60286	67300	0	177	0	177	0	180	7	165	CLR/COL/POL
ISS_142EN_PLMHPHR001_PIE	2010-354	1	588137	158137	158137	0	222	0	222	0	222	7	27	CLR
ISS_142EN_NITEMAP001_CIRS	2010-354	13	97411	105777	105777	0	280	0	281	0	281	7	79	CLR

TABLE A1. (continued)

Observation	Date YYYY-DDD	No. of images	Range to Enceladus (km)			Subspacecraft Point						Subsolar Point		Filter Notes
						Start		C/A		End				
			Start	C/A	End	Lat	Lon	Lat	Lon	Lat	Lon	Lat	Lon	
ISS_142EN_PLMHPHR002_PIE	2010-354	36	52578	26082	26082	0	304	0	315	0	315	7	116	CLR/COL
ISS_142EN_ORSCA001_PIE	2010-355	31	24897	24897	48265	1	160	1	160	0	171	7	150	CLR/COL
ISS_142EN_ENCEL001_VIMS	2010-355	16	54909	54909	102458	0	174	0	174	0	194	7	163	CLR/COL
ISS_144EN_PLMHPHR002_PIE	2011-031	29	92151	68838	68838	–1	295	–1	287	–1	287	8	142	CLR/COL
ISS_144EN_ENCEL001_PIE	2011-031	37	51322	60479	80639	–1	275	–1	270	–1	241	8	162	CLR/COL/POL
ISS_153EN_ENCEL001_PIE	2011-256	38	44032	42483	51713	–1	343	–1	1	–1	43	11	52	CLR/COL/POL
ISS_154EN_PLMHPHR001_PIE	2011-274	4	164629	82547	82547	0	153	0	176	0	176	11	317	CLR
ISS_154EN_FP1SECLX001_CIRS	2011-274	11	13604	23604	31853	–1	37	–1	37	–1	40	11	13	CLR/COL/POL
ISS_154EN_ENCELADUS001_PRIME	2011-274	26	7190	1190	158630	0	64	0	64	0	76	11	56	CLR/COL/POL
ISS_155EN_PLMHPHR001_PIE	2011-292	2	148820	127212	127212	0	156	0	161	0	161	12	313	CLR
ISS_155EN_ENCEL002_PIE	2011-292	30	35045	21081	21081	0	177	–1	195	–1	195	12	337	CLR/COL/POL
ISS_155EN_ENCELADUS001_CIRS	2011-292	40	90085	40085	56141	–1	44	–1	44	–1	49	12	20	CLR
ISS_156EN_ENCELADUS001_CIRS	2011-310	8	947234	144505	144505	0	157	0	158	0	158	12	311	CLR
ISS_156EN_ENCEL001_PRIME	2011-310	23	67485	77485	149734	–1	54	–1	54	0	73	12	38	CLR/COL/POL
ISS_158EN_ENCEL001_PRIME	2011-346	32	21544	21544	28952	–2	78	–2	78	–1	110	12	100	CLR/COL
ISS_161EN_PLMHPMR001_PIE	2012-051	1	133961	133961	133961	0	136	0	136	0	136	13	322	CLR
ISS_163EN_PLMHPHR001_PIE	2012-087	1	111954	111954	111954	1	166	1	166	1	166	14	333	CLR
ISS_163EN_ENCELADUS001_PIE	2012-087	20	77338	27338	63620	–1	38	–1	38	–1	49	14	31	CLR/COL
ISS_164EN_PLMHPMR001_PIE	2012-105	26	165929	1475	119726	0	152	0	157	0	164	14	324	CLR/COL
ISS_165EN_PLMHPHR002_PRIME	2012-123	6	342281	138321	138321	0	158	0	159	0	159	14	324	COL
ISS_165EN_ENCELADUS002_CIRS	2012-123	11	75650	75650	79978	–1	52	–1	52	–1	53	14	42	CLR/COL
ISS_219EN_ENCEL001_CIRS	2015-208	30	113438	1131	111964	1	262	1	262	1	262	25	258	CLR/COL/POL
ISS_220EN_ENCELNP001_PRIME	2015-230	19	59880	59880	63238	2	154	2	154	2	156	25	252	CLR/COL/POL
ISS_220EN_PLMHPMR001_PIE	2015-230	29	114961	114961	158194	1	189	1	189	0	206	25	306	CLR/COL
ISS_221DI_REGMAP001_PIE	2015-251	12	364783	355405	355405	0	142	0	141	0	141	25	148	CLR/COL
ISS_221EN_ENCELADUS001_PIE	2015-251	38	129178	128133	129884	1	103	1	102	1	102	25	207	CLR/COL/POL
ISS_223EN_ENCEL001_PIE	2015-287	47	121425	5840	73296	1	160	20	192	1	41	25	178	CLR/COL/POL
ISS_224EN_ENCEL001_PRIME	2015-301	40	342201	60925	60925	1	315	1	331	1	331	25	282	CLR/COL/POL
ISS_228EN_ENCELINB001_CIRS	2015-353	13	316966	92681	92681	–1	328	–1	332	–1	332	26	292	CLR/COL/POL
ISS_228EN_ENCELOUTB001_CIRS	2015-353	43	15006	9436	40764	–19	352	–32	354	–9	189	26	330	CLR/COL/POL
ISS_230EN_ENCEL001_PIE	2016-014	16	76351	74169	79334	4	96	4	90	3	69	26	153	CLR/COL/POL
ISS_232EN_ENCELADUS001_VIMS	2016-046	25	86153	86153	97239	–58	269	–58	269	–59	244	26	305	CLR/COL/POL
ISS_250EN_ENCELNPOL001_PIE	2016-332	45	77455	32227	32227	79	33	62	344	62	344	27	201	CLR/COL/POL
ISS_250EN_ENCELSPOL001_CIRS	2016-332	6	130005	130005	140551	–64	335	–64	335	–65	337	27	238	CLR/COL

Part 2:

Enceladus Plumes and the E Ring

Postberg F., Clark R. N., Hansen C. J., Coates A. J., Dalle Ore C. M., Scipioni F., Hedman M. M., Waite J. H. (2018) Plume and surface composition of Enceladus. In *Enceladus and the Icy Moons of Saturn* (P. M. Schenk et al., eds.), pp. 129–162. Univ. of Arizona, Tucson, DOI: 10.2458/azu_uapress_9780816537075-ch007.

Plume and Surface Composition of Enceladus

Frank Postberg
University of Heidelberg, Free University of Berlin

Roger N. Clark and Candice J. Hansen
Planetary Science Institute

Andrew J. Coates
University College London

Cristina M. Dalle Ore and Francesca Scipioni
NASA Ames Research Center

Matthew M. Hedman
University of Idaho

J. Hunter Waite
Southwest Research Institute

This chapter provides a comprehensive review of Enceladus' plume and surface composition as determined by the end of the Cassini mission. The enceladean plume is composed of three different phases: Gas, solids (dust), and ions. In all three phases, water is by far the most abundant constituent, but other chemical species are also present. In the gas, H_2 and NH_3 and CO_2 are present with volume mixing ratios of at least fractions of a percent. Methane is the most abundant organic compound (≈ 0.2%), but several, yet unspecified, C_2 — and probably C_3 — species are present. The D/H ratio in the plume is much higher than in Saturn's atmosphere. Macroscopic (r > 0.2 μm) plume ice grains appear to be composed of ice in a primarily crystalline state. The main non-icy compounds in these grains are sodium salts and organic material. These materials are heterogeneously distributed over three compositionally diverse ice grain populations and can reach percent-level abundance in individual grains. Ice grains carrying salts or organics are larger than pure water ice grains and are found at higher frequencies in the plume than in the E ring. Organics in ice grains are generally more refractory than in plume gas. In some ice grains, complex macromolecules with atomic masses above 200 u have been detected. Oxygen- and nitrogen-bearing organics are likely present in both gas and ice grains. The plume also hosts at least two kinds of nanograins. One, probably icy, population is dispersed inside the plume. The other is dominated by SiO_2 and appears to be embedded in larger ice grains and is only released later in the E ring. The possible origins of the different constituents in Enceladus' interior are discussed. There are more negatively than positively charged water ions in the plume. Cations have the form H_nO^+ (n = 0–3) and respective dimers, anions are dissociated water molecules (OH, H^-, O^-) or cluster of the form $(H_2O)_nOH^-$ (n = 1–3). Saturn's magnetosphere hosts an abundance of non-water cations that likely originate from plume constituents: NH^+ is the most abundant, with a mixing ratio of a few percent, followed by N^+, C^+, and cations with masses of 28 u. Vertical compositional stratification of ice grains in the plume has been clearly documented, but there are also hints of compositional variations in ice grains emerging from different tiger stripe fractures. In cones of supersonic gas, heavier molecular species (e.g., CO_2) have a narrower lateral spread than light species (e.g., H_2). Other spatial variations in the gas are likely, but could not be observed by Cassini's instruments. Although the plume shows clear activity variations over time, currently no compositional fluctuation could be linked to these variations. Enceladus' surface is subject to constant resurfacing by deposition of plume ices and exhibits the cleanest water ice surface in the solar system. From infrared observations, CO_2 is present in these ices in the south polar terrain (SPT); deposition of aliphatic organics is also indicated there. From disk-averaged ultraviolet (UV) observations, NH_3 might be generally present in surface ices. The best candidates for an additional UV absorber are "tholins" or iron-rich nanograins. Predictions for plume deposition rates can be reasonably well matched with observations in various wavelengths, and indicate strong variations in grain size, but possibly also compositional variations. From plume composition, the SPT should be enriched in salts but these have not yet been detected with remote sensing. In the SPT, crystalline ice from plume deposits seems to be predominant. It is currently unclear whether amorphous ice exists on Enceladus' surface.

1. INTRODUCTION

The plume of Enceladus can be seen as the defining feature of Enceladus' uniqueness because no other icy body in the solar system is currently known to exhibit such continuous and large-scale activity. The composition of the plume immediately became one of the highest priorities of the Cassini mission because it was suspected that compositional information would yield unique insights into interior processes, including (at the time putative) subsurface liquid water. And indeed, the current knowledge of the moon's interior exceeds that of most other planetary bodies (see the chapters in this volume by Glein et al., Spencer et al., and Hemingway et al.).

The diverse and flexible payload of the Cassini-Huygens flagship mission turned out to be a huge advantage because it enabled immediate follow-up investigations of the plume's composition without designing and launching an entirely new mission. Cassini's instruments were able to measure the composition of emitted gas, solid material (dust), and charged particles with both *in situ* and remote sensing techniques. During Cassini's extended mission (2008–2017), multiple close Enceladus flybys were incorporated into Cassini's tour around Saturn. During these flybys the spacecraft flew directly through the plume in order to allow Cassini's *in situ* instruments to investigate fresh samples from the enceladean subsurface. Other flybys allowed high-resolution imaging at ultraviolet (UV), visible, and infrared wavelengths to observe the ice grains from the plume and the surface composition that resulted from the outflowing plume materials. A greater part of the ejected micrometer-sized and submicrometer-sized ice grains falls back onto the moon (*Porco et al.,* 2006; *Kempf et al.,* 2008; *Ingersoll and Ewald,* 2011) and Enceladus' surface is constantly exposed to ice particle deposition from the plume (*Kempf et al.,* 2010; *Schenk et al.,* 2011). Therefore the composition of the moon's surface is closely linked to the composition of the plume.

However, material emerging from Enceladus is not only apparent close to the moon: Cassini's measurements have shown that a greater part of all matter residing in the vast space between the orbits of the main rings and Titan is dominated by compounds that once were part of Enceladus. The most prominent witness to this fact is Saturn's diffuse E ring, which consists of ice grains that, after ejection into the plume, escape Enceladus' gravitational domain (see the chapter by Kempf et al. in this volume). In fact, a substantial part of our current knowledge about the plume's composition has been inferred by analyzing the material that escaped the moon's gravity. Neutral gas is ejected at such high velocities that it almost entirely escapes from Enceladus (*Hansen et al.,* 2008). This is also true for charged particles that are quickly coupled to Saturn's magnetosphere.

In this chapter the word "jet" is used for individual, collimated sources that emerge from Enceladus' south polar fractures. "Plume" is used to refer to the entire south polar emission composed of all jets and diffuse sources along the fractures. We first discuss the composition of the plume's three distinct phases: Gas, micrometer-sized grains, and ionized particles (sections 2–4). We then try to link the identified compounds to possible subsurface sources (section 5). We review the current state of knowledge regarding Enceladus' surface composition (section 6) and then explore relationships between the compositions of the plume and the surface (section 7). The chapter concludes with an integrated summary and a presentation of major open questions (section 8).

2. COMPOSITION OF THE GAS PHASE

Two Cassini instruments have measured the gaseous component of Enceladus' plume: the Ion and Neutral Mass Spectrometer (INMS) and the Ultraviolet Imaging Spectrograph (UVIS). These instruments observe the plume from quite different and complementary perspectives. While INMS is an *in situ* detector that measures the composition along the flight path of the spacecraft through the plume, UVIS is a remote sensing instrument that observes the plume gas from a large distance, thereby integrating along the line of sight between the spacecraft and the star or Sun that is being occulted.

The analytical method of INMS is mass spectrometry. In its "closed source mode" neutral gas is collected in an antechamber, and is subsequently transferred through a tube into an ionization chamber. There, the gas is ionized by electron impact from electron guns. The atomic and molecular masses of the cations that form from this electron bombardment are then inferred by a quadrupole mass spectrometer with integer mass resolution ($m/\Delta m \sim$ unity). The mass range of the instrument is 2 u–99 u (u = atomic mass unit). The INMS also provides an open source mode where the neutral gas is ionized "on the fly" and directly enters the quadrupole mass selection unit without interaction with the walls of the instrument's interior. For details about the INMS instrument, see *Waite et al.* (2004).

UVIS identifies neutral gases in the plume by absorption of star- or sunlight at UV wavelengths (110–190 nm for stellar occultations, 55–110 nm for the solar occultation). Certain absorption features are specific to the molecules in the gas. The spectrum of starlight transmitted through an absorbing gas will be attenuated at different wavelengths in a manner that is diagnostic of the composition of the gas. The extinction due to absorption at a given wavelength for a particular gas is generally given as a cross-section. Then, to estimate the column density, the spectrum of the transmitted signal is compared to the spectral absorption features of a specific gas calculated from these cross-sections as a function of wavelength. For details about the UVIS instrument, see *Esposito et al.* (2004).

To fully understand the current view of the plume gas composition, it is helpful to review Cassini's exploration in chronological order. For that reason, this section starts with the "early results" (section 2.1). Those readers that are just interested in the current state of the art might directly jump to section 2.2.

2.1. Early Cassini Results

The first measurements of the gas composition of the plume of Enceladus by INMS and UVIS were both obtained on July 14, 2005. Since the exact location of the plume was not known at that time, the measurements were serendipitous for both instruments.

It turned out that Cassini just touched the fringe of the plume during the July 2005 encounter (E2). Nevertheless, INMS mass spectra identified the main gas components of the plume. The data indicated that the atmospheric plume and coma was dominated by water, with significant amounts of carbon dioxide (~3%), an unidentified species with a mass of 28 u [~4% at that time reported to be either carbon monoxide (CO) or molecular nitrogen (N_2)], and methane (CH_4). Ammonia was detected at a level that did not exceed 0.5%. Trace quantities of acetylene and propane were also reported (*Waite et al.,* 2006).

The UVIS occultation of gamma Orionis, observed on the same day as the INMS measurement, showed that water is a clear match for all absorption features observed with adequate signal to noise ratio in the spectra. The best fit for the column density was given as 1.5×10^{16} H_2O molecules cm^{-2}. From that column density a total gaseous water emission rate of 150–350 kg s^{-1} could be inferred (*Hansen et al.,* 2006). This number varied only slightly through subsequent occultations. The occultations of zeta Orionis (October 2007), the Sun (May 2010), epsilon and zeta Orionis (October 2011), and epsilon Orionis (March 2016) yielded similar emission rates ranging from 170 kg s^{-1} to 250 kg s^{-1}, assuming the same typical gas velocities (*Hansen et al.,* 2011, 2017). Data shown in Fig. 1 are from the occultation of zeta Orionis observed in 2007, compared to the theoretical water vapor spectrum calculation (Fig. 1).

In all individual occultations UVIS does not detect components other than water. However, tight upper limits for a number of constituents could be constrained and will be discussed in section 2.2.

The next opportunities to measure the plume of Enceladus *in situ* were the E3 and E5 encounters by Cassini in March and October 2008. As in the case of E2, the trajectories were highly inclined but the relative speed was much higher than at E2, with E5 providing the highest flyby speed of all Enceladus encounters (17.7 km s^{-1} compared to 8.2 km s^{-1} for E2). The higher flyby speed and the closer distance to the sources of the plume provided a substantial increase in signal-to-noise ratio of E5 compared to E2 and E3.

The E5 flyby data indicated inferential evidence for a liquid water ocean based on ^{40}Ar and ammonia detection in the plume (*Waite et al.,* 2009). However, ^{40}Ar has not been reproduced in subsequent measurements. The plume composition measurements, shown in Fig. 2 and Table 1 [reproduced from Table 1 and Fig. 1 of *Waite et al.* (2009)] indicate similar values to those of the earlier E2 flyby with two notable exceptions: (1) the number and concentration of organic compounds, especially above 50 u, were significantly enhanced, and (2) there was a substantially increased

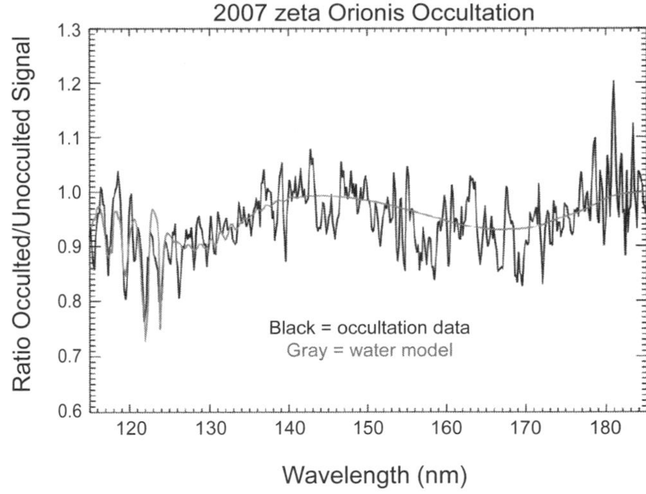

Fig. 1. The spectrum from zeta Orionis shows narrow absorption features between 115 and 130 nm diagnostic of water vapor and a broad absorption at ~155–175 nm. The smooth curve compares water vapor to the spectrum for the best-fit column density of 1.4×10^{16} cm^{-2}.

H_2 component in the plume. These anomalies were later explained as follows:

1. *Excess organic compounds were fragmentation products:* A view from the broader perspective of subsequent flybys (E14, E17, and E18; <8 km s^{-1} flyby, see below) suggests that the higher flyby speed of E5 (17.7 km s^{-1}) led to significant fragmentation of organic compounds with heavy molecular weight, outside the INMS mass range, that are found in the ice grains (*Postberg et al.,* 2018) (see also sections 3.1 and 5.1). Ice grains inevitably enter the INMS antechamber during plume traversals. In addition, fragmentation of gaseous molecules that hit the antechamber walls at these high speeds also contribute to the ambiguity of the spectrum. Consequently, most of the abundances depicted in Fig. 2 and Table 1 and concentrations given therein do not reflect the intrinsic gas composition (see footnote of Table 1 for details). This is especially true for species with masses greater than 27 u.

2. *Excess hydrogen:* The hydrogen excess was explained by the vaporization of raw titanium from the walls of the titanium antechamber by hypervelocity impacts of ice grains. Subsequent titanium oxidation reactions (TiO, TiO_2) led to dissociation of H_2O vapor, yielding gaseous H_2 that was then measured by the mass spectrometer. Verification of this hypothesis using ballistic impact modeling has been published by *Walker et al.* (2015). Later low levels of native H_2 have been detected in the plume (section 2.3).

The serendipitous occurrence of the dissociation reaction described in point 2 above allowed the determination of the D/H ratio in H_2O from the relatively low-mass-resolution measurements provided by INMS. The value of 2.9 (+1.5/–0.7) $\times 10^{-4}$ is in the mid range of observed cometary D/H values and very similar to the values measured at Comet Halley (*Balsiger et al.,* 1995), as opposed to the order of

TABLE 1. Apparent volume mixing ratios based on analysis of the E5 data presented in *Waite et al.* (2009).

Group	Species	E5 Volume Mixing Ratio (%)	Fragmentation Class
1	H_2O^*	90 ± 0.01	
	H_2^*	[0.39]	II
	CO_2^*	0.053 ± 0.01	
	CO^*	[0.044]	II
	CH_4	0.009 ± 0.005	I
	NH_3	0.0082 ± 0.0002	
2	C_2H_2	$(3.3 \pm 2) \times 10^{-3}$	II
	C_2H_4	<0.012	I
	C_2H_6	$<1.7 \times 10^{-3}$	I
	HCN	$<7.4 \times 10^{-3}$	I
	N_2	<0.011	I
	H_2CO	$(3.1 \pm 1) \times 10^{-3}$	I
	CH_3OH	$(1.5 \pm 0.6) \times 10^{-4}$	I
	H_2S	$(2.1 \pm 1) \times 10^{-5}$	I+
3	$(^{40}Ar)^\dagger$	$(3.1 \pm 0.3) \times 10^{-4}$	II
	C_3H_4	$<1.1 \times 10^{-4}$	I
	C_3H_6	$(1.4 \pm 0.3) \times 10^{-3}$	II
	C_3H_8	$<1.1 \times 10^{-4}$	II
	C_2H_4O	$<7.0 \times 10^{-4}$	I
	C_2H_6O	$<3.0 \times 10^{-4}$	I
4	C_4H_2	$(3.7 \pm 0.8) \times 10^{-5}$	I+
	C_4H_4	$(1.5 \pm 0.6) \times 10^{-5}$	I+
	C_4H_6	$(5.7 \pm 3) \times 10^{-5}$	I+
	C_4H_8	$(2.3 \pm 0.3) \times 10^{-4}$	II
	C_4H_{10}	$<7.2 \times 10^{-4}$	II
	C_5H_6	$<2.7 \times 10^{-6}$	I+
	C_5H_{12}	$<6.2 \times 10^{-5}$	I+
	C_6H_6	$(8.1 \pm 1) \times 10^{-5}$	I+

* The mixing ratios for H_2 in brackets have been included in the mixing ratio for H_2O as it is believed the vast majority of H_2 is produced from interaction of hypervelocity ice grains on the INMS antechamber (see item "II" above). The fragmentation class assigned to H_2 is with respect to H_2O. The mixing ratio for CO in brackets has likewise been included in the mixing ratio for CO_2 due to indications of a similar hypervelocity-induced dissociation process. However, from the low abundance of CO_2 during slow flybys (see Table 2), it seems that CO is rather an organic fragmentation product and thus the value for CO_2 in Table 1 and *Waite et al.* (2009) is substantially overestimated by the addition of CO (4.4%) to the E5 CO_2 raw signal of just 0.9%.

† ^{40}Ar abundance was originally based upon the lack of fit at mass channel 40 from other potentially contributing species such as C_3H_4 and C_3H_6. However, subsequent analysis of the "slow" Enceladus flybys indicates a large reduction in the abundance of species at mass 40 (and neighbors), as can be seen by the fragmentation class of II. ^{40}Ar cannot be a product of heavy organic fragmentation, so it is much more likely that the ^{40}Ar signal originally reported for the E5 flyby is rather due to some mixture of organic fragments not yet fully understood.

Abundances cover the range of accepted composition models for ionization and fragmentation by INMS's electron guns that adequately fit the E5 mass spectrum. Insights from later flybys show that the given values mostly do not reflect the intrinsic composition of plume gas. In particular, most values are heavily influenced by molecular fragmentation from high velocity wall impacts, of mostly heavy organics probably residing in ice grains (*Postberg et al.*, 2018), that are responsible for a greater part of the organic species in the spectrum. Species listed with upper limits (gray color) are present in some INMS ionization models but absent from others and are potentially present rather than definitive detections. "Fragmentation Class" indicates the apparent contribution from heavy organic fragmentation by high-velocity wall impacts upon the listed abundances observed during the fast E5 flyby (~17.7 km s^{-1}). It is based upon the increase in abundance compared to "slow" (<8 km s^{-1}) flybys (E14, E17, E18, E21). Class I indicates a species with a substantial contribution from fragmentation (factor of 2–20 larger than at "slow" flybys); Class II indicates species that are almost exclusively due to fragmentation (>factor of 20 abundance compared to slow flybys). Class "I+" indicates a species that has not been detected on slow flybys and thus Class I is only a lower limit for the degree of contributions from fragmentation. It is very likely that the abundances for these species are primarily or exclusively due to heavy organic fragmentation. Group 1 indicates major species, group 2 represents the "C2 region" of the spectrum (masses 24–34), group 3 represents the "C3 region" (masses 36–46) and group 4 represents the "C4+ region" (masses 48–80) of the spectrum. See Tables 2 and 3 for mixing ratios of intrinsic plume gas.

magnitude lower values found in the atmosphere of Saturn (*Pierel et al.*, 2017). Based on this finding, *Waite et al.* (2009) hypothesized that Enceladus might not have formed by subnebula condensation processes during the cooling of the subnebula but was formed or captured late in the saturnian subnebula formation process.

The E5 flyby could not resolve the determination and quantification of the species at 28 u because CO was abundantly produced by dissociation of larger CO-bearing species by molecular or ice grain hypervelocity impacts on, and reaction with, the walls of the INMS antechamber. *Waite et al.* (2009) estimated that up to 80% of the signal at mass

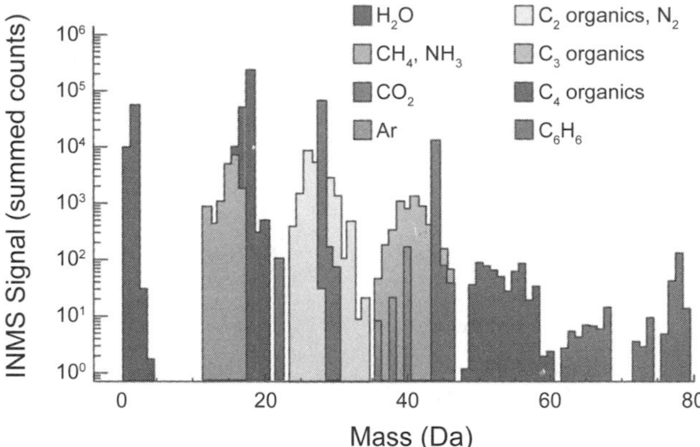

Fig. 2. See Plate 12 for color version. Volume mixing ratios based on analysis of the E5 data presented in *Waite et al.* (2009). Most organic species do not reflect the intrinsic gas composition. Due to the high velocity of the flyby, the spectrum is dominated by species from molecular fragmentation of high mass organic compounds probably residing in plume ice grains (Table 1). The presence of argon could not be confirmed by later measurements. Published with permission of Nature Publishing Group.

28 u was produced this way. The residual (~20%) of the mass 28 signal was attributed to N_2 or C_2H_4 (ethylene) or a combination of both with an upper limit of 1.2% (volume mixing ratio) for each substance. A "small contribution" from intrinsic CO was also possible.

2.2. Refined Cassini Results

UVIS plume occultations between 2007 and 2011 helped place strong constraints on the possibilities for the ambiguous species at mass 28 (CO, N_2, C_2H_4). An analysis of deviations from a pure water vapor spectrum during stellar occultations yielded an upper limit of 0.9% for CO (*Hansen et al.,* 2006, 2017). A solar occultation by the plume in 2010 presented the unique opportunity to use UVIS' extreme UV channel (55–110 nm) to constrain the N_2 abundance in the plume because this wavelength range includes N_2 absorption features. Although the solar occultation did not reveal any absorptions by N_2, the non-detection of such features set an upper limit of 0.5% N_2 in the plume. These upper limits further reduced the options left from the INMS E5 data (*Waite et al.,* 2009). Recently, the UVIS team summed many extremely long UVIS integrations of UV light reflected by the plume and produced a multiply scattered spectrum with features associated with those of hydrocarbon absorbers, primarily C_2H_4 (*Shemansky et al.,* 2016). Although this is the first unambiguous detection of a 28 u gas species, a mixing ratio has not yet been inferred.

The UVIS data have been tested for the presence of both methanol (CH_3OH) and ammonia (NH_3). Fits of the spectrum improve when methanol is added to the pure water absorption spectrum; however, there are no spectral features with adequate signal-to-noise ratios to allow unambiguous identification or even an upper limit of methanol as a constituent. As was the case for methanol, adding NH_3 to the model plume composition improved the overall fit by increasing absorption at short wavelengths. In the case of NH_3 there are definitive spectral features that should show up in the spectrum. However, these features are not detectable at the 0.4–1.3% level reported by INMS (*Waite et al.,* 2017) (see Table 2). UVIS data are thus consistent with, but cannot be used to independently verify, the INMS NH_3 estimate.

INMS plume spectra obtained from the E14, E17, and E18 flybys in 2011 and 2012 provided a much more consistent picture of the gas composition of the plume than previous INMS data. The flybys all occurred with ~7.5 km s^{-1} relative speed horizontally over the south polar region with closest approaches (CA) ranging from 75 to 100 km in altitude above the vent surface. This configuration allowed for good signal to noise ratios, but avoided effective molecular breakup from wall impacts (evident by the lack of species above 50 u). As can be seen in Fig. 3, the mass spectra were remarkably similar, allowing a deconvolution analysis of the compositional data (Magee et al., in preparation). This allowed a more confident determination of the major volatiles (Table 2) with the exception of the abundance of native H_2 (due to interference with H_2 from H_2O dissociation; see item 2 in section 2.1).

Therefore, the efforts of E14, E17, and E18 were complemented in October 2015 by the measurements of the E21 flyby, a horizontal south polar flyby with a CA of only 50 km and a relative speed of 8.5 km s^{-1}. Here, the INMS for the first time used its open source mode in the enceladean plume. Although the open source mode is a factor of 400 less sensitive than the closed source mode and comes with strict pointing requirements, it allows for mitigation of the effects from titanium reactions of the closed source antechamber since the material is ionized without wall interaction. The open source concentrations of major volatiles measured during E21 agree with the numbers inferred during E14, E17, and E18 with the exception of mass 28 (see discussion below). The E21 measurements most importantly enabled the detection and quantification of the mixing ratio of H_2 in the plume (*Waite et al.,* 2017). All resulting major volatiles are shown in Table 2.

Surprisingly, a species with a mass of 28 u was not seen in the open source data from E21, suggesting that the respective signal seen in the closed source is largely due to a fragmentation product (CO or C_2H_4) from heavier, maybe organic, molecules. The results exclude any 28 u intrinsic species in the plume gas at a mixing ratio of 0.1% (*Waite et al.,* 2017). It is currently unclear if this low fraction of a native plume volatile at 28 u is sufficient to be in agreement with the tentative detection of weak C_2H_4 reflection features by UVIS (*Shemansky et al.,* 2016).

Analysis of the organic compounds via mass deconvolution for both organic compounds carrying 2 or 3 carbon atoms (C_2 and C_3 species) obtained at E14, E17, and E18 leads to a host of ambiguities (Magee et al., in preparation) that can only be resolved with higher-mass-resolution mass

TABLE 2. Final volume mixing ratios of all confirmed neutral gas compounds in Enceladus' plume from Cassini INMS measurements [reproduced from *Waite et al.* (2017), Table 1, with permission from *Science*].

H_2O	CO_2	CH_4	NH_3	H_2
96–99%	0.3–0.8%	0.1–0.3%	0.4–1.3%	0.4–1.4%

spectrometers on future missions. These compounds with unresolved ambiguities that might be present in the plume are shown in Table 3. Organic molecular species with 3 or more carbon atoms or other species above 50 u were not detected on the low-speed flybys. These compounds are not present in the plume gas at mixing ratios accessible to INMS and the concentrations given in Table 3 can be seen as upper limits. The detection of organic species with high molecular masses in detectable concentration on high-speed flybys, at E3 and E5 (*Waite et al.,* 2009), was due to fragmentation of organic molecules (Fig. 2 and Table 1) above the mass range of INMS, likely residing in ice grains (*Postberg et al.,* 2018). See also sections 3.1 and 5.1.

2.3. Compositional Variation in Space and Time

Horizontal variation of the neutral gas density within the plume is substantial. The gas plume seems to consist of supersonic collimated high-velocity components ("jets") and emissions from much slower outgassing all along the south polar fractures (*Hansen et al.,* 2008; *Teolis et al.,* 2017). This idea is in agreement with the observed stratified ice grain emission (*Postberg et al.,* 2011a; *Porco et al.,* 2014; *Spitale et al.,* 2015) (see section 3.2). It is currently unclear if a compositional variation is linked to this emission because of the low spatial resolution of the published UVIS and INMS compositional results. However, INMS measurements indicate that in supersonic jets, emissions of heavy molecular species (e.g., CO_2, 44 u) are subject to a smaller lateral spread than lighter species (e.g., H_2, H_2O) (*Yeoh et al.,* 2015; *Perry et al.,* 2015). Modeling of these effects shows that the relative abundance of CO_2 and H_2O at altitude can vary more than 30% from the center to the edge of a jet (*Hurley et al.,* 2015). This scenario is true for individual sources (Fig. 4) but probably also generally affects the south polar plume as a whole, which is a superposition of these sources.

In contrast to ice grain emission, variations in time of the integrated gas emission rate in the entire south polar plume appear to be mild (*Hansen et al.,* 2011, 2017). However, emission rate variations of individual gas jets are substantial (*Hansen et al.,* 2017; *Teolis et al.,* 2017). It is not known if these variations over time correlate with variation in the composition of the gas phase. Limited compositional variations in the gas of individual jet sources over time are likely but have not yet been identified in the Cassini data although major compositional changes in the overall plume gas composition were not observed (*Waite et al.,* 2017; Magee et al., in preparation).

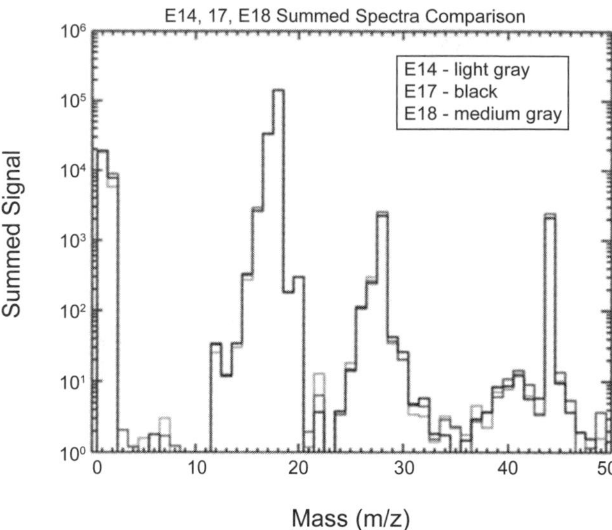

Fig. 3. The mass spectra from flybys E14 (light gray), E17 (black), and E18 (medium gray) show the reproducibility of the gas composition from these lower-velocity flybys (~7.5 km s^{-1}). The summed signal amplitude of each spectrum is set relative to the noise floor, such that the minimum value on the y-scale represents unit signal-to-noise ratio.

3. COMPOSITION OF THE SOLID PHASE

Three Cassini instruments have assessed the chemical composition of the icy component of Enceladus' plume: the Cosmic Dust Analyzer (CDA) (*Srama et al.,* 2004), the Cassini Plasma Spectrometer (CAPS) (*Young et al.,* 2004), and the Visible and Infrared Mapping Spectrometer (VIMS) (*Brown et al.,* 2004). The former two are *in situ* detectors that measure the particles along the flight path of the spacecraft through the plume. While the CDA's time of flight-mass spectrometer (TOF-MS) subsystem is sensitive to ice grains with radii of about 0.2–2 μm, CAPS observes much smaller grains with sizes of up to about 0.003 μm. VIMS is a remote sensing instrument that observes the plume ice grains from a large distance at infrared wavelengths (1–5 μm), thereby integrating along its line of sight. On a few occasions the observation geometry allowed VIMS to acquire spatially resolved spectra of the plume.

It is important to remember that a small fraction of plume particles (about 5–10% by mass) are launched fast enough to escape Enceladus' gravity and populate the E ring, while the rest of the icy grains fall back onto the moon's surface (*Porco et al.,* 2006; *Kempf et al.,* 2008, 2010; *Hedman et al.,* 2009; *Ingersoll and Ewald,* 2011). Hence the composition of E-ring grains, as well as plume particles, can provide information about Enceladus' interior.

We will first discuss implications for the plume composition derived from the CDA's E-ring spectra (section 3.1) and then the results from plume traversals (section 3.2). We address the nanodust population in the plume observed by CAPS (section 3.3) and finally the remote sensing results from VIMS (section 3.4).

TABLE 3. Ambiguous plume gas constituents and their possible concentrations from the deconvolution analysis of INMS spectra obtained at "slow" flybys (E14, E17, E18, and E21); molecular masses of the main isotopes are given in parentheses.

Minor Species I* Moderate Ambiguity >100 ppm <0.2%	Minor Species II† High Ambiguity <100 ppm				
	Hydrocarbons	N-bearing	O-bearing	NO-bearing	Others
C_2H_2 (26)	C_3H_4 (40)	CH_5N (31)	O_2 (32)	C_2H_7NO (61)	H_2S (34)
HCN (27)	C_3H_6 (42)	C_2H_3N (41)	CH_3OH (32)	$C_2H_5NO_2$ (75)	PH_3 (34)
C_2H_4 (28)	C_3H_8 (44)	C_2H_7N (45)	C_2H_2O (42)	$C_3H_7NO_2$ (89)	Ar (36,40)
CO (28)	C_4H_8 (56)	$C_2H_6N_2$ (58)	C_2H_4O (44)		C_3H_5Cl (76)
N_2 (28)	C_4H_{10} (58)	C_4H_9N (71)	C_2H_6O (46)		
C_2H_6 (30)	C_5H_{10} (70)	$C_4H_8N_2$ (84)	C_3H_6O (58)		
CH_2O (30)	C_5H_{12} (72)	$C_6H_{12}N_4$ (140)	C_3H_8O (60)		
NO (30)	C_8H_{18} (114)		$C_2H_4O_2$ (60)		
			$C_2H_6O_2$ (62)		
			$C_4H_{10}O$ (74)		
			$C_4H_6O_2$ (86)		

* Some combination of at least four of the listed species at concentration >100 ppm is required to match the INMS spectra. These species dominate the C2 region of the spectrum (masses 24–34).

† Many possible combinations of the listed species at low concentrations may match the INMS spectra, but at least some of these species are necessary to do so. Isomers are possible in some cases. These species are primarily used to fit the C3 region (masses 36–46).

Modified from *Magee and Waite* (2017).

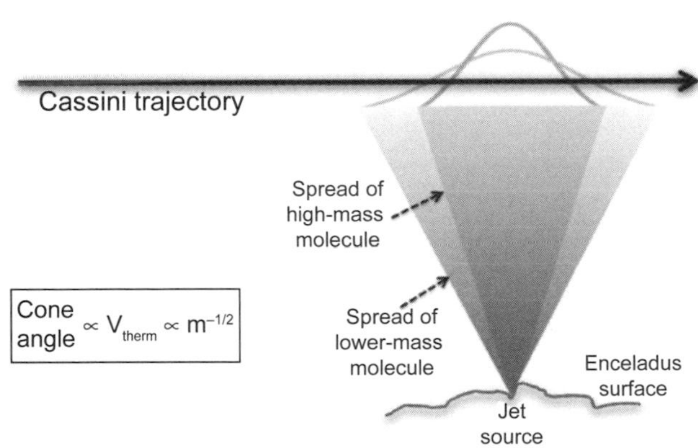

Fig. 4. Illustration of the mass-dependent behavior of high-velocity molecules emitted by the jets. For molecules emitted at the same supersonic velocity and in thermal equilibrium with each other at the time they are emitted, the cone angle or spreading of the molecules depends on the molecular mass. This behavior causes differences in spatial composition that are measured by the INMS. Depending on the temperature, bulk velocity, and mass, spreading angles vary from 10° to 45°. Cassini's CA during horizontal south polar flybys typically was between 50 and 100 km. From *Perry et al.* (2015).

3.1. Cosmic Dust Analyzer Measurements in the E Ring

In practice, most of the knowledge of the composition of ice grains emitted by Enceladus comes from the CDA, an impact ionization detector. When dust or ice grains strike the detector's metal target plate with speeds in excess of 1 km s^{-1}, a part of the impactor is ionized. CDA then produces time-of-flight mass spectra of the cations present in each individual impact cloud with a transmission cadence of up to 1 spectrum every 2 s. The mass resolution of the instrument is relatively low: $m/\Delta m$ increases from ≈10 at 1 u up to ≈50 at the upper end of the CDA mass range at about 200 u (*Postberg et al.,* 2006). From 2004 to 2017 CDA obtained tens of thousands of mass spectra of individual ice grains. Most of them were recorded in the E ring and only a few hundred directly in the plume. As CDA was not built to operate in a dense dusty environment like the enceladean plume, the instrument settings needed to be tweaked to allow plume measurements. These instrument settings always compromised the quality of plume spectra. However, in the E ring, CDA could be operated nominally, and therefore the E-ring spectra provide both the highest-quality spectra and the better statistics, whereas measurements in the plume provide insights into its spatial compositional structure.

The impact ionization process yields different cation abundances of identically composed particles at different impact speeds (*Postberg et al.,* 2008, 2011b). Impact speeds of ice grains during the Cassini mission vary between 3.5 km s^{-1} to 20 km s^{-1}. It often is a challenge to disentangle speed effects on the spectra from spectral variations that are due to actual compositional variations.

CDA measurements obtained during Cassini's first E-ring crossing in October 2004 quickly confirmed that the particles in the E ring are mostly composed of water ice (*Hillier et al.,* 2007). Over 99% of particles detected in the entire E ring are dominated by water ice. However, CDA also found that there are significant differences in the compositions of these ice grains, and that about 95% of all E-ring mass spectra from the CDA can be categorized into three major distinct families (*Postberg et al.,* 2008, 2009a). The abundance of each compositional type given below is based on the evaluation of about 10,000 E-ring ice grain spectra. These families are also present in the plume itself, although in different proportions (see section 3.2). Although interplanetary and interstellar dust (*Altobelli et al.,* 2016) has been detected in the E ring, it does not have an E-ring origin and therefore is not discussed here.

Type I particles: About 65% of all E-ring spectra belong to this group, increasing to even higher fractions with decreasing particle size. These grains appear to be composed of nearly pure water ice because their spectra are dominated by mass lines caused by water-cluster cations $(H_2O)_n(H_3O)^+$, (n = 0–15; see Fig. 5a). Na^+ and K^+ and their respective water cluster ions $(H_2O)_n(Na, K)^+$ are often present and form the only non-water mass lines. These lines imply mostly very low concentrations of alkali salts in the ice grains with Na/H_2O ratios on the order of 10^{-7} (*Postberg et al.,* 2009a).

Type II particles: Type II particles on average produce higher total ion yields upon impact, implying they are larger on average than Type I particles. Type II spectra represent the second most abundant E-ring family (≈25%, increasing with increasing grain size) and in most cases show the same characteristic as Type I with an additional distinct feature at mass 27 u to 31 u and/or 39 u to 45 u, with each of these mostly organic features representing more than one ion species (*Postberg et al.,* 2008). In some cases additional non-water signatures, indicative of additional organic compounds, appear. The fraction of organics and the composition of the organic species can vary dramatically among the different grains (Figs. 4c,d). Furthermore, while organic signatures are the most prominent non-water species, sometimes contributions from silicates and salts may be present. Most Type II spectra are salt-poor, similar to Type I.

Type III particles: This family, comprising about 10% of E-ring spectra, exhibits a totally different pattern of mass lines (Fig. 5b) than the other two. In contrast to Type I and II, the water cluster peaks $(H_2O)_n(H_3O)^+$ are absent or barely recognizable. The characterizing mass lines are of the form $(NaOH)_n(Na)^+$ (n = 0–4), indicating a Na/H_2O mole ratio well above 10^{-3}. Frequent mass lines of $NaCl\text{-}Na^+$ and $Na_2CO_3\text{-}Na^+$ reveal NaCl and $NaHCO_3$ and/or Na_2CO_3 as the main sodium-bearing compounds. Ground experiments with analog material indicate an average concentration of 0.5–2% sodium and potassium salts by mass, with K compounds being less abundant by far (*Postberg et al.,* 2009a,b). Impacts of Type III particles have an average ion yield that is several times higher than of Type I particles, implying a considerably larger size (*Postberg et al.,* 2011a).

In a few impact spectra, a combination of spectral features from different types (e.g., Type II and Type III) are found. While Type I and Type III grains are fairly homogenous within their compositional family, the organic-bearing Type II grains are quite compositionally diverse, with the concentrations of organic species varying from traces up to the percent level (*Postberg et al.,* 2018). Most Type II grains show one or two groups of organic mass lines between 26 u and 31 u and 39 u and 45 u respectively. These cations are indicative of C_2 and C_3 hydrocarbons respectively but could also contain oxygen- and nitrogen-bearing species, e.g., CH_xO^+, x = 1–3 or $CH_2NH_2^+$ between 29 and 31 u or respective C_2 species between 41 u and 45 u. In general these organic species are in agreement with volatile organics observed by INMS (Table 3).

However, a small fraction of Type II grains exhibits strong organic mass lines at masses in excess of 70 u up to the end of the CDA mass range at 200 u (*Postberg et al.,* 2018). These high mass organic cations (HMOC) indicative of concentrations on the percent level stem from refractory organic inclusions in the ice grains (section 5.1). The HMOC-type grains show aromatic and aliphatic constituents with functional groups containing oxygen and likely nitrogen. Most aromatic constituents are attached to non-carbon functional groups or dehydrogenated C-atoms. It is possible that all these constituents originate from cross-linked or polymerized macromolecules where mostly small aromatic structures are connected by short aliphatics chains (*Postberg et al.,* 2018). These high-mass organic species residing in ice grains might have also been observed by the INMS in the plume during high-speed flybys, where the high impact velocity disintegrated large molecules to organic fragments small enough to show up in INMS limited mass range (<100 u) (*Postberg et al.,* 2018) (see section 2.1 and Fig. 2 and Table 1 therein). The observation of unspecified high mass molecular species in the plume by CAPS (*Coates et al.,* 2010a,b) might also be due to large fragment ions from these organic species (see section 4.1).

Another dust population observed by the CDA instrument that can provide information about the E ring's composition, and thus Enceladus' plume, are the so-called "stream particles." These are high-speed, nanometer-sized dust particles that are not gravitationally bound to the saturnian system and were seen for the first time well before Cassini reached Saturn (*Kempf et al.,* 2005b; *Hsu et al.,* 2010). These tiny grains of dust, once charged, gain sufficient kinetic energy from Saturn's magnetic field to be thrown out of orbit into interplanetary space (*Grün et al.,* 1993; *Hamilton and Burns,* 1993; *Horanyi et al.,* 1993; *Kempf et al.,* 2005b; *Hsu et al.,* 2010, 2011, 2012). Numerical modeling of their trajectories indicates that the majority of Saturn's stream particles were once part of the E ring before they were ejected into the streams. Moreover, from their composition and dynamical modeling, these particles are thought to be inclusions released from much larger E-ring ice grains by the magnetospheric plasma erosion (*Hsu et al.,* 2011, 2015).

CDA can only detect these tiny grains because they hit the detector with extraordinary high speed, acquired by their magnetospheric interaction, typically exceeding 100 km s^{-1} (*Hsu et al.,* 2010). Still, only the largest of stream particles (≈20%) produce a signal that is strong enough to allow a rough characterization of their composition with CDA (*Hsu*

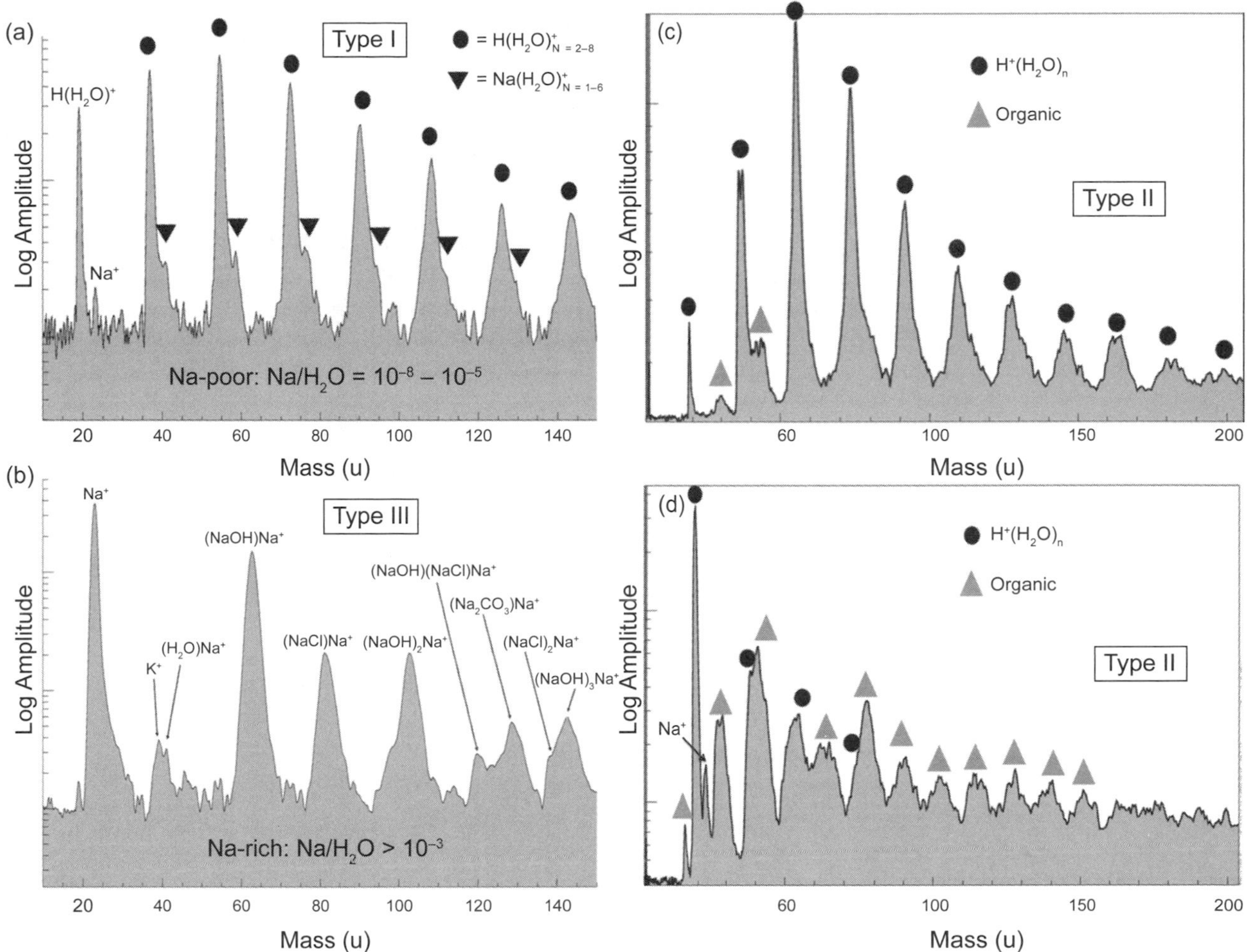

Fig. 5. Different compositional types in representative CDA mass spectra of E-ring ice grains. **(a)** Type I spectra show mostly water with only traces of sodium. From *Postberg et al.* (2009a). (b) Type III spectra exhibit strong mass lines from different sodium salts. From *Postberg et al.* (2009a). **(c),(d)** Type II spectra, showing a wide variety within their group. Many are very similar to Type I with the addition of subtle mass lines in agreement with molecular organic cations carrying only 1–3 carbon atoms [**(c)**]. A few Type II spectra show abundant organic cations, some of them extending over the entire CDA mass range up to ≈200 u [**(d)**].

et al., 2011). These particles provide unique information about the plume's composition because, unlike E-ring grains, many of these largest saturnian stream particles have silicon as a major constituent and are depleted in water ice (*Kempf et al.,* 2005a; *Hsu et al.,* 2011). Co-adding of the weaker signals shows that at least a part of the grains that show no individual particle signal (≈80%) have a similar silicon-rich composition (*Hsu et al.,* 2011). The strongest stream particle spectra nearly all show a silicon mass line and have been used by *Hsu et al.* (2015) for more detailed compositional analysis. They find that they are almost metal-free with a composition in agreement with pure silica (SiO_2) rather than typical rock-forming silicates (e.g., olivine or pyroxene). A size estimate derived from their dynamical properties by numerical modeling (*Hsu et al.,* 2011) agrees with the sizes inferred from the spectra signal of the silca grains on a very confined size range with radii ranging 2–9 nm (*Hsu et al.,* 2015). A rough quantitative estimate by *Hsu et al.* (2015) gives a silica/water ice mixing ratio of 150–3500 ppm in the material ejected from the plume into the E ring. This number is based on the assumption that all nanograins detected by CDA are made of silica, although the composition can only definitely be assessed for a fraction of them. In this sense, the mixing ratio given above represents an upper limit. Note that this population is different from the nanograins observed directly in the plume by CAPS (section 3.4).

Tables 4 and 5 give an overview on the inner E-ring composition near Enceladus estimated from CDA data. Table 5 also shows how concentrations inferred in the E ring might be extrapolated to plume composition. The entries in Table 5 do come with some caveats, as described in the caption. An additional ambiguity is introduced, but not considered in Table 5, because CDA is not sensitive to grains below ≈0.2 μm in the E ring or the plume, yet CAPS measurements show that these small grains are present (section 3.3). This situation becomes even more complicated when assessing the overall composition including the vapor, because it is important to consider the different gas/solid mass ratios in the plume and in the E ring. In the plume this ratio is about 10 (section 8.1). However, only ≈10% of the grains escape the plume (*Porco et al.,* 2006, 2017; *Spahn et al.,* 2006; *Schmidt et al.,* 2008; *Kempf et al.,* 2010; *Ingersoll and Ewald,* 2011), increasing the gas/dust ratio injected into the E ring to ≈100.

3.2. Cosmic Dust Analyzer Measurements Inside the Plume

The compositional types identified in E-ring ice grains are also found in the plume. Modeling of the plume indicates that the plume is stratified in grain size (*Schmidt et al.,* 2008), which was observed by VIMS (*Hedman et al.,* 2009) and the HRD subsystem of CDA (*Kempf et al.,* 2008). So it is naturally interesting to see if this dynamical stratification comes along with a compositional stratification. Normally the CDA's maximum cadence of spectra transmission (<0.6 s^{-1}) does not allow measurements with high spatial resolution in the plume. In 2008 a special modification to the CDA's processing software was made to allow a spectra transmission rate of up to 5 s^{-1} for a short time. However, the mass resolution and the mass range of the CDA had to be reduced to accommodate this increased transmission rate. The new mode was successfully executed during the highly inclined E5 flyby with a relative speed of 17.7 km s^{-1}. The result is shown in Fig. 6. Each data point of the profile can be compared with the highly inclined spacecraft trajectory shown in Fig. 7.

Figure 6 shows that on E5 near CA to Enceladus (21 km) the proportions of the three main types exhibited significant variations: (1) a steep increase in salt-rich Type III grains from near zero shortly before CA to a maximum of >40% a few seconds after CA and a subsequent shallower decrease toward the dense plume; (2) a corresponding simultaneous decrease in the Type I grain proportion with respect to the E-ring background shortly before CA; and (3) a less-pronounced increase in the proportion of Type II grains after CA with a subsequent sharp maximum between +45 s to +51 s.

The most plausible explanation for the simultaneous increase of Type III and decrease of Type I proportions in the fringe region of the plume (Fig. 7) is that salt-rich grains are ejected at slower speeds than salt-poor grains: The slow Type III grains are dominant at low altitudes, whereas the faster Type I grains are enriched at higher altitudes and in the E ring but depleted close to Enceladus. In contrast, the increased ratio of organic-containing Type II ice grains in the core region of the plume does not seem to depend on altitude and thus ejection speed.

TABLE 4. Abundances of main ice grain types as identified by CDA in the E ring and their non-water constituents (*Hillier et al.,* 2007; *Postberg et al.,* 2008, 2009a,b, 2011a, 2018; *Hsu et al.,* 2015).

	Type I	Type II	Type III	Stream Particle Nanograins
Number fraction	60–70%	20–30%	≈10%	—
Main non-water constituent (MNWC)	Na, K	Organic	Na and K salts	SiO_2
Typical MNWC concentration in individual grains	<0.0001%	0.000001–10%	0.5–2%	High

Valid for particle radii for which the composition can be assessed by CDA (≈0.2–2 µm). The percentage given in the table is size dependent because in larger grains Type I become less frequent while the other two become more frequent. The abundances of the different types thus also depend on the minimum size threshold detectable by CDA, which again depends on impact speed and instrument settings. Stream particle nanograins actually do not belong to the E ring in a dynamical sense. However, their E-ring origin was indirectly determined by *Hsu et al.* (2011, 2012).

TABLE 5. Integrated abundances of constituents found by CDA in the inner E ring.

	Water Ice	Organic Material	Na and K Salts	SiO_2
Concentration in E-ring solids	99.0%–99.9%	0.01–0.3%	0.1–0.4%	0.015%–0.35% (upper limit)
Concentration in plume solids	Decreasing	Increasing	Increasing	Similar (?)

These estimates require some assumptions. It is assumed that all compositional types have a similar ion yield when impinging CDA and that Type II and III grains are on average more massive than Type I grains by a factor of 5. The organic fraction is calculated based on the assumption that most of the organics emitted into the E ring reside in a specific subpopulation that show extraordinary high organic concentrations up to the percent level (*Postberg et al.,* 2018). An average organic concentration of 0.5–5% in these grains was assumed. The actual detection frequency of this grain subtype in the E ring varies between 1% and 3% (*Postberg et al.,* 2018). These two factors determine the uncertainty for organics given in the table. The number for SiO_2 from stream particles is taken from *Hsu et al.* (2015) and constitutes an upper limit as explained in the text. The water abundance is inferred from the abundances of the non-water constituents and neglects the possibility that there might be further compounds not seen by CDA. In this sense the value is an upper limit.

Postberg et al. (2011a) suggest that different size distributions of Type I and III grains are responsible for the different ejection speeds implied by the measurements. As a consequence of the radius-dependent friction force that a grain experiences when accelerated by a gas in a subsurface ice vent (*Schmidt et al.,* 2008), ejection speeds of particles are size dependent. This size dependence leads to the observed tendency of large particles preferentially populating the lower regions of the plume (*Kempf et al.,* 2008; *Hedman et al.,* 2009). Figure 6d shows the model fits to the E5 data, assuming size-dependent ejection speeds in a uniform particle flux emerging from all four tiger stripes. It qualitatively reproduces the rise in the fraction of salt-rich grains around CA. An even better match could be achieved if the eight faster and more collimated jet-like particle sources known at that time (*Spitale and Porco,* 2007; *Hansen et al.,* 2008) were added to the uniform flux. These jets are modeled with a steeper particle size distribution, and therefore are richer in small grains compared to the slower uniform particle flux, and, in this model, are preferably salt poor (*Postberg et al.,* 2011a). Figure 7 shows a graphical representation of the modeled plume including the jets and the E-ring background.

The profile of organic-bearing Type II grains does not seem to follow a trend where speed and size are linked to composition as in the case of Type I and III grains. Their proportion is slightly higher in the dense plume compared to the plume fringe region, in which the Type III maximum of E5 lies (Fig. 6a–c). This implies a general enrichment of ice grains containing organic material that could be associated with fast, collimated jets (*Postberg et al.,* 2011a). A second significant increase of Type II grains between 45 and 51 s after CA coincides with the passage of jet source III identified by *Spitale and Porco* (2007). This event also coincides with the time where Cassini's ground track lies over the tiger stripe fissure called Damascus Sulcus and indicates a passage through a region significantly enriched in organic-bearing ice grains. The short timing of this Type II increase only agrees with a very collimated jet source with an opening angle of about 10° because the spacecraft was already 600 km above the surface at that point (Fig. 7). The implied

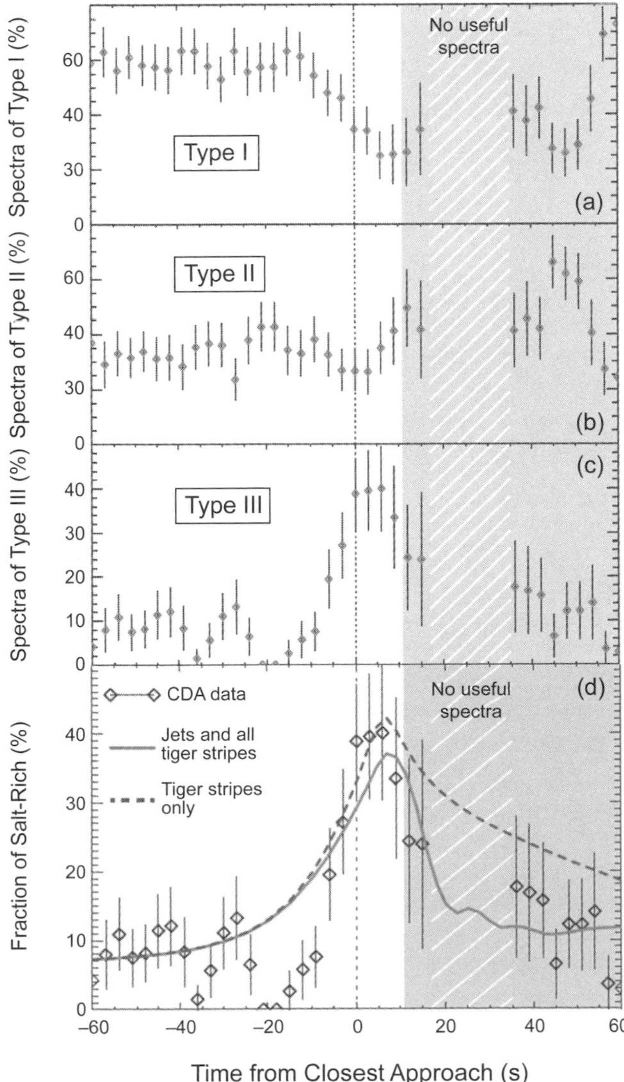

Fig. 6. The compositional plume profile during E5 (modified from *Postberg et al.,* 2011a). **(a)–(c)** The relative frequencies of the compositional grain types (I, II, and III) are plotted. Each data point represents an interval of ±4 s and includes ~40 spectra (≈5/s) before CA and slightly less afterward. Error bars are 1 σ from the mean derived from counting statistics. CDA continuously recorded at its maximum rate, therefore the measured frequencies reflect proportions and not absolute abundances. During the period of highest impact rate between ~18 s and ~35 s after CA, too few useful spectra were obtained (hatched region) due to the overload of the instrument. Similarly, unspecified selection effects may have led to fluctuations in the statistics starting from ~11 s after CA. Although the data obtained during this time interval (gray) may therefore have been affected by instrument performance issues, they exhibit a stable trend matching the model predictions of *Postberg et al.* (2011a) shown on the right panel. **(d)** Compositional profile of Type III grains with overlaying contours obtained from two models. The dashed line shows a modeled uniform particle flux emerging from all four tiger stripes. The solid line shows a model including eight faster and more collimated jet-like particle sources observed by *Spitale and Porco* (2007). The contribution from the small jet particles helps fit the rapid decrease of salt-rich grains when the spacecraft was entering the densest part of the plume, starting about 5 s after CA. The model fit including fast sources is also considerably better in matching the relatively low level of Type III grains after 40 s from CA, when the spacecraft was still within range of the jets. The background flux of E-ring grains is also part of the model. It dominates until about 10 s before CA.

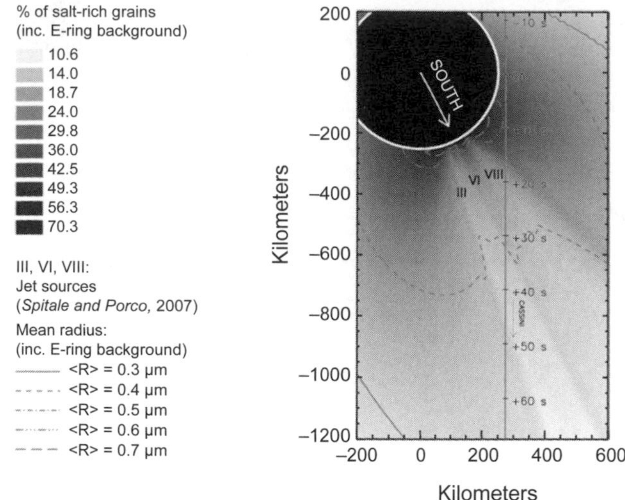

Fig. 7. Graphical representation of the model plume, including the E-ring background, as derived from the E5 flyby data (modified from *Postberg et al.,* 2011a). The shades shown b on the left show the modeled proportion of salt-rich grains (Type III). Pure water ice (Type I) and organic-bearing grains (Type II) are subsumed as salt-poor in this model. Overlaid are contours of constant mean particle radius obtained from the model. The projection used is in the plane of the E5 spacecraft trajectory (solid black line, shown with 10-s intervals). It is expected to see both the largest particles and the highest fraction of salt-rich grains, a few seconds after CA to Enceladus. Structures of the three most relevant localized supersonic jets for this projection are clearly visible in both the compositional profile and size contours. Note that the model only considers particles with radii above the estimated instrument's detection threshold ($r \geq 0.2$ μm).

organic-rich emission from Damascus Sulcus is supported by VIMS observations of surface deposits that show the strongest organic absorption at 3.44 μm (*Brown et al.,* 2006) on Enceladus around Damascus Sulcus (section 6). Moreover, plume grains emerging from Baghdad Sulcus and Damascus Sulcus show IR features in VIMS spectra not observed in emissions of the other two fractures (section 3.4).

The high-rate spectra recording mode employed during E5 turned out to be a risk to the CDA instrument's health and safety and therefore could not be used again. Despite the limits in detection rate, other attempts to map the plume stratification were done during E17: E18 in 2012 and E21 in 2015. Here the maximum spectral recording rate was limited to ~0.6 s^{-1}, about 10× lower than during E5. Unlike E5 the trajectory of all three flybys were not inclined and led Cassini almost horizontally over the SPT. The flyby speeds were less than half of E5: 7.5 km s^{-1} on E17 and E18 and 8.5 km s^{-1} on E21. In terms of spatial resolution, the lower speed partially compensated for the lower detection rate.

In contrast to E5, where the spacecraft trajectory was perpendicular to the tiger stripe fractures, Cassini flew almost parallel to these surface features during E17 and E18 (Fig. 8). With a mean anomaly of 146° (E17) and 153° (E18) the position in Enceladus' orbit was also very similar. Of these three low-velocity flybys, E17 yielded the dataset with the highest quality. The compositional profile of E17 (*Khawaja et al.,* 2017) is shown in Fig. 8. When entering the plume from the E-ring background, the proportion of Type I goes down, whereas the proportions of the other two groups go up. However, the increase in the proportion of Type III grains is much less pronounced than in E5, reaching ≈18% around the CA (≈7% in the E ring), whereas at the same time the Type II proportion reaches 55% (starting from ≈30% in the E ring).

The domination of Type II grains in the plume on E17 might only be reconciled with the E5 findings when a general enrichment of these organic-bearing grains inside a large number of fast jets is assumed, which, at the same time, suppress the Type III proportion that are ejected from slower sources. Compositional plume variations in space — during E5 the closest approach was near Alexandria Sulcus whereas in E17 it was near Cairo and Bagdhad Sulci — and in time — between E5 (in 2008) and in E17 (in 2012) — might also be a factor. Efforts are ongoing to modify the E5 model in a way that it can be reconciled with the E17 data.

3.3. Cassini Plasma Spectrometer Measurements on Nanograins in the Plume

Two subsystems of CAPS, the Ion Mass Spectrometer (IMS) and the Electron Spectrometer (ELS) (*Young et al.,* 2004), frequently observed a population of charged nanograins in the Enceladus plume (*Jones et al.,* 2009). The principle of operation of ELS allows the detection of negatively charged nanograins, their energy per charge (as ELS uses an electrostatic analyzer), and direction, using a microchannel plate (MCP). For the Enceladus encounters, the flyby speed may be used to determine the mass per charge via a conversion involving that speed. In the case of IMS, there is an additional linear electric field time of flight (TOF) determination of mass, beyond the input electrostatic analyzer, that operates similarly to ELS. Most of the results

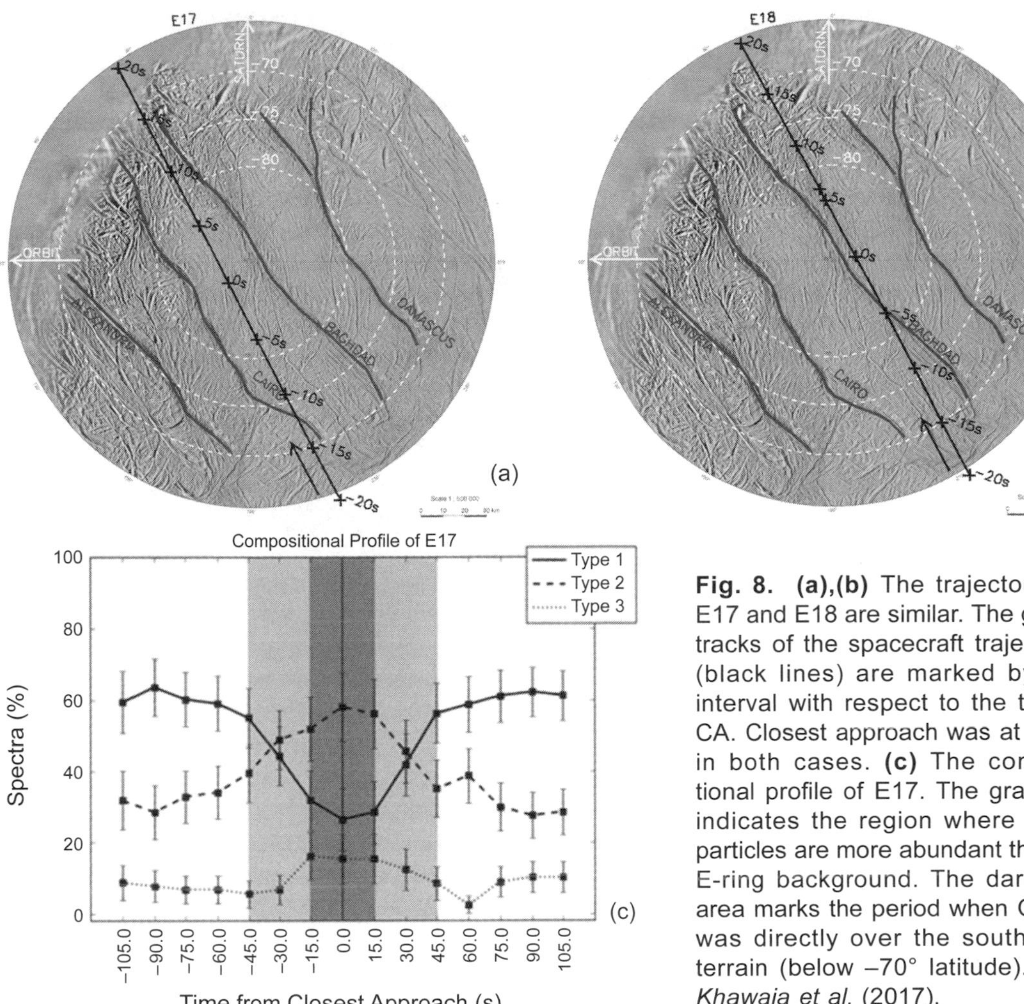

Fig. 8. (a),(b) The trajectories of E17 and E18 are similar. The ground tracks of the spacecraft trajectories (black lines) are marked by time interval with respect to the time of CA. Closest approach was at 76 km in both cases. **(c)** The compositional profile of E17. The gray area indicates the region where plume particles are more abundant than the E-ring background. The dark gray area marks the period when Cassini was directly over the south polar terrain (below –70° latitude). From *Khawaja et al.* (2017).

here refer to the analysis of "START" MCP pulses beyond the IMS electrostatic section.

With a mass to charge ratio m/q ~ 10^3–10^4 u/q the detected grains are inside the mass gap between neutral and charged molecular species (sections 2 and 4) and larger, macroscopic (r > 0.2 μm) solid ice grains observed by CDA (section 3.1 and 3.2) and VIMS (section 3.4) (Figs. 9 and 13b). During plume traversals the CDA is not sensitive enough to detect such small grains and CAPS is the only instrument capable of observing this subpopulation in the plume. In contrast to the hypervelocity SiO_2 nanograins observed by CDA in Saturn's magnetosphere, the nanograins in the plume are generally assumed to be of icy composition by comparison with the compositional analysis of gas (*Hansen et al.,* 2008, 2011; *Waite et al.,* 2017) (section 2), ions (*Coates et al.,* 2010a,b) (section 4) and larger grains observed by the CDA (*Postberg et al.,* 2011a) (section 3.2). Thus the icy composition of nanograins is mainly based on plausibility arguments but not on direct measurements.

The nanoparticles are charged both negatively and positively and dominate the energy spectra at ~1 keV and higher in the regions in which they are seen. *Jones et al.* (2009) suggested that triboelectric charging during the plume emission process could provide the charging mechanism. Detailed comparison of the observed fluxes with known jet emission sites gave a good correspondence, and the oppositely charged species were seen to deflect in Saturn's electric and magnetic field.

The charged nanograin analysis of these heavy species (up to about 30,000 amu/q, corresponding to ~2000 H_2O molecules) was further pursued by *Hill et al.* (2012). On this basis they suggest that most of the grain charging occurs in the plume itself via electron impact. They also argued for single charge on the grains. The measured density of negative and positive nanograins near Enceladus is shown in Fig. 9d from *Hill et al.* (2012) and indicates that negative nanograins are clearly dominant near Enceladus.

Charged nanograins can easily escape Enceladus' gravity field on trajectories bent by Saturn's co-rotating electromagnetic field (*Hill et al.,* 2012; *Dong et al.,* 2015) and can provide a significant source of material for the Enceladus torus and for the E ring. If one uses the (uncertain) negative and positive nanograin densities from *Hill et al.* (2012) and assumes a grain speed of ~500 m s^{-1} over an area πR_E^2 (where R_E is Enceladus radii) and an average mass of 10,000 amu, then an approximate mass flux of negative

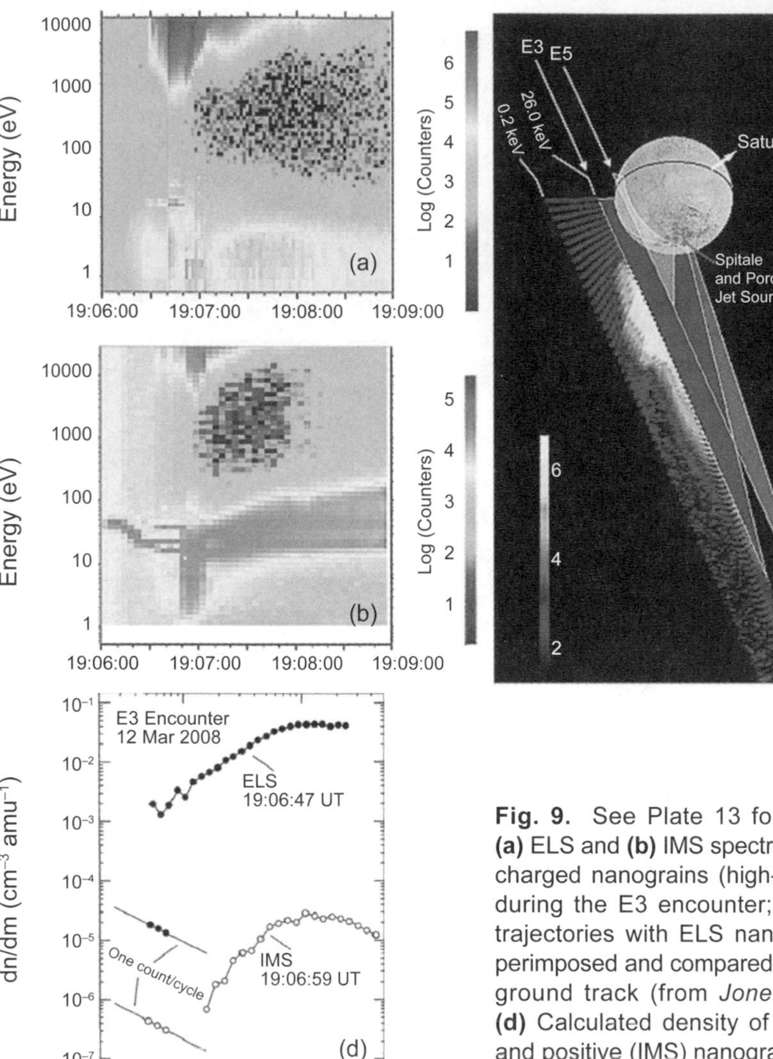

Fig. 9. See Plate 13 for color version. **(a)** ELS and **(b)** IMS spectrograms showing charged nanograins (high-energy signals) during the E3 encounter; **(c)** E3 and E5 trajectories with ELS nanograin data superimposed and compared with the Cassini ground track (from *Jones et al.,* 2009). **(d)** Calculated density of negative (ELS) and positive (IMS) nanograins showing the dominance of negatively charged grains (from *Hill et al.,* 2012).

and positive charged nanophase dust grains of 5.6 kg s^{-1} can be estimated, which is a mass flux comparable to those of grains with r > 0.6 μm (see the chapter by Kempf et al. in this volume). *Dong et al.* (2015) further analyzed the CAPS nanograin data and combined them with CDA and RPWS observations to provide a composite size distribution. This model distribution was based on fitting a composite dust nanograin size distribution peaking at ~2 nm to the other observations. From these fits, the total mass production rate of all grains was found to be ~20% of the INMS water vapor mass density at ~15–65 kg s^{-1}, the majority of which resides in grains with radii below 100 nm. However, this estimate is controversial because it assumes a continuous size distribution between grains of a few nanometers in size up to micrometer-sized grains, which is ambiguous as Cassini's instruments do not well constrain the size distribution for radii between 4 nm and about 1 μm (*Dong et al.,* 2015). Recent calibration experiments indicate that *Dong et al.* (2015) drastically underestimated the nanograin detection efficiency of the CAPS sensor. If this underestimate is verified, the flux would go down by a factor of 10–20 to not more than a few kilograms per second. In any case, the mass flux of nanograins could be close to the mass flux of macroscopic ice grains escaping into the E ring and would therefore be an important source of matter into the E ring and Saturn's magnetosphere. For a detailed discussion, see the chapter by Kempf et al. in this volume.

3.4. Visible and Infrared Mapping Spectrometer Measurements of the Plume

Information about the plume ice particle composition can also be derived from remote sensing spectral data. Challenges associated with these sorts of measurements are that the plume has a low optical depth and the plume particles are strongly forward scattering. The former means that the plume spectra have low signal-to-noise ratios, while the latter means that the observations with the highest signals do not typically exhibit strong absorption signals. Indeed, the only clear spectral feature in near-infrared plume spectra is a dip at 3 µm that is due to the extremely strong fundamental water-ice absorption band. Thus far, no other component than water has been securely detected in near-infrared plume spectra, although efforts are ongoing.

While the CDA data contains much more information about the chemical composition of plume particles, the VIMS spectral data provide important constraints on the physical structure of the ice grains. For example, the position of the band minimum depends on whether the ice is in an amorphous or crystalline state, and the observed spectra indicate that the plume particles consist primarily of crystalline water ice (Fig. 10). This crystalline state implies that the grains formed at temperatures above 130 K, which is consistent with other evidence that the plume sources are rather warm. Further studies of the spectral and photometric properties of the plume particles could also reveal whether the plume particles are compact grains or loose aggregates of smaller particles.

VIMS solar occultation data obtained in 2010 together with UVIS measurements show that the plume material above Baghdad and Damascus Sulci has a dust-to-gas mass ratio that is roughly an order of magnitude higher than the material above Alexandria and Cairo Sulci (*Hedman et al.,* 2018). The highest-resolution near-infrared spectral data obtained by VIMS can resolve material coming from three of the fissures, allowing spatial variations in plume particle properties to be detected (Fig. 11). The ice grains emerging from Baghdad, Cairo, and Damascus Sulci all show a strong 3-µm water-ice absorption band with a band minimum position consistent with primarily crystalline water ice. However, the detailed shape of this band, as well as the spectral slope at shorter wavelengths, does differ from fissure to fissure. In particular, the spectra of the Cairo material seem to be distinct from those of the material emerging from Baghdad and Damascus. This observation almost certainly reflects differences in the particle size distributions of the material erupted from the different fractures, but it could also imply variations in the structure and compositions of the grains emerging from the different sources (*Dhingra et al.,* 2017). The high amount of organic-bearing grains observed by CDA when flying over Damascus during the E5 flyby (*Postberg et al.,* 2011a) (section 3.2) might also reflect these compositional differences.

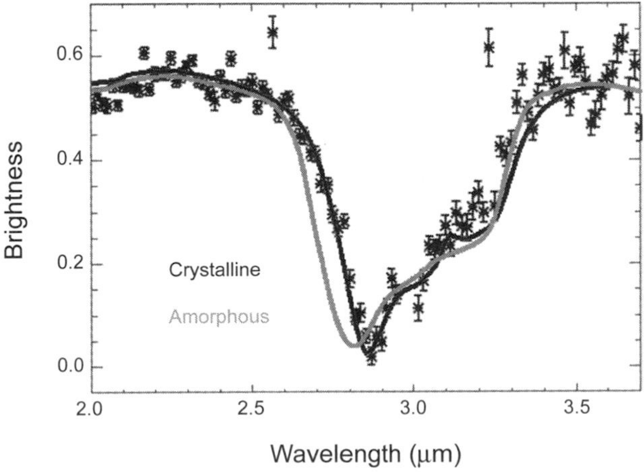

Fig. 10. Character of ice in Enceladus' plume. Mie-theory based model spectra for crystalline and amorphous ice compared with plume spectra collected with the VIMS instrument onboard Cassini. The VIMS spectra are best fit with a crystalline water ice spectrum. Adapted from *Dhingra et al.* (2017).

4. COMPOSITION OF CHARGED PARTICLES

Two of Cassini's instruments assessed the composition of charged particles emitted by Enceladus. The positive ion composition was determined by the CAPS IMS (*Young et al.,* 2004), which has a TOF subsystem. The principle of operation of ELS allows the detection of negatively charged species, their energy per charge (as ELS uses an electrostatic analyzer), and direction, using a microchannel plate (MCP). Measurements have been performed from inside the plume as well as from escaping material in Saturn's magnetosphere. The analysis of charged species in the magnetosphere has also been pursued using the Charge Energy Mass Spectrometer (CHEMS) sensor of the Magnetospheric Imaging Instrument (MIMI).

In general, the sensitivity of Cassini instruments to minor species in ionized form is lower than for neutral molecules or macroscopic ice grains. In fact, only the measurements in Saturn's magnetosphere allowed for long enough integration times (months to years) to unambiguously identify non-water constituents. In some cases, quantification was possible, nicely complementing the neutral gas and solid grains compositional measurements of Enceladus' plume material.

4.1. Cassini Plasma Spectrometer Measurements of Charged Molecules in the Plume

The discovery of the Enceladus plume via the magnetic field deflection (*Dougherty et al.,* 2006) and subsequent Cassini measurements (*Spahn et al.,* 2006; *Hansen et al.,* 2006; *Porco et al.,* 2006; *Waite et al.,* 2006) provided the impetus to determine the plasma and neutral interaction and composition of the plume. The first ion measurements were presented by *Tokar et al.* (2006), who analyzed the plasma flow around Enceladus. The measured deflection was initially compared to models developed for Io (*Hill and Pontius,* 1998). The initial estimate for the total plasma mass loading rate was ~3×10^{27} H_2O s^{-1}, corresponding to ~100 kg s^{-1}. The positive ion composition near the plume was found to be dominated by water group ions including O^+, OH^+, H_2O^+, and H_3O^+. The presence of H_3O^+ shows that ion-neutral chemistry occurs in the plume as charge exchange is required for its formation and is velocity dependent with slower velocities favoring charge exchange.

The Radio and Plasma Wave Spectrometer (RPWS) and CAPS ELS observe a substantial plasma density increase near Enceladus (*Morooka et al.,* 2011; *Coates et al.,* 2013), again indicating that the main mass-loading process is charge

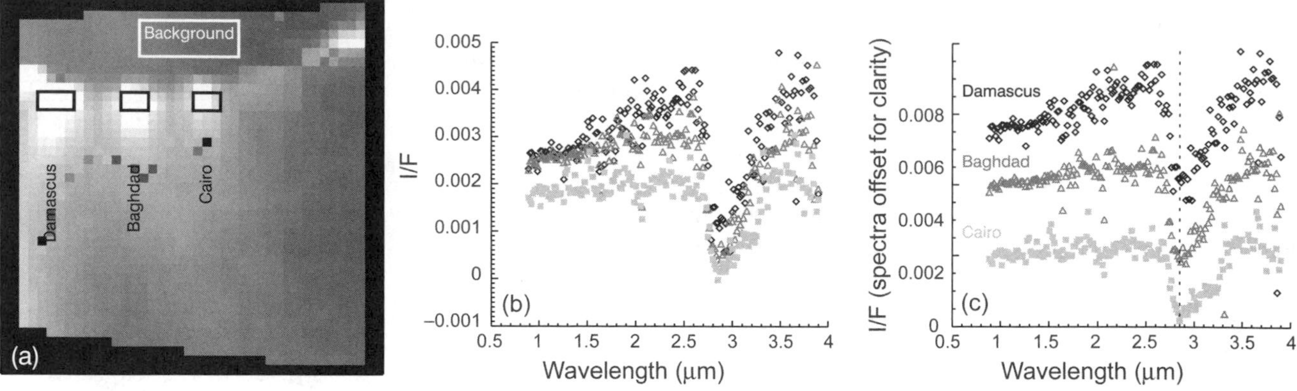

Fig. 11. Spatial variability in the near-IR spectral properties of Enceladus' plume. **(a)** Spatially-resolved VIMS observation of Enceladus' plume from the E10 encounter (cube: V1652853941). Rectangles indicate sampled regions in the plume and correspond to eruptions along Cairo, Baghdad, and Damascus tiger stripes. **(b)** Spectral character of eruptions along individual tiger stripes. Note the differences in the spectral slope between 1 and 2.5 μm. **(c)** Same spectra as in **(b)** but vertically offset for clarity. Note the spectral asymmetry (bump) present in the Cairo spectrum (indicated by the arrow), which is not observed in spectra from Baghdad and Damascus. Similarly, a spectral bump around 2.6 μm is only apparent on Baghdad and Damascus. The dotted line indicates the band minimum position, which is the same for all three spectra and indicates crystalline water ice grains in the plume. Uncertainties on individual data points are not plotted for the sake of clarity. Sizes of error bars are typically comparable to those of the symbols, but they vary with wavelength. See *Dhingra et al.* (2017) for further details on the uncertainties associated with these spectral observations.

exchange. Additional ionization processes include electron impact ionization and photoionization. Charge exchange provides energetic neutrals that create an expansion of the Enceladus-related neutral cloud seen as a large-scale OH cloud from the Hubble Space Telescope (*Shemansky et al.,* 1993) and later from Cassini (*Esposito et al.,* 2005).

Further analysis of positive ions in the plume itself was presented by *Tokar et al.* (2009) using two close Enceladus encounters in 2008, E3 (52 km CA) and E5 (25 km). Cold (<10 eV) ions were observed in the ram direction, indicating an almost stagnant plume ionosphere produced from the plume's neutral exosphere. Slowing of the plasma was observed north of Enceladus at some 4–6 R_E away, while south of Enceladus signatures were seen up to 22 R_E away. The composition of the plume ionosphere was again water group (O^+, OH^+, H_2O^+, and H_3O^+) ions, and in addition, heavier water dimer positive ions were found [$(H_xO_2)^+$] with x = 1–4. These heavier ions, predicted by *Johnson et al.* (1989), may be formed by charge exchange with a neutral dimer from the plume or via ion-molecule interactions in the stagnant plasma. Figure 12a shows a mass spectrum of positive ions in the Enceladus plume ionosphere.

The cold ions in the plume and the low relative speed between the ions and neutrals here indicate that the initial ions are converted to fresh pickup ions via ion-molecule interactions, consistent with the presence of H_3O^+. The ambient ions from the magnetosphere nearby (O^+, OH^+, H_2O^+, and H_3O^+) interact with H_2O in the plume itself, giving H_2O^+ and H_3O^+, in a similar process of H_3O^+ production to comets (*Cravens et al.,* 2011). These initially stagnant ions are gradually accelerated and move into the magnetospheric wake. They become the principal source of H_3O^+ for the saturnian magnetosphere and contribute to the ambient plasma torus, which interacts via charge exchange with Saturn's extended neutral cloud (*Tokar et al.,* 2006) and becomes redistributed through the magnetosphere (*Johnson et al.,* 2006).

In addition to the positive water dimer ions, another remarkable discovery was that of negative water cluster ions in the plume (*Coates et al.,* 2010a). These cold ions were also seen in the spacecraft ram direction and form part of the plume ionosphere. The ions have a short lifetime and were inferred to be constantly produced from H_2O or ice grains in the plume itself. Enceladus thus joins Earth, Comets Halley and Churyumov-Gerasimenko, and Titan as locations where negative ions have been detected. Figure 12b shows a mass spectrum of negative ions in the Enceladus plume ionosphere, while Fig. 9a–c shows overviews of the CAPS observations.

Water-associated negative ions (e.g., OH^-, O^-, H^-) can be produced from H_2O by dissociative electron attachment. Peaks are visible in mass groups 9–27, 27–45, 45–70, 70–300, and 300–500 u/q, which may be $(OH)_n^-$ or perhaps $(OH^-)(H_2O)_n$ with n = 1,2,3,4 . . . 30. Considering the very limited mass resolution of the measurement, the first three peaks are well centered around masses of negative clusters (17 u, 35 u, 53 u). However, the peaks visible in the higher-mass groups may include not only water-related clusters, but perhaps more complex carbon-based species such as seen by CDA and INMS (see sections 3 and 5). The density of the negative ions decreased with mass and can reach 50% of the total density of ambient electrons. However, there is also evidence that the highest density of negative species is in the charged dust grains in this region (see section 3.3). A further discussion of the negative ions and comparison with Titan can be found in *Coates et al.* (2010b).

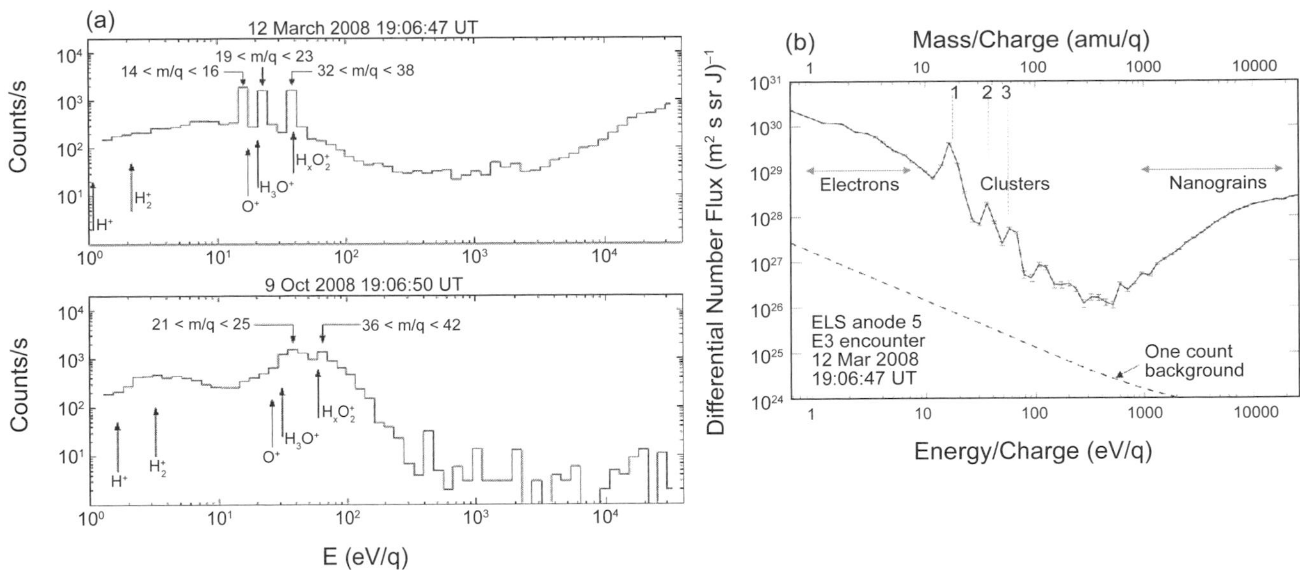

Fig. 12. **(a)** Positive ion spectra in the Enceladus plume ionosphere measured during the E3 (top) and E5 (bottom) encounters, showing water group and dimer (with x = 4) ion masses (*Tokar et al.,* 2009). **(b)** Negative ion spectrum measured during the E3 encounter showing multiples of m = 18 (adapted from *Coates et al.,* 2010a). Nanograins can be seen at the high mass end of the spectrum.

The complexity of the negatively charged population outside the plume but near Enceladus is presented by *Coates et al.* (2013). The population includes cold magnetospheric electrons, negative and positive water clusters, charged nanograins, "magnetospheric photoelectrons" produced from ionization of neutrals throughout the magnetosphere near Enceladus, and "plume photoelectrons" from photoionization in the plume. The plume and magnetospheric photoelectrons provide a source of warm electrons with energy >20 eV, which can cause electron impact ionization (>13 eV is needed for this ionization). These warm electrons increase the importance of this process in the region near Enceladus and probably throughout Saturn's inner magnetosphere.

The variability of water molecule production rate in the plume was studied using models and Cassini CAPS, MIMI, and INMS data by *Smith et al.* (2010), who found at least a factor of 4 variation in the production rate of Enceladus in a seven-month period covering encounters E2, E3, and E5. The results are consistent with variability based on orbital location (*Hedman et al.,* 2013), as mentioned by *Blanc et al.* (2015) (see also the chapter by Smith et al. in this volume).

4.2. Compositional Cassini Plasma Spectrometer and Magnetospheric Imaging Instrument Measurements of Ionized Plume Material in Saturn's Magnetosphere

The first measurements by CAPS at Saturn (*Young et al.,* 2005) revealed that the dominant species in the magnetosphere is water group ions (W^+, corresponding to combinations of O^+, OH^+, H_2O^+, H_3O^+). They also reported small concentrations of N^+ between ~3.5 and 8 R_S indicating that, unexpectedly, something in the inner magnetosphere (rather than Titan, at 20 R_S) was producing the N^+.

Further analysis of the water group population has been performed. The dominance of the W^+ density implied a dominant plasma source within 5.5 R_S (*Wilson et al.,* 2008). A more comprehensive survey of plasma parameters including regions inside Enceladus' orbit (*Thomsen et al.,* 2010) over 4.5 yr showed that (1) the ratio of the density of H_2^+ to H^+ is higher near Titan's orbit, indicating Titan as a source of H_2^+, and (2) W^+ ions dominate in the inner magnetosphere within ~3 R_s of the equatorial plane. *Arridge et al.* (2011) reviewed the published plasma data and concluded that the inner magnetosphere is dominated by low-energy electrons and water group ions sourced from Enceladus. Figure 13 shows the densities of various electron and ions species, collected by *Arridge et al.* (2011).

Pickup water group ions near Enceladus' orbit were studied by *Tokar et al.* (2008). They suggest that the ions are formed by charge exchange near Enceladus between water group neutrals (O, OH, H_2O) and thermal ions co-rotating with Saturn. The velocity space distribution of the pickup ions, assumed to be OH^+, is ring-like in velocity space (see section 4.1), indicating that they are relatively new pickup ions. Their density corresponded to ~8% of the total ion density between 4 and 4.5 R_S. Related ion cyclotron waves were studied by *Leisner et al.* (2006), who found that waves produced by W^+ (O^+, OH^+, H_2O^+, H_3O^+) and also O_2^+ were visible throughout the E-ring region.

Further analysis of the nitrogen population has also been pursued by *Smith et al.* (2005, 2007), who confirmed a source in the inner magnetosphere, although the molecular source (N_2 or NH_3) has not yet been determined. Subsequent work by *Smith et al.* (2008) using a combination of data analysis and modeling showed that the most likely source is NH_x^+, likely from ammonia, representing a fraction of a few

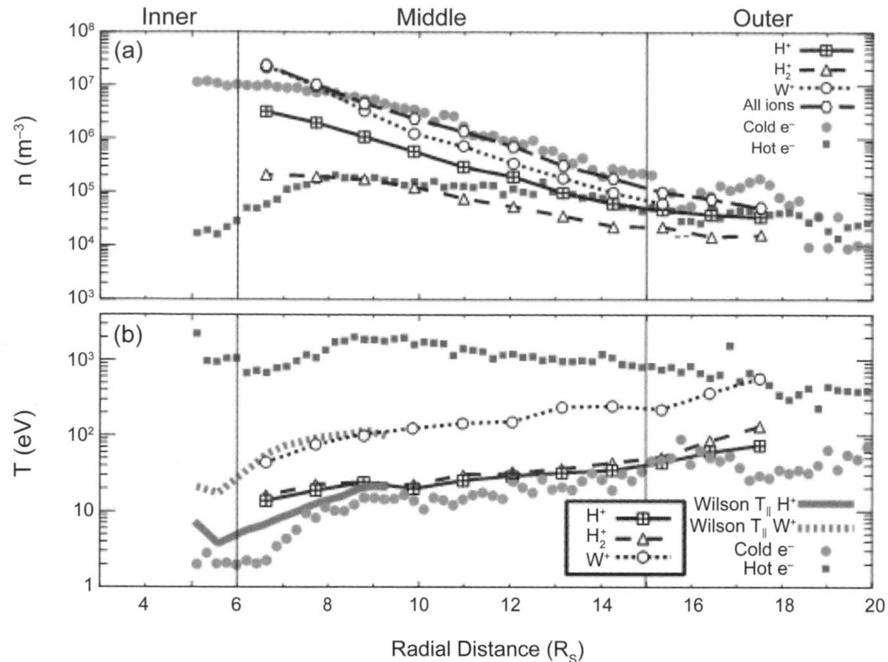

Fig. 13. Density and temperature in Saturn's magnetosphere from various sources collected by *Arridge et al.* (2011): **(a)** number densities of hot and cold electrons (*Schippers et al.*, 2008) and thermal ions (*Thomsen et al.*, 2010); **(b)** plasma temperatures of hot and cold electrons (*Schippers et al.*, 2008) and thermal ions (*Thomsen et al.*, 2010; *Wilson et al.*, 2008).

percent compared to water ions (see Fig. 14). An N_2^+ origin may additionally be present. However, the non-detection by UVIS in the neutral plume gas restricts the N_2 abundance to be below 0.5% of the emitted water vapor (*Hansen et al.*, 2011), and the INMS detects NH_3 in the plume with a volume mixing ratio of 0.4%–1.3% (see section 2.2). It is currently unclear why charged nitrogen species appear to be more abundant in the magnetosphere compared to what can be supplied by the known nitrogen-bearing neutral species. One possibility may be that NH_x^+ species have longer lifetimes than water ions.

The analysis of minor ions compared to W⁺ in the magnetosphere with integration times of several years (Fig. 15) has also been pursued using the CHEMS sensor of the

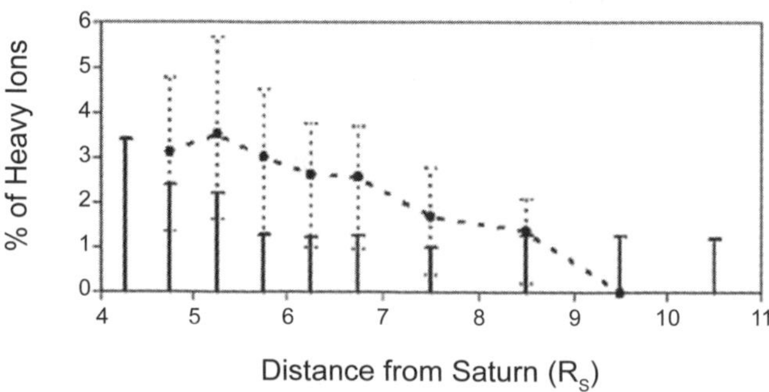

Fig. 14. Concentrations for N_2^+ (solid black lines) and NH_x^+ (dotted line) from CAPS IMS data. Percentages of all heavy ions are shown as a function of radial distance from Saturn (R_S). For N_2^+ only an upper limit could be inferred. From *Smith et al.* (2008).

MIMI instrument (*Krimigis et al.*, 2004), which detects energetic particles in the range 83–167 keV (*Christon et al.*, 2013, 2014). The presence of N⁺ (~2% compared to W⁺) confirmed the CAPS detection of nitrogen-bearing species. Furthermore, C⁺ indicates dissociation of abundant organic material in agreement with the organic species found in the plume (section 2 and 3). In particular, *Christon et al.* (2013, 2014) studied the $^{28}M^+$ (<1% compared to W⁺) and O_2^+ (~2%). The main nearby source for the O_2^+ is photolysis of Saturn's main rings, based on the observed seasonal variation, which favors a ring-related source. For $^{28}M^+$ the source may be Enceladus or the main rings, and $C_2H_5^+$, HCNH⁺, N_2^+, Si⁺, and CO⁺ were suggested as possible species. Further seasonal variations are under study. *Christon et al.* (2015) studied an ion with mass 56 u with an abundance of ~10^{-4} compared with W⁺, identifying it as Fe⁺. However, they suggested that this ion may be produced from meteor ablation near Saturn's mesosphere-ionosphere boundary, or perhaps from impacted interplanetary dust particles in the main rings, and that Enceladus is probably not the source in this case.

5. SOURCES OF THE DIVERSE PLUME CONSTITUENTS

5.1. Solid Compounds

The distinct compositional ice grain families probably have very different origins and/or generation mechanisms. The composition of Type III particles matches the composition expected for liquid water within Enceladus (*Zolotov*, 2007), which has washed out salts from primordial rock

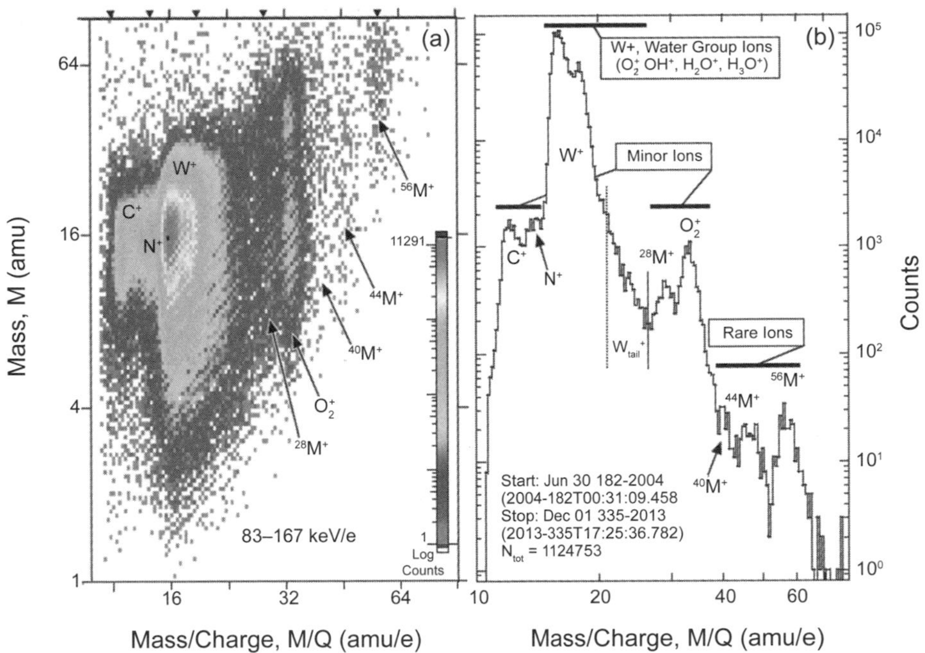

Fig. 15. See Plate 14 for color version. **(a)** Triple coincidence pulse height analysis events by MIMI-CHEMS measured in Saturn's near-equatorial magnetosphere under certain selection criteria (see *Christon et al.,* 2015) inside of 20 R_S or the magnetopause, whichever is closer to Saturn, from 2004 to 2013 are displayed in a mass (M) vs. mass-per-charge (M/Q), color spectrogram. The water group (W^+) (mostly O^+ at 16 amu/e, followed by roughly equal amounts of OH^+ and H_2O^+ at 17 and 18 amu/e, respectively, at about half the O^+ abundance, along with a little H_3O^+) and the minor and rare heavy ion species are identified. C^+ and N^+ are the most abundant non-water species. **(b)** Histogram of the data plotted in **(a)**. The histogram permits clearer identification of the minor ions like ^{28}M and $O_2{}^+$ and the "rare group" ions of $CO_2{}^+$ (44 M^+) and iron (56 M^+) and facilitates qualitative visual comparisons to W^+ and the minor ions. From *Christon et al.* (2015).

inside the moon's potentially porous (*McKinnon,* 2015) core (*Postberg et al.,* 2009a, 2011a; *Hsu et al.,* 2015). *Postberg et al.* (2009a) suggest that a spray of droplets is inevitably generated when bubbles reaching the water table of the ocean burst (e.g., *Lhuissier and Villermaux,* 2012). The bubbles can be formed either from water evaporating close to its triple point or upwelling volatile gases (*Matson et al.,* 2012) such as CO_2, CH_4, or H_2. If the spray droplets are sufficiently small, they will not fall back onto the water table but will be carried by vapor (emerging from the evaporating water) and follow the pressure gradient upward through the cracks and vents into space (*Postberg et al.,* 2009a). Following this model, Type III grains would be direct samples from the water table of the ocean. From their composition, the ocean salinity would be above 0.5%, with NaCl as the most abundant dissolved component followed by about half the amount of $NaHCO_3$ and/or $NaCO_3$ (see the chapter by Glein et al. in this volume).

By contrast, the salt-poor (or salt-free) Type I grains cannot be generated from ocean spray. Most of these grains are probably produced from vapor condensation (*Schmidt et al.,* 2008; *Yeoh et al.,* 2015). Whether the vapor stems from evaporating ocean-water or sublimated ice makes no noticeable difference in their composition, and both mechanisms are likely to contribute. However, most Type I spectra show traces of sodium ($Na/H_2O \approx 10^{-7}$) that are in good agreement with the traces of salts that one expects to find in the gas phase of evaporating salt water (*Postberg et al.,* 2009a). The observation of predominately crystalline ice grains in the plume (*Dhingra et al.,* 2017 (see Fig. 10, section 3.4)) implies that the grains formed at temperatures well above 130 K, which is consistent with other evidence that the conditions in the ice vents are rather warm (see the chapter by Goldstein et al. in this volume).

Vapor that rapidly moves upward inside the ice vents condenses to ice grains when narrow passages in the ice channels cause local supersaturation. This process naturally forms smaller grains than the freezing of ocean spray. Compared to the latter, these smaller salt-poor grains are accelerated to higher average speeds (for any given density and speed of the carrier gas) (*Schmidt et al.,* 2008; *Postberg et al.,* 2011a). Homogeneous vapor condensation can also occur during adiabatic cooling after the vapor left the vents. This condensation is limited to altitudes equivalent to 10–100 vent diameters (after which the gas becomes collisionless) and produces small grains from nanometer scales up to radii clearly below 1 μm for plausible vent diameters and gas velocities (*Yeoh et al.,* 2015; see also the chapter by Goldstein et al. in this volume). This scenario is again consistent with salt-poor vapor condensates being preferentially smaller than grains formed from salt-rich ocean spray.

The origin of the organic-enriched Type II grains is the least constrained of the three types. However, their great abundance and frequent detection during Cassini crossings of the enceladean plume again suggest that Enceladus is their main source (*Postberg et al.,* 2008, 2011a; *Khawaja et al.,* 2017; *Postberg et al.,* 2018). Most of them are salt-poor, in agreement with a formation from condensing vapor. In this case, besides water, initially volatile organic compounds may have condensed onto ice grains as the vapor cooled on its way upward through the icy channels. A small fraction of grains shows spectral features of Type II as well as Type III. These grains could be frozen salt-water droplets that may have incorporated organic compounds from the enceladean ocean. Alternatively, organics that initially were in the gas phase could have condensed onto the salty ice grains in the vents.

About 3% of Type II grains exhibit mass lines that stem from complex organic parent species with molecular masses in excess of 200 u in particularly high concentration, which probably originate from Enceladus' hydrothermally active core (*Postberg et al.,* 2018). These organic species are too massive to be in the gas phase at plausible physical conditions above the evaporating water table (T ≤ 0°C), where the water is inevitably in contact with the ice crust. The refractory organic material has been detected mostly in salt-poor ice grains and thus they did not form from the salty ocean spray, which preserves the liquid composition. Consequently, the organic material was not dissolved in the ocean water when the ice grain formed. According to *Postberg et al.* (2018), the most plausible way to generate these ice grains containing abundant high-mass-organics is if the organic material exists as a separate phase, such as a thin film or layer of mostly refractory, insoluble organic species floating on top of the water table (Fig. 16). When bubbles burst in such a scenario, they tear apart the organic film and, besides salty water droplets, throw up droplets or flakes rich in hydrophobic organic material (see also the chapter by Glein et al. in this volume). They will then serve as efficient nucleation cores for ice condensation: Droplets ascending in the icy vents become coated by water ice condensing from the vapor carrying the grains (*Postberg et al.,* 2018). These high mass organic species have also been potentially observed by INMS in the plume during high-speed flybys, where the high impact velocity of organic-bearing ice grains disintegrated large molecules to organic fragments small enough to show up in INMS limited mass range (<100 u) (*Postberg et al.,* 2018) (see section 2). The observation of unspecified high mass molecular species in the plume by CAPS (*Coates et al.,* 2010a,b) might also be due to large fragment ions from these organic species (see section 4.1 and Fig. 12).

The nanophase silica (SiO_2) grains emitted from the E ring can be interpreted as silica colloids with radii of 2–9 nm that formed during the cooling of hydrothermal waters in the subsurface ocean of Enceladus (*Hsu et al.,* 2015). Combined with long-term laboratory experiments, the composition and narrow size distribution of these grains argue for ongoing hydrothermal activities within Enceladus and place constraints on the temperature, alkalinity, and salinity of Enceladus' subsurface waters (*Hsu et al.,* 2015; *Sekine et al.,* 2015). For further details see the chapter by Glein et al. in this volume. These nanoparticles would be

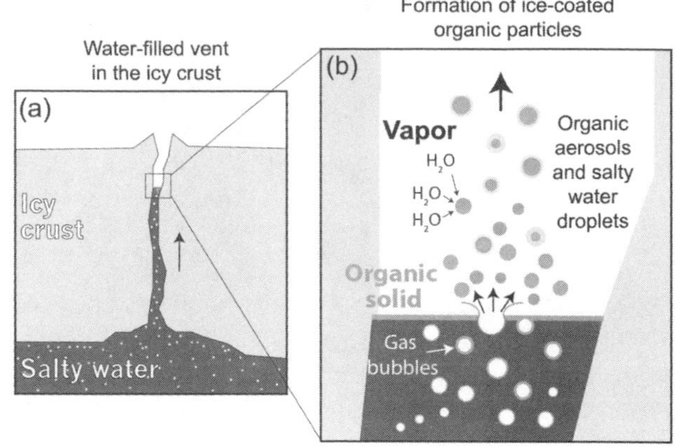

Fig 16. Schematic on the formation of ice grains from heterogenous nucleation (not to scale). From *Postberg et al.* (2018) with permission from *Nature.* **(a)** Ascending gas bubbles in the ocean efficiently transport organic material into water-filled cracks in the south polar ice crust. **(b)** Organics ultimately concentrate in a thin organic layer on top of the water table, located inside the icy vents. When gas bubbles burst, they form aerosols made of insoluble organic material that later serve as efficient condensation cores for the production of an icy crust from water vapor, thereby forming HMOC-type particles. Another effect of the bubble bursting is that larger, pure saltwater droplets form, which freeze and are later detected as salt-rich type-3 ice particles in the plume and the E ring. The figure implies the parallel formation of both organic and saltwater spray, but their formation could actually be separated in space (e.g., at different tiger stripes cracks) or time (e.g., dependent on the varying tidal stresses working on the cracks) (*Hedman et al.,* 2013; *Kite and Rubin,* 2016).

transported from hydrothermal reaction sites, probably located inside Enceladus' porous rocky core to the water table (*Hsu et al.,* 2015; *Choblet et al.,* 2017) and then naturally become inclusions of salty ocean spray from which Type III ice grains form. Nanosilica might also be hovering in the vapor phase above the water table, dragged into ice vents and then becoming condensation cores for the formation of salt-poor ice grains of Types I and II. Interestingly, MIMI-CHEMS observes a species with a mass of 28 u in Saturn's magnetosphere with Enceladus being a possible source (section 3). Besides volatile gases (section 5.2), Si^+ from eroded nanosilica grains might be a source.

Other nanophase species measured in the plume are charged grains, presumably made of water ice, observed by CAPS (*Jones et al.,* 2009; *Hill et al.,* 2012). These tiny grains inevitably form when the water vapor cools during adiabatic expansion into space (*Yeoh et al.,* 2015) (see also the Goldstein et al. chapter in this volume). They quickly become mostly negatively charged after leaving the vents (e.g., *Hill et al.,* 2012) (see section 3.3 for details). The CAPS measurements indicate a size of 1–3 nm and hence are even smaller than the silica nanograins. Although these grains are invisible in Cassini images, they might be as abundant by mass as macroscopic ice grains (see section 3.3).

5.2. Volatile Compounds

With a mixing ratio above 95%, INMS and UVIS measurements clearly rank H_2O vapor as the most abundant gaseous plume constituent (section 2). A large part likely comes from evaporation of ocean water, close to its triple point, that has hydrostatically ascended inside the "tiger stripe" cracks through most of the south polar ice crust. With an ice shell thickness below 5 km at the south pole (*Cadek et al.,* 2016; *Le Gall et al.,* 2017), the evaporating water table lies less than 1000 m below the surface (*Postberg et al.,* 2016) (see the Spencer et al. chapter in this volume). Ice sublimation will cause a currently unspecified but substantial contribution to the observed water vapor. Above the water table, ice in the walls of the vents is warmed by the ascending gas. Even at the outlets, where the vents reach the moon's surface, ice temperatures reach almost 200 K, causing a non-negligible vapor pressure from sublimating ice (*Goguen et al.,* 2013).

Probably the most remarkable volatile plume constituent is H_2, which during E21 has been detected in concentrations exceeding 0.3%. Its extreme volatility precludes "storage" over geological timescales on a small body like Enceladus and suggests that it is currently (or has been very recently) produced inside the moon. Its detection is highly suggestive of ongoing serpentinization reactions in hydrothermal systems within the ocean of Enceladus (*Waite et al.,* 2017). Most importantly, the data in Table 2 allowed *Waite et al.* (2017) to calculate the chemical viability of H_2 as the chemical energy source in the reaction $4H_2 + CO_2 \rightarrow CH_4 + H_2O$ — a reaction that expresses methanogenic metabolism in Earth's hydrothermal systems. At moderate alkaline pH values, the chemical affinity is positive, thus verifying the habitability of the interior ocean (see the chapter by McKay et al. in this volume).

Ammonia (NH_3) is detected in the plume in similar mixing ratios (0.4–1.3%) as H_2. It is a very reproducible constituent because it has been detected at all occasions when the INMS acquired plume composition, and its presence also enhances the spectral fit for UVIS plume spectra (see section 2). The nitrogen-bearing ion species observed in Saturn's magnetosphere (section 4.2) indicate even higher concentrations there. Possible sources for NH_3 are numerous and are poorly constrained from Cassini measurements. For example, it can (1) form from the dissolving gases in the ocean, (2) be a product of a chemical reaction in the ocean or hydrothermal sites, or (3) be released from clathrates that might reside deep in the icy crust (*Kieffer et al.,* 2006).

Similar considerations are true for methane (CH_4), which has been frequently measured in the plume at concentrations of about 0.2 % (section 2.2). *Bouquet et al.* (2015) discuss methane contributions from clathrate decomposition as well as hydrothermal production and conclude that both scenarios are viable. *Waite et al.* (2017) suggest hydrothermal scenarios such as Sabatier- or Fischer-Tropsch-like processes as well as thermogenesis.

Carbon dioxide with a mixing ratio of about 0.5% likely is released when pressurized water saturated in CO_2 ascends from the depth of the ocean. *Matson et al.* (2012) present a model where dissolved gases, mostly CO_2, exsolves as the water moves toward the surface inside conduits in the south polar ice crust. Bubbles formed by exsolution can decrease the bulk density of the vertical column of water enough that the pressure at the bottom of the column is less than that at the top of the ocean. It is suggested that this pressure difference drives ocean water into and up the conduit toward the surface. CO_2-saturated ocean water would be in good agreement with the substantial concentrations of carbonate salts found in the salt-rich ice grains that are suggested to resemble ocean water composition (*Postberg et al.,* 2009a), with more dissolved CO_2 for a lower ocean pH (see the Glein et al. chapter in this volume).

The most controversial Cassini observation in the plume gas is the species with a mass of 28 u, which could be attributed to CO, N_2, C_2H_4, or even Si (as a cationic species; see below). Both INMS and UVIS measurements now agree that the abundance of CO and N_2 lies below 0.5%. Whereas the latest INMS results exclude any intrinsic plume gas with mass 28 u at a level of 0.1% (*Waite et al.,* 2017), UVIS results indicate at least traces of ethylene (C_2H_4) to be present (*Shemansky et al.,* 2016). These constraints are particularly interesting in the context of a 28 u cation species observed by MIMI-CHEMS in Saturn's magnetosphere (*Christon et al.,* 2014) (section 4.2). However, it cannot be differentiated if this species stems from a gaseous plume compound or dissociated silicates, like silica nanograins or interplanetary dust (with the latter obviously not being of Enceladus origin).

6. SURFACE COMPOSITION

6.1. Infrared Observations

Earth-based telescopic spectra of Enceladus already indicated the presence of water ice (*Grundy et al.,* 1999; *Cruikshank et al.,* 2005; *Emery et al.,* 2005; *Verbiscer et al.,* 2006). At opposition, the icy surface reflects more than 130% of the visible sunlight [geometric albedo = 1.375 ± 0.008 (*Verbiscer et al.,* 2007), Bond albedo = 0.85 ± 0.11 (*Pitman et al.,* 2010)], which makes Enceladus the body with the highest visible geometric albedo of all bodies in the solar system. The Bond albedo of the trailing hemisphere is higher (0.93 ± 0.11) than that of the leading hemisphere (0.77 ± 0.09) (*Pitman et al.,* 2010). *Grundy et al.* (1999), *Emery et al.* (2005), and *Verbiscer et al.* (2006) reported detecting a weak absorption in the 2.2 to 2.4 μm region, indicating the possible presence of NH_3 or NH_3 hydrate, but *Cruikshank et al.* (2005) did not detect the feature, and so far, it has not been definitively detected in Cassini's VIMS data with the latest instrument calibration (*Clark et al.,* 2016).

Early studies by Cassini VIMS confirmed dominant water ice on Enceladus' surface (Figs. 17a,b). Trapped CO_2 was

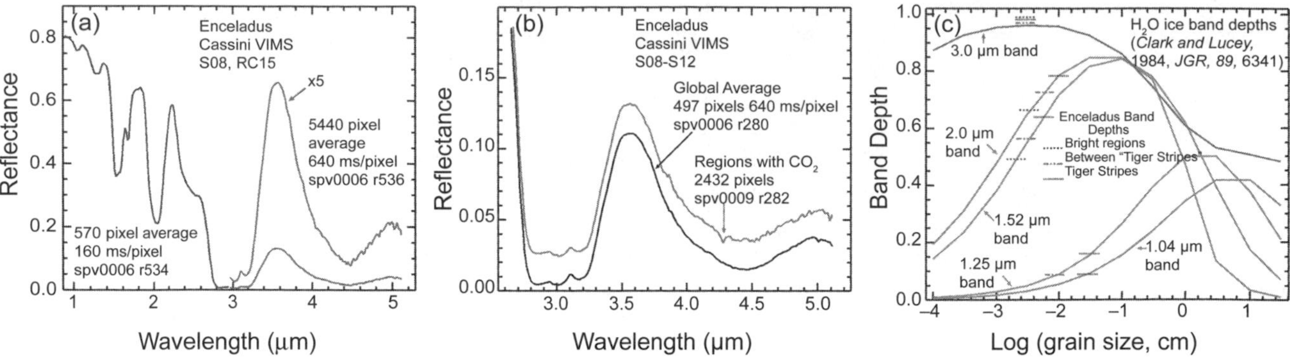

Fig. 17. (a) VIMS average spectra of Enceladus show the signatures of crystalline water ice at 1.65 μm and 3.1 μm. **(b)** VIMS spectra of Enceladus showing crystalline water ice (3.1 μm) and a CO_2 signature in some locations, predominantly in the tiger stripe region. **(c)** Ice band depth as a function of grain diameter with band depths observed on Enceladus. This graph does not include photometric effects or the effects of submicrometer ice grains, which will show enhanced band depths as the abundance of such grains increases.

found in most locations, including the tiger stripe region (*Brown et al.,* 2006). Brown et al. also reported amorphous ice. However, this early work did not include the effects of diffraction by submicrometer ice grains, which are common in the E ring and in the plume (e.g., *Kempf et al.,* 2008; *Hedman et al.,* 2009; *Postberg et al.,* 2011a) (see section 3.4.). The signatures of submicrometer ice grains, probably from E-ring deposition in most cases, can be observed throughout the Saturn system (*Clark et al.,* 2012). Clark et al. showed that a radiative transfer model that included diffraction from submicrometer particles modified the spectral structure in a unique way, changing relative band depths and shifting band shapes toward longer wavelengths, and verified the effects with lab spectra of small ice grains. Amorphous ice shifts the bands to shorter wavelengths and changes the shapes differently than submicrometer grains. For further discussion on crystalline and amorphous surface ice, see section 7.2.

Jaumann et al. (2008) mapped the grain size of ice across Enceladus' surface and found that the observed ice absorption strengths in Cassini VIMS spectra could be explained by pure crystalline ice of varying grain sizes. They found the largest grain sizes (~0.2 mm) in the south polar tiger stripe region. Jaumann et al. also found that the particle diameter of water ice grains increases toward younger tectonically altered surface units and the largest particles are found in relatively "fresh" surface material. The smallest ice grains were generally found in old, densely cratered terrains. They also found that the ice grain diameters are strongly correlated with geologic features and surface ages, indicating a stratigraphic evolution of the surface that is caused by plume deposition and distribution of materials from cratering events. A complicating factor in deriving grain sizes in the presence of submicrometer ice grains is that, as tiny ice grains become more abundant, the surface looks more like a block of ice and the absorption band depth increases. Fortunately, the presence of small grains is revealed by the modification of the shapes of the absorption bands (*Clark et al.,* 2012), but a more sophisticated analysis of the spectral properties is needed than a simple calculation of band depth (Fig. 17c). *Jaumann et al.* (2008) completed their study before these effects were known. Some of the areas they found with larger ice grain sizes, especially those subject to large amounts of plume "snow," are definitely affected by submicrometer ice grains and thus need to be reevaluated. *Scipioni et al.* (2017) have begun that evaluation and have generally found much smaller grain sizes than *Jaumann et al.* (2008) and a correlation with plume deposition (see section 7.1 for a details). However, the highest-spatial-resolution data have yet to be analyzed at the native resolution.

Additional work on the relation between band depth and grain size has been accomplished by *Verbiscer et al.* (2006), who demonstrated the effects of the photometric or scattering properties of surface particles on absorption band depths. Such analysis is further complicated by the relationship between band depth and phase angle (e.g., *Pitman et al.,* 2017), which needs to be included in any modeling effort.

Derivation of compositional abundances from reflectance spectroscopy requires the identification of the components making up the surface, and a simultaneous solution of the grain sizes and abundances of each component. While water ice dominates Enceladus' surface, *Brown et al.* (2006) found trace organic compounds in the tiger stripe region (Fig. 18). The band position of the organics is consistent with aliphatic hydrocarbons, but the signature is weak and a more specific identification could not be made. Detecting the presence of the organic absorptions is made more complex by the spectral curvature in the strong signatures of ice. A more sophisticated analysis than that performed by Brown et al., with the latest VIMS calibration (*Clark et al.,* 2016), needs to be done and might reveal more information on the nature of the organics. However, the signatures of organic trace compounds (*Brown et al.,* 2006) are consistent with the detection of organic-bearing ice grains and methane in the plume by CDA (*Postberg et al.,* 2008, 2018) and INMS, respectively (*Waite et al.,* 2006, 2017), and are expected to be deposited primarily near the tiger stripe region (section 7.1). *Brown et al.* (2006) also set an upper limit of 2% for solid NH_3 global surface deposits. *Hodyss et al.* (2009)

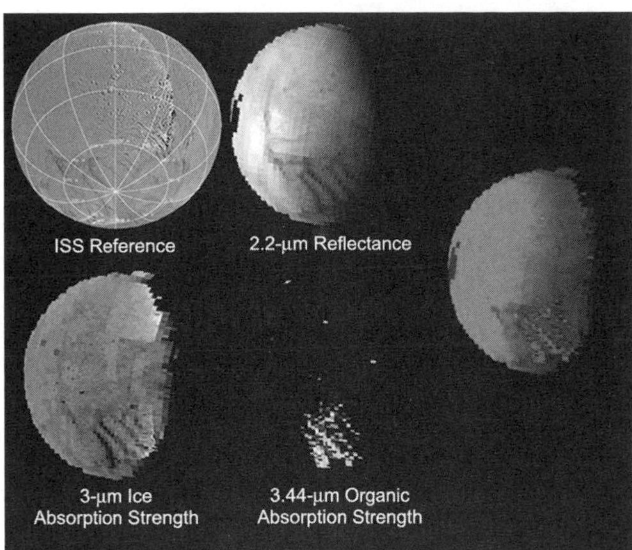

Fig. 18. See Plate 15 for color version. Cassini VIMS surface composition from *Brown et al.* (2006).

reported detecting methanol in VIMS spectra from Enceladus' surface, but it has not been verified with the latest VIMS calibration.

Sodium salts (NaCl and $NaHCO_3/Na_2CO_3$) have been reported in the ice grains in Enceladus' plume at the 0.1–1% level (*Postberg et al.,* 2009a, 2011a) (section 3.1, 3.2). It is possible that these salts could be detected by optical remote sensing. Carbonate salts have strong absorptions in the 2+-μm region due to C-O stretch-bend combinations. Sodium chloride, NaCl, is a naturally occurring mineral, halite. Halite is transparent in the VIMS spectral range unless the halite contains water. Adsorbed water absorption is shifted to shorter wavelengths than ice absorptions, but a high signal-to-noise ratio is needed to detect adsorbed water in small abundances. Halite shows a narrow 0.27-μm absorption (*Clark et al.,* 2007), but this spectral region is not covered by any instrument on Cassini. Halite has not been measured in the deeper UV covered by the UVIS instrument. At present, the published VIMS spectra of the plumes have insufficient signal-to-noise ratios to constrain salt abundance. By averaging all VIMS spectra of the surface or the plume obtained over the entire mission might produce a high enough signal-to-noise ratio to constrain the salt abundance in the future.

6.2. Ultraviolet and Visible Light Observations

At far ultraviolet (FUV) and middle-ultraviolet (MUV) wavelengths, Enceladus' reflectance drops precipitously. Figure 19a shows an Enceladus spectrum acquired by UVIS (*Hendrix et al.,* 2010). Water ice has an absorption edge between 165 and 180 nm that is diagnostic of grain size. A combination of grain sizes can be found to fit the absorption edge of water ice to the UVIS spectrum, consistent with the water ice composition established by longer wavelength spectra. *Anderson and van Dishoeck* (2008) show that the ice-absorption edge, in the UVIS spectral region, shifts to shorter wavelengths in amorphous ice relative to crystalline ice. While we do not have amorphous ice optical constants in the UVIS range at the temperatures of Enceladus, the crystalline optical constants provide a good match if the grain size is on the order of 10 μm (*Hendrix et al.,* 2010), consistent with the UV indicating the dominance of crystalline ice. For further discussion on crystalline and amorphous surface ice see section 7.2. There is an additional absorber(s) longward of 175 nm that reduces the reflectivity in the UVIS spectral range (Fig. 19a).

Figure 19b shows the spectrum of Enceladus when UVIS data are combined with the data at visible and near-infrared wavelengths from *Verbiscer et al.* (2005) and *Verbiscer et al.* (2006), as illustrated in *Hendrix et al.* (2010). With an I/F of >1 Enceladus is very reflective at visible wavelengths >400 nm. At wavelengths shorter than 400 nm, the disk-averaged I/F drops to 80% at 275 nm (*Verbiscer et al.,* 2005), then to <40% at 190 nm. The UVIS FUV spectrum exhibits a "ledge" from 175 to 185 nm with ~30% reflectivity, with an upturn between 185 and 190 nm. The final drop-off short of 175 nm is due to the presence of water ice.

The spectrum of Enceladus at FUV and MUV wavelengths is not consistent with pure water ice. Some additional contaminant(s) must be present to darken the surface at wavelengths from 175 nm to 400 nm. The effort to identify this component(s) from the spectrum alone is hampered by a severe lack of laboratory data and optical constants in this wavelength range. There is, however, a fairly good idea of the constituents in the plume from INMS, UVIS, and CDA data. While gases escape (*Hansen et al.,* 2006), the larger plume particles preferentially fall back to the surface of Enceladus (*Kempf et al.,* 2010; *Postberg et al.,* 2011a). Smaller plume particles tend to go into orbit forming the E ring, but they can be reaccreted on Enceladus' surface (for details see the chapter by Kempf et al. in this volume). A logical approach is to model non-water species in the plume as the contaminants darkening Enceladus' surface.

INMS has identified 0.4–1.3% NH_3 in the plume gas (*Waite et al.,* 2017) and, although the CDA has not reported a detection of ammonia in the ice grains, it is possible that nitrogen-bearing compounds are also emitted in the solid phase as minor species. The addition of ~1% NH_3 to water ice with larger grain size in the model surface spectra computed by *Hendrix et al.* (2010) reproduces the ledge seen in the UVIS spectrum from 175 to 185 nm and the upturn at ~185 nm (Fig. 20). This finding was confirmed by UV observations with the Hubble Space Telescope (*Zastrow et al.,* 2012). It is also in agreement with the upper limit of 2% NH_3 inferred from VIMS observations (*Brown et al.,* 2006) (see also section 6.1). Although NH_3 is not expected to be stable to photolysis and radiolysis on Enceladus' surface, ammonia hydrate may be, and is also consistent with groundbased near-infrared spectra (*Verbiscer et al.,* 2006). Moreover, NH_3 may not need to be stable over long timescales since it is probably constantly replenished.

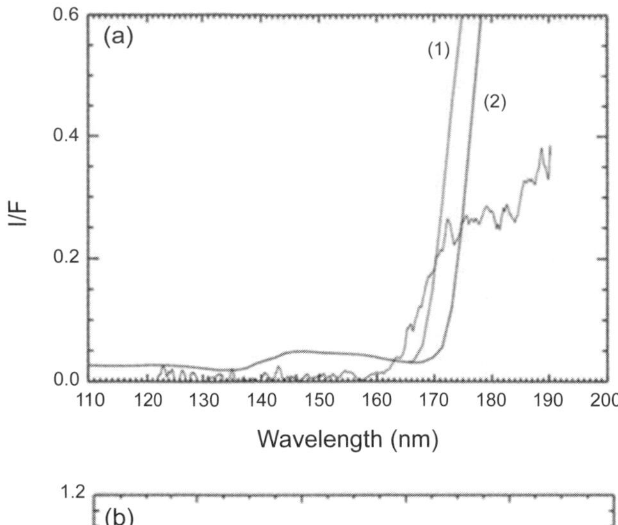

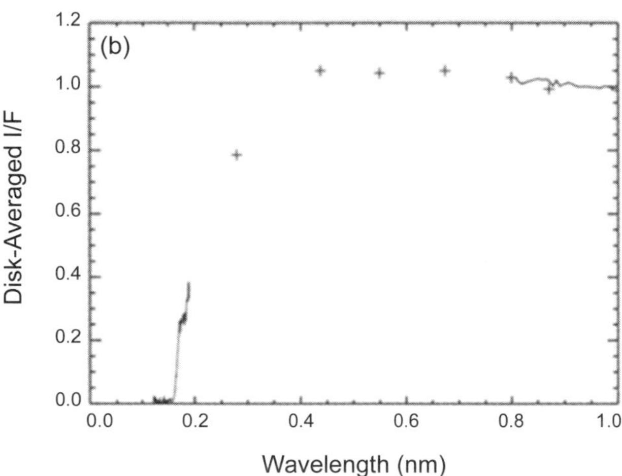

Fig. 19. **(a)** This UVIS spectrum (black) was acquired on May 27, 2007, at a range of 620,000 km at a solar phase angle of 2°. Two model spectra of crystalline water ice with different grain sizes are shown [curve (1): areal mixture of 70% 45-μm and 30% 400-μm grains, curve (2): areal mixture of two intimate ice grain mixtures of the two sizes]. The mixture of grain sizes in the ice determines the precise location of the absorption edge between 165 and 180 nm. An additional absorber longward of ≈175 nm is required to fit the Enceladus spectrum. From *Hendrix et al.* (2010). **(b)** UVIS FUV data are combined with a single MUV data point from HST at 275 nm (*Verbiscer et al.,* 2005), the visible spectrum (plus signs) from *Verbiscer et al.* (2005), and data from 800 to 1000 nm from *Verbiscer et al.* (2006). The visible data were acquired at larger phase angles and thus do not reach the absolute albedo maximum of 1.375 ± 0.008 (*Verbiscer et al.,* 2007).

At wavelengths longer than 190 nm NH_3 is no longer an absorber toward the visible portion of the spectrum and an additional component must be present to explain the drop-off in reflectivity between 190 and 400 nm.

The aliphatic hydrocarbons detected by *Brown et al.* (2006) (see section 6.1) generally do not have UV absorbers that can explain Enceladus' UV spectrum. However, it has been shown that mixtures of H_2O + CH_4 + NH_3 ices that have been irradiated in the lab produce tholins (*Thompson et al.,* 1987). Moreover, more complex organics have been detected in the plume by CDA (*Postberg et al.,* 2008, 2018) and INMS (*Waite et al.,* 2009) and these may be processed on Enceladus' surface to form such tholins. Unfortunately, very limited spectral data have been published; however, it does appear that some type of tholin material could be responsible for the darkening of Enceladus' surface at MUV wavelengths (*Hendrix et al.,* 2010).

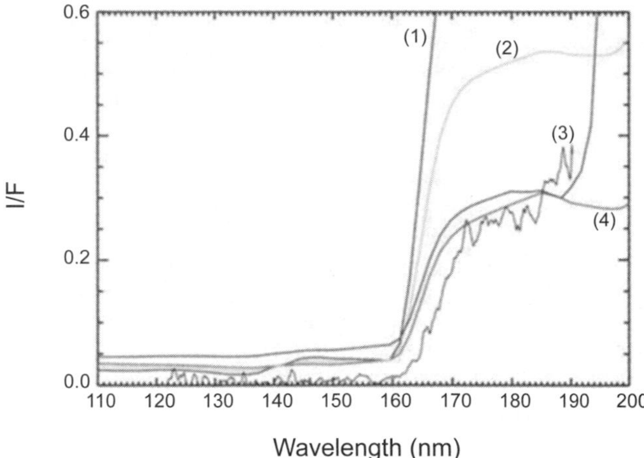

Fig. 20. Intimate mixtures of 99% H_2O (grain size 1 μm) and 1% NH_3 [varying grain sizes shown by numbered curves: (1) 1 μm, (2) 15 μm, (3) and (4) 30 μm]. Models used NH_3 data from *Dawes et al.* (2007) except (4), which used NH_3 data from *Martonchik et al.* (1984). Figure from *Hendrix et al.* (2010).

A plausible alternative to an organic UV absorber would be nanophase iron in space-weathered meteoritic dust. *Clark et al.* (2012) showed common spectral characteristics of the visible to UV absorber at some locations on Iapetus with other icy satellite surfaces in the Saturn system including Enceladus, and in the Cassini Division in Saturn's rings. On Iapetus, the deposits of dark material are relatively pure, and the VIMS data definitively reject tholins, leaving the nano-phase metallic iron and iron oxides as the best explanation. However, at present on Enceladus, the available spectral data are insufficient to definitively distinguish between the tholin vs. nano-iron plus nano-iron oxide explanation because the UV absorber signatures are relatively weak (see the chapter by Hendrix et al. in this volume).

7. PLUME/SURFACE INTERACTION

7.1. Constraints from Grain Size

Water ice is the main component observed in the plume ejected from Enceladus' tiger stripes. Most of the icy dust particles and gas erupted from this region redeposit on the surface at a rate that decreases with increasing distance from the surface fractures. The deposition rate had been simulated by *Kempf et al.* (2010) and *Southworth et al.* (2018) to be 0.5 mm yr^{-1} close to the vents, and 10 μm yr^{-1} at certain regions north of the equator assuming compact ice deposition (density ≈ 0.9 g cm^{-3}). The plumes' deposits are broad below 45°S, then, due to interactions with Saturn's gravity, they split into two patterns centered at ~45°W and ~225°W, respectively (Fig. 21).

Scipioni et al. (2017) investigated whether the plume material that accumulates on the surface leaves a spectral signature that can be observed by analyzing data returned by VIMS between 0.88 and 5.12 μm (Fig. 22). VIMS-IR spectra are sampled in 256 spectral channels, with an average spectral sampling of 16.6 nm (*Brown et al.,* 2004). In general, the depth of absorption features is influenced by the variation of water ice abundance and grain size. Larger grains generally cause a deeper absorption, whereas if water ice is mixed with a contaminant material, band depths become shallower (*Hapke et al.,* 1978; *Clark,* 1981a,b). However, micrometer and submicrometer ice grains show enhanced band depths as the abundance of such grains increases (Fig. 17c) (*Clark et al.,* 2012). Therefore, plume ice grain deposition, the sizes of which predominately lie in such a small size regime (*Schmidt et al.,* 2008; *Kempf et al.,* 2010; *Postberg et al.,* 2011a), might increase band depth because of reduced scattering in closely packed small particles when the particles are much smaller than the wavelength (*Clark et al.,* 2012). Although the grain sizes of up to ~0.2 mm inferred by *Jaumann et al.* (2008) are probably drastically overestimated due to the aforementioned effect, the general finding of Jaumann et al. that larger grains are observed closer to the plume sources is in agreement with the observed size stratification of the plume (e.g., *Hedman et al.,* 2009) (see section 3.2 and the chapter by Kempf et al. in this volume).

To have a comprehensive view of the distribution of the abundance of the water ice and/or of the variation of the grain size across the surface, *Scipioni et al.* (2017) created spatially-resolved, cylindrically-projected maps of the selected water ice band depths and of the reflectance peak. From the comparison between model-predicted ice deposits and water ice distribution maps, the observation of deeper absorptions is expected where the plume deposition rate increases.

Water ice spectral signatures vary little across the surface of Enceladus (Fig. 23), and water ice band depths only have subtle variations across Enceladus on average. The most pronounced difference in band depths values involves the SPT. Indeed, the tiger stripes, and the terrains surrounding them, up to about –60° latitude, show by far the deepest water ice absorption bands, and the smallest value of the 3.6-μm reflectance peak. Elsewhere on the surface, the band depths and the reflection peak show a longitudinal variation. The terrains with the lowest band depth values are located between 0°W and 45°W, between 315°W and 360°W, and around 180°W, and they have almost constant band depths across the latitudinal direction. A regional bright spot shows up in the leading side, centered at about 90°W and 30°N. The near-infrared reflectance (Fig. 23d) of the bright spot is relatively high, while this spectral index decreases to background levels moving toward 0°W and 180°W.

To visualize the redeposition processes taking place on the surface of Enceladus, the extracted level curves from the modeled deposition rate (Fig. 21) are plotted on top of VIMS-derived maps in Fig. 23. The water ice distribution maps show overall a good agreement with the predicted ejecta deposits in the SPT and in the eastern portion of the trailing hemisphere. The ice deposits along ~225°W predicted by the model (Fig. 21) are in fact reproduced by color changes observed in the VIMS maps. From 205°W to 360°W, there is a qualitative match between the model and the data, but the "wedged" shape of the deposition map is not well reproduced. On parts of the leading side (0°W–135°W), the map and the plume deposit predictions diverge. Although both the ejecta deposition rate and the water ice band depths show a longitudinal trend on the leading hemisphere, their positions in latitude do not exactly overlap. This divergence is at least partially caused by a regional bright spot centered at 30°N, 90°W. At this location, the deposition model predicts a rate below 10 μm yr^{-1} (*Kempf et al.,* 2010).

The location of this bright spot on the leading hemisphere matches that of a microwave scattering anomaly (*Ries and Jansen,* 2015) found by Cassini's RADAR instrument. The feature correlates with a tectonized terrain with very few craters, indicative of a recent (<100 m.y.) resurfacing event, maybe caused by an ice diapir (*Ries and Jansen,* 2015). It is

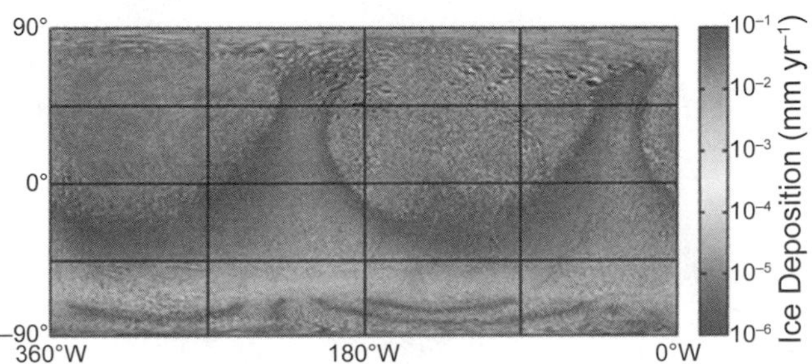

Fig. 21. See Plate 16 for color version. Plume deposition on Enceladus surface as modeled by *Southworth et al.* (2018). The model assumes a homogenous distribution of sources all along the tiger stripe fractures.

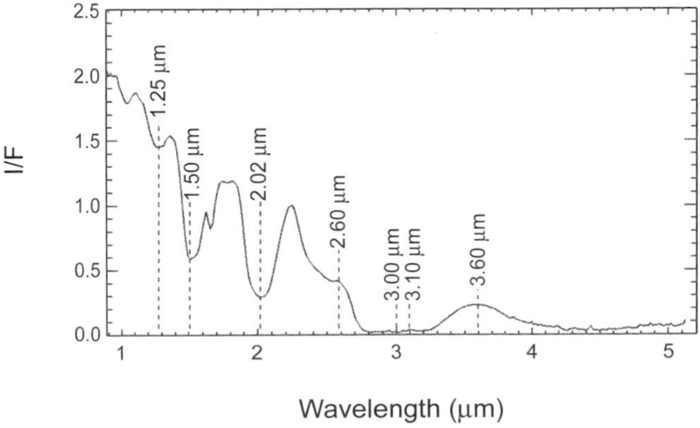

Fig. 22. VIMS Enceladus spectrum showing water ice absorption and reflection features in the near-infrared. The main water ice overtones and combinations in the near-infrared range are located at 1.04, 1.25, 1.5, 2.0, and 3.0 μm, while a reflectance peak arises at 3.6 μm.

possible that this caused accumulation of fresh water ice, or annealing to bigger grain sizes, that would explain the deeper band depth in the near-infrared observations and would mask the faint signature of the plumes' deposits in this region.

The near-infrared data can be compared to four global, high-spatial-resolution color ratio maps from Cassini's Imaging Science Subsystem (ISS) produced by *Schenk et al.* (2011) by cylindrically projecting and mosaicking ISS data in the IR3 (0.930 μm), GRN (0.586 μm), and UV3 (0.338 μm) filters.

Except for the IR3/GRN ratio map, the maps displayed in Fig. 24 agree with the plume deposition model very well. The position — as well as the shapes — of the features in the ratio maps matches the deposition model. In contrast to the near-infrared maps, the match is also good on the leading hemisphere and in general the match is more accurate for these visible maps (Fig. 24) compared to the near-infrared maps of Fig. 23. The reason for this mismatch might be that in visible light, water ice is more transparent than in the near-infrared, leading to a more apparent optical effect produced by a trace abundance of a non-ice component (discussed above), or due to a grain size difference between the non-ice components and the ice. The bright region in the leading hemisphere, observed in the near-infrared water

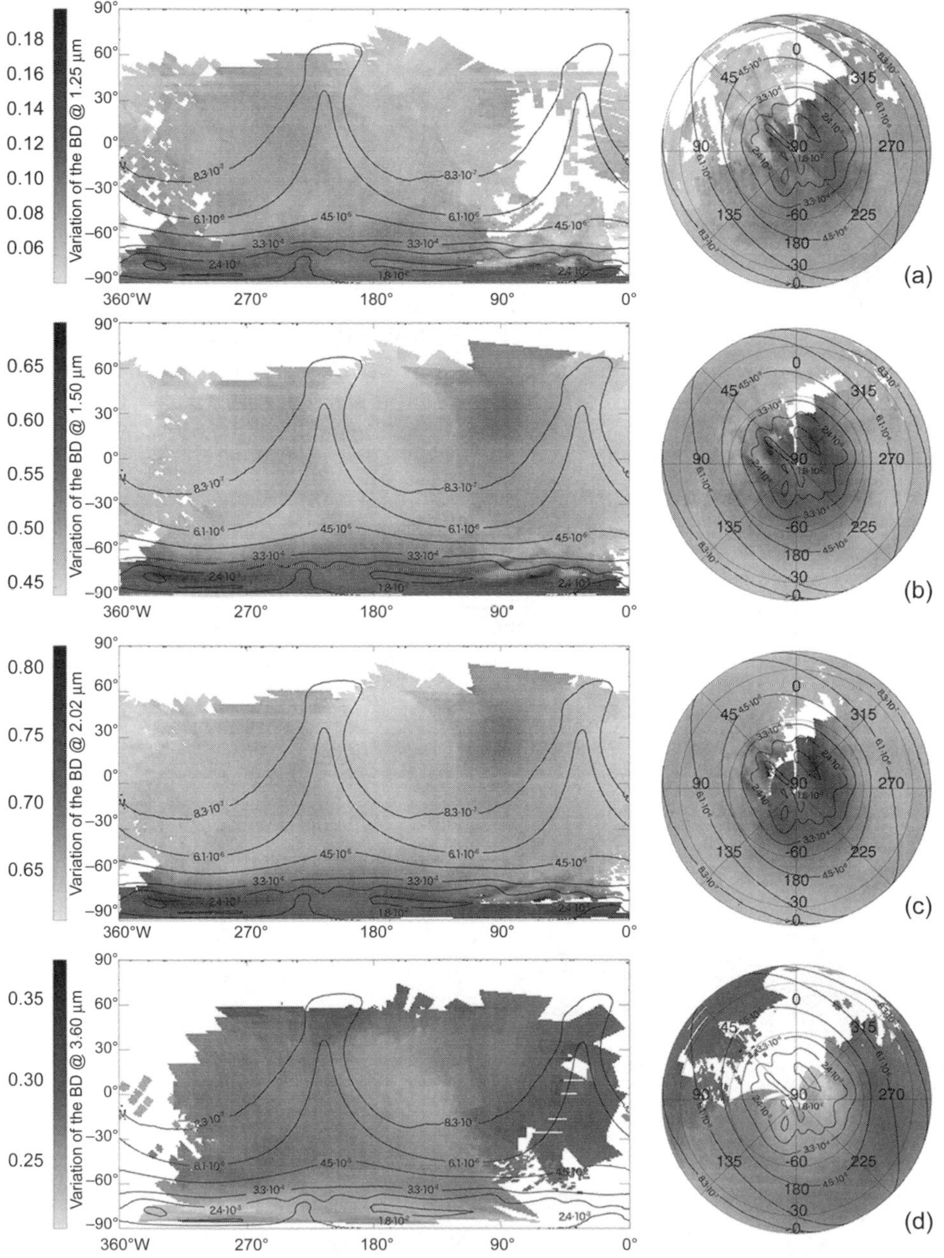

Fig. 23. See Plate 18 for color version. **(a),(b),(c)** Panels map the 1.25-, 1.5-, and 2-μm band depths, respectively; (d) strength of the 3.6-μm reflectance peak. Contour lines correspond to yearly plume deposition rate in millimeters (*Southworth et al.,* 2018) under the unrealistic assumption of compact ice deposition with a density of 0.9 g cm^{-3}.

ice band depth maps, is only apparent in the GRN/UV3 and not as obvious in the IR3/UV3 map.

7.2. Constraints from Ice Crystallinity

Mapping the crystallinity of surface ices on Enceladus provides an alternative method to find indicators of plume deposition on its surface. At the average surface temperature on the icy saturnian satellites, ~80 K, amorphous ice is stable against thermal recrystallization for long timescales (*Mastrapa et al.,* 2013). Temperature enhancements beyond 135 K are known to cause a phase change in H_2O ice from amorphous to crystalline; H_2O ice can then reamorphize over time when subjected to ion bombardment, as shown by *Mastrapa and Brown* (2006), or by micrometeorite bombardment. Changes in H_2O ice phase can thus be used to track

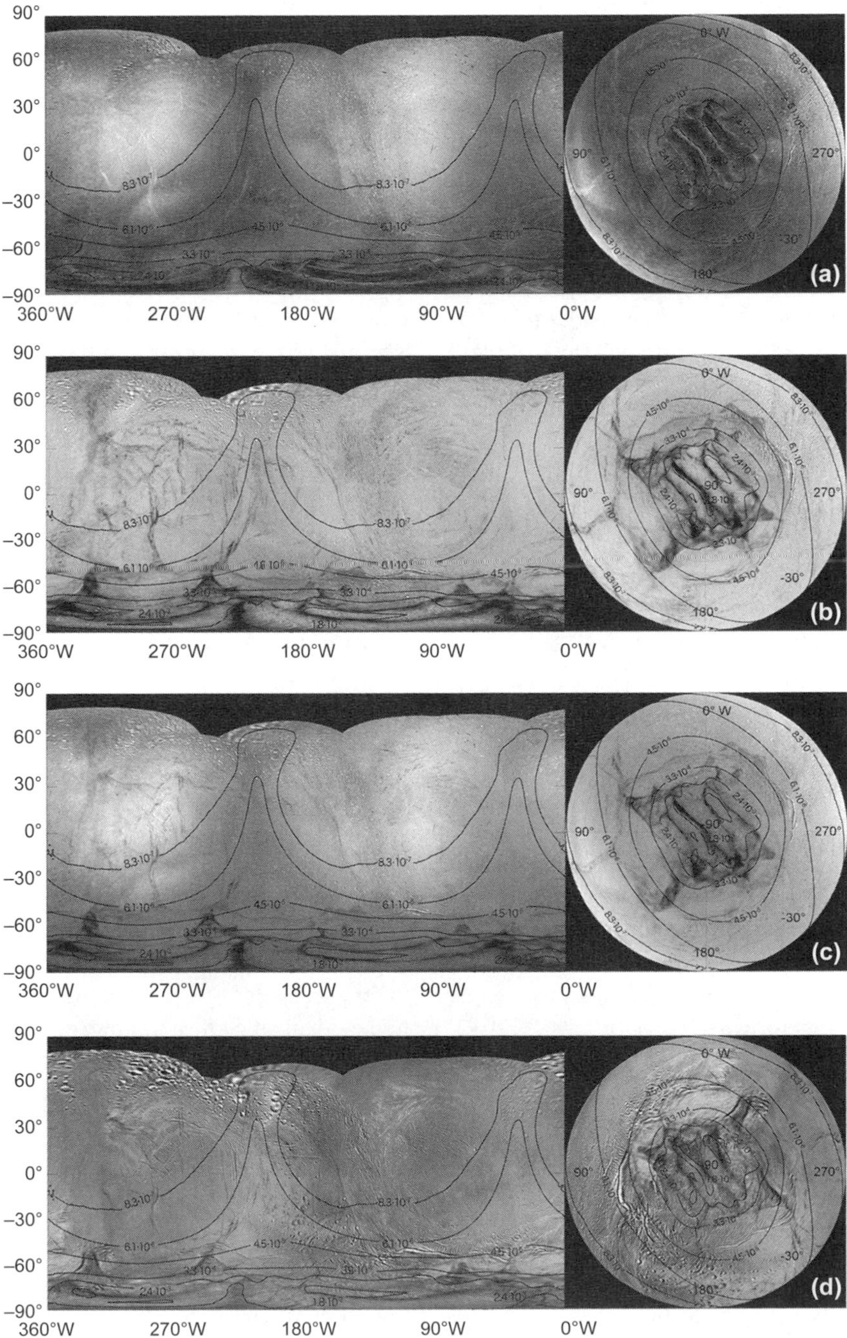

Fig. 24. The first three panels show three Cassini ISS ratio maps: **(a)** GRN/UV3, **(b)** IR3/GRN, and **(c)** IR3/UV3 (*Schenk et al.,* 2011). Bright areas are associated with a positive slope relative to the ratioed bands. **(d)** See Plate 17 for color version. RGB combination of the three filters. The bright regions in the **(a)**, **(c)**, and **(d)** combination resemble well the plume fallout outlined by deposit models (*Southworth et al.,* 2018). The IR3/GRN map shown in **(b)** is smoother and does not resemble ejecta deposits.

variations in temperature and physical conditions on the surface, ideal to reveal past and/or present processes on the surface. In the case of Enceladus, H_2O plume vapor deposited on a cold surface (≤130 K) might produce amorphous ice (*Baragiola et al.,* 2008), while icy plume deposits are expected to be crystalline (*Dhingra et al.,* 2017) (see also section 3.4). Consequently, amorphous H_2O ice at the SPT might mark the presence of a vapor-deposition mechanism, whereas crystalline ice would be indicative of deposition of plume ice grains as the dominant process. Moreover, amorphous ice further away from the SPT could indicate an "old" surface where secondary effects, like space weathering, are more efficient than plume deposition.

Although early Cassini results by *Brown et al.* (2006) indicated amorphous ice, later work by *Clark et al.* (2012) indicated that the effects of diffraction from submicrometer ice grains might explain the observed VIMS spectra of Enceladus rather than amorphous ice. *Brown et al.* (2006) based amorphous ice detection on a decrease in strength of absorption at 1.65 μm, and a decrease in intensity of the 3.1-μm Fresnel peak. *Clark et al.* (2012) showed both of these effects are also caused by diffraction from submicrometer ice particles in the surface.

The analysis of VIMS data in principle allows spatially resolved measurements but no results on the detection and distribution of amorphous ice have been published yet. However, various analyses to constrain the effects of plume deposition with a "crystallinity map" are currently ongoing [e.g., with the method established by *Dalle Ore et al.* (2015) but modified for the effects of submicrometer ice grains].

8. CONCLUSIONS AND OPEN QUESTIONS

8.1. Plume Composition

The enceladean plume is composed of three different phases: gas, solids (dust), and ions. Neutral gas is the most abundant component and is emitted with an average rate of 170–250 kg s^{-1} (*Hansen et al.,* 2017). The estimates for the emitted solid material vary much more. This estimate only partially reflects the orbital variation in emitted dust, which appears to be larger than for gas (*Hedman et al.,* 2013; *Nimmo et al.,* 2014; *Ingersoll and Ewald,* 2017; *Hansen et al.,* 2017), but also reflects the larger uncertainty of the dust flux estimates. Estimates range from about 3–5 kg s^{-1} (*Schmidt et al.,* 2008; *Kempf et al.,* 2010) to 50 kg s^{-1} (*Ingersoll and Ewald,* 2011) or 15–65 kg s^{-1} (*Dong et al.,* 2015). The arguably most robust value is given in the chapter by Kempf et al. in this volume, with about 20 kg s^{-1}, and yields a dust to gas ratio about 10% (for a detailed discussion see the chapter by Kempf et al.). The ejection speed of the gas is much larger than that of the dust grains (*Schmidt et al.,* 2008; *Hedman et al.,* 2009), therefore the gas almost completely escapes into space. The ionic component is picked up by Saturn's magnetosphere and also escapes Enceladus' gravitational influence. The abundant nanometer-sized grains observed by CAPS (*Hill et al.,* 2012; *Dong et al.,* 2015) probably also almost completely escape. However, only a fraction (5–10%) of the solid material larger than ≈0.1 μm in the plume escapes into the E ring, whereas the greater part is falling back to the surface (*Porco et al.,* 2006; *Spahn et al.,* 2006; *Schmidt et al.,* 2008; *Kempf et al.,* 2010; *Ingersoll and Ewald,* 2011), a fact that strongly couples plume composition and surface composition (see section 8.3). The probability of escape is coupled to the grain size: Larger grains are ejected at lower speeds, leading to the observed tendency of large particles preferentially populating the lower regions of the plume and smaller grains preferentially escaping into the E ring (*Kempf et al.,* 2008; *Hedman et al.,* 2009). This size dependence also causes a compositional plume stratification (*Postberg et al.,* 2011a; *Khawaja et al.,* 2017) (see section 8.2).

All three phases (gas, dust, and ionic) are primarily composed of water. For the neutral gas the water abundance is larger than 96%, and for dust grains with $r > 0.2$ μm it is on the order of 99%, whereas the water proportion in the ionic phase, although dominant, is less well constrained. The majority of ice grains in the plume are in a crystalline (and not amorphous) state (*Dhingra et al.,* 2017), indicative of formation temperatures above 130 K. Although the general interpretation that these grains are made of water ice is justified (e.g., *Hill et al.,* 2012), the composition of the predominately negatively charged nanograins in the plume is actually not known (see open questions in section 8.4).

The main non-icy compounds in the solid phase ($r > 0.2$ μm) are sodium salts and organic material. Salts and organics are heterogeneously distributed over three compositional diverse main ice grain populations and can both reach percent level abundance in individual grains (*Postberg et al.,* 2009a, 2011a, 2018). SiO_2 is another important constituent (>100 ppm) but was indirectly inferred by measurements in the outer saturnian system (*Hsu et al.,* 2011, 2015). The most abundant volatiles are ammonia, molecular hydrogen, carbon dioxide, and organics. Each of these compounds is present in mixing ratios of fractions of a percent with upper limits of H_2 and NH_3 slightly above 1% (*Waite et al.,* 2017). The D/H ratio is 2.9 (+1.5/–0.7) × 10^{-4} and is approximately in the mid-range of observed cometary D/H values and much higher than in Saturn's atmosphere (*Waite et al.,* 2009). Methane (≈0.2%) is the most abundant organic compound in the gas phase, with less abundant C_2 and maybe C_3 species (*Magee and Waite,* 2017). In comparison, the ice grains carry more refractory organic material with atomic masses ranging from about 28 u up to at least 200 u (*Postberg et al.,* 2008, 2018). Oxygen- and nitrogen-bearing organic species might be present in both neutral gas and ice grains (*Magee and Waite,* 2017; *Postberg et al.,* 2018). The tentative detection of ammonia in the surface ice of Enceladus (*Emery et al.,* 2005; *Verbiscer et al.,* 2006; *Hendrix et al.,* 2010) (see section 4.2) supports the idea of nitrogen-bearing ice grains being emitted by the plume.

In general, the sensitivity of Cassini instruments to minor species in ionized form is lower than for neutral molecules or macroscopic ice grains. Ions directly measured in the plume are composed almost exclusively of water and water products. Detected cations have the form H_nO^+ (n = 0–3) and respective dimers. Detected anions are dissociated water molecules (OH, H^-, O^-) or a cluster of the form $(H_2O)_nOH^-$ (n = 1–3) (*Tokar et al.,* 2009; *Coates et al.,* 2010a,b). The only indication for ions other than from water are high-mass anions (>200 u) possibly in agreement with complex organics (*Coates et al.,* 2010a,b). In contrast, there is ample evidence for non-water ions from integrated measurements in Saturn's magnetosphere that likely were emitted by Enceladus' plume. The most apparent are nitrogen-bearing cations, in particular NH^+ detected by CAPS, present on a level of a few percent (*Smith et al.,* 2008), a proportion that seems to be difficult to reconcile with the lower abundance of nitrogen-bearing species in the plume (see section 8.4 for further discussion). The detection of N^+ by the MIMI-CHEMS sensor supports this result. The instrument also finds C^+ in similar abundance (≈1%), indicative of dissociation of organic plume constituents. Another ion species of possible plume origin detected in trace abundance are cations with a mass of 28 u that could be $C_2H_5^+$, $HCNH^+$, N_2^+, Si^+, and CO^+ (*Christon et al.,* 2014, 2015). See section 8.4 for further discussion.

Possible origins of the different constituents in Enceladus' interior are discussed in section 5 and in the chapter by Glein et al. this volume.

8.2. Variability of the Plume Composition in Space and Time

There is clear indication of spatial variation in the plume's composition. Most apparent is the compositional stratification of the ice grains with r > 0.2 μm measured by the CDA. Ice grains containing significant amounts of sodium and potassium salts are significantly more abundant at low altitudes. This stratification is indicative of lower ejection speeds, caused either by slower ejection speed of the carrier gas or by the larger size of these grains (*Postberg et al.,* 2011a). In turn the almost pure water ice grains are more abundant at high altitudes and in the fraction that escapes into the E ring. Organic-bearing grains appear to be most abundant in confined high-velocity jets at all altitudes (*Postberg et al.,* 2011a; *Khawaja et al.,* 2017). The stratification outlined above is based on proportions; obviously the absolute abundance of ice grains of any composition increases with lower altitudes. The plume material above Baghdad and Damascus Sulci has a dust-to-gas mass ratio that is roughly an order of magnitude higher than the material above Alexandria and Cairo Sulci (*Hedman et al.,* 2018). This difference suggests, but does not prove, that there is a lateral variation in the ice grain composition (*Postberg et al.,* 2011a; *Dhingra et al.,* 2017). Organic-bearing grains are possibly more abundant in emissions from Damascus Sulcus than from the other sulci (*Postberg et al.,* 2011a; *Brown et al.,* 2006).

The spatial density variation of the neutral gas inside the plume is substantial (*Hansen et al.,* 2008; *Teolis et al.,* 2017). However, it is not known if a compositional variation is linked to this spatial variation. The only clue from Cassini is INMS measurements, which indicate that heavier molecular species (e.g., CO_2, 44 u) have a narrower lateral spread than lighter species (e.g., H_2, 2 u) in a supersonic jet cone (*Perry et al.,* 2015).

Spatial variations in the composition of the ionized phase have not been reported. However, charged molecules and charged nanograins both will have very different trajectories compared to the ballistic trajectories of the neutral gas and larger grains. The plasma environment is dominated by the dynamics of the interactions between the plume neutral gas and the co-rotational plasma in Saturn's magnetosphere (see the chapter by Smith et al. in this volume). Because the charge-to-mass ratios of nanograins are sufficiently large, their dynamics are strongly influenced by the Lorentz forces and thus sensitive to the local electromagnetic environment resulting from the plume-magnetosphere interactions. Grains with radii of a few nanometer with both polarities are found in the plume (*Jones et al.,* 2009), and these grains are likely responsible for several features observed in the Cassini magnetic field and thermal plasma data (*Simon et al.,* 2011; *Hill et al.,* 2012; *Meier et al.,* 2014, 2015; *Dong et al.,* 2015). Fluxes of both plasma ions and charged nanograins emitted by Enceladus will strongly vary with magnetic field and co-rotational plasma conditions and could potentially be linked to a compositional variation.

The plume is known to be time variable. The most apparent activity cycle is coupled to the moon's orbital period (*Hedman et al.,* 2013; *Nimmo et al.,* 2014; *Ingersoll and Ewald,* 2017). The plume brightness from emitted ice grains varies by about a factor of 4 between apocenter (maximum) and pericenter (minimum). This variation can be explained either by a change in total mass flux or by a drastic change in the grain size distribution or by a superposition of both effects (*Hedman et al.,* 2013). A variation by a factor of 4 has also been observed for the nanograin flux (*Smith et al.,* 2010; *Blanc et al,* 2015) (for details, see the chapter by Smith et al. in this volume). There is also an indication for brightness variations on the order of years (*Hedman et al.,* 2013; Ingersoll and Ewald, 2017). By contrast, the variations in the integrated emitted gas flux over time seem to be milder (*Hansen et al.,* 2017; *Teolis et al.,* 2017). Currently no compositional variation can be linked to either these variations (see section 8.4.).

8.3. Surface Composition

Enceladus exhibits the cleanest water ice surface with the highest visible albedo known in the solar system (*Verbiscer et al.,* 2007). The only non-icy component that has been measured with certainty is CO_2 that is trapped in surface water ice. CO_2 appears to be most abundant in the SPT (*Brown et al.,* 2006). An absorption in the FUV consistent with NH_3 mixed in water ice has been identified in disk-averaged UVIS images (*Hendrix et al.,* 2010), but is only in agreement with its absence in near-infrared spectra at concentrations of less than 2% (*Emery et al.,* 2005; *Verbiscer et al.,* 2006; *Brown et al.,* 2006). Infrared bands consistent with aliphatic hydrocarbons have been identified along the tiger stripe fissures (*Brown et al.,* 2006). Suitable candidates for an absorption in the MUV, also observed on other icy moons of Saturn (see the chapter by Hendrix et al. in this volume), might be some sort of tholins or traces of a mixture of nanophase iron and iron oxide.

Deposition of plume material, especially micrometer-sized ice grains, happens on most of Enceladus' surface. If E-ring deposition is also taken into account, the entire surface is exposed to plume material (see the chapter by Kempf et al. in this volume), although deposition rates vary with surface location. The SPT is subject to intense resurfacing by icy plume deposits and the composition there should be largely governed by the composition of the emitted ice grains. From deposition maps (*Schenk et al.,* 2011; *Scipioni et al.,* 2017) and plume dynamics (see the chapter by Goldstein et al. in this volume), the largest grains fall back onto the surface closest to their tiger stripe sources. CDA data indicate that these large grains are particularly rich in non-icy components, which make it likely that sodium salts, in particular chloride and carbonate/bicarbonate, and organic compounds are present in substantial quantities of up to ≈1% close to the tiger stripe fissures.

At present, the published VIMS spectra of the plumes have insufficient signal-to-noise ratios to constrain the salt abundance. Organic material is likely present near the tiger stripes (*Brown et al.,* 2006) and "tholins" are tentatively indicated (*Hendrix et al.,* 2010). If the latter are present, they could be the result of space weathering of organic precursors in the plume and can form on geological timescales. This process might be more efficient at regions further away from the SPT, which are not subject to the highest deposition rates of fresh plume material. Although the deposition of crystalline ice grains from the plume might argue for mostly crystalline surface ice, at least near the SPT, the question of whether amorphous ice is present on Enceladus' surface has not been fully resolved.

8.4. Major Open Questions

The only reduced carbon plume compound that can be identified with certainty is methane. Although there are definite measurements of further (unspecified) organic species with higher masses in the ice grains (*Postberg et al.,* 2008, 2018) and the neutral plume gas (*Waite et al.,* 2009; *Magee and Waite,* 2017), the actual organic compounds are poorly constrained qualitatively and quantitatively. In the plume gas, this uncertainty concerns unspecified species with up to three C atoms, whereas in the ice grains the organic molecules, at least occasionally, are much more complex (*Postberg et al.,* 2018). It is also likely that the limited mass resolution of the CDA not only limits the detection and specification of organic compounds but also the detection of inorganic compounds, like minerals and salts other than the most abundant species reported by *Postberg et al.* (2009a). In particular, ammonia or other nitrogen-bearing species would be in agreement with observations of Enceladus' surface (*Hendrix et al.,* 2010). One of the major compositional uncertainties is the unidentified species at 28 u. There is no such species detected by the INMS in the plume with an upper limit below 0.1% (*Waite et al.,* 2017). However, UVIS reports traces of organic species, attributed mostly to C_2H_4 (28 u), in the plume (*Shemansky et al.,* 2016) and an unspecified ionic species with 28 u is detected in Saturn's magnetosphere (*Christon et al.,* 2013, 2014). Obviously, a better understanding of the organic as well as the inorganic plume inventory is a critical element for the investigation of Enceladus' subsurface ocean by future missions (see the chapter by Lunine et al. in this volume).

Cassini's instruments could not directly measure the composition of nanograins in the plume. Although it is a plausible assumption that they are mostly or exclusively composed of water ice, the actual composition is in principle unknown. Nanograins play an important role in the overall emission of solid plume material (*Hill et al.,* 2012), and they are in contrast to the larger grains ($r > 0.1$ μm) observed by CDA that completely escape into the saturnian system. At least some of the emitted nanograins are composed of SiO_2. Dynamical modeling indicates that these are different from the nanograins directly observed in the plume (*Hsu et al.,* 2011, 2015; *Hill et al.,* 2012). SiO_2 grains might leave the vents primarily incorporated into macroscopic ice grains and are only released later by plasma sputter erosion of their carriers (*Hsu et al.,* 2015) in the E ring, but it is actually unclear whether SiO_2 nanograins also exist as individual particles in the plume.

The overabundance of nitrogen-bearing ions (in particular NH_x^+) from Enceladus in Saturn's magnetosphere (*Smith et al.,* 2008) compared to the apparent plume composition is a long-standing unresolved question. Possible explanations could be (1) NH_x^+ has a long lifetime compared with water ions and is therefore overrepresented in the ionic phase; (2) abundant nitrogen-bearing species are emitted in the form of organic molecules (in the gas and in ice grains) that, after their dissociation, feed NH_x^+ into the magnetosphere; or (3) nanograins are effective carriers of nitrogen-bearing species that are later released as ions into Saturn's magnetosphere. The tentative detection of ammonia in the surface ices of Enceladus (*Emery et al.,* 2005; *Verbiscer et al.,* 2006; *Hendrix et al.,* 2010) supports the idea of nitrogen-bearing ice grains emitted by the plume.

Except for the compositional stratification of the macroscopic ice grains in the plume (*Postberg et al.,* 2011a), variations of the plume's composition in space and time are not yet well constrained by Cassini measurements. However, given the rich dynamical fine structure (*Hansen et al.,* 2008, 2017; *Porco et al.,* 2014; *Teolis et al.,* 2017; *Southworth et al.,* 2018) and the quite drastic diurnal orbital variation in the plume brightness (*Hedman et al.,* 2013; *Nimmo et al.,* 2014; *Ingersoll and Ewald,* 2017), it is highly likely that yet-unknown compositional variations coincide with the variations in activity.

Sodium salts and organic compounds are currently poorly constrained on the enceladean surface although their presence in quantities up to the percent level on the SPT is highly likely. The effects of submicrometer grains on VIMS spectra have been underestimated at first and a new VIMS calibration was required for observations that occurred later in the mission (*Clark et al.,* 2016). Therefore, a more sophisticated analysis of the VIMS data might reveal more information on organics in the near future. Ultimately, co-adding all VIMS spectra of the surface obtained over the entire Cassini mission may produce a spectrum with a sufficiently high signal-to-noise ratio to constrain salt abundances in the near future.

There is a non-water MUV absorber on Enceladus. It seems to have similar properties to impurities observed on other icy moons (see the chapter by Hendrix et al. in this volume) and the main rings (e.g., *Clark et al.,* 2012; *Filacchione et al.,* 2012). At present, the available spectral data are insufficient to definitively distinguish between the two best candidates: tholins or a mixture of nanophase iron and iron oxide. The latter could potentially indicate impact gardening by space-weathered meteoritic dust.

The detection of amorphous water ice on Enceladus could reveal more of its surface history by identifying regions where ice condensed from vapor at low temperatures or locations where previous surface modification by particle bombardment is/was more efficient than the deposition of crystalline plume ice grains. Amorphous water ice may still be detected on the surface of Enceladus in Cassini VIMS spectra if the latest VIMS calibration (*Clark et al.,* 2016) is applied, but such analyses have not been done at the full spatial resolution of the VIMS dataset.

REFERENCES

Altobelli N., Postberg F., Fiege K., Trieloff M., and 11 colleagues (2016) Flux and composition of interstellar dust at Saturn from Cassini's Cosmic Dust Analyser. *Science, 352,* 312–318.

Anderson S. and van Dishoeck E. F. (2008) Photodesorption of water ice. *Astron. Astrophys., 91,* 907–916.

Arridge C. S., André N., McAndrews H. J., Bunce E. J., Burger M. H., Hansen K. C., Hsu H.-W., Johnson R. E., Jones G. H., Kempf S., Khurana K. K., Krupp N., Kurth W. S., Leisner J. S., Paranicas C., Roussos E., Russell C. T., Schippers P., Sittler E. C., Smith H. T., Thomsen M. F., and Dougherty M. K. (2011) Mapping magnetospheric equatorial regions at Saturn from Cassini Prime Mission observations. *Space Sci. Rev., 164,* 1–83.

Balsiger H., Altwegg K., and Geiss J. (1995) D/H and $^{18}O/^{16}O$ ratio in the hydronium ion and neutral water from *in situ* ion measurements in Comet Halley. *J. Geophys. Res., 100,* 5827–5834.

Baragiola R. A., Famá M., Loeffler M. J., Raut U., and Shi J. (2008) Radiation effects in ice: New results. *Nucl. Instr. Meth. Phys. Res. B, 266,* 3057–3062.

Blanc M., Andrews D. J., Coates A. J., Hamilton D. C., Jackman C., Jia X., Kotova A., Morooka M., Smith H. T., and Westlake J. H. (2015) Saturn plasma sources and associated transport processes. *Space Sci. Rev., 192,* 237–283.

Bouquet A., Mousis O., Waite J. H., and Picaud S. (2015) Possible evidence for a methane source in Enceladus' ocean. *Geo. Res. Lett., 42,* 1334–1339.

Brown R. H. and 21 colleagues (2004) The Cassini Visual and Infrared Mapping Spectrometer (VIMS) investigations. *Space Sci. Rev., 115,* 111–168.

Brown R. H. and 24 colleagues (2006) Composition and physical properties of Enceladus' surface. *Science, 311,* 1425–1428.

Cadek O., Tobie G., van Hoolst T., Masse M., Choblet G., Lefevre A., Mitri G., Baland R.-M., Behounkova M., Bourgeois O., and Trinh A. (2016) Enceladus's internal ocean and ice shell constrained from Cassini gravity, shape, and libration data. *Geophys. Res. Lett., 43(11),* 5653–5660.

Choblet G., Tobie G., Sotin C., Běhounková M., Čadek O., Postberg F., and Souček O. (2017) Powering prolonged hydrothermal activity inside Enceladus. *Nature Astron., 1,* 841–847.

Christon S. P., Hamilton D. C., DiFabio R. D., Mitchell D. G., Krimigis S. M., and Jontof-Hutter D. S. (2013) Saturn suprathermal O_2^+ and mass-28^+ molecular ions: Long-term seasonal and solar variation. *J. Geophys. Res., 118,* 3446–3462.

Christon S. P., Hamilton D. C., Mitchell D. G., DiFabio R. D., and Krimigis S. M. (2014) Suprathermal magnetospheric minor ions heavier than water at Saturn: Discovery of $^{28}M^+$ seasonal variations. *J. Geophys. Res., 119,* 5662–5673.

Christon S. P., Hamilton D. C., Plane J. M. C., Mitchell D. G., DiFabio R. D., and Krimigis S. M. (2015) Discovery of suprathermal Fe^+ in Saturn's magnetosphere. *J. Geophys. Res., 120,* 2720–2738.

Clark R. N. (1981a) The spectral reflectance of water-mineral mixtures at low temperatures. *J. Geophys. Res., 86,* 3074–3086.

Clark R. N. (1981b) Water frost and ice: The near-infrared spectral reflectance 0.65–2.5 μm. *J. Geophys. Res., 86,* 3087–3096.

Clark R. N., Swayze G. A., Wise R., Livo E., Hoefen T., Kokaly R., and Sutley S. J. (2007) *USGS Digital Spectral Library splib06a.* U.S. Geological Survey, Data Series 231, *http://speclab.cr.usgs.gov/spectral-lib.html.*

Clark R. N., Cruikshank D. P., Jaumann R., Brown R. H., Stephan K., Dalle Ore C. M., Livo K. E., Pearson N., Curchin J. M., Hoefen T. M., Buratti B. J., Filacchione G., Baines K. H., and Nicholson P. D. (2012) The composition of Iapetus: Mapping results from Cassini VIMS. *Icarus, 218,* 831–860.

Clark R. N., Brown R. H., and Lytle D. M. (2016) *The VIMS Wavelength and Radiometric Calibration.* NASA Planetary Data System, The Planetary Atmospheres Node, *http://atmos.nmsu.edu/data_and_services/atmospheres_data/Cassini/vims.html.* 23 pp.

Coates A. J., Jones G. H., Lewis G. R., Wellbrock A., Young D. T., Crary F. J., Johnson R. E., Cassidy T. A., and Hill T. W. (2010a) Negative ions in the Enceladus plume. *Icarus, 206,* 618–622.

Coates A. J., Wellbrock A., Lewis G. R., Jones G. H., Young D. T., Crary F. J., Waite J. H., Johnson R. E., Hill T. W., and Sittler E. C. Jr. (2010b) Negative ions at Titan and Enceladus: Recent results. *Faraday Discussions, 147,* 293–305.

Coates A. J., Wellbrock A., Jones G. H., Waite J. H., Schippers P., Thomsen M. F., Arridge C. S., and Tokar R. L. (2013) Photoelectrons in the Enceladus plume. *J. Geophys. Res., 118,* 5099–5108.

Cravens T. E., Ozak N., Richard M. S., Campbell M. E., Robertson I. P., Perry M., and Rymer A. M. (2011) Electron energetics in the Enceladus torus. *J. Geophys. Res., 116,* A09205, DOI: 10.1029/2011JA016498.

Cruikshank D. P., Owen T. C., Dalle Ore C., Geballe T. R., Roush T. L., de Bergh C., Sandford S. A., Poulet F., Benedix G. K., and Emery J. P. (2005) A spectroscopic study of the surfaces of Saturn's large satellites: H_2O ice, tholins, and minor constituents. *Icarus, 175,* 268–283.

Dalle Ore C. M., Cruikshank D. P., Mastrapa R. M. E., Lewis E., and White O. L. (2015) Impact craters: An ice study on Rhea. *Icarus, 261,* 80–90.

Dawes A., Mukerji R. J., Davis M. P., Holtom P. D., Webb S. M., Sivaraman B., Hoffmann S. V., Shaw D. A., and Mason N. J. (2007) Morphological study into the temperature dependence of solid ammonia under astrochemical conditions using vacuum ultraviolet and Fourier-transform infrared spectroscopy. *J. Chem. Phys., 126,* 244711-1 to 244711-12.

Dhingra D., Hedman M. M., Clark R. N., and Nicholson P. D. (2017) Spatially resolved near infrared observations of Enceladus' tiger stripe eruptions from Cassini VIMS. *Icarus, 292,* 1–12.

Dong Y., Hill T. W., and Ye S.-Y. (2015) Characteristics of ice grains in the Enceladus plume from Cassini observations. *J. Geophys. Res.–Space Physics, 120,* 915–937.

Dougherty M. K., Khurana K. K., Neubauer F. M., Russell C. T., Saur J., Leisner J. S., and Burton M. E. (2006) Identification of a dynamic atmosphere at Enceladus with the Cassini magnetometer. *Science, 311,* 1406–1409.

Emery J. P., Burr D. M., Cruikshank D. P., Brown R. H., and Dalton J. B. (2005) Near-infrared (0.8–4.0 micron) spectroscopy of Mimas, Enceladus, Tethys, and Rhea. *Astron. Astrophys., 435,* 353–362, DOI: 10.1051/0004- 6361:20042482.

Esposito L. W. and 19 colleagues (2004) The Cassini Ultraviolet Imaging Spectrograph investigation. *Space Sci. Rev., 115, 299–361.*

Esposito L. W., Colwell J. E., Larsen K., McClintock W. E., Stewart A. I. F., Hallett J. T., Shemansky D. E., Ajello J. M., Hansen C. J., Hendrix A. R., West R. A., Keller H. U., Korth A., Pryor W. R., Reulke R., and Yung Y. L. (2005) Ultraviolet imaging spectroscopy shows an active saturnian system. *Science, 307,* 1251–1255.

Filacchione G. and 18 colleagues (2012) Saturn's icy satellites and rings investigated by Cassini-VIMS. III. Radial compositional variability. *Icarus, 22(2),* 1064–1096.

Goguen J. D. et al. (2013) The temperature and width of an active fissure on Enceladus measured with Cassini VIMS during the 14 April 2012 south pole flyover. *Icarus, 226,* 1128–1137.

Grundy W. M., Buie M. W., Stansberry J. A., Spencer J. R., and Schmitt B. (1999) Near-infrared spectra of icy outer solar system surfaces: Remote determination of H_2O ice temperatures. *Icarus, 142,* 536–549.

Grün E. and 23 colleagues (1993) Discovery of jovian dust streams and interstellar grains by the Ulysses spacecraft. *Nature, 362,* 428–430.

Hamilton D. and Burns J. (1993) Ejection of dust from Jupiter's gossamer ring. *Nature, 364,* 695–699.

Hansen C. J., Esposito L., Stewart A. I. F., Colwell J., Hendrix A., Pryor W., Shemansky D., and West R. (2006) Enceladus' water vapor plume. *Science, 311,* 1423–1425.

Hansen C. J., Esposito L. W., Stewart A. I. F., Meinke B., Wallis B., Colwell J. E., Hendrix A. R., Larsen K., Pryor W., and Tian F. (2008) Water vapour jets inside the plume of gas leaving Enceladus. *Nature, 456,* 477–479.

Hansen C. J., Shemansky D. E., Esposito L. W., Stewart I. A. F., Lewis B. R., Colwell J. E., Hendrix A. R., West R. A., Waite J. H. Jr., Teolis B., and Magee B. A. (2011) The composition and structure of the Enceladus plume. *Geophys. Res. Lett., 38,* L11202.

Hansen C. J., Esposito L. W., Aye K.-M., Colwell J. E., Hendrix A. R., Portyankina G., and Shemansky D (2017) Investigation of diurnal variability of water vapor in Enceladus' plume by the Cassini ultraviolet imaging spectrograph. *Geophys. Res. Lett., 44,* 672–677, DOI: 10.1002/2016GL071853.

Hapke B., Wagner J., Cohen A., and Partlow W. (1978) Reflectance measurements of lunar materials in the vacuum ultraviolet. *Proc. Lunar Planet. Sci. Conf. 9th,* pp. 2935–2947.

Hedman M. M., Nicholson P. D., Showalter M. R., Brown R. H., Buratti B. J., and Clark R. N. (2009) Spectral observations of the Enceladus plume with Cassini-VIMS. *Astrophys. J., 693,* 1749–1762.

Hedman M. M., Gosmeyer C. M., Nicholson P. D., Sotin C., Brown R. H., Clark R. N., Baines K. H., Buratti B. J., and Showalter M. R. (2013) An observed correlation between plume activity and tidal stresses on Enceladus. *Nature, 500,* 182–184.

Hedman M. M., Dhingra D., Nicholson P. D., Hansen C. J., Portyankina G., Ye S., and Dong Y. (2018) Spatial variations on the dust-to-gas ratio of Enceladus' plume. *Icarus, 305,* 123–138.

Hendrix A. R., Hansen C. J., and Holsclaw G. M. (2010) The ultraviolet reflectance of Enceladus: Implications for surface composition. *Icarus, 206,* 608–617.

Hill T. W. and Pontius D. (1998) Plasma injection near Io. *J. Geophys. Res., 103,* 19879–19886.

Hill T. W., Thomsen M. F., Tokar R. L., Coates A. J., Lewis G. R., Young D. T., Crary F. J., Baragiola R. A., Johnson R. E., Dong Y., Wilson R. J., Jones G. H., Wahlund J.-E., Mitchell D. G., and Horányi M. (2012) Charged nanograins in the Enceladus plume. *J. Geophys. Res., 117,* A05209.

Hillier J. K., Green S. F., McBride N., Schwanethal J. P., Postberg F., Srama R., Kempf S., Moragas-Klostermeyer G., McDonnell J. A. M., and Grün E. (2007) The composition of Saturn's E ring. *Mon. Not. R. Astron. Soc., 377,* 1588–1596.

Hodyss R., Parkinson C. D., Johnson P. V., Stern J. V., Goguen J. D., Yung Y. L., and Kanik I. (2009) Methanol on Enceladus. *Geophys. Res. Lett., 36,* L17103, DOI: 10.1029/2009GL039336.

Horányi M., Morfill G., and Grün E. (1993) Mechanism for the acceleration and ejection of dust grains from Jupiter's magnetosphere. *Nature, 363,* 144–146.

Hsu H. W., Kempf S., Postberg F., Srama R., Jackman C. M., Moragas-Klostermeyer G., Helfert S., and Grün E. (2010) Interaction of the solar wind and stream particles, results from the Cassini dust detector. In *Twelfth International Solar Wind Conference* (M. Maksimovic et al., eds.), pp. 510–513. AIP Conf. Proc. 1216, American Institute of Physics, Melville, New York.

Hsu H.-W., Postberg F., Kempf S., Trieloff M., Burton M., Roy M., Moragas-Klostermeyer G., and Srama R. (2011) Stream particles as the probe of the dust-plasma-magnetosphere interaction at Saturn. *J. Geophys. Res., 116, A9,* A09215.

Hsu H. W., Krüger H., and Postberg F. (2012) Dynamics, composition, and origin of jovian and saturnian dust stream particles. In *Nanodust in the Solar System: Discoveries and Interpretations* (I. Mann et al., eds.), pp. 77–117. Astrophysics and Space Science Library 385, Springer-Verlag, New York.

Hsu H. W., Postberg F., Sekine Y., and 12 colleagues (2015) Ongoing hydrothermal acitivities within Enceladus. *Nature, 519,* 207–210.

Hurley D. M., Perry M. E., and Waite J. H. (2015) Modeling insights into the locations of density enhancements from the Enceladus water vapor jets. *J. Geophys. Res.–Planets, 120,* 1763–1773, DOI: 10.1002/ 2015JE004872.

Ingersoll A. P. and Ewald S. P. (2011) Total particulate mass in Enceladus plumes and mass of Saturn's E ring inferred from Cassini ISS images. *Icarus, 216,* 492–506.

Ingersoll A. P. and Ewald S. P. (2017) Decadal timescale variability of the Enceladus plumes inferred from Cassini images. *Icarus, 282,* 260–275.

Jaumann R. and 16 colleagues (2008) Distribution of icy particles across Enceladus' surface as derived from Cassini-VIMS measurements. *Icarus, 193,* 407–419, DOI: 10.1016/j.icarus.2007.09.013.

Johnson R. E., Pospieszalska M. K., Sittler E. C. Jr., Cheng A. F., Lanzerotti L. J., and Sieveka E. M. (1989) The neutral cloud and heavy ion inner torus at Saturn. *Icarus, 77,* 311–329.

Johnson R. E., Smith H. T., Tucker O. J., Liu M., Burger M. H., Sittler E. C., and Tokar R. L. (2006) The Enceladus and OH tori at Saturn. *Astrophys. J. Lett., 644,* L137–L139.

Jones G. H., Arridge C. S., Coates A. J., Lewis G. R., Kanani S., Wellbrock A., Young D. T., Crary F. J., Tokar R. L., Wilson R. J., Hill T. W., Johnson R. E., Mitchell D. G., Schmidt J., Kempf S., Beckmann U., Russell C. T., Jia Y. D., Dougherty M. K., Waite J. H. Jr., and Magee B. (2009) Fine jet structure of electrically-charged grains in Enceladus' plume. *Geophys. Res. Lett., 36,* L16204.

Kempf S., Srama R., Horányi M., Burton M., Helfert S., Moragas-Klostermeyer G., Roy M., and Grün E. (2005a) High-velocity streams of dust originating from Saturn. *Nature, 433,* 289–291.

Kempf S., Srama R., Postberg F., Burton M., Green S. F., Helfert S., Hillier J. K., McBride N., McDonnell J. A. M., Moragas-Klostermeyer G., Roy M., and Grün E. (2005b) Composition of saturnian stream particles. *Science, 307,* 1274–1276.

Kempf S., Beckmann U., Moragas-Klostermeyer G., Postberg F., Srama R., Economou T., Schmidt J., Spahn F., and Grün E. (2008) The E ring in the vicinity of Enceladus I: Spatial distribution and properties of the ring particles. *Icarus, 193,* 420–437.

Kempf S., Beckmann U., and Schmidt J. (2010) How the Enceladus dust plume feeds Saturn's E ring. *Icarus, 206,* 446–457.

Khawaja N., Postberg F., and Schmidt J. (2017) The compositional profile of the enceladian ice plume from the latest Cassini flybys. *Lunar Planet. Sci. XLVIII,* Abstract #2005. Lunar and Planetary Institute, Houston.

Kieffer S. W. et al. (2006) A clathrate reservoir hypothesis for Enceladus' south polar plume. *Science, 314,* 1764–1766.

Kite E. S. and Rubin A. M. (2016) Sustained eruptions on Enceladus explained by turbulent dissipation in tiger stripes. *Proc. Natl. Acad. Sci., 113(15),* 3972–3975.

Krimigis S. M., Mitchell D. G., Hamilton D. C., et al. (2004) Magnetosphere Imaging Instrument (MIMI) on the Cassini mission to Saturn/Titan. *Space Sci. Rev., 114,* 233–329, DOI: 10.1007/ s11214-004-1410-8.

Le Gall A. and 17 colleagues (2017) Thermally anomalous features in the subsurface of Enceladus's south polar terrain. *Nature Astron., 1,* 0063.

Leisner J. S., Russell C. T., Dougherty M. K., Blanco-Cano X., Strangeway R. J., and Bertucci C. (2006) Ion cyclotron waves in Saturn's E ring: Initial Cassini observations. *Geophys. Res. Lett., 33,* L11101.

Lhuissier H. and Villermaux E. (2012) Bursting bubble aerosols. *J. Fluid Mech., 696,* 5–44, DOI: 10.1017/jfm.2011.418.

Magee B. A. and Waite J. H. (2017) Neutral gas composition of Enceladus' plume — Model parameter insights from Cassini-INMS. *Lunar Planet. Sci. XVLIII,* Abstract #2974. Lunar and Planetary Institute, Houston.

Martonchik J. V., Orton G. S., and Appleby J. F. (1984) Optical

properties of NH_3 ice from the far infrared to the near ultraviolet. *Appl. Opt., 23,* 541–547.
Mastrapa R. M. E. and Brown R. H. (2006) Ion irradiation of crystalline H_2O-ice: Effect on the 1.65-μm band. *Icarus, 183,* 207–214.
Mastrapa R. M. E., Grundy W. M., and Gudipati M. S. (2013) Amorphous and crystalline H_2O-Ice. In *The Science of Solar System Ices* (M. Gudipati and J. Castillo-Rogez, eds.), pp. 371–408. Springer-Verlag, New York.
Matson D. L., Castillo-Rogez J. C., Davies A. G., and Johnson T. V. (2012) Enceladus: A hypothesis for bringing both heat and chemicals to the surface. *Icarus, 221,* 53–62.
McKinnon W. B. (2015) Effect of Enceladus's rapid synchronous spin on interpretation of Cassini gravity. *Geophys. Res. Lett., 42(7),* 2137–2143.
Meier P., Kriegel H., Motschmann U., Schmidt J., Spahn F., Hill T. W., Dong Y., and Jones G. H. (2014) A model of the spatial and size distribution of Enceladus' dust plume. *Planet. Space Sci., 104,* 216–233.
Meier P., Motschmann U., Schmidt J., Spahn F., Hill T. W., Dong Y., Jones G. H., and Kriegel H. (2015) Modeling the total dust production of Enceladus from stochastic charge equilibrium and simulations. *Planet. Space Sci., 119,* 208–221.
Morooka M. W., Wahlund J.-E., Eriksson A. I., Farrell W. M., Gurnett D. A., Kurth W. S., Persson A. M., Shafiq M., Andre M., and Holmberg M. K. G. (2011) Dusty plasma in the vicinity of Enceladus. *J. Geophys. Res., 116,* A12221.
Nimmo F., Porco C., and Mitchel C. (2014) Tidally modulated eruptions on Enceladus: Cassini ISS observations and models. *Astron. J., 148,* 46.
Perry M. E. et al. (2015) Cassini INMS measurements of Enceladus plume density. *Icarus, 257,* 139–162.
Pierel J. D. R, Nixon C. A., Lellouch E., Fletcher L. N., Bjoraker G. L., Achterberg R. K., Bézard B., Hesman B. E., Irwin P. G. J., and Flasar F. M. (2017) D/H ratios on Saturn and Jupiter from Cassini CIRS. *Astron. J., 154,* 154–178.
Pitman K. M., Buratti B. J., and Mosher J. A. (2010) Disk-integrated bolometric Bond albedos and rotational light curves of saturnian satellites from Cassini Visual and Infrared Mapping Spectrometer. *Icarus, 206,* 537–560.
Pitman K. M., Kolokova L., Verbiscer A. J., Mackowski D. W., and Joseph E. C. S. (2017) Coherent backscattering effect in spectra of icy satellites and its modeling using multi-sphere T-matrix (MSTM) code for layers of particles. *Planet. Space Sci., 149,* 23–31.
Porco C. C., Helfenstein P., Thomas P. C., Ingersoll A. P., Wisdom J., West R., Neukum G., Denk T., Wagner R., Roatsch T., Kieffer S., Turtle E., McEwen A., Johnson T. V., Rathbun J., Veverka J., Wilson D., Perry J., Spitale J., Brahic A., Burns J. A., DelGenio A. D., Dones L., Murray C. D., and Squyres S. (2006) Cassini observes the active south pole of Enceladus. *Science, 311,* 1393–1401.
Porco C., DiNino D., and Nimmo F. (2014) How the geysers, tidal stresses, and thermal emission across the south polar terrain of Enceladus are related. *Astron. J., 148,* 45.
Porco C., Dones L., and Mitchell C. (2017) Could it be snowing microbes on Enceladus? Asessing conditions in its plume and implications for future missions. *Astrobiology, 17(9),* 876–912.
Postberg F., Kemp S., Sram R., Gree S. F., Hillie J. K., McBride N., and Grün E. (2006) Composition of jovian dust stream particles. *Icarus, 183,* 122–134.
Postberg F, Hillier J. K., Kempf S., Srama R., Green S. F., McBride N., and Grün E. (2008) The E-ring in the vicinity of Enceladus II. Probing the moon's interior — the composition of E-ring particles. *Icarus, 193,* 438–454.
Postberg F., Kempf S., Schmidt J., Brillantov N., Beinsen A., Abel B., Buck U., and Srama R. (2009a) Sodium salts in E ring ice grains from an ocean below the surface of Enceladus. *Nature, 459,* 1098–1101.
Postberg F., Kempf S., Rost D., Stephan T., Srama R., Trieloff M., Mocker A., and Goerlich M. (2009b) Discriminating contamination from particle components in spectra of Cassini's dust detector CDA. *Planet. Space Sci., 57,* 1359–1374.
Postberg F., Schmidt J., Hillier J. K., Kempf S., and Srama R. (2011a) A salt-water reservoir as the source of a compositionally stratified plume on Enceladus. *Nature, 474(7353),* 620–622.
Postberg F., Grün E., Horanyi M., Kempf S., Krüger H., Srama R., Sternovsky Z., and Trieloff M. (2011b) Compositional mapping of planetary moons by mass spectrometry of dust ejecta. *Planet. Space Sci., 59,* 1815–1825.
Postberg F., Tobie G., and Dambeck T. (2016) Under the sea of Enceladus. *Sci. Am., 315(4),* 38–45.
Postberg F., Khawaja N., and 18 colleagues (2018) Macromolecular organic compounds from the depths of Enceladus. *Nature, 558,* 564–568.
Ries P. A. and Jansen M (2015) A large-scale anomaly in Enceladus' microwave emission. *Icarus, 257,* 88–102.
Schenk P., Hamilton D., Johnson R., McKinnon W., Paranicas C., Schmidt J., and Showalter M. (2011) Plasma, plumes and rings: Saturn system dynamics as recorded in global color patterns on its midsize icy satellites. *Icarus, 211,* 740–757.
Schippers P., Blanc M., Andre N., Dandouras I., Lewis G. R., Gilbert L. K., Persoon A. M., Krupp N., Gurnett D. A., Coates A. J., Krimigis S. M., Young D. T., and Dougherty M. K. (2008) Multi-instrument analysis of electron populations in Saturn's magnetosphere. *J. Geophys. Res., 113,* A07208.
Schmidt J., Brillantov N., Spahn F., and Kempf S. (2008) Slow dust in Enceladus' plume from condensation and wall collisions in tiger stripe fractures. *Nature, 451,* 685–688.
Scipioni F., Schenk P., Tosi F., D'Aversa E., Clark R., Combe J.-Ph., Dalle Ore C. M. (2017) Deciphering sub-micron ice particles on Enceladus surface. *Icarus, 290,* 183–200.
Sekine Y., Shibuya T., Postberg F., Hsu H. W., Suzuki K., Masaki Y., Kuwatani T., Mori M., Hong P. K., Yoshizaki M., Tachibana S., and Sirono S. (2015) High-temperature water-rock interactions and hydrothermal environments in the chondrite-like core of Enceladus. *Nature Commun., 6,* 8604, DOI: 10.1038/ncomms9604.
Shemansky D. E., Matheson P., Hall D.T., Hu H.-Y., and Tripp T. M. (1993) Detection of the hydroxyl radical in the Saturn magnetosphere. *Nature, 363,* 329–331.
Shemansky D. E., Yoshii J., Hansen C., Hendrix A. R., Liu X., and Yung Y. (2016) The complex highly structured near Enceladus environment: Analysis of Cassini UVIS image cube vectors in the FUV. *AOGS 2016,* #PS06-A025.
Simon S., Saur J., Kriegel H., Neubauer F. M., Motschmann U., and Dougherty M. K. (2011) Influence of negatively charged plume grains and hemisphere coupling currents on the structure of Enceladus' Alfven wings: Analytical modeling of Cassini magnetometer observations. *J. Geophys. Res., 116,* A04221.
Smith H. T., Shappirio M., Sittler E. C., Reisenfeld D., Johnson R. E., Baragiola R. A., Crary F. J., McComas D. J., and Young D. T. (2005) Discovery of nitrogen in Saturn's inner magnetosphere. *Geophys. Res. Lett., 32,* L14S03.
Smith H. T., Johnson R. E., Sittler E. C., Shappirio M., Reisenfeld D., Tucker O. J., Burger M., Crary F. J., McComas D. J., and Young D. T. (2007) Enceladus: The likely dominant nitrogen source in Saturn's magnetosphere. *Icarus, 188,* 356–366.
Smith H. T., Shappirio M., Johnson R. E., Reisenfeld D., Sittler E. C., Crary F. J., McComas D. J., and Young D. T. (2008) Enceladus: A potential source of ammonia products and molecular nitrogen for Saturn's magnetosphere. *J. Geophys. Res., 113,* A11206.
Smith H. T., Johnson R. E., Perry M. E., Mitchell D. G., McNutt R. L., and Young D. T. (2010) Enceladus plume variability and the neutral gas densities in Saturn's magnetosphere. *J. Geophys. Res., 115,* A10252.
Southworth B. S., Kempf S., and Spitale J. (2018) Surface deposition of the Enceladus plume and the zenith angle of emissions. *Icarus,* in press.
Spahn F. and 15 colleagues (2006) Cassini dust measurements at Enceladus and implications for the origin of the E ring. *Science, 311,* 1416–1418.
Spitale J. N. and Porco C. C. (2007) Association of jets on Enceladus with the warmest regions on its south-polar fractures. *Nature, 449,* 695–697.
Spitale J. N., Hurford T. A., Rhoden A. R., Berkson E. E., and Platts S. S. (2015) Curtain eruptions from Enceladus' south-polar terrain. *Nature, 521(7550),* 57–60.
Srama R. and 43 colleagues (2004) The Cassini Cosmic Dust Analyser. *Space Sci. Rev., 114,* 465–518.
Teolis B. D., Perry M. E., Hansen C. J., Waite J. H., Porco C. C., Spencer J. R., and Howett C. J. A. (2017) Enceladus plume structure and temporal variability: Comparison of Cassini observations. *Astrobiology, 19(7),* DOI: 10.1089/ast.2017.1647.
Thompson W. R., Murray B. G. J. P. T., Khare B. N., and Sagan C. (1987) Coloration and darkening of methane clathrate and other ices by charged particle irradiation: Applications to the outer solar

system. *J. Geophys. Res., 92,* 14933–14947.

Thomsen M. F., Reisenfeld D. B., Delapp D. M., Tokar R. L., Young D. T., Crary F. J., Sittler E. C., McGraw M. A., and Williams J. D. (2010) Survey of ion plasma parameters in Saturn's magnetosphere. *J. Geophys. Res., 115,* A10220.

Tokar R. L., Johnson R. E., Hill T. W., Pontius T. D. H., Kurth W. S., Crary F. J., Young D. T., Thomsen M. F., Reisenfeld D. B., Coates A. J., Lewis G. R., Sittler E. C., and Gurnett D. A. (2006) The interaction of the atmosphere of Enceladus with Saturn's plasma. *Science, 311,* 1409–1412.

Tokar R. L., Wilson R. J., Johnson R. E., Henderson M. G., Thomsen M. F., Cowee M. M., Sittler E. C. Jr., Young D. T., Crary F. J., McAndrews H. J., and Smith H. T. (2008) Cassini detection of water-group pick-up ions in the Enceladus torus. *Geophys. Res. Lett., 35,* L14202.

Tokar R. L., Johnson R. E., Thomsen M. F., Wilson R. J., Young D. T., Crary F. J., Coates A. J., Jones G. H., and Paty C. S. (2009) Cassini detection of Enceladus' cold water-group plume ionosphere. *Geophys. Res. Lett., 36,* L13203.

Verbiscer A. J., French R. G., and McGhee C. A. (2005) The opposition surge of Enceladus: HST observations 338–1022 nm. *Icarus, 173,* 66–83.

Verbiscer A. J., Peterson D. E., Skrutskie M. F., Cushing M., Helfenstein P., Nelson M. J., Smith J. D., and Wilson J. C. (2006) Near-infrared spectra of the leading and trailing hemispheres of Enceladus. *Icarus, 182,* 211–223.

Verbiscer A., French R., Showalter M., and Helfenstein P. (2007) Enceladus: Cosmic graffiti artist caught in the act. *Science, 315(5813),* 815.

Waite J. H. and 15 colleagues (2004) The Cassini Ion and Neutral Mass Spectrometer (INMS) investigation. *Space Sci. Rev., 114,* 113–231.

Waite J. H., Combi M. R., Ip W.-H., Cravens T. E., McNutt R. L., Kasprzak W., Yelle R., Luhmann J., Niemann H., Gell D., Magee B., Fletcher G., Lunine J., and Tseng W.-L. (2006) Cassini Ion and Neutral Mass Spectrometer: Enceladus plume composition and structure. *Science, 311,* 1419–1422.

Waite J. H., Lewis W. S., Magee B. A., Lunine J. I., McKinnon W. B., Glein C. R., Mousis O., Young D. T., Brockwell T., Westlake J., Nguyen M.-J., Teolis B., Niemann H., McNutt R., Perry M., and Ip W.-H. (2009) Liquid water on Enceladus from observations of ammonia and ^{40}Ar in the plume. *Nature, 460,* 487–490.

Waite J. H. and 12 colleagues (2017) Cassini finds molecular hydrogen in the Enceladus plume: Evidence for hydrothermal processes. *Science, 356(6334),* 155–159.

Walker J. D., Chocron S., Waite J. H., and Brockwell T. (2015) The vaporization threshold: Hypervelocity impacts of ice grains into a titanium Cassini spacecraft instrument chamber. *Procedia Engineering, 103,* 628–635, DOI: 10.1016/j.proeng.2015.04.081.

Wilson R. J., Tokar R. L., Henderson M. G., Hill T. W., Thomsen M. F., and Pontius D. H. Jr. (2008) Cassini plasma spectrometer thermal ion measurements in Saturn's inner magnetosphere. *J. Geophys. Res., 113,* A12218.

Yeoh S. K., Chapman T. A., Goldstein D. B., Varghese P. L., and Trafton L. M. (2015) On understanding the physics of the Enceladus south polar plume via numerical simulation. *Icarus, 253,* 205–222.

Young D. T., Berthelier J.-J., Blanc M., Burch J. L., Coates A. J., Goldstein R., Grande M., Hill T. W., Johnson R. E., Kelha V., McComas D. J., Sittler E. C., Svenes K. R., Szegö K., Tanskanen P., Ahola K., Anderson D., Bakshi S., Baragiola R. A., Barraclough B. L., Black R., Bolton S., Booker T., Bowman R., Casey P., Crary J., Delapp D., Dirks G., Eaker N., Funsten H., Furman J. D., Gosling J. T., Hannula H., Holmlund C., Huomo H., Illiano J.–M., Jensen P., Johnson M. A., Linder D., Luntama T., Maurice S., McCabe K., Mursula K., Narheim B. T., Nordholt J. E., Preece A., Rutzki J., Ruitberg A., Smith K., Szalai S., Thomsen M. F., Viherkanto K., Vilppola J., Vollmer T., Wahl T. E., Wüest M., Ylikorpi T., and Zinsmeyer C. (2004) Cassini Plasma Spectrometer investigation. *Space Sci. Rev., 114,* 1–112.

Young D. T., Berthelier J.-J., Blanc M., Burch J. L., Bolton S., Coates A. J., Crary F. J., Goldstein R., Grande M., Hill T. W., Johnson R. E., Baragiola R. A., Kelha V., McComas D. J., Mursula, K., Sittler E. C., Svenes K. R., Szegö K., Tanskanen P., Thomsen M. F., Bakshi S., Barraclough B. L., Bebesi D., Delapp D., Dunlop M. W., Gosling J. T., Furman J. D., Gilbert L. K., Glenn D., Holmlund C., Illiano J.-M., Lewis G. R., Linder D. R., Maurice S., McAndrews H. J., Narheim B. T., Pallier E., Reisenfeld D., Rymer A. M., Smith H. T., Tokar R. L.,Vilppola J., and Zinsmeyer C. (2005) Composition and dynamics of plasma in Saturn's magnetosphere. *Science, 307,* 1262–1265.

Zastrow M., Clarke J. T., Hendrix A. R., and Noll K. S. (2012) UV spectrum of Enceladus. *Icarus, 220,* 29–35.

Zolotov M. Y. (2007) An oceanic composition on early and today's Enceladus. *Geophys. Res. Lett., 34,* L23203.

Spencer J. R., Nimmo F., Ingersoll A. P., Hurford T. A., Kite E. S., Rhoden A. R., Schmidt J., and Howett C. J. A. (2018) Plume origins and plumbing (ocean to surface). In *Enceladus and the Icy Moons of Saturn* (P. M. Schenk et al., eds.), pp. 163–174. Univ. of Arizona, Tucson, DOI: 10.2458/azu_uapress_9780816537075-ch008.

Plume Origins and Plumbing: From Ocean to Surface

J. R. Spencer
Southwest Research Institute

F. Nimmo
University of California, Santa Cruz

A. P. Ingersoll
California Institute of Technology

T. A. Hurford
NASA Goddard Space Flight Center

E. S. Kite
University of Chicago

A. R. Rhoden
Southwest Research Institute

J. Schmidt
University of Oulu

C. J. A. Howett
Southwest Research Institute

The plume of Enceladus provides a unique window into subsurface processes in the ice shell and ocean of an icy world. Thanks to a decade of observations and modeling, a coherent picture is emerging of a thin ice shell extending across the south polar region, cut through by fractures directly connected to the underlying ocean, and at least partially filled with water. The plume jets emerging from the fractures directly sample this water reservoir. The shell undergoes daily tidal flexing, which modulates plume activity by opening and closing the fractures. Dissipation in the ice and conduit water components due to this flexing is likely to generate the several gigawatts of observed power that are lost from the south pole as infrared radiation and plume latent heat.

1. INTRODUCTION

Since the discovery of activity at the south pole of Enceladus, enormous progress has been made in interpreting the rich data from the multiple instruments on the Cassini spacecraft to understand what is happening below the visible surface. In this chapter, we review the visible and thermal observations of the surface expression of the plume fractures. We then review the current state of understanding of the processes that connect the surface to the underlying ocean, constrained by these observations and by the properties of the plume itself, which are described in more detail in the chapter in this volume by Goldstein et al.

2. SURFACE OBSERVATIONAL CONSTRAINTS

2.1. Plume Source and Jet Morphology

The plume emanates from four prominent parallel fractures, informally named the "tiger stripes," and associated branches. These fractures are approximately 130 km long and spaced 35 km apart, aligned approximately 30° westward from the direction to Saturn. The major tiger stripe fractures have raised margins enclosing a central trough approximately 2 km wide and 0.5 km deep, although some plume jets and thermal emission emanate from simple open fractures without raised margins or complex structure. These troughs are the source of both the thermal emission (*Spencer et al.,* 2006; *Spencer and Nimmo,* 2013) and the plume jets (*Spitale and Porco,* 2007; *Porco et al.,* 2014; *Spitale et al.,* 2015; *Helfenstein and Porco,* 2015). The fractures themselves are unresolved at the ~10-m/pixel resolution of the best Cassini ISS images, appearing only as dark linear features at the bottom of narrow trenches within the tiger stripes. In some locations there are multiple trenches, separated by medial ridges, within the tiger stripe trough (*Helfenstein and Porco,* 2015) (see also Fig. 1f). The ice-particle plume consists of a combination of continuous curtains (*Spitale et al.,* 2015) and at least 100 discrete jets (*Porco et al.,* 2014). The source locations on the surface of the discrete jets are not in general obvious in the available ISS images of the tiger stripes, although some appear to be associated with radial grooves or cross-cutting fractures (*Helfenstein and Porco,* 2015). Individual jets are time-variable in ways that are not obviously related to the tidal modulation of the plume as a whole (*Porco et al.,* 2014). Discrete jets of water vapor are also seen in the UV stellar occultation data (*Hansen et al.,* 2011, 2017).

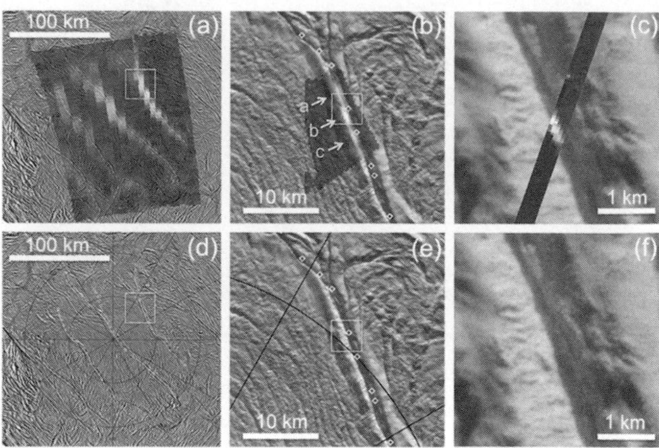

Fig. 1. See Plate 19 for color version. The appearance of the tiger stripes in the visible (Cassini ISS) and thermal infrared [Cassini CIRS, 7–16 μm, **(a)–(c)**] at a variety of scales. **(a)** The entire tiger stripe system with the exception of part of Alexandria Sulcus (left), which is known from other observations to be relatively faint. CIRS 9–16-μm data from March 2008. Dashed lines mark the location of the active tiger stripes. The square shows the location of **(b)**. **(b)** Closeup of the most active part of Damascus Sulcus (a combination of 9–16-μm and 7–9-μm images), mapped by CIRS in August 2010. Diamonds show the location of ISS plume jets reported by *Helfenstein and Porco* (2015). Arrows mark the locations of three discrete hot spots, a, b, and c (also shown in Fig. 2a), two of which correspond well to plume jet locations. The square shows the location of **(c)**. **(c)** The highest-resolution CIRS image of Damascus Sulcus emission, taken by direct sampling of the CIRS 7–9-μm interferograms from October 2015, showing dominant emission from a single fissure <100 m wide, and isolated hot spots elsewhere within the central trough (*Gorius et al.,* 2015). Placement relative to the terrain is approximate. **(d)–(f)** The same images without the CIRS data overlay, to show the terrain more clearly.

2.2. Spatial Distribution of Thermal Emission

At 10 km resolution thermal emission is present along the entire length of the tiger stripes where plume activity is seen (*Howett et al.,* 2011) (see also Fig. 1). One striking example is a branch of Baghdad Sulcus that is much narrower and less conspicuous in ISS images than some of its inactive neighbors, but is a site of both thermal emission and plume activity (points 75–80 in Fig. 3). However, emission strength varies greatly along the fractures, being particularly strong toward the Saturn-facing ends of Damascus and Baghdad Sulci and to a lesser extent at the anti-Saturn end of Alexandria Sulcus, where some fractures perpendicular to the main tiger stripes are also active. At higher spatial resolution, larger local spatial variations in emission strength are seen. Figure 1 shows the appearance of the tiger stripes at visible and thermal wavelengths at a variety of scales. On 1-km scales, thermal emission shows peaks that, in the few places where data of sufficient quality exist, correspond in some locations (but not all) to discrete plume jets seen by ISS (Fig. 1b, Fig. 2a). Similar discrete peaks of thermal emission, corresponding to plume jet locations, are seen by Cassini's VIMS instrument at 5 μm (*Goguen et al.,* 2013, 2016). Excess 2.2-cm thermal emission has also been reported via passive radiometry and radar-derived emissivities from the Cassini RADAR instrument near 60°S, 240°W and 65°S, 340°W, just north of the tiger stripe region (*Le Gall et al.,* 2017).

2.3. Tiger Stripe Temperatures

Peak temperatures along the tiger stripes have been measured both by CIRS and VIMS by fitting blackbody curves to tiger stripe emission spectra, with similar results. The width of the emitting region can also be derived by assuming a linear source parallel to the fractures, which varies greatly from place to place. *Goguen et al.* (2013) obtained a best-fit isothermal temperature of 197 ± 20 K , with a best-fit width of 9 m, from 4–5-μm spectra of one location on Baghdad Sulcus (Fig. 2c), and *Spencer et al.* (2011) obtained a temperature of 176.7 ± 1.3 K, with a best-fit width of 147 m, from 7–9-μm spectra of the brightest point on Damascus Sulcus (Fig. 2b). The fact that longer wavelengths tend to produce fits with lower temperatures (e.g., the best fit of 167 ± 0.7 K to the full 600–1400-cm^{-1} wavelength range in Fig. 2b) suggests that the surface near the tiger stripes has (not surprisingly) a range of temperatures. The fact that temperatures higher than 200 K have not been seen, despite evidence for liquid water at depth, is plausibly due to rapid sublimation and evaporative cooling of the water ice surface at the vents (*Goguen et al.,* 2013), which would tend to buffer water ice surface temperatures due to the strong dependence of sublimation rate on temperature. In theory, lag deposits or other impermeable coatings of non-volatile materials, such as salts, could supress sublimation cooling and allow higher temperatures; the lack of observed higher temperatures may indicate that such coatings are uncommon.

The spatial distribution of emission as a function of distance from the vents has been compared to simple models of conduction away from a warm, isothermal, vertical fracture by *Abramov and Spencer* (2009). The intensity and wavelength distribution of the thermal emission spectrum of the warmest part of Damascus, taken in August 2010, can be matched quite well with such models, assuming a fracture temperature of 225 K, except that up to six parallel fractures within the ~1-km width of the CIRS footprint are required (Fig. 2b). Forty-meter resolution imaging of a nearby region in October 2015 (Fig. 1c) shows extended thermal emission consistent with the presence of multiple active fractures, but emission is dominated by a single fracture. This region may not correspond precisely to the region of the spectrum in Fig. 2b, or emission may be time-variable.

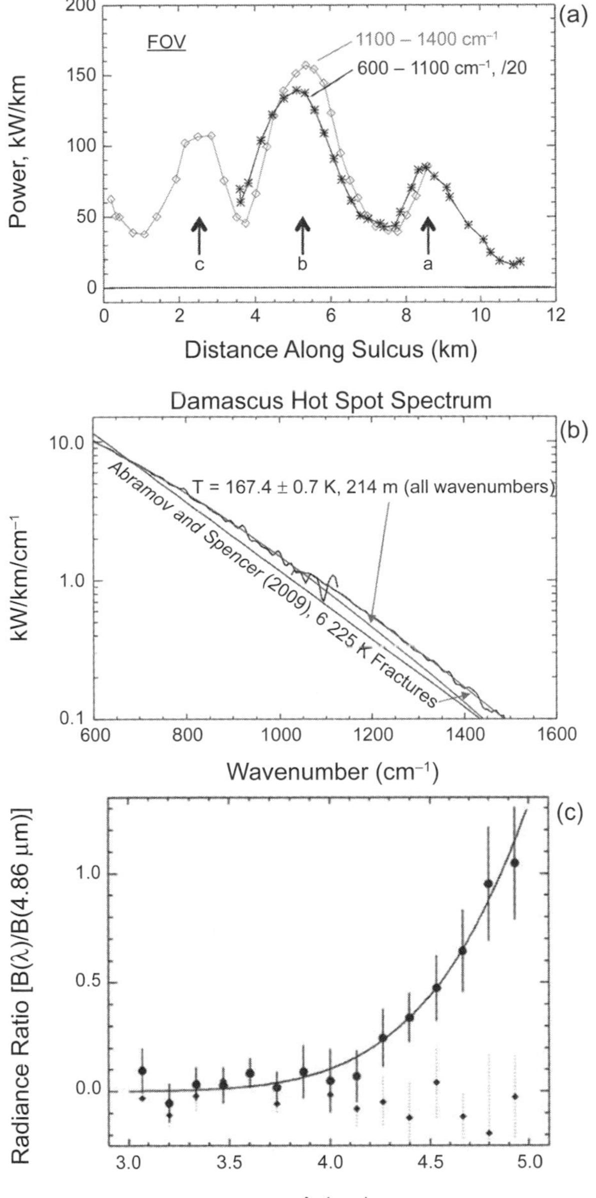

Fig. 2. (a) Power profile along Damascus sulcus in two CIRS wavebands, showing peaks associated (in two cases) with ISS jets, from the data shown in Fig. 1b. Arrows identify the same three discrete hot spots, a, b, and c. **(b)** CIRS spectrum of the brightest part of Damascus Sulcus [peak near 5 km (*Spencer et al.,* 2011), compared to blackbody fits and the model of *Abramov and Spencer* (2009), multiplied sixfold to approximate the signature of six adjacent fractures. **(c)** VIMS spectrum of Baghdad Sulcus, compared to blackbody fit with temperature 197 K (*Goguen et al.,* 2013). The diamonds show the mean of background measurements obtained on either side of the Sulcus.

The thermal observations discussed above are at wavelengths shorter than 16.7 μm (600 cm^{-1}), where both CIRS and VIMS have relatively high spatial resolution. However, most of the thermal emission from the fractures is radiated at longer wavelengths. Understanding this long-wavelength emission is important to determine both the lower-temperature emission from the stripes and their total heat flow. Unfortunately, the long-wavelength (10–600 cm^{-1}) detector of the CIRS instrument has much lower spatial resolution than the short-wavelength detectors, and can resolve the tiger stripes only for brief intervals during the closest Cassini flybys. Models are thus required to extrapolate these limited long-wavelength resolved observations to the entire tiger stripe system. *Spencer et al.* (2013) found that the available high-resolution 10–600-cm^{-1} observations could be matched by adding a low-temperature emission component, with a temperature of 80 K, to the higher-temperature emission inferred at shorter wavelengths. Adjusting the local width of this low-temperature component to match the observations, and extrapolating to the rest of the tiger stripe system by assuming a constant 80 K temperature, and assuming that the spatial variation in the radiated power for the long-wavelength emission was similar to that of the well-mapped short-wavelength emission (Fig. 1a), *Spencer et al.* (2013) derived a complete model of observed emission from the tiger stripes, updated slightly here, that is consistent with, although certainly not uniquely constrained by, the available data (Fig. 3).

2.4. Total Heat Flow

Total heat flow from the tiger stripe system, an important geophysical parameter, has been difficult to determine precisely. Blackbody fits to the first 600–1100 cm^{-1} integrated CIRS spectra of the entire tiger stripe system in 2005 produced a heat flow of 5.8 ± 1.9 GW (*Spencer et al.,* 2006). Subsequent integrated 10–600-cm^{-1} observations of the south polar terrain (SPT) indicated much higher heat flow (15.8 ± 3.1 GW), but required model-dependent subtraction of the "passive" contribution from re-radiated sunlight (*Howett et al.,* 2011). Because it is highly likely that the emission associated with the tiger stripes fractures themselves is endogenic, fracture emission provides a lower limit to the total endogenic component (this is a lower limit because there may also be endogenic emission from between the tiger stripes, either from smaller unresolved fractures or from conduction through the ice shell that raises the background temperature). The model of *Spencer et al.* (2013), described above (Fig. 3), produces a tiger stripe radiated heat flow of 4.2 GW. Note that this number includes a passive component due to solar heating of the radiating regions [the so-called "thermal pedestal effect" (*Veeder et al.* 1994)], which when subtracted might reduce the total by a few tenths of a gigawatt. To this must be added the latent heat content of the water vapor plume [~0.5 GW (*Ingersoll and Pankine*, 2010)], and any additional

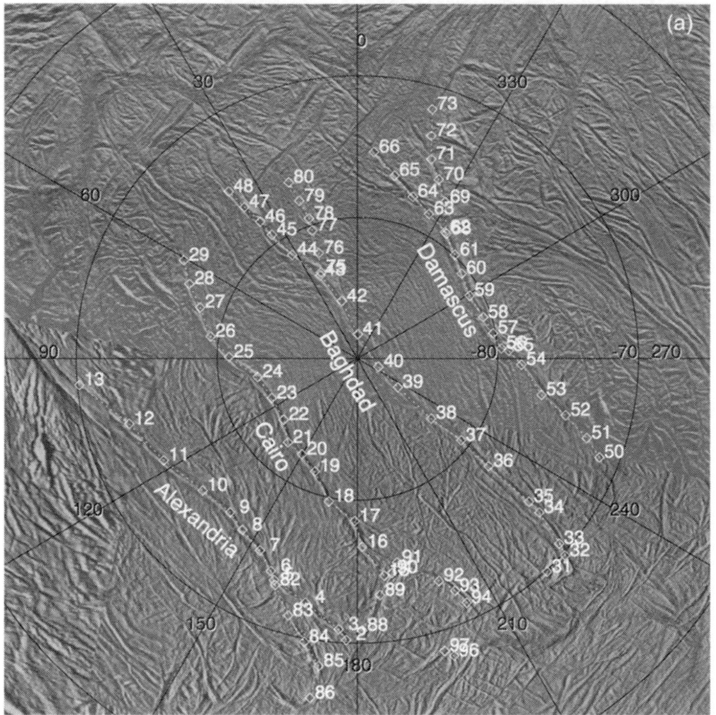

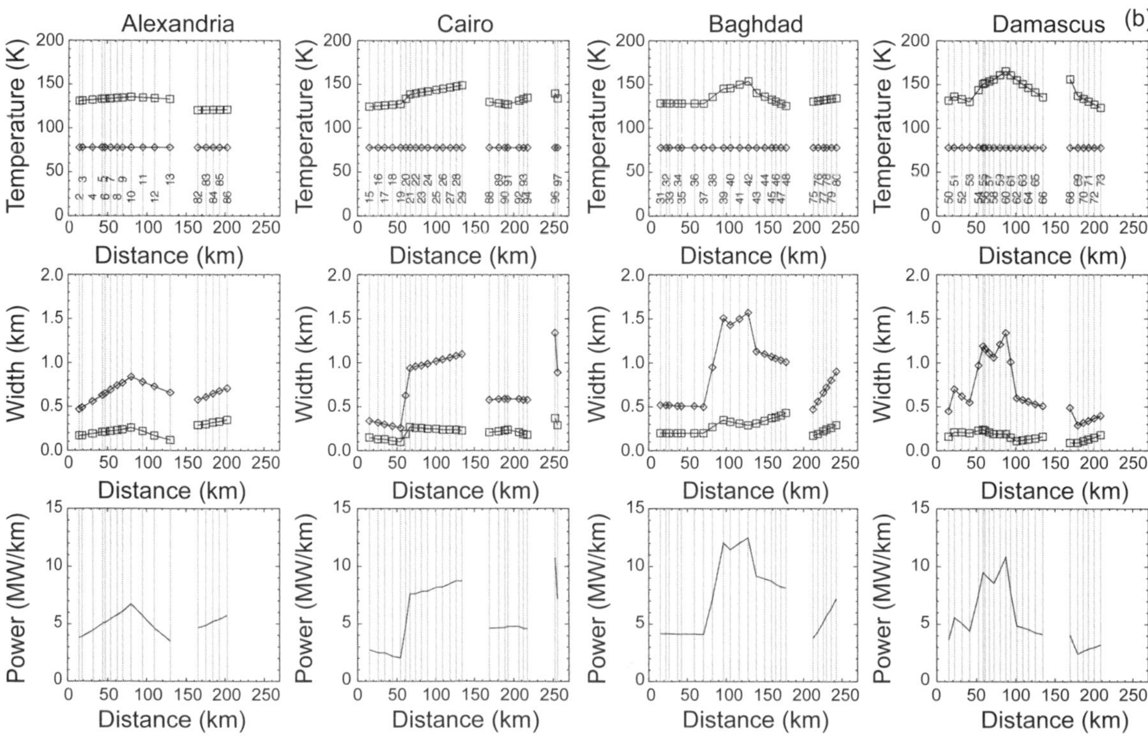

Fig. 3. Model of thermal emission from the tiger stripes, matched to multiple CIRS datasets, updated from *Spencer et al.* (2013). **(a)** Locations of the warm material along the fractures. Numbered points correspond locations in the model where the model parameters are defined, and correspond to numbered vertical lines in **(b)**. **(b)** Model temperatures, emission widths, and power per unit stripe length at each of those points along the four named tiger stripes and their branches. Inflection points in the model parameters are locations where the model parameters were fit to the multiple datasets; intermediate points are linearly interpolated. This model produces a total radiated heat flow of 4.2 GW.

broadly distributed endogenic radiation from the surface between and surrounding the tiger stripes. The discrepancy between the >13-GW estimates for the total heat flow and the ~4.2-GW tiger stripe heat flow suggests the existence of this broad component, but it is difficult to measure. For instance, the heat conducted up from the ocean through 2 km of ice with thermal conductivity 3.5 W m^{-1} K^{-1} would raise the temperature of the surface from 73 K to 77 K, which would be difficult to detect without a well-constrained model of passive thermal emission. If this heat flux were constant from the pole to 75°S latitude, the total endogenic power would be an additional ~4.7 GW.

3. THE NATURE OF THE PLUME SOURCE

Figure 4 summarizes our current picture of the subsurface configuration of the tiger stripe fractures, and the principle processes operating there. These are discussed in more detail below.

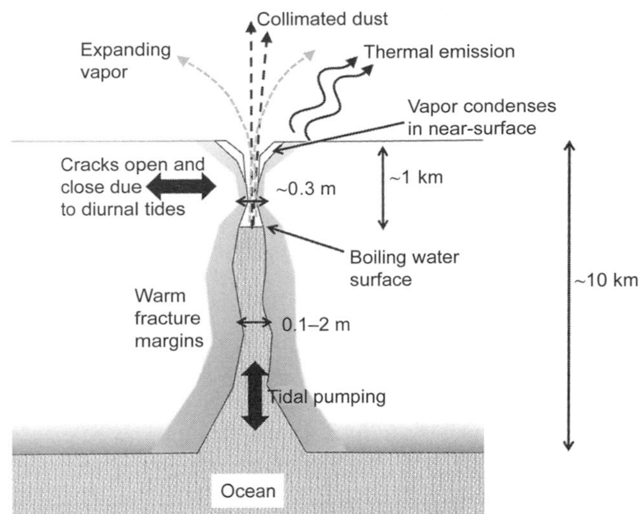

Fig. 4. Schematic of the major processes, discussed in detail in this chapter, thought to be taking place in the plumbing connecting the surface tiger stripes to the underlying ocean.

3.1. Is the Plume Source Solid or Liquid?

The detection of plumes emanating from fractures in the SPT of Enceladus (*Dougherty et al.,* 2006; *Porco et al.,* 2006; *Spencer et al.,* 2006; *Spitale and Porco,* 2007) led to multiple physical models of the eruptions. The earliest models invoked either a rapidly boiling, near-surface liquid water reservoir (*Porco et al.,* 2006), degassing of clathrates (*Kieffer et al.,* 2006), or direct sublimation from warm ice (*Nimmo et al.,* 2007). Over time, measurements of the composition, structure, and density of the plumes have favored a liquid water source, as described in detail below.

Enceladus' plumes are composed of solid particles that are mainly water ice, along with a gaseous component that is dominated by water vapor (see the chapter in this volume by Goldstein et al.). Minor components of the plume gases include CO_2, NH_3, CH_4, H_2, heavier hydrocarbons, and other species (*Waite et al.,* 2009; *Hansen et al.,* 2011; *Waite et al.,* 2017). These minor gas species are suggestive of hydrothermal cycling between the rocky interior and cryosphere (e.g., *Matson et al.,* 2007; *Glein et al.,* 2008; see the chapters in this volume by Glein et al. and Postberg et al.).

The presence of sodium and potassium salts in the plume and E-ring particles (*Postberg et al.,* 2009), especially in larger, lower-speed particles found preferentially near the plume source (*Postberg et al.,* 2011), is strong evidence of a liquid source. If the particles were formed by condensation from the vapor, which had sublimated from ice, they would be practically salt-free. Hence, salt-bearing plume particles would not be expected in "dry" models (e.g., *Kieffer et al.,* 2006; *Loeffler et al.,* 2006; *Nimmo et al.,* 2007; *Cooper et al.,* 2009). Liquid water interacting with silicates will result in dissolved salts; as the water freezes slowly, the salts become progressively more concentrated in the liquid, while the ice remains salt-free. Frozen droplets derived from the liquid will retain the salt content. The salty grains imply direct connections between the liquid ocean and the vacuum of space.

Cassini's Cosmic Dust Analyzer (CDA) also identified nanometer-sized grains rich in silicon (*Hsu et al.,* 2015) that are most likely SiO_2. Their size and composition suggest formation in a hydrothermal system within Enceladus, further supporting a liquid water source for the plumes that is in contact with rock (*Glein et al.,* 2008; *Hsu et al.,* 2015). Hydrothermal SiO_2 particles will grow to micrometer-sized, larger than is observed, over thousands of years, which may be evidence that the production of nanosilica grains is very recent or even ongoing. The nanosilica particles further constrain the deep ocean conditions to moderate salinity, an alkaline pH, and relatively warm temperature (*Hsu et al.,* 2015; *Sekine et al.,* 2015). However, the exact values depend on many assumptions, including the composition of Enceladus' rocky core and whether the system is chemically open or closed (*Sekine et al.,* 2015; *Glein et al.,* 2015).

The total particulate mass within the plumes and the ratio of ice to vapor both provide clues as to the source of the plume material. Early attempts to constrain the relative abundances of ice to vapor resulted in quite disparate values, with *Porco et al.* (2006) reporting a high abundance that favored a liquid source and *Kieffer et al.* (2009) finding a low value that suggested a sublimation-driven process of plume formation. Estimates using high-phase-angle Cassini images suggested a large solid/gas ratio of 0.35–0.7 if particles are spherical (*Ingersoll and Ewald,* 2011), more consistent with an evaporating liquid source, but much lower ratios, 0.07 ± 0.01, if most particles are aggregates (*Gao et al.,* 2016). As discussed below, this picture may be complicated by the presence of different eruptive sources, with different solid/gas ratios. For instance, there may be a "background" component of vapor flux with very little solid material (*Postberg et al.,* 2009, 2011). Nonetheless, the compositional measurements suggest that, overall, the liquid source is dominant.

3.2. Plume Plumbing and Source

3.2.1. The plumbing system problem. The plumes, and thus their plumbing system, appear to be long-lived phenomena (see the chapter by Goldstein et al. in this volume). In order to reach the surface as a plume of ice grains and vapor, Enceladus ocean water faces an uphill struggle. First, ocean-to-surface conduits must penetrate the ice shell; second, the conduits must be kept open in spite of ice creep; third, water must flow through the shell without freezing; and fourth, water must overcome its greater density relative to the surrounding ice. These challenges define the plumbing-system problem for eruptions on Enceladus.

Ocean-to-surface conduits thread through the cold near-surface layer of the ice shell, where ice creep is negligible but conduits may seal off by freezing shut. Conduits also pass through the warmer near-ocean layer, where conduits are vulnerable to closure by ice creep. Near the surface, magmatic eruptions are the best terrestrial analogs (e.g., *Rubin* 1995). Near the ocean, englacial channels are the best terrestrial analogs (*Cuffey and Patterson,* 2010). Ice shell thickness (D) at the SPT is <45 km but is more likely <10 km (*Iess et al.,* 2014; *Čadek et al.,* 2016; Hemingway et al., this volume). By simple buoyancy arguments, the time-averaged water table depth for a non-overpressured, volatile-free ocean is a distance 0.1 D below the surface. Ocean overpressure, diurnal tides, or siphoning by bubble exsolution, could all spill water onto the surface, although there is no evidence from geomorphology or IR data, where observed temperatures peak at ~200 K as discussed above, for liquid water reaching the surface today.

Forming an ocean-to-surface conduit from a crack that initiates at the surface would require tensile stresses that exceed overburden pressure (*Crawford and Stevenson,* 1988). Therefore, if the only source of stress is diurnal eccentricity-tide stresses of ~10^5 Pa (*Nimmo et al.,* 2007), surface-initiated cracks on Enceladus have zero width below a depth of 1 km. Stresses $\gg 10^5$ Pa are possible with non-synchronous rotation, polar wander or ice shell freezing, but these stresses develop over long timescales and so will relax away in the ductile, lower part of the shell (*Crawford and Stevenson,* 1988). Despite this relaxation effect, fractures that initiate at the surface can propagate under these larger stresses to the ocean for D < 25 km and ice tensile strength 3 MPa (*Rudolph and Manga,* 2009). Water-filled cracks that initiate at the ocean can puncture the shell more readily than cracks that initiate at the surface, and ice can fracture even close to its melting point (*Schulson,* 2001). That is because the crack-closing stress is the pressure-head difference between the ice column and the water column, which is more favorable than the pressure-head difference between ice and vacuum (*Crawford and Stevenson,* 1988; *Matson et al.,* 2012). Gas exsolution can contribute to the buoyancy of the rising water column (*Crawford and Stevenson,* 1988). Furthermore, the ocean may be overpressured (*Manga and Wang,* 2007), favoring crack propagation. If D $\leq$ 10 km, conduits could form at the current orbital eccentricity. Alternatively, conduits may have formed during a higher-eccentricity period in Enceladus' past, when stresses were higher.

A pre-existing ocean to surface conduit can be sealed by ice creep, driven by the pressure difference, ΔP, between the water column and the surrounding lithostatic pressure. For a cylindrical pipe of radius R, the pipe-closing rate is (*Nye,* 1953; *Cuffey and Patterson,* 2010)

$$\frac{1}{R}\frac{\partial R}{\partial t} = A\left[\frac{\Delta P}{n}\right]^n \quad (1)$$

with n = 3 (Glen's law) and A ~ 2 × 10^{-24} s^{-1} Pa^{-3} for ice at the melting point (*Cuffey and Patterson,* 2010). The maximum value of ΔP is (D/10 km) × 10^5 Pa if the ocean is not overpressured. Melt-back by turbulent dissipation can balance inflow and lead to a steady-state conduit aperture, but only if the water flows swiftly (*Röthlisberger,* 1972; *Weertman,* 1972; *Kite and Rubin,* 2016). At steady state, the power consumed in meltback (per unit length pipe) is $2\pi R(\partial R/\partial t)\rho_i L$ (*Cuffey and Patterson,* 2010), where ρ_i is the density of ice and L is the latent heat of melting. These models assume that water is at the local pressure-melting temperature of ice, and that frictional heat is absorbed locally and instantaneously (*Cuffey and Patterson,* 2010). Currents driven by flushing of the fractures in response to tidal flexing, given by 2π(D)/(1.37 days), where 1.37 days is the orbital period of Enceladus, are roughly 1 m s^{-1} for D = 20 km, which is sufficient to keep the fractures from closing (*Kite and Rubin,* 2016).

If the ice shell is decreasing in volume globally over sufficiently fast timescales, the ocean will be underpressured. However, if the ocean is freezing, then it will be overpressured, to the extent that the ice shell can contain the pressure: The maximum ocean overpressure is the tensile strength of ice, 1–3 MPa (*Rudolph and Manga,* 2009). If the rock core is rigid and impermeable, overpressure of >10^6 Pa can be generated (*Manga and Wang,* 2007). Ocean overpressure/underpressure is important for conduit initiation and conduit stability. For ocean overpressure > (D/10 km) × 10^5 Pa, conduits will expand by shouldering aside adjacent warm ice, and water can reach the surface even without gas exsolution.

3.2.2. Geometry of the plumbing system. Enceladus' plumbing system could consist of focused flow in slots or channels, distributed porous flow, or something in between. A narrow slot undergoing unidirectional flow is unstable to the formation of discrete pipes (*Walder,* 1982), but horizontal water transport during tidally pumped flow might suppress this instability (*Kite and Rubin,* 2016). Pipes may be unstable to elongation perpendicular to the flow direction (*Dallaston and Hewitt,* 2014). It is difficult for pipes or slots to pinch off at a particular depth because any turbulent dissipation (and thus melt-back) would be focused at the pinch-point. The presence of localized dust jets (*Porco et al.,* 2014), localized gas jets (*Hansen et al.,* 2017; *Yeoh et al.,* 2017), and localized ~100-m-scale hot spots [some of which correlate with apparent jets in visible images; see *Goguen et al.* (2016) and section 2.2 above] suggest that pipes within the fissure system may be supplying a significant fraction of the overall heat and mass flux, at least above the water table, where partial blocking by vapor condensation (considered later) may play a role (*Nakajima and Ingersoll,* 2016).

Closed plumbing systems reaching to just below the surface have also been proposed (*Matson et al.,* 2012). In this model, sensible heat is transferred to the ice and conducted to the surface where it is radiated to space. Bubbles of exsolved CO_2 provide buoyancy on the way up from the ocean, a 2° temperature difference between the warm ocean and the freezing salt water provides the sensible heat, and the gas loss provides the density gain that causes the liquid to sink back to the ocean. The details remain uncertain, but it is clear that a lot of water is involved. The sensible heat associated with a 2° change in temperature is only about 0.3% of the latent heat associated with vaporization. Also, the closed system might be unstable: The pipes might burst and the system could revert to a vapor deposition system.

3.2.3. Sustainability. Models of the Enceladus plumbing system usually assume that it is long-lived. Enceladus' eruptions have been ongoing since at least 2005 (*Hedman et al.,* 2013; *Porco et al.,* 2014), although individual jets do appear to turn on and off (*Nimmo et al.,* 2014) and activity has decreased over the course of the Cassini observations (*Ingersoll and Ewald,* 2017). The persistence of the E ring, whose grains have a lifetime of decades against sputtering (*Jurac et al.,* 2001), suggests that the current eruption rate is similar to the average over the last ~100 years. The widths of the warm belts bracketing the tiger stripes is ~1 km, as described above. If these belts are kept warm by conductive heating from central fissures, then thermal timescales require the fissures to have been warm for ~10^4 yr . Finally, complete removal of craters from the SPT probably requires $\gg 10^3$ yr of activity. Taken together, the data suggest that the eruptions are sustained but cannot rule out system-wide periods of repose. In either event, the energy transfer from the ocean to the ice shell is so large that it would have a major effect on ice-shell thermal structure.

3.3. Tidal Dissipation and Flexing in the Fracture System

3.3.1. Energy sources. While the distribution of tidal heating within Enceladus is not well established, it is only directly detectable in the south polar region, and is likely to be focused there. Enceladus' internal ocean enhances tidal dissipation by increasing Enceladus' response to gravitation tidal forces. A locally thinner and/or weaker shell may further concentrate heating in the SPT (*Souček et al.,* 2016; *Běhounková et al.,* 2015). Solid ice near its melting point has a viscoelastic response (Maxwell) time comparable to the orbital period of Enceladus, which makes it especially susceptible to tidal heating. This heating will be most effective near the base of the ice shell, where viscosity is lowest.

For some geometries, dissipation of tidally driven flow in the plumbing system could produce power output equal to the observed power output (section 2.4) (see also *Kite and Rubin,* 2016). This is analogous to tidal dissipation on Earth, which happens mostly in shallow seas and narrow channels rather than the open ocean or the solid Earth (*Egbert and Ray,* 2003). The generated power staves off crack freeze-out and helps maintain the ocean against freezing. *Kite and Rubin*'s (2016) model gives a width of 2 m for the water-filled portion of the cracks: In this model a crack with a full width less than 1 m freezes shut. This width is significantly greater than the 0.1-m estimate for cracks in the top tens of meters below the surface derived from the ratio of latent and radiated heat (section 4.2) (see also *Ingersoll and Pankine,* 2010; *Nakajima and Ingersoll,* 2016), suggesting that the cracks may narrow toward the surface.

3.3.2. Tidal flexing. The presence of an ocean enhances tidal flexing and generates large tidal stresses on Enceladus' surface. The tiger stripes are approximately aligned with one of the two directions of maximum tidal tensile stresses (*Nimmo et al.,* 2007), suggesting that they may have formed in response to tidal stress. Once formed, these stresses can rework the fractures and control the eruptive activity observed from them.

The strongest direct evidence for tidal control of the eruptive behavior comes from the fact that, as predicted by tidal flexing models (*Hurford et al.,* 2007), the overall plume is brighter at apoapse than at periapse, as seen both by VIMS (*Hedman et al.,* 2013) and ISS (*Nimmo et al.,* 2014; *Ingersoll and Ewald,* 2017) (Fig. 5). This response is presumably because the opening and closing of cracks modulates the solid mass flux. Plume activity does not cease

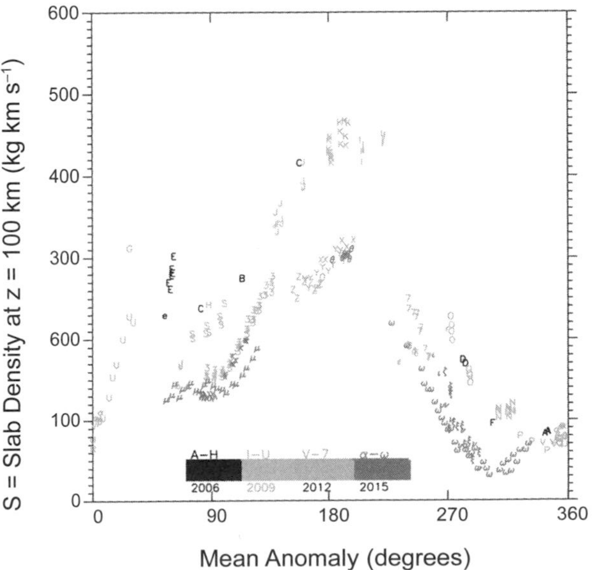

Fig. 5. See Plate 20 for color version. Orbital and long-term variability of the Enceladus dust plume brightness at visible wavelengths, from *Ingersoll and Ewald* (2017). Different symbols identify different datasets. Mean anomaly describes the orbital position of Enceladus relative to periapse, while the Y axis is a measure of integrated plume brightness at 100 km altitude.

even at periapse, which must indicate either that the cracks do not close entirely (thus placing a lower bound on crack width, if it is uniform), or — less likely — that part of the (solid) plume material emanates from between the tiger stripes. *Nimmo et al.* (2014) conclude that the velocity of the solid material does not appear to change significantly over the orbital period, while *Hedman et al.* (2013) and *Ingersoll and Ewald* (2017) infer a reduction of launch speed around periapse. In contrast, UVIS observations of erupted gas, while sparser, suggest only very small variations in flux with orbital position, but also suggest that the gas velocities may be higher at apoapse (*Hansen et al.,* 2017). Thus, the details of tidal control on eruptive behavior are not yet clear. A further intriguing observation is that the diurnal peak in the eruptive plume lags the predicted peak response of an elastic shell, which is just before apocenter (*Hurford et al.,* 2007, 2012) by several hours (*Nimmo et al.,* 2014; *Ingersoll and Ewald,* 2017). This could be because the shell is responding in a viscoelastic fashion (*Běhounková et al.,* 2015), or perhaps because of the response of water-filled cracks (*Kite and Rubin,* 2016).

On a decadal timescale, the plume appears to be growing fainter (*Hedman et al.,* 2013; *Ingersoll and Ewald,* 2017) (Fig. 5), which suggests a decreasing solid mass flux. This could be due to some change in the plumbing, but may also be an effect of subtle 11-year periodic variations in orbital parameters (*Horanyi et al.,* 1992; *Vienne and Duriez,* 1992) or (less likely) seasonal insolation-driven changes in the near-surface temperature. To date, no such long-period variation is evident in the vapor flux estimates (*Hansen et al.,* 2017).

3.3.3. Energy balance within the fractures. Given the apparently low shell thickness at the SPT (*Čadek et al.,* 2016), the possible broad regional heat flow between the tiger stripes, hinted at in the CIRS data (section 2.4), is probably transferred to the surface via conduction. However, the highly localized heat fluxes along the tiger stripes indicate that there must be additional processes operating. An isothermal crack at temperatures above ~200 K will transfer heat to the surroundings, generating a local thermal anomaly similar to that observed (*Abramov and Spencer,* 2009). Transfer of heat from the water-filled or vapor-filled fracture to the surroundings occurs primarily by deposition of latent heat (*Ingersoll and Pankine,* 2010). This is because, as heat flows away by conduction from the ice fracture walls to balance radiation at the surface, any reduction in wall temperature will immediately result in condensation or freezing of water onto the walls to maintain isothermality. In the absence of competing processes, such as local dissipation and melt-back, this deposition inexorably narrows the fractures. Below the water surface, the heat comes from the latent heat of fusion as water freezes onto the fracture walls. *Ingersoll and Nakajima* (2016) parameterize the heat conduction using the formula $F = 4k_i\Delta T/(\pi z)$, where F is the heat flux into the wall, k_i is the thermal conductivity of ice, ΔT is the temperature difference between the warm wall and the cold surface, and z is the depth. The formula is valid for an isothermal crack and an isothermal surface. The upward speed of the water column is assumed to match that required to balance condensation of vapor near the surface, determined by the total radiated power (assumed in this paper to be ~4.7 GW) divided by the 500-km length of the tiger stripes and by the latent heat of vaporization. The result is that the water loses slightly more than half its mass in going from the ocean to the evaporating surface. It also means the build-up of ice is closing the crack at a rate of 5.4 cm yr^{-1}, about 10% of the rate above the water surface (section 4.2). The flushing that accompanies the opening and closing of the cracks will counteract this freezing (*Kite and Rubin,* 2016).

4. PHYSICS OF THE PLUME VENTS AT AND ABOVE THE PRESUMED LIQUID INTERFACE

4.1. The Water/Vapor Interface

Much of the important physics in the plume plumbing occurs at the water surface within the fractures. At least some of the plume particles are probably created here, giving us direct samples from this location, and the fact that this interface appears to resist freezing over, at least on short timescales, provides key constraints on plume physics.

Evaporative cooling will tend to rapidly freeze the water surface, which would shut down the eruptions. Ice forms when liquid water at 0°C evaporates, producing 7.5 g of ice for every 1 g of vapor, based on the heats of vaporization and fusion. Fresh water has a density maximum at 4.3°C, so the freezing water accumulates at the surface — it is not replaced by warmer water from below (water with salinity greater than 24.7 g kg^{-1} does not have this problem — the densest water is at the freezing point). Several solutions to the evaporative cooling problem have been proposed. Each involves exchange with a deeper thermal bath, such as the ocean:

1. If the surface area that is evaporating is much larger than the vent cross-sectional area, then the evaporative power (W m^{-2} of evaporating area) is modest due to the vapor pressure above the water surface, and can be balanced by convective exchange with a deeper reservoir (*Postberg et al.,* 2009). This requires large vapor chambers above the liquid that narrow to the vent channels. However, *Ingersoll and Nakajima* (2016) and *Nakajima and Ingersoll* (2016) point out that even if the walls are parallel, friction with the walls of a long, narrow channel produces a backpressure that is almost as large as the saturation vapor pressure of the liquid. A stable equilibrium develops — too much flow causes the backpressure to increase, and that reduces the evaporation rate. Too little flow and the backpressure will decrease, causing faster evaporation.

2. Evaporative cooling at the top of a constant-width slot can be balanced by convective overturn, if the water has salinity >16.2 g kg^{-1} and is stirred near the water table by exsolving bubbles (*Ingersoll and Nakajima,* 2016). In this model, evaporation of salty water leaves colder, saltier, and thus denser water behind, which sinks and brings warmer, fresher water to the surface. Since 20 g kg^{-1} is the upper limit

of the measured range (*Postberg et al.,* 2009), *Ingersoll and Nakajima* (2016) concluded that narrow conduits filled with salty liquid could support the plumes observed at the surface.

3. Dissipation of kinetic energy due to tidal pumping will delay and may prevent icing over (*Kite and Rubin,* 2016).

4. Siphoning driven by the exsolution of exsolved gases could bring warm gas-charged fluid from the ocean to the water table, with cold degassed water sinking back to the ocean (*Matson et al.,* 2012).

Approaches 1–3 above share the requirement that conduit width below the water table is ≥0.1 m; the results of *Nakajima and Ingersoll* (2016) suggest that the slot width above the water must be less than that below. Approaches 2–4 share the feature that water circulates between the ice shell and the ocean. If water that has changed its temperature and composition during its passage through the ice shell returns to the ocean, this will affect ocean-top temperature and composition (*Melosh et al.,* 2004).

As discussed in section 3.1, the characteristics of the plumes' particles and gases are clues to the physical and chemical processes at the ocean-gas interface. Early estimates of a high solids-to-vapor ratio in the plume led to comparisons to a Yellowstone geyser, where a bubbly liquid erupts into a low-pressure environment (*Porco et al.,* 2006). The low solubility in water of methane and other alkanes led to theories of explosive dissociation of clathrate hydrates containing gases physically trapped in the lattice (*Kieffer et al.,* 2006; *Gioia et al.,* 2007). The geyser model suffers because the particles don't go high enough (*Brilliantov et al.,* 2008), and is also ruled out by the lack of the sodium vapor that would be expected from explosive evaporation of water containing dissolved sodium salts (*Schneider et al.,* 2009; *Postberg et al.,* 2011). The clathrate model suffers because water vapor is a minor component of clathrate decomposition — a substantial part of the water vapor in the plumes would come from sublimation of ice grains in a flow driven by the volatile-entrapped gases, including CO_2, N_2, CO, and CH_4 (*Postberg et al.,* 2011). *Postberg et al.* (2009, 2011) argue that the liquid water reservoir is wide and close to the surface, such that droplets from the ocean spray can freeze and be ejected as plume particles. Additional models of the plume generation process are still being actively developed (e.g., *Matson et al.,* 2012; *Kite and Rubin,* 2016).

Micrometer-sized particles can be launched from the ocean and entrained into the plume gas by controlled boiling involving bubbles breaking at the surface, which generates spray (*Ingersoll and Nakajima,* 2016; *Porco et al.,* 2017). There are two mechanisms for launching droplets. Film droplets form when the thin upper surface of a bubble, protruding into the air, shatters into hundreds of particles with radii ranging between ~0.01 μm and 1–2 μm. Jet droplets form when the bubble cavity collapses; several droplets are formed in a vertical column with radii in the ~1–50-μm range (*de Leeuw et al.,* 2011; *Veron,* 2015). These sizes overlap with those of particles in the Enceladus plumes (*Hedman et al.,* 2009; *Ingersoll and Ewald,* 2011; *Postberg et al.,* 2011).

There is abundant evidence for plume jets from multiple sources. The CDA data imply three compositional types (*Postberg et al.,* 2011). Type I grains are almost pure water ice. Type II grains contain organic compounds and/or silicates. Type III grains are rich in sodium and potassium salts, are more massive, and are concentrated close to the surface, suggesting they are launched at speeds less than the escape velocity. UVIS and INMS data imply both distributed sources along the tiger stripes and narrow jets with Mach numbers >5 (*Yeoh et al.,* 2015, 2017). These patterns may imply that sources extending to the ocean have large particles rich in salt and leave the vent at slower speeds. Or it is possible that all the sources are the same and the large particles intrinsically have slower speeds? *Schmidt et al.* (2008) and *Postberg et al.* (2011) pursue the latter assumption. In their model, the particles are constantly having their speeds reset to zero by collisions with the walls of the conduits, which are assumed to have variable cross-section. The collision rate is the same for all particles, but the time to reaccelerate back to the speed of the gas is greater for the larger particles. The authors get a good fit to the particle size distribution and the preponderance of larger, slower, particles close to the surface with a distance between collisions of about 0.1 m. This is comparable to the widths of the conduits, according to some models (*Ingersoll and Pankine,* 2010; *Nakajima and Ingersoll,* 2016).

4.2. Escape to the Surface

There are two pathways by which heat generated inside Enceladus is lost to space (*Ingersoll and Pankine,* 2010). One is infrared radiation emitted at the surface, discussed above. The other is the latent heat of vaporization associated with the ~220 kg s^{-1} of water vapor escaping from the vents (*Hansen et al.,* 2011, 2017). Assuming a heat of vaporization of 2.5 × 10^6 J kg^{-1}, the latent heat power is 0.55 GW, about 13% of the tiger stripe infrared power of ~4.2 GW. The ratio of these two pathways provides a useful constraint on the width of the conduit. Assuming that the heat radiated to space comes from the freezing and condensation on the walls, the infrared power will not change with crack width but the latent heat associated with the escaping vapor will. Using a fluid dynamical model of vapor-filled cracks, *Ingersoll and Pankine* (2010) and *Nakajima and Ingersoll* (2016) derived a crack width of ~0.1 m extending over the 500-km cumulative length of the tiger stripes in order to match the observed ratio. A larger width would give a larger ratio of latent heat to infrared emission, and that was not observed. However, very close to the surface, the vents appear to flare, probably due to sublimation of the water ice: *Goguen et al.* (2013) estimate a width of 9 m for the fracture at a hot spot on Baghdad Sulcus.

The vapor-filled vent will self-seal rapidly due to deposition of latent heat as discussed above (*Ingersoll and Pankine,* 2010). Figure 3 shows that the hot component of the model fit to the Cassini CIRS data, which will preferentially be radiated close to the vent and thus from shallow depths, has a typical temperature of about 140 K and emission width of about 200 m (i.e., 100 m either side of the fractures, if symmetrical). If this radiated power were supplied by deposition of latent heat in the uppermost 100 m of the fracture, the deposition rate would be 0.3 m per year. If this were the only process, a crack 0.1 m wide would close shut in a few months. Condensation near the top is likely because thermal conduction through the ice to the surface is greatest there, and the walls are likely to be colder down to a depth that is comparable to the radiating strip. Ice flow at depth may contribute to keeping the near-surface fractures open, as discussed in section 5.1. Mechanical erosion due to differential tidal motion of the fracture walls may also play a role.

5. BROADER IMPLICATIONS

5.1. Tectonic Setting

We do not currently know how processes at the tiger stripes relate to the larger-scale tectonics of the south polar region (see the chapter in this volume by Patterson et al.). The overall energy budget of the SPT is strongly influenced by advection: ascent of warm water through cracks and shallow deposition of ice releasing latent heat. But there is also likely to be a significant background contribution from heat conducted across the ice shell. Such a regime would be intermediate between the tectonics of Earth (energy loss dominated by conduction) and the tectonics of Io (energy loss dominated by magmatic advection).

Feedbacks between cryovolcanism and tectonics are likely. For example, ice inflow into the base of the conduits, if sustained for $>10^6$ yr, would cause regional subsidence (*Kite and Rubin,* 2016). This subsidence may indirectly create accommodation space near the fractures, perhaps keeping them open in the face of ice condensation (Fig. 6). This model would predict negligible heat flow between the tiger stripes, due to the subsidence.

Intermittent tectonic resurfacing (*O'Neill and Nimmo,* 2010) could both cause and be caused by cryovolcanic regime shifts. If the tiger stripe thermal emission is carried by latent heat deposited by condensation on the walls of the tiger stripe (*Ingersoll and Pankine,* 2010; *Porco et al.,* 2014), the flow of mass from the water source to the topmost kilometer of the ice shell, $\sim 2 \times 10^3$ kg s^{-1}, is (if not transmitted downward by subsidence) equivalent to an expansive strain distributed across the whole tiger stripe terrain of $\sim 10^{-13}$ s^{-1}. Such a strain would double the width of the SPT in ~3 m.y. If this situation is maintained long term, cryovolcanic mass and energy fluxes are comparable to tectonic mass and energy fluxes, so that cryovolcanism and tectonics are strongly coupled. Three-dimensional models are now being applied to the ductile (*Běhounková et al.,* 2015) and brittle (*Souček et al.,* 2016) behaviors of the solid ice shell, but these three-dimensional models have not yet been combined with each other, nor with feedbacks on the liquid-water plumbing system.

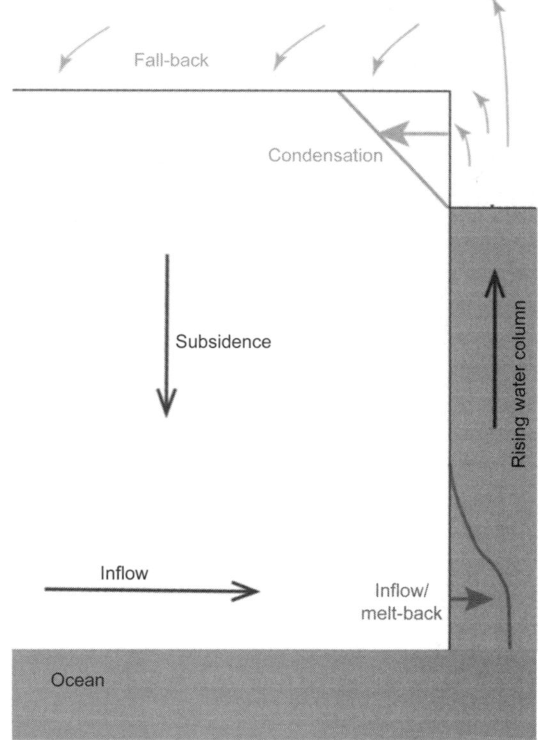

Fig. 6. Schematic of possible flow in the ice shell surrounding the tiger stripes, adapted from *Kite and Rubin* (2016). In this model, melt-back due to turbulent dissipation in the water-filled portion of the tiger stripes (right) causes subsidence of the ice shell, which opens the vapor-filled portion of the fractures near the surface, perhaps preventing them from self-sealing due to ice condensation.

5.2. Astrobiology

For astrobiology, the message from plumbing-system modeling is that the salt-rich jets of Enceladus represent a fresh sample of the ocean. There is no reason to expect that water sits in cryomagma chambers inside the ice shell for long timescales that might lead to the breakdown of biomarkers. Progress on the volcano-tectonic connection will also constrain the rate at which oxidants produced at Enceladus' surface by radiation-driven chemistry are recycled into Enceladus' ocean: potentially a source of energy for life at Europa (*Vance et al.,* 2016) and perhaps at Enceladus too (*Parkinson et al.,* 2008). Bubbles near the water-vapor interface may concentrate biological material, if any, and carry it into the plume (*Porco et al.,* 2017).

6. SUMMARY AND OPEN QUESTIONS

We have learned a lot about the plume fractures and vents in the dozen years since the discovery of activity. Fractures appear to cut through the entire thickness of the ice shell, and are likely mostly filled with liquid water, connected directly to the ocean. Water circulation at speeds of about 1 m s^{-1}, driven by tidal flexing in fractures roughly

1 m wide, may generate enough dissipation to match much of the observed power radiated by the tiger stripes, and the resulting melt-back may combat fracture narrowing by viscous creep and the expected freezing of water onto the fracture walls required to balance conductive heat loss. At the water surface, expected from buoyancy arguments to be situated ~10% of the way down to the ocean, observed plume particles may be generated by bursting of bubbles. Freezing of the water surface by evaporative cooling may be avoided by water vapor pressure buildup above the water surface, due to a combination of narrowing of the fractures toward the surface and viscous throttling of the escaping vapor in the fractures, and also by salinity-driven overturn or dissipative heating of the water. The upper, vapor-filled, part of the fractures may have typical widths of 0.1 m. However, rapid condensation of water vapor onto the fracture walls is expected to narrow the fractures by ~0.3 m yr^{-1}, raising unanswered questions about their long-term stability. This condensation transfers several Gigawatts of latent heat to the fracture walls, where it is conducted to the surface and radiated to space, while a smaller fraction escapes directly as latent heat. Near-surface temperatures approach 200 K, probably limited by sublimation cooling.

However, many outstanding questions remain about the plume origins and plumbing. One is the spatial distribution of heat, solid, and vapor sources. For instance, are the solids lofted mainly in jets or in curtains? Are there vapor sources that do not produce significant solids? How continuous is the emission of heat, solids, and vapor along the tiger stripes? And in particular, what is the magnitude of the likely background-distributed heat source between the tiger stripes?

Space-time variability of the plume and its jets is not fully understood. While individual jets do appear to turn on and off (*Porco et al.,* 2014), unlike the plume as a whole, they do not do so in a manner explicable by simple tidal models; instead, choking of conduits by ice deposition in the near-surface, and subsequent opening of new channels, may be the culprit. Are vapor and solids modulated by tides in the same way? It is not clear why the plume persists at periapse when tidal forces should close the fractures, or why there is a time lag in the orbital response. Also, the origin of the secular decrease in plume brightness seen during the Cassini mission (*Ingersoll and Ewald,* 2017) remains a mystery.

We do not understand the width and shape of the cracks. To keep them from freezing, widths of 1 m are best for the water-filled parts of the channels (*Ingersoll and Nakajima,* 2016; *Kite and Rubin,* 2016). Widths of 0.1 m are best for the vapor-filled parts of the channels to match the latent heat to radiation ratio (*Nakajima and Ingersoll,* 2016) and to match the particle sorting with altitude (*Schmidt et al.,* 2008; *Postberg et al.,* 2011). But these are model-dependent results with limited observational confirmation to date.

While continued analysis of the rich legacy of Cassini observations may answer many of these questions, some must await the next mission to Enceladus, which we hope will occur within the next few decades.

REFERENCES

Abramov O. and Spencer J. R. (2009) Endogenic heat from Enceladus' south polar fractures: New observations and models of conductive surface heating. *Icarus, 199,* 189–196.

Běhounková M., Tobie G., Čadek O., Choblet G., Porco C., and Nimmo F. (2015) Timing of water plume eruptions on Enceladus explained by interior viscosity structure. *Nature Geosci., 8,* 601–604.

Brilliantov N. V., Schmidt J., and Spahn F. (2008) Geysers of Enceladus: Quantitative analysis of qualitative models. *Planet. Space Sci., 56,* 1596–1606.

Čadek O. and 10 colleagues (2016) Enceladus's internal ocean and ice shell constrained from Cassini gravity, shape, and libration data. *Geophy. Res. Lett., 43,* 5653–5660.

Cooper J. F., Cooper P. D., Sittler E. C., Sturner S. J., and Rymer A. M. (2009) Old Faithful model for radiolytic gas-driven cryovolcanism at Enceladus. *Planet. Space Sci., 57,* 1607–1620.

Crawford G. D. and Stevenson D. J. (1988) Gas-driven water volcanism in the resurfacing of Europa. *Icarus, 73,* 66–79.

Cuffey K. and Patterson W. (2010) *The Physics of Glaciers, 4th edition.* Elsevier (Academic), New York.

Dallaston M. C. and Hewitt I. J. (2014) Free-boundary models of a meltwater conduit. *Phys. Fluids, 26,* article ID 083101.

de Leeuw G., Andreas E. L., Anguelova M. D., Fairall C. W., Lewis E. R., O'Dowd C., Schulz M., and Schwartz S. E. (2011) Production flux of sea spray aerosol. *Rev. Geophys., 49,* RG2001, DOI: 10.1029/2010RG000349.

Dougherty M. K., Khurana K. K., Neubaur F. M., Russell C. T., Saur J., Leisner J. S., and Burton M. E. (2006) Identification of a dynamic atmosphere at Enceladus with the Cassini magnetometer. *Science, 311,* 1406–1409.

Egbert G. D. and Ray R. D. (2003) Semi-diurnal and diurnal tidal dissipation from TOPEX/Poseidon altimetry. *Geophys. Res. Lett., 30,* 1907.

Gao P., Kopparla P., Zhang X., and Ingersoll A. P. (2016) Aggregate particles in the plumes of Enceladus. *Icarus, 264,* 227–238.

Gioia G., Chakrobarty P., Marshak S., and Kieffer S. W. (2007) Unified model of tectonics and heat transport in a frigid Enceladus. *Proc. Natl. Acad. Sci., 104,* 13578–13581.

Glein C. R., Zolotov M. Y., and Shock E. L. (2008) The oxidation state of hydrothermal systems on early Enceladus. *Icarus, 197,* 157–163.

Glein C. R., Baross J. A., and Waite J. H. (2015) The pH of Enceladus' ocean. *Geochim. Cosmochim. Acta, 162,* 202–219.

Goguen J. D. and 12 colleagues (2013) The temperature and width of an active fissure on Enceladus measured with Cassini VIMS during the 14 April 2012 south pole flyover. *Icarus, 226,* 1128–1137.

Goguen J. D., Buratti B. J., and the Cassini VIMS Team (2016) Cassini VIMS spectra of the thermal emission from hot spots along Enceladus south pole fissures. *AAS/Division for Planetary Sciences Meeting Abstracts, 48,* 214.10.

Gorius N., Howett C., Spencer J., Albright S., Jennings D., Hurford T., Romani P., Segura M., and Verbiscer A. (2015) Resolving Enceladus thermal emission at the 10s of meters scale along Baghdad Sulcus using Cassini CIRS. *AAS/Division for Planetary Sciences Meeting Abstracts, 47,* 410.01.

Hansen C. J. and 10 colleagues (2011) The composition and structure of the Enceladus plume. *Geophys. Res. Lett., 38,* L11202.

Hansen C. J., Esposito L. W., Aye K.-M., Colwell J. E., Hendrix A. R., Portyankina G., and Shemansky D. (2017) Investigation of diurnal variability of water vapor in Enceladus' plume by the Cassini ultraviolet imaging spectrograph. *Geophys. Res. Lett., 44,* 672–677.

Hedman M. M., Nicholson P. D., Showalter M. R., Brown R. H., Buratti B. J., and Clark R. N. (2009) Spectral observations of the Enceladus plume with Cassini-VIMS. *Astrophys. J., 693,* 1749–1762.

Hedman M. M., Gosmeyer C. M., Nicholson P. D., Sotin C., Brown R. H., Clark R. N., Baines K. H., Buratti B. J., and Showalter M. R. (2013) An observed correlation between plume activity and tidal stresses on Enceladus. *Nature, 500,* 182–184.

Helfenstein P. and Porco C. C. (2015) Enceladus' geysers: Relation to geological features. *Astron. J., 150,* 96.

Horanyi M., Burns J. A., and Hamilton D. P. (1992) The dynamics of Saturn's E ring particles. *Icarus, 97,* 248–259.

Howett C. J. A., Spencer J. R., Pearl J., and Segura M. (2011) High heat flow from Enceladus' south polar region measured using 10–600 cm^{-1} Cassini/CIRS data. *J. Geophys. Res.–Planets, 116,* E03003.

Hsu H.-W. and 14 colleagues (2015) Ongoing hydrothermal activities within Enceladus. *Nature, 519,* 207–210.

Hurford T. A., Helfenstein P., Hoppa G. V., Greenberg R., and Bills B. G. (2007) Eruptions arising from tidally controlled periodic openings of rifts on Enceladus. *Nature, 447,* 292–294.

Hurford T. A., Helfenstein P., and Spitale J. N. (2012) Tidal control of jet eruptions on Enceladus as observed by Cassini ISS between 2005 and 2007. *Icarus, 220,* 896–903.

Iess L. and 10 colleagues (2014) The gravity field and interior structure of Enceladus. *Science, 344,* 78–80.

Ingersoll A. P. and Ewald S. P. (2011) Total particulate mass in Enceladus plumes and mass of Saturn's E ring inferred from Cassini ISS images. *Icarus, 216,* 492–506.

Ingersoll A. P. and Ewald S. P. (2017) Decadal timescale variability of the Enceladus plumes inferred from Cassini images. *Icarus, 282,* 260–275.

Ingersoll A. P. and Nakajima M. (2016) Controlled boiling on Enceladus. 2. Model of the liquid-filled cracks. *Icarus, 272,* 319–326.

Ingersoll A. P. and Pankine A. A. (2010) Subsurface heat transfer on Enceladus: Conditions under which melting occurs. *Icarus, 206,* 594–607.

Jurac S., Johnson R. E., and Richardson J. D. (2001) Saturn's E ring and production of the neutral torus. *Icarus, 149,* 384–396.

Kieffer S. W., Lu X., Bethke C. M., Spencer J. R., Marshak S., and Navrotsky A. (2006) A clathrate reservoir hypothesis for Enceladus' south polar plume. *Science, 314,* 1764.

Kieffer S. W., Lu X., McFarquhar G., and Wohletz K. H. (2009) A redetermination of the ice/vapor ratio of Enceladus' plumes: Implications for sublimation and the lack of a liquid water reservoir. *Icarus, 203,* 238–241.

Kite E. S. and Rubin A. M. (2016) Sustained eruptions on Enceladus explained by turbulent dissipation in tiger stripes. *Proc. Natl. Acad. Sci., 113,* 3972–3975.

Le Gall A. and 17 colleagues (2017) Thermally anomalous features in the subsurface of Enceladus's south polar terrain. *Nature Astron., 1,* 0063.

Loeffler M. J., Raut U., and Baragiola R. A. (2006) Enceladus: A source of nitrogen and an explanation for the water vapor plume observed by Cassini. *Astrophys. J., 649,* L133–L136.

Manga M. and Wang C.-Y. (2007) Pressurized oceans and the eruption of liquid water on Europa and Enceladus. *Geophys. Res. Lett., 34,* L07202.

Matson D. L., Castillo J. C., Lunine J., and Johnson T. V. (2007) Enceladus' plume: Compositional evidence for a hot interior. *Icarus, 187,* 569–573.

Matson D. L., Castillo-Rogez J. C., Davies A. G., and Johnson T. V. (2012) Enceladus: A hypothesis for bringing both heat and chemicals to the surface. *Icarus, 221,* 53–62.

Melosh H. J., Ekhol A. G., Showman A. P., and Lorenz R. D. (2004) The temperature of Europa's subsurface water ocean. *Icarus, 168,* 498–502.

Nakajima M. and Ingersoll A. P. (2016) Controlled boiling on Enceladus. 1. Model of the vapor-driven jets. *Icarus, 272,* 309–318.

Nimmo F., Spencer J. R., Pappalardo R. T., and Mullen M. E. (2007) Shear heating as the origin of the plumes and heat flux on Enceladus. *Nature, 447,* 289–291.

Nimmo F., Porco C. P., and Mitchell C. (2014) Tidally modulated eruptions on Enceladus: Cassini ISS observations and models. *Astron. J., 148,* DOI: 10.1088/0004-6256/148/3/46.

Nye J. F. (1953) The flow law of ice from measurements in glacier tunnels, laboratory experiments and the Jungfraufirn borehole experiment. *Proc. R. Soc. London, 1139,* 477–489.

O'Neill C. and Nimmo F. (2010) The role of episodic overturn in generating the surface geology and heat flow on Enceladus. *Nature Geosci., 3,* 88–91.

Parkinson C. D., Liang M.-C., Yung Y. L., and Kirschivnk J. L. (2008) Habitability of Enceladus: Planetary conditions for life. *Origins Life Evol. Biosph., 38,* 355–369.

Porco C. C., Helfenstein P., Thomas P. C., Ingersoll A. P., Wisdom J., et al. (2006) Cassini observes the active south pole of Enceladus. *Science, 311,* 1393–1401.

Porco C., Di Nino D., and Nimmo F. (2014) How the geysers, tidal stresses, and thermal emission across the south polar terrain of Enceladus are related. *Astron. J., 148,* 45.

Porco C. C., Dones L., and Mitchell C. (2017) Could it be snowing microbes on Enceladus? Assessing conditions in its plume and implications for future missions. *Astrobiology, 17(9),* DOI: 10.1089/ast.2017.1665.

Postberg F., Kempf S., Schmidt J., Brilliantov N., Beinsen A., Abel B., Buck U., and Srama R. (2009) Sodium salts in E-ring ice grains from an ocean below the surface of Enceladus. *Nature, 459,* 1098–1101.

Postberg F., Schmidt J., Hillier J., Kempf S., and Srama R. (2011) A salt-water reservoir as the source of a compositionally stratified plume on Enceladus. *Nature, 474,* 620–622.

Röthlisberger H. (1972) Water pressure in intra- and subglacial channels. *J. Glaciol., 11,* 177–203.

Rubin A. M. (1995) Propagation of magma-filled cracks. *Annu. Rev. Earth Planet. Sci., 23,* 287–336.

Rudolph M. L. and Manga M. (2009) Fracture penetration in planetary ice shells. *Icarus, 199,* 536–541.

Schmidt J., Brilliantov N., Spahn F., and Kempf S. (2008) Slow dust in Enceladus' plume from condensation and wall collisions in tiger stripe fractures. *Nature, 451,* 685–688.

Schneider N. M., Burger M. H., Schaller E. L., Brown M. E., Johnson R. E., et al. (2009) No sodium in the vapour plumes of Enceladus. *Nature, 459,*1102–1104.

Schulson E.M. (2001) Brittle failure of ice. *Eng. Fracture Mech., 68,* 1839–1887.

Sekine Y. and 11 colleagues (2015) High-temperature water-rock interactions and hydrothermal environments in the chondrite-like core of Enceladus. *Nature Commun., 6,* 8604.

Souček O., Hron J., Běhounková M., and Čadek O. (2016) Effect of the tiger stripes on the deformation of Saturn's moon Enceladus. *Geophys. Res. Lett., 43,* 7417–7423.

Spencer J. R. and Nimmo F. (2013) Enceladus: An active ice world in the Saturn system. *Annu. Rev. Earth Planet. Sci. 41,* 693–717.

Spencer J. R., Pearl J. C., Segura M., Flasar F. M., Mamoutkine A., Romani P., Buratti B. J., Hendrix A. R., Spilker L. J., and Lopes R. M. C. (2006) Cassini encounters Enceladus: Background and the discovery of a south polar hot spot. *Science, 311,* 1401–1405.

Spencer J. R., Howett C. J. A., Verbiscer A. J., Hurford T. A., Segura M. E., and Pearl J. C. (2011) Observations of thermal emission from the south pole of Enceladus in August 2010. *EPSC-DPS Joint Meeting 2011,* 1630.

Spencer J. R., Howett C. J. A., Verbiscer A., Hurford T. A., Segura M., and Spencer D. C. (2013) Enceladus heat flow from high spatial resolution thermal emission observations. *EPSC 2013 Abstracts,* 8–13 September, London, UK.

Spitale J. N. and Porco C. C. (2007) Association of the jets of Enceladus with the warmest regions on its south-polar fractures. *Nature, 449,* 695–697.

Spitale J. N., Hurford T. A., Rhoden A. R., Berkson E. E., and Platts S. S. (2015) Curtain eruptions from Enceladus' south-polar terrain. *Nature, 521,* 57–60.

Vance S. D., Hand K. P., and Pappalardo R. T. (2016) Geophysical controls of chemical disequilibria in Europa. *Geophys. Res. Lett., 43,* 4871–4879.

Veeder G. J., Matson D. L., Johnson T. V., Blaney D. L., and Goguen J. D. (1994) Io's heat flow from infrared radiometry: 1983–1993. *J. Geophys. Res., 99,* 17095–17162.

Veron F. (2015) Ocean spray. *Annu. Rev. Fluid Mech., 47,* 507–538.

Vienne A. and Duriez L. (1992) A general theory of motion for the eight major satellites of Saturn III. Long-period perturbations. *Astron. Astrophys., 257,* 331–352.

Waite J. H. Jr. and 15 colleagues (2009) Liquid water on Enceladus from observations of ammonia and ^{40}Ar in the plume. *Nature, 460,* 487–490.

Waite J. H. and 12 colleagues (2017) Cassini finds molecular hydrogen in the Enceladus plume: Evidence for hydrothermal processes. *Science, 356,* 155–159.

Walder J. (1982) Stability of sheet flow of water beneath temperate glaciers and implications for glacier surging. *J. Glaciol., 28,* 273–293.

Weertman J. (1972) General theory of water flow at the base of a glacier or ice sheet. *Rev. Geophys. Space Phys., 10,* 287–333.

Yeoh S. K., Chapman T. A., Goldstein D. B., Varghese P. L., and Trafton L. M. (2015) On understanding the physics of the Enceladus south polar plume via numerical simulation. *Icarus, 253,* 205–222.

Yeoh S. K., Li Z., Goldstein D. B., Varghese P. L., Levin D. A., and Trafton L. M. (2017) Constraining the Enceladus plume using numerical simulation and Cassini data. *Icarus, 281,* 357–378.

Goldstein D. B., Hedman M., Manga M., Perry M., Spitale J., and Teolis B. (2018) Enceladus plume dynamics. In *Enceladus and the Icy Moons of Saturn* (P. M. Schenk et al., eds.), pp. 175–194. Univ. of Arizona, Tucson, DOI: 10.2458/azu_uapress_9780816537075-ch005.

Enceladus Plume Dynamics: From Surface to Space

David B. Goldstein
The University of Texas at Austin

Matthew Hedman
University of Idaho

Michael Manga
University of California Berkeley

Mark Perry
Johns Hopkins University Applied Physics Laboratory

Joseph Spitale
Planetary Sciences Institute

Benjamin Teolis
Southwest Research Institute

The vapor and particulate plume rising out of Enceladus' south polar region discovered by Cassini is a dramatic and active geologic feature. This plume not only feeds the E ring but also carries information about the satellite's interior. The plume contains both jets and more distributed emissions that emerge from various fractures in the icy crust. The particle flux is tidally modulated over the course of each Enceladus day, but also appears to vary on longer and perhaps shorter timescales. Dynamically, the plume provides a thus-far-unique example of a cool multi-phase (particulates and vapor) flow into a weak gravity, vacuum environment. In this chapter we describe the plume dynamics from the continuum flow in the near-vent region, through the gas dynamic expansion process, out to the far field where vapor and particulate motions uncouple, where intermolecular collisions cease, and where the Cassini observations were made. It is ultimately those Cassini observations, interpreted through dynamical models, that will enable us to understand the source conditions for the plume.

1. INTRODUCTION

The geysers comprised of largely water vapor and icy particles feeding the broader plume over the south pole of Enceladus constitute a unique opportunity. That flow of material, moderated by several thermal and physical processes along its passage from a presumed liquid source all the way out to Cassini's orbit and beyond, provides unique access to conditions deep within the satellite. Multiple Cassini instruments [Ion and Neutral Mass Spectrometer (INMS), Cosmic Dust Analyzer (CDA), Cassini Plasma Spectrometer (CAPS), Magnetometer (MAG), Ultraviolet Imaging Spectrograph (UVIS), Visible and Infrared Imaging Spectrometer (VIMS), Imaging Science Subsystem (ISS), and Composite Infrared Spectrometer (CIRS)] characterized the gas-particle plume over and near the warm tiger stripe region. Numerous observations also exist of the near-vent regions in the visible and the IR. As described in the chapter in this volume by Spencer et al., the most likely source for these extensive eruptions is a subsurface liquid reservoir of saline water and other volatiles boiling and then escaping through crevasse-like conduits into the vacuum of space. Realistic analysis of the observations of Enceladus' plume phenomena is especially critical for making successful inferences of subsurface conditions (Spencer et al., this volume), notably concerning habitability (McKay et al. and Lunine et al., this volume). Moreover, beyond Enceladus itself, the plume injects gas and particles into Saturn's E ring and the Saturn system as a whole, influences Saturn's magnetosphere, and coats the surfaces of Enceladus and other satellites with "snow."

Both ISS and VIMS images observed Enceladus' plume via sunlight scattered by particles that are either entrained in the vapor or moving ballistically on their own (Fig. 1). The plume, as traced by the particles, clearly emanates from the tiger stripes and has both a diffuse component and many dense, narrow components, labeled "jets" or "geysers." The scattered light intensity is greatest near Enceladus' surface and decreases with altitude over distances of hundreds of kilometers (*Hedman et al.,* 2013; *Nimmo et al.,* 2014; *Ingersoll and Ewald,* 2017). Four stellar/solar occultations observed by UVIS remotely measured the vapor, finding that the jets are of high Mach number, and can be four times denser than the diffuse components, which extend laterally only slightly beyond the region of the tiger stripes. *In situ* measurements of the gas molecules by INMS (Appendix Table A1) show densities and spatial variations consistent with UVIS. Vapor densities over the south-polar region peaked at a few times 10^8 molecules cm^{-3} at 100 km altitude and depend approximately on the altitude squared. The gas is predominately water vapor, with 1–4% other volatiles (*Waite et al.,* 2017; Postberg et al., this volume). CAPS and INMS measured ions in the plume, but, at less than 100 ions cm^{-3}, the ion density is too low to affect plume dynamics.

Individual jets and the plume as a whole show temporal variability on decadal, diurnal, and shorter timescales, presumably reflecting variations in the mass flux out of the satellite. Analyses of UVIS, INMS, and MAG data produce estimates of 100 to more than 1500 kg s^{-1} for the mass flux in vapor (although this high value is perhaps an outlier event). There is some indication in UVIS and INMS data that much of the variability occurs in the jets, with less variation in the diffuse components.

The narrowness of the vapor jets (which implies highly supersonic vapor flow) indicates that at least some of the surface vents act as converging-diverging nozzles, with vapor passing through a constriction and then rapidly expanding. Near the surface vents, the two-component flow of vapor and particles is well represented as a continuum and has well-defined pressures, densities, and species mass fractions. Within tens of meters (*Yeoh et al.,* 2015), as the plume expands to low densities and temperatures, the particles and gas no longer influence each other. At still greater distances, as the expansion continues, the gas flow transitions to the free-molecular regime, where intermolecular collisions become uncommon and both molecules and the remaining particles move ballistically, influenced predominantly by the gravity fields of Enceladus and then Saturn. Finally, at distances greater than a few hundred kilometers from Enceladus, electromagnetic effects become important; these are described in the chapter by Kempf et al. in this volume.

Fig. 1. See Plate 21 for color version. An ISS color-stretched image acquired on November 27, 2005, illustrating overall plume morphology threaded by many finely resolved jets or geysers.

The particulates are primarily ice grains, and vary in size, composition, and formation process. The ice grains that are seen by VIMS, ISS and CDA are on the order of 1 μm in radius and probably originate either from salty-ocean spray or from vapor condensation within the subsurface fissures (see the chapter in this volume by Postberg et al.). The CAPS instrument has measured much smaller nanometer-sized grains that presumably condense from the vapor during expansion and cooling processes. CIRS finds that the surface temperatures increase with proximity to the tiger stripes and reach 170 K at the hottest regions, but the expansion cools the plume vapor to less than 30 K above the surface (*Yeoh et al.,* 2015).

The suggested Europa south polar plume (*Roth et al.,* 2013; *Sparks et al.,* 2016) may be similar to Enceladus in all but the gravity field and net mass flow. At the other end of the spectrum, cometary plumes are similar to the Enceladus plume in terms of their gas/particulate dynamics, constituents, mass flow, velocities, and exhaust into vacuum, but differ in their driving source mechanisms.

One of the most intriguing aspects of the Enceladus plume is its potential link to extraterrestrial habitability and the search for life elsewhere in the solar system. If the plume constituents ultimately arise from a liquid water source, the plume may present an exceptional and unique opportunity to examine and sample liquid water constituents from deep within a surface-frozen ocean world. Important questions then relate to how those constituents, whether molecular indicators or even possibly entire organisms, were processed below the surface and within the plume on the way to their examination (see the chapters in this volume by McKay et al. and Lunine et al.).

Below, we first describe the visible and measurable aspect of the plumes and their source — the tiger stripes — in section 2. Section 3 describes the observed temporal variability and some of the possible controlling mechanisms. Section 4 discusses the processes encountered as the subsurface water becomes the vapor and the ice grains that escape from the surface vents. Once we have begun to detail the physics of the Enceladus plume, we take a brief diversion in section 5 to set this plume into the context of other observed related plumes in the solar system. Sections 6 and 7 examine the physical properties of the emitted particles and vapor from the surface to hundreds of kilometers, where they transition to the environment external to Enceladus.

2. VENT LOCATIONS: TIGER STRIPES, STRONG JETS, DISTRIBUTED SOURCES, JET RELATIONSHIPS TO THERMAL OUTPUT

2.1. Analysis of Initial Observations

While the material emerging from the south pole has been observed with multiple instruments, the most detailed information about the source locations comes from the ISS observations, which include highly resolved observations of the particle-rich jets from the tiger stripe fissures. In 2005, ISS observations revealed large rifts in the crust, informally called "tiger stripes" (*Porco et al.,* 2006). These fractures (each of which comprises numerous branches) are named Alexandria, Cairo, Baghdad, and Damascus (Fig. 2), and were hypothesized to be the sources of the observed jets. That hypothesis was supported by Cassini CIRS observations (*Spencer et al.,* 2006), which detected anomalously high temperatures (~85 K) south of 65°S latitude, and identified several localized regions (labeled A–F in Fig. 2) of particularly high temperatures (~145 K) along the tiger stripe fractures.

Triangulating ISS observations of jets taken between 2005 and 2007, *Spitale and Porco* (2007) inferred the positions of eight major jet sources. Those triangulations were performed by representing jets in images as line segments and identifying clusters of intersections that simultaneously yielded common geographic source locations and direction vectors. Many of the resulting source locations, labeled I through VIII (Fig. 2), coincided with the *Spencer et al.* (2006) CIRS hot spots, and all of them were consistent with emission from the tiger stripe fractures.

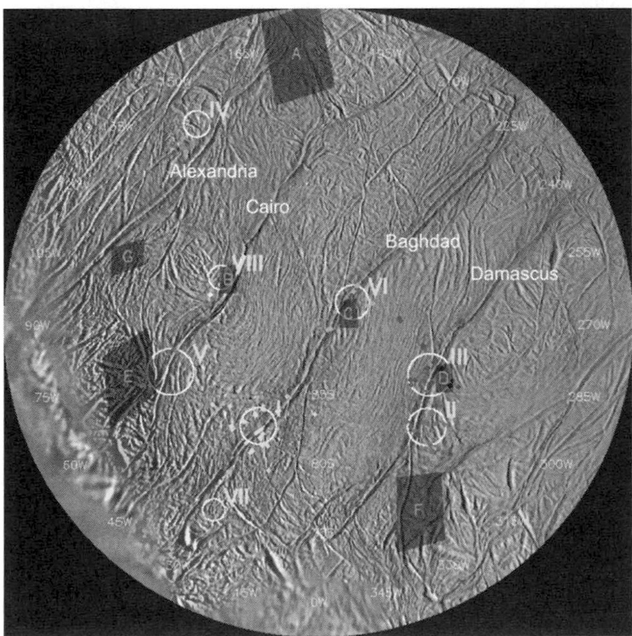

Fig. 2. See Plate 22 for color version. Polar stereographic projection of Enceladus' south polar region showing the eight source locations inferred by *Spitale and Porco* (2007) using triangulation. Locations of CIRS hot spots from *Spencer et al.* (2006) are shown as red quadrilaterals.

In *Spitale and Porco* (2007), the observations that contributed to each solution were mostly taken from large distances, and occurred over multiple Cassini orbits. As a result, that approach characterized the prominent collimated jets (or concentrations of unresolved fine jets) that could be seen from large distances, and which remained reasonably static during the intervals between those observations, which were typically months or more. Because of the large distances, the precision of the inferred source locations was typically no better than 10 to 20 km, sufficient to distinguish which tiger stripe was involved, but not necessarily which branch.

2.2. Analysis of High-Resolution Observations

Understanding the time variability of the jetting activity over the south polar terrain (important in determining the connection with tides; see section 3 below) required significantly better spatial and temporal resolution than provided by the 2004–2006 images used in *Spitale and Porco* (2007). To that end, a new series of observations was conducted at distances that enabled resolution of the fine structure of the jets and provided sufficient parallax to perform triangulation during a single encounter.

Porco et al. (2014) analyzed those new observations (Fig. 3 is an example) using an updated triangulation approach. Approximately 100 sources were inferred, and maps of jetting activity at various times were produced (Fig. 4). The geographical distribution of the inferred jetting activity was shown to be consistent with CIRS temperatures (*Howett et al.,* 2011) and with localized hot spots seen in high-resolution Cassini VIMS observations (*Goguen et al.,* 2013). The possibility of lower-speed eruptions forming sheets of material was left as an open question by *Porco et al.* (2014). Many of the inferred jets were observed be in various states of activity in different observations, although a tidally controlled relationship (*Hurford et al.,* 2007) was not apparent.

Spitale et al. (2015) analyzed several of those same datasets using a different approach. The motivation for developing a different approach was that much of the activity (per unit length of tiger stripe fracture) appears as a continuous glow rather than as discrete jets that can be represented as simple lines in an image. Moreover, much of the fine discrete structure that does appear is difficult to reliably identify among successive images. In this approach, active regions were located by comparing a *simulated curtain* of material with the emission seen in an image. The simulated curtain could emerge at any zenith angle, and linear spreading with altitude could also be included. Because those observations were obtained during the onset of southern winter, most of the source locations were in shadow, and the shadow of

Fig. 3. Cassini ISS image showing Enceladus' shadow cast on the planes of material associated with each tiger stripe.

Enceladus was cast across the emerging jets (Fig. 3). In many cases, those shadows allowed for the unique identification of the active fracture associated with a given curtain of observed material.

Spitale et al. (2015) produced two notable results: (1) Maps of activity at five different times (Fig. 5) spanning about a year showed that most of the fracture system is active at all times. (2) It was noted that jet-like artifacts (referred to as phantom jets) may appear in a simulated image due to a fortuitous combination of the viewing angle relative to the fracture geometry (Fig. 6).

Results corroborate the geographic link between eruptive activity and elevated CIRS temperatures that was first established by *Spitale and Porco* (2007) and expanded by *Porco et al.* (2014). Moreover, activity was detected on fractures for which CIRS has not shown elevated temperatures; in particular, the fracture system branching from Baghdad Sulcus at Source VI was seen to be active on day 2010-225.

The *Spitale et al.* (2015) result (point 2 above) may explain why many fine features are difficult to reliably identify throughout an imaging sequence as the geometry varies. *Spitale et al.* (2015) also suggested that some of the *Porco et al.* (2014) jets may not be real jets but instead ripples in a curtain of emission. On the other hand, *Porco et al.* (2015) have pointed out that 90 to 95 of the 98 jets were triangulated on the basis of images from a wide range of viewing angles at >45° (*Porco et al.,* 2014) to the tiger stripes, lending confidence to their identification as real jets. Indeed, *Helfenstein and Porco* (2015) have identified a number of surface features corresponding to *Porco et al.* (2014) solutions that support the eruption of discrete jets at those locations (see below). Moreover, some discrete jets may have been missed because they spread out and blend with their neighbors before emerging into the sunlight, appearing as continuous curtains.

2.3. Jets and Curtains

It is likely that both jet and curtain eruptive styles coexist, may be end members of the same phenomenon, and reflect different effects of the geology and the dynamics of the Enceladus system. Presumably, there exists a distribution of vent aspect ratios (length:width) with round holes having a value of 1.0 and long slots having a value approaching infinity. Due to their great heights, the features measured by *Spitale and Porco* (2007) are almost certainly supersonic jets with embedded particles that feed the E ring and do not fall out directly onto the surface of Enceladus. Many of the most convincing *Porco et al.* (2014) jets are bright to high altitudes and also likely fall into this category. Indeed, *Mitchell et al.* (2015) were able to reproduce the "tendril" structures in Enceladus' E ring using the fastest particles launched from the most active *Porco et al.* (2014) jets. The

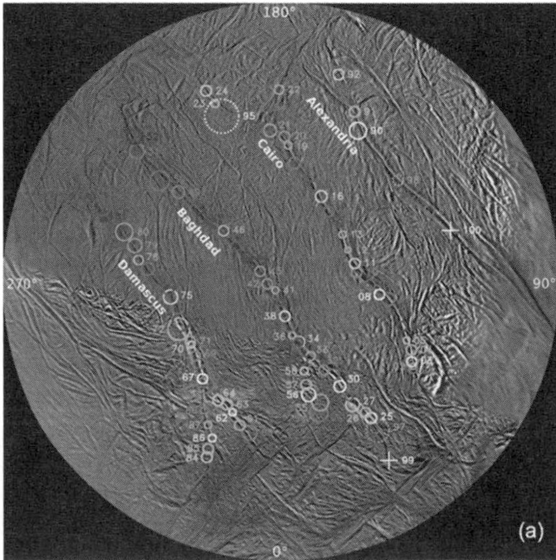

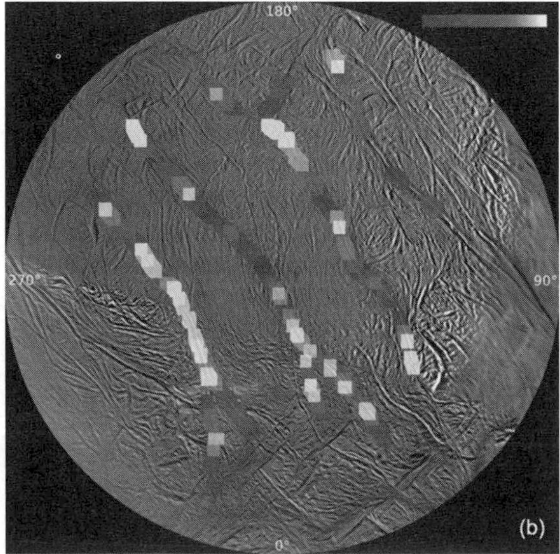

Fig. 4. See Plate 23 for color version. **(a)** Locations of the jets. **(b)** Jet activity derived from ISS observations. There is strong correlation with the jet activity and the tiger-stripe temperatures measured by CIRS and VIMS. From *Porco et al.* (2014).

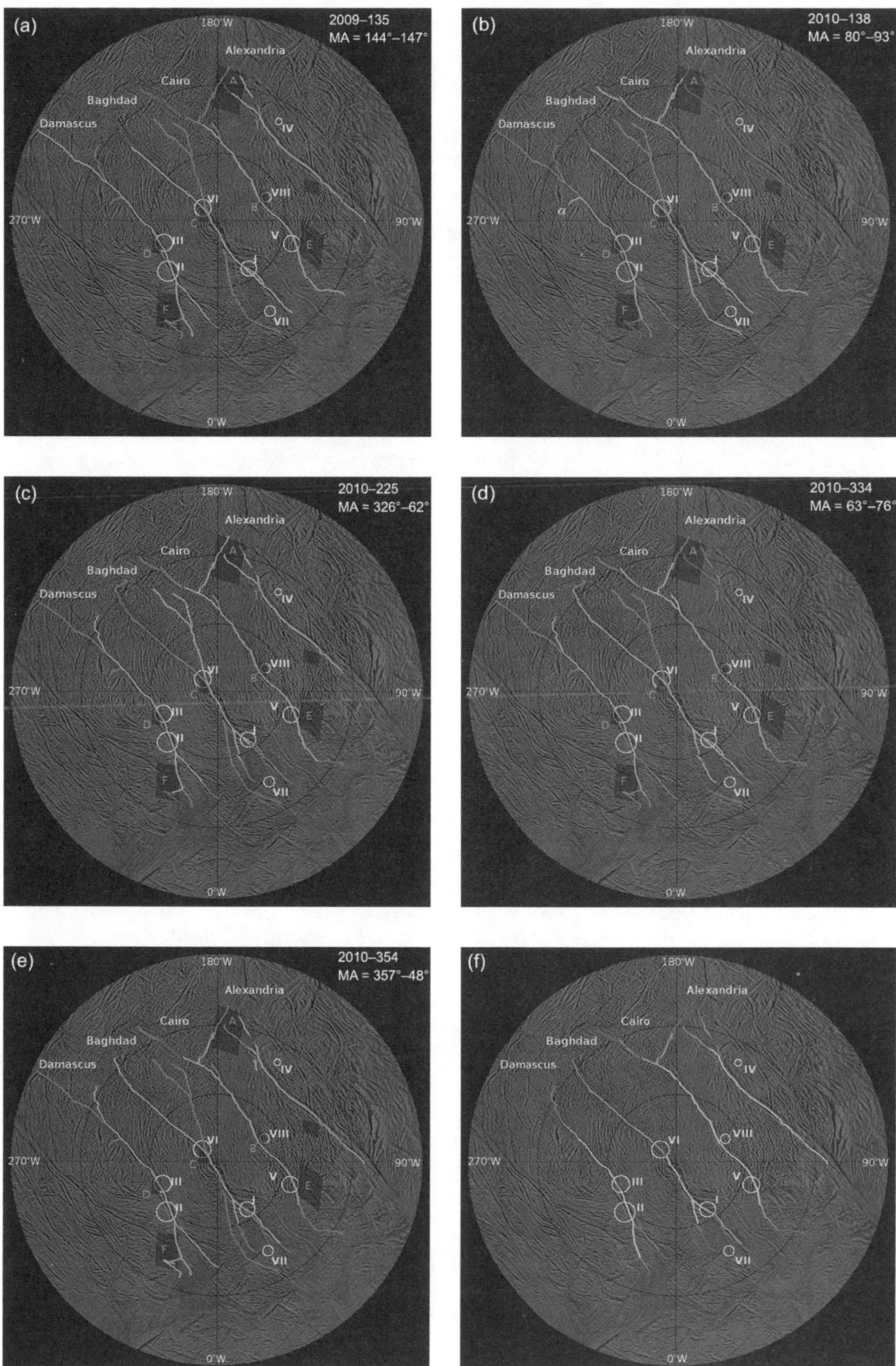

Fig. 5. See Plate 24 for color version. Curtain activity at five different mean anomalies (MA), and the average (f). Green areas are active, red areas are inactive, and blue areas are undetermined.

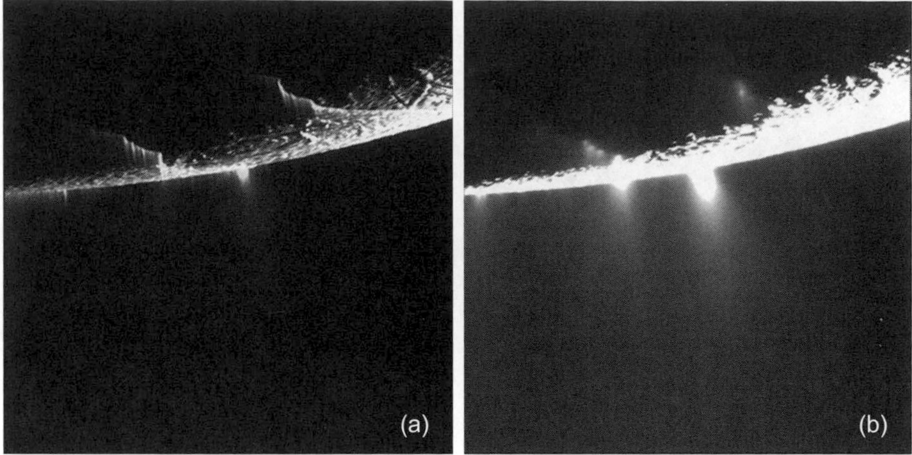

Fig. 6. Curtain simulation overlain on image N1637461416. **(a)** Image is displayed with no overlay, and stretched to make the erupted material visible. **(b)** Image is displayed with no stretch, with simulated curtains overlain. The simulated curtains are sampled *uniformly* at 250-m intervals along each fracture, and the ground-level intensities are also uniform. "Phantom" jets appear at locations where the line of sight intersects the curtain at a shallower angle than in the immediate surroundings. Every relatively bright feature in the simulated curtain is a phantom; note the correspondence with apparent "jets" in the Cassini image. Nightward of the terminator, the bottom edge of the curtains are defined by the shadow of Enceladus, allowing the unique determination of the source fractures.

curtain-like sprays, on the other hand, tend to have heights closer to 10 km and likely consist of material launched at lower velocities. That hypothesis is supported by simulations (*Kempf et al.,* 2010) comparing the fallout patterns for initial conditions corresponding to *Porco et al.* (2014) jets and those for *Spitale et al.* (2015) curtains to the observed albedo pattern on Enceladus' surface.

In addition to feeding the E ring, the supersonic jets identified in *Spitale and Porco* (2007) and *Porco et al.* (2014) likely account for most of the tidally-modulated emission noted by *Hedman et al.* (2013) (section 3 below), as those observations covered altitudes from 50 to 450 km. At lower altitudes, where the curtains may dominate, a tidal signature has yet to be discovered in the ISS data.

2.4. Correlation with Geology and Morphology

The broad correlation between jetting activity and prominent fractures in the south polar terrain has been thoroughly established based on imaging of the region at moderate resolutions (*Porco et al.,* 2015). Coverage of the region at the very high resolutions needed to associate jetting activity with local morphological features is incomplete, although some intriguing associations have emerged. *Helfenstein and Porco* (2015) examined the very-high-resolution Cassini images and found that (1) azimuths of the jets align preferentially with the main fractures, local cross-cutting fractures, or a local tectonic fabric; and (2) some *Porco et al.* (2014) source locations correspond to local features suggestive of ballistic fallout, scouring, or other processes that create radial striations and/or buildups centered at locations on the primary fractures [see Fig. 7, as well as Fig. 16c of *Helfenstein and Porco* (2015)].

One difficulty with result 1 above is that multiple directions of cross-cutting are apparent in some surface images, and other fractures may exist that would only be discerned at higher resolutions than those available, reducing the statistical significance of any directional correlation. However, if a variety of orientations are available to respond to the rotating stress field at a particular locale, vapor may be emitted with different orientations at different times, and it might explain why low-altitude activity is seen at some level throughout the tidal cycle.

Approximately 20% of the jets that do not appear to align preferentially with fractures or tectonic fabric may align with unresolved fractures, or their existence may argue against the proposed correlation. They also may be evidence for phantoms in the *Porco et al.* (2014) solutions, although the 57 jets examined in the *Helfenstein and Porco* (2015) study were selected for their prominence and consistent non-zero zenith angles, conditions that appear to favor real discrete jets. Those conditions also favor the supersonic jets that likely contribute to the high-altitude plume, which is tidally modulated. The strong correlation of the density of the high-altitude plume with mean anomaly implies a response to a consistent fracture orientation, potentially in conflict with the idea of a significant contribution from fractures not aligned with primary fracture system.

The association of morphologic features with *Porco et al.* (2014) jet solutions in result 2 above (Fig. 7) provides strong confirmation for those features being physical jets, and provides a window into the jetting mechanism at the surface. The appearance of these features, particularly the radial striations, does not lend itself to other obvious explanations, but a systematic analysis of the correlation between jet solutions and morphologic features has yet to be performed.

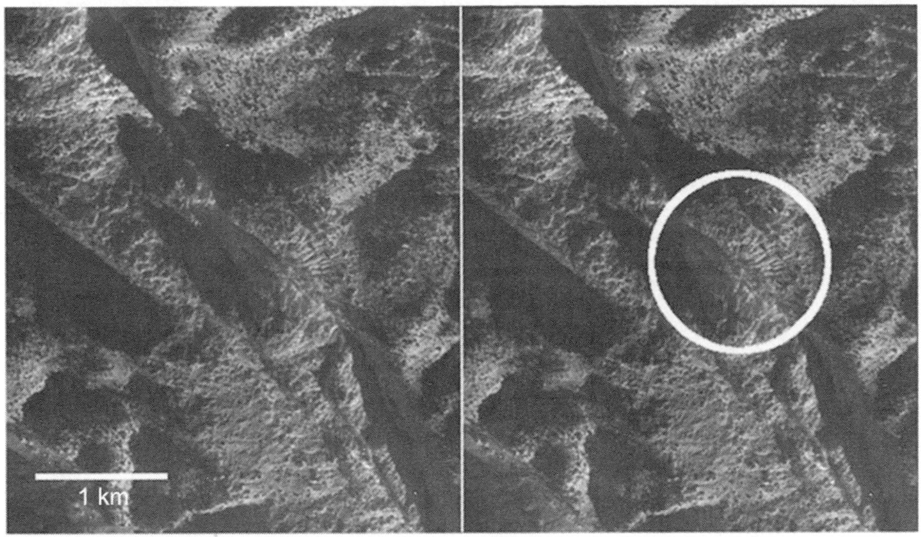

Fig. 7. Radial striations emanating from Baghdad Sulcus suggesting localized emission. From *Helfenstein and Porco* (2015).

3. TIME VARIABILITY OF JETS AND THE PLUME: PRESUMED PROCESSES AND THEIR RELATION TO THE ORBITAL DYNAMICS OF ENCELADUS

The sources of Enceladus' plume material not only have a complex distribution in space, they also vary significantly with time over periods ranging from days to years. Thus far, the best-documented temporal variations are periodic changes in the plume's total particle output that are correlated with the satellite's location on its orbit around Saturn. Enceladus' orbit is not a perfect circle, so both the distance between Enceladus and Saturn and Enceladus' orbital velocity oscillate back and forth slightly during the satellite's 33-hour orbit. As discussed in the chapters in Part 1 of this volume, these small oscillations change the tidal stresses across the south polar terrain, altering the connections between the plume vents and the subsurface and thus influencing the satellite's plume activity.

Two instruments onboard the Cassini spacecraft have detected particle-output variations that are clearly correlated with Enceladus' location along its eccentric orbit: the cameras of ISS (*Nimmo et al.,* 2014; *Porco et al.,* 2014; *Ingersoll and Ewald,* 2017) and VIMS (*Hedman et al.,* 2013). Both instruments obtained images of the micrometer-sized particles erupting from the satellite over a wide range of times and observing conditions. The apparent brightness of the plume in these datasets depends upon both the number density of plume particles and the viewing geometry, particularly the phase angle (i.e., the Sun-plume-camera angle). This angle is important because the particles in the plume scatter light more efficiently at higher phase angles, and so the same material appears brighter when the phase angle is larger. Fortunately, these viewing-geometry-dependent variations can be determined by comparing observations at the same phase angle and/or by applying appropriate corrections based on light-scattering models.

In practice, different authors have used different methods of estimating the dependence of the plume's brightness on phase angle. *Hedman et al.* (2013) used a simple correction where the plume's brightness was a power-law function of the scattering angle, which was based on the observed variations in the plume's brightness at different orbital phases. *Nimmo et al.* (2014) used a more complex phase function that was computed using Mie theory and the particle-size distributions of the plume particles derived from VIMS spectra (*Hedman et al.,* 2009). Finally, *Ingersoll and Ewald* (2017) used several different phase functions derived by *Gao et al.* (2016) that were extrapolated from an earlier study of extremely high-phase plume observations (*Ingersoll and Ewald,* 2011). The *Hedman et al.* (2013) approach has the advantage of being simple, but since the assumed phase function was derived from the VIMS data by assuming that the plume always had the same brightness at a given orbital phase, and since different phase angles were observed at different times, there is a chance that this method would suppress long-term trends in VIMS measurements of the plume's brightness. By contrast, the other methods assume that measurements of the plume's spectra and/or phase function measured at one time can be extrapolated and applied to all the other observations. A potential problem with the latter approach is that it assumes that the physics of the flow/particles remains unchanged — it is just that the mass flow changes between observations. While all these methods do yield similar results, i.e., the relative variation with mean anomaly is well established, there are differences in the detailed trends that will probably require more indepth investigations of the plume's photometry before they can be resolved.

Figure 8 shows the ISS and VIMS estimates of the plume's ice-grain output — corrected for viewing geometry — as functions of the satellite's position along its orbit. Both sets of data clearly show that the plume's ice-particle output is roughly four times higher when the satellite is furthest from Saturn than it is at other points in its orbit. This trend is repeatable among different subsets of the data and is not linked to other phenomena such as shadowing by Saturn, and therefore appears to be a persistent feature of Enceladus' geological activity for the duration of the Cassini mission.

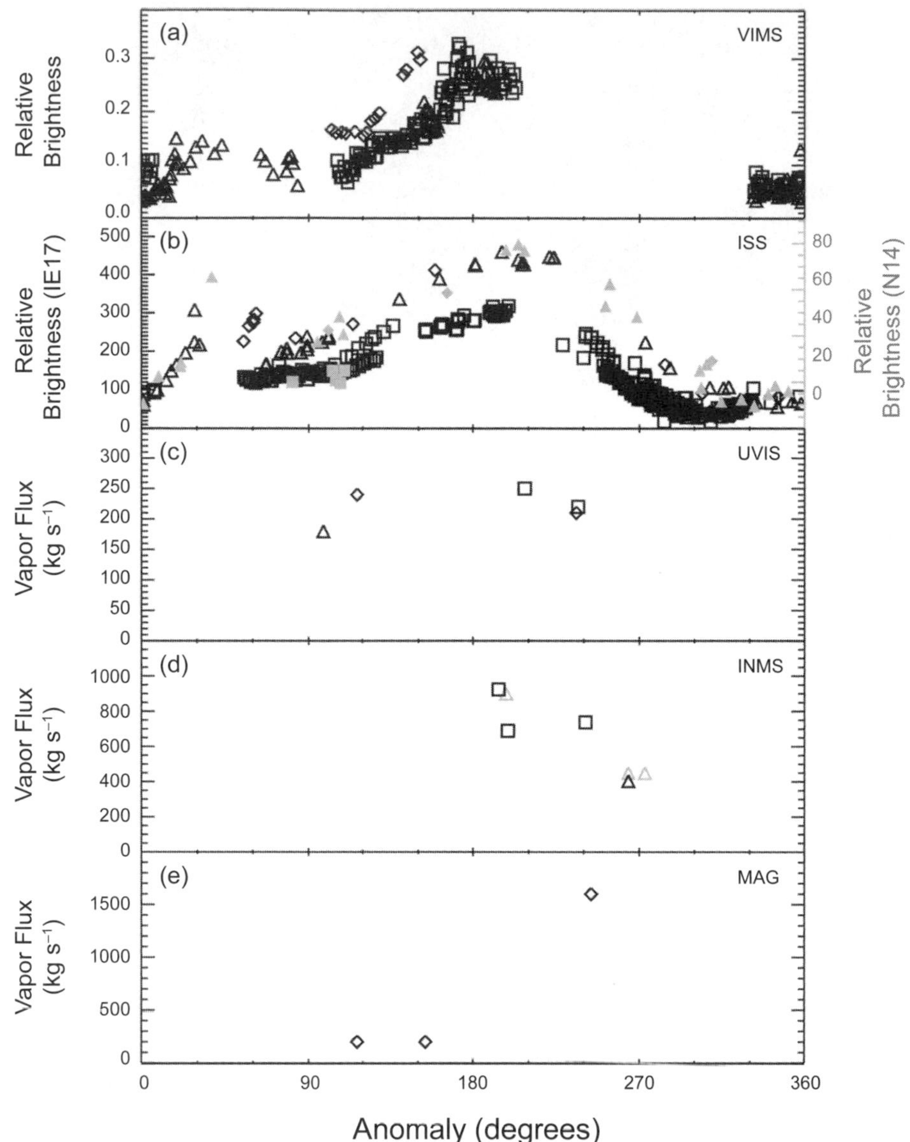

Fig. 8. Temporal variations in Enceladus' plume activity as observed by various Cassini instruments. All panels show a measure of plume activity as a function of the satellite's orbital anomaly (an anomaly of 0° corresponds to when the satellite is closest to Saturn, and 180° is when the satellite is farthest from the planet). **(a)** VIMS estimates of the plume's brightness at 85 km altitude and 1.2 μm, corrected for phase angle (*Hedman et al.,* 2013). **(b)** Estimates of the plumes' brightness derived from ISS images by *Nimmo et al.* (2014) (gray) and *Ingersoll and Ewald* (2017) (black). **(c)** Estimates of the gas flux derived from UVIS occultations using the UVIS-derived Mach number for vapor spreading with a temperature of 170 K (*Hansen et al.,* 2017). **(d)** Gas fluxes derived from INMS data obtained during several flybys by *Yeoh et al.* (2017) (E3, E5, and E7 flybys in gray) and *Teolis et al.* (2017) (E7, E14, E17, and E18 "adiabatic" estimates in black). **(e)** Estimates of the gas flux derived from MAG data from the E0, E1, and E2 flybys by *Saur et al.* (2008). In all five panels, different symbols correspond to different observation times: Measurements marked with diamonds were obtained before 2008, those marked with triangles were obtained in 2008 through 2010, and those marked with squares were obtained in 2011 through 2016. The VIMS measurements and the two ISS analyses show clear increases in the plume's brightness as the satellite moves through anomalies of 180°. There is also evidence that the plume's overall activity level has decreased over time. Differences in the shapes and magnitudes of these trends may in part reflect differences in how the VIMS and ISS groups corrected for variations in the observed phase angle. While the UVIS data show comparatively little variation in the gas flux, the MAG and INMS analyses could indicate higher gas fluxes when the satellite is near apoapsis.

This implies that changing tidal stresses are producing repeatable physical changes below the surface of Enceladus.

Interestingly, while the total flux of particles from Enceladus changes dramatically as the satellite moves around Saturn, the vertical structure of the plume does not change nearly as much. Again, direct comparisons of the published analyses are challenging because different authors used different models for the plume's vertical structure. For example, *Nimmo et al.* (2014) fit an exponential vertical profile to integrated brightness observations and reported a nearly constant scale height parameter. *Hedman et al.* (2013), by contrast, found a linear trend better described the particle velocity distribution and found the effective maximum velocity is 10% smaller when the satellite is near its orbital apoapsis and the particle flux is highest. Finally, *Ingersoll and Ewald* (2017) argued that the plume's brightness decayed with altitude like a power law, and that the power-law index was larger (~–0.4) when Enceladus was near its orbital apoapsis and smaller (~–0.2) at other orbital phases. Note that despite their different ways of parameterizing the trends, both the *Hedman et al.* (2013) and *Ingersoll and Ewald* (2017) analyses suggest that the plume's brightness variation with altitude, as derived from ISS data, declines more quickly when Enceladus' activity level is high. Since the visible plume particles follow nearly ballistic trajectories (see below), this implies that the launch velocity distribution of the plume particles is steeper when Enceladus is near its orbital apoapsis. However, the velocity distribution of the particles does not change as dramatically as the particle output. These trends probably reflect changes in the fissures driven by the changing tidal forces (see the chapters in Part 1 of this volume), but the subtlety of the velocity distribution changes also places stringent limits on the degree to which tidal forces can change the geometry of source vents (see section 6).

The gas flux also varies, but its behavior is less well characterized than that of the particles. The first and largest variations in gas flux were described by *Saur et al.* (2008), who attributed variations in the magnetometer data from the first three Enceladus encounters to flux decrease from 1600 kg s^{-1} for E0 (orbit phase of 230°) to 200 kg s^{-1} for E1 (orbit phase of 160°) and E2 (orbit phase of 110°). This early analysis did not incorporate some of the later knowledge of the plume, including the large component of charged dust, and that may explain the E0 flux, which is larger than other measurements. The first INMS data that were sufficient for mass-flow analyses were E3 (270°) and E5 (200°), which showed more-modest variations in vapor flow [*Smith et al.* (2010) found a factor of 4 difference between E3 and E5; *Dong et al.* (2011) and *Yeoh et al.* (2017) found a factor of 2] that were consistent with the particle-flux orbit-dependent variations. [Analyses of later INMS data, particularly E14, E17, and E18, which had parallel trajectories, also showed variations that correlated to orbit phase (*Perry et al.,* 2015).] These INMS total-inferred flow rates depend heavily on modeling, as INMS only measures gas density along a single instantaneous line through the plume. In contrast, the UV occultations provide an integrated sampling across nearly the entire plume, providing column density, but assume a gas velocity to obtain a gas mass flux. The first three UV occultations, which all occurred during the low-grain-output phase of Enceladus' orbit, showed little variation in the plume gas flux output (*Hansen et al.,* 2011). The fourth and last occultation (*Hansen et al.,* 2017) occurred closer to Enceladus' apoapsis. Although the total column density was similar to previous UVIS measurements, there is indication that the stronger, narrow jets were denser than during previous measurements (*Hansen et al.,* 2017). A tentative explanation is that the diurnal variations of the prominent jets are greater than those of the broader plume. Since vapor velocity was not measured directly in these analyses, most of them [*Saur et al.* (2008) is an exception] required an assumed temperature to convert the derived Mach number to total flux. As discussed in the chapter by Smith et al. in this volume, the total mass of the E-ring neutral torus provides an additional constraint on the time-averaged vapor flux from Enceladus.

The changes in the plume's overall particle output on orbital timescales may be the best-understood variations in Enceladus' geological activity. Furthermore, the observed activities of particular sources do not show any apparent correlation to orbital phase or tidal stress state (*Hurford et al.,* 2012). This suggests that while the overall particle flux from the plume can exhibit very regular behavior associated with orbital position over several years, individual sources and vents can turn on and off in a much more stochastic manner. In this context, it is worth noting that *Roddier et al.* (1998) described a bright streak in the E ring in 1995 telescopic observations that could potentially represent a major outburst of material from Enceladus, the likes of which has never been observed by Cassini.

On longer timescales, it appears that the plume's activity level, as measured by the particle flux, has decreased over the course of the Cassini mission, being between 50% and 100% higher prior to 2008 than it has been since that time (*Hedman et al.,* 2013; *Ingersoll and Ewald,* 2017). This could represent a slow decrease in Enceladus' activity driven by the progressive choking off of various vents, but the magnitude of the decrease makes this idea problematic because it would imply that Enceladus will stop venting material entirely in a few more years, and the persistence of the E ring suggests that Enceladus' output has persisted in some form for decades. Instead, it seems more likely that the slow decrease is part of some long-term cycle in plume activity. Two mechanisms that might be involved in this are the 11-year variations in Enceladus' orbital eccentricity (which may influence tidal stresses) (*Ingersoll and Ewald,* 2017) and seasonal changes in the solar illumination at the south pole (which may affect thermal stresses in the near-surface crust). Note that the eccentricity variations are very small, and so it is unclear if they can significantly affect plume activity, and no one has yet evaluated whether solar heating can significantly affect the plume output given that the relevant thermal wave can only penetrate a few meters below the surface.

4. TRANSITION TO/FROM WATER TO VAPOR/GRAINS ABOVE THE SURFACE: NOZZLE FLOWS, PARTICLE DISTRIBUTIONS NEAR THE JETS, AND THERMODYNAMICS

We now begin a discussion of the gas- and grain-dynamics. To provide context, Fig. 9 illustrates features and dominant physics expected in different regions in the eruption. At region A, which is well below the surface, there is liquid, yielding droplets and vapor that make their way to the surface as described in the chapter by Spencer et al. in this

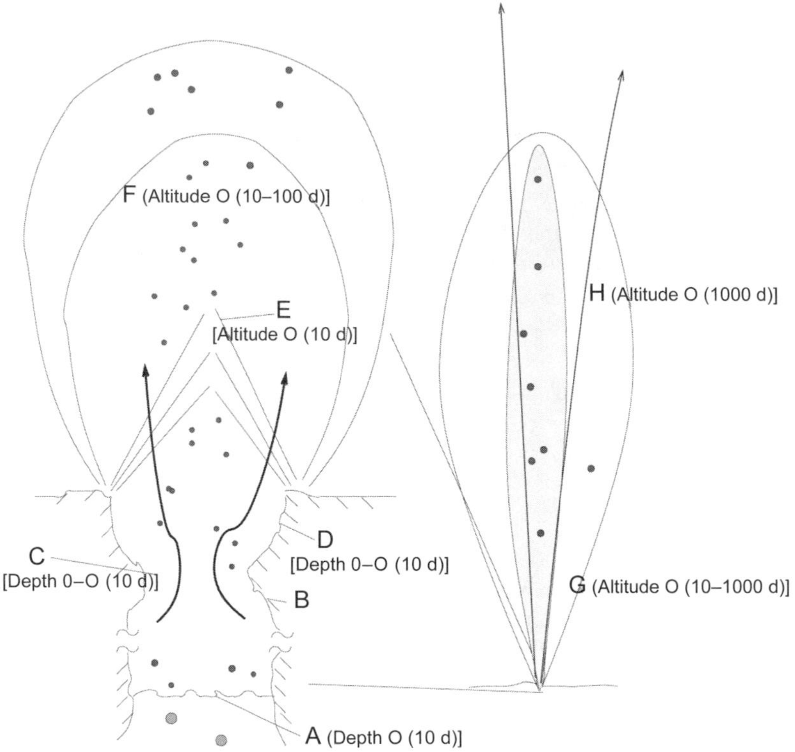

Fig. 9. Schematic drawing of the features and regions of dominant physics (A–H described in the text) of a geyser. The depth or altitude of the region is given in terms of an order of magnitude multiple of vent diameters (d).

volume. As the mixture of vapor and particles approaches the surface (region B), it has been moving slowly but can accelerate through a constriction or gas-dynamic throat and continue expanding and accelerating below the surfacc (region C) due to an increase in conduit cross section or perhaps due to condensation on the sidewalls. In region D (discussed in this section), the flow may still be interacting with the icy conduit via heat, mass, and momentum exchange. Once above the surface (section 6), the gas/particulate flow will pass through expansion waves (region E). There will be a continuing conversion of thermal energy to directed energy as the gas accelerates upward from the surface and cools. As the above-surface geyser continues to expand, the gas and particulate densities drop (region F). There is no background atmosphere to interfere with the plume expansion into vacuum. The particulate motion will decouple from the vapor motion in region G (*Yeoh et al.,* 2015), and some particulates can fall back to the surface. Even further above the vent (region H; see section 7), the mean free path length for intermolecular collisions increases to tens of kilometers and molecules and particles thereafter follow ballistic trajectories subject to only gravitational and, when ionized, electromagnetic forces. All Cassini actual observations of aloft gas and particles occurred near and above region G where the particles and vapor had already parted ways.

As the subsurface two-phase flow approaches and broaches the satellite's surface it will expand rapidly (Fig. 9, region D). Due to the rapidity of the final stage of the expansion, it is a good approximation that the process is adiabatic — there is no energy exchange via conduction or radiation between the flow and the surrounding conduit sidewalls. If the flow is sufficiently rapid, the time is too short for equilibration of vapor pressure with the sidewalls. It is presently unclear if the process of vapor pressure equilibration would be causing condensation or sublimation of sidewall ice right near the exit — it depends on the exact details of surface temperature, vent geometry, and gas dynamics and detailed simulations coupling all of the relevant physics have not been done (see discussions in *Ingersoll and Nakajima,* 2016; *Nakajima and Ingersoll,* 2016).

If the flow *is* adiabatic and mass exchange between the gas and the sidewalls at the exit is negligible, the vent will act as a nozzle through which the flow expands to supersonic speeds, as has been suggested by the narrowness of the jets (*Hansen et al.,* 2011). In an ideally designed rocket nozzle meant for operation with gas flows into the vacuum of space, the flow is *sonic* at the narrowest area (the throat). Sonic conditions mean the flow speed is equal to the local speed of sound, $\sqrt{\gamma RT}$ if the flow is of an ideal gas, where γ is the ratio of specific heats and R the gas constant [and perhaps less if it contains even a modest mass fraction of particles; e.g., if the particulate mass loading is 10% and the grains move with the gas in complete gas-particle equilibrium, the speed of sound is reduced by ~8% (*Kieffer,* 1982; *Kilegel,* 1966)]. Once past, the throat flow expands to high speed at the exit while moving straight out of the nozzle along its axis. In a less-than-ideal nozzle, there are viscous losses of flow momentum that may cause shock waves, flow separation from the nozzle walls, and condensation. Also, if the length of the nozzle is insufficient to accelerate the flow to near zero temperature (at which point all thermal energy in the flow has been converted into directed kinetic energy), the exit plane Mach number will be finite and the gas (but not the particles) will continue to expand laterally and axially even after it leaves the nozzle. Presumably, the actual Enceladus vents are not ideal and the finite exit Mach number flow will continue to expand above the surface. Since the processes that cause temporal variability in particle output have less than a 10% effect on particle speed (Fig. 10), those processes may occur below any gas dynamic throat because the exit-plane velocity and Mach number are determined by the throat-to-exit area ratio.

The phase diagram of water for low temperatures and entropies, Fig. 11, provides insight into some of the processes controlling the mass fractions of ice and vapor that enter the plume. If the plume is supplied by liquid water transported in near local thermodynamic equilibrium directly from an ocean at close to 273 K (red circle in Fig. 11), only a few percent of the mass of the plume would be vapor, assuming decompression following either the red or green curves in Fig. 11 ends at ~190 K due to kinetic limitations when temperature and density are low and gas velocity is high (p. 459 in *Lu and Kieffer,* 2009). But the *observed* mass fraction of vapor appears to be 90–95% (e.g., *Gao et al.,* 2016), although this value is poorly constrained and is certainly dependent on exactly where one looks within the plume. Still, one-way flow of ocean water into a thermodynamically equilibrium plume is incompatible with the observed ice/water ratio derived from Cassini measurements, and there must therefore be some process of circulation or overturning in the plumbing system to recycle liquid that then does not erupt as vapor (e.g., *Postberg et al.,* 2011; *Kite and Rubin,* 2016; *Ingersoll and Nakajima,* 2016) and/or removal combined with ballistic return of solid ice in the lower portions of the jets and plume. These general conclusions should not differ much if the water is salty because the dissolved salts do not change the thermodynamic properties of water significantly. However, the presence of salt does change the temperature-dependence of liquid density, which in turn affects convection and freezing of liquid in the vent (*Ingersoll and Nakajima,* 2016).

The height to which particles are ejected and hence how plume brightness decays with altitude, the scale height, depends on the particle-size distribution and the square root of the eruption speed of particles at the vent. If that scale height does not change over time (*Hedman et al.,* 2009; *Nimmo et al.,* 2014; *Ingersoll and Ewald,* 2016) or vary with integrated brightness for numerous different Cassini images, the implications are that there is a constant velocity distribution for visible particles and a constant particle-size distribution. This observation is consistent with the hypothesis that the flow emerging from the vents is choked to the speed of sound of the gas-and-solid mixture. One interpretation is that the integrated brightness is proportional to the cross-sectional area of the throats of the vents feeding the eruptions and

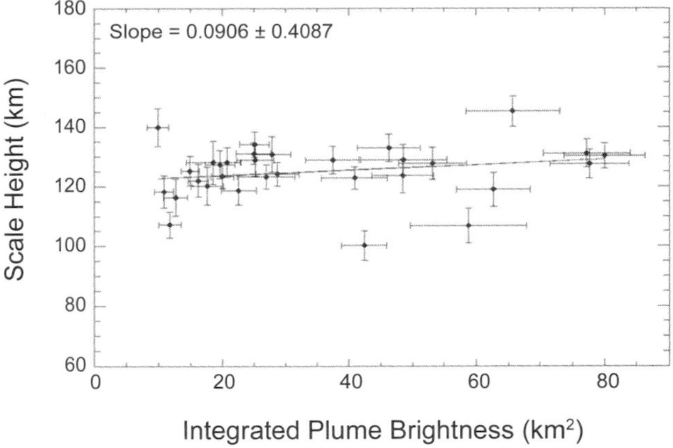

Fig. 10. Plume scale height, obtained assuming brightness decays exponentially with height, vs. integrated plume brightness from many ISS images (*Nimmo et al.,* 2014). The analysis is based on I/F measurements for altitudes from 50 km to 500 km. An early analysis (*Porco et al.,* 2006) found a scale height of 30 km when analyzing data from the surface to 50 km. This difference could be due to the larger, slower, bound particles that dominate the lower altitudes (*Postberg et al.,* 2011; *Hedman et al.,* 2009) and that brightness does not decay exponentially with height (*Hedman et al.,* 2013; *Ingersoll and Ewald,* 2017).

the Mach number is very high: Since brightness varies by a factor of ~4 throughout the orbital cycle, the open throat area thereby also varies by a factor of ~4.

5. COMPARISON WITH GEYSERS AND VOLCANIC ERUPTIONS ON EARTH

Several other bodies in the solar system have active plumes. Earth's largest plumes arise from volcanos and are similar to Enceladus only in that they are gas + particulate flows. But Earth's water geysers are perhaps the closest analog to those on Enceladus. However, there are significant differences: Terrestrial geysers are hot (the vapor + droplet emanating flow was not recently in contact with ice), they exhaust into a dense background atmosphere, and gravity dominates the large-scale motions. Entrainment of ambient air and heat exchange with this air dominate the ascent of volcanic plumes (e.g., *Woods,* 1995) and modifies the height and shape of geyser jets (*Karlstrom et al.,* 2013). Enceladus' jets expand into a vacuum and particles should follow ballistic trajectories once they separate from the vapor. A volcanic plume on Io, in contrast, gains a certain relation to the Enceladus plume in that there is no background atmosphere. But ionian plumes also arise from hot rock or lava, sometimes impinging on ice, and are ultimately gravity-dominated. Triton's plumes, though only slightly explored, also rise into a background atmosphere. The suggested Europa south polar plume (*Roth et al.,* 2014; *Sparks et al.,* 2016) may be similar to Enceladus in all but the gravity field and net mass flow. At another end of the spectrum, cometary plumes are similar to the Enceladus plume in terms of their

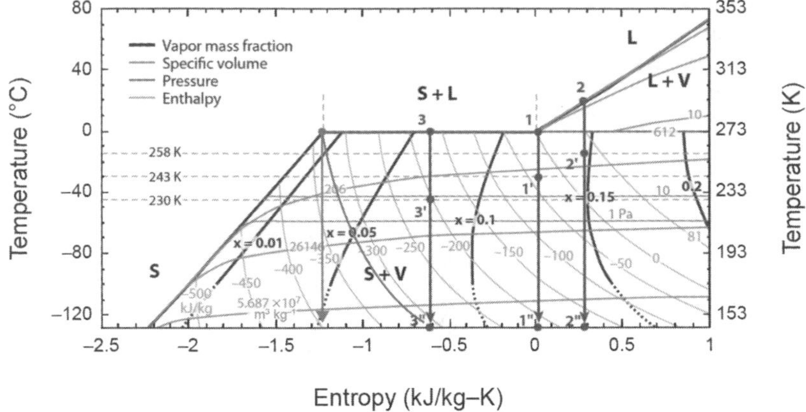

Fig. 11. See Plate 25 for color version. Phase diagram for water at low temperatures and entropies (from *Lu and Kieffer,* 2009). From the red circle to point 1 is liquid water in contact with ice. Isenthalpic (green curve) or isentropic (red line and other vertical lines) decompression increases the mass fraction of vapor (burgundy curves), but only leads to a mass fraction of several percent vapor. In contrast, the plume has an observed vapor mass fraction of 90–95% (e.g., *Gao et al.,* 2016).

gas/particulate dynamics, constituents (*Waite et al.,* 2009), mass flow, velocities, and exhaust into vacuum, but differ in their driving source mechanisms as comets do not have a reservoir of liquid water. The high Mach number narrow jets present on Enceladus and comets indicate that both have vapor expanding through a throat or nozzle. This might be due to the competing effects of condensation and freezing that tend to close a vent opening and the vapor pressure that keeps the vent open or clear.

In terrestrial instances of geologic plumes, eruption speed is usually assumed to be limited to the speed of sound of the gas + particle mixture at the vent (e.g., *Mastin,* 1995) — flow is choked. Observational evidence for choked flow is lacking for magmatic volcanos because the masses of gas and particles are difficult to constrain. Choking has been inferred from measured velocities and vapor-liquid ratios at geysers (*Karlstrom et al.,* 2013; *Munoz-Saez et al.,* 2015), but the uncertainties are large. Establishing that flow is choked is "a notoriously difficult problem" (*Kieffer,* 1989). The observations of high-Mach number vapor (sections 6 and 7) and the collimated particle flow (Fig. 3) on Enceladus, a case in which there is no atmospheric backpressure, are perhaps the strongest observations for choking in natural multiphase eruptions. Important questions are *where* they choke, and whether the flow above that point is reasonably modeled as adiabatic and without much mass exchanged with the conduit walls. If the Enceladus plume nozzles were simply converging nozzles, flow would be sonic at the surface and there would be much expansion of the gas above the surface. The fact that we see particulate beams suggests that that size range of particles we see were beam-formed below the surface. This might have been in a diverging section of the conduit near the surface and the particles became collimated like the gas (see discussion in *Yeoh et al.,* 2015). Or it could have occurred by some process that selectively skimmed laterally moving particles that touched the side walls, irrespective of the gas motion.

The eruptions of magmatic volcanos and geysers on Earth are driven by the volume expansion of gas produced during decompression. Enceladus' eruptions are more similar to geysers, which are one-component systems with mass exchange between liquid, solid, and vapor (Fig. 11), whereas magmatic volcanos only exsolve dissolved gases. Non-condensable gases, such as N_2 and CO_2, have been proposed to play a role in initiating the eruptions of terrestrial geysers (*Lu and Kieffer,* 2009; *Ladd and Ryan,* 2016; *Hurwitz et al.,* 2016), and there are episodically erupting cold geysers driven by CO_2 exsolution (e.g, *Watson et al.,* 2014). The importance of the other gases for boiling geysers, other than slightly changing the liquid stability field, remains unclear (*Hurwitz and Manga,* 2017). The role of observed non-condensable gases [such as CO_2, CH_4, NH_3 and H_2 (*Waite et al.,* 2014)] in eruption processes on Enceladus is also unclear, including their role in driving ascent and sustaining flow within conduits (*Matson et al.,* 2012). But compared to hot geysers on Earth, components dissolved in the cold Enceladus geysers other than H_2O appear to be a larger fraction of the erupted materials.

6. EXPANSION AND ACCELERATION AT THE VENT: GRAIN VELOCITIES, TEMPERATURES, NUCLEATION, AND GROWTH

The observed erupted grains can be separated into three regimes based on their ejected velocity. The slowest group rises out of vents and falls out of the plume in close proximity to the tiger stripes. These particles tend to be large (*Degruyter and Manga,* 2011). Another, faster-moving group, distributes itself broadly over the surface (*Kempf et al.,* 2010). A third group has velocities greater than the escape speed of Enceladus and leaves the satellite altogether. The size and speed relationships of the different populations serve as a diagnostic of the physics of their origins. In this section we discuss the first and second groups of particles that return to the Enceladus surface; in section 7 we concentrate on the escaping group.

The vapor in a high-Mach-number geyser leaving the surface (region E and beyond, Fig. 9) will be highly supersaturated and will condense on existing ice grains and water droplets (heterogeneous nucleation) to increase their size. There exists a competition, however, between the rapid decrease of gas density that results in a decrease of the rate of molecules colliding with particles (a kinetic limitation), and the growth of the particulate size. *Yeoh et al.* (2015) show that given an assumption of meter-scale vent openings and sonic exit conditions (the throat being at the surface in their discussion), particulates can only grow by at most ~1 μm before the collision rate drops enough to stifle grain growth within several vent diameters of the exit. Condensation of the supersaturated gas may also produce new grains, which range from clusters of a few molecules to several hundred nanometers in size. These nanograins, some of which acquired a charge and were measured by CAPS, are discussed in section 7.

Analyses of CDA *in situ* measurements of grains near Enceladus and in the E ring provide crucial constraints on the size distribution, the velocity distribution, and the sources of the particles. In the E ring, 0.5–2% of the ice grains are rich in sodium salts (*Postberg et al.,* 2009) and have compositions (see the chapter in this volume by Postberg et al.) that confirm Enceladus harbors a subsurface ocean in contact with a rocky core. Near Enceladus, the sodium-rich grains are larger and slower (*Postberg et al.,* 2011), and comprise 99% of the particulate mass ejected from the surface vents (Fig. 12). This is incompatible with non-liquid models for the plume source, and is expected for ice grains that are frozen aerosols or sprays working their way from a liquid ocean to the surface (Schmidt *et al.,* 2008; Spencer et al., this volume). Combining the near-Enceladus and E-ring CDA measurements, *Kempf et al.* (2010) find that the same particle size-and-velocity distributions can explain the particle-size distribution measured over the surface by VIMS (*Jaumann et al.,* 2008), the plume measurements by VIMS and CDA, and the scale height and distribution of particles in the E ring.

Near-infrared spectra of the plume obtained by VIMS provide relatively direct measurements of the size distribution for particles between 1 and 4 μm in radius. The VIMS plume observations are obtained at high phase angles where the signal is primarily due to light diffracting around individual particles. In this limit, a particle of a given radius s observed at a phase angle α scatters light most efficiently at wavelengths $\lambda \sim s/(\pi-\alpha)$. Hence there is a fairly direct mapping between the plume's brightness variations with wavelength and the shape of the particle size distribution. Indeed, *Hedman et al.* (2009) were able to use the VIMS spectra to determine the relative numbers of particles with radii of 1, 2, and 3 μm over a range of altitudes within the plume. This work demonstrated that the number density of 3-μm particles falls much more rapidly with altitude than the number density of 1-μm particles. This implies that the typical launch velocity of 3-μm particles is lower than that of 1-μm particles.

VIMS data can also indicate grain sizes of the large particles on the surface next to the tiger stripes (*Jaumann et al.,* 2008), providing additional constraints on the size of particles that erupt. Deposited ice particles can grow from sputtering (e.g., *Clark et al.,* 1983) or sintering (e.g., *Kaempfer and Schneebeli,* 2007), but at cold surface temperatures these processes may be slow (*Spencer et al.,* 2006) compared to particle deposition times (*Kempf et al.,* 2010). Using the gas flow model of *Ingersoll and Pankine* (2010) and the assumption that the acceleration of particles is limited by the distance they travel between collisions with conduit walls (*Schmidt et al.,* 2008), *Degruyter and Manga* (2011) modeled the acceleration of particles within the conduit and their ballistic transport once they exited the vent — large particles achieve lower exit speeds and hence are transported to lower altitudes and are deposited closer to the vents. *Degruyter and Manga* (2011) fit the predicted transport to the VIMS surface observations of grain sizes in the range of 20–30 μm at distances of 5–10 km from cracks (*Jaumann et al.,* 2008), to determine the relationship between the distance over which particles are accelerated in the conduits and the gas temperature (although grain growth

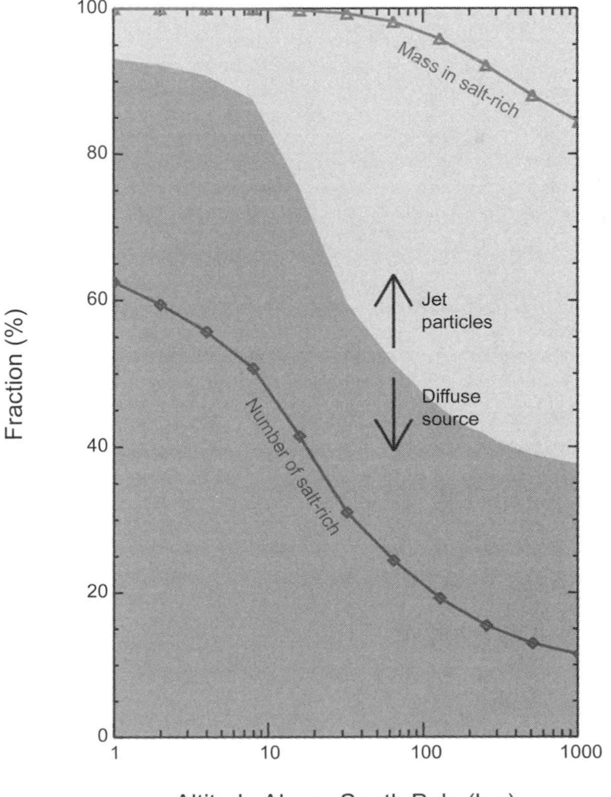

Fig. 12. The fraction of salt-rich grains in the plume decreases with increasing altitude as the larger, slower, salt-rich grains fall back to the surface. Most of the mass of the grains that emerge from the vents is in the larger, salt-rich grains, but most of the grains that escape from Enceladus are salt-poor. From *Postberg et al.* (2011).

through sintering could contribute as well). Figure 13 shows the predicted relationship between particle size and altitude. While the model is constrained only by the surface observations, it is consistent with the high-altitude CDA, VIMS, and E-ring measurements, suggesting that the surface and high-altitude particle sizes originate from the same eruptions but that larger particles move slower than small ones.

Particle-size distributions derived from ISS (visible) images depend on the topology of the particles as smooth spheres interact with light differently than rough-surfaced aggregates of the same size and mass. The *Gao et al.* (2016) analyses of high-phase-angle ISS data show that aggregate models produce grain/vapor ratios of 0.07 ± 0.01, explaining some of the discrepancy between the larger ratios derived from smooth-sphere models (*Ingersoll et al.,* 2010) and the ratios derived from CDA data.

There are two implications of the model and observations of surface grain size. First, particles greater than tens of micrometers erupt, and this is a size greater than what would simply condense from vapor (*Schmidt et al.,* 2008; *Yeoh et al.,* 2015). These large particles may be frozen droplets from the ocean, or mechanically produced particles from the conduit walls. Second, still larger particles may be erupting, but these will remain close to the surface and be deposited close to the surface vents.

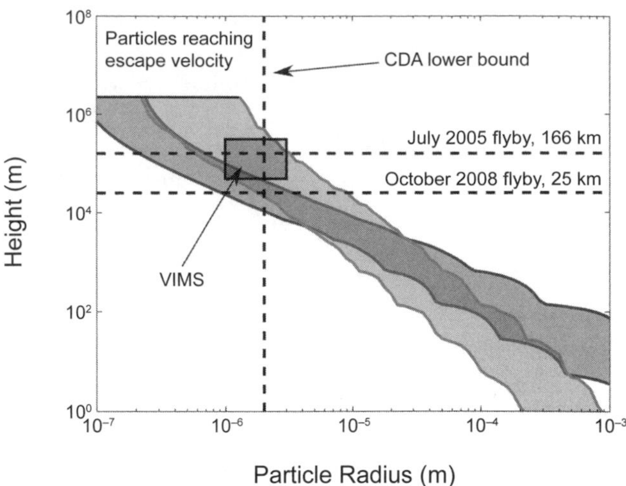

Fig. 13. Maximum particle height as a function of particle size for two models, both constrained by VIMS inferences of particles sizes adjacent to the tiger stripes (*Degruyter and Manga,* 2011). The gray band, which at 1 m height spans particle sizes of 2×10^{-4} to 10^{-3} m, indicates acceleration is limited by particle collisions with walls, while in the other band acceleration is limited by gas drag. The region upper and lower bounds for each color are for gas temperatures between 190 K and 273 K. The CDA lower bound is from *Spahn et al.* (2006); the VIMS box is from *Hedman et al.* (2009). The wiggles in the curves arise from the form of the drag model for particles as it transitions from free molecular to transitional to slip flow to continuum flow (*Crowe et al.,* 1997).

7. THE COLLISIONLESS REGIME: SEPARATION OF MOLECULE AND GRAIN MOTION AND MODEL FITS TO IDENTIFY VENT EXHAUST PROPERTIES

CAPS observations of charged nanometer grains in the plume (*Jones et al.,* 2009; *Dong et al.,* 2015) compliment the observations of larger grains discussed in the previous section. CAPS data show a size distribution consistent with condensation during rapid expansion at the surface vent (*Yeoh et al.,* 2015). The observed grains range in size from clusters of a few water molecules to radii of 3 nm, with a peak number density at approximately 2 nm. Particles with sizes between the upper limit (4 nm) of CAPS and the lower limit (0.1 μm) of CDA during Enceladus encounters are likely, but those are unmeasured by Cassini instruments [see *Dong et al.* (2015) for a discussion of the possible distributions between the two size regimes]. The nanometer-sized grains are entrained in the vapor, and accelerate to near-gas velocities as they leave the vent. Most escape Enceladus, but comprise only a small mass fraction of the particulate supply to the E ring. They acquire a charge quickly, and are influenced by Saturn's magnetic field more than the larger particles (see the chapter by Kempf et al. in this volume). Estimates of the total mass of the nanograins range from 1% to 20% of the vapor (*Hill et al.,* 2012; *Dong et al.,* 2015); the large range is due to uncertainties in grain charging and in the CAPS response to nanograins.

The expansion process just above the vents involves a transition from the collisional gas flow regime to a collisionless one. *Yeoh et al.* (2017) showed that if the gas is sufficiently collisional at the vent, the subsequent gas dynamics adiabatic expansion will allow the nearly complete conversion of thermal motion to directed motion and the gas will reach nearly the ultimate speed of $\sqrt{2\gamma RT_0/(\gamma - 1)}$. Here, T_0 represents the stagnation temperature of the gas when it was last in thermal equilibrium with liquid or walls of the conduit. If the vent Knudsen number — the ratio of gas mean free path to vent diameter — is too high (greater than ~0.1), residual thermal (random) motion will remain. Hence molecules should reach an ultimate speed limited to about 1005 m s^{-1} by the water triple point temperature of 273 K. Whether the particles follow the gas flow will depend on the local Stokes number, which represents a ratio of grain response time to flow field changes compared to the rate of change of the flow field density along mean streamlines. As described by *Yeoh et al.* (2015), the particle and gas flows decouple at a height of up to 1000 vent diameters for nanometer-sized particles but only perhaps 100 diameters for micrometer-sized particles. Hence, the decoupling happens well below the height at which Cassini actually observed the gas or particulates.

Cassini observed the plume vapor and particulates when both were in the "free molecular" regime — when intermolecular collisions and molecule-particle collisions had ceased to affect molecule and particle trajectories. Hence, the observed characteristics of the plume were imprinted by lower-altitude physics. Those observations may thus be used to infer the unseen physics and conditions at the vent. In this section, we discuss INMS measurements of the plume vapor distribution, including observations of the broad vapor cloud and discrete gas jets, obtained during these flybys. These *in situ* data, in concert with UVIS stellar and solar occultations of the gas jets, and imaging of the grain jets, provide constraints on the properties (locations, magnitudes, and gas velocity) and time variability of the plume surface sources. Early observations of the gas jets by UVIS, during a plume occultation of the star zeta Orionis on October 24, 2007 (*Hansen et al.,* 2006), showed fine structure in the water vapor density on the scale of a few kilometers in the plume, suggesting the presence of supersonic gas jets having a ratio of bulk velocity to thermal velocity of 1.5 ± 0.2. In the subsequent years, UVIS would acquire several more stellar occultations, and one solar occultation during the May 18, 2010, E10 flyby with exceptionally good signal-to-noise (*Hansen et al.,* 2011), enabling multiple narrow, supersonic jets to be discerned. Additionally, during this time, INMS detection of the individual gas jets required several attempts, over multiple flybys, during which a number of measurement and instrumental challenges were overcome.

The E3 and E5 flybys on March 12, 2008, and October 9, 2008, were the earliest flybys through the plume for which INMS was aimed toward the spacecraft direction of motion to sample and measure the gas density (Fig. 14) and composition (see the chapter in this volume by Postberg et al., as well as Appendix Table A1). These two flybys both took place along similar north-to-south trajectories, and thereby encountered the plume after closest approach, sampling the plume density and composition as the spacecraft moved away from the south polar region. In their analysis of these two flybys, *Teolis et al.* (2010) modeled the adsorption of water vapor on the walls of the INMS gas inlet

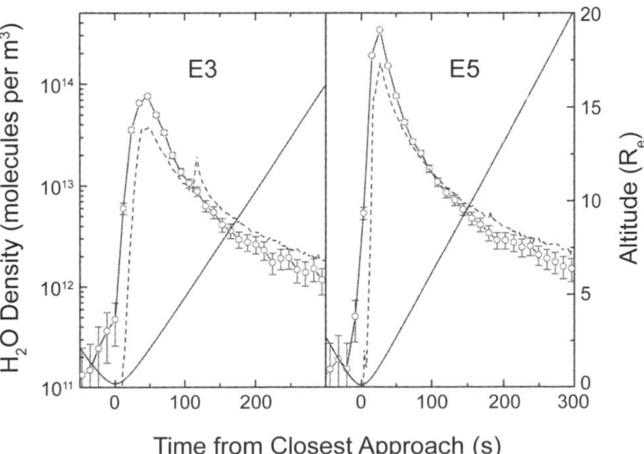

Fig. 14. Plume density (black line, left-axis scale) from models of the INMS measurements of H_2O at mass 18 u (dashed line) vs. time from closest approach for E3 and E5. The solid line shows the altitude (right scale) in Enceladus radii (252 km). Since water is adsorbed to the walls of the INMS inlet aperture, INMS response is delayed and extended compared to the plume density. From *Teolis et al.* (2010).

thermalization antechamber, and found that such sticking introduced a time delay and distortion in the INMS H_2O data. They determined that other plume volatiles including CO_2 vapor provided a more accurate representation of the plume density vs. position along Cassini's trajectory. The CO_2 E3 and E5 data showed that the plume vapor density decays approximately as the inverse square of the distance from the south polar terrain, consistent with collisionless vapor expansion from Enceladus well above the escape speed. Expanding on early UVIS based modeling, *Tian et al.* (2007), *Smith et al.* (2010), *Tenishev et al.* (2010) (both using E3 and E5 INMS data), *Dong et al.* (2011), *Tenishev et al.* (2014), and *Yeoh et al.* (2017) (using UVIS and E3, E5, and E7 INMS data and distributed tiger-stripe sources) applied analytical and Monte Carlo modeling, exemplified in Fig. 15, to estimate plume source properties, i.e., source rate, temperature, and gas velocity, by fitting the eight major grain jets identified from Cassini imagining by *Spitale and Porco* (2007) to these data.

Beginning with the 91-km E7 flyby on November 2, 2009, the Cassini spacecraft carried out a series of low-altitude (<100 km) traversals over the south polar terrain, directly through the plume and sufficiently close to the tiger stripes to observe the detailed spatial distribution or structure of vapor in the gas jets. During these flybys, only the most abundant plume non-sticky species, CO_2 [with a 5% mixing ratio (*Waite et al.,* 2009)], had sufficient signal-to-noise in INMS data to enable accurate measurements of local density variations due to jets along Cassini's trajectory. For the 7.7-km s^{-1} flyby speed and sample cadence of 1.5 s for E7, the spatial resolution of CO_2 measurements was 12 km, which provided only poor resolution of the jets in the E7 data (*Perry et al.,* 2015). To improve resolution in later encounters, the INMS team adjusted the measurement strategy, concentrating the INMS mass scans on the 44-µm CO_2 channels, which yielded CO_2 density data at a higher, 0.25-s temporal and 1.9-km spatial, resolution. As shown in Fig. 16, CO_2 data from E14, E17, and E18 clearly resolved density variations, indicative of gas jets, along Cassini's trajectories. In Fig. 17, we show a three-dimensional projection of these data over the Enceladus south polar terrain, to illustrate how the jet structure observed by INMS is spatially distributed relative to the tiger stripes. Using the E14, E17, and E18 INMS data, *Hurley et al.* (2015a) suggested on the basis of Monte Carlo models that the plume source may be continuously distributed, albeit variable, along the tiger stripes. The complete plume three-dimensional structure is difficult to uniquely constrain solely on the basis of the few INMS flybys as multiple combinations of jet pointing directions and intensities can fit the data, and the jets may be time variable (section 3).

Several groups have modeled the gas density and velocity distributions as the sums of many non-interacting jets emanating from the *Spitale and Porco* (2007) or *Porco et al.* (2014) jets and tiger-stripe fissures. *Teolis et al.* (2017) (with UVIS and INMS data) and *Portyankina et al.* (2016) (using UVIS data) have modeled the plume with all 98 *Porco et al.* (2014)

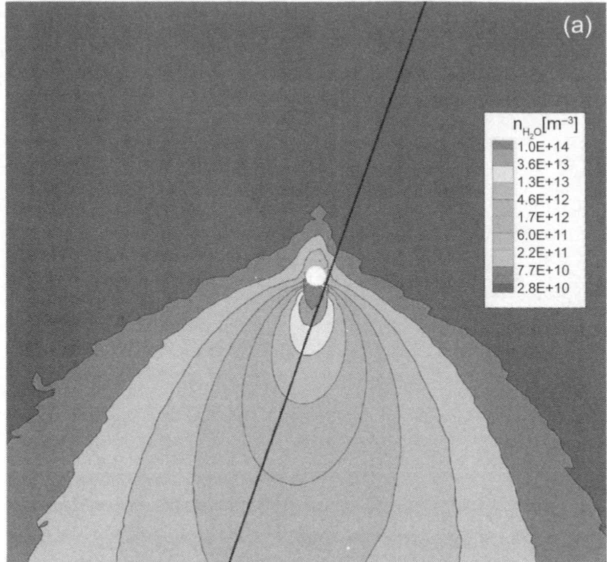

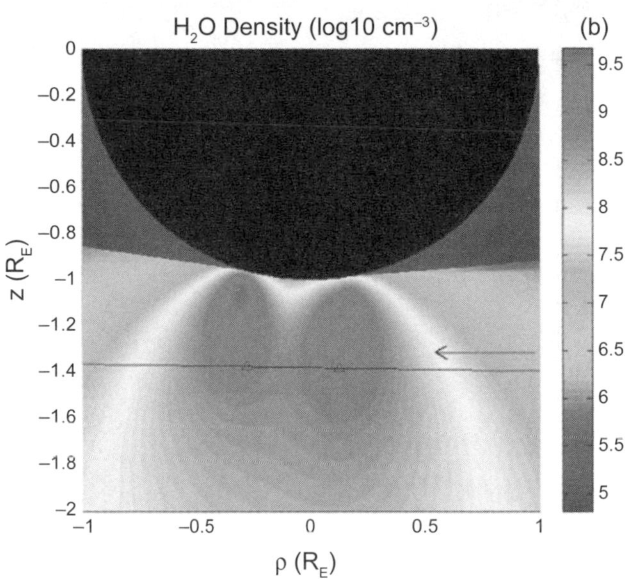

Fig. 15. See Plate 26 for color version. **(a)** Monte Carlo [*Tenishev et al.* (2010) E5 flyby shown] and **(b)** analytical [*Dong et al.* (2011) E7 flyby shown] plume water vapor density models, from fits to INMS E3 and E5 (both models), and E7 [*Dong et al.* (2011) only], with the *Spitale and Porco* (2007) sources as the constraint.

jets. *Teolis et al.* (2017) assumed a "drifted Maxwellian" velocity distribution: a random (thermal) spreading superimposed on a bulk velocity. They summed the inputs from separate sources assuming non-interacting (i.e., collisionless) jets, a justified approximation since the jet spacings of a few kilometers are below the ~10-km molecular mean free path at the maximum ~3×10^{14} m^{-3} H_2O plume gas densities observed by INMS. *Yeoh et al.* (2017) accounted for intermolecular collisions using well-resolved DSMC simulations for the individual *Spitale and Porco* (2007) jets, including a solution directly coupled to a detailed subsurface simulation.

Modeling results from the INMS team (*Teolis et al.*, 2017) are shown in Fig. 18. Their modeling considers two possible plume contributions: (1) an upward-directed gas

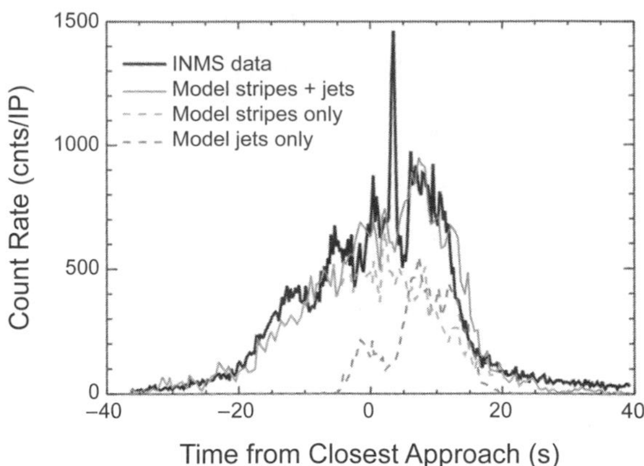

Fig. 16. See Plate 27 for color version. INMS measurements of mass 44 u species during the E17 flyby are shown in black. Model (*Hurley et al.*, 2015) predictions using constant emission along the tiger stripes at 500 m s^{-1} and 270 K (gray dashed line) are selected to match the rise and fall on the outskirts of the plume. The jet model using 1500 m s^{-1} and 270 K are included (blue dashed line) to reproduce the overall enhancement near closest approach. The sum of the two models, shown in red, reproduces the overall structure of the plume but misses some of the fine structure.

source continuously distributed along the tiger stripes, and (2) multiple jets at discrete tiger stripe locations. The jets are given a four-point Mach number distribution ranging from 0 (gas at rest) to 12 (the fast component), as required to best fit the shapes of the features in the six best INMS datasets and two UVIS occultations. Although the fits are not unique, there are only a few families of solutions, and the distribution of Mach numbers is similar among them, with jets with Mach numbers of four or greater required. The fit to the INMS densities and to the UVIS column densities yields broad agreement with the continuous emission model, but the jet model more successfully captures local variations and peaks. The results also suggest stochastic time variability in the plume sources along the tiger stripes. A pattern of systematic variation in specific jets with mean anomaly cannot be presently discerned, probably due to details in the geometry of individual fissures beyond detectability.

Other approaches used a combination of localized jets and continuous emission from the tiger stripes. As an illustration of the quality of the various model fits and to highlight that the models appear to encompass much of the physics to explain the bulk of the observations, we highlight the following: *Hurley et al.* (2015) (Fig. 18) fit the tiger-stripe model to the broad shape of the INMS measurements and used the jets to fit the fine structure. They also investigated the potential for different-mass molecules to have different behavior based on their thermal velocities. *Tenishev et al.* (2014) (Fig. 19) found that simulating the UVIS 2007 occultation data required both jets and continuous tiger stripe emission. *Yeoh et al.* (2017) (Fig. 20) also finds that the best fits require the combination of strong jets and a more-diffuse source from the tiger stripes. Analyses from all these groups used only the *Spitale and Porco* (2008) jets.

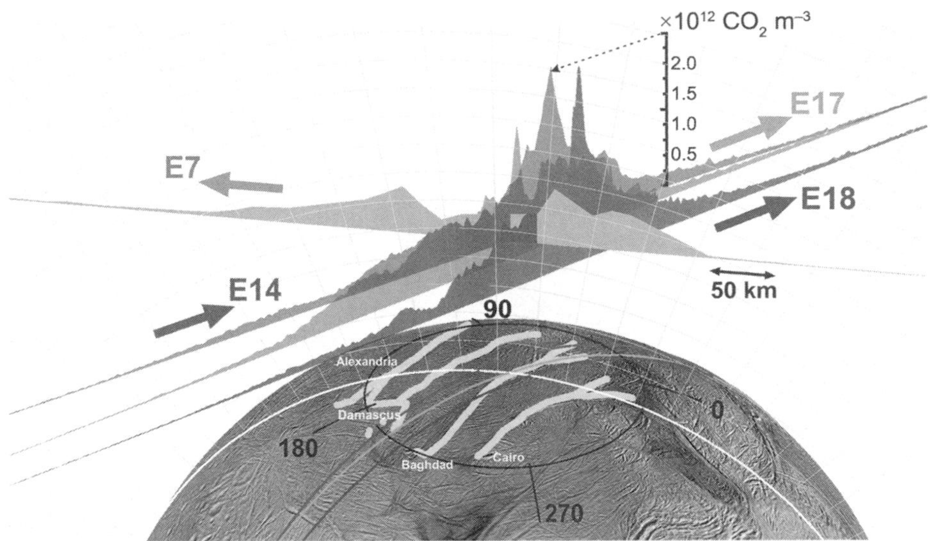

Fig. 17. See Plate 28 for color version. To-scale three-dimensional representation of the E14, E17, E18, and (lower-resolution) E7 INMS data, with vertical areas representing (in linear scale) the density, and the flat base of the areas corresponding to the Cassini trajectories.

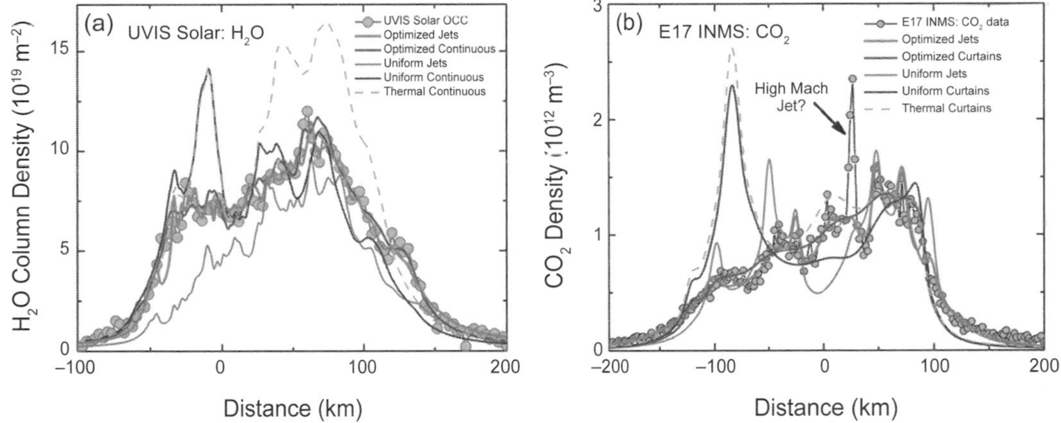

Fig. 18. See Plate 29 for color version. **(a)** Enceladus plume water vapor column density measurement from the UVIS 2010 solar occultation (dotted line), plotted vs. distance across the plume along the occultation line of sight minimum ray height. Thin lines: modeling solutions (*Teolis et al.,* 2017) assuming continuous emission along the tiger stripes. Thick lines: Solutions assuming the *Porco et al.* (2014) jets. **(b)** INMS CO_2 density measurement (dotted line) along the E17 flyby trajectory, showing peaks suggestive of discrete plume sources. Lines: Solutions with *Porco et al.* (2014) jets as the constraint.

During four encounters, INMS used its alternate mode for observing neutrals, the Open Source Neutral Beam (OSNB) mode. During one of encounters, E8, Cassini passes through a jet while rotating so that INMS sampled the water-molecule velocities relative to Enceladus (*Perry et al.,* 2016). The measurements show a bulk velocity of 1.1 ± 0.2 km s^{-1}, which is consistent with the 1.0-km s^{-1} ultimate velocity of adiabatic expansion. During another encounter, E11, similar measurements showed velocities of 600 m s^{-1} outside the strong jets. Combined, these results support the multiple-velocity models of the plume and supply the velocities necessary to interpret the Mach numbers derived from those models. Also, since the different groups did not use a single speed or jet spreading angle for the gas in their models, it is not straightforward to say whether or not simply a difference in assumed/modeled surface exit flow properties is what leads to the different fitted mass fluxes.

8. CONCLUSIONS AND OPEN ISSUES

We list several issues concerning the plume that appear to be resolved by the Cassini observations and modeling but whose resolution itself opens up other questions.

Is Enceladus the dominant source of the E ring? Yes, by ice grains entrained in the plume, not by excavation of grains by micrometeorites, the mechanism proposed before the plumes were discovered (see the chapter in this volume by Dougherty et al.).

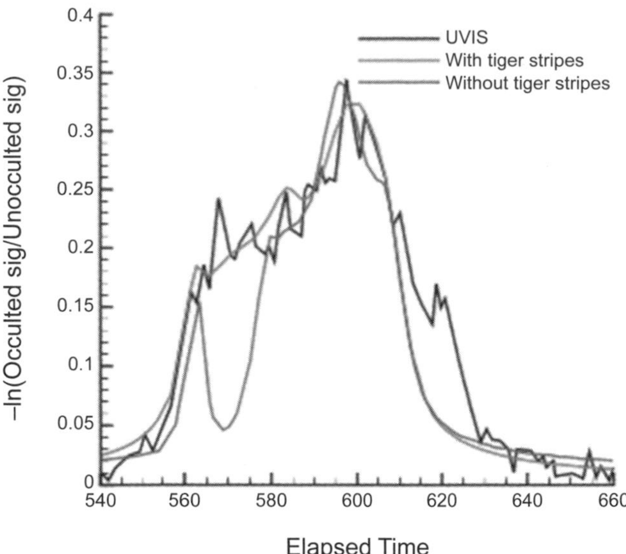

Fig. 19. Comparison of the *Tenishev et al.* (2014) results with the 2010 UVIS solar occultation observation of *Hansen et al.* (2011). Without gas production distributed along the tiger stripes, the model does not reproduce the UVIS scan at the elapsed time of 448.9 s. Gray line with big dip at 570 s: Only the *Spitale and Porco* (2007) sources were used, and the model does not reproduce all the peaks in the UVIS data. Higher-lying gray line: The addition of a tiger stripe gas source to the model is required to fit the data, suggesting additional sources distributed along the tiger stripes as later identified by *Porco et al.* (2014).

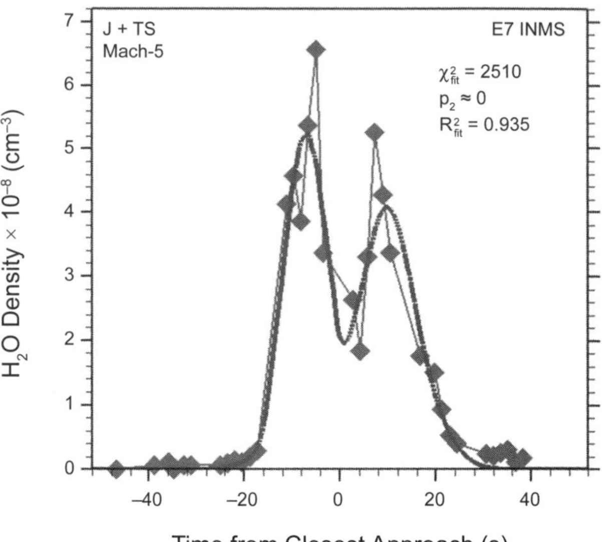

Fig. 20. Best fit of the *Yeoh et al.* (2017) model using Mach 5 exit conditions to the INMS E7 data using both jet and tiger-stripe sources.

Are clathrates the dominant source of the plumes? No. The CDA detection of frozen salty water, combined with gravity data and libration analyses, all point to a subsurface ocean.

Where does the plume originate? From the tiger stripe fractures in the south polar region. Geysers and diffuse sources appear to rise from vents that can be either highly elongated or compact. There remain important questions about the distribution of orifice aspect ratios among the vents.

What is the net mass flow in the plume? Most analyses estimate a few hundred kilograms per second, but there is a factor-of-2 uncertainty in some of the analyses and uncertainty due to the plume variability.

Is the plume just water vapor and ice grains? No, it contains other volatiles and ice particles with several distinct compositional variations. There remain questions about the relative fractions of the constituents and the nature of yet to be identified constituents (see the chapter by Postberg et al. in this volume).

Do the plumes vary, and, if so, over what timescales? Diurnal variations have been observed in the ice-grain flux by VIMS and ISS, and gravitational stress has been identified as the likely cause. ISS imaging and INMS data also give evidence for short-term stochastic variability in discrete sources/jets. But many details remain unknown, for example, the motion of the channel walls and vents that control the flux, the variability of the vapor, and the difference between jet and diffuse variability. The nature of shorter- and longer-term variability also still needs to be fully quantified.

APPENDIX

TABLE A1. Parameters of the Enceladus encounters discussed in this chapter.

Encounter	Date	Minimum Altitude	Features
E0	17 Feb 2005	1259	Magnetometer observations indicated an atmosphere; distant ISS images of plume
E1	9 Mar 2005	497	Imaging, magnetometer, and plasma measurements
E2	14 Jul 2005	168	Relatively far from ejected vapor; minimal data useful for plume structure; first INMS and UVIS identification of H_2O; first CDA measurements of dust; CAPS measurement of plume ions
	24 Oct 2007		UVIS stellar occultation; measurement of high-Mach jets, source rate
E3	12 Mar 2008	51	INMS pass; steeply inclined, fast (14 km s^{-1}) north-south trajectories, following the plume at altitudes more than 250 km
E5	9 Oct 2009	25	INMS pass; steeply inclined, fast (17 km s^{-1}) north-south trajectories, following the plume at altitudes more than 250 km
E7	2 Nov 2009	100	INMS pass; horizontal, slow pass (7 km s^{-1}), low, perpendicular to stripes, outbound from Saturn

TABLE A1 (continued)

Encounter	Date	Minimum Altitude	Features
E8	21 Nov 2009	950	Horizontal, slow pass south of Enceladus 1200 km below the equatorial plane; includes vapor velocity measurements from INMS OSNB data
	18 May 2010		UVIS solar occultation; high-Mach jets; measured source rate
E11	13 Aug 2011		
E14	1 Oct 2011	100	Horizontal, slow pass at low altitude with high-resolution INMS data, parallel to stripes, inbound to Saturn
E16	6 Nov 2011		
E17	27 Mar 2012	75	Horizontal, slow pass at low altitude, high-resolution INMS data, parallel to stripes, inbound to Saturn
E18	14 Apr 2012	75	Horizontal, slow pass at low altitude, high-resolution INMS data, parallel to stripes, inbound to Saturn; last CAPS data
E21	28 Oct 2015 2015	49	Horizontal, slow pass at lowest altitude through plumes, parallel to stripes, INMS OSNB measurements of H_2

REFERENCES

Clark R., Fanale F., and Zent A. (1983) Frost grain size metamorphism: Implications for remote sensing of planetary surfaces. *Icarus, 56,* 233–245.

Crowe C., Sommerfeld M., and Tsuji Y. (1997) *Multiphase Flows with Droplets and Particles.* CRC, Boca Raton, Florida.

Degruyter W. and Manga M. (2011) Cryoclastic origin of particles on the surface of Enceladus. *Geophys. Res. Lett., 38,* L16201, DOI: 10.1029/2011GL048235.

Dong Y., Hill T. W., Teolis B. D., Magee B. A., and Waite J. H. (2011) The water vapor plumes of Enceladus. *J. Geophys. Res., 116,* A10204.

Dong Y., Hill T. W., and Ye S.-Y. (2015) Characteristics of ice grains in the Enceladus plume from Cassini observations. *J. Geophys. Res.–Space Physics, 120,* 915–937, DOI: 10.1002/2014JA020288.

Gao P., Kopparla P., Zhang X., and Ingersoll A. (2016) Aggregate particles in the plumes of Enceladus. *Icarus, 264,* 227–238.

Goguen J. D., Buratti B. J., Brown R. H., Clark R. N., Nicholson P. D., Hedman M. M., Howell R. R., Sotin C., Cruikshank D. P., Baines K. H., Lawrence K. J., Spencer J. R., and Blackburn D. G. (2013) The temperature and width of an active fissure on Enceladus measured with Cassini VIMS during the 14 April 2012 south pole flyover. *Icarus, 226,* 1128–1137.

Hansen C. J., Esposito L., Stewart A. I. F., Colwell J., Hendrix A. P., Shemansky W. D., and West R. (2006) Enceladus' water vapor plume. *Science, 311(5766),* 1422.

Hansen C. J. et al. (2011) The composition and structure of the Enceladus plume. *Geophys. Res. Lett., 38,* L11202.

Hansen C. J., Esposito L. W., Aye K.-M., Colwell J. E., Hendrix A. R., Portyankina G., and Shemansky D. (2017) Investigation of diurnal variability of water vapor in Enceladus' plume by the Cassini ultraviolet imaging spectrograph. *Geophys. Res. Lett., 44,* DOI: 10.1002/2016GL071853.

Hedman M., Nicholson P. D., Showalter M. R., Brown R. H., Buratti B. J., and Clark R. N. (2009) Spectral observations of the Enceladus plume with Cassini-VIMS. *Astrophys. J., 693,* 1749–1762.

Hedman M. M., Gosmeyer C. M., Nicholson P. D., Sotin C., Brown R. H., Clark R. N., Baines K. H., Buratti B. J., and Showalter M. R. (2013) An observed correlation between plume activity and tidal stresses on Enceladus. *Nature, 500,* 182.

Helfenstein P. and Porco C. (2015) Enceladus' geysers: Relation to geological features. *Astron. J., 150,* 96.

Hill T. W., Thomsen M. F., Tokar R. L., Coates A. J., Lewis G. R., Young D. T., Crary F. J., Baragiola R. A., Johnson R. E., Dong Y., Wilson R. J., Jones G. H., Wahlund J.-E., Mitchell D. G., and Horányi M. (2012) Charged nanograins in the Enceladus plume. *J. Geophys. Res., 117,* A05209, DOI: 10.1029/2011JA017218.

Howett C. J. A., Spencer J. R., Pearl J., and Segura M. (2011) High heat flow from Enceladus' south polar region measured using 10–600 cm^{-1} Cassini/CIRS data. *J. Geophys. Res., 116,* E03003.

Hurford T. A., Helfenstein P., Hoppa G. V., Greenberg R., and Bills B. G. (2007) Eruptions arising from tidally controlled periodic openings of rifts on Enceladus. *Nature, 447,* 292–294.

Hurford T. A., Helfenstein P., and Spitale J. N. (2012) Tidal control of jet eruptions on Enceladus as observed by Cassini ISS between 2005 and 2007. *Icarus, 220,* 896–903.

Hurley D. A., Perry M. E., and Waite J. H. (2015) Modeling insights into the locations of density enhancements from the Enceladus water vapor jets. *J. Geophys. Res.–Planets, 120,* 1763.

Hurwitz S. and Manga M. (2017) The fascinating and complex dynamics of geyser eruptions. *Annu. Rev. Earth Planet. Sci., 45,* 31–59.

Hurwitz S., Clor L., McCleskey R. B., Nordstrom D. K., Hune A., and Evans W. (2016) *Dissolved gases in hydrothermal (phreatic) and geyser eruptions at Yellowstone National Park, USA. Geology, 44,* 235. DOI: 10.1130/G37478.1.

Kite E. and Rubin A. (2016) Sustained eruptions on Enceladus explained by turbulent dissipation in tiger stripes. *Proc. Natl. Acad. Sci., 113(15),* 3972–3975.

Ingersoll A. and Pankine A. (2010) Subsurface heat transfer on Enceladus: Conditions under which melting occurs. *Icarus, 206,* 594–607.

Ingersoll A. P. and Nakajima M. (2016) Controlled boiling on Enceladus. 2. Model of the liquid-filled cracks. *Icarus, 272,* 319–326.

Ingersoll A. P. and Ewald S. P. (2011) Total particulate mass in Enceladus plumes and mass of Saturn's E ring inferred from Cassini ISS images. *Icarus, 216(2),* 492–506.

Ingersoll A. P. and Ewald S. P. (2017) Decadal timescale variability of the Enceladus plumes inferred from Cassini images. *Icarus, 282,* 260–275.

Jaumann R. et al. (2008) Distribution of icy particles across Enceladus' surface as derived from Cassini-VIMS measurements. *Icarus, 193,* 407–419.

Jones G. H. et al. (2009) Fine jet structure of electrically charged grains in Enceladus' plume. *Geophys. Res. Lett., 36,* L16204, DOI: 10.1029/2009GL038284.

Kaempfer T. U., and Schneebeli M. (2007) Observations of isothermal metamorphism of new snow and interpretation as a sintering process. *J. Geophys. Res., 112,* D24101, DOI: 10.1029/2007JD009047.

Karlstrom L., Hurwitz S., Sohn R., Vandemeulebrouck J., Murphy F., Rudolph M., Johnston M., Manga M., and McCleskey R. B. (2013) Eruptions at Lone Star Geyser, Yellowstone National Park, USA; 1. Energetics and eruption dynamics. *J. Geophys. Res.–Solid Earth, 118,* 4048-4062.

Kempf S., Beckmann U., and Schmidt J. (2010) How Enceladus dust plume feeds Saturn's E ring. *Icarus, 206,* 446–457.

Kieffer S. W. (1982) Dynamics and thermodynamics of volcanic eruptions — Implications for the plumes on Io. In *Satellites of Jupiter* (D. Morrison, ed.), pp. 647–723. Univ. of Arizona, Tucson.

Kieffer S. W. (1989) Geologic nozzles. *Rev. Geophys., 27(1),* 3–38.

Kliegel J. R. (1963) Gas particle nozzle flow. *Intl. Symposium on Combustion, 9(1),* 811–826.

Ladd B. S. and Ryan M. C. (2016) Can CO_2 trigger a thermal geyser eruption? *Geology, 44(4),* DOI: 10.1130/G37588.1.

Lu X. and Kieffer S. W. (2009) Multicomponent liquid and ice systems on the planets: Thermodynamics and fluid dynamics. *Annu. Rev. Earth Planet. Sci., 37,* 449–477.

Mastin L. G. (1995) *A Numerical Program for Steady-State Flow of Hawaiian Magma-Gas Mixtures Through Vertical Eruptive Conduits.* USGS Open File Report 95-756.

Matson D. L., Castillo-Rogez J. C., Davies A. G., and Johnson T. V. (2012) Enceladus: A hypothesis for bringing both heat and chemicals to the surface. *Icarus, 221,* 53–62.

Mitchell C. J., Porco C. C., and Weiss J. W. (2015) Tracking the geysers of Enceladus into Saturn's E ring. *Astron. J., 149,* 156.

Munoz-Saez C., Manga M., Hurwitz S., Rudolph M., Namiki A., and Wang C-Y. (2015) Dynamics within geyser conduits, and sensitivity to environmental perturbations: Insights from a periodic geyser in the El Tatio geyser field, Atacama Desert, Chile. *J. Volc. Geotherm. Res., 292,* 41–55.

Nakajima N. and Ingersoll A. P. (2016) Controlled boiling on Enceladus. 1. Model of the vapor-driven jets. *Icarus, 272,* 309–318.

Nimmo F., Porco C., and Mitchell C. (2014) Tidally modulated eruptions on Enceladus: Cassini ISS observations and models. *Astron. J., 148,* 46.

Perry M. E., Teolis B. D., Hurley D. M., Magee B. A., Waite J. H., Brockwell T. G., Perryman R. S., and McNutt R. L. (2015) Cassini INMS measurements of Enceladus plume density. *Icarus, 257,* 136.

Perry M. E., Teolis B. D., Grimes J., Miller G. P., Hurley D. M., Waite J. H. Jr., Perryman R. S., and McNutt R. L. Jr. (2016) Direct measurement of the velocity of the Enceladus vapor plumes. *Lunar and Planetary Science XLVII,* Abstract #2846. Lunar and Planetary Institute, Houston, Texas.

Porco C. C., Helfenstein P., Thomas P. C., Ingersoll A. P., Wisdom J., West R., Neukum G., Denk T., Wagner R., Roatsch T., Kieffer S., Turtle E., McEwen A., Johnson T. V., Rathbun J., Veverka J., Wilson D., Perry J., Spitale J., Brahic A., Burns J. A., Del Genio A. D., Dones L., Murray C. D., and Squyres S. (2006) Cassini observes the active south pole of Enceladus. *Science, 311,* 1393–1401.

Porco C., DiNino D., and Nimmo F. (2014) How the geysers, tidal stresses, and thermal emission across the south polar terrain of Enceladus are related. *Astron. J., 148,* 45.

Porco C., DiNino D., and Nimmo F. (2015) Enceladus' 101 geysers: Phantoms? Hardly! Abstract P13A-2118 presented at the 2015 Fall Meeting, AGU, San Francisco, California, December 14–18.

Portyankina G., Esposito L. W., Ali A., and Hansen C. J. (2016) Modeling of the Enceladus water vapor jets for interpreting UVIS star and solar occultation observations. *Lunar and Planetary Science XLVII,* Abstract #2600. Lunar and Planetary Institute, Houston, Texas.

Postberg F., Kempf S., Schmidt J., Brilliantov N., Beinsen A., Abel B., Buck U. and Srama R. (2009) Sodium salts in E-ring ice grains from an ocean below the surface of Enceladus. *Nature, 459,* 1098–1101, DOI: 10.1038/nature08046.

Postberg F., Schmidt J., Hillier S., Kempf S., and Srama R. (2011) A salt-water reservoir as the source of a compositionally stratified plume on Enceladus. *Nature, 474,* 620–622, DOI: 10.1038/nature10175.

Roddier C., Roddier F., Graves J., and Northcott M. (1998) Discovery of an arc of particles near Enceladus' orbit: A possible key to the origin of the E ring. *Icarus, 136(1),* 50–59.

Roth L., Saur J., Retherford K. D., et al. (2014) Transient water vapor at Europa's south pole. *Science, 343(6167),* 171–174.

Saur J., Schilling N., Neubauer F. M., Strobel D. F., Simon S., Dougherty M. K., Russell C. T., and Pappalardo R. T. (2008) Evidence for temporal variability of Enceladus' gas jets: Modeling of Cassini observations. *Geophys. Res. Lett., 35,* L20105, DOI: 10.1029/2008GL035811.

Schmidt J., Brilliantov N., Spahn F., and Kempf S. (2008) Slow dust in Enceladus' plume from condensation and wall collisions in tiger stripe fractures. *Nature, 457,* 685–688.

Smith H. T., Johnson R. E., Perry M. E., Mitchell D. G., McNutt R. L., and Young D. T. (2010) Enceladus plume variability and the neutral gas densities in Saturn's magnetosphere. *J. Geophys. Res., 115,* A10252.

Spahn F. et al. (2006) Cassini dust measurements at Enceladus and implications for the origin of the E ring. *Science, 311,* 1416–1418.

Sparks W. B., Hand K. P., McGrath M. A., Bergeron E., Cracraft M., and Deustua S. (2016) Probing for evidence of plumes on Europa with HST/STIS. *Astrophys. J., 829(2),* 121, DOI: 10.3847/0004-637X/829/2/121.

Spencer J. R. et al. (2006) Cassini encounters Enceladus: Background and the discovery of a south polar hot spot. *Science, 311,* 1401–1405.

Spitale J. N. and Porco C. C. (2007) Association of the jets of Enceladus with the warmest regions on its south-polar fractures. *Nature, 449,* 695.

Spitale J. N., Hurford T. A., Rhoden A. R., Berkson E. E., and Symeon S. P. (2015) Curtain eruptions from Enceladus' south-polar terrain. *Nature, 521,* 57.

Tenishev V., Combi M. R., Teolis B. D., and Waite J. H. (2010) An approach to numerical simulation of the gas distribution in the atmosphere of Enceladus. *J. Geophys. Res., 115,* A09302.

Tenishev V., Ozturk O. C. S., Combi M. R., Rubin M., Hunter J. H., and Perry M. E. (2014) Effect of the tiger stripes on the water vapor distribution in Enceladus' exosphere. *J. Geophys. Res.–Planets, 119,* 2658.

Teolis B. D., Perry M. E., Magee B. A., Westlake J., and Waite J. H. (2010) Detection and measurement of ice grains and gas distribution in the Enceladus plume by Cassini's Ion Neutral Mass Spectrometer. *J. Geophys. Res., 115,* A09222.

Teolis B. D., Perry M. E., Hansen C. J., Waite J. H., Porco C. C., Spencer J. R., and Howett C. J. A. (2017) Enceladus plume structure and time variability: Comparison of Cassini observations. *Astrobiology, 17(9),* DOI: 10.1089/ast.2017.1647.

Tian F., Stewart A. I. F., Toon O. B., Larsen K. W., and Esposito L. W. (2007) Monte Carlo simulations of the water vapor plumes on Enceladus. *Icarus, 188,* 154.

Waite J. H. et al. (2009) Liquid water on Enceladus from observations of ammonia and ^{40}Ar in the plume. *Nature, 460,* 487.

Waite J. H., Glein C. R., Perryman R. S., Teolis B. D., Magee B. A., Miller G., Grimes J., Perry M. E., Miller K. E., Bouquet A., Lunine J. I., Brockwell T., and Bolton S. J. (2017) Cassini finds molecular hydrogen in the Enceladus plume: Evidence for hydrothermal processes. *Science, 356(6334),* 155–159, DOI: 10.1126/science.aai8703.

Watson Z. T., Han W. S., Keating E. H., Jung N. H., and Lu M. (2014) Eruption dynamics of CO_2 driven cold-water geysers: Crystal, Tenmile geysers in Utah and Chimayó geyser in New Mexico. *Earth Planet. Sci. Lett., 408,* 272–284.

Woods A.W. (1995) The dynamics of explosive volcanic eruptions. *Rev. Geophys., 33,* 495–530.

Yeoh S. K., Chapman T., Goldstein D., Varghese P., and Trafton L. (2015) Understanding the physics of Enceladus south polar plume via direct numerical simulation. *Icarus, 253,* 205–222, DOI: *10.1016/j.icarus.2015.02.020.*

Yeoh S. K., Li Z., Goldstein D. B., Varghese P. L., Levin D. A., and Trafton L. M. (2017) Constraining the Enceladus plume using numerical simulation and Cassini data. *Icarus, 281,* 357.

Kempf S., Horányi M., Hsu H.-W., Hill T. W., Juhász A., and Smith H. T. (2018) Saturn's diffuse E ring and its connection with Enceladus. In *Enceladus and the Icy Moons of Saturn* (P. M. Schenk et al., eds.), pp. 195–210. Univ. of Arizona, Tucson, DOI: 10.2458/azu_uapress_9780816537075-ch010.

Saturn's Diffuse E Ring and Its Connection with Enceladus

Sascha Kempf, Mihály Horányi, and Hsian-Wen Hsu
University of Colorado Boulder

Thomas W. Hill
Rice University

Antal Juhász
Wigner Research Centre for Physics

H. Todd Smith
Johns Hopkins University

Saturn's E ring is the second largest planetary ring in the solar system, encompassing the icy satellites Mimas (r_M = 3.07 R_S), Enceladus (r_E = 3.95 R_S), Tethys (r_{Th} = 4.88 R_S), Dione (r_D = 6.25 R_S), Rhea (r_R = 8.73 R_S), as well as the more distant moon Titan (r_{Ti} = 20.25 R_S). Enceladus was proposed early on as the dominant source of ring particles, since the edge-on brightness profile peaks near the moon's mean orbital distance. We now know that the cryovolcanos at the moon's south-polar terrain are the main sources for the E-ring particles. Many features such as the ring's blue color and the local ring structure can be explained by the dynamics of the plume particle ejection. The large-scale structure of the ring is due to the concurrent orbital evolution of the particles due to plasma drag and their mass loss due to sputtering. Particles originating in the plumes of Enceladus can supply the entire E ring extending beyond the orbit of Titan.

1. INTRODUCTION

There is an intimate connection between Enceladus and Saturn's diffuse E ring, which envelops the ice moons Mimas, Enceladus, Tethys, Dione, Rhea, and Titan. The ring is composed of predominantly water-ice grains (*Hillier et al.,* 2007) smaller than 10 μm (*Nicholson et al.,* 1996; *Kempf et al.,* 2008). Studies of the ring particle dynamics suggest particle lifetimes of less than 200 years (*Haff et al.,* 1983; *Morfill et al.,* 1983; *Jurac,* 2001; *Juhász et al.,* 2007; *Beckmann,* 2008), implying that a mechanism resupplying the ring with fresh dust must exist. The discovery of the Enceladus plume by various Cassini instruments (*Dougherty et al.,* 2006; *Hansen et al.,* 2006; *Porco et al.,* 2006; *Spahn et al.,* 2006b; *Waite et al.,* 2006) dramatically changed the picture, with a lot of the missing puzzle pieces falling into place. The current view is that Enceladus is by far the strongest, and probably the only, source of E-ring particles.

We begin with a review of the history of the ring's discovery and early attempts to model the ring formation and evolution. We then review the observational evidence, which is mostly coming from Cassini observations. Finally, we will review the current understanding of the ring-particle dynamics.

1.1. History of the Ring Discovery

In the first half of the twentieth century, there were numerous unconfirmed and often conflicting reports of visual telescopic sightings of what we now call the E ring (Hill, 1984). The E ring was first documented photographically during the 1966–1967 ring-plane crossing (*Feibelman,* 1967). Advances in CCD detector technology produced much sharper Earth-based images during the next ring-plane crossing in 1980 (*Baum et al.,* 1981). Sample results from that paper are shown in Fig. 1.

These 1980 results already reveal at least three prominent (and now known to be durable) features of the E ring: (1) its optical depth peaks sharply at the orbital distance of Enceladus at 3.95 R_S (Saturn's equatorial radius $R_S \approx$ 60,300 km) with a value $\tau \sim 10^{-6}$, (2) it is detectable (even from Earth) out to at least 9 R_S, and (3) its thickness perpendicular to the ring plane is a significant fraction of 1 R_S, vastly exceeding that of any other known planetary ring. These three results, in turn, seem to imply (*Baum et al.,* 1981) that (1) Enceladus is the principle source of E-ring particles; (2) there is radial particle transport away from Enceladus' orbit, largely but not entirely outward, that determines the radial distribution of

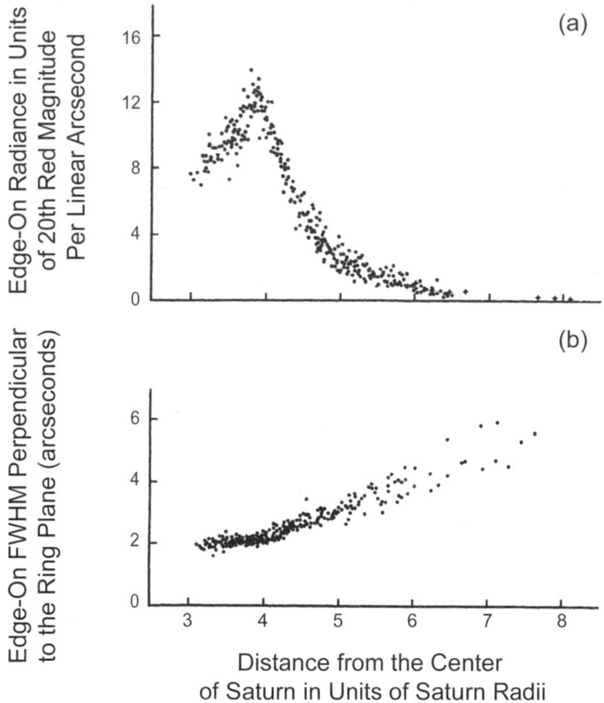

Fig. 1. Results from CCD images taken during the 1980 Earth crossing of Saturn's ring plane, reproduced from *Baum et al.* (1981). **(a)** E-ring brightness (in astronomers' units) vs. equatorial distance from Saturn's spin axis in units of R_S. **(b)** E-ring thickness (FWHM of observed brightness distribution away from the ring plane), vs. distance. For this epoch, 10 arcsec ≈ 1 R_S.

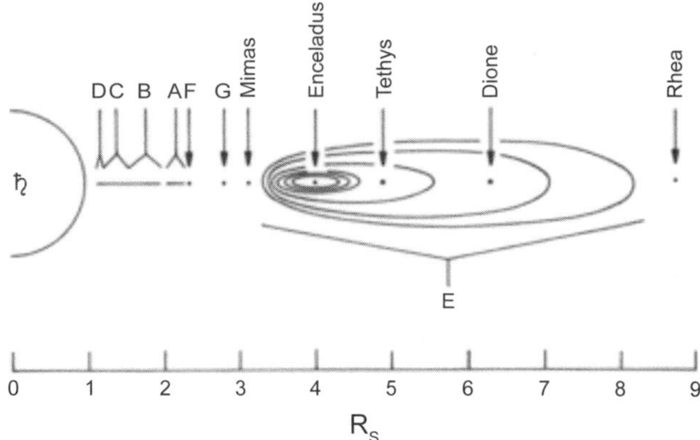

Fig. 2. Qualitative meridian-plane location and two-dimensional structure of the E ring, reproduced from *Hill* (1984).

E-ring particles; and (3) the lifetime of the E-ring structure must be much shorter than the timescales ~10^{5-6} yr that characterize the collisional collapse of most planetary rings to a very thin equatorial disk.

Infrared (*Terrile and Tokunaga,* 1980) and optical (*Dollfus and Brunier,* 1981) reflectance spectra indicated that at least some E-ring particles have radii ≤ a few micrometers, consistent with their detection in forward-scattered but not back-scattered sunlight by the Voyager 1 visible-light imager (*Smith et al.,* 1981). Figure 2 shows an early attempt to summarize these observational findings in the form of a qualitative map of E-ring density in a ρ–z plane, where (ρ, ϕ, z) is a cylindrical coordinate system centered on Saturn's spin axis (*Hill,* 1984). The classical rings A–D, F, and G are razor-thin on this scale. All the icy moons are within the detectable E ring, but Enceladus is clearly its main source.

Independently, and more or less concurrently, *in situ* charged-particle signatures from the brief Saturn encounters by Pioneer 11 (September 1, 1979) and Voyagers 1 and 2 (November 12, 1980, and August 25, 1981) provided further information about the structure and grain properties of the E ring. *Thomsen and van Allen* (1979) introduced the concept of using charged-particle absorption signatures to probe the rings. Several subsequent studies employed similar approaches with complementary results and sometimes conflicting conclusions (*McDonald et al.,* 1980; *van Allen et al.,* 1980; *Bastian et al.,* 1980; *Sittler et al.,* 1981; *Krimigis and Armstrong,* 1982; *Carbary et al.,* 1983; *Hamilton et al.,* 1983; *Hood,* 1983; *Schardt and McDonald,* 1983).

The larger extent of the E ring, compared to groundbased observations, was also indicated by the spurious electric field pulses recorded by the Voyager plasma wave instruments that were attributed to dust impacts on the spacecraft during the ring-plane crossings (*Gurnett et al.,* 1983).

Two dust production mechanisms have been postulated for Enceladus. Interestingly, geyser-like processes were proposed early (*Haff et al.,* 1983; *Pang et al.,* 1984), although in later years most attention was paid to meteoroidal impact ejection (*Juhász and Horányi,* 2002), or ejecta generation by the E-ring particles themselves colliding with their own parent bodies (*Hamilton and Burns,* 1994). This change of focus was probably due to the fact that the structure of Jupiter's diffuse gossamer rings could be explained by collisional ejecta originating from the ring moons Almathea and Thebe, whose orbital extermes coincide with the outer boundaries of the rings (*Burns et al.,* 1999).

To observe the faint E ring from Earth, the Sun and Earth have to lie in the same plane as Saturn's rings, which is a rather rare event. The first opportunity for such an Earth-bound telescopic observation after the Voyager flybys was in August 1995, when *de Pater et al.* (2004) found in telescopic infrared edge-on images small density enhancements around the orbits of Tethys and Dione, suggesting that all ring moons may contribute to the ring-particle population. Surprisingly, the de Pater et al. finding has not been confirmed by any of the Cassini instruments (*Horányi et al.,* 2009).

1.2. Pre-Cassini Ring Models

Two observations have intrigued scientists trying to model the ring from early on. First, the E ring has an unusual blue color, implying a narrow size distribution centered around 1 μm (*Nicholson et al.,* 1996), while rings resupplied with collisional ejecta would be expected to have a broad size distribution. Second, the ring's vertical scale height is not bound by the orbital extremes of Enceladus but increases with the radial distance to the moon (*Showalter et al.,* 1991). This implies that either the grains' ejection speeds are much larger than the satellite's escape speed of 207 m s^{-1}, contradicting laboratory data for impact ejecta (*Stöffler et al.,* 1975; *Hartmann,* 1985), or the inclinations of the ring particles have to grow rapidly after their escape into the ring.

The unique properties of the E ring stimulated the development of detailed dynamical models. *Horányi et al.* (1992) investigated the evolution of ring particles subject to perturbations by the planet's oblate gravity field, electromagnetic forces, and solar radiation pressure. They demonstrated that for a certain particle size the precession of the particle orbit due to the planet's oblateness and electromagnetic forces cancel each other. These particles swiftly acquire large eccentricities induced by the solar radiation pressure and spread out until ~7 R_S. *Hamilton* (1993) reproduced these results using orbital-averaged perturbation equations. These models predict highly eccentric particle orbits resulting in short lifetimes due to collisions with the main ring, and do not reproduce the radial extent of the ring. *Dikarev and Krivov* (1998) and *Dikarev* (1999) recognized that drag forces by the ambient plasma cause the ring particles to slowly migrate outward, allowing grains in less-eccentric orbits to cover the visible radial range of the ring. *Juhász and Horányi* (2002, 2004) computed self-consistently the spatial distribution of E-ring particles using an innovative library technique. This model closely reproduced the ring brightness profile and blue color as reported by *Showalter et al.* (1991),as well as the outward displacement of the densest ring point perceived in Earth-bound observations (*de Pater et al.,* 2004). Thus, at Cassini's arrival, not many researchers expected big surprises about the E ring, and we couldn't have been more wrong.

2. OBSERVATIONS

The optically visible E ring is populated with particles in excess of about 100 nm. Those are the particles we usually call ring particles. However,there is a second dust population in Saturn's E ring with distinct characteristics — the so-called nanograins. Because of their small sizes (<10 nm), these particles are invisible in the images and can only be detected by *in situ* instruments. Nanograins associated with the E ring have been observed at two different locations in the saturnian system: in the plume of Enceladus by the Cassini Plasma Spectrometer (CAPS) (*Young et al.,* 2004; *Jones et al.,* 2009), and at the outer magnetosphere and in interplanetary space by the Cosmic Dust Analyzer (CDA) (*Kempf et al.,* 2005a).

2.1. Spatial Distribution

Cassini has two *in situ* detectors capable of detecting ring particles: the CDA, which is designed to measure the size, composition, charge, and number density of grains ≤10 μm (*Srama et al.,* 2004), and the Radio and Plasma Wave Science (RPWS) investigation (*Gurnett et al.,* 2004), which senses dust impacts via voltage pulses in its antenna signals induced by dust impacts (*Ye et al.,* 2014, 2016a). The RPWS instrument has similar sensitivity to the High Rate Detector (HRD), a CDA subsystem designed for dust-dense regions, but is more robust with regard to spacecraft attitude. However, it is not specifically calibrated for dust detection, and detailed studies of the recollection of impact-generated charges onto the spacecraft and/or antennae as well as the comparisons with CDA/HRD measurements have only recently begun (*Ye et al.,* 2014, 2016b). Hereafter we will focus mostly on the *in situ* measurements by CDA.

Remote sensing observations by the Cassini Imaging Science Subsystem (ISS) (*Porco et al.,* 2004) provide information about the global structure of the ring as well as about the ring's column density of larger grains and the particle size distribution.

Already, the first *in situ* dust number density measurements by CDA revealed the E ring to be very different from what had been perceived from pre-Cassini measurements. The ring was found to extend at least until Titan's orbital distance (r_{Ti} = 20.25 R_S), with its density decreasing smoothly outside of Enceladus' orbital distance (*Srama et al.,* 2006), while previous Earth-bound telescopic observations (*Nicholson et al.,* 1996; *de Pater et al.,* 2004) could only follow it to a much smaller outer radius due to the ring's small optical depth.

The radial distribution of the ring particles is well described by a pair of power laws centered at the densest point within the ring (*Showalter et al.,* 1991), which is displaced outward from Enceladus' orbit by at least 0.05 R_S (*Kempf et al.,* 2008). The ring's vertical full width half maximum (FWHM) depends on the distance to Saturn's rotation axis ρ, has its minimum of ~4300 km at Enceladus and is ~5400 km at Mimas. Outside Enceladus' orbit the FWHM increases linearly with ρ and is ~5500 km at Tethys and ~6300 km at 5.01 R_S (*Kempf et al.,* 2008). The empirical E-ring density profile for grains ≥900 nm derived from early CDA measurements in 2005

$$n(\rho, z) = n_0 \exp\left(-\frac{(z - z_0(\rho))^2}{2\sigma(\rho)^2}\right)\begin{cases}(\rho/\rho_c)^{+50} & \rho \le \rho_c \\ (\rho/\rho_c)^{-20} & \rho > \rho_c\end{cases} \tag{1}$$

and

$$\sigma(\rho) = \sigma_c + (\rho - \rho_c)\begin{cases}\frac{\sigma_i - \sigma_c}{\rho_i - \rho_c} & \rho \le \rho_c \\ \frac{\sigma_0 - \sigma_c}{\rho_0 - \rho_c} & \rho > \rho_c\end{cases}$$

with

$$z_0(\rho) = \begin{cases} z_0(\rho_i)\frac{\rho - \rho_c}{\rho_i - \rho_c} & \rho \le \rho_c \\ 0 & \rho > \rho_c\end{cases}$$

Kempf et al. (2008) reproduces most of the Cassini *in situ* sets within roughly Dione's orbit data sufficiently well. The best-fit model parameters are $\sigma_i = 2293$ km, $\rho_i = 3.16\ R_S$, $\rho_c = 3.98\ R_S$, $\rho_0 = 4.75\ R_S$, $\sigma_0 = 2336$ km, and $z_0(\rho_i) = -1220$ km. Simultaneous measurements by Cassini RPWS found a slightly thicker ring (*Kurth et al.,* 2006), which can most likely be attributed to a lower size detection threshold of RPWS. The 2005 CDA data indicated that the E ring's symmetry plane inside of Enceladus' orbit is displaced southward by ~1000 km at Mimas' orbital distance and is aligned with Saturn's equatorial plane outward of Enceladus, while *Hedman et al.* (2012) and similarly *Ye et al.* (2016a) reported about a southward displacement of the rings' peak brightness of 1000–2000 km at the orbit of Mimas and a northward displacement of 1000 km near the Tethys orbit. Outside Dione's orbit the vertical ring profile exhibits pronounced seasonal variations (e.g., *Juhász and Horányi,* 2004) and deviates strongly from a simple shifted Gaussian.

The peak number density n_0 derived from *in situ* measurements has been found to range from $16 \times 10^{-2}\ m^{-3}$ to $21 \times 10^{-2}\ m^{-3}$. Remote sensing and *in situ* observations obtained by Cassini revealed further that the radial density profile of the E ring depends on the longitude relative to the Sun (*Hedman et al.,* 2012; *Kempf et al.,* 2012).

The brightness profile of the ring shows a day-night asymmetry, with the brightness peak on the dayside being brighter and located closer to Saturn and vice versa for the nightside. The ring's brightness along the morning ansa also appears bluer in color with a sharp inner edge, in comparing with the evening ansa. It is worthy to note that the aforementioned vertical warp appears to be independent of the hour angle, meaning that the E ring is symmetric and asymmetric at the same time.

Hedman et al. (2010) showed that a diffuse ring can be "heliotropic" (i.e., the ring particles' apoapses are directed toward the Sun) due to the combined effect from the directional perturbation of solar radiation pressure and grain's orbital precession. Considering the fast precession rate of micrometer-sized E-ring grains (one round per four years), *Hedman et al.* (2012) argued that some of the observed asymmetries of the E ring may be attributed to the perturbation of the solar radiation pressure received on grains' trajectories caused by the seasonal variation of Saturn's shadow coverage on the E ring. *Hsu et al.* (2016) attributed the day-night-asymmetry to the recently discovered moon-to-midnight electric field in Saturn's magnetosphere (*Andriopoulou et al.,* 2012, 2014; *Wilson et al.,* 2013). With grains charged negatively (*Kempf et al.,* 2006), this electric field could hinder the radiation pressure and result in a dichotomous, size-dependent distribution of the grains' orbital elements. At this stage, the E-ring asymmetry has not been fully understood and is not included in the empirical ring profile (equation (1)).

ISS has obtained the highest resolution images of the E ring so far. Between 2005 and 2006, ISS obtained global views of the ring at fairly large solar illumination angles of 15.8°–19.8° (*Hedman et al.,* 2012). The vertical brightness and thickness does not depend on the longitudinal distance to Enceladus, which is in contrast to the pronounced dependence of the ring thickness on the longitudinal distance to the Sun: inside Enceladus' orbit the ring is thicker at midnight, while exterior to Enceladus' orbit the ring is thicker at noon (*Hedman et al.,* 2012). The vertical optical brightness profile does not peak at the ring plane (i.e., $z = 0\ R_S$) but shows clearly two bands at $z\pm = 1000$ km (*Hedman et al.,* 2012). Remarkably, the double-banded structure is not apparent in the CDA data, but appears weakly in RPWS vertical profiles (*Kurth et al.,* 2006) (most likely enabled by the higher signal-to-noise of the RPWS measurements).

The large number of Cassini traversals through the inner E ring allowed the Cassini *in situ* instruments, mainly RPWS and CDA, to monitor the temporal variation of the ring particle density. *Kurth et al.* (2006) reported on one equatorial and five vertical passages through the inner E ring during which RPWS measured ring particle impacts, while *Kempf et al.* (2008) published CDA dust density data for grains ≥0.9 μm recorded during two equatorial and six vertical Cassini ring traversals. The data of the latter team provides evidence for a variation in the ring's peak density of 20% within 37 days. It is not clear whether the observed small variations stem from azimuthal variations within the dust torus or from variable plume activity. The Cassini Ultraviolet Imaging Spectrograph (UVIS) also observed only small variations of the plume's water vapor production (*Hansen et al.,* 2017). The CDA and UVIS data are in contrast to plume brightness variations by a factor of about 3 by, e.g., VIMS (*Hedman et al.,* 2013), which are most likely caused by tidal flexing (see the chapter by Spencer et al. in this volume). Nevertheless, the *in situ* ring data indicate that the Enceladus dust source supplies the ring with fresh dust at a surprisingly constant rate.

The vertical ring profile is not the only imprint of the localized dust injection by Enceladus' south pole plume. The radial brightness profiles derived from Earthbound observations indicated that the densest point of the E ring lies outside of Enceladus' orbit, displaced by 10^4 km (*de Pater et al.,* 2004), while *in situ* measurements by CDA found the displacement to be >3000 km (*Kempf et al.,* 2008). Numerical simulations before the discovery of the Enceladus plume did not reproduce this observation. *Juhász et al.* (2007) showed that the initial inclinations of the plume particles decrease the likelihood of the particles recolliding with Enceladus, allowing them to migrate outward by drag forces exerted by the ambient plasma, explaining the outward displacement of the ring densest point. The imprint of the dust jets emerging from the plumes on Enceladus are also visible in Cassini images of the E ring (*Mitchell et al.,* 2015) (Fig. 3).

2.2. Particle Sizes

*2.2.1. **Ring particles.*** CDA measurements indicate that the differential size distribution of ring particles $a_d \geq 0.9$ μm within Rhea's orbit is described best by power laws $n(a_d),d\ a_d \sim a_d^q$ with slopes q between –5 and –4 (*Kempf et al.,* 2008). RPWS measurements indicate for ring particles <10 μm a slope of –4 at Enceladus and of –3 at 3 R_S (*Ye et al.,* 2014).

*2.2.2. **Nanograins.*** The south-polar geyser plumes of Enceladus contain singly-charged nanometer-size water-ice

Fig. 3. Cassini image of the densest region of the E ring taken at a distance of about 2.1 × 6 km from Enceladus (NASA/JPL/SCI PIA 08321). The moon is within a dust torus fed by the Enceladus ice jets. The strands emerging from the moon form the various microsignatures apparent in the ring's vertical (see, e.g., Fig. 6) and radial profiles obtained by Cassini *in situ* instruments such as RPWS (*Kurth et al.,* 2006) or CDA (*Kempf et al.,* 2008). Tethys is visible to the left of Enceladus.

grains discovered by CAPS (*Jones et al.,* 2009), in addition to their predominantly H_2O vapor content (*Waite et al.,* 2006) and the micrometer-sized water-ice grains (*Kempf et al.,* 2008) that make the plumes visible in ISS images (*Porco et al.,* 2006). The nanograins, which are formed by nucleation in the ascending water vapor (*Schmidt et al.,* 2008), must be charged in order to be detected by the CAPS electrostatic analyzers. Their charge is acquired by electron attachment from the cold but dense plasma in the plume (*Farrell et al.,* 2010; *Morooka et al.,* 2011; *Shafiq et al.,* 2011; *Hill et al.,* 2012). Negatively charged nanograins vastly outnumber the positive ones, as expected from the large ratio of electron thermal flux to positive-ion thermal flux (*Horányi,* 1996).

The most likely nanograin charge is a single excess electron per grain because a second incident electron would be strongly repelled by the negative potential of the first electron for such a small grain (*Hill et al.,* 2012; *Meier et al.,* 2014, 2015). CAPS analyzes the energy per charge ratio E/q of incident particles. When Cassini encounters the Enceladus plumes, the incident particle energy $E = \frac{1}{2} mv^2$ is dominated by the ram speed $v \sim 10$ km s^{-1} of Cassini relative to the plume, so $E \propto m$. The grain mass m can in turn be converted to grain radius a_d with the conventional assumption of spherical water-ice grains. This results in the size spectra illustrated in Fig. 4 for four of the five Cassini plume encounters for which CAPS had ram pointing.

2.3. Grain Composition

2.3.1. Ring particles. Even before Cassini's arrival it was suspected that water plays an important role in the E ring's composition. However, Cassini revealed that the region outside the main rings is dominated by water-ice particles (*Hillier et al.,* 2007), as well as water molecules and ions that form from water dissociation (*Young et al.,* 2005; *Esposito et al.,* 2005; *Melin et al.,* 2009).

For the first time, CDA enabled the *in situ* characterization of the ring particle composition. The instrument records time-of-flight (TOF) mass spectra from cations and cationic aggregates of the plasma generated by single grain impacts onto its rhodium target (*Srama et al.,* 2004). CDA carried out compositional measurements of both E-ring particles (*Hillier et al.,* 2007; *Postberg et al.,* 2008, 2009) and freshly ejected particles during plume crossings (*Postberg et al.,* 2011). Furthermore, CDA obtained compositional information about saturnian stream particles (*Kempf et al.,* 2005a,b; *Hsu et al.,* 2010, 2011a,b), which are the end members of the E-ring particle population.

CDA identified at least three distinct families of E-ring particles, all of which are dominated by water-ice (*Postberg et al.,* 2008, 2009). Type I particles are composed of almost pure water-ice and constitute about 65% of the ring-particle population CDA is sensitive to. The low abundance of Na^+ and K^+ in the spectra corresponds to an Na/H_2O ratio of about 10^{-7} (*Postberg et al.,* 2009). Spectra of Type II particles show, in addition to the dominating Type I mass lines, a distinct feature between mass 27 and 31 u, as well as peaks that can be most likely attributed to organic compounds and/or a silicate (*Postberg et al.,* 2009). Spectra of about 6% of the ring particles exhibit the Type III characteristics: the water cluster ion lines $(H_2O)H_n^+$ predominant in Type I and II spectra are replaced by $(NaOH)_nNa^+$ cluster lines. The frequent appearance of $NaCl–Na^+$ and $Na_2CO_3–Na^+$ mass lines implies that the most probable Na-bearing compounds are NaCl and/or Na_2CO_3. Type III spectra can be reproduced in laser-induced liquid beam ionization/desorption (LILBID) laboratory experiments (*Charvat et al.,* 2004), which imply an Na concentration in Type III ring particles of about 0.5–2% (*Postberg et al.,* 2009).

2.3.2. Nanograins. Mass spectra of saturnian stream particles have already been obtained by CDA in 2004 during Cassini's approach to Saturn (*Kempf et al.,* 2005a,b). The initial analysis of CDA impact spectra indicated little, if any, water-ice and a major siliceous component indicated by the presence of Si^+, SiO_2^+, and $SiRh^+$ mass lines (*Kempf et al.,* 2005b). A dynamical analysis by *Hsu et al.* (2011b) demonstrated convincingly that a stream particle is in fact the refractory remnants of sputtered icy E-ring grains in which they were initially embedded. A thorough reanalysis of stream particle spectra identified silica (SiO_2) as the main component of most stream particles. The most plausible explanation for this finding is the abundant formation of nanocolloidal silica in Enceladus' subsurface ocean, which is then partly integrated in ice grains expelled by the plumes (*Hsu et al.,* 2015).

2.4. Electrostatic Grain Charge

The surfaces of E-ring grains are continually bombarded by ambient plasma particles and photons, causing a charge

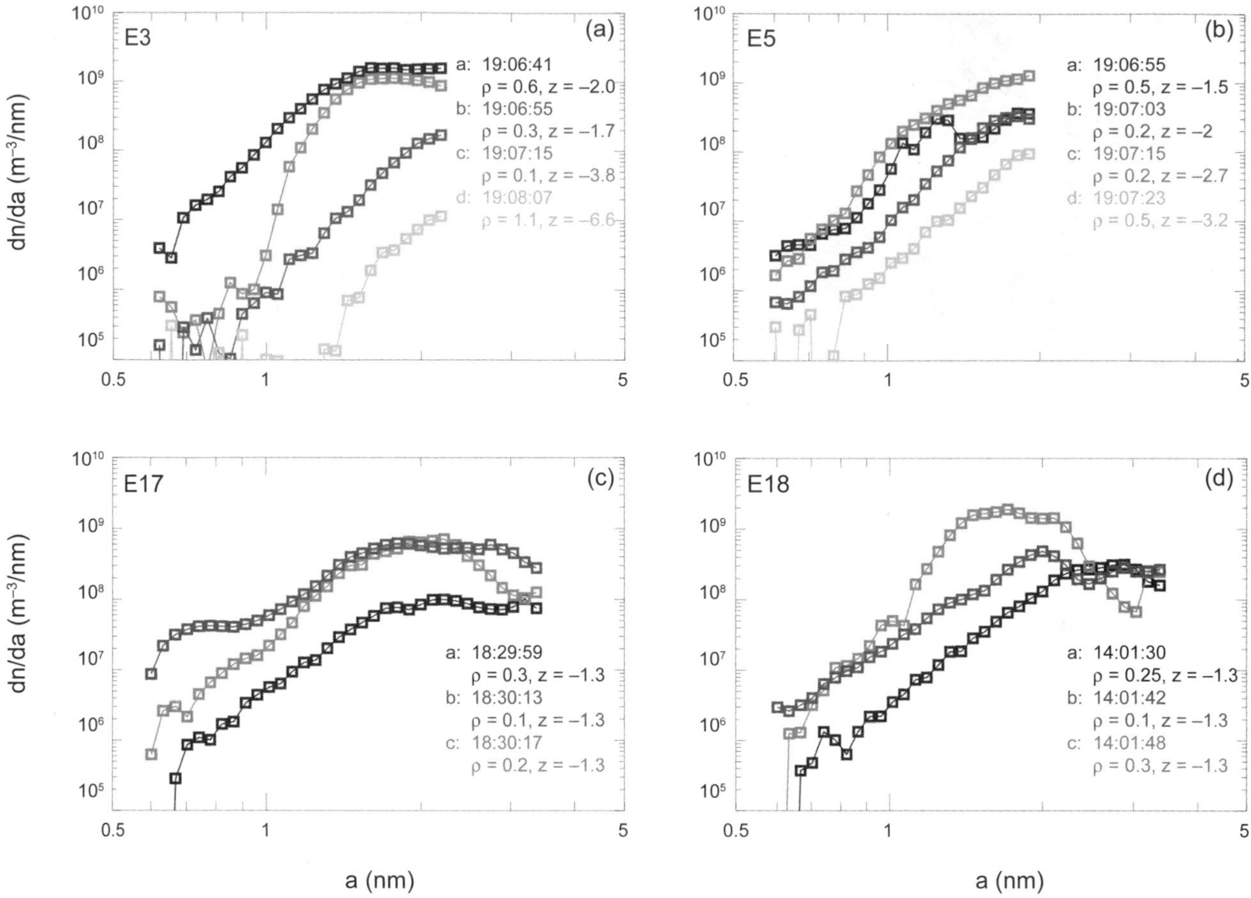

Fig. 4. CAPS nanograin size distributions for four encounters of Cassini with the Enceladus plume, at various times and Enceladus-centered cylindrical coordinates. From *Dong et al.* (2015).

transfer between the dust particle and its environment (*Horányi,* 1996). As a consequence, the ring particles carry an electrostatic charge $Q_d = 4\pi\varepsilon_0 a_d \phi_d$, where equilibrium potential ϕ_d is, to a good approximation, independent of the grain size and depends only on the plasma environment. Smaller grains have larger charge-to-mass ratios $Q_d/m_d \sim a_d^{-2}$, and hence are more susceptible to electromagnetic forces. Cassini's CDA detector is equipped with a charge-sensitive grid system (*Auer et al.,* 2002), which enables measurements of the charge carried by larger ring particles before they hit the detector. Particles detected approximately within Rhea's orbital distance have been found to be negatively charged with $\phi_d \sim -2$ V (*Kempf et al.,* 2006), which is consistent with RPWS Langmuir Probe (LP) potentials ranging between −2 and −3 V (*Wahlund et al.,* 2005). Outside Rhea's orbit the grain charging is dominated by photoionization due to solar UV and grains carry positive charges with $\phi_d \sim +3$ V (*Kempf et al.,* 2006).

3. ENCELADUS AS THE MAIN SOURCE OF THE RING

Despite the remarkable progress since the discovery of the Enceladus plume, there are many phenomena still not understood. One important open question is whether the Enceladus geysers are really the only source of E ring dust. An early analysis of ISS plume images suggested a dust escape to the ring of about 0.04 kg s^{-1} (*Porco et al.,* 2006), while *in situ* data imply an escape rate of 0.2 kg s^{-1} (*Spahn et al.,* 2006b), which is sufficient to maintain a steady-state E ring. However, it is difficult to understand why the influx of interplanetary meteoroids does not seem to produce noticeable amounts of collisional ejecta at any of the saturnian moons. Our expectations were thwarted for several possible reasons. The ejecta production by dust impacts from either external origin, like in the case of the icy Galilean satellites (*Krivov et al.,* 2003; *Sremčević et al.,* 2003), or the E ring itself (*Hamilton and Burns,* 1994) depends on the product of the bombarding flux and the yield, and on the ratio of outgoing over the incoming mass of the dust particles, which is a sensitive function of the impact speed. The flux of interplanetary particles, hence their size and speed distributions, remain poorly constrained in the outer solar system. Similarly, the makeup and the temperature of their icy surfaces could greatly affect their yield. As for now, impact-generated ejecta particles have only been identified near the tiny moons Methone and Anthe (*Sun et al.,* 2017), Rhea (*Jones et al.,* 2008), and maybe Enceladus

(*Spahn et al.,* 2006b; *Kempf et al.,* 2010). In the following we review the two possible mechanisms for feeding the E ring: the escape of plume particles and the impactor ejecta mechanism. Both mechanisms differ with respect to the size and speed distribution of the ejected particles, which leads to observable effects.

3.1. Ejecta Injection Into the Ring

During close flybys by the Galileo spacecraft of the jovian moon Ganymede, the onboard dust detector discovered that the moon is wrapped within a faint dust cloud (*Krüger et al.,* 1999), which has been found to be populated by ejecta dust of a broad size range generated by collisions of micrometeoroids of interplanetary origin with the moon's surface (*Krüger et al.,* 2000). This phenomenon has turned out to be quite common in the solar system: the Galilean moons Europa and Callisto *(Krüger et al.,* 2002), as well as Earth's Moon (*Horányi et al.,* 2005), are engulfed in dust atmospheres. Moons with dust envelopes are effective dust sources since ejecta launched faster than the moon's escape speed are injected into the moon's orbit and may replenish tenuous dust rings. The impactor-ejecta mechanism has been studied in great detail by *Krivov et al.* (2003) and *Sremčević et al.* (2003) and will be briefly summarized next.

The surfaces of saturnian moons are eroded by E-ring particle impacts and by collisions with interplanetary dust particles (IDPs) (*Spahn et al.,* 2006a). Because of their higher impact speed and larger size, the ejecta mass production is dominated by IDP projectiles. Hypervelocity impacts of large projectiles (tens of micrometers) generate ejecta particles, whose mass and speed distributions are assumed to be (1) given by power laws and (2) independent of each other, which is supported by experimental data (see references in *Krivov et al.,* 2003). For Enceladus ejecta, *Spahn et al.* (2006a) suggest a differential mass distribution with a power-law exponent of $-\gamma = 12/5$, and a differential speed distribution with an exponent $-\beta$ between 2 (hard surface) and 3 (regolith). The independence of the speed distribution on the ejecta implies that γ is also the slope of the mass distribution of escaping ejecta. Thus, rings dominantly populated by surface ejecta such as Jupiter's gossamer rings (*Burns et al.,* 1999) are characterized by rather broad size distributions and have a reddish tint.

The number of ejecta $>a_d$ escaping from the moon's gravity to the ring is then

$$N_{esc}(> a_d) = \left(\frac{v_0}{v_{esc}}\right)^{\beta-1} \left(\frac{a_{max}}{a_d}\right)^{\gamma} \frac{\pi R_E^2}{m_{max}} \frac{3-\gamma}{\gamma} F_{imp} Y \quad (2)$$

(*Krivov et al.,* 2003; *Spahn et al.,* 2006a), where v_0 is the speed of the slowest ejecta particle. The dependence of equation (2) on the mass m_{max} of the largest ejecta is rather weak, which justifies us to replace m_{max} by the mean projectile mass m_{imp}. The yield $Y(m_{imp}, v_{imp})$ is the total ejected mass per projectile mass and depends on the surface properties and on the mass and speed v_{imp} of the impactor (e.g., *Koschny and Grün,* 2001). The impactor mass flux F_{imp} and the mean impactor speed $\langle v_{imp}\rangle$ at Saturn are only roughly known. *Spahn et al.* (2006a) considered F_{imp} = 1.8×10^{16} kg m^{-2} s^{-1} and $\langle v_{imp}\rangle = 9.5$ km s^{-1}, while new CDA measurements indicate a 10× lower flux and $\langle v_{imp}\rangle$ = 4 km s^{-1} (*Kempf et al.,* 2015). The large uncertainties of the model parameters imply an order of magnitude uncertainty for numerical values of N_{esc}. For an in-depth discussion of the impactor-ejecta mechanism, see *Krivov et al.* (2003).

Spahn et al. (2006a) reevaluated the expected ejecta production rates of the E-ring moons in the light of the new Cassini data and concluded that IDPs produce ~0.1 kg s^{-1}, while E-ring impactors produce ~10 kg s^{-1} of fresh dust. If this were the case, collisional ejecta would still be the dominating E-ring dust source. However, there is only circumstantial evidence for ejecta in the Cassini data. The best fit of the CDA plume density profile of grains ≥1.6 μm obtained during the first close Enceladus flyby in 2005 (*Spahn et al.,* 2006b; *Kempf et al.,* 2010) to numerical plume models is obtained by assuming an additional ejecta source with a density of 0.01 m^{-3} of grains ≥1.6 μm at ~171 km altitude (see Fig. 5). It should also be noted that no density enhancements along the orbits of the other ice moons have been identified. This apparent lack of impact induced ejecta remains an unexplained observation.

3.2. Plume Particle Injection Into the Ring

The speed distribution of emerging plume particles depends on the grain size a_d and is given by

$$p(v_d \mid a_d) = \left(1 + \frac{a_d}{a_c}\right)\frac{a_d}{a_c} \cdot \frac{v_d}{v_{gas}^2}\left(1 - \frac{v_d}{v_{gas}}\right)^{\frac{a_d}{a_c}-1} \quad (3)$$

with $\int_0^{v_{gas}} p(v_d|a_d)dv = 1$ (*Schmidt et al.,* 2008). Because plume particles are assumed to be accelerated by the emerging gas, the gas speed v_{gas} sets an upper bound on the particle speed distribution. The critical radius a_c is a measure of the reacceleration rate for particles between their last subsurface collision with a wall of the vent and ejection from the surface of the moon. Particles with $a_d < a_c$ will be reaccelerated to velocities approaching gas velocity, while larger particles $a_d > a_c$ will be ejected at slower speeds. Occultation data by UVIS constrain v_{gas} to be 300–1000 m s^{-1} (*Tian et al.,* 2007; *Hansen et al.,* 2011), while $a_c \sim 0.8$ μm is the best fit to the CDA plume density profile (*Schmidt et al.,* 2008). The size-dependent speed distribution (equation (3)) is consistent with observations by Cassini's Visiual and Infrared Mapping Spectrometer (VIMS) (*Brown et al.,* 2004; *Hedman et al.,* 2009) and implies that the plume is stratified with respect to grain size. CDA data obtained during Cassini plume traversals indicate a power-law distribution with a slope of ~−3.5 for the plume particle size distribution at the Enceladus surface (*Kempf et al.,* 2016).

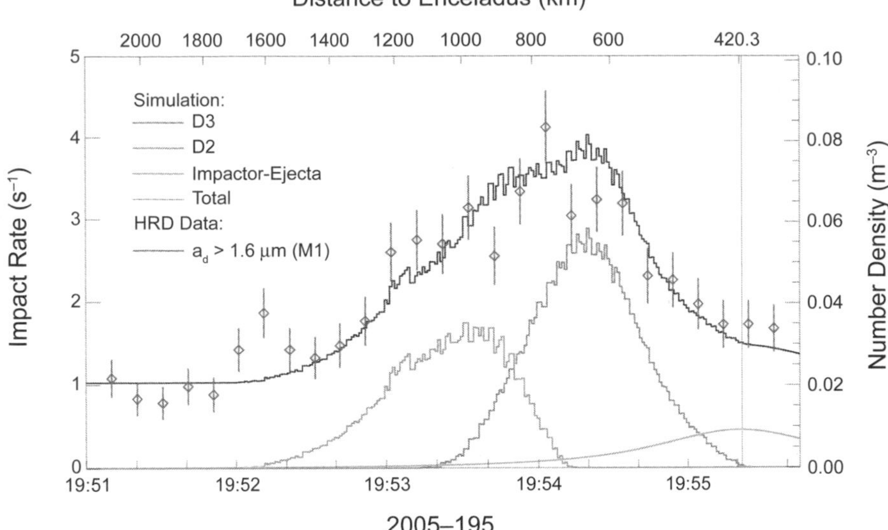

Fig. 5. Comparison of count rates predicted by simulations to CDA dust measurements of grains >1.6 μm during the Cassini flyby at Enceladus in 2005 indicates a weak ejecta contribution of 0.005 m^{-3} to the moon's dust exosphere of 0.06 m^{-3} at ~350 km altitude. From *Kempf et al.* (2010).

The injection of dust grains from localized sources in the south polar terrain of Enceladus explains many aspects of the ring's peculiar vertical profile. Since the grains' ejection speed is much slower than Enceladus' orbital speed, v_E = 12.4 km s^{-1}, the orbital elements of fresh ring particles and the moon are almost identical. However, because plume particles are ejected roughly orthogonal to the ring plane, the minimum inclination of the ejected grains is $i_{min} = R_H/r_E = 0.11°$, where R_H = 948 km is the Hill radius of Enceladus, characterizing the extent of the moon's gravitational influence. Particles ejected faster than the three-body escape speed from Enceladus of v_{esc} = 207 m s^{-1} may escape to the ring and have an initial orbital inclination of $i^2(v_e) \approx i^2_{min} + (v^2_e - v^2_{esc}/v^2_E)$, where v_E is Enceladus' orbital speed. Numerical studies by *Kempf et al.* (2010) showed that, in order to escape to the ring, plume particles need be launched faster than 224 m s^{-1}, as slower grains recollide with Enceladus during their first orbit. Furthermore, the initial inclination of grains launched slower than 244 m s^{-1} will be altered by three-body effects, because these grains traverse the Enceladus Hill sphere during their first ring-plane crossings. Vertical ring profiles obtained by tracing the trajectories of plume particles match CDA ring data well (*Kempf et al.*, 2010). In fact, individual dust jets appear as kinks in the ring's vertical density profile (Fig. 6). In ISS images of the E ring in the vicinity of Enceladus, the localized dust injection from the jets in the plume is clearly visible as long sinusoidal strands (dubbed "tendrils") along the moon's orbit (see Fig. 3). While the numerical simulations by *Mitchell et al.* (2015) do not fully reproduce all the fine structure seen in the ISS images, they do approximately capture the large-scale morphology, and a cascade of finer and finer structures emerging in the vicinity of Enceladus. The strands can be associated with many of the 99 jets recently identified by *Porco et al.* (2014) (see Fig. 7).

Because larger plume particles are ejected at lower speeds, predominantly smaller grains escape to the ring. This is in contrast to ejecta particles and explains the unique blue color of the E ring.

4. RING DYNAMICS AND EVOLUTION

4.1. Particle Evolution

Due to the low particle number density the E ring is collision free and the particle dynamics is due to the competing effects of gravity, radiation pressure, drag, and electromagnetic forces. The resulting orbital evolution involves a large span of timescales. The characteristic charging time of micrometer-sized dust particle near Enceladus is on the order of minutes (*Kempf et al.*, 2006; *Juhász et al.*, 2007), and the orbital evolution due to the combined effects of planetary oblateness, radiation pressure, drag and the Lorentz force becomes noticeable in just a few hours (*Juhász et al.*, 2007; *Kempf et al.*, 2010). The morphology of the ring develops in weeks to months, but its large structure changes with the orbital motion of Saturn about the Sun of 30 years (*Juhász and Horányi*, 2004). The characteristic lifetime of a micrometer-sized particle against sputtering would be about 50 years if it were to stay near Enceladus (*Jurac*, 2001); however, in a few centuries a continually mass losing and migrating particle can reach even beyond the orbit of Titan due to the combined effects of sputtering and plasma drag (*Horányi et al.*, 2008; *Kempf and Beckmann*, 2017).

Saturn's oblateness leads to a prograde orbital precession, while Saturn's outward pointing co-rotational field causes a

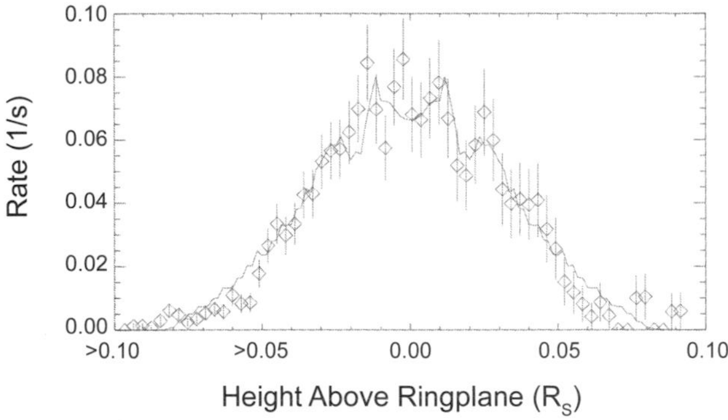

Fig. 6. Vertical profile of the E ring at Enceladus' orbital distance as measured by CDA during a steep ring plane crossing at 3.93 R_S (diamonds). The detector was sensitive to grains >0.9 μm. The solid line compares the Cassini data with model calculations for a ring fed by the eight Enceladus dust jets identified by *Spitale and Porco* (2007). Each pronounced spike in the vertical profile is associated with a dust jet emerging from the moon's south polar terrain (*Kempf et al.,* 2010).

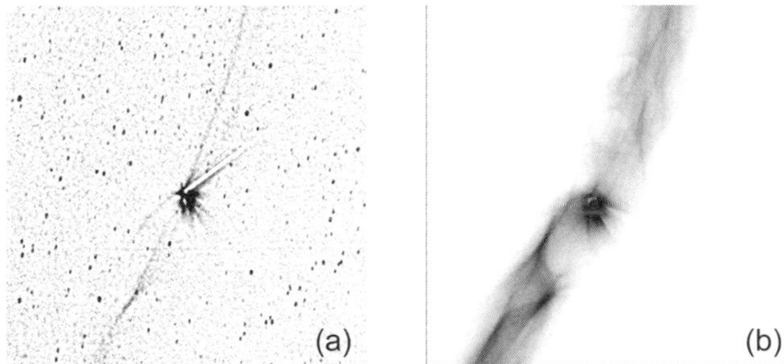

Fig. 7. Numerical simulations match the dust strands emerging from the Enceladus south-polar terrain apparent in ISS images (*Mitchell et al.,* 2015). **(a)** Median filtered high-phase ISS image of E ring with Enceladus at the center. **(b)** Synthetic image using numerical simulations of a subset of jets identified by *Porco et al.* (2014) approximately reproduces the large-scale morphology, and some of the fine structures that can be identified in the real image on the left.

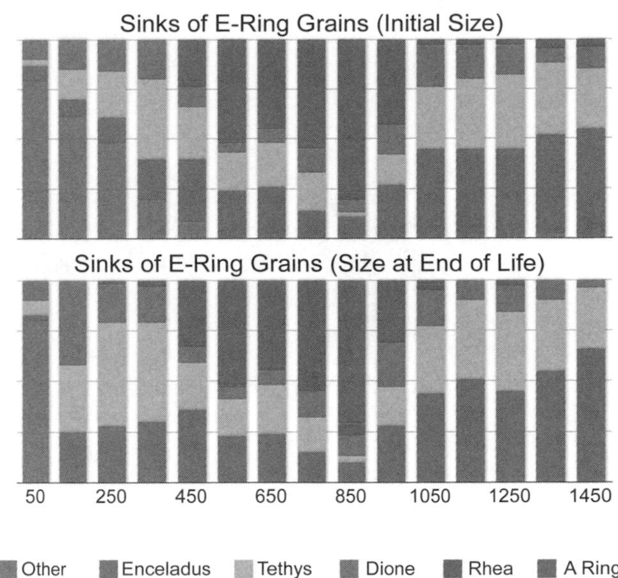

Fig. 8. Sinks of E-ring particles as function of the initial grain size (top panel) and of the size at the end of life (bottom panel). From *Kempf and Beckmann* (2017).

retrograde orbital precession for negatively charged grains (see section 2.4). Because the grain charge depends on a_d there is a particle size range for which the prograde and retrograde orbital precessions cancel each other out, allowing the solar radiation pressure to induce large orbital eccentricities (*Horányi et al.,* 1992; *Hamilton,* 1993). The rapid decrease of the pericenters of such particles puts them onto a collision course with Saturn's dense A ring (see Fig. 8), which reduces their lifetime to a few tens of years. Early A-ring collisions can only be avoided if the grains' semimajor axis grows fast enough under the action of plasma drag forces (*Morfill et al.,* 1983; *Havnes et al.,* 1992; *Dikarev,* 1999).

The life cycle of an E ring particle depends on its initial size and can be summarized as follows (see Figs. 8 and 9) (*Horányi et al.,* 2009; *Kempf and Beckmann,* 2017). Small grains (≤400 nm) experience a rather rapid growth of their semimajor axes to about 7 R_S, while their eccentricities increase only moderately. As a consequence, their orbits' pericenter lies safely outside the A ring, implying a rather long lifetime and sputtering as the dominating loss mechanism. Grains with initial sizes between 400 nm and ~1 μm are affected by the "locking mechanism" described above and get dominantly lost by A-ring collisions. Grains bigger than that do not collide with the main rings owing to the slow growth of their eccentricities and are predominantly lost due to collisions with the ring moons.

Sputtering erosion turns out to be of particular relevance for understanding the life cycle of E-ring particles. Sputtering erosion of E-ring grains by magnetospheric ions was first considered to explain the required water production rate of 10^{27} molecules s^{-1} in the inner saturnian system inferred from the Hubble Space Telescope observations more than a decade prior to Cassini's arrival at the saturnian system (*Jurac,* 2001; *Jurac et al.,* 2002). While the water production from sputtering of E-ring ice grains was found insufficient (and we now know that Enceladus' activity was responsible), these works showed that the grain sputtering lifetime is comparable to the dynamical lifetime, suggesting the erosion process plays a significant role in shaping the E ring. The grain size is a key parameter in the ring-particle dynamical evolution. The short sputtering lifetime [~50 yr for a micrometer-sized grain (*Jurac,* 2001)] implies that the grain size decreases substantially during its dynamical evolution. The corresponding forces and their ratios therefore vary with a grain's age. Meanwhile, grains also experience acceleration by plasma drag forces that increases their orbital energy, i.e., their semimajor axis, and slowly drift outward (*Dikarev,* 1999). As a result of that, the outer E ring is composed of older and smaller

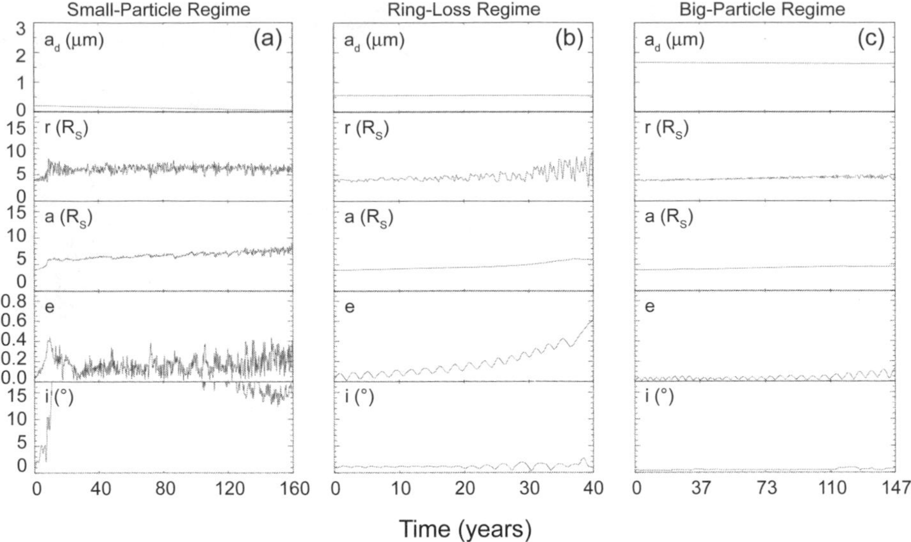

Fig. 9. Dynamical evolution of E-ring particles launched at Enceladus. **(a)** Evolution of a 0.4-μm grain; **(b)** typical example of a "ring-loss" regime particle (a_d = 0.6 μm); **(c)** evolution of a 1.7-μm grain characteristic of the "big particle regime." Each panel illustrates the evolution of the grain size a_d (top row), distance to Saturn r (second row), semimajor axis a (third row), eccentricity e (fourth row), and inclination i (bottom row). From *Kempf and Beckmann* (2017).

grains. However, since both processes are proportional to the grains' surface area and the plasma density, the slope of the size distribution remains unchanged across the E ring as the entire distribution shifts toward smaller sizes at larger radial distances (*Horányi et al.,* 2008). The radial extent of the E ring provides in fact a lower limit on the ring age as well as on the duration of Enceladus' activity of at least a few hundred years (*Horányi et al.,* 2008).

Grains that do not collide with one of the moons or the A ring will eventually be eroded by plasma sputtering. The sputtering timescale is proportional to the grain's mass-to-surface ratio, i.e., a 1-μm grain will survive sputtering roughly 10× longer than a 100-nm grain. The lifetime of nanometer-sized particles is even shorter because of the larger surface curvature (*Nietiadi et al.,* 2014). Because the sputtering efficiency strongly depends on the material properties, sputtering erosion is an important process for the E ring as water ice is relatively "soft" and prone to sputtering erosion compared to other materials, such as silicates. The sputtering lifetime of nanometer-sized ice particles is expected to be on the order of days (shorter if heat spike sputtering is considered), meaning that most negatively charged nanograins detected in the plume are consumed by sputtering in the dense plasma torus around Enceladus' orbit.

4.2. Mass Deposition on Ring Moons

All the moons engulfed in the E ring are continually bombarded by the particles comprising this ring, leading to an ongoing "contamination" by matter originating from the plumes, hence, the undersurface ocean of Enceladus. The influx of E-ring dust onto the surfaces of the other moons has been suggested to contribute to their observed albedo and color features (*Cruikshank et al.,* 2005; *Verbiscer et al.,* 2007; *Schenk et al.,* 2011). These features are attributed to the combined effects of magnetospheric plasma-induced space weathering and dust impacts. In particular, *Verbiscer et al.* (2007) noted a strong correlation between the visual geometric albedo of the icy moons and the E ring's pole-on reflectance at the moons' orbital distances (Fig. 10), which may indicate a major role of Enceladus dust for their surface properties. An indepth discussion of the moons' surface alteration is provided in the chapters by Howett et al. and Verbiscer et al. in this volume. The surface composition is discussed in the chapter by Hendrix et al.

It is still an open question of whether dust impacts contribute to the generation or the removal of albedo features by delivering salt-laden ice particles, possibly even organic mat-

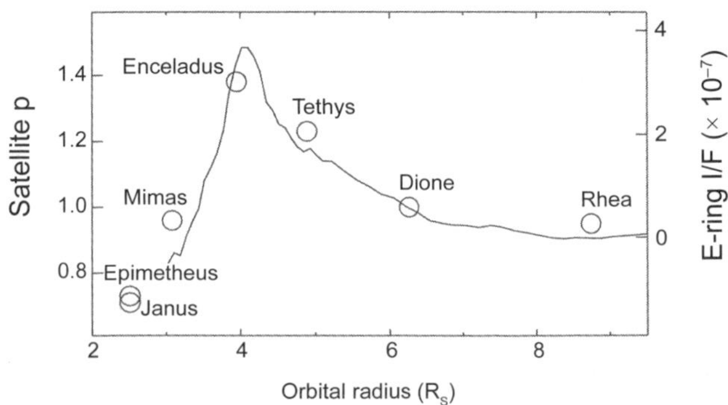

Fig. 10. The mean visual geometric albedo of the E-ring moons vs. the ring's I/F brightness. From *Verbiscer et al.* (2007).

ter, from the plumes of Enceladus, or by exposing a fresh icy surface below these markings, respectively. Plasma-induced space weathering patterns due to the co-rotating magnetospheric ions and low-energy electrons impact all the moons on their trailing side; however, more energetic populations of magnetospheric electrons can drift in the opposite direction, impacting the moons on their leading hemispheres (*Khurana et al.,* 2008). Interestingly, the complex orbital evolution of small charged dust particles also leads to a deposition pattern that is different for each moon. These patterns can be calculated by following the trajectories (*Juhász and Horányi,* 2002; *Juhász et al.,* 2007; *Horányi et al.,* 2008; *Kempf et al.,* 2010; *Kempf and Beckmann,* 2017) of a large number of particles from the plumes of Enceladus with the appropriate initial velocity and size distributions (*Spahn et al.,* 2006b; *Schmidt et al.,* 2008; *Kempf et al.,* 2016) until they intersect the orbit of a moon, and using their velocity and size distributions at that distance to calculate the total flux reaching the surface.

4.2.1. Plume particle deposition on Enceladus. The mass deposition on Enceladus is dominated by plume particles ejected too slowly to escape from the moon's gravity. *Kempf et al.* (2010) studied numerically the snowfall pattern of the plume material (Fig. 11b) and estimated annual deposition rates of 0.5 mm in the immediate vicinity of the tiger stripes and about 10^{-5} mm at the equator. Between −45° and 45° latitude the snowfall pattern forms two broad bands along the 45°W meridian and the 205°W meridian resulting from the three-body dynamics of the plume material. There is a pronounced dependence of the mass deposition on the particle size — the deposits in the southern terrain are dominantly due to big plume particles, while the deposits in the northern hemisphere are formed mostly by small grains (*Southworth and Kempf,* 2017). Remarkably, the predicted snowfall pattern is clearly visible in the IR/UV color ratio map of the moon (*Schenk et al.,* 2011) (Fig. 11a).

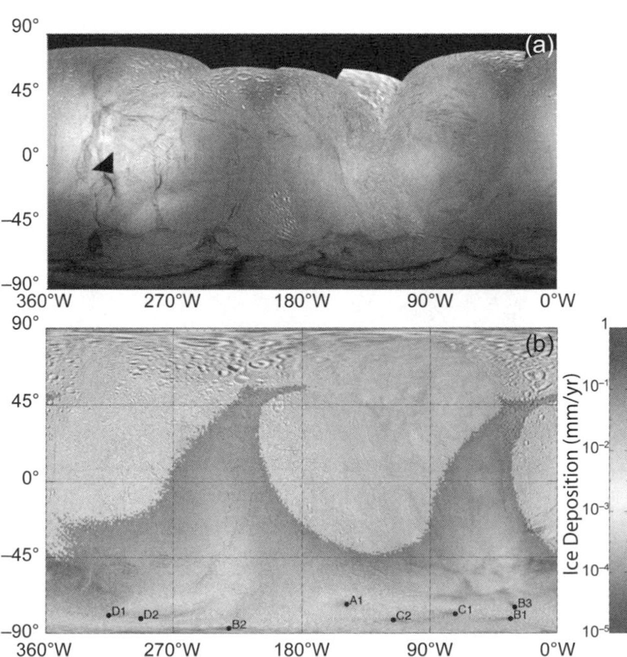

Fig. 11. See Plate 30 for color version. **(a)** Global IR/UV color ratio map of Enceladus (*Schenk et al.,* 2011). **(b)** Snowfall pattern of plume particles between 500 nm and 5 μm on the Enceladus surface (*Kempf et al.,* 2010).

4.2.2. Deposition of ring particles. Figure 12 shows deposition maps of E-ring material on the ring moons. For Mimas, Enceladus, Tethys, Dione, and Rhea, these maps indicate the influx of E-ring particles bombarding their surface. In the case of Titan, however, its nitrogen-dominated atmosphere is thick enough to ablate the incoming dust particles, which could possibly contribute to haze formation (*Sittler et al.,* 2009).

The maps indicate a systematic shift of the location where most of the E-ring particles reach the surfaces of these moons, from the trailing hemisphere of the innermost Mimas toward the leading hemisphere of the outermost moon Titan, intimately linked to the orbital evolution of the E-ring particles with increasing eccentricities and inclinations due to the combined effects of gravity, electromagnetic forces, radiation pressure, and plasma drag, while continually losing mass due to sputtering.

5. NANOGRAINS ORIGIN AND EVOLUTION

Nanograins associated with the E ring have been observed in the plume of Enceladus by CAPS (*Young et al.,* 2004; *Jones et al.,* 2009) and at the outer magnetosphere and in interplanetary space by CDA (*Kempf et al.,* 2005a). These detections are serendipitous and strongly biased by the detection methods and instrument limitations. It is fair to state that our current knowledge of nanograins in the E ring is far from complete. Yet, by constraining their origins and interactions with the surroundings, they have proven very useful to probe Enceladus and its vicinity in ways that other means cannot.

5.1. Nanograins in the Enceladus Plumes

The nanograins observed by CAPS in the Enceladus plume are formed from the cooling plume vapor during its adiabatic expansion into space (*Yeoh et al.,* 2015; see also the chapter by Goldstein et al. in this volume). To place the CAPS observations of nanograins in the plume (*Jones et al.,* 2009; *Hill et al.,* 2012) in context with the much larger water-ice grains measured by CDA (*Spahn et al.,* 2006b) and the submicrometer grains inferred from RPWS impacts (*Ye et al.,* 2014), *Dong et al.* (2015) produced the composite grain-size distributions shown in Fig. 13. The solid curve is an analytical and continuous fit to both CAPS and CDA datasets, while the dashed curves are revisions of the fits suggested by the requirement of charge neutrality between the (mostly negative) nanograins and the (mostly positive) plume plasma reported from the RPWS Langmuir probe measurements (*Morooka et al.,* 2011). The fitted curves have no theoretical motivation; they simply show that a

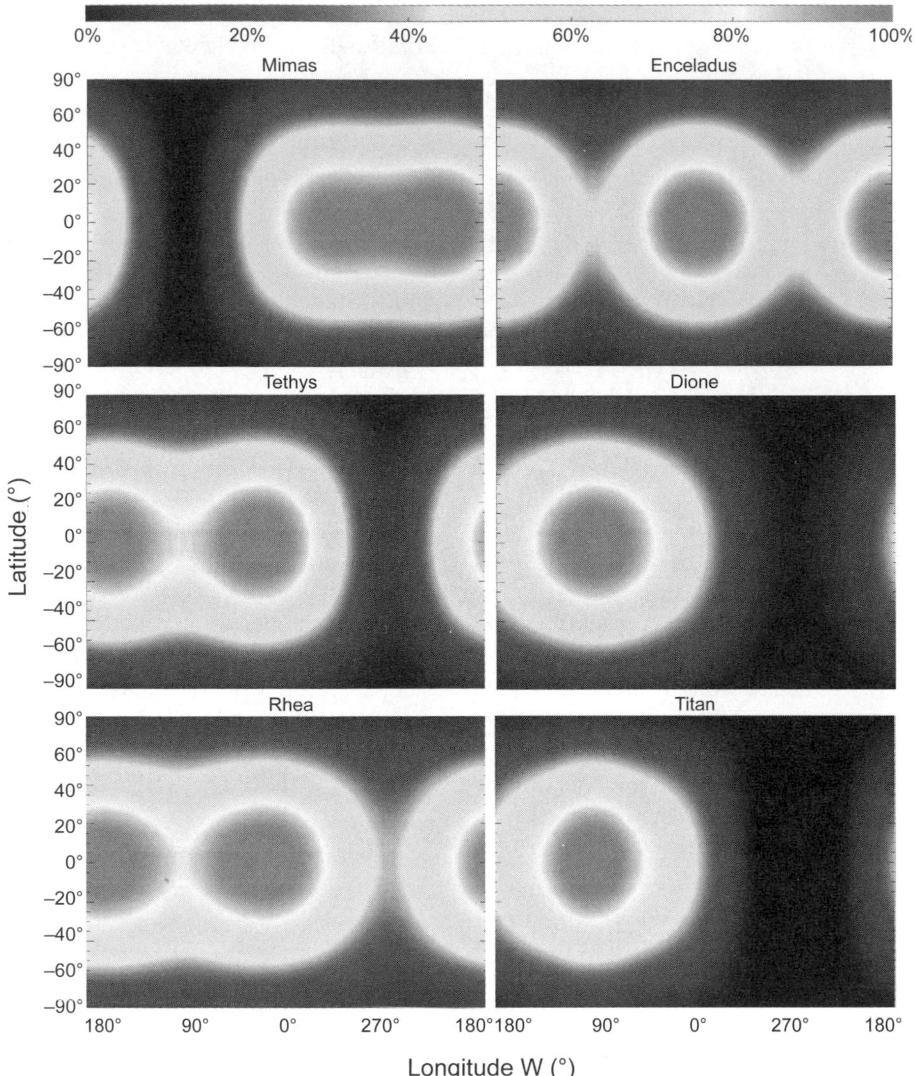

Fig. 12. The mass influx map for the moons Mimas, Enceladus, Tethys, Dione, Rhea, and Titan. In each case the color code represents the mass influx normalized to its maximum value. 90°W and 270°W are on the leading and trailing sides of the moons, respectively.

bimodal size distribution, while allowed by the available observations, is not required by those observations. Both the observations and the fitted curves are subject to significant (factor ~2) uncertainties.

The fate of these nanograins after leaving the Enceladus plumes is unknown. Unlike the larger micrometer and submicrometer grains, due to their small sizes and related long charging times, the charge state of nanograins remains insensitive to changes in the local plasma parameters and will remain a stochastic variable (*Yaroshenko et al.,* 2014; *Yaroshenko and Lühr,* 2014; *Meier et al.,* 2015). Their contribution to the mass production rate for the much broader E ring is quite uncertain owing to both experimental and data-analysis uncertainties, but may well be at least comparable to that of the micrograins (*Dong et al.,* 2015). However, due to their short lifetime, plume nanograins cannot contribute significantly to the overall ring particle population.

5.2. Saturnian Stream Particles

In contrast to the nanograins in the Enceladus plume, the nanometer-sized species observed by CDA are characterized by their high speeds (>70 km s^{-1}). These fast, tiny particles from Saturn resemble so-called "stream particles" emerging from the inner jovian system into interplanetary space (*Grün et al.,* 1993, 1996). Such high speeds, resulting from their interactions with the co-rotational electric field in Saturn's magnetosphere (*Horányi et al.,* 1993a,b; *Horányi,* 2000), indicate that they are in fact not gravitationally bound to Saturn or even to the Sun (*Zook et al.,* 1996). Yet grains as small as those measured by CAPS have charge-to-mass ratios large enough to be tied to the saturnian magnetic field. In addition, the magnetic field configuration of Saturn allows only positively charged grains to gain energy from Saturn's co-rotation electric field and propagate outward.

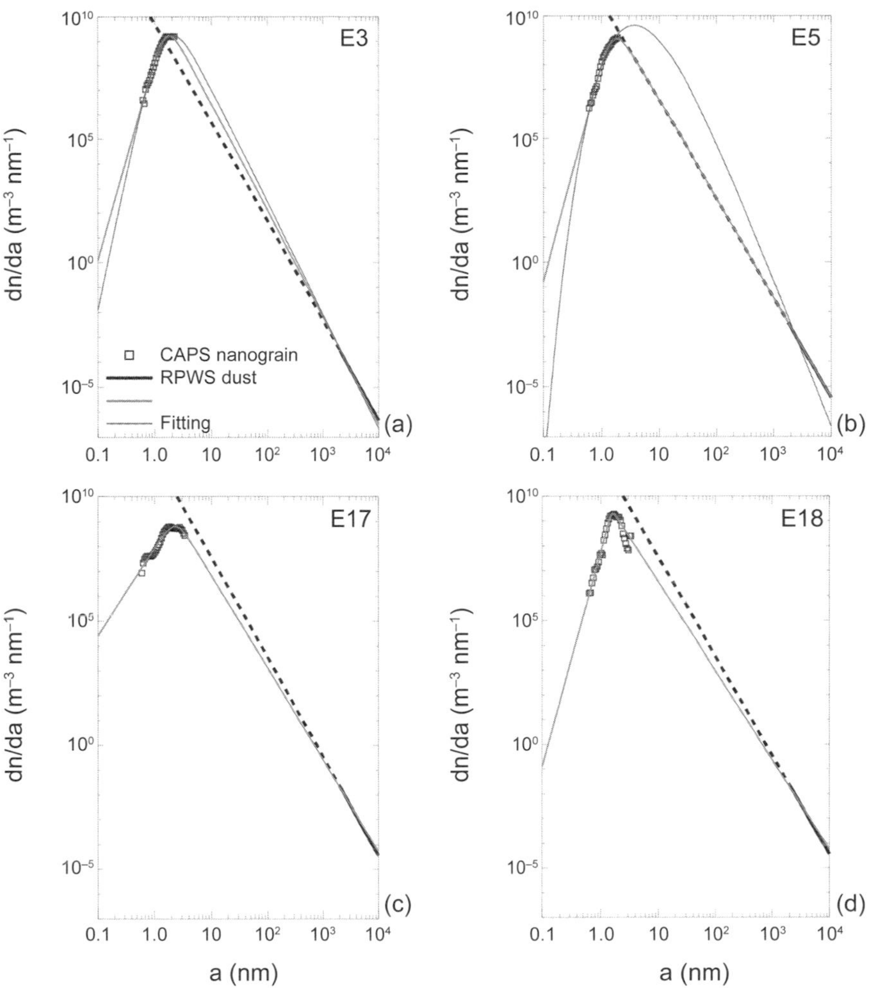

Fig. 13. Composite size distributions proposed by *Dong et al.* (2015) for the same four Cassini plume encounters shown in Fig. 4. The black squares below a_d ~ 3 nm are CAPS measurements. The solid black lines above a_d ~ 1 μm represent the CDA measurements. The dashed black lines are extrapolations of the CDA power-law curve for submicrometer impacts inferred from RPWS measurements.

This means that, from the grain dynamics point-of-view, the nanograins observed by CAPS in the plume is not likely to be the source of saturnian dust streams observed by CDA (*Hsu et al.,* 2011b).

Grain composition provides another important constraint on their origin. Silicon is consistently found in the saturnian stream particles (*Kempf et al.,* 2005b; *Hsu et al.,* 2015), which stands out from the water-ice dominated worlds of Saturn and excludes the nanograins detected in the plume as the source. Silicious stream particles are more sputtering resistant than water ice (*Tielens et al.,* 1994) and are suggested to be sputtered residuals of E-ring grains. The dynamical properties inferred from the interaction with the interplanetary magnetic field (IMF) suggest that they originate from the 6–10 R_S region (*Hsu et al.,* 2011a), the region where E-ring grains are trapped and sputtered because of the grain potential gradient (*Horányi,* 1996; *Beckmann,* 2008; *Kempf and Beckmann,* 2017). This allows extensive sputtering to release nanometer-sized silica particles from the eroded ice matrix of micrometer-sized E-ring grains (*Hsu et al.,* 2011b, 2015).

In contrast to the cold, dense plasma near Enceladus' orbit, dust grains located at the outer part of the E ring are charged positively mainly due to higher secondary electron emission in the hotter plasma environment. E-ring ice grains thus provide a means to relocate these silicious grains to a more dynamically favorable location for them to be ejected outward, implying that the saturnian stream particles also originate from Enceladus.

Stream particles may serve as condensation cores in subsurface vents, or simply be carried within frozen droplets from the subsurface plume sources to become integrated into the E-ring grains. The nano-silica mass fraction relative to water ice is in the range of 100–1000 ppm, too low to be detected in the E-ring grain mass spectra (*Hsu et al.,* 2015).

Silica is a known indicator of hydrothermal reactions and the presence of nano-phase silica in Enceladus ice grains indicates ongoing high-temperature (>90°C) hydrothermal interactions in the subsurface alkaline ocean of this geologically active icy moon (*Hsu et al.,* 2015; *Sekine et al.,* 2015). More details can be found in the chapters in this volume by Postberg et al., McKay et al., and Hemingway et al..

6. SUMMARY

Twelve years of Cassini measurements have provided us with the data needed to study Enceladus and the E ring as a system. We now understand the connection between the cryovolcanos on the moon and the properties of the ring. The E ring is an important means for the erosion of, as well as for the mass deposition on, the surfaces of all ring moons.

The ring particles turn out to be carriers of valuable information about ongoing geophysical processes in the interior of the moon as well as about the Enceladus ice plume itself. The dynamics and evolution of the ring particles are still not fully understood. Cassini findings such as the pronounced noon-midnight asymmetry of the ring have so far escaped a satisfying explanation.

Acknowledgments. This work was partially supported by NASA/JPL.

REFERENCES

Andriopoulou M., Roussos E., Krupp N., Paranicas C., Thomsen M., Krimigis S., Dougherty M. K., and Glassmeier K.-H. (2012) A noon-to-midnight electric field and nightside dynamics in Saturn's inner magnetosphere, using microsignature observations. *Icarus, 220,* 503–513.

Andriopoulou M., Roussos E., Krupp N., Paranicas C., Thomsen M., Krimigis S., Dougherty M. K., and Glassmeier K.-H. (2014) Spatial and temporal dependence of the convective electric field in Saturn's inner magnetosphere. *Icarus, 229,* 57–70.

Auer S., Grün E., Srama R., Kempf S., and Auer R. (2002) The charge and velocity detector of the Cosmic Dust Analyser on Cassini. *Planet. Space Sci., 50,* 773–779.

Bastian T. S., Chenette D. L., and Simpson J. A. (1980) Charged particle anisotropics in Saturn's magnetosphere. *J. Geophys. Res., 85,* 5763–5771.

Baum W. A., Kreidl T., Westphal J. A., Danielson G. E., Seidelmann P. K., Pascu D., and Currie D. G. (1981) Saturn's E ring. *Icarus, 47,* 84–96.

Beckmann U. (2008) Dynamik von Staubteilchen in Saturn's E-Ring. Ph.D. thesis, Universität Heidelberg, Germany.

Brown R. H., Baines K. H., Bellucci G., et al. (2004) The Cassini Visual and Infrared Mapping Spectrometer (VIMS) investigation. *Space Sci. Rev., 115,* 111–168.

Burns J. A., Showalter M. R., Hamilton D. P., Nicholson P. D., de Pater I., Ockert-Bell M. E., and Thomas P. C. (1999) The formation of Jupiter's faint rings. *Science, 284,* 1146–1149.

Carbary J. F., Krimigis S. M., and Ip W. H. (1983) Energetic particle microsignatures of Saturn's satellites. *J. Geophys. Res., 88,* 8947–8958.

Charvat A., Lugovoj E., Faubel M., and Abel B. (2004) New design for a time-of-flight mass spectrometer with a liquid beam laser desorption ion source for the analysis of biomolecules. *Rev. Sci. Instr., 75,* 1209–1218.

Cruikshank D. P., Owen T. C., Ore C.D., Geballe T. R., Roush T. L., de Bergh C., Sandford S. A., Poulet F., Benedix G. K., and Emery J. P. (2005) A spectroscopic study of the surfaces of Saturn's large satellites: H_2O ice, tholins, and minor constituents. *Icarus, 175,* 268–283.

de Pater I., Martin S. C., and Showalter M. R. (2004) Keck near-infrared observations of Saturn's E and G rings during Earth's ring plane crossing in August 1995. *Icarus, 172,* 446–454.

Dikarev V. V. (1999) Dynamics of particles in Saturn's E ring: Effects of charge variations and the plasma drag force. *Astron. Astrophys., 346,* 1011–1019.

Dikarev V. V. and Krivov A. V. (1998) Dynamics and spatial distribution of particles in Saturn's E ring. *Solar System Res., 32,* 128–143.

Dollfus A. and Brunier S. (1981) Photometry of the outer ring E of Saturn. *Bull. Am. Astron. Soc., 13,* 727.

Dong Y., Hill T. W., and Ye S. Y. (2015) Characteristics of ice grains in the Enceladus plume from Cassini observations. *J. Geophys. Res., 120,* 915–937.

Dougherty M. K., Khurana K. K., Neubauer F. M., Russell C. T., Saur J., Leisner J. S., and Burton M. E. (2006) Identification of a dynamic atmosphere at Enceladus with the Cassini magnetometer. *Science, 311,* 1406–1409.

Esposito L. W., Colwell J. E., Larsen K., et al. (2005) Ultraviolet imaging spectroscopy shows an active saturnian system. *Science, 307,* 1251–1255.

Farrell W. M., Kurth W. S., Tokar R. L., et al. (2010) Modification of the plasma in the near-vicinity of Enceladus by the enveloping dust. *Geophys. Res. Lett., 37,* 20202.

Feibelman W. (1967) Concerning the "D" ring of Saturn. *Nature, 214,* 793–794.

Grün E., Zook H., Baguhl M., et al. (1993) Discovery of jovian dust streams and interstellar grains by the Ulysses spacecraft. *Nature, 362,* 428–430.

Grün E., Baguhl M., Hamilton D., et al. (1996) Constraints from Galileo observations on the origin of jovian dust streams. *Nature, 381,* 395–398.

Gurnett D. A., Grün E., Gallagher D., Kurth W. S., and Scarf F. L. (1983) Micron-sized particles detected near Saturn by the Voyager plasma wave instrument. *Icarus, 53,* 236–254.

Gurnett D. A., Kurth W. S., Kirchner D. L., et al. (2004) The Cassini Radio and Plasma Wave Investigation. *Space Sci. Rev., 114,* 395–463.

Haff P. K., Siscoe G. L., and Eviatar A. (1983) Ring and plasma — The enigmae of Enceladus. *Icarus, 56,* 426–438.

Hamilton D. C., Brown D. C., Gloeckler G., and Axford W. I. (1983) Energetic atomic and molecular ions in Saturn's magnetosphere. *J. Geophys. Res., 88,* 8905–8922.

Hamilton D. P. (1993) Motion of dust in a planetary magnetosphere — Orbit-averaged equations for oblateness, electromagnetic, and radiation forces with application to Saturn's E ring. *Icarus, 101,* 244–264.

Hamilton D. P. and Burns J. A. (1994) Origin of Saturn's E ring: Self-sustained, naturally. *Science, 267,* 550–553.

Hansen C. J., Esposito L., Stewart A. I. F., Colwell J., Hendrix A., Pryor W., Shemansky D., and West R. (2006) Enceladus' water vapor plume. *Science, 311,* 1422–1425.

Hansen C. J., Shemansky D. E., Esposito L. W., et al. (2011) The composition and structure of the Enceladus plume. *Geophys. Res. Lett., 38,* 11202.

Hansen C. J., Esposito L. W., Aye K.-M., Colwell J. E., Hendrix A. R., Portyankina G., and Shemansky D. (2017) Investigation of diurnal variability of water vapor in Enceladus' plume by the Cassini ultraviolet imaging spectrograph. *Geophys. Res. Lett., 44,* 672–677.

Hartmann W. K. (1985) Impact experiments. I — Ejecta velocity distributions and related results from regolith targets. *Icarus, 63,* 69–98.

Havnes O., Morfill G. E., and Melandso F. (1992) Effects of electromagnetic and plasma drag forces on the orbit evolution of dust in planetary magnetospheres. *Icarus, 98,* 141–150.

Hedman M. M., Nicholson P. D., Showalter M. R., Brown R. H., Buratti B. J., and Clark R. N. (2009) Spectral observations of the Enceladus plume with Cassini-VIMS. *Astron. J., 693,* 1749–1762.

Hedman M. M., Burt J. A., Burns J. A., and Tiscareno M. S. (2010) The shape and dynamics of a heliotropic dusty ringlet in the Cassini division. *Icarus, 210,* 284–297.

Hedman M. M., Burns J. A., Hamilton D. P., and Showalter M. R. (2012) The three-dimensional structure of Saturn's E ring. *Icarus, 217,* 322–338.

Hedman M. M., Gosmeyer C. M., Nicholson P. D., Sotin C., Brown R. H., Clark R. N., Baines K. H., Buratti B. J., and Showalter M. R. (2013) An observed correlation between plume activity and tidal stresses on Enceladus. *Nature, 500,* 182–184.

Hill T. W. (1984) Saturn's E ring. *Adv. Space Res., 4,* 149–157.

Hill T. W., Thomsen M. F., Tokar R. L., et al. (2012) Charged nanograins in the Enceladus plume. *J. Geophys. Res., 117,* A05209.

Hillier J. K., Green S. F., McBride N., Schwanethal J. P., Postberg F., Srama R., Kempf S., Moragas-Klostermeyer G., McDonnell J. A. M., and Grün E. (2007) The composition of Saturn's E ring. *Mon. Not. R. Astron. Soc., 377,* 1588–1596.

Hood L. L. (1983) Radial diffusion in Saturn's radiation belts — A modeling analysis assuming satellite and ring E absorption. *J. Geophys. Res., 88,* 808–818.

Horányi M. (1996) Charged dust dynamics in the solar system. *Annu. Rev. Astron. Astrophys., 34,* 383–418.

Horányi M. (2000) Dust streams from Jupiter and Saturn. *Phys. Plasmas, 7,* 3847–3850.

Horányi M., Burns J. A., and Hamilton D. (1992) The dynamics of Saturn's E ring particles. *Icarus, 97,* 248–259.

Horányi M., Morfill G. E., and Grün E. (1993a) The dusty ballerina skirt of Jupiter. *J. Geophys. Res., 98,* 21245–21251.

Horányi M., Morfill G., and Grün E. (1993b) Mechanism for the acceleration and ejection of dust grains from Jupiter's magnetosphere. *Nature, 363,* 144–146.

Horányi M., Juhász A., and Morfill G. E. (2008) Large-scale structure of Saturn's E-ring. *Geophys. Res. Lett., 35,* 4203.

Horányi M., Burns J. A., Hedman M. M., Jones G. H., and Kempf S. (2009) Diffuse rings. In *Saturn from Cassini-Huygens* (M. K. Dougherty et al., eds.), pp. 511–536. Springer, Dordrecht.

Horányi M., Szalay J. R., Kempf S., Schmidt J., Grün E., Srama R., and Sternovsky Z. (2015) A permanent, asymmetric dust cloud around the Moon. *Nature, 522,* 324–326.

Hsu H. W., Kempf S., Postberg F., Srama R., Jackman C. M., Moragas-Klostermeyer G., Helfert S., and Grün E. (2010) Interaction of the solar wind and stream particles, results from the Cassini dust detector. In *Twelfth International Solar Wind Conference* (M. Maksimovic et al., eds.), pp. 510–513. AIP Conf. Proc. 1216, American Institute of Physics, Melville, New York.

Hsu H. W., Kempf S., Postberg F., Trieloff M., Burton M., Roy M., Moragas-Klostermeyer G., and Srama R. (2011a) Cassini dust stream particle measurements during the first three orbits at Saturn. *J. Geophys. Res., 116,* 8213.

Hsu H. W., Postberg F., Kempf S., Trieloff M., Burton M., Roy M., Moragas-Klostermeyer G., and Srama R. (2011b) Stream particles as the probe of the dust-plasma magnetosphere interaction at Saturn. *J. Geophys. Res., 116,* 9215.

Hsu H. W., Postberg F., Sekine Y., et al. (2015) Ongoing hydrothermal activities within Enceladus. *Nature, 519,* 207–210.

Hsu S., Horányi M., Juhász A., Kempf S., Sternovsky Z., and Ye S. (2016) Understanding the E-ring puzzle. Abstract P33E-01 presented at 2016 Fall Meeting, AGU, San Francisco, California, 11–15 December.

Jones G. H., Roussos E., Krupp N., et al. (2008) The dust halo of Saturn's largest icy moon, Rhea. *Science, 319,* 1380.

Jones G. H., Arridge C. S., Coates A. J., et al. (2009) Fine jet structure of electrically charged grains in Enceladus' plume. *Geophys. Res. Lett., 36,* 16204.

Juhász A. and Horányi M. (2002) Saturn's E ring: A dynamical approach. *J. Geophys. Res., 107,* 1–10.

Juhász A. and Horányi M. (2004) Seasonal variations in Saturn's E-ring. *Geophys. Res. Lett., 31,* 19703.

Juhász A., Horányi M., and Morfill G. E. (2007) Signatures of Enceladus in Saturn's E ring. *Geophys. Res. Lett., 34,* 9104.

Jurac S. (2001) Saturn's E ring and the production of the neutral torus. *Icarus, 149,* 384–396.

Jurac S., McGrath M. A., Johnson R. E., Richardson J. D., Vasyliunas V. M., and Eviatar A. (2002) Saturn: Search for a missing water source. *Geophys. Res. Lett., 29,* 2172–2175.

Kempf S. and Beckmann U. (2017) Dynamics and long-term evolution of Saturn's E ring particles. *Icarus,* in press.

Kempf S., Srama R., Horányi M., Burton M., Helfert S., Moragas-Klostermeyer G., Roy M., and Grün E. (2005a) High-velocity streams of dust originating from Saturn. *Nature, 433,* 289–291.

Kempf S., Srama R., Postberg F., et al. (2005b) Composition of saturnian stream particles. *Science, 307,* 1274–1276.

Kempf S., Beckmann U., Srama R. Horányi M., Auer S., and Grün E. (2006) The electrostatic potential of E ring particles. *Planet. Space Sci., 54,* 999–1006.

Kempf S., Beckmann U., Moragas-Klostermeyer G., Postberg F., Srama R., Economou T., Schmidt J., Spahn F., and Grün E. (2008) The E ring in the vicinity of Enceladus. I. Spatial distribution and properties of the ring particles. *Icarus, 193,* 420–437.

Kempf S., Beckmann U., and Schmidt J. (2010) How the Enceladus dust plume feeds Saturn's E ring. *Icarus, 206,* 446–457.

Kempf S., Horányi M., Juhász A., Cruz A., Srama R., Postberg F., Spahn F., and Schmidt J. (2012) The 3-dimensional structure of Saturn's E ring inferred from Cassini CDA observations. *EPSC Abstracts*, EPSC2012-701.

Kempf S., Horányi M., Srama R., and Altobelli N. (2015) Exogenous dust delivery into the saturnian system and the age of Saturn's rings. *EPSC Abstracts 2015,* EPSC2015-411.

Kempf S., Southworth B., Schmidt J., Srama R., and Postberg F. (2016) How much dust does Enceladus eject? In *EGU General Assembly Conference Abstracts,* 18041.

Khurana K. K., Russell C. T., and Dougherty M. K. (2008) Magnetic portraits of Tethys and Rhea. *Icarus, 193,* 465–474.

Koschny D. and Grün E. (2001) Impacts into ice-silicate mixtures: Ejecta mass and size distributions. *Icarus, 154,* 402–411.

Krimigis S. M. and Armstrong T. P. (1982) Two-component proton spectra in the inner Saturnian magnetosphere. *Geophys. Res. Lett., 9,* 1143–1146.

Krivov A. V., Sremčević M., Spahn F., Dikarev V. V., and Kholshevnikov K. V. (2003) Impact-generated dust clouds around planetary satellites: Spherically symmetric case. *Planet. Space Sci., 51,* 251–269.

Krüger H., Krivov A., Hamilton D., and Grün E. (1999) Detection of an impact-generated dust cloud around Ganymede. *Nature, 399,* 558–560.

Krüger H., Krivov A. V., and Grün E. (2000) A dust cloud of Ganymede maintained by hypervelocity impacts of interplanetary micrometeoroids. *Planet. Space Sci., 48,* 1457–1471.

Krüger H., Horányi M., Krivov A. V., and Graps A. (2002) Jovian dust: Streams, clouds and rings. In *Jupiter: The Planet, Satellites and Magnetosphere* (F. Bagenal et al., eds.), pp. 219–240. Cambridge Univ., Cambridge.

Kurth W. S., Averkamp T. F., Gurnett D. A., and Wang Z. (2006) Cassini RPWS observations of dust in Saturn's E ring. *Planet. Space Sci., 54,* 988–998.

McDonald F. B., Schardt A. W., and Trainor J. H. (1980) If you've seen one magnetosphere, you haven't seen them all — Energetic particle observations in the Saturn magnetosphere. *J. Geophys. Res., 85,* 5813–5830.

Meier P., Kriegel H., Motschmann U., Schmidt J., Spahn F., Hill T. W., Dong Y., and Jones G. H. (2014) A model of the spatial and size distribution of Enceladus' dust plume. *Planet. Space Sci., 104,* 216–233.

Meier P., Motschmann U., Schmidt J., Spahn F., Hill T. W., Dong Y., Jones G. H., and Kriegel H. (2015) Modeling the total dust production of Enceladus from stochastic charge equilibrium and simulations. *Planet. Space Sci., 119,* 208–221.

Melin H., Shemansky D. E., and Liu X. (2009) The distribution of atomic hydrogen and oxygen in the magnetosphere of Saturn. *Planet. Space Sci., 57,* 1743–1753.

Mitchell C. J., Porco C. C., and Weiss J. W. (2015) Tracking the geysers of Enceladus into Saturn's E ring. *Astron. J., 149,* 156.

Morfill G. E., Grün E., and Johnson T. V. (1983) Saturn's E, G, and F rings — Modulated by the plasma sheet? *J. Geophys. Res., 88,* 5573–5579.

Morooka M. W., Wahlund J. E., Eriksson A. I., Farrell W. M., Gurnett D. A., Kurth W. S., Persoon A. M., Shafiq M., André M., and Holmberg M. K. G. (2011) Dusty plasma in the vicinity of Enceladus. *J. Geophys. Res., 116,* A12221.

Nicholson P. D., Showalter M. R., and Dones L. (1996) Observations of Saturn's ring-plane crossing in August and November. *Science, 272,* 509–516.

Nietiadi M. L., Sandoval L., Urbassek H. M., and Möller W. (2014) Sputtering of Si nanospheres. *Phys. Rev. B, 90,* 045417.

Pang K. D., Voge C. C., Rhoads J. W., and Ajello J. M. (1984) The E ring of Saturn and satellite Enceladus. *J. Geophys. Res., 89,* 9459–9470.

Porco C. C., West R. A., Squyres S., et al. (2004) Cassini Imaging Science: Instrument characteristics and anticipated scientific investigations at Saturn. *Space Sci. Rev., 115,* 363–497.

Porco C. C., Helfenstein P., Thomas P. C., et al. (2006) Cassini observes the active south pole of Enceladus. *Science, 311,* 1393–1401.

Porco C. C., DiNino D., and Nimmo F. (2014) How the geysers, tidal stresses, and thermal emission across the south polar terrain of Enceladus are related. *Astron. J., 148,* 45.

Postberg F., Kempf S., Hillier J. K., Srama R., Green S. F., McBride N., and Grün E. (2008) The E-ring in the vicinity of Enceladus. II. Probing the moon's interior — The composition of E-ring particles. *Icarus, 193,* 438–454.

Postberg F., Kempf S., Schmidt J., Brilliantov N., Beinsen A., Abel B., Buck U., and Srama R. (2009) Sodium salts in E-ring ice grains from an ocean below the surface of Enceladus. *Nature, 459,* 1098–1101.

Postberg F., Schmidt J., Hillier J. K., Kempf S., and Srama R. (2011) A salt-water reservoir as the source of a compositionally stratified plume on Enceladus. *Nature, 474,* 620–622.

Schardt A.W. and McDonald F. B. (1983) The flux and source of energetic protons in Saturn's inner magnetosphere. *J. Geophys. Res., 88,* 8923–8935.

Schenk P., Hamilton D. P., Johnson R. E., McKinnon W. B., Paranicas C., Schmidt J., and Showalter M. R. (2011) Plasma, plumes and rings: Saturn system dynamics as recorded in global color patterns on its midsize icy satellites. *Icarus, 211,* 740–757.

Schmidt J., Brilliantov N., Spahn F., and Kempf S. (2008) Slow dust in Enceladus' plume from condensation and wall collisions in tiger stripe fractures. *Nature, 451,* 685–688.

Sekine Y., Shibuya T., Postberg F., et al. (2015) High temperature water-rock interactions and hydrothermal environments in the chondrite-like core of Enceladus. *Nature Commun., 6,* 8604.

Shafiq M., Wahlund J. E., Morooka M. W., Kurth W. S., and Farrell W. M. (2011) Characteristics of the dust-plasma interaction near Enceladus' south pole. *Planet. Space Sci., 59,* 17–25.

Showalter M. R., Cuzzi J. N., and Larson S. M. (1991) Structure and particle properties of Saturn's E ring. *Icarus, 94,* 451–473.

Sittler E. C., Scudder J. D., and Bridge H. S. (1981) Distribution of neutral gas and dust near Saturn. *Nature, 292,* 711–714.

Sittler E. C., Ali A., Cooper J. F., Hartle R. E., Johnson R. E., Coates A. J., and Young D. T. (2009) Heavy ion formation in Titan's ionosphere: Magnetospheric introduction of free oxygen and a source of Titan's aerosols? *Planet. Space Sci., 57,* 1547–1557.

Smith B. A., Soderblom L., Beebe R. F., et al. (1981) Encounter with Saturn — Voyager 1 imaging science results. *Science, 212,* 163–191.

Southworth B. and Kempf S. (2017) Surface deposition maps for the Enceladus dust plume. *Icarus,* in press.

Spahn F., Albers N., Hörning M., Kempf S., Krivov A. V., Makuch M., Schmidt J., Seiß M., and Sremčević M. (2006a) E ring dust sources: Implications from Cassini's dust measurements. *Planet. Space Sci., 54,* 1024–1032.

Spahn F., Schmidt J., Albers N., et al. (2006b) Cassini dust measurements at Enceladus and implications for the origin of the E ring. *Science, 311,* 1416–1418.

Spitale J. N. and Porco C. C. (2007) Association of the jets of Enceladus with the warmest regions on its south-polar fractures. *Nature, 449,* 695–697.

Srama R., Ahrens T. J., Altobelli N., et al. (2004) The Cassini Cosmic Dust Analyzer. *Space Sci. Rev., 114,* 465–518.

Srama R., Kempf S., Moragas-Klostermeye, G., et al. (2006) *In situ* dust measurements in the inner saturnian system. *Planet. Space Sci., 54,* 967–987.

Sremčević M., Krivov A. V., and Spahn F. (2003) Impact generated dust clouds around planetary satellites: Asymmetry effects. *Planet. Space Sci., 51,* 455–471.

Stöffler D., Gault D. E., Wedekind J., and Polkowski G. (1975) Experimental hypervelocity impact into quartz sand — Distribution and shock metamorphism of ejecta. *J. Geophys. Res., 80,* 4062–4077.

Sun K.-L., Seiß M., Hedman M. M., and Spahn F. (2017) Dust in the arcs of Methone and Anthe. *Icarus, 284,* 206–215.

Terrile R. J. and Tokunaga A. (1980) Infrared photometry of Saturn's E-ring. *Bull. Am. Astron. Soc., 12,* 701.

Thomsen M. F. and van Allen J. A. (1979) On the inference of properties of Saturn's ring E from energetic charged particle observations. *Geophys. Res. Lett., 6,* 893–896.

Tian F., Stewart A. I. F., Toon O. B., Larsen K. W., and Esposito L. W. (2007) Monte Carlo simulations of the water vapor plumes on Enceladus. *Icarus, 188,* 154–161.

Tielens A. G. G. M., McKee C. F., Seab C. G., and Hollenbach D. J. (1994) The physics of grain-grain collisions and gas-grain sputtering in interstellar shocks. *Astrophys. J., 431,* 321–340.

van Allen J. A., Randall B. A., and Thomsen M. F. (1980) Sources and sinks of energetic electrons and protons in Saturn's magnetosphere. *J. Geophys. Res., 85,* 5679–5694.

Verbiscer A., French R., Showalter M., and Helfenstein P. (2007) Enceladus: Cosmic graffiti artist caught in the act. *Science, 315,* 815.

Wahlund J. E., Boström R., Gustafsson G., et al. (2005) The inner magnetosphere of Saturn: Cassini RPWS cold plasma results from the first encounter. *Geophys. Res. Lett., 32,* L20S09.

Waite J. H., Combi M. R., Ip W. H., et al. (2006) Cassini Ion and Neutral Mass Spectrometer: Enceladus plume composition and structure. *Science, 311,* 1419–1422.

Wilson R. J., Bagenal F., Delamere P. A., Desroche M., Fleshman B. L., and Dols V. (2013) Evidence from radial velocity measurements of a global electric field in Saturn's inner magnetosphere. *J. Geophys. Res., 118,* 2122–2132.

Yaroshenko V. V. and Lühr H. (2014) Random dust charge fluctuations in the near-Enceladus plasma. *J. Geophys. Res., 119,* 6190–6198.

Yaroshenko V. V., Lühr H., and Miloch W. J. (2014) Dust charging in the Enceladus torus. *J. Geophys. Res., 119,* 221–236.

Ye S. Y., Gurnett D. A., Kurth W. S., Averkamp T. F., Kempf S., Hsu H. W., Srama R., and Grün E. (2014) Properties of dust particles near Saturn inferred from voltage pulses induced by dust impacts on Cassini spacecraft. *J. Geophys. Res., 119,* 6294–6312.

Ye S. Y., Kurth W. S., Hospodarsky G. B., Averkamp T. F., and Gurnett D. A. (2016a) Dust detection in space using the monopole and dipole electric field antennas. *J. Geophys. Res., 121,* 11964–11972.

Ye S. Y., Gurnett D. A., and Kurth W. S. (2016b) *In-situ* measurements of Saturn's dusty rings based on dust impact signals detected by Cassini RPWS. *Icarus, 279,* 51–61.

Yeoh S. K, Chapman T. A., Goldstein D. B., Varghese P. L., and Trafton L. M. (2015) On understanding the physics of the Enceladus south polar plume via numerical simulation. *Icarus, 253,* 205–222.

Young D. T., Berthelier J. J., Blanc M., et al. (2004) Cassini Plasma Spectrometer investigation. *Space Sci. Rev., 114,* 1–112.

Young D. T., Berthelier J. J., Blanc M., et al. (2005) Composition and dynamics of plasma in Saturn's magnetosphere. *Science, 307,* 1262–1266.

Zook H., Grün E., Baguhl M., Hamilton D., Linkert G., Liou J. C., Forsyth R., and Phillips J. (1996) Solar wind magnetic field bending of jovian dust trajectories. *Science, 274,* 1501–1503.

Smith H. T., Crary F. J., Dougherty M. K., Perry M. E., Roussos E., Simon S., and Tokar R. L. (2018) Enceladus and its influence on Saturn's magnetosphere. In *Enceladus and the Icy Moons of Saturn* (P. M. Schenk et al., eds.), pp. 211–234. Univ. of Arizona, Tucson, DOI: 10.2458/azu_uapress_9780816537075-ch011.

Enceladus and Its Influence on Saturn's Magnetosphere

H. T. Smith
Johns Hopkins University Applied Physics Laboratory

F. J. Crary
University of Colorado Boulder

M. K. Dougherty
Imperial College London

M. E. Perry
Johns Hopkins University Applied Physics Laboratory

E. Roussos
Max Planck Institute

S. Simon
Georgia Institute of Technology

R. L. Tokar
Planetary Science Institute

The discovery and study of active water plumes emanating from the southern pole of the Enceladus has provided key information for understanding particle sources and dynamics in Saturn's magnetosphere. Prior to 2004, knowledge of the saturnian system was limited to Earth-based observations and only three *in situ* flybys. The subsequent arrival of the Cassini spacecraft rapidly changed what we know about the influence of this moon on the saturnian system, causing dramatic revisions to our understanding of the physical processes occurring in Saturn's magnetosphere. The very dense, relatively unprotected atmosphere of the large moon Titan, which was once believed to be a dominant source of plasma, has taken a back seat to its much smaller cousin, Enceladus. This relatively small moon serves as the primary source of particles in the magnetosphere and thus has a significant impact on both the near-Enceladus environment as well as the entire magnetosphere. Here we present an overview of the current understanding of these impacts.

1. INTRODUCTION/BACKGROUND

Saturn contains a very diverse and distinctive magnetosphere. While Jupiter's magnetosphere is dominated by plasma, neutral particles in Saturn's magnetosphere can outnumber charged particles by over 1 to 2 orders of magnitude. This interesting phenomenon, as well other key characteristics, are caused by Saturn's largest source of heavy particles, Enceladus.

Similar to Jupiter, Saturn's magnetosphere is a result of the planet's magnetic field, which is generated by its internal rotation. Interestingly, Saturn's magnetic field is basically aligned with the planetary rotational axis (<1° tilt) and rotates with a period of less than 11 hours. The size of Saturn's magnetosphere increases and decreases based on solar wind pressure but is relatively large. At the shortest point between Saturn and the Sun, *Kanani et al.* (2010) determine the standoff distance is generally 21–27 Saturn radii (R_S, or approximately 60,268 km).

Prior to Cassini arrival at Saturn in 2004, *in situ* observations of the saturnian system were limited to Pioneer 11 and Voyagers 1 and 2. These missions helped to show that Saturn's plasma consists of hydrogen, water-group, and/or nitrogen ions. However, of the neutral particle populations, only the H torus was directly observed (*Broadfoot et al.,* 1981; *Bridge et al.,* 1982). Early neutral cloud and plasma modeling efforts were able to reproduce ion observations with an equatorial bound neutral population of only ~30 cm^{-3} in the vicinity of the icy satellites (*Ip,* 1984; *Johnson et al.,* 1989; *Pospieszalska and Johnson,* 1991).

Additionally, plasma transport and chemistry models were able to simulate these observations assuming icy satellite and ring-sputtering sources (*Richardson and Sittler,* 1990; *Richardson,* 1995).

Voyager UVS H observations (*Shemansky and Hall,* 1992) and subsequent multiple modeling efforts led to the search for, and detection of, OH (*Shemansky et al.,* 1993) with the Hubble Space Telescope (HST). These observations dramatically changed the perspective of Saturn's magnetosphere. Follow-on modeling showed much higher OH densities than anticipated, leading to the discovery that neutral particles greatly outnumber charged particles. *Jurac and Richardson* (2005) went on to infer the presence of a source in the vicinity of Enceladus' orbit and that source rates were much larger than could be produced by ion sputtering on icy satellites alone.

Following Cassini's arrival at Saturn, the understanding of neutrals, plasma, and their interactions became even more complex. Titan's dense nitrogen atmosphere, once thought to be a significant source of neutrals in Saturn's magnetosphere (*Barbosa,* 1987; *Ip,* 1997; *Smith et al.,* 2004) turned out be a negligible net source in that most of the species ejected from Titan do not appear to accumulate. Unfortunately, predicted neutral nitrogen densities are too low for direct detection, so nitrogen ionization products must be observed to infer the presence of neutral nitrogen. However, the majority of the nitrogen ions detected appeared to originate from the tiny icy satellite Enceladus (*Smith et al.,* 2005, 2007). This discovery was significant because it is unexpected that such a small inner icy moon could be a much more significant magnetospheric source than the very dense, unprotected atmosphere of an outer moon, Titan, which is larger than the planet Mercury.

The discovery of the active neutral water gas and dust plumes emanating from the south pole of Enceladus (see volume 311 of *Science,* 2006) proved this small moon is the major source of neutral molecules and heavy ions in Saturn's magnetosphere. This dominance of Enceladus over Titan is even evident in the outer magnetosphere near Titan, where ionization products from Enceladus (primarily O^+) are the dominant heavy species in the magnetosphere up to as close as several satellite radii from Titan, where Titan-originated species (primarily hydrocarbon ions) then become visible. Figure 1 shows the current understanding of Saturn's primary particle source rates, illustrating how Enceladus is a dominant magnetospheric source. Enceladus' role as a primary magnetospheric source has large implications for significantly impacting Saturn's local as well as global magnetosphere; this is discussed in detail below.

2. INFLUENCE ON LOCAL MAGNETOSPHERIC ENVIRONMENT

As the primary source of heavy particles in Saturn's magnetosphere, the impact of Enceladus is first experienced in the local environment near the moon. As these particles interact with the local magnetic field, plasma environment, and dust, as well as solar photons, a feedback system is established between these components that generates an interesting local magnetospheric environment.

2.1. Magnetic Field

The initial discovery of the interaction of Saturn's magnetic field and plasma with an atmospheric plume at Enceladus was carried out by the magnetometer instrument onboard Cassini (*Dougherty et al.,* 2006; see also the chapter in this volume by Dougherty et al.) during three flybys of the moon in 2005. Clear magnetic field perturbations in the vicinity of the moon revealed that the saturnian plasma and magnetic field were being deflected and slowed down by Enceladus. In addition, increased ion cyclotron wave activity at water group ion gyro-frequencies underpinned that the moon was acting as a major source of ions. Numerous subsequent flybys of Enceladus confirmed the venting of water vapor, dust, and organic material from the cracks/tiger stripes at the south pole of Enceladus.

The nearly corotating plasma of Saturn's magnetosphere is traveling at a relative speed of ~26 km s^{-1} when it reaches Enceladus, with the magnetic field of Saturn on the order of 320 nT. The sonic and Alfvenic speeds are on the order of 20 and 150 km s^{-1} respectively (*Khurana et al.,* 2007), hence the resultant interaction of Enceladus with Saturn's magnetospheric plasma is both sub-Alfvenic and sub-magnetosonic.

The obstacle to Saturn's corotational plasma is provided by the water vapor plumes emanating from the south pole of Enceladus. New ions are generated mainly through charge exchange with the incident magnetospheric plasma, and are also due to electron impacts and solar UV radiation. These newly produced ions are added to the plasma flow, which extracts momentum from the flow, slowing it down. The expected interaction of a conducting or mass-loading object in magnetized flowing plasma has long been understood, with the schematic in Fig. 2 (*Khurana et al.,* 2007) revealing the expected behavior of the magnetic field lines in the vicinity of the object. The coordinate system shown here is a Cartesian system known as the Enceladus Interaction System (ENIS), where the x-component is in the direction of co-rotating flow (flowing at a velocity of v), the y-axis points toward Saturn, and the z-axis is along the direction of Saturn's rotation axis. Thermal and magnetic pressure gradients are generated by the slowed plasma in the mass-loaded region and a magnetic field curvature diverts and accelerates the plasma flow around the mass-loading region. Alfven wings both above and below the obstacle (and shown in Fig. 2) are generated by the resulting currents of the interaction (*Neubauer,* 1980, 1998), with the current system closing in Saturn's ionosphere, carried by the Alfven wings.

A special characteristic of Enceladus' magnetospheric interaction is the pronounced north-south asymmetry of the obstacle to the plasma flow. The Alfven waves generated in the local interaction region cannot propagate unimpeded into Saturn's northern hemisphere, but they are partially blocked at the icy crust of the moon. As shown by *Saur et al.* (2007),

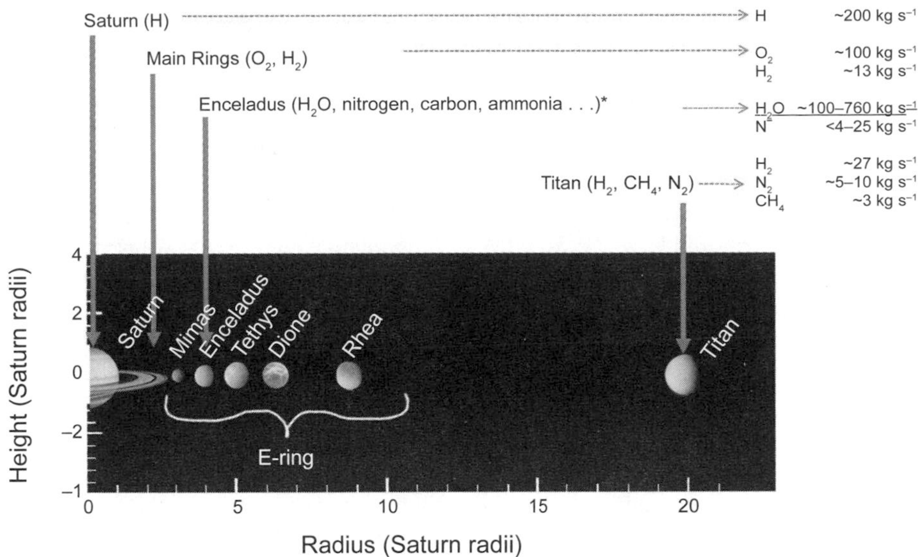

Fig. 1. Current estimates of Saturn's primary magnetopsheric particle source rates: H from Saturn (*Tseng et al.,* 2013a), maximum O_2 and H_2 from main rings (*Tseng et al.,* 2013b, divided by 100 for equinox), H_2O from Enceladus (*Hansen et al.,* 2006; *Burger et al.,* 2007; *Tian et al.,* 2007; *Smith et al.,* 2010; *Dong et al.,* 2011), N from Enceladus (*Smith et al.,* 2007), H_2 from Titan (*Cui et al.,* 2008), N_2 and CH_4 from Titan (*DeLaHaye et al.,* 2007).

this partial reflection of the Alfven wing gives rise to a system of "hemisphere coupling currents" that are tangential to the surface of Enceladus and directly connect Saturn's northern and southern polar ionospheres. These surface current systems generate rotational discontinuities in the magnetic field, which were observed by Cassini during numerous crossings of the Enceladus flux tube (*Simon et al.,* 2014). An auroral footprint within Saturn's ionosphere is a manifestation of the moon's plasma interaction process (*Pryor et al.,* 2011). Such an electrodynamic coupling between a mass-loading moon and its parent planet is not unknown, with Jupiter and its volcanic moon Io being a well-known example (*Kivelson et al.,* 2001).

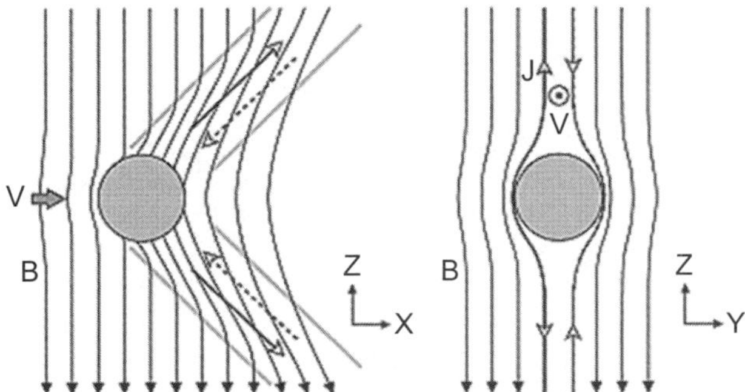

Fig. 2. The interaction of a conducting or mass-loading object with a magnetized flowing plasma (*Khurana et al.,* 2007). The inflowing corotating Saturn plasma flows at a velocity v, and the resultant draped magnetic field lines are denoted by B. The left and right figures show the (X,Z) and (Y,Z) planes of the ENIS coordinate system respectively.

Analytical work by *Simon et al.* (2011) has revealed that electron absorption by dust grains within the plume can cause a reversal in the sign of the Hall conductivity (known as the anti-Hall effect). This changes the sign of the B_y component in the Alfven wings, as has been observed during all of the Enceladus flybys, with hybrid simulations (*Kriegel et al.,* 2011) able to reproduce this result. Extensive modeling of the interaction of the magnetosphere with Enceladus and its plume has been carried out (e.g., *Jia et al.,* 2010; *Kriegel et al.,* 2009, 2011; *Simon et al.,* 2011), and all conclude that the presence of charged dust within the plume has a strong influence on the resulting electromagnetic interaction. It was also demonstrated that the observed magnetic field perturbations contain detailed information on the charging process and the motion of nanograins in the interaction region (*Kriegel et al.,* 2014).

Clear field-aligned current signatures are observed on each side of the moon forming the Alfven wing structure (*Engelhardt et al.,* 2015), and these are associated with whistler mode emissions similar to terrestrial auroral hiss (*Gurnett et al.,* 2011; *Leisner et al.,* 2013) and magnetic field aligned electron beams. The typical intensities of the currents are consistent with the expected size of the perpendicular current coupling the two field aligned currents across the ionosphere.

Magnetic field observations have also revealed ion cyclotron waves produced by water-group (O^+, OH^+, H_2O^+, H_3O^+) and O_2^+ ions at nearly all radial distances and local times within the E ring (*Leisner et al.,* 2006; *Meeks et al.,* 2016). The waves are observed to extend in a radial sense from just inside the orbit of Enceladus to outside that of Dione's orbit. In latitudinal direction, the ion cyclotron waves appear to be confined to a thin layer around Saturn's magnetic equatorial plane (*Leisner et al.,* 2011). These waves are generated by the pickup ions that are created from the ionization of the neutral exosphere around the E-ring material and they grow from the free energy of the highly anisotropic distribution of the pickup ions. The energy flux of the waves is related to the energy that is added to the ions when they are accelerated and can therefore be used to estimate the loss rate from the exosphere of the ring, which is about 2 orders of magnitude less than the mass production arising in the Io torus.

2.2. Local Environment Key Interaction Processes

Particle interaction rates help to provide critical insight into how Enceladus influences the local magnetospheric environment. The local particle environment is impacted by the local magnetic field environment but also serves to impact the local magnetic field. Thus, a feedback mechanism is created that serves to enhance some interaction processes over others.

These interactions are initiated with the Enceladus plume water source. The exact plume generation mechanisms are still being debated (see the chapter in this volume by Goldstein et al.), but there is general agreement that most of the mass in the Enceladus plumes is exiting as neutral water molecules (see the chapter in this volume by Kempf et al.). These water particles are ejected through lower-speed gas outflow as well as higher-speed jets, but the bulk velocity is on the order of <1.0 km s^{-1} (*Smith et al.,* 2010; *Hansen et al.,* 2011). This is a key parameter because the particle escape velocity is much lower than the ~12.6 km s^{-1} orbital speed of Enceladus. Thus, when a neutral water molecule is escaping from Enceladus, its primary velocity component comes from Enceladus itself. From the perspective of the Saturn inertial reference frame, the particle's trajectory appears very similar to the Enceladus orbit but with a slight orbit adjustment. The net result is a particle that moves relatively slowly with respect to Enceladus until it interacts with something else.

Over time, this small velocity difference will cause the particles to drift further upstream and downstream of Enceladus, and with enough time, they form a narrow torus of co-rotating water molecules. However, the H_2O molecules that are perturbed by interactions start to form different local particle populations. The vertical scale height of this water is observed to be ~0.4 R_S. In general, the source water molecules can interact with electrons, ions, and solar photons or simply impact the surface of Enceladus (*Hartogh et al.,* 2011).

When an existing electron from the ambient plasma (or energetic particle population) interacts with a water molecule, the result is an electron-impact "collision" that modifies the original source water population distribution. This interaction could simply liberate an electron from the molecule, producing a water (positively charged) ion with this liberated electron, adding to the local ambient electron population. Additionally, the interaction could dissociate the molecule into a combination of neutral and/or ion daughter species. Any resulting daughter neutral species receives a small increase in kinetic energy from this interaction, which serves to move the particles further away from Enceladus. The interaction rates of these processes are dependent on the density of the ambient electron population and its energy distribution as well the energy-dependent cross-section for that particular interaction. This water molecule electron-impact cross section is a measured and/or calculated distribution function that identifies how likely an electron of a particular energy will interact with a water molecule for each possible resulting ion/neutral particle combination. It is important to note that the resulting OH, O, and H particles can also experience electron-impact interactions.

Solar photons can also interact with water molecules. In the case of Enceladus, photon interaction, also referred to as photolysis, is a chemical reaction where photons from the Sun collide with a water molecule. Similar to electron-impact, these interactions can simply ionize the water molecule (with a resulting electron) or they can dissociate the target into combination of ions and/or neutral components. Photons of all solar wavelengths can participate in these interactions; however, photons with wavelengths shorter than visible light tend to dominate because a photon's energy is inversely proportional to wavelength. During this process, energy is transferred to the resulting particles, which also serve to move particles further away from Enceladus. Thus, these interactions can be considered similar to electron impact regarding their impact to the particle distribution around Enceladus. However, the probability (or interaction rate) of such processes is dependent on distance from the Sun and solar conditions rather than an interacting particle population. Also similar to electron-impact, the resulting OH, O, and H particles can also experience photon interactions. *Heubner et al.* (2015) provides a comprehensive collection of solar-condition-based photo interaction rates at 1 AU from the Sun, which can be scaled for Saturn's distance from the Sun.

Neutral source water particles can also interact with ions either through a simple collision or one where charge exchange occurs (effectively ionizing a neutral and neutralizing an ion). This process is much more efficient as plasma and neutral particle velocities approach each other. The net result of charge exchange in general is that while a fresh ion is created, an existing ion has become an energetic neutral particle so there is usually no net increase in the ion population. This process dramatically spreads out the neutral clouds and enhances particle escape from the magnetosphere. Similar to electron impact, the ion-collision interaction rates are dependent on the density of the ambient ion population and its energy distribution as well the energy-dependent cross-section for that particular interaction. Also similar to electron-impact, the resulting OH, O, and H particles can also experience ion-collision interactions that are also dependent on the relative velocity of the neutral particle and the ion.

The above processes can then be compared to trace the interaction pathways and ultimate fate of source neutral water molecules emanating from the Enceladus plumes in the local magnetospheric environment. Figure 3 shows the main interactions for Enceladus plume-generated water molecules

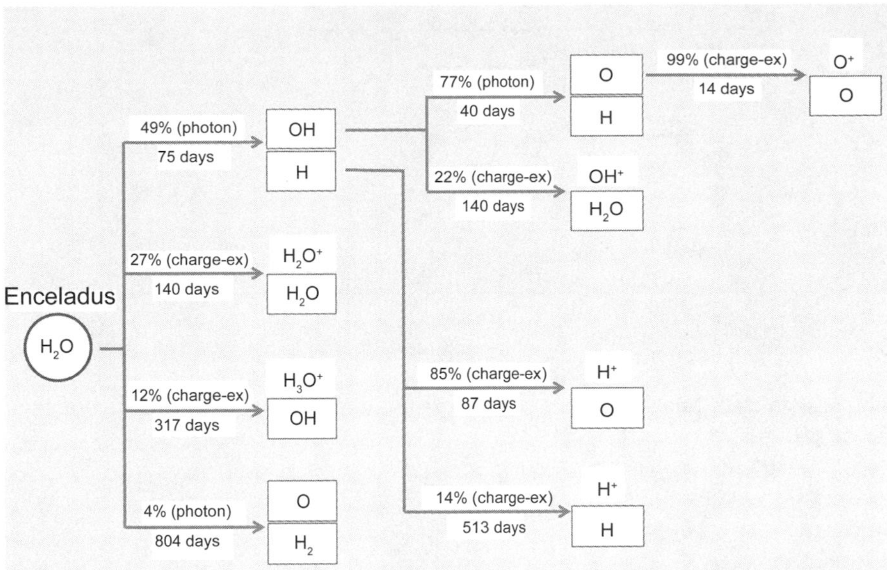

Fig. 3. Interaction flow chart for neutral water source molecules ejected via the Enceladus plumes. Rates are calculated using average plasma densities and temperature for 4 R_S from Saturn. Interactions expressed by process type (photolysis, charge-exchange, and ionization), interaction process probability, and interaction time (lifetime in days).

(and the derivative neutral species), along with estimated timescales and relative probability compared to competing processes based on average plasma and solar conditions. These calculations are based on *Smith et al.* (2010) but are also consistent with *Cassidy and Johnson* (2010) and *Fleshman et al.* (2010). This figure provides some key insight into what is actually occurring near Enceladus. About half of the water is dissociated into O and OH by photons in less than 3 months. Additionally, about a quarter of the water interactions experience symmetric charge exchange, resulting in no net impact to total particle populations but with noticeable changes to particle energy and velocity. About three-fourths of the resulting OH is also mainly dissociated by photons into O and H in less than two months, with the remaining OH experiencing charge exchange. The resulting O and H mostly experiences charge exchange in the near-Enceladus region in two weeks and less than three months, respectively. The dominance of photons and ions over electron impact is because the bulk of the electron population near Enceladus is at energies less than the electron-impact ionization cutoff energies (generally <13.6 eV), so most electron-impact interactions are caused by the much smaller electron population that is above this minimum energy (i.e., "hotter" electrons). Additionally, the relatively low plasma flow velocities (<38 km s^{-1}) in the inner magnetosphere enhance charge exchange interactions as compared to the much larger plasma flow velocities in the middle and outer magnetosphere.

2.3. Plasma Slowing/Stagnation Region

During close encounters of Cassini with Enceladus, the Cassini Plasma Spectrometer (CAPS) instrument obtained *in situ* measurements that strongly suggest the presence of water group ions freshly produced in the dense plume. The dominant species are light ions (H^+, H_2^+), water-group ions (O^+, OH^+, H_2O^+, H_3O^+), and single water cluster ions, ($H_2O*H_2O^+$), all observed close to and nearly due south of Enceladus. The ions have kinetic energies in the CAPS frame roughly equal to ions that are at rest with respect to Enceladus and rammed into the CAPS instrument roughly at the Cassini spacecraft speed. This is the signature of freshly produced ions in the plume due to (e.g., generally caused by) charge exchange interactions of incoming magnetospheric ions and neutral plume gas. Figure 4 shows an example of these ions detected by CAPS during the Cassini E3 Enceladus encounter. The high concentration of ions close to the water group ion and charged water cluster instrument ram energies (denoted by vertical arrows) are clearly visible. Note also the presence of positive nanograins (discussed below in section 2.5). Further details of these data are discussed in *Tokar et al.* (2009).

The Cassini E7 Enceladus encounter on November 2, 2009, provided additional observations of the plume stagnation region with Cassini's motion directly through the plume region as depicted in Fig. 5. The strong plume interaction region in this figure extends from about 07:41:40 UT to 07:43:00 UT, the region in which the CAPS detector, which was sensitive to rammed ions (Fig. 6), observed stagnation and fresh ions at the ram energies.

The E7 encounter also exhibited the close correspondence between the CAPS-observed ion slowing and the Ion Neutral Mass Spectrometer (INMS)-observed entry into the dense plume water vapor. This is depicted in Fig. 7, which shows the CAPS ion flow speed and the INMS atomic mass 44 counts, a proxy for plume water vapor concentration. The E7 closest approach is at ~07:41:58 UT, and CAPS observes

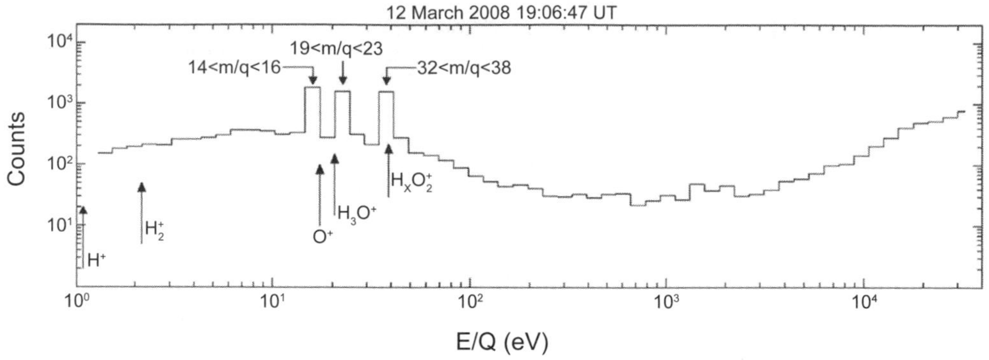

Fig. 4. CAPS IMS ion counts as a function energy per charge in eV for the Cassini E3 Enceladus encounter on March 12, 2008. From *Tokar et al.* (2009).

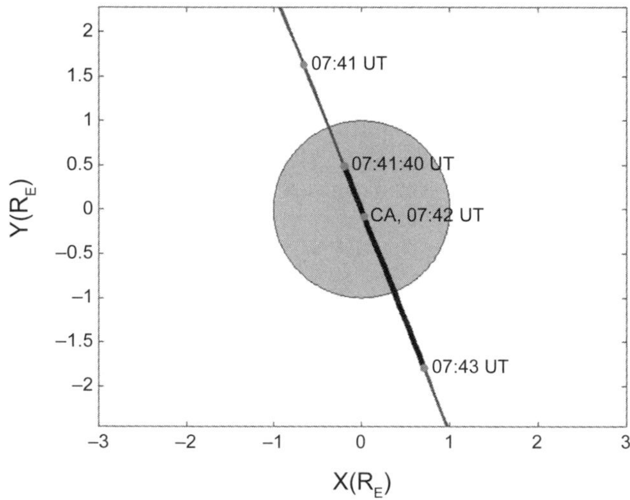

Fig. 5. Cassini trajectory in the XY plane (in Enceladus radii) for the Cassini E7 Enceladus encounter. Closest approach is indicated by CA and the plume stagnation region is indicated by the gray circle.

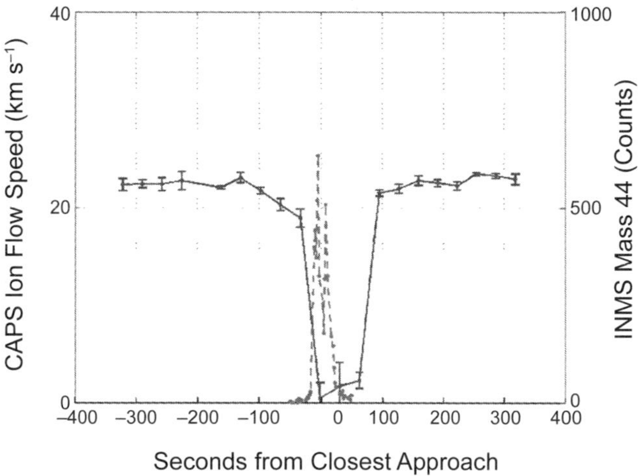

Fig. 7. Comparison of CAPS measured ion flow velocity (kilometers per second) to INMS mass 44 amu counts as a function of time (in seconds) relative to closest approach during the E7 Enceladus encounter. Flow velocity is the solid line associated with the left Y-axis. INMS counts are the dotted line associated with the right Y-axis.

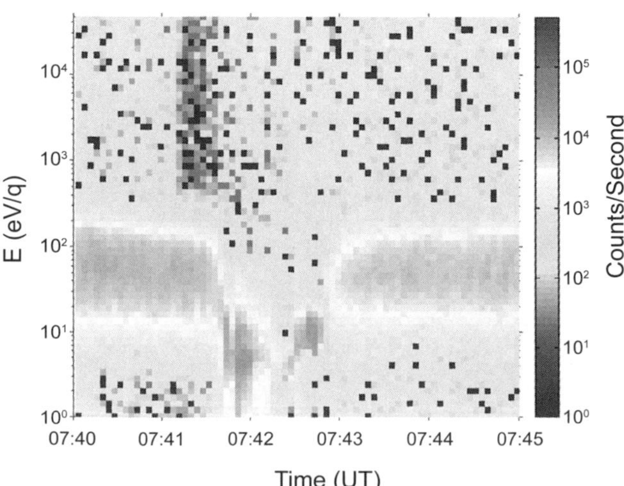

Fig. 6. CAPS IMS ion spectra (counts per second) as a function of energy (energy per charge in eV) and time (UT) for the Cassini E7 Enceladus encounter. Closest approach is at 07:42 UT.

a rapid decrease in the ion flow speed ~20 s before closest approach (CA). Similarly, INMS observes an increase from 29 to 399 particle counts in the mass 44 counts from 17.5 s to 11.3 s before closest approach (*Perry et al.,* 2015). The data suggest that magnetospheric plasma enters the dense plume, leading to charge exchange with plume water vapor and subsequent pickup into the corotating magnetospheric plasma. The signature of this process is sharp on the Saturnward side of the plume as the new ions are picked up and gyrate away from Saturn. The transition out of the plume opposite Saturn for E7 is more extended due to a number of factors, e.g., ion drift velocity and variable ion gyroradii for the various masses created. Note in Fig. 7 that the CAPS-observed ion flow speeds from the count distributions in the ram direction imply speeds as low as a few kilometers per second in the Enceladus frame.

2.4. Energetic Particles

Almost all local interaction signatures of Enceladus' interaction with high-energy particles are seen as electrons. Charge exchange with the extended neutral gas cloud of Saturn's magnetosphere depletes all ambient ions with energies above few kiloelectronvolts before they reach the magnetic L-shell of the moon (*Paranicas et al.,* 2008). As a result, this section (and section 3.3. below) reviews mostly signatures of Enceladus in energetic electrons, unless otherwise stated. Most observations come from Cassini's Magnetospheric Imaging Instrument (MIMI), which measured charged particles above several kiloelectronvolts and into the megaelectron-volt energy range using three detectors: the Low-Energy Magnetospheric Measurement System (LEMMS), the Charge-Energy-Mass Spectrometer (CHEMS), and the Ion and Neutral Camera (INCA) (*Krimigis et al.,* 2004).

The subsonic interaction of moons with the plasma in the jovian magnetosphere is known to drive field-aligned electron acceleration (*Williams and Thorne,* 2003; *Williams,* 2004; *Hess et al.,* 2011). Relevant signatures near Enceladus have been observed with the CAPS Electron Spectrometer (ELS) instrument; together with a kiloelectron-volt proton beam seen by MIMI/INCA, this is the only energetic ion feature resolved so far near Enceladus (*Pryor et al.,* 2011). The energies of the accelerated electrons seen with CAPS/ELS did not extend much above 1 keV, meaning that all electrons observed with Cassini's LEMMS instrument near Enceladus (>18 keV) cannot have their origin in the same local acceleration source but belong to the ambient environment. Furthermore, these energetic electrons have a negligible feedback to the dynamics of Enceladus' interaction with the magnetosphere: Their low impact-ionization cross-sections and fluxes (*Kollmann et al.,* 2011) mean that they cannot contribute significantly to the ionization of neutrals that surround the moon or to subsequent processes like mass-loading.

What is typically observed in LEMMS electrons are energetic electron flux dropouts. While at most saturnian moons these dropouts reflect the loss of these particles on the moon surfaces (*Roussos et al.,* 2007; *Simon et al.,* 2015), at Enceladus the interpretation remains controversial. That is because the dropouts can be highly structured. They can have a variety of spatial scales compared to the size of the moon and characteristics that change strongly as a function of energy, the LEMMS instrument pointing (pitch angle), the type of Cassini's flyby trajectory (highly inclined, equatorial etc.), and the flyby date (*Krupp et al.,* 2012). Figure 8 includes all the observed electron dropout signature types, which are also described in more detail in *Simon et al.* (2015) using a different set of LEMMS and Cassini Magnetometer (MAG) flyby observations. In brief, the Enceladus interaction signatures include:

- *Deep, moon-centered flux dropouts (Fig. 8, all flybys, dropout near closest approach)*: These typically occur when Cassini crosses field lines that map onto the surface of Enceladus or the expected location of the moon's wake. The wake is downstream for electrons below several hundred kiloelectronvolts and upstream for higher energies (*Jones et al.,* 2006). This is the only feature that can be attributed with little doubt to the absorption of electrons at the moon's surface and that appears consistently in all close flybys. These dropouts can be separated from additional flux reductions at the moon's vicinity on the basis of their large depth, the sharpness of their boundaries, and the fact that they are sometimes contained within the field-aligned current signatures that mark the boundaries of Enceladus' flux tube (*Saur et al.,* 2007; *Simon et al.,* 2014).
- *"Ramps" (Fig. 8, e.g., flyby E3 around 19:06, top panel):* These are partial flux dropouts seen at sub-megaelectronvolt electrons. When visible, they are detected at the flanks of moon-generated absorption regions that we described above, they are usually asymmetric at ingress and egress, and they appear to coincide with regions of interaction-driven, magnetic field gradients. The latter characteristic will drive a local modification of gradient and curvature electron drifts that in turn could lead to diverted electron trajectories and a partial exclusion of energetic electrons from those regions. This could imply a lossless flux dropout mechanism (e.g., transport), as has been suggested to explain small flux dropout features near Rhea (*Roussos et al.,* 2012). On the other hand, ramps are also collocated with enhanced densities of gas and dust that originate from Enceladus' plumes. It remains unclear if and how much the energy losses and scattering due to the propagation of electrons through that medium contribute to the observed flux reductions (*Tadokoro and Katoh,* 2014).
- *"Spikes" (Fig. 8, e.g., flyby E3 at 19:06, second panel):* The transition from the kiloelectronvolt to the megaelectronvolt energy range sometimes reveals flux dropouts at the location of the ramps that have the form of a narrow spike. When present, the spike matches the location of an equally narrow magnetic field perturbation. The theory is that the ramps and spikes represent zones of electron exclusion due to magnetic field perturbations: If electrons at those regions were lost to dust or neutrals, it would have been challenging to explain the observed energy dependencies in the profile and depth of these features.
- *Broad megaelectronvolt electron dropouts (Fig. 8, e.g., flyby E3, E5, second panels):* Besides the spikes, megaelectronvolt electron measurements feature flux reductions that may extend for many radii away from the moon's flux tube. The energy dependence is not standardized. In some cases, the profile is similar for all energies (see, e.g., *Simon et al.,* 2015). On the other hand, in the indicated

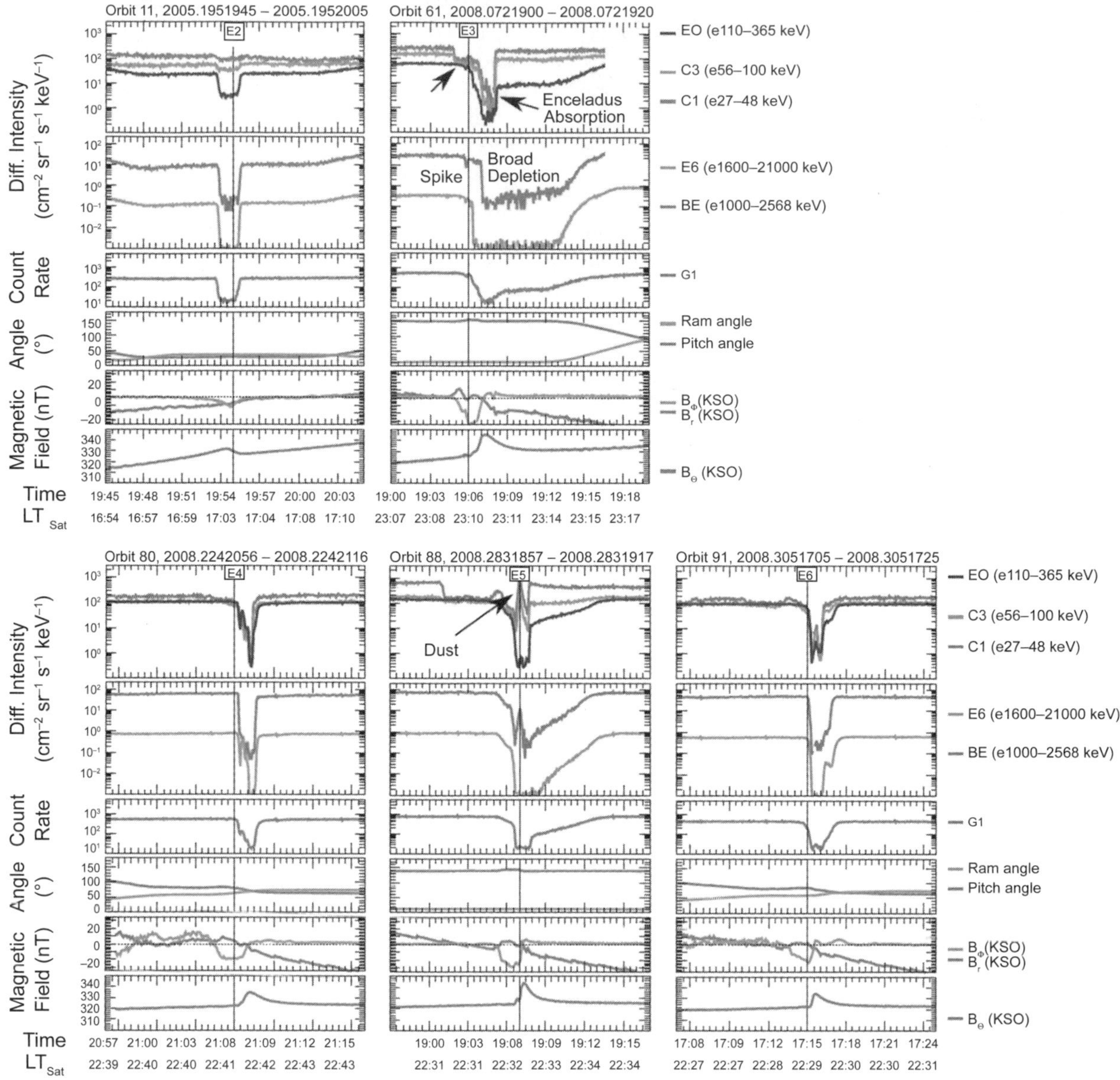

Fig. 8. LEMMS measurements during several highly inclined Enceladus flybys of Cassini (E2, E3, E4, E5, and E6). The time of closest approach to the moon is marked by a solid vertical line. The properties of the energy channels are mentioned on the legends to the right of the frames. Notice that LEMMS E-channels can have similar notations as Enceladus flybys. From *Krupp et al.* (2012).

Cassini E3 and E5 Enceladus encounters of Fig. 1, where the interaction region was crossed from the north to the south, the width of the flux depletions exceeded 10 Enceladus radii (R_{Enc}, ~252 km), typically increasing in size with electron energy. The scale size of the flux dropout is much larger than the size of the interaction region in terms of magnetic field perturbation or the extent of the dense part of the neutral and gas cloud of the moon. Although *Krupp et al.* (2012) calculated that signatures of megaelectronvolt electron interactions may extend far from the moon, it was suggested that this can occur even along the equatorial plane of the interaction, something that is not clearly seen during equatorial flybys. *Meier et al.* (2014) showed ray-like structures of charged dust extending more than 10 R_{Enc} downstream that may lead to electron losses further from the moon. However, it still remains unclear why such dropouts would appear only at the inclined orbits. An alternative is that electrons can be lost due to perturbations propagating away from Enceladus, such as waves; Whistler waves have been attributed as drivers of equal broad losses at Saturn's moon Rhea (*Santolik et al.,* 2011).

- *Distant dropouts (microsignatures) (Fig. 9)*: Losses at the Enceladus environment can be sustained along a large longitudinal section of the moon's L shell (*Jones et al.,* 2006; *Andriopoulou et al.,* 2014). That is because the primary process for refilling electron wakes in the kiloelectronvolt and megaelectronvolt range is the slow radial diffusion. Due to radial magnetospheric drifts, these distant wakes (termed microsignatures) can be seen approximately between L = 3.85 and 4.2. That defines the L-shell extent in the magnetosphere where the immediate effects of Enceladus can be realized. It is also worth noting that the majority of Enceladus' microsignatures are not highly structured, as we see at close flybys. They resemble the absorption by a single obstacle, similar to what has been seen at other saturnian moons (*Roussos et al.,* 2007). That may indicate that the additional dropout features seen at close flybys correspond to lossless disturbances in the electron distribution function that are visible only in the vicinity of the moon, where the electromagnetic fields are strongly modified.

Overall, LEMMS electron observations demonstrate that the moon itself is a sink for magnetospheric electrons, despite the presence of magnetic field and the gas/dust enhancements that are observed near Enceladus that could potentially shield its surface from electron precipitation. The role of the gas, dust, and magnetic field perturbations in modifying the electron distribution function near the moon still lacks a quantitative assessment. That requires tracing electrons in the output of a magneto-hydrodynamic (MHD) or hybrid simulation of the interaction region, a task that poses several challenges, primarily due to the small scales of electron motion (*Roussos et al.,* 2012). Despite these limitations, it is clear that energetic electron observations hold key information for the field topology of the local interaction region and the properties of gas and dust populations in Enceladus' plumes.

2.5. Plume Interaction

The interaction between the Enceladus plume and Saturn's magnetosphere is unlike that of other satellites, either of Saturn or Jupiter, due to the high abundance of dust in the plume. The dust acquires an electric charge and electron densities are significantly depleted. However, the details of this dusty interaction are still poorly understood. While the data from multiple instruments and model calculations are qualitatively consistent, there are large, quantitative discrepancies.

The first indications of this unusual plasma environment came from observations of electron density during the E3 encounter on March 12, 2008 (DOY 72), which was the first encounter to pass through the plume itself. During this encounter, the Cassini Radio and Plasma Wave System (RPWS) instrument observed an abrupt drop in the electron density (*Farrell et al.,* 2009), while the ion density remained unaffected or even increased. The electron density

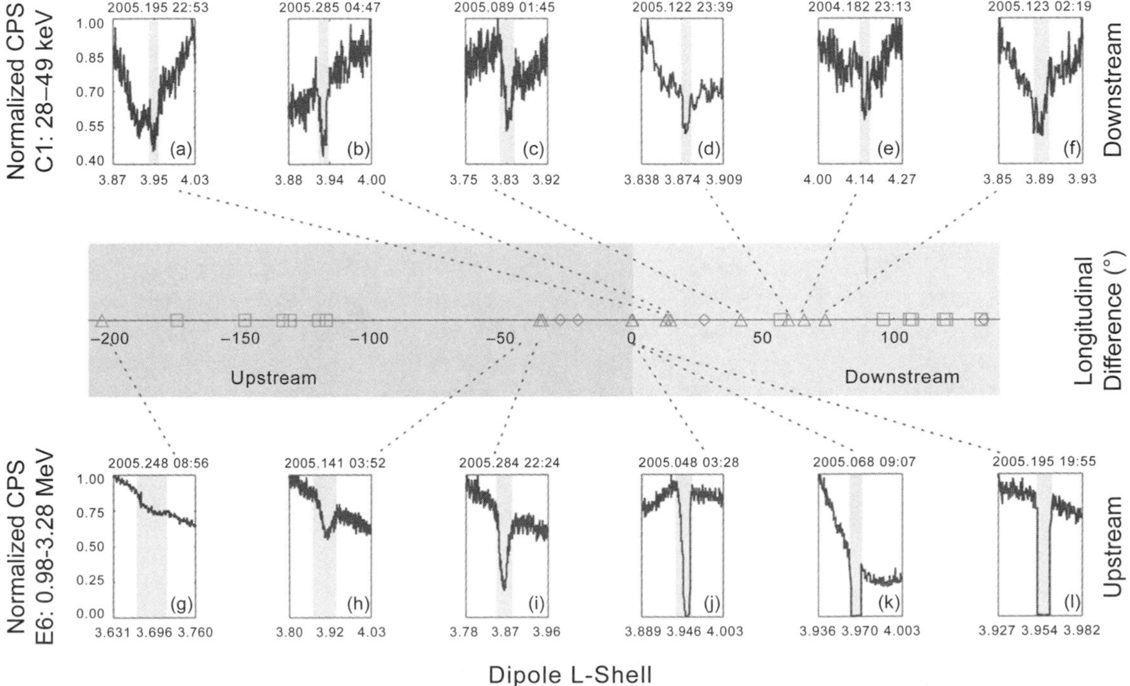

Fig. 9. Electron microsignatures of Enceladus. **(a)–(f)** Microsignatures in keV electrons, downstream of Enceladus in the corotational flow. **(g)–(l)** Upstream microsignatures [**(j)** through **(l)** correspond to three close flybys; **(g)** may not come from Enceladus (*Roussos et al.,* 2016)]. From *Jones et al.* (2006).

is measured by identifying emission at the upper hybrid frequency (*Gurnett and Bhattacharjee,* 2005).

$$f_{uh} = \frac{\sqrt{\dfrac{e^2B^2}{m_e^2} + \dfrac{n_e e^2}{m_e \varepsilon_o}}}{2\pi} = \sqrt{\left(28\text{Hz}\left(\frac{B}{1\text{nt}}\right)\right)^2 + \left(8.98\text{kHz}\left(\frac{n_e}{1\text{cm}^{-3}}\right)\right)^2} \quad (1)$$

For this equation, e is the unit charge (in coulombs), n_e is the electron density (cm^3), m_e is the mass of an electron (kg), and ε_0 is the permittivity of free space. Since the magnetic field is independently measured, the electron density can be directly calculated from the upper hybrid frequency.

Figure 10 shows the RPWS data from this encounter. Away from the plume, the electron density was between 80 and 90 cm^{-3}. As the spacecraft entered the plume, the density dropped abruptly to below 20 cm^{-3}. Within the plume itself, dust impacts produced noise (see below), which precluded the identification of the upper hybrid line. On subsequent, plume-crossing encounters, these RPWS data were similar [e.g., the E6 encounter, October 31, 2008 (DOY 305) (*Engelhardt et al.,* 2015)].

Electron density depletion within the plumes has also been observed by the RPWS Langmuir probe. This sensor operates in two modes. The most informative measurement is a current-voltage sweep. The probe is biased relative to the spacecraft for 256 voltages covering ±32 V. At each voltage, the instrument measures net current onto the probe (from ions, electrons, photoelectrons, etc.). This resulting curve may be fit to determine electron density and temperature, spacecraft potential, and ion density and speed. Since the ion current is much lower than the electron current, the uncertainty in ion measurements is significantly larger. Unfortunately, sweeps are performed only once every 24 s (typically providing only a few sweeps during a plume crossing).

Between current-voltage sweeps, the probe operates in its second mode. In this mode, the spacecraft-probe voltage is fixed at 11.5 V and the current sampled at 20 Hz. This provides one of the highest time-resolution datasets, which is critical since the spacecraft may cross small structures in the plumes in under 1 s. This 20-Hz mode does not provide unique information on plasma conditions, but is, to good approximation, proportional to $n_e\sqrt{T_e}[1 + (U_{bias} + U_{SC})/T_e]$. This has been used as a proxy for electron density, by assuming electron temperature and spacecraft potential are either constant or may be estimated from sweep data. [See *Engelhardt et al.* (2015), *Morooka et al.* (2011), *Wahlund et al.* (2009), *Yaroshenko et al.* (2009), and *Shafiq et al.* (2011) for specific details of the Cassini RPWS/LP analysis.]

Figure 11 shows an example of the Langmuir probe results. Overall, the data typically show a decrease in electron density on the edges of the plume, by a factor of a few to five. Inside the plume, the electron densities are elevated relative to the background plasma, reaching densities of a few hundred to as much as 10^4 cm^{-3} [in the case of the E5 encounter,

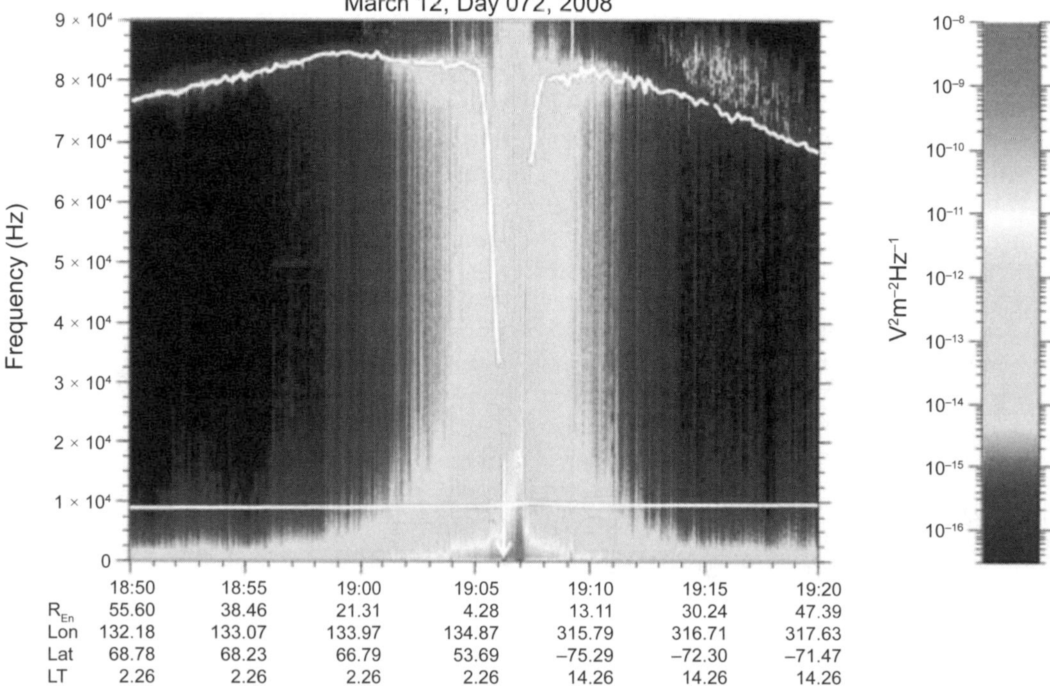

Fig. 10. RPWS electric wave spectra from the E3 Enceladus encounter (March 12, 2008.) The upper hybrid line, from which electron densities are calculated, is shown in white. The broadband noise seen around closest approach is due to dust impacts on the spacecraft. From *Farrell et al.* (2009).

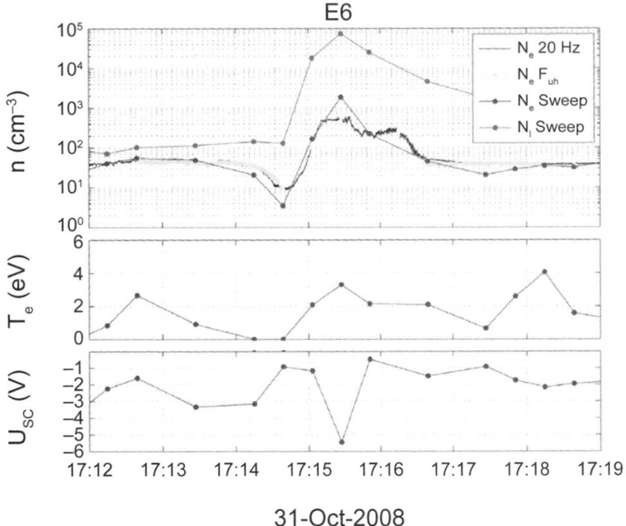

Fig. 11. Electron and ion densities, electron temperature, and spacecraft potential derived from Langmuir probe data from the E6 Enceladus encounter (October 31, 2008.) Electron density derived from the upper hybrid line is also shown in the top panel. From *Engelhardt et al.* (2015).

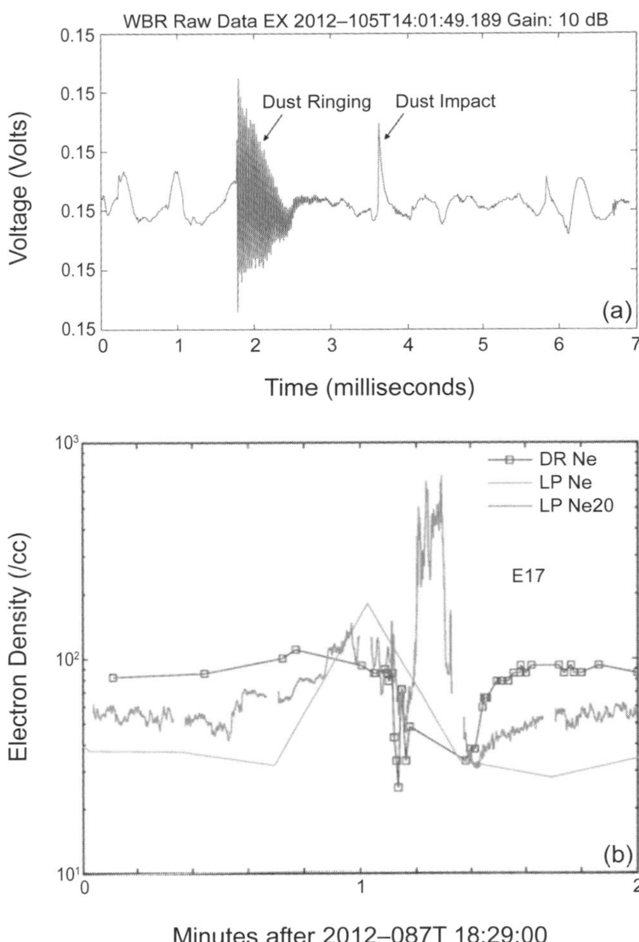

Fig. 12. (a) An example of a "ringing" dust impact. **(b)** Derived electron density (DR) shown for the E17 encounter compared with the Langmuir probe results (20 Hz shown by middle jagged line, sweeps by bottom solid line).

October 9, 2008, (DOY 283)]. In many cases, electron density derived from 20-Hz data differ from near-simultaneous sweep data. This may indicate that the assumptions about electron temperature or spacecraft potential are not reliable.

The third and final method used to determine electron densities in the plume is a result of dust impacts on the RPWS antennas themselves. *Ye et al.* (2014) showed that such impacts produce a ringing or damped oscillation in the wideband (voltage vs. time) RPWS measurements. This is interpreted as an oscillation at the upper hybrid frequency, and used to calculate the electron density. However, it is not clear if this is the unperturbed electron density, or if the spacecraft's potential and/or photoelectrons have altered the local density near the RPWS antennas. Figure 12 shows an example of a ringing impact in the raw wideband data and the derived electron densities for the Cassini E17 Enceladus encounter.

At some times, all three electron density determinations agree reasonably well. At other times, however, the agreement is better on some encounters and worse on others. Ideally, a fourth method of determining electron densities would resolve this. In a different plasma environment, this might be possible, since data from the Cassini CAPS/ELS could be used. In the vicinity of Enceladus, however, the spacecraft is charged to a negative potential of a few volts. As a result, electrons with energies below a few electron-volts are repelled and cannot be measured by ELS. While providing good data on the nonthermal tail of the electron distribution, ELS data cannot resolve the discrepancies between the other three determinations of electron density.

While showing a complex plasma interaction, the electron densities in the plume are even more interesting when compared to ion density measurements. The primary instrument for ion density determinations, the CAPS Ion Mass Spectrometer (IMS), produced ambiguous results during Enceladus encounters due to the short duration of plume crossings (<150 s). This instrument measures ion flux as a function of energy (per charge) and direction. Energies are sampled over a 4-s scan from ~1 eV to ~32 keV. In one plane, the particle direction is measured simultaneously in eight 8° × 20° pixels. The remaining angular dimension is sampled by mechanically rotating the instrument, thereby providing coverage of velocity space. Since the time required for this rotation is over 52 s (for the minimum, 28° scan), the instrument was held at a constant pointing, which contained the ram direction on five plume-crossing encounters. This provided two-dimensional cuts through velocity space. Without knowing that the plasma flow is in that two-dimensional plane, the CAPS/IMS spectra cannot be uniquely used to compute ion density. They may, however, be used as a consistency check, either for other measurements of ion density or for model calculations. An additional concern is the 4-s time resolution, which may result in time aliasing if plasma conditions are changing on 4-s or shorter timescales (30 km or larger scales along track).

The available ion density determinations were made using Langmuir probe data. These data are produced by fitting the negative voltage portion of a current-voltage sweep. The results are relatively uncertain since the total current in this portion of the sweep are relatively low. Fits to the Langmuir probe sweeps (*Shafiq et al.,* 2011; *Morooka et al.,* 2011; *Engelhardt et al.,* 2015) show a dramatic increase in ion density within the plume. Figure 11 shows the ion density profile for the Cassini E6 Enceladus encounter, peaking at 8×10^4 cm^{-3}. On the six plume-crossing encounters reported by these authors, the peak ion densities were between 2×10^4 and 10^5 cm^{-3}. The simultaneous measurements of electron densities were one to two orders of magnitude lower, and *Morooka et al.* (2011) report an electron-to-ion density ratio dropping below 0.01 during the Cassini E3 Enceladus encounter.

An explanation for the unusual inequality between ion and electron densities was suggested by *Farrell et al.* (2009), based solely on the drop in electron density (i.e., prior to the reports of enhanced ion densities). The plume contains a dense dust population. Such dust would acquire a negative electric charge. With sufficient dust density and charge, the free electrons would be largely absent and the dust would be the primary negative charge carrier. This interpretation has been confirmed by measurements of charged dust and by theoretical calculations of dust charging. However, the extreme values of free electron depletion (e.g., 1% of the ion density or lower) are difficulty to reconcile with theoretical calculations (*Yaroshenko et al.,* 2009, 2015; *Meyer-Vernet,* 2013) and models are unable to account for the high ion densities reported by the Langmuir probe (*Kreigel et al.,* 2014). Although unable to uniquely determine ion density, the CAPS/IMS sensor did observe, on all ram-pointed plume crossings, maximum phase space densities around 10^{-15} s^3 cm^{-6}. This is consistent with 250 H_2O^+ ions per cm^{-3} at a temperature of 1 eV and with a flow velocity in the instrument's field of view. While instrument pointing and ion temperature could a produce a lower observed phase space density in CAPS, a factor of at least 2 or 3 reduction in the Langmuir probe ion density would be required for consistency with CAPS. This discrepancy is unresolved.

Unlike the observations of ion and electron densities in the plume, observations of dust are less ambiguous and more consistent. Dust particles of ~2 μm and larger were observed by both the Cosmic Dust Analyzer (CDA) (*Spahn,* 2006), which was designed to do so, and by the RPWS instrument, which was not designed for such observations (*Kurth et al.,* 2006; *Ye et al.,* 2016). The RPWS detections are a result of dust impacts on the spacecraft producing voltage transients and there are uncertainties in its absolute calibration, while the CDA measurements are direct measurements of particle impacts on the instrument's high rate detector (HRD). The size of the dust particles may be determined either from a histogram of peak voltage of spikes in the RPWS wideband data or from the relative rates in five HRD detectors of different sensitivities. For this mass, size distributions are assumed to be a power law, $\frac{dn}{dr} \propto r^{\mu}$, and described by the exponent μ. For particles of this size, μ is always negative (fewer large particles than small ones). Figure 13 shows examples of dust density and size distribution from the E7 encounter. Dust densities of 5–10 m^3 have were observed in the center of the plums, compared to the ~0.1 m^{-3} of the E ring near the orbit of Enceladus. However, the density varies by a factor of a few within the plume, probably due to small-scale jets embedded within the plume. The size distribution, while variable, was typically around μ ~ –4. CDA can also measure the charge of larger dust particles (*Kempf et al.,* 2006) and their composition, using the main dust analyzer sensor. This sensor reaches its maximum sampling rate during Enceladus encounters and therefore the HRD is used for dust density profiles. The observed charge implies dust particles with a charge between –1 and –2 V. Measurements of composition are discussed in the chapter in this volume by Kempf et al.

In addition to the measurements by CDA and RPWS, micrometer-sized dust was also observed by the INMS and MIMI instruments (*Teolis et al.,* 2010; D. G. Mitchell, personal communication, 2016). The response of these instruments to dust impacts is uncertain, whereas the signatures seen by RPWS have also been observed on many spacecraft, going back to the Voyager Saturn encounter (*Gurnett et al.,* 1983), and their interpretation is relatively well-understood (*Collette et al.,* 2015, and references therein).

The CAPS instrument provided a surprising measurement of nanometer-sized particles in the plume (*Jones et al.,* 2009; *Hill et al.,* 2012; *Dong et al.,* 2015). Although designed to measure ions and electrons, the sensors are sensitive to any particles with the appropriate energy-to-charge ratio (1 V

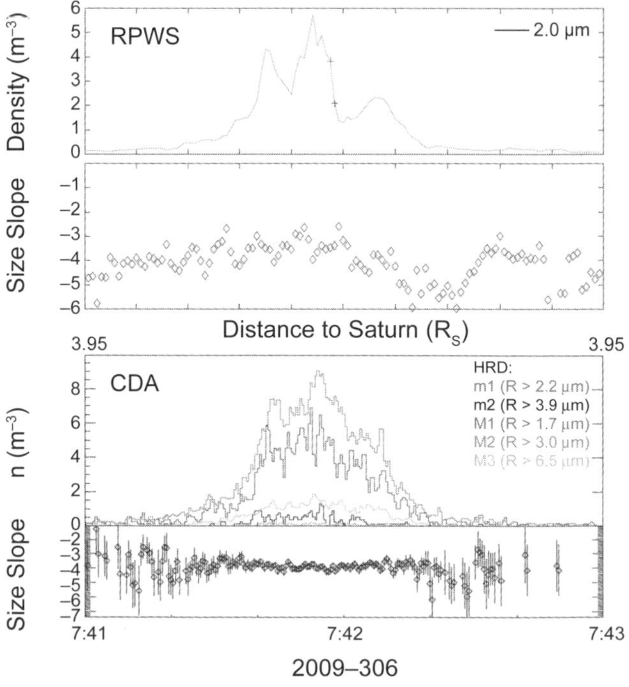

Fig. 13. Dust density and mass distribution during the E7 encounter. Derived from RPWS spectra (top) and CDA High Rate Detector (bottom). From *Ye et al.* (2014).

to 33 kV for ions and –1 V to –26 kV for electrons.) The nanograin dust is observed at energies above ~1 keV and confined to the ram direction. The identification of these particles as nanograin dust is based on this directionality: 1-keV electrons and water-group ions have a velocity of 18,750 and 100 km s^{-1}, respectively, and their apparent direction would not be significantly influenced by the spacecraft's 7.5 to 17.7 km s^{-1} speed.

Assuming these particles are a narrow beam from the ram direction, the density of these nanograins can be calculated from their observed flux, the spacecraft's velocity, the effective area of the sensors, and detector efficiency. The first three of these quantities are well known. The detector efficiency is much less certain, since neither sensor was calibrated with dust particles. The detection efficiency of a 10-keV electron striking a microchannel plate is very different from that of a 10-keV dust particle. *Hill et al.* (2012) assumed the efficiency for the electron sensor was 0.05, based on laboratory work with large molecules and similar detectors. Hill et al. also noted that this was uncertain by a factor of 2 to 3. For the ion sensor, an efficiency of 1 was assumed. The ion sensor, as part of a time-of-flight measurement, sends particles through an ultrathin carbon foil before they reach the detector. Traversing the foil was expected to fragment the dust particle, and the resulting shower of impacts on the detector was believed to have a very high probability of detection. The negatively charged nanograin dust densities peak at 1–2 × 10^3 cm^{-3} on all encounters except E7, while the positive nanograin densities are approximately 3 orders of magnitude lower with peak densities of ~0.5 cm^{-3}.

In all encounters, there were multiple peaks and/or fine structure within the plume. These may be associated with the known, small-scale jets within the plume (*Jones et al.,* 2009), and sometimes match peaks seen in other datasets such as micrometer-sized dust density or INMS measurements of neutral gas (*Dong et al.,* 2015). It is also interesting to note that the peaks in positive and negative nanograin density do not occur at the same time during the E3 and E5 encounters (*Jones et al.,* 2009). This may be due to electromagnetic forces accelerating the positive and negative particles in opposite directions. If so, these offsets would be expected to be smaller during the E17 and E18 encounters, since they were much closer to Enceladus and the source. Such a comparison has not yet been performed.

The nanograin flux was much lower on the E7 encounter. *Hill et al.* (2012) interpreted this in terms of the dust charging process (below), but the subsequent encounters suggest this is an observational artifact. Plume material has a southward velocity of ~0.5–1.0 km s^{-1}. On the E3 and E5 encounters, the spacecraft had a high speed relative to Enceladus (14.4 and 17.7 km s^{-1}) with a strong southward component. On the other three encounters — E7, E17, and E18 — the speed was lower (7.7, 7.5, and 7.5 km s^{-1}) and had no appreciable southward component. The southward motion of the plume material would have caused a shift in the apparent ram direction. This would have been a negligible ~1° shift on E3 and E5, but a 4°–7° shift on the later encounters (compared with the 8° field of view of the instrument). During E3, E5, and E7, the instrument's actuator was fixed to observe the ram direction. During E17 and E18, it was oriented to point 4° north of the ram direction, compensating for the southward motion of the plume material. As a result, E7 was the only encounter where this aberration was significant and not compensated for. It was also the only encounter with nanograin dust fluxes well below 1000 cm^{-3}.

The size distribution of the nanograin dust may be calculated from the CAPS energy spectra. Since the particles are moving at the ram velocity, this observed energy per charge may be converted to a mass per charge. For E3, E5, E7, E17, and E18, this conversion factor is 1.1, 1.6, 0.31, 0.29, and 0.29 eV/amu. The charge state of the nanograins is not measured, but several different estimates suggest that all, or almost all, of these particles are singly charged (*Hill et al.,* 2012; *Yaroshenko et al.,* 2014; *Dong et al.,* 2015). Over the various encounters and over the ~1–32-keV range observed by CAPS, this covers particles with a mass of roughly 600 to 90,000 amu. If the particles had a density of 1000 kg m^{-3} and were spherical, this would correspond to a radius of 0.6 to 3.3 nm. In practice, the particles are almost certainly larger, since it is unlikely that they are either spherical or zero porosity. It is important to note that, except for these size calculations of nanograins and the similar calculations for micrometer-sized particles by CDA and RPWS, the radius is a derived quantity. The direct measurements are of kinetic energy, which is converted to mass based on a measured speed or a reasonable estimate of it. The inferred radius is much less certain, since neither the shape nor the density is well-constrained.

Given the observations of observations of nanograin and micrometer-sized dust particles, it is possible to estimate a complete size distribution (dn/da). *Dong et al.* (2015) has done so using a four-parameter functional form of

$$\frac{dn}{da} = Ca^{\alpha}\left[1+\frac{\alpha}{\kappa}\left(\frac{a}{a_0}\right)^{\beta}\right]^{-\frac{\alpha+\kappa}{\beta}} \tag{2}$$

selected so that a_0 is the peak grain radius (in nm), n is the density (in m^3), and the distribution is a power law proportional to a^{α} for small radii and a power law proportional to $a^{-\kappa}$ at large radii. For the four encounters they fit, shown in Fig. 14, they obtained $\alpha \sim 8$ and $\beta \sim 5$ (two dimensionless parameters that control the shape of the curve), $\kappa \sim 4$ [the power-law exponent from *Ye et al.* (2014)], and $a_0 \sim 2$ nm.

3. INFLUENCE ON GLOBAL MAGNETOSPHERIC ENVIRONMENT

While the most immediate impacts of Enceladus are observed in the local magnetospheric environment, this moon has a much larger impact. In particular, the relatively large source rate combined with the particular chemical composition of this dominate source serve to shape composition, interactions, and dynamics throughout the entire magnetosphere.

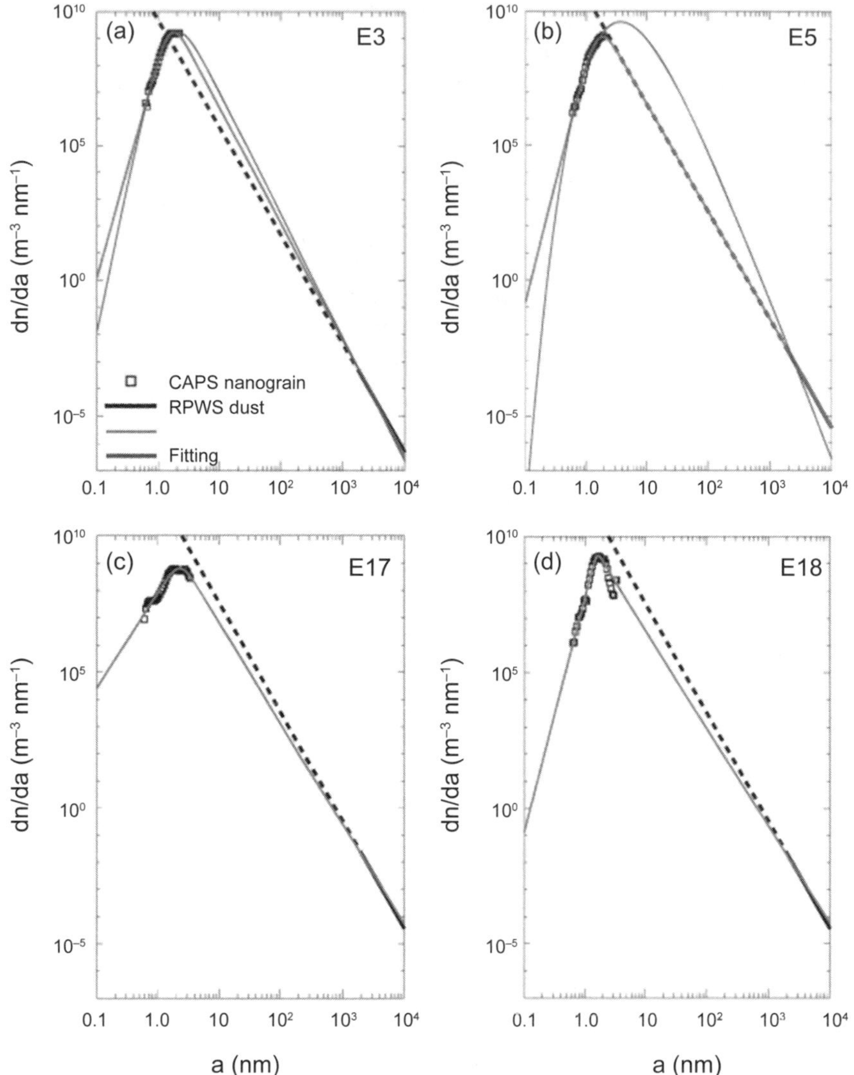

Fig. 14. Dust-sized distributions calculated by combining measurements of both nanometer- and micrometer-sized dust by CAPS and RPWS. From *Dong et al.* (2015).

3.1. Global Environment Key Interaction Processes

Similar to the previous section, particle interaction rates also provide essential insight into how Enceladus impacts the global magnetosphere. In particular, interactions with electrons, photons, and other ions create and redistribute neutral particles. Figure 15 shows the relative water-group interaction rates as a function of distance from Saturn for the primary magnetospheric species originating from Enceladus-generated water. These lifetimes are derived using the same method as applied in the previous section. It is important to consider these rates as a function of distance from Saturn because their relative impacts change in different regions of the magnetosphere. By examining each process in more detail, their relative roles and feedback on the global magnetosphere becomes evident.

Near Enceladus, water molecules (and OH) tend to dissociate into daughter neutral species from photon interactions more often (and faster) than the other interaction process. The net result is a slightly more extended water distribution, with production of new OH and O particle distributions (or tori). One might assume that photon interactions would therefore also dominate throughout the magnetosphere, because such interaction rates are constant, as they are a function of distance from the Sun and solar activity. However, Fig. 15 shows that in some regions, electron impact interactions can actually dominate.

The relative importance of this interaction process changes as a function of distance from Saturn because the plasma population is also changing, as shown in Fig. 16. This figure shows the average equatorial water group temperatures and densities based on *Schippers et al.* (2008), *McAndrews et al.* (2009), and *Thomsen et al.* (2010). Near Enceladus, the average temperature of the core electron population is below electron-impact ionization thresholds, which is why such interaction rates are so low in the inner magnetosphere. Almost

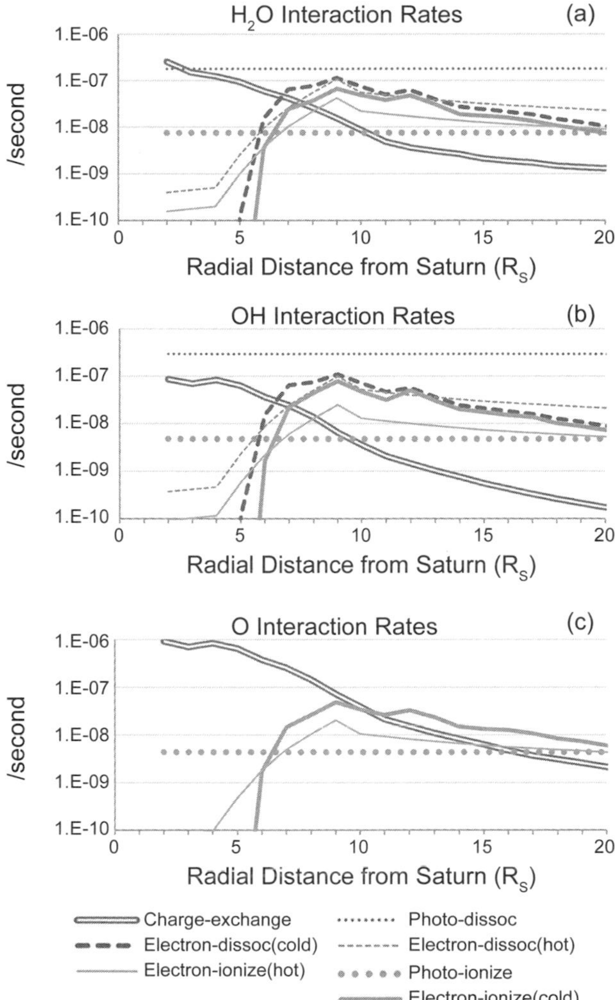

Fig. 15. Relative average water group equatorial interaction rates (per second) as a function of distance from Saturn (R_S) for **(a)** water, **(b)** OH, and **(c)** O. Rates are for charge exchange (double line), photo-dissociation (thin dotted line), photo-ionization (thick dotted line), core electron dissociation (thick dashed line), hot electron dissociation (thin dashed line), core electron ionization (thick line), and hot electron ionization (thin line).

all of the interactions are caused by the hotter, much less dense population of electrons. Figure 15 separates electron interaction rates into these two populations and shows how virtually all such interactions are caused by hotter electrons near Enceladus. However, the average electron temperature increases with distance from Saturn, so the electron interaction rate also increases with such distance, because a greater percentage of the electron population can interact. Notice that the ionization rate peaks in the middle magnetosphere at approximately 6–9 Saturn radii (R_S) from Saturn. This occurs because the energy-dependent cross sections and electron densities are most optimized in this region, causing the electron impact interactions to become dominant. Thus, more ions are created in this region (through ionization), causing ~6–9 R_S to be the primary plasma mass loading region. This conclusion is supported by the fact that plasma flux tubes are loaded with more plasma in this region than any other in Saturn's magnetosphere (*Sittler et al.,* 2008; *Chen et al.,* 2010). This result is interesting because one might assume that most plasma is produced near the primary particle source of Enceladus. However, Saturn's unique environment impacts interaction rates such that the primary plasma source region is actually in a different location than the neutral particle source.

The relatively low electron interaction rates near Saturn allow for other processes to dominate near Enceladus. In addition to photon processes, Fig. 16 shows that ion interactions (primarily charge exchange) are significant in this region. As mentioned in the previous section, charge exchange interaction cross sections are more favorable for lower relative velocities. Near Enceladus, the neutral orbital speed is ~12.6 km s^{-1} and the plasma speed is <38 km s^{-1}. However, at 9 R_S, the orbital speed is only ~8.4 km s^{-1}, while the rigid plasma co-rotation speed is <88 km s^{-1}. This larger speed difference serves to reduce charge exchange interaction rates and make them less significant with greater distance from Enceladus' orbit. Thus, charge exchange interactions tend to have the strongest impact in the inner magnetosphere, but can still play a noticeable role in Enceladus' global magnetospheric impact because a significant portion of the water (and OH and O) near Enceladus experiences charge exchange interactions. In general, these neutral particles

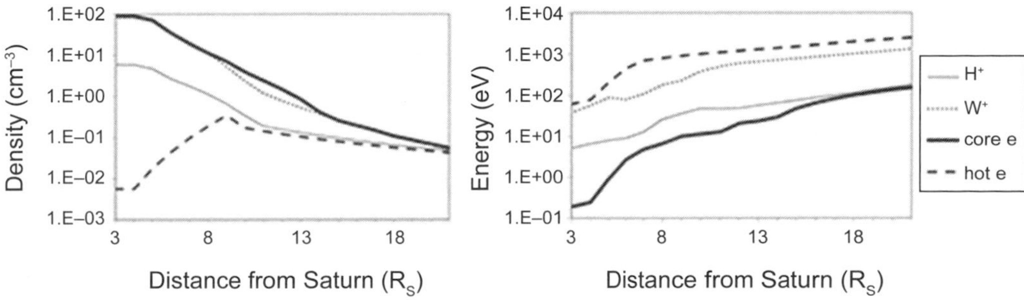

Fig. 16. Average equatorial plasma/electron temperatures (eV) and densities (cm^{-3}) based on *Schippers et al.* (2008), *McAndrews et al.* (2009) and *Thomsen et al.* (2010) as a function of radial distance from Saturn (R_S) for light ions ("H+"), water group ions ("W+") the core electron population ("e"), and the hotter electron population ("hot e").

are converted into fresh pickup ions (usually a water group ion) and the ions are converted into neutral particles. These newly created neutral particles retain the plasma co-rotation velocity, which is much faster than the local orbital speed. Additionally, these particles are no longer controlled by Saturn's magnetic field, resulting in an energetic neutral particle that tends to escape the Saturn system on a ballistic trajectory. The newly created ions, however, are influenced by Saturn's magnetic field, but are traveling much slower than the local plasma. Saturn's magnetic field must therefore exert work to increase particle velocities (or "pickup") to match the local plasma velocity. This process is known as "momentum loading" because in general, the net number of ions does not increase but the momentum is impacted.

The prevalence of charge exchange in Saturn's magnetosphere allows for many neutral particles originating from Enceladus to actually escape Saturn's magnetosphere before creating any new ions. Thus, the total Enceladus neutral particle source rate does not directly equate into a plasma source rate. *Cassidy et al.* (2010) and *Fleshman et al.* (2010) determine that 31–66% of Enceladus plume particles actually escape Saturn's magnetosphere, with only 11–26% experiencing ionization (the remaining particles collide with Saturn, the main rings, or satellites).

3.2. Torus Formation Processes

As the neutral H_2O, OH, O, and H begin to spread out, these particles are bound by Saturn's gravity, causing them to co-orbit the planet along with Enceladus and form toriodal clouds (tori) in Saturn's magnetosphere. Additionally, the energetic neutral particles that are on escaping ballistic trajectories produce a nearly constant outflow of these particles. Interestingly, even though Enceladus is relatively small compared to Titan and orbits in the inner magnetosphere, Enceladus-generated tori are impacting Saturn's magnetosphere so much that their influence dominates over Titan even in the outer magnetosphere. Figure 17 shows Cassini CAPS ion detections in the region near Titan with the dominant species being oxygen (from Enceladus) until within 7 Titan radii of the moon. This dominance of water-group ions as far out as the outer magnetosphere provides insight into the large size of Enceladus-generated tori. The OH torus has been directly observed by the Hubble Space Telescope (*Shemansky et al.,* 1993), the O torus using Cassini UVIS (*Shemansky et al.,* 2009), and the H torus by Voyager (*Shemansky and Hall,* 1992) and Cassini UVIS (*Melin et al.,* 2009), all of which reveal tori that greatly extend beyond the inner magnetosphere. Subsequent modeling work constrained by these observations help to support the pervasiveness of Enceladus-generated water-group tori throughout Saturn's magnetosphere (*Jurac and Richardson,* 2005; *Johnson et al.,* 2006).

These tori are a result of the global particle interactions previously listed, with each process contributing in a different way to the formation of these remarkable features.

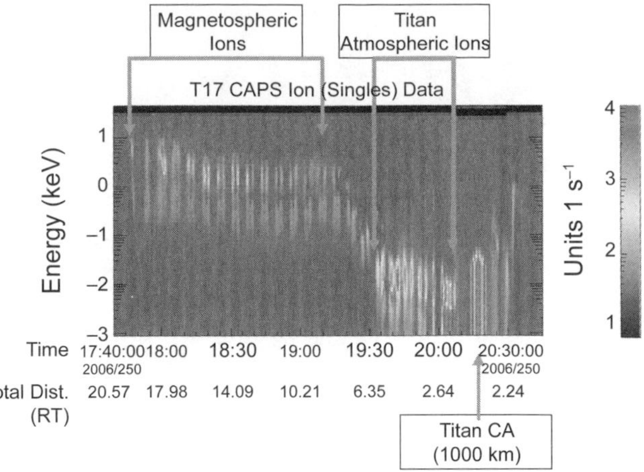

Fig. 17. Cassini CAPS ion detection spectra as a function of energy (keV) and distance from Titan (Titan radii or ~2575 km) for DOY 250 in 2006. Higher-energy magnetospheric ions and ions originating from Titan's atmosphere as well as the Cassini closest approach to Titan are annotated.

The generation of the initial tori starts with the Enceladus water source itself. The relatively slow escape velocities of these particles keep the spatial distribution relatively narrow in the radial direction. However, as shown in Fig. 15 above, the lifetimes of most of these particles are on the order of months or longer, which allows for them to spread out in the orbital direction and form a more complete ring (torus) around Saturn until they interact with other particles (*Johnson et al.,* 2006).

This inner, narrow torus of water particles then becomes the seed mechanism for the larger, more extended tori. As Fig. 15 shows, many of these water particles are dissociated by photons into OH and H. The energy transferred to the OH molecule causes these particles to form a more extended torus than the source water, while the energy imparted on the lighter H atom causes these particles to dramatically spread out to start forming a very large hydrogen torus. On average, this process takes less than three months. Subsequently, this OH torus is then subject to additional dissociation in O and H. The resulting H adds to the existing extended hydrogen torus, while the oxygen forms an additional torus that is larger than the water and OH tori. It is important to note that electron-impact interactions also dissociate molecules. However, these processes tend to take longer, except in the 6–9 R_S region, as discussed in the previous section. Thus, the dissociation interactions serve to generate all the tori as well as slightly enlarge the water-group tori while dramatically generating and enlarging the hydrogen tori.

After these tori are generated, they can spread out more through interactions with other neutral particles. In general, neutral particle distributions are not dense enough to facilitate direct collisions between such neutral particles because their mean free paths are too long. However, the dipole particles, H_2O and OH, allow for an interesting phenomenon. In the densest region of the neutral tori, these particles can react to each other through enhanced dipole interactions at greater distances (*Farmer,* 2009; *Cassidy and Johnson,* 2010). The trajectories of these particles are slightly altered because of these interactions. While the impact of a single interaction is relatively small and only occurs in the densest region, H_2O and OH can orbit for several months before they experience their first interaction with photons and/or charged particles, which is much longer than the ~33-hour orbital period. Thus, many neutral-neutral interactions can occur in the dense torus region, causing a cumulative effect that enlarges the water and OH tori and then indirectly expands the O torus.

Interactions with ions serve to extend the apparent tori even more. As mentioned above, interaction with ions can be in the form of a simple collision or one where charge exchange occurs (effectively ionizing a neutral and neutralizing an ion). A simple collision will spread out the tori somewhat as their trajectories are altered. However, in the case of charge exchange, the apparent tori are dramatically extended by the neutralized ion, which can contribute to the H_2O, OH, O, and/or H tori. These newly formed neutral particles are traveling at the co-rotating plasma velocity but no longer bound by Saturn's magnetic field. Near Enceladus' orbit, the plasma velocity is on the order of ~26 km s^{-1} larger than the orbital velocity; however, the velocity difference increases with increasing distance from Saturn. Thus, depending on where the charge exchange interaction occurs, the resulting neutral particle can experience significant trajectory impacts. Some of these particles will remain bound to Saturn on very large orbits while many others will simply exit the system on ballistic trajectories because their velocities are too large to remain orbitally bound to Saturn. Interestingly, these escaping particles can also appear to increase the size of Saturn tori. Although escaping, there is a nearly constant flux of these escaping particles (most likely oxygen), which appears to increase the density of outer magnetospheric distributions. This occurs because as one of these particles departs a region of space, it is replaced by another.

Ionization was not mentioned, as it should not generally generate or expand the neutral tori. These interaction processes instead serve to limit and/or reduce the tori through the loss of neutral particles by ionization. In theory, this effect should be greatest in the 6–9 R_S region where ionization is enhanced; however, this impact has not been directly observed. Figure 18 shows theoretical three-dimensional modeling results that help to summarize the impact of each interaction process on the oxygen torus. The results were derived from the *Smith et al.* (2004, 2007, 2008, 2010) models. Figure 18a shows a scenario where the only interaction process is dissociation, illustrating a slightly expanded torus. Figure 18b expands on the scenario in Fig. 18a by adding neutral collisions with other neutrals, which further extends the torus. Finally, Fig. 18c includes charge exchange, which serves to dramatically expand the oxygen torus. Therefore, all the above processes serve to produce the extensive and complex system of tori in Saturn's magnetosphere.

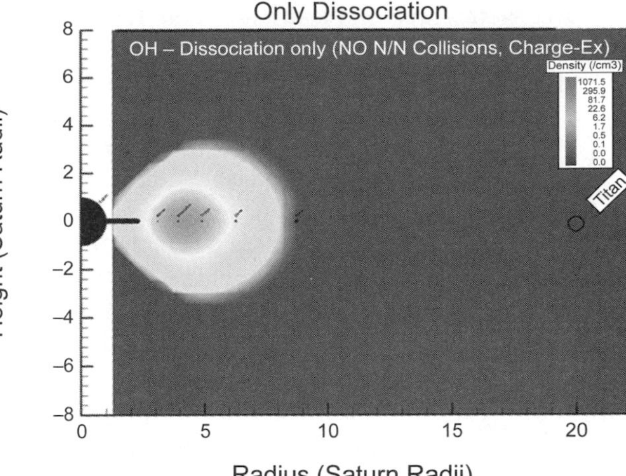

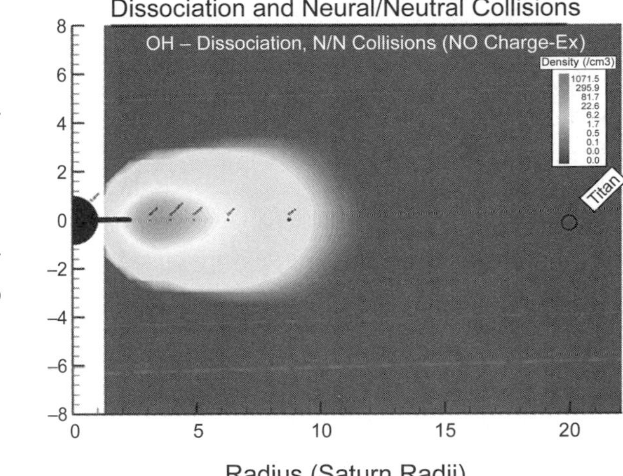

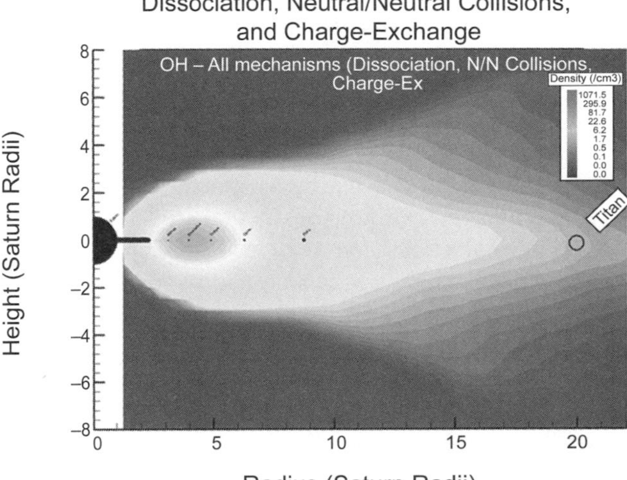

Fig. 18. Three-dimensional theoretical neutral oxygen model density (cm^{-3}) results derived from *Smith et al.* (2004, 2005, 2007, 2010) as a function of Z (y-axis on each panel) and radial distance (x-axis on each panel) in Saturn radii. Model cases show with **(a)** only dissociation processes allowed, **(b)** only dissociation and neutral/neutral interactions allowed, and **(c)** charge-exchange also allowed.

3.3. Energetic Particle Production and Transport

Enceladus is a dominant source of magnetospheric plasma for Saturn's magnetosphere and it is naturally expected to be a key driver in shaping, directly or indirectly, the global energetic charged particle environment of Saturn. Water-group ions, at least up to about 100 keV/q, for instance, constitute a large fraction of the ion composition (*DiFabio et al.,* 2011).

The effects of mass loading by Enceladus on the global circulation of plasma and energetic particles are briefly discussed by *Kivelson* (2006) and *Simon et al.* (2015). Large-scale injections in the middle and outer magnetosphere are attributed to a global dipolarization process that results from the release of flux tubes loaded with heavy, water group ions from Enceladus toward the magnetotail. Such injections, which are visible with the MIMI/INCA ENA camera, are likely a dominant source of high-energy electrons and ions for Saturn's ring current. Short-duration flux enhancements detected between the inner and middle magnetosphere, called interchange-injections, are also an indirect signature of Enceladus' presence (*Mauk et al.,* 2005; *Chen and Hill,* 2008). These injections, which sustain Saturn's radiation belts (*Paranicas et al.,* 2008, 2010; *Thomsen et al.,* 2016), result from a centrifugal-interchange instability that develops in that region, also driven primarily by mass-loading of the magnetosphere with water-group ions.

Interchange injections are a fast inward transport process and are the most effective way of supplying the inner magnetosphere with energetic particles from the ring current. These particles will otherwise undergo heavy losses if they cross the extended neutral gas cloud generated by Enceladus through a slow diffusive transport. Losses are more severe for ions that have short lifetimes against charge-exchange, a process not affecting electrons (*Dialynas et al.,* 2009). Energy-time spectrograms with LEMMS and CHEMS reveal that inside of about L = 7 energetic ions below 100–200 keV are at instrumental background levels, excluding few isolated interchange injections that rarely penetrate inside L = 6 (Fig. 19). Energetic electron observations also feature many interchange injections below a few hundred kiloelectronvolts, superimposed on a background population that can still reach the orbit of Enceladus, despite the attenuation by neutrals and dust (*Kollmann et al.,* 2011). Electrons and ions above several hundred kiloelectronvolts are less affected by Enceladus' gas and dust cloud.

The heavy losses of the kiloelectronvolt energetic particle population inside L = 7 has several implications. *Mauk and Fox* (2010) and *Mauk* (2015) demonstrated that Saturn's radiation belts are among the weakest in the solar system. That also results in relatively weak sputtering rates of the inner, large saturnian moons, including Mimas, Tethys, and Dione, that cannot sustain dense exospheres (see the chapter by Teolis et al. in this volume) and/or compete with Enceladus as magnetospheric ion sources (*Saur and Strobel,* 2005; *Roussos et al.,* 2008).

At higher energies, Enceladus has a profound effect on megaelectronvolt ions. *Jones et al.* (2006) and many follow-up studies demonstrated that Enceladus is emptying its L shell of these ions, forming a "macrosignature." A macrosignature is absorption of charged particles sustained at all magnetospheric longitudes. Radial transport of megaelectronvolt ions across such macrosignatures is prevented by Enceladus' absorption effects. This is, however, a feature not associated with Enceladus' plumes. The plume material is transparent to megaelectronvolt ions. Losses of these ions occur at the surface of Enceladus. Macrosignatures are also not unique to Enceladus. They are seen also at the L shells of the Janus/Epimetheus pair, Mimas and Tethys. They can form due to a combination of effects involving those moons' sizes and the properties of energetic ion motion and the slow magnetospheric diffusion at these low L shells. This is discussed in detail in *Simon et al.* (2015) and *Roussos et al.* (2016). In the same studies it is explained why macrosignatures are not visible in energetic electrons, at least below few megaelectronvolts. It is worth noting that the shape and/or the temporal evolution such absorption signatures (including the microsignatures discussed in section 2.4 above) have been used to deduce fundamental properties about particle transport in Saturn's magnetosphere and Enceladus' distance, such as diffusion rates and co-rotational and radial plasma flows (*Jones et al.,* 2006; *Paranicas et al.,* 2005; *Roussos et al.,* 2007; *Andriopoulou et al.,* 2012, 2014; *Kollmann et al.,* 2013).

3.4. Water-Group Ion Ratios

The relative fractions of the ions that comprise the water group, H_3O^+, H_2O^+, OH^+, and O^+, provide valuable information on the conditions in Saturn's magnetosphere and on the fate of the Enceladus neutrals that contribute the bulk of the plasma. Moreover, these fractions, which are sensitive probes of the source, transport, and loss mechanisms, supply essential calibration data for the magnetosphere models, enabling higher confidence in the many conclusions drawn from those models.

One source of ion data in Saturn's magnetosphere is the INMS, a quadrupole mass spectrometer with an open source ion (OSI) mode that measures ions with masses less than 99 u and energies less than 42 eV (21 km s^{-1} for H_2O^+). The resolution of the mass filter is 1/8 amu, so the identification of each water-group ion species is unambiguous. In the OSI mode, each 31-ms integration period (IP) counts a specified mass within a specified narrow energy band. Combined with a velocity resolution of approximately 1 km s^{-1}, the 2°–4° field of view of INMS results in only a small volume of velocity space measured during each IP. The low sensitivity of the INMS OSI mode, which was designed for the dense Titan atmosphere, requires the binning of hundreds or thousands of inner magnetosphere (IM) measurements to achieve a useful signal-to-noise ratio. Only the water-group ions between 3.5 and 7 R_S have densities that are sufficiently above the radiation background for OSI measurements in Saturn's IM. The count rates for all other species, including H^+, H_2^+, and N^+, are indistinguishable from the noise level.

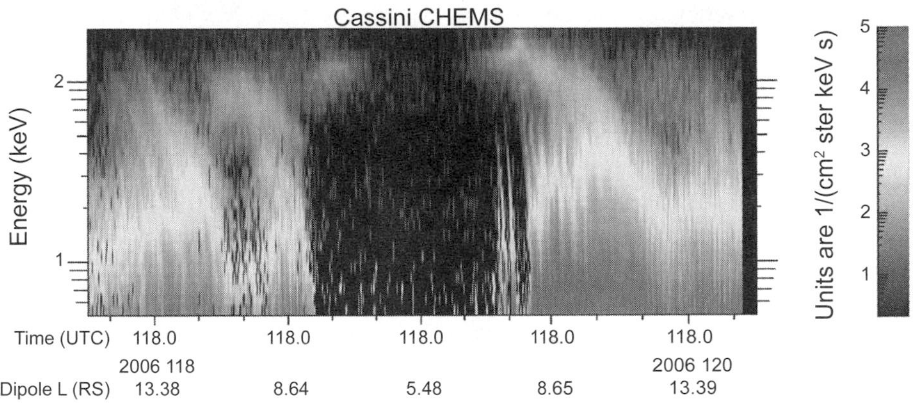

Fig. 19. Proton differential flux energy-time spectrogram for a periapsis pass on day 2006/118, using MIMI/CHEMS data. Note the intensity gap at periapsis (due to heavy charge-exchange losses) and the sporadic interchange injections (at fractional day 119.3). From *Paranicas et al.* (2008).

INMS data (Fig. 20) were acquired over 19 Cassini orbits with sufficient pointing to enable ion measurements for at least a portion of the orbit. Only measurements that were linked in time, velocity space, and location were binned, so measurements from different orbits are not combined. The measurements span a wide range of locations in velocity space, ranging from deep in the core of the velocity distribution, which is centered in a frame co-rotating with Saturn's magnetic field, to greater than 60 km s^{-1} (Fig. 21). The uncertainty in each measurement is determined statistically based on the signal level, so the measurements near the core have lower uncertainties.

H_2O^+ comprises the bulk of the ions near 4.0 R_S, and the H_2O^+ fraction decreases with increasing distance from 4.0 R_S, the source of neutral water at Enceladus. At 4.0 R_S, the fraction of H_2O^+ ranges from 60% to 100%. At 6.5 R_S, the three main water-group constituents, H_2O^+, OH^+, and O^+, are equal within the measurement errors. H_3O^+, which

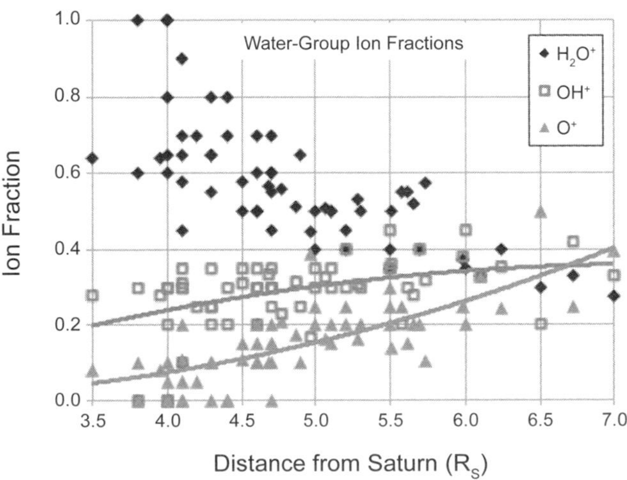

Fig. 20. The relative abundance of the dominant water group ions, H_2O^+, OH^+, and O^+, varies with distance from 4 R_S, the orbit radius of Enceladus. All statistically significant results are shown. Uncertainties range from 10% to 50%, depending on the signal level. H_3O^+ is usually inseparable from the noise floor but it can be as high as 10% (not shown).

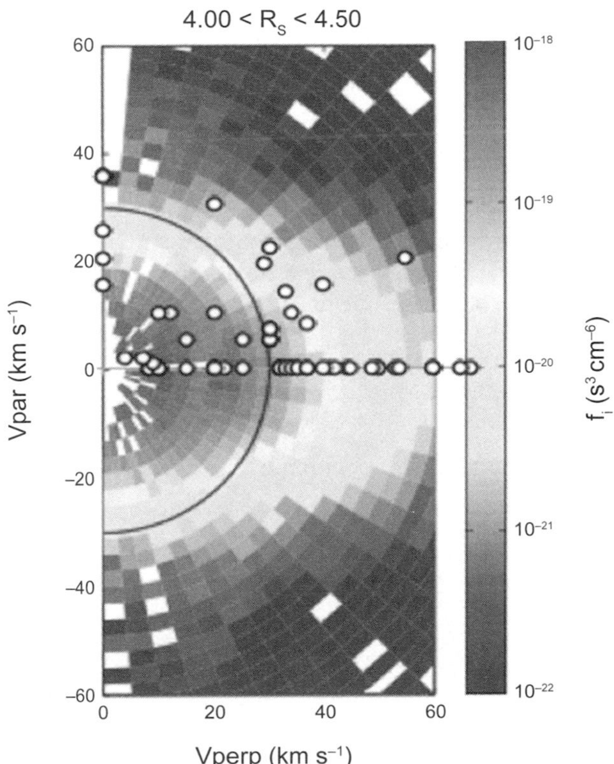

Fig. 21. The small white circles show the location in velocity space of each of the INMS ion measurement. The velocities are given in the reference frame of the rotating magnetic field. The vertical axis shows the velocity of the ions along the magnetic field lines and the horizontal axis shows the velocity perpendicular to the field, which is nominally aligned with Saturn's rotation axis. The size of the white circles reflects the extent of the INMS FOV in velocity space. The background plot is from *Tokar et al.* (2008) and shows the distribution of the total water-group ions near 4 R_S.

dominates the water-group ion fractions in the Enceladus plume, is 10% or less in Saturn's magnetosphere outside the plume.

Particularly near 4 R_S, the relative ion fractions have variations that are greater than the uncertainty in the values. The variations are not clearly linked to any of the studied parameters, including ion velocity, location in velocity space, density, injection events, and the orbit-phase of Enceladus. Although INMS samples a wide variety of these parameters, including many locations in Saturn's magnetosphere and in velocity space, all the parameters were sparsely covered, and correlations between the parameters could be lost due to overlapping dependencies or to statistical variation. Comparing the data to modeling that includes these factors may lead to explanations for the observed variations.

The radial dependence of the INMS measurements of H_2O^+ and OH^+ fractions, and the very low abundance of H_3O^+, are similar to early (*Richardson and Jurac,* 2004; *Jurac and Richardson,* 2005) and some later (*Smith et al.,* 2010) models of Saturn's magnetosphere. The models predicted a lower fraction of O^+ than INMS observations, which found O^+ reaching the same concentration as the other species near 7 R_S. Later models by *Fleshman et al.* (2013) differ in showing lower fractions of H_2O^+ and much higher abundance of H_3O^+, which is 70% of the water group for some results. Additionally, *Wilson et al.* (2015) calculated water-group relative ion ratios using CAPS data; however, separation of individual water-group ion species is much more difficult and uncertain than INMS observations. They also showed the fraction of H_3O^+ in significant amounts at least as far as the middle magnetosphere. However, *Cravens et al.* (2009) analyzed INMS data from the Cassini E3 Enceladus encounter and found H_3O^+ only present inside the Enceladus plume. The relative abundance of H_3O^+ is significant because this species is generated and disposed of under relative low energy environments and is thus a critical marker for understanding the magnetospheric environment. This inconsistency is not currently resolved.

3.5. Plume Variability

The discovery and characterization of the Enceladus plumes was quickly followed by the logical question as to what is the water gas source rate, and does it change over time. This topic was originally quite controversial as multiple studies of this phenomenon appeared to be contradictory, with some showing the plume source rates were very stable, while others reported noticeable variability. However, research now appears to be converging on the conclusion that these plumes are variable.

The first hints that the plume might be variable started to emerge shortly after their discovery. Cassini ISS visible observations of the plumes showed many little jets and the observations of the local environment appeared variable (*Esposito et al.,* 2005; *Waite et al.,* 2006; *Jones et al.,* 2006; *Gurnett et al.,* 2007). However, these ISS images are of the ice grains, and the local environment could be impacted by factors other than plume source variability. The plume rate controversy was exacerbated by the limited number of available direct observations. Thus, modeling studies began to emerge to overcome this research limitation. One early effort involved using magnetic field data to constrain a two-fluid plasma model with a fixed neutral particle distribution to analyze the plume gas source rate. Using this technique, *Saur et al.* (2008) deduce that the plumes are variable because they determine that the E0 Enceladus encounter has a source rate of ~1600 kg s^{-1} of H_2O, while the E1 and E2 encounters were only producing water at ~200 kg s^{-1}. Unfortunately, Cassini did not fly through the plume region during these encounters, so these observations were not optimal for studying plume variability.

After additional Enceladus encounter observations were available, more exhaustive studies of plume variability became possible. *Smith et al.* (2010) applied modeling efforts with Cassini INMS neutral particle observations to constrain plume geometry, determine plume velocity, and estimate the net plume source rate for the E2, E3, and E5 Enceladus encounters. These observations involved detection of *in situ* water density along the Cassin trajectories. Using their three-dimensional Monte Carlo particle-tracking, multispecies computational model, they forward modeled these parameters and determine a bulk velocity of ~720 km s^{-1} with variable plume H_2O bulk source rates of ~72 kg s^{-1} for E2, ~190 kg s^{-1} for E3, and ~750 kg s^{-1} for E5 [E7 tentatively ~275 kg s^{-1} (*Blanc et al.,* 2015)]. However, they also determined that the E2 flyby trajectory barely skimmed the edge of the plume region and therefore was not a reliable estimate. Thus, they reported that the plumes source rate likely varies by a factor of ~3.5.

Additional studies followed that also reported plume variability. *Tenishev et al.* (2010) applied a three-dimensional Monte Carlo test particle model that was also constrained by Cassini neutral particle (INMS) observations; however, they modeled eight variable plume source locations to determine the relative outflow from the different tiger stripe regions. They simultaneously fit the E3 and E5 encounters and reported the source rate for E5 was within a factor of 2 greater than the E3 source rate. *Dong et al.* (2011) next used Cassini INMS neutral observations of the E3 and E5 encounters to constrain their research. However, they used an analytical model, as well as data from the E7 Enceladus encounter, and reported source rates ranging from ~450 kg s^{-1} to ~1000 kg s^{-1}.

However, results of plume variability are contradicted by results based on Cassini UVIS plume occultation observations. *Hansen et al.* (2011) studied extreme ultraviolet solar plume occultations of the H_2O gas plumes to analyze plume characteristics during observations in 2005, 2007, and 2010. Not only did they report evident of high-speed jets (>1 km s^{-1}), but they also reported highly stable plume source rates. More specifically, they estimated that the plume source rates varied by less than 20% of ~200 kg s^{-1} and concluded there was no evidence of short-term plume variability. *Hansen et al.* (2017) theorize this discrepancy may be the result of the increased water gas source rate actually being caused by high-speed jets rather than the slower bulk escape process.

This apparent contradiction remained unresolved until *Hedman et al.* (2013) conducted an extensive study of Cassini VIMS infrared observations. Although these observations involve escaping plume dust particles, variability in these observations can still provide a good proxy of H_2O rate variability. They examined a much larger dataset than previous studies, with 252 images collected from 2005 until 2012. This study also covered a much wider range of spatial locations. Based on this research, they reported that the Enceladus plume source rate varies by a factor of ~3, consistent with many previous studies. Interestingly, they also determine that this variability appears to be modulated by Enceladus' orbital location (mean anomaly). The smallest source rates were observed when Enceladus is closest to Saturn, with the tiger stripes under the least strain. This is consistent with the theory that the plume source rate is enhanced during periods of enhanced surface tension.

Figure 22 shows the *Hedman et al.* (2013) results. Additionally, this figure contains the superimposed, normalized *Smith et al.* (2010) results with the respective locations in the orbit, showing a very good agreement and further supporting the existence of Enceladus plume variability, helping prove that the plume water gas source rates are variable, similar to the plume grain rate. Thus, with the exception of *Hansen et al.* (2011), results indicate the plumes are variable on the order of about a factor of ~3. While this variability should likely impact the local magnetospheric environment, it has yet to be determined if such variability could impact the global magnetospheric environment.

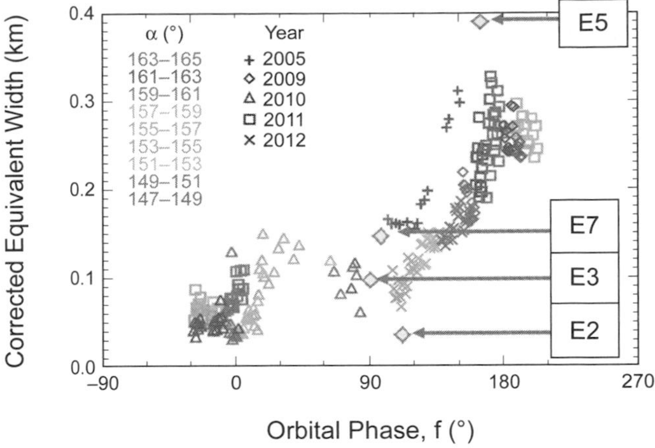

Fig. 22. Figure from *Hedman et al.* (2013) showing plume activity as a function of Enceladus mean anomaly (width vs. orbital phase). *Smith et al.* (2010) scaled source rate values for the E2, E3, E5, and E7 encounters are also plotted (assuming orbital symmetry).

4. SUMMARY AND OPEN QUESTIONS

Research based on over 13 years of Cassini observations at Saturn has provided greater insight into the significance of Enceladus for Saturn's magnetosphere. Water plumes from this small inner icy moon are actually the dominate source of particles for the entire system; interestingly, they are also a much more significant source than the large moon, Titan. As these particles exit the plumes, they interact with other particles and the local magnetic field to create an unusual slow-moving environment around the moon. This environment then interacts with the newly produced particles to create a feedback system that further impacts the local environment. Over time, these particles spread out and populate the entire magnetosphere, thus having a significant impact on the global environment as well.

While many questions have been answered by recent research, many more questions have emerged. Some significant unanswered questions include:

- Why is Enceladus a larger source than Titan?
- What is causing the large difference in positive and negative charged particles in the plume?
- Is Enceladus responsible for Saturn's neutral-dominated magnetosphere?
- Is plume variability impacting the magnetosphere?
- What interactions are occurring inside the plume?
- What are the relative water group ion abundances?
- What is causing the energetic electron flux "dropouts" at Enceladus?

Thus, the last 13 years of discovery at Saturn has shown just how significant a little icy moon can be for a large planetary magnetosphere. However, many more years of exciting discoveries regarding the impact of Enceladus still lay ahead.

REFERENCES

Andriopoulou M., Roussos E., Krupp N., et al. (2012) A noon-to-midnight electric field and nightside dynamics in Saturn's inner magnetosphere, using microsignature observations. *Icarus, 220,* 503–513.

Andriopoulou M., Roussos E., Krupp N., et al. (2014) Spatial and temporal dependence of the convective electric field in Saturn's inner magnetosphere. *Icarus, 229,* 57–70.

Barbosa D. D. (1987) Titan's atomic nitrogen torus: Inferred properties and consequences for the Saturnian aurora. *Icarus, 72,* 53–61.

Blanc M., Andrews D. J., Coates A. J., et al. (2015) Saturn plasma sources and associated transport processes. *Space Sci. Rev., 192,* 209–236, DOI: 10.1007/s11214-015-0184-5.

Bridge H. S., Belcher J. W., Lazarus A. J., et al. (1982) Plasma observations near Saturn — Initial results from Voyager 1. *Science, 212,* 217–224.

Broadfoot A. L., Sandel B. R., Shemansky D. E., et al. (1981) Extreme ultraviolet observations from Voyager 1 encounter with Saturn. *Science, 212,* 206–211.

Burger M. H., Sittler E. C., Johnson R. E., Smith H. T., and Tucker O. J. (2007) Understanding the escape of water from Enceladus. *J. Geophys. Res., 112,* A06219, DOI: 10.1029/2006JA012086.

Cassidy T. A. and Johnson R. E. (2010) Collisional spreading of Enceladus' neutral cloud. *Icarus, 209(2),* 696–703, DOI: 10.1016/j.icarus.2010.04.010.

Chen Y. and Hill T. W. (2008) Statistical analysis of injection/dispersion events in Saturn's inner magnetosphere. *J. Geophys. Res., 113,* A12311, DOI: 10.1029/2008JA013529.

Chen Y., Hill T. W., Rymer A. M., and Wilson R. J. (2010) Rate of radial transport of plasma in Saturn's inner magnetosphere. *J. Geophys. Res., 115,* A10211, DOI: 10.1029/2010JA015412.

Collette A., Meyer G., Malaspina D., and Sternovsky Z. (2015) Laboratory investigation of antenna signals from dust impacts on spacecraft. *J. Geophys. Res, 120(7),* 5298–5305. DOI: 10.1002/2015JA021198.

Cravens T. E., McNutt R. L. Jr., Waite J. H. Jr., et al. (2009) Plume ionosphere of Enceladus as seen by the Cassini ion and neutral mass spectrometer. *Geophys. Res. Lett., 36,* L08106, DOI: 10.1029/2009GL037811.

Cui J., Yelle R. V., and Volk K. (2008) Distribution and escape of molecular hydrogen in Titan's thermosphere and exosphere. *J. Geophys. Res., 113,* E10004, DOI: 10.1029/2007JE003032.

De La Haye V., Waite J. H. Jr., Cravens T. E., et al. (2007) Titan's corona: The contribution of exothermic chemistry. *Icarus, 19(1),* 236–250.

Dialynas K., Krimigis S. M., Mitchell D. G., et al. (2009) Energetic ion spectral characteristics in the saturnian magnetosphere using Cassini/MIMI measurements. *J. Geophys. Res., 114,* A01212.

DiFabio R. D., Hamilton D. C., Krimigis S. M., and Mitchell D. G. (2011) Long term time variations of the suprathermal ions in Saturn's magnetosphere. *Geophys. Res. Lett, 38,* L18103.

Dong Y., Hill T. W., Teolis B. D., Magee B. A., and Waite J. H. (2011) The water vapor plumes of Enceladus. *J. Geophys. Res., 116,* A10204, DOI: 10.1029/2011JA016693.

Dong Y., Hill T. W., and Ye S.-Y. (2015) Characteristics of ice grains in the Enceladus plume from Cassini observations. *J. Geophys. Res., 120,* 915–937, DOI: 10.1002/2014JA020288.

Dougherty M. K., Khurana K. K., Neubauer F. M., et al. (2006) Identification of a dynamic atmosphere at Enceladus with the Cassini magnetometer. *Science, 311(5766),* 1406–1409, DOI: 10.1126/science.1120985.

Engelhardt I. A. D., Wahlund J.-E., Andrews D. J., et al. (2015) Plasma regions, charged dust and field-aligned currents near Enceladus. *Planet. Space Sci., 117,* 453–469, DOI: 10.1016/j.pss.2015.09.010.

Esposito L. W., Colwell J. E., et al. (2005) Ultraviolet imaging spectroscopy shows an active saturnian system. *Science, 307(5713),* 1251–1255.

Farmer A. J. (2009) Saturn in hot water: Viscous evolution of the Enceladus torus. *Icarus, 202(1),* 280–286, DOI: 10.1016/j.icarus.2009.02.031.

Farrell W. M., Kurth W. S., Gurnett D. A., et al. (2009) Electron density dropout near Enceladus in the context of water-vapor and water-ice. *Geophys. Res. Lett., 36,* L10203, DOI: 10.1029/2008GL037108.

Fleshman B. L., Delamere P. A., and Bagenal F. (2010) Modeling the Enceladus plume-plasma interaction. *Geophys. Res. Lett., 37,* L03202, DOI: 10.1029/2009GL041613.

Fleshman B. L., Delamere P. A., Bagenal F., et al. (2013) A 1-D model of physical chemistry in Saturn's inner magnetosphere. *J. Geophys. Res., 118,* 1567–1581, DOI: 10.1002/jgre.20106.

Gurnett D. A. and Bhattacharjee A. (2005) *Introduction to Plasma Physics: With Space and Laboratory Applications*. Cambridge Univ., Cambridge. 464 pp.

Gurnett D. A., Grün E., Gallagher D., et al. (1983) Micron-sized particles detected near Saturn by the Voyager plasma wave instrument. *Icarus, 53(2),* 236–254.

Gurnett D. A., Persoon A. M., Kurth W. S., et al. (2007) The variable rotation period of the inner region of Saturn's plasma disk. *Science, 316(5823),* 442–445.

Gurnett D. A., Averkamp T. F., Schippers P., et al. (2011) Auroral hiss, electron beams and standing Alfvén wave currents near Saturn's moon Enceladus. *Geophys. Res. Lett., 38,* L06102, DOI: 10.1029/2011GL046854.

Hansen C. J., Esposito L., Stewart A. I. F., et al. (2006) Enceladus' water vapor plume. *Science, 311(5766),* 1422–1425.

Hansen C. J. et al. (2011) The composition and structure of the Enceladus plume. *Geophys. Res. Lett., 38,* L11202, DOI: 10.1029/2011GL047415.

Hansen C. J., Esposito L. W., Aye K.-M., et al. (2017) Investigation of diurnal variability of water vapor in Enceladus' plume by the Cassini ultraviolet imaging spectrograph. *Geophys. Res. Lett., 44,* 672–677, DOI: 10.1002/2016GL071853.

Hartogh P., Lellouch E., Moreno R., et al. (2011) Direct detection of the Enceladus water torus with Herschel. *Astron. Astrophys., 532,* L2, DOI: 10.1051/0004-6361/201117377.

Hedman M. M., Gosmeyer C. M., Nicholson P. D., et al. (2013) An observed correlation between plume activity and tidal stresses on Enceladus. *Nature, 500,* 182–184, DOI: 10.1038/nature12371.

Hess S. L. G., Delamere P. A., Dols V., and Ray L. C. (2011) Comparative study of the power transferred from satellite-magnetosphere interactions to auroral emissions. *J. Geophys. Res., 116,* A01202.

Hill T. W., Thomsen M. F., Tokar R. L., et al. (2012) Charged nanograins in the Enceladus plume. *J. Geophys. Res., 117,* A05209, DOI: 10.1029/2011JA017218.

Huebner W. F. and Mukherjee J. (2015) Photoionization and photodissociation rates in solar and blackbody radiation fields. *Planet. Space Sci., 106,* 11–45.

Ip W.-H. (1984) On the equatorial confinement of thermal plasma generated in the vicinity of the rings of Saturn. *J. Geophys. Res., 89(A1),* 395–398, DOI: 10.1029/JA089iA01p00395.

Ip W. (1997) On neutral cloud distributions in the saturnian magnetosphere. *Icarus, 97,* 42–47.

Jia Y.-D, Russell C. T., Khusrana K. K., Leisner J. S., et al. (2010). Time-varying magnetospheric environment near Enceladus as seen by the Cassini magnetometer. *Geophys. Res. Lett., 37,* L09203, DOI: 10.1029/2010GL042948.

Johnson R. E., Pospieszalska M. K., Sittler E. C., et al. (1989) The neutral cloud and heavy ion inner torus at Saturn. *Icarus, 77,* 311–329.

Johnson R. E., Smith H. T., Tucker O. J., et al. (2006) The Enceladus and OH tori at Saturn. *Astrophys. J. Lett., 644,* L137–L139.

Jones G. H., Roussos E., Krupp N., et al. (2006) Enceladus' varying imprint on the magnetosphere of Saturn. *Science, 311,* 1412–1415.

Jones G. H., Arridge C. S., Coates A. J., et al. (2009) Fine jet structure of electrically charged grains in Enceladus' plume. *Geophys. Res. Lett., 36,* L16204, DOI: 10.1029/2009GL038284.

Jurac S. and Richardson J. D. (2005) A self-consistent model of plasma and neutrals at Saturn: Neutral cloud morphology. *J. Geophys. Res., 110,* A09220.

Kanani S. J. et al. (2010) A new form of Saturn's magnetopause using a dynamic pressure balance model, based on *in situ*, multi-instrument Cassini measurements. *J. Geophys. Res., 115,* A06207, DOI: 10.1029/2009JA014262.

Kempf S., Beckmann U., Srama R., et al. (2006) The electrostatic potential of E ring particles. *Planet. Space Sci., 54(9–10),* 999–1006.

Kivelson M. G. (2006) Does Enceladus govern magnetospheric dynamics at Saturn? *Science, 311,* 1391–1392.

Khurana K. K., Dougherty M. K., Russell C. T., and Leisner J. S. (2007) Mass loading of Saturn's magnetosphere near Enceladus. *J. Geophys. Res., 112(A8),* A08203, DOI: 10.1029/2006JA012110.

Kivelson M. G., Khurana K. K., Russell, C. T., et al. (2001) Magnetized or unmagnetized: Ambiguity persists following Galileo's encounters with Io in 1999 and 2000. *J. Geophys. Res., 106,* 26121, DOI: 10.1029/2000JA002510.

Kollmann P., Roussos E., Paranicas C., et al. (2011) Energetic particle phase space densities at Saturn: Cassini observations and interpretations. *J. Geophys. Res., 116,* A05222.

Kollmann P., Roussos E., Paranicas C., et al. (2013) Processes forming and sustaining Saturn's proton radiation belts. *Icarus, 222,* 323–341.

Kriegel H., Simon S., Mueller, J., et al. (2009) The plasma interaction of Enceladus: 3D hybrid simulations and comparison with Cassini MAG data. *Planet. Space Sci., 57(14–15),* 2113–2122, DOI: 10.1016/j.pss.2009.09.025.

Kriegel H., Simon S., Motschmann, U., et al. (2011) Influence of negatively charged plume grains on the structure of Enceladus' Alfvén wings: Hybrid simulations versus Cassini magnetometer data. *J. Geophys. Res., 116,* A1022319, DOI: 10.1029/2011JA016842.

Kriegel H., Simon S., Meier P., et al. (2014) Ion densities and magnetic signatures of dust pickup at Enceladus. *J. Geophys. Res., 119,* 2740–2774, DOI: 10.1002/2013JA019440.

Krimigis S. M., Mitchell D. G., Hamilton D. C., et al. (2004) Magnetosphere Imaging Instrument (MIMI) on the Cassini Mission to Saturn/Titan. *Space Sci. Rev., 114,* 233–329.

Krupp N., Roussos E., Kollmann P., et al. (2012) The Cassini Enceladus encounters 2005–2010 in the view of energetic electron measurements. *Icarus, 218,* 433–447.

Kurth W. S., Averkamp T. F., Gurnett D. A., et al. (2006) Cassini RPWS observations of dust in Saturn's E ring. *Planet. Space Sci., 54(9–10),* 988–998.

Leisner J. S., Russell C. T., Dougherty K. K., et al. (2006) Ion cyclotron waves in Saturn's E ring: Initial Cassini observations. *Geophys. Res. Lett., 33,* L11101, DOI: 10.1029/2005GL024875.

Leisner J. S., Russell C. T., Wei H. Y., and Dougherty M. K. (2011) Probing Saturn's ion cyclotron waves on high-inclination orbits: Lessons for wave generation. *J. Geophys. Res., 116,* A09235, DOI: 10.1029/2011JA016555.

Leisner J. S., Hospodarsky G. B., and Gurnett D. A. (2013) Enceladus auroral hiss observations: Implications for electron beam locations. *J. Geophys. Res, 118,* 160–166, DOI: 10.1029/2012JA018213.

Mauk B. H. (2015) Comparative investigation of the energetic ion spectra comprising the magnetospheric ring currents of the solar system. *J. Geophys. Res.–Space Phys., 119,* 9729–9746, DOI: 10.1002/2014JA020392.

Mauk B. H. and Fox N. J. (2010) Electron radiation belts of the solar system. *J. Geophys. Res., 115,* A12220, DOI: 10.1029/2010JA015660.

Mauk B. H., Saur J., Mitchell D. G., et al. (2005) Energetic particle injections in Saturn's magnetosphere. *Geophys. Res. Lett., 32,* L14S05.

McAndrews H. J. et al. (2009) Plasma in Saturn's nightside magnetosphere and the implications for global circulation. *Planet. Space Sci., 57(14–15*), 1714–1722.

Meeks Z., Simon S., and Kabanovic S. (2016) A comprehensive analysis of ion cyclotron waves in the equatorial magnetosphere of Saturn. *Planet. Space Sci., 129,* 47–60.

Meier P., Kriegel H., Motschmann U., et al. (2014) A model of the spatial and size distribution of Enceladus' dust plume. *Planet. Space Sci., 104,* 216–233.

Melin H., Shemansky D. E., and Liu X. (2009) The distribution of atomic hydrogen and oxygen in the magnetosphere of Saturn. *Planet. Space Sci., 57(14–15),* 1743–1753.

Meyer-Vernet N. (2013) On the charge of nanograins in cold environments and Enceladus dust. *Icarus, 226(1),* 583–590.

Morooka M. W., Wahlund J.-E., Eriksson A. I., et al. (2011) Dusty plasma in the vicinity of Enceladus. *J. Geophys. Res., 116,* A12221, DOI: 10.1029/2011JA017038.

Neubauer F. M. (1980) Nonlinear standing Alfvén wave current system at Io: Theory. *J. Geophys. Res., 85,* 1171–1178, DOI: 10.1029/JA085iA03p01171.

Neubauer F. M. (1998) The sub-Alfvénic interaction of the Galilean satellites with the jovian magnetosphere. *J. Geophys. Res., 103(E9),* 19843–19866, DOI: 10.1029/97JE03370.

Paranicas C., Mitchell D. G., Livi S., et al. (2005) Evidence of Enceladus and Tethys microsignatures. *Geophys. Res. Lett., 32,* L20101.

Paranicas C., Mitchell D. G., Krimigis S. M., et al. (2008) Sources and losses of energetic protons in Saturn's magnetosphere. *Icarus, 197,* 519–525.

Paranicas C., Mitchell D. G., Roussos E., et al. (2010) Transport of energetic electrons into Saturn's inner magnetosphere. *J. Geophys. Res., 115,* A09214.

Perry M. E., Teolis B. D., Hurley D. M., et al. (2015) Cassini INMS measurements of Enceladus plume density. *Icarus, 257,* 139–162.

Pospieszalska M. K. and Johnson R. E. (1991) Micrometeorite erosion of the main rings as a source of plasma in the inner saturnian plasma torus. *Icarus, 93,* 45–52.

Pryor W. R., Rymer A. M., Mitchell D. G., et al. (2011) The auroral footprint of Enceladus on Saturn. *Nature, 472,* 331–333.

Richardson J. D. (1995) An extended plasma model for Saturn. *Geophys. Res. Lett., 22(10),* 1177–1118, DOI: 10.1029/95GL01018.

Richardson J. D. and Jurac S. (2004) A self-consistent model of plasma and neutrals at Saturn: The ion tori. *Geophys. Res. Lett., 31(24),* L24803.

Richardson J. D. and Sittler E. C. Jr. (1990) A plasma density model for Saturn based on Voyager observations. *J. Geophys. Res., 95(A8),* 12019–12031, DOI: 10.1029/JA095iA08p12019.

Roussos E., Jones G. H., Krupp N., et al. (2007) Electron microdiffusion in the saturnian radiation belts: Cassini MIMI/LEMMS observations of energetic electron absorption by the icy moons. *J. Geophys. Res., 112,* A06214.

Roussos E., Krupp N., Armstrong T. P., et al. (2008) Discovery of a transient radiation belt at Saturn. *Geophys. Res. Lett., 35,* L22106.

Roussos E., Kollmann P., Krupp N., et al. (2012) Energetic electron observations of Rhea's magnetospheric interaction. *Icarus, 221,* 116–134.

Roussos E., Krupp N., Kollmann P., et al. (2016) Evidence for dust-driven, radial plasma transport in Saturn's inner radiation belts. *Icarus, 274,* 272–283.

Santolik O., Gurnett D. A., Jones G. H., et al. (2011) Intense plasma wave emissions associated with Saturn's moon Rhea. *Geophys. Res. Lett., 38,* L19204.

Saur J. and Strobel D. F. (2005) Atmospheres and plasma interactions at Saturn's largest inner icy satellites. *Astrophys. J. Lett., 620,* L115–L118.

Saur J., Neubauer F. M., and Schilling N. (2007) Hemisphere coupling in Enceladus' asymmetric plasma interaction. *J. Geophys. Res., 112,* A11209.

Saur J., Schilling N., Neubauer F. M., et al. (2008) Evidence for temporal variability of Enceladus' gas jets: Modeling of Cassini observations. *Geophys. Res. Lett., 35(20),* L20105.

Schippers P., Blanc M., André N., et al. (2008) Multi-instrument analysis of electron populations in Saturn's magnetosphere. *J. Geophys. Res., 113,* A07208.

Shafiq M., Wahlund J.-E., Morooka M. W., et al. (2011) Characteristics of the dust-plasma interaction near Enceladus' south pole. *Planet. Space Sci., 59(1),* 17–25.

Shemansky D. E. and Hall D. T. (1992) The distribution of atomic hydrogen in the magnetosphere of Saturn. *J. Geophys. Res., 97,* 4143–4161.

Shemansky D. E., Matherson P., Hall D. T., et al. (1993) Detection of the hydroxyl radical in the Saturn magnetosphere. *Nature, 363,* 329–332.

Shemansky D. E., Liu X., and Melin H. (2009) The Saturn hydrogen plume. *Planet. Space Sci., 57(14–15),* 1659–1670.

Simon S., Saur J., Kriegel H., et al. (2011) Influence of negatively charged plume grains and hemisphere coupling currents on the structure of Enceladus' Alfvén wings: Analytical modeling of Cassini magnetometer observations. *J. Geophys. Res., 116,* A04221, DOI: 10.1029/2010JA016338.

Simon S., Saur J., van Treeck S. C., Kriegel H., and Dougherty M. K. (2014) Discontinuities in the magnetic field near Enceladus. *Geophys. Res. Lett., 41,* 3359–3366, DOI: 10.1002/2014GL060081.

Simon S., Roussos E., and Paty C. S. (2015) The interaction between Saturn's moons and their plasma environments. *Phys. Rept., 602,* 1–65.

Sittler E. C., Andre N., Blanc M., et al. (2008) Ion and neutral sources and sinks within Saturn's inner magnetosphere: Cassini results. *Planet. Space Sci., 56(1),* 3–18, DOI: 10.1016/j.pss.2007.06.006.

Smith H. T., Johnson R. E., and Shematovich V. I. (2004) Titan's atomic and molecular nitrogen tori. *Geophys. Res. Lett., 31(16),* L16804, DOI: 10.1029/2004GL020580.

Smith H. T. et al. (2005) Discovery of nitrogen in Saturn's inner magnetosphere. *Geophys. Res. Lett., 32(14),* L14S03.

Smith H. T., Johnson R. E., Sittler E. C., et al. (2007) Enceladus: The likely dominant nitrogen source in Saturn's magnetosphere. *Icarus, 188(2),* 356–366.

Smith H. T., Shappirio M., Johnson R. E., et al. (2008) Enceladus: A potential source of ammonia products and molecular nitrogen for Saturn's magnetosphere. *J. Geophys. Res., 113,* A11206, DOI: 10.1029/2008JA013352.

Smith H. T. et al. (2010) Enceladus plume variability and the neutral gas densities in Saturn's magnetosphere. *J. Geophys. Res., 115,* A10252, DOI: 10.1029/2009JA015184.

Spahn F., Schmidt J., Albers N., et al. (2006) Cassini dust measurements at Enceladus and implications for the origin of the E ring. *Science, 311(5766),* 1416–1418.

Tadokoro H. and Katoh Y. (2014) Test-particle simulation of energetic electron-H_2O elastic collision along Saturn's magnetic field line around Enceladus. *J. Geophys. Res., 119,* 8971–8978.

Tenishev V., Combi M. R., Teolis B. D., and Waite J. H. (2010) An approach to numerical simulation of the gas distribution in the atmosphere of Enceladus. *J. Geophys. Res., 115,* A09302, DOI: 10.1029/2009JA015223.

Teolis B. D., Perry M. E., Magee B. A., et al. (2010) Detection and measurement of ice grains and gas distribution in the Enceladus plume by Cassini's ion neutral mass spectrometer. *J. Geophys. Res., 115,* A09222, DOI: 10.1029/2009JA015192.

Thomsen M. F., Reisenfeld D. B., Delapp D. M., et al. (2010) Survey of ion plasma parameters in Saturn's magnetosphere. *J. Geophys. Res., 115,* A10220, DOI: 10.1029/2010JA015267.

Thomsen M. F., Coates A. J., Roussos E., et al. (2016) Suprathermal electron penetration into the inner magnetosphere of Saturn. *J. Geophys. Res., 121,* 5436–5448.

Tian F., Stewart A. I. F., Toon O. B., et al. (2007) Monte Carlo simulations of the water vapor plumes on Enceladus. *Icarus, 188(1),* 154–161.

Tokar R. L., Wilson R. J., Johnson, R. E., et al. (2008) Cassini detection of water-group pick-up ions in the Enceladus torus. *Geophys. Res. Lett., 35,* L14202, DOI: 10.1029/2008GL034749.

Tokar R. L., Johnson R. E., Thomsen M. F., et al. (2009) Cassini detection of Enceladus' cold water-group plume ionosphere. *Geophys. Res. Lett., 36,* L13203, DOI: 10.1029/2009GL038923.

Tseng W.-L., Johnson R. E., and Ip W.-H. (2013a) The atomic hydrogen cloud in the saturnian system. *Planet. Space Sci., 85,*164–174.

Tseng W.-L., Johnson R. E., and Elrod M. K. (2013b) Modeling the seasonal variability of the plasma environment in Saturn's magnetosphere between main rings and Mimas. *Planet. Space Sci., 77,* 126–135, DOI: 10.1016/j.pss.2012.05.001.

Wahlund J.-E., André M., Eriksson A. I. E., et al. (2009) Detection of dusty plasma near the E-ring of Saturn. *Planet. Space Sci., 57(14–15),* 1795–1806.

Waite J. H. and 13 colleagues (2006) Cassini ion and neutral mass spectrometer: Enceladus plume composition and structure. *Science, 311(5766),* 1419–1422.

Williams D. J. (2004) Energetic electron beams in Ganymede's magnetosphere. *J. Geophys. Res., 109,* A09211.

Williams D. J. and Thorne R. M. (2003) Energetic particles over Io's polar caps. *J. Geophys. Res., 108(A11),* 1397.

Wilson R. J., Bagenal F., Cassidy T., et al. (2015) The relative proportions of water group ions in Saturn's inner magnetosphere: A preliminary study. *J. Geophys. Res., 120,* 6624–6632, DOI: 10.1002/2014JA020557.

Yaroshenko V. V., Ratynskaia S., Olson J., et al. (2009) Characteristics of charged dust inferred from the Cassini RPWS measurements in the vicinity of Enceladus. *Planet. Space Sci., 57(14–15),* 1807–1812.

Yaroshenko V. V., Lühr H., and Miloch W. J. (2014) Dust charging in the Enceladus torus. *J. Geophys. Res., 119,* 221–236, DOI: 10.1002/2013JA019213.

Yaroshenko V. V., Miloch W. J., and Lühr H. (2015) Particle-in-cell simulation of spacecraft/plasma interactions in the vicinity of Enceladus. *Icarus, 257,* 1–8.

Ye S.-Y., Gurnett D. A., Kurth W. S., et al. (2014) Electron density inside Enceladus plume inferred from plasma oscillations excited by dust impacts. *J. Geophys. Res., 119,* 3373–3380, DOI: 10.1002/2014JA019861.

Ye S.-Y., Gurnett D. A., and Kurth W.S. (2016) *In-situ* measurements of Saturn's dusty rings based on dust impact signals detected by Cassini RPWS. *Icarus, 279,* 51–61.

Part 3:

Saturn's Icy Moons

Schenk P., White O. L., Byrne P. K., and Moore J. M. (2018) Saturn's other icy moons: Geologically complex worlds in their own right. In *Enceladus and the Icy Moons of Saturn* (P. M. Schenk et al., eds.), pp. 237–265. Univ. of Arizona, Tucson, DOI: 10.2458/azu_uapress_9780816537075-ch012.

Saturn's Other Icy Moons: Geologically Complex Worlds in Their Own Right

Paul Schenk
Lunar and Planetary Institute

Oliver L. White
SETI Institute

Paul K. Byrne
North Carolina State University

Jeffrey M. Moore
NASA Ames Research Center

Global mapping of the mid-sized icy moons of Saturn (besides Enceladus) at multiple wavelengths by the Cassini orbiter has revealed geologically complex worlds. These bodies exhibit diverse geological histories, much of which was not recognized by the Voyager reconnaissance. All the moons, except perhaps Iapetus, have been tectonically deformed and all but Iapetus reveal indications of higher heat flow in the past by virtue of viscously relaxed impact and tectonic features, or by relatively low-amplitude spheroidal global shapes that contrast sharply with cold and lumpy Iapetus. The tectonic signatures of these moons include relatively pristine and so presumably late-forming or reactivated rift systems on Tethys, Dione, and Rhea. These moons, as well as Mimas, also show more degraded and so likely older tectonic systems, indicating complex and protracted thermal histories. Impact craters dominate and are characterized by bowl shapes, hummocky floors, or prominent central peaks. Central pits are not observed (except probably at Odysseus on Tethys), and extensive ponded impact melt deposits are not evident. Giant Odysseus might have been responsible for Tethys' Ithaca Chasma rift system, although independent origins are not yet precluded. A cryptic red-stained fracture system on Tethys appears to be part of a long history of tectonism on that ice-rich moon. Putative volcanic resurfacing in the form of (relatively) smooth plains is limited in scope but appears to have affected nearly half of Dione and parts of Tethys. Any such terrains on the other moons have been erased by impact cratering. The origins of these terrains are as yet uncertain, but the erasure of older craters, flat-lying topography, and formation of two large irregular scarp-enclosed depressions in the center of smooth terrains on Dione tend to support a volcanic origin. An alternative explanation is that these terrains are remnants of an old apex-antapex cratering asymmetry that have been rotated 180° of longitude. The extent of smooth plains, relaxed craters, and tectonic patterns of various ages indicate that Dione has the most complex and perhaps most protracted thermal evolution of any of these moons, after Enceladus. The factor of 5 lower topographic amplitudes of the five inner moons indicates that a major heating event erased the ancient large basins and deep relief evident on outer moon Iapetus. Two satellites, Rhea and Iapetus, bear equatorial scars of the reaccretion of circum-satellite debris rings, in the form of discolorations on promontories and a large discontinuous equatorial ridge. Whether Saturn's satellite system is very young relative to the age of the solar system, as has been speculated, remains unclear, but the geologic record of these mid-sized bodies does not rule out this possibility.

1. INTRODUCTION

Among Saturn's 60+ icy moons, organic-rich Titan and geologically active Enceladus have been a primary focus of observation and study during Cassini's 13 years in Saturn orbit. Yet these bodies do not exist in a metaphorical vacuum, likely forming contemporaneously with six other major icy saturnian moons and interacting with them and the Saturn environment ever since. These mid-sized bodies are members of a class of ice-rich bodies between ~400 and ~1000 km in diameter orbiting Saturn, Uranus, and Pluto. They are intermediate between the smaller, lumpy, irregularly shaped ring shepherd and captured moons (see the chapters by Thomas et al. and Denk et al. in this volume), and the larger "planet-sized" icy moons such as Ganymede, Titan, and Callisto. (Hyperion is considered here as an irregular object and is discussed in the chapter in this volume by Thomas et al.) These mid-sized worlds — Mimas, Tethys, Dione, Rhea, and Iapetus (Fig. 1; see also Table 1 in the chapter by Castillo-Rogez et al. in this volume) —are the focus of this chapter.

Geologic investigations of these bodies focus on the search for physical evidence of heat production in these bodies, which elucidates their dynamical and geologic history and that of the satellite system itself. Geologic activity on these mid-sized moons will relate to the degree to which tidal and radiogenic heating has reshaped these worlds, and potentially also the heat source(s) powering Enceladus' ongoing activity (see the chapters in this volume by Patterson et al., Nimmo et al., Spencer et al., and Goldstein et al.). Hence the roles of tectonics and volcanism on these other moons are particularly important. These mid-sized icy satellites represent a natural laboratory for many processes, including volcanism on small bodies, impact on low gravity and icy bodies, ring formation at small bodies, and charged particle bombardment of airless bodies.

2. CASSINI MAPPING OF THE ICY MOONS

Incomplete mapping by Voyager at resolutions of 0.5 to 30 km/pixel (*Smith et al.,* 1981, 1982) suggested that the mid-sized satellites were heavily cratered and scarred by occasional fracture sets, but were not overly geologically complex. *Jaumann et al.* (2009) provided a summary of Voyager-related studies and an initial review of Cassini's first wave of discovery from the targeted encounters with these bodies in the period 2004–2007, based on new Cassini data at better than 500 m/pixel over roughly 50% of their surfaces. Complemented by numerous additional targeted and untargeted observations between 2008 and 2016, global mapping (Fig. 1) of ~100% of each body is now complete (the exception being Iapetus, where the two targeted encounters provide mapping at better than 500 m/pixel for ~75% of the surface). These data also allowed for global color (Fig. 1, Plate 31) (*Schenk et al.,* 2011) and topography (this chapter) maps of the surfaces of these satellites to be generated.

Encounter geometries and velocities, target sizes, and pointing capabilities of the Cassini orbiter (e.g., *Porco et al.,* 2004) dictated the default mapping resolution of each

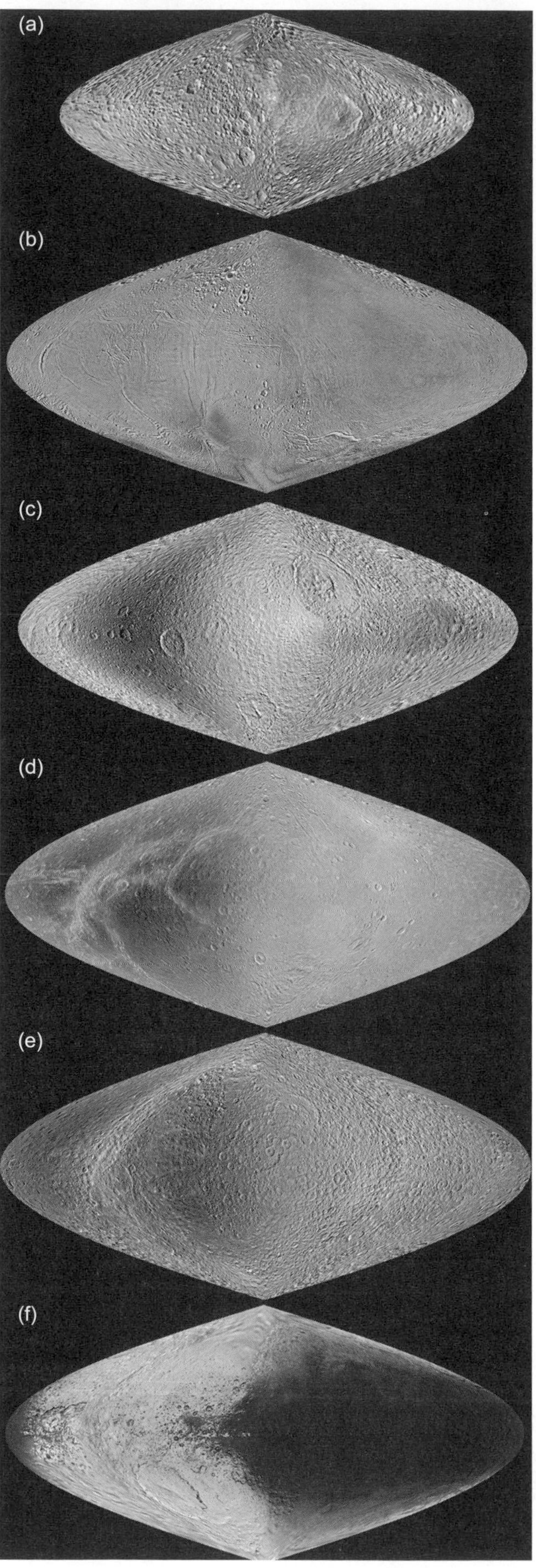

Fig. 1. See Plate 31 for color version. Global map views of the six classical mid-sized icy satellites of Saturn. Sinusoidal projections are centered on longitude 180° and are from the three-color global maps of *Schenk* (2014). Brightnesses are representative but not to scale and are optimized for each view to provide best overall contrast. Maps are to scale in each pair [**(a)** and **(b)**, **(c)** and **(d)**, **(e)** and **(f)**]. Base resolutions are **(a)** 200 m for Mimas, **(b)** 100 m for Enceladus, **(c)** 250 m for Tethys, **(d)** 250 m for Dione, **(e)** 400 m for Rhea, and **(f)** 400 m for Iapetus.

satellite. Mapping was accomplished with clear filter ISS panchromatic imaging and supporting color mapping. Dione and Tethys are most effectively mapped at 250 m/pixel, while Rhea and Iapetus are mapped at 400 m/pixel. Concurrent color imaging using 930-, 560-, and 338-μm-centered filters was usually acquired in summation mode at 2× lower resolution than concurrent panchromatic clear filter mosaics, and subsequently merged with the higher-resolution panchromatic images to create global mapping products (see Plate 31) with cartographic control and approximately correct albedo representation.

Cassini targeted selected areas for dedicated high-resolution images and mosaics at resolutions of tens of meters per pixel down to several meters per pixel. Many of these were serendipitous targets dictated by specific encounter geometry and solar illumination, and of value for small crater statistics or regolith characterization. Among those targeted for geologic focus were the bright ray crater Inktomi on Rhea; fracture networks on Tethys, Dione, and Rhea; and equatorial spots on Rhea, all discussed below.

Stereo imaging was comprehensively acquired for these satellites (and ~50–70% for Iapetus). Stereo images improve geologic interpretability and allow for production of topographic maps [digital elevation models (DEMs); see examples published by *Giese et al.* (2007), *Schenk and McKinnon* (2009), *Dombard et al.* (2012), and *White et al.* (2013, 2017), among others]. Low-resolution maps of global shape derived from limb profiles are described by *Thomas et al.* (2007), *Thomas* (2010), and *Nimmo et al.* (2010, 2011). Global integrated stereogrammetric topographic maps created by P. Schenk are shown and discussed in this chapter.

Visible and Infrared Mapping Spectrometer (VIMS) global multispectral and absorption band maps [taking advantage of updated calibration procedures (e.g., *Scipioni et al.,* 2017); see also the chapters in this volume by Hendrix et al. and Verbiscer et al.] can be produced at resolutions of ca. 10 km/pixel, although a few mapping sites were acquired at 1 km resolution or better. Ultraviolet Imaging Spectrograph (UVIS) maps of ultraviolet spectral properties (e.g., *Hendrix and Hansen,* 2008; *Hendrix et al.,* 2010; Hendrix et al., this volume) can also be produced at similar low resolution. Composite Infrared Spectrometer (CIRS) maps of thermal properties have also been produced (*Howett et al.,* 2010, 2011, 2012, 2014, 2016), although these mostly show variations in thermal inertia, the most prominent of which are related to the equatorial thermal and color anomalies attributed to high-energy electrons (*Schenk et al.,* 2011; *Howett et al.,* 2011, 2012). Integrated maps of spectral and geologic properties have been produced for Dione, Tethys, and Rhea (e.g., *Stephan et al.,* 2010, 2012, 2016).

Cassini's new global mapping products have reinforced the discovery that these bodies are geologically complex. They have also revealed new, unanticipated geologic processes and activity on these mid-sized icy satellites. Here, after the end of the Cassini mission, we survey our understanding of these unique worlds, their geologic histories, and the geologic processes modifying their surfaces, with an emphasis on observed geologic constraints on internal processes.

3. CRATER MORPHOLOGIES

From casual visual inspection, the icy moons of Saturn appear to be dominated by impact craters and we focus on this phenomenon first. These multitudes of impact events have reshaped their surfaces violently, but also reveal aspects of satellite thermal history through post-impact modification. Initial crater morphologies are also a function of the impact parameters and target properties. High-resolution observations of pristine unmodified craters were obtained in only a few instances, however.

Almost all craters on these moons are bowl-shaped simple craters or central peak craters (Fig. 2). Interesting features include the presence of secondary craters in a handful of fresh craters in the >50-km size range (Fig. 3), which all form at >1 crater diameter from the rim and are consistent with gravity scaling of ejecta processes on these low-gravity worlds (e.g., *Housen et al.,* 1983). Examples include Sagaris (Fig. 3), Telemachus, Inktomi, Yu-ti, and Odysseus (*Schenk et al.,* 2017). We also observe the presence of pancake (or pedestal) ejecta styles in several cases. These form a low annular plateau adjacent to the rim with an outward-facing scarp a few hundred meters high and are similar to those seen on Ganymede (*Horner and Greeley,* 1982; *Schenk and Ridolfi,* 2002; *Schenk et al.,* 2004) and in some craters on Mars (*Head and Roth,* 1976; *Carr et al.,* 1977; *Wohletz and Sheridan,* 1983). Examples in the Saturn system include Sagaris (Fig. 3), Sabinus, and Evander on Dione, as well as indications at other craters on Tethys and Rhea and at Herschel on Mimas. Subtle ejecta features are most easily recognized in fresh craters where degradation has not erased their signatures and where they formed on smoother units and are thus more recognizable. Given the low surface temperatures and lack of broad, thick impact melt deposits on these bodies, formation of this type of ejecta suggests that liquid water is likely not responsible for these features but are more likely a form of debris flow-type emplacement component of the ballistic ejecta process. Otherwise, there are few unusual characteristics of impact cratering in the Saturn system.

One surprise is that central pit craters so prevalent in craters >25 km across on Ganymede and Callisto (e.g., *Schenk,* 1993; *Bray et al.,* 2012) are almost nonexistent on the mid-sized moons. Inverse gravity scaling of this peak-to-pit transition diameter (e.g., *Pike,* 1980; *Schenk et al.,* 2004) indicates that craters larger than 75 to 100 km on the saturnian moons (depending on target) should be central pits, but all craters up to at least 350 km across are dominated by large conical central peaks extending across one-third or more of the crater width (Figs. 2 and 3). The lone exception is Odysseus, a 425-km-wide well-preserved impact basin on Tethys (Fig. 4), which features an oval central depression 60–80 km wide and 4 km deep (relative to the crater floor), surrounded by broken massifs. This is the closest to a central

Fig. 2. Oblique Cassini ISS images of typical simple (bowl-shaped) and complex (central-peak) craters on the icy moons of Saturn. **(a)** Simple crater (D ~10 km) at 30°N, 199°W on Rhea. **(b)** Central peak crater Telemachus (D ~92 km; 55°N, 337°W) on Tethys. The knobby floor and prominent central peak typical of complex craters on these satellites are well expressed at Telemachus.

pit type feature in the Saturn system, but occurs well above the predicted diameter for such landforms. Lack of central pits at Saturn is exacerbated by the discovery of central pit craters on Ceres (*Schenk et al.,* 2018a), a low-density, ice-rich dwarf planet similar in size to Dione, at crater diameters of >75 km, consistent with predictions from gravity scaling of Ganymede central pits.

The confirmation of central pits on Ceres and their conspicuous lack in the Saturn system potentially places new constraints on the pit-formation process and the role of volatiles in particular (cf. *Schenk et al.,* 2018a). A consensus may be building that central pits form as a result of melted ice occurring in the region normally associated with central peak uplifts and subsequent drainage of water into the

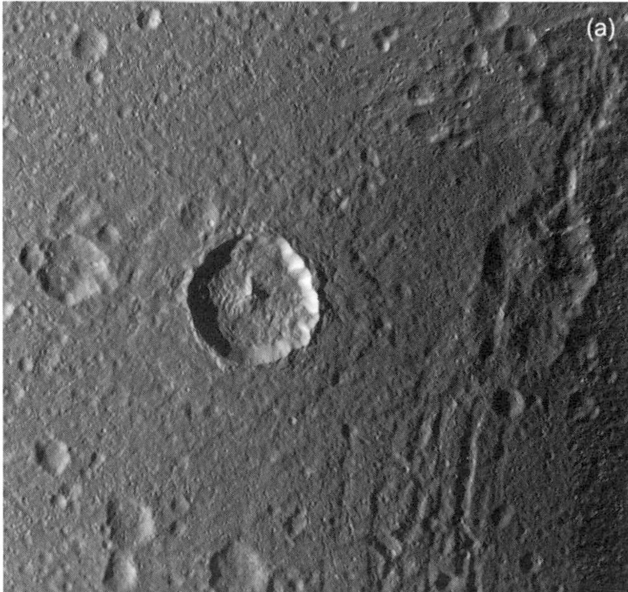

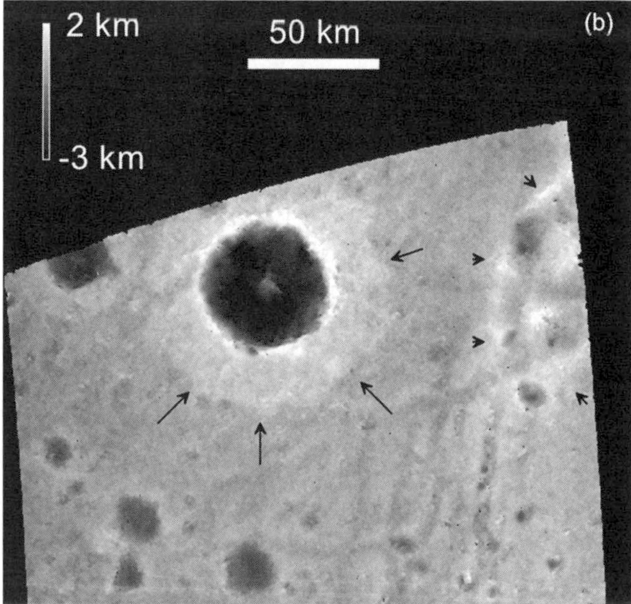

Fig. 3. Central peak crater Sagaris (D ~48 km; 5°N, 104°W) on Dione, showing well-preserved ejecta deposit (in the form of a annular plateau or "pancake" deposit, long arrows), and secondary fields on these moons. **(a)** View from global 250-m-resolution panchromatic mosaic. **(b)** View from stereo-derived DEM of this region showing relief of ±2 km. Also visible at right and upper right (short arrows) are the irregular craters Metiscus and Murranus, respectively, described in section 5 showing their shallow topography, as well as the Fidena Fossae fracture zone at lower right.

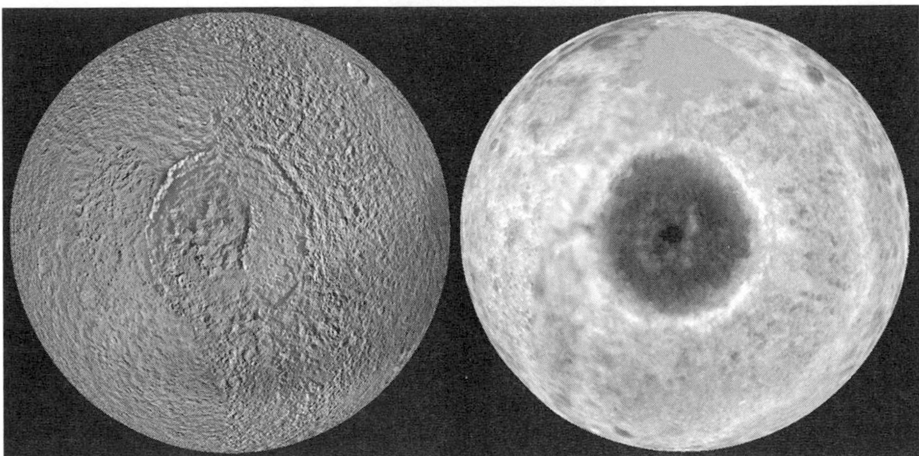

Fig. 4. Orthographic projections of Tethys centered on 425-km-wide central-pit crater Odysseus. View at left is from global 250-m-resolution panchromatic and color mosaic; at right from global stereo-derived DEM. Washboard terrains are visible at top and bottom. Visible in the topography is the partially rimmed central pit of Odysseus as well as the arcuate ridge and wide ejecta scour zone external to the rim to the east (right). A highly relaxed smaller impact basin is visible to the upper left of the basin rim. Total relief shown in view is ~10 km.

subsurface (see discussions in *Schenk et al.,* 2004, 2018a; *Bray et al.,* 2012; *Elder et al.,* 2012). If correct, the melt model for pits would indicate that the lack of central pits is consistent with the observational evidence that extended impact melt deposits are also lacking. Of course, resolution may play a role as many impact melt deposits on Mars and the Moon required meter-scale imaging or better to resolve (e.g., *Denevi et al.,* 2012), something we have very little of in the Saturn system. Central pits on ice-rich bodies indicate a role for volatiles. Impact velocities are comparable in the jovian and saturnian system, and while thermal conditions in the saturnian moons are likely colder than those in the jovian system or on Ceres, independent evidence (below) points to elevated heat flows in the past, raising the question of why there are no ancient pit craters in the Saturn system either. A final resolution of the origins of central pits and the role of composition may have to await a return to the Jupiter system, given the current lack of high-resolution (<100-m-scale) imaging from Ganymede and Callisto.

Several craters on Dione are relatively shallow and feature prominent central peaks rising 3–5 km above the crater rim and the surrounding plains (*White et al.,* 2017). The craters are quite circular and the shapes resemble regular impact craters that have undergone significant viscous relaxation. Other craters also show evidence of relaxation. The nature and implications of these craters is discussed below.

Despite the evidence for higher heat flow and speculation regarding subsurface oceans in one or more of these moons (*Hammond et al.,* 2013; *Tajeddine et al.,* 2014; *Thomas et al.,* 2016; *Beuthe et al.,* 2016), we do not recognize any of the unusual crater landforms seen on icy worlds with confirmed oceans, specifically the multiring basins of Ganymede, Europa, and Callisto (*McKinnon and Melosh,* 1980; *Moore et al.,* 1998; cf. *Schenk et al.,* 2004) and the shallow distorted Manannan and Pwyll craters on Europa (cf. *Schenk and Turtle,* 2009). Although the absence of these types of features does not preclude oceans, it does suggest they were deep enough to not influence crater formation. We return to the issue of internal oceans later.

Three impact features are worth further discussion. Bright and dark rayed craters are a measure of the current flux of impactors. A few dozen such craters have been identified on all six of these moons and display a dearth of craters of small craters on all the satellites from Mimas to Iapetus (*Schenk and Murphy,* 2011). (Whether preferential erasure of smaller rays occurs has not been determined.) None were observed on these moons by Voyager due to a combination of limited mapping coverage and image resolution (and a bit of bad luck on image location), but the confirmed presence of rayed craters by Cassini demonstrates that familiar ejecta processes occur on these lower-gravity worlds.

The largest of the bright ray craters is Inktomi, a 49-km-diameter flat-floored crater on Rhea with a ray system radiating several hundred kilometers from the crater rim (Fig. 5). This was the target of a campaign in 2007 that imaged the crater in three colors at resolutions as good as 32 m/pixel and in stereo, and by VIMS. The mosaics are the best for a larger pristine crater in the Saturn system and reveal a rugged landscape. A few small slump features formed on the inner rimwall, but most of the floor is dominated by rolling hills and scarps related to floor uplift and rim failure (Fig. 5b) that are typical of older craters in the 10–50-km size range on these moons. The ejecta deposit is well preserved and shows a smooth textured surface mantling the rugged preexisting cratered landscape. Most of the ejecta and floor is essentially free of small craters, except for an ovoid region over the central eastern half of the crater, which is anomalously heavily cratered. These have been interpreted as auto-secondary craters that formed on the rim and floor, due to the lack of any other possible source crater (*Schenk et al.,* 2017).

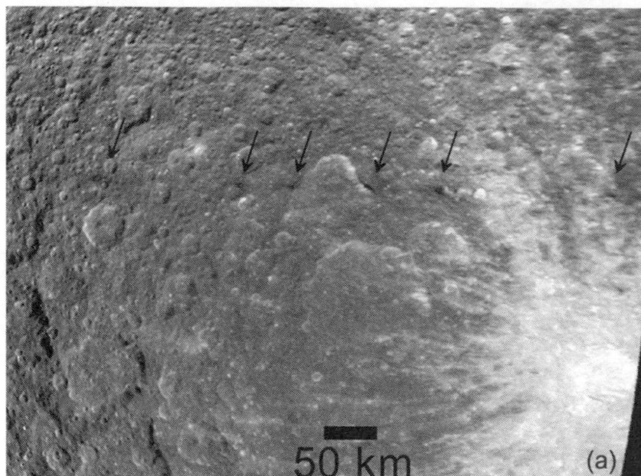

Fig. 5. (a) Cassini three-color mosaic of bright-rayed 50-km-wide crater Inktomi, Rhea (11°S, 112°W) and areas to the west and north. Inktomi is the bright crater at lower right. Bright rays extend to the north (and south) but are less prominent to the west (and east), indicating possible oblique impact trajectory. Note also chain of dark spots across center of images (arrows): the "blue pearls" discussed in the text. Image acquired at ~490-m/pixel scales in three colors (IR3-GRN-UV3). **(b)** High-resolution two-frame mosaic of Inktomi crater and parts of the eastern ejecta deposit on Rhea acquired at ~32-m/pixel scales. Most of the crater floor is composed of hummocky debris related to complex crater collapse, but several very small landslides are visible along the base of the inner rimwall at upper right and a field of numerous small craters covers the central eastern crater floor. North is up in this orthographic map.

Curiously, flat-floor deposits analogous to large "melt sheets" of large craters on the Moon and Mercury are lacking at all crater sizes in the Saturn system. Melt (i.e., water) can drain into a fractured crater floor, but some local ponding must occur naturally, and their absence at 32-m/pixel scales (Fig. 5), coupled with the lack of central pits (if the melted central peak model is correct; see above discussion), suggests that large volumes of melt do not form to begin with on these bodies.

The 425-km-wide Odysseus basin (Fig. 4) was also observed by Cassini on multiple opportunities, the best in three colors and stereo at 250 m/pixel and the western half obliquely at 230 m/pixel, the best for any well-preserved larger basin in the Saturn system. Aside from the rimmed central depression, which resembles a form of central pit, these images also reveal a complex surface on the floor of the basin with numerous scarps and irregular mounds. Odysseus is ~6 km deep relative to surrounding terrains and ~9 km relative to rim (with the deeper central pit adding 4 km to these totals), easily the deepest feature on any of the icy moons. Recent work on crater shapes on Dione and Tethys (*White et al.,* 2013, 2017) suggests that Odysseus may be up to 20% relaxed, despite its deep profile. These fractures are likely evidence of post-impact uplift of the floor of the basin.

A widely distributed network of narrow sinuous fractures crosses the floor of Odysseus (Fig. 6). These fractures are little more than a kilometer across but up to 100 km long. They appear to be relatively recent, crossing all other features. They also have a relatively bluish color, consistent with the color of younger crater rims and other recently formed steep walls on these moons (*Schenk et al.,* 2011). A sinuous ridge ~2 km high forms an arc ~360 km from the eastern rim of Odysseus (Fig. 4). This is the approximate location where one might expect the distal edge of an ejecta deposit and this ridge could be the outer margin of such a deposit. The ridge also coincides approximately but not exactly with the outer limit of a zone of small densely spaced craters that surround Odysseus in this same quadrant.

The 345-km-wide Evander basin (Fig. 7), the largest preserved on Dione, is similar in scale to Odysseus but is in some ways its opposite. Unlike Odysseus, Evander features a prominent central peak ~4 km high, surrounded by a nearly complete but relatively narrow ridge, or peak

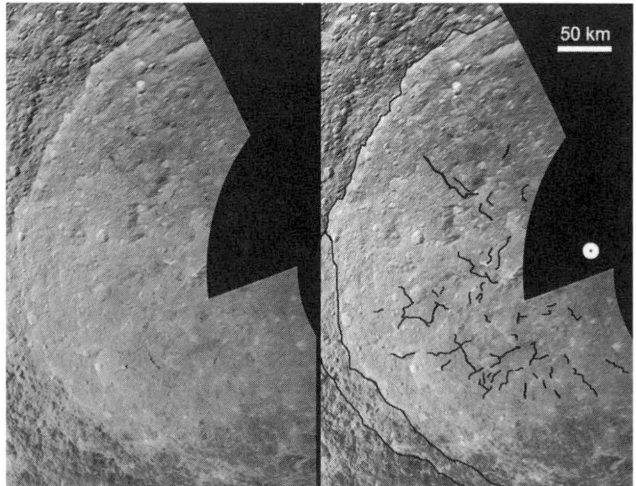

Fig. 6. Oblique high-Sun view of western Odysseus impact basin, Tethys. Annotated version at right shows locations of prominent floor fractures. Dot is the center of the basin. See Fig. 4 for hemispheric view centered on Odysseus.

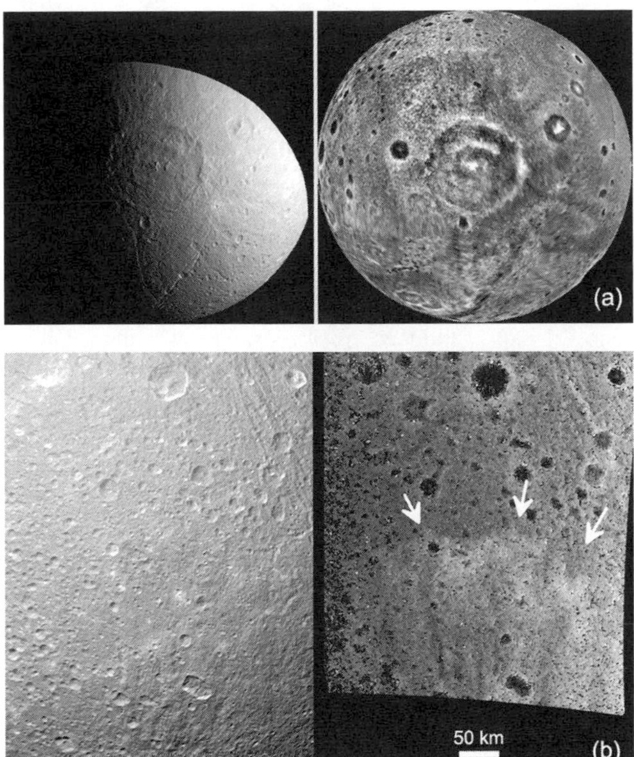

Fig. 7. (a) Orthographic projections of Dione centered on 345-km-wide central-peak crater Evander (62°S, 146°W). View at left is from global 650-m/pixel resolution oblique image; at right from global stereo-derived DEM, showing preservation of prominent structures (rim scarp, inner ring, and central peak), but also extensive relaxation of overall basin depth. Note undulating topography to the north and east of Evander, corresponding to the ejecta deposit of Evander. Compare with relief across Odysseus in Fig. 4. Total relief shown in topography map is ~5 km. **(b)** View of northern ejecta deposit of Evander, Dione. View at left is from 250-m-resolution panchromatic base mosaic; view at right is steregrammetric DEM showing lobate ejecta deposits on the cratered plains north of the rim (arrows, bottom half of view). Arrows indicate partial marginal ridge along outer edge of ejecta deposit. Area centered at ~24°S, 148°W.

ring. Unlike Odysseus, Evander is strongly relaxed (Fig. 7a), with the floor nearly elevated up to the ground level (e.g., *White et al.,* 2013, 2017). Despite this, relief of up to 4 km is preserved within the basin. The peak/peak ring complex is also seen in the large relaxed basin Telamus on Tethys and could be an artifact of the viscous relaxation of the outer edge of the broad central peak producing an artificial inner ring. Alternatively, this unusual central morphology could indicate the presence of an anomalous layer at depth on Dione not present elsewhere in the Saturn system.

Outside the rim of Evander, low topographic lobes radiate out from Evander approximately one crater radius from the rim (Fig. 7a,b). These are likely ejecta deposits, a small portion of which was observed by Voyager (*Plescia,* 1983) and interpreted as something other than ejecta due to Evander being unresolved in the Voyager maps. A ridge-like rise broadly similar to the Odysseus ridge (Fig. 4) is evident in the topography at the outer edge of this deposit (Fig. 7). Although broader and more subtle in form, both ridges bear some resemblance in relief and placement to ridge-like ramparts observed around martian crater ejecta, primarily those referred to as single-layer ejecta craters (e.g., *Barlow,* 2006). We note that such morphologies are not as yet recognized in smaller craters, only these two large younger basins. We do not propose that any liquid is involved in the formation of these terminal ridges, as is sometimes concluded for ejecta in the rock-ice mixtures of Mars, merely that they are an aspect of ballistic emplacement of large amounts of debris in the form of a mobile debris slide radiating from crater center.

4. TECTONICS

Although not displaying the same pervasive level of tectonic deformation as Enceladus (see the chapter by Patterson et al. in this volume), all the other mid-sized saturnian satellites other than perhaps Iapetus show evidence of some tectonic deformation. This activity has overwhelmingly yielded landforms interpreted as extensional structures, with little by way of ridges or other features that imply crustal shortening. The most heavily deformed body of the group is Dione, which displays multiple generations of rifting across much of its surface. Tethys and Rhea are tectonized to lesser extents than Dione, but similarities in how they are deformed lead us to group these three bodies below. Iapetus and Mimas are the least deformed of this group, and so are discussed separately.

4.1. Dione

Aside from Enceladus, Dione is perhaps the most tectonically complex of the saturnian satellites. Voyager observations of Dione indicated the presence of ridges, scarps, troughs, and coalesced pit crater chains (*Smith et al.,* 1981; *Plescia,* 1983; *Moore,* 1984). The enigmatic network of bright, curvilinear "wispy" markings observed on Dione's trailing hemisphere (Fig. 1d) (*Smith et al.,* 1981) had been interpreted as high-albedo, surficial pyroclastic deposits (*Stevenson,* 1982; *Plescia,* 1983), possibly associated with nearby tectonic deformation (*Moore,* 1984). Some combination of volume change (from interior heating and/or cooling), tidal spindown, and orbital recession was invoked as a means by which these structures collectively may have formed (*Moore,* 1984).

Dione's most dramatic tectonic features are a distributed network of linear walled depressions (Fig. 8) interpreted as normal-fault-bound graben and half graben (e.g., *Wagner et al.,* 2006; *Goff-Pochat and Collins,* 2009; *Collins et al.,* 2010; *Tarlow and Collins,* 2010). The generally well preserved walls of these landforms have morphological properties characteristic of normal fault scarps on terrestrial worlds (*Dawers and Anders,* 1995; *Willemse et al.,* 1996; *Willemse,* 1997), including displacements that taper toward their tips, areas of under- and overlap as well as evidence of linkage, variations in strike along their length, and even relay structures (some of which appear breached). Fault dip angles may vary considerably across Dione, although some fault scarps show evidence of shallowing by viscous relaxation (*Beddingfield et al.,* 2015). Most of these structures and hence most of the extensional strain are concentrated in the trailing hemispheres. The most prominent such zones have roughly north-south trends, notably Palatine, Eurotas, and Padua Chasmata. Collectively, these rifts extend ~1300 km, subtend 133° of arc, and are 40–130 km in width.

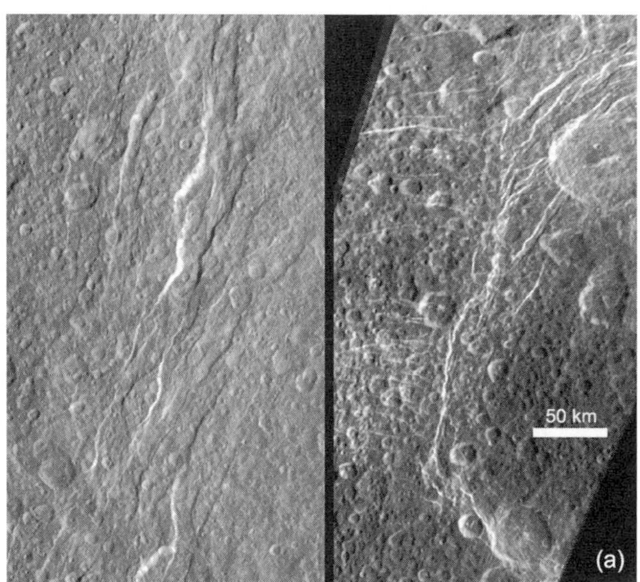

The so-called "wispy terrain," associated directly with these extensional tectonic structures, may reflect the exposure of clean water ice along normal fault scarps (*Porco et al.,* 2005; *Stephan et al.,* 2010; *Beddingfield et al.,* 2015; *Harita,* 2016); the appearance of this terrain could also reflect yet more flanking fractures, such as joints, that remain too small to be seen even with Cassini ISS images, and/or by outgassing and deposition of water or other volatile vapors. Nonetheless, that many of the fine-scale structures in the wispy terrain are visible today, and show little evidence of impact induced erosion or mass wasting (*Harita,* 2016), attests to their relative youth (*Stephan et al.,* 2010; *Kirchoff and Schenk,* 2015). Model ages for the wispy terrain range from ~2.5–1.0 G.y. (*Wagner et al.,* 2006; *Kirchoff and Schenk,* 2015), although the fractures themselves may have slipped as recently as 300 Ma (*Harita,* 2016). Indeed, extensional structures on Dione in general are rarely superposed by impact craters, even though the faults cut through

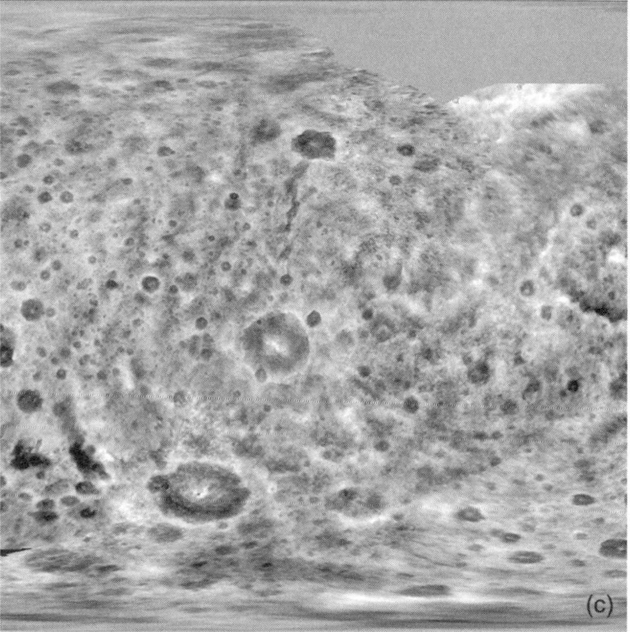

Fig. 8. (a) Cassini three-color mosaics of well-preserved late-stage graben systems on Rhea (left) and Dione (right). Image resolutions are ~200 m/pixel in each case. North is up in these simple cylindrical projections. Rhea scene centered at 25°N, 280°W; Dione scene at 20°N, 240°W. **(b)** Cassini ISS mosaic of Rhea showing cryptic ridge and trough features near 20°S, 260°W, extending from upper left to center. Image resolution is ~315 m/pixel, north is to top. Relief across structures is ~2–3 km. **(c)** Stereogrammetric DEM of entire trailing hemisphere of Rhea, showing irregular mounds and ridges of indeterminate but possible compressional origin across the western half of this hemisphere. Other prominent features include the large craters Powehiwehi (D ~270 km) and Izanagi (D ~240 km) at near center and lower left, the older ridges and troughs shown in **(b)** as curvilinear depressions visible at lower right center, and the deepest of the graben systems shown in **(a)** at upper center.

many hundreds of preexisting such features. Most of Dione's vast assemblage of normal faults is therefore geologically relatively recent — or, at least, the last increment of strain accommodated by these structures is so.

Faint bright lineaments cross the sub- and anti-Saturn points (*Collins et al.,* 2010); these narrow linear features may be joints — fractures with opening displacements only and no associated vertical motion. Older troughs include Drepanum Chasma, a 50-km-wide degraded/relaxed structure on the southern leading hemisphere that records an earlier episode of extension (*Wagner et al.,* 2009). There are fewer extensional structures on Dione's leading hemisphere, although several of those present also strike north-south, such as Fidena Fossae and Tibur Chasmata. There may also be a population of currently unrecognized fractures distributed across Dione (*Beddingfield et al.,* 2015), apparent only by the control they have exerted over polygonal crater rims (*Moore,* 1984).

The strains accommodated by numerous faults on Dione are on the order of a few percent (*Tarlow and Collins,* 2010), corresponding to less than 1% expansion of the moon's surface (*Collins et al.,* 2010). Together, the orientations and strains of these structures are most consistent with having formed from non-synchronous rotation of Dione's icy shell relative to its interior, perhaps coupled with global volume change from interior cooling (*Collins et al.,* 2010).

Enigmatic, subtle parallel lineaments trending northeast-southwest near the center of Dione's leading hemisphere, and with amplitudes of a few hundred meters and characteristic wavelengths of ~10 km, give this region a washboard-like texture (visible in part in Fig. 3). Nonetheless, it is not clear if these lineaments are tectonic in origin and, if so, whether they represent extensional or shortening strains. These are among the few features on the other moons that resemble features on Enceladus, specifically the "washboard" texture on the older cratered terrains (near 0°N, 180°W) of that moon.

There is a notable dearth of positive-relief landforms on Dione that are plausibly shortening tectonic structures; the most obvious such landform is Janiculum Dorsa, a single ridge ~900 km long that trends approximately north-south in the moon's leading hemisphere. Most other positive-relief landforms are resolvably related to impact craters (e.g., *Wagner et al.,* 2006). A linear depression flanking Janiculum Dorsa interpreted as a flexural trough indicates that the elastic thickness of the icy shell under this ridge is about 2 km thick (*Hammond et al.,* 2013), comparable with the estimated elastic thickness under a portion of Palatine Chasmata and with estimates from limb profiles (*Nimmo et al.,* 2011; *Hammond et al.,* 2013).

4.2. Rhea

Observations of Rhea by Voyager indicated the possible presence of narrow, linear troughs and grooves in the body's trailing hemisphere, as well as bright streaks similar to the wispy terrain on Dione (*Smith et al.,* 1981). These linear landforms, interpreted as tectonic in nature, were further classified by *Moore et al.* (1985) into groups of both extensional and shortening structures that included ridges, scarps, "megascarps," troughs, and coalesced pit crater chains. *Moore et al.* (1985) found the Rhean tectonic landforms consistent with a geological history that included at least one phase of global expansion and possibly a period of global contraction. *Thomas and Squyres* (1988) concluded that distortions induced by changes in the moon's rotational period and/or its orbital semimajor axis, coupled with global expansion (and perhaps global contraction), could account for those landforms' distributions and orientations.

Cassini ISS data revealed many of the tectonic lineaments on Rhea to be normal faults and graben (Fig. 8) (*Wagner et al.,* 2007, 2010), many of which reveal bright ice along their scarp faces (*Stephan et al.,* 2012). Most of this deformation is concentrated in the trailing hemisphere, manifest as two major rift zones (Galunlati and Yasmi Chasmata) that trend roughly northeast-southwest and that are up to 3 km deep (*Hammond et al.,* 2011); these rifts are morphologically similar to those on Dione (Fig. 8). Galunlati is the longer of two, extending for ~1500 km and subtending 115° of arc. Both systems vary in width, from ~40 to 90 km, and are characterized by normal fault segments that display minor local variations in strike, relay ramps, and numerous stepover regions tens of kilometers long. Weak flexural deformation associated with these rifts on Rhea indicates that local elastic thicknesses are on the order of at least 10 km (*Hammond et al.,* 2011). Similar to Dione, the scarps are not heavily degraded by subsequent impacts, and so the most recent slip events along these structures may have been geologically recent.

A set of more degraded ridge-and-trough landforms in the center of the trailing hemisphere Rhea (Fig. 8b) may be extensional, but an element of horizontal lithospheric shortening is also possible. These form along a great circle together with the relatively old Puchou Catenae trough observed by Voyager on the other side of Rhea, extending for at least 180° of circumference. The high degree of cratering of much of Rhea's surface challenges unequivocal interpretations of older linear landforms that may be tectonic in origin. For example, there are no shortening structures on Rhea as clearly resolvable as such as Dione's Janiculum Dorsa, and the "megascarps" proposed by *Moore et al.* (1985) and discussed by *Thomas and Squyres* (1988) are difficult to distinguish from basin rims with Cassini data. Irregular positive-relief undulations across Rhea across the trailing hemispheric (Fig. 8c), evident only in the topography, may reflect shortening strains, but are ancient and of uncertain origins. Whatever the origins of these various features, the thermal and tectonic history of Rhea is more complicated than implied by the dense cratering record, and may have included a period in which a regional or global compressive stress state prevailed.

4.3. Tethys

Like Dione, the tectonics of Tethys are primarily extensional in nature. The dominant feature is Ithaca Chasma, a giant rift zone first observed by Voyager (*Smith et al.,* 1981, 1982), 1800 km in length and subtending at least 270° of arc, between 70 km and 110 km in width, and 2–5 km deep (*Moore and Ahern,* 1983; *Moore et al.,* 2004). Ithaca Chasma hosts many individual closely spaced fault segments along its course that together form a complex, branching map pattern (Fig. 9). Topographic profiles across the central section indicate a local elastic thickness of 5–7 km (*Giese et al.,* 2007).

Curiously, Ithaca Chasma lies along a warped great circle ~20° from the axis centered on the Odysseus basin. This spatial relationship motivated *Moore and Ahern* (1983) to propose that the formation of this rift zone was causally linked to the impact event that produced Odysseus, and specifically that a dampened, whole-body oscillation of Tethys resulting from that impact led to tensional stresses that exceeded the yield strength of the satellite's icy lithosphere (*Moore et al.,* 2004). This hypothesis is discussed further below.

Numerous additional linear structures were noted on Tethys (Fig. 9b,c), including those subparallel to (but not part of) Ithaca Chasma (*Moore and Ahern,* 1983). Other landforms, such as narrow fractures and pit chains, are also broadly distributed across Tethys. Their origins are not well understood, however, and further analysis of these landforms, and the role they played in the tectonic and thermal evolution of the moon, is warranted.

The most enigmatic linear landforms on Tethys (and perhaps the entire Saturn system) are the "red streaks" that form a symmetric pattern centered on the Saturn-centered tidal axis (*Schenk et al.,* 2015). These lineaments take the

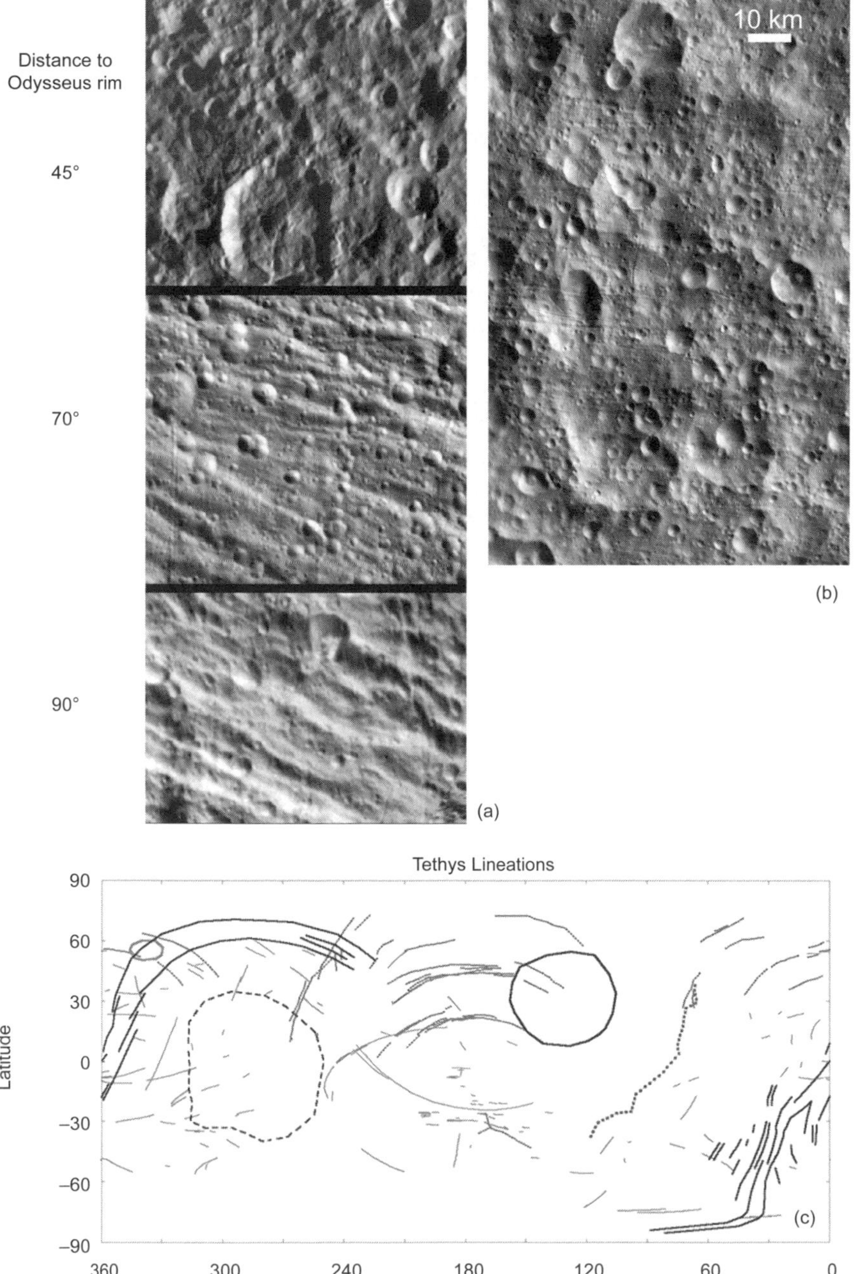

Fig. 9. (a) Three views of Ithaca Chasma, Tethys, at comparable resolutions (~250 m) showing changes in morphology and crater density along its track. Top view is northern end of the structure closest to Odysseus, center view near equator, and bottom view is along southern section furthest from Odysseus. Largest crater in top view is 42 km across. **(b)** Cassini 100-m-resolution mosaic of equatorial region of Tethys (centered at 1180°W) showing dense network of east-west lineations. Additional randomly oriented lienations are visible as well. Scene ~90 km across and centered at 6°S, 181°W. **(c)** Global cylindrical map of Tethys showing locations of known linear features, including troughs and pit chains (solid lines) and red streaks (dashed curved lines), as well as several of the largest craters (circles). Ithaca Chasma is the prominent continuous trough at lower right and upper left.

form of faintly darkened, arcuate markings with an enhanced near-infrared (NIR) color signature, an unusual color for features on Saturn's icy satellites, and are most apparent under high-solar-incidence and low-emission viewing angles. These features show no surface deformation even with image resolutions of 60 m/pixel, although there are indications of grain-size variations or the presence of organic compounds in VIMS spectra. Without resolvable tectonic deformation, differences in grain size or color must be related to localized precipitation or molecular redeposition processes. Nonetheless, these streaks are geologically young and may reflect incipient deformation of Tethys' icy shell (*Schenk et al.,* 2015; Schenk et al., in preparation).

4.4. Similarities in Faulting Across Dione, Tethys, and Rhea

Despite differences in size, density, internal structure, impact cratering history, and dynamical environment, fundamental similarities exist in the style and distribution of large-scale extensional deformation on these three icy moons. Rifting is most concentrated within or near each body's trailing hemisphere, and shows a preference for ~north-south orientations and a systematic increase in complexity from Rhea, to Tethys, to Dione.

Sections of each moon's chasma or chasmata appear to lie along substantial portions of great circles (*Moore et al.,* 2004). Furthermore, tidally induced stresses alone tend to result in degree-two patterns in strain, in contrast to the hemispherical dichotomy in the localization of strain observed on these three icy satellites (and on Rhea and Dione in particular). For example, hemispherical differences in ice shell thickness, arising from spatially heterogeneous uneven heat flow from the interior, may account for this "degree-one" dichotomy (*Byrne et al.,* 2016). Thus, these large-scale rifts may have formed under a scenario in which tidal processes contributed at most a secondary component of stress, with Rhea representing the least, and Dione the most, advanced stage of deformation. Of note, the uranian moon Titania hosts two north-south-trending chasmata: Messina and Belmont Chasma are 20–50 km across and are 2–5 km deep (*Smith et al.,* 1986), and at least parts of both are situated near the sub-Uranus point.

4.5. Mimas

Both Voyager and Cassini observed widely distributed linear grooves across Mimas that are putatively of tectonic origin (*Smith et al.,* 1981). The most conspicuous grooves are tens of kilometers long, a few to as many as 10 km across, and ~2–3 km deep (Fig. 10). *Smith et al.* (1981) suggested that at least some of these grooves may be causally linked to the formation of the then-unnamed Herschel impact basin, 140 km in diameter. Some sections are arrayed in an en echelon manner, and some even show evidence for fault linkage via the presence of relay ramps. Alternatively, stresses arising from changes in Mimas' orbital or rotational periods may have played a role in the formation of these structures (e.g., *Moore et al.,* 2004). No single stress regime unequivocally matches the deformation patterns as yet, however, a point we return to in the discussion in section 10.

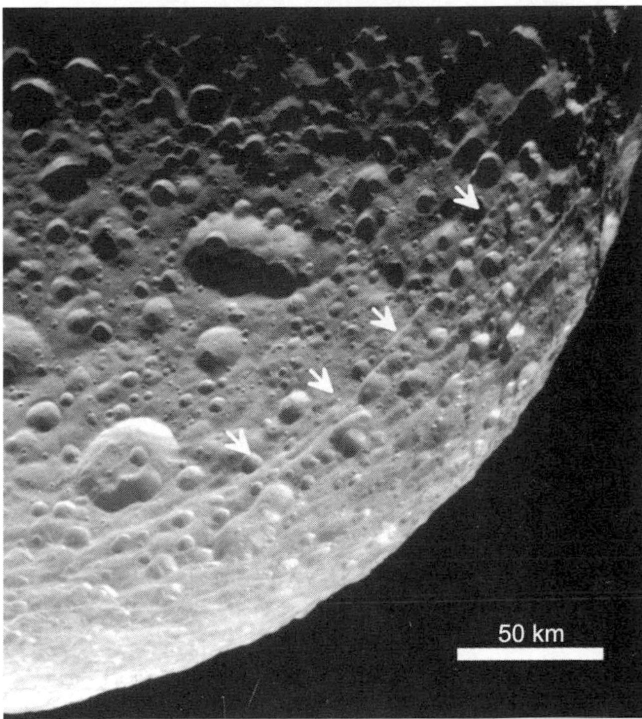

Fig. 10. Cassini image showing troughs on Mimas (arrows), trending diagonally from lower left to upper right. North polar terrains are to the upper left. Image resolution is 300 m/pixel, and is centered near 45°N, 45°W.

4.6. Iapetus

Cassini revealed few unequivocal tectonic landforms on Iapetus, and certainly far fewer than those documented on Dione, Tethys, and Rhea. *Singer and McKinnon* (2011) surveyed Iapetus for likely tectonic landforms, and concluded that a tidal spindown pattern such as that predicted by *Pechmann and Melosh* (1979) was not present (or at least not preserved) on the moon. These authors classified mapped linear landforms into four groups [large troughs (Fig. 11), linear crater rim wall segments, non-crater-related lineaments, and linear central peaks (*Singer and McKinnon,* 2011)]. Although the linear troughs appear to be secondary crater chains related to large basins, it is not yet clear what process(es) are responsible for all these structures.

Iapetus does boast an enigmatic ridge situated almost exactly at its equator (*Porco et al.,* 2005). In places, this ridge is up to 20 km high and 70 km across, and spans almost 75% of the moon's circumference (*Porco et al.,* 2005; *Giese et al.,* 2008; *Singer and McKinnon,* 2011; *Dombard et al.,* 2012; *Lopez Garcia et al.,* 2014; *Kuchta et al.,* 2015), but it is topographically discontinuous. Defying any obvious

Fig. 11. Cassini mosaic of cratered plains centered on the equator of Iapetus at longitude 310°. The dark linear features are crater chains that are floored by dark material otherwise observed on most of the leading hemisphere. These linear features are traceable to large impact basins and are likely secondary crater chains. The dark floored crater near center is ~90 km across, and the rugged terrains at bottom are ejecta deposits from nearby 500-km central-peak Engelier basin. North is to the top; black bar is 50 km.

explanation, numerous endogenic and exogenic formation mechanisms have been proposed for the ridge. Endogenic processes include tidal spindown (*Porco et al.,* 2005; *Castillo-Rogez et al.,* 2007; *Sandwell and Schubert,* 2010), convection (*Czechowski and Leliwa-Kopystyński,* 2008), despinning coupled with global volume change (*Beuthe,* 2010), global contraction (*Sandwell and Schubert,* 2010), or even intrusion (*Melosh and Nimmo,* 2009), although none of these mechanisms explains fully the location and morphology of the ridge.

An exogenic origin for the ridge has been attributed to infall of a former iapetian ring remnant (*Ip,* 2006; *Levison et al.,* 2011) formed either by a destroyed subsatellite (*Dombard et al.,* 2012), or a giant impact coincident with despinning (*Kuchta et al.,* 2015). The ring infall hypothesis was challenged by *Giese et al.* (2008), although formation of the ridge by some infall scenario was found plausible by *Lopez Garcia et al.* (2014). Considering only endogenous ridge-formation mechanisms, *Singer and McKinnon* (2011) did not find a tectonic (or volcanic) origin for this curious landform likely. The ring hypothesis, and unexpected support from blue patches on Rhea (*Schenk et al.,* 2011), is discussed more fully in section 8.

5. VOLCANISM

A key question dating back to Voyager observations in 1980–1981 is whether volcanism has ever occurred on any of the icy moons of Saturn. By "volcanism" we refer to that process which is also known as cryovolcanism. Cryovolcanism can be defined as "eruptions... that consist of liquid or vapor phases of materials that freeze at temperatures of icy satellite surfaces" (*Prockter,* 2004). The term "cryovolcanism" can also include the possibility of solid-state (diapiric) emplacement of materials from deeper inside a body. Evidence for volcanism would be direct evidence of heat sufficient to melt or otherwise vertically and/or horizontally mobilize water ice or other ice mixtures. Studies of crater density variations on Rhea hinting at possible ancient resurfacing (*Lissauer et al.,* 1988) have not yet been tested with the new Cassini data. Smooth plains were observed on Dione and Tethys (*Smith et al.,* 1981, 1982; *Moore and Ahern,* 1983; *Moore,* 1984; *Plescia,* 1983), and Cassini mapping confirms that these are the only moons on which such plains are observed (e.g., *Jaumann et al.,* 2009). No similar candidate volcanic units have been observed on Mimas, Rhea, or Iapetus.

The smooth plains unit on Tethys is ~600 km across and is centered close to the center of the trailing hemisphere. Although heavily cratered, the unit is characterized by a lack of the dense large craters that populate other regions (Fig. 12) and mostly features craters <50 km across. This is evident in the global DEM, which shows a profound difference in the ruggedness of the two terrains (Fig. 12). Whether or not this unit forms an embayment contact along its margin with heavily cratered terrains (as does the lunar mare) is obscured by the heavy cratering on both terrains.

The origin of this terrain on Tethys is cryptic. *Moore et al.* (2004) considered the possibility it could be an antipodal terrain (e.g., *Bruesch and Asphaug,* 2004) formed by seismic shaking induced by Odysseus. This terrain is centered near the antipode of the Odysseus impact but is offset ~40° from it, rendering a connection unclear. A severe amount of shaking would be necessary to remove the ancient deeply cratered topography seen elsewhere on Tethys. Moreover, seismic shaking should progressively diminish radially away from the Odysseus antipode, a hypothesis not supported by the topography of the plains unit. *Smith et al.* (1981) and *Moore and Ahern* (1983) suggested these units could be volcanic, formed by a low-viscosity fluid that did not form major edifices. Any evidence for flow fronts or source vents has been either removed by dense cratering or was not resolved.

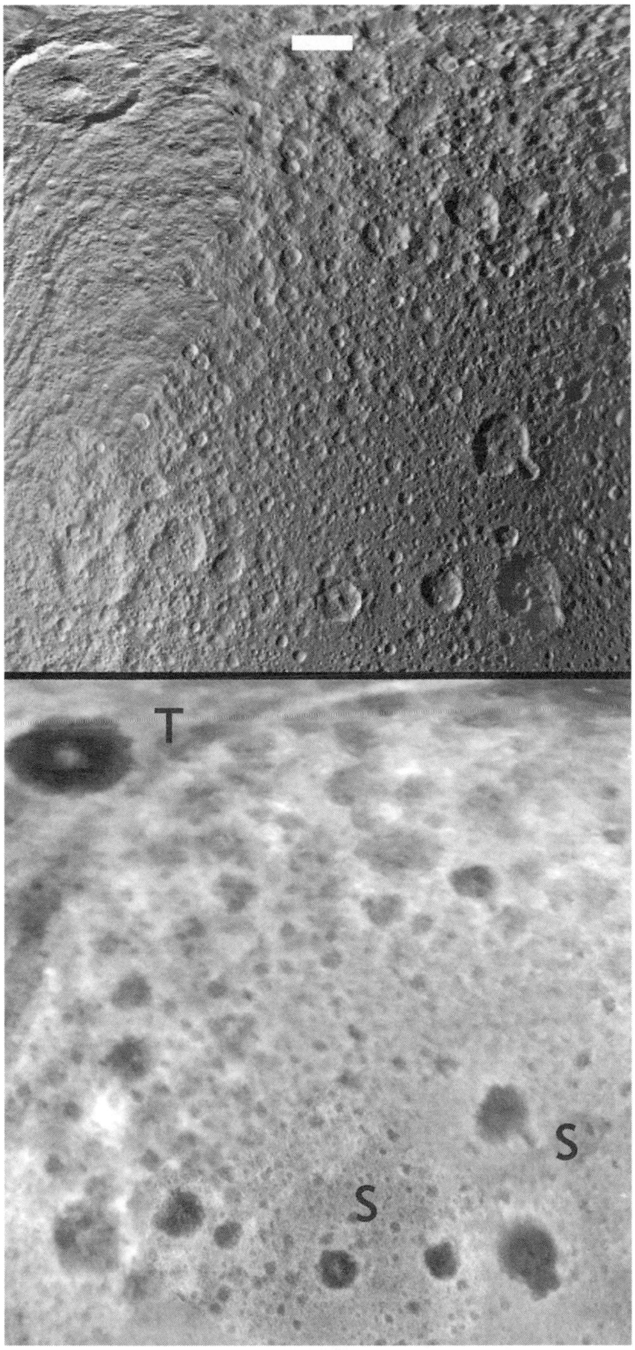

The case for volcanism on Dione is more interesting. Cassini mapping revealed that the smooth plains observed by Voyager extend across most of the leading hemisphere of Dione (Figs. 1 and 13 and Plate 31d) and are thus a key part of that moon's geologic story. The plains are mostly low-lying and only moderately cratered, although the southern plains are partly obscured by Evander crater and its ejecta (Fig. 7). A number of structures cross these plains, including extensional landforms such as troughs as well as the possible shortening ridge Janinulum Dorsa (*Hammond et al.,* 2013).

At the center of the plains is a pair of oblong depressions (Fig. 13), Murranus and Metiscus. These scarp-enclosed depressions are 45 to 70 km across, but only ~1 km deep,

Fig. 12. Portion of "smooth plains" on Tethys (lower right quadrant of scene). Top is portion of global cylindrical map centered on trailing hemisphere at 0°N, 270°W, and bottom is topographic map of same region showing elevations (dark is low). Change from deeply cratered highland at upper left to smoother cratered plains at lower right is evident. Ithaca Chasma trough system and 90-km Telemachus crater are prominent at upper left. Total topographic range shown is 5 km. Simple cylindrical map projection centered near 30°N, 310°W; scale bar is ~100 km.

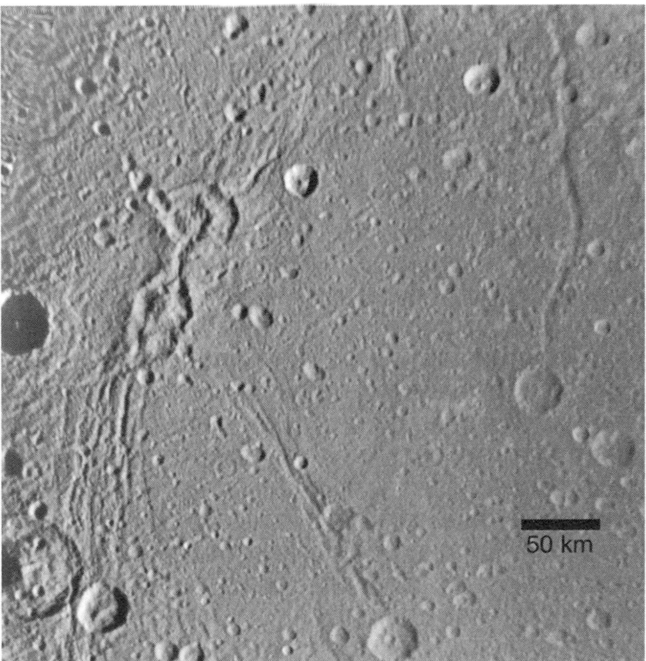

Fig. 13. Center of leading hemisphere of Dione showing central region of smooth plains at the center of the leading hemisphere. The two shallow oblong craters are Murranus (45 km wide, top center) and Metiscus (40 × 70 km wide, bottom center) are distinct and interpreted to be volcanic in origin. Sagaris crater and its pancake ejecta are on the left edge. View is 300 km across. Note fractures radiating northward and southward from these walled depressions. Topographic map of center left area shown in Fig. 3.

shallow by saturnian impact crater standards (Fig. 3). The craters are also joined by a deep groove, presumably part of the tectonic pattern that also radiates from the southern and northern ends of these two depressions (i.e., Fidena and Petelia Fossae). Several irregular mounds occupy the centers of these structures; the northern feature is also a nested pair of craters. The irregular shapes of these features and the central mounds do not resemble circular relaxed impact craters elsewhere on Dione (*White et al.,* 2017) (see section 6). *Moore and Schenk* (2007) suggested that Murranus and Metiscus might be volcanic structures, based on their unusual shapes (Fig. 13) and locations in the centers of both the smooth plains and several of the sets of fractures within the plains. These putative volcanic features are also reminiscent of the irregular shapes of walled depressions on Ganymede such as Rum Patera (*Schenk et al.,* 2001; *Schenk,* 2010; *Spaun et al.,* 2001), which are not impact craters and have been interpreted as volcanic. If volcanic, these two irregular depressions are the only evidence for explosive or collapse-forming volcanism in the Saturn system.

The age(s) of these putative volcanic features on Dione is indeterminate. Murranus and Metiscus are partly obscured by extensive fields of small superposed craters (Fig. 3), most of which are secondaries from the nearby 53-km-diameter fresh crater Sagaris, located only 80 km to the west. Global crater counts suggest the age of the smooth plains is likely several gigayears (*Kirchoff and Schenk,* 2015; see the chapter by Kirchoff et al. in this volume). If the caldera-like features described above formed late in the smooth plains formation process, as seems likely, then they are likely of similar age and are now quiescent. Searches for ongoing activity at Dione, however, have thus far proved futile.

6. INFERRING THERMAL HISTORIES FROM LANDFORM MODIFICATION

Relaxation of topography has been observed on many of the saturnian mid-sized icy satellites by the Voyager and Cassini spacecraft, and attributed to viscous relaxation of the icy crusts of these bodies (e.g., *Passey,* 1983; *Schenk,* 1989; *Moore et al.,* 2004). The scale of viscous relaxation is a function of the ductile rheological properties of the satellite (which are themselves sensitive to the crustal heat flux), as well as the timescale over which relaxation has occurred. Examination of relaxed impact craters (which are abundant and have predictable initial shapes) can therefore be exploited to probe the thermal histories of these icy satellites (e.g., *Dombard and McKinnon,* 2006). We refer the reader to the chapter in this volume by Castillo-Rogez et al. for further discussion of the thermal evolution of Saturn's mid-sized icy satellites.

The apparent depth relaxation fraction (RF) of a crater is defined as $1–d_a(t)/d_a(0)$ (*Dombard and McKinnon,* 2006), where $d_a(0)$ is the depth of the initial, unrelaxed crater (from the ground plane) and $d_a(t)$ is the depth of the relaxed crater after time, t, has elapsed. Figure 14 shows d/D plots for Mimas, Tethys, Dione, Rhea, and Iapetus, measured using stereo DEMs derived from Cassini imaging (*White et al.,* 2013, 2017). Best-fit lines can be applied to complex craters/basins across all diameters for both the Iapetus and Mimas plots, but the Rhea complex crater trend transitions to a shallower slope above ~100 km diameter. These plots indicate that relaxation has affected craters on Mimas or Iapetus to only a negligible degree at most, while relaxation has appreciably affected impact basins exceeding ~100 km diameter on Rhea. The Dione and Tethys plots exhibit a much greater range of depths for complex craters and impact basins than those of the other three satellites, indicating that relaxation on these two satellites has affected craters down to lower diameters than is seen on the other three. An independent study of crater relaxation on Rhea and Dione (*Phillips et al.,* 2012, 2013) also found that large craters on these satellites with diameters ranging from ~150 to 500 km have generally high degrees of relaxation, with large craters on Dione being more relaxed than large craters on Rhea. *White et al.* (2017) noted that, whereas no significant geographical control on relaxation could be discerned across the surfaces of Rhea and Tethys, relaxed craters on Dione tended to concentrate in a zone encompassing much of the southern hemisphere, as well as equatorial regions centered on the leading and

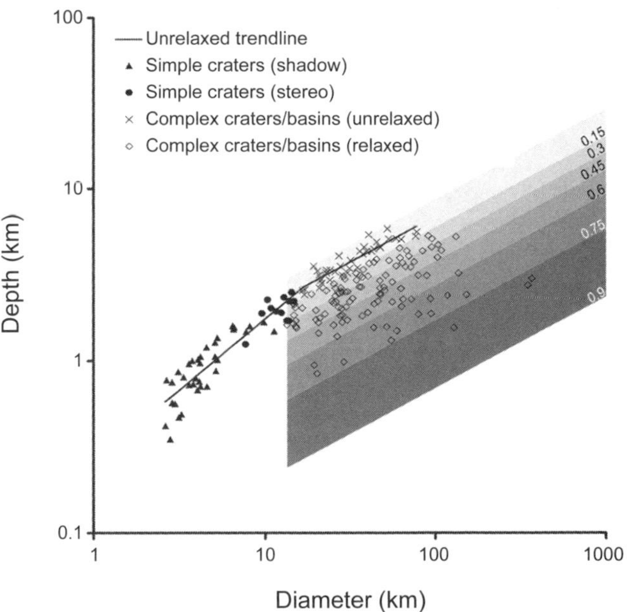

Fig. 14. Depth/diameter plot for craters on Dione, reproduced from *White et al.* (2017). On saturnian satellites, simple craters are bowl- or cone-shaped, complex craters display flat floors and (sometimes) central peaks, and impact basins are essentially an extension of complex crater morphology, but can also display a central ring complex (*Pike,* 1974, 1980; *Williams and Zuber,* 1998). For complex craters and impact basins, unrelaxed and relaxed bins are represented by the shaded zones oriented parallel to the least-squares slope of unrelaxed craters. Unrelaxed craters fill the topmost bin, and the relaxation fraction increases for each bin are as indicated.

trailing hemispheres. This area correlates well to a region of low crater spatial density as mapped by *Kirchoff and Schenk* (2015), and also resembles the configuration of the dichotomy between the heavily cratered northern terrain and the southern, less cratered, heavily tectonized terrain on Enceladus (*Crow-Willard and Pappalardo,* 2015).

White et al. (2013, 2017) used measurements of topographic profiles of relaxed and unrelaxed craters on Rhea, Dione, and Tethys to estimate historical heat flux across their surfaces through viscoelastic relaxation simulations and crater age dating of craters for which simulations were performed. The simulations input initial topographic conditions of the relaxed craters, which are represented by unrelaxed craters of comparable diameter that have been size-scaled (and gravity-scaled, if the unrelaxed crater exists on a different satellite to the relaxed crater). A range of thermal profiles was entered to find which combination of heat flux, relaxation timescale, and ice grain size produces a final, relaxed crater profile that is most consistent with observation. Figure 15 presents simulation results for two craters, 350-km-diameter Evander on Dione (Fig. 7) and 397-km-diameter Tirawa on Rhea, which display RFs of 0.80 and 0.58, respectively.

On Rhea, where recognizable relaxation is mostly confined to craters 100 km in diameter or greater, heat fluxes reaching 30 mW m^{-2} are necessary to achieve the observed relaxation. Relaxation is seen to affect smaller craters on Dione and Tethys than on Rhea, and higher heat fluxes, which frequently measure in the vicinity of 50 to 60 mW m^{-2}, are necessary to relax these craters (*White et al.,* 2017), due to relaxation being less effective for shorter-wavelength topography. Unrelaxed craters have not experienced heat fluxes exceeding radiogenic levels [estimated at a few milliwatts per square meter at ~4.5 Ga (*Schubert et al.,* 1986; *Turcotte and Schubert,* 2002)] since their formation. In all cases, the relaxation process is mostly complete after several million years, with relatively little relaxation occurring thereafter, meaning that only a very short period of elevated heat flux is necessary to facilitate the required relaxation. The high heat fluxes derived by these simulations that are necessary to achieve the very relaxed states of some craters on these satellites have strong implications for the potential past existence of subsurface oceans. The thermo-viscoelastic model of *Phillips et al.* (2012, 2013) underpredicted the observed relaxation on Rhea and Dione, causing them to conclude that the interior temperature necessary to achieve the observed relaxation must have caused melting of the water ice at depth. Alternatively, convection in the ice could drive a melt interface deeper, possibly negating the need for localized oceans forming underneath these craters (*Dombard et al.,* 2007).

This method of deducing historical heat flux based on crater relaxation indicates that satellites such as Mimas and Iapetus have not experienced any periods of heat flux elevated above such levels since formation of their craters. This interpretation is consistent with the low rock/ice mass fractions that have been inferred for these satellites (*Schubert et al.,* 2007; *Castillo-Rogez et al.,* 2007), as well as their dynamic histories, which are not regarded as having ever been affected by tidal resonances that could contribute to interior heating (*Castillo-Rogez et al.,* 2007, 2011).

The relaxed state of large craters on Rhea, which date from Rhea's immediate post-accretion era (*White et al.,* 2013), indicate that global elevated heat fluxes stemming from Rhea's accretion and radioisotopes prevailed in its early

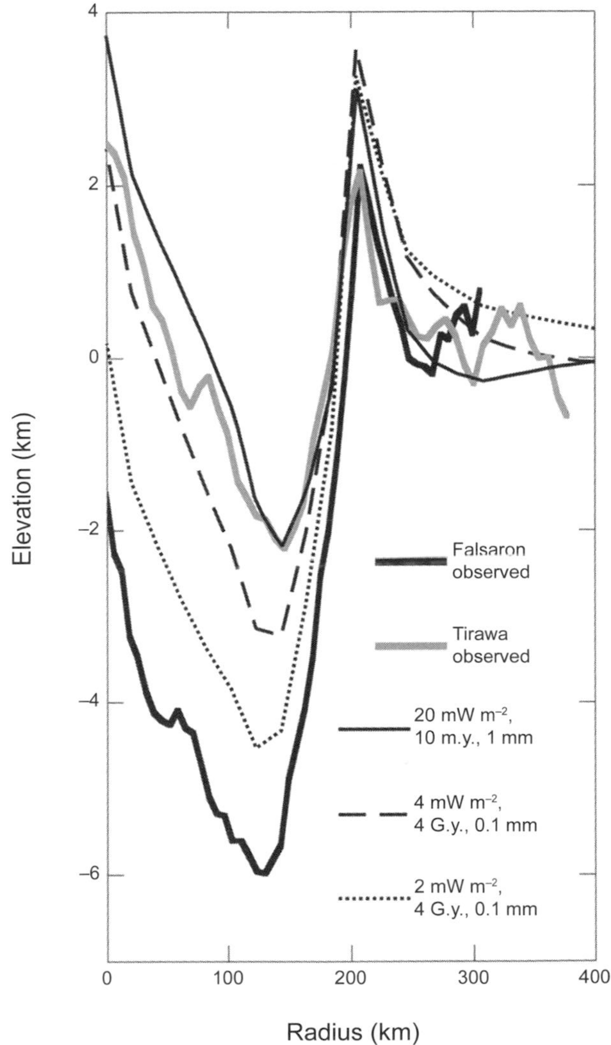

Fig. 15. Radial topographic profiles resulting from viscoelastic simulations of relaxation of Tirawa impact basin on Rhea, assuming the non-Newtonian rheology of ice (*White et al.,* 2013). The unrelaxed, 422-km-diameter crater Falsaron, on Iapetus, has been size- and gravity-scaled in order to represent the initial topography of Tirawa. Simulations are performed for a variety of heat fluxes, relaxation timescales, and ice grain sizes. The simulations indicate that a heat flux of 20 mW m^{-2} is necessary to relax Tirawa to its present morphology, and that such relaxation can occur on a timescale of only several million years and for a large ice grain size. In contrast, relaxation under radiogenic heat flux conditions of a few milliwatts per meters squared is insufficient to reproduce the present morphology even for a relaxation timescale of billions of years and a small ice grain size.

history, but did not persist long enough to cause relaxation of smaller craters that formed later, whereas Dione displays many relaxed craters down to small diameters (~20 km). Dione relaxed craters also exhibit a range of ages based on crater counting (*White et al.,* 2017), and therefore has experienced protracted heat fluxes in excess of those on Tethys, Rhea, or Mimas. In addition, relaxation on Dione has a strong geographical dependency, forming a double-sided concentration on the leading and trailing hemisphere through the south polar region and thus the resemblance of the baseball-stitching-shaped geologic dichotomy between heavily cratered, less relaxed terrains, and lightly cratered, more relaxed terrains on Enceladus (see the chapter in this volume by Patterson et al.) and on Miranda. This crude similarity suggests that similar heating mechanisms may have affected these two satellites. However, heating for Enceladus has persisted to the present day and powered large-scale tectonic deformation of the crust, whereas that for Dione has not persisted, and has only been sufficient to relax, rather than erase, impact craters. The evidence for resurfacing via crustal heating on both of these satellites can be explained by their high rock/ice mass ratios (*Schubert et al.,* 2007; *Thomas et al.,* 2007) and the fact that they are currently trapped in a 2:1 mean-motion orbital resonance with each other, the tidal effects of which are currently instrumental in heating Enceladus, and which may have affected Dione to a greater degree in the past than they are at present (*Ojakangas and Stevenson,* 1986; *Shoji et al.,* 2013, 2014).

The presence of heavily relaxed craters on Tethys is more difficult to reconcile with potential heat sources. Its small mass and very small rock/ice mass fraction (*Thomas et al.,* 2007) means that heat sources such as radiogenic or accretional heating are less likely to provide the necessary heat flux. Tethys currently has a negligible orbital eccentricity of 1×10^{-4} and is locked in a 1:2 orbital resonance with Mimas. Although this resonance does not excite eccentricity, it does excite inclination, which in principle can lead to obliquity tidal heating (*Chen et al.,* 2014). There is also the possibility that Tethys has passed through a 3:2 resonance with Dione, which would have resulted in an increased eccentricity and elevated heating (*Chen and Nimmo,* 2008). Assessing the relaxation state of Tethys is complicated by the comparatively recent Odysseus impact (*Kirchoff and Schenk,* 2010), the crater and ejecta of which have likely erased or obscured most of the large, preexisting craters on the leading hemisphere (Fig. 4).

In addition to crater relaxation, measurements of the morphologies of fractures and ridges on icy satellites have been used to assess lithospheric thickness and heat flux based on considerations of flexural support. Such topographic features also impose loads on the lithosphere, causing the surface to bend in response, with the magnitude and wavelength of bending controlled by the effective elastic thickness (e.g., *Hammond et al.,* 2013). Thus, gauging the shape of the flexed topography at these features allows modeling of the effective elastic thickness of the lithosphere and, together with an assumed lithospheric strength envelope, provides estimates of its mechanical thickness and the heat flux (*Giese et al.,* 2007). Other than Enceladus, this technique is most applicable to Dione and Tethys in the saturnian system, as they display the most well-developed, global-scale rift zones.

Giese et al. (2007) used Cassini stereo-derived topography of Ithaca Chasma on Tethys to identify apparent flexural deformation along the concave flanks of the chasma, which are upraised by up to 6 km above the surrounding terrain in some areas. By fitting a broken elastic plate model to profiles of the eastern flank of Ithaca Chasma, *Giese et al.* (2007) estimated the effective elastic thickness to have been between 7.2 km and 4.9 km, which translates to heat fluxes in the range of 18.1 to 30 mW m^{-2} for adopted geological strain rates of 10^{-17}–10^{-14} s^{-1}. *Hammond et al.* (2013) estimated a local lithospheric thickness from stereo-generated topography of the Janiculum Dorsa ridge on Dione of 3.5 ± 1 km [falling within the 1.5–5 km range determined by *Nimmo et al.* (2011) based on long-wavelength topography], with a corresponding heat flux range of 25–60 mW m^{-2} for strain rates between 10^{-18} and 10^{-16} s^{-1}. As with the crater relaxation modeling of *White et al.* (2013, 2017) and *Phillips et al.* (2012, 2013), the results of these investigations indicate that portions of the crusts of these satellites have been subject to heat fluxes in excess of radiogenic levels since the formation of Ithaca Chasma and Janiculum Dorsa. Based on the heat flux values obtained by the modeling of *Giese et al.* (2007), *Chen and Nimmo* (2008) inferred that eccentricity-induced tidal heating, brought on by a resonance passage with Dione, is likely to have affected Tethys at some point in its history. Whether this thermal record is coincident with that recorded by the impact craters is unclear.

7. GLOBAL TOPOGRAPHY AND HEAT

Global topographic maps are now essentially complete for the five moons discussed here (Fig. 16), as well as for 35–40% of the uranian moons Titania, Miranda, and Ariel (e.g., *Peterson et al.,* 2015). With the exception of Ariel, topography of these moons as a group is dominated by widespread impact cratering and localized (if extensive) tectonic structures described above. In a comparative sense, the overwhelming finding is that the global topographic character of the four inner moons (relative to their mean triaxial shapes) contrasts sharply with that of Iapetus (Fig. 16). The topography of Iapetus features large local and regional deviations from the triaxial shape dominated by impact basins of various sizes. The 2σ histogram of topography on Iapetus ranges from +18 to –18 km (Fig. 17). The topographic range for each of the four inner moons (leaving aside Enceladus for the moment) is much lower, only +5 to –5 km, including all basins and depressions. Among mid-sized icy bodies mapped topographically to date (Figs. 16 and 17), including Charon (*Schenk et al.,* 2018b), the four inner saturnian satellites have the lower topographic amplitudes known, despite their rugged cratered surfaces.

Despite (or more likely because of) the active resurfacing, the global topographic amplitude of Enceladus is the

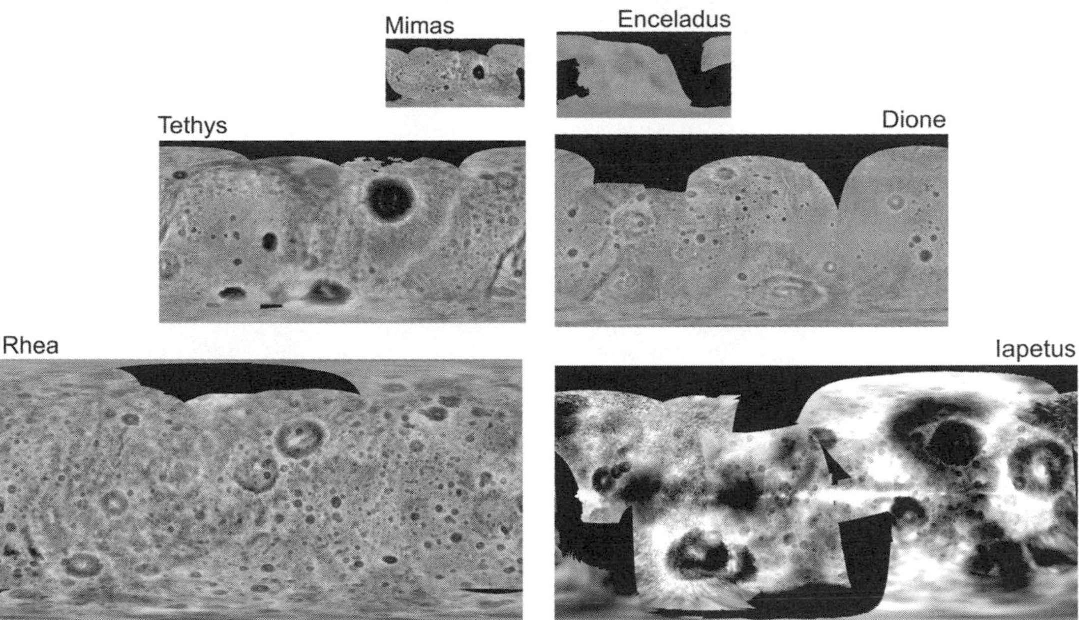

Fig. 16. Global cylindrical maps of the topography of the six classical icy saturnian satellites. Maps are scaled in brightness to show topography of ±5 km relative to global spheroid, highlighting how low in amplitude topography is on the five inner moons compared to Iapetus. The large deep depression on Tethys is Odysseus and the large ringed basin on Dione is Evander. Maps are shown at identical physical scale. All data produced by the lead author.

flattest of any of the mid-sized bodies. We presume Europa will prove to be similarly smooth, but global topographic datasets do not yet exist for the Galilean satellites. Of the remaining mid-sized saturnian or uranian satellites, Dione is the next flattest (Figs. 16 and 17), consistent with it having the greatest geologic complexity of the other saturnian moons and the best evidence for resurfacing (i.e., the smooth plains of the leading hemisphere; Figs. 1 and 13). Tethys and Rhea are the more rugged of the Saturn set, but still less rugged than Charon (*Schenk et al.,* 2018b) and the three

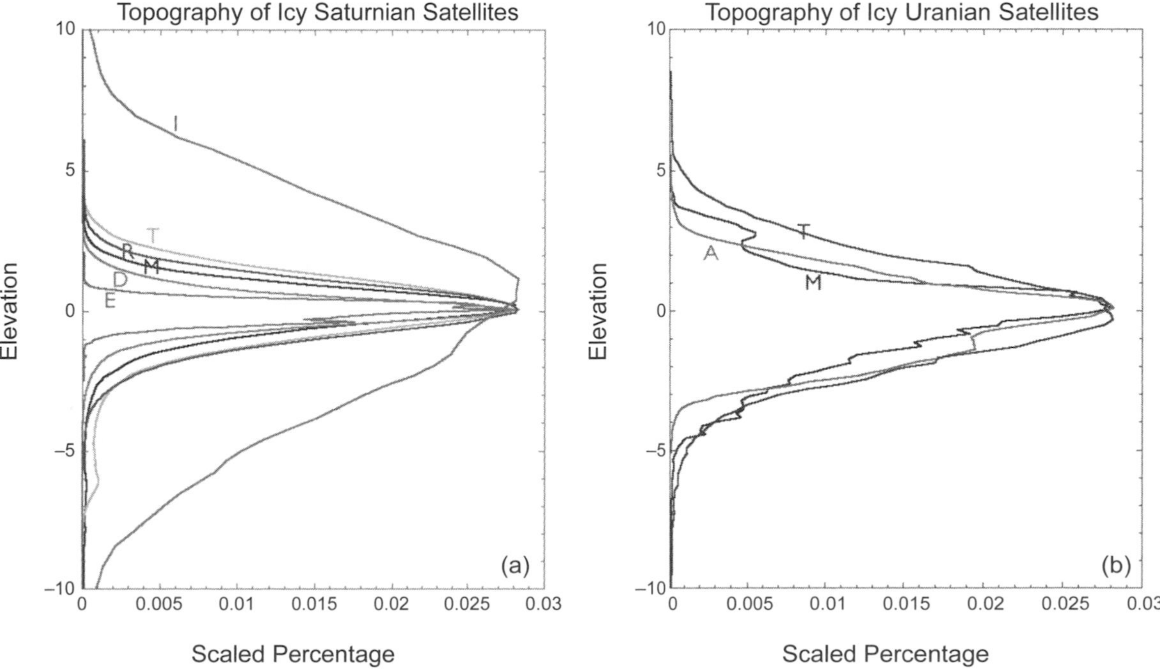

Fig. 17. Hypsogram, in the form of a cumulative histogram, of elevations for icy bodies measured to date, showing **(a)** saturnian and **(b)** uranian moons separately for clarity. Letters refer to first letter in each moon's name. Note that data for uranian satellites are for only ~40% of the surfaces in each case, and subject to improvement from future data sources.

uranian moons for which we have extensive data.

The low topographic range for the four inner icy moons, despite the heavy cratering record, is a surprise. If Iapetus originated within the Saturn system (see the chapter by Castillo-Rogez et al. in this volume) and was not captured, the inner moons should not have escaped the projectile flux that is recorded on Iapetus, and the record of large impacts and deep topographic variations on Iapetus should have in principle occurred on the inner moons as well. Gravitational focusing (*Zahnle et al.,* 2001) should lead to even larger and more numerous basins on the inner satellites. A total of 10 basins larger than 300 km have been identified in both the topographic and base mosaics of Iapetus (despite the lack of resolved imaging in some locations), but only four on Rhea, which has approximately the same surface area (Dione, Tethys, and Mimas are similarly depleted when scaled for orbit and size). The inference from these observations is that a major event, presumably global heating, removed the record of large projectile bombardment observed on Iapetus. Thus the inner moons either escaped this record of heavy bombardment (by forming later perhaps) or, more likely, that major heating episodes severely heated each moon, erasing the topographic record of large early impacts observed on Iapetus. If so, these moons would have had to cool sufficiently and quickly enough to record the most recent basin-forming events and the currently observed crater populations. Elevated heat flow recorded by relaxed craters and ridges (section 6) occurred after these global episodes. If Iapetus is a captured object, interpretation of its topography, geology, and cratering record hinge on whether capture occurred very early (in which case Iapetus behaved as if it were always in the Saturn system) or later.

8. SATELLITE RING SYSTEMS

One of the most stunning of Cassini's many surprises at Saturn was the discovery of equatorial great circle features on Iapetus and Rhea, and their probable origin as fallback deposits of circumsatellite debris rings (e.g., *Ip,* 2006; *Schenk et al.,* 2011). Ring systems were not thought to be very stable around smaller planetary bodies, especially satellites with low surface gravity, and this idea was not taken very seriously, at least initially. The first evidence came in the form of the giant equatorial ridge on Iapetus (*Porco et al.,* 2005) rising a few to almost 20 km in height (Fig. 18), a feature not observed on any other body (*Ip,* 2006). The difficulty in explaining the observed discontinuous great circle ridge segments on Iapetus by endogenic means such as global contraction or tidal stress fields, which tend to produce spatially distributed deformation, left reaccretion of a circum-Iapetus debris ring (*Ip,* 2006), perhaps due to impact-generated orbital decay of a temporary late-accretional moon (*Dombard et al.,* 2012; *Levison et al.,* 2011; *Lopez Garcia et al.,* 2014; *Damptz et al.,* 2018), increasingly plausible.

The discovery of the "blue pearls" (*Schenk et al.,* 2011) on Rhea, a similar yet different equatorial deposit (Fig. 18), and of physical, although sparse, ring systems around the small icy centaur and Kuiper belt objects Chariklo (*Braga-Ribas et al.,* 2014; *Duffard et al.,* 2014) and Haumea (*Ortiz et al.,* 2017) strengthens the case that equatorial debris ring systems are (or have been) possible if not common around mid-sized planetary bodies. The blue pearls of Rhea are a series of NIR-dark irregular patches located at the crests of the highest ridges or massifs located along the equator of Rhea (*Schenk et al.,* 2011). Although the Rhea blue pearls lack constructional relief in and of themselves (other than forming on preexisting highs), they share the same irregularly spaced great-circle interrupted staccato pattern of discrete occurrences on high points along the equator, separated by intervals of ordinary cratered terrain, observed in the topographic expression of the Iapetus ridge (Fig. 18). Neither feature is associated with any tectonic structures along their lengths or in near proximity. Furthermore, both the blue pearls and the Iapetus ridge are discontinuous, with the pearls on Rhea forming only on high points along the equator, and ridge segments being highest and longest on the leading hemisphere and correlating with regions where flanking topography is also higher (Fig. 18). We note also the relative paucity of ridge segments on Iapetus and blue pearls on Rhea on the centers of their trailing hemisphere, which could reflect either regional lows or an offset of the center of figure from the center of mass, leading to more ridge or pearl accumulation on one hemisphere than another.

Observing two such discontinuous equatorial features on the larger mid-sized icy satellites that orbit at greater distances from Saturn (and hence less disturbed by Saturn's large gravity field) strengthens the case that both are related to infalling or collapse of an orbiting debris ring. The mechanism involves impact of debris spiraling in from orbit and striking high-standing terrain along the equator; continued deposition will accumulate a ridge as on Iapetus. Less or no accumulation would occur on low areas due to accretion of orbiting debris at higher topographic obstacles upstream. The difference at Rhea is that its ring system must have either been much lower mass or very little of its

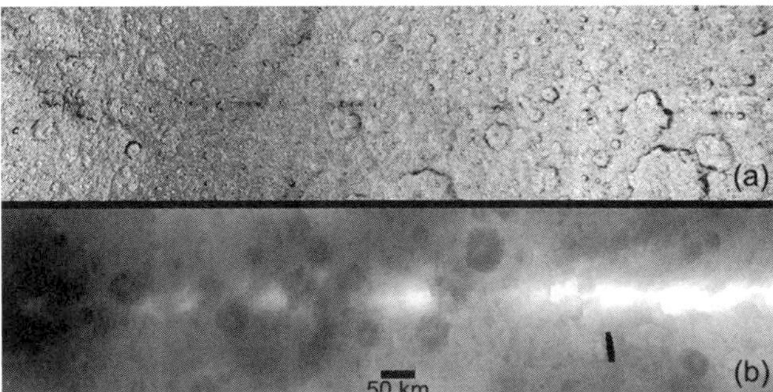

Fig. 18. Comparison of the equatorial regions of **(a)** Rhea and **(b)** Iapetus. Rhea image is an IR/UV ratio map showing irregular distribution of IR-dark "blue pearls" (see also Fig. 5a for another view). Iapetus image is from the global DEM map of topography (bright is high) showing the similarly discontinuous nature of the ridge along the equator. Both maps are to scale and north is up.

material was forced to the surface [as proposed for Iapetus (*Levison et al.,* 2011)].

Numerical models suggest that rapid despinning (e.g., *Kuchta et al.,* 2015) could indeed produce an equatorial ridge on Iapetus under ideal circumstances, but inspection of the maps indicates that the discontinuities in the ridge are not associated with post-impact-cratering scars, which suggests that the ridge originally formed in a discontinuous manner more consistent with reaccreting ring debris on topographic highs than an internal mechanism. Thus, although another mechanism may be viable, the weight of evidence currently favors ring reaccretion for the equatorial features on both Rhea and Iapetus. No evidence for additional equatorial features has been observed or reported elsewhere in the Saturn or Pluto systems (e.g., *Moore et al.,* 2016). No definitive equatorial features have yet been reported for uranian moons, but the case for such landforms on these moons remains open as the polar illumination during the Voyager 2 encounter offered poor views of the equatorial regions of those moons; nonetheless, because of its distance from Uranus, the outermost moon Oberon is the most likely candidate to host such a feature. Although none have been reported in the jovian system either, the case for such occurrences there remains open until high-resolution mapping is complete. Whether formed by ring reaccretion or another mechanism, study of these features will provide key insights into a type of astrophysical and/or geologic process not observed before in the solar system.

9. MASS WASTING

Mass wasting in the Saturn system can be divided into two categories. The first is the movement of thin surface layers, such as downslope creep of regolith and shallow failures as avalanches and debris flow. These are usually modeled as continuum processes. Mass wasting on the surfaces of the mid-sized icy satellites of Saturn is most widely expressed as "linear creep" in which relief is reduced and small features destroyed by erosion of topographic highs, such as crater rims, and the infilling of topographic lows, such as crater floors (e.g., *Fassett and Thomson,* 2014). Linear creep is induced either directly by small impacts and impact gardening, or by seismic shaking associated with more distant large impacts. The steep inner rimwall slopes of craters on cratered terrains of most mid-sized icy satellites appear brighter and "bluer" than their surroundings in regional-scale high-Sun imaging, probably due to the freshening of steep slopes by mass wasting. In general, however, this form of mass wasting is difficult to document.

The second category of mass wasting is deep-seated failures (slumps, landslides, flows). These are found in scattered locations (e.g., crater wall interiors, tectonic scarps) on most planets and satellites. Discrete landslides have been observed extending from well-preserved crater rims on Rhea, Tethys, and Mimas (Fig. 19). The association of these slides with crater rims on steep topography is similar to that observed in many similar landslides on Ceres (*Buczkowski et al.,* 2016)

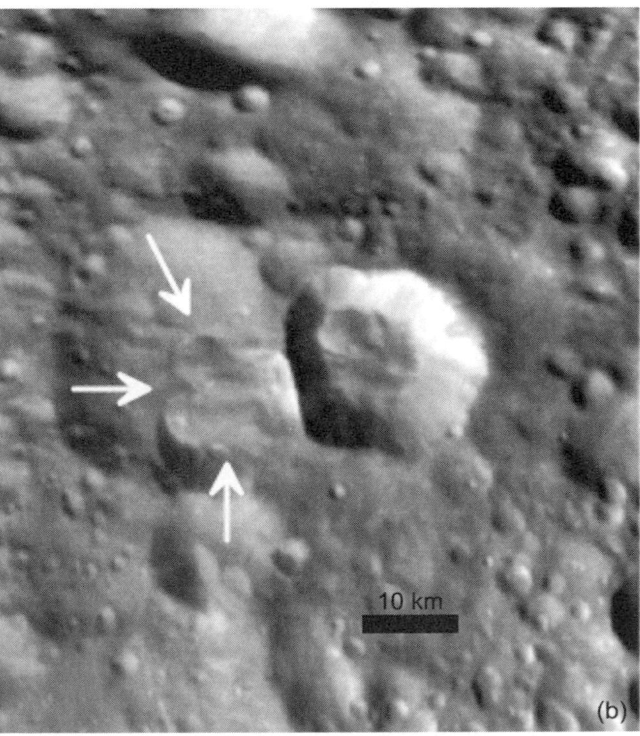

Fig. 19. Landslides (arrows) associated with recent craters on **(a)** Tethys and **(b)** Mimas. Tethys crater at upper center is ~12 km across and located at 7°N, 186°W; on Mimas the originating crater is 27 km across and is located at 47°N, 165°W.

and suggests that the impact event both triggers the slide and perhaps imparts lateral momentum to the debris. This style of failure is lacking on Vesta, however, despite similar surface gravity on all these objects. Vesta is different from the others by way of its very low water ice content, consistent with the idea that water ice is relatively weak and materials containing them will be more susceptible to collapse.

An interesting variant within this category are the long-runout landslides on Iapetus and Rhea. *Singer et al.* (2012) investigated these and concluded that they formed as a result of localized frictional heating in ice rubble such that sliding surfaces became slippery. The great heights from which the landslides started were thought to contribute to the energy needed to initiate this frictional heating. Several types of mass wasting are attributed to the formation of "honeycomb" topography of Hyperion, which was interpreted by *Howard et al.* (2012) as a product of impact cratering with partial ejecta redeposition, loss of bedrock strength by sublimation, and diffusive mass wasting.

10. DISCUSSION

In this section we examine each moon and how the geologic processes described above come together to elucidate their geologic histories. We also discuss possible origins for major features and their relationships.

10.1. Iapetus

Iapetus serves as a benchmark for mid-sized icy worlds removed from the proximal influence of their powerful parent planets. Initial Cassini studies (e.g., *Castillo-Rogez et al.,* 2007, 2011) suggested that Iapetus has undergone no tidal heating and cooled sufficiently quickly such that a "primordial" 10-hour rotational shape was frozen in to produce the oblate spheroid observed today. The origins of Iapetus are not well understood. It is often assumed that Iapetus formed in the outer Saturn system or was ejected into a larger orbit early on (see the chapter in this volume by Castillo-Rogez et al.). This would have subjected it to Saturn-focused incoming impactors but removed it from the strongest flux, as well as placing it outside the magnetosphere and away from Enceladus. A capture origin for Iapetus is also plausible but capture is usually associated with significant tidal heating during the adjustment of the orbit to its current state, as in the case of Triton (e.g., *Nimmo and Spencer,* 2014). The fate of a captured Iapetus is unclear.

Iapetus has essentially been quietly weathering in space since the fossil rotation shape was frozen in. Large impact basins were accumulated in this time, also producing long linear impact crater chains (Fig. 11) (*Singer and McKinnon,* 2011). Although internal mechanisms remain plausible, evidence currently indicates that the great equatorial ridge (Fig. 20) likely formed by redeposition of a circum-Iapetan ring of debris (e.g., *Dombard et al.,* 2012; *Rivera-Valentin et al.,* 2014) early enough to be ancient, but late enough such that it has not been blasted apart by heavy cratering. Phoebe dust has likely been painting the surface, which then evolved under the influence of insulation-driven volatile migration into the stark bright and dark patterns (Figs. 1 and 20, Plate 31) we see today (*Denk et al.,* 2010; *Spencer and Denk,* 2010), troubled only by the occasional landslide (*Singer et al.,* 2012) and impact crater. Iapetus thus escaped the complex histories observed on moons like Dione and Tethys.

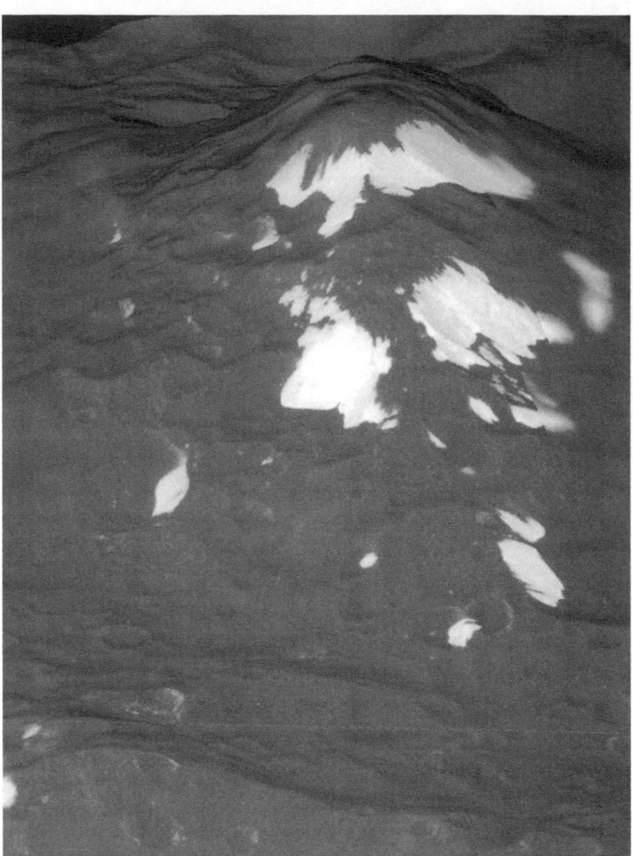

Fig. 20. Oblique perspective view looking west along a segment of the Iapetus equatorial ridge deposit. The main massif at upper center is 25 km wide and 5 km high. Patchwork of bright ice and dark material is also evident. View is centered on equator near longitude 180° near the boundary between the bright and dark hemispheres.

10.2. Rhea

The geologic history of Rhea is also mostly dominated by impact cratering but there remain several intriguing features that betray a more complex past. There is no direct evidence for obvious extensive resurfacing on Rhea, except perhaps the as yet unconfirmed Voyager-based crater erasure record of *Lissauer et al.* (1988), which could be related to basin erasure, volcanism, or viscous relaxation. There are several styles of tectonic deformation, however, the most significant being an older ridge and trough system (Figs. 8b,c) and the late-stage formation or reactivation of the graben sets across the trailing hemisphere (Fig. 8a). The geologically recent fracturing evidenced by the graben system discovered on the trailing hemisphere is similar in style to the more extensive system on Dione (also discussed in the next section) and remains the most recent non-impact endogenic geologic process to have occurred on the surface of Rhea. [Rare high-resolution views of Rhea at <20 m/pixel (e.g., ISS frame s01741547885) reveal occasional isolated fault scarps that may be more common globally than might be guessed from lower-resolution global imaging.] Their probable extensional morphology indicates that Rhea underwent global expansion relatively late in its history, perhaps due to freezing of an ocean or simple secular cooling of an ice mantle (e.g., *Moore et al.,* 1985).

The enigmatic, more degraded, and hence older linear features of Rhea (Fig. 8b) suggest that this moon underwent additional episodes of cooling and heating. The degraded state of these features makes it difficult to ascertain if they are extensional or shortening in nature, although initial examination suggests they are parallel extensional fractures. We also have the highly degraded irregular mound pattern of unknown origin visible in the topography of the trailing hemisphere (Fig. 8b). This evidence for a more complex tectonic and thermal record on Rhea is consistent with evidence for higher heat flows in the past from the crater relaxation record (e.g., *White et al.,* 2013). Finally we have the apparent global-scale early phase heating event that erased the large basins that dominate Iapetus. Why Rhea would have undergone fracturing at several different times and later only in the trailing hemisphere remain open questions.

A number of isolated non-tectonic crater chains occur across the surface and initial studies (*Johnston and White,* 2013) suggest that many of them are secondaries associated with larger craters on Rhea; further study is required. The "blue pearls" (Fig. 18) described above are likely related to infalling ring debris and are similar in distribution and probably in origin to the equatorial ridge on Iapetus (see section 8). The more pertinent fact here is their existence. They form a chain of surficial features within 2° of the equator and span ~270° of longitude, and are thus highly unlikely to be related to randomly oriented exogenic factors such as secondary or sesquinary impact cratering from primary craters on Rhea or other moons. The lack of any constructional artifacts associated with these color patterns on Rhea implies that they are due to regolith disruption (*Schenk et al.,* 2011), and color imaging elsewhere on Rhea shows that these "bluish" colors are associated with steep slopes (e.g., crater rims) and fade with time. Thus we would expect that a color signature due to ring reaccretion of this type would also fade with time, implying that this is a geologically recent phenomenon. A more recent origin would seem to argue against an ancient origin involving formation of a proto-moon formation or debris thereof early in its history, unlike at Iapetus, where the event forming the ridge appears to be older (e.g., *Ip,* 2006; *Levison et al.,* 2011; *Dombard et al.,* 2012). One scenario that might explain this is that of a recent crater formed on Rhea and launching a small amount of debris into orbit, much smaller than that responsible for the ridge deposit on Iapetus. Inktomi (Fig. 5) is a candidate crater and likely struck obliquely given the butterfly ray pattern, but is only 49 km across and it remains to be determined whether such an impact would produce any orbital debris.

10.3. Dione

The relatively pristine and presumably recent fracture networks (once referred to as "wispy terrains") on the trailing hemisphere of Dione are similar in morphology to, if more extensively distributed than, those on Rhea (Fig. 8). The origins of both networks remain the subject of ongoing investigations. No definitive stress origins or histories have emerged except that the youngest fractures form discrete, graben-like fault blocks almost exclusively on the trailing hemispheres of both Dione and Rhea. The reason for this distribution is unknown, but if global stress fields are involved this could imply substantial differences in lithospheric strength on the two hemispheres (*Byrne et al.,* 2016). The observation that (displacements on) these features are the most recent non-impact events on the surfaces of both bodies also has significant if unknown implications for thermal histories, either secular cooling or recently imposed stress fields on these bodies.

The other central question on Dione remains the origins of the smoother materials covering most of the leading hemisphere (Fig. 13). The leading hypothesis remains that these terrains were emplaced due to some variant of volcanic resurfacing by water, warm mobilized ice, or other water-volatile mixtures, the most commonly invoked (although not yet detected) candidate being ammonia-water fluids and soft ices (e.g., *Jankowski and Squyres,* 1988; *Schenk,* 1991). The low viscosities for such mixtures could explain the lack of high-relief features across these plains. *Schenk and Moore* (2007) hypothesized that the linear troughs and the oddly shaped double non-impact scarp-enclosed depressions Murranus and Metiscus within the smooth terrains (Fig. 13) are relicts of volcanic resurfacing followed by late-stage outgassing at two central vents and radial fracturing perhaps due to broad uplift or other dynamic stress regimes.

The crater relaxation mapping of *White et al.* (2017) indicates that much of Dione experienced elevated heat flows (by tens of milliwatts per square meter) at some point in its history. However, the mapping pattern of relaxation does not correspond to that of the smooth plains [overlapping its southern half but extending onto the trailing hemisphere and south polar terrains as well (*White et al.,* 2017)], and could have post-dated or outlasted smooth plains formation. The relatively low topographic amplitude of Dione compared to the other mid-sized non-Enceladus satellites (Fig. 17) also speaks to a process that has erased much of Dione's relief early on, in part due to the low relief of the smooth plains.

Although the volcanic hypothesis for smooth plains emplacement remains viable, given the central location of the oddly-shaped central scarp-enclosed depressions (Fig. 13) and spatially associated fracture patterns within the smooth deposits, many questions remain. Cassini did not resolve flow fronts on the smooth plains unit, perhaps due to the moderate but pervasive post-emplacement cratering. Also, the margins of the smooth terrain boundary are not sharp but instead grade into heavily cratered highlands over a significant distance (*Kirchoff and Schenk,* 2015), implying that there may be no recognizable discrete contact (boundary) between the two terrains or that it was erased by cratering. Whether the smooth plains could instead be the result of crater erasure by extremely high heat flow is uncertain, but the lack of any cryptic rings representing nearly flattened impact craters within these terrains suggests either near melting of the near-surface layers or returns us to the volcanic resurfacing hypothesis.

An alternative hypothesis for the smooth plains (on Dione at least) is suggested by the near-hemispheric scale of this geologic terrain and its gradational boundary with heavily cratered terrains. Heliocentric impactor fluxes are predicted to produce highly asymmetric crater densities on synchronous satellites orbiting Saturn (e.g., *Zahnle et al.,* 2001), perhaps as high as 40 to 1 from leading to trailing hemisphere. On Dione we see rather the opposite, where the trailing hemisphere (except in the vicinity of a large relict impact basin and the most densely tectonized zones) is more densely cratered than the leading (*Kirchoff and Schenk,* 2015). The more lightly cratered terrains on the leading hemisphere could be a relict of this heliocentric cratering asymmetry, but would require that this hemisphere had been the original lightly cratered trailing hemisphere and that Dione "flipped" hemispheres ~180° to move this hemisphere to its current leading orientation. Under this scenario, the irregular central structures would still be volcanic in origin but date to a much earlier epoch in Dione's history rather than relatively late. Why Dione would have reoriented 180° about its polar axis is unclear, but we note that the Evander basin, which is superposed on and younger than the smooth plains, and currently estimated to have an age of <2.5 Ga (*Kirchoff and Schenk,* 2015), could possibly have triggered a reorientation event (*Nimmo and Matsuyama,* 2007). While such hypotheses may be plausible, we consider volcanic origins for the smooth plains and the walled depressions Murranus and Metiscus as the most likely for their formation.

The possibility of an ocean within Dione has been raised based on gravity and topography (*Beuthe et al.,* 2016). Geologically, an ocean is difficult to confirm. If the outer shell is on the order of 100 km thick as suggested, it may be difficult to fracture in a way that would betray or invoke the presence of an ocean beneath. Solid-state diapirism, perhaps as a large single plume under the smooth plains, could be an effect of an ocean but would require detailed thermal calculations to ascertain. The largest impact feature, Evander (Fig. 7), is unusual in morphology and degree of relaxation, and could be related to a large impact into a floating ice shell, but many other craters are also relaxed (*White et al.,* 2017) and only require elevated heat flows. These craters are only tens of kilometers across and unlikely to have been affected by a deep ocean. The case for an ocean within Dione thus remains provocative but not yet conclusively demonstrated by geology or geophysics.

10.4. Tethys

The geologic complexity of Tethys, with possible resurfacing indicated by smooth plains (Fig. 12) and multiple tectonic episodes (Fig. 9c), is puzzling given its relatively low ice-like density (see Table 1 in the chapter by Castillo-Rogez et al., this volume) and thus the inferred low abundance of radiogenic nuclides. As discussed above, global topography of Tethys (Figs. 16 and 17) apparently experienced the same erasure of a presumed ancient irregular surface and large basin population otherwise recorded by Iapetus. Mapping of relaxed impact craters shows that despite its negligible non-ice composition, Tethys also experienced a prolonged period of elevated heat flow (*White et al.,* 2017), sufficient to largely erase the relief, if not the rim and peak structures, of several large preserved basins. Odysseus was the last of the large basins to form on Tethys [estimated at 400 Ma to 1 Ga (*Kirchoff and Schenk,* 2010)] and displays a relaxation fraction of 0.66 (*White et al.,* 2017) despite its 9-km depth. Odysseus may have partially relaxed despite its younger age, which also implies that high heat flow conditions have affected Tethys comparatively recently. The observed fracture network on the floor (Fig. 6) is consistent with post-impact uplift.

The formation of smooth plains (Fig. 12) on Tethys remains cryptic. Elevated heat flows recorded by relaxed impact craters could have allowed for limited partial melting, but the composition of resulting melts is unknown and the origin of these units have not been the focus of detailed study. They are close to both the antapex of motion (and thus the area of lowest heliocentric cratering flux as was discussed for Dione) and to the antipodal region of Odysseus (although smoothing is not necessarily an expected outcome, as discussed above).

Aside from the origin of smooth plains, a central question on Tethys remains the postulated link between the giant fracture system Ithaca Chasma and the large basin Odysseus (*Smith et al.,* 1982; *Moore and Ahern,* 1983; *Moore et al.,* 2004). The trough system forms a confined band of closely spaced linear fractures unique in the Saturn system that runs crudely concentric to Odysseus for ~270° of its circumference. The spatial relationship to Odysseus, the largest crater in the system with respect to the diameter of the target body, is complex. As shown in Fig. 21, one half of Ithaca Chasma is concentric to Odysseus at a distance of ~120° from the center of the crater. This is also the portion that features the raised rim studied by *Giese et al.* (2007). This section also runs parallel to the unique sinuous ridge described above, which lies ~350 km from the rim. Between the ridge and the crater rim, the rugged topography of overlapping craters typical of most of Tethys is erased and replaced with innumerable overlapping small craters, chains, and scour marks (Fig. 4). This terrain marks what appears to be an ejecta deposit directly related to Odysseus formation. The trace of the sinuous ridge corresponds closely although not exactly with the limit of secondary craters in this region. This observation raises two possibilities. The ridges around

Odysseus and Evander (Figs. 4 and 7) may be rampart-style ejecta deposits in which ballistically emplaced ejecta piles up at the outer edge of the deposit, or on Tethys some sort of shortening ridge related to shock waves in the interior from the Odysseus event (see below), subsequently run over by the edge of the secondary crater field.

The other half of Ithaca Chasma, nearer the current north pole, runs at a steeper angle to a concentric trace to Odysseus and approaches within ~400 km of the rim (Fig. 21). This section has no raised rim and the sinuous ridge becomes indistinct or nonexistent in this region. This section is also more heavily cratered, consistent with its location nearer the rim of Odysseus, and appears to deflect around the outline of the smooth plains occurrence (Fig. 21).

Understanding the age of Ithaca Chasma relative to Odysseus is crucial, but unfortunately is unclear. Crater counts of Ithaca Chasma by *Giese et al.* (2007) and discussed by *Stephan et al.* (2016) that indicate an older age are suspect for two reasons. One is that the counts date the floor of the fracture system, and in a graben system the original cratered surface will be mostly preserved as the intervening crustal blocks are downdropped. Thus the reported cratering "age" of Ithaca Chasma may date the original surface more than it dates the fracturing event(s). Second, the density of craters apparently increases as Ithaca Chasma tracks closer to the Odysseus rim (Fig. 9), suggesting that secondaries may contaminate the crater counts across the fracture system, and perhaps over much of Tethys itself. Detailed reanalysis of the cratering record to search for a secondary cratering signature on Tethys and determine the actual age of faulting is recommended.

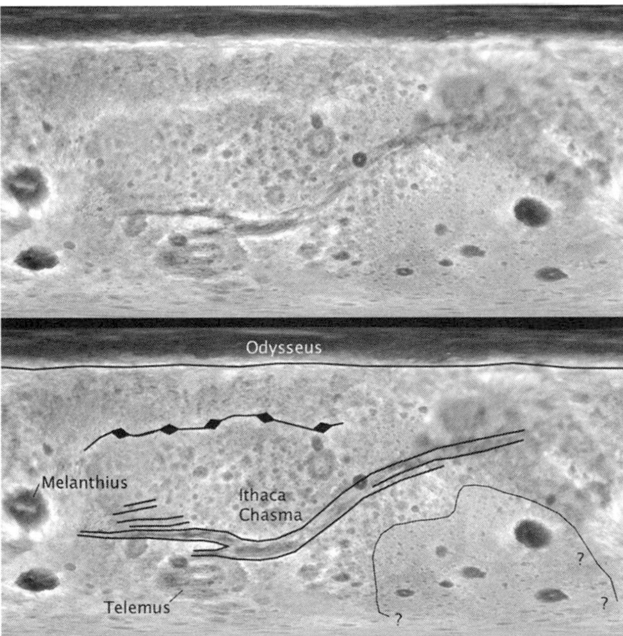

Fig. 21. Cylindrical map of global topography of Tethys, rotated such that the large basin Odysseus is at the paleopole, and illustrating spatial relationships of major structural features to the basin. Features concentric to Odysseus form horizontal lines, features radial to Odysseus form vertical lines. Bottom view is annotated to highlight major features, including the arcuate ridge (diamond line), Ithaca Chasma, and outline of smooth plains (thin curved line). Compare with current global base mosaics in section 2 and topography map in Fig. 16.

To explain the unique characteristics of Tethys, we propose that the large Odysseus impact produced seismic waves across the interior, together with possible internal mass redistribution as the impact basin was excavated and the floor uplifted. This promptly produced a ring of fractures roughly circumferential to Odysseus (e.g., *Moore and Ahern,* 1983), followed by later-arriving ejecta that strongly modified the near-rim environment and also lightly cratered the newly formed fracture system. Horizontal shortening may or may not have been involved in the formation of the sinuous ridge. The offset and asymmetric characteristics of Ithaca Chasma could be related to oblique impact or variations in internal strength. Of course, it cannot be ruled out that the trough system may be completely unrelated to Odysseus, in which case formation of a giant crater near the center of the great circle of the troughs is coincidental. The fact that ejecta scouring, a large concentric ridge, and the rift of Ithaca Chasma all appear only on the same side of Odysseus (Fig. 21) would seem to strengthen a connection between these features, and may even require an oblique impact event.

The equatorial Divalia Fossae trough system on Vesta (*Buczkowski et al.,* 2012) may strengthen the case for a genetic link between a large impact and tectonism on Tethys. This trough system is likely a direct effect of seismic waves generated by the giant Rheasilvia impact basin at Vesta's south pole (*Ivanov and Melosh,* 2013; *Bowling et al.,* 2013). These seismic waves produce a concentration of extensional stresses in an annulus around the widest parts of Vesta ~90° from the center of impact. The Vesta fractures are also asymmetric and their center is offset 10°–20° from the basin center. The fossae are also more heavily cratered than Rheasilvia, due to the fact that the seismic waves and fracturing outrun the slower moving ejecta curtain and secondary projectiles (*Buczkowski et al.,* 2012), which land on top of the newly formed features and possibly explain its apparent "older" appearance. Moreover, a similarity to the Vesta troughs may also help explain why Ithaca Chasma is more narrowly concentrated and not as broadly distributed as the Dione and Rhea fracture systems. Detailed crater counts, and numerical impact models of large impacts in small bodies, of the type employed for Vesta (e.g., *Ivanov and Melosh,* 2013; *Bowling et al.,* 2013) should help test these proposals for Tethys.

The spatial associations outlined above allow the hypothesis that the Odysseus impact and fracturing at Ithaca Chasma are directly related but do not prove it to be so. Alternative origins for Ithaca Chasma include freeze expansion of the interior (*Moore and Ahern,* 1983) or tidal stresses (e.g., *Chen and Nimmo,* 2008). Whatever the origin of the chasma, determining the timing of its formation is key to either confirming an impact origin or to placing it within Tethys' thermal history.

The origin of the late-stage red streaks on Tethys (*Schenk et al.,* 2015) remains cryptic as of this writing, but they postdate all other structures, crossing small craters and the giant Odysseus basin unaffected. The pattern is centered on the current tidal axis with Saturn and is clearly related to global stress fields of some sort. Reorientation of Tethys due to the inferred negative mass load of the giant crater is possible (*Nimmo and Matusyama,* 2007), but Odysseus is not at any of the preferred a, b, or c primary axial locations for reorientation. If any reorientation has occurred it was not very large, likely due the prominent triaxial shape and the difficulty of reorienting such a cold, stiff, oblate body. Nonetheless, even a modest amount of reorientation could produce the fracturing observed. To date, polar wander provides a less satisfactory match to the observed pattern than do other mechanisms such as non-synchronous rotation, although no match is completely satisfying.

10.5. Mimas

Mimas is a touchstone in the Saturn system; broadly similar in size and density to active Enceladus and the closest mid-sized icy satellite to Saturn itself, yet strangely devoid of any but the most rudimentary deformation (*Schenk,* 2011, and figures therein). The origin of Mimas' trough system (Fig. 10) is unclear. Although apparently extensional, their orientations (Fig. 22) are not obviously or at least exclusively related to a single tidal or other global stress field, and whereas many are radial to Herschel, some are not (*Schenk,* 2011). Recent models of the librational wobble of Mimas (*Tajeddine et al.,* 2014) suggest that either Mimas has a core that is also irregular in shape or (less likely) that a deep subsurface ocean exists within Mimas. There is no obvious geological evidence of an ocean, except perhaps the troughs. If unrelated to Herschel-induced fracturing, the global trough system could indicate a significant period of global expansion, in turn indicating freezing of an ocean (e.g., *Rhoden et al.,* 2017) or perhaps simple secular cooling. The troughs are sufficiently preserved to indicate they are not extremely ancient (*Schenk,* 2011). They predate Herschel, but by how much is also undetermined; fracturing may have occurred during impact but before the arrival of any ejecta that might or might not have formed from Herschel, as likely happened on Vesta (*Buczkowski et al.,* 2012; *Bowling et al.,* 2013). The modest degree of degradation of the troughs thus could be related to ejecta or seismic disruption directly linked to Herschel itself. Whether freezing of a water ocean could produce such an episode also remains unclear (*Rhoden et al.,* 2017). The primary alternatives include global tidal or rotational forces likely greater than they are today, or global fracturing due to Herschel.

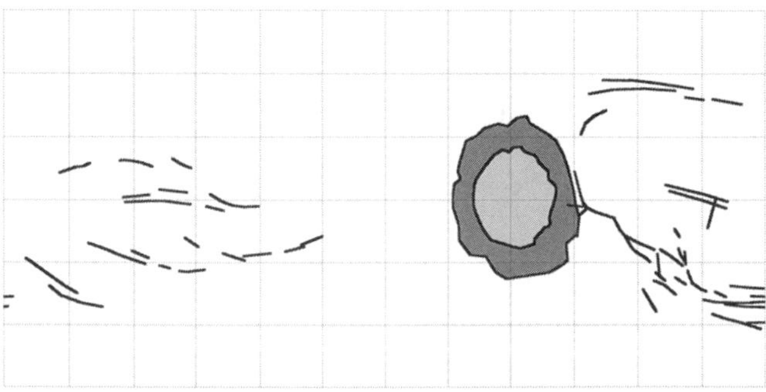

Fig. 22. Global map showing current distribution of grooves and troughs (dark lines) on Mimas. Double circle is Herschel impact crater and its inner ejecta blanket, which is visible in the global topography and as an elevated plateau in the global mosaic.

A thick ice shell over an ocean on a small world could be resistant to fracturing and volcanism, with the ocean thus remaining hidden, but it would not be resistant to impact cratering. Herschel (Fig. 23) was a significant event on Mimas and formation of a crater this size relative to its target would likely be influenced by a subsurface ocean in some way, as craters are shallow or form multi-ring structures under the influence of the ocean under Europa's surface (see *Schenk and Turtle,* 2009). Herschel is a typical central peak crater with an essentially unrelaxed and unmodified depth of ~11 km, betraying no evidence of alteration by or influence of a subsurface liquid layer.

Although searches for tectonic disruption at the antipodes to large impacts in the Saturn system, akin to that proposed for Mercury (*Schultz and Gault,* 1975; *Fleitout and Thomas,* 1982; *Watts et al.,* 1991; *Bruesch and Asphaug,* 2004; *Lü et al.,* 2011), have returned ambiguous results, some evidence for this effect may have been found on Mimas. An area ~125 km across located almost exactly at the Herschel antipode appears to be more dissected by irregular troughs and knobs (Fig. 23). If this is disrupted antipodal terrain, it could indicate the presence of a core on Mimas that would focus seismic energy as it does on Mercury. Again, large impact simulations for Mimas may be warranted.

A prominent, lens-shaped equatorial color (*Schenk et al.,* 2011) and thermal (*Howett et al.,* 2011) anomaly straddles the equator of the leading hemisphere of Mimas due to ice grain alteration by high-energy electrons. A similar such feature is observed on Tethys but not on Enceladus because of ongoing redeposition of plume particulates (*Schenk et al.,* 2011). These and other continuing exogenic processes are discussed in the chapters in this volume by Howett et al. and Hendrix et al., but it is worth noting that the intensity of the color anomaly is less on the floor of Herschel crater, which lies near the center of the lens. This is probably related to the brecciated or possibly partly melted properties of impact crater floor material.

11. SUMMARY AND OPEN QUESTIONS

In a process begun by Voyager, the diverse instruments onboard the Cassini orbiter have revealed the mid-sized icy moons of Saturn to be geologically complex worlds.

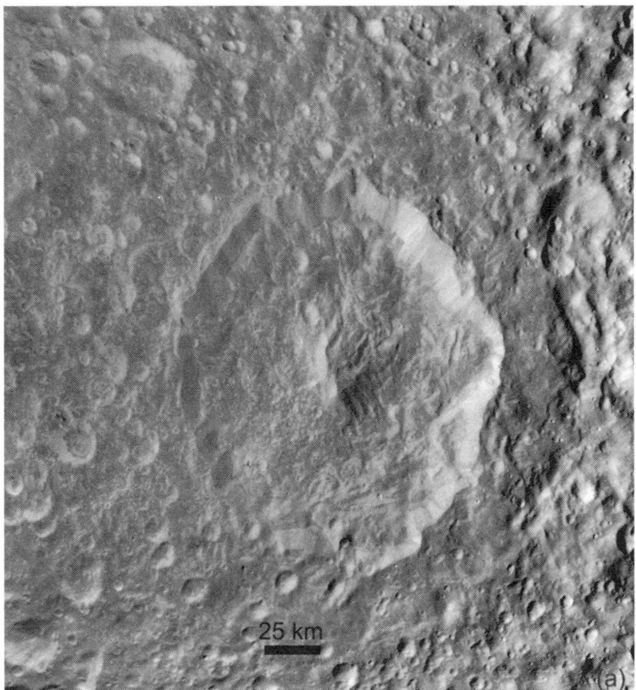

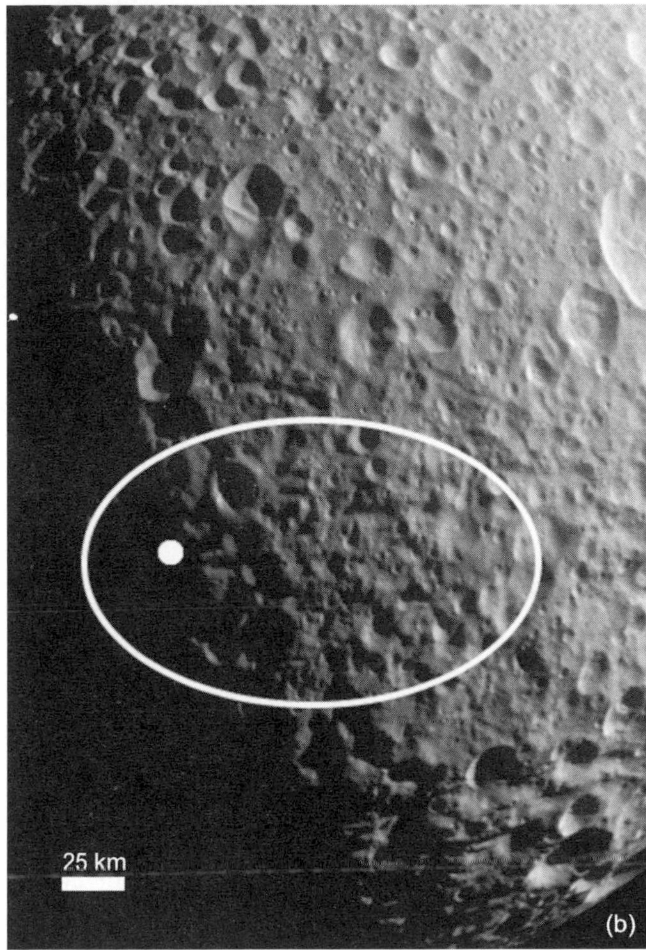

Fig. 23. Best Cassini views of **(a)** 125-km-diameter Mimas crater Herschel located at 0°N, 115°W, and **(b)** region directly antipode to Herschel, at 0°N, 305°W. Rugged putative antipodal terrain in center of right view (approximately outlined in white oval) includes the antipodal point (dot) and is more chaotic than typical cratered terrains to the north or south. North is top in each view.

The Cassini observations have also revealed the nature of cryptic Voyager features and physical and geological processes unimagined before visits by spacecraft. Setting aside the ongoing activity on Enceladus, which is covered in numerous other chapters, these discoveries include regolith alteration by electrons, circumsatellite ring systems, and putative volcanic features on Dione and perhaps Tethys. These discoveries also raise as-yet new and fundamental questions about these bodies and their evolution.

Outstanding questions include: How did the smooth plains become smooth? Was resurfacing in smooth plains the consequence of low-viscosity molten materials, warm plastically deforming and flowing ice, or some other lower-energy process? If heat flows inferred from relaxation were so high, why was volcanism not more extensive? Why did rift systems form when they did (late in their geologic history)? How does Tethys, with its apparently negligible rock fraction and internal heating, manifest evidence of past high heat flow and the recently formed red-streak fracture system? Why did these fractures form where they did? Were global directional stress fields (e.g., tidal) involved, or is relatively simple more uniform global expansion due to cooling a sufficient explanation? Are heat flow estimates consistent or even reliable? What past or current heating sources ever modified these bodies?

Other questions include: Did impact melting occur? Why do we see no central pits in craters on these bodies, when we do see them on Ceres? What is the origin(s) of crater chains on Rhea, Dione, Tethys, and Iapetus? Are they mostly secondaries or do split comet crater chains also form at Saturn? Are there any unambiguous signatures of subsurface oceans on any of these bodies?

Key to all investigations regarding volcanism, tectonism, and cratering (both the event and subsequent modification) is the timing of these events. Reliable ages for tectonic features, both absolute and relative, are difficult to measure. Landform degradation and the contamination from secondary craters (Fig. 3) (*Schenk et al.,* 2017) can also affect the reliability of crater counts. Of course, the assumption of heliocentric projectile origin lies behind most efforts to determine absolute ages and these have large uncertainties (see the chapter by Kirchoff et al. in this volume).

One of the more interesting questions raised by studies of the Cassini mission data is the now apparent youthful age of the ring system (e.g., *Kempf et al.,* 2017; *Iess et al.,* 2017), and, by association, (possibly) the age of the satellite system (e.g., *Charnoz et al.,* 2011; *Ćuk et al.,* 2016). If the rings are ~100 Ma in age, then the inner satellites at least could be of similar age. Testing for such a young age is not easy. Ages derived from the observed crater population (see the chapter by Kirchoff et al.) assume a heliocentric source population, and the possibility that some or most of these craters are from planetocentric debris (and thus derived ages irrelevant) is discussed in that chapter. The formation of satellites during a recent major event is a dynamical problem (not addressed here) in terms of accretion, and a fragmentation problem in terms of what sort of debris would be left to crater the newly formed moons. Whether such debris would have a diagnostic signature is not known but worth considering.

From a geologic perspective, it is not clear what signatures might be manifest, except perhaps for preservation of distinctly different compositional remnants in the regional geology of these moons. No such signatures are known. The apparent erasure of ancient topography and large basins on the inner moons could be related to a late formation event but would require heat flows sufficient to erase even cryptic signatures of such large structures, which are not apparent in the topography. On the other hand, the paucity of the same large impact basins as found on Iapetus could indicate they never formed on these inner moons, and thus point to a young age. Significant heating from tidal interactions during the satellite (re)formation process are possible (*Charnoz et al.,* 2011). The observed cratering record could thus be the record of sweep-up of system-wide debris mixed with a subsequent heliocentric flux. The interpretation of the global topography and impactor records rests in part on the assumption that Iapetus also formed within the Saturn system (see the chapter in this volume by Castillo-Rogez et al.) and not elsewhere, a conclusion difficult to demonstrate with certainty. Distinguishing this scenario from an ancient formation event or from other early heating events may be difficult. One might ask whether it is likely that the complex geologic records we observe on these moons, including rapid cooling of higher heat flows, could form in such a short period of time. We point to Europa, which has been completely resurfaced within the past 60–120 m.y. (*Bierhaus et al.,* 2009) and suggests that a recent origin for these moons, although not compelling from a geological perspective, is at least not implausible.

Acknowledgments. The portion of this work performed by P.S. was carried out at the Lunar and Planetary Institute, which is managed by the Universities Space Research Association under a cooperative agreement with the National Aeronautics and Space Administration. This paper is LPI Contribution No. 2104. We also thank the Cassini Project, and especially Tilmann Denk, Paul Helfenstein, and Todd Ansty, for their supreme efforts in planning and acquiring these data. We thank the Cassini Data Analysis and Participating Scientist Programs for support.

REFERENCES

Barlow N. (2006) Impact craters in the northern hemisphere of Mars: Layered ejecta and central pit characteristics. *Meteoritics & Planet. Sci., 41,* 1425–1436.

Beddingfield C. B., Burr D. M., and Dunne W. M. (2015) Shallow normal fault slopes on saturnian icy satellites. *J. Geophys. Res.–Planets, 120,* 2053–2083.

Beuthe M. (2010) East-west faults due to planetary contraction. *Icarus, 209,* 795–817.

Beuthe M., Rivaldi A., and Trihn A. (2016) Enceladus's and Dione's floating ice shells supported by minimum stress isostasy. *Geophys. Res. Lett., 43,* 10,088–10,096.

Bierhaus E., Zahnle K., and Chapman C. (2009) Europa's crater distributions and surface ages. In *Europa* (R. T. Pappalardo et al., eds.), pp. 161–180. Univ. of Arizona, Tucson.

Bowling T. J., Johnson B. C., Melosh H. J., Ivanov B. A., O'Brien D. P., Gaskell R., and Marchi S. (2013) Antipodal terrains created by the Rheasilvia basin forming impact on asteroid 4 Vesta. *J. Geophys. Res., 118,* 1821–1834, DOI: 10.1002/jgre.20123.

Braga-Ribas F., et al. (2014) A ring system detected around the Centaur (10199) Chariklo. *Nature, 508,* 72–75, DOI: 10.1038/nature13155.

Bray V., Schenk P. M., Melosh H. J., Morgan J. V., and Collins, G. S. (2012) Ganymede crater dimensions – Implications for central peak and central pit formation and development. *Icarus, 217,* 115–129, DOI: 10.1016/j.icarus.2011.10.004.

Bruesch L. S. and Asphaug E. (2004) Modeling global impact effects on mid-sized icy bodies: Applications to Saturn's moons. *Icarus, 168,* 457–466, DOI: 10.1016/j.icarus.2003.11.007.

Buczkowski D. L. et al. (2012) Large-scale troughs on Vesta: A signature of planetary tectonics. *Geophys. Res. Lett., 39,* L18205, DOI: 10.1029/2012GL052959.

Buczkowski D. L. and 22 colleagues (2016) The geomorphology of Ceres. *Science, 353,* DOI: 10.1126/science.aaf4332.

Byrne P. K., Schenk P. M., McGovern P. J., and Collins G. C. (2016) Hemispheric-scale rift zones on Rhea, Tethys, and Dione. In *Enceladus and the Icy Moons of Saturn,* Abstract #3020. LPI Contribution No. 1927, Lunar and Planetary Institute, Houston.

Carr M. H., Crumpler L. S., Cutts J. A., Greeley R., Guest J. E., and Masursky H. (1977) Martian impact craters and emplacement of ejecta by surface flow. *J. Geophys. Res., 82,* 4055–4065, DOI: 10.1029/JS082i028p04055.

Castillo-Rogez J. C., Matson D. L., Sotin C., Johnson T. V., Lunine J. I., and Thomas P. C. (2007) Iapetus' geophysics: Rotation rate, shape and equatorial ridge. *Icarus, 190,* 179–202, DOI: 10.1016/j.icarus.2007.02.018.

Castillo-Rogez J. C., Efroimsky M., and Lainey V. (2011) The tidal history of Iapetus: Spin dynamics in the light of a refined dissipation model. *J. Geophys. Res., 116,* E09008, DOI: 10.1029/2010JE003664.

Charnoz S., Crida A., Castillo-Rogez J., Lainey V., Dones L., Karatekin O., Tobie G., Mathis S., Le Poncin-Lafitte C., and Salmon J. (2011) Accretion of Saturn's mid-sized moons during the viscous spreading of young massive rings: Solving the paradox of silicate-poor rings versus silicate-rich moons. *Icarus, 216,* 535–550.

Chen, E. M. A. and Nimmo F. (2008) Implications from Ithaca Chasma for the thermal and orbital history of Tethys. *Geophys. Res. Lett., 35,* L19203, DOI: 10.1029/2008GL035402.

Chen E. M. A., Nimmo F., and Glatzmaier G. A. (2014) Tidal heating in icy satellite oceans. *Icarus, 229,* 11–30, DOI: 10.1016/j.icarus.2013.10.024.

Collins G. C., McKinnon W. B., Moore J. M., Nimmo F., Pappalardo R. T., Prockter L. M., and Schenk P. M. (2010) Tectonics of the outer planet satellites. In *Planetary Tectonics* (R. A. Schultz and T. R. Watters, eds.), pp. 264–350. Cambridge Univ., Leiden.

Crow-Willard E. N. and Pappalardo R. T. (2015) Structural mapping of Enceladus and implications for formation of tectonized regimes. *J. Geophys. Res., 120,* 928–950, DOI: 10.1002/2015JE004818.

Ćuk M., Dones L., and Nesvorny D. (2016) Dynamical evidence for a late formation of Saturn's moons. *Astrophys. J., 820,* 97, DOI: 10.3847/0004-637X/820/2/97.

Czechowski L. and Leliwa-Kopystyński J. (2008) The Iapetus's ridge: Possible explanations of its origin. *Adv. Space Sci. Res., 42,* 61–69.

Damptz A., Dombard A., and Kirchoff M. (2018) Testing the formation of the equatorial ridge on Iapetus via crater counting. *Icarus, 302,* 134–144.

Dawers N. H. and Anders M. H. (1995) Displacement-length scaling and fault linkage. *J. Structural Geol., 17,* 607–614.

Denevi B. and 10 colleagues (2012) Physical constraints on impact melt properties from Lunar Reconnaissance Orbiter Camera images. *Icarus, 219,* 665–675.

Denk T., Neukum G., Roatsch T., Porco C. C., Burns J. A., Galuba G. G., Schmedemann N., Helfenstein P., Thomas P. C., Wagner R. J., and West R. A. (2010) Iapetus: Unique surface properties and a global color dichotomy from Cassini imaging. *Science, 327,* 435–439, DOI: 10.1126/science.1177088.

Dombard A. J. and McKinnon W.B. (2006) Elastoviscoplastic relaxation of impact crater topography with application to Ganymede and Callisto. *J. Geophys. Res., 111,* E01001, DOI: 10.1029/2005JE002445.

Dombard A. J., Bray V. J., Collins G. S., Schenk P. M., and Turtle E. P. (2007) Relaxation and the formation of prominent central peaks in large craters on the icy satellites of Saturn. *Bull. Am. Astron. Soc., 39,* 429.

Dombard A. J., Cheng A. F., McKinnon W. B., and Kay J. P. (2012) Delayed formation of the equatorial ridge on Iapetus from a subsatellite created in a giant impact. *J. Geophys. Res., 117,* E03002, DOI: 10.1029/2011JE004010.

Duffard R., Pinilla-Alonso N., Ortiz J. L., Alvarez-Candal A., Sicardy B., Santos-Sanz P., Morales N., Colazo C., Fernández-Valenzuela E., and Braga-Ribas F. (2014) Photometric and spectroscopic evidence for a dense ring system around Centaur Chariklo. *Astron. Astrophys., 568,* A79.

Elder C., Bray V., and Melosh H. (2012) The theoretical plausibility of central pit crater formation via melt drainage. *Icarus, 221,* 831–843.

Fassett C. I. and Thomson B. J. (2014) Crater degradation on the lunar maria: Topographic diffusion and the rate of erosion on the Moon. *J. Geophys. Res.–Planets, 119,* 2255–2271, DOI: 10.1002/2014JE004698.

Fleitout L. and Thomas P. G. (1982) Far-field tectonics associated with a large impact basin: Applications to Caloris on Mercury and Imbrium on the Moon. *Earth. Planet. Sci. Lett., 58,* 1, 104–115.

Giese B., Wagner R., Neukum G., Helfenstein P., and Thomas P. C. (2007) Tethys: Lithospheric thickness and heat flux from flexurally supported topography at Ithaca Chasma. *Geophys. Res. Lett., 34,* L21203, DOI: 10.1029/2007GL031467.

Giese B., Denk T., Neukum G., Roatsch T., Helfenstein P., Thomas P. C., Turtle E. P., McEwan A., and Porco C. C. (2008) The topography of Iapetus' leading side. *Icarus, 193,* 359–371.

Goff-Pochat N. and Collins G. C. (2009) Strain measurement across fault scarps on Dione. *Lunar Planet. Sci. XL,* Abstract #2111. Lunar and Planetary Institute, Houston.

Hammond N., Phillips C., Nimmo F., and Kattenhorn S. (2011) Stereo topography of fault systems and crater-fault interactions on Rhea. Abstract P43D-1714 presented at 2011 Fall Meeting, AGU, San Francisco, California, Dec. 5–9.

Hammond N. P., Phillips C. B., Nimmo F., and Kattenhorn S. A. (2013) Flexure on Dione: Investigating subsurface structure and thermal history. *Icarus, 223,* 418–422, DOI: 10.1016/j.icarus.2012.12.021.

Head J. W. and Roth R. (1976) Mars pedestal crater escarpments: Evidence for ejecta-related emplacement. In *Abstracts of Papers Presented to the Symposium on Planetary Cratering Mechanics,* pp. 50–52. LSI Contribution No. 259, Lunar Science Institute, Houston.

Hendrix A. R. and Hansen C. J. (2008) The albedo dichotomy of Iapetus measured at UV wavelengths. *Icarus, 193,* 344–351, DOI: 10.1016/j.icarus.2007.07.025.

Hendrix A. R., Hansen C. J., and Holsclaw G. M. (2010) The ultraviolet reflectance of Enceladus: Implications for surface composition. *Icarus, 206,* 608–617, DOI: 10.1016/j.icarus.2009.11.007.

Hirata N. (2016) Timing of the faulting on the wispy terrain of Dione based on stratigraphic relationships with impact craters. *J. Geophys. Res.–Planets, 121,* 2325–2334.

Horner V. M. and Greeley R. (1982) Pedestal craters on Ganymede. *Icarus, 51,* 549–562.

Housen K. R., Schmidt R. M., and Holsapple K. A. (1983) Crater ejecta scaling laws: Fundamental forms based on dimensional analysis. *J. Geophys. Res., 88,* 2485–2499.

Howard A., Moore J., Schenk P., White O., and Spencer J. (2012) Sublimation-driven erosion on Hyperion: Topographic analysis and landform simulation model tests. *Icarus, 220,* 268–276.

Howett C. J. A., Spencer J. R., Pearl J., and Segura M. (2010) Thermal inertia and bolometric Bond albedo values for Mimas, Enceladus, Tethys, Dione, Rhea and Iapetus as derived from Cassini/CIRS measurements. *Icarus, 206,* 573–593. DOI: 10.1016/j.icarus.2009.07.016.

Howett C. J. A., Spencer J. R., Schenk P., Johnson R. E., Paranicas C., Hurford T. A., Verbiscer A., and Segura M. (2011) A high-amplitude thermal inertia anomaly of probable magnetospheric origin on Saturn's moon Mimas. *Icarus, 216,* 221–226, DOI: 10.1016/j.icarus.2011.09.007.

Howett C. J. A., Spencer J. R., Hurford T., Verbiscer A., and Segura M. (2012) PacMan returns: An electron-generated thermal anomaly on Tethys. *Icarus, 221,* 1084–1088, DOI: 10.1016/j.icarus.2012.10.013.

Howett C. J. A., Spencer J. R., Hurford T., Verbiscer A., and Segura M. (2014) Thermophysical property variations across Dione and Rhea. *Icarus, 241,* 239–247, DOI: 10.1016/j.icarus.2014.05.047.

Howett C. J. A., Spencer J. R., Hurford T., Verbiscer A., and Segura M. (2016) Thermal properties of Rhea's poles: Evidence for a meter-deep unconsolidated subsurface layer. *Icarus, 272,* 140–148, DOI: 10.1016/j.icarus.2016.02.033.

Iess L. and 10 colleagues (2017) The dark side of Saturn's gravity. Abstract U22A-03 presented at 2017 Fall Meeting, AGU, New Orleans, Louisiana, Dec. 11–15.

Ip W.-H. (2006) On a ring origin of the equatorial ridge of Iapetus. *Geophys. Res. Lett., 33,* L16203, DOI: 10.1029/2005GL025386.

Ivanov B. A. and Melosh H. J. (2013) Two-dimensional numerical modeling of the Rheasilvia impact formation. *J. Geophys. Res., 118,* 1545–1557, DOI: 10.1002/jgre.20108.

Jankowski D. G. and Squyres S.W. (1988) Solid-state ice volcanism on the satellites of Uranus. *Science, 241,* 1322–1325.

Jaumann R. et al. (2009) Icy satellites: Geological evolution and surface processes. In *Saturn from Cassini-Huygens* (M. K. Dougherty et al., eds.), pp. 637–781. Springer, New York.

Johnston R. and White O. (2013) Crater chain classification and origins on Rhea. *Lunar Planet. Sci. XLIV,* Abstract #2581. Lunar and Planetary Institute, Houston.

Kempf S., Altobelli N., Srama R., Cuzzi J., and Estrada P. (2017) The age of Saturn's rings constrained by the meteoroid flux into the system. Abstract P34A-05 presented at the 2017 Fall Meeting, AGU, New Orleans, Louisiana, Dec. 11–15.

Kirchoff M. R. and Schenk P. M. (2010) Impact cratering records of the mid-sized, icy saturnian satellites. *Icarus, 206,* 485–497, DOI: 10.1016/j.icarus.2009.12.007.

Kirchoff M. R. and Schenk P. M. (2015) Dione's resurfacing history as determined from a global impact crater database. *Icarus, 256,* 78–89, DOI: 10.1016/j.icarus.2015.04.010.

Kuchta M., Gabriel T., Miljković K., Běhounková M., Souček O., Choblet G., and Čadek O. (2015) Despinning and shape evolution of Saturn's moon Iapetus triggered by a giant impact. *Icarus 252,* 454–465.

Levison H. F., Walsh K. J., Barr A. C., and Dones L. (2011) Ridge formation and de-spinning of Iapetus via an impact-generated satellite. *Icarus, 214,* 773–778, DOI: 10.1016/j.icarus.2011.05.031.

Lissauer J. J., Squyres S. W., and Hartmann W. K. (1988) Bombardment history of the Saturn system. *J. Geophys. Res., 93,* 13776–13804, DOI: 10.1029/JB093iB11p13776.

Lopez Garcia E., Rivera-Valentin E., Schenk P., Hammond N., and Barr A. (2014) Topographic constraints on the origin of the equatorial ridge on Iapetus. *Icarus, 237,* 419–421.

Lü J., Sun Y., Toksöz M. N., Zheng Y., and Zuber M. T. (2011) Seismic effects of the Caloris basin impact Mercury. *Planet. Space Sci., 59,* 1981–1991, DOI: 10.1016/j.pss.2011.07.013.

McKinnon W. B. and Melosh H. J. (1980) Evolution of planetary lithospheres: Evidence from multiringed structures on Ganymede and Callisto. *Icarus, 44,* 454–471.

Melosh H. J. and Nimmo F. (2009) An intrusive dike origin for Iapetus' enigmatic ridge? *Lunar Planet. Sci. XL,* Abstract #2478. Lunar and Planetary Institute, Houston.

Moore J. M. (1984) The tectonic and volcanic history of Dione. *Icarus, 59,* 205–220.

Moore J. M. and Ahern J. L. (1983) The geology of Tethys. *J. Geophys. Res., 88,* 577–584, DOI: 10.1029/JB088iS02p0A577.

Moore J. M. and Schenk P. M. (2007) Topography and endogenic features on Saturnian mid-sized satellites. *Lunar Planet. Sci. XXXVIII, Abstract #2136. Lunar and Planetary Institute, Houston.*

Moore J. M., Horner V. M., and Greeley R. (1985) The geomorphology of Rhea: Implications for geologic history and surface processes. *J. Geophys. Res., 90,* C785–C795.

Moore J. M., Schenk P. M., Bruesch L. S., Asphaug E., and McKinnon W. B. (2004) Large impact features on mid-sized icy satellites. *Icarus, 171,* 421–443, DOI: 10.1016/j.icarus.2004.05.009.

Moore J. and 42 colleagues (2016) The geology of Pluto and Charon through the eyes of New Horizons. *Science, 351,* 1284–1293.

Nimmo F. and Matsuyama I. (2007) Reorientation of icy satellites by impact basins. *Geophys. Res. Lett., 34,* L19203, DOI: 10.1029/2007GL030798.

Nimmo F. and Spencer J. (2014) Powering Triton's recent geological activity by obliquity tides: Implications for Pluto geology. *Icarus, 246,* 2–10.

Nimmo F., Bills B. G., Thomas P. C., and Asmar S. W. (2010) Geophysical implications of the long-wavelength topography of Rhea. *J. Geophys. Res., 115,* E10008, DOI: 10.1029/2010JE003604.

Nimmo F., Bills B. G., and Thomas P. C. (2011) Geophysical implications of the long-wavelength topography of the saturnian satellites. *J. Geophys. Res., 116,* E11001, DOI: 10.1029/2011JE003835.

Ojakangas G. W. and Stevenson D. J. (1986) Episodic volcanism of tidally heated satellites with application to Io. *Icarus, 66,* 341–358, DOI: 10.1016/0019-1035(83)90125-286)90163-6.

Ortiz J. and 72 colleagues (2017) The size, shape, density and ring of the dwarf planet Haumea from a stellar occultation. *Nature, 550,* 219–223.

Passey Q. R. (1983) Viscosity of the lithosphere of Enceladus. *Icarus, 53,* 105–120, DOI: 10.1016/0019-1035(83)90024-6.

Pechman J. B. and Melosh H. J. (1979) Global fracture patterns of a despun planet: Application to Mercury. *Icarus, 38,* 243–250.

Peterson G., Nimmo F., and Schenk P. (2015) Elastic thickness and heat flux estimates for the uranian satellite Ariel. *Icarus, 250,* 116–122.

Phillips C. B., Hammond N. P., Robuchon G., Nimmo F., Beyer R., and Roberts J. (2012) Stereo imaging, crater relaxation, and thermal histories of Rhea and Dione. *Lunar Planet. Sci. XLIII,* Abstract #2571. Lunar and Planetary Institute, Houston.

Phillips C. B., Hammond N. P., Roberts J. H., Nimmo F., Beyer R. A., and Kattenhorn S. (2013) Stereo topography and subsurface thermal properties on icy satellites of Saturn. *Lunar Planet. Sci. XLIV,* Abstract #2766. Lunar and Planetary Institute, Houston.

Pike R. J. (1974) Depth/diameter relations of fresh lunar craters: Revision from spacecraft data. *Geophys. Res. Lett., 1,* 291–294, DOI: 10.1029/GL001i007p00291.

Pike R .J. (1980) Control of crater morphology by gravity and target type: Mars, Earth, Moon. *Proc. Lunar Planet. Sci. Conf. 11th,* pp. 2159–2189. Lunar and Planetary Institute, Houston.

Plescia J. B. (1983) The geology of Dione. *Icarus, 56,* 255–277.

Porco C. and 19 colleagues (2004) Cassini imaging science: Instrument characteristics and anticipated scientific investigations at Saturn. *Space Sci. Rev., 115,* 363–497.

Porco C. and 40 colleagues (2005) Cassini imaging science: Initial results on Phoebe and Iapetus. *Science, 307,* 1237–1242, DOI: 10.1126/science.1107981.

Prockter L. M. (2004) Ice volcanism on Jupiter's moons and beyond. In *Volcanic Worlds: Exploring The Solar System's Volcanoes* (R. M. C. Lopes and T. K. P. Gregg, eds.), pp. 145–177. Springer-Verlag, Berlin.

Rhoden A., Henning W., Hurford T., Patthoff A., and Tajeddine R. (2017) The implications of tides on the Mimas ocean hypothesis. *J. Geophys. Res., 122,* 400–410.

Rivera-Valentin E. G., Barr A. C., Lopez Garcia E. J., Kirchoff M. R., and Schenk P. M. (2014) Constraints on planetesimal disk mass from the cratering record and equatorial ridge on Iapetus. *Astrophys. J., 792(2),* 127, DOI: 10.1088/0004-637X/792/2/127.

Sandwell D. and Schubert G. (2010) A contraction model for the flattening and equatorial ridge of Iapetus. *Icarus, 210,* 817–822.

Schenk P. M. (1989) Crater formation and modification on the icy satellites of Uranus and Saturn: Depth/diameter and central peak occurrence. *J. Geophys. Res., 94,* 3813–3832, DOI: 10.1029/JB094iB04p03813.

Schenk P. M. (1991) Fluid volcanism on Miranda and Ariel: Flow morphology and composition. *J. Geophys. Res., 96,* 1887–1906, DOI: 10.1029/90JB01604.

Schenk P. M. (1993) Central pit and dome craters: Exposing the interiors of Ganymede and Callisto. *J. Geophys. Res., 98,* 7475–7498.

Schenk P. (2010) *Atlas of Galilean Satellites.* Cambridge Univ., Cambridge. 406 pp.

Schenk P. (2011) Geology of Mimas? *Lunar Planet. Sci. XLII,* Abstract #2729. Lunar and Planetary Institute, Houston.

Schenk P. (2014) *Icy Moons. https://www.lpi.usra.edu/icy_moons/* [accessed July 9, 2018], Lunar and Planetary Institute, Houston.

Schenk P. and McKinnon W. (2009) One-hundred-km-scale basins on Enceladus: Evidence for an active ice shell. *Geophys. Res. Lett., 36,* L16202.

Schenk P. M. and Moore J. M. (2007) Impact crater topography and morphology on saturnian mid-sized satellites. *Lunar Planet. Sci. XXXVIII,* Abstract #2305. Lunar and Planetary Institute, Houston.

Schenk P. and Murphy S. (2011) The rayed craters of Saturn's icy satellites (including Iapetus): Current impactor populations and origins. *Lunar Planet. Sci. XLII,* Abstract #2098. Lunar and Planetary Institute, Houston.

Schenk P. M. and Turtle E. P. (2009) Europa's impact craters: Probes of the icy shell. In *Europa* (R. T. Pappalardo et al., eds.), pp. 181–198. Univ. of Arizona, Tucson.

Schenk P. M., McKinnon W. B., Gwynn D., and Moore J. M. (2001) Flooding of Ganymede's bright terrains by low-viscosity water-ice lavas. *Nature, 410,* 57–60.

Schenk P., Chapman C., Zahnle K., and Moore J. (2004) Ages and interiors, the cratering record of the Galilean satellites. In *Jupiter: The Planets, Satellites, and Magnetospheres* (F. Bagenal et al., eds.), pp. 427–456. Cambridge Univ., New York.

Schenk P. M., Hamilton D. P., Johnson R. E., McKinnon W. B., Paranicas C., Schmidt J., and Showalter M. R. (2011) Plasma, plumes and rings: Saturn system dynamics as recorded in global color patterns on its mid-sized icy satellites. *Icarus, 211,* 740–757, DOI: 10.1016/j.icarus.2010.08.016.

Schenk P. M., Buratti B. J., Byrne P. K., McKinnon W. B., Nimmo F., and Scipioni F. (2015) "Blood stains" on Tethys: Evidence for recent activity? *European Planet. Sci. Congr., 2015,* EPSC2015-893.

Schenk P., Hoogenboom T., and Kirchoff M. (2017) Auto-secondaries on a mid-sized icy moon: Bright rayed crater Inktomi (Rhea). *Lunar Planet. Sci. XLVIII,* Abstract #2686. Lunar and Planetary Institute, Houston.

Schenk P. and 12 colleagues (2018a) The central pit and dome at Cerealia facula bright deposit and floor deposits in Occator Crater, Ceres: Morphology, comparisons and formation. *Icarus,* in press.

Schenk P. and 15 colleagues (2018b) Canyons, craters, and volcanism: Global cartography and topography of Pluto's moon Charon from New Horizons. *Icarus,* in press.

Schubert G., Spohn T., and Reynolds R. T. (1986) Thermal histories, compositions and internal structures of the moons of the solar system. In *Satellites* (J. A. Burns and M. S. Matthews, eds.), pp. 224–292. Univ. of Arizona, Tucson.

Schubert G., Anderson J. D., Travis B. J., and Palguta J. (2007) Enceladus: Present internal structure and differentiation by early and long-term radiogenic heating. *Icarus, 188,* 345–355, DOI: 10.1016/j.icarus.2006.12.012.

Schultz P. H. and Gault D. E. (1975) Seismic effects from major basin formations on the Moon and Mercury. *Earth Moon Planets, 12(2),* 159–177.

Scipioni F., Schenk P., Tosi F., D'Aversa E., Clark R., Combe J.-Ph., and Dalle Ore C. M. (2017) Deciphering sub-micron ice particles on Enceladus surface. *Icarus, 290,* 183–200, DOI: 10.1016/j.icarus.2017.02.012.

Shoji D., Hussmann H., Kurita K., and Sohl F. (2013) Ice rheology and tidal heating of Enceladus. *Icarus, 226,* 10–19, DOI: 10.1016/j.icarus.2013.05.004.

Shoji D., Hussmann H., Sohl F., and Kurita K. (2014) Non-steady state tidal heating of Enceladus. *Icarus, 235,* 75–85, DOI: 10.1016/j.icarus.2014.03.006.

Singer K. and McKinnon W. (2011) Tectonics on Iapetus: Despinning, respinning, or something completely different? *Icarus, 216,* 198–211.

Singer K. N., McKinnon W. B., Schenk P. M., and Moore J. M. (2012) Massive ice avalanches on Iapetus mobilized by friction reduction during flash heating. *Nature Geosci., 5,* 574–578, DOI: 10.1038/ngeo1526.

Smith B. A. et al. (1981) Encounter with Saturn: Voyager 1 imaging science results. *Science, 212,* 163–191, DOI: 10.1126/science.212.4491.163.

Smith B. A. et al. (1982) A new look at the Saturn system: The Voyager 2 images. *Science, 215,* 505–537, DOI: 10.1126/science.215.4532.504.

Smith B. A. and the Voyager Imaging Team (1986) Voyager 2 in the uranian system: Imaging science results. *Science, 233,* 43–64.

Spaun N. A., Head J. W., Pappalardo R. T., and the Galileo SSI Team (2001) Scalloped depressions on Ganymede from Galileo (G28) very high resolution imaging. *Lunar Planet. Sci. XXXI,* Abstract #1448. Lunar and Planetary Institute, Houston.

Spencer J. R. and Denk T. (2010) Formation of Iapetus' extreme albedo dichotomy by exogenically triggered thermal ice migration. *Science, 327,* 432–435, DOI: 10.1126/science.1177132.

Stephan K. and 15 colleagues (2010) Dione's spectral and geological properties. *Icarus, 206,* 631–652.

Stephan K. and 19 colleagues (2012) The saturnine moon Rhea as seen by Cassini VIMS. *Planet. Space Sci., 61,* 142–160.

Stephan K. and 15 colleagues (2016) Cassini's geological and compositional view of Tethys. *Icarus, 274,* 1–22.

Tajeddine R., Rambaux N., Laine V., Charno S., Richard A., Rivoldin A., and Noyelle B. (2014) Constraints on Mimas' interior from Cassini ISS libration measurements. *Science, 346,* 322–324, DOI: 10.1126/science.1255299.

Tarlow S. and Collins G. C. (2010) Fault scarp offsets and fault population analysis on Dione. Abstract P21B-1602 presented at the 2010 Fall Meeting, AGU, San Francisco, California, Dec. 13–17.

Thomas P. C. (2010) Sizes, shapes, and derived properties of the saturnian satellites after the Cassini nominal mission. *Icarus, 208,* 395–401, DOI: 10.1016/j.icarus.2010.01.025.

Thomas P. J. and Squyres S. W. (1988) Relaxation of impact basins on icy satellites. *J. Geophys. Res., 93,* 14919–14932.

Thomas P. C., Burns J. A., Helfenstein P., Squyres S., Veverka J., Porco C., Turtle E. P., McEwen A., Denk T., Giese B., Roatsch T. V., and Jacobson R. A. (2007) Shapes of the saturnian icy satellites and their significance. *Icarus, 190,* 573–584, DOI: 10.1016/j.icarus.2007.03.012.

Thomas P. C., Tajeddine R., Tiscareno M. S., Burns J. A., Joseph J., Loredo T. J., Helfenstein P., and Porco C. (2016) Enceladus's measured physical libration requires a global subsurface ocean. *Icarus, 264,* 37–47, DOI: 10.1016/j.icarus.2015.08.037.

Turcotte D. L. and Schubert G (2002) *Geodynamics*. Wiley, New York. 456 pp.

Wagner R. J., Neukum G., Giese B., Roatsch T., Wolf U., Denk T., and the Cassini ISS Team (2006) Geology, ages and topography of Saturn's satellite Dione observed by the Cassini ISS camera. *Lunar Planet. Sci. XXXVII,* Abstract #1805. Lunar and Planetary Institute, Houston.

Wagner R. J., Neukum G., Giese B., Roatsch T., and Wolf U. (2007) Fault scarp offsets and fault population analysis on Dione. *Lunar Planet. Sci. XXXVIII,* Abstract #1958. Lunar and Planetary Institute, Houston.

Wagner R. J., Neukum G., Stephan K., Roatsch T., Wolf U., and Porco C. C. (2009) Stratigraphy of tectonic features on Saturn's satellite Dione derived from Cassini ISS camera data. *Lunar Planet. Sci. XL,* Abstract #2142. Lunar and Planetary Institute, Houston.

Wagner R. J., Neukum G., Giese B., Roatsch T., Denk T., Wolf U., and Porco C. C. 2010. The geology of Rhea: A first look at the ISS camera data from orbit 121 (Nov. 21, 2009) in Cassini's Extended Mission. *Lunar Planet. Sci. Conf. XLI,* Abstract #1672. Lunar and Planetary Institute, Houston.

Watts A. W., Greeley R., and Melosh H. J. (1991) The formation of terrains antipodal to major impacts. *Icarus, 93,* 159–168.

White O. L., Schenk P. M., and Dombard A. J. (2013) Impact basin relaxation on Rhea and Iapetus and relation to past heat flow. *Icarus, 223,* 699–709, DOI: 10.1016/j.icarus.2013.01.013.

White O. L., Schenk P. M., Bellagamba A. W., Grimm A. M., Dombard A. J., and Bray V. J. (2017) Impact crater relaxation on Dione and Tethys and relation to past heat flow. *Icarus, 288,* 37–52, DOI: 10.1016/j.icarus.2017.01.025.

Williams K. K. and Zuber M. T. (1998) Measurement and analysis of lunar basin depths from Clementine altimetry. *Icarus, 131,* 107–122.

Wohletz K. H. and Sheridan M. F. (1983) Martian rampart crater ejecta: Experiments and analysis of melt-water interaction. *Icarus, 56,* 15–37, DOI: 10.1016/0019-1035(83)90125-2.

Zahnle K., Schenk P., Sobieszczyk S., Dones L., and Levison H. (2001) Differential cratering of synchronously rotating satellites by ecliptic comets. *Icarus, 153,* 111–129.

Kirchoff M. R., Bierhaus E. B., Dones L., Robbins S. J., Singer K. N., Wagner R. J., and Zahnle K. J. (2018) Cratering histories in the saturnian system. In *Enceladus and the Icy Moons of Saturn* (P. M. Schenk et al., eds.), pp. 267–284. Univ. of Arizona, Tucson, DOI: 10.2458/azu_uapress_9780816537075-ch013.

Cratering Histories in the Saturnian System

Michelle R. Kirchoff
Southwest Research Institute

Edward B. Bierhaus
Lockheed Martin

Luke Dones, Stuart J. Robbins, and Kelsi N. Singer
Southwest Research Institute

Roland J. Wagner
German Aerospace Center–DLR

Kevin J. Zahnle
NASA Ames Research Center

We review the state of knowledge on the cratering histories of the saturnian satellites Mimas, Enceladus, Tethys, Dione, Rhea, and Iapetus. The Voyager spacecraft yielded the first look at crater populations of these satellites, and the Cassini mission has provided extensive, new data to better understand these populations. Meanwhile, continued observations of outer solar system small-body populations and improvements in dynamical models have increased our knowledge of the impactor populations, which is also vital to understanding cratering histories. The surfaces of Mimas, Rhea, and Iapetus appear to be composed of old, densely cratered terrain that likely date from an early era of heavy bombardment. Along with having areas of this old terrain, Enceladus, Dione, and Tethys also have large regions of younger terrain indicated by their lower crater densities. Absolute model ages of these terrains range from ~4 Ga to being currently active (south polar terrain on Enceladus).

1. INTRODUCTION

Impact craters are the primary tool that planetary scientists have to understand the population of impactors that created them, and they are a key tool to assign relative ages to the formation of different surfaces across the solar system. Small-body populations are only observable to a minimum size limited by object brightness and detector technology, and given the outer solar system's distance from Earth, the smallest known bodies would produce enormous craters if they struck the moons of the outer planets. Craters therefore probe the small, unobserved population of bodies in the outer solar system. Reliable radiometric ages of surfaces are limited to Earth and the Moon, and therefore the buildup of craters over time is the only method available on all other solid solar system bodies to derive absolute model ages. To use this incredibly valuable and useful probe requires an interplay between observations and modeling.

Observationally, the craters must first be identified and measured. This requires sufficient image resolution (typically, limits of crater identification are diameters corresponding to ≈6–10 pixels across), sufficient solar incidence (typically the Sun more than 50° away from noon, with the best incidence angles ≈60–80° relative to noon Sun), sufficient exposure duration (such that signal-to-noise is reasonable for the identification of features), and sufficient emission angle (such that foreshortening is not too extreme to hinder both identification and accurate measurement). While the Voyager 1 and 2 spacecraft returned nearly 2000 images of Saturn's moons, the vast majority were for astrometric measurements, Titan's clouds, or full-disk images that permitted little detailed crater investigation relative to the >43,000 satellite images returned by Cassini as of this report.

Most solar system bodies — including many moons of Saturn — display varied geology that requires at least a basic level of geologic mapping. This separates the terrain by type, such that potentially different crater populations may be better distinguished in analysis. Once the crater population has been established by the researcher, it can be compared with models of the impactor populations to constrain the models.

In the inner solar system, the impactors are predominantly asteroids (*Neukum et al.,* 2001; *Bottke et al.,* 2002; *Strom et al.,* 2005; *Marchi et al.,* 2009) and the asteroid population is well known to magnitude +18, corresponding to ≈1-km-diameter bodies (*Harris and D'Abramo,* 2015). Additionally, the bodies in the inner solar system have very little exchange of material between themselves. This means that impacts are limited almost exclusively to heliocentric bodies (mostly asteroids) and ejecta fragments, which can form secondary impact craters (e.g., *McEwen and Bierhaus,* 2006; *Singer et al.,* 2014; *Bierhaus et al.,* 2018).

In the outer solar system, the impactors are primarily "comets," and the size distributions of their sources (Kuiper belt objects and Oort cloud bodies) are significantly less well constrained (see section 3 for details). Additionally, these heliocentric impactors can also have a significantly different velocity distribution than asteroids [e.g., compare *Yue et al.* (2013), for asteroids striking the Moon, with *Zahnle et al.* (1993) for comets striking Ganymede, Callisto, and Titan]. In the outer solar system, ejecta fragments can form both proximal and distal secondary craters, but can also be ejected to escape velocity and return to impact the parent satellites, or may impact a different satellite entirely. These planetocentric impactors would then form what are termed "sesquinary" impact craters (from a prefix meaning "one-and-a-half"), indicating that they are intermediate in character between primary and secondary craters.

In addition to this more complex impactor population, the flux of impactors over time may have changed, as may have the heliocentric population's size distribution. The Nice model (*Tsiganis et al.,* 2005; *Morbidelli et al.,* 2005; *Gomes et al.,* 2005) invokes an instability in the orbits of the giant planets that would have unleashed a flood of small bodies from the primordial Kuiper belt. For tens of millions of years, the impact rate on Saturn's moons (and other solar system bodies) would have been orders of magnitude higher than it is at present [see, e.g., *Rivera-Valentin et al.* (2014) and *Movshovitz et al.* (2015); see *Dones et al.* (2015) for a review of more recent versions of the Nice model]. Finally, the populations of objects themselves may have changed over time as they collide to produce a collisionally evolved population, but this may be significantly slowed in the outer solar system relative to its rate in the asteroid belt (e.g., *Nesvorný et al.,* 2011; *Campo Bagatin and Benavidez,* 2012).

All these different factors act together to create the observed crater population: a changing impactor flux, a changing impactor size distribution, heliocentric vs. planetocentric impactors, and potential secondary craters on larger bodies. However, the relative importance of each of these at Saturn, and as a function of distance from Saturn, is unknown. While they can be informed by very limited observations of the present-day impact flux, the predominant tool to understand them is models. By comparing these models to the observed crater population on each body, we can better constrain the roles of each component.

Therefore, the crater population on the saturnian satellites is an important constraint on these models. In turn, by better understanding each component of the impactor population, a better model for the overall impact history of the system can be determined. With a model for the impact history in hand, the crater spatial density itself can be used for estimation of absolute model ages.

2. CRATER DISTRIBUTIONS

2.1. Crater Size-Frequency Distributions

Analysis of impact crater size-frequency distributions (SFDs) on the mid-sized saturnian satellites, which for this chapter are defined as Mimas, Enceladus, Tethys, Dione, Rhea, and Iapetus, began with the return of images from Voyager 1 and 2 and were reported by *Smith et al.* (1981, 1982). Several researchers built on this initial analysis in the following years and continued to refine or revise interpretations (*Shoemaker and Wolfe,* 1981; *Strom,* 1981; *Strom and Woronow,* 1982; *Plescia and Boyce,* 1982, 1983, 1985; *Plescia,* 1983; *Moore,* 1984; *Horedt and Neukum,* 1984; *Morrison et al.,* 1984, 1986; *Chapman and McKinnon,* 1986; *Lissauer et al.,* 1988). One primary interpretation from these studies was that two crater populations were indicated by the data (*Shoemaker and Wolfe,* 1981; *Smith et al.,* 1981, 1982; *Strom and Woronow,* 1982; *Horedt and Neukum,* 1984; *Plescia and Boyce,* 1985; *Chapman and McKinnon,* 1986). Neutral names were assigned by these authors: "Population I" and "Population II". Population I craters looked similar to those observed on the galilean satellites, with shallow SFD slopes (ratio of small to large craters is smaller than average for crater populations throughout the solar system) at small crater diameters [diameter (D) < 20 km] and a multitude of large craters (D > 20 km). Population I is thought to be most clearly expressed on Rhea. Population II was characterized by a lack of larger craters (D > 20 km) and a higher density of smaller craters (D < 20 km). It appeared to be strongly exhibited on Mimas and the young terrains of Dione, Tethys, and Enceladus. Population I was considered to be harder to observe, since it was likely obscured by Population II, which was thought to be younger. Furthermore, one could not appear to identify a particular crater by its appearance as belonging either to Population I or Population II.

Meanwhile, *Lissauer et al.* (1988) proposed an alternative — that there was only one crater population and differences in crater SFDs were due to crater saturation (see section 2.3). However, imaging at the time was insufficient to determine if small craters are in saturation equilibrium.

In 2004, Cassini went into orbit around Saturn and soon started sending back higher-resolution images of the mid-sized saturnian satellites of terrains not seen before, along with those previously imaged by the Voyager missions. After 12 years of the Cassini mission, we now have near-global views of all the satellites, at pixel scales ranging from 100 to 500 m/pixel (*Roatsch et al.,* 2009, 2012, 2013). These global views have not only allowed a better understanding of the crater SFDs across the satellites' surfaces on a variety of terrains, but the the extension of crater SFDs to smaller diameters (D < 5–10 km). Multiple researchers have utilized these images to obtain crater SFDs for all the mid-sized saturnian satellites (*Neukum et al.,* 2005, 2006; *Porco et al.,*

2005, 2006; *Wagner et al.,* 2006, 2007, 2008; *Schmedemann et al.,* 2008, 2009; *Kirchoff and Schenk,* 2009, 2010; *Stephan et al.,* 2010, 2012; *Jaumann et al.,* 2011; *Schmedemann and Neukum,* 2011; *Bierhaus et al.,* 2012; *Robbins et al.,* 2014, 2015; *Damptz et al.,* 2017). There is good agreement in crater SFDs among these researchers, as demonstrated by comparing data from Mimas in Fig. 1. Furthermore, crater SFDs from Voyager-era analyses, where diameter ranges overlap, are also similar. Only one significant difference (i.e., outside of error) is indicated: a lower density for craters D = 7–20 km for *Bierhaus et al.* (2012). This may be explained by variation in the age of the terrain examined. *Bierhaus et al.* (2012) concentrated on the region around Herschel basin, which is less densely cratered than the terrains examined by other researchers. However, the shape of the crater SFD is generally similar to that found by other researchers. We also note that similar crater SFD shapes for D < 20 km are found on the inner, small satellites of Saturn (see the chapter by Thomas et al. in this volume).

Cassini crater SFDs of densely cratered terrains still appear to indicate two crater populations (Fig. 2), but they may not have the exact same characteristics as described from the Voyager data. Overall, the crater SFDs for Mimas, Tethys, and Dione have a different shape than those for Rhea and Iapetus. All satellite crater SFDs appear to have increasing R-values with increasing D, and thus shallow slopes, for small diameters. However, the diameter at which the slope changes is different for Mimas, Dione, and Tethys compared to Rhea and Iapetus. Furthermore, the change in the slope is different for these two groups. Mimas, Dione, and Tethys crater SFDs start to have steeper slopes (a higher ratio of small to large craters than average) at D = 10–20 km (until D ≈ 200 km). Meanwhile, the crater SFDs for Rhea and Iapetus turn over to relatively flat R values at D = 7–10 km (until D ≈ 100 km), such that the ratio of small to large craters is close to average for the inner solar system. All satellites with the exception of Mimas, which has poor statistics for D > 100 km (only Herschel is that large), have steep slopes from D ≈ 100–200 km and shallow slopes for D > 200 km. Finally, SFDs for densely cratered terrains on Enceladus show a similar shape trend to Mimas, Dione, and Tethys; however, the change from a shallow to a steep slope occurs at smaller diameters (D = 3–6 km).

Although the variation in diameter of where the slope changes occur might be explained by different impact speeds on these satellites [assuming heliocentric impactors, impact speeds will be higher closer to Saturn, which produces larger craters for a given impactor (*Zahnle et al.,* 2003)], there are three other potential explanations for the differences in *slope*: different impactor populations, crater saturation, and surface geologic activity. The first is discussed further in section 4 and the second in section 2.2. Here we focus on the possibility of surface geologic activity modifying craters to produce the different slopes. These processes would need to remove smaller craters (D < 20 km) on Iapetus and Rhea and D > 20 km on Mimas, Enceladus, Tethys, and Dione. Identified processes on the saturnian satellites that remove craters are cryovolcanism, tectonism, and viscous relaxation (*Parmentier and Head,* 1981; *Smith et al.,* 1981, 1982; *Plescia,* 1983; *Moore,* 1984; *Plescia and Boyce,* 1985; *Porco et al.,* 2006;

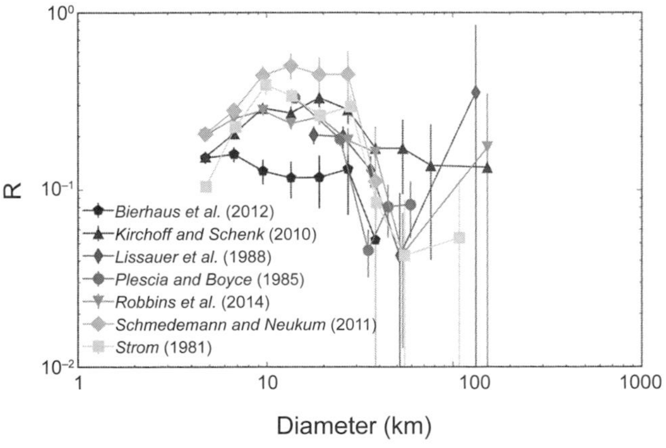

Fig. 1. Relative (R) plot (*Crater Analysis Techniques Working Group,* 1979) of crater SFDs for Mimas from different researchers using Voyager and Cassini imaging. R-value is plotted vs. geometric mean of a crater diameter bin (D_{bin}). R-value is computed by dividing the differential density of craters within a diameter bin by D_{bin}^{-3}. Bins are standard $2^{1/2}$D, except for those of *Plescia and Boyce* (1985), which are 10 km in size. In some cases bin size is doubled for large craters where data is sparse. Error bars are $N^{1/2}$, where N is the number of craters in a bin. Data for *Plescia and Boyce* (1985) are from their "sp" terrain.

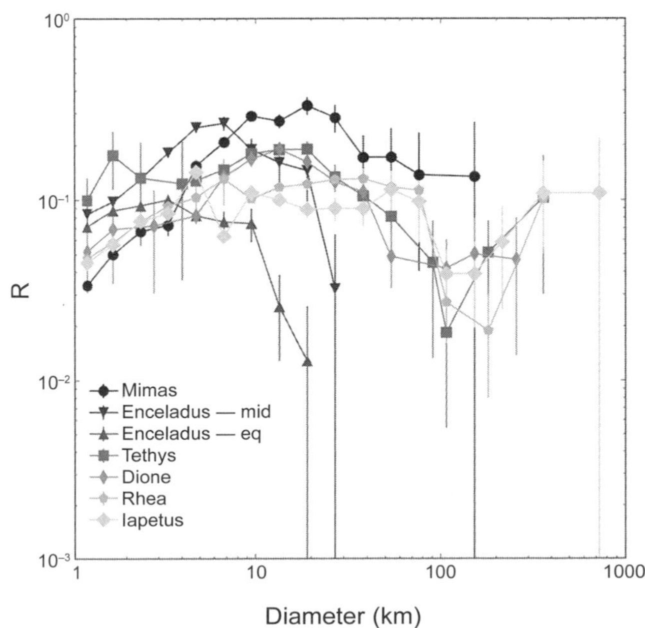

Fig. 2. Comparison of relative (R) crater SFDs from densely cratered terrains of the mid-sized saturnian satellites for D ≥ 1 km. Data compiled from *Kirchoff and Schenk* (2009, 2010, 2015) and *Robbins et al.* (2014). Diameters are not scaled to a common impact velocity.

Wagner et al., 2006; *Bray et al.,* 2007; *Giese et al.,* 2007; *Smith et al.,* 2007; *Chen and Nimmo,* 2008; *Smith-Konter and Pappalardo,* 2008; *Kirchoff and Schenk,* 2009, 2015; *Schenk and Moore,* 2009; *Hammond et al.,* 2011; *Bland et al.,* 2012; *Phillips et al.,* 2012; *White et al.,* 2013, 2017; *Yin and Pappalardo,* 2015; *Martin,* 2016; see also the chapters by Patterson et al. and Schenk et al. in this volume).

In principle, cryovolcanic flows can be an efficient way to erase craters [it is unclear how efficient cryovolcanic plumes, such as those on Enceladus, are at erasing craters (see *Bland et al.,* 2012; *Hirata et al.,* 2014; Patterson et al., this volume)]. In general, higher flow volumes result in more and larger craters removed. Thus, cryovolcanism could potentially explain the lower number of smaller craters on Iapetus and Rhea, except that no evidence for cryovolcanism has been found on these satellites. Meanwhile, cryovolcanism never covers only large craters, and therefore, it cannot likely explain the reduced number of large craters in the densely cratered terrains of Mimas, Enceladus, Tethys, and Dione. Dione has the strongest evidence for cryovolcanic flows with its smooth plains and enigmatic crater-like features ["depression complexes" (*Schenk and Moore,* 2009; *Kirchoff and Schenk,* 2015; Schenk et al., this volume)], but the reduced crater density at all observed diameters in the smooth plains indicates that both small and large craters were erased (Fig. 3).

A similar argument for tectonism can be made. The formation of graben and ridges more efficiently erases smaller craters than large, which is particularly evident on

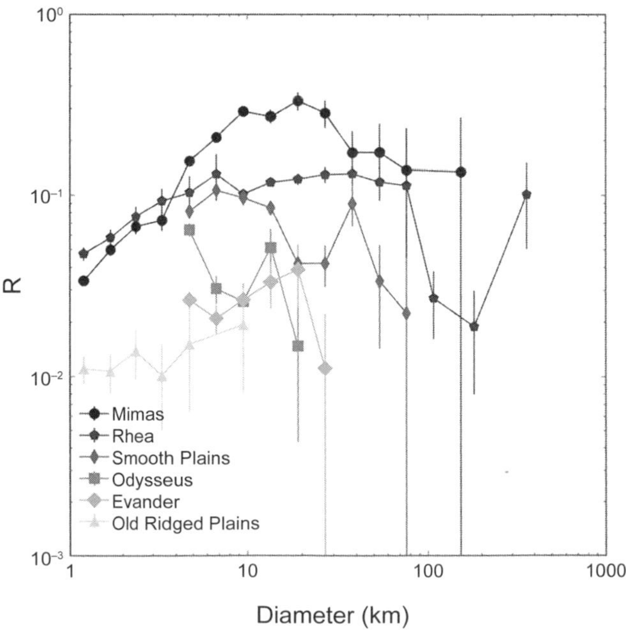

Fig. 3. Comparison of relative (R) crater SFDs from densely cratered terrains to young terrains. Mimas and Rhea are plotted to represent the two end-member distributions of densely cratered terrains. To represent younger terrains we plot the smooth plains and Evander basin on Dione (*Kirchoff and Schenk,* 2015), Odysseus on Tethys (*Kirchoff and Schenk,* 2010), and old ridged plains on Enceladus [rp6 from *Kirchoff and Schenk* (2009); 20°N, 225°W]. Diameters are not scaled to a common impact velocity.

Enceladus, where remnants of large craters are observed in the tectonized plains of the leading and trailing hemispheres, but few smaller craters are found except those formed after tectonic processes were complete (e.g., see the chapter by Patterson et al. in this volume). Therefore, tectonism could explain the lack of smaller craters on Iapetus and Rhea, except there are no large, cohesive areas of tectonic features (e.g., Schenk et al., this volume). However, it cannot explain the lack of larger craters in the densely cratered regions of Mimas, Enceladus, Tethys, and Dione, while smaller craters remain, even though large tectonized areas are observed on all except Mimas (e.g., Schenk et al., this volume).

Finally, viscous relaxation has been shown to be an important process in modifying craters on all icy satellites. Furthermore, it is a wavelength-dependent process and is more effective at modifying larger craters than small. Thus, viscous relaxation could potentially be responsible for removing large craters on Mimas, Enceladus, Tethys, and Dione. Partially relaxed craters have been observed and higher heat flows inferred in localized areas on Enceladus, Tethys, and Dione (*Bray et al.,* 2007; *Smith et al.,* 2007; *Bland et al.,* 2012; *Phillips et al.,* 2012; *White et al.,* 2017; Patterson et al., this volume; Schenk et al., this volume). However, viscous relaxation does not tend to completely remove a crater, and therefore this process alone cannot account for fewer large craters on these satellites, such as has been suggested for Enceladus, where viscously relaxed craters may be buried by plume particles (*Kirchoff and Schenk,* 2009; *Hirata et al.,* 2014). In addition, since smaller craters are less affected, viscous relaxation is not plausible for erasing smaller craters on Rhea and Iapetus. Overall, crater erasure by geological processes does not seem to be a promising explanation for the differences observed in the crater SFD slopes.

Now we compare the end-member crater SFDs for the heavily cratered terrains (represented by Mimas and Rhea) to those for some younger terrains (Fig. 3). For most of the satellites, the youngest terrains are within large basins (such as Herschel on Mimas or Odysseus on Tethys). Enceladus and Dione, however, have large areas of less densely cratered plains (*Smith et al.,* 1981, 1982; *Plescia,* 1983; *Plescia and Boyce,* 1983; *Moore,* 1984; *Porco et al.,* 2006; *Wagner et al.,* 2006; *Kirchoff and Schenk,* 2009, 2015). Here we examine the smooth plains and Evander basin on Dione, older ridged plains on Enceladus [unit rp6 from *Kirchoff and Schenk* (2009)], and Odysseus on Tethys. Roughly, slope trends seem to reflect those seen for the densely cratered terrains of Mimas, Enceladus, Tethys, and Dione, with shallow slopes for D < 20 km and steep slopes for D > 20 km. However, we do not see the shallow slopes for Odysseus or Dione's smooth plains; their SFDs remain steep to smaller diameters. In contrast, the steep slope is not observed for Evander or Enceladus' old ridged plains. In both cases, this may be more due to lack of data, and not an indication of actual trends. More high-resolution data of young terrains could potentially help refine our knowledge of how crater populations changed with time, but we are ultimately limited by the area of young terrains available.

Finally, we compare saturnian satellite crater SFDs to those of other solar system worlds (Fig. 4). We again use Rhea and Mimas to represent Population I and II of the Saturn system, and compare these to the Neukum lunar crater production function [derived using equation (3) and Table 1 in *Neukum et al.* (2001)], a Callisto crater SFD (*Schenk et al.,* 2004), and a Charon crater SFD for Vulcan Planum [informal name (*Stern et al.,* 2015; *Moore et al.,* 2016; *Robbins et al.,* 2017; *Singer et al.,* 2017)]. Some of the same slope trends are observed, but they occur at different diameters depending on the impact speed inherent to the body. An important similarity is the shallow slopes observed at D ≲ 10–20 km for outer solar system worlds, which is different from the inner solar system, whose worlds have a steep slope. Even if the potential influence of secondary craters on the Moon, which have a steeper slope, is accounted for (*McEwen and Bierhaus,* 2006; *Robbins and Hynek,* 2014), the inner solar system crater population is steeper at these sizes (e.g., *Chapman and McKinnon,* 1986). This relative lack of small craters in the outer solar system appears to be an inherent difference with the inner solar system.

2.2. Apex-Antapex Asymmetry

When a satellite's rotational period is synchronously locked with its orbital period such that the same hemisphere always faces the planet, a crater density asymmetry is expected to develop between the leading and trailing hemispheres (*Shoemaker and Wolfe,* 1981; *Horedt and Neukum,* 1984; *Zahnle et al.,* 1998, 2001; *Valsecchi et al.,* 2014; *Hirata,* 2016) if the main impactor population is heliocentric. Specifically, the leading hemisphere is predicted to have a higher crater density than the trailing. The difference in crater density is dependent on several factors, including characteristics of the impactor population, and it is estimated that saturnian satellites should have 16–110× greater crater spatial density within 30° of the apex than within 30° of the antapex, where lower values are for ecliptic comet impactor populations with shallower size distributions, i.e., relatively fewer small bodies (*Zahnle et al.,* 2001). Several studies have searched for this asymmetry on satellites that appear to be unaffected by internal geological processes (Mimas, Rhea, Iapetus, and inner, small satellites) and in general little (<factor of 4 difference) to no asymmetry is found (*Horedt and Neukum,* 1984; *Zahnle et al.,* 2001; *Kirchoff and Schenk,* 2010; *Leliwa-Kopystyński et al.,* 2012; *Hirata,* 2016; see also the chapter by Thomas et al. in this volume). On the contrary, a couple of studies have potentially found a weak opposite asymmetry on Mimas (*Leliwa-Kopystyński et al.,* 2012; *Alvarellos et al.,* 2017).

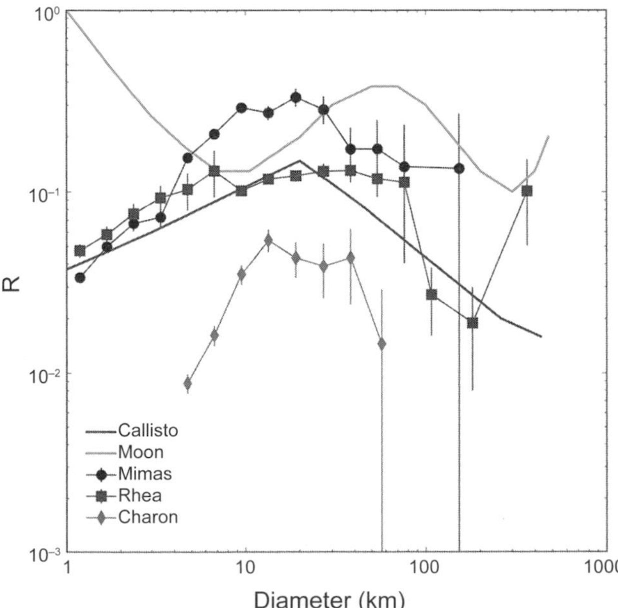

Fig. 4. R-plot of crater SFDs from Mimas and Rhea compared to densely cratered terrains on Callisto, Earth's Moon, and Pluto's moon Charon. Callisto data are from *Schenk et al.* (2004) and have a shape similar to the other outer solar system worlds. Lunar data from *Neukum et al.* (2001) and have much higher densities and steeper distribution at D < 10 km. Charon data from *Moore et al.* (2016). Diameters are not scaled to a common impact velocity.

The mismatch of observed asymmetry densities to those predicted has been hypothesized to result from a dominant nonheliocentric impactor population, impact crater saturation, or nonsynchronous rotation. The possibility of a dominant nonheliocentric impactor population is discussed in section 4 and impact crater saturation is discussed in the next section. Here we focus on nonsynchronous rotation as the potential cause of a reduced asymmetry.

Nonsynchronous rotation may occur for a satellite that either has its outer ice crust separated from the core by an ocean or if a large impact has temporarily caused the satellite to rotate faster or slower (*Lissauer,* 1985; *Zahnle et al.,* 2001; *Wieczorek and Le Feuvre,* 2009; *Goldreich and Mitchell,* 2010). An asymmetry is not expected to develop for nonsynchronous rotation because the ice shell rotates through the leading and trailing hemispheres, causing crater densities to be more uniformly distributed. There is little evidence that Rhea and Iapetus currently have an ocean or ever had one [Mimas has an unexpectedly large physical libration; one possible explanation is that Mimas has an ocean (*Tajeddine et al.,* 2014; although see *Rhoden et al.,* 2017)]. Furthermore, breaking synchronous rotation by large impacts is difficult (*Lissauer,* 1985; *Wieczorek and Le Feuvre,* 2009). Therefore, it seems unlikely that the reduced asymmetry on the saturnian satellites has been caused by nonsynchronous rotation.

2.3. Crater Saturation

Impact crater saturation is defined as when crater density has reached a point at which the formation of new craters erases the same area of older craters, and crater density has reached equilibrium (*Gault,* 1970). It is a concern when trying to use saturated crater distributions to derive characteristics of impactor populations, impact rates, and cratering histories because information has been lost.

There is general agreement in the literature that at least some, if not all, of the heavily cratered terrains of the mid-sized saturnian satellites are likely saturated (*Hartmann,* 1984; *Chapman and McKinnon,* 1986; *Strom,* 1987; *Lissauer et al.,* 1988; *Squyres et al.,* 1997; *Kirchoff and Schenk,* 2010; *Kirchoff,* 2018). In particular, crater SFDs between 5 and 30 km on Mimas, Enceladus, Tethys, and Dione do show high spatial densities often taken to represent crater saturation (R = 0.1–0.3). However, analysis of saturation equilibrium in the saturnian system is somewhat complicated because saturated crater SFDs for shallow sloped crater populations likely look different than saturated crater SFDs of steeper sloped crater populations of the inner solar system (*Chapman and McKinnon,* 1986; *Richardson,* 2009; *Hirabayashi et al.,* 2017). The crater SFDs likely do not reach true equilibrium density, but reach a quasi-equilibrium that is dominated by the formation of large craters. Moreover, crater SFDs do not evolve to a single slope as is common for the inner solar system (*Gault,* 1970), but have been shown to retain the changes in slope observed for a nonsaturated crater population (*Chapman and McKinnon,* 1986; *Richardson,* 2009; *Hirabayashi et al.,* 2017). These aspects can make it difficult to determine which saturnian system crater distributions are saturated, especially using crater SFDs alone. Thus, there has been work taking a different approach that combines analysis of the crater SFDs with studies of the spatial statistics of the crater distributions (*Lissauer et al.,* 1988; *Squyres et al.,* 1997; *Kirchoff,* 2018). For instance, the density of craters in different parts of a saturated surface varies less from region to region than would be expected for a random (Poisson) distribution. This work further supports the conclusion that densely cratered terrains on all the mid-sized satellites are likely saturated.

Evidence for saturation has also been inferred from the observations of the two crater populations (*Lissauer et al.,* 1988). In particular, saturation may be reached for different diameters at different times, which theoretically may result in changes to the crater SFD shapes (*Gault,* 1970; *Chapman and McKinnon,* 1986; *Lissauer et al.,* 1988). However, it has not yet been shown that the observed differences between Population I and II could be produced by these types of changes. Furthermore, saturation has been used to potentially explain the lack of an apex-antapex asymmetry (*Zahnle et al.,* 2001; *Kirchoff and Schenk,* 2010; *Hirata,* 2016). If the crater density has reached a high enough value to be in equilibrium, then it has also reached a value where any asymmetry would be difficult to discern. Nevertheless, these last two arguments have a viable alternative, as there may be more than one impactor population (see section 4).

3. PRESENT-DAY IMPACTOR POPULATIONS

3.1. Types of Craters: Primary, Secondary, Sesquinary

Impact craters on the saturnian satellites can be primary, secondary, or sesquinary (*Bierhaus et al.,* 2012). Primary craters are made by the direct impact of small bodies, typically comets or Centaurs (small bodies orbiting the Sun in the region of the giant planets). Secondary craters are those made by ejecta launched in a primary impact that never escape the gravity of the impacted body. The slowest ejecta land closest to the primary crater and create an ejecta blanket. Faster ejecta make secondary craters. Because of the low surface gravities of Saturn's moons (except Titan), much of the ejecta could fly far across a moon, so secondaries likely would not cluster near the primary crater as much as they do on larger bodies such as Mars, Mercury, the Moon, and Jupiter's Galilean satellites (*Schultz and Singer,* 1970; *McEwen and Bierhaus,* 2006, *Bierhaus et al.,* 2012, 2017), although recent observations may indicate secondary craters can form just over one crater diameter from a primary crater's rim (see the chapter by Schenk et al. in this volume). The fastest, "sesquinary" ejecta escape the satellite after a primary cratering event and go into orbit around Saturn [other short-lived planetocentric bodies, e.g., those produced in collisions between moons (*Ćuk et al.,* 2016), are not included in this category]. Most sesquinary ejecta are swept up by the source moon, but the orbits of some fragments can be perturbed enough [in particular, most Hyperion ejecta strike Titan (*Dobrovolskis and Lissauer,* 2004)], or the original ejection velocity can be big enough, that the ejecta impact another satellite (*Alvarellos et al.,* 2002, 2005, 2008, 2017; *Zahnle et al.,* 2008; *Singer et al.,* 2013). For Mimas and possibly Enceladus, the escape velocities are so small that there may be no traditional secondaries at all, but only sesquinaries (*Bierhaus et al.,* 2012). The very fastest could in principle escape the Saturn system entirely for outer satellites such as Iapetus, but this would be a tiny fraction of all ejecta.

3.2. Ecliptic Comets/Centaurs

By the mid-nineteenth century, three categories of cometary orbits were emerging. The largest group consisted of comets whose orbits were so vast that they could be fit with parabolas, or at least had semimajor axes (a) larger than Neptune's (30 AU). These comets had roughly equal numbers of prograde (in the same sense as the planets) and retrograde orbits. The next group had orbits with semimajor axes slightly interior to Uranus's orbit (19 AU) and a wide range of orbital inclinations, although most were prograde. The third group had much smaller orbits ($a \approx 3$ AU) and, in most cases, small inclinations to the ecliptic. These groups roughly match what we now call, respectively, long-period, Halley-type, and Jupiter-family comets (JFCs). The formation and evolution of the cometary reservoirs is discussed in detail by *Dones et al.* (2015).

The Oort cloud was proposed to be a long-lived reservoir for comets, possibly including the JFCs (*Oort,* 1950). However, it was not clear whether the round Oort cloud could supply enough JFCs, given the strong concentration of JFCs toward the ecliptic. *Joss* (1973) concluded that the Oort cloud was *not* an adequate source of JFCs, and *Fernández* (1980) inferred that a flattened "comet belt" beyond Neptune was likely to be the reservoir of the JFCs [called "short-period comets" by *Fernández* (1980)]. *Duncan et al.* (1988) bolstered this conclusion and proposed that the comet belt be called the Kuiper belt. Perturbations by Neptune and/or hypothetical massive bodies in the belt would perturb some objects in the Kuiper belt onto Neptune-crossing orbits. Some of those bodies, in turn, would in time reach orbits that crossed the orbits of the other giant planets, with those that became Jupiter-crossing being identified with the JFCs. The first Kuiper belt object (KBO) was discovered in 1992 (*Jewitt and Luu,* 1993).

The Kuiper belt is now known to have a complex structure (*Gladman et al.,* 2008), with dynamically distinct populations including a "classical" belt (*Petit et al.,* 2011): bodies such as Pluto, which completes two orbits for every three of Neptune, in mean-motion resonances with Neptune (*Gladman et al.,* 2012; *Pike et al.,* 2015; *Volk et al.,* 2016), and "scattered disk" objects (*Luu et al.,* 1997), which follow highly eccentric orbits with perihelion distances (q) generally between 30 and 40 AU. Scattered disk objects (SDOs) with $q \leq 37$ AU are perturbed strongly enough by Neptune that the orbits of most become Neptune-crossing during the age of the solar system (*Elliot et al.,* 2005; *Lykawka and Mukai,* 2007). Because SDOs have less stable orbits than other components of the Kuiper belt, they are a plausible source for the Centaurs and JFCs (*Duncan and Levison,* 1997; *Tiscareno and Malhotra,* 2003; *Volk and Malhotra,* 2008).

The main difficulty in estimating impact rates on saturnian moons is that the size distribution of Centaurs is unknown because discovery is very incomplete, except for the largest bodies. The crude arguments used to fill the gaps between JFCs and the Kuiper belt is presented in *Zahnle et al.* (2003), *Dones et al.* (2009), and section 4. As an example of the vast extrapolation required, *Brasser and Wang* (2015) infer from a set of 406 catalogued JFCs that the scattered disk contains $N_{SD} = 5.9\ (+2.2,-5.1) \times 10^9$ bodies with diameters >2.3 km, while *Nesvorný et al.* (2017) find $(3-9) \times 10^8$ such bodies — a factor of 10 fewer than *Brasser and Wang*'s (2015) nominal value, but within their error bars.These studies involve numerical integrations of bodies that escape the scattered disk and become comets that pass within 2 or 2.5 AU of the Sun. The biggest uncertainty in N_{SD} arises because comets are prone to vanish, often by falling apart spectacularly, as seen recently for Comet 332P/Ikeya-Murakami (*Jewitt et al.,* 2016). Therefore Brasser and Wang, Nesvorný et al., and other modelers (e.g., *Di Sisto et al.,* 2009) must invoke a "fading law," which quantifies cometary splitting, fading, and development of a crust that reduces outgassing. In addition, the sizes of the nuclei of most JFCs are inferred from the brightness of their comae, rather than from any certain knowledge of the solid chunk at the heart of the comet.

3.3. Long-Period and Halley-Type Comets

Long-period and Halley-type comets (LPCs and HTCs, respectively) have often been defined as comets with orbital periods in excess of 200 years and between 20 and 200 years, respectively. These correspond to semimajor axes >34 AU and 7–34 AU, respectively. A more modern taxonomy relies on a comet's Tisserand parameter, T, with respect to Jupiter, with comets with $T > 2$ corresponding to ecliptic comets and comets with $T < 2$ being "nearly isotropic" comets (LPCs and HTCs) (*Levison,* 1996) [see *Gladman et al.* (2008), *Tancredi* (2014), and *Jewitt et al.* (2015) for revised versions of Levison's classification scheme]. Most LPCs and HTCs are thought to originate in the Oort cloud (*Oort,* 1950; *Duncan et al.,* 1987; *Heisler,* 1990; *Wiegert and Tremaine,* 1999; *Levison et al.,* 2001, 2002; *Kaib and Quinn,* 2009; *Brasser and Morbidelli,* 2013; *Wang and Brasser,* 2014). A tightly bound "inner inner" Oort cloud containing bodies like Sedna may have formed early while the Sun was a member of a cluster (*Brasser,* 2008; *Brasser et al.,* 2008). After the Sun escaped the cluster, the vaster "classical" Oort cloud would form, its population peaking about 1 b.y. later.

Brasser and Morbidelli (2013) find that the Oort cloud contains $(7.6 \pm 3.3) \times 10^{10}$ comets with diameters (d) > 2.3 km. [*Nesvorný et al.* (2017) infer about 10^{10} Oort cloud comets of this size.] Using the estimate from *Brasser and Morbidelli* (2013), on the order of 10^8 Oort cloud comets are likely to be on Saturn-crossing orbits. However, because of their long periods of revolution around the Sun, LPCs and HTCs have fewer opportunities to traverse the Saturn system than do Centaurs. In addition, because they approach the Saturn system at higher velocities than do Centaurs, Saturn gravitationally focuses LPCs and HTCs much less than it does Centaurs. For these reasons, the impact flux due to LPCs on Saturn's moons interior to Titan's orbit is estimated to be only ~1% of the flux due to Centaurs (*Zahnle et al.,* 2003).

3.4. Main-Belt Asteroids

The census of asteroids is much more complete than that of comets. There are some 2×10^5 main-belt asteroids with $d > 2.3$ km (*Bottke et al.,* 2005; *Sheppard and Trujillo,* 2010). Although some asteroids leak out of the main belt and traverse the region of the giant planets, they are not expected to be an important source of impactors on saturnian moons, given their small numbers. Neukum and colleagues proposed that asteroids could be captured onto long-lived, low-to-moderate eccentricity orbits around Jupiter and Saturn (*Horedt and Neukum,* 1984), but have not suggested a dynamical mechanism for circularizing the orbits.

3.5. Other Possible Impactors

Members of several other classes of small bodies are capable, in principle, of striking Saturn's moons if they escape their source regions. These include the recently discovered main-belt comets (*Hsieh and Jewitt,* 2006) [also called "active asteroids" (*Jewitt et al.,* 2015)], Hildas in 3:2 resonance with Jupiter, Trojans in 1:1 resonance with Jupiter, and Trojans in 1:1 resonance with Neptune. Generally speaking, these bodies have low-to-moderate inclinations and should behave dynamically like Centaurs if they become Saturn-crossing. However, there are likely to be few such escaped objects, compared with the number of Centaurs of the same size, so they can be neglected when calculating cratering rates.

Like the other giant planets, Saturn has a retinue of distant "irregular" satellites (*Gladman et al.,* 2001), the biggest of which is Phoebe. These moons can collide with each other (*Nesvorný et al.,* 2003), and a much larger early population of irregular satellites may have been an important source of impactors for Iapetus (*Bottke et al.,* 2010). However, irregular satellites are not expected to be an important source of impactors for the moons within Titan's orbit. However, catastrophic breakup of small moons within Titan's orbit could be a significant additional source of impactors for mid-sized moons from Mimas to Rhea (*Movshovitz et al.,* 2015).

3.6. Impact Ejecta

Many craters on the Moon have rays [fine ejecta from the impact (*Pieters et al.,* 1985)], along with "associated gouges" (*Shoemaker,* 1962). Copernicus crater is the type example. *Shoemaker* (1962) interpreted these gouges as "secondary impact craters formed by individual large fragments or clusters of large fragments ejected from Copernicus." Secondary craters have subsequently been found on many other worlds, including Mars, Mercury, Europa, Ganymede, Vesta, Ceres, and Rhea. The largest secondary craters are 4–8% the diameter of their primary for a wide range of primary sizes on different bodies (*Allen,* 1979; *Robbins and Hynek,* 2011; *Singer et al.,* 2013, 2015). Several investigators have used the observed sizes of secondary craters as a function of distance from the primary to try to infer the sizes of the fragments that made the secondaries (e.g., *Vickery,* 1986; *Hirata and Nakamura,* 2006; *Singer et al.,* 2013, 2014). These studies assume that the fragments follow ballistic trajectories, so ejecta launched fastest fly farthest, and use a variety of crater-scaling relationships. They find a relationship $d_{max} \propto v^{-\beta}$, where d_{max} is the diameter of the biggest fragment ejected at velocity v, and β is an exponent that is found to have values between 0.2 and 3 (*Singer et al.,* 2015).

Bierhaus et al. (2012, 2018) have modeled the role of crater ejecta on Saturn's mid-sized moons, using impact rates from *Zahnle et al.* (2003) and crater-scaling relationships from *Housen and Holsapple* (2011). *Alvarellos et al.* (2005, 2017) investigated the orbital evolution of ejecta that escaped after large impacts on Mimas, Enceladus, Tethys, Dione, and Rhea to quantify the formation of sesquinary craters when the ejecta struck moons. As an example, for ejecta from the ~400-km-diameter Odysseus basin on Tethys, *Alvarellos et al.* (2005) found that the biggest sesquinary crater that could form would have a diameter of 19 km if the fragments were spalls (i.e., pieces that flew off the free surface of the target moon) or 10 km if the fragments were rubble. However, there is no firm evidence as yet that the moons of Jupiter or Saturn harbor a significant population of sesquinary craters (*Zahnle et al.,* 2008; *Bierhaus et al.,* 2012). The ejecta might break up while in orbit into much smaller bodies that then make craters too small to be detectable.

4. IMPACT RATES

4.1. Crater Chronologies

Recent cratering rates in the inner solar system can be determined by counting the craters on surfaces with radiometrically determined ages, recording observations of newly forming craters (*Daubar et al.,* 2013; *Speyerer et al.,* 2016), or by assessing the size, number, and orbital distributions of the current observable populations of asteroids and comets that can strike the terrestrial planets, as constrained by the observed crater populations; however, only the first technique can provide information about past cratering rates. The first method — termed the "historical record" here — uses the populations of craters that are observed on surfaces of known age estimated by other methods, such as radiometric dating. This is currently possible *only* for Earth and the Moon. For Earth, the cratering record is severely affected by erosion, so that it is probably limited to assessing only the largest craters that are most likely to be preserved. For the Moon, the record is dominated by craters on lunar maria and rocks that have been returned from the Moon and dated radiometrically. The historical record indicates that lunar cratering rates were significantly higher 3.8 b.y. ago than they have been since 3.2 Ga (e.g., *Hartmann et al.,* 2000; *Neukum et al.,* 2001; *Robbins,* 2014). Very little can be said of the lunar impact rate after 3.0 Ga other than to determine an average rate for the duration (e.g., *Neukum et al.,* 2001; *Kirchoff et al.,* 2013; *Robbins,* 2014).

In the second method, repeated imaging is used to find where new, small craters have formed. This technique has been applied on Mars (*Daubar et al.,* 2013) and the Moon (*Speyerer et al.,* 2016). However, this technique requires very-high-resolution imaging due to the relatively short timescales and low impact probabilities in the imaging baseline, which would require designated satellite orbiters or other detailed observations, such as impact flashes.

In the third method — termed "dead reckoning" here — the probability that any two objects will collide is estimated from orbital mechanics. The mutual velocity can be computed, although where an impact takes place, at what velocity, and at what incidence angle all have to be reckoned probabilistically. Impact probabilities are then summed over the known or estimated populations of potential impactors. At best, this method provides only the current rate. Then, with considerable extrapolation and comparison to observed crater SFDs, past rates could be estimated with significant uncertainty. Even now, with the current population of possible Earth-impacting asteroids rather well characterized, there is enough uncertainty that *Bottke et al.* (2016) have been willing to argue that the current cratering rate on Venus is a factor of 5 higher than it is nominally assumed (*Korycansky and Zahnle,* 2005).

Dead reckoning is the only approach currently available for the Saturn system. Furthermore, it is even more uncertain than for the inner solar system. There are four basic ingredients: how often a given satellite is struck by a heliocentric impactor, the physical properties of the impactors and their targets, the crater scaling rule one uses to predict the size of the craters made by these impactors in these targets, and estimation of the shape of the crater production function using the observed crater records.

How often large, heliocentric comets hit Saturn's satellites can be assessed from the observed population of big bodies in the outer solar system (either Centaurs in the orbital space bracketing Saturn, or from the historical record of comets encountering Jupiter), coupled to a model that relates the relative population of Jupiter-encountering comets to the population of Saturn-encountering comets. The known Centaurs, such as Chiron, are so large that their impact rates with Saturn's satellites are effectively zero. The population of largest Centaurs is therefore extrapolated to smaller sizes, over several orders of magnitude, to the <5-km-sized comets that matter most for cratering. The extrapolation to these small comets is guided by the SFDs of craters on lightly cratered terrains of Ganymede and Europa and by the observed current rate that kilometer-sized comets closely encounter and/or hit Jupiter (*Zahnle et al.,* 2003; *Hueso et al.,* 2013). Finally, the impact probabilities are mapped to a cratering rate using an impact crater scaling formula (e.g., *Housen and Holsapple,* 2011).

The shape (size-frequency distribution) of the crater production function is estimated from the observed crater records. It is necessary that the observed crater population be formed by heliocentric impactors, since the impact probabilities have been derived for these populations. However, whether the observed crater populations on the mid-sized saturnian satellites are formed primarily by heliocentric or planetocentric (a combination of impactors orbiting Saturn produced by catastrophic breakup of small satellites and sesquinary impact ejecta; see sections 3.5 and 3.6) impactors is yet unknown. The fact that the predicted large apex-antapex asymmetry from heliocentric impactors is not seen on any satellites (see section 2.2) could be a result of a mostly planetocentric impactor population, which will not have a preferred impact hemisphere (*Zahnle et al.,* 2001). Furthermore, if some satellites are more frequently struck by planetocentric debris, that could explain the two crater populations discussed in section 2. For example, it has become increasingly apparent over the last 15 years that the SFD of heliocentric comets is markedly deficient in small bodies compared with a collisional distribution (*Bierhaus et al.,* 2001; *O'Brien and Greenberg,* 2003; *Moore et al.,* 2016; *Robbins et al.,* 2017; *Singer et al.,* 2017), which could explain the shallow slopes of Population I for smaller craters. Furthermore, planetocentric impactors could be responsible for the steeper slope seen at smaller diameters for Population II.

The first estimate of current cratering rates in the outer solar system was developed by E. Shoemaker with the Voyager imaging team in the first set of real data from Jupiter and Saturn (*Smith et al.,* 1981, 1982; *Shoemaker and Wolfe,* 1981). They considered short-period comets, long-period comets, and asteroids as impactors of Jupiter's galilean satellites (see section 3). Asteroids were found to be currently relatively unimportant because there are few of them in Jupiter-crossing orbits compared to comets, and the long-period comets, which mostly have orbits highly inclined to the ecliptic, are relatively unlikely to encounter Jupiter and experience relatively little gravitational focusing. By contrast, short-period comets — especially those known as JFCs — are concentrated by Jupiter's gravity and are by far the greatest source of craters.

For Saturn, *Smith et al.* (1982) extrapolated the concept of JFCs to a hypothetical subclass of short-period comets they called the "Saturn family" that they calibrated to the one object then known, Chiron. [It has since turned out that Saturn-family comets are at most a small peak, much less than the concentration found at Jupiter, in the radial distribution of what are now called "ecliptic comets" (*Levison and Duncan,* 1997; *Tiscareno and Malhotra,* 2003; *Di Sisto and Brunini,* 2007).] *Smith et al.* (1982) made several other assumptions that differ to a greater or lesser extent from what might be chosen today; withal, their estimates for current cratering rates of the mid-sized icy satellites of Saturn were roughly a third of what would be recommended today (*Dones et al.,* 2009, Case A). They ignored the contributions of asteroids because it is difficult for conventional asteroids to pass Jupiter, where they are already a relatively small population, to reach Saturn. *Lissauer et al.* (1988) used similar arguments and obtained generally similar results. The biggest discrepancies are with Iapetus and Phoebe (for which modern estimates are lower) and for tiny moons [where estimates in *Dones et al.* (2009) are higher].

Zahnle et al. (2003) and *Dones et al.* (2009) considered cratering rates at Saturn by comets from two different reservoirs, one with an abundance of small comets (Case B) and the other depleted as the population is at Jupiter (Case A). However, the anecdotal argument that Cassini has seen no sign of a comet striking Saturn or its rings [although see *Tiscareno et al.* (2013) for evidence of much smaller — perhaps up to several-meter-sized — bodies striking Saturn's rings] and an apparent near absence of primary impact craters on the very lightly cratered terrains of Enceladus (*Porco et al.,* 2006; *Kirchoff and Schenk,* 2009; *Bierhaus et al.,* 2012) together argue that small comet (tens of meters to a kilometer) impacts are potentially currently rare at Saturn. More recently, crater measurements on Pluto and Charon suggest that the paucity of these small comets extends to the orbit of Pluto and to much deeper timescales than can be established by craters on Ganymede or Europa (*Moore et al.,* 2016; *Robbins et al.,* 2017; *Singer et al.,* 2017). It is possible that a lack of these small bodies, relative to a canonical collisional distribution (*Dohnanyi,* 1969; *O'Brien and Greenberg,* 2003), may be a property of the Kuiper belt.

Most recently, *Di Sisto and Zanardi* (2013, 2016) have estimated crater rates on the saturnian satellites from the Centaur population. They combine a detailed numerical model of the dynamical evolution of the Centaurs with observations of crater populations. They also consider two extrapolations of the Centaur population below an impactor diameter of 60 km, one with a shallow slope and the other with a steep slope. They found that the steep slope better matches the observed crater populations until crater diameters ~30 km, where the shallow slope is a better match. Computed current crater rates are, overall, similar to those computed by *Zahnle et al.* (2003), if slightly lower.

According to any of these cometary impact rate estimates, current cratering falls far short of reproducing the observed number of craters on Rhea and Iapetus in the age of the solar system. It therefore follows that impact rates were once much higher than they are now. The higher impact rates are often associated with activity during accretion and/or the late heavy bombardment (e.g., *Gomes et al.,* 2005; *Dones et al.,* 2009), but there may be other possibilities not yet conceived.

Meanwhile, *Neukum et al.* (2006) chose to work with the assumption that a majority of the impactors were asteroids from the main belt. They derived a production function polynomial for each mid-sized saturnian satellite empirically from the lunar polynomial through a lateral shift of the lunar polynomial in log-D to account for different crater scaling. They also assume a lunar-like bombardment scenario with an exponentially declining impact rate prior to ~3.5 Ga. The production function is fitted to a crater measurement and the cumulative frequency for $D \geq 1$ km obtained from the curve fit is used to derive a cratering model age. The polynomials were shown to fit measured crater distributions by *Wagner et al.* (2006, 2007) and *Stephan et al.* (2010, 2012) quite well. However, other researchers (*Kirchoff and Schenk,* 2009, 2010; *Robbins et al.,* 2015) do not find a match between their crater SFDs and the extrapolated Neukum production function. Furthermore, an explanation why the lunar production function and asteroids as a preferential impactor population is applicable for the moons of Saturn still remains an open issue.

Another thing that has changed as a result of Cassini is that we now have several other "clocks" apart from cratering to consider. These include (1) the geysering rate of Enceladus, which can be compared to tectonic disruption of, or burial of, craters and cratered terrain (*Kirchoff and Schenk,* 2009; *Hirata et al.,* 2014; *Bland et al.,* 2012; *Kempf et al.,* 2010; chapter by Kempf et al. in this volume), and may be linked with apex-antapex asymmetric coloring of Mimas (*Schenk et al.,* 2011); (2) observation from the Cassini Visual and Infrared Mapping Spectrometer (VIMS) of decreasing size of ice particles with increasing surface age on Enceladus due to space weathering [50–100 μm down to 10–20 μm on the oldest terrains (*Jaumann et al.,* 2008)]; (3) the conversion of crystalline ice back to amorphous ice after an impact (*Dalle Ore et al.,* 2015); (4) the photochemical destruction of a limited methane reservoir on Titan at a rate controlled by photons from the Sun (*Lunine et al.,* 1983; *Yung et al.,* 1984; *Atreya et al.,* 2009); (5) the (limited) cumulative production of photochemical products on the surface of Titan, also controlled by sunlight (*Lorenz et al.,* 2008); (6) the dark side of Iapetus, which is a cumulative deposit of photochemical or radiolytic products from Phoebe and other irregular satellites (*Soter,* 1974; *Tamayo et al.,* 2011; chapter by Hendrix et al. in this volume); (7) cumulative coloring of Saturn's rings (*Cuzzi and Estrada,* 1998); and (8) the Q of Saturn, which determines the rate of evolution and the time between resonance encounters between satellites, and thus provides a background timescale to the system. Of these, the one that has the largest effect on understanding the cratering histories of Saturn's satellites is the value and evolution of Saturn's Q.

4.2. A Young Satellite System?

It has recently been claimed, based on (1) the last century's telescopic record of the evolution of Mimas, Enceladus, Tethys, Dione, and Rhea's orbits (*Lainey et al.,* 2012) and (2) 10 years of more precise Cassini observations that do not contradict the hypothesis (*Lainey et al.,* 2017), that Saturn's effective tidal Q is very small, on the order of 1500, a value that is too small by at least an order of magnitude to have been maintained for the age of the solar system with the present system of satellites (*Goldreich,* 1965). It follows that, if Q is as small as has been derived, the architecture of the current system of satellites has been in place for only a small fraction of the age of the solar system. *Ćuk et al.* (2016) have quantified this fraction by bounding the tidal evolution of the system between orbital resonances among satellites that appear to have occurred in the past and earlier orbital resonances that appear not to have occurred. From this they deduced that with Q of 1500 the current system of satellites is ~100 m.y. old. They then speculate that the current system of satellites was built from the dispersed debris of an earlier system of satellites, generally resembling the present system but differing in its particulars, that had encountered fatal resonances in its tidal evolution. *Fuller et al.* (2016) present an alternative model in which Q ~ 1500 does *not* imply recent origin of the moons. In Fuller's scenario, the rate of outward migration of satellites due to the tides they excite in Saturn does not depend strongly on distance from the planet as it does in conventional tidal models.

The hypothesis that the surfaces of the current satellites are all roughly 100 m.y. old is surprisingly hard to test against the cratering record due to the uncertainty in the craters' origins from heliocentric or planetocentric impactors. Planetocentric impactors record an event or events that took place within Saturn's satellite system but are silent about when that event took place. Test particle models in the current Saturn system (from Mimas to Rhea) show that the lifetime of planetocentric debris is typically on the order of 100 years or less (*Alvarellos et al.,* 2005, 2017), although some debris may last for millions of years or more in protected resonant orbits (e.g., co-orbitals). In other words, planetocentric impactors, if significant, may be able to produce a large number of craters relatively rapidly. For further detailed discussion of Saturn's small Q and implications for the tidal evolutions and ages of the satellites, particularly Enceladus, see the chapter by Nimmo et al. in this volume.

Perhaps the best argument against young satellites is to be found in the big craters Herschel on Mimas and Odysseus on Tethys. Herschel is ≈130 km in diameter, near the apex of motion, and likely <1 Ga (*Kirchoff and Schenk,*

2008) (see section 5.2). Its location, size, and relative youth make it a good candidate for a crater made by a comet, which would have been about 5 km in diameter — typical of comets — and striking at about 30 km s^{-1}. If heliocentric cometary impact rates were higher during the past billion years, Herschel would be commensurately younger, but there is little support for a recent (<100 m.y.) origin of Mimas. To fit Herschel into the past 100 m.y. for the young satellite hypothesis would probably require that it be made by Mimas sweeping up a small moon. Such a moon would have struck at roughly 1.6 km s^{-1} (the difference in orbital velocities between Mimas and Enceladus, which seems like it should be a good bound for how fast a destabilized co-orbital or nearby formerly resonant moon might move) and would have been about 25 km in diameter. Even an impact at this speed could be hypervelocity for an icy moon and could produce a crater that looks like Hershel, but the probability of such an impact is unknown.

Odysseus on Tethys poses the same sorts of problems as Herschel on Mimas, but amplified. Odysseus is also on the leading hemisphere, as expected of a heliocentric comet, and it is ≈440 km in diameter. If made by a comet, the comet would have been roughly 20 or 25 km in diameter, a size object that would only have a very small probability of impacting in the last 100 m.y. [That is, the probability, p, of such an event in time t is $1 - \exp(-t/20\ \text{G.y.}) \sim t/20$ G.y. for $t \ll 20$ G.y., so $p \sim 0.5\%$ for t = 100 m.y.] Furthermore, Odysseus is not a shallow crater (*White et al.,* 2017; Schenk et al., this volume), which suggests that it formed on a rather cool moon. This is no problem for convention, which would assign to Odysseus an age on the order of 3.5 Ga (*Giese et al.,* 2007; *Dones et al.,* 2009) (see section 5.2), nor would such an ancient age conflict with our uncertain knowledge of comet impact rates. Therefore, the depth of Odysseus presents a challenge to a recent origin, because a recent origin implies that Tethys was recently warm, and so there needs to be time enough for Tethys to cool before Odysseus formed. The specific energy released when accumulating Tethys's mantle during reaccretion is on the order of $v_{esc}^2 \sim 1.6 \times 10^9$ ergs g^{-1}; this estimate takes into account Tethys' gravity and the random velocities of the accreting debris. Reaccretion is very fast [~100–1000 years (*Burns and Gladman,* 1998)] and there is not time enough for radiative cooling to remove significant energy, and as Tethys is made of water ice, all the accretional energy goes into heating water ice. For Tethys, this means that the mantle is heated by about 120 K, which would leave the newly remade tethyian mantle at a temperature of around 240 K. At 240 K, Odysseus should viscously relax on a timescale on the order of 1 year (*Parmentier and Head,* 1981; *Lunine and Stevenson,* 1987; *Dombard and McKinnon,* 2006; *White et al.,* 2017). Because it takes about 100 m.y. for conductive cooling to penetrate to 100 km and more than a billion years to penetrate to 400 km, it seems unlikely that Odysseus could have survived as a deep basin unless it formed after Tethys had cooled for several hundred million years.

In the confines of the recent origins hypothesis, we would again have to require that Odysseus was formed by the impact of another moon. Taking an impact velocity of 1.3 km s^{-1} (the difference between Tethys' orbital velocity and those of Enceladus on one side and Dione on the other), we estimate that the moon would have been on the order of 120 km in diameter.

5. CRATERING HISTORIES

5.1. Previous Work Based on Voyager Images

In the first detailed images of the surface of Enceladus returned from the flyby of Voyager 2 at a spatial resolution of approximately 1 km/pixel, *Smith et al.* (1982) identified five types of terrain (geologic units), including densely cratered plains and sparsely cratered, smooth or ridged, plains indicative of tectonic and/or cryovolcanic resurfacing. Using cratering rates derived in *Smith et al.* (1982), they estimated that the tectonically altered units could have a maximum age of ~1 Ga. Meanwhile, *Plescia and Boyce* (1985) used an extrapolation of the lunar crater chronology (see section 4.1) and estimated the absolute model age of densely cratered regions to be on the order of 3.6 Ga, while younger, tectonically resurfaced terrains could be as old as 1–2 Ga. They concluded that Enceladus' surface could have been endogenically active throughout most of its history. Finally, *Kargel and Pozio* (1996) carried out a thorough study of the geologic evolution of Enceladus using various cratering chronology models to estimate absolute surface ages. Using constant impact rates given by *Smith et al.* (1982), they found the younger tectonically resurfaced units are on the order of several hundred million years old. *Kargel and Pozio* (1996) then assumed that the constant cratering rate was equal to the crater frequency in the most densely cratered regions divided by the age of the solar system and found that the tectonically altered regions could be considerably younger, on the order of several tens of millions of years. Lastly, they assumed an exponentially declining flux of projectiles with a half-life of 700 m.y. In this scenario, densely cratered plains have model ages >4 Ga, while the tectonized regions could be as young as several million years old, indicating that Enceladus could be active today, which was confirmed by Cassini.

Although not as extensive or recent as Enceladus' activity, Dione and Tethys also have terrains with varying crater histories. From Voyager imaging, smooth and "wispy" terrains were observed on Dione on the leading and trailing hemisphere, respectively (*Smith et al.,* 1981; *Strom,* 1981; *Plescia and Boyce,* 1982, 1985; *Plescia,* 1983; *Moore,* 1984; *Morrison et al.,* 1984, 1986). Both appeared to have lower crater densities than other observed areas. *Plescia and Boyce* (1985) estimated an age of ~2.2 Ga for the smooth plains, but were not able to derive an age for the wispy terrain due to the oblique viewing geometry. On Tethys, Ithaca Chasma, a large tectonic canyon, was recognized to have a lower crater density than the cratered plains (*Smith et al.,* 1981, 1982; *Plescia and Boyce,* 1983, 1985; *Morrison et al.,* 1984, 1986), and *Plescia and Boyce* (1985) derived a model age of ~3.7 Ga. Otherwise, it was recognized that the majority of the terrains on Mimas, Dione, Tethys, Rhea, and Iapetus were densely cratered. The absolute model ages, which are highly model dependent and assume heliocentric impactors dominate, of these units were estimated to be likely older than 3.9 Ga (*Smith et al.,* 1982; *Plescia and Boyce,* 1983, 1985).

5.2. Cassini Data

In the first three close flybys in 2005 (Cassini orbits 3, 4 and 11), a significant portion of Enceladus' major surface units was geologically mapped at a resolution a factor of ~5–10 better than Voyager. These units range from densely cratered plains to the tectonically resurfaced south polar terrain (SPT), which has been pbserved to be cryovolcanically active (*Porco et al.,* 2006). These geologic key units are shown in Fig. 5 and reviewed in detail in the chapter by Patterson et al. in this volume. Here, the unit designations used in the global geologic map of Enceladus by *Crow-Willard and Pappalardo* (2015) were adopted. Surface feature names used in association with geologic units were either defined prior to the Cassini mission, or new names based on Cassini images have been confirmed by the International Astronomical Union (IAU) (*Roatsch et al.,* 2013). Cratering model ages based upon crater SFDs derived from these images and those that followed were computed using either the outer solar system chronology of *Zahnle et al.* (2003) or a modern extrapolation of the lunar crater chronology (*Neukum et al.,* 2006; *Michael et al.,* 2016).

Figure 6 shows that crater frequencies from the oldest to the youngest unit vary by about 3 orders of magnitude. The crater frequencies correspond to cratering model ages ranging from ~4 Ga down to units formed recently or at present. The oldest unit, designated as cp1 by *Crow-Willard and Pappalardo* (2015), was formed at ~2–4 Ga (*Porco et al.,* 2006; *Jaumann et al.,* 2008, 2011; *Kirchoff and Schenk,*

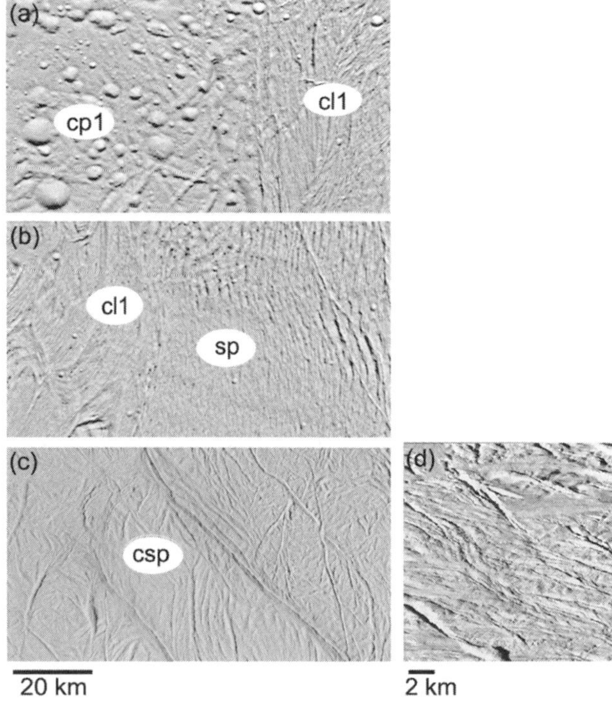

Fig. 5. Geologic key units in image data from Cassini orbits 003, 004, and 011. Unit designations from *Crow-Willard and Pappalardo* (2011, 2015). **(a)** Cratered plains (cp1) and trailing hemisphere curvilinear terrain (cl1) of Samarkand Sulcus (110 m/pxl). **(b)** Curvilinear terrain (cl1) of Samarkand Sulcus and striated plains (sp) of Sarandib Planitia (110 m/pxl). **(c)** Unit central south plains (csp) of the south polar terrain (SPT) (110 m/pxl). **(d)** WAC frame W1500063766 from orbit 011 with 33 m/pxl located within unit csp.

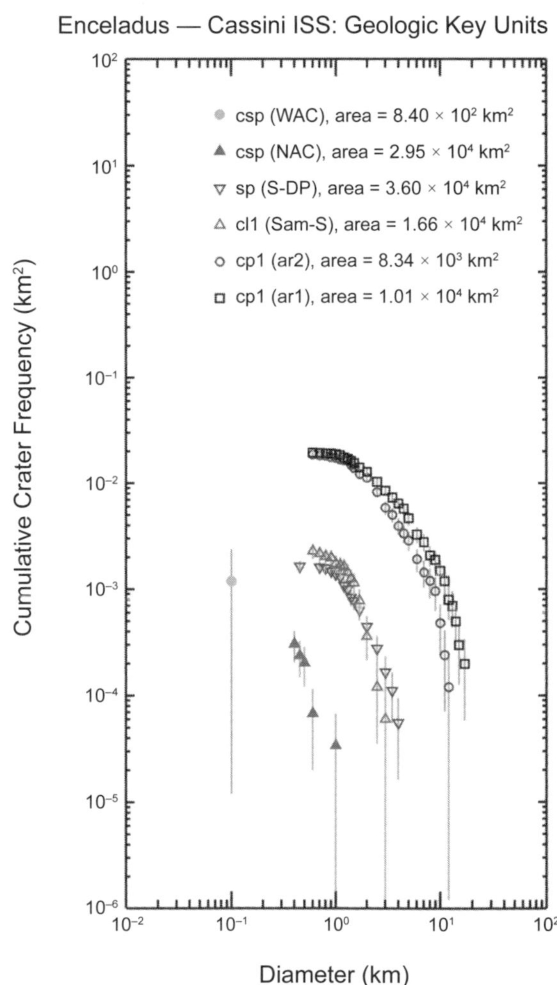

Fig. 6. Cumulative crater size-frequency diagram of the geologic units shown in Fig. 5. Geologic unit designations are from *Crow-Willard and Pappalardo* (2015). Measurements from oldest to youngest: unit cp1 (two areas); unit sp (Sarandib and Diyar Planitia), and unit cl1 (Samarkand Sulcus). Unit csp in the south polar terrain: filled triangles denote measurements from 110-m/pxl global basemap (*Roatsch et al.,* 2008) and filled circles denote estimated maximum crater frequency in WAC frame W1500063766. Units cp1, sp, and cl1 are from measurements by *Jaumann et al.* (2011); crater statistics of unit csp are taken from *Porco et al.* (2006).

2009). The youngest unit is the cryovolcanically active SPT [unit csp, central south polar (*Crow-Willard and Pappalardo,* 2015)]. While most of this terrain has model ages on the order of 1–100 Ma (*Porco et al.,* 2006), a maximum cratering model age of <~1 Ma could be extracted for a portion of unit csp covered by a single 33-m/pixel wide angle camera (WAC) image, by using the method described in detail by *Michael et al.* (2016) for areas that are devoid of impact craters (unit indicated by the yellow data point in Fig. 6). This reflects that parts of the SPT are being resurfaced by cryovolcanism at the present time. Crater model ages for other tectonized units fall in between the cp1 unit and SPT (*Porco et al.,* 2006; *Jaumann et al.,* 2008, 2011; *Kirchoff and Schenk,* 2009).

Cassini has also considerably extended our view of the cratering histories of Mimas, Tethys, Dione, Rhea, and Iapetus (*Neukum et al.,* 2005, 2006; *Wagner et al.,* 2006, 2007, 2008; *Giese et al.,* 2007; *Schmedemann et al.,* 2008, 2009; *Dones et al.,* 2009; *Kirchoff and Schenk,* 2010, 2015; *Schmedemann and Neukum,* 2011; *Di Sisto and Zanardi,* 2016). First, it has been revealed that Mimas, Rhea, and Iapetus do not have significant areas of lightly cratered surfaces and that their densely cratered surfaces have a crater model age >3.9 Ga (*Neukum et al.,* 2005, 2006; *Wagner et al.,* 2007, 2008; *Giese et al.,* 2007; *Schmedemann et al.,* 2008, 2009; *Dones et al.,* 2009; *Schmedemann and Neukum,* 2011; *Di Sisto and Zanardi,* 2016). The exception is Herschel crater on Mimas, which could be <1 Ga (*Kirchoff and Schenk,* 2008). For Dione and Tethys, no new, large areas of lightly cratered terrains were discovered, but the smooth plains, "wispy" plains, and Ithaca Chasma could be better characterized. Crater model ages for the smooth plains seem to be ~2 Ga (*Dones et al.,* 2009; *Kirchoff and Schenk,* 2015). The "wispy" plains are now known to be a dense set of tectonic features with ages possibly as young as 1–2 Ga, but could be >3.7 Ga depending on crater size and chronology used (*Wagner et al.,* 2006; *Kirchoff and Schenk,* 2015). Ithaca Chasma appears to be quite old, likely >3.3 Ga (*Giese et al.,* 2007). Finally, Dione and Tethys both have one large, young basin: Evander on Dione and Odysseus on Tethys. The crater model age for Evander, which is <1 Ga using the outer solar system chronology (*Wagner et al.,* 2006; *Kirchoff and Schenk,* 2015), indicates it is likely younger than Odysseus, which is 1–3 Ga using the outer solar system chronology (*Giese et al.,* 2007; *Dones et al.,* 2009).

6. REMAINING ISSUES

The above sections summarize the state of research, but more work is yet needed to bring various lines of evidence and modeling into concordance. Open questions relate to determining the source of impactors forming the craters on the saturnian satellites and deconvolving the influence of various preservation/degradation processes. Some questions in need of future work include:

- *Are the satellites old or young?* As described in section 4, recent dynamical modeling of the saturnian satellite orbital resonances has led to the proposal that the mid-sized satellites may have formed only ~100 m.y. ago (*Ćuk et al.,* 2016). In this scenario, the current crater population would be primarily the result of planetocentric debris. Can any empirical observations of the geology or cratering records on the saturnian satellites be found to support this hypothesis?
- *What causes the differences in the crater populations among the satellites?*
  - *What is the contribution of planetocentric debris, sesquinaries, secondaries, and captured asteroids?* Is there a size limit for the influence of these populations? What are the dynamical restrictions on their ability to impact the saturnian satellites?
  - *Does crater preservation state or saturation possibly obscure the apex-antapex asymmetry for small craters?* Although many studies have delved into this topic, new information is needed to help us understand the origin of the system impactors, and whether they can be traced to a heliocentric or a planetocentric source.
  - *How can Pluto and Charon data help further constrain the contribution of all populations?* Data recently returned from the New Horizons mission provide clues to the distant outer solar system impactor flux (*Robbins et al.,* 2017; *Singer et al.,* 2017). How can Pluto system data be related to the crater SFDs on giant planet satellites given the different dynamical conditions? Some saturnian satellite terrain SFDs share a feature with the Pluto system SFDs: There is a break to a shallower SFD slope for craters below ~10 km in diameter. Does this imply that at least some portion of the saturnian system impactors are heliocentric?
  - *Have subtle resurfacing events been missed?* Variation in crater densities occur even within relatively well-defined geologic units. Some differences may be attributed simply to stochastic variation in cratering, but others may be from nonobvious resurfacing events (e.g., a degraded, relatively ancient surface flow with no obvious margin). If the latter is true, researchers must carefully consider what geologic event (or combination of events) is being dated.
- *What are the impact rates in the outer solar system?*
  - *How young is young?* Young terrains are notoriously difficult to date, generally due to the low numbers of craters found there (small number statistics) and the fact that impact flux is not well constrained at small impactor sizes. Given the current uncertainties, terrain age estimates for some younger regions (e.g., on Enceladus) can differ by up to several billion years.

- *How should any early bombardment events be incorporated into the rates?* If the satellites are indeed old (≥4 Ga, as opposed to a few hundred million years or younger), are the crater populations reflecting any portion of a much higher impact rate in the distant past? Can a higher impact rate in the past explain any features we see in the crater SFDs?

Future work may elucidate answers to these questions, but these issues challenge us to consider nontraditional interpretations of the saturnian satellites' origin and evolution. Further dynamical modeling, geophysical modeling, mapping, and measurements can contribute information in the near-term. Future spacecraft missions, possibly even those bringing back samples, will be vital in the future to make new measurements relevant to these topics. Higher-resolution imaging, along with spectral, thermal, and gravity data, can yield important constraints on the surface and interior evolution of saturnian satellites.

7. SUMMARY

The two crater populations — Population I and Population II — discovered in crater analyses using Voyager imaging still seem to be evident in analyses based upon Cassini imaging. However, the characteristics may be slightly different than originally reported. Overall, Population II appears to be most clearly expressed on Mimas, Dione, and Tethys, where we find that crater SFDs have steeper slopes at D = 10–20 km until D ≈ 200 km, while the crater SFDs for Rhea and Iapetus remain shallow, possibly a representation of Population I. The processes behind the production of two crater populations is still unknown, but has been hypothesized to be due to either different impactor populations, crater saturation, or surface geologic activity. The last explanation seems to be the most unlikely, as the differences observed in the crater SFDs are not predicted by any known geological process for the saturnian satellites.

A potential clue comes from the fact that a significant apex-antapex asymmetry has not yet been observed for the saturnian satellites least affected by geologic activity (Mimas, Rhea, and Iapetus). An asymmetry in crater density, with a higher density on the apex (or leading hemisphere) than the antapex (or trailing hemisphere), is expected to develop for heliocentric impactors bombarding synchronously rotating satellites, such as Mimas, Rhea, and Iapetus. Since there is little evidence these satellites have ever experienced nonsynchronous rotation, a lack of asymmetry could be explained by a nonheliocentric impactor population or crater saturation. Unfortunately, the data are not yet sufficient to determine which of these is the case for these satellites. However, there is other observational evidence that densely cratered satellite surfaces in the saturnian system may be saturated, such as their high crater density and uniform spatial crater distribution.

Recent numerical simulations have shed light on means of generating a considerable planetocentric impactor population in the Saturn system. There are three important groups of impactors for Saturn's satellites: Centaurs/ecliptic comets, debris from the catastrophic breakup of small satellites, and impact ejecta. The first is the main heliocentric impactor population, while the last two make up the planetocentric population. Dynamical calculations show that a large number of planetocentric impactors can be created when a satellite is disrupted or a large primary crater forms on Mimas, Enceladus, Tethys, or Dione. Most of this planetocentric debris is found to impact the satellite on which they were generated, and therefore this population may explain the different crater SFDs seen on Mimas, Enceladus, Tethys, and Dione.

When we compare saturnian satellite crater SFDs to the Moon crater production function, Callisto's crater SFD, and Charon's crater SFD for Vulcan Planum (informal name), we find similarities and differences. An important observation is that outer solar system worlds all seem to have shallow slopes for D ≲ 10–20 km, while inner solar system worlds have a very steep slope. This lack of small craters appears to be an inherent difference between the inner and outer solar system, and is likely caused by the different heliocentric impactor populations that dominate.

Once the main impactor sources are generally understood, crater chronologies for the saturnian satellites can be developed by assessing the size, number, and orbital distributions of the current observable populations of heliocentric comets and converting that population to crater distributions through scaling laws. Since observations of heliocentric comets in the outer solar system are currently limited to much larger sizes than would form the majority of craters observed, most of the crater production functions are currently derived by extrapolation using crater SFDs. The first crater chronologies were developed using crater SFDs compiled from Voyager images and a rudimentary understanding of the Centaur (or ecliptic comet) population. As understanding of the Centaur population has improved over the last 30 years and with new observations of the crater populations from Cassini, impact rates have continued to be refined. However, until we more fully understand the Centaur population, especially at small impactor sizes, or obtain samples for radiometric dating, there will be little further improvement in these rates.

In the meantime, new determinations of Saturn's Q value have indicated that all the mid-sized satellites might be quite young, ~100 Ma. If true, this would dramatically alter our understanding of the impact rates. Due to the uncertainty in the contribution of planetocentric impactors, it is difficult to use the crater populations to disprove these young ages. However, the presence of two large, deep (mostly unrelaxed) basins, Herschel on Mimas and Odysseus on Tethys, provide strong evidence against such youthful satellites. One hundred million years would provide little time for a just-formed moon to cool enough to maintain an unrelaxed basin.

Enceladus shows the widest range in crater frequencies of all mid-sized saturnian satellites. Densely cratered units are stratigraphically the oldest, while sparsely cratered units, or units devoid of craters at specific image resolution, are the youngest. The south polar terrain has been confirmed to be geologically active at the present time. Dione and Tethys have some variation in crater density, while Mimas, Rhea, and Iapetus appear to all be simply densely cratered surfaces.

Due to the uncertainties in impact cratering chronology models, absolute model ages can only be roughly estimated. Model ages for the densely cratered units on all mid-sized saturnian satellites are on the order of 4 Ga. If a lunar chronology is assumed (which we do not recommend), Enceladus has been endogenically active since the earliest times at ~4 Ga until the present day. Whether this has been ongoing geologic activity or whether there have been highly active periods alternating with less active, or even inactive, episodes is difficult to clarify based on crater measurements. For the outer solar system chronology, Enceladus' geologic history seems to be compressed into the last ~2 Ga, while the cratering record prior to that time would be considered lost. Dione's smooth and tectonized terrains appear to be ~2–3 Ga, while Ithaca Chasma on Tethys is >3 Ga.

REFERENCES

Allen C. C. (1979) Large lunar secondary craters — Size-range relationships. *Geophys. Res. Lett., 6,* 51–54.

Alvarellos J. L., Zahnle K. J., Dobrovolskis A. R., and Hamill P. (2002) Orbital evolution of impact ejecta from Ganymede. *Icarus, 160,* 108–123.

Alvarellos J. L., Zahnle K. J., Dobrovolskis A. R., and Hamill P. (2005) Fates of satellite ejecta in the Saturn system. *Icarus, 178,* 104–123.

Alvarellos J. L., Zahnle K. J., Dobrovolskis A. R., and Hamill P. (2008) Transfer of mass from Io to Europa and beyond due to cometary impacts. *Icarus, 194,* 636–646.

Alvarellos J. L., Dobrovolskis A., Zahnle K. J., Hamill P., Dones L., and Robbins S. J. (2017) Fates of satellite ejecta in the Saturn system, II. *Icarus, 284,* 70–89.

Atreya S. K., Niemann H. B., Encrenaz T., and Owen T. C. (2009) Formation of Jupiter and Saturn and the origin of their atmospheres: Current constraints and future prospects. In *EGU General Assembly Conference Abstracts,* Abstract #6324.

Bierhaus E. B., Chapman C. R., Merline W. J., Brooks S. M., and Asphaug E. (2001) Pwyll secondaries and other small craters on Europa. *Icarus, 153,* 264–276.

Bierhaus E. B., Dones L., Alvarellos J. L., and Zahnle K. (2012) The role of ejecta in the small crater populations on the mid-sized Saturnian satellites. *Icarus, 218,* 602–621.

Bierhaus E. B., McEwen A. S., Robbins S. J., Singer K. N., Dones L., Kirchoff M. R., and Williams J.-P. (2018) Secondary craters and ejecta across the solar system: Populations and effects on impact-crater–based chronologies. *Meteoritics & Planet. Sci., 53(4),* 638–671.

Bland M. T., Singer K. N., McKinnon W. B., and Schenk P. M. (2012) Enceladus' extreme heat flux as revealed by its relaxed craters. *Geophys. Res. Lett., 39,* L17204.

Bottke W. F., Morbidelli A., Jedicke R., Petit J.-M., Levison H. F., Michel P., and Metcalfe T. S. (2002) Debiased orbital and absolute magnitude distribution of the near-Earth objects. *Icarus, 156,* 399–433.

Bottke W. F., Durda D. D., Nesvorný D., Jedicke R., Morbidelli A., Vokrouhlický D., and Levison H. F. (2005) The fossilized size distribution of the main asteroid velt. *Icarus, 175,* 111–140.

Bottke W. F., Nesvorný D., Vokrouhlický D., and Morbidelli A. (2010) The irregular satellites: The most collisionally evolved populations in the solar system. *Astron. J., 139,* 994–1014.

Bottke W. F., Vokrouhlicky D., Ghent B., Mazrouei S., Robbins S. J., and Marchi S. (2016) On asteroid impacts, crater scaling laws, and a proposed younger surface age for Venus. In *Lunar and Planetary Science XLVII,* Abstract #2036. Lunar and Planetary Institute, Houston.

Brasser R. (2008) A two-stage formation process for the Oort comet cloud and its implications. *Astron. Astrophys., 492,* 251–255.

Brasser R. and Morbidelli A. (2013) Oort cloud and scattered disk formation during a late dynamical instability in the solar system. *Icarus, 225,* 40–49.

Brasser R. and Wang J. H. (2015) An updated estimate of the number of Jupiter-family comets using a simple fading law. *Astron. Astrophys., 573,* A102.

Brasser R., Duncan M. J., and Levison H. F. (2008) Embedded star clusters and the formation of the Oort cloud. III. Evolution of the inner cloud during the galactic phase. *Icarus, 196,* 274–284.

Bray V .J., Smith D. E., Turtle E. P., Perry J. E., Rathbun J. A., Barnash A. N., Helfenstein P., and Porco C. C. (2007) Impact crater morphology variations on Enceladus. In *Lunar and Planetary Science XXXVIII,* Abstract #1873. Lunar and Planetary Institute, Houston.

Burns J. A. and Gladman B. J. (1998) Dynamically depleted zones for Cassini's safe passage beyond Saturn's rings. *Planet. Space Sci., 46,* 1401–1407.

Campo Bagatin A. and Benavidez P. G. (2012) Collisional evolution of trans-neptunian object populations in a Nice model environment. *Mon. Not. R. Astron. Soc., 423,* 1254–1266.

Chapman C. R. and McKinnon W. B. (1986) Cratering of planetary satellites. In *Satellites* (J. A. Burns and M. S. Matthews, eds.), pp. 492–580. Univ. of Arizona, Tucson.

Chen E. M. A. and Nimmo F. (2008) Implications from Ithaca Chasma for the thermal and orbital history of Tethys. *Geophys. Res. Lett., 35,* L19203.

Crater Analysis Techniques Working Group (1979) Standard techniques for presentation and analysis of crater size-frequency data. *Icarus, 37,* 467–74.

Crow-Willard E. N. and Pappalardo R. T. (2011) Global geologic mapping of Enceladus. In *EPSC Abstracts, Vol. 6,* EPSC-DPS2011-635.

Crow-Willard E. N. and Pappalardo R. T. (2015) Structural mapping of Enceladus and implications for formation of tectonized regions. *J. Geophys. Res., 120,* 928–950.

Ćuk M., Dones H. L., and Nesvorný D. (2016) Dynamical evidence for a late formation of Saturn's moons. *Astrophys. J., 820,* article ID 97.

Cuzzi J. N. and Estrada P. R. (1998) Compositional evolution of Saturn's rings due to meteoroid bombardment. *Icarus, 132,* 1–35.

Dalle Ore C. M., Cruikshank D. P., Mastrapa R. M. E., Lewis E., and White O. L. (2015) Impact craters: An ice study on Rhea. *Icarus, 261,* 80–90.

Damptz A. L., Dombard A. J., and Kirchoff M. R. (2017) Testing models for the formation of the equatorial ridge on Iapetus via crater counting. *Icarus, 302,* 134–144, DOI: 10.1016/j.icarus.2017.10.049.

Daubar I. J., McEwen A. S., Byrne S., Kennedy M. R., and Ivanov B. (2013) The current martian cratering rate. *Icarus, 225,* 506–516.

Di Sisto R. P. and Brunini A. (2007) The origin and distribution of the Centaur population. *Icarus, 190,* 224–235.

Di Sisto R. P., Fernández J. A., and Brunini A. (2009) On the population, physical decay and orbital distribution of Jupiter family comets: Numerical simulations. *Icarus, 203,* 140–154.

Di Sisto R. P. and Zanardi M. (2013) The production of craters on the mid-sized saturnian satellites by Centaur objects. *Astron. Astrophys., 553,* article ID A79.

Di Sisto R. P. and Zanardi M. (2016) Surface ages of mid-size saturnian satellites. *Icarus, 264,* 90–101.

Dobrovolskis A. R. and Lissauer J. J. (2004) The fate of ejecta from Hyperion. *Icarus, 169,* 462–473.

Dohnanyi J. S. (1969) Collisional model of asteroids and their debris. *J. Geophys. Res., 74,* 2531–2554.

Dombard A. J. and McKinnon W. B. (2006) Elastoviscoplastic relaxation of impact crater topography with application to Ganymede and Callisto. *J. Geophys. Res., 111,* E01001.

Dones L., Chapman C. R., McKinnon W. B., Melosh H. J., Kirchoff M. R., Neukum G., and Zahnle K. J. (2009) Icy satellites of Saturn: Impact cratering and age determination. In *Saturn from Cassini-Huygens* (M. K. Dougherty et al., eds.), pp. 613–636. Springer, Dordrecht.

Dones L., Brasser R., Kaib N., and Rickman H. (2015) Origin and evolution of the cometary reservoirs. *Space Sci. Rev., 197,* 191–269.

Duncan M. J. and Levison H. F. (1997) A disk of scattered icy objects and the origin of Jupiter-family comets. *Science, 276,* 1670–1672.

Duncan M., Quinn T., and Tremaine S. (1987) The formation and extent of the solar system comet cloud. *Astron. J., 94,* 1330–1338.

Duncan M., Quinn T., and Tremaine S. (1988) The origin of short-period

comets. *Astrophys. J. Lett., 328,* L69–L73.
Elliot J. L., Kern S. D., Clancy K. B., Gulbis A. A. S., Millis R. L., Buie M. W., Wasserman L. H., Chiang E. I., Jordan A. B., Trilling D. E., and Meech K. J. (2005) The deep ecliptic survey: A search for Kuiper belt objects and Centaurs. II. Dynamical classification, the Kuiper belt plane, and the core population. *Astron. J., 129,* 1117–1162.
Fernández J. A. (1980) On the existence of a comet belt beyond Neptune. *Mon. Not. R. Astron. Soc., 192,* 481–491.
Fuller J., Luan J., and Quataert E. (2016) Resonance locking as the source of rapid tidal migration in the Jupiter and Saturn moon systems. *Mon. Not. R. Astron. Soc., 458,* 3867–3879.
Gault D. E. (1970) Saturation and equilibrium conditions for impact cratering on the lunar surface: Criteria and implications. *Radio Sci., 5,* 273–291.
Giese B., Wagner R., Neukum G., Helfenstein P., and Thomas P. C. (2007) Tethys: Lithospheric thickness and heat flux from flexurally supported topography at Ithaca Chasma. *Geophys. Res. Lett., 34,* L21203.
Gladman B. and 10 colleagues (2001) Discovery of 12 satellites of Saturn exhibiting orbital clustering. *Nature, 412,* 163–166.
Gladman B., Marsden B. G., and VanLaerhoven C. (2008) Nomenclature in the outer solar system. In *The Solar System Beyond Neptune* (M. A. Barucci et al., eds.), pp. 43–57. Univ. of Arizona, Tucson.
Gladman B., Lawler S. M., Petit J.-M., Kavelaars J., Jones R. L., Parker J. Wm., van Laerhoven C., Nicholson P., Rousselot P., Bieryla A., and Ashby M. L. N. (2012) The resonant trans-neptunian populations. *Astron. J., 144,* article ID 23.
Goldreich P. (1965) An explanation of the frequent occurrence of commensurable mean motions in the solar system. *Mon. Not. R. Astron. Soc., 130,* 159–181.
Goldreich P. M. and Mitchell J. L. (2010) Elastic ice shells of synchronous moons: Implications for cracks on Europa and non-synchronous rotation of Titan. *Icarus, 209,* 631–638.
Gomes R., Levison H. F., Tsiganis K., and Morbidelli A. (2005) Origin of the cataclysmic late heavy bombardment period of the terrestrial planets. *Nature, 435,* 466–469.
Hammond N. P., Phillips C. B., Robuchon G., Beyer R., Nimmo F., and Roberts J. (2011) Crater relaxation and stereo imaging of Rhea. In *Lunar and Planetary Science XLI,* Abstract #2633. Lunar and Planetary Institute, Houston.
Harris A. W. and D'Abramo G. (2015) The population of near-Earth asteroids. *Icarus, 257,* 302–312.
Hartmann W. K. (1984) Does crater "saturation equilibrium" occur in the solar system? *Icarus, 60,* 56–74.
Hartmann W.K., Ryder G., Dones L., and Grinspoon D. (2000) The time-dependent intense bombardment of the primordial Earth/Moon system. In *Origin of the Earth and Moon* (R. M. Canup and K. Righter, eds.), pp. 493–512. Univ. of Arizona, Tucson.
Heisler J. (1990) Monte Carlo simulation of the Oort comet cloud. *Icarus, 88,* 104–121.
Hirabayashi M., Minton D. A., and Fassett C. I. (2017) An analytical model of crater count equilibrium. *Icarus, 289,* 134–143.
Hirata N. (2016) Differential impact cratering of Saturn's satellites by heliocentric impactors. *J. Geophys. Res., 121,* 111–117.
Hirata N. and Nakamura A. M. (2006) Secondary craters of Tycho: Size-frequency distributions and estimated fragment size-velocity relationships. *J. Geophys. Res., 111,* E03005.
Hirata N., Miyamoto H., and Showman A. P. (2014) Particle deposition on the saturnian satellites from ephemeral cryovolcanism on Enceladus. *Geophys. Res. Lett., 41,* 4135–4141.
Horedt G. P. and Neukum G. (1984) Planetocentric versus heliocentric impacts in the jovian and saturnian satellite system. *J. Geophys. Res., 89,* 10405–10410.
Housen K. R. and Holsapple K. A. (2011) Ejecta from impact craters. *Icarus, 211,* 856–875.
Hsieh H. H. and Jewitt D. (2006) A population of comets in the main asteroid belt. *Science, 312,* 561–563.
Hueso R., Pérez-Hoyos S., Sánchez-Lavega A., et al. (2013) Impact flux on Jupiter: From superbolides to large-scale collisions. *Astron. Astrophys., 560,* article ID A55.
Jaumann R., Stephan K., Hansen G. B., et al. (2008) Distribution of icy particles across Enceladus' surface as derived from Cassini-VIMS measurements. *Icarus, 193,* 407–419.
Jaumann R., Stephan K., Brown R. H., et al. (2011) Enceladus: Correlation of surface particle distribution and geology. *EPSC Abstracts, Vol. 6,* EPSC-DPS2011-435-1.
Jewitt D. and Luu J. (1993) Discovery of the candidate Kuiper belt object 1992 QB1. *Nature, 362,* 730–732.
Jewitt D., Hsieh H., and Agarwal J. (2015) The active asteroids. In *Asteroids IV* (P. Michel et al., eds.), pp. 221–241. Univ. of Arizona, Tucson.
Jewitt D., Mutchler M., Weaver H., Hui M.-T., Agarwal J., Ishiguro M., Kleyna J., Li J., Meech K., Micheli M., Wainscoat R., and Weryk R. (2016) Fragmentation kinematics in Comet 332P/Ikeya-Murakami. *Astrophys. J. Lett., 829,* article ID L8.
Joss P. C. (1973) On the origin of short-period comets. *Astron. Astrophys., 25,* 271–273.
Kargel J. S. and Pozio S. (1996) The volcanic and tectonic history of Enceladus. *Icarus, 119,* 385–404.
Kaib N. A. and Quinn T. (2009) Reassessing the source of long-period comets. *Science, 325,* 1234–1236.
Kempf S., Beckmann U., Schmidt J. (2010) How the Enceladus dust plume feeds Saturn's E ring. *Icarus, 206(2),* 446–457, DOI: 10.1016/j.icarus.2009.09.016.
Kirchoff M. R. (2018) Can spatial statistics help decipher impact crater saturation? *Meteoritics & Planet. Sci., 53(4),* 874–890, DOI: 10.1111/maps.13014.
Kirchoff M. R. and Schenk P. M. (2008) Bombardment history of the saturnian satellites. In *Workshop on the Early Solar System Impact Bombardment,* Abstract #3023. LPI Contribution No. 1439, Lunar and Planetary Institute, Houston.
Kirchoff M. R. and Schenk P. (2009) Impactor populations in the saturnian system: Constraints from the cratering records. In *Lunar and Planetary Science XL,* Abstract #2067. Lunar and Planetary Institute, Houston.
Kirchoff M. R. and Schenk P. (2010) Global impact cratering record of Saturn's moon Dione: Constraining the geological history. In *Lunar and Planetary Science XLI,* Abstract #1455. Lunar and Planetary Institute, Houston.
Kirchoff M. R. and Schenk P. (2015) Dione's resurfacing history as determined from a global impact crater database. *Icarus, 256,* 78–89.
Kirchoff M. R., Chapman C. R., Marchi S., Curtis K. M., Enke B., and Bottke W. F. (2013) Ages of large lunar impact craters and implications for bombardment during the Moon's middle age. *Icarus, 225,* 325–341.
Korycansky D. G. and Zahnle K. J. (2005) Modeling crater populations on Venus and Titan. *Planet. Space Sci., 53,* 695–710.
Lainey V., Karatekin Ö., Desmars J., Charnoz S., Arlot J.-E., Emelyanov N., Le Poncin-Lafitte C., Mathis S., Remus F., Tobie G., and Zahn J.-P. (2012) Strong tidal dissipation in Saturn and constraints on Enceladus' thermal state from astrometry. *Astrophys. J., 752,* article ID 14.
Lainey V., Jacobson R. A., Tajeddine R., Cooper N. J., Murray C., Robert V., Tobie G., Guillot T., Mathis S., Remus F., Desmars J., Arlot J.-E., De Cuyper J.-P., Dehant V., Pascu D., Thuillot W., Le Poncin-Lafitte C., and Zahn J.- P. (2017) New constraints on Saturn's interior from Cassini astrometric data. *Icarus, 281,* 286–296.
Leliwa-Kopystyński J., Banaszek M., and Wlodarczyk I. 2012. Longitudinal asymmetry of craters' density distributions on the icy satellites. *Planet. Space Sci., 60,* 181–192.
Levison H. F. (1996) Comet taxonomy. In *Completing the Inventory of the Solar System* (T. W. Rettig and J. M. Hahn, eds.), pp. 173–191. ASP Conf. Ser. 107, Astronomical Society of the Pacific, San Francisco.
Levison H. F. and Duncan M. J. (1997) From the Kuiper belt to Jupiter-family comets: The spatial distribution of ecliptic comets. *Icarus, 127,* 13–32.
Levison H. F., Dones L., Chapman C. R., Stern S. A., Duncan M. J., and Zahnle K. (2001) Could the lunar 'late heavy bombardment' have been triggered by the formation of Uranus and Neptune? *Icarus, 151,* 286–306.
Levison H. F., Morbidelli A., Dones L., Jedicke R., Wiegert P. A., and Bottke W. F. (2002) The mass disruption of Oort cloud comets. *Science, 296,* 2212–2215.
Lissauer J. J. (1985) Can cometary bombardment disrupt synchronous rotation of planetary satellites? *J. Geophys. Res., 90,* 11289–11293.
Lissauer J. J., Squyres S. W., and Hartmann W. K. (1988) Bombardment history of the Saturn system. *J. Geophys. Res., 93,* 13776–13804.
Lorenz R. D., Mitchell K. L., Kirk R. L., et al. (2008) Titan's inventory of organic surface materials. *Geophys. Res. Lett., 35,* L02206.

Lunine J. I. and Stevenson D. J. (1987) Clathrate and ammonia hydrates at high pressure — Application to the origin of methane on Titan. *Icarus, 70,* 61–77.

Lunine J. I., Stevenson D. J., and Yung Y. L. (1983) Ethane ocean on Titan. *Science, 222,* 1229–1230.

Luu J., Marsden B. G., Jewitt D., Trujillo C. A., Hergenrother C. W., Chen J., and Offutt W. B. (1997) A new dynamical class of object in the outer solar system. *Nature, 387,* 573–575.

Lykawka P. S. and Mukai T. (2007) Dynamical classification of trans-neptunian objects: Probing their origin, evolution, and interrelation. *Icarus, 189,* 213–232.

Marchi S., Mottola S., Cremonese G., Massironi M., and Martellato E. (2009) A new chronology for the Moon and Mercury. *Astron. J., 137,* 4936–4948.

Martin E. S. (2016) The distribution and characterization of strike-slip faults on Enceladus. *Geophys. Res. Lett., 43,* 2456–2464.

McEwen A. S. and Bierhaus E. B. (2006) The importance of secondary cratering to age constraints on planetary surfaces. *Annu. Rev. Earth Planet. Sci., 34,* 535–567.

Michael G. G., Kneissl T., and Neesemann A. (2016) Planetary surface dating from crater size-frequency distribution measurements: Poisson timing analysis. *Icarus, 277,* 279–285.

Moore J. M. (1984) The tectonic and volcanic history of Dione. *Icarus, 59,* 205–220.

Moore J. M., McKinnon W. B., Spencer J. R., et al. (2016) The geology of Pluto and Charon through the eyes of New Horizons. *Science, 351,* 1284–1293.

Morbidelli A., Levison H. F., Tsiganis K., and Gomes R. (2005) Chaotic capture of Jupiter's Trojan asteroids in the early solar system. *Nature, 435,* 462–465.

Morrison D., Johnson T., Shoemaker E., Soderblom L., Thomas P., Veverka J., and Smith B. (1984) Satellites of Saturn: Geological perspective. In *Saturn* (D. Morrison, ed), pp. 609–639. Univ. of Arizona, Tucson.

Morrison D., Owen T., and Soderblom L. A. (1986) The satellites of Saturn. In *Satellites* (J. A. Burns and M. S. Matthews, eds.), pp. 764–801. Univ. of Arizona, Tucson.

Movshovitz N., Nimmo F., Korycansky D. G., Asphaug E., and Owen J. M. (2015) Disruption and reaccretion of midsized moons during an outer solar system late heavy bombardment. *Geophys. Res. Lett., 42,* 256–263.

Nesvorný D., Alvarellos J. L., Dones L., and Levison H. L. (2003) Orbital and collisional evolution of the irregular satellites. *Astron. J., 126,* 398–429.

Nesvorný D., Vokrouhlický D., Bottke W. F., Noll K., and Levison H. F. (2011) Observed binary fraction sets limits on the extent of collisional grinding in the Kuiper belt. *Astron. J., 141,* article ID 159.

Nesvorný D., Vokrouhlický D., Dones L., Levison H. F., Kaib N., and Morbidelli A. (2017) Origin and evolution of short-period comets. *Astrophys. J., 845(27),* 25 pp.

Neukum G., Ivanov B. A., and Hartmann W. K. (2001) Cratering records in the inner solar system in relation to the lunar reference system. *Space Sci. Rev., 96,* 55–86.

Neukum G., Wagner R., Denk T., Porco C. C., and Cassini ISS Team (2005) The cratering record of the saturnian satellites Phoebe, Tethys, Dione and Iapetus in comparison: First results from analysis of the Cassini ISS imaging data. In *Lunar and Planetary Science XXXVI,* Abstract #2034. Lunar and Planetary Institute, Houston.

Neukum G., Wagner R., Wolf U., and Denk T. (2006) The cratering record and cratering chronologies of the saturnian satellites and the origin of impactors: Results from Cassini ISS data. *EPSC Abstracts, Vol. 1,* Abstract #610.

O'Brien D. P. and Greenberg R. (2003) Steady-state size distributions for collisional populations: Analytical solution with size-dependent strength. *Icarus, 164,* 334–345.

Oort J. H. (1950) The structure of the cloud of comets surrounding the solar system and a hypothesis concerning its origin. *Bull. Astron. Inst. Neth., 11,* 91–110.

Parmentier E. M. and Head J. W. (1981) Viscous relaxation of impact craters on icy planetary surfaces: Determination of viscosity variation with depth. *Icarus, 47,* 100–111.

Petit J.-M., Kavelaars J. J., Gladman B. J., Jones R. L., Parker J. Wm., van Laerhoven C., Nicholson P., Mars G., Rousselot P., Mousis O., Marsden B., Bieryla A., Taylor M., Ashby M. L. N., Benavidez P., Campo Bagatin A., and Bernabeu G. (2011) The Canada-France Ecliptic Plane Survey — Full Data Release: The orbital structure of the Kuiper belt. *Astron. J., 142,* article ID 131.

Phillips C. B., Hammond N. P., Robuchon G., Nimmo F., Beyer R. A., and Roberts J. (2012) Stereo imaging, crater relaxation, and thermal histories of Rhea and Dione. In *Lunar and Planetary Science XLIII,* Abstract #2571. Lunar and Planetary Institute, Houston.

Pieters C. M., Adams J. B., Smith M. O., Mouginis-Mark P. J., and Zisk S. H. (1985) The nature of crater rays — The Copernicus example. *J. Geophys. Res., 90,* 12392–12413.

Pike R. E., Kavelaars J. J., Petit J. M., Gladman B. J., Alexandersen M., Volk K., and Shankman C. J. (2015) The 5:1 Neptune resonance as probed by CFEPS: Dynamics and population. *Astron. J., 149,* article ID 202.

Plescia J. B. (1983) The geology of Dione. *Icarus, 56,* 255–277.

Plescia J. B. and Boyce J. M. (1982) Crater densities and geological histories of Rhea, Dione, Mimas and Tethys. *Nature, 295,* 285–290.

Plescia J. B. and Boyce J. M. (1983) Crater numbers and geological histories of Iapetus, Enceladus, Tethys and Hyperion. *Nature, 301,* 666–670.

Plescia J. B. and Boyce J. M. (1985) Impact cratering history of the saturnian satellites. *J. Geophys. Res., 90,* 2029–2037.

Porco C. C., Baker E., Barbara J., et al. (2005) Cassini imaging science: Initial results on Phoebe and Iapetus. *Science, 307,* 1237–1242.

Porco C. C., Helfenstein P., Thomas P. C., et al. (2006) Cassini observes the active south pole of Enceladus. *Science, 311,* 1393–1401.

Rhoden A. R., Henning W., Hurford T. A., Patthoff D. A., and Tajeddine R. (2017) The implications of tides on the Mimas ocean hypothesis. *J. Geophys. Res.–Planets, 122,* 400–410.

Richardson J. E. (2009) Cratering saturation and equilibrium: A new model looks at an old problem. *Icarus, 204,* 697–715.

Rivera-Valentin E. G., Barr A. C., Lopez Garcia E. J., Kirchoff M. R., and Schenk P. M. (2014) Constraints on planetesimal disk mass from the cratering record and equatorial ridge on Iapetus. *Astrophys. J., 792,* article ID 127.

Roatsch Th., Wählisch M., Hoffmeister A., et al. (2009) High-resolution atlases of Mimas, Tethys, and Iapetus derived from Cassini-ISS images. *Planet. Space Sci., 57,* 83–92.

Roatsch Th., Kersten E., Wählisch M., et al. (2012) High-resolution atlas of Rhea derived from Cassini-ISS images. *Planet. Space Sci., 61,* 135–141.

Roatsch T., Kersten E., Hoffmeister A., Wählisch M., Matz K.-D., and Porco C. C. (2013) Recent improvements of the saturnian satellites atlases: Mimas, Enceladus, and Dione. *Planet. Space Sci., 77,* 118–125.

Robbins S. J. (2014) New crater calibrations for the lunar crater-age chronology. *Earth Planet. Sci. Lett., 403,* 188–198.

Robbins S. J. and Hynek B. M. (2011) Secondary crater fields from 24 large primary craters on Mars: Insights into nearby secondary crater production. *J. Geophys. Res., 116,* E10003.

Robbins S. J and Hynek B. M. (2014) The secondary crater population of Mars. *Earth Planet. Sci. Lett., 400,* 66–76.

Robbins S. J., Bierhaus E. B., and Dones L. (2014) Craters of the saturnian satellite system: I. Mimas. In *Planetary Crater Consortium 5,* Abstract #1411.

Robbins S. J., Bierhaus E. B., and Dones L. (2015) Craters of the saturnian satellite system: II. Mimas and Rhea. In *Lunar and Planetary Science XLVI,* Abstract #1654. Lunar and Planetary Institute, Houston.

Robbins S. J., Singer K. N., Bray V. J., et al. (2017) Craters of the Pluto-Charon system. *Icarus, 287,* 187–206.

Schenk P. M. and Moore J. M. (2009) Eruptive volcanism on Saturn's icy moon Dione. In *Lunar and Planetary Science XL,* Abstract #2465. Lunar and Planetary Institute, Houston.

Schenk P. M., Chapman C. R., Zahnle K., and Moore J. M. (2004) Ages and interiors: The cratering record of the Galilean satellites. In *Jupiter: The Planet, Satellites and Magnetosphere* (F. Bagenal et al., eds.), pp. 427–456. Cambridge Univ., Cambridge.

Schenk P. M., Hamilton D. P., Johnson R. E., McKinnon W. B., Paranicas C., Schmidt J., and Showalter M. R. (2011) Plasma, plumes and rings: Saturn system dynamics as recorded in global color patterns on its midsize icy satellites. *Icarus, 211,* 740–757.

Schmedemann N. and Neukum G. (2011) Impact crater size-frequency distribution (SFD) and surface ages on Mimas. In *Lunar and Planetary Science XLII,* Abstract #2772. Lunar and Planetary Institute, Houston.

Schmedemann N., Neukum G., Denk T., and Wagner R. J. (2008)

Stratigraphy and surface ages on Iapetus and other saturnian satellites. In *Lunar and Planetary Science XXXIX,* Abstract #2070. Lunar and Planetary Institute, Houston.

Schmedemann N., Neukum G., Denk T., and Wagner R. (2009) Impact crater size-frequency distribution (SFD) on saturnian satellites and comparison with other solar-system bodies. In *Lunar and Planetary Science XL,* Abstract #1941. Lunar and Planetary Institute, Houston.

Schultz P. H. and Singer J. (1970) A comparison of secondary craters on the Moon, Mercury, and Mars. *Proc. Lunar Planet. Sci. Conf. 11th,* pp. 2243–2259.

Sheppard S. S. and Trujillo C. A. (2010) The size distribution of the Neptune Trojans and the missing intermediate-sized planetesimals. *Astrophys. J. Lett., 723,* L233–L237.

Shoemaker E. M. (1962) Interpretation of lunar craters. In *Physics and Astronomy of the Moon* (Z. Kopal, ed.), pp. 283–359. Academic, London.

Shoemaker E. M. and Wolfe R. F. (1981) Evolution of the saturnian satellites: The role of impact. In *Lunar and Planetary Science XII, Supplement A, Satellites of Saturn,* pp. 1–3. LPI Contribution No. 428, Lunar and Planetary Institute, Houston.

Singer K. N., McKinnon W. B., and Nowicki L. T. (2013) Secondary craters from large impacts on Europa and Ganymede: Ejecta size-velocity distributions on icy worlds, and the scaling of ejected blocks. *Icarus, 226,* 865–884.

Singer K. N., Jolliff B. L., and McKinnon W. B. (2014) Lunar secondary craters measured using LROC imagery: Size-velocity distributions of ejected fragments. In *Lunar and Planetary Science XLV,* Abstract #1162. Lunar and Planetary Institute, Houston.

Singer K. N., McKinnon W. B., Jolliff B. L., and Plescia J. B. (2015) Icy satellite and lunar ejecta from mapping of secondary craters: Implications for sesquinary forming fragments. In *Workshop on Issues in Crater Studies and the Dating of Planetary Surfaces,* Abstract #9034. LPI Contribution No. 1841, Lunar and Planetary Institute, Houston.

Singer K. N., McKinnon W. B., Greenstreet S., Gladman B., Bierhaus E. B., Stern S. A., Parker A. H., et al. (2017) Impact craters on Pluto and Charon indicate a deficit of small Kuiper belt objects. *Asteroids Comets Meteors,* Abstract #Parallel3.c.3.

Smith B. A., Soderblom L., Beebe R., et al. (1981) Encounter with Saturn: Voyager 1 imaging science results. *Science, 212,* 163–191.

Smith B. A., Soderblom L., Batson R., et al. (1982) A new look at the Saturn system: The Voyager 2 images. *Science, 215,* 504–537.

Smith D. E., Turtle E. P., Melosh H. J., and Bray V. J. (2007) Viscous relaxation of craters on Enceladus. In *Lunar and Planetary Science XXXVIII,* Abstract #2237. Lunar and Planetary Institute, Houston.

Smith-Konter B. and Pappalardo R. T. (2008) Tidally driven stress accumulation and shear failure of Enceladus's tiger stripes. *Icarus, 198,* 435–451.

Soter S. (1974) The brightness asymmetry of Iapetus. Paper presented at IAU Colloquium 28, Cornell University, Ithaca.

Speyerer E. J., Povilaitis R. Z., Robinson M. S., Thomas P. C., and Wagner R. V. (2016) Quantifying crater production and regolith overturn on the Moon with temporal imaging. *Nature, 538,* 215–218.

Squyres S. W., Howell C., Liu M. C., and Lissauer J. J. (1997) Investigation of crater "saturation" using spatial statistics. *Icarus, 125,* 67–82.

Stephan K. and 14 colleagues (2010) Dione's spectral and geological properties. *Icarus, 206,* 631–652.

Stephan K., Jaumann R., Wagner R., et al. (2012) The saturnian satellite Rhea as seen by Cassini VIMS. *Planet. Space Sci., 61,* 142–160.

Stern S. A., Bagenal F., Ennico K., et al. (2015) The Pluto system: Initial results from its exploration by New Horizons. *Science, 350(6258),* aad1815.

Strom R. G. (1981) Crater populations on Mimas, Dione and Rhea. In *Lunar and Planetary Science XII,* pp. 7–9. Lunar and Planetary Institute, Houston.

Strom R. G. (1987) The solar system cratering record: Voyager 2 results at Uranus and implications for the origin of impacting objects. *Icarus, 70,* 517–35.

Strom R. G. and Woronow A. (1982) Solar system cratering populations. In *Lunar and Planetary Science XIII,* pp. 782–783. Lunar and Planetary Institute, Houston.

Strom R. G., Malhotra R., Ito T., Yoshida F., and Kring D. A. (2005) The origin of planetary impactors in the inner solar system. *Science, 309,* 1847–1850.

Tajeddine R., Rambaux N., Lainey V., Charnoz S., Richard A., Rivoldini A., and Noyelles B. (2014) Constraints on Mimas' interior from Cassini ISS libration measurements. *Science, 346,* 322–324.

Tamayo D., Burns J. A., Hamilton D. P., and Hedman M. M. (2011) Finding the trigger to Iapetus' odd global albedo pattern: Dynamics of dust from Saturn's irregular satellites. *Icarus, 215,* 260–278.

Tancredi G. (2014) A criterion to classify asteroids and comets based on the orbital parameters. *Icarus, 234,* 66–80.

Tiscareno M. S. and Malhotra R. (2003) The dynamics of known Centaurs. *Astron. J., 126,* 3122–3131.

Tiscareno M. S., Mitchell C. J., Murray C. D., Di Nino D., Hedman M. M., Schmidt J., Burns J. A., Cuzzi J. N., Porco C. C., Beurle K., and Evans M. W. (2013) Observations of ejecta clouds produced by impacts onto Saturn's rings. *Science, 340,* 460–464.

Tsiganis K., Gomes R., Morbidelli A., and Levison H. F. (2005) Origin of the orbital architecture of the giant planets of the solar system. *Nature, 435,* 459–461.

Valsecchi G. B., Alessi E. M., and Rossi A. (2014) The geometry of impacts on a synchronous planetary satellite. *Cel. Mech. Dyn. Astron., 119,* 257–270.

Vickery A. M. (1986) Size-velocity distribution of large ejecta fragments. *Icarus, 67,* 224–236.

Volk K. and Malhotra R. (2008) The scattered disk as the source of the Jupiter family comets. *Astron. J., 687,* 714–725.

Volk K. and 12 colleagues (2016) OSSOS III — Resonant trans-neptunian populations: Constraints from the first quarter of the Outer Solar System Origins Survey. *Astron. J., 152,* article ID 23.

Wagner R. J., Neukum G., Giese B., Roatsch T., Wolf U., and Denk T. (2006) Geology, ages and topography of Saturn's satellite Dione observed by the Cassini ISS camera. In *Lunar and Planetary Science XXXVII,* Abstract #1805. Lunar and Planetary Institute, Houston.

Wagner R. J., Neukum G., Giese B., Roatsch T., and Wolf U. (2007) The global geology of Rhea: Preliminary implications from the Cassini ISS data. In *Lunar and Planetary Science XXXVIII,* Abstract #1958. Lunar and Planetary Institute, Houston.

Wagner R. J., Neukum G., Giese B., Roatsch T., Denk T., Wolf U., and Porco C. C. (2008) Geology of Saturn's satellite Rhea on the basis of the high-resolution images from the targeted flyby 049 on Aug. 30, 2007. In *Lunar and Planetary Science XXXIX,* Abstract #1930. Lunar and Planetary Institute, Houston.

Wang J. H. and Brasser R. (2014) An Oort cloud origin of the Halley-type comets. *Astron. Astrophys., 563,* article ID A122.

White O. L., Schenk P. M., and Dombard A. J. (2013) Impact basin relaxation on Rhea and Iapetus and relation to past heat flow. *Icarus, 223,* 699–709.

White O. L., Schenk P. M., Bellagamba A. W., Grimm A. M., Dombard A. J., and Bray V. J. (2017) Impact crater relaxation on Dione and Tethys and relation to past heat flow. *Icarus, 288,* 37–52.

Wieczorek M. A. and Le Feuvre M. (2009) Did a large impact reorient the Moon? *Icarus, 200,* 358–366.

Wiegert P. and Tremaine S. (1999) The evolution of long-period comets. *Icarus, 137,* 84–121.

Yin A. and Pappalardo R. T. (2015) Gravitational spreading, bookshelf faulting, and tectonic evolution of the south polar terrain of Saturn's moon Enceladus. *Icarus, 260,* 409–439.

Yue Z., Johnson B. C., Minton D. A., Melosh H. J., Di K., Hu W., and Liu Y. (2013) Projectile remnants in central peaks of lunar impact craters. *Nature Geosci., 6,* 435–437.

Yung Y. L., Allen M., and Pinto J. P. (1984) Photochemistry of the atmosphere of Titan — Comparison between model and observations. *Astrophys. J. Suppl., 55,* 465–506.

Zahnle K., Mac Low M.-M., and Chyba C. F. (1993) Some consequences of the collision of a comet and Jupiter. In *Division for Planetary Sciences Meeting Abstracts,* Abstract #1043.

Zahnle K., Dones L., and Levison H. F. (1998) Cratering rates on the Galilean satellites. *Icarus, 136,* 202–222.

Zahnle K., Schenk P., Sobieszczyk S., Dones L., and Levison H. F. (2001) Differential cratering of synchronously rotating satellites by ecliptic comets. *Icarus, 153,* 111–129.

Zahnle K., Schenk P., Levison H. F., and Dones L. (2003) Cratering rates in the outer solar system. *Icarus, 163,* 263–289.

Zahnle K., Alvarellos J. L., Dobrovolskis A., and Hamill P. (2008) Secondary and sesquinary craters on Europa. *Icarus, 194,* 660–674.

Castillo-Rogez J. C., Hemingway D., Rhoden A., Tobie G., and McKinnon W. B. (2018) Origin and evolution of Saturn's mid-sized moons. In *Enceladus and the Icy Moons of Saturn* (P. M. Schenk et al., eds.), pp. 285–305. Univ. of Arizona, Tucson, DOI: 10.2458/azu_uapress_9780816537075-ch014.

Origin and Evolution of Saturn's Mid-Sized Moons

J. C. Castillo-Rogez
Jet Propulsion Laboratory, California Institute of Technology

D. Hemingway
University of California Berkeley

Alyssa Rhoden
Southwest Research Institute

G. Tobie
Universite de Nantes

W. B. McKinnon
Washington University in Saint Louis

This chapter focuses on the origin(s) and evolution of the mid-sized moons of Saturn: Mimas, Tethys, Enceladus, Dione, Rhea, and Iapetus. Enceladus will also be considered for comparative planetology. We describe the contribution of the Cassini-Huygens mission to the knowledge of the moons' internal evolution. Geophysical constraints remain scarce despite several dedicated flybys. Hence our overall understanding of the moons interiors remains primarily based on modeling. Besides Enceladus, observational evidence exists for the occurrence of deep oceans in Tethys and Dione in the past in Mimas at present, which is difficult to reconcile with their low heat budgets. The diverse physical properties encountered across the mid-sized satellites raise important questions about the origin of the system and leaves open many questions to be addressed by future missions.

1. INTRODUCTION

The saturnian satellite system is the most complex across the solar system with four distinct types of satellites: (1) Titan, a large moon akin to the Galilean satellites in terms of mass and radius; (2) the inner, mid-sized moons located within Titan's orbit; (3) moonlets embedded in the rings; and (4) irregular satellites that are likely captured (*Nesvorny et al.,* 2007). Beyond Titan's orbit lies Iapetus, almost 60 R_S (saturnian radii) away, whose relationship to the other regular satellites remains to be clarified. Other oddities in the saturnian satellite system include the irregular Hyperion, which could be a fragment of a disrupted precursor (*Farinella et al.,* 1983), and the co-orbitals of Tethys and Dione. Figure 1 shows the distribution of saturnian moons covered in this chapter.

Most of the mid-sized satellites were discovered during the seventeenth and eighteenth centuries by Giovanni Domenico Cassini (Tethys, Dione, Iapetus, and Rhea) and William Herschel (Mimas, Enceladus). These astronomers played a crucial role in charting the saturnian system. *In situ* reconnaissance of the moons started with the Pioneer 11 mission in 1979 with flybys of the inner moons from between 100,000 and 300,000 km away. A sharper picture of Saturn's system was delivered by the twin Voyager spacecraft in the 1980s. They revealed a complex suite of objects in terms of physical and geological properties. These missions provided estimates of the moons' mean radii and densities, which led to the first interior evolution models of these bodies (e.g., *Ellsworth and Schubert,* 1983). Models at the time put emphasis on two properties: the physics of ice-rock mixtures, and compaction in bodies believed to have accreted with significant porosity (see *Eluszkiewicz et al.,* 1998, for a review). Finally, all the mid-sized moons were found in spin-orbit resonance with Saturn.

The Cassini-Huygens mission arrived at the saturnian system in 2004 and started its exploration of the mid-sized moons with a >100,000-km flyby of Iapetus on December 30. The list of Cassini-Huygens flybys targeted at a range of <10,000 km is found in Fig. 2. The activities of the Cassini-Huygens mission were centered around Saturn, Titan, and Enceladus. The mid-sized moons were targeted on a best effort basis and included one close flyby for Mimas, three for Tethys, and five each for Dione and Rhea. Additional "Voyager-class" flybys at ~>100,000 km contributed to increasing spatial coverage. There have been several recent

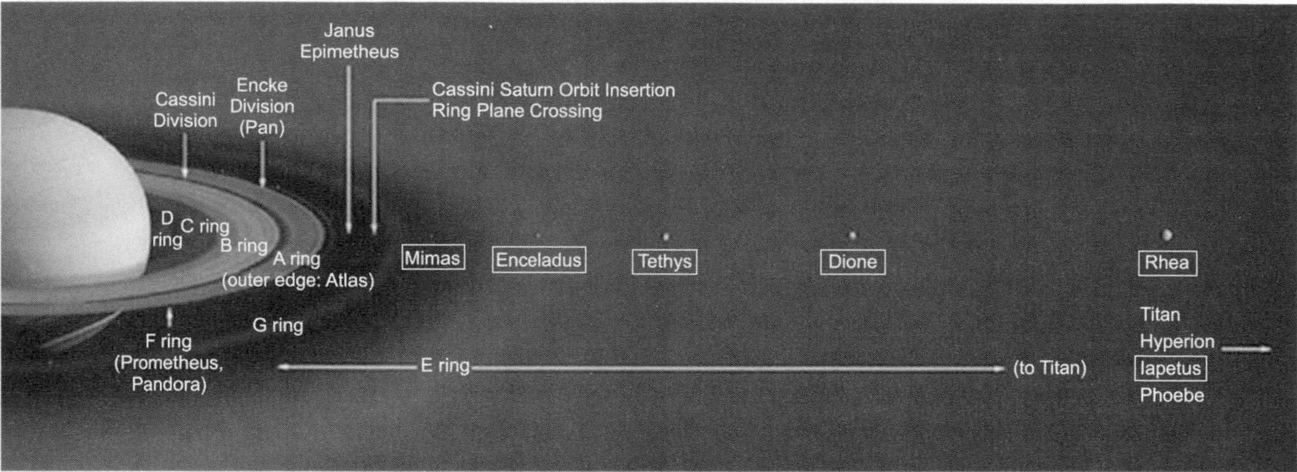

Fig. 1. This sketch of the saturnian system configuration shows the distribution of the moons covered in this chapter (boxed). Credit: NASA/JPL.

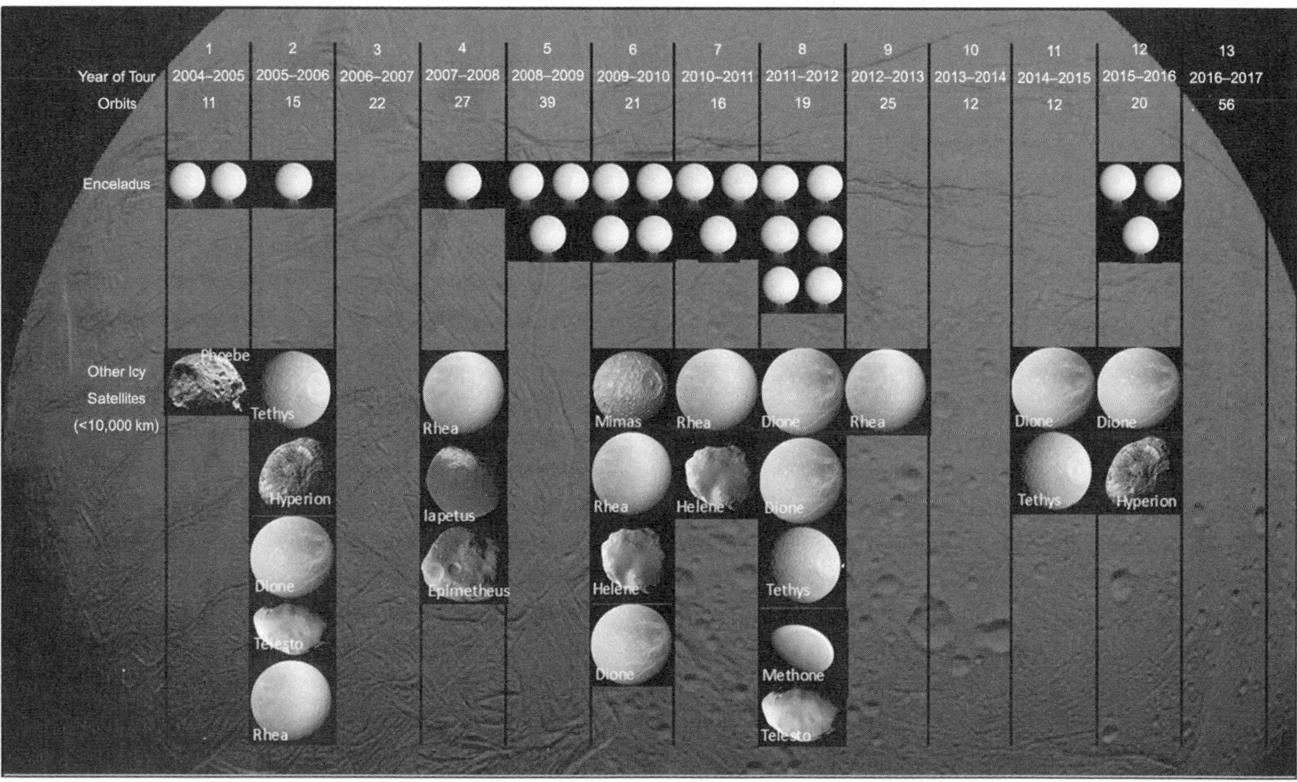

Fig. 2. Summary of the flybys dedicated to Saturn's moons (besides Titan) during the course of the Cassini-Huygens mission. Credit: NASA/JPL.

reviews of the saturnian satellite system, focusing on their thermophysical properties and thermal evolution (*Matson et al.,* 2009; *Collins et al.,* 2009; *Jaumann et al.,* 2009; *Schubert et al.,* 2010; *McCarthy and Castillo-Rogez,* 2012; *Hussmann et al.,* 2015), and we will also refer to other chapters in this volume (e.g., Kirchoff et al., Nimmo et al., Schenk et al.). Hence the purpose of the present chapter is not to revisit well-known considerations on thermal evolution. Instead we focus on the state of understanding of the mid-sized satellites as a system in space and time — going back to their origin — as Cassini-Huygens has completed its mission.

The current state of knowledge of each mid-sized satellite is detailed in section 2. While these observations led to refined estimates of the satellites' physical parameters, key properties of these bodies, especially pertaining to their interior structure, are missing. On the other hand, Cassini-Huygens revealed the complex relationships between the moons and rings, which led to the introduction of new concepts for their origin, potentially shifting the paradigm of an old satellite system formed in Saturn's nebula to bodies that were born in and emerged sequentially from the rings (section 3). Different origins determine different types of evolution, although the many parameters driving internal evolution make it difficult to favor a particular formation scenario (section 4). This review closes with a summary of research gaps and major open questions that may be tested by future missions to the outer solar system and especially to the uranian system, which shares many similarities with Saturn's.

2. STATE OF KNOWLEDGE OF ICY SATELLITE PROPERTIES

Close-proximity observations by the Cassini-Huygens mission have refined the moons' physical properties. Exquisite details of surface geology, surface composition, gravity science, and the detection of magnetic signature at some of these bodies have led to deeper insights into their internal properties. The Cassini-Huygens observations have yielded refined GM and mean radii for all the mid-sized satellites. Triaxial shapes were obtained with typical uncertainties 0.2–0.5 km (*Thomas,* 2010), insufficient in most cases for deriving meaningful moment of inertia estimates, especially with the additional uncertainty introduced by non-hydrostatic anomalies. As an extreme case, Iapetus exhibits a highly non-hydrostatic, fossil shape as a witness of an early period of faster rotation. Degree-two gravity harmonics were inferred from radio science and shape observations of Enceladus (*Iess et al.,* 2014), Rhea (*Tortora et al.,* 2016), and Dione (*Hemingway et al.,* 2016). Librational amplitudes were derived at Enceladus and Mimas from imaging (*Thomas et al.,* 2016; *Tajeddine et al.,* 2014). Geological studies led to constraints on the heat flux of the satellites over time (see chapter by Schenk et al., this volume). Oceans have been suggested in smaller bodies, such as Tethys (*Chen and Nimmo,* 2008) and Mimas (*Tajeddine et al.,* 2014), although observations do not strictly require oceans. Furthermore, the development and long-term preservation of deep oceans in these energy-starved bodies remain to be modeled. The case for a past, and possibly also present, deep ocean in Dione has been made based on combined studies of gravity and topography (*Hemingway et al.,* 2016; *Beuthe et al.,* 2016). Groundbased astrometric observations recorded for more than a century have also contributed to better understanding the orbital evolution of these moons, which eventually led to constraints on the tidal dissipation of Saturn (*Lainey et al.,* 2012, 2017).

We summarize below the current knowledge of the mid-sized moons (see also the chapter by Schenk et al. in this volume). Key physical and orbital properties are gathered in Table 1.

2.1. Mimas

Mimas is ~185,520 km from the center of Saturn (i.e., ~3 R_S) and it is currently in a 4:2 nearly commensurable, *inclination*-type resonance with Tethys (*Dermott,* 1971). It is likely that Mimas' initial eccentricity was somewhat larger than its present value (e.g., *Peale et al.,* 1980; *Lissauer and Safronov,* 1991) and that it damped as a result of tidal dissipation within the satellite. However, the critical dynamical property to be understood is the current high value of Mimas' *free* eccentricity, e = 0.0196. Such a high eccentricity is a strong constraint for models because Mimas is so close to Saturn and tidal dissipation is expected to damp free eccentricities to smaller values, such as those exhibited by Enceladus (e = 10^{-3}) and Rhea (e = 10^{-5}). It is also puzzling that Mimas shares this peculiarity with more distant and massive Titan and Iapetus, which have also preserved free eccentricities greater than 0.02, although in these cases this feature can be explained by the distances of these bodies from Saturn, which substantially reduces the effects of tides, lengthening the timescales over which their orbits would circularize. Thus, modeling Mimas' eccentricity may lead to a broader understanding of the way tidal dissipation evolves satellites' orbits. Also, Mimas, in the course of its evolution, could have been affected by resonances with other satellites. Then the satellite's eccentricity would have slowly decreased to its current value. As an alternative, assuming Mimas accreted in the rings of Saturn, it could have gained its high eccentricity following large impacts incurred when it left the rings (*Charnoz et al.,* 2011).

Mimas has a mean radius of 198.2 km. Its surface is covered with craters and appears to be ancient (*Plescia and Boyce,* 1982). Like most other mid-sized moons, it displays a large basin called Herschel (Fig. 3). Its density

Fig. 3. Mimas as seen from the Cassini orbiter on February 13, 2010, from about 9500 km. This picture highlights Mimas' heavily cratered surface and the ~130-km Herschel crater. Credit: NASA/JPL-Caltech/Space Science Institute (PIA12570).

TABLE 1. Basic characteristics of Saturn's mid-sized satellites.

Property	Mimas	Enceladus	Tethys	Dione	Rhea	Titan	Hyperion	Iapetus	Phoebe
Mean radius (km)	198.2 ± 0.4	252.10 ± 0.2	531.0 ± 0.6	561.4 ± 0.4	763.5 ± 0.6	2574.73 ± 0.09	135.0 ± 4.0	734.3 ± 2.8	106.5 ± 0.7
a (equatorial) (km)	207.8 ± 0.5	256.6 ± 0.3	538.4 ± 0.3	563.4 ± 0.6	765.0 ± 0.7	2574.91 ± 0.11	180.1 ± 2.0	746.7 ± 2.9	109.4 ± 1.4
b (equatorial) (km)	196.7 ± 0.5	251.4 ± 0.2	528.3 ± 1.1	561.3 ± 0.5	763.1 ± 0.6	(equat. average)	133.0 ± 4.5	745.7 ± 2.9	108.5 ± 0.6
c (polar) (km)	190.6 ± 0.3	248.3 ± 0.2	526.3 ± 0.6	559.6 ± 0.4	762.4 ± 0.6	2574.32 ± 0.05 (north pole)	102.7 ± 4.5	712.1 ± 1.6	101.8 ± 0.3
GM ($km^3 s^{-2}$)	2.5026 ± 0.0006	7.2027 ± 0.0125	41.2067 ± 0.0038	73.1146 ± 0.0015	153.9426 ± 0.0037	8978.138 ± 1.334	0.375 ± 0.003	120.5038 ± 0.0080	0.553 ± 0.001
Density (kg m^{-3})	1149 ± 7	1609 ± 5	985 ± 3	1478 ± 3	1237 ± 3	1879.8 ± 4.4	544 ± 50	1088 ± 13	1638 ± 33
Heat flow estimates (discrete features) (mW m^{-2})	Low	60–220	20–60+	50–60+	20–30	~7	Low	Low	Low
Central pressure (MPa)	~7	~23	~37	~95	~124	~3500	~1	~89	~4
Rotation period (days)	0.942 (synchronous)	1.370 (synchronous)	1.888 (synchronous)	2.737 (synchronous)	4.518 (synchronous)	15.95 (synchronous)	chaotic	79.33 (synchronous)	0.387
Semimajor axis (km)	185539	238042	294672	377415	527068	1221865	1500933	3560854	12947918
Orbital period (days)	0.942	1.370	1.888	2.737	4.518	15.95	21.28	79.33	548.02
Eccentricity	0.0196	0.0047	0.0001	0.0022	0.0002	0.0288	0.0232	0.0293	0.1634
Obliquity (deg.)	1.574	0.003	1.091	0.028	0.333	0.3	variable	8.298	152.14
Geometric albedo	0.962 ± 0.004	1.375 ± 0.008	1.229 ± 0.005	0.998 ± 0.004	0.949 ± 0.003	0.22	0.3	0.02–0.7	0.06

The a, b, c radii refer to the moon ellipsoidal shape axes [*Thomas* (2010), except in the case of Titan, *Zebker et al.* (2009)]; heat flux estimates come from *White et al.* (2017) for Tethys and Dione in the past and from the chapter by Schenk et al. in this volume for Mimas, Rhea, and Iapetus. The heat flux range provided for Enceladus is based on estimates for the south polar region that are uncertain by a factor of 3 (e.g., *Howett et al.*, 2011). The heat flow estimated for Titan is based on theoretical modeling (*Tobie et al.*, 2005).

of ~1150 kg m^{-3} corresponds to a rock volume fraction of about 9%, assuming no porosity, which means Mimas is most likely a heat-starved body due to its limited potential for radiogenic heating. Mimas' large libration amplitude (*Tajeddine et al.,* 2014) has been attributed to the decoupling of an ice shell from the deep interior via a liquid layer, or by the presence of a highly non-hydrostatic core. The latter explanation is more likely: Mimas does not show evidence of intense internal heating in its geology and the absence of obvious tectonic features argues against the presence of a deep ocean (*Rhoden et al.,* 2017; Schenk et al., this volume). Mimas' librations also imply that Mimas is differentiated, which further complicates models of its thermal-orbital evolution. The lack of tidally-driven geologic activity (such as the fracture systems on Europa and Enceladus), the lack of relaxed craters or other evidence of high heat flows, and the current high eccentricity of Mimas all require that Mimas avoid significant recent tidal dissipation. *Neveu and Rhoden* (2017) modeled the coupled thermal-orbital evolution of Mimas and identified multiple pathways by which Mimas could appear differentiated today while remaining consistent with these observations (see section 4).

2.2. Tethys

Tethys is located about 4 planet radii from Saturn. This moon has an inclination of 1° owing to an inclination resonance with Mimas. Its eccentricity is close to zero.

Tethys' radius is 531 km. Its mean density is 985 kg m^{-3} (*Thomas,* 2010), suggesting it is almost entirely ice. Assuming no porosity, the rock mass fraction is only ~7 wt.%. However, an object of that size could preserve significant porosity following accretion, up to 40% at shallow depth, depending on thermal evolution (*Matson et al.,* 2009), which would imply a rock fraction of 15–20 wt.%. Tethys' ellipsoidal shape departs from hydrostatic equilibrium by about 2.6 km (a–c, assumes a homogeneous interior) (*Thomas, 2010*).

One of Tethys' most prominent features is the large Ithaca Chasma, a rift that is about 1000 km long and 2–3 km deep. It is 100 km wide in the north and becomes two narrower branches in the south. Its flanks are raised by up to 6 km above its surroundings (*Giese et al.,* 2007). Crater counting suggests it formed approximately 4 G.y. ago (*Giese et al.,* 2007). Previous work has suggested that the formation of Ithaca Chasma may be related to the 400 km Odysseus impact basin (Fig. 4) or expansion due to the freezing of Tethys' interior (*Smith et al.,* 1982; *Moore et al.,* 2004). More recent work involving crater counts suggests that Ithaca Chasma pre-dates Odysseus (*Giese et al.,* 2007; *Stephan et al.,* 2016). However, Schenk et al. (this volume) argue otherwise based on geological considerations; hence the formation of Ithaca Chasma is still debated. Flexural modeling of Ithaca Chasma suggests that the surface heat flux was 18–30 mW m^{-2} when it formed (*Giese et al.,* 2007) and analysis of large crater morphology yields heat fluxes in excess of 60 mW m^{-2} (*White et al.,* 2017). These geological observations imply a past history of tidal heating that could have occurred if Tethys passed through a 3:2 resonance with Dione (*Chen and Nimmo,* 2008). Owing to the very low eccentricity and the inclination resonance with Mimas, tidal dissipation in Tethys comes primarily from obliquity tides, at least at present and in the recent past (*Chen et al.,* 2014). However, the total tidal heating budget of Tethys is one of the lowest among mid-sized icy moons (uranian satellites included). Hence the inferred heat flux likely required the eccentricity to be high for a prolonged period of time, for example, excited by a resonance. This could have led to melting of the ice layer, which would increase tidal heat production and further thin the ice shell (*Chen and Nimmo,* 2008). Thermal evolution models (excluding tidal heating) predict that if convection ever occurred, it ceased early in solar system history (*Multhaup and Spohn,* 2007).

Fig. 4. Tethys as seen from the Cassini orbiter on April 11, 2015, from 190,000 km. This picture highlights the ~445-km basin Odysseus and partial resurfacing due to the basin ejecta. Credit: NASA/JPL-Caltech/Space Science Institute (PIA18317).

2.3. Dione

Dione is similar in size to Tethys but much denser (Table 1). It has a rock fraction of roughly 50% by mass and a surface that shows signs of an intriguing history of endogenic activity, suggestive of strong internal heating. Dione's participation in the 2:1 mean-motion resonance with Enceladus maintains its present orbital eccentricity of 0.0022.

Dione's predominantly water ice surface is characterized by several different terrain types, spanning a broad range of ages, and suggests extensive tectonic activity and multiple episodes of large-scale resurfacing (*Kirchoff and Schenk,* 2015) (Fig. 5). A province of densely cratered terrain covers the northern portion of the anti-saturnian face; the leading

Fig. 5. Dione as seen from the Cassini orbiter on September 30, 2007, from about 45,000 km. Dione's surface is one of the most tectonized among mid-sized icy moons as indicated by the extensive set of fractures (wispy terrain) highlighted in this picture. Credit: NASA/JPL/Space Science Institute (PIA09764).

hemisphere is dominated by a large province of significantly smoother plains that suggests an event of resurfacing; and much of the sub- and anti-saturnian faces are covered by an intermediately cratered terrain, indicating one or more major resurfacing events prior to the one associated with the smooth terrain (*Kirchoff and Schenk,* 2015). Perhaps most strikingly, the trailing hemisphere is characterized by a significantly darker surface (likely the result of bombardment by particles embedded in Saturn's fast-rotating magnetosphere (e.g., *Clark et al.,* 2008) crosscut by a network of bright features, historically known as "wispy" terrain, comprising an extensive series of steep fault scarps (e.g., *Stephan et al.,* 2010). These were first discovered by the Voyager missions (*Smith et al.,* 1981; *Plescia,* 1983) and have been better imaged by the Cassini-Huygens mission (e.g., *Stephan et al.,* 2010). Unfortunately, the age of these features is not well constrained (*Kirchoff and Schenk,* 2015), but their crosscutting relationships suggest they are among the youngest features on the surface (*Martin et al.,* 2016). The wispy terrain structures appear to be extensional in origin (*Jaumann et al.,* 2009), but the moon also contains ridges that point to a compressional history (*Collins et al.,* 2009). Finally, several large impact craters, some of which may date to <1 G.y., appear to be substantially relaxed, suggesting high heat fluxes even late in Dione's history (e.g., *Phillips et al.,* 2012; *White et al.,* 2017).

Although Dione is large enough to have relaxed to near spherical symmetry, it is considerably more flattened than expected for a perfectly hydrostatic body. Its triaxial ellipsoidal figure (Table 1) yields (a–c)/(b–c) ≈ 2.2, far from the ratio close to 4 expected for a hydrostatic body and Dione's distance from Saturn. Equivalently, the unnormalized nonzero degree-2 spherical harmonic shape coefficients are $H_{20} = -1920 \pm 160$ m and $H_{22} = 370 \pm 30$ m, having the ratio $-H_{20}/H_{22} = 5.2 \pm 0.6$ (1σ), far from the nearly hydrostatic expectation of 3.3 (*Hemingway et al.,* 2016), and indicating a substantial amount of global scale non-hydrostatic topography. Analysis of Dione's topographic variance spectrum has been interpreted as showing a transition from isostatic to flexural support that is consistent with a globally averaged effective lithospheric elastic thickness of 1.5–5 km (*Nimmo et al.,* 2011). In complementary work, *Hammond et al.* (2013) used stereotopography to show that a prominent ridge known as Janiculum Dorsum, standing more than 1 km above the local elevation, appears to be supported in part by lithospheric flexure. The degree of flexure suggests an effective elastic thickness of 2.5–4.5 km at the time of the ridge's formation (*Hammond et al.,* 2013), in agreement with results from global-scale topography (*Nimmo et al.,* 2011), and corresponding to a heat flux of 25–60 mW m^{-2}. *White et al.* (2017) also suggested a heat flow as high as 60 mW m^{-2} to explain Dione's relaxed crater morphology, especially in its southern hemisphere. This leads these authors to draw analogies with Enceladus, invoking concentrated, resonance-induced tidal heating. Such a large heat flux is argued to be possible only in the presence of a subsurface liquid layer that decouples the ice shell from the deeper interior, permitting more significant tidal flexing and therefore greater tidal heating than would be possible otherwise (*Hammond et al.,* 2013). It is possible the high heat flow is evidence for past convection. It is uncertain whether such convection continues today (*Zhang and Nimmo,* 2009), and thus little is known about the possible transport of material between the surface and deeper layers at present.

Recently, combined analysis of gravity and topography has ruled out the possibility of an undifferentiated interior and has provided further evidence for a subsurface liquid water ocean (*Hemingway et al.,* 2016; *Beuthe et al.,* 2016). Doppler tracking of the Cassini orbiter during a series of dedicated gravity flybys, the final two of which took place in 2015, allowed for the determination of Dione's quadrupole (degree-2) gravity field, with the measured zonal and (zero phase) sectorial terms being $J_2 = 1454 \pm 16$ and $C_{22} = 363 \pm 2$, in units of dimensionless gravitational potential $\times 10^6$. The ratio $J_2/C_{22} = 4.00 \pm 0.06$ indicates a statistically significant departure from the hydrostatic expectation of close to 3.31 (*Tricarico,* 2014). While this ratio is large, it is considerably smaller than the corresponding ratio for the shape (5.2 ± 0.6), an indication that Dione's global scale non-hydrostatic topography is substantially compensated (*Hemingway et al.,* 2016).

Possible mechanisms for compensation include lateral variations in the density of the icy mantle (i.e., Pratt isostasy), lateral variations in the thickness of an ice shell overlying a subsurface liquid water ocean (i.e., Airy isostasy), or some combination of the two. Gravity-based interior structure models are inherently non-unique and depend on several assumptions including the ice shell and ocean densities, the degree of elastic support, and the precise definition of isostatic equilibrium (see the Hemingway et al. chapter in this volume). However, if Dione's degree-2 topography is compensated in an Airy sense (i.e., it is supported isostatically in a subsurface liquid ocean), and neglecting the small contribution of long wavelength elastic support, the mean ice shell and ocean thicknesses are 76–122 km and 35–95 km, respectively, leaving a core radius of roughly 400 km and a core density around 2400 kg m^{-3} (*Hemingway et al.,* 2016; *Beuthe et al.,* 2016); the corresponding moment of inertia factor is approximately 0.33.

It should, however, be stressed that alternative interpretations of the gravity observations are possible, especially given the considerable uncertainties in the shape and the fact that a pure Airy compensation model accounts for only one endmember scenario. Future measurements that are better able to constrain the topography, gravity field, and rotation state of Dione will permit more precise interior modeling. Such results, in combination with future geodynamical modeling efforts, would lead us to a better understanding of Dione's evolution.

Fig. 6. Rhea as seen from the Cassini orbiter on November 26, 2005, from about 60,000 km distance. The large crater on the upper right of the picture is Rhea's largest crater Tirawa (~397 km). Credit: NASA/JPL/Space Science Institute (PIA07763).

2.4. Rhea

Rhea is Saturn's second largest moon and the outermost of the five mid-sized satellites interior to Titan (Table 1). Rhea is similar in size to Iapetus, but denser, with a rock fraction of roughly 25–35% by mass, depending on assumed rock and ice densities (*Castillo-Rogez,* 2006; *Anderson and Schubert,* 2007). Rhea is currently not in a mean motion resonance.

Spectroscopic observations from both Voyager and Cassini indicate a nearly pure water ice surface, down to at least the base of the deepest (~5 km) impact craters and with only superficial deposits of CO_2 and other contaminants (*Clark and Owensby,* 1981; *Stephan et al.,* 2012; *Scipioni et al.,* 2013). Rhea's surface resembles that of its neighbor, Dione, in that it is dominated by water ice and its darkened trailing hemisphere (perhaps due to bombardment by particles co-rotating with Saturn's magnetosphere) is interrupted by a series of bright structures that appear to be of tectonic origin (*Wagner et al.,* 2010; *Stephan et al.,* 2012) (Fig. 6). Compared with Dione, however, the extent of tectonic features on Rhea is limited, with the surface being more heavily cratered and showing somewhat less evidence of resurfacing. Although Rhea is much larger than Dione, meaning greater capacity to retain heat, its potential for prolonged endogenic activity is limited by the reduced tidal heating it experiences due to the much larger semimajor axis of its orbit and its small orbital eccentricity of $\sim 10^{-3}$.

The largest impact structure on Rhea's surface is the 360-km-diameter, 5-km deep, Tirawa basin on the anti-saturnian hemisphere; the youngest major crater is Inktomi (47 km), centered on the leading (western) hemisphere. Although Rhea's surface is rough, analysis of depth-to-diameter ratios for the largest impact craters (>100 km diameter) indicates considerable relaxation compared with similarly-sized craters on Iapetus (*White et al.,* 2013), suggesting higher heat fluxes (perhaps tens of milliwatts per square meter) early in Rhea's history. Rhea's topographic variance spectrum shows a break in slope that could be another indication of relaxation of long wavelength structures — although a transition from isostatic to flexural support at shorter wavelengths could also account for this signal, and has been used to obtain an estimated lithospheric elastic thickness of ~5 km (*Nimmo et al.,* 2010, 2011).

Rhea's approximately triaxial ellipsoidal figure has semi-axes a = 765.7 ± 0.5 km, b = 763.6 ± 0.5 km, c = 762.9 ± 0.3 km (2σ) (*Thomas,* 2010; *Tortora et al.,* 2016). Although the absolute uncertainties here are similar to those of the other saturnian satellites, Rhea's relatively small tidal and rotational deformations mean that it is more difficult to assess whether or not its shape conforms to the expectation for a body in hydrostatic equilibrium. The unnormalized non-zero degree-2 spherical harmonic shape coefficients are $H_{20} = -1190 \pm 150$ m and $H_{22} = 350 \pm 60$ m, assuming a reference radius of 764.0 km, and having the ratio $-H_{20}/H_{22} = 3.4 \pm 1.0$ (1σ) [although we note that the uncertainties reported in *Tortora et al.* (2016) were overly conservative]. While this ratio is compatible with the hydrostatic expectation of 10/3, the uncertainty is far too large to permit any strong conclusions. Early models suggested that long-lived radioisotope decay heat was sufficient to promote the formation of a deep ocean in Rhea (*Hussmann et al.,* 2006).

However, this feature would have most likely been short-lived and is not expected based on available observations.

Doppler radio tracking during Cassini's first targeted flyby of Rhea in 2005 yielded a series of initial estimates of Rhea's quadrupole (degree-2) gravity field, with the large values of the gravity moments indicating that Rhea's interior is unlikely to be strongly differentiated (*Anderson and Schubert,* 2007, 2010; *Iess et al.,* 2007; *Mackenzie et al.,* 2008). An additional, high-inclination, dedicated gravity flyby in 2013 finally permitted independent estimates of $J_2 = 946 \pm 14$ and $C_{22} = 242 \pm 4$, in units of dimensionless gravitational potential $\times 10^6$ (*Tortora et al.,* 2016), with the ratio $J_2/C_{22} = 3.91 \pm 0.10$ indicating a statistically significant departure from the hydrostatic expectation of ~10/3 (uncertainties here are 1σ).

Interior modeling suffers from a non-uniqueness problem that is made worse by the fact that hydrostatic equilibrium cannot be assumed in this case. Nevertheless, if the shape model is accurate, then the high J_2/C_{22} ratio indicates some excess internal mass in the equatorial region. *Tortora et al.* (2016) have suggested that Rhea consists of an envelope of clean water ice surrounding a low-density, hydrated silicate core. If the degree-2 exterior shape was established early in Rhea's history, when heating and relaxation were appreciable (*White et al.,* 2013), the relatively weaker icy exterior could conform closely to the expected hydrostatic figure even as the partly silicate interior supports some degree of excess flattening compared to the hydrostatic expectation, accounting for the large J_2 (*Tortora et al.,* 2016). The thickness of the rock-free water ice envelope in such a model, however, is not well constrained. The shape and gravity model 1σ uncertainties permit icy envelope thicknesses anywhere from ~5 km (the depth of Tirawa crater) up to ~200 km, corresponding to moment of inertia factors anywhere from 0.399 to 0.353 (*Tortora et al.,* 2016). The upper bound on the shell thickness is not certain, but making the core smaller than about 600 km moves the model results outside the 1σ uncertainties in the observations (dominated by the uncertain shape). Hence, in absence of more accurate results, a wide range of interior structures, and therefore formation and evolutionary histories, are possible (*Castillo-Rogez,* 2006; *Barr and Canup,* 2008).

2.5. Iapetus

Located at about 60 R_S (semimajor axis of 3.56×10^6 km), Iapetus is the farthest regular satellite of Saturn. Compared to the other mid-sized satellites, Iapetus has an orbit significantly inclined [7.5° with respect to the Laplace surface (*Mosqueira et al.,* 2010)]. Its large eccentricity (0.03) and spin axis inclination (~8.3°) reflect limited tidal dissipation owing to the large distance of Iapetus to its primary.

Iapetus' mean radius is about 735 km. Iapetus is similar in size to Rhea, but much less dense (Table 1), with a rock fraction of about 20–30% by mass, depending on the assumed porosity (*Castillo-Rogez et al.,* 2007; *Robuchon et al.,* 2010). The low silicate content combined with very weak tidal forcing from Saturn imply a very small heat budget for this satellite.

Iapetus distinguishes itself from the other moons by an exceptionally large flattening and a narrow equatorial ridge reaching heights up to 20 km above the surrounding terrain. The large flattening implies that Iapetus was rotating much faster in its early past, and then slowed down to is present-day rotation rate (*Castillo-Rogez et al.,* 2007; *Robuchon et al.,* 2010). The difference between the polar radius (712.1 ± 1.6 km) and the equatorial one (745.7 ± 2.9 km) (*Thomas,* 2010) would be consistent with a rotational period of 16 h for a homogenous body or 15 h for a differentiated body (*Castillo-Rogez et al.,* 2007; *Thomas,* 2010), while the present-day rate is about 79 d. However, taking into account the large semimajor axis of Iapetus, the despinning of Iapetus due to tidal interactions with distant Saturn is problematic. The time for a moon to reach tidal locking (i.e., synchronous rotation) is proportional to a^6 (e.g., *Gladman et al.,* 1996) and depends on the ability of the moon to dissipate rotational energy by tidal friction. *Castillo-Rogez et al.* (2007) showed that Iapetus could reach synchronous rotation if the interior was hot due to early warming by short-lived radiogenic elements and possessed a low viscosity during a sufficiently long period of time (about 1 G.y.). However, for such a weak interior, it is difficult to explain how the strong flattening was preserved. Subsequently, *Robuchon et al.* (2010) reevaluated the scenario of early warming and despinning proposed by *Castillo-Rogez et al.* (2007) by solving consistently heat transfer by thermal convection, viscoelastic tidal friction, and viscous relaxation of the body shape. They showed that the present-day 34-km flattening could be the result of an ancient larger flattening ranging between 45 and 80 km, which requires an initial rotation period between 8.5 h and 10 h and large amounts of short-lived radiogenic elements, implying very early satellite accretion, <4 m.y. after the condensation of calcium-aluminum-rich inclusions (CAIs). More recently, by considering more realistic rheological properties for the icy interior of Iapetus, *Kuchta et al.* (2015) argued that, in the absence of additional external interactions, despinning of a fast rotating Iapetus due to internal friction is impossible even for warm initial conditions (T > 250 K). Alternatively, it has been proposed that the despinning is largely the result of a giant impact, either as a consequence of direct momentum exchange upon impact (*Kuchta et al.,* 2015), or owing to interactions with a subsatellite formed by the impacts (*Levison et al.,* 2011; *Dombard et al.,* 2012).

A very ancient origin of Iapetus is indicated by the cratering record. Its surface is significantly older than Rhea and seems to have been hit by a different impactor population (*Dones et al.,* 2009) [and see discussions in *Greenstreet et al.* (2015), their Fig. 7]. *Schmedemann et al.* (2008) inferred a surface age of 4.3 G.y., and *Giese et al.* (2008) and *Kirchoff and Schenk* (2010) identified large basins, one up to 800 km across, likely signatures of an early history of intense bombardment. Schenk et al. (this volume) pointed out that Iapetus has more basins than Rhea (10 > 300 km vs. 4, respectively), while gravitational focusing would imply that Rhea experienced more impacts than Iapetus. Iapetus' very old surface is consistent with thermal models that predicted

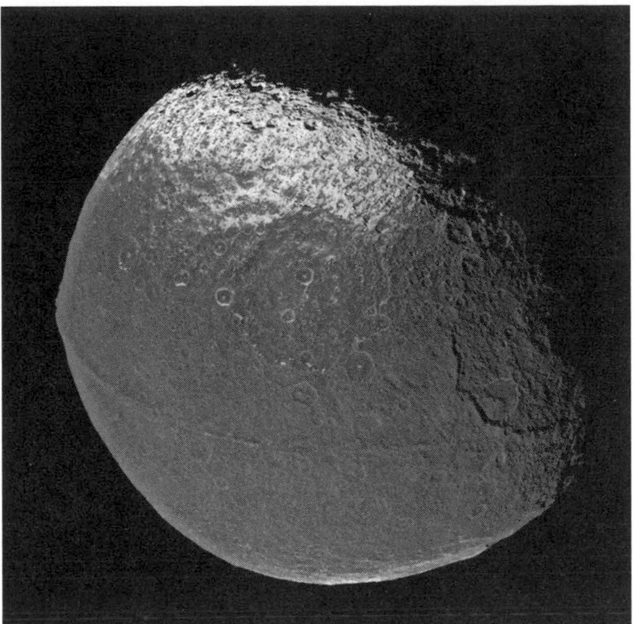

Fig. 7. Iapetus as seen from the Cassini orbiter on December 31, 2004, from about 172,400 km. This view shows some of Iapetus' key features: the equatorial ridge, the large basins, and sharp color contrasts. Credit: NASA/JPL/Space Science Institute (PIA06166).

internal temperatures could not reach the water ice melting point. If Iapetus accreted some ammonia, as suggested by cosmochemical models, then its interior could have been subject to partial melting and separation of the rock from the volatile phase at cold temperatures (well below the water ice eutectic) (*Castillo-Rogez et al.,* 2007; *Robuchon et al.,* 2010). That state could have lasted a few tens of millions of years; otherwise Iapetus likely remained frozen for most of its history. This scenario is supported by the analysis of crater morphology (i.e., relaxation), implying that the main heat source throughout Iapetus' history has been long-lived radioisotope decay (*White et al.,* 2013). Magnetometer observations by Cassini-Huygens did identify a magnetic anomaly that could be explained by interaction between Iapetus' intrinsic magnetic field and solar wind (*Leisner et al.,* 2008), but this observation remains to be confirmed.

The narrow equatorial ridge runs ≥75% of the satellite circumference, segmented in several discontinuous portions (*Giese et al.,* 2008; *Singer and McKinnon,* 2011; *Dombard et al.,* 2012; *Lopez Garcia et al.,* 2014). The high crater density indicates that it is an ancient feature (*Denk et al.,* 2010) that has been subsequently modified by cratering processes and landsliding (*Singer et al.,* 2012). Its location on the top of equatorial bulge suggests a causal link with the oblate shape; however, the mechanism explaining their common origin still remains debated and the most recent studies are in favor of an exogenic origin (for a review, see the chapter in this volume by Schenk et al.).

2.6. Looking at the System as a Whole

The mid-sized moons show broad diversity in their physical properties. Their radii range from ~200 to ~800 km and their densities range from 0.98 g cm^{-3} at Tethys up to 1.61 g cm^{-3} at Enceladus. The lack of pattern in the density and size of Saturn's mid-sized satellites is a key open question that must relate to their origin(s), as discussed in section 3. Iapetus' separation distance of ~50 R_S with respect to the inner satellites likely points to variations in accretional environments across the saturnian system. This situation is even more peculiar as Titan is located between Iapetus and the rest of the mid-sized moons.

Crater distributions also show peculiarities across the system. First, crater densities and size frequency distributions (SFD) suggest the satellites may not have formed in the same timeframe (*Kirchoff and Schenk,* 2010). Similarly, Kirchoff et al. (this volume) propose that variations in the slope of the crater size frequency distribution may be due to the occurrence of different populations of impactors (planetocentric and heliocentric) in the system. Furthermore, Kirchoff et al. point out that the apex/anti-apex asymmetry in the distribution of craters formed by heliocentric impactors expected for moons locked in orbital resonance is weak (zero to less than a factor of 4 difference) on Mimas, Rhea, and Iapetus. These moons are taken as a reference because they have been subject to little geological activity that could have erased the cratering record. Kirchoff et al. interpret this observation as being due to crater saturation or to a dominant non-heliocentric impactor population [e.g., crater ejecta exchanged among moons (*Alvarellos et al.,* 2005)]. Further analysis of Rhea and Iapetus by *Harita* (2016) indicates an apex-antiapex asymmetry in the distribution of large craters (tens of kilometers), believed to be primarily of heliocentric origin, whereas no asymmetry was found for smaller craters. *Harita* (2016) suggests the latter may either reflect a planetocentric source of impactors or saturation in small size craters.

Surface composition suggests that these objects have rather clean ice shells, except in the case of Iapetus' surface, which may be contaminated by dust ejected from Phoebe's surface and ring (*Spencer and Denk,* 2010). This characteristic is a potential clue to the state of differentiation of these bodies with the caveat that icy dust ejected from Enceladus forms the E ring and coats the surfaces of Tethys and other nearby satellites (*Schenk et al.,* 2011).

High heat fluxes inferred at Tethys, Dione, and Rhea suggests the occurrence of enhanced tidal dissipation at some point in the history of these satellites, which appear to require the presence of a liquid layer decoupling the icy shell as well as resonances occurring for at least part of their histories.

The interior structure inferred for each of the moons is summarized in Fig. 8. Geophysical measurements provide firm evidence for Enceladus' differentiation and current existence of a deep ocean. In contrast, Iapetus' very old surface suggests that it remained undifferentiated, consistent with its poor heat budget, and likely preserved some

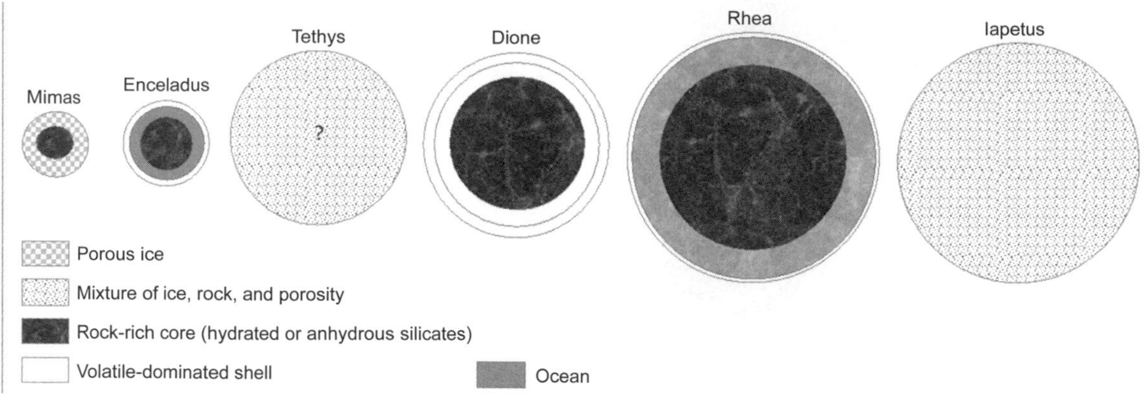

Fig. 8. Summary of the state of knowledge of Saturn's mid-sized moons from observations obtained with the Cassini orbiter. Many uncertainties remain. Semi-transparent textures reflect the uncertainties in layer thicknesses. For example, estimates on Rhea's core range from ~600 to 759 km and Dione's ocean may be absent or up to ~95 km thick.

porosity (*Castillo-Rogez et al.,* 2007). As discussed above, topographic relaxation suggests high heat flow on Dione in the geologically recent past. In the case of a conductive ice shell, with the temperature increasing approximately linearly with depth, this suggests the presence of a deep ocean. If solid-state convection occurred in the icy mantle, however, a deep ocean is not required to account for the past high surface heat flow on Dione. Libration measurements imply that Mimas is differentiated in some manner (*Tajeddine et al.,* 2014), and measurements of flexure (*Giese et al.,* 2007) and crater relaxation (*White et al.,* 2013, 2017) are suggestive of high heat flows, and possibly oceans, within Tethys and Rhea. However, it is not possible with the information in hand to further elaborate on the interiors of Mimas, Tethys, or Rhea.

3. SCENARIOS FOR MID-SIZED SATELLITE ORIGINS

A variety of scenarios and their variants have been suggested so far for the origin of Saturn's moons and are described in detail in the chapter by McKinnon et al. in this volume. Accretion in Saturn's circumplanetary disk, fed from a nearby planetesimal and nebular reservoir, has been the leading hypothesis until recently (e.g., *Canup and Ward,* 2006). In that model, satellite densities might be expected to increase toward the planet, like in the case of the Galilean satellites. However, the apparently random variations in density observed across Saturn's and Uranus' regular satellites challenge that model. These objects are organized as a function of their respective masses (Fig. 9), not their densities. This has led to the introduction of an alternative model in which mid-sized satellites form within the rings and migrate as a result of both tidal interaction between the planet and the satellites, and resonant interactions between the rings and the satellites (*Canup,* 2010; *Charnoz et al.,* 2011). In this framework, the satellites' distribution is determined by tidal interactions, with the most massive objects moving outward at a faster rate. In this case, Iapetus may be the survivor of the early stage of satellite accretion (*Mosqueira et al.,* 2010), as Titan would be as well. In this sense, this ring-based model does not imply there was no circumplanetary disk, but that the process(es) that formed the rings may have led to the inner mid-sized satellites. This model has gained momentum with multiple approaches to dating the rings and the possibility for multiple generations of ring-born satellites. We also review additional formation scenarios.

3.1. Formation in Saturn's Subnebula

The regular icy satellites of Saturn may have formed in a circumplanetary accretion disk (CPD) that surrounded the growing gas giant at the end of its formation (*Peale and Canup,* 2015). Proto-Saturn would have been sufficiently massive to open a gap in the solar nebula (*Morbidelli and Crida,* 2007), although likely not as deep a gap as at Jupiter (*Fung and Chiang,* 2016). Accretion of gas and entrained particles (now colloquially described as "pebbles") would then proceed through streamers, before coalescing and forming a prograde-rotating CPD (e.g., *Ayliffe and Bate,* 2009; *Tanigawa et al.,* 2014). Accretion of pebbles could build satellitesimals, which could in turn create mid-sized moons as well as bodies as large as Titan. All of these bodies would be subject to gas drag, and in the case of one or several Titan-sized bodies, Type I drift (*Canup and Ward,* 2006; *Canup,* 2010), which would drive them all inward toward the bloated proto-Saturn. It is only the dispersal of the solar nebula and any protosatellite nebular gas with it that halts this inward drift, with the implication that the satellites we see are the survivors of the most recently formed sets of satellites.

Early in the protosatellite nebular phase, temperature and pressure conditions in the CPD, in principle, may have been high enough to induce gas phase chemistry and to devolatilize solids captured from the solar nebula. Depending on the conditions in the CPD, we might expect a gradient in satellite density, however slight, decreasing with distance

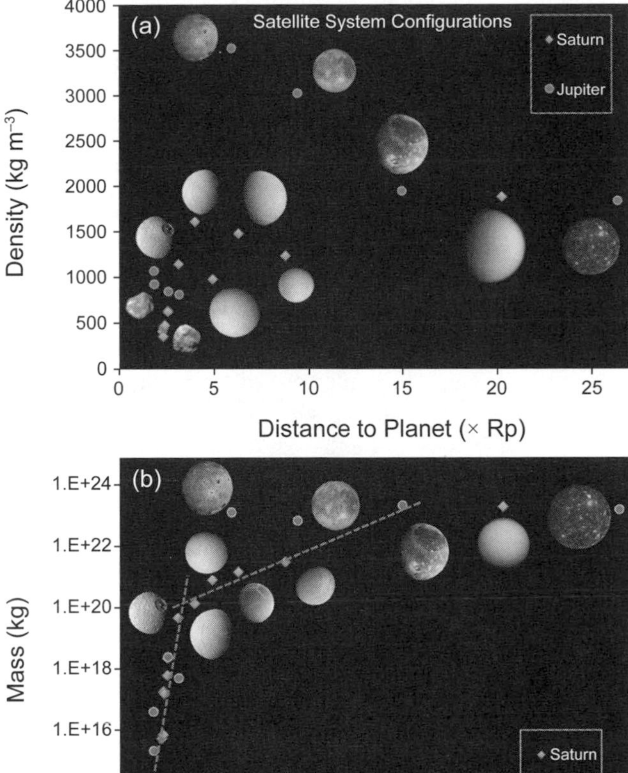

Fig. 9. Comparison of Jupiter's and Saturn's moon distribution as a function of distance to their primary and their **(a)** densities and **(b)** masses. The Galilean system shows a decrease in density as a function of distance, whereas Saturn's inner moons are organized as a function of mass.

from Saturn, as volatile ices would be progressively inhibited from condensing at closer distances (see the chapter by McKinnon et al.). As the solar nebula dissipated, however, the CPD would have progressively attenuated and cooled, enough so that gas phase chemistry would have been kinetically inhibited and that the compositions of solids captured into the protosatellite disk would have remained largely unaltered. However, the lack of any such compositional gradient at all, as mentioned above, and indeed the iciness of the mid-sized satellites in general, point to cooler and less-gas-rich nebular conditions when these moons (or their precursors; see below) formed.

3.2. Origin of Iapetus

Compared to the other mid-sized satellites, Iapetus has an orbit significantly inclined, 7.5° with respect to the Laplace plane (*Mosqueira et al.,* 2010). This characteristic, together with its position beyond Titan's orbit, suggests an origin distinct from the inner mid-sized satellites. This is also supported by unusually large number of basins across Iapetus (Schenk et al., this volume). Iapetus may be the survivor of the early stage of satellite accretion (*Mosqueira et al.,* 2010). Its low rock mass fraction has been explained by a mechanism of ablation (i.e., erosion) of heliocentric planetesimals supplied from the Jupiter-Saturn feeding zone (*Mosqueira et al.,* 2010). These authors suggested that mechanism based on the premise that heliocentric planetesimals formed early in Saturn's feeding zone could be partially differentiated by ^{26}Al decay heat. However, heating and differentiation in the 10-km large planetesimals assumed in that study was not modeled and the very large porosity expected in objects of that size, similar to comets, was not accounted for. Hence Iapetus' relatively low ice to rock ratio, similar to that of the inner mid-sized satellites, lacks a firm explanation.

3.3. Formation from Saturn's Rings

The idea that planetary satellites could form from ring material created by an impact or disruption of a large object entering the Roche's zone of a planet was introduced for the Moon (*Hartmann and Davis,* 1975; *Cameron and Ward,* 1976) and applied to the formation of Charon by *McKinnon* (1989) and *Canup* (2005). In the latter case, *Canup* (2005) suggested that Charon formed quasi-intact from the collision of Pluto with an interloper that contained about 30–50% of Pluto's mass. Alternatively, Charon could have reaccreted from a disk of material formed as a consequence of that impact (*Canup,* 2011). *Bromley and Kenyon* (2015) suggested that that disk of material is the likely accretional environment for Nix, Hydra, and other small satellites.

The relevance of that scenario to Saturn's inner moons was suggested following the discovery of kilometer-sized "propeller moons" embedded in Saturn's rings (*Porco et al.,* 2005) and a menagerie of small satellites within Mimas' orbit and as co-orbitals of Tethys and Dione (*Charnoz et al.,* 2010). Based on the accretion models initially proposed for the smallest moons interior to Mimas' orbit (*Charnoz et al.,* 2010), *Charnoz et al.* (2011) introduce the possibility that the entire inner moon system formed from Saturn's rings. This was also motivated by the estimation of dissipation function of Saturn (*Lainey et al.,* 2012), suggesting fast tidal expansion and hence much later satellite formation than initially anticipated.

Following the suggestion that the current Galilean satellites represent the last generation of satellites formed in the jovian nebula and migrated by Type I migration (*Canup and Ward,* 2006; *Ward and Canup*, 2010), *Canup* (2010) also suggested that the saturnian rings formed from an early Titan-sized satellite. In this model, the core of that large satellite was lost within Saturn; hence the satellites born from the rings would be mostly icy, consistent with Cassini's observations. *Charnoz et al.* (2011) then suggested that the ring progenitor could have been partially differentiated, allowing for the distribution of a few large shards of silicates that eventually formed the core of the newly formed satellites via accretion of ice within their Hill spheres. *Charnoz et al.* (2011) were able to reproduce the overall architecture of the mid-sized moon system.

The underlying idea is that ring material spreads over time and starts reaccreting when it crosses the giant planet Roche limit (about 2.5 planet radii in the case of Saturn). Multiple observations of small satellites embedded in Saturn's rings are direct evidence of that phenomenon. *Crida and Charnoz* (2012) generalized this scenario to Saturn, Uranus, and Neptune and suggested that many regular satellites could have formed this way.

Inside Mimas' orbit protosatellites exchange angular momentum with the rings, leading to a progressive orbital expansion of the satellite's orbit. Hence satellite dynamical evolution is coupled to the rings' structure and large-scale evolution. Simulations by *Charnoz et al.* (2011) estimated that about 50% of the ring mass is sourced to form the moons. The current mass of Saturn's rings has been estimated on the order of one to several Mimas masses (~4 × 10^{19} kg) (see *Charnoz et al.,* 2009a). Hence it is difficult to bound the original mass of the rings. Even with a good estimate of the current mass of the rings derived from the radio science measurements that took place during the Cassini Grand Finale (*Iess et al.,* 2017), assessing the initial mass of the rings would rely on several factors that are not well constrained, in particular which satellites formed from the rings [only Mimas-Tethys in the Salmon and Canup model or the five mid-sized satellites as in *Charnoz et al.* (2011)], and on what timescales. Viscous spreading plays a major role in the evolution of the rings' mass, and *Salmon et al.* (2010) showed that the current estimated mass could be accommodated by initial masses varying by an order of magnitude.

Cuk et al. (2016) went further by inferring from the dynamical properties of the system that the current inner moon system likely formed very recently, possibly within the last 100 m.y., from the reaccretion of debris from a previous generation of moons born from the rings that went into unstable orbits. However, *Hyodo and Charnoz* (2017) invalidated the premise for the *Cuk et al.* (2016) scenario by showing the catastrophic collision of two moons is followed by their rapid reaccretion. That is, the spreading of the debris is too slow to explain Saturn's rings.

Salmon and Canup (2017) proposed a "hybrid" model where Mimas, Enceladus, and Tethys formed from ring material, whereas Dione and Rhea formed in the saturnian subnebula with Titan. Indeed, Saturn's rings appear to be dominated by ice with no evidence for the rocky "shards" suggested by *Charnoz et al.* (2011). Hence the rocky component of the satellites need to be supplied from an external source. Assuming an early formation, the satellites could have gained their rocky component, as well as additional volatiles, from comets migrated during the late heavy bombardment (LHB) (*Salmon and Canup,* 2017). However, this model cannot reproduce the rock mass fractions of Dione and Rhea and suggests an origin separate from Mimas, Enceladus, and Tethys. In this framework, the rocky component would presumably be distributed homogeneously in the satellites. This contrasts with the *Charnoz et al.* (2011) model in which the satellites accreted already differentiated in a porous ice shell surrounding an irregular chunk of rock.

Numerical simulations by *Movshovitz et al.* (2015) suggested that Mimas, Enceladus, and Tethys should have suffered at least one catastrophic impact during the LHB. Large high-velocity impacts, characteristics of LHB, are expected to lead to a complete disruption of the target (e.g., *Monteux et al.,* 2016). While these bodies are expected to reaccrete, catastrophic disruption would lead to homogenization of their interiors and would represent a major cooling event. If the moons held a deep ocean at the time of the LHB, a reasonable assumption in the case of Enceladus, then such an event might have led to a significant loss of volatiles.

3.4. Additional Formation Scenarios

A recent alternative approach to the diversity question was introduced by *Asphaug and Reufer* (2013), who suggested that the satellites formed from the debris of large satellites that merged, resulting in the formation of Titan. The combined mass of the mid-sized moons represents just a few percent of Titan. Their statistical modeling shows that the variety of debris' properties can match the observed physical properties of the moons. However, this model does not track the long-term dynamical evolution of the fragments and the fraction that eventually ends up in stable planetocentric orbits or within Saturn itself. Like for a ring origin, the timing of that impact determines the heat budget available to the satellites. While promising, little attention has been devoted to that scenario to evaluate how it can explain the overall properties of the moons (e.g., cratering history, geophysical evolution, and, most importantly, semimajor axes).

A variant scenario forms icy moons as debris generated from the collisions between large proto-Titans migrating within Saturn's circumplanetary disk and moons growing in the inner part of the disk (*Sekine and Genda,* 2012). These authors find that the collisions yield mid-sized moons with a broad range of rock fractions. They further suggest that the heat generated upon impacting could kickstart tidal dissipation in Enceladus.

3.5. Summary and Implications

Formation conditions bear on the long-term evolution of the satellites, in terms of initial conditions and especially of heat budget. These aspects are compared in Table 2 for the two leading models. A formation within Saturn's rings provide a way to explain the variations in rock mass fraction among the satellites, although this model is somewhat ad hoc: It requires a partially differentiated precursor and then selective accretion of ice onto a few shards of various sizes interspersed in the rings. This also implies that the satellites accreted with irregular cores, hence already differentiated. The cores may evolve and relax as a result of thermal evolution, at least in those bodies that contain >30 wt% rock, e.g., Dione and Enceladus. On the other hand, satellites formed

TABLE 2. Expected original state and main evolution traits of Saturn's inner mid-sized moons for the two main origin scenarios.

Property	Formation in Subnebula	Formation in Saturn's Rings
Diversity in rock:ice mass ratio	Requires special processes (e.g., impacts) to depart from rock:ice gradient with distance	Rock:ice ratio is not deterministic, but plausibly low
Heat budget	Moons benefited from full long-lived radioisotope heat, possibly short-lived radioisotopes, and several resonance crossing	No short-lived radioisotope decay heat; long-lived radioisotope budget and resonance history depend on time of formation
Current interior structure	Accretion of ice-rock mixture, differentiation determined by internal heating history (see section 4)	Differentiated core with porous ice shell (*Charnoz et al.,* 2011); injection of rock, presumably homogeneous (*Salmon and Canup,* 2017)
Surface age	Dominated by heliocentric impactors with additional secondary impacts from debris	Two populations of impactors (planetocentric and heliocentric) in the *Charnoz et al.* (2011) model; dominated by heliocentric impactors in *Salmon and Canup* (2017)

in the saturnian subnebula as a homogeneous mixture of ice and rock may not differentiate owing to a limited heat budget and the possibility for effective convective heat transfer (*Ellsworth and Schubert,* 1983) (see section 4).

In the subnebula origin model, all the moons accreted within the same timeframe, whereas the formation in Saturn's rings implies different ages for each of the moons. This carries important implications on the heat budgets of the moons and thus their long-term evolution, as described in more detail in section 4. In the rings' origin scenario model, the absolute satellites' ages are tied to the rings' emplacement time and then as a function of the distance to the ring. Hence, Rhea would be the oldest satellite, while Mimas would be the last satellite to emerge from the rings and therefore the youngest. The ring formation has been suggested to range from 4.5 G.y. (*Canup,* 2010) to the LHB period (*Hyodo et al.,* 2017). The current inner moon system itself could have been produced as late as 100 m.y. ago (*Cuk et al.,* 2016), with the caveats noted above regarding the assumptions made by the latter authors on Saturn's dissipation factor. In the latter model the moons we see today escaped the LHB, and most cometary impacts altogether. In the *Salmon and Canup* (2017) model the moons formed very early and thus most of their cratering should be ascribed to cometary, i.e., heliocentric, impacts. In all these scenarios, a major unknown concerns orbital evolution due to tides raised on Saturn (the tidal Q question), and whether each satellite has its own unique and time-varying effective Q (*Fuller et al.,* 2016); see below and the chapter in this volume by Nimmo et al. for a detailed discussion.

Key constraints on the formation of the moons should come from their crater densities and morphologies. However, as described in the chapter by Kirchoff et al. (this volume) and above, the record available does not allow us to draw firm conclusions. It is possible inner moons were subject to several populations of impactors. The environment in which Mimas acquired its heavily cratered surface could involve protomoons emerging from the ring with, or following, Mimas. *Charnoz et al.* (2011) suggested that the relatively large craters and basins encountered at all mid-sized satellites within Titan's orbit points to the occurrence of more ring-formed moons than reflected in the current system. On their side, *Dones et al.* (2016), building on *Alvarellos et al.* (2005), investigated the possibility that the small craters on the moons were created by planetocentric debris impacts, with low relative velocities, whereas the large craters were produced by heliocentric (cometary) impactors.

Applied to the *Cuk et al.* (2016) model, this scenario would imply that most of the moon's cratering record was created from planetocentric debris. However, Kirchoff et al. (this volume) question an age as young as 100 m.y. as being inconsistent with the large basins observed on most mid-sized moons in the saturnian system, unless such basins result from collisions with the remaining large planetocentric debris from the late accretion phase.

Another potential key difference in the initial conditions of the mid-sized moons between these two main origin models is their volatile complement. Moons formed in the rings should reflect the composition of the ring progenitor, whether it is a differentiated Titan-sized body (*Canup,* 2010) or a large transneptunian object (*Charnoz et al.,* 2009b; *Hyodo et al.,* 2017) that likely brought in a large fraction of non-icy volatiles as well as organics. On the other hand, satellites formed within Saturn's original protosatellite nebula could reflect a volatile composition enriched, for example, in CH_4 compared with CO and a D/H ratio much closer to the lower, protosolar value (see the chapter by McKinnon et al. in this volume).

The implications of these initial conditions on the long-term evolution of the moons and comparison with available observations are addressed in section 4.

4. DRIVERS OF GEOPHYSICAL EVOLUTION

The heat budget available to the mid-sized satellites depends in part on their accretional environment, which determined their formation timeframe and timescale and in turn the amount of accreted radioisotopes. Internal evolution depends on the balance between heat produced from radioisotopes and tidal dissipation and heat lost via conduction and convection. Depending on their heat budgets, some of the satellites may have undergone partial melting of their volatile phases with the potential to promote significant tidal dissipation and drive endogenic activity.

The past decade has seen major progress in the way internal processes are modeled. This is in part thanks to an effort to better understand the microphysics involved in the deformation of ice in response to stress via theoretical and experimental research.

This chapter does not delve into the equations or the mechanisms involved in mid-sized moon evolution because these can be found in several recent reviews (*Hussmann et al.,* 2015). Instead, we assess the possibility for key processes, and their associated impacts, to occur in the various satellites in order to explain the diversity of geological surfaces observed across the inner system (Fig. 10).

4.1. Radioisotope Heat Budget

Accretional and potential energy associated with differentiation is small (*Leliwa-Kopystynki and Kossacki,* 2000; *Matson et al.,* 2009), commensurate with the small size of these bodies. Accretion in a disk or rings is at low relative speed and involves little to modest heating (*Squyres et al.,* 1988). However, certain scenarios might permit conditions where accretional energy was significant, for example, the energy delivered by the Sekine/Asphaug-style impacts (see section 3.4).

Radioisotope decay heat comes from two types of sources: short-lived radioisotopes with aluminum-26 (^{26}Al) as the most energetic and iron-60 as a possible secondary source. Long-lived radioisotopes are potassium-40, uranium-235 and -238, and thorium-232. Short-lived radioisotopes might have been active if the moons formed in the saturnian nebula within a few million years after a reference time determined by the original amount of ^{26}Al measured in CAIs in chondrites. Indeed, theoretical modeling of Saturn's accretion and

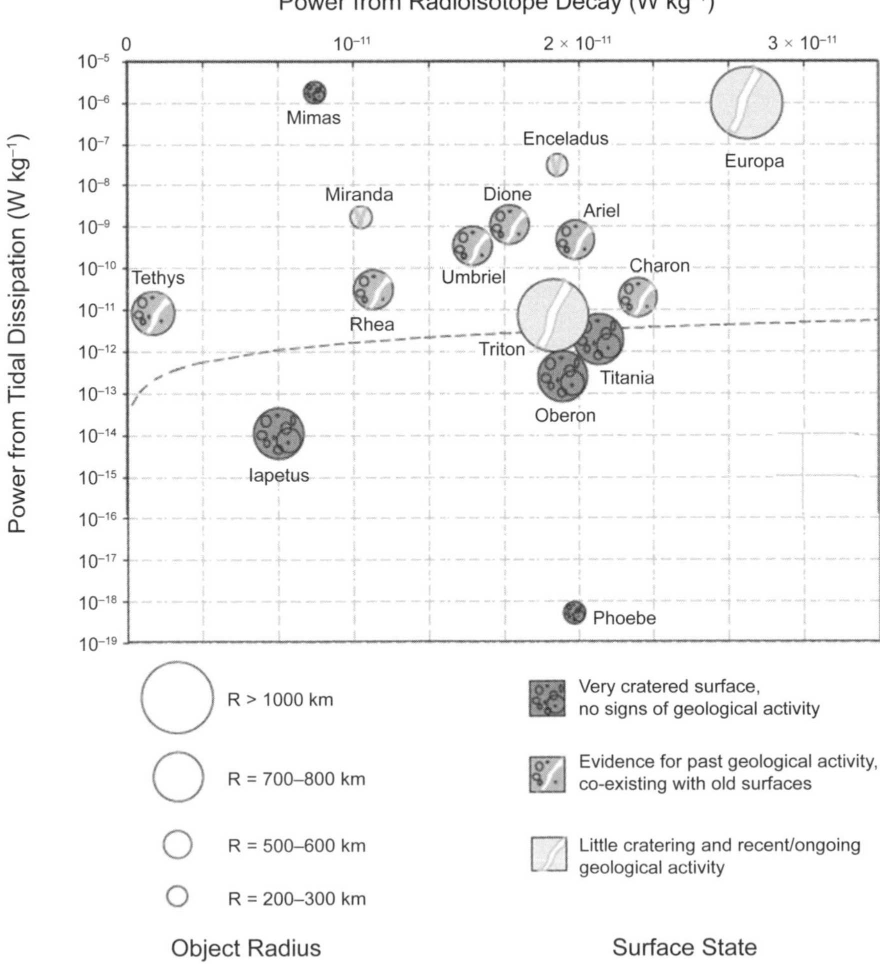

Fig. 10. Mapping of giant planet icy moons against the magnitudes of their energy sources: radioisotopes decay heat on the x-axis and eccentricity-driven tidal dissipation on the y-axis. The latter parameter assumes $k_2/Q = 1$. Updated from *Castillo-Rogez and Lunine* (2012).

system evolution suggests that accretion in about 3 m.y. after the beginning of the solar system was likely (e.g., *Dodson-Robinson et al.,* 2008).

In theory, it might be possible to infer whether a moon accreted short-lived radioisotopes if it is otherwise heat starved owing to a small rock fraction and absence of tidal dissipation. *Castillo-Rogez et al.* (2007) suggested this might be the case for Iapetus since its silicate mass fraction is too low otherwise to trigger sufficient internal heating for the interior to experience geophysically significant tidal forcing and despinning (see section 2). We note, however, that even with significant short-lived radioisotopes, it appears difficult to produce full despinning of Iapetus (see section 2) (*Kuchta et al.,* 2015). In most other cases, it is difficult to prove the past occurrence of ^{26}Al decay heat considering the many parameters involved in modeling icy moons close to their primary.

In the accretional framework suggested by *Charnoz et al.* (2011), the satellites accreted without short-lived radioisotopes. However, if the ring's progenitor were a Titan-sized satellite, as suggested by *Canup* (2010), then it is possible that body was subject to early melting and aqueous activity resulting from accretional heating and ^{26}Al decay. In that case the rock phase could have been hydrated, at least partially (*Monteux et al.,* 2016), and even more so if the moons accreted early with significant amounts of ^{26}Al. More generally, the heat budget of satellites formed in the rings depends on the timing of the ring formation and could vary significantly depending on whether the rings formed from a proto-Titan or a transneptunian object.

Long-lived radioisotope abundances used in geophysical modeling are generally based on CI chondrite composition, which reflects solar abundances. This assumption may not apply to icy moons accreted in the rings. If the ring progenitor was subject to melting and chemical fractionation, then radioisotopes could have been redistributed within the rock phase and the liquid layer in that body. However, it is not possible to elaborate further with the information in hand.

4.2. Tidal Dissipation

Tidal deformation and dissipation of planets and satellites is a fundamental driver of orbital and thermal evolution (see the chapter by Nimmo et al. in this volume for more detail on the basics of tidal heating and orbital evolution). Tidal forcing exerted by a planet on its moons produces a torque resulting in an exchange of angular momentum between the planet's rotation and the orbital motions of the moons. The consequences of that tidal forcing takes two forms: the evolution of the satellite's orbit (e.g., variation in semimajor axis, eccentricity, inclination) and the production of internal heat in regions of the moons that are not purely elastic. The tide-raising satellite evolves further from its primary (as long as it is outside the synchronous orbit) and decelerates due to the dissipation in the primary, whereas loss of energy in the satellite decreases its orbital energy leading to acceleration of the orbital motion. The dynamical evolution of the saturnian satellite systems is difficult to decipher owing to the many resonances that could develop in the course of the satellite evolution.

The contribution of tidal dissipation to internal evolution is demonstrated in the case of Enceladus but is less evident for the other satellites. The amplitude of tidal forcing decreases by 5 orders of magnitude between Enceladus and Iapetus. The impact of tidal heating in the course of a satellite's history is determined by its dynamical evolution and, conversely, tidal dissipation drives orbital evolution. Current inclination (i-) and eccentricity (e-) resonances among the mid-sized moons suggest that resonances have affected the evolution of Mimas, Enceladus, Tethys, Dione, and Rhea (e.g., *Zhang and Nimmo,* 2009). If satellites formed in Saturn's rings, then their dynamical history is more uncertain. *Charnoz et al.* (2011) suggested the protosatellites could have been subject to extensive impacting upon exiting the rings that could have led to increased eccentricities, up to 0.2, and enhanced tidal dissipation that could be responsible for the high flux inferred at Tethys and Dione. However, this scenario remains to be modeled. On the other hand, the *Cuk et al.* (2016) model circumvents resonances among satellites, which may not be consistent with the high heat flow inferred from geologic observations. In particular, rock-poor Tethys needs a heat source that necessarily has to be tidal and that requires an ancient resonance crossing since Tethys' current eccentricity is close to zero.

Tidal effects are also quantified by two parameters describing the response of a moon to tidal forcing: the tidal Love number k_2 and the dissipation factor Q. The former characterizes the amplitude of the response of the body subject to tidal forcing, while the latter characterizes the amount of energy dissipated by friction over a tidal cycle. Both parameters are functions of the density and viscoelastic properties, and thus the thermal evolution, of the moons. Ice dissipation shows a strong dependence on temperature (see reviews by, e.g., *Durham et al.,* 2001; *McCarthy and Castillo-Rogez,* 2012). A fully frozen moon is characterized by a k_2 number that is $\sim<10^{-2}$. The presence of a deep ocean leads to a decoupling of the outer shell from the deep interior, which translates into a k_2 at least 1 order of magnitude greater than the frozen case. In highly porous material, dissipation is determined by friction at the grain boundaries (see *McCarthy and Castillo-Rogez,* 2012) and thus by the lithostatic pressure. An object that formed cold might not become dissipative without the help of radioisotope decay heat [e.g., discussion in *Castillo-Rogez et al.* (2007) for Iapetus].

Finally, the extent of tidal dissipation and orbital evolution depend in part on the dissipative properties of the primary. Tides within the central planet expand the moon's orbit, although the intensity of these tides has been debated. Independent of the moon's origin, the long-time assumed value of Saturn's dissipation factor Q ~ 18,000 (*Sinclair,* 1983) has made it difficult to reproduce the current orbital architecture of Saturn's satellite system. That value assumes that Mimas formed at the synchronous orbit [at which the moon's orbital period matches the planet's rotation period (see *Goldreich and Soter,* 1966)] and moved toward its current position in 4.5 G.y. The recent reevaluation of Saturn's Q by *Lainey et al.* (2012, 2017) suggests at least three times as much dissipation as previously thought (during the current epoch), leading to stronger tidal interaction with the satellites. This revised value allows one to reproduce, at least to first order, the spatial organization of Saturn's moons formed in the rings. Furthermore, using astrometric data of the mid-sized moons, *Lainey et al.* (2017) show that the dissipation factor Q_s of Saturn is a function of the orbital periods of the moons (e.g., Q_s at Rhea's orbital period is much less than Q_s at Tethys' period). Accounting for the strong frequency dependence of Saturn's dissipation factor, *Fuller et al.* (2016) showed that the outward evolution of the moons orbits may be modest and had to develop over several billion years. Nimmo et al. (this volume, their Fig. 3) model the orbital evolution of the moons for constant-Q and time-variable-Q assumptions, demonstrating that the latter can produce resonances among the satellites in the course of their evolution without requiring the satellites to be young.

4.3. Differentiation

The extent of differentiation of mid-sized icy satellites is poorly constrained, except in the case of Enceladus (Hemingway et al., this volume) and Dione (*Hemingway et al.,* 2016). Hence, at this point, our understanding is limited to modeling considerations. In the case of a formation in the rings, *Charnoz et al.* (2011) suggested that the inner moons acquired their rock from shards dispersed throughout the rings that gathered porous ring material in their Hill sphere. This implies that the satellites formed already differentiated. The possible occurrence of non-hydrostatic cores in Enceladus (*McKinnon,* 2013) and Mimas (*Tajeddine et al.,* 2014) support this idea, although *Monteux et al.* (2016) showed that irregular cores in small icy moons, like for instance Enceladus, could also reflect deformations incurred by large impacts. *Salmon and Canup* (2017) also suggested some of the satellites (Mimas, Tethys) could form mostly icy and acquire their rock content from impactors of heliocentric origin.

Differentiation following melting of the deep interior has turned out to be a process less straightforward than presented in early studies. These assumed that satellites would differentiate into a clean core and an icy shell, independent of their size (e.g., *Hussmann et al.,* 2006). However, observations of Ceres by the Dawn mission suggest that separation of the rock from the volatile component in a body with a gravity 2 or more orders of magnitude lower than Earth) is only partial, as indicated by its mean moment of inertia (*Park et al.,* 2016). The homogeneous distribution of hydrated materials on Ceres' surface at the global scale (*Ammannito et al.,* 2016) suggests the dwarf planet hosted a large ocean for some period of time where pervasive aqueous alteration or rocky material occurred on a global scale. In that ocean, large particles should sink while the smaller (<30 μm) particles might remain entrained in hydrothermal convection, as suggested by *Kirk and Stevenson* (1987) in the case of Ganymede and modeled by *Bland and Travis* (2017) for a broad range of water-rich bodies. In these conditions, the core did not end up as a monolith but instead as a mudball or porous aggregate. This model was also suggested to explain part of Enceladus' large heat flow (*Roberts,* 2014; *Travis and Schubert,* 2015; *Choblet et al.,* 2017). Furthermore, a phase of aqueous alteration drives chemical differentiation via leaching of major elements, such as alkali and alkaline earth metals, from the rock to the liquid phase. This is consistent with the occurrence of various salts found in Enceladus' plumes (*Postberg et al., 2011*). Potassium is very mobile in hydrothermal environments (e.g., *Neveu et al.,* 2017), especially in the presence of ammonium (*Engel et al.,* 1994), which it exchanges with in clay interlayers. The redistribution of ^{40}K from the rock to a volatile-rich shell can result in a cooler core evolution, thus limiting internal differentiation (*Castillo-Rogez and Lunine, 2010*).

4.4. Heat Transfer

Heat transfer and loss to space can take two forms in mid-sized moons: diffusion and subsolidus convection. The dominance of one over the other depends on the temperature gradient across the body, its size, its material thermophysical properties, and heat budget.

Ice at the temperatures expected in icy satellites is very conductive, up to 4.2 W m^{-1} K^{-1} at 100 K, and it decreases with increasing temperature. Thus, it takes a lot of heat for ice to reach a temperature threshold where it becomes dissipative and convective. As a result, if a moon formed cold, for example without the heat pulse from short-lived radioisotope decay heat, then it might be difficult to bring its interior to temperatures suitable for convection to begin. Assuming formation in Saturn's circumplanetary disk, *Hussmann et al.* (2006) showed that convection is not expected to occur in most mid-sized satellites because they are too small and/or too low density. *Multhaup and Spohn* (2007) came to the opposite conclusion, where all the mid-sized satellites could reach conditions propitious to stagnant-lid convection in undifferentiated interiors. The latter paper resorted to ammonia hydrates as a means to promote low-temperature endogenic activity. However, the low peritectic temperature of ammonia hydrates implies the water ice viscosity would actually be very high, limiting the prospect for convection. Dione and Rhea could reach conditions for ice melting without the need for ammonia, which might be confirmed at Dione (*Beuthe et al.,* 2016). This is consistent with earlier assessment by *Forni et al.* (1991) who also found out that Dione could be convective. On the other hand, *Matson et al.* (2009) and *Hussmann et al.* (2006) agreed that the smaller and less dense satellites, such as Mimas and Tethys, were unlikely to be convective at any point during their history. Finally, *Hussmann et al.* (2006) assumed that the largest of the moons, Rhea and Iapetus, would end up differentiated in a rocky core and ice shell. This contrasts with other studies that suggest Iapetus could not reach conditions for global melting of its interior (*Castillo-Rogez et al.,* 2007).

Few thermal evolution models have looked into the evolution of mid-sized moons formed from Saturn's ring material. This is in part because the key scenario parameters driving thermal evolution are still being defined. For example, ice accreted in this way is very porous, whereas interior evolution models generally have assumed solid interiors. End-to-end models accounting for the feedback between porosity and thermal evolution remain to be developed for this scenario.

4.5. Application to Saturn's Moons

The two leading formation scenarios imply very different heat budgets with drastic implications for long-term internal evolution of the moons. The occurrence of resonances, whose timing and duration are unknown, further complicate the story. Whether the moons formed at once or emerged one after the other from the rings determines their dynamical evolution and the prospect for resonances to occur. Proximity to the primary is not a sufficient condition for a moon to experience tidal dissipation bearing significant geophysical implications. Mimas shows all the characteristics of a frozen body despite its proximity to Saturn. Mimas serves as a reference for quantifying dissipation in another ice-dominated body, Tethys, which is poorer in rock by a factor of 5. Yet, Tethys is suspected to have been subject to extensive tidal dissipation in the past based on heat flow estimates from geomorphology (*Giese et al.,* 2007). This situation is yet another discrepancy observed in the mid-sized moon system. How tidal dissipation gets kickstarted in bodies that are starved in radioisotopes is unclear. This leads to the question of why Mimas and Enceladus followed such different evolutionary pathways if they formed together in Saturn's rings. Enceladus' rock volume fraction is three times that of Mimas and could have provided the energy kick required for Enceladus to start dissipating. However, for long-lived radioisotopes to play a role in warming up ice, the satellites must have emerged before or shortly after the heat peak due to ^{40}K decay was reached. Uranium and thorium radioisotopes have half-lives too long to play a significant role over a timescale of a few hundred million years. Hence, Enceladus seems to have reached a threshold with enough rock-based radioisotopes to warm up the ice to the point where it became dissipative. Mimas, on the other hand, could not reach that state. *Neveu and Rhoden* (2017) showed that if Mimas melts and differentiates through some combination of short-lived radioisotopes and tidal heating, it cannot end up with its present orbital eccentricity unless it is driven by a resonant passage. A more favorable scenario is that Mimas was produced from ring material within the past billion years, and potentially even more recently, consistent with the models of *Charnoz et al.* (2011) and *Salmon and Canup* (2017). Hence, Mimas emerged from the rings too late for heat from radioisotope decay to build up. In this model, Mimas begins its life as a rocky ring fragment and accretes ice from the rings as its orbit expands. Mimas' interior is layered without ever heating up enough to lose its orbital eccentricity. It is also possible that Mimas acquired its eccentricity recently through a resonance passage (e.g., Nimmo et al., this volume). The conditions for rock-poor Tethys to become dissipative to the point of being subject to partial melting remain to be elucidated. As an alternative, *Sekine and Genda* (2012) suggested that heat generated by large impacts could have increased temperatures in Tethys, Enceladus, and Dione to the point that these moons became dissipative. However, the details of that chain of events remain to be modeled.

The end-to-end journey of the moons from the rings to their current positions has been studied by *Charnoz et al.* (2011), and *Salmon and Canup* (2017) describe in detail the orbital evolution of the satellites as they emerge from the rings and evolve away due to tides and ring torques. On the other hand, in the *Cuk et al.* (2016) model, the moons did not migrate over long distances. In this context, the conditions for Enceladus and Tethys to become dissipative and melt remains to be modeled. Cuk et al. find that Tethys and Dione likely did not cross their 3:2 resonance as expected for a system that would have more time to evolve dynamically. On the other hand, Cuk et al. identified a possible 5:3 Dione-Rhea resonance and a Tethys-Dione secular resonance over the past 100 m.y. This model assumes a constant (i.e., non-frequency dependent) dissipation factor for Saturn, which is at odds with the current understanding of giant planet internal processes (e.g., *Ogilvie,* 2014). Instead, using astrometric data of the mid-sized moons, *Lainey et al.* (2017) show that the dissipation factor Q_s of Saturn is a function of the orbital periods of the moons (e.g., Q_s at Rhea's orbital period is much less than Q_s at Tethys' period). *Fuller et al.* (2016) have shown that the outward evolution of the moons' orbits is modest because of the dissipation factor's strong dependence on frequency (see also the chapter by Nimmo et al. in this volume). Much work is needed to derive a geophysical and dynamical evolution framework consistent throughout the icy moon system.

5. CONCLUSIONS AND OPEN QUESTIONS

Cassini-Huygens brought a trove of new data about Saturn's satellites but also opened new questions. Taken in the broader context of Saturn's system, these observations shed new light on accretion processes with applicability to the early solar system evolution. The diversity in physical properties and geological evolution observed across that satellite system offers a laboratory for testing fundamental physical processes: response to tidal stress, feedback between dynamical and geophysical evolution, onset of convection, and the physics of differentiation. Considered in the broader context of icy moons observed across the outer solar system, these bodies help test the boundaries of ocean worlds. Finally, the mid-sized satellites offer context for comprehending the peculiar evolutionary path followed by Enceladus. Open questions include, but are not limited to, reconciling the diversity of physical properties observed across the saturnian system; understanding the relationship, or lack thereof, between Iapetus and the satellites inside Titan's orbit; reconciling the cratering record with the increasingly young origin proposed for the satellites; the place of the co-orbital moons of Dione and Tethys in the overall formation picture; and the divergent evolutionary paths followed by Enceladus and Mimas.

An area of primary importance in addressing these questions is deciphering the information contained in the cratering record, including the apex/antiapex observations (see the chapter by Kirchoff et al. in this volume for more extensive discussion). Chronology inferred from crater counting is muddled by the possibility for multiple sources of impactors in the saturnian system: heliocentric impactors (comets, Centaurs) and planetocentric impactors from small satellites formed in the rings and ejecta from impacts on neighboring bodies (*Dones et al.,* 2016).

The possibility for extensive oceans at some point in the history of ice-dominated moons (e.g., Tethys), is another aspect that requires detailed modeling. Deep oceans may be created from an episode of intensive tidal heating, for example, promoted by resonance crossing. However, the pathway for tidal runaway in objects that are frigid remains to be demonstrated. This is even more problematic if the objects formed late (i.e., with little remaining live radioisotopes) and considering that solid ice thermal conductivity is high (~4 W m^{-1} K^{-1}) at the temperatures under consideration. Many unconstrained or not yet explored parameters and processes complicate internal modeling: remnant porosity can have a significant impact on material thermophysical properties; mechanisms driving tidal dissipation in mixtures are not well understood or supported by experimental measurements; and passage through resonances could have resulted in erasing information on early epochs as a result of increased heating. Limited constraints on the state of differentiation of the mid-sized moons limit interpretation, and non-hydrostatic anomalies add confusion. There is a compelling need for more experimental measurements to gain insight into the material thermophysical properties at relevant conditions of pressure, temperature, and composition; analog simulations of material interaction with particles; study of the response of multi-phase systems, especially in the presence of water or large lateral variations in viscous properties that call for fine-scale three-dimensional modeling that may be accessible with today's supercomputer capabilities; and experimental study of the physical process of differentiation in reduced gravity simulations.

Finally, the study of Saturn's mid-sized moons may be used to make sense of the observations of the uranian mid-sized satellites observed by Voyager and to help prepare for their future exploration. These bodies also exhibit a diversity of physical properties but also a pattern of increasing mass with distance to the planet (Fig. 11). A future mission to Uranus should search for geophysical evidence of the satellites' possible accretion in Uranus' rings or chemical markers for an origin in Uranus' subnebula.

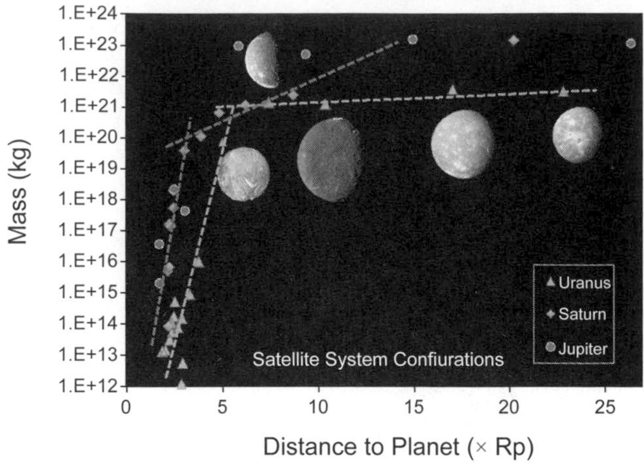

Fig. 11. Same as Fig. 9, now including the uranian moons.

Acknowledgments. The authors are thankful to J. Salmon and P. Schenk for their thorough reviews. Part of this work was carried out at the Jet Propulsion Laboratory, California Institute of Technology under contract to NASA.

REFERENCES

Alvarellos J. L., Zahnle K. J., Dobrovolskis A. R., and Hamill P. (2005) Fates of satellite ejecta in the Saturn system. *Icarus, 178,* 104–123.

Ammannito E., De Sanctis M. C., Ciarniello M., Frigeri A., Carrozzo F. G., Combe J.-Ph., Ehlmann B. E., et al. (2016) Distribution of ammoniated magnesium phyllosilicates on Ceres. *Science, 353,* aaf4279.

Anderson J. D. and Schubert G. (2007) Saturn's satellite Rhea is a homogeneous mix of rock and ice. *Geophys. Res. Lett., 34(2),* L02202, DOI: 10.1029/2006GL028100.

Anderson J. D. and G. Schubert (2010) Rhea's gravitational field and interior structure inferred from archival data files of the 2005 Cassini flyby. *Phys. Earth Planet. Inter., 178,* 176–182, DOI: 10.1016/j.pepi.2009.09.003.

Asphaug E. and Reufer A. (2013) Late origin of the Saturn system. *Icarus, 223,* 544–565.

Ayliffe B. A. and Bate M. R. (2009) Gas accretion on to planetary cores: Three-dimensional self-gravitating radiation hydrodynamical calculations. *Mon. Not. R. Astron. Soc., 393,* 49–64.

Barr A. C. and Canup R. M. (2008) Constraints on gas giant satellite formation from the interior states of partially differentiated satellites. *Icarus, 198(1),* 163–177, DOI: 10.1016/j.icarus.2008.07.004.

Beuthe M., Rivoldini A., and Trinh A. (2016) Enceladus' and Dione's floating ice shells supported by minimum stress isostasy. *Geophys. Res. Lett., 43,* 1–9, DOI: 10.1002/2016GL070650.

Bland P. A. and Travis B. J. (2017) Giant convecting mud balls of the early solar system. *Sci. Adv., 3,* e1602514.

Bromley B. C. and Kenyon S. J. (2015) Evolution of a ring around the Pluto-Charon binary. *Astrophys. J., 809,* 88.

Cameron A. G. W. and Ward W. B. (1976) The origin of the Moon. *Lunar Planet. Sci. VII,* p. 120. Lunar and Planetary Institute, Houston.

Canup R. M. (2005) A giant impact origin of Pluto-Charon. *Science, 307,* 546–550.

Canup R. M. (2010) Origin of Saturn's rings and inner moons by mass removal from a lost Titan-sized satellite. *Nature, 468,* 943–946.

Canup R. M. (2011) On a giant impact origin of Charon, Nix, and Hydra. *Astron. J., 141,* 35.

Canup R. M. and Ward W. R. (2006) A common mass scaling for satellite systems of gaseous planets. *Nature, 441,* 834–839.

Castillo-Rogez J. (2006) The internal structure of Rhea. *J. Geophys. Res., 111,* E11005, DOI: 10.1029/2004JE002379.

Castillo-Rogez J. C. and Lunine J. I. (2010) Evolution of Titan's rocky core constrained by Cassini observations. *Geophys. Res. Lett., 37,* L20205, DOI: 10.1029/2010GL044398.

Castillo-Rogez J. C. and Lunine J. I. (2012) Small worlds habitability. In *Frontiers of Astrobiology* (C. Impey et al., eds.), pp. 201–228. Cambridge Univ., Cambridge.

Castillo-Rogez J. C. et al. (2007) Iapetus' geophysics: Rotation rate, shape, and equatorial ridge. *Icarus, 190,* 179–202.

Charnoz S., Dones L., Esposito L. W., Estrada P. R., and Hedman M. M. (2009a) Origin and evolution of Saturn's ring system. In *Saturn from Cassini-Huygens* (M. K. Dougherty et al., eds.), pp. 537–576. Springer, Berlin.

Charnoz S., Morbidelli A., Dones L., and Salmon J. (2009b) Did Saturn's rings form during the late heavy bombardment? *Icarus, 199,* 413–428.

Charnoz S., Salmon J., and Crida A. (2010) The recent formation of Saturn's moonlets from viscous spreading of the main rings. *Nature, 465,* 752–774.

Charnoz S., Crida A., Castillo-Rogez J. C., Lainey V., Dones L., Karatekin O., Tobie G., Mathis S., Le Poncin-Lafitte C., and Salmon J. (2011) Accretion of Saturn's mid-sized moons during the viscous spreading of young massive rings: Solving the paradox of silicate-poor rings versus silicate-rich moons. *Icarus, 216,* 535–550.

Chen E. M. A. and Nimmo F. (2008) Implications from Ithaca Chasma for the thermal and orbital history of Tethys. *Geophys. Res. Lett., 35,* L19203, DOI: 10.1029/2008GL035402.

Chen E. M. A., Nimmo F., and Glatzmaier G. A. (2014) Tidal heating in icy satellite oceans. *Icarus, 229,* 11–30.

Choblet G., Tobie G., Sotin C., Behounkova M., Cadek O., and Postberg F. (2017) Powering prolonged activity inside Enceladus. *Nature Astron., 1,* 841–847.

Clark R. N. and Owensby P. D. (1981) The infrared spectrum of Rhea. *Icarus, 46(3),* 354–360, DOI: 10.1016/0019-1035(81)90138-X.

Clark R. N. et al. (2008) Compositional mapping of Saturn's satellite Dione with Cassini VIMS and implications of dark material in the Saturn system. *Icarus, 193(2),* 372–386, DOI: 10.1016/j.icarus.2007.08.035.

Collins G. C., McKinnon W. B., Moore J. M., Nimmo F., Pappalardo R. T., Prockter L. M., and Schenk P. (2009) Tectonics of the outer planet satellites. In *Planetary Tectonics* (T. R. Watters and R. A. Schultz, eds.), pp. 264–350. Cambridge Univ., Cambridge.

Crida A. and Charnoz S. (2012) Formation of regular satellites from ancient massive rings in the solar system. *Science, 338,* 1196–1199.

Cuk M., Dones L., and Nesvorny D. (2016) Dynamical evidence for a late formation of Saturn's moons. *Astrophys. J., 820,* 97.

Denk T. et al. (2010) Iapetus: Unique surface properties and a global color dichotomy from Cassini imaging. *Science, 327,* 435–439.

Dermott S. F. (1971) The Mimas-Tethys resonance formation problem. *Mon. Not. R. Astron. Soc., 153,* 83–96.

Dodson-Robinson S. E., Bodenheime P., Laughli G., Willac K., Turne N. J., and Beichma C. A. (2008) Saturn forms by core accretion in 3.4 Myr. *Astrophys. J. Lett., 688,* L99.

Dombard A.J. et al. (2012) Delayed formation of the equatorial ridge on Iapetus from a subsatellite created in a giant impact. *J. Geophys. Res.–Planets, 117,* E03002.

Dones L., Chapman C. R., McKinnon W. B., Melosh H. J., Kirchoff M. R., Neukum G., and Zahnle K. H. (2009) Icy satellites of Saturn: Impact cratering and age determination. In *Saturn from Cassini-Huygens* (M. K. Dougherty et al., eds.), pp. 613–635. Springer, Berlin.

Dones H. C. L., Alvarellos J., Bierhaus E. B., Bottke W., Cuk M., Hamill P., Nesvorny D., Robbins S., and Zahnle K. (2016) Could the craters on the mid-sized moons of Saturn have been made by satellite debris? In *AAS/DPS Meeting Abstracts #48,* 518.07.

Durham W. B. and Stern L. A. (2001) Rheological properties of water ice — Applications to satellites of the outer planets. *Annu. Rev. Earth Planet. Sci., 29,* 295–330.

Ellsworth K. and Schubert G. (1983) Saturn's icy satellites — Thermal and structural models. *Icarus, 54,* 490–510.

Eluszkiewicz J., Leliwa-Kopystynski J., and Kossacki K. J. (1998) Metamorphism of solar system ices. In *Solar System Ices* (B. Schmitt et al., eds.), pp. 119–138. Kluwer, Dordrecht.

Engel S., Lunine J. I., and Norton D. L. (1994) Silicate interactions with ammonia-water fluids on early Titan. *J. Geophys. Res., 99(E2),* 3745–3752.

Farinella P., Milani A., Nobili A., Paolicchi P., and Zappala V. (1983) Hyperion: Collisional disruption of a resonant satellite. *Icarus, 54,* 353–360.

Forni O., Coradini A., and Federico C. (1991) Convection and lithospheric strength in Dione, an icy satellite of Saturn. *Icarus, 94,* 232–245.

Fuller J., Luan J., and Quataert E. (2016) Resonance locking as the source of rapid tidal migration in the Jupiter and Saturn moon systems. *Mon. Not. R. Astron. Soc., 458,* 3867–3879.

Fung J. and Chiang E. (2016) Gap opening in 3D: Single-planet gaps. *Astrophys. J., 832,* 105.

Giese B., Wagner R., Neukum G., Helfenstein P., and Thomas P. C. (2007) Tethys: Lithospheric thickness and heat flux from flexurally supported topography at Ithaca Chasma. *Geophys. Res. Lett., 34,* DOI: 10.1029/2007GL031467.

Giese B., Denk T., Neukum G., Roatsch T., Helfenstein P., Thomas P. C., Turtle E. P., McEwen A., and Porco C. C. (2008) The topography of Iapetus' leading side. *Icarus, 193,* 359–371.

Gladman B. et al. (1996) Synchronous locking of tidally evolving satellites. *Icarus, 122,* 166–192.

Goldreich P. and Soter S. (1966) Q in the solar system. *Icarus, 5,* 375–389.

Greenstreet S., Gladman B., and McKinnon W. B. (2015) Impact and cratering rates onto Pluto. *Icarus, 258,* 267–288.

Hammond N. P., Phillips C. B., Nimmo F., and Kattenhorn S. A. (2013) Flexure on Dione: Investigating subsurface structure and thermal history. *Icarus, 223(1),* 418–422, DOI: 10.1016/j.icarus.2012.12.021.

Harita N. (2016) Differential impact cratering of Saturn's satellites by heliocentric impactors. *J. Geophys. Res., 121,* 111–117.

Hartmann W. K. and Davis D. R. (1975) Satellite-sized planetesimals and lunar origin. *Icarus, 24,* 504–515.

Hemingway D. J., Zannoni M., Tortora P., Nimmo F., and Asmar S. W. (2016) Dione's internal structure inferred from Cassini gravity and topography. *Lunar Planet. Sci. XLVII,* Abstract #1314. Lunar and Planetary Institute, Houston.

Howett C. J. A., Spencer J. R., Pearl J., and Segura M. (2011) High heat flow from Enceladus' south polar region measured using 10–600 cm^{-1} Cassini/CIRS data. *J. Geophys. Res., 116,* E03003.

Hussmann H., Sohl F., and Spohn T. (2006) Subsurface oceans and deep interiors of medium-sized outer planet satellites and large trans-Neptunian objects. *Icarus, 185,* 258–273.

Hussmann H., Sotin C., and Lunine J. I. (2015) Interiors and evolution of icy satellites. In *Treatise in Geophysics, Second Edition, Vol. 10: Physics of Terrestrial Planets and Moons* (G. Schubert, ed.), pp. 605–635. Elsevier, Amsterdam. DOI: 10.1016/B978-0-444-53802-4.00178-0.

Hyodo R. and Charnoz S. (2017) Dynamical evolution of the debris disk after a satellite catastrophic disruption around Saturn. *Astron. J., 154,* 34.

Hyodo R., Charnoz S., Ohtsuki K., and Genda H. (2017) Ring formation around giant planets by tidal disruption of a single passing large Kuiper belt object. *Icarus, 282,* 195–213.

Iess L., Rappaport N. J., Tortora P., Lunine J., Armstrong J. W., Asmar S. W., Somenzi L., and Zingoni F. (2007) Gravity field and interior of Rhea from Cassini data analysis. *Icarus, 190(2),* 585–593, DOI: 10.1016/j.icarus.2007.03.027.

Iess L., Stevenson D. J., Parisi M., Hemingway D., Jacobson R. A., Lunine J. I., Nimmo F., Armstrong J. W., Asmar S. W., Ducci M., and Tortora P. (2014) The gravity field and interior structure of Enceladus. *Science, 344,* 78–80.

Iess L., Racioppa P., Durante D., Mariani M. Jr., Anabtawi A., Armstrong J. W., Gomez Casajus L., Tortora P., and Zannoni M. (2017) The dark side of Saturn's gravity. Abstract U22A-03 presented at the 2017 Fall Meeting, AGU, San Francisco, California, Dec. 11–15.

Jaumann R., Clark R. N., Nimmo F., Hendrix A. R., Burrati B. J., Denk T., Moore J. M., Schenk P. M., Ostro S. J., and Srama R. (2009) Icy satellites: Geological evolution and surface processes. In *Saturn from Cassini-Huygens* (M. K. Dougherty et al., eds.), pp. 577–612. Springer, Berlin.

Kirchoff M. R. and Schenk P. (2010) Impact cratering records of the mid-sized, icy saturnian satellites. *Icarus, 206,* 485–497.

Kirchoff M. R. and Schenk P. (2015) Dione's resurfacing history as determined from a global impact crater database. *Icarus, 256,* 78–89, DOI: 10.1016/j.icarus.2015.04.010.

Kirk R. L. and Stevenson D. J. (1987) Thermal evolution of a differentiated Ganymede and implications for surface features. *Icarus, 69,* 91–134.

Kuchta M. et al. (2015) Despinning and shape evolution of Saturn's moon Iapetus triggered by a giant impact. *Icarus, 252,* 454–465.

Lainey V., Karatekin O., Desmars J., Charnoz S., Arlot J.-E., Emelyanov N., Le Poncin-Lafitte C., Mathis S., Remus F., and Tobie G. (2012) Strong tidal dissipation in Saturn and constraints on Enceladus' thermal state from astrometry. *Astrophys. J., 752,* 14.

Lainey V., Jacobson R. A., Tajeddine R., Cooper N. J., Murray C., Robert V., Tobie G., Guillot T., Mathis S., and Remus F. (2017) New constraints on Saturn's interior from Cassini astrometric data. *Icarus, 281,* 286–296.

Leisner J. S., Russell C. T., Strangeway R. J., Omidi N., Dougherty M. K., and Kurth W. S. (2008) The interior of Iapetus: Constraints provided by the solar wind interaction. Abstract # P31C-08 presented at the 2008 Fall Meeting, AGU, San Francisco, Dec. 15–19.

Leliwa-Kopystynski J. and Kossacki K. J. (2000) Evolution of porosity in small icy bodies. *Planet. Space Sci., 48,* 727–745.

Levison H. F. et al. (2011) Ridge formation and de-spinning of Iapetus via an impact-generated satellite. *Icarus, 214,* 773–778.

Lissauer J. J. and Safronov V. S. (1991) The random component of planetary rotation. *Icarus, 93,* 288–297.

Lopez Garcia E.J. et al. (2014) Topographic constraints on the origin of the equatorial ridge on Iapetus. *Icarus, 237,* 419–421.

Mackenzie R. A., Iess L., Tortora P., and Rappaport N. J. (2008) A non-hydrostatic Rhea. *Geophys. Res. Lett., 35,* L05204, DOI: 10.1029/2007GL032898.

Martin E. S., Patthoff D. A., and Watters T. R. (2016) Mysterious linear virgae across the icy satellites. *Lunar Planet. Sci. XLVII,* Abstract #2958. Lunar and Planetary Institute, Houston.

Matson D., Castillo-Rogez J. C., Schubert G., Sotin C., and McKinnon W. B. (2009) The thermal evolution and internal structure of Saturn's mid-sized icy satellites. In *Saturn from Cassini-Huygens* (M. K. Dougherty et al., eds.), pp. 577–612. Springer, Berlin.

McCarthy C. M. and Castillo-Rogez J. C. (2012) Planetary ices attenuation properties. In *The Science of Solar System Ices* (M. S. Gudipati and J. Castillo-Rogez, eds.), pp. 183–225. Astrophysics and Space Science Library, Vol. 356, Springer, Berlin.

McKinnon W. B. (1989) On the origin of the Pluto-Charon binary. *Astrophys. J. Lett., 344,* L41–L44.

McKinnon W. B. (2013) The shape of Enceladus as explained by an irregular core: Implications for gravity, libration, and survival of its subsurface ocean. *J. Geophys. Res., 118,* 1775–1788.

Monteux J., Collins G. R., Tobie G., and Choblet G. (2016) Consequences of large impacts on Enceladus' core shape. *Icarus, 264,* 300–310.

Moore J. M., Schenk P. M., Bruesch L. S, Asphaug E., and McKinnon W. B. (2004) Large impact features. *Icarus, 171,* 421–443.

Morbidelli A. and Crida A. (2007) The dynamics of Jupiter and Saturn in the gaseous protoplanetary disk. *Icarus, 191,* 158–171.

Mosqueira I., Estrada P. R., and Charnoz S. (2010) Deciphering the origin of the regular satellites of gaseous giants — Iapetus: The Rosetta ice-moon. *Icarus, 207,* 448–460.

Movshovitz N., Nimmo F., Korycansky D. G., Asphaug E. A., and Owen J. M. (2015) Disruption and reaccretion of midsized moons during an outer solar system late heavy bombardment. *Geophys. Res. Lett., 42,* 256–263.

Multhaup K. and Spohn T. (2007) Stagnant lid convection in the mid-sized icy satellites of Saturn. *Icarus, 186,* 420–435.

Nesvorny D., Vokrouhlicky D., and Morbidelli A. (2007) Capture of irregular satellites during planetary encounters. *Astron. J., 133,* 1962.

Neveu M. and Rhoden A. R. (2017) The origin and evolution of a differentiated Mimas. *Icarus, 296,* 183–196.

Neveu M., Desch S. J., and Castillo-Rogez J. C. (2017) Aqueous geochemistry in icy world interiors: Equilibrium fluid, rock, and gas compositions, and fate of antifreezes and radionuclides. *Geochim. Cosmochim. Acta, 212,* 324–371.

Nimmo F., Bills B. G., Thomas P. C., and Asmar S. W. (2010) Geophysical implications of the long-wavelength topography of Rhea. *J. Geophys. Res.–Planets, 115(10),* 1–11, DOI: 10.1029/2010JE003604.

Nimmo F., Bills B. G., and Thomas P. C. (2011) Geophysical implications of the long-wavelength topography of the saturnian satellites. *J. Geophys. Res., 116(E11),* E11001, DOI: 10.1029/2011JE003835.

Ogilvie G. I. (2014) Tidal dissipation in stars and giant planets. *Annu. Rev. Astron. Astrophys., 52,* 171–210.

Park R. S., Konopliv A. S., Bills B. G., Rambaux N., Castillo-Rogez J. C., Raymond C. A., Vaughan A. T., Ermakov A. I., Zuber M. T., Fu R. R., Toplis M. J., Russell C. T., Nathues A., and Preusker F. (2016) A partially differentiated interior for (1) Ceres deduced from its gravity field and shape. *Nature, 537,* 515–517.

Peale S. J. and Canup R. M. (2015) Deciphering the motions of planets and moons. *Proc. Natl. Acad. Sci., 112,* DOI: 10.1073/pnas.1512536112.

Peale S. J., Cassen P., and Reynolds R. T. (1980) Tidal dissipation, orbital evolution, and the nature of Saturn's inner satellites. *Icarus, 43,* 65–72.

Phillips C. B., Hammond N. P., Robuchon G., Nimmo F., Beyer R., and Roberts J. (2012) Stereo imaging, crater relaxation, and thermal histories of Rhea and Dione. *Lunar Planet. Sci. XLIII,* Abstract #2571. Lunar and Planetary Institute, Houston.

Plescia J. B. (1983) The geology of Dione. *Icarus, 56,* 255–277.

Plescia J. B. and Boyce J. M. (1982) Crater densities and geological histories of Rhea, Dione, Mimas and Tethys. *Nature, 295,* 285–290.

Porco C. C. et al. (2005) Cassini imaging science: Initial results on Phoebe and Iapetus. *Science, 307,* 1237–1242.

Postberg F., Schmidt J., Hillier J., Kempf S., and Srama R. (2011) A salt-water reservoir as a source of compositionally stratified plume on Enceladus. *Nature, 474,* 620–622.

Rhoden A. R., Henning W., Hurford T. A., Patthoff D. A., and Tajeddine R. (2017) The implications of tides on the Mimas ocean hypothesis. *J. Geophys. Res., 122,* 400–410.

Roberts J. H. (2014) The fluffy core of Enceladus. *Icarus, 258,* 54–66.

Robuchon G. et al. (2010) Coupling of thermal evolution and despinning of early Iapetus. *Icarus, 207,* 959–971.

Salmon J. and Canup R. M. (2017) Accretion of Saturn's inner mid-sized moons from a massive primordial ice ring. *Astrophys. J., 836,* 109.

Salmon J., Charnoz S., Crida A., and Brahic A. (2010) Long-term and large-scale viscous evolution of dense planetary rings. *Icarus, 209,* 771–785.

Schenk P. M., Hamilton D. P., Johnson R. E., McKinnon W. B., Paranicas C., Schmidt J., and Showalter M. R. (2011) Plasma, plumes and rings: Saturn system dynamics as recorded in global color patterns on its midsize icy satellites. *Icarus, 211,* 740–757.

Schmedemann N., Neukum G., Denk T., and Wagner R. (2008) Stratigraphy and surface ages on Iapetus and other saturnian satellites. *EGU General Assembly Abstracts, 10,* EGU2008-A-08745.

Schubert G., Hussmann H., Lainey V., Matson D. L., McKinnon W. B., Sohl F., Sotin C., Tobie G., Turrini D., and Van Hoolst T. (2010) Evolution of icy satellites. *Space Sci. Rev., 153,* 447–484, DOI: 10.1007/s11214-010-9635-1.

Scipioni F., Tosi F., Stephan K., Filacchione G., Ciarniello M., Capaccioni F., and Cerroni P. (2013) Spectroscopic classification of icy satellites of Saturn I: Identification of terrain units on Dione. *Icarus, 226(2),* 1331–1349, DOI: 10.1016/j.icarus.2013.08.008.

Sekine Y. and Genda H. (2012) Giant impacts in the saturnian system: A possible origin of diversity in the inner mid-sized satellites. *Planet. Space Sci., 63,* 133–138.

Sinclair A. T. (1983) A re-consideration of the evolution hypothesis of the origin of the resonances among Saturn's satellites. In *Dynamical Trapping and Evolution in the Solar System* (V. V. Markellos and Y. Kozai, eds.), pp. 19–25. Reidel, Dordrecht.

Singer K. N. and McKinnon W. B. (2011) Tectonics on Iapetus: Despinning, respinning, or something completely different? *Icarus, 216,* 198–211.

Singer K. N. et al. (2012) Massive ice avalanches on Iapetus mobilized by friction reduction during flash heating. *Nature Geosci., 5,* 574–578.

Smith B. A., Soderblom L., Beebe R., Boyce J., et al. (1981) Encounter with Saturn: Voyager 1 imaging science results. *Science, 212,* 163–191.

Smith B. A., Soderblom L., Batson R., Bridges P., Inge J., et al. (1982) A new look at the Saturn system: The Voyager 2 images. *Science, 215,* 504–537.

Spencer J. R. and Denk T. (2010) Formation of Iapetus' extreme albedo dichotomy by exogenically triggered thermal ice migration. *Science, 327,* 432–435, DOI: 10.1126/science1177132.

Squyres S. W., Reynolds R. T., Summers A. L., and Shung F. (1988) Accretional heating of the satellites of Saturn and Uranus. *J. Geophys. Res., 93,* 8779–8794.

Stephan K. et al. (2010) Dione's spectral and geological properties. *Icarus, 206(2),* 631–652, DOI: 10.1016/j.icarus.2009.07.036.

Stephan K. et al. (2012) The saturnian satellite Rhea as seen by Cassini VIMS. *Planet. Space Sci., 61(1),* 142–160, DOI: 10.1016/j.pss.2011.07.019.

Stephan K., Wagner R., Jaumann R., Clark R. N., Cruikshank D. P., Brown R. H., Giese B., Roatsch T., Filacchione G., Matson D., Ore C. D., Capaccioni F., Baines K. H., Rodriguez S., Krupp N., Buratti B. J., and Nicholson P. D. (2016) Cassini's geological and compositional view of Tethys. *Icarus, 274,* 1–22.

Tajeddine R., Rambaux N., Lainey V., Charnoz S., Richard A., Rivoldini A., and Noyelles B. (2014) Constraints on Mimas' interior from Cassini ISS libration measurements. *Science, 346,* 322–324.

Tanigawa T., Maruta A., and Machida M. N. (2014) Accretion of solid materials onto circumplanetary disks from protoplanetary disks. *Astrophys. J., 784,* 109.

Thomas P. C. (2010) Sizes, shapes, and derived properties of the saturnian satellites after the Cassini nominal mission. *Icarus, 208(1),* 395–401, DOI: 10.1016/j.icarus.2010.01.025.

Thomas P. C., Tajeddine R., Tiscareno M. S., Burns J. A., Joseph J., Loredo T. J., Helfenstein P., and Porco C. (2016) Enceladus's measured physical libration requires a global subsurface ocean. *Icarus, 264,* 37–47, DOI: 10.1016/j.icarus.2015.08.037.

Tobie G., Mocquet A., and Sotin C. (2005) Tidal dissipation within large icy satellites: Applications to Europa and Titan. *Icarus, 177,* 534–549.

Tortora P., Zannoni M., Hemingway D., Nimmo F., Jacobson R. A., Iess L., and Parisi M. (2016) Rhea gravity field and interior modeling from Cassini data analysis. *Icarus, 264,* 264–273, DOI: 10.1016/j.icarus.2015.09.022.

Travis B. J. and Schubert G. (2015) Keeping Enceladus warm. *Icarus, 250,* 32–42.

Tricarico P. (2014) Multi-layer hydrostatic equilibrium of planets and synchronous moons: Theory and application to Ceres and to solar system moons. *Astrophys. J., 782(2),* 99, DOI: 10.1088/0004-637X/782/2/99.

Wagner R. J., Neukum G., Giese B., Roatsch T., Denk T., Wolf U., and Porco C. C. (2010) The geology of Rhea: A first look at the ISS camera data from orbit 121 (Nov 21, 2009) in Cassini's extended mission. *Lunar Planet. Sci. XLI,* Abstract #1672. Lunar and Planetary Institute, Houston.

Ward W. R. and Canup R. M. (2010) Circumplanetary disk formation. *Astron. J., 140,* 1168.

White O. L., Schenk P. M., and Dombard A. J. (2013) Impact basin relaxation on Rhea and Iapetus and relation to past heat flow. *Icarus, 223(2),* 699–709, DOI: 10.1016/j.icarus.2013.01.013.

White O. L., Schenk P. M., Bellagamba A. W., Grimm A. M., Dombard A. J., and Bray V. J. (2017) Impact crater relaxation on Dione and Tethys and relation to past heat flow. *Icarus, 288,* 37–52.

Zebker H. A., Stiles B., Hensley S., Lorenz R., Kirk R. L., and Lunine J. (2009) Size and shape of Saturn's Moon Titan. *Science, 324,* 921–923.

Zhang K. and Nimmo F. (2009) Recent orbital evolution and the internal structures of Enceladus and Dione. *Icarus, 204,* 597–609.

Hendrix A. R., Buratti B. J., Cruikshank D. P., Clark R. N., Scipioni F., and Howett C. J. A. (2018) Surface composition of Saturn's icy moons. In *Enceladus and the Icy Moons of Saturn* (P. M. Schenk et al., eds.), pp. 307–322. Univ. of Arizona, Tucson, DOI: 10.2458/azu_uapress_9780816537075-ch015.

Surface Composition of Saturn's Icy Moons

Amanda R. Hendrix
Planetary Science Institute

Bonnie J. Buratti
Jet Propulsion Laboratory

Dale P. Cruikshank
NASA Ames Research Center

Roger N. Clark
Planetary Science Institute

Francesca Scipioni
NASA Ames Research Center

Carly J. A. Howett
Southwest Research Institute

The surface compositions of Saturn's moons are largely controlled by their environments. At the inner mid-sized moons, gradients in water-ice abundance, as well as hemispheric albedo dichotomies, with distance from Enceladus demonstrate the key role of the E ring in driving surface composition. A variation with distance from Enceladus is also seen in a reddish coloring agent in the system, whose identification is still under discussion. In the outer part of the system, Hyperion and the leading hemisphere of Iapetus are colored by dark, organic-rich dust originating from Phoebe.

1. INTRODUCTION

In this chapter, we discuss what is known of the surface compositions of the moons of Saturn, focusing on Mimas, Tethys, Dione, Rhea, Iapetus, Hyperion, Phoebe and the small moons. Discussion of the surface composition of Enceladus is handled in the chapter by Postberg *et al.* in this volume.

The icy moons of Saturn orbit the planet in a dynamic environment where grain bombardment by ring material (namely E ring or Phoebe ring) and radiolytic effects from charged particles play critical roles in influencing surface composition (e.g., *Schenk et al.,* 2011; see also the chapter by Howett et al. in this volume). As we discuss here, the surface compositions of the moons are thus driven largely by their settings within the system. It is these exogenous contamination sources that determine the current surface composition of Saturn's moons rather than a temperature gradient in the region around the proto-Saturn. Their bulk composition is dominated by H_2O ice (e.g., see the chapter by Castillo-Rogez et al. in this volume), and although water ice is a primary constituent of their surface compositions [water ice was found to be the dominant surface component of Tethys, Dione, Rhea, and Iapetus by *Fink et al.* (1976) and was subsequently found by various investigators on Hyperion, Enceladus, Mimas, and Phoebe], much of the H_2O ice observed on the surfaces of the inner moons (Mimas, Enceladus, Tethys, Dione, Rhea) derives from the E ring (and ultimately Enceladus's plume). Native H_2O ice is found primarily at fresh impact craters and tectonic fractures (e.g., *Clark et al.,* 2008; *Stephan et al.,* 2010, 2012, 2016). In this chapter, we discuss variations in surface compositions that are largely controlled by the exogenic processes of E-ring grain and radiolytic modifications, as well as thermal and seasonal variations.

All of Saturn's major satellites have prograde, synchronous orbits. A direct consequence of these orbital characteristics is an observed albedo asymmetry between the surfaces of their leading and trailing hemispheres (e.g., *Buratti et al.,* 1998); in this chapter we also discuss hemispheric compositional variations that are a result of the orbital characteristics. Of the satellites discussed here, Phoebe and Hyperion are the exceptions: Phoebe is in a retrograde, nonsynchronous orbit and Hyperion rotates chaotically.

At many moons elsewhere in the solar system, albedo, color, and compositional units can be correlated with geologic units. In general, more ancient, highly cratered surfaces are darker, and for a given icy moon, albedo correlates with both crater counts and terrain age. For example, the grooved terrain of Ganymede, which underwent geologic processing after the formation of the satellite, is brighter than the primordial cratered terrain (*Squyres and Veverka,* 1981). Another example is the dwarf planet Pluto, which has a low-albedo cratered equatorial region (Cthulhu Regio) and a very high-albedo, geologically active feature centered near the equator on the anti-Charon hemisphere (known as Sputnik Planitia) (*Buratti et al.,* 2017).

In contrast, the surfaces of the moons of Saturn are dominated by exogenous effects, controlling the albedo and to a large extent the composition. The key driver is the ring system, and because it is diverse and extends over vast reaches of Saturn's system, its effects are complex. The orbits of the inner mid-sized moons (Mimas, Enceladus, Tethys, Dione, Rhea) within the E ring result in bombardment and/or "sandblasting" of the surfaces by E-ring grains, affecting the composition of the uppermost layers of the icy regolith. In the outer part of the Saturn system, material from Phoebe's giant ring is significant at Iapetus, as well as at Hyperion. Meteoritic bombardment also "gardens" the surfaces of the satellites, to brighten and "fluff up" the upper layer of the regolith, with a more pronounced effect on the leading side of the moons (*Buratti et al.,* 1990), and can also darken it by adding dark meteoritic material to the surfaces (e.g., *McCord et al.,* 2012). Charged particle bombardment (both within the Saturn magnetosphere and in the solar wind) can drive chemical modifications within the regoliths of these moons (see also the chapter by Howett et al. in this volume). High-energy particles in the saturnian magnetosphere (e.g., *Paranicas et al.,* 2014) bombard and anneal the surfaces of the inner moons and cause morphological changes (e.g., *Howett et al.,* 2011). Cold plasma co-rotating with Saturn's magnetosphere (e.g., *Pospieszalska and Johnson,* 1989) may be responsible for the lower albedos on the trailing hemispheres of Tethys, Dione, and Rhea, and at Dione, evidence of implantation of small, dark grains (*Clark et al.,* 2008) has been found on the trailing hemisphere. Combinations of these processes can lead to leading-trailing hemisphere grain size differences.

2. INNER SYSTEM MID-SIZED SATELLITES

2.1. Water Ice and Variations with Distance from Enceladus

Water ice is the dominant surface component of the icy saturnian moons Mimas, Enceladus, Tethys, Dione, and Rhea, as first observed (on Tethys, Dione, and Rhea) by *Fink et al.* (1976). Groundbased measurements (e.g., *Verbiscer et al.,* 2006) also show the infrared absorption characteristics of water ice, as do Cassini Visual and Infrared Mapping Spectrometer (VIMS) measurements (*Brown et al.,* 2006a,b; *Filacchione et al.,* 2007; *Clark et al.* 2008, 2012). Furthermore, Cassini Ultraviolet Imaging Spectrograph (UVIS) measurements demonstrate that the UV H_2O ice absorption edge (near 165 nm) is ubiquitous in the Saturn system (e.g., *Hendrix et al.,* 2010, 2012).

Voyager 1 and 2 measurements found high visible albedos consistent with water ice and showed an almost complete lack of correlation between spectrophotometric properties and geologic units for the saturnian system. Enceladus was found to exhibit terrains with widely varying ages, dating from the tail end of the late heavy bombardment to possibly geologically active, crater-free terrains (*Buratti,* 1988). The other moons were found to have high albedos, regardless of their geologic age as derived from crater counts. The apparent pattern showed that Tethys, Rhea, and Dione were brighter on their leading sides, while Mimas and Enceladus seemed to be somewhat brighter on their trailing sides (*Buratti and Veverka,* 1984; *Verbiscer and Veverka,* 1992, 1994). An analysis of the Voyager-derived albedos of these moons with distances from Saturn showed that their albedo decreased with distance from Enceladus (*Pang et al.,* 1984); this was also shown using Hubble Space Telescope (HST)-measured geometric albedos obtained at true opposition (*Verbiscer et al.,* 2007) (Fig. 1). Because the plume of Enceladus is the source of the E ring (e.g., *Hansen et al.,* 2006), this pattern was interpreted to mean that the ice particles from the ring were bombarding and/or sandblasting the moons, with fine-grained water-ice-rich particles being enriched on those moons closest to Enceladus (*Verbiscer et al.,* 2007). Moreover, a model predicting E-ring grain bombardment on the moons (*Hamilton and Burns,* 1994) matches the observation that the leading sides of the moons exterior to Enceladus are brighter than their trailing hemispheres, while the trailing hemisphere of Mimas, interior to Enceladus, has a slightly brighter trailing side (*Buratti and Veverka,* 1984; *Verbiscer and Veverka,* 1992; *Buratti et al.,* 1998). An updated model of E-ring grain bombardment is discussed in the chapter by Kempf et al. in this volume.

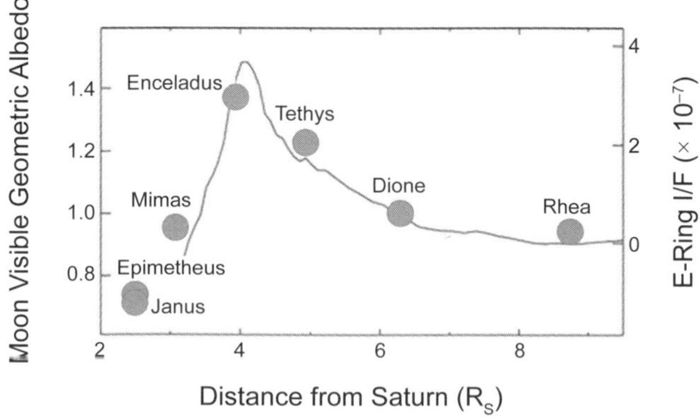

Fig. 1. The visible geometric albedos of the medium-sized saturnian moons as a function of distance from the planet. The intensity of the E-ring as a function of distance from Saturn is superimposed on the observations. Figure based on information in *Verbiscer et al.* (2007).

Cassini VIMS data also show longitudinal variations consistent with E-ring grain bombardment, in terms of water-ice absorption band depths and grain sizes (*Filacchione et al.,* 2013). The main water-ice overtones and combinations in the near infrared range occur at 1.04, 1.25, 1.5, 2.0, and 3.0 μm, along with a reflectance peak at 3.6 μm (Fig. 2). Water-ice band depth maps for Dione and Rhea are shown in Fig. 3 in the form of cylindrically-projected maps. (As this chapter is written, similar maps for Mimas and Tethys are in progress and are not included here.) The band depth values were calculated following the technique of *Clark and Roush* (1984)

$$BD = 1-(Rb/Rc)$$

where Rb is the reflectance value at the band bottom and Rc is the spectral continuum value measured at the same wavelength. Rc is found through a linear fit between the left and right wings of each band. Because illumination conditions can influence the depth of the absorption bands (e.g., *Kolokolova et al.,* 2010), only VIMS cubes at phase angles between 10° and 50° were used in creating the maps of Fig. 3, and spectra were normalized at the continuum value of 2.23 μm to ensure minimization of photometric effects, as demonstrated by *Scipioni et al.* (2013, 2014).

In general, the depth of absorption features is influenced by the variation of water-ice abundance and grain size. It decreases if water ice is mixed with a contaminant material, or if the size of the grains decreases (*Hapke et al.,* 1978; *Clark et al.,* 2011, and references therein). A higher concentration of contaminants, or smaller water-ice grains, makes the water-ice absorption bands shallower (e.g., *Clark and Lucey,* 1984). However, submicrometer ice grains have been shown to exhibit increased diffraction effects, distorting the shape of the ice absorptions and changing the relationship of band depth with grain size, making band depth increase with submicrometer grains (*Clark et al.,* 2012). *Verbiscer et al.* (2006) also point out that the depth of absorption features can be influenced by particles that are relatively small and/or microstructurally complex.

The band depth maps of Fig. 3 demonstrate the marked asymmetry between the leading and the trailing hemispheres, for both Dione and Rhea, as shown by the increase in the depth of the three water-ice absorption bands on the leading side, with maximum band depths around 90°W. The darkest terrains on Dione's and Rhea's surface are observed on the trailing hemisphere. On Dione, the trailing hemisphere dark region is a well-defined shape, centered at 270°W, 0°N, extending to roughly 60° in latitude and more than 90° in longitude, crossing into the leading hemisphere. Rhea's dark trailing hemisphere terrain is more extended than Dione's dark region. In these regions, the depth of the water-ice absorption bands has the lowest value. (The lowest values of the absorption bands are, for Dione, 0.0 for the 1.25-μm band, 0.15 for the 1.5-μm band, and 0.34 for the 2.0-μm band; for Rhea, 0.016 for the 1.25-μm band, 0.33 for the 1.5-μm band, and 0.53 for the 2.0-μm band.) This band depth distribution is consistent with the leading hemispheres being constantly refreshed by water-ice particles from the E ring. Furthermore, for both Dione and Rhea, the leading hemisphere has a smoother, more uniform appearance (Fig. 3). Similarly, *Verbiscer et al.* (2006) found that the water-ice absorption bands were stronger on the leading side of Enceladus than the trailing side, which they suggested could be related to the preferential erosion of particles on the trailing side by charged particles within Saturn's magnetosphere.

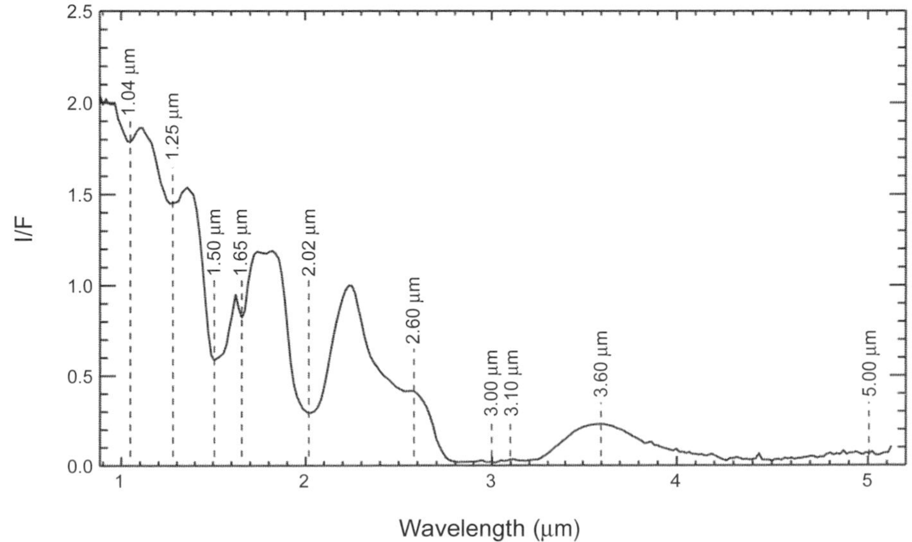

Fig. 2. Example of VIMS spectrum of an icy saturnian satellite demonstrating the characteristic water-ice infrared features. While this is a spectrum of Enceladus, it is used here as an example because the indicated water-ice absorption features also appear on the other moons. Spectra have been normalized at the continuum wavelength of 2.23 μm to minimize photometric effects induced by different illumination conditions.

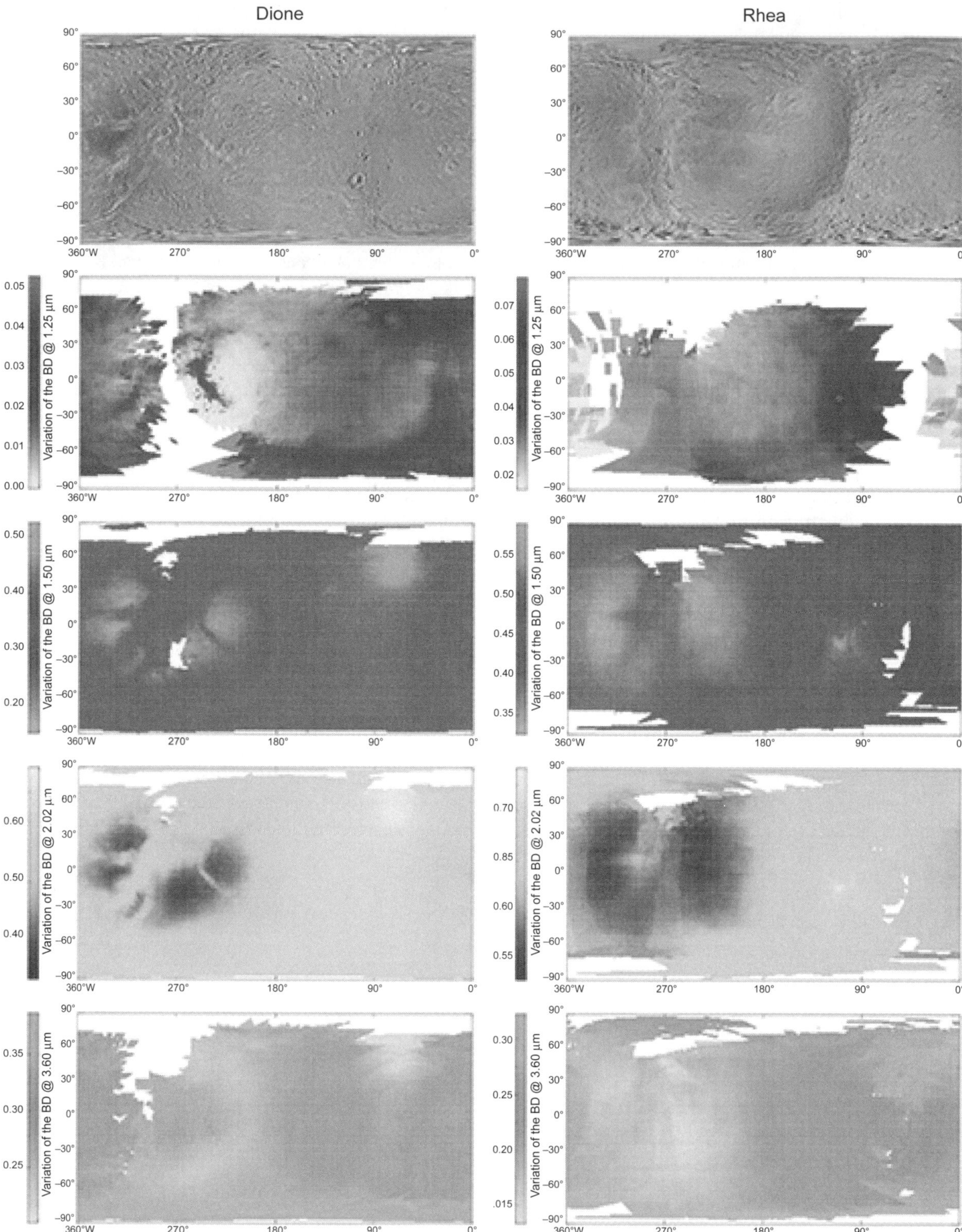

Fig. 3. See Plate 32 for color version. Maps of variation of the water-ice absorption bands at 1.25, 1.5, and 2.0 µm, and of the reflectance maximum at 3.6 µm, for Dione (left column) and Rhea (right column). The variation of each spectral indicator is represented by a unique color code. The associated color bar is on the left of each image, and values increase from bottom to top. The maps were produced following the method described in detail in *Scipioni et al.* (2017). Each map is sampled by using a fixed-resolution grid with angular bins of 1° latitude × 1° longitude. Within each bin, the values of the spectral indices computed on the basis of all VIMS data covering that particular bin had been averaged (for details, see *Scipioni et al.,* 2017). Base maps are provided for comparison in the top row (maps courtesy of NASA/JPL-Caltech/Space Science Institute).

Besides the E-ring-related water ice discussed above, *native* water ice is found at the inner saturnian moons at fresh craters and tectonic fractures (*Jaumann et al.,* 2008; *Stephan et al.,* 2010, 2012, 2016). At Dione, the crater Creusa is recognized as a bright spot with deep water-ice absorption bands (Fig. 3, left column). Rhea (Fig. 3, right column) hosts a fresh, bright crater (Inktomi) on its leading hemisphere. It is located close to the leading hemisphere's center, and it shows the deepest water-ice band depths across the whole surface. Deep water-ice absorption bands are also found corresponding to the "wispy" tectonic terrains of Rhea and Dione, although the wispy terrain water-ice band depths are more suppressed than the leading hemisphere's terrains and Dione's Creusa crater region.

Two phases of H_2O ice that are thermodynamically stable in the conditions on the surfaces of Saturn's satellites are the hexagonal crystalline phase and an amorphous phase in which the molecular arrangement is disordered. H_2O frozen from a vapor at $T < \sim 135$ K assumes the disordered state, which spontaneously anneals through a metastable cubic and then a hexagonal crystalline structure as the temperature reaches ~170 K (*Mastrapa et al.,* 2008). Lowering the temperature of crystalline ice does not cause it to revert to the amorphous state, but the crystalline structure can be disrupted by charged particle bombardment (e.g., *Baragiola et al.* 2013). Large-scale impacts on a cold, crystalline ice surface are expected to have the dual effects of widespread melting at the impact site and the formation of amorphous ice in some of the ejecta that is either mechanically disrupted or is caused to condense rapidly from the vapor produced by the energy of the impact.

Although there are competing time- and temperature-dependent processes to amorphize and anneal the ice, the general sense is that an old impact crater on an icy planetary surface will have a different pattern of distribution of the two phases of water ice than a younger crater of comparable size. *Dalle Ore et al.* (2015) demonstrated this difference with the craters Inktomi and Obatala on the leading and trailing (respectively) hemispheres of Rhea using a spectroscopic technique based on the difference in shape of the 2.0-μm H_2O ice band in the amorphous and crystalline phases. Using estimates of annealing times and resurfacing rates by E-ring particle accretion, they estimate the age of Obatala as ~450 Ma. Although the distribution pattern of amorphous and crystalline phases around Inktomi is clear, *Dalle Ore et al.* (2015) concluded that the age is indeterminate by this technique because the leading hemisphere is not subjected to the same amorphizing effects of charged-particle bombardment as the trailing hemisphere. Amorphous water ice on Enceladus is also discussed in the chapter by Postberg et al. in this volume.

2.2. Non-Water-Ice Contaminants

2.2.1. Reddish material. Like water-ice abundances, non-water-ice contaminants on Saturn's moons also tend to vary with distance from Enceladus. In addition to discrete diagnostic absorption bands, the trend of the intensity of scattered sunlight with wavelength can be broadly indicative of the presence of ill-defined molecular complexes. Specifically, the trend toward higher reflectance with increasing wavelength over the UV-visible spectral region translates to shades of yellow or red coloration of a surface. This color trend is important, especially on icy planetary surfaces, because ices have little or no intrinsic color, and the coloring agent is therefore something different. One of the primary non-ice species appearing on the inner icy moons of Saturn is exemplified by a red spectral slope in the ~0.2–0.5-μm range (e.g., *Noll et al.,* 1997; *Hendrix et al.,* 2010; *Filacchione et al.,* 2012). *Filacchione et al.* (2013) used Cassini VIMS data to show that the spectral slope (0.35–0.55 μm) increases (becomes redder) with distance from Enceladus (Fig. 4). The visibly reddish material is present on both the leading and trailing hemispheres, with increased abundances on the trailing hemispheres and with distance from Enceladus as shown using both VIMS and HST data (*Filacchione et al.,* 2012; *Hendrix et al.,* 2017). Also of note is that radar albedos have been found to decrease with distance from Enceladus, attributed to increasing abundances of contaminants including possibly ammonia, silicates, metallic oxides, and polar organics (*Ostro et al.,* 2006).

The nature of the red-sloped material is not entirely clear. Several materials can impart colors ranging from yellow to red to surfaces where ices have been identified. Nano-sized particles of neutral or oxidized iron have this effect, as noted by *Clark et al.* (2008, 2012) and supported by laboratory measurements and scattering model calculations referenced in those papers; these small particles may also contribute to an upturn in the spectral reflectance by Rayleigh scattering at wavelengths shorter than ~0.4 μm. Nanophase neutral and oxidized iron derived from Fe-bearing minerals in igneous rock grains and redeposited on the grain surfaces are considered to be responsible for the general reddening of the lunar surface (*Pieters et al.* 2000). However, as discussed in the chapter by Postberg et al. in this volume, very little evidence of iron-rich silicates is present in the saturnian system, so if iron is responsible for the reddish character, it would likely need to derive from interplanetary/interstellar material. Salts are also known to redden with weathering (e.g., *Hand and Carlson,* 2015) and are present in the E ring (see the chapter by Postberg et al.), so these are a possible source of the red material in the system.

Another reddening material consists of a macromolecular carbonaceous tholin defined as a highly disordered network of small aromatic units (moieties) linked by short and branched aliphatic chains. Nitrogen atoms can be substituted for carbon in both the aromatic and aliphatic components. Tholins came into prominence as the probable source of the yellow-brown color of Titan's atmosphere when colored refractory solids were created in the laboratory by exposing a Titan-mixture of gases ($N_2 + CH_4$ in proportions 100:10) to electrical spark, UV light, or corona discharge (e.g., *Khare et al.,* 1984; *Imanaka et al.,* 2004). This work was an outgrowth of the Miller-Urey experiments (*Miller,* 1953), in which

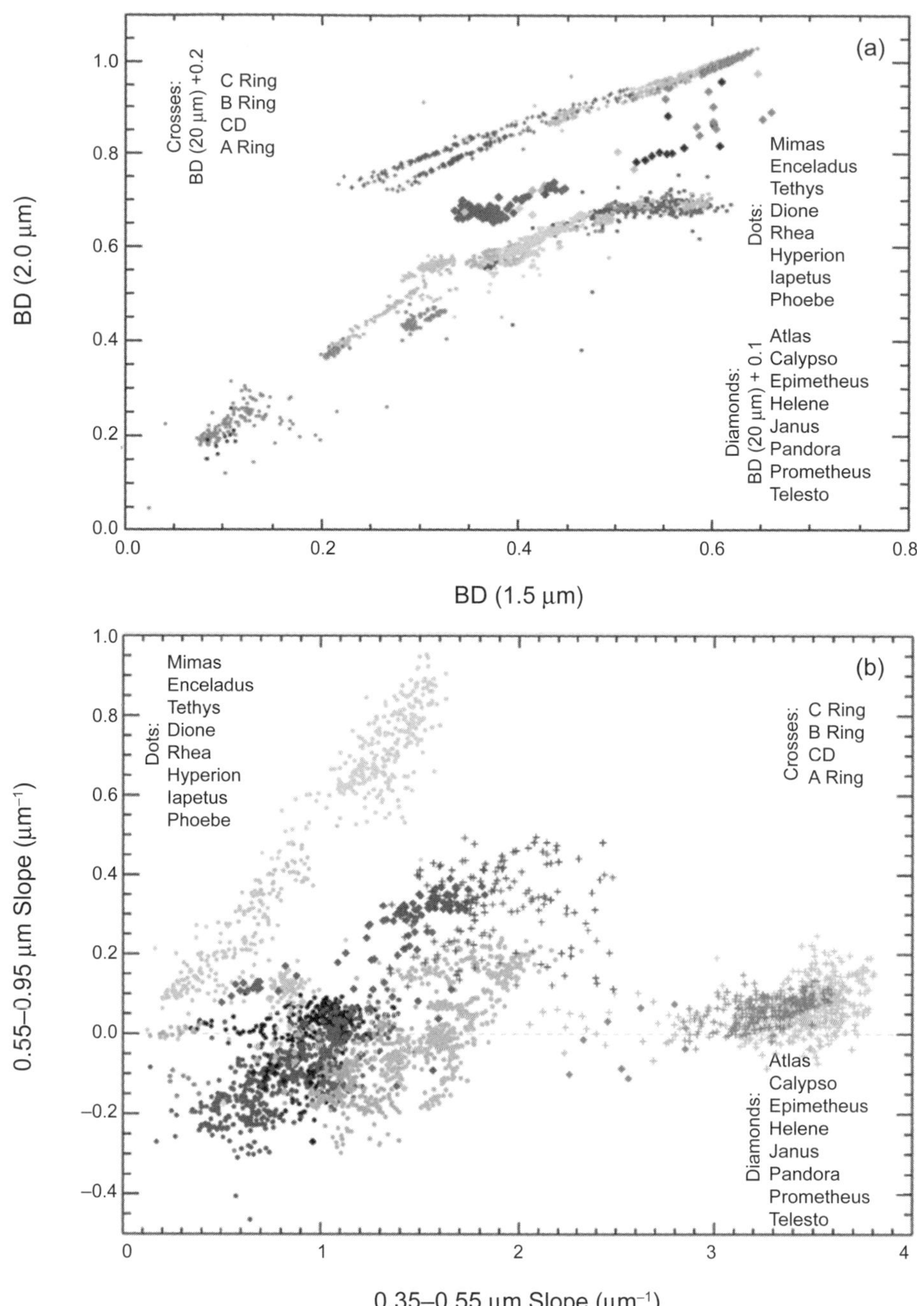

Fig. 4. See Plate 33 for color version. **(a)** Water-ice band-depths of the moons of Saturn and its rings, showing the trends with distance from Saturn. **(b)** Spectral slopes of the moons of Saturn and its rings, showing the trends with distance from Enceladus. After *Filacchione et al.* (2012).

colored solids were an additional result of the well-known production of amino acids in a simulated primitive terrestrial atmosphere. The Titan tholin produced in the works cited and in several other laboratories is widely regarded as comprising Titan's atmospheric aerosol particles sampled by the Cassini INMS instrument (*Waite et al.*, 2009a) and the material precipitated to the surface that contributes to the yellow-brown coloration of the region observed by the Huygens probe (*Karkoschka and Schroder,* 2016). The exact composition of the tholin in Titan's atmosphere and on the surface, the chemical pathway(s) by which it forms, and its relevance to natural processes are subjects of continuing discussion and laboratory research.

Tholins can be created in ices as well as in gases, provided that carbon is present; *Thompson et al.* (1987) produced colored solids by irradiation of a mixture of CH_4 and H_2O ice, with the degree of coloration and the albedo of the solid dependent on radiation dose. Yellow- and red-colored organic refractory residues were created and analyzed from a mixture of $N_2 + CH_4 + CO$ (100:1:1) irradiated both with UV photons and 1.2-keV electrons in separate experiments (*Materese et al.* 2014, 2015). In the latter experiments there is a high degree of nitrogen incorporation into the resulting residues (N/C in the range 0.5–0.9). There have been a few other tholin experiments in ices (e.g., *McDonald et al.,* 1996).

A type of tholin or complex organic species has been supported by studies as the darkening/reddening agent on the surfaces of the inner moons of Saturn, especially Rhea, Tethys and Dione (e.g., *Noll et al.,* 1997; *Hendrix et al.,* 2010, 2017, 2018). *Filacchione et al.* (2012) has suggested that the source of the non-ice contaminant(s) is the main rings of Saturn, a chromophore whose reddening effects are offset by the bluing effect of E-ring grains (e.g., at Enceladus). [Reddish color is also found in Saturn's rings; tholins have been proposed as the source, as well as polycyclic aromatic hydrocarbons (PAH) and nanophase iron or iron oxide (*Cuzzi et al.,* 2009).] *Hendrix et al.* (2018) showed that the amount of reddening generally increases with distance from Enceladus; on the trailing hemispheres the reddening is correlated with the electron intensity, and Hendrix et al. pointed out that radiolytic processing of organics causes the products to become spectrally redder. They suggested that such processing occurs while organic-rich grains are in the E ring (*Postberg et al.,* 2008), and increases with exposure time in the E ring, such that grains interacting with Rhea are redder (more processed) than those impacting moons closer to Enceladus.

2.2.2. Other non-ice species. A CO_2 absorption at 4.25 μm is present in VIMS data of all of the inner icy moons (*Clark et al.,* 2008) and has been shown to appear primarily on the trailing hemispheres of Dione (*Stephan et al.,* 2010) and Rhea (*Stephan et al.,* 2012), attributed to an exogenic process such as plasma bombardment. Furthermore, VIMS data of Dione show a pattern of bombardment of fine, sub-0.5-μm diameter particles impacting the satellite from the trailing side direction (*Clark et al.,* 2008). *Clark et al.* (2012) report that this dark material "is composed of metallic iron, nano-size iron oxide (hematite), CO_2, H_2O ice, and possible signatures of ammonia, bound water, H_2 or OH-bearing minerals, trace organics, and as yet unidentified materials."

Another absorption feature is found on the satellites, centered near 0.26 μm. *Noll et al.* (1997) first observed this feature in HST observations of Rhea; the feature was attributed to radiolytically-produced ozone (O_3). Further HST observations confirmed the presence of this feature at nearly all satellites in the inner system, on both leading and trailing hemispheres but slightly stronger on the trailing hemispheres of Tethys, Dione, and Rhea (particularly strong on the Tethys trailing hemisphere); *Hendrix et al.* (2017) suggested it could be related to organics such as those found in the E ring (*Postberg et al.,* 2008) and point out that benzene, naphthalene, and other PAHs exhibit absorptions near 0.26 μm. They suggested that cold plasma bombardment of E-ring-derived organics may contribute to radiolytic production of the 0.26-μm absorption, along with the dark, reddish species.

Consistent with an increase in non-ice contaminant away from Enceladus, the water-ice band depths at 1.5 and 2.0 μm tend to decrease with increasing distance from Enceladus (*Filacchione et al.,* 2012, 2013). As discussed above, the suppression of the depth of the water-ice bands in the dark terrains is the result of the presence of contaminants mixed with the water ice, and of the reduced size of the ice grains. Along these lines, the water-ice band depth maps of Dione and Rhea (Fig. 3) show decreased band depths on the trailing hemispheres of these bodies, consistent with lower abundances of pure water ice, increased contaminant abundances, and/or smaller grain sizes. The smaller grain sizes may be due to effects of charged particle bombardment on the trailing hemispheres of those moons (*Verbiscer et al.,* 2006), although sputtering has been found to preferentially remove small grains (*Cassidy et al.,* 2013). At Mimas, the VIMS-measured water-ice band depths are slightly shallower on the trailing hemisphere than on the leading hemisphere, which could be consistent with preferential deposition of fine-grained E-ring particles there and/or energetic electron annealing of leading hemisphere grains, and/or radiolytic processing of the trailing hemisphere grains by cold plasma (*Hendrix et al.,* 2012).

Perhaps the most unusual and unique features discovered during the Cassini mission are the red streaks on Tethys (*Schenk,* 2015). Up to 250 km long and only 3–4 km wide, they seem to form large arcs across the surface and are unlike any features found on other icy moons, and their origin remains unknown. Comparison of VIMS spectra between the red streaks and surrounding regions shows substantial differences, with deeper water-ice bands at 1.6 and 2.0 μm, implying some type of fresh processing (*Buratti et al.,* 2018), consistent with the observation that the streaks appear to be superimposed on underlying geologic features. There is also evidence in the VIMS spectra for organic compounds in the region of the red streaks (*Buratti et al.,* 2018). Alternatively, the spectral signatures of the streaks would be due to particle-size effects, or to the recent sublimation of water ice (*Brown et al.,* 2012), which would also imply some type of recent or ongoing activity.

Finally, ammonia (or ammonia hydrate) may be present in small amounts (~1%) on the surface of Enceladus as measured using groundbased near-IR (2.235- and 2.21-μm features) and UV spectroscopy (*Emery et al.,* 2005; *Verbiscer et al.,* 2006; *Hendrix et al.,* 2010), and may also be present on the Tethys trailing hemisphere (*Verbiscer et al.,* 2008; *Hendrix et al.,* 2015). These features have not been detected in Cassini VIMS spectra; however, VIMS may have found trace signatures of ammonia as the N-H stretch fundamental at 2.97 μm (*Clark et al.,* 2008, 2012). It is likely that the NH_3 in the Enceladus plume (*Waite et al.,* 2009b) is continually deposited onto the surfaces of Enceladus and perhaps Tethys.

2.3. Seasonal Variations

For a planet in a circular orbit, an inclined spin axis with respect to the ecliptic can cause seasons and substantial surface volatile transport. Similarly, seasons can occur on a moon if it also possesses a tilted axis with respect to incoming solar radiation. Indirect evidence for the possible seasonal changes on the surface of Dione and Rhea is provided by observation of their exospheres (*Teolis and Waite,* 2016). Both CO_2 and O_2 have been observed to produce seasonal exospheres that are consistent with sublimation or desorption at the moons' polar regions during the equinoxes and condensation or absorption with advancing winter (see the chapter by Teolis et al. in this volume). A further seasonal variation is found at Mimas and Tethys, where Cassini UVIS observes the likely presence of photolytically-produced hydrogen peroxide (H_2O_2) (*Hendrix et al.,* 2012). The peroxide, an efficient UV absorber, is readily produced via photolysis and was proposed to explain the latitudinal variation in brightness observed at far-UV wavelengths on Mimas; during southern summer the southern hemisphere was observed to be UV-darker than the northern hemisphere. A similar observation was made at Tethys; later in the Tethys year, as the Sun moved northward, the northern hemisphere became darker and the southern hemisphere became brighter, consistent with photolytically-produced H_2O_2 in the summer hemisphere.

2.4. Summary of Each Inner Mid-Sized Moon

2.4.1. Mimas. Water ice is the dominant species, with smaller grains found on the trailing hemisphere, consistent with higher fluxes of E-ring grains there; it may also be consistent with cold plasma radiolytic processing, creating contaminants in the trailing hemisphere ice, and/or annealing of the leading hemisphere grains due to energetic electron bombardment. The visible spectrum is slightly redder than at Enceladus, although with no strong leading-trailing asymmetry in absorber. Photolytically-produced hydrogen peroxide may be present in the top layers of the regolith, and appears to be seasonally variable.

2.4.2. Tethys. Water ice is the dominant species, with deeper H_2O absorption bands on the leading hemisphere, which could be due to emplacement of water-ice-rich E-ring grains there. A strong leading-trailing hemisphere asymmetry in brightness and color is present, with the trailing hemisphere being dominated by reddish material that may be related to cold plasma bombardment. Ammonia and/or ammonia hydrate may be present in small amounts on the trailing hemisphere, and may have the Enceladus plume as a source. Photolytically-produced hydrogen peroxide may be present in the top layers of the regolith, and appears to be seasonally variable. The origin of the bizarre red streaks remains inconclusive.

2.4.3. Dione. Water ice is the dominant species, with deeper H_2O absorption bands on the leading hemisphere, which could be due to emplacement of water-ice-rich E-ring grains there. CO_2 is present in the dark regions, perhaps radiolytically produced. There may be infall of dark material on the trailing hemisphere from an unknown source. The strong leading-trailing hemisphere albedo dichotomy is likely due to a combination of E-ring grain emplacement/bombardment on the leading hemisphere and radiolytic darkening on the trailing hemisphere. Relatively fresh water ice is found at fresh craters and tectonic cracks. A 0.26-μm absorption feature, present at all of the satellites, is the strongest in the system at Dione and Rhea, particularly on the trailing hemisphere.

2.4.4. Rhea. Water-ice band depths are found to be stronger on the leading hemisphere (where E-ring grain bombardment dominates) and in tectonic cracks. The leading-trailing hemisphere albedo dichotomy is likely due to a combination of E-ring grain emplacement/bombardment on the leading hemisphere and radiolytic darkening on the trailing hemisphere. Relatively fresh water ice is found at fresh craters and tectonic cracks.

3. OUTER SYSTEM SATELLITES: HYPERION, IAPETUS, PHOEBE

Hyperion, Iapetus, and Phoebe have long been known to exhibit different spectral and albedo properties from the inner moons of Saturn discussed above. Their environments are different from the inner moons, as they orbit Saturn in regions unaffected by the E ring and in a much more benign magnetospheric environment. Phoebe, with its distant retrograde orbit, has long been thought to be a captured body, likely from the Kuiper belt (*Johnson and Lunine,* 2005), and orbits well outside Saturn's magnetosphere, in the solar wind. Iapetus, with its famous hemispheric albedo dichotomy, also orbits in the solar wind. Hyperion, known for its chaotic rotation, reddish appearance and non-spherical shape, spends part of its orbit within Saturn's magnetosphere and part of the time in the solar wind. The Phoebe ring (*Verbiscer et al.,* 2009), an enormous swath of dust emanating from the retrograde moon, has a significant impact on the surface compositions of Iapetus and Hyperion.

3.1. Compositional Clues from Ultraviolet and Visible-Near-Infrared Spectroscopy

Since its discovery in 1899, Phoebe's great distance from Saturn (215 R_S) and highly inclined orbit (i = 173°) suggested that it was a captured object (*Pollack et al.,* 1979; *Ćuk and Burns,* 2004). Scientific curiosity centered around the question of Phoebe's origin: Is it an errant asteroid, or was it formed deep in the outer solar system? Prior to the Cassini flyby of Phoebe, Voyager and groundbased observations revealed that Phoebe was primarily composed of dark carbonaceous material, leading to comparisons with C-type asteroids (*Degewij et al.,* 1980; *Tholen and Zellner,* 1983; *Jarvis et al.,* 1997). However, several lines of evidence pointed to an origin for Phoebe in the distant outer solar system. With the detection of water ice on the surface (*Owen et al.,* 1999), the main asteroid belt was largely ruled out as a source for Phoebe. *Johnson and Lunine* (2005) showed that density measurements of Phoebe from Cassini are consistent with a composition similar to that measured in Kuiper belt objects.

Phoebe's surface composition is dominated by crystalline H_2O and CO_2 ices in addition to a substantial amount of low-albedo material, presumably consisting of a combination of rock and organic-bearing components. Cassini VIMS measured ferrous-iron-bearing minerals, bound water, trapped CO_2, probable phyllosilicates, organics, and nitriles (*Clark et al.,* 2005), while Cassini UVIS observed water ice mixed with a dark material of a likely carbonaceous nature (*Hendrix and Hansen,* 2008a). The presence of nitriles was later discounted using improvements in the VIMS calibration (*Clark et al.,* 2008, 2012); the "nitrile" band at 2.42 μm was shown to be due to trapped H_2 (*Clark et al.,* 2012). Details on the organic component of Phoebe's surface composition are discussed in the next section. An extremely fine-grained material, possibly nanophase iron or iron oxide, may contribute to an upturn in the spectral reflectance at wavelengths shorter than about 0.7 μm, attributed to Rayleigh scattering. The CO_2 is trapped primarily in the dark material, but can also be complexed, possibly as a clathrate, with H_2O. The $^{13}CO_2$ band at 4.367 μm on Phoebe is strong in comparison with other satellites where CO_2 is present.

Iapetus has intrigued planetary scientists for centuries, primarily due to its striking hemispheric albedo dichotomy. The leading hemisphere (centered on 90°W) is very dark, reflecting just ~4% of the visible light that hits it, while the trailing hemisphere (centered on 270°W), is relatively quite bright and has a visible albedo of ~60% (*Squyres et al.,* 1984). The signature of water ice on the trailing hemisphere was first noted by *Fink et al.* (1976). A long-standing question has been whether Iapetus' leading hemisphere dark terrain was created through exogenic processes (*Cook and Franklin,* 1970), or whether geologic activity emplaced dark material from within Iapetus (*Smith et al.,* 1981, 1982). Voyager images of dark-floored craters within the bright terrain pointed to an endogenic source; they also suggested that the bright-dark boundary is too irregular to be consistent with infalling dust (*Smith et al.,* 1981, 1982). Researchers theorized (*Soter,* 1974) that the dark material is exogenically emplaced on Iapetus' leading hemisphere as material is lost from the moon Phoebe (*Burns et al.,* 1979). Retrograde Phoebe dust from 215 R_S would travel inward and impact the leading hemisphere of Iapetus, orbiting at 59 R_S. However, Phoebe is spectrally gray at visible wavelengths, while the Iapetus dark material is reddish (*Cruikshank et al.,* 1983; *Squyres et al.,* 1984). Phoebe material coating Iapetus was thought to undergo some sort of chemistry or impact volatilization to change the color and darken the material (*Cruikshank et al.,* 1983; *Buratti and Mosher,* 1995). Furthermore, Iapetus' dark, red spectrum is suggestive of primitive, unaltered material (e.g., *Jarvis et al.,* 2000); the relative freshness of Iapetus' dark, red material was noted to be consistent with the idea that organic-containing surfaces exposed to radiation eventually become darker and blacker (*Andronico et al.,* 1987).

The exact composition of Iapetus' dark material was studied using groundbased observations prior to Cassini; the presence of a deep 3.0-μm absorption feature led to comparisons with C-type asteroids (*Lebofsky et al.,* 1982) and primitive meteorite-type material (*Bell et al.,* 1984), while its red visible–near-infrared (VNIR) spectrum has been compared with organic material (*Cruikshank et al.,* 1983; *Vilas et al.,* 1996). Groundbased Iapetus data at VNIR wavelengths (e.g., *Cruikshank et al.,* 1983; *Vilas et al.,* 1996; *Jarvis et al.,* 2000; *Buratti et al.,* 2002) and longer IR wavelengths (*Owen et al.,* 2001) were used in spectral mixture models and compared with asteroids and Phoebe spectra in attempts to understand the surface composition. Models including water ice, amorphous carbon, clay silicates, and tholins provided adequate fits. *Vilas et al.* (1996) pointed out the existence of a 0.67-μm absorption feature, likely indicating aqueously-altered silicates in the dark material; VIMS later also observed the 0.67-μm absorption and analyses found it to be consistent with nanophase iron oxide (*Clark et al.,* 2012).

Groundbased radar observations at 13 cm (*Black et al.,* 2004) and Cassini radar data at 2.2 cm (*Ostro et al.,* 2006) indicate that the dark terrain must be quite thin (one to several decimeters); an ammonia–water-ice mixture may be present below several decimeters of the surface on both the leading and trailing hemispheres of Iapetus. The radar results appear to rule out the theories of a thick dark material layer (*Matthews,* 1992; *Wilson and Sagan,* 1996).

Cassini and Spitzer Space Telescope observations helped to solve many of the mysteries about Iapetus. Albedo patterns observed by Cassini cameras in late 2004 suggested external emplacement of material (e.g., dark material on ram-facing crater walls at high latitudes) (*Porco et al.,* 2005), consistent with an external source of dark material. Cassini UVIS results (*Hendrix and Hansen,* 2008b) suggested that the darkening process may be a recent or ongoing activity. UVIS spectra of the lowest latitudes of the dark terrain display the diagnostic UV water-ice absorption feature; water-ice amounts increase within the dark material away from the apex (at 90°W longitude, the center of the dark leading hemisphere), consistent with thermal segregation of water ice. The water ice in the darkest, warmest low-latitude regions is not expected to be stable and was noted as a sign of ongoing or recent emplacement of the dark material from an exogenic source (*Hendrix and Hansen,* 2008b). This is also supported by images of large basins and craters in the dark terrain, but with no bright material exposed (*Porco et al.,* 2005), along with the radar evidence that the dark material is a thin coating.

Initial analysis of Cassini VIMS data of Iapetus (*Buratti et al.,* 2005) showed that carbon dioxide was present in the low-albedo material, probably as a photochemically produced molecule trapped in H_2O ice or in some mineral or complex organic solid. The spectrum of the low-albedo hemisphere was modeled with a combination of organic tholin, poly-HCN, and small amounts of H_2O ice and Fe_2O_3. The high-albedo hemisphere was modeled with H_2O ice slightly darkened with tholin. The detection of CO_2 in the low-albedo material on the leading hemisphere supports the contention that it is carbon-bearing material from an external source that has been swept up by the satellite's orbital motion. Further compositional mapping and radiative transfer modeling of VIMS data (*Clark et al.,* 2012) showed that the dark material is composed of metallic iron, nano-sized iron oxide (hematite), CO_2, H_2O ice, and possible signatures of ammonia, bound water, H_2- or OH-bearing minerals, trace organics, and as yet unidentified materials. The CO_2 band depth increased in strength from the leading side apex to the transition zone to the icy trailing side. A Rayleigh scattering peak in the visible part of the spectrum indicates the dark material has a large component of fine, sub-0.5-μm-diameter particles consistent with nanophase hematite and nanophase iron. $^{13}CO_2$ is detected on Iapetus, but is weaker than on Phoebe.

The discovery of the Phoebe ring using Spitzer Space Telescope observations (*Verbiscer et al.,* 2009), a large envelope of dust emanating from Phoebe, provided the missing link between Phoebe and Iapetus. (Further discussion of the relationship between Phoebe and Iapetus is included in the following section.) *Spencer and Denk* (2010) showed that, in addition to an exogenic source of dust, thermal migration of volatiles is critical for determining the large-scale and small-scale albedo patterns on Iapetus. Temperatures in the dark material, given the slow rotation of Iapetus (and thus long daytime), increase to the point where volatile materials are unstable and migrate to colder regions. Volatiles become cold-trapped in locally bright and cold regions, such as the polar regions of the leading hemisphere.

Hyperion is spectrally red at visible wavelengths, similar to D-type asteroids, and has been compared with organic material (*Jarvis et al.,* 2000). *Buratti et al.* (2002) found that Hyperion and Iapetus dark material can be fit with similar models using reddish, D-type material, similar to the result obtained by *Jarvis et al.* (2000). Water ice is also present at the surface, as first identified in groundbased observations (*Cruikshank et al.,* 1980) and later further studied using far-UV and infrared Cassini observations (e.g., *Cruikshank et al.,* 2007). Hyperion's chaotic rotation, irregular shape, and very low bulk density, as well as the pattern of geologic structures on its surface, challenge theories of its origin and evolution. Much of the surface is covered with crater-like depressions that have been sculpted by the evaporation of the H_2O ice that dominates the infrared spectral properties, forming surface features termed suncups. CO_2 ice or frost is found, appearing to favor the elevated rims of the suncups, possibly representing condensation of CO_2 evaporating from warmer regions. Many of the suncups have deposits of dark material on their floors, thought to be accreted dust from the Phoebe ring that was not swept up by Iapetus. A hydrocarbon spectral signature is found in both the high- and low-albedo material across Hyperion's surface.

Although organics have been invoked to explain the reddish nature of the Iapetus dark material at visible wavelengths, *Clark et al.* (2012) showed how differing amounts of dust mixed in the ice can explain both the visible wavelength spectra of Phoebe and Iapetus using the nano-iron plus nano-iron oxide model. This model can also include small amounts of organics and, as discussed in section 2, can also explain the reddish nature of regions of the inner moons of Saturn. On Iapetus, the abundance of the dark red material is much higher than on the inner icy moons of Saturn. *Clark et al.* (2012) found that a carbon-tholin model is not compatible with spectra of the dark material on Iapetus and noted further that (1) spectra of tholins produced in terrestrial laboratories have a 3-μm absorption shifted from the position observed in VIMS spectra of Iapetus and the tholin spectra have different shapes, and (2) known tholins show near-infrared absorption features not seen in spectra of Iapetus. Furthermore, *Clark et al.* (2012) showed that the spectral mixtures that match Iapetus features mimic observed spectral features of dark material in the inner Saturn system, including Dione (which has the strongest signatures of dark material on the inner satellites) and in the Cassini Division. This suggests that the reddening agents in the outer and inner Saturn system could have similar compositions. *Clark et al.* (2012) modeled the spectrum of Hyperion, showing that it too could be explained with the nano-iron/nano-iron oxide model.

3.2. Compositional Clues from Cassini Infrared Spectroscopy (2.5–5 μm): Organic Molecules

The spectral region accessible to groundbased telescopes and its extension to about 5 μm afforded by spacecraft instrumentation such as Cassini VIMS contains absorption bands diagnostic of certain classes of organic molecules. The spectral region 2.5–5 μm encompasses the stretching fundamentals of C-H in both aliphatic and aromatic molecules at approximately 3.38–3.51 μm and 3.25–3.30 μm, respectively (see tables in *Cruikshank et al.,* 2014), as well as other carbon-bearing molecules of interest such as CO, CO_2, HCN, and alcohols (specifically CH_3OH methanol). While absorption bands of CH_4, C_2H_6, and CH_3OH have been found in the spectra of several small, icy bodies in the solar system, they have not yet been detected on the satellites or rings of Saturn. However, the stretching modes of C-H in aromatic and aliphatic hydrocarbons have been found in the Cassini VIMS spectra of Phoebe, Iapetus, and Hyperion (*Clark et al.,* 2005; *Dalle Ore et al.,* 2012; *Cruikshank et al.,* 2014) and in the C ring and Cassini Division (*Filacchione et al.,* 2014). The analysis of the spectrum of Iapetus by *Cruikshank et al.* (2014) has been the most extensive in the exploration of the aromatic and aliphatic hydrocarbon bands, but as noted below is in the process of revision because of improved processing of the VIMS data upon which it was based.

The stretching fundamentals of C-H in both aliphatic and aromatic molecules at approximately 3.51–3.38 μm and 3.25–3.30 μm, respectively, occur on a region of the spectra that is steeply sloped upward toward longer wavelengths. The slope arises from the presence of H_2O ice and possible OH in other constituents of the surface materials. The interpretation of the organic molecules on Iapetus depends critically on the fitting of this slope and the extraction of the very weak residuals attributed to hydrocarbon bands. The fitting can be done with a model that accounts for the H_2O spectrum, but the model is sensitive to the complex refractive indices of the ice, which are in turn sensitive to the temperature of the ice. A second, more subjective technique is to apply a multi-point spline across the 3.5–3.8-μm spectral region and then extract the residuals. Both techniques have been applied, and they achieve the same basic result, showing residuals that are on the order of several σ above the intrinsic noise in the data, although the depths and widths of the residuals are different.

Figure 5a shows the smoothed and normalized spectrum of the low-albedo region of the leading hemisphere of Iapetus derived from data taken at the 2007 flyby of the satellite and processed with data reduction pipeline (RC19) of the VIMS data that is superior to previous pipeline versions. In Fig. 5b, the absorption profile is fitted with a suite of gaussian curves representing absorption by both aromatic and aliphatic C-H functional groups. The central wavelengths of the absorption bands of aromatic C-H, and aliphatic $-CH_2-$ (symmetric and asymmetric) and $-CH_3$ (symmetric and asymmetric), are taken from the literature of molecular theory and laboratory spectroscopy (*Ricca et al.,* 2012; *Wexler,* 1967; *Pendleton et al.,* 1994). An additional gaussian representing methanol (CH_3OH) is included to improve the fit, although methanol, while a plausible component, has not been identified by a discrete band in the spectrum of Iapetus.

As in previous analyses of this and similar VIMS datasets for Iapetus, the aromatic band at 3.28 μm is prominent, and the envelope of absorption attributed to the aliphatic bands (3.35–3.6 μm) is clearly seen. An estimate of the abundance of aromatic C-H relative to aliphatic C-H can be made from the strengths of the individual components of the bands defined by the gaussian curves representing the functional groups, with knowledge of the intrinsic strengths (A values) derived from the references cited above and given in *Cruikshank et al.* (2014). In the new reduction of the Iapetus spectrum presented here, the ratio of aromatic C-H to aliphatic C-H is on the order of 1, indicating the presence of a significant component of aromatic rings, possibly in the form of PAHs. Alternatively, the aromatic rings may be single or clusters of two or three rings, linked together in a network of aliphatic chains, comparable to the notional structure of organic molecular assemblages in the atmosphere of Titan (Fig. 4 of *Lebreton et al.,* 2009) and the interstellar medium (*Kwok and Zhang,* 2011).

The prominence of the aromatic band is notable because it is seen in VIMS spectra of Phoebe and Hyperion (*Clark et al.,* 2005; *Dalle Ore et al.,* 2012; *Cruikshank et al.,* 2014), but is rare in other solar system materials. The appearance of this band links the organic-bearing material among these three Saturn satellites, and appears to be fully consistent with the scenario in which dust particles removed from Phoebe coat the leading hemisphere of Iapetus as the dust spirals inward toward Saturn by Poynting-Robertson drag. The discovery of the dust ring in Phoebe's orbit (*Verbiscer et al.,* 2009) and an elucidation of the dynamical means for the transport of that dust inward toward Saturn (*Tamayo et al.,* 2011) argue strongly for Phoebe as the source of the hydrocarbon-bearing dust. The color of Phoebe is different from that of Iapetus, but this can be explained by the idea that impacts remove dust from the *interior* of Phoebe, forming the Phoebe ring, whereas the *surface* of Phoebe is exposed to the space environment and processes of space weathering, carbonizing the surface (e.g., *Strazzulla,* 1986).

A complication to this scenario lies in the color difference of the low-albedo materials on the leading and trailing hemispheres of Iapetus (*Denk et al.,* 2010). This difference may result from the accretion of two different kinds of dust, and possibly two or more different epochs of dust accretion, or as noted above, the effects of thermal migration of ice in the scenario developed by *Spencer and Denk* (2010). As *Tamayo et al.* (2011) have noted, dust removed from small, irregular satellites of Saturn other than Phoebe also spirals inward toward Saturn, and may help explain the color difference. The difference in color might be explained by different abundances of ice and nanophase iron, as well as grain-size distribution. As seen in the earlier analysis

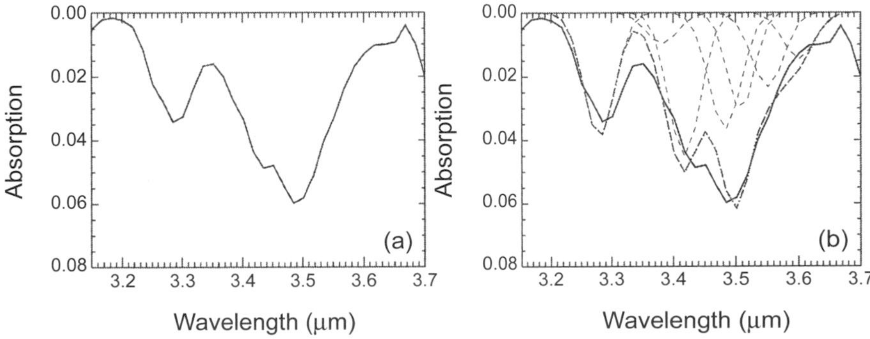

Fig. 5. VIMS spectra of Iapetus dark material. The modeled continuum has been divided out (for model details, see *Cruikshank et al.,* 2014). **(a)** Absorption bands in the spectrum of the low-albedo material of Iapetus, 3.15–3.65 μm, smoothed and normalized. **(b)** The Iapetus absorption profile is fitted with a suite of gaussian curves centered at 3.280 μm (aromatic C-H), 3.381 μm (aliphatic $-CH_3$ asymmetric), 3.418 μm (aliphatic $-CH_2-$ asymmetric), 3.483 μm (aliphatic $-CH_3$ symmetric), 3.509 μm (aliphatic $-CH_2-$ symmetric), and 3.537 μm (CH_3OH). The heavy dashed line shows the sum of the gaussians that give the best fit to the Iapetus spectrum using the modified gaussian method (MGM) of *Sunshine et al.* (1993). The intrinsic absorption strength (A value) of the aromatic band is ~8–10× weaker than the equivalent for the aliphatic bands, resulting in similar abundances of aromatic and aliphatic C-H groups.

by *Cruikshank et al.* (2014), both varieties of low-albedo material exhibit an aromatic hydrocarbon spectral signature. The variety of dust favored for the leading hemisphere is the dust precipitated from the Phoebe ring, a process that appears to be ongoing, as determined by its shallow depth of a few tens of centimeters (*Ostro et al.,* 2006), and the paucity of impact craters that penetrate the accreted layer to reveal the icy bedrock below.

In summary, we note that aromatic hydrocarbon is present in detectable amounts on Phoebe, Iapetus, and Hyperion, and a new analysis of the abundance of C-H bonds on Iapetus indicates that the ratio of aromatic to aliphatic bonds is about 1. This is a revision to the abundance published by *Cruikshank et al.* (2014); a new analysis of the Phoebe and Hyperion abundance is in progress. The detectability of the aromatic material in proportion to the aliphatic component is in contrast to the much smaller abundance ratio determined from spectra of this type for both interstellar dust and most solar system materials observed telescopically or measured in the laboratory from meteorite, comet, and interplanetary dust samples (*Cruikshank et al.,* 2014).

4. SMALL MOONS

4.1. Inner Moons and Ring Moons

The small inner moons of Saturn represent a variety of worlds, from highly battered objects such as the co-orbital satellites Janus and Epimetheus, to smooth Methone, to Atlas encircled with an apron. Helene, which is nearly spherical, is covered with intriguing branching patterns that indicate some type of downslope transport. These moons offer important clues to the evolution of the saturnian system, including collisional processes and history, the origin of the ring systems, and exogenous alteration by Saturn's magnetosphere and dust environment. Two moons — Pan and Daphnis, as well as several moonlets — are embedded in Saturn's main ring system. The geology and dynamics of these moons is discussed in *Thomas et al.* (2013); see also the chapter by Thomas et al. in this volume.

Eleven of the small inner moons of Saturn — Calypso, Atlas, Epimetheus, Helene, Janus, Methone, Pallene, Pan, Pandora, Prometheus, and Telesto — were observed by Cassini data at sufficient signal to noise to allow VIMS spectra to be extracted. So far spectra have been analyzed for all these objects except Methone (*Buratti et al.,* 2010; *Filacchione et al.,* 2012). Figure 6 is a composite of Imaging Science Subsystem (ISS) images of these six moons, and Table 1 is a summary of their physical properties and distance from Saturn. All these moons are in synchronous rotation.

Figure 7 shows spectra for these six moons from VIMS (*Buratti et al.,* 2010), supplemented with ISS data in the 0.35–0.9-μm region. None of the small moons shows absorption bands other than those for water ice. However, their spectra show evidence for darker contaminants, and they have been modeled with organic-rich material including Triton tholin, Iapetus low-albedo material, and Murchison meteorite, a CM2-type primitive object (*Buratti et al.,* 2010). The moons exhibit important color differences among themselves. The inner "shepherd" moons, Atlas and Pandora, tend to be redder than the co-orbitals and the Tethys Lagrangians in the visible, possibly due to contamination by the ring system of Saturn (*Filacchione et al.,* 2012) ("shepherd" is a convenient but obsolete term, as we now know the edges of the rings are maintained by resonances). These moons appear to possess the same reddish material observed on the medium-sized moons (*Buratti et al.,* 2010; *Filacchione et al.,* 2010; *Thomas et al.,* 2013) (see Fig. 7). The small moons more distant from Saturn are bluer in the visible, which could be due to contamination by the E ring (*Filacchione et al.,* 2012).

The moons may also exhibit color changes between their leading and trailing sides, as the main moons do (*Noland et al.,* 1974; *Buratti et al.,* 1990). However, the bottom cell on Fig. 7 shows that Calypso, which was measured on its leading side, is redder than Telesto, which was measured on its trailing side. Recent additional measurements gathered by the Cassini spacecraft but not yet analyzed may resolve the question of hemispheric color differences on these small moons.

The small inner moons also show color and possible compositional differences within themselves. A few have sufficiently resolved morphology to correlate geologic units with spectral differences. The upper and lower surface areas of Helene exhibit distinctly different colors. The low-lying unit, which is filled with downsloped material, is substantially bluer, with an ISS IR3/UV3 (0.93 μm/0.34 μm) ratio that is almost 15% lower than the higher-elevation materials (*Thomas et al.,* 2013). One explanation for this color difference is that the lower particles are smaller. Telesto and Calypso also show color changes in the visible and near-IR, but they are not as distinct as those of Helene.

Fig. 6. The six inner saturnian moons for which VIMS spectra between 0.40 and 5.1 μm have been derived. The images were obtained with the Cassini ISS Camera. The sizes are not to scale; radii are listed in Table 1. Individual images courtesy of NASA/JPL-Caltech/Space Science Institute.

4.2. Outer Irregular Moons

A large family of small irregular moons of Saturn orbits the planet at great distances (~0.1 AU), falling into four dynamical families. Very little is known about the composition of these objects, except that it must be diverse. Their colors range from gray to red, although they are not as a red as Kuiper belt objects, which have V-R indices approaching 0.9, compared to a maximum of 0.65 for the saturnian moons (*Jewitt and Haghighipur,* 2007; *Grav and Bauer,* 2007). *Grav and Bauer* (2007) showed that the colors of the dynamical families were correlated in at least three of four cases, offering further proof that the families may have originated from a common body.

TABLE 1. Physical properties of the inner small moons with published spectra.

Name	Mean Distance from Saturn (km)	Mean Radius (km)	Density (kg m^{-3})	Visible Albedo (10° phase)	Type
Atlas	137,774	14.9 ± 0.8	473 ± 83	0.49 ± 0.08	Shepherd*
Pandora	141,810	40.6 ± 1.5	487 ± 56	0.62 ± 0.08	Shepherd*
Janus	151,450	89.0 ± 0.4	643 ± 9	0.37 ± 0.03	Coorbital
Epimetheus	151,450	58.2 ± 0.7	639 ± 22	0.41 ± 0.04	Coorbital
Telesto	294,720	12.2 ± 0.3		0.61 ± 0.04	Tethys Lagrangian
Calypso	294,721	9.5 ± 0.6		0.69 ± 0.14	Tethys Lagrangian

* Shepherd nomenclature is historical; ring edges are caused by resonances rather than by shepherding by adjacent moons.

Data from the JPL horizon database (*http://ssd.jpl.nasa.gov/?satellites*) and the chapter by Thomas et al. in this volume.

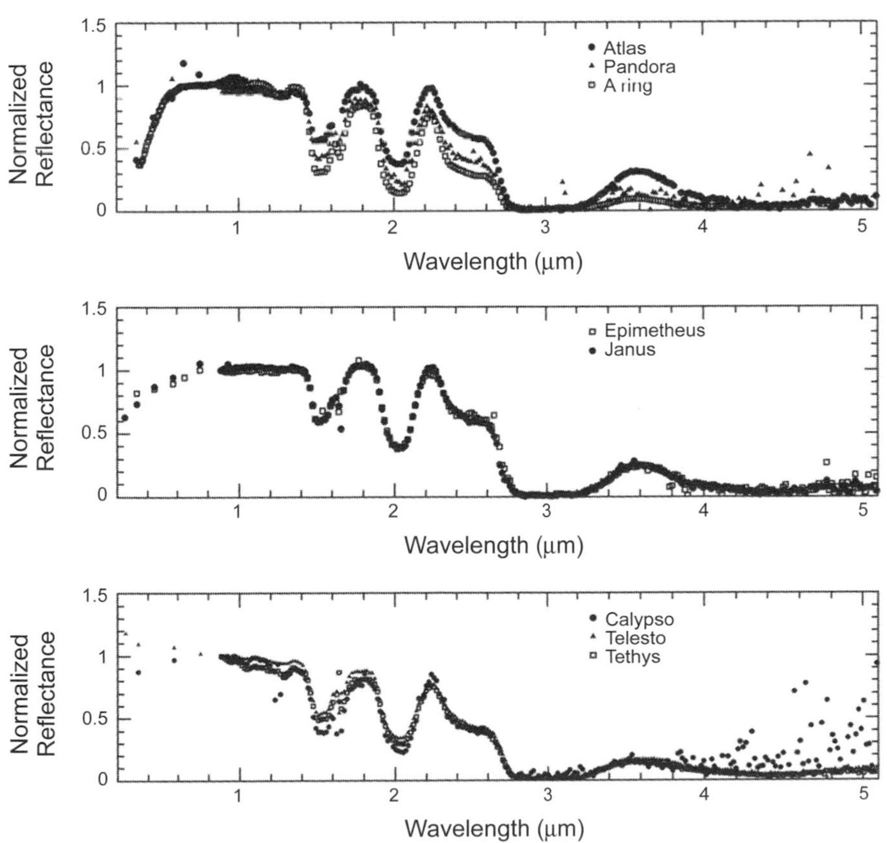

Fig. 7. Spectra of six small inner saturnian satellites based on VIMS and ISS data. The most inner moons, Atlas and Pandora, are shown with the A ring of Saturn to illustrate that their spectra are redder shortward of 1 μm, and very similar to the spectrum of the outer (A) ring of Saturn in this spectral region. The spectra of the coorbitals Janus and Epimetheus are very similar. The spectra of the Lagrangians of Tethys, Calypso and Telesto, are very similar to that of Tethys. However, Calypso is redder in the visible region of the spectrum, possible because its spectrum represents the trailing side while Telesto's leading hemisphere is shown. The scatter in the small moons' spectra is greater at wavelengths beyond 4 μm. Figure adapted from *Buratti et al.* (2010).

5. FUTURE AREAS OF STUDY/OPEN QUESTIONS

Several issues surrounding the surface compositions of Saturn's icy moons remain in the post-Cassini era. While some species have been detected definitively (e.g., H_2O, CO_2), some infrared spectral identifications remain uncertain. A lingering open issue is the reddish component on the surfaces of the mid-sized inner moons of Saturn and possible contributions from carbon-bearing, salty, and/or metallic species, and whether there is a relationship between the reddish material of Iapetus and Hyperion and that on the inner moons. If so, how is such material transported? If nano-iron particles are significant, are there enough silicates in the Saturn system to be space-weathered, resulting in nano-iron particles, or are the silicates brought in from other regions of the solar system? (Since the reddish species are seen on the surface they must be continually emplaced, competing with E-ring grain coating.) Continued study of Cassini Cosmic Dust Analyzer (CDA) data of E-ring and interplanetary grains (and their organic and silicate components), along with laboratory studies of organics and their spectral evolution to processing, will help address these issues, as will continued study of Cassini VIMS, ISS, UVIS, and Composite Infrared Spectrometer (CIRS) datasets.

An additional remaining open issue regards the red streaks on Tethys. How are they formed? Are they related to an active (or recently active) internal global stress field? They appear to be relatively young, overlying existing craters and other geologic features. Continued study of Tethys and the other icy moons of Saturn and their evolutionary processes and modeling of interiors may aid in understanding these mysterious features.

Acknowledgments. The authors are grateful to G. Filacchione, A. Verbiscer, and an anonymous reviewer for careful reviews of this chapter, and to the many people around the world who made the Cassini-Huygens mission possible.

REFERENCES

Andronico G., Baratta G. A., Spinella F., and Strazzulla G. (1987) Optical evolution of laboratory-produced organics: Applications to Phoebe, Iapetus, outer belt asteroids and cometary nuclei. *Astron. Astrophys., 184,* 333–336.

Baragiola R. A., Famá M. A., Loeffler M. J., Palumbo M. E., Raut U., Shi J., and Strazzulla G. (2013) Radiation effects in water ice in the outer solar system. In *The Science of Solar System Ices* (M. S. Gudipati and J. Castillo-Rogez, eds.), pp. 527–549, Springer, Berlin.

Bell J. F., Cruikshank D. P., and Gaffey M. J. (1984) The composition and origin of the Iapetus dark material. *Icarus, 61,* 192–207.

Black G. J., Campbell D. B., Carter L. M., and Ostro S. J. (2004) Radar detection of Iapetus. *Science, 304,* 553.

Brown R. H., Baines K. H., Bellucci G., Buratti B. J., Capaccioni F., Cerroni P., Clark R. N., Coradini A., Cruikshank D. P., Drossart P., Formisano V., Jaumann R., Langevin Y., Matson D. L., McCord T. B., Mennella V., Nelson R. M., Nicholson P. D., Hansen G. B., Hibbitts C. A., Momary T. W., and Showalter M. R. (2006a) Observations in the Saturn system during approach and orbital insertion, with Cassini's Visual and Infrared Mapping Spectrometer (VIMS). *Astron. Astrophys., 446,* 707–716, DOI: 10.1051/0004-6361:20053054.

Brown R. H., Clark R. N., Buratti B. J., Cruikshank D. P., Barnes J. W., Mastrapa R. M. E., Bauer J., Newman S., Momary T., Baines K. H., Bellucci G., Capaccioni F., Cerroni P., Combes M., Coradini A., Drossart P., Formisano V., Jaumann R., LangevinY., Matson D. L., McCord T. B., Nelson R. M., Nicholson P. D., Sicardy B., and Sotin C. (2006b) Composition and physical properties of Enceladus' surface. *Science, 311,* 1425–1428.

Brown R. H., Lauretta D. S., Schmidt B., and Moores J. (2012) Experimental and theoretical simulations of ice sublimation with implications for the chemical, isotopic, and physical evolution of icy objects. *Planet. Space Sci., 60,* 166–180.

Buratti B. J. (1988) Enceladus: Implications of its unusual photometric properties. *Icarus, 75,* 113–126.

Buratti B. J. and Mosher J. A. (1995) The dark side of Iapetus: Additional evidence for an exogenous origin. *Icarus, 115,* 219–227.

Buratti B. J. and Veverka J. (1984) Voyager photometry of Rhea, Dione, Tethys, Enceladus and Mimas. *Icarus, 58,* 254–264.

Buratti B. J., Mosher J. A., and Johnson T. V. (1990) Albedo and color maps of the saturnian satellites. *Icarus, 87,* 339–357.

Buratti B. J., Mosher J. A., Nicholson P. D., McGhee C., and French R. (1998) Photometry of the saturnian satellites during ring plane crossing. *Icarus, 136,* 223–231.

Buratti B. J., Hicks M. D., Tryka K. A., Sittig M. S., and Newburn R. L. (2002) High-resolution 0.33–0.92 μm spectra of Iapetus, Hyperion, Phoebe, Rhea, Dione and D-type asteroids: How are they related? *Icarus, 155,* 375–381.

Buratti B. J. and 28 colleagues (2005) Cassini Visual and Infrared Mapping Spectrometer observations of Iapetus: Detection of CO_2. *Astrophys. J. Lett., 622,* L149–L152.

Buratti B. J. et al. (2010) Cassini spectra and photometry 0.25–5.1 μm of the small inner satellites of Saturn. *Icarus, 206,* 524–536.

Buratti B. J. and 16 co-authors (2017) Global albedos of Pluto and Charon from LORRI New Horizons observations. *Icarus, 287,* 207.

Buratti B. J., Brown R. H., Clark R. N., Cruikshank D., and Filacchione G. (2018) Spectral analyses of Saturn's moons using Cassini-VIMS. In *Remote Sensing* (J. Bishop, ed.), in press. Cambridge Univ., Cambridge.

Burns J. A., Lamy P. L., and Soter S. (1979) Radiation forces on small particles in the solar system. *Icarus, 40,* 1–48.

Cassidy T. A., Paranicas C. P., Shirley J. H., Dalton J. B., Teolis B. T., Johnson R. E., Kamp L., and Hendrix A. R. (2013) Magnetospheric ion sputtering and water ice grain size at Europa. *Planet. Space Sci., 77,* 64–73.

Clark R. and Lucey P. (1984) Spectral properties of ice-particulate mixtures and implications for remote sensing — 1. Intimate mixtures. *J. Geophys. Res., 80,* 6341–6348.

Clark R. and Roush T. (1984) Reflectance spectroscopy — Quantitative analysis techniques for remote sensing applications. *J. Geophys. Res., 89,* 6329–6340.

Clark R. N., Brown R. H., Jaumann R., Cruikshank D. P., Nelson R. M., Buratti B. J., McCord T. B., Lunine J., Hoefen T. M., Curchin J. M., Hansen G., Hibbitts C., Matz KD., Baines K. H., Bellucci G., Bibring J.-P., Capaccione F., Cerroni P., Coradini A., Formisano V., Langevin Y., Matson D. L., MennellaV., Nicholson P. D., Sicardy B., and Sotin C. (2005) Compositional mapping of Saturn's moon Phoebe with imaging spectroscopy. *Nature, 435,* 66–69.

Clark R. N., Brown R. H., Jaumann R., Cruikshank D. P., Buratti B., Baines K. H., Nelson R. M., Nicholson P. D., Moore J. M., Curchin J. M., Hoefen T., and Stephan K. (2008) Compositional mapping of Saturn's satellite Dione with Cassini VIMS and the implications of dark material in the Saturn system. *Icarus, 193,* 372–386.

Clark R., Carlson R., Grundy W., and Noll K. (2011) Observed ices in the solar system. In *The Science of Solar System Ices* (M. S. Gudipati and J. Castillo-Rogez, eds.), pp. 3–46. Springer, Berlin.

Clark R. N., Cruikshank D. P., Jaumann R., Brown R. H., Stephan K., Dalle Ore C. M., Livi K. E., Pearson N., Curchin J. M., Hoefen T. M., Buratti B. J., Filacchione G., Baines K. H., and Nicholson P. D. (2012) The surface composition of Iapetus: Mapping results from Cassini VIMS. *Icarus, 218,* 831–860.

Cook A. F. and Franklin F. (1970) An explanation for the light curve of Iapetus. *Icarus, 13,* 282–291.

Cruikshank D. P. (1980) Near infrared studies of the satellites of Saturn and Uranus. *Icarus, 41,* 246–258.

Cruikshank D. P., Bell J. F., Gaffey M. J., Brown R. H., Howell R., Beerman C., and Rognstad M. (1983) The dark side of Iapetus. *Icarus, 53,* 90–104.

Cruikshank D. P., Dalton J. B., Dalle Ore C. M., Bauer J., Stephan K., Filacchione G., Hendrix A. R., Hansen C. J., Coradini A., Cerroni P.,

Tosi F., Capaccioni F., Jaumann R., Buratti B. J., Clark R. N., Brown R. H., Nelson R. M., McCord T. B., Baines K. H., Nicholson P. D., Sotin C., Meyer A. W., Bellucci G., Combes M., Bibring J.-P., Langevin Y., Sicardy B., Matson D. L., Formisano V., Drossart P., and Mennella V. (2007) Surface composition of Hyperion. *Nature, 448,* 54–56, DOI: 10.1038/nature05948.

Cruikshank D. P., Dalle Ore C. M., Clark R. N., and Pendleton Y. J. (2014) Aromatic and aliphatic organic materials on Iapetus: Analysis of Cassini VIMS data. *Icarus, 233,* 306–315.

Ćuk M. and Burns J. A. (2004) Gas-drag-assisted capture of Himalia's family. *Icarus, 167,* 369–381.

Cuzzi J., Clark R., Filaccione G., French R., Johnson R., Marouf E., and Spilker L. (2009) Ring particle composition and size distribution. In *Saturn from Cassini-Huygens* (M. K. Dougherty et al., eds.), pp. 459–509. Springer, Berlin.

Dalle Ore C. M., Cruikshank D. P., and Clark R. N. (2012) Infrared spectroscopic characterization of the low-albedo materials on Iapetus. *Icarus, 221,* 735–743.

Dalle Ore C. M., Cruikshank D. P., Mastrapa R. M. E., Lewis E., and White O. L. (2015) Impact craters: An ice study on Rhea. *Icarus, 261,* 80–90.

Degewij J., Cruikshank D. P., and Hartmann W. K. (1980) Near-infrared colorimetry of J6 Himalia and S9 Phoebe: A summary of 0.3–2.2 μm reflectances. *Icarus, 44,* 541–547.

Denk T., Neukum G., Roatsch Th., Porco C. C., Burns J. A., Galuba G. G., Schmedemann N., Helfenstein P., Thomas P. C., Wagner R. J., and West R. A. (2010) Iapetus: Unique surface properties and a global color dichotomy from Cassini Imaging. *Science, 327,* 435–439.

Emery J. P., Burr D. M., Cruikshank D. P., Brown R. H., and Dalton J. B. (2005) Near-infrared (0.8–4.0 μm) spectroscopy of Mimas, Enceladus, Tethys, and Rhea. *Astron. Astrophys., 435,* 353–362.

Filacchione G., Capaccioni F., McCord T. B., Coradini A., Cerroni P., Bellucci G., Tosi F., D'Aversa E., Formisano V., Brown R. H., Baines K. H., Bibring J. P., Buratti B. J., Clark R. N., Combes M., Cruikshank D. P., Drossart P., Jaumann R., Langevin Y., Matson D. L., Mennella V., Nelson R. M., Nicholson P. D., Sicardy B., Sotin C., Hansen G., Hibbitts K., Showalter M., and Newman S. (2007) Saturn's icy satellites investigated by Cassini-VIMS — I. Full-disk properties: 350–5100 nm reflectance spectra and phase curves. *Icarus, 186(1),* 259–290.

Filacchione G., Capaccioni F., Clark R. N., Cuzzi J. N., Cruikshank D. P., Coradini A., Cerroni P., Nicholson P. D., McCord T. B., Brown R. H., Buratti B. J., Tosi F., Nelson R. M., Jaumann R., and Stephan K. (2010) Saturn's icy satellites investigated by Cassini-VIMS. II. Results at the end of nominal mission. *Icarus, 206,* 507–523.

Filacchione G. et al. (2012) Saturn's icy satellites and rings investigated by Cassini-VIMS. III —Radial compositional variability. *Icarus, 220,* 1064–1096.

Filacchione G. et al. (2013) The radial distribution of water ice and chromophores across Saturn's system. *Astrophys. J., 766,* 76–80.

Fink U., Larson H. P., Gautier T. N. III, and Treffers R. R. (1976) Infrared spectra of the satellites of Saturn — Identification of water ice on Iapetus, Rhea, Dione, and Tethys. *Astrophys. J. Lett., 207,* L63–L67.

Grav T. and Bauer J. (2007) A deeper look at the colors of the saturnian irregular satellites. *Icarus, 191,* 267–285.

Hamilton D. P. and Burns J. A. (1994) Origin of Saturn's E ring: Self-sustained, naturally. *Science, 264,* 550–553.

Hand K. and Carlson R. W. (2015) Europa's surface color suggests an ocean rich with sodium chloride. *Geophys. Res. Lett., 42,* DOI: 10.1002/ 2015GL063559.

Hansen C. J., Esposito L., Stewart A. I. F., Colwell J., Hendrix A., Pryor W., Shemansky D., and West R. (2006) Enceladus's water vapor plume. *Science, 311,* 1422–1425.

Hapke B., Wagner J., Cohen A., and Partlow W. (1978) Reflectance measurements of lunar materials in the vacuum ultraviolet. *Lunar and Planetary Science IX,* pp. 456–458. Lunar and Planetary Institute, Houston.

Hendrix A. R. and Hansen C. J. (2008a) Ultraviolet observations of Phoebe from Cassini UVIS. *Icarus, 193,* 323–333.

Hendrix A. R. and Hansen C. J. (2008b) The albedo dichotomy of Iapetus measured at UV wavelengths. *Icarus, 193,* 344–351.

Hendrix A. R., Hansen C. J., and Holsclaw G. M. (2010) The ultraviolet reflectance of Enceladus: Implications for surface composition. *Icarus, 206,* 608–617.

Hendrix A. R., Cassidy T. A., Buratti B. J., Paranicas C., Hansen C. J., Teolis B., Roussos E., Bradley E. T., Kollmann P., and Johnson R. E. (2012) Mimas' far-UV albedo: Spatial variations. *Icarus, 220,* 922–931.

Hendrix A. R., Noll K. S., and Spencer J. R. (2015) Investigating Saturn's icy moons using HST/STIS. Abstract P31B-2072 presented at 2015 Fall Meeting, AGU, San Francisco, California, 14–18 December.

Hendrix A. R., Noll K. S., and Spencer J. R. (2017) Icy Saturnian satellites: Using UV-Vis data to study surface composition and surface processing. *AAS/DPS Meeting Abstracts #49,* 210.03.

Hendrix A. R., Filacchione G., Paranicas C., Schenk P., and Scipioni F. (2018) Icy Saturnian satellites: Disk-integrated UV-IR spectral characteristics and links to exogenic processes. *Icarus, 300,* 103–114.

Howett C. J. A., Spencer J. R., Schenk P., Johnson R. E., Paranicas C., Hurford T. A., Verbiscer A., and Segura M. (2011) A high-amplitude thermal inertia anomaly of probable magnetospheric origin on Saturn's moon Mimas. *Icarus, 216,* 221–226.

Imanaka H., Khare B. N., Elsila J. E., Bakes E. L. O., McKay C. P., Cruikshank D. P., Sugita S., Matsui T., and Zare R. N. (2004) Laboratory experiments of Titan tholin formed in cold plasma at various pressures: Implications for nitrogen-containing polycyclic aromatic compounds in Titan haze. *Icarus, 168,* 344–366.

Jarvis K. S., Vilas F., Larson S. M., and Gaffey M. J. (1997) S4 Hyperion and S9 Phoebe — Testing a link with Iapetus. *Lunar and Planetary Science XXVIII,* Abstract #1745. Lunar and Planetary Institute, Houston.

Jarvis K. S., Vilas F., Larson S. M., and Gaffey M. J. (2000) Are Hyperion and Phoebe linked to Iapetus? *Icarus, 146,* 125–132.

Jaumann R. et al. (2008) Distribution of icy particles across Enceladus' surface as derived from Cassini-VIMS measurements. *Icarus, 193,* 407–419.

Jewitt D. and Haghighipur N. (2007) Irregular satellites of the planets: Products of capture in the early solar system. *Annu. Rev. Astron. Astrophys., 45,* 261–95.

Johnson T. V. and Lunine J. I. (2005) Saturn's moon Phoebe as a captured body from the outer solar system. *Nature, 435,* 69–71.

Karkoschka E. and Schroder S. E. (2016) The DISR imaging mosaic of Titan's surface and its dependence on emission angle. *Icarus, 270,* 307–325.

Khare B. N., Sagan C., Arakawa E. T, Suits F., Callcott T. A., and Williams M. W. (1984) Optical constants of organic tholins produced in a simulated titanian atmosphjere: From soft X-rays to microwave frequencies. *Icarus, 60,* 127–137.

Kolokolova L., Buratti B., and Tishkovets V. (2010) Impact of coherent backscattering on the spectra of icy satellites of Saturn and the implications of its effects for remote sensing. *Astrophys. J. Lett., 711,* L71–L74.

Kwok S. and Zhang Y. (2011) Mixed aromatic-aliphatic organic nanoparticles as carriers of unidentified infrared emission features. *Nature, 479,* 80–83.

Lebreton J.-P., Coustenis A., Lunine J., Raulin F., Owen T., and Strobel D. (2009) Results from the Huygens probe on Titan. *Astronomy Astrophys. Rev., 17,* 149–179.

Lebofsky L. A., Feierberg M. A., and Tokunaga A. T. (1982) Infrared observations of the dark side of Iapetus. *Icarus, 49,* 382–386.

Mastrapa R. M., Bernstein M. P., Sandford S. A., Roush T. L., Cruikshank D. P., and Dalle Ore C. M. (2008) Optical constants of amorphous and crystalline H_2O-ice in the near infrared from 1.1 to 2.6 μm. *Icarus, 197,* 307–320.

Materese C. K., Cruikshank D. P., Sandford S. A., Imanaka H., Nuevo M., and White D. (2014) Ice chemistry on outer solar system bodies: Carboxylic acids, nitriles, and urea detected in refractory residues produced from the UV photolysis of N_2:CH_4:CO containing ices. *Astrophys. J., 788,* 111, DOI: 10.1088/0004-637X/788/2/111.

Materese C. K., Cruikshank D. P., Sandford S. A., Imanaka H., and Nuevo M. (2015) Ice chemistry on outer solar system bodies: Electron radiolysis of N_2-, CH_4-, and CO-containing ices. *Astrophys. J., 812,* 150, DOI: 10.1088/0004-637X/812/2/150.

Matthews R. A. J. (1992) The darkening of Iapetus and the origin of Hyperion. *Q. J. R. Astron. Soc., 33,* 253–258.

McCord T. B. et al. (2012) Dark material on Vesta from the infall of carbonaceous volatile-rich material. *Nature, 491,* 83–86.

McDonald G. et al. (1996) Production and chemical analysis of cometary ice tholins. *Icarus, 122,* 107–117.

Miller S. (1953) A production of amino acids under possible primitive Earth conditions. *Science, 117,* 528–529.

Noland M. et al. (1974) Six-color photometry of Iapetus, Titan, Rhea, Dione and Tethys. *Icarus, 23,* 334–354.

Noll K. S., Roush T. L., Cruikshank D. P., Johnson R. E., and Pendleton Y. E. (1997) Detection of ozone on Saturn's satellites Rhea and Dione. *Nature, 388,* 45–47.

Ostro S. et al. (2006) Cassini RADAR observations of Enceladus, Tethys, Dione, Rhea, Iapetus, Hyperion, and Phoebe. *Icarus, 183,* 479–490.

Owen T. C., Cruikshank D. P., Dalle Ore C. M., Geballe T. R., Roush T. L., and de Bergh C. (1999) Detection of water ice on Saturn's satellite Phoebe. *Icarus, 140,* 379–382.

Owen T. C., Cruikshank D. P., Dalle Ore C. M., Geballe T. R., Roush T. L., de Bergh C., Meier R., Pendleton Y. J., and Khare B. N. (2001) Decoding the domino: The dark side of Iapetus. *Icarus, 149,* 160–172.

Pang K. D., Voge C., Rhoads J., and Ajello J. (1984) The E ring of Saturn and satellite Enceladus. *J. Geophys. Res., 89,* 9459–9470.

Paranicas C., Roussos E., Decker R. B., Johnson R. E., Hendrix A. R., Schenk P., Kollmann P., Cassidy T., Dalton J. B., Patterson W., Hand K., Nordheim T., Howett C. J. A., Krupp N., and Mitchell D. G. (2014) The lens feature on the saturnian satellites. *Icarus, 234,* 155–161.

Pendleton Y. J., Sandford S. A., Allamandola L. J., Tielens A. G. G. M., and Sellgren K. (1994) Near-infrared absorption spectroscopy of interstellar hydrocarbon grains. *Astrophys. J., 437,* 683–696.

Pieters C. M., Taylor L. A., Noble S. K., Keller L. P., Hapke B., Morris R. V., Allen C. C., McKay D. S., and Wentworth S. (2000) Space weathering on airless bodies: Resolving a mystery with lunar samples. *Meteoritics & Planet. Sci., 35,* 1101–1107.

Pollack J. B., Burns J. A., and Tauber M. E. (1979) Gas drag in primordial circumplanetary envelopes: A mechanism for satellite capture. *Icarus, 37,* 587–611.

Porco C. C. and 34 colleagues (2005) Cassini imaging science: Initial results on Phoebe and Iapetus. *Science, 307,* 1237–1242.

Pospieszalska M. K. and Johnson R. E. (1989) Magnetospheric ion bombardment profiles of satellites: Europa and Dione. *Icarus, 78,* 1–13.

Postberg F., Kempf S., Hillier J. K., Srama R., Green S. F., McBride N., and Grün E. (2008) The E-ring in the vicinity of Enceladus. II. Probing the moon's interior — The composition of E-ring particles. *Icarus, 193,* 438–454.

Ricca A., Bauschlicher C. W. Jr., Boersma C., Tielens A. G. G. M., and Allamandola L. J. (2012) The infrared spectroscopy of compact polycyclic aromatic hydrocarbons containing up to 384 carbons. *Astrophys. J., 754,* 75.

Schenk P. (2015) "Blood Stains" on Tethys: Evidence for recent activity? Abstract P21B-02 presented at 2015 Fall Meeting, AGU, San Francisco, California, 14–18 December.

Schenk P., Hamilton D. P., Johnson R. E., McKinnon W. B., Paranicas C., Schmidt J., and Showalter M. R. (2011) Plasma, plumes and rings: Saturn system dynamics as recorded in global color patterns on its midsize icy satellites. *Icarus, 211,*740–757.

Scipioni F., Tosi F., Stephan K., Filacchione G., Ciarniello M., Capaccioni F., Cerroni P., and the VIMS Team (2013) Spectroscopic classification of icy satellites of Saturn I: Identification of terrain units on Dione. *Icarus, 226,* 1331–1349.

Scipioni F., Tosi F., Stephan K., Filacchione G., Ciarniello M., Capaccioni F., Cerroni P., and the VIMS Team (2014) Spectroscopic classification of icy satellites of Saturn II: Identification of terrain units on Rhea. *Icarus, 243,* 1–16.

Scipioni F., Schenk P., Tosi F., D'Aversa E., Clark R., Combe J. Ph., and Dalle Ore C. M. (2017) Deciphering sub-micron ice particles on Enceladus surface. *Icarus, 290,* 183–200.

Smith B. A., Soderblom L., Beebe R., Boyce J., Briggs G., Bunker A., Collins S., Hansen C., Johnson T., Mitchell J., Terrile R., Carr M., Cook A., Cuzzi J., Pollack J., Danielson G., Ingersoll A., Davies M., Hunt G., Masursky H., Shoemaker E., Morrison D., Owen T., Sagan C., Veverka J., Strom R., and Suomi V. (1981) Encounter with Saturn: Voyager 1 imaging science results. *Science, 212,* 163–191.

Smith B., Soderblom L., Batson R., Bridges P., Inge J., Masursky H., Shoemaker E., Beebe R., Boyce J. M., Briggs G., Bunker A., Collins S., Hansen C., Johnson T., Mitchell J., Terrile R., Cook A., Cuzzi J., Pollack J., Danielson G., Ingersoll A., Davies M., Hunt G., Morrison D., Owen T., Sagan C., Veverka J., Strom R., and Suomi V. (1982) A new look at the Saturn system: The Voyager 2 images. *Science, 215,* 504–537.

Soter S. (1974) Brightness of Iapetus. Presented at Planetary Satellites, August 18–21, Cornell University, Ithaca, New York, IAU Colloquium 28.

Spencer J. R. and Denk T. (2010) Formation of Iapetus' extreme albedo dichotomy by exogenically triggered thermal ice migration. *Science, 327,* 432.

Squyres S. W. and Veverka J. (1981) Voyager photometry of surface features on Ganymede and Callisto. *Icarus, 46,* 137–155.

Squyres S. W., Buratti B., Veverka J., and Sagan C. (1984) Voyager photometry of Iapetus. *Icarus, 59,* 426–435.

Stephan K., Jaumann R., Wagner R., Clark R., Cruikshank D., Hibbitts C., Roatsch T., Hoffmann H., Brown R., Filacchione G., Buratti B., Hansen G., McCord T., Nicholson P., and Baines K. (2010) Dione's spectral and geological properties. *Icarus, 206,* 631–652.

Stephan K., Jaumann R., Wagner R., Clark R. N., Cruikshank D. P., Hibbitts C. A., Roatch T., Matz K.-D., Giese B., Brown R. H., Filacchione G., Cappacioni F., Buratti B. J., Hansen G. B., Nicholson P. D., Baines K. H., Nelson R. M., and Matson D. L. (2012) The saturnian satellite Rhea as seen by Cassini VIMS. *Planet. Space Sci., 61,* 142–160.

Stephan K., Wagner R., Jaumann R., Clark R. N., Cruikshank D. P., Brown R. H., Giese B., Roatsch T., Matson D., Dalle Ore C., Filacchione G., Capaccione F., Baines K., Rodriguez S., Buratti B. J., and Nicholson P. D. (2016) Cassini's geological and compositional view of Tethys. *Icarus, 274,* 1–22.

Strazzulla G. (1986) Organic material from Phoebe to Iapetus. *Icarus, 66,* 397–400.

Sunshine J. M. and Pieters C. M. (1993) Estimating modal abundances from the spectra of natural and laboratory pyroxene mixtures using the modified Gaussian model. *J. Geophys. Res., 98,* 9075–9087.

Tamayo D., Burns J. A., Hamilton D. P., and Hedman M. M. (2011) Finding the trigger to Iapetus' odd global albedo pattern: Dynamics of dust from Saturn's irregular satellites. *Icarus, 215,* 260–278.

Teolis B. D. and Waite J. H. (2016) Dione and Rhea seasonal exospheres revealed by Cassini CAPS and INMS. *Icarus, 272,* 277–289.

Tholen D. J. and Zellner B. (1983) Eight-color photometry of Hyperion, Iapetus and Phoebe. *Icarus, 53,* 341–347.

Thomas P. C., Burns J. A., Hedman M., Helfenstein P., Morrison S., Tiscareno M. S., and Veverka J. (2013) The inner small satellites of Saturn: A variety of worlds. *Icarus, 226,* 999–1019.

Thompson W. R., Murray B. G. J. P. T., Khare B. N., and Sagan C. (1987) Coloration and darkening of methane clathrate and other ices by charged particle irradiation: Applications to the outer solar system. *J. Geophys. Res., 92,* 14933–14947.

Verbiscer A. J. and Veverka J. (1992) Mimas: Photometric roughness and albedo map. *Icarus, 99,* 63–69.

Verbiscer A. J. and Veverka J. (1994) A photometric study of Enceladus. *Icarus, 110,* 155–164.

Verbiscer A. J. et al. (2006) Near-infrared spectra of the leading and trailing hemispheres of Enceladus. *Icarus, 182,* 211–223.

Verbiscer A. J. et al. (2007) Enceladus: Cosmic graffiti artist caught in the act. *Science, 315,* 815–817.

Verbiscer A. J., Peterson D. E., Skrutskie M. F., Cushing M., Helfenstein P., Nelson M. J., Smith J. D., and Wilson J. C. (2008) Ammonia hydrate on Tethys' trailing hemisphere. *The Science of Solar System Ices: A Cross-Disciplinary Workshop,* Abstract #9064. Lunar and Planetary Institute, Houston.

Verbiscer A. J., Skrutskie M. F., and Hamilton D. P. (2009) Saturn's largest ring. *Nature, 461,* 1098–1100.

Vilas F., Larson S. M., Stockstill K. R., and Gaffey J. J. (1996) Unraveling the zebra: Clues to the Iapetus dark material composition. *Icarus, 124,* 262–267.

Waite J. H. Jr., Young D. T., Westlake J. H., Lunine J. I., McKay C. P., and Lewis W. S. (2009a) High-altitude production of Titan's aerosols. In *Titan from Cassini-Huygens* (R. H. Brown et al., eds.), pp. 201–214. Springer, Berlin.

Waite J. H. Jr. and 15 colleagues (2009b) Liquid water on Enceladus from observations of ammonia and ^{40}Ar in the plume. *Nature, 460,* 487–490.

Wilson P. D. and Sagan C. (1996) Spectrophotometry and organic matter on Iapetus. 2. Models of interhemispheric asymmetry. *Icarus, 122,* 92–106.

Wexler A. S. (1967) Integrated intensities of absorption bands in infrared spectroscopy. *Appl. Spect. Rev., 1,* 29–98.

Verbiscer A. J., Helfenstein P., Buratti B. J., and Royer E. (2018) Surface properties of Saturn's icy moons from optical remote sensing. In *Enceladus and the Icy Moons of Saturn* (P. M. Schenk et al., eds.), pp. 323–341. Univ. of Arizona, Tucson, DOI: 10.2458/azu_uapress_9780816537075-ch016.

Surface Properties of Saturn's Icy Moons from Optical Remote Sensing

Anne J. Verbiscer
University of Virginia

Paul Helfenstein
Cornell University

Bonnie J. Buratti
Jet Propulsion Laboratory

Emilie Royer
University of Colorado

Much of our knowledge of physical surface properties of the icy satellites of Saturn comes from optical remote sensing, thermal infrared data, and radar. Among the regolith properties that can be characterized from optical remote sensing are the composition, state of compaction, grain sizes, and thermophysical properties such as bolometric albedo and thermal inertia. The spectrophotometric behavior of the surfaces of Saturn's satellites provides a record of the processes that have shaped their regolith formation and evolution. Here we examine the known surface properties of the satellites and their terrains from the far-ultraviolet to the near-infrared and how they depend on geological setting and placement in the saturnian system.

1. INTRODUCTION

Prior to the advent of spacecraft missions like Voyager and Cassini, nearly all information about the icy satellites of Saturn was obtained exclusively from whole-disk telescopic observations. Fundamental thermophysical quantities such as geometric albedos and Bond albedos could be estimated by extrapolating observed reflectances to zero phase; however, accurate measurements of these quantities are enabled only by spacecraft observations that provide satellite sizes and observations during node crossings in which satellites align with the Sun and the detector. Even with new high-resolution close-up observations from spacecraft, modern revised estimates of these quantities are still important for characterizing the average global and even hemispheric (leading- and trailing-side) thermophysical properties.

High-quality disk-resolved observations of the icy satellites provide a much more detailed picture of how thermophysical properties vary with terrain type and geological setting. Global mapping of wavelength-dependent normal reflectances and its correlation with spatial variations in thermal inertia, surface roughness, surface porosity, and other regionally variable physical quantities are leading to new insights about surface physical processes and how they relate to the Saturn environment including ring-satellite interactions and the effects of magnetospheric interactions.

Modeling of radiative transfer in planetary regoliths is critical to the interpretation of optical spectrophotometric observations in the context of the surface physical properties that they imply. In general and as discussed in detail below, the wavelength-dependent light-scattering behavior of a surface is affected in predictable ways by different regolith properties, such as constituent particle albedos, grain microstructure, compaction state, and macroscopic surface relief. The ability to estimate and accurately constrain these properties critically hinges on the availability of observations covering a wide range of photometric geometries, specifically phase angle, α, incidence angle, i, and emission or emergence angle e. An accurate, unambiguous representation of the spectrophotometric characteristics of each satellite surface cannot presently be obtained without reliance on both whole-disk and disk-resolved photometric observations. When heavy reliance is needed on whole-disk phase curves, it is important to correct the phase curve for systematic rotational and latitudinal variations in brightness due to geographic albedo variegations. Even when phase coverage to very high phase angles is available, care should be taken to account for how the signal/noise quality of the data may degrade as the satellites reflect progressively less light as the solar phase increases toward a narrow crescent. In the Saturn environment and especially for Saturn's high-albedo satellites, it may be necessary to subtract the effects of

Saturnshine from spacecraft observations of the Saturn-facing side of the satellites — especially at large phase angles. Recent summaries of modern spectrophotometric models used for planetary surface analysis can be found in *Li et al.* (2015) and *Verbiscer et al.* (2013). Although we will discuss different models below, in the present chapter, we will primarily focus on results from the application of the most widely used model of *Hapke* (1981, 1984, 1986, 2002, 2008, 2012a,b) and its variants. Table 1 provides a summary of the Hapke model parameters and their physical interpretations. These include the single-scattering albedo, the amplitude and width of the shadow hiding opposition effect (SHOE) and the coherent backscatter opposition effect (CBOE), macroscopic surface roughness, the single particle phase function (SPPF), and a porosity coefficient. Further details of these parameters are provided in specific discussions below.

2. ALBEDOS AND ALBEDO VARIATIONS

2.1. Geometric Albedos

The geometric albedo is the ratio of the reflectance of a planetary body viewed at opposition to the reflectance of a flat disk of the same size that scatters incident light equally in all directions. Prior to 2005, visible mean geometric albedos p_V of satellites in the Saturn system (Table 2) were estimated by extrapolating their solar phase curves to the reflectance at zero phase. However, the January 14, 2005, node crossing of the Saturn system enabled the *measurement* of the geometric albedos of the icy saturnian satellites from Earth-based telescopes, revealing that they span the widest range of any of the giant planet systems. Geometric albedos obtained with the Hubble Space Telescope at opposition are available for most of Saturn's large icy satellites; however, geometric albedos for some of the smaller and ring-embedded moons must still be estimated from Cassini spacecraft observations. Thomas et al. (this volume) estimated whole-disk albedos of these smaller moons at 10° phase. Figure 1 compares known geometric albedos with the reflectance at 10° phase. With the exception of Helene and Calypso, most geometric albedos fall along the empirical trend shown; however, it should be noted that the geometric albedos of Helene and Calypso are provisional only since they have not been corrected for any variations in their rotational lightcurves. We have used this relationship in Fig. 1 to estimate geometric albedos in cases where opposition measurements are not available. With a mean geometric albedo of $p_V = 1.4$, Enceladus is not only the most reflective satellite in the Saturn system, it is the most reflective body in the solar system (*Verbiscer et al.,* 2007). One of Saturn's darkest satellites, Phoebe (*Miller et al.,* 2011) has $p_V = 0.086$, a value consistent with its presumed outer solar system origin (*Johnson and Lunine,* 2005). Visible reflectances generally correlate with the dominant surface alteration processes that are active in the local environment. Albedos of the satellites embedded within the E ring are the highest in the solar system owing to Enceladus' geologically active plumes; however, despite the fact that it is embedded within the G ring, the geometric albedo of Aegaeon ($p_V <$ 0.15) may be close to that of Phoebe (*Hedman et al.,* 2011), meaning Aegaeon may also be one of Saturn's darkest moons.

Far-ultraviolet (FUV) geometric albedos of saturnian satellites have been derived from Cassini Ultraviolet Imaging Spectrometer (UVIS) observations by extrapolating the FUV solar phase curves of Mimas, Tethys, and Dione at 180 nm to their reflectance at zero phase. These satellites are much darker at FUV wavelengths than in the visible, mainly due to the presence of a water ice absorption band around 165 nm. Mimas' leading and trailing hemispheres as well as Tethys' leading hemisphere all exhibit geometric albedos of about 0.48 at 180 nm, while the trailing hemisphere of Tethys and the leading hemisphere of Dione have geometric albedos of about 0.32. Dione's trailing hemisphere is the darkest of these satellites with a geometric albedo of about 0.18 (*Royer and Hendrix,* 2014). Geometric albedos in the FUV generally correlate with exogenic processes affecting the surfaces and their proximity to Enceladus.

TABLE 1. Summary of *Hapke*'s (2012b) model parameters.

Hapke Term	Description
ϖ_o	Average particle single-scattering albedo: A measure of the efficiency of a typical regolith grain to scatter and absorb light. Related to the composition, optical constants, size, and mechanical structure of the regolith grains.
b,c	Parameters of the two-parameter Henyey-Greenstein (2PHG) particle phase function. The phase function is a linear combination of two single-term Henyey-Greenstein functions in which b describes the assumed equal angular narrowness of the backscattering and forward-scattering lobes and c is a partition factor that determines the relative amplitude of each lobe. Describes mechanical structure of grains (*McGuire and Hapke,* 1995).
h_C, $B_{0,C}$	Angular half-width (radians) and amplitude, respectively, of the coherent backscatter opposition effect (CBOE). Depends on density and size of small (typically subgrain) scatterers and the transparency of the medium (mean optical path length of a photon), and porosity.
h_S, $B_{0,S}$	Angular half-width (radians) and amplitude, respectively, of the shadow hiding opposition effect (SHOE). h_S depends on regolith porosity and grain-size distribution. $B_{0,S}$ measures grain transparency.
$\bar{\theta}$	Average macroscopic slope angle of subresolution-scale surface relief.
K	*Hapke*'s (2008, 2012b) porosity coefficient; corrects for grain packing in regolith.

TABLE 2. Visible geometric albedos, phase integrals, and spherical albedos of Saturn's icy satellites.

Satellite	Geometric Albedo	Phase Integral	Spherical Albedo	Reference
Pan	0.86 ± 0.07			[1]
Daphnis	0.91 ± 0.07			[1]
Atlas	0.82 ± 0.13			[1]
Prometheus	1.05 ± 0.11			[1]
Pandora	0.99 ± 0.13			[1]
Epimetheus	0.73 ± 0.03			[2]
Janus	0.71 ± 0.02			[2]
Aegaeon	<0.15			[3]
Mimas	0.962 ± 0.004	0.62	0.60	[2]
Methone	0.72 ± 0.09			[1]
Pallene	0.52 ± 0.20			[4]
Enceladus	1.375 ± 0.008	0.68	0.93	[2]
Tethys	1.229 ± 0.005	0.52	0.64	[2]
Calypso	1.34 ± 0.10			[2]
Telesto	0.97 ± 0.04			[1]
Dione	0.998 ± 0.004	0.70	0.70	[2]
Polydeuces	1.24 ± 0.09			[1]
Helene	1.67 ± 0.20			[2]
Rhea	0.949 ± 0.003	0.63	0.60	[2]
Hyperion	0.44 ± 0.13			[1]
Iapetus leading hemisphere	0.04	0.26	0.01	[5]
Iapetus trailing hemisphere	0.45	0.61	0.27	[5]
Phoebe	0.0856 ± 0.0023	0.32	0.03	[6]
Irregular satellites	0.06			[5]

References: [1] Values using Fig. 1 and Thomas et al. (this volume), [2] *Verbiscer et al.* (2007), [3] *Hedman et al.* (2011), [4] *Hedman et al.* (2010), [5] *Blackburn et al.* (2010), [6] *Miller et al.* (2011).

3. REGOLITH STRUCTURAL PROPERTIES

3.1. Regolith Porosity: Shadow Hiding Opposition Effect (SHOE) Width

The reliability of porosity estimates based upon the SHOE angular width strongly depends on being able to distinguish the CBOE from the SHOE in the dataset. While this can be reliably done only with circular polarization data (cf. *Hapke*, 2012a, and references therein), meaningful estimates have been obtained (cf. *Helfenstein and Shepard,* 2011) from imperfect coverage that begins at phase angles just beyond the most well-defined portion of the CBOE spike, but well within the broader SHOE (see Fig. 6). However, if coverage extends into the realm of the CBOE, then the contributions of both SHOE and CBOE must be disentangled. This requires having phase coverage over the full opposition effect with data points collected in small increments of phase angle especially near zero phase, and preferably with similar coverage in circular polarization.

In the Hapke model, the porosity of the uppermost regolith layer can be estimated from the angular width parameter, h_S, of the SHOE and also from *Hapke*'s (2008, 2012b) compaction correction, K. The porosity, P = (1–ϕ) is related to ϕ, known as the packing factor, and

$$K = \frac{-\ln\left(1-1.209\phi^{2/3}\right)}{1.209\phi^{2/3}} \tag{1}$$

The relationship between h_S and P can be complicated depending upon the rate at which the regolith compacts with depth near the surface and the size-distribution of regolith particles (see equation (3)). In practice, it is generally assumed that the porosity is effectively constant over the shallow regolith depths to which optical reflectance is measured and that the particle size distribution is reasonably represented by an average grain diameter. Then

$$h_S = \frac{3}{8}K\phi = -\frac{3}{8}\frac{\phi\ln\left(1-1.209\phi^{2/3}\right)}{1.209\phi^{2/3}} = -0.31\phi^{1/3}\ln\left(1-1.209\phi^{2/3}\right) \tag{2}$$

where ϕ = (1–P). Equations (1) and (2) are considered valid for porosities in the range between 24.8% and 100%

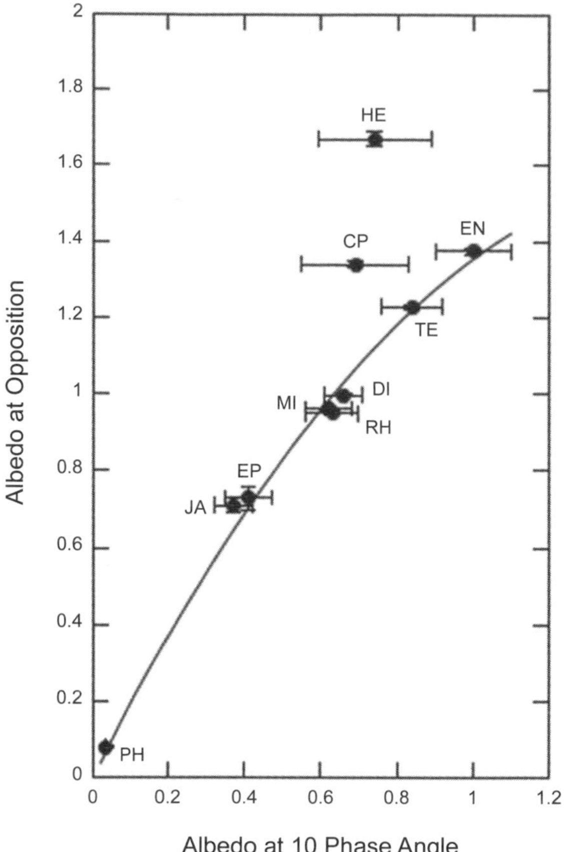

Fig. 1. Comparison of visible geometric albedos to whole-disk albedos for Janus (JA), Epimetheus (EP), Mimas (MI), Enceladus (EN), Tethys (TE), Calypso (CP), Dione (DI), Helene (HE), and Rhea (RH) obtained at 10° phase angle (data from the chapter by Thomas et al. in this volume). Curved line is an empirical least-squares fit of polynomial $p_0 = 1.97\ p_{10} - 0.61\ p_{10}^2$ to all data points except outliers CP and HE. Here, p_0 and p_{10} are the geometric albedos measured at opposition and at 10° phase angle, respectively.

(*Helfenstein and Shepard,* 2011). The corresponding ranges of $1.00 \leq K < 8.43$ and $0.0 < h_s < 2.4$. *Hapke* (2008) notes that the validity for porosities less than about 50% is not yet known because of coherent effects that are not treated in this model, but which must increasingly contribute as the grain spacing approaches the wavelength of light. However, the HWHM of SHOE is $2h_s$ so that its theoretical upper limit (~2.4) corresponds to a HWHM that is unrealistically large (~275°). In reality, it is unlikely that any detectable SHOE HWHM would be greater than 90°, so that in practice, h_s should be less than ~0.8. Such a large value might arise if regolith compacts at shallow depths in a peculiar way (cf. *Hapke,* 1986) or if the particle size distribution and size scale of photometrically expressed regolith porosity is comparable to the size scales of detectable macroscopic roughness (see below). It is also important to note that, especially in the case of bright icy satellites, the SHOE can be obscured by multiply-scattered light, so that when the SHOE cannot be reliably detected the porosity must be estimated from the parameter K alone using equation (1).

Using Enceladus as a representative example, $h_s = 0.20$ (Helfenstein et al., in preparation, 2018) with equation (2) yields a porosity of P = 65 ± 10%, a value that is about 20% more compacted than the recent lunar regolith estimate of 83 ± 3% (*Hapke and Sato,* 2016). For the Rhea value of $h_s = 0.08 \pm 0.02$ (*Verbiscer and Veverka,* 1989) the corresponding porosity P = 83 ± 4% is remarkably lunar-like.

A widely used earlier approach assumes a non-uniform grain size lunar-like distribution and relates the h_s parameter to porosity P

$$h_S = -\left(\frac{3}{8}\right) * \ln(P) * Y, = -\left(\frac{3}{8}\right) * \ln\left(1 - \phi\right) \qquad (3)$$

where Y = sqrt(3)/($\ln(r_l/r_s)$). r_l/r_s is the ratio of the effective radius of the largest grain to the effective radius of the smallest grain (*Hapke,* 1986; *Helfenstein and Veverka,* 1987).

Table 3 lists photometric estimates of porosity for a wide variety of saturnian satellites derived using equation (2). Porosities range from 9% to almost 100% with a mean of 71 ± 25%. Of the values given in Table 3, the three representing the most extremely porous surfaces are probably overestimates because the retrieved values of h_s appear to characterize the narrow CBOE rather than the broader SHOE (see Fig. 6).

In at least one case, a significant difference in porosity between the leading hemisphere and trailing hemisphere of the body was found in the FUV. Applying equation (3) to Cassini UVIS data, *Royer and Hendrix* (2014) derived porosities of 25% on the leading hemisphere of Tethys and 49% on its trailing hemisphere. Mimas and Dione have very low porosities (<5%) as seen at 180 nm. Note that these porosity estimates are made under the assumption of a lunar-like grain distribution, with significant uncertainties on the h_s term. This calculation did not take into account the CBOE, based on preliminary results showing that adding the CBOE contribution does not significantly improve the fit of phase curves on these satellites in the FUV. Equation (2) confirms the same relative trends but with somewhat larger porosities. For example, Mimas and Dione have more compact regoliths with porosities of 31 ± 4% and 42 ± 7%, respectively, while the leading and trailing hemispheres of Tethys have porosities of 74 ± 1% and 86 ± 2%, respectively.

3.2. Macroscopic Surface Texture: Photometric Roughness

Depending on the representative size scale at which it is measured and the way it is defined, macroscopic surface roughness can provide insights into the kinds of alteration processes, both endogenic and exogenic, that are at work on the surfaces of Saturn's icy moons. For example, on typical scales from submeter to many kilometers, infilling of topographic depressions by particulate deposits from cryovolcanos or exogenously placed dust tends to mute

TABLE 3. Aggregate textural properties of saturnian satellite regoliths.

Satellite	SHOE			$\bar{\theta}$(°)	Reference
	HWHM (°)	h_S	Porosity (%)*		
Janus	55 ± 6	0.48 ± 0.05	42 ± 3	24 ± 5	Helfenstein et al. (in preparation, 2018)
Mimas	9 ± 1	0.077 ± 0.006	84 ± 1	30 ± 1	*Verbiscer and Veverka* (1992)
	94 ± 10	0.82 ± 0.18	31 ± 4	[20]‡	*Royer and Hendrix* (2014)‡
(LH)	73 ± 11	0.64 ± 0.05	35 ± 1	[20]‡	*Royer and Hendrix* (2014)‡
(TH)	115 ± 57	1.0 ± 0.5	$28 \pm ^{13}_{3}$	[20]‡	*Royer and Hendrix* (2014)‡
Enceladus	23 ± 6	0.20 ± 0.05	65 ± 5	21 ± 5	Helfenstein et al. (in preparation, 2018)
	18 ± 1	0.16 ± 0.02	70 ± 5	NA	*Verbiscer et al.* (2005)
Tethys	18 ± 2	0.16 ± 0.03	70 ± 5	23 ± 5	*Elder et al.* (2007)
	11 ± 4	0.10 ± 0.04†	80 ± 7	[20]‡	*Royer and Hendrix* (2014)‡
(LH)	15 ± 6	0.13 ± 0.05	75 ± 7	[20]‡	*Royer and Hendrix* (2014)‡
(TH)	$7 \pm ^{8}_{7}$	$0.06 \pm ^{0.07}_{0.06}$	$87 \pm ^{13}_{12}$	[20]‡	*Royer and Hendrix* (2014)‡
Dione	55 ± 4	0.48 ± 0.03†	42 ± 7	[20]‡	*Royer and Hendrix* (2014)‡
(LH)	53 ± 8	0.46 ± 0.07	43 ± 5	[20]‡	*Royer and Hendrix* (2014)‡
(TH)	58 ± 11	0.51 ± 0.10	40 ± 7	[20]‡	*Royer and Hendrix* (2014)‡
Rhea	9 ± 2	0.08 ± 0.02	83 ± 3	13 ± 5	*Verbiscer and Veverka* (1989)
	0.8 ± 1.1	0.007±0.001¶	98 ± 1	16 ± 2	*Domingue et al.* (1997)§
	0.05	0.0004¶	>99	33	*Ciarniello et al.* (2011)
(LH)	[9 ± 2]	[0.08 ± 0.02]	[83 ± 3]	6 ± 5	*Verbiscer and Veverka* (1989)
	0.8 ± 1.1	0.007 ± 0.001¶	98 ± 1	15 ± 2	*Domingue et al.* (1997)
(TH)	[9 ± 2]	[0.08 ± 0.02]	[83 ± 3]	10 ± 5	*Verbiscer and Veverka* (1989)
	0.8 ± 1.1	0.007 ± 0.001¶	98 ± 1	16 ± 2	*Domingue et al.* (1997)
Phoebe	4.4 ± 1.3	0.038 ± 0.011	91 ± 2	31 ± 4	*Simonelli et al.* (1999)
	7.1 ± 0.4	0.062 ± 0.003	86 ± 1	NA	*Miller et al.* (2011)
	4.6 ± 0.1	0.04 ± 0.01	91 ± 2	33 ± 3	*Buratti et al.* (2008)
Iapetus					
Dark terrain	NA	NA	NA	6	*Lee et al.* (2010)
Bright terrain	NA	NA	NA	NA	*Lee et al.* (2010)

* Porosity computed using equation (2).
† Value in brackets was assumed in model fit.
‡ Results obtained in FUV at λ =180 nm.
§ Mean of separately retrieved leading hemisphere and trailing hemisphere parameters.
¶ Values most likely represents CBOE angular width.

topographic relief and hence reduce roughness. Fluid flows from cryovolcanic events would similarly cause smooth uniform surfaces at these scales. In contrast, heavily cratered surfaces should preserve a relatively high degree of roughness. Differential measures of surface roughness can be used to derive relative surface chronology, delineate geologic units, and determine the areal extent over which specific physical and geological processes have operated.

In the context of photometrically detected roughness, topographic surface relief alters the photometric behavior of otherwise smooth planetary surfaces in fundamental ways. At any given phase angle, the reflectance of a surface of uniform albedo varies with incidence and emission angle across the disk. The rate at which the surface darkens as the emission angle increases to 90° at the limb (i.e., "limb-darkening") is significantly affected by surface roughness. At the same time, the rate at which shadows projected by topographic features appear and lengthen with increasing incidence angle toward the terminator is also affected by surface roughness. These effects are visible over many phase angles, but are most pronounced at high phase where the apparent separation of the limb and terminator across the illuminated crescent is very narrow (i.e., especially at large phase angles where the limb and terminator are close together on the thin crescent).

Geological features that may affect photometric behavior encompass size scales from mountains, craters, and scarps down to millimeter-sized clumps of regolith grains. Rough features change the amount of reflected radiation by altering the local radiance incident and emission angles, by casting shadows, and by obscuring structures behind them. In addition, the extent to which shadows cast by topographic relief are diluted by multiply-reflected light from Sun-facing surface facets plays a role in determining the extent and accuracy to which photometric measures of roughness can be determined. Surfaces composed of highly reflective materials may effectively camouflage the presence of surface relief. These factors must be carefully modeled and accounted for in any complete photometric model. Perhaps most importantly, roughness modeling enables the determination of average roughness *below the resolution limit of the detector*. Roughness models are commonly scale-invariant, and studies of at least the Moon have shown that small facets — mainly clumps of particles — dominate on the natural regolith surfaces that are likely widespread over most airless planetary bodies (*Helfenstein and Shepard,* 1999; *Cord et al.,* 2005).

Two commonly used models have been developed to describe photometrically detectable macroscopic roughness: Hapke's mean slope model (*Hapke,* 1984) and a crater roughness model (*Buratti and Veverka,* 1985). The problem of calculating the photometric effects of macroscopic roughness entails both computing the local incidence and emission angles on a surface covered with rough features, and computing the fraction of illuminated surface that is lost to facets blocking solar radiation as well as accounting for positive relief that blocks from view the surfaces behind it. Ideally, the effects of multiply-reflected light should be treated especially for surfaces composed of high-albedo materials. The resulting function can be fit through an iterative procedure to the specific intensity at each solar phase, incidence, and emission angle to derive the mean topographic slope angle (for the case of Hapke's model), or to scans of specific intensity vs. emission angle to derive the mean depth to radius in the case of the crater roughness model. The models yield similar results in describing ideal, rough features: Hapke's features are statistically represented as Gaussian distributions of slope angles, while the model of *Buratti and Veverka* (1985) covers a planetary surface with paraboloidal craters defined by a depth-to-radius ratio.

In practice, the retrieval of macroscopic roughness from photometric observations of a planetary surface or terrains is achieved while simultaneously solving for best-fit values of other components of the photometric model, such as the average particle single-scattering albedo, the opposition effect parameters, and the average particle single-scattering phase function. The ability to constrain all the functional components uniquely depends strongly on the available coverage of photometric geometry. With regard to best constraints on macroscopic roughness, for disk-integrated solar phase curves the effects of roughness are greatest at large solar phase angles (*Helfenstein et al.,* 1988). But even with a broad range in solar phase angles, the effects of roughness are convolved with the single-particle phase function. The most effective fits of photometric roughness are obtained by employing a disk-resolved form of the model, because the function form of I/F with respect to the emission angle is unique for a given value of the roughness parameter and assumed average particle albedo.

The estimation of photometric roughness from whole-disk phase curves has several limitations. Even with full coverage in phase angle, the results may not be unique, and the results may not accurately describe all the relevant physical characteristics of the surface [*Shepard and Helfenstein* (2007) and *Hapke*'s (2008) response; *Helfenstein and Shepard* (2011), and *Souchon et al.* (2011)]. The first criticism can be overcome by fitting a disk-resolved model, as described above. Global and hemispheric mean values of the parameters are best constrained by simultaneously fitting whole-disk and disk-resolved datasets. Reliable fits to the disk-resolved model require a good range in emission angles (see the discussion on Enceladus in the next section). Another serious concern is that current roughness models do not account for multiply-scattered photons and assume that all shadows are primary. For the high-albedo icy saturnian moons, most of which are bright at visible wavelengths, that assumption is not valid. In their assessment of when multiple scattering becomes important, *Buratti and Veverka* (1985) show that for icy surfaces with normal reflectances (when the emission, incident, and solar phase angle are all zero) less than 0.6, current models are sufficient. Therefore, for Phoebe, one of Saturn's darkest moons in the visible with normal reflectances less than 0.6 at wavelengths <0.98 μm, the measurement of mean topographic roughness of 33° ± 3° (*Buratti et al.,* 2008) is valid. For higher albedos, only comparisons of the roughness of surfaces of similar albedos is valid. One way around this problem is to fit data at wavelengths at which the albedo is low, such as within the water ice absorption bands in the near-infrared. Although, in principle, photometric roughness should be independent of wavelength within the geometric optics limit, it can indirectly be affected when the albedo of the surface varies strongly with wavelength. Then, the illumination of shadows from multiply-reflected light can dilute shadows at wavelengths where the surface albedo is high in comparison to wavelengths where the surface albedos and the shadows are dark. The expected result would be a correlation of decreasing photometric roughness with increasing wavelength-dependent albedo.

Cassini observations enabled a whole new level of study of macroscopic roughness, not only because there were so many more observations at large solar phase angles, but images at higher resolution enabled studies of separate geologic terrains. In addition, the accumulation of data by Cassini's Visual and Infrared Mapping Spectrometer (VIMS) and UVIS at wavelengths where water ice is dark, eliminated the problem of accounting for multiple scattering.

Recent analyses of Cassini VIMS observations show that Phoebe's macroscopic scale roughness is substantially greater than what was estimated from fits of the Hapke photometric roughness model to Voyager ISS imaging data (*Simonelli et al.,* 1999; *Buratti et al.,* 2008). Phoebe's roughness is consistent with a violent history of impacts with the other outer saturnian moons, as well as a long collisional history in the Kuiper belt, from which it is believed to originate (*Johnson and Lunine,* 2005).

Differences between roughness on leading and trailing hemispheres offer insights into which processes are at work on each. For example, annealing and possible "smoothing" of surfaces by high-energy electrons should dominate on the leading sides of the saturnian moons. This process appears to occur on Mimas and Tethys and to a lesser extent on Dione, producing lens-shaped regions of anomalously high thermal inertia (*Howett et al.,* 2011, 2012, 2014) that are not as rough as the non-thermally anomalous terrains (*Annex et al.,* 2013; *Verbiscer et al.,* 2014). The abundance of small rough features may explain why high-energy particles in the Saturn system are so effective at changing the roughness:

They would smooth mainly small asperities, which dominate the regolith (*Rozitis and Green,* 2011).

Lee et al. (2010) fit the *Buratti and Veverka* (1985) roughness model to high- and low-albedo regions of Iapetus and found that the dark, leading hemisphere is much less rough than the high-albedo regions on the trailing hemisphere (Fig. 2). The best fits to the low-albedo terrain yield depth-to-radius values of only 0.12, equivalent to a Hapke mean slope angle of 9°, while fits to the high-albedo regions yield a roughness equivalent to a mean slope angle of 30°–40°. Infill by dust originating from the Phoebe ring (*Verbiscer et al.,* 2009) has likely obliterated shallow craters and facets on Iapetus's dark, leading hemisphere.

Measurements of roughness on Enceladus are of particular interest, not only because roughness would help to tell the story of how small micrometer-sized particles might coat the moon's south polar terrain (SPT), but also because roughness that is characteristic of a terrain where plume deposits are known to exist could be used as "ground-truth" to seek out similar active regions on other icy satellites such as Europa and perhaps Dione. However, the SPT is difficult to model, mainly because the limited range in emission angle means the parameters are not well-constrained. Figure 3 shows several roughness fits to a scan of I/F extracted from Alexandria Sulcus. Formal fits yield depth-to-radius values of 0.1 (*Chang et al.,* 2009).

Table 3 lists estimates of Hapke's macroscopic roughness parameter $\overline{\theta}$ for various saturnian satellites. Values range from 6° to 33° with a mean of 22° ± 10°. Most values fall in the range that is typical of heavily cratered airless planetary bodies. As described earlier, multiply-reflected light between relief on high-albedo surfaces is a potential source of significant error. However, the three highest-albedo objects in Table 3 (Enceladus, Mimas, and Tethys) do not have atypically low values that might be expected if multiply-reflected light were attenuating dark shadows. Neither do the darkest surfaces (Phoebe and dark terrain on Iapetus) exhibit exceptionally high roughnesses.

The lunar regolith is our only *in situ* reference for the subdecimeter size scales that are represented by visual wavelength photometric studies (*Helfenstein and Shepard,* 1999). Because of the shadow-brightening effects of multiply-reflected light on high-albedo surfaces, it is possible that the most representative size scale that is detected as photometric roughness may change with albedo, with larger scales being detected on brighter surfaces and smaller scales being filtered out (cf. *Shepard and Campbell,* 1998). The mechanical architecture of regolith is very much fractal-like (cf. *Helfenstein and Shepard,* 1999; *Shkuratov,* 1995; *Shkuratov and Helfenstein,* 2001) and, if the size scale of photometric roughness is not much larger than the fairy-castle structure that is characteristic of such high-porosity, fluffy aggregations of particles, then the porosity and photometric roughness might be expected to correlate; however, data in Table 3 show no clear correlation between h_S and macroscopic roughness.

4. REGOLITH PARTICLE STRUCTURE AND TRANSPARENCY

The lunar surface remains our best source of information about the near-surface architecture of regolith that covers the airless planetary bodies. As a generic analog at optical wavelengths, lunar regolith is typically an aggregate composed largely of powdery, fine grains that clump to form a complex fairy-castle structure at subdecimeter size scales

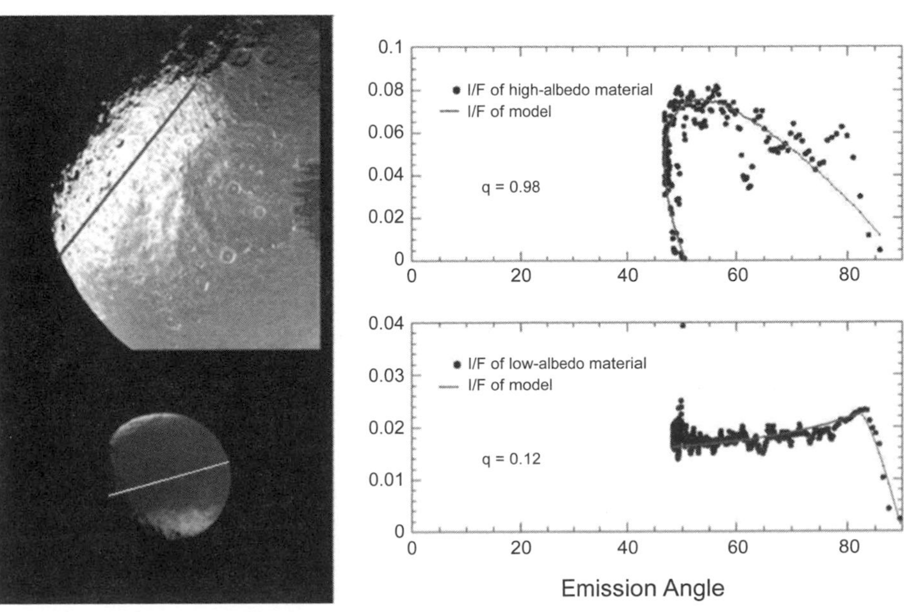

Fig. 2. Scans of I/F (black dots) extracted from both high-albedo and low-albedo regions of Iapetus and model fit (line) to the crater roughness model (*Lee et al.,* 2010).

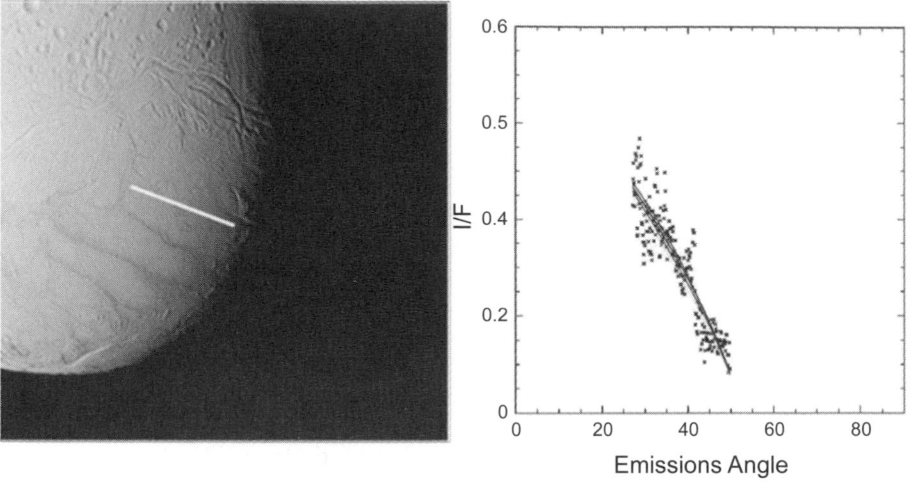

Fig. 3. From *Chang et al.* (2009), best fit of the crater roughness model to Alexandria Sulcus on Enceladus with a depth to diameter ratio of 0.10, showing substantial infill by small particles in the region. The various lines correspond to different amounts of multiply scattered radiation, which is approximated by a Lambert (isotropic) surface, ranging from 30% to 70%. The model does not account for partial illumination of shadows by multiply scattered photons.

(cf. *Helfenstein and Shepard,* 1999). Remotely-sensed information about the physical structure and transparency of regolith particles derive from two aspects of photometric behavior: the characteristic directional scattering of light over all phase angles by average grains in the regolith, which is represented by the average particle SPPF (see Appendix A), and the regolith's characteristic opposition effect. The grain structure of the regolith is inferred mostly from the average particle phase function, but it can be further constrained by testing the consistency of the interpretation against the amplitude of the SHOE and the angular width of the CBOE.

4.1. Regolith Grain Structure: Single Particle Phase Function (SPPF)

For regolith grains in mutual contact that are large compared to a wavelength of light, a qualitative assessment of the average regolith grain shape and sometimes its internal structure can be determined from the phase angle dependence of singly-scattered light from the regolith. However, the reliability of this approach depends strongly on the availability of high-quality observations that cover many phase angles from opposition to the forward-scattering direction, preferably much greater than $\alpha = 90°$. When excellent, high-quality coverage is available, it is possible to detect both the forward- and backward-scattering lobes of the SPPF. However, past experience with laboratory and planetary datasets have shown that the lobes are often asymmetrical in width, with the forward-scattering lobe being narrower than the backscattering lobe (*Hillier,* 1993; *McGuire and Hapke,* 1995; *Kamei and Nakamura,* 2002; *Helfenstein and Shepard,* 2011) and phase coverage to very large phase angles may be needed to detect accurately the detailed geometry of forward-scattering.

Earth-based observations of Saturn's satellites are limited to phase angles less than about 6°. However, with the advent of planetary space probes like Voyager and Cassini, phase coverage over a wide range has become a reality. The Voyager flyby missions provided generally non-uniform phase coverage of many of the mid-sized saturnian satellites to phase angles greater than 90°, but most of the best-quality images were disproportionately obtained at smaller phase angles. Voyager-based analyses of the surfaces of Saturn's icy moons indicated that their regolith grains generally scatter incident visible radiation primarily in the backward direction (*Verbiscer et al.,* 1990).

Cassini has provided greatly improved coverage at all phase angles, in some cases even out to $\alpha \cong 155°$. Recent studies have shown that saturnian satellite SPPFs do have a forward-scattering component (e.g., *Ciarniello et al.,* 2011). This result was not entirely unexpected because even the typical SPPF of lunar regolith, on average a comparatively dark material, exhibits a broad forward-scattering lobe (cf. *Hapke,* 1981). An important question that can be answered with Cassini is how strongly forward-scattering are the icy satellite regolith grains in comparison to the backscattering behavior. *Annex et al.* (2013) and *Verbiscer et al.* (2014) measured the directional scattering properties of Mimas using Cassini ISS images at 340 nm and *Royer and Hendrix* (2014) did the same using Cassini UVIS observations at 180 nm. Both studies found that Mimas is predominantly backscattering, although, as Fig. 4 shows, *Royer and Hendrix* (2014) found no difference in the directional scattering properties of Mimas' leading and trailing hemispheres

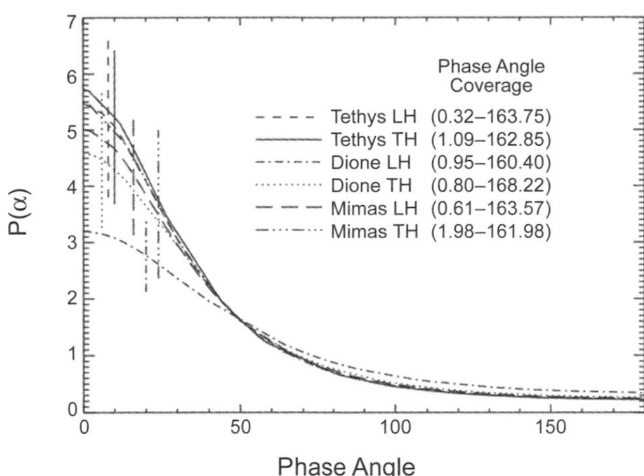

Fig. 4. Two-parameter Henyey-Greenstein phase function at 180 nm, where P(α) is the 2PHG from *Domingue et al.* (1991) (Appendix A, equation (A4)). The vertical bars represent the root mean square (rms) values for each hemisphere of each satellite. The phase angle coverage (in degrees) used to determine the P(α) values is given for each curve. Phase curves are plotted from results of *Royer and Hendrix* (2014).

at 180 nm, whereas *Annex et al.* (2013) and *Verbiscer et al.* (2014) showed that surface particles on Mimas' thermally anomalous terrain (found only on the moon's leading hemisphere) are more forward scattering at 340 nm than those elsewhere on the satellite. In the FUV, *Royer and Hendrix* (2014) found that Mimas' scattering properties are homogenous, whereas in the visible *Annex et al.* (2013) and *Verbiscer et al.* (2014) found that they are distinct. Although a slight difference between both leading and trailing hemispheres of Mimas was expected in the FUV, the quality of the UVIS datasets on Mimas and the poor coverage at very low phase angle on the trailing hemisphere do not allow for the detection of such an asymmetry. *Royer and Hendrix* (2014) also found that the surfaces of Tethys and Dione were backscattering at 180 nm, and while they also did not detect any difference between the directional scattering properties of the leading and trailing hemispheres of Tethys, as Fig. 4 shows, they did find that particles on the leading hemisphere of Dione are more forward-scattering than those on the trailing hemisphere.

As indicated in Table 4, in their analyses of Rhea from whole-disk phase curves, *Domingue et al.* (1997) and, more recently, *Ciarniello et al.* (2011) interpreted Rhea's regolith grains as being predominantly forward-scattering, in contrast to previous analyses by *Verbiscer and Veverka* (1989), who used both whole-disk and disk-resolved data. We discuss Rhea in greater detail below.

A useful analysis tool for interpreting the structural characteristics of regolith particles is an empirical relationship between the two model parameters (b,c) of the McGuire-Hapke (*McGuire and Hapke,* 1995) double Henyey-Greenstein representation of the average particle phase function, and the structural characteristics of corresponding well-characterized particles in the laboratory (see *McGuire and Hapke,* 1995; *Hapke,* 2012a,b; *Souchon et al.,* 2011; *Shepard and Helfenstein,* 2007). The two parameters of this function are b, which describes the (assumed equal) angular widths of forward- and backward-scattering lobes of the phase function, and c, which measures the relative amplitude of each lobe. The physical characteristics of different particle types map into specific regions of the McGuire-Hapke b-c plot, also known as the "hockey stick" diagram (*Hapke,* 2012a) because of the empirical shape of a mapped domain of b,c parameters and corresponding particle types. The mapped domain does not represent a physical limit and indeed it does not allow for particles that have forward- and backward-scattering lobes of different angular widths. As we will show later, an occasional pitfall of the McGuire-Hapke function can occur when observations are almost entirely restricted to the backscattering direction with some only sampling in the forward direction.

Figure 5 provides a representative sampling of saturnian satellite average global SPPF values. Figure 5a shows that most examples appear within a narrow portion of the diagram that is characterized by rough particles with relatively high-densities of internal scatterers. This region of the plot corresponds to particles that are dominantly backscattering, consistent with the early findings of *Verbiscer and Veverka* (1990). *Verbiscer and Veverka* (1990) and *Domingue et al.* (1997) showed that terrestrial snow surfaces exhibit particle phase functions that are dominantly forward-scattering and much different from those of icy regolith surfaces on airless icy planetary bodies. Figure 5a indicates that the primary difference is due to the more euhedral shapes and absence of significant internal scatterers in terrestrial snow grains. *Jost et al.* (2013) investigated the scattering properties of surfaces composed of micrometer-scale ice particles and monitored their change in scattering behavior at regular time intervals while the grains were allowed to sinter. Fresh particles were found to be strongly backscattering, and indeed, many of the icy satellites plotted in Fig. 5 fall within or near the trends defined by Jost et al.'s ice-dust aggregates. Lunar regolith appears to be intermediate between terrestrial snow and most saturnian satellites in the detectability of internal scatterers, but particle shape irregularities and surface roughness of lunar and typical saturnian satellite regoliths are greater than for terrestrial snow particles. Also, while Janus' regolith particles are not unusual for their density of internal scatterers, they appear to be somewhat less irregular or rough in shape than regolith grains on the other icy satellites.

Figure 5b identifies terrain-dependent differences in regolith particle types on Enceladus and Rhea. On Enceladus, the leading hemisphere appears to have particles that are slightly less irregular in shape but which contain a higher density of internal scatterers than average global examples. Anomalously scattering (dark) features on Enceladus' leading hemisphere (*Helfenstein,* 2012) exhibit particles that have an unusually high density of internal scatterers compared to other Enceladus features, like bright crater ejecta, which are most similar to average global values. Values for

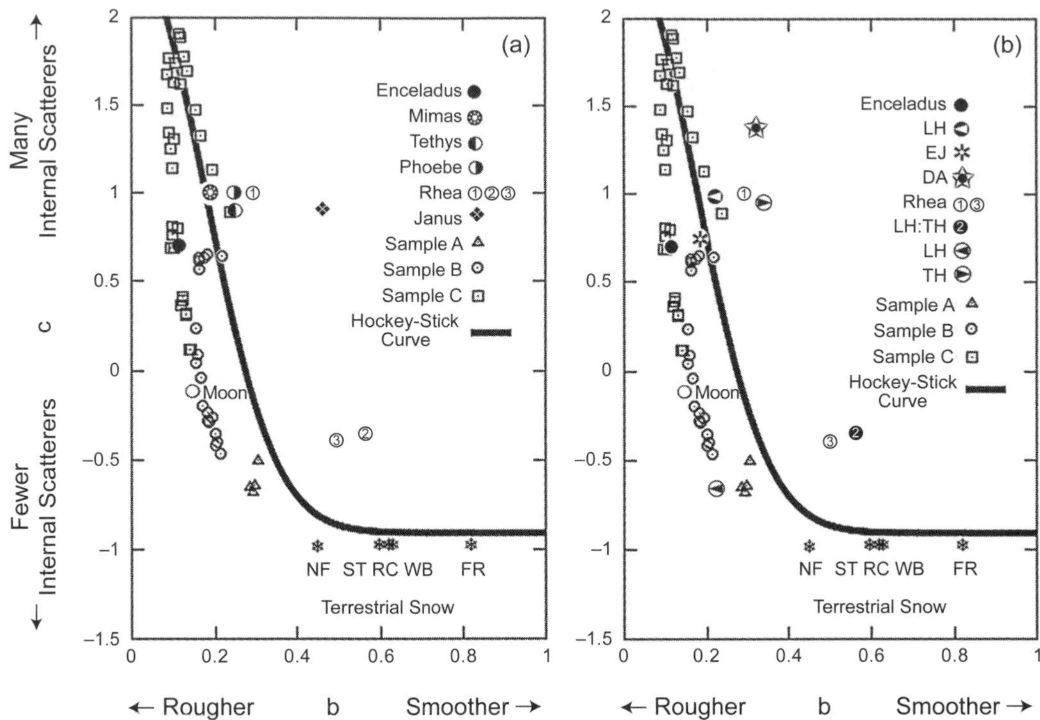

Fig. 5. (a) McGuire-Hapke "hockey stick" empirical relation (*Hapke,* 2012b) showing global average visual wavelength (~0.55 μm) particle phase function (SPPF) parameters b and c derived from a representative collection of saturnian satellites, the Moon (*Helfenstein et al.,* 1997), different terrestrial snow surfaces (*Domingue et al.,* 1997; *Verbiscer and Veverka,* 1990), and laboratory samples of micrometer-scale ice particles and aggregates of them (*Jost et al.,* 2013). Terrestrial snow subtypes are newly fallen (NF), settling (ST), rain crusted (RC), and frost (FR). Samples A, B, and C are the experimental ice-dust regolith analogs (*Jost et al.,* 2013) for which aggregates of micrometer-scale ice grains were observed at regular time intervals while being allowed to sinter. Sample A was observed with a 750-nm optical filter, samples B and C with a 650-nm filter. Saturnian satellites are Enceladus (*Helfenstein,* 2012), Tethys (*Elder et al.,* 2007), Mimas (*Verbiscer and Veverka,* 1989), Phoebe (*Simonelli et al.,* 1999), Rhea (① *Verbiscer and Veverka,* 1990; ② *Domingue et al.,* 1997; ③ *Ciarniello et al.,* 2011), and Janus (Helfenstein et al., in preparation, 2018). **(b)** Hockey stick plot that identifies Enceladus leading hemisphere (LH) and trailing hemisphere (TH) parameters, as well as dark anomaly (DA) and bright crater ejecta (EJ) parameters. The Rhea data point labeled ❷ identifies LH, TH, and global average b,c values, all of which coincide with the data point location. *Note:* McGuire-Hapke double Henyey-Greenstein b,c parameter values for Enceladus, Mimas, Tethys, Janus, Rhea [3PHG parameters are from *Domingue et al.* (1997)] and the Moon have been converted from original three-parameter double Henyey-Greenstein function parameters (section A1.2). For terrestrial snows and Rhea [2PHG parameters of *Domingue et al.* (1997)] the McGuire-Hapke b,c parameter values have been converted from the original b,c parameters of the Domingue double Henyey-Greenstein function (section A1.4).

Rhea differ greatly among different studies and depending on the range of data that were used to constrain the parameters.

In our earlier discussion (see Table 4), we noted that *Domingue et al.* (1997) and *Ciarniello et al.* (2011) obtained similarly forward-scattering SPPFs from fits of the Hapke model to whole-disk phase curve observations. In strong contrast, *Verbiscer and Veverka* (1989) found a dominantly backscattering SPPF from independent fits of Hapke's model to whole-disk and disk-resolved observations. The forward scattering SPPF results are intuitively attractive because they fall between lunar regolith grains, which are darker composed of lower-albedo materials (and thus would be expected to be less transparent) and terrestrial snow, which is typically composed of clean, pure water-ice with few internal crystal imperfections and internal scatterers and would be expected to be more transparent. However, these results were derived from whole-disk observations alone and they are vulnerable to ambiguities and complicated parameter couplings that can generally be eliminated only by constraining the Hapke model both with whole-disk and disk-resolved observations. Table 4 and Fig. 6 show that *Ciarniello et al.*'s (2011) phase curves provide good coverage of the backscattering lobe of the SPPF, but the forward-scattering direction is starkly sampled — in fact, only about 20° beyond α = 90°. *Domingue et al.* (1997) obtained a similar forward-scattering SPPF with more complete forward phase-curve coverage (to over 135°), but their solution trades off with a much lower surface roughness. In both of these cases, the angular width of the forward-scattering lobe is being almost entirely controlled by the fitted shape of the backscattering

TABLE 4. Hapke parameters for Rhea at 0.55 μm.

	ϖ_o	SHOE $B_{0,S}$	SHOE h_S	SPPF g_1 or b	SPPF g_2 or c	SPPF f	$\bar{\theta}$(°)	Photometric Angle Coverage (°)
Global								
Ve89*	0.86 ± 0.01	0.66 ± 0.02	0.08 ± 0.02	–0.29 ± 0.01 [b = 0.29]	0.0 [c = +1.0]	0.0	13 ± 5	1.8–135.3 (whole-disk + disk-resolved)
Do97†	0.97 ± 0.03	0.45 ± 0.02	0.0073 ± 0.001	0.57 ± 0.01	–0.35 ± 0.01	NA	16 ± 2	0.3–135.6 (whole-disk only)
Ci11‡	0.989 ± 0.001	1.8 ± 0.1	0.0004	0.5 ± 0.1	–0.4 ± 0.1	NA	33 ± 1	0.08–109.8 (whole-disk only)
Leading Hemisphere (Do97†)								
2PHG	0.986 ± 0.010	0.46 ± 0.01	0.0071 ± 0.001	0.58 ± 0.01	–0.34 ± 0.01	NA	16 ± 2	0.3–135.6
3PHG	0.996 ± 0.010	0.44 ± 0.01	0.0071 ± 0.001	0.41 ± 0.01 [b=0.23]	–0.65 ± 0.01 [c = –0.66]	0.171 ± 0.005	15 ± 2	0.3–135.6
Trailing Hemisphere (Do97†)								
2PHG	0.95 ± 0.01	0.44 ± 0.01	0.0074 ± 0.001	0.57 ± 0.01	–0.35 ± 0.01	NA	16 ± 2	1.8–16.6
3PHG	0.92 ± 0.01	0.57 ± 0.01	0.0074 ± 0.001	0.024 ± 0.005 [b = 0.33]	–0.35 ± 0.01 [c = 0.93]	0.964 ± 0.005	16 ± 2	1.8–16.6

* *Verbiscer and Veverka* (1989).
† *Domingue et al.* (1997).
‡ *Ciarniello et al.* (2011).

lobe. In the solution from *Verbiscer and Veverka* (1989), the forward-scattering lobe has an amplitude of zero and the extrapolation beyond the upper limit of coverage does not vary dramatically as in the previous two cases. None of the three studies sought to fit the narrow CBOE, and instead used only a SHOE model in their fits. The angular width of the CBOE for Rhea is known from *Verbiscer et al.* (2007) to have an angular HWHM of only 0.16°. Consequently, two forward-scattering SPPF solutions have corresponding SHOE solutions that better describe CBOE than the actual SHOE. However, Fig. 5b shows that, in the forward-scattering examples, the observations extend to small enough phase angles to detect the CBOE. However, in the study of *Verbiscer and Veverka* (1989), phase coverage does not extend to small enough phase angles to detect the CBOE. Consequently, Verbiscer and Veverka's fits to the SHOE cover an appropriate range of phase angles for that contribution. Verbiscer and Veverka's dominantly backscattering SPPF solution describes a reasonably broad SHOE that is similar in angular width to that for lunar regolith. However, the two forward-scattering solutions of the other studies effectively fit the CBOE with the SHOE function and fail to accurately detect the SHOE at larger phase angles.

4.2. Particle Transparency: Shadow Hiding Opposition Effect (SHOE) Amplitude

Hapke (1986, 2012a) provides a model parameter, B_0, to represent the amplitude of the SHOE and which may be interpreted as a measure of the transparency of an average regolith particle. Specifically

$$B_0 \cong \frac{S_0}{\varpi_0 P(0)} \tag{4}$$

where ϖ_0 is the average particle single-scattering albedo, P(0) is the value of the average particle phase function evaluated at zero phase, and S_0 represents the contribution of singly-scattered light that reflects from on or near the top surface of the particle. For a perfectly euhedral particle, S_0 is simply the Fresnel reflectance. For an opaque particle, all the light comes from the first surface and $B_0 = 1$ and B_0 decreases toward zero with increasing transparency of grains. When this parameter is well-determined it provides a consistency check with the particle structure implied by the hockey stick diagram. Observational constraints on the shape of SHOE generally require coverage at multiple phase angles less than a few tens of degrees and in small increments of phase angle near opposition. Modeling of FUV observations retrieve values for $B_{0,s}$ above 0.78 for both hemispheres on Mimas, Tethys, and Dione showing rather opaque particles.

However, with regard to icy satellites, multiply-scattered photons in high-albedo regoliths may either completely obscure the SHOE or otherwise increase the size-scale of detectable shadows to those between aggregate particles that approach the sizes of shadows that define the macroscopic scale roughness (cf. *Helfenstein and Shepard,* 2011). Then in the latter case, the amplitude of the SHOE no longer represents

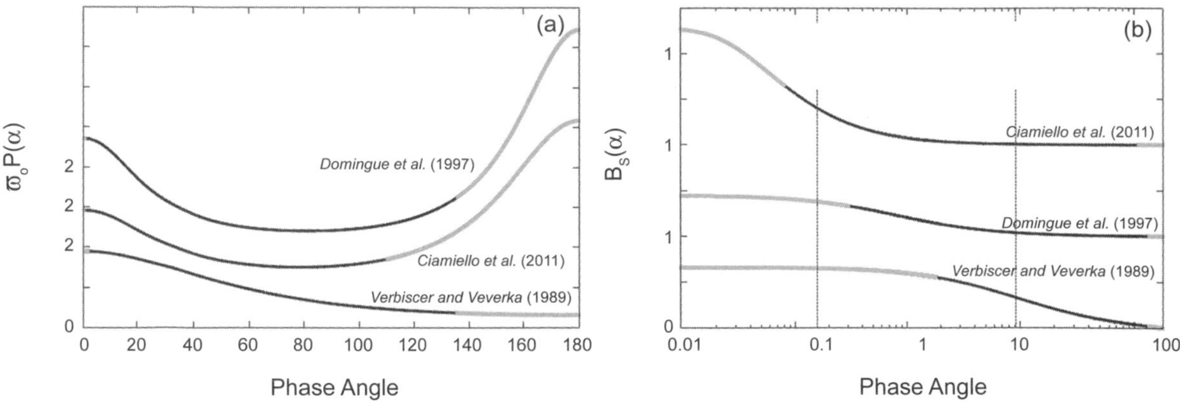

Fig. 6. Comparison of SPPF and SHOE curves for Table 4 fits of the Hapke model to Rhea. For clarity, all curves have been vertically offset by 1.0. Dark segments of each curve show the functions over the range of phase angles over which phase angle data were available to constrain the fits. Light gray segments represent extrapolations of the functions into the range where there are no data to constrain the fits. **(a)** Fits that show strong forward scattering are poorly constrained by available observations. Instead, the angular width of the forward-scattering lobe is almost entirely determined by the shape of the backscattering lobe. In the dominantly backscattering curve of *Verbiscer and Veverka* (1989), the amplitude of the forward-scattering lobe is zero and there is no ambiguity between the forward- and backward scattering lobes. **(b)** Plots of the best-fit opposition surge function for the three cases, using a logarithmic scale for phase angle to show details at small phase angles. Vertical bars identify the HWHM for the CBOE (from *Verbiscer et al.,* 2007) and from the best-fit of *Verbiscer and Veverka* (1989), respectively. The fits of *Ciarniello et al.* (2011) and *Domingue et al.* (1997) primarily describe the narrow CBOE, which is poorly constrained by the phase curve coverage used. The absence of a broader contribution forces a tradeoff in which the SPPF has a relatively narrow backscattering lobe. The *Verbiscer and Veverka* (1989) fit describes a broad SHOE that is wide enough to be well-constrained by the phase curve coverage and leads to a broad, unimodal SPPF that exhibits only weak forward scattering.

the characteristics of individual regolith particles, but rather clumps of particles or particle aggregate assemblages that are large enough to be opaque. The detectability of regolith-particle-scale SHOE amplitude can be complicated by the relatively high-transparency of water (and other) ices — so even for a relatively euhedral grain, there must be a very high density of internal scatterers for $B_{0,S}$ to be observable on the particle scales. The Fresnel reflectance of pure water ice at $\lambda = 0.56$ μm is ~0.2 (*Warren and Brandt,* 2008), which would approximate S_0 for a euhedral water ice crystal with no internal scatterers. Photometrically determined values for Enceladus from Cassini clear-filter imaging are $B_{0,S} = 0.53$ and $\varpi_0P(0) = 2.1$ (cf. Helfenstein et al., in preparation, 2018), consistent with the interpretation that high-albedo icy Enceladus regolith grains are neither opaque nor perfectly transparent. The measured values yield $S_0 = 1.1$, a factor greater than 5 over the theoretical value for a euhedral ice crystal lacking internal scatterers, consistent with the high density of internal scatterers indicated in Fig. 5. The Rhea value of $B_{0,S} = 0.66 \pm 0.02$ and $\varpi_0P(0) = 1.9 \pm 0.1$ yields $S_0 = 1.3 \pm 1$, not significantly different from Enceladus.

4.3. Grain Microstructure: Coherent-Backscatter Opposition Effect (CBOE)

The CBOE is generally observed at very small phase angles (less than a couple of degrees) near opposition. The detectability of this narrow contribution depends on the presence of small (approximately wavelength-sized or smaller) scatterers. While these scatterers can be whole regolith grains or minute dust particles coating larger regolith grains, they can also be structural imperfections on and within regolith grains. Examples of such structural imperfections include microscopic pits and surface asperities, and internal cracks, crystal defects, inclusions, bubbles, or other voids.

In *Hapke*'s (2002, 2012b) model, the angular-width of the CBOE is characterized by a parameter $h_C = \lambda/(4\pi\Lambda_T)$, where λ is the wavelength of light and Λ_T is the transport mean free optical path length of a photon. The expression predicts that, at any given wavelength, the angular width of the CBOE narrows as the Λ_T increases. The Λ_T itself depends on a variety of factors, including the packing density of the scatterers, their albedos, and the extent to which they are forward- or backward-scattering. The Λ_T is proportional to the spatial density and scattering cross-section of the scatterers and it is proportional to $(1-\xi)$, where $-1 \leq \xi \leq +1$ is known as the asymmetry factor. It is negative for dominantly backscattering and positive for forward-scattering.

As mentioned earlier, in the absence of circular polarization data, there is no absolutely reliable way to separate the contributions of CBOE and SHOE from a photometric phase curve. Although CBOE theory is well-established and observationally verified for well-separated particles smaller than the wavelength, this is not true for media of large particles in contact, as they are in regolith-covered surfaces. In particular, the predicted dependency on wavelength does not agree with observations. Consequently, attempts to derive the quantitative measures of regolith grain microstructure from the width of the opposition effect cannot be given much weight.

Moreover, it is difficult to interpret the structural properties of regolith grains uniquely from the angular width of the CBOE alone because there are many and varied configurations that can produce similar angular widths at a single wavelength. For example, if the scatterers that produce the observed CBOE are subwavelength-scale scatterers within individual grains, a more complex particle architecture would be suggested (cf. *Shkuratov and Helfenstein,* 2001), in comparison to the case for which the scatterers are mostly minute individual dust or frost grains. Thus, the interpretation of the angular width of CBOE is best used in concert with the constraints above and at best, as a qualitative consistency check.

Verbiscer et al. (2007) obtained observations of the CBOE for several saturnian satellites at $\lambda = 0.55$ μm (Table 5). Using Enceladus as an example, $h_C = 0.0013$ at $\lambda = 0.55$ μm, which gives $\Lambda_T = 34$ μm. This would be consistent with the presence of relatively transparent ice crystals containing a low density of internal, backward-scatterers, less-transparent crystals, but with a higher density of internal forward-scatterers. The latter is most consistent with the predictions of SHOE. Rhea's value is not significantly different from Enceladus'. Of the seven saturnian satellites for which *Verbiscer et al.* (2007) measured h_C, Tethys has the narrowest with $h_C = 0.0004$ with a $\Lambda_T = 109$ μm and Epimetheus has the broadest with $h_C = 0.015$ and $\Lambda_T = 3$ μm. For Tethys, the implication is that the individual regolith particles have a much lower density of internal scatterers than Enceladus. However, Fig. 5 suggests that the internal density of scatterers is not much different than for average Enceladus. One possibility is that the internal scatterers behave differently than those on Enceladus, for example, that the dominant scatterers may be complex grain-surface irregularities rather than internal scatterers and that the Λ_T for Tethys is more indicative of the separation distance of particles in the regolith It is interesting that, of the eight satellites in Table 5, the co-orbiting ring satellites, Janus and Epimetheus, similarly have the lowest estimated Λ_T values (7 μm and 3 μm, respectively). The scatterers for these objects may be close aggregates of roughly wavelength-sized grains themselves, perhaps captured from interactions with the F and G rings. We note that Hapke's model also includes a coherent backscatter amplitude coefficient $B_{0,C}$ that is related to the particle albedo, but at present its relationship to particle structure is not understood (*Hapke,* 2012b).

TABLE 5. Coherent backscatter opposition effect (CBOE) properties.

	Amplitude (mag.)	HWHM (°)	$B_{0,C}$	h_C	Λ_T (μm)
Tethys*	0.33	0.047	0.32	0.00041	109
Enceladus*	0.32	0.145	0.35	0.0013	33
Rhea*	0.47	0.158	0.33	0.0014	31
Mimas*	0.42	0.262	0.31	0.0023	19
Dione*	0.42	0.457	0.32	0.0040	11
Janus*	0.58	0.709	0.29	0.0062	7
Epimetheus*	0.51	1.705	0.27	0.0149	3
Phoebe†	0.32	0.41	0.20	0.0036	12

* *Verbiscer et al.* (2007).
† *Miller et al.* (2011).

In the near-infrared, the water ice absorption bands offer opportunities to probe regolith grain structure and transparency via the SHOE and CBOE. Within the water ice absorption band at 2 μm where grains are darker, Cassini VIMS observations (*Buratti et al.,* 2009) demonstrate that the shapes of the opposition surge of the leading hemisphere of Rhea are distinct: the opposition surge (both CBOE and SHOE) is broader and has a greater amplitude within the absorption band than outside the band (at 0.9 and 2.23 μm) (Fig. 7). Despite the fact that the CBOE is a multiple-scattering effect, it clearly is more pronounced within the band where Rhea is darker than outside the band where Rhea is brighter.

5. EFFECTS OF BOMBARDMENT BY RING PARTICLES

Light scattering in the FUV is particularly sensitive to relatively small amounts of surface weathering (*Hapke,* 2001; *Hendrix et al.,* 2003), but it is largely insensitive to effects of energetic charged particle bombardment due to the shallower sensing depths. Probing the uppermost layers of the icy satellite surfaces, FUV wavelengths are the most suited for the analysis of the alteration of icy surfaces of airless bodies by exogenic processes. Bombardment by E-ring grains, charged particles, and plasma are common processes acting in the saturnian system. Ion bombardment of ices is known to produce defects in the ice, creating voids and bubbles that affect the light-scattering properties of the surface. Exogenic processes can also change the chemistry of the surface by trapping gases, which can produce spectral absorption features (*Johnson,* 1997; *Johnson and Quickenden,* 1997; *Kouchi and Kuroda,* 1990; *Sack et al.,* 1992). Heavy (less-penetrating) damaging ions are known to brighten surfaces in the visible (*Sack et al.,* 1992). Bombardment by charged particles is responsible for several phenomena: implantation of new chemical species, chemical reactions and species creation, alteration of grain size and other microstructures, and even sputtering away of the surface.

Competing ring-interaction and magnetospheric sweeping processes act on the surface reflectance. As a general rule, E-ring grain bombardment brightens the surface on the leading hemispheres of satellites at visible wavelengths, while plasma bombardment darkens the surface on the trailing side. No asymmetry has been observed in the FUV between the saturnian and anti-saturnian hemispheres of those satellites, indicating that exogenic processes act primarily on the leading and trailing hemispheres (see the chapter by Howett et al. in this volume).

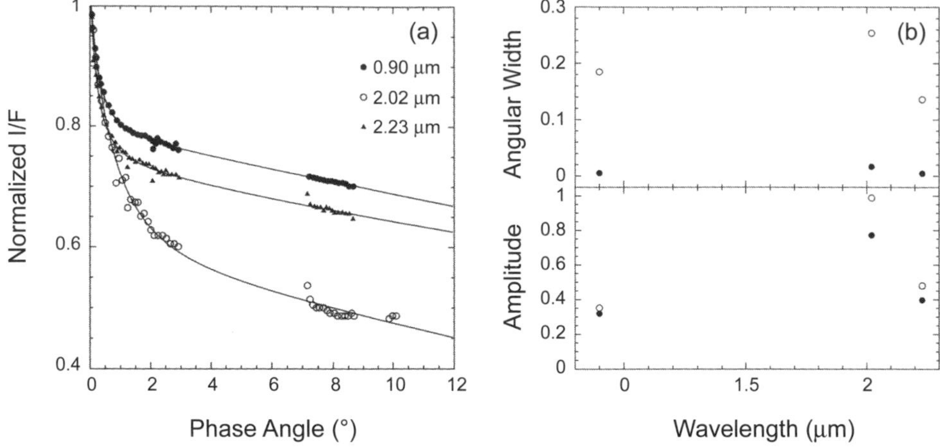

Fig. 7. (a) Normalized, near-infrared disk-integrated phase curves of the leading hemisphere of Rhea from Cassini VIMS observations (from *Buratti et al.,* 2009) illustrate morphological changes in the opposition surge inside (at 2.02 μm) and outside (at 0.90 and 2.23 μm) the 2-μm water ice absorption band. **(b)** Opposition surge parameters derived from fits of the *Hapke* (2012b) equation to the phase curves shown on the left. Filled circles are SHOE parameters and open circles are the CBOE parameters. Both opposition surges due to SHOE and CBOE are broader and have higher amplitude within the water ice absorption band at 2 μm, where the albedo is low.

Tethys and Dione have brighter leading hemispheres than trailing hemispheres at FUV wavelengths, while Mimas exhibits a more uniform, global FUV reflectance. Mimas' global uniformity can be explained by E-ring grains that are preferentially directed toward its trailing hemisphere (*Hamilton and Burns,* 1994). In this specific case, competing E-ring grain brightening and plasma bombardment darkening can explain Mimas' trailing hemisphere reflectance at 180 nm comparable to its leading hemisphere, impacted by fewer E-ring grains than on other satellites.

FUV retrieval of the SHOE parameters shows that Tethys exhibits a different behavior than Mimas and Dione, displaying a narrower opposition effect ($h_s = 0.134 \pm 0.05$ and $h_s = 0.064 \pm 0.07$ on the leading and trailing hemispheres respectively) at phase angles less than 20°. In comparison, FUV global average values of h_s for Mimas and Dione (Table 3) are 0.82 ± 0.18 and 0.48 ± 0.03, respectively. This result suggests a different microstructure and granularity of the regolith on Tethys, compared to that on Mimas and Dione. The orbital position of each satellite relative to Enceladus's plume, the source of the E-ring grains, is the most probable explanation, suggesting a more intense bombardment by the E-ring grains at the orbit of Tethys.

Because the icy satellites are relatively dark in the FUV compared to visible wavelengths, the contribution of multiply-scattered light between regolith particles is also lower and in turn, the amplitude of the SHOE is not as strongly attenuated by it. So FUV estimates of Hapke's B_0 parameter are especially useful for determining the transparency of surface grains. The $\varpi_0P(0)$ term of equation (4) relates to the total (internal + surface) incoherent backscattering from single particles, while the S_0 term relates only to the amount of particle surface backscattering. As described in section 4.2, the B_0 ratio is thus a measure of particle transparency.

Based on a study of the Galilean satellites, *Hendrix et al.* (2005) linked the $\varpi_0P(0)$ value to the amount of bombardment experienced by the satellite surfaces in the FUV, and results from Cassini support this interpretation. The single-scattering albedo ϖ_0 directly correlates with surface composition. (See the chapter by Hendrix et al. in this volume for a detailed discussion of satellite surface composition.) As seen in the FUV, Mimas' trailing and leading hemispheres should thus have roughly similar composition, while the trailing hemispheres of Tethys and Dione include an additional absorbing species. As shown in Fig. 8, Tethys and Dione have higher values of $\varpi_0P(0)$ on the leading hemisphere than on the trailing hemisphere. This result is consistent with the fact that more bombardment by E-ring grains on their leading side is expected. Dione also has lower values than Tethys, consistent with the fact that Dione is further from Enceladus, the source of the E-ring bombardment. Dione is expected to experience a lower amount of bombardment by E-ring grains than Tethys and Mimas and thus should exhibit less fresh bright water-ice on its surface. In addition, an exogenic process acting only on the trailing side, such as a darkening agent coming from outside the Saturn system [suggested by *Clark et al.* (2008)], could explain the low photometric values of its trailing hemisphere. Mimas displays a slightly higher value on the trailing hemisphere, within large uncertainties. Mimas is expected to experience more E-ring grain bombardment on its trailing side (*Hamilton and Burns,* 1994).

Figure 8 provides a straightforward explanation for these observations. Values of S_0 for both hemispheres of Dione

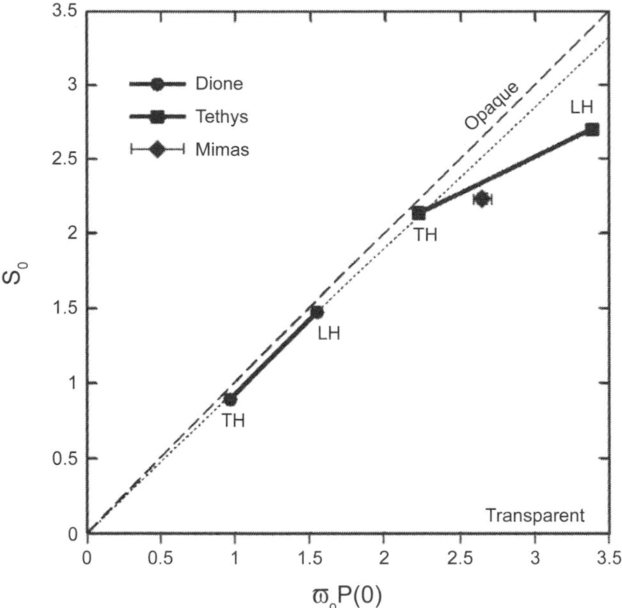

Fig. 8. Systematic LH-TH differences in the properties controlling the amplitude of the SHOE for Mimas, Tethys, and Dione as a measure of the amount of received particle bombardment (modified from *Royer and Hendrix,* 2014). Rearranging equation (4), $S_0 = B_0\ [\varpi_0 P(0)]$ so that the plot of a straight line intersecting the origin would have a slope of constant B_0. The short dashed line represents a line of constant $B_0 = 0.95$, the mean global value for Dione. The long-dashed line represents perfectly opaque particles (i.e., $B_0 = 1$). Ideally, transparent particles would plot along the horizontal line $S_0 = 0$. Data point for Mimas is the global mean with 1σ error bars.

and the trailing hemisphere of Tethys plot along Dione's mean global opposition amplitude line (corresponding to $B_0 = 0.95$). The particles that are represented by this trend are all almost perfectly opaque to the same extent despite their differences in composition. This is an indication that the grains have evolved to the same high degree of structural damage and mechanical complexity as a result of magnetic sweeping (especially in the case of THs) and damage from radiation and high-energy particle bombardment over time. However, the LH of Tethys and both hemispheres of Mimas fall significantly below this opacity trend in the direction that represents particles that are more transparent. Since Mimas and the LH of Tethys are the most heavily bombarded by the E ring, it is clear that a primary effect of E-ring bombardment is to emplace or otherwise facilitate the surface growth of clean ice grains that are youthful and less damaged by radiation and ion bombardment than the highly-damaged particles represented in the other cases. However, given the optical constants of pure ice at λ = 180 nm [1.4358 + 2.8255 × 10^{-6}i (*Warren and Brandt,* 2008)], the Fresnel reflectance of clean ice at normal incidence would give $S_0 = 0.032$ for clear, euhedral ice crystals. Thus, even the E-ring-affected regolith grains have considerable microstructure, perhaps as a result of microscopic-scale impact welding and aggregation into larger particles.

6. REMAINING QUESTIONS, FUTURE WORK

For the immediate future, it will be important to complete a survey of the surface properties of the satellites and their terrains. Because of Saturn's unique ring/satellite environment, it will be important to characterize how ring/satellite interactions affect the surface evolution of these bodies. With the identification of thermally anomalous terrains on Mimas and Tethys, it will also be important to define the range of the observable regolith physical properties that distinguish the anomalous terrains from the unaffected materials and tie these differences to the high-energy magnetospheric interactions that are believed to be the cause.

In the longer term, detailed spectrophotometric mapping at spatial resolutions that can be tied to specific classes of geological features and terrains will provide a useful framework for understanding how the surfaces of the objects have evolved over time through impact cratering, tidal tectonics, icy volcanism, thermal relaxation of topography, mass wasting, and thermal segregation of materials.

By far, Cassini ISS images and optical spectrophotometric data from the UVIS, VIMS, and Cassini Infrared Spectrometer (CIRS) instruments will be the focus of these studies. A key point when dealing with imaging photometry and optical spectrophotometry is that, ideally, reliable constraints on any of the surface physical properties should be obtained from observations at many different viewing and illumination geometries and that coverage over all geographic locations. Even though the successful Cassini mission has come to an end, we are still obtaining new coverage, especially of the small saturnian satellites. Even with the extensive coverage now at hand, the coverage especially at high spatial resolutions is seldom uniform over all geographic locations on any specific object. Consequently, novel strategies will need to be applied to optimize the reliability of surface physical properties that are measured with our collection of optical remote sensing data and provide rigorous estimates of uncertainty. An often overlooked critical need is for the reporting of realistic absolute confidence limits for derived model parameters and corresponding implied surface physical properties. Formal error bars related to goodness-of-fit of a model to different subsets of a particular collection of observations are useful for sorting the best-fit from the poorest ones, but they do not place results in a useful context for comparison to results of other studies. In the context of providing realistic uncertainty estimates, laboratory and field studies that test the reliability of photometric model interpretation to retrieve estimates of real surface physical properties, such as soil porosity and particle microstructure, will continue to be important. Such studies need to be as complete as possible in their physical characterization of the geological site and/or analog samples and their structural and optical properties from decimeter to microscopic size scales.

Analysis of the opposition effects of the satellites has turned out to be especially useful in providing estimates of surface physical properties of the satellite regoliths. Until a future space mission is launched to the Saturn system, new observations of the opposition effects in cases where we lack this information must come from telescopic observing. Ongoing improvements in remote sensing technology and instrumentation may soon be able to extend the quality of data and the range of observable satellites. Existing telescopic observations of the narrow coherent backscatter opposition effect that are available for some of the satellites have turned out to be a critically valuable aid in efforts to separate the contribution of CBOE from SHOE in Cassini observations. While the opportunities to observe saturnian satellites at exact opposition from Earth are rare, improving the existing telescopic dataset even for near-opposition apparitions can be valuable.

The importance of ring:satellite interactions has become a useful framework for investigating the surface properties of the satellites (*Buratti et al.,* 1990; *Verbiscer et al.,* 2007; *Lopes et al.,* 2008; *Schenk et al.,* 2011). *Thomas et al.* (2013) (see also the Thomas et al. chapter in this volume) demonstrate that the airless saturnian satellites fall into distinct morphological and compositional classes that coincide with their distances from Saturn and the ring environment in which each is embedded. The germane questions that remote sensing can help answer involve how ring:satellite interactions change the physical state of exposed surface materials over time. With regard to particle structure, what differences can be found among the satellites between native comminuted regolith, accreted ring material, Enceladus jet fallout, thermally segregated materials and contaminants, and thermally anomalous materials on Tethys and Mimas? How are the compaction and/or lithological states of regolith observed on the icy satellites affected by gravity differences, geological age, and surface processes such as mass wasting and thermal degradation?

Topics of special interest vary with Thomas et al.'s satellite groupings, and comparative analyses of regolith properties among all of them is an important direction for ongoing and future work. For the inner small ring shepherds, Pan, Atlas, and Daphnis, a focus will be on identifying the physical properties that distinguish their peculiar smooth-appearing equatorial ridges from the structurally distinct rougher terrain at higher latitudes. The F-ring shepherds, Prometheus and Pandora, are more heavily cratered objects with notable surface morphological variations that are likely accompanied by variations in regolith properties. Pandora exhibits a variety of crater forms, some relatively shallow and filled with ejecta and/or accumulations of F-ring materials, others that are relatively fresh-appearing with relatively bright materials exposed on the crater walls and darker floor deposits. Prometheus may be an example of a partially delaminated body that exposes a core that is morphologically distinct from the more typical cratered surface (see the chapter by Thomas et al. in this volume). The co-orbital moons, Janus and Epimetheus, are also heavily cratered objects that orbit further out between the F ring and the G ring. Their placement suggests that they should be less affected by ring-particle effects than Prometheus and Pandora. The arc-embedded or Alkyonide moons (Aegaeon, Pallene, Methone, and Anthe) are peculiar for their smooth, egg shapes. Methone is the only example for which we have relatively high-resolution coverage; however, whole-disk phase curves may identify the extent to which their surface materials are distinct from those on other icy saturnian satellites.

Moons that are or expected to be affected by the E ring include four major mid-sized moons (Mimas, Enceladus, Tethys, and Dione) as well as Trojan satellites of Tethys (Calypso and Telesto) and Dione (Helene and Polydeuces). As discussed earlier, identifying leading- and trailing-hemisphere surficial differences are key to understanding the different roles of E-ring bombardment on the leading hemispheres and magnetic sweeping of the trailing hemispheres. Rhea, which orbits just outside of the influence of the E ring, will serve as a good comparison body. Mimas and Tethys are both heavily cratered objects exhibiting similar geographic patterns of thermally anomalous terrain (see the chapter in this volume by Howett et al.) that are expected to be correlated with differences in surface physical properties. Tethys also has peculiar narrow reddish streaks (*Schenk,* 2015) as well as a mysterious equatorial band of relatively darker material. The Trojan moons all have large craters that show evidence of filling by particulate debris. However, they are especially peculiar because they exhibit dendritic patterns of topography and albedo that resemble drainage basins that carry very smooth looking materials in comparison to the coarser-textured mesas that border them. Because of its active geyser activity that feeds the E ring, Enceladus is a special focus of attention. Preliminary work already demonstrates global variations in surface albedo related to reaccumulation of E-ring materials and plume deposits. The surface of Enceladus exhibits a wide range of cratered, tectonically-disrupted, young, actively deposited volcano-tectonic terrains and thermally-modified features. The possible importance of sintering to regolith evolution on Enceladus is suggested by Fig. 5 and widespread examples of thermally-modified terrains that range from relatively ancient features such as viscously relaxed impact craters (*Bland et al.,* 2012), flexurally supported topography (*Giese et al.,* 2008), and unstable tectonic extension of ridge-and-trough features *(Bland et al.*, 2007) to structures such as pinnacle ridges that appear to be sculpted by recently active geysers in the south polar region (cf. *Helfenstein and Porco,* 2015). The surface physical properties that characterize these varied features will help to define how they and Enceladus as a whole have evolved through time.

Beyond the orbit of Titan are the Phoebe ring moons and the small, dark outer satellites of Saturn. The geometry of the Phoebe ring is consistent with dust ejected from low-albedo Phoebe (see Table 2) feeding this enormous diffuse ring that engulfs the outer small moons as well as extending to Iapetus and Hyperion closer to Saturn. Accumulations of dark Phoebe-dust on Iapetus has led to runaway thermal segregation of large regions of nearly black surfaces and bright accumulations of cold-trapped frost that subdivides the heavily cratered surface of Iapetus (cf. *Spencer and Denk,* 2010). Relatively dark deposits on material on the heavily cratered surface of the irregular moon Hyperion tend to accumulate on the flat, topographically low crater floor, while the sloping walls of the craters are significantly brighter. Thermal segregation has not evolved on Hyperion as it has on Iapetus. An important aim of photometric analysis of these objects will be to compare the physical properties of their dark and bright deposits in the context of surface evolution. We have no surface-resolved images of the dark, outermost moons of Saturn, so any analysis of their surface properties must be derived by whole-disk spectrophotometry, possibly from observations made by the James Webb Space Telescope (JWST). Nevertheless, comparing what can be ascertained about their surfaces, especially with regard to Phoebe's surface properties, will help to paint a more complete picture of their possible role in contributing to the Phoebe ring, or as targets of bombardment by Phoebe dust.

Acknowledgments. The authors gratefully acknowledge support from NASA's Cassini Data Analysis Program and the Planetary Geology and Geophysics Program.

REFERENCES

Annex A. M., Verbiscer A. J., Helfenstein P., Howett C. J. A., and Schenk P. (2013) Photometric properties of thermally anomalous terrains on icy saturnian satellites. *DPS Meeting #45,* 417.02.

Blackburn D. G., Buratti B. J., Ulrich R., and Mosher J. (2010) Solar phase curves and phase integrals for the leading and trailing hemispheres of Iapetus from the Cassini Visual Infrared Mapping Spectrometer. *Icarus, 209*, 738–744.

Bland M. T., Beyer R. A., and Showman A. P. (2007) Unstable extension of Enceladus lithosphere. *Icarus, 192,* 92–105.

Bland M. T. et al. (2012) Enceladus' extreme heat flux as revealed by its relaxed craters. *Geophys. Res. Lett., 39,* L17204.

Buratti B. J. and Veverka J. (1985) Photometry of rough planetary surfaces — The role of multiple scattering. *Icarus, 64,* 320–328.

Buratti B. J., Mosher J. A., and Johnson T. V. (1990) Albedo and color maps of the saturnian satellites. *Icarus, 87(2),* 339–357.

Buratti B. J. and 10 colleagues (2008) Infrared (0.83–5 μm) photometry of Phoebe from the Cassini Visual and Infrared Mapping Spectrometer. *Icarus, 193,* 309–322.

Buratti B. J. and 8 colleagues (2009) Opposition surges of the satellites of Saturn from the Cassini Visual Infrared Mapping Spectrometer (VIMS). In *Lunar and Planetary Science XL,* Abstract #1738. Lunar and Planetary Institute, Houston.

Chang J. P., Buratti B. J., Hicks M., Mosher J., and Landry B. (2009) Surface texture analysis of Enceladus' south polar region. Abstract P51A-1120 presented at 2009 Fall Meeting, AGU, San Francisco, Calif., 3–7 Dec.

Ciarniello M. and 10 colleagues (2011) Hapke modeling of Rhea surface properties through Cassini VIMS spectra. *Icarus, 214,* 541–555.

Clark R. and 11 colleagues (2008) Compositional mapping of Saturn's satellite Dione with Cassini VIMS and implications of dark material in the Saturn system. *Icarus, 193,* 372–386.

Cord A. M., Pinet P. C., Daydou Y., and Chevrel S. (2005) Experimental determination of the surface photometric contribution in the spectral reflectance deconvolution processes for a simulated martian crater-like regolith target. *Icarus, 175,* 78–91.

Domingue D., Hapke B., Lockwood G., and Thompson D. (1991) Europa's phase curve: Implications for surface structure. *Icarus, 90,* 30–42.

Domingue D., Hartman B., and Verbiscer A. (1997) The scattering properties of natural terrestrial snows versus icy satellite surfaces. *Icarus, 128,* 28–48.

Elder C., Helfenstein P., Thomas P. C., Veverka J., Burns J. A., Denk T., and Porco C. (2007) Tethys' mysterious equatorial band. *DPS Meeting #39,* 11.06.

Giese B. and six colleagues (2008) Enceladus: An estimate of heat flux and lithospheric thickness from flexurally supported topography. *Geophys. Res. Lett., 35,* L24204.

Hamilton D. P. and Burns J. A. (1994) Origin of Saturn's E ring: Self-sustained naturally. *Science, 264,* 550–553.

Hapke B. (1981) Bidirectional reflectance spectroscopy. 1. Theory. *J. Geophys. Res., 86,* 4571–4586.

Hapke B. (1984) Bidirectional reflectance spectroscopy. III — Correction for macroscopic roughness. *Icarus, 59,* 41–59.

Hapke B. (1986) Bidirectional reflectance spectroscopy. IV — The extinction coefficient and the opposition effect. *Icarus, 67,* 264–280.

Hapke B. (2001) Space weathering from Mercury to the asteroid belt. *J. Geophys. Res., 106,* 10039–10073.

Hapke B. (2002) Bidirectional reflectance spectroscopy. 5. The coherent backscatter opposition effect and anisotropic scattering. *Icarus, 157,* 523–534.

Hapke B. (2008) Bidirectional reflectance spectroscopy. 6. Effects of porosity. *Icarus, 195,* 918–926.

Hapke B. (2012a) Bidirectional reflectance spectroscopy. 7. The single particle phase function hockey stick relation. *Icarus, 221,* 1079–1083.

Hapke B. (2012b) *Theory of Reflectance and Emittance Spectroscopy,* 2nd edition. Cambridge Univ., Cambridge. 513 pp.

Hapke B. and Sato H. (2016) The porosity of the upper lunar regolith. *Icarus, 273,* 75–83.

Hedman M. M., Cooper N. J., Murray C. D., Beurle K., Evans M. W., Tiscareno M. S., and Burns J. A. (2010) Aegaeon (Saturn LIII), a G-ring object. *Icarus, 207,* 433–447.

Hedman M. M., Burns J. A., Thomas P. C., Tiscareno M. S., and Evans M. W. (2011) Physical properties of the small moon Aegaeon (Saturn LIII). *EPSC Abstracts, Vol. 6,* EPSC-DPS2011-531-2.

Helfenstein P. (2012) Seeing through frost on Enceladus. Abstract P32A-09 presented at 2012 Fall Meeting, AGU, San Francisco, Calif., 3–7 Dec.

Helfenstein P. and Porco C. C. (2015) Enceladus geysers: Relation to geological features. *Astron. J., 150,* 96–129.

Helfenstein P. and Shepard M. K. (1999) Submillimeter-scale topography of the lunar regolith. *Icarus, 141,* 107–131.

Helfenstein P. and Shepard M. K. (2011) Testing the Hapke photometric model: Improved inversion and the porosity correction. *Icarus, 215,* 83–100.

Helfenstein P. and Veverka J. (1987) Photometric properties of lunar terrains derived from Hapke's equations. *Icarus, 72,* 342–357.

Helfenstein P., Veverka J., and Hillier J. (1997) The lunar opposition effect: A test of alternative models. *Icarus, 128,* 2–14.

Helfenstein P., Veverka J., and Thomas P. C. (1988) Uranus satellites: Hapke parameters from Voyager disk-integrated photometry. *Icarus, 74,* 231–239.

Hendrix A., Vilas F., and Festou M. (2003) Vesta's UV lightcurve: Hemispheric variation in brightness and spectral reversal. *Icarus, 162,* 1–9.

Hendrix A. R., Domingue D. L., and Kimberly K. (2005) The icy Galilean satellites: Ultraviolet phase curve analysis. *Icarus, 173,* 29–49.

Henyey L. G. and Greenstein J. L. (1941) Diffuse radiation in the galaxy. *Astrophys. J., 93,* 70–83.

Hillier J. K. (1993) Voyager photometry of Triton. Ph.D. thesis, Cornell Univ., Ithaca.

Howett C. J. A., Spencer J. R., Schenk P., Johnson R. E., Paranicas C., Hurford T. A., Verbiscer A., and Segura M. (2011) A high amplitude thermal inertia anomaly of probable magnetospheric origin on Saturn's moon Mimas. *Icarus, 216,* 221–226.

Howett C. J. A., Spencer, J. R., Hurford T., Verbiscer A., and Segura M. (2012) Pac-Man returns: An electron-generated thermal anomaly on Tethys. *Icarus, 221,* 1084–1088.

Howett C. J. A., Spencer J. R., Hurford T., Verbiscer A., and Segura M. (2014) Thermophysical property variations on Dione and Rhea. *Icarus, 241,* 239–247.

Johnson R. (1997) Polar caps on Ganymede and Io revisited. *Icarus, 128,* 469–471.

Johnson R. and Quickenden T. (1997) Photolysis and radiolysis of water ice on outer solar system bodies. *J. Geophys. Res., 102,* 10985–10996.

Johnson T. V. and Lunine J. I. (2005) Saturn's moon Phoebe as a captured object from the outer solar system. *Nature, 435,* 69–71.

Jost B., Gundlach B., Pommerol A., Oesert J., Gorb S. N., Blum J., and Thomas N. (2013) Micrometer-sized ice particles for planetary-science experiments — II. Bidirectional reflectance. *Icarus, 225,* 352–366.

Kamei A. and Nakamura A. M. (2002) Laboratory study of the bidirectional reflectance of powdered surfaces: On the asymmetry parameter of asteroid photometric data. *Icarus, 156,* 551–561.

Kouchi A. and Kuroda T. (1990) Amorphization of cubic ice by ultraviolet irradiation. *Nature, 344,* 134–135.

Lee J. S., Buratti B. J., Hicks M., and Mosher J. (2010) The roughness of the dark side of Iapetus from the 2004 to 2005 flyby. *Icarus, 206,* 623–630.

Li J. Y., Helfenstein P., Buratti B. J., Takir D., and Clark B. E. (2015) Asteroid photometry. In *Asteroids IV* (P. Michel et al., eds.), pp. 129–150. Univ. of Arizona, Tucson.

Lopes R. M. C., Buratti B. J., and Hendrix A. R. (2008) The Saturn system's icy satellites: New results from Cassini. *Icarus, 193,* 305–308.

McGuire A. F. and Hapke B. W. (1995) An experimental study of light scattering by large, irregular particles. *Icarus, 113,* 134–155.

Miller C., Verbiscer A. J., Chanover N. J., Holtzman J. A., and Helfenstein P. (2011) Comparing Phoebe's 2005 opposition surge in four visible light filters. *Icarus, 212,* 819–834.

Royer E. M. and Hendrix A. R. (2014) First far-ultraviolet disk-integrated phase curve analysis of Mimas, Tethys, and Dione from the Cassini-UVIS data sets. *Icarus, 242,* 158–171.

Rozitis B. and Green S. (2011) Directional characteristics of thermal-infrared beaming from atmosphere-less planetary surfaces — a new thermophysical model. *Mon. Not. R. Astron. Soc., 415,* 2042–2062.

Sack N., Johnson R., Boring, J., and Baragiola R. (1992) The effect of magnetospheric ion bombardment on the reflectance of Europa's surface. *Icarus, 100,* 534–540.

Schenk P. (2015) "Blood stains" on Tethys: Evidence for recent activity? Abstract P21B-02 presented at Fall Meeting, AGU, San Francisco, Calif., 15–18 Dec.

Schenk P. and six colleagues (2011) Plasma, plumes and rings: Saturn system dynamics as recorded in global color patterns on its midsize icy satellites. *Icarus, 211,* 740–757.

Shepard M. K. and Campbell B. A. (1998) Shadows on a planetary surface and implications for photometric roughness. *Icarus, 134,* 279–291.

Shepard M. K. and Helfenstein P. (2007) A test of the Hapke photometric model. *J. Geophys. Res., 112,* E03001.

Simonelli D. P., Kay J., Adinolfi D., Veverka J., Thomas P. C., and Helfenstein P. (1999) Phoebe: Albedo map and photometric properties. *Icarus, 138,* 249–258.

Shkuratov Y. (1995) Fractoids and the photometry of solid surfaces of terrestrial bodies. *Solar System Res., 29,* 421–432.

Shkuratov Y. and Helfenstein P. (2001) The opposition effect and the quasi-fractal structure of regolith: I. Theory. *Icarus, 152,* 96–116.

Souchon A. L. and seven colleagues (2011) An experimental study of Hapke's modeling of natural granular surface samples. *Icarus, 215,* 313–331.

Spencer J. R. and Denk T. (2010) Formation of Iapetus' extreme albedo dichotomy by exogenically triggered thermal ice migration. *Science, 327,* 432.

Thomas P. C., Burns J. A., Hedman M., Helfenstein P., Morrison S., Tiscareno M. S., and Veverka J. (2013) The inner small satellites of Saturn: A variety of worlds. *Icarus, 226,* 999–1019.

Verbiscer A. J. and Veverka J. (1989) Albedo dichotomy of Rhea: Hapke analysis of Voyager photometry. *Icarus, 82,* 336–353.

Verbiscer A. J. and Veverka J. (1990) Scattering properties of natural snow and frost: Comparison with icy satellite photometry. *Icarus, 88,* 418–428.

Verbiscer A. J. and Veverka J. (1992) Mimas: Photometric roughness and albedo map. *Icarus, 99,* 63–69.

Verbiscer A. J., Helfenstein P., and Veverka J. (1990) Backscattering from frost on icy surfaces in the outer solar system. *Nature, 347,* 162–164.

Verbiscer A. J., French R. G., and McGhee C. A. (2005) The opposition surge of Enceladus: HST observations 338–1022 nm. *Icarus, 173,* 66–83.

Verbiscer A., French R., Showalter M., and Helfenstein P. (2007) Enceladus: Cosmic graffiti artist caught in the act. *Science, 315,* 815.

Verbiscer A. J., Skrutskie M. F., and Hamilton D. P. (2009) Saturn's largest ring. *Nature, 461,* 1098–1100.

Verbiscer A. J., Helfenstein P., and Buratti B. J. (2013) Photometric properties of solar system ices. In *The Science of Solar System Ices* (M. S. Gudipati and J. Castillo-Rogez, eds.), pp. 47–72. Astrophysics and Space Science Library, Vol. 356, Springer, New York.

Verbiscer A. J., Helfenstein P., Howett C., Annex A., and Schenk P. (2014) Photometric properties of thermally anomalous terrain on Mimas. DPS Meeting #46, 502.08.

Warren S. G. and Brandt R. E. (2008) Optical constants of ice from the ultraviolet to the microwave: A revised compilation. *J. Geophys. Res., 113,* D14220, DOI: 10.1029/2007JD009744.

APPENDIX A: VARIANTS OF THE HENYEY-GREENSTEIN PARTICLE PHASE FUNCTION

Over the 35 years since *Hapke*'s (1981) photometric model was introduced, a variety of different average particle SPPFs have been adopted for use in the Hapke model. These include Legendre polynomial representations, Mie functions, and most frequently, variants of the *Henyey and Greenstein* (1941) particle phase function. In this chapter, we summarized results of Hapke model fits that originally differed in terms of the SPPF that each worker adopted. However, to interpret the physical significance of the fits in Fig. 5 and Table 3, we expressed all results in terms of the model parameters of the McGuire-Hapke double Henyey-Greenstein function.

We present the variants below, and Table A1 lists conversions between the model parameters of each SPPF variant. It is important to note that exact conversions exist between some model pairs, and for others, only effective (i.e., non-equivalent) conversions exist. All the models can be exactly expressed in terms of the three-parameter double Henyey-Greenstein function (3PHG). As Table A1 shows, there are also exact conversions between both two-parameter Henyey-Greenstein functions (2PHG) and conversions from the single Henyey-Greenstein (1PHG) function to the two 2PHG variants.

The inexact, *effective* parameter conversions cannot generally be used to accurately model the SPPF phase curves equivalent to the predictions of the original source model parameters, except in special cases where the relations simplify. For example, an exact conversion from the 3PHG ⇨ 1PHG exists only for the case where the 3PHG partition coefficient $f = 0$ or $f = 1$, which in either case reduces the 3PHG to the 1PHG function. The effective parameters are often useful for presenting the results of fits in a context for physical interpretation (see Fig. 5).

A1. Single Henyey-Greenstein Function (1PHG)

The classical Henyey-Greenstein function is an empirical model that uses only one model parameter, the asymmetry parameter $g = \langle\cos(\alpha)\rangle$, where the angle brackets indicate the SPPF weighted mean of the cosine of the phase angle. The relation is normalized such that $g = 0$ for isotropically scattering particles, $g < 0$ for dominantly backscattering particles, and $g > 0$ for dominantly forward-scattering particles. Expressed terms of phase angle, α,

TABLE A1. Conversions between parameters of different particle phase functions.

Convert to	Convert from	Conversion
Equivalent Conversions		
3PHG: g_1, g_2, f	HG: g	$g_1 = g$, $f = 0$
3PHG: g_1, g_2, f	McGuire-Hapke: b, c	$g_1 = -b$, $g_2 = +b$, $f = (1-c)/2$
3PHG: g_1, g_2, f	Domingue: b′, c′	$g_1 = -b'$, $g_2 = +b'$, $f = (1-c')$
McGuire-Hapke: b, c	Domingue: b′, c′	$b = b'$, $c = 2c'-1$
Domingue: b′, c′	McGuire-Hapke: b, c	$b' = b$, $c' = (1 + c)/2$
Effective Conversions (Non-Equivalent)		
HG: g	3PHG: g_1, g_2, f	$g_{eff} = (1-f)\, g_1 + f\, g_2$
HG: g	McGuire-Hapke: b,c	$g_{eff} = -(b)(c)$
HG: g	Domingue: b′, c′	$g_{eff} = -(b')(2c'-1)$
McGuire-Hapke: b,c	3PHG: g_1 , g_2 , f	$b_{eff} = \| (1-f)\, g_1 + f\, g_2 \|$ $c_{eff} = 1-2f$
Domingue: b′, c′	3PHG: g_1, g_2, f	$b'_{eff} = \| (1-f)\, g_1 + f\, g_2 \|$ $c'_{eff} = (1-f)$

$$P_{HG}(\alpha; g) = \frac{\left(1-g^2\right)}{\left(1+2g\cos(\alpha)+g^2\right)^{3/2}} \tag{A1}$$

A2. Three-Parameter Double Henyey-Greenstein Function (3PHG)

The 3PHG is a simple linear combination of the 1PHG, in which there are two asymmetry parameters, g_1 and g_2, respectively, and a partition coefficient f

$$P_{3PHG}(\alpha; g_1, g_2, f) = \frac{(1-f)\left(1-g_1^2\right)}{\left(1+2g_1\cos(\alpha)+g_1^2\right)^{3/2}} + \frac{f\left(1-g_2^2\right)}{\left(1+2g_2\cos(\alpha)+g_2^2\right)^{3/2}} \tag{A2}$$

The separate asymmetry parameters allow the model to describe a SPPF that has a forward-scattering lobe that is different in angular width than the backscattering lobe. In most cases, the symmetry parameters have different signs, but if both have the same sign, it can describe a single lobe with a complex shape.

A3. McGuire-Hapke Double Henyey-Greenstein Function (2PHG)

The 2PHG function is a simplification of the 3PHG in which the angular widths of the forward- and backward-scattering lobes of the SPPF are assumed to be equal

$$P_{MH}(\alpha; b, c) = \frac{1+c}{2}\frac{\left(1-b^2\right)}{\left(1-2b\cos(\alpha)+b^2\right)^{3/2}} + \frac{1-c}{2}\frac{\left(1-b^2\right)}{\left(1+2b\cos(\alpha)+b^2\right)^{3/2}} \tag{A3}$$

The asymmetry parameter, $b \geq 0$, and the use of a single positive-valued asymmetry parameter to represent both forward- and backscattering is achieved by making use of the relation that $\cos(\pi-\alpha) = -\cos(\alpha)$ to substitute a negative sign for the positive sign in the denominator in one of the 1PHG contributions in equation (A2).

A4. Domingue Double Henyey-Greenstein Function (2PHG)

The Domingue 2PHG function is almost identical to the McGuire-Hapke 2PHG, except for a difference in the way that the forward- and backward-scattering components are partitioned

$$P_{Do}(\alpha; b', c') = \frac{c'\left(1-b'^2\right)}{\left(1-2b'\cos(\alpha)+b'^2\right)^{3/2}} + \frac{(1-c')\left(1-b'^2\right)}{\left(1+2b'\cos(\alpha)+b'^2\right)^{3/2}} \tag{A4}$$

Color Section

Plate 1. Io-like water plumes on Enceladus drawn in a popular article published in 1981 (*Gore,* 1981). Credit: Lloyd K. Townsend Jr./National Geographic Creative.

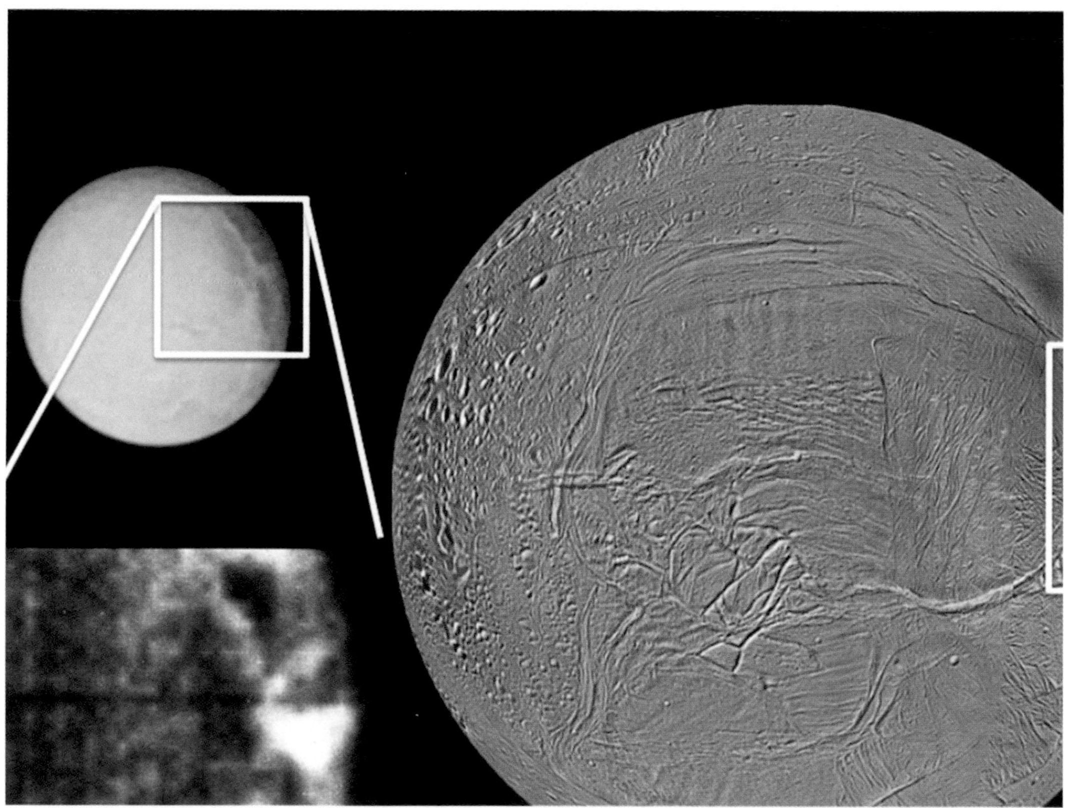

Plate 2. *Left:* An enhanced image of Enceladus from frame FDS 3493157 obtained by Voyager 1 of the southern hemisphere of Enceladus (whose limb is visible on the righthand side of the image) showing high-albedo markings on its trailing side at southern latitudes around 50°. *Right:* The Cassini map of Enceladus shows that this region is near the edge of the tiger stripe region. In each image the south pole is just off to the right of each image. Credit: NASA/JPL/Caltech; Cassini map is PIA18435.

Plates accompany chapter by Dougherty et al. (pp. 3–16).

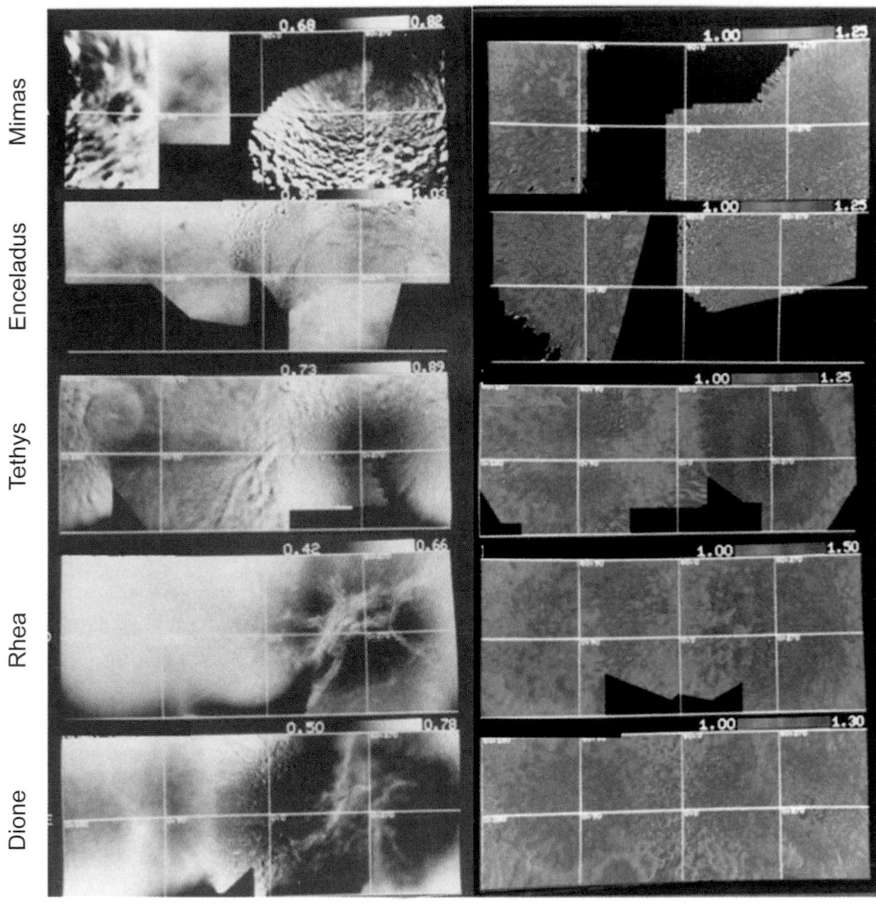

Plate 3. Voyager maps of the main inner moons of Saturn. The maps on the left depict the normal reflectance in the Voyager clear filter (0.47 µm), corrected for all the effects of viewing geometry (current values are higher because of a better knowledge of their opposition surges). The right set of maps are color ratios of the Voyager orange filter (0.59 µm; in the case of Tethys, the green filter at 0.56 µm) to the Voyager violet filter (0.41 µm). Adapted from *Buratti et al.* (1990).

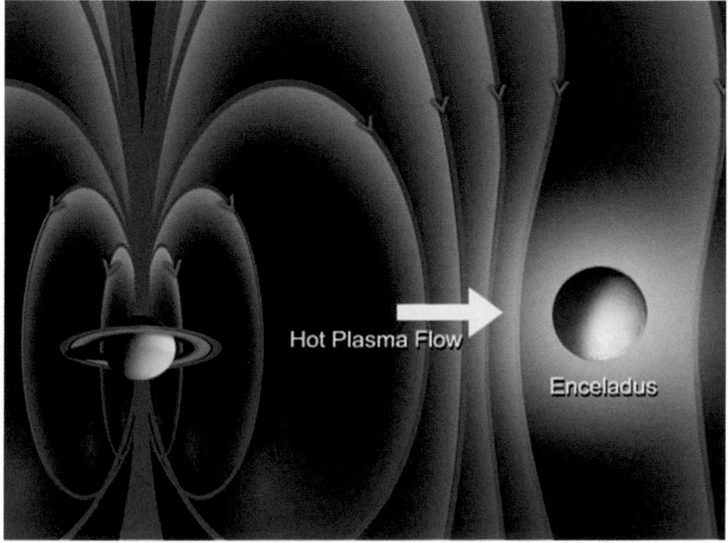

Plate 4. A schematic (where Saturn and Enceladus are not to scale) showing the co-rotating Saturn magnetic field and plasma being draped ahead of Enceladus by a diffuse extended atmosphere.

Plates accompany chapter by Dougherty et al. (pp. 3–16).

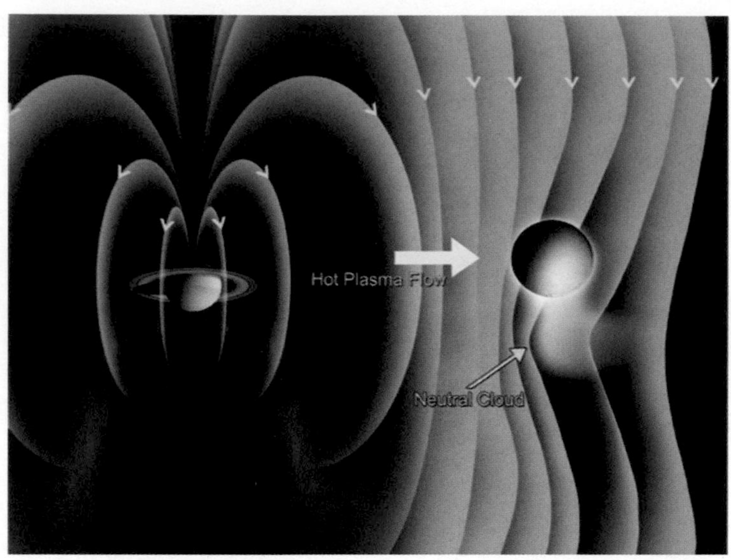

Plate 5. A schematic (where Saturn and Enceladus are not to scale) showing the co-rotating Saturn magnetic field and plasma being perturbed by the polar plume of water vapor generated at the south pole of Enceladus.

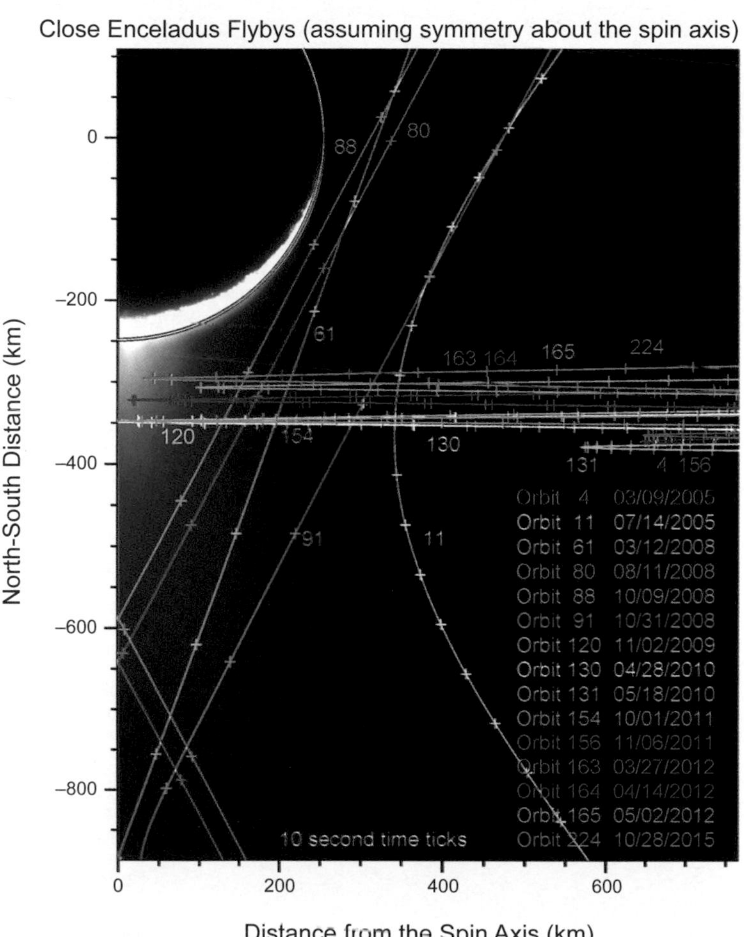

Plate 6. Schematic revealing the geometries of the targeted flybys relative to Enceladus, where the different colors show the different flybys and the date of their closest approach.

Plates accompany chapter by Dougherty et al. (pp. 3–16).

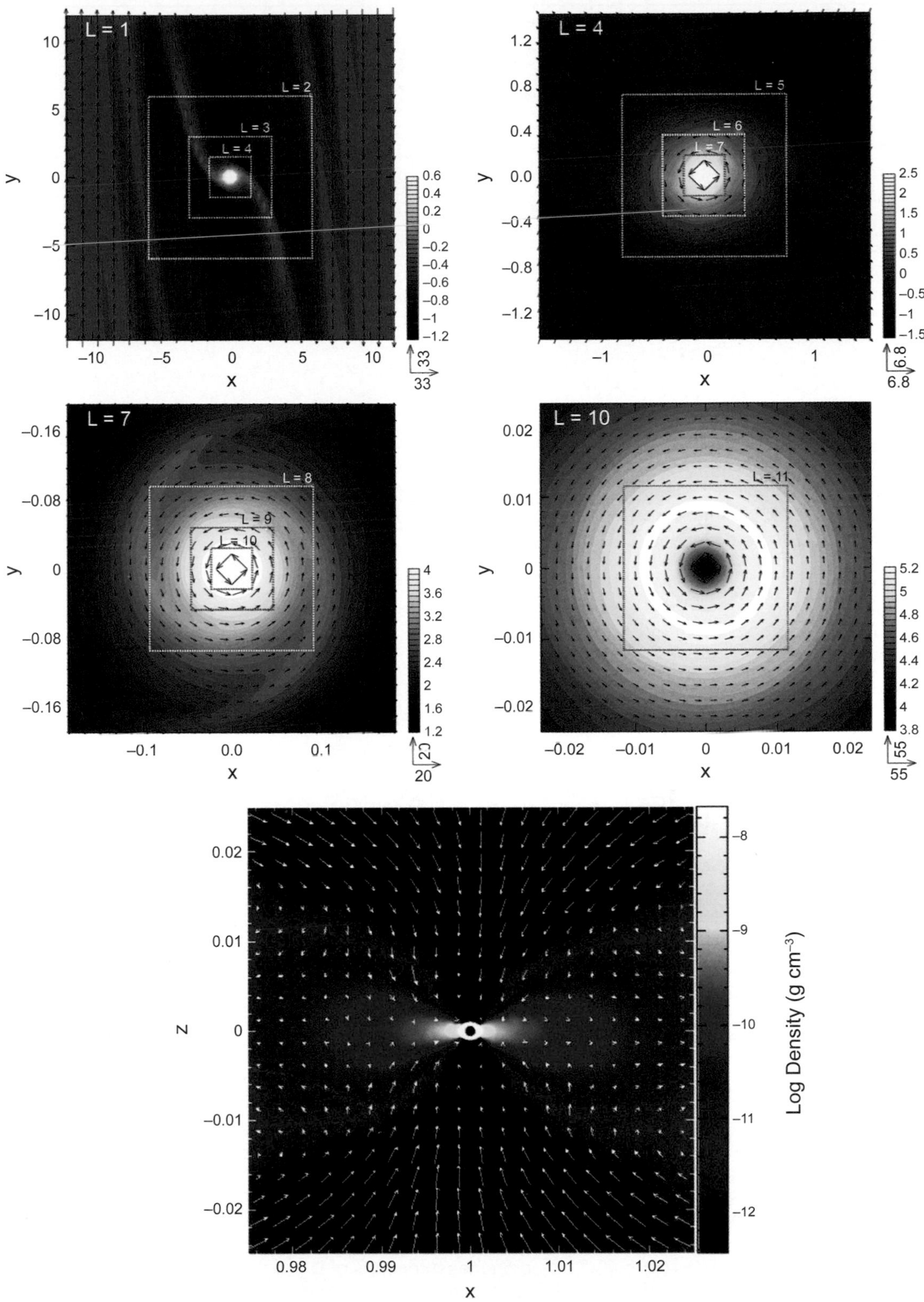

Plate 7. *Top:* Velocity (arrows) and log density (color) (both non-dimensional) of the flow around an accreting "Jupiter" in the protosolar nebular midplane, and at nested spatial scales (each factor of L is a power of 2); x and y are distances from the planet in units of the planet's Hill radius. From three-dimensional hydrodynamic calculations of *Tanigawa et al.* (2012). *Bottom:* Similar calculation for an accreting "Jupiter" from *Ayliffe and Bate* (2009), but for a vertical slice, showing rain of accreting gas and entrained solids onto surface of a circumplanetary accretion disk (CPD); x is normalized distance from the Sun, z is vertical to the mid-plane.

Plate accompanies chapter by McKinnon et al. (pp. 17–38).

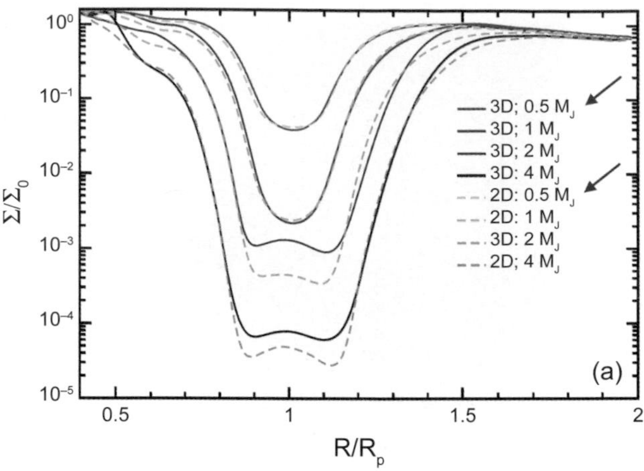

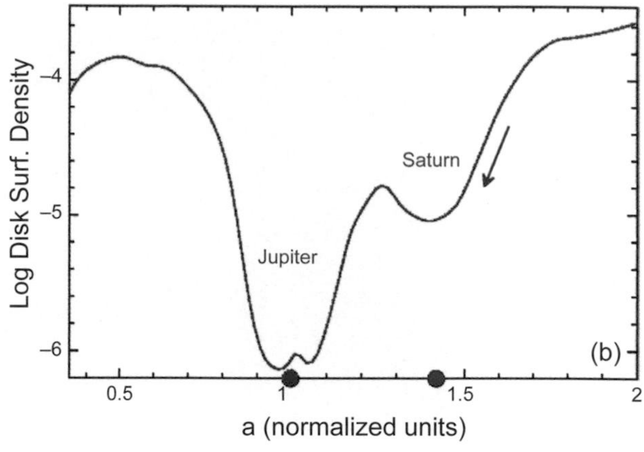

Plate 8. (a) Protosolar nebular surface density profiles, as a function of normalized radius from the Sun, averaged azimuthally except for a small area around a giant planet and averaged in time after steady-state conditions are achieved (after 1000 orbits); modified from *Fung and Chiang* (2016). **(b)** Nebular surface density with an embedded Jupiter and Saturn near but outside their 3:2 mean-motion resonance (A. Morbidelli, personal communication, 2016, and see text). Local density peak near Jupiter indicates formation of a circumplanetary accretion disk whereas the lack of a similar peak near Saturn signifies a more distended, circumplanetary envelope.

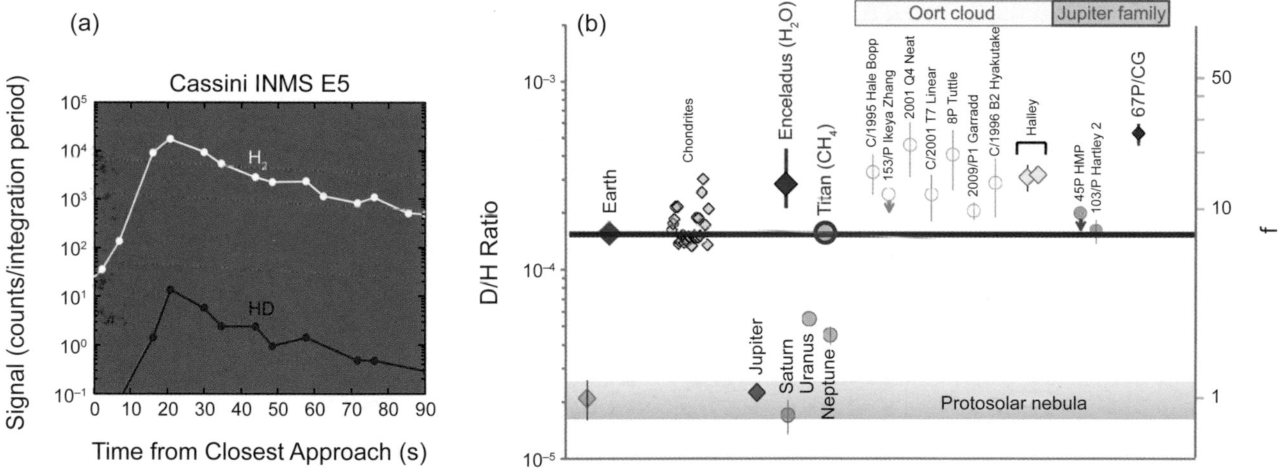

Plate 9. (a) The bulk H_2 signal seen during the E5 flyby, shown here, is representative of plume H_2O. During the time of maximum plume material influx into the INMS, signal-representing HD rose above noise levels for approximately 60 seconds, allowing a direct comparison of H_2 (black) to HD (red) to obtain the D/H ratio in water shown in Table 2. **(b)** D/H ratios in different objects of the solar system. Diamonds represent data obtained by means of *in situ* mass spectrometry measurements, and circles refer to data obtained with astronomical methods. The protosolar nebula (PSN) D/H value (lower left) is estimated to be 2.1 (± 0.5) × 10^{-5} based on measurements of D/H and $^3He/^4He$ in the atmosphere of Jupiter (*Mahaffy et al.,* 1998) and $^3He/^4He$ in the solar wind and meteorites (*Geiss and Gloeckler,* 1998). The fractionation factor f is defined as $[D/H]_{object}/[D/H]_{PSN}$. Modified from *Altwegg et al.* (2015).

Plates accompany chapter by McKinnon et al. (pp. 17–38).

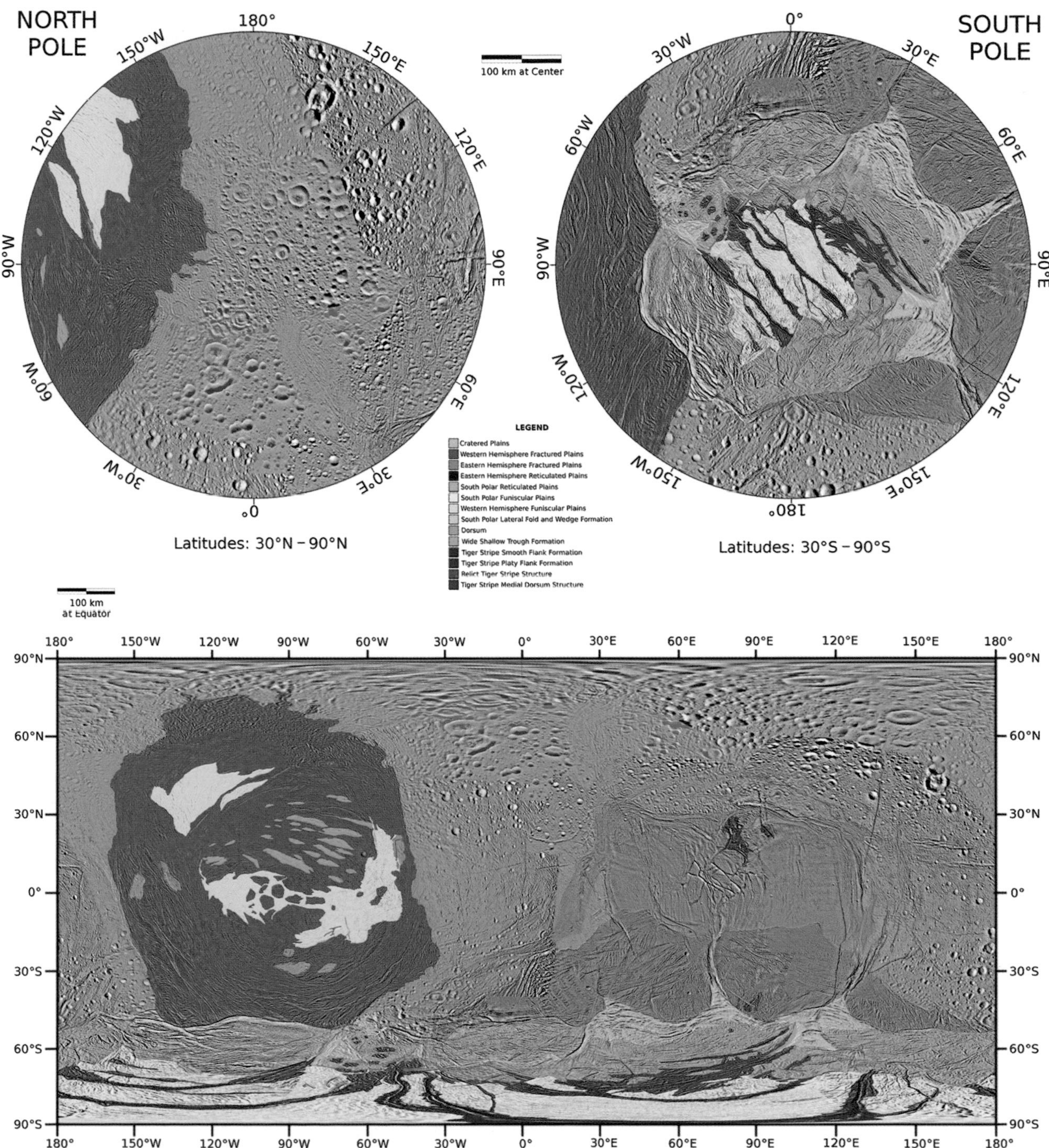

Plate 10. Polar stereographic (above) and simple cylindrical (below) image mosaic (*Becker et al.*, 2016) with superposed terrain unit map from *Spencer et al.* (2009); updated according to *Helfenstein et al.* (2010a,b).

Accompanies chapter by Patterson et al. (pp. 95–125).

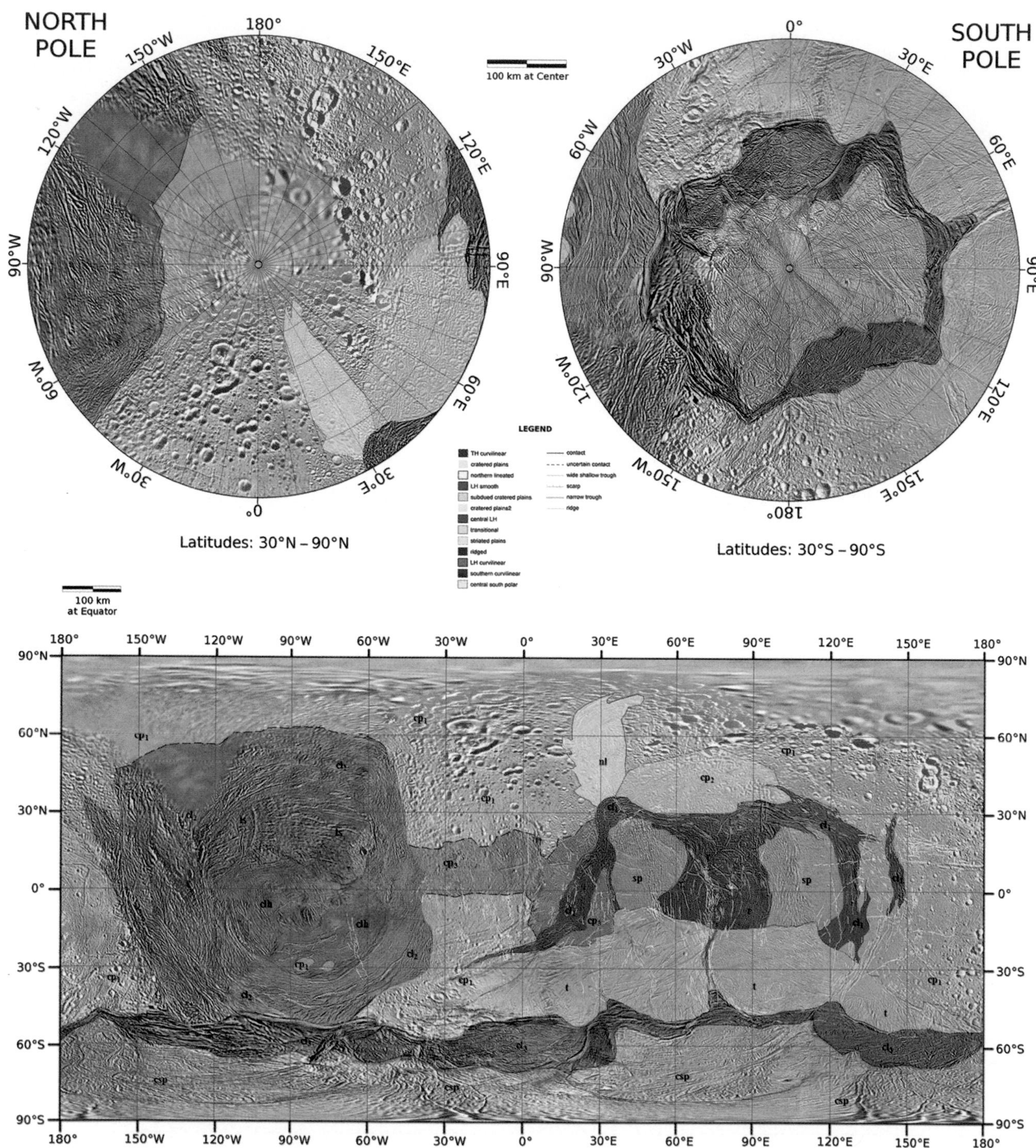

Plate 11. Polar stereographic (above) and simple cylindrical (below) image mosaic (*Becker et al.,* 2016) with superposed terrain unit map from *Crow-Willard and Pappalardo* (2015).

Accompanies chapter by Patterson et al. (pp. 95–125).

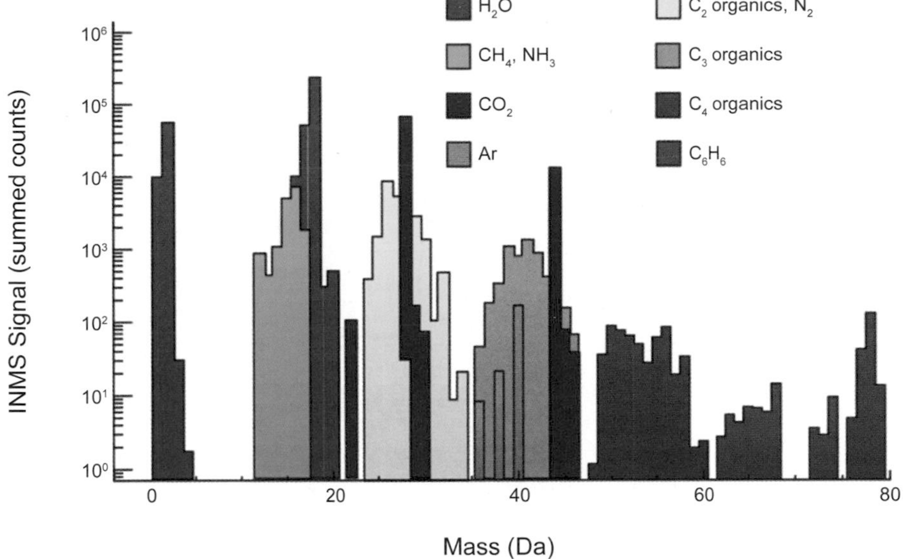

Plate 12. Volume mixing ratios based on analysis of the E5 data presented in *Waite et al.* (2009). Most organic species do not reflect the intrinsic gas composition. Due to the high velocity of the flyby, the spectrum is dominated by species from molecular fragmentation of high mass organic compounds probably residing in plume ice grains (Table 1). The presence of argon could not be confirmed by later measurements. Published with permission of Nature Publishing Group.

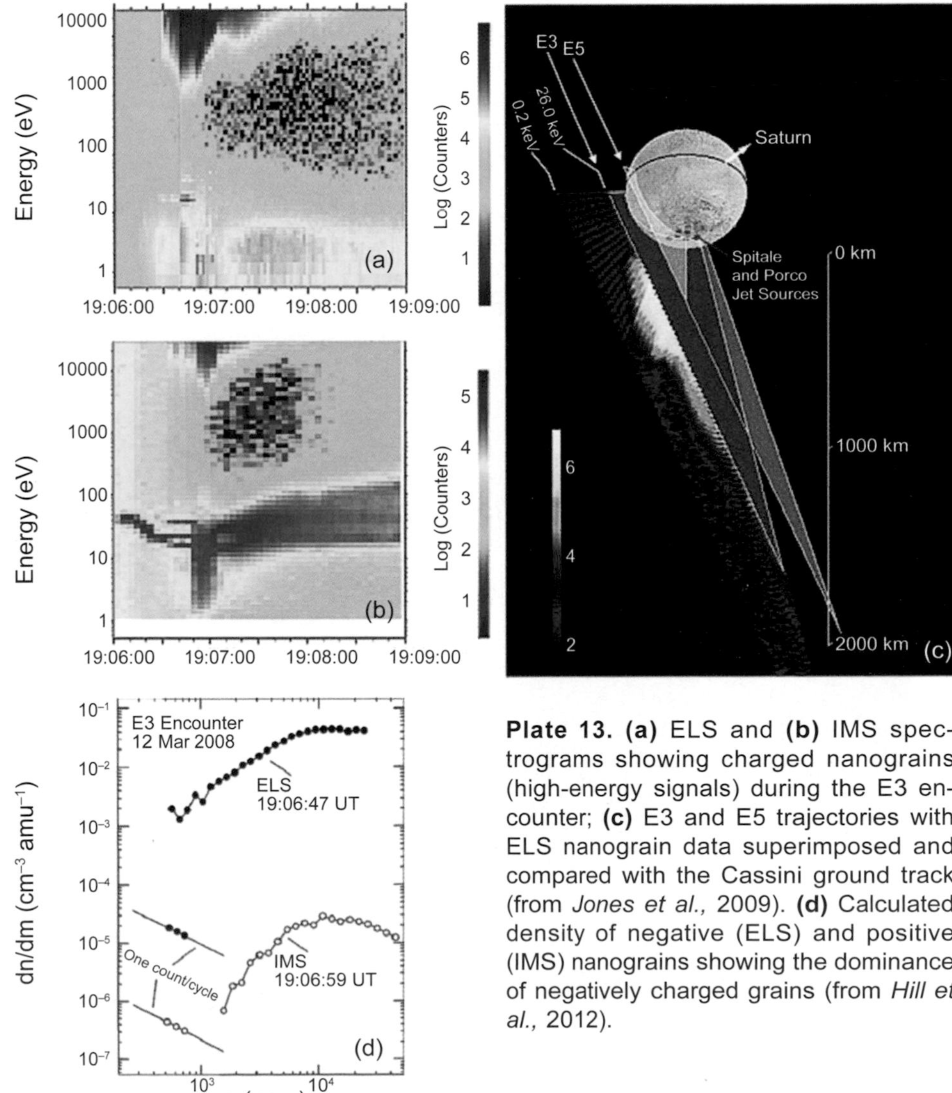

Plate 13. (a) ELS and **(b)** IMS spectrograms showing charged nanograins (high-energy signals) during the E3 encounter; **(c)** E3 and E5 trajectories with ELS nanograin data superimposed and compared with the Cassini ground track (from *Jones et al.*, 2009). **(d)** Calculated density of negative (ELS) and positive (IMS) nanograins showing the dominance of negatively charged grains (from *Hill et al.*, 2012).

Plates accompany chapter by Postberg et al. (pp. 129–162).

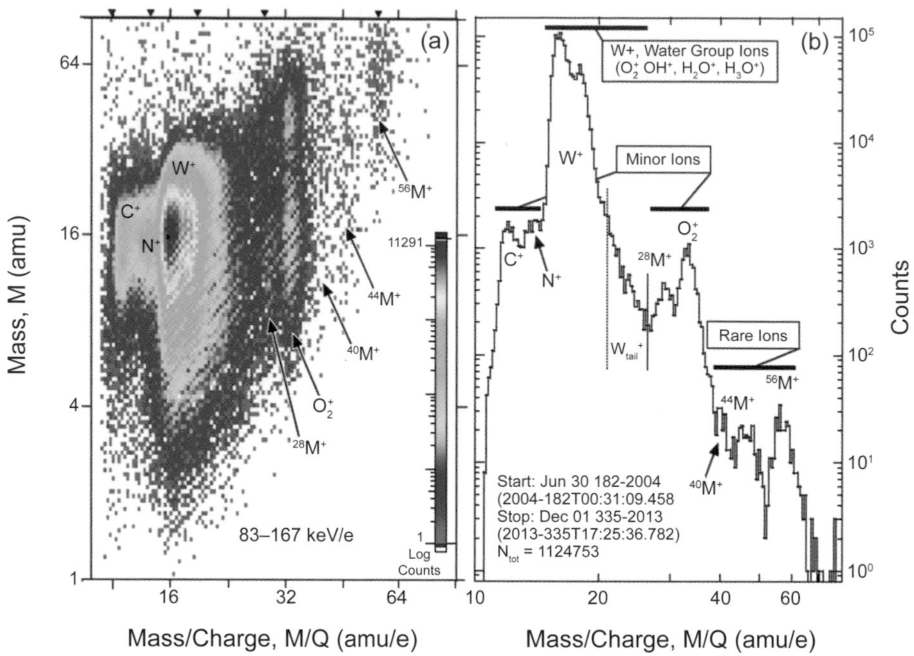

Plate 14. (a) Triple coincidence pulse height analysis events by MIMI-CHEMS measured in Saturn's near-equatorial magnetosphere under certain selection criteria (see *Christon et al.,* 2015) inside of 20 R_S or the magnetopause, whichever is closer to Saturn, from 2004 to 2013 are displayed in a mass (M) vs. mass-per-charge (M/Q), color spectrogram. The water group (W^+) (mostly O^+ at 16 amu/e, followed by roughly equal amounts of OH^+ and H_2O^+ at 17 and 18 amu/e, respectively, at about half the O^+ abundance, along with a little H_3O^+) and the minor and rare heavy ion species are identified. C^+ and N^+ are the most abundant non-water species. **(b)** Histogram of the data plotted in **(a)**. The histogram permits clearer identification of the minor ions like ^{28}M and O_2^+ and the "rare group" ions of CO_2^+ (44 M^+) and iron (56 M^+) and facilitates qualitative visual comparisons to W^+ and the minor ions. From *Christon et al.* (2015).

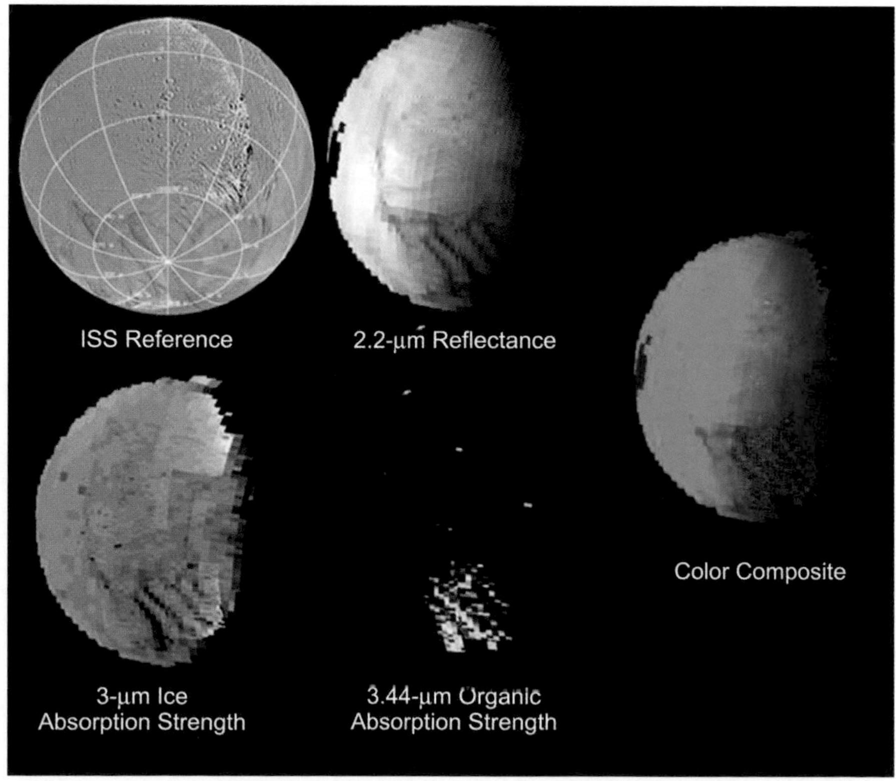

Plate 15. Cassini VIMS surface composition from *Brown et al.* (2006).

Plates accompany chapter by Postberg et al. (pp. 129–162).

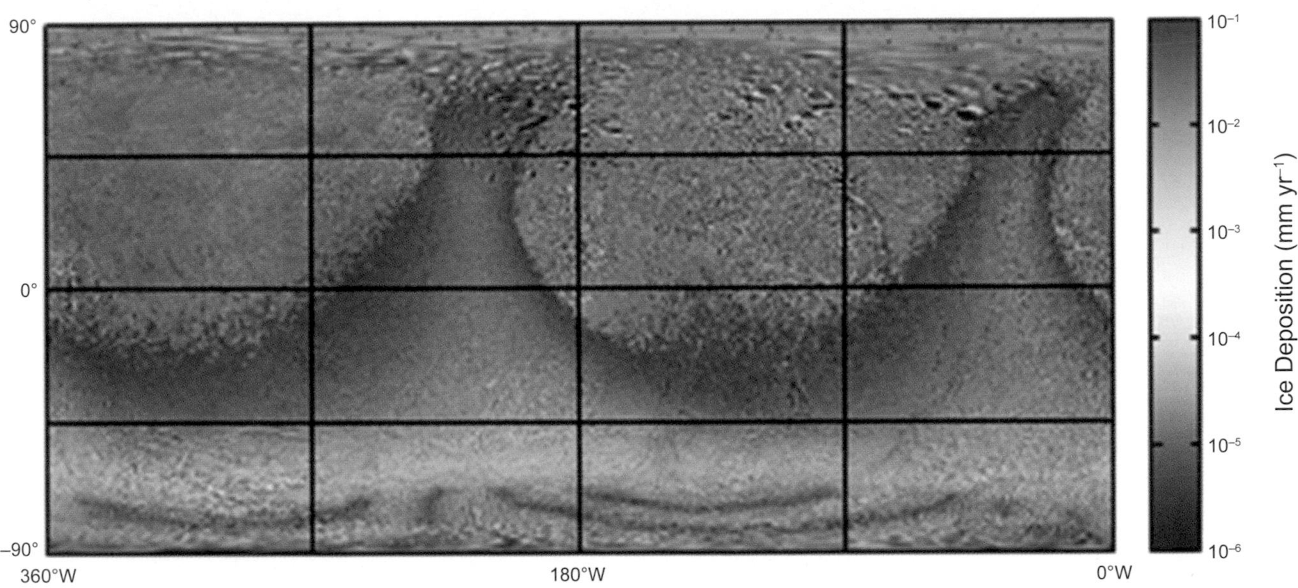

Plate 16. Plume deposition on Enceladus surface as modeled by *Southworth et al.* (2018). The model assumes a homogenous distribution of sources all along the tiger stripe fractures.

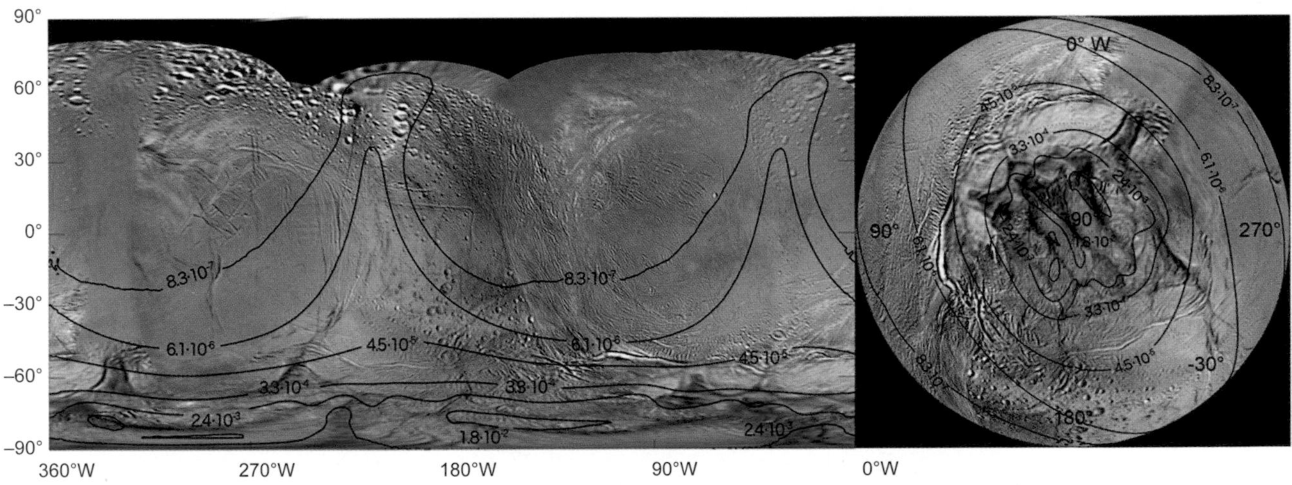

Plate 17. RGB combination of ratio maps from three Cassini ISS filters: GRN/UV3, IR3/GRN, and IR3/UV3 (*Schenk et al.,* 2011). The bright regions resemble well the plume fallout outlined by deposit models(*Southworth et al.,* 2018).

Plates accompany chapter by Postberg et al. (pp. 129–162).

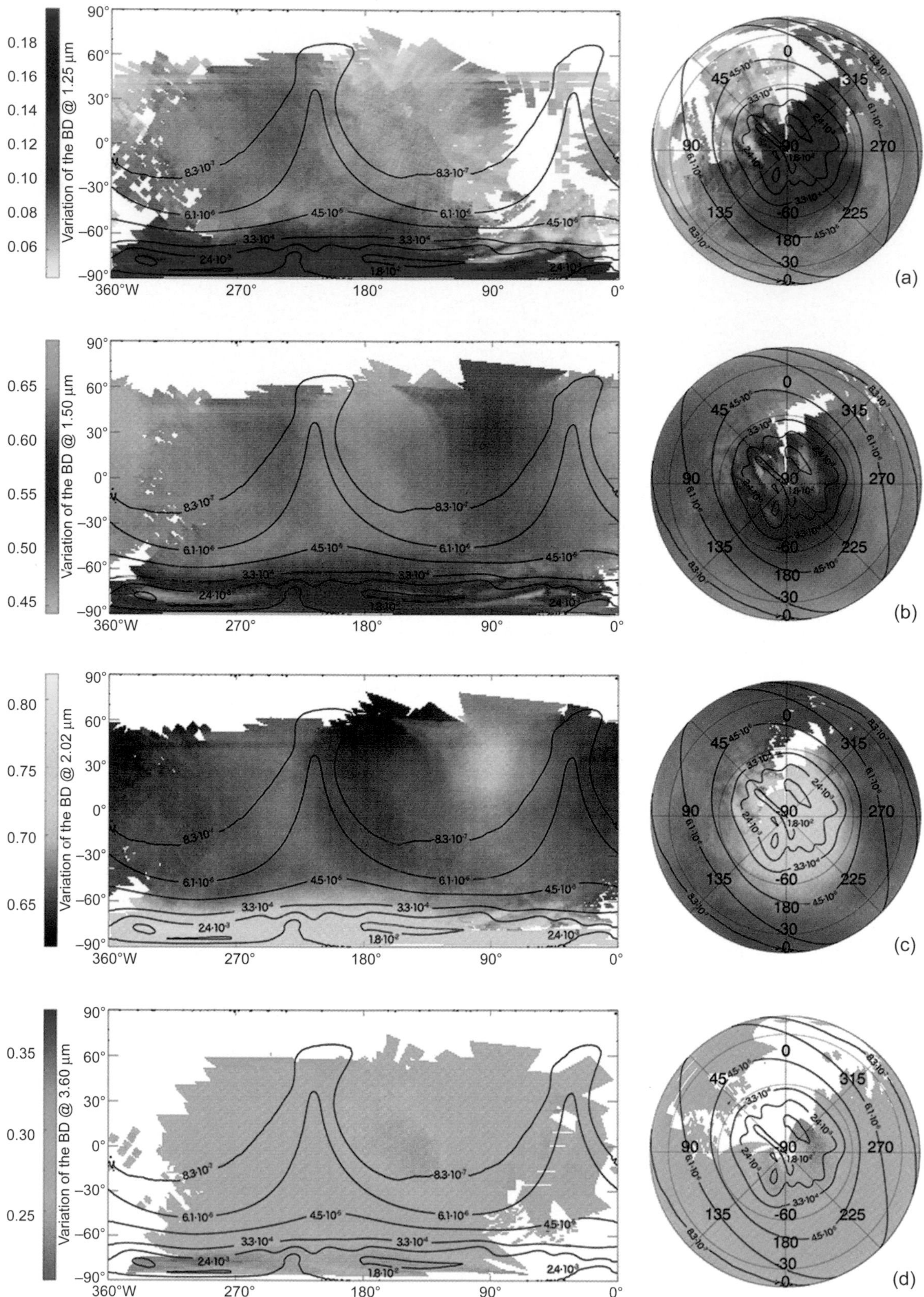

Plate 18. (a),(b),(c) Panels map the 1.25-, 1.5-, and 2-µm band depths, respectively; **(d)** strength of the 3.6-µm reflectance peak. Contour lines correspond to yearly plume deposition rate in millimeters (*Southworth et al.,* 2018) under the unrealistic assumption of compact ice deposition with a density of 0.9 g cm^{-3}.

Plates accompany chapter by Postberg et al. (pp. 129–162).

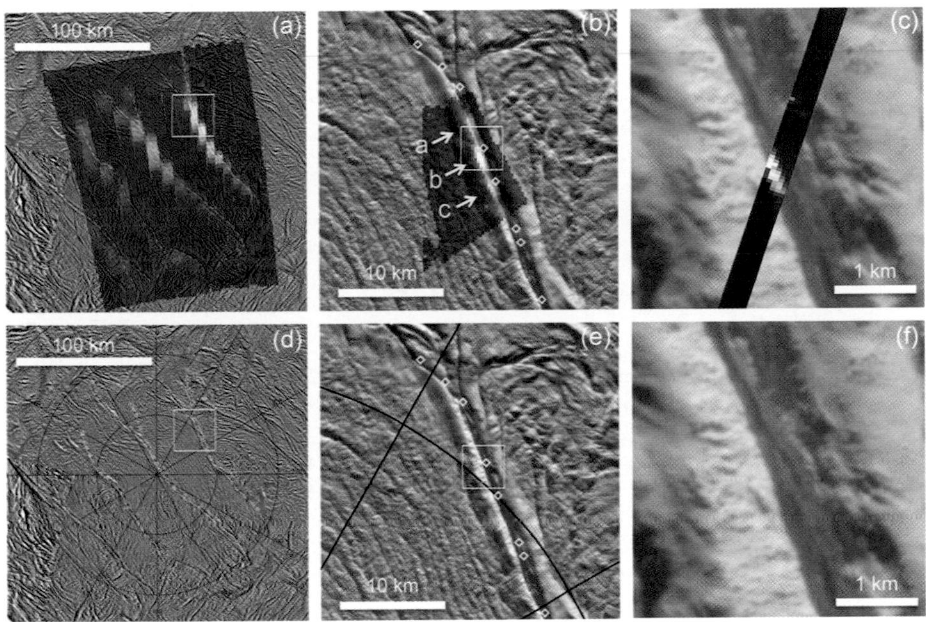

Plate 19. The appearance of the tiger stripes in the visible (Cassini ISS) and thermal infrared [Cassini CIRS, 7–16 μm, **(a)**–**(c)**] at a variety of scales. **(a)** The entire tiger stripe system with the exception of part of Alexandria Sulcus (left), which is known from other observations to be relatively faint. CIRS 9–16-μm data from March 2008. Dashed lines mark the location of the active tiger stripes. The square shows the location of **(b)**. **(b)** Closeup of the most active part of Damascus Sulcus (a combination of 9–16-μm and 7–9-μm images), mapped by CIRS in August 2010. Diamonds show the location of ISS plume jets reported by *Helfenstein and Porco* (2015). Arrows mark the locations of three discrete hot spots, a, b, and c (also shown in Fig. 2a), two of which correspond well to plume jet locations. The square shows the location of **(c)**. **(c)** The highest-resolution CIRS image of Damascus Sulcus emission, taken by direct sampling of the CIRS 7–9-μm interferograms from October 2015, showing dominant emission from a single fissure <100 m wide, and isolated hot spots elsewhere within the central trough (*Gorius et al.,* 2015). Placement relative to the terrain is approximate. **(d)**–**(f)** The same images without the CIRS data overlay, to show the terrain more clearly.

Accompanies chapter by Spencer et al. (pp. 163–174).

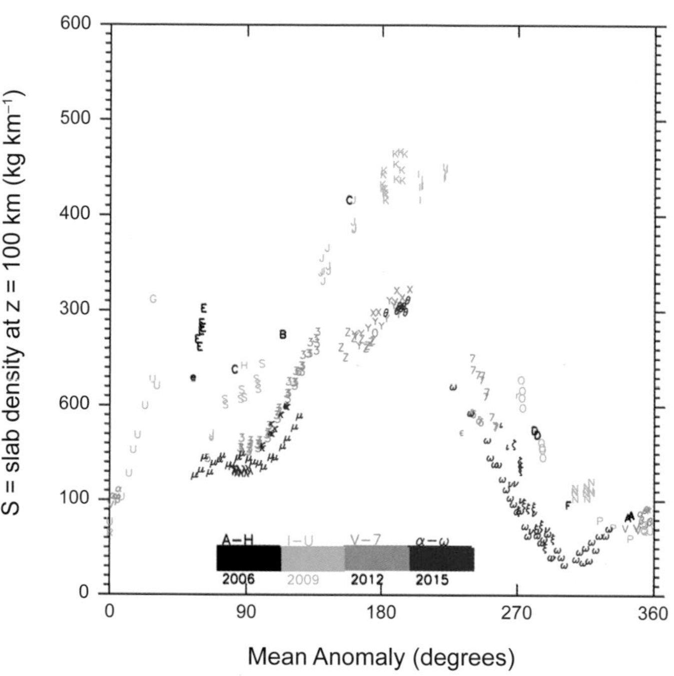

Plate 20. Orbital and long-term variability of the Enceladus dust plume brightness at visible wavelengths, from *Ingersoll and Ewald* (2017). Different symbols identify different datasets. Mean anomaly describes the orbital position of Enceladus relative to periapse, while the Y axis is a measure of integrated plume brightness at 100 km altitude.

Accompanies chapter by Spencer et al. (pp. 163–174).

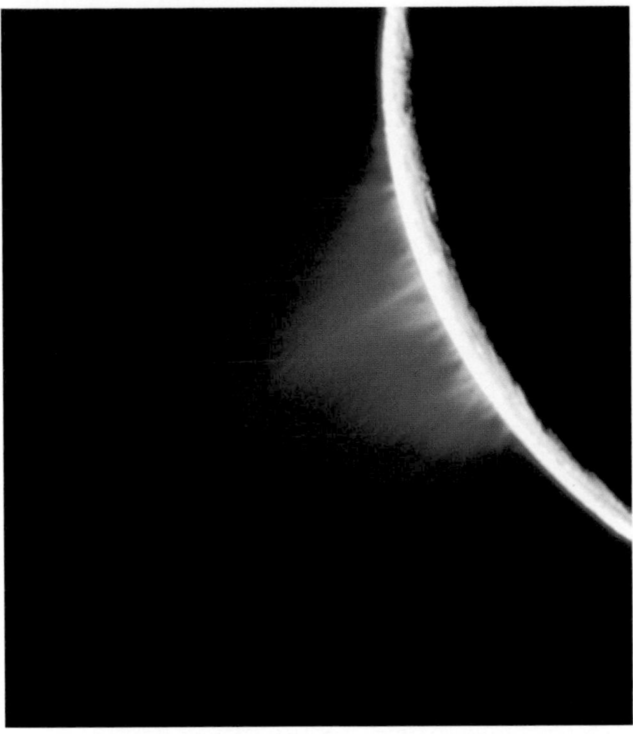

Plate 21. An ISS color-stretched image acquired on November 27, 2005, illustrating overall plume morphology threaded by many finely resolved jets or geysers.

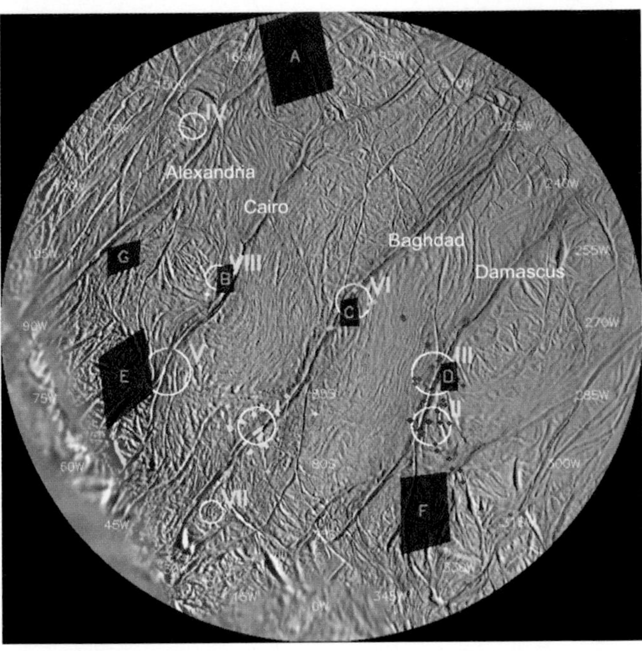

Plate 22. Polar stereographic projection of Enceladus' south polar region showing the eight source locations inferred by *Spitale and Porco* (2007) using triangulation. Locations of CIRS hot spots from *Spencer et al.* (2006) are shown as red quadrilaterals.

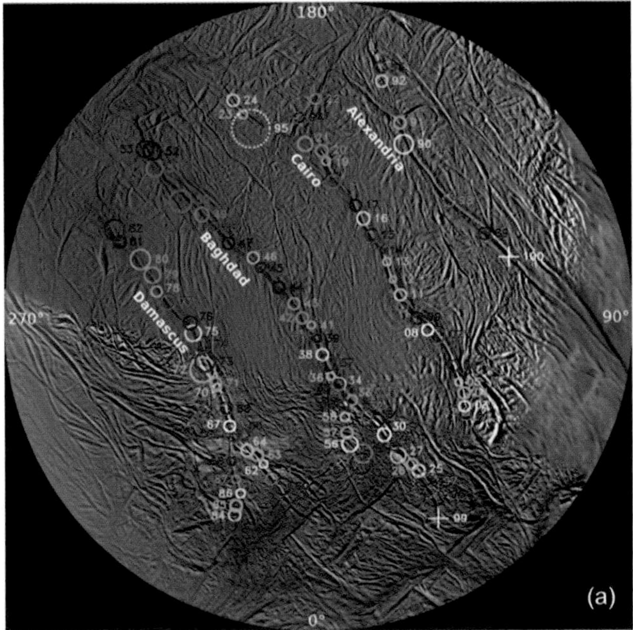

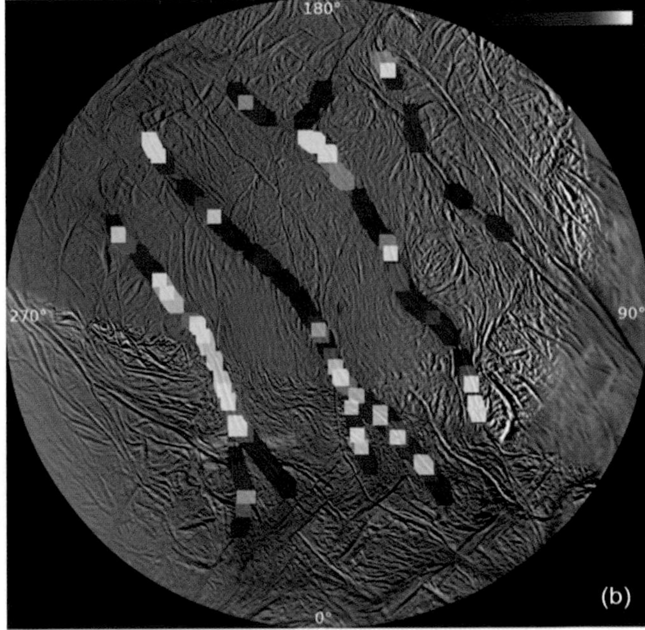

Plate 23. (a) Locations of the jets. **(b)** Jet activity derived from ISS observations. There is strong correlation with the jet activity and the tiger-stripe temperatures measured by CIRS and VIMS. From *Porco et al.* (2014).

Plates accompany chapter by Goldstein et al. (pp. 175–194).

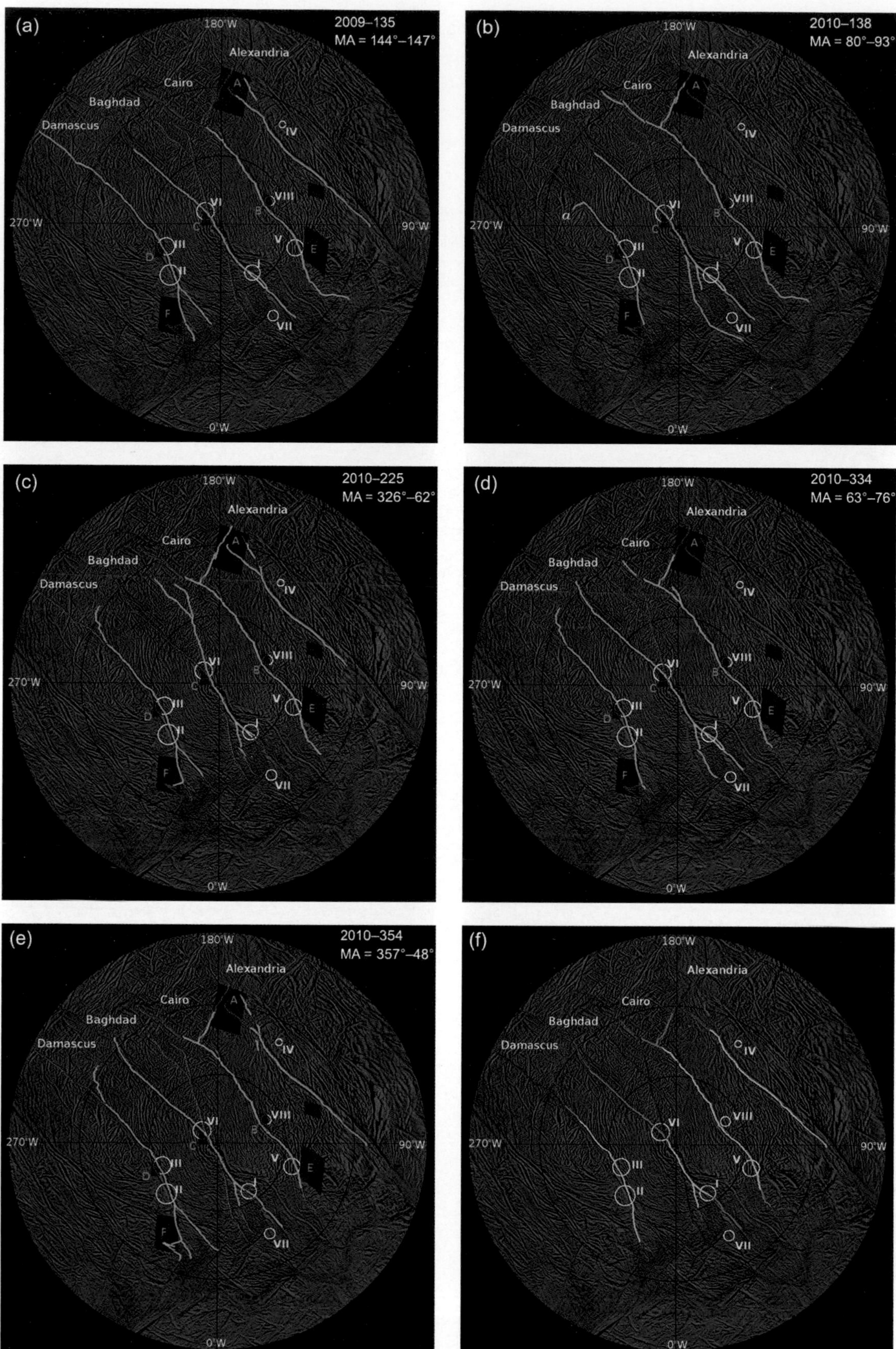

Plate 24. Curtain activity at five different mean anomalies (MA), and the average (f). Green areas are active, red areas are inactive, and blue areas are undetermined.

Plates accompany chapter by Goldstein et al. (pp. 175–194).

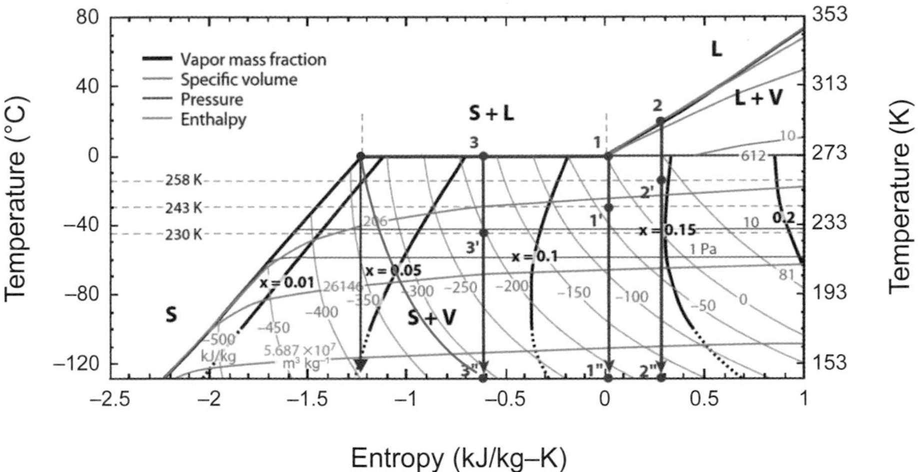

Plate 25. Phase diagram for water at low temperatures and entropies (from *Lu and Kieffer,* 2009). From the red circle to point 1 is liquid water in contact with ice. Isenthalpic (green curve) or isentropic (red line and other vertical lines) decompression increases the mass fraction of vapor (burgundy curves), but only leads to a mass fraction of several percent vapor. In contrast, the plume has an observed vapor mass fraction of 90–95% (e.g., *Gao et al.,* 2016).

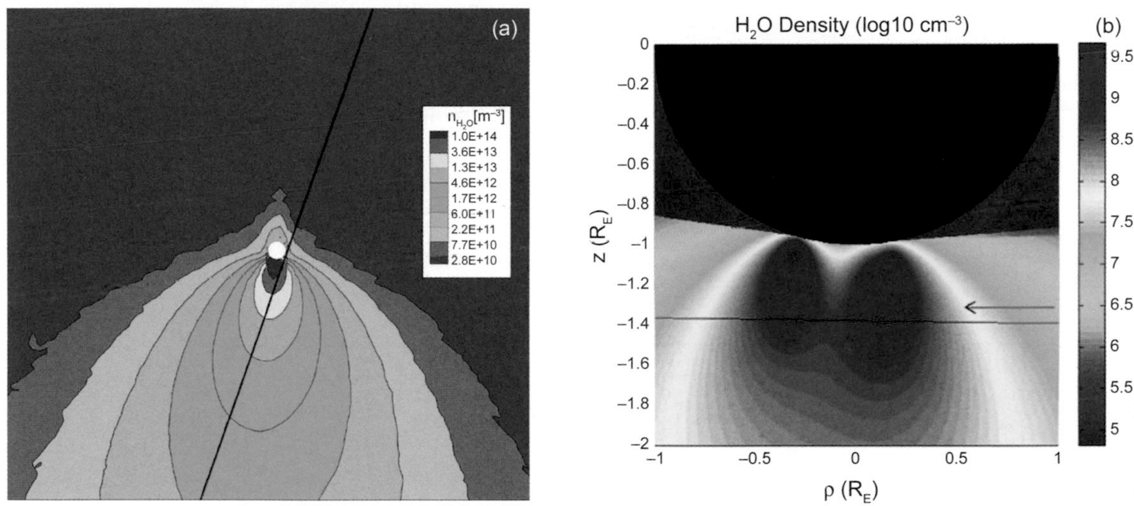

Plate 26. (a) Monte Carlo [*Tenishev et al.* (2010) E5 flyby shown] and **(b)** analytical [*Dong et al.* (2011) E7 flyby shown] plume water vapor density models, from fits to INMS E3 and E5 (both models), and E7 [*Dong et al.* (2011) only], with the *Spitale and Porco* (2007) sources as the constraint.

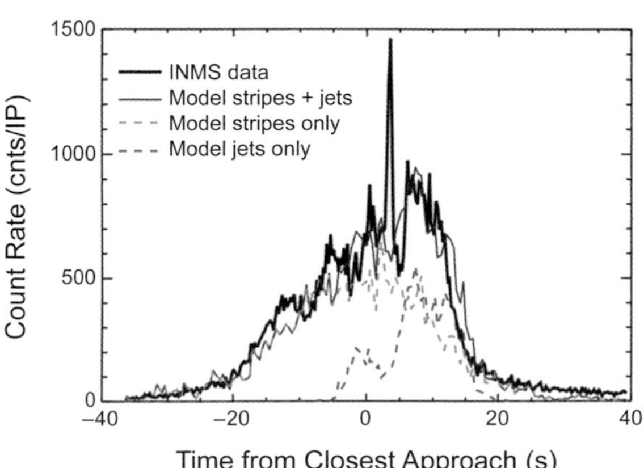

Plate 27. INMS measurements of mass 44 u species during the E17 flyby are shown in black. Model (*Hurley et al.,* 2015) predictions using constant emission along the tiger stripes at 500 m s^{-1} and 270 K (gray dashed line) are selected to match the rise and fall on the outskirts of the plume. The jet model using 1500 m s^{-1} and 270 K are included (blue dashed line) to reproduce the overall enhancement near closest approach. The sum of the two models, shown in red, reproduces the overall structure of the plume but misses some of the fine structure.

Plates accompany chapter by Goldstein et al. (pp. 175–194).

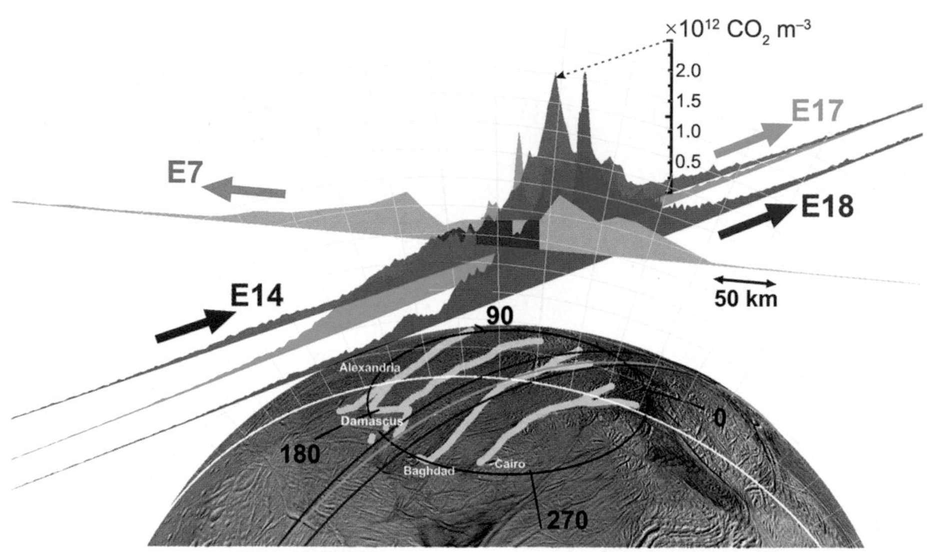

Plate 28. To-scale three-dimensional representation of the E14, E17, E18, and (lower-resolution) E7 INMS data, with vertical areas representing (in linear scale) the density, and the flat base of the areas corresponding to the Cassini trajectories.

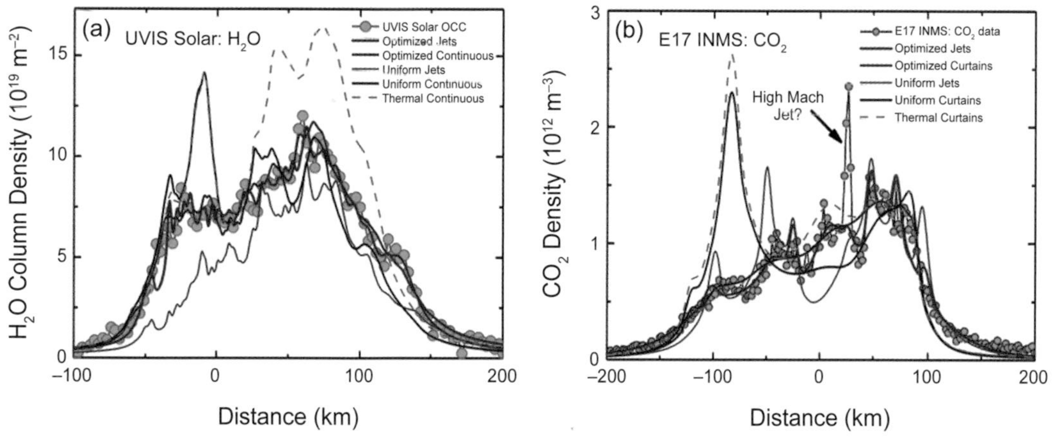

Plate 29. (a) Enceladus plume water vapor column density measurement from the UVIS 2010 solar occultation (dotted line), plotted vs. distance across the plume along the occultation line of sight minimum ray height. Thin lines: modeling solutions (*Teolis et al.,* 2017) assuming continuous emission along the tiger stripes. Thick lines: Solutions assuming the *Porco et al.* (2014) jets. **(b)** INMS CO_2 density measurement (dotted line) along the E17 flyby trajectory, showing peaks suggestive of discrete plume sources. Lines: Solutions with *Porco et al.* (2014) jets as the constraint.

Plates accompany chapter by Goldstein et al. (pp. 175–194).

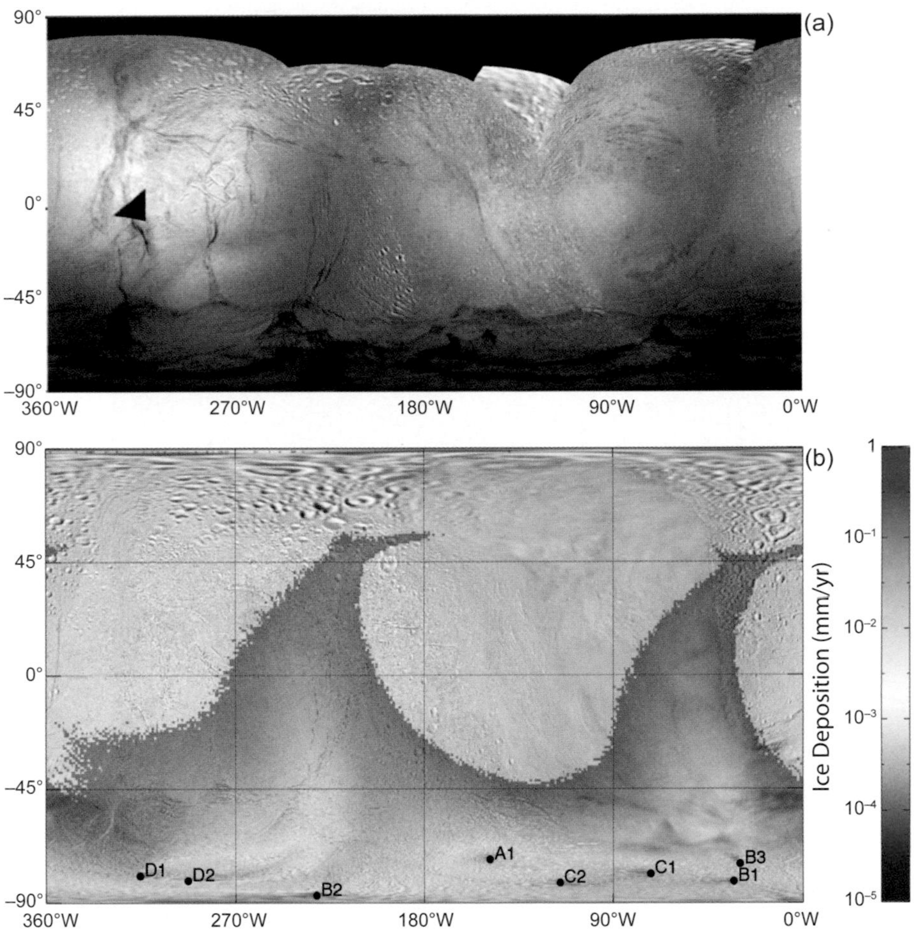

Plate 30. (a) Global IR/UV color ratio map of Enceladus (*Schenk et al.,* 2011). **(b)** Snowfall pattern of plume particles between 500 nm and 5 μm on the Enceladus surface (*Kempf et al.,* 2010).

Accompanies chapter by Kempf et al. (pp. 195–210).

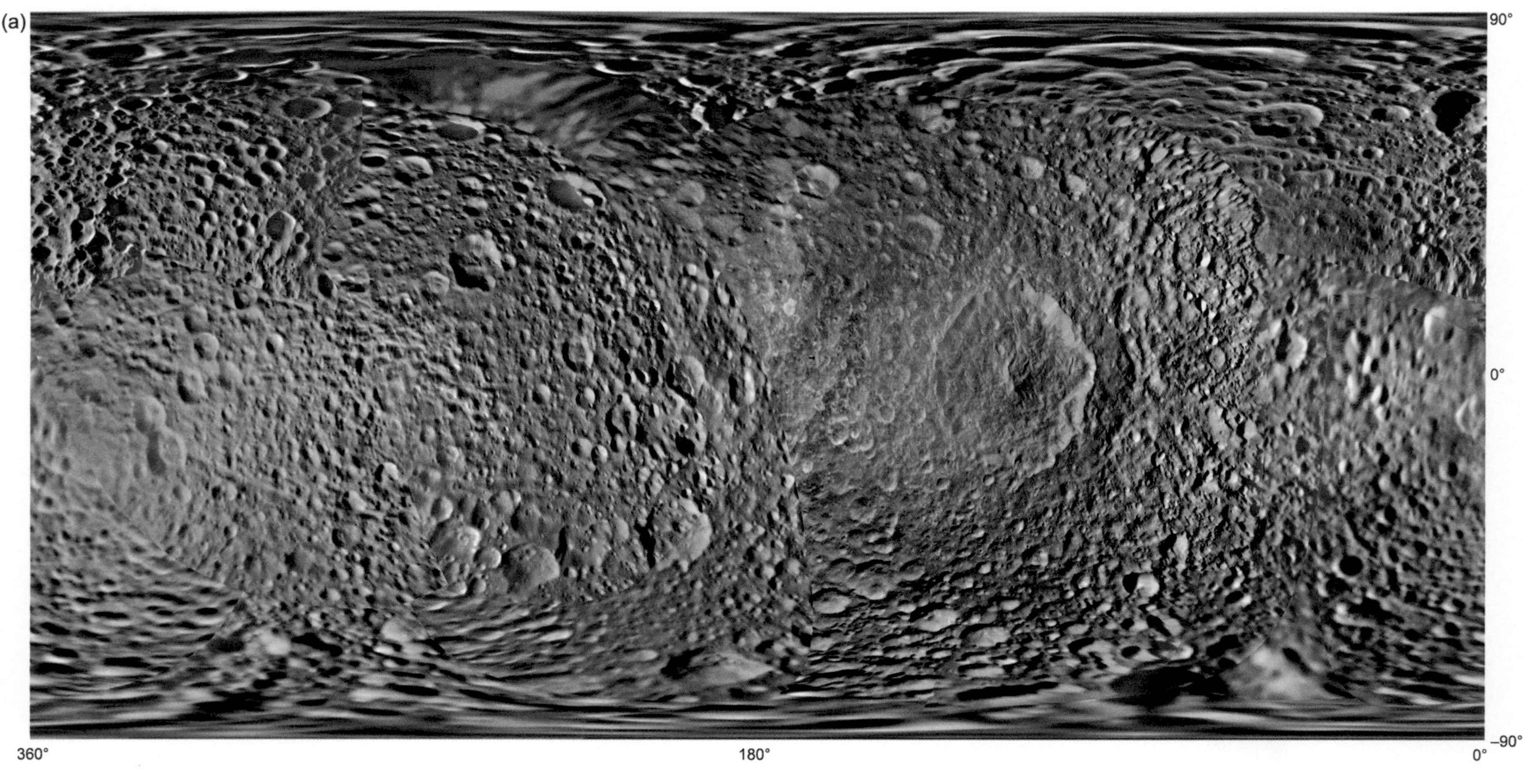

Plate 31. (a) Global map view of Mimas. Sinusoidal projection is centered on longitude 180° and is from the three-color global maps of *Schenk* (2014). Brightness is representative but not to scale and is optimized to provide best overall contrast. Base resolution is 200 m.

Accompanies chapter by Schenk et al. (pp. 237–265).

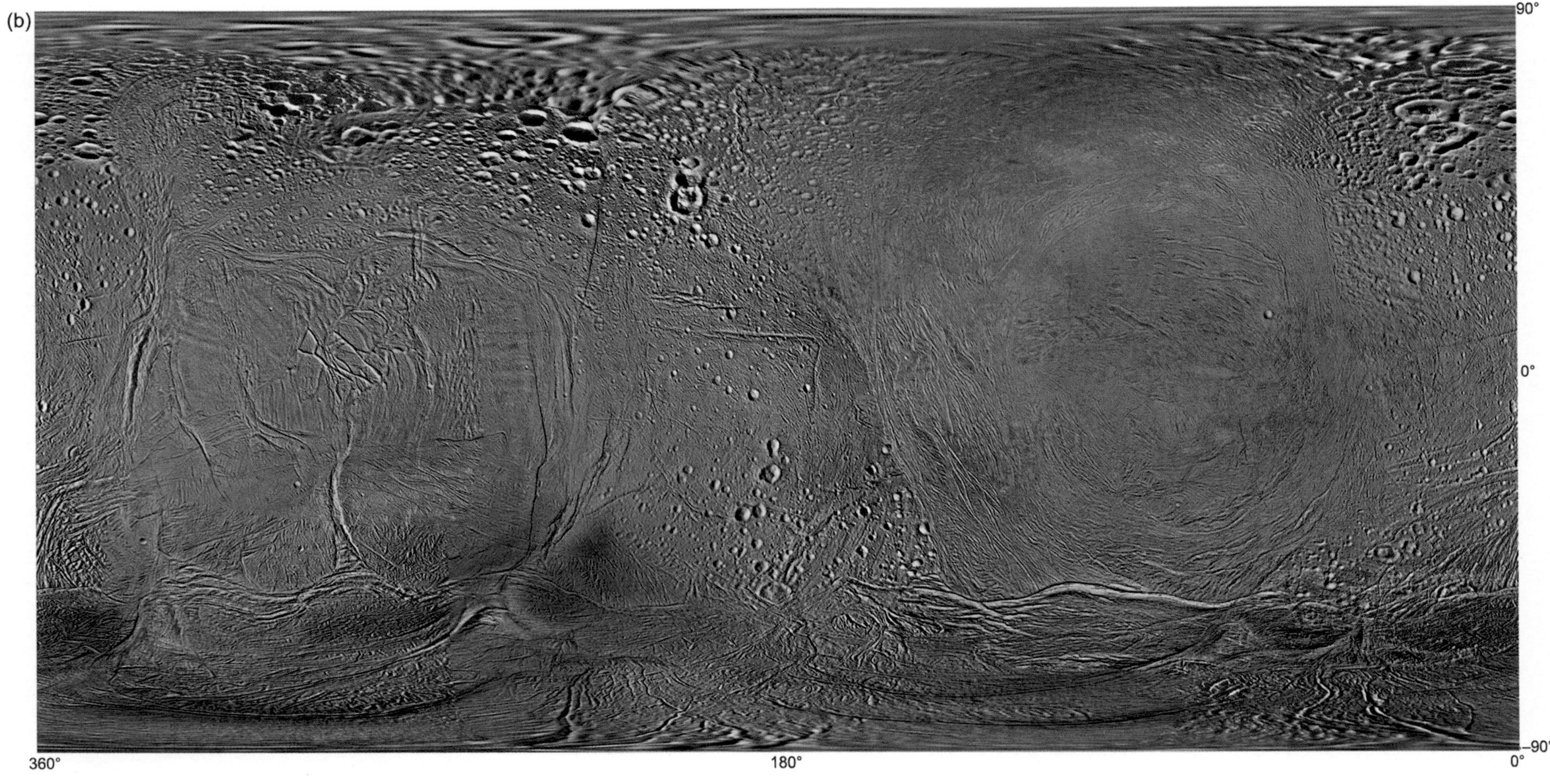

Plate 31. (b) Global map view of Enceladus. Sinusoidal projection is centered on longitude 180° and is from the three-color global maps of *Schenk* (2014). Brightness is representative but not to scale and is optimized to provide best overall contrast. Base resolution is 100 m.

Accompanies chapter by Schenk et al. (pp. 237–265).

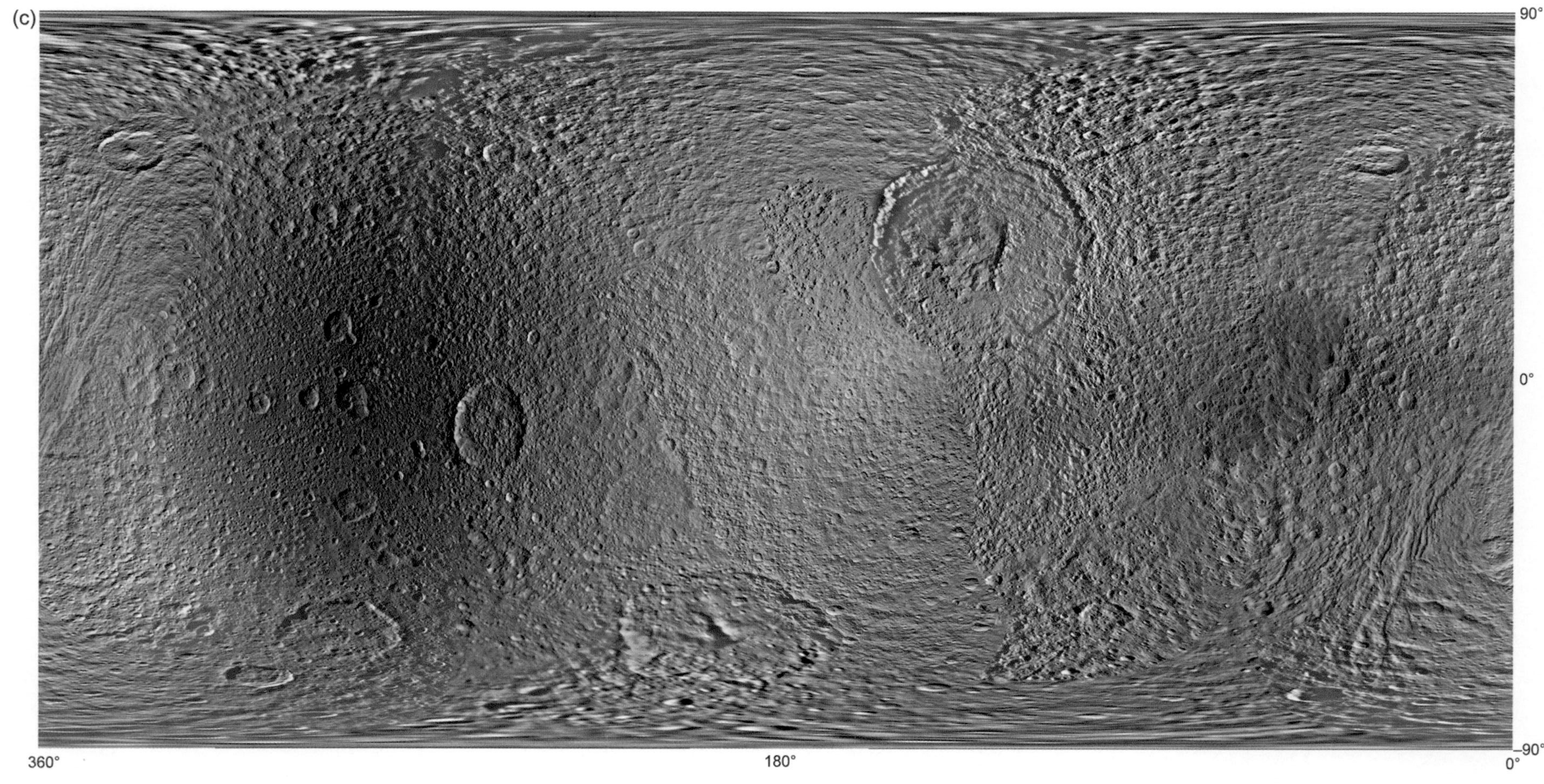

Plate 31. (c) Global map view of Tethys. Sinusoidal projection is centered on longitude 180° and is from the three-color global maps of *Schenk* (2014). Brightness is representative but not to scale and is optimized to provide best overall contrast. Base resolution is 250 m.

Accompanies chapter by Schenk et al. (pp. 237–265).

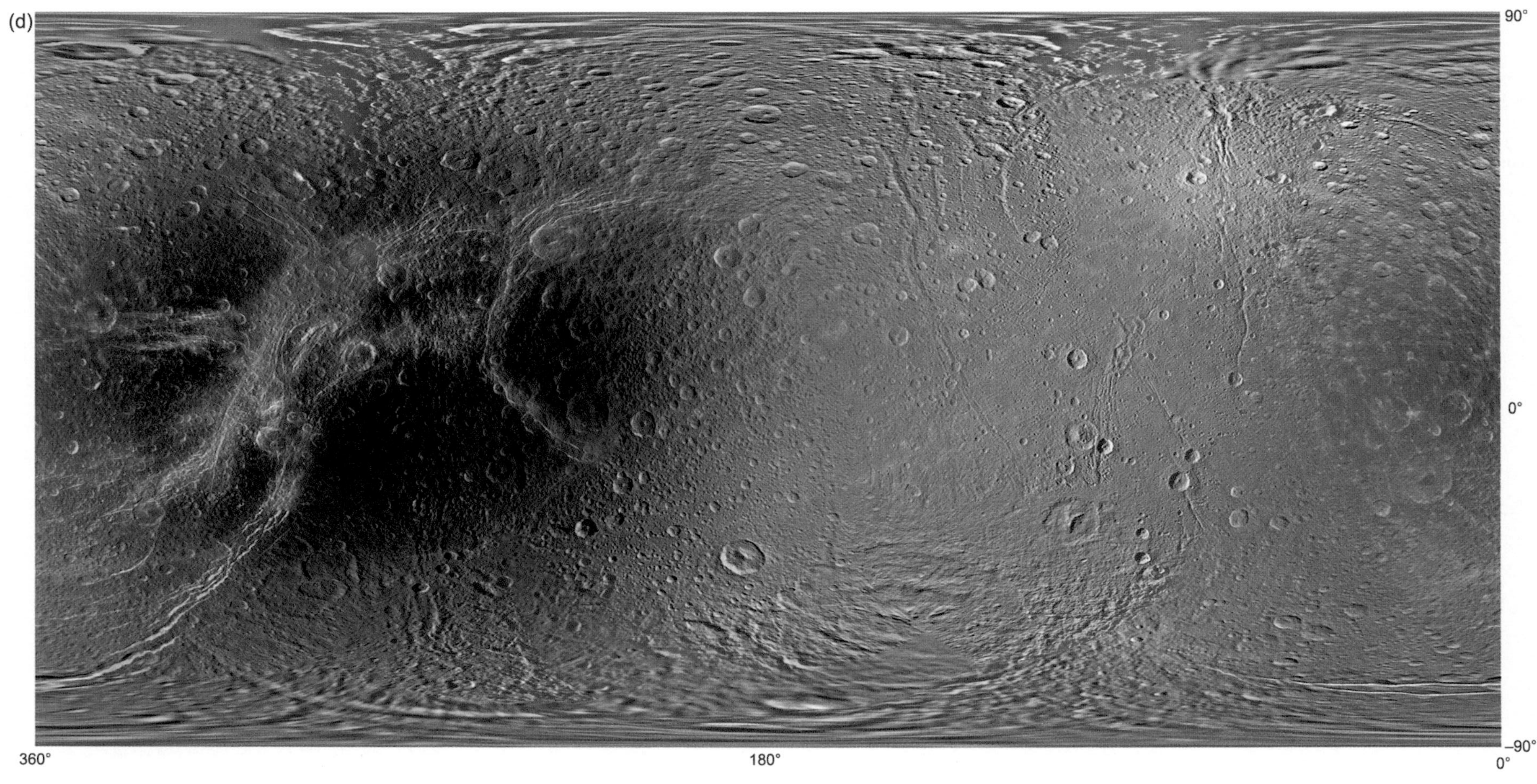

Plate 31. (d) Global map view of Dione. Sinusoidal projection is centered on longitude 180° and is from the three-color global maps of *Schenk* (2014). Brightness is representative but not to scale and is optimized to provide best overall contrast. Base resolution is 250 m.

Accompanies chapter by Schenk et al. (pp. 237–265).

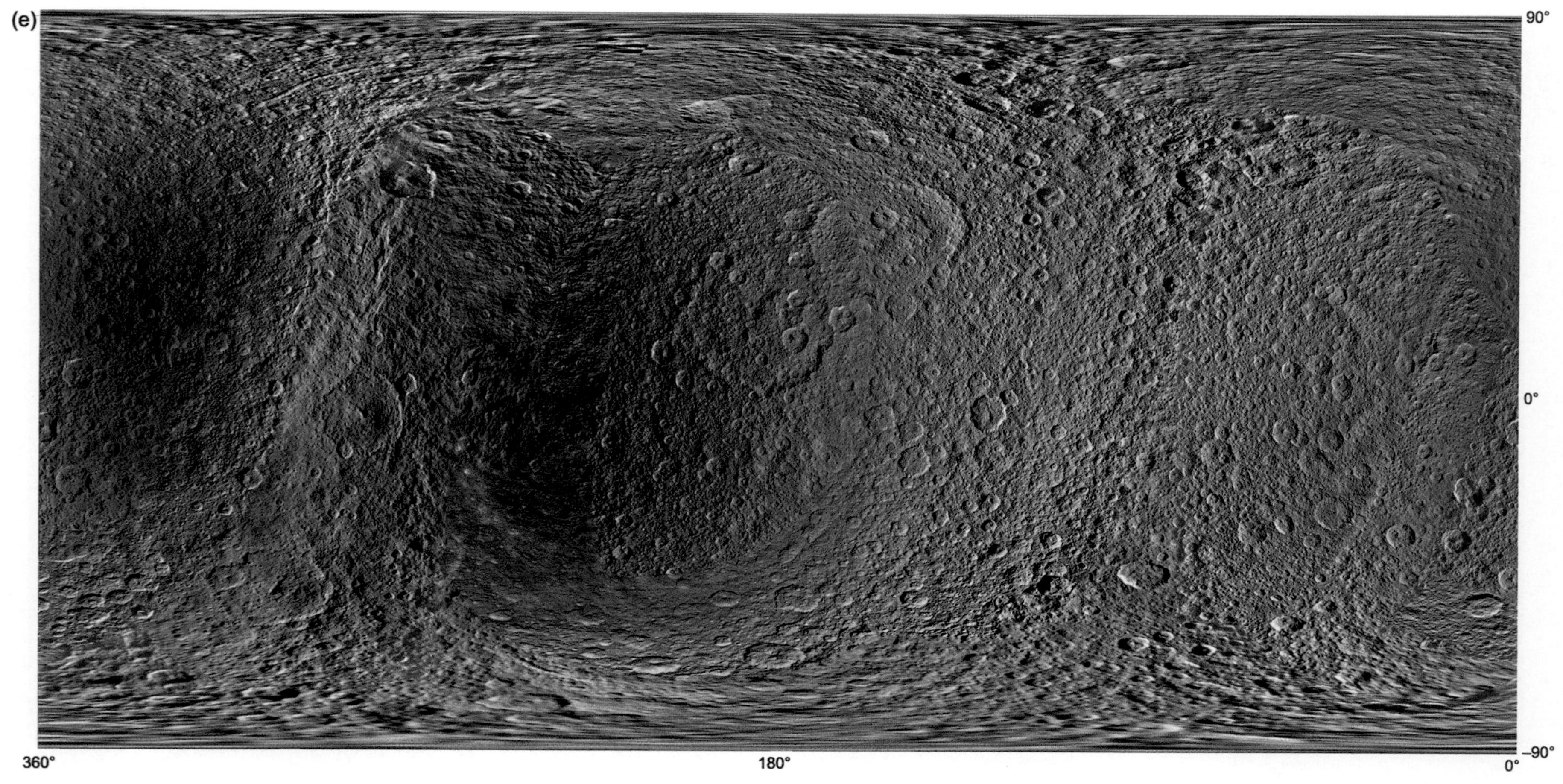

Plate 31. **(e)** Global map views of Rhea. Sinusoidal projection is centered on longitude 180° and is from the three-color global maps of *Schenk* (2014). Brightness is representative but not to scale and is optimized to provide best overall contrast. Base resolution is 400 m.

Accompanies chapter by Schenk et al. (pp. 237–265).

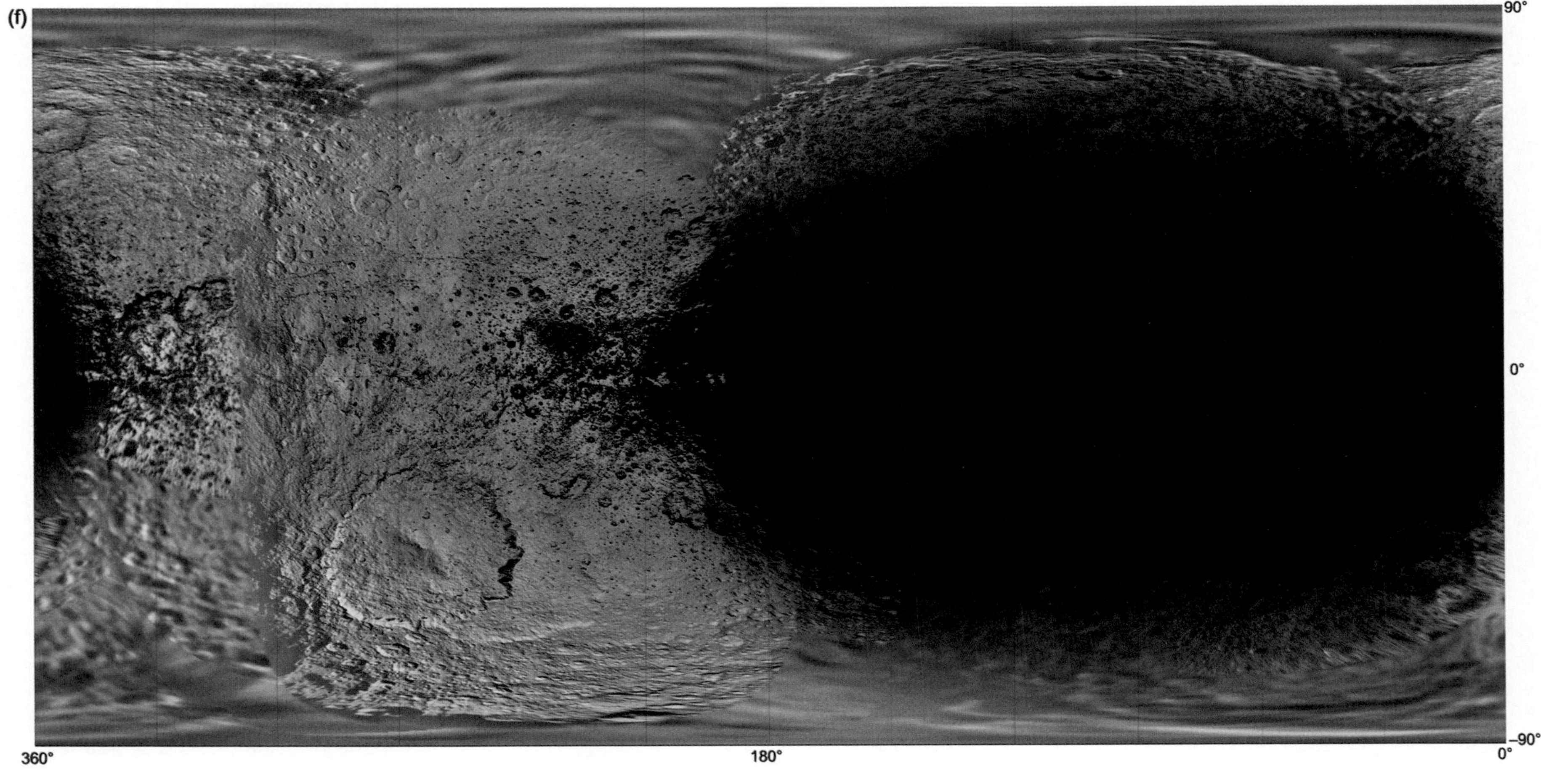

Plate 31. **(f)** Global map views of Iapetus. Sinusoidal projection is centered on longitude 180° and is from the three-color global maps of *Schenk* (2014). Brightness is representative but not to scale and is optimized to provide best overall contrast. Base resolution is 400 m.

Accompanies chapter by Schenk et al. (pp. 237–265).

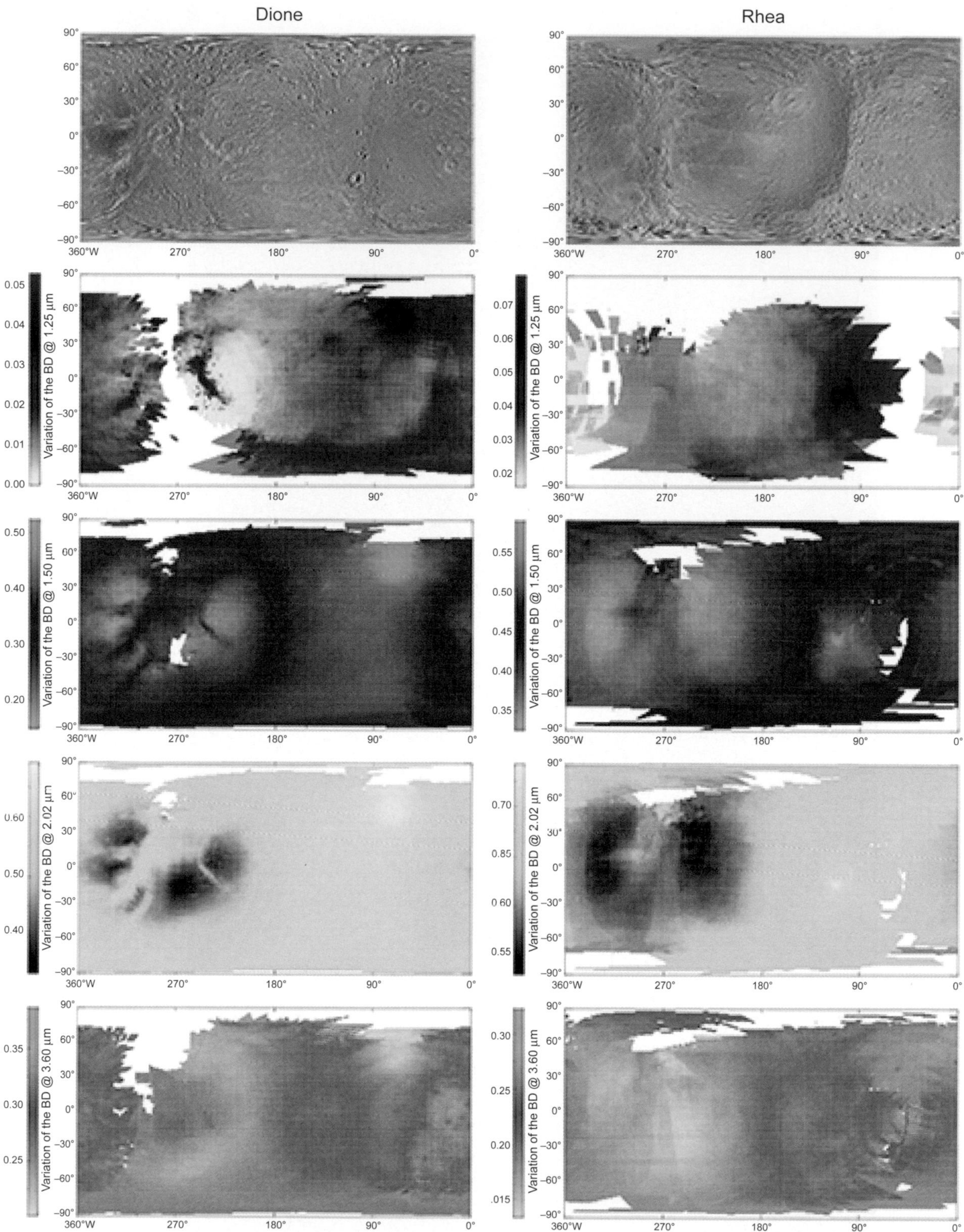

Plate 32. Maps of variation of the water-ice absorption bands at 1.25, 1.5, and 2.0 μm, and of the reflectance maximum at 3.6 μm, for Dione (left column) and Rhea (right column). The variation of each spectral indicator is represented by a unique color code. The associated color bar is on the left of each image, and values increase from bottom to top. The maps were produced following the method described in detail in *Scipioni et al.* (2017). Each map is sampled by using a fixed-resolution grid with angular bins of 1° latitude × 1° longitude. Within each bin, the values of the spectral indices computed on the basis of all VIMS data covering that particular bin had been averaged (for details, see *Scipioni et al.,* 2017). Base maps are provided for comparison in the top row (maps courtesy of NASA/JPL-Caltech/Space Science Institute).

Accompanies chapter by Hendrix et al. (pp. 307–322).

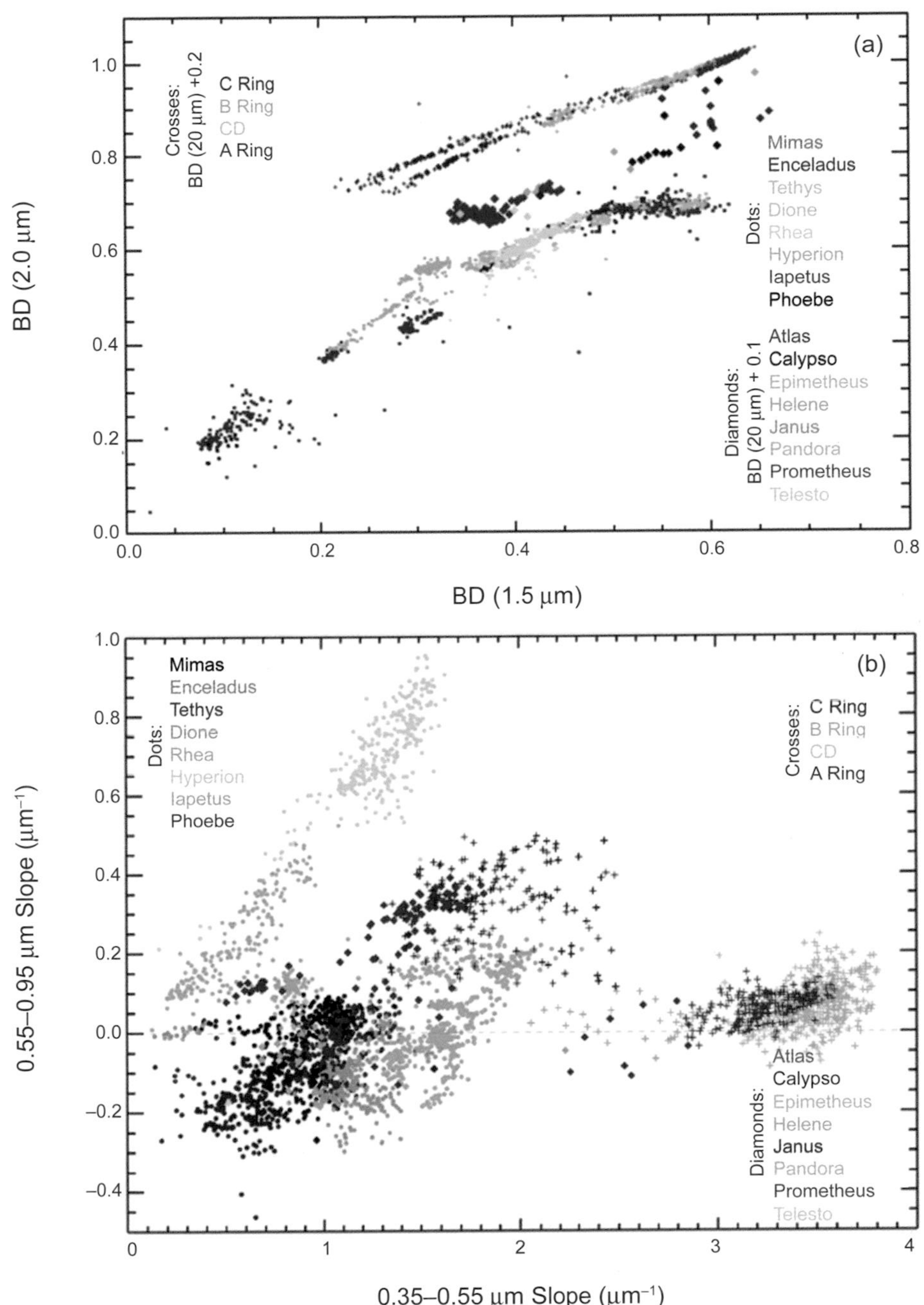

Plate 33. (a) Water-ice band-depths of the moons of Saturn and its rings, showing the trends with distance from Saturn. **(b)** Spectral slopes of the moons of Saturn and its rings, showing the trends with distance from Enceladus. After *Filacchione et al.* (2012).

Accompanies chapter by Hendrix et al. (pp. 307–322).

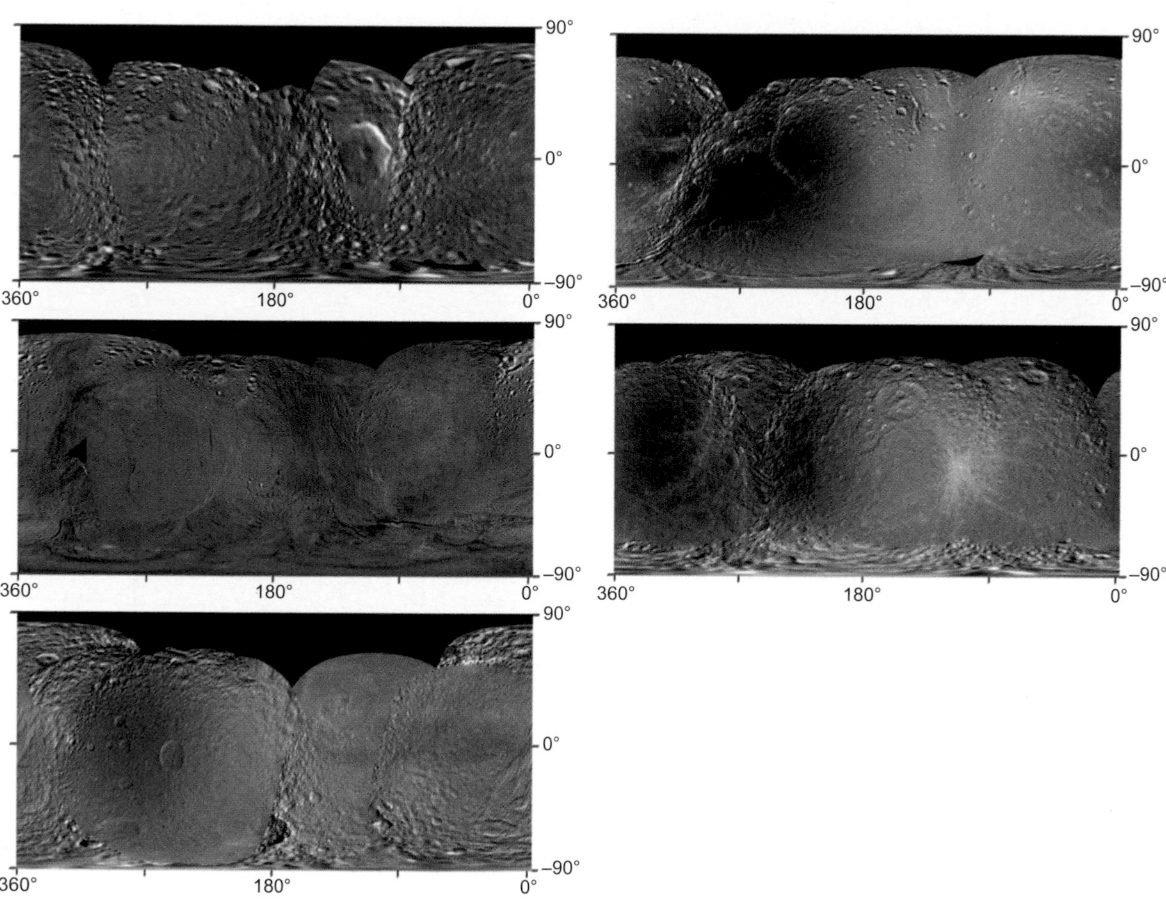

Plate 34. Enhanced Cassini three-color (IR-green-UV filter) global maps of the five inner midsized icy satellites of Saturn (from top to bottom: Mimas, Enceladus, Tethys, Dione, Rhea). Maps are in simple cylindrical projection from 90°S to 90°N and from –2°W to 360°W. From *Schenk et al.* (2011).

Accompanies chapter by Howett et al. (pp. 343–360).

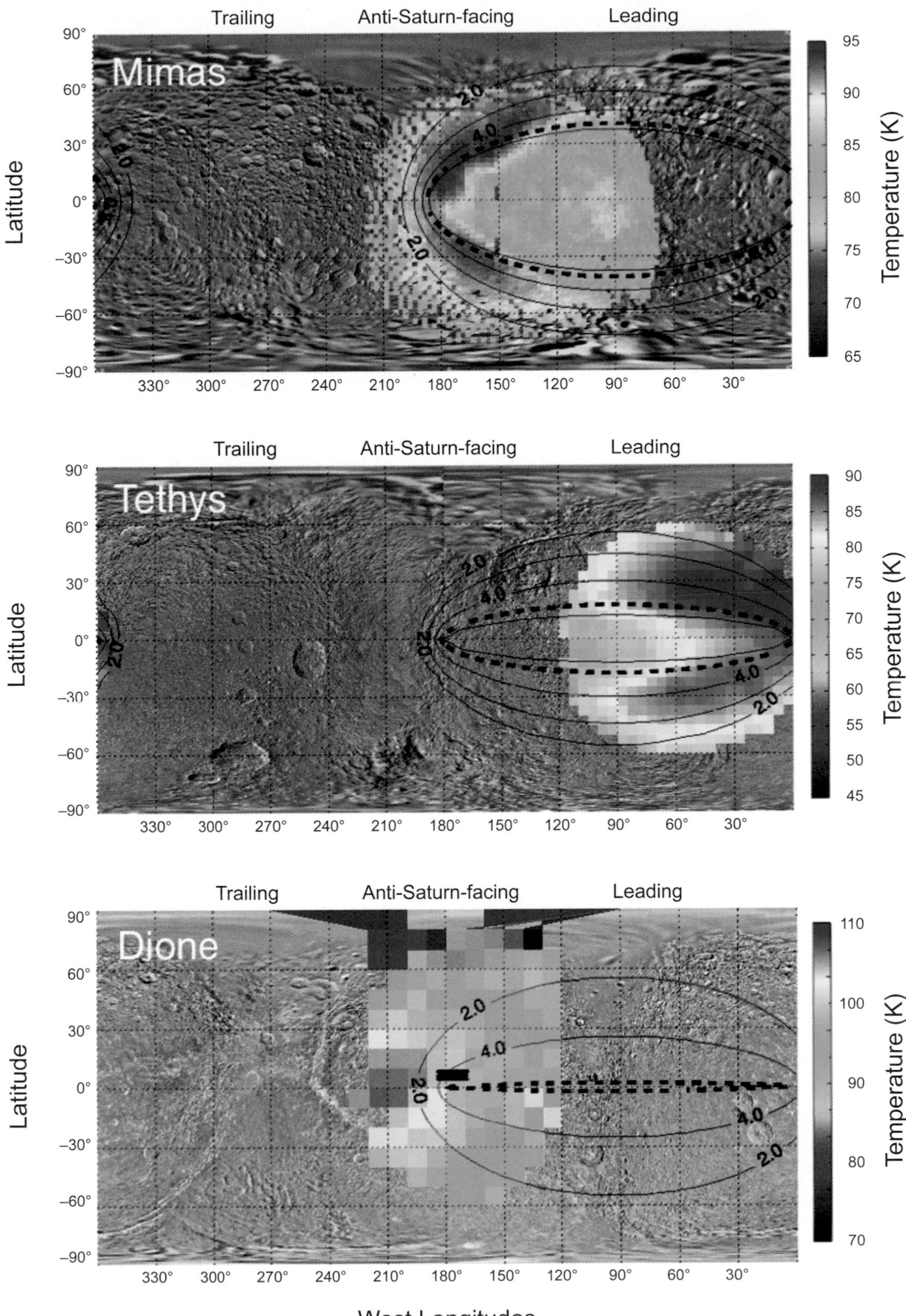

Plate 35. Daytime temperature observations of Mimas, Tethys, and Dione. Overlaid are contours of energetic electron power deposited into the surface per unit area, Q (log10 MeV cm^{-2} s^{-1}), determined using updated results from Cassini's Magnetospheric Imaging Instrument (MIMI). The best-fitting contour to the Mimas (Tethys) color and thermal inertia anomaly boundary (cf. *Howett et al.,* 2011, 2012) is given by the dashed line at 5.6 (1.8) × 10^4 MeV cm^{-2} s^{-1}.

Accompanies chapter by Howett et al. (pp. 343–360).

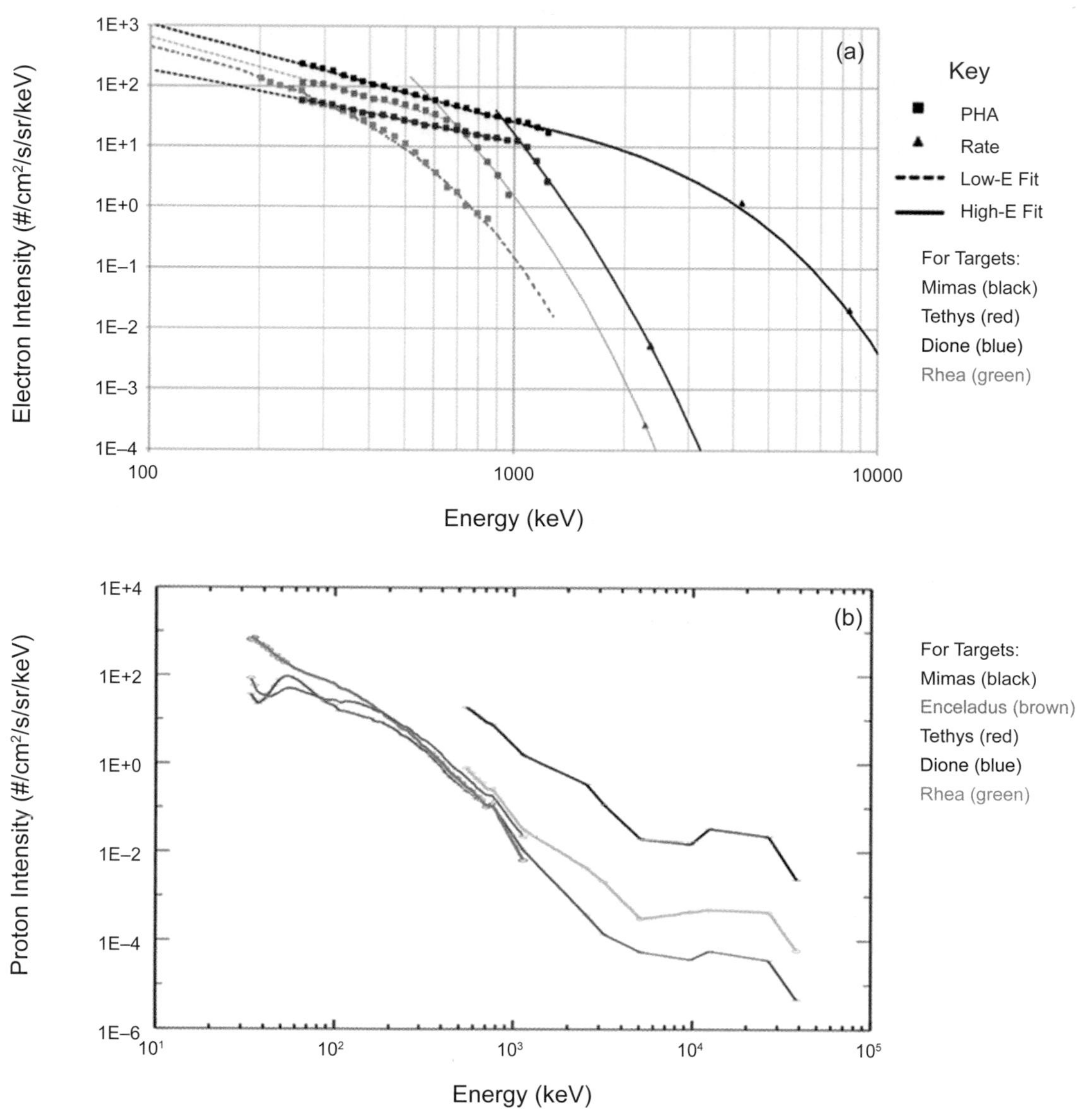

Plate 36. **(a)** Differential energy spectra (electrons per cm^2 s sr keV) at the L shells of four inner satellites of Saturn, adapted from *Paranicas et al.* (2014). Cassini MIMI pulse height analysis (PHA) and (higher-energy) rate data are shown with different plot symbols. Separate fits are created for each moon and energy range (see key for details). **(b)** Differential energy spectra (protons per cm^2 s sr keV) at the L shells of five inner satellites of Saturn.

Accompanies chapter by Howett et al. (pp. 343–360).

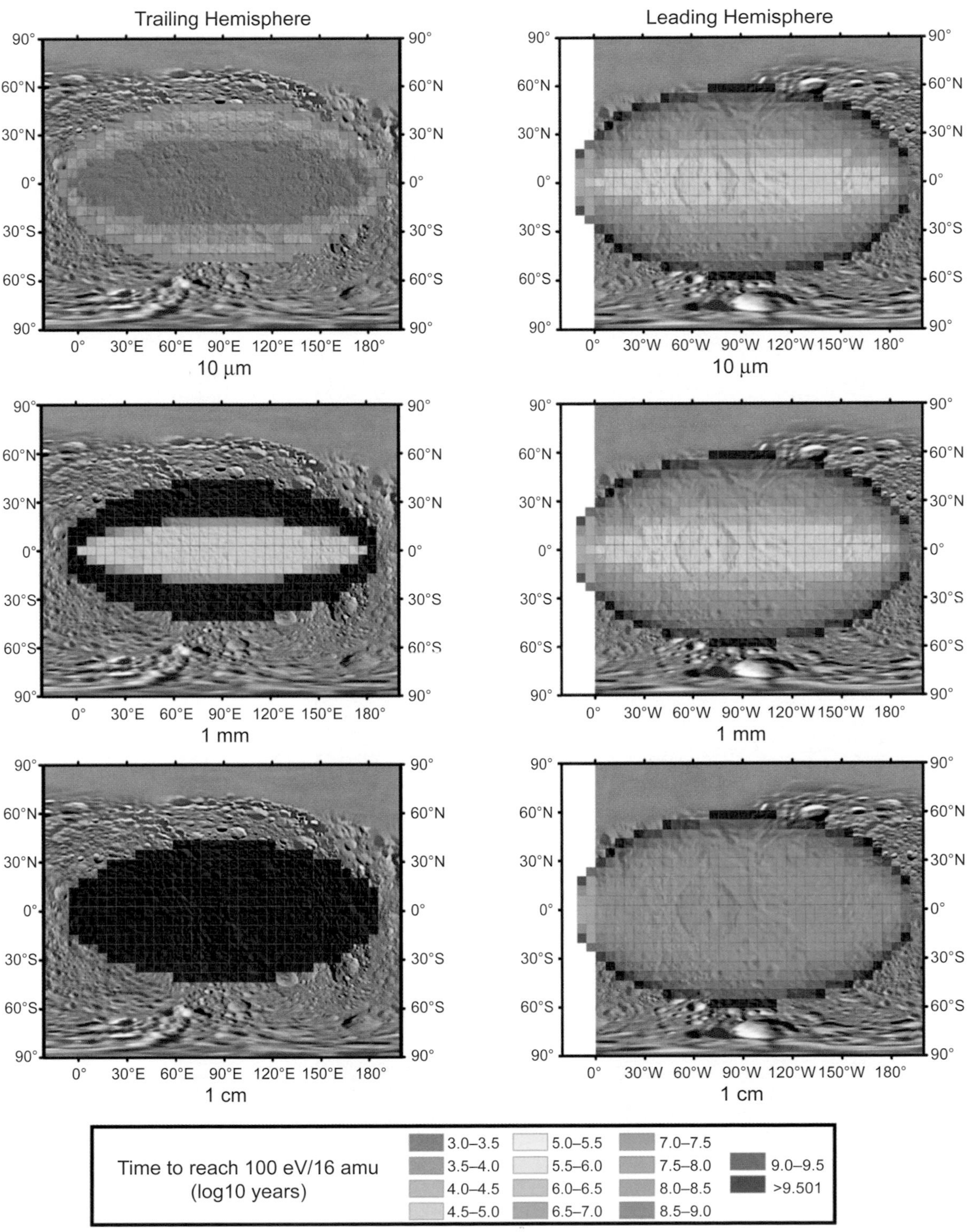

Plate 37. Energy deposition maps at different depths on the trailing hemisphere of Mimas. The energetic electron dose is given in terms of years to reach a significant dose of 100 eV/16 amu, which is equal to a dose of 60.3 G_{rad}. From *Nordheim et al.* (2017).

Accompanies chapter by Howett et al. (pp. 343–360).

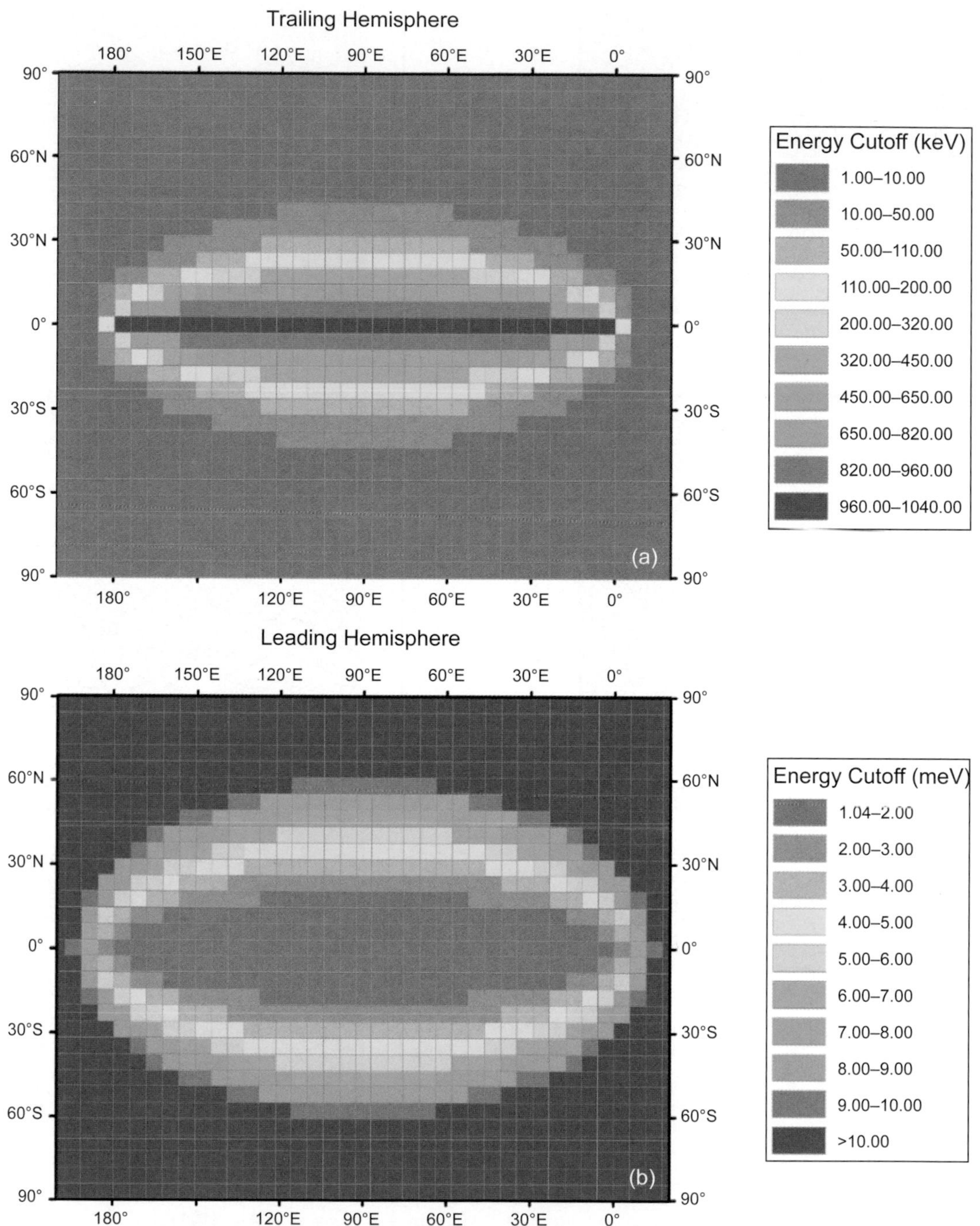

Plate 38. Predicted cutoff energy vs. surface location for Mimas. (a) On the trailing hemisphere, the cutoff energy represents the highest-energy electrons capable of accessing that point on the surface. **(b)** On the leading hemisphere, the cutoff energy represents the lowest-energy electrons capable of accessing that point. From *Nordheim et al.* (2017).

Accompanies chapter by Howett et al. (pp. 343–360).

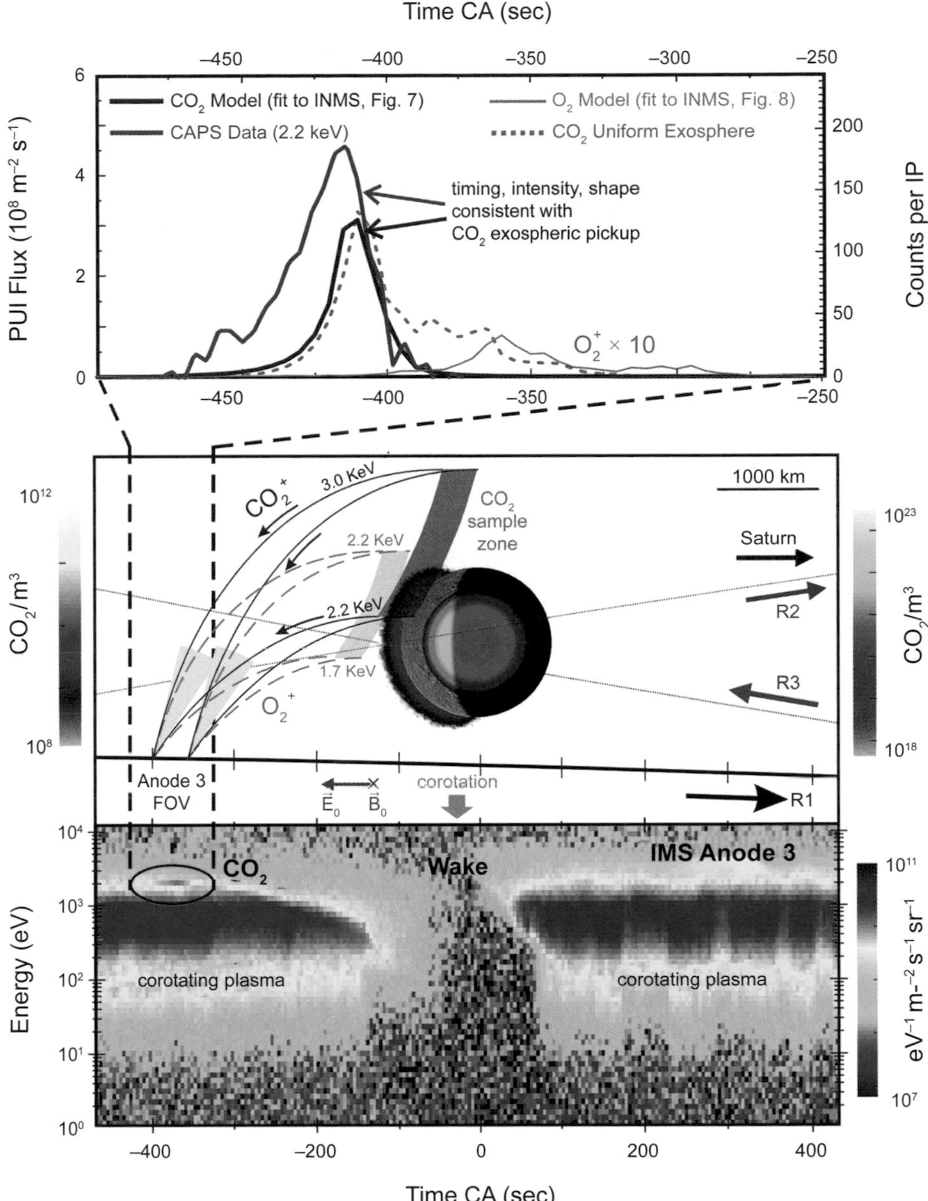

Plate 39. From *Teolis and Waite* (2016). Comparison of the Rhea pickup ion flux and distribution detected by CAPS IMS anode 3 on the R1 flyby, to that anticipated by the model CO_2 and O_2 exospheres, fit to the later INMS R2 and R3 data and projected back to the R1 flyby date and time. (*Bottom*) Anode 3 ion flux energy spectrogram with the CO_2^+ pickup ion signature circled. (*Center*) R1 flyby configuration showing the Rhea exospheric CO_2 density cross section, Rhea surface illumination and adsorbed column density, the region (purple) sampled by anode 3 during the pickup ion detection, the anode 3 FOV at the time and location of the pickup ion detection along the R1 trajectory, and the limiting O_2^+ and CO_2^+ energies (on arrival at Cassini) and trajectories accepted into anode 3. The R1 sampling region is approximated as two-dimensional in the flyby's equatorial plane. (*Top*) Anticipated (dark blue: CO_2^+, orange: O_2^+ scaled × 10) and measured (red) total pickup ion signal vs. time from closest approach, compared to the expected profile (dashed line) for a uniform CO_2 exosphere. As shown the timing, intensity and shape of the signal is not consistent with O_2^+ (nor with fragments such as CO^{n+}, O^{n+}, C^{n+}), but is consistent with CO_2^+. The absence of a shoulder after –350 s is consistent with a dayside exosphere, as anticipated by the CO_2 exosphere model and later observed by INMS.

Accompanies chapter by Teolis et al. (pp. 361–386).

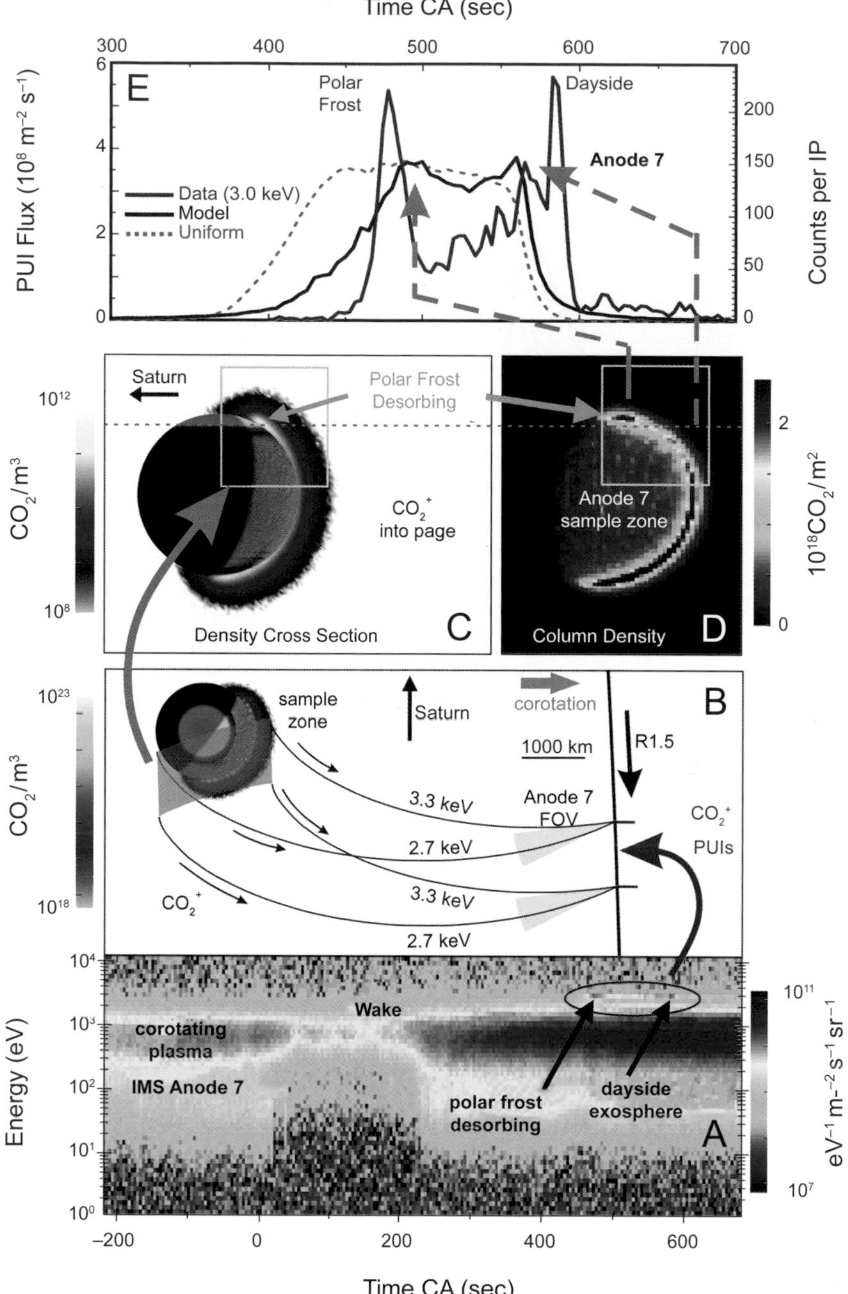

Plate 40. From *Teolis and Waite* (2016). Comparison of the Rhea pickup ion flux and distribution detected by CAPS IMS anode 7 on the R1.5 flyby, to that anticipated by the model CO_2 exosphere (fit to the later INMS R2 and R3 data) projected back to the R1.5 flyby date and time. (A) Anode 7 ion flux energy spectrogram with the CO_2^+ pickup ion signature circled. (B) R1.5 flyby configuration showing Rhea with the model exospheric CO_2 density cross section and adsorbed column density, the surface illumination, the region (purple box) sampled by anode 7 during the pickup ion detection, the anode 7 FOV at the time and location of the pickup ion detection along the R1.5 trajectory, and the limiting CO_2^+ energies (on arrival at Cassini) and trajectories accepted into anode 7. (C) Equatorial view of the model exosphere viewed through the sample box, showing the box's approximate north-south extent if the anode 8.5° FWHM elevation angular acceptance is projected back to Rhea. (D) Exospheric column density from the orientation of (C) as sampled by anode 7, showing the intense feature from polar frost desorption. (E) Anticipated total anode 7 pickup ion signal vs. time from closest approach (red), with arrows showing the correspondence of the double peaked structure to the polar frost desorption, and dayside exospheric signatures seen in (D). As at R1 (Fig. 2), the timing, intensity, and shape of the signal is not consistent with O_2^+ (not shown) nor fragment species, but is consistent with CO_2^+. The observed pickup ion signature (red) shows a similar (even more intense) double peaked structure, which implies that CAPS is observing a CO_2 exosphere with an abundance and spatial structure consistent with INMS based modeling. Also shown for comparison (top, dashed line) is the expected pickup ion profile for a uniform CO_2 exosphere.

Accompanies chapter by Teolis et al. (pp. 361–386).

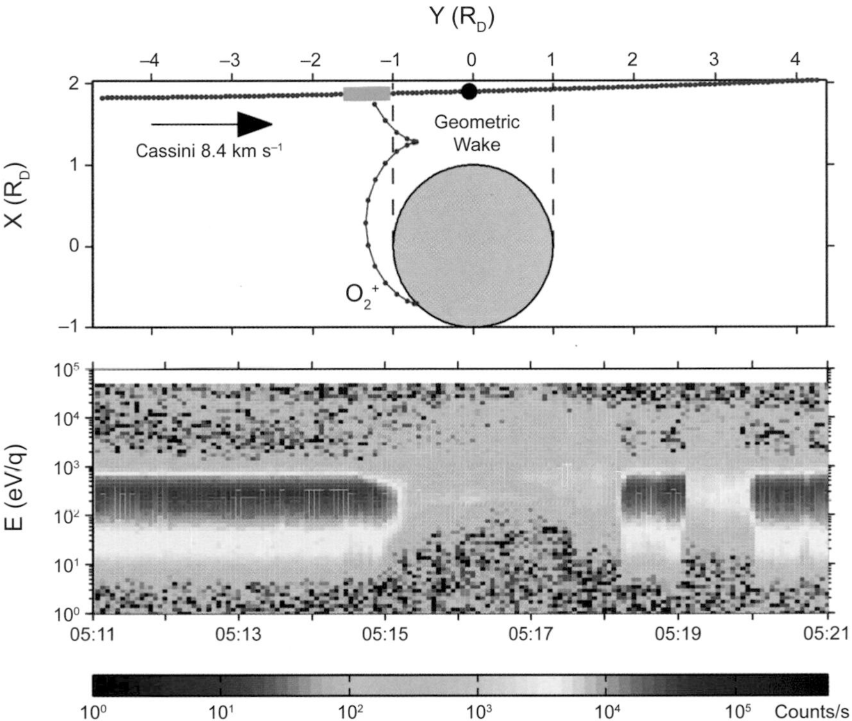

Plate 41. From *Tokar et al.* (2012). (*Bottom*) CAPS ion count rate at 4 s resolution as a function of time and energy per charge. The measurements are for anode 4 of CAPS, sensing flow predominantly in the co-rotation direction. (*Top*) Cassini trajectory during D2 and Dione's geometric wake, with a sample pick-up O_2^+ ion trajectory in the ambient flow plotted at 2 s resolution. The pickup ions are detected by CAPS just before entering the wake, in the green shaded region on the trajectory.

Accompanies chapter by Teolis et al. (pp. 361–386).

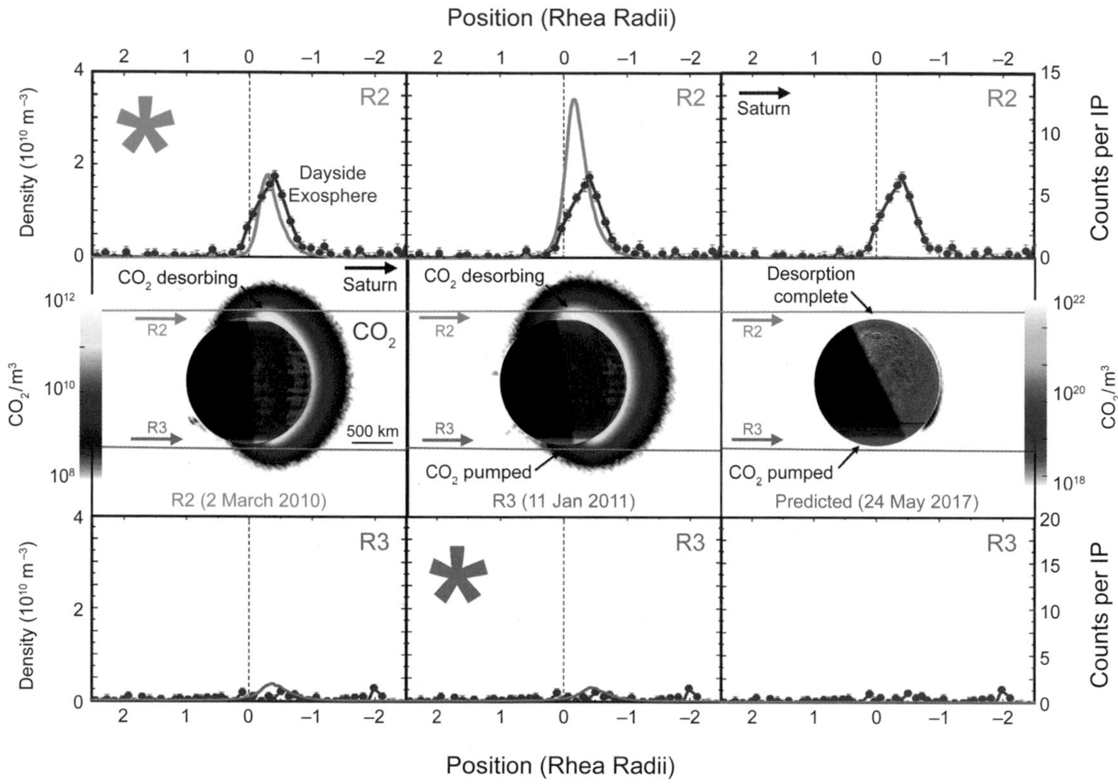

Plate 42. From *Teolis and Waite* (2016). Rhea CO_2 exospheric model results, showing (center row) the surface solar illumination (dawn terminator as viewed over the equator), and predicted gas density cross sections and surface frost column density at the R2 (left) and R3 (center column) flyby dates and times, and prediction (right) for the 2017 Saturn solstice. (*Top*) Comparison of the observed (red points, data binned) and predicted (green line) CO_2 densities vs. position from closest approach along the R2 trajectory (x-axis, in Rhea radii, dashed line: closest approach). (*Bottom*) Same for the R3 flyby trajectory. Asterisks denote model-data comparisons with the same flyby date; all other plots show model projections to different dates. The model assumed a 3.7×10^{21} CO_2 s^{-1} source rate as required to match the observed peak density on both flybys. The observation of CO_2 concentrated on the dayside is consistent with the modeled exospheric structure due to nightside and polar cryopumping. CO_2 was desorbing from the north, and cryopumping onto the southern polar surface during the flybys. Accordingly, as anticipated by the model (bottom center), CO_2 was undetectable in the south on R3. As shown (right), exhaustion of the northern frost cap should have resulted in the collapse of the CO_2 exosphere by the time of the 2017 solstice.

Accompanies chapter by Teolis et al. (pp. 361–386).

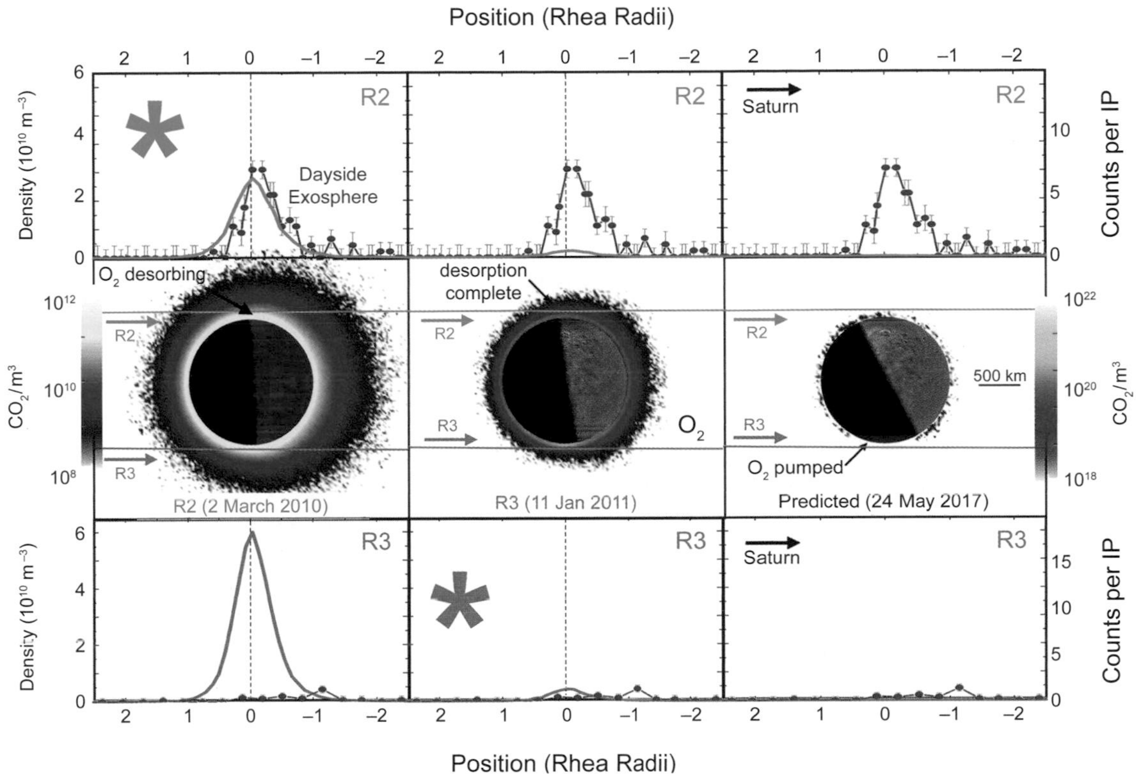

Plate 43. From *Teolis and Waite* (2016). Rhea O_2 exospheric model results, showing (center row) the surface solar illumination (dawn terminator as viewed over the equator), and predicted gas density cross sections and surface frost column density at the R2 (left) and R3 (center) flyby dates and times, and prediction (right) for the 2017 Saturn solstice. (*Top*) Comparison of the observed (data binned) and predicted O_2 densities vs. position from closest approach along the R2 trajectory (x-axis, in Rhea radii, dashed line: closest approach). (*Bottom*) Same for the R3 flyby trajectory. Asterisks denote model-data comparisons with the same flyby date; all other plots show model projections to different dates. The model assumed a 7.2×10^{21} O_2 s^{-1} source rate as required to match the observed peak density on both flybys. An O_2 tail is detected on the dayside, consistent with the anticipated greater dayside scale height. Unlike CO_2, the southern latitudes are not yet cold enough to cryopump O_2 from the southern exosphere on the R2 and R3 flybys dates. However, the modeled O_2 frost cap is exhausted between the R2 and R3 flybys, resulting in an anticipated collapse of the global exosphere, consistent with the reduced O_2 detection on R3.

Accompanies chapter by Teolis et al. (pp. 361–386).

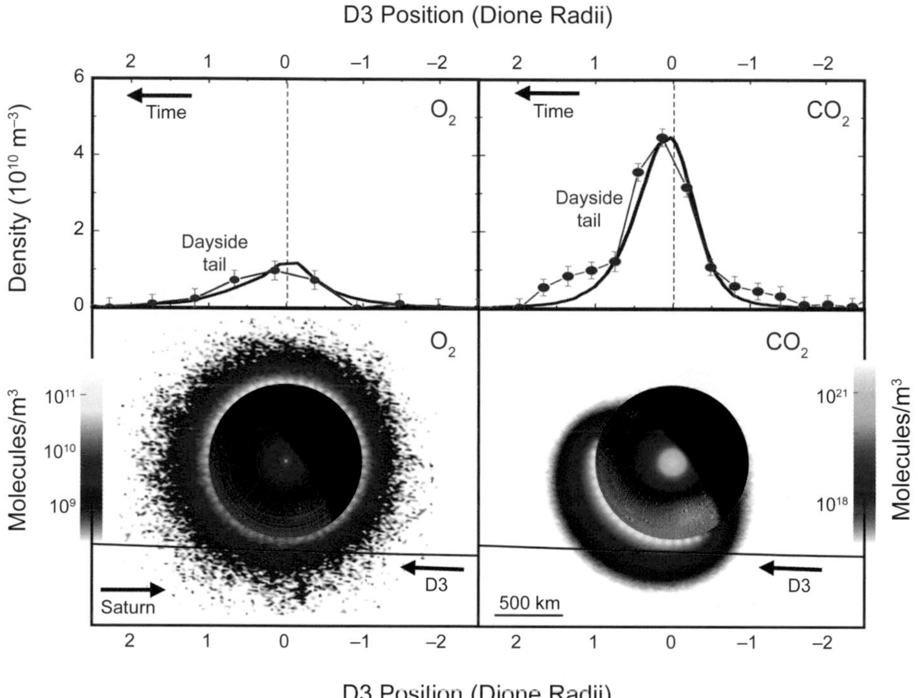

Plate 44. From *Teolis and Waite* (2016). Dione O_2 (left) and CO_2 (right) exospheric model results, showing (bottom) the surface solar illumination (north polar view), and predicted equatorial gas density cross sections and surface frost column density at the December 12, 2011, D3 flyby date and time. (*Top*) Comparison of the observed (blue points: O_2, red points: CO_2, data binned) and predicted (black lines) O_2 and CO_2 densities vs. position from closest approach (dotted lines) along the D3 trajectory (x-axis, in Dione radii). The model assumed 45 and 1.6×10^{21} s^{-1} O_2 and CO_2 source rates, as required to match the observed peak density. CO_2 and, to a lesser degree, O_2 are concentrated over the dayside as predicted by the model. As observed at Rhea, the Dione exospheric model predicts seasonal exospheric variability.

Accompanies chapter by Teolis et al. (pp. 361–386).

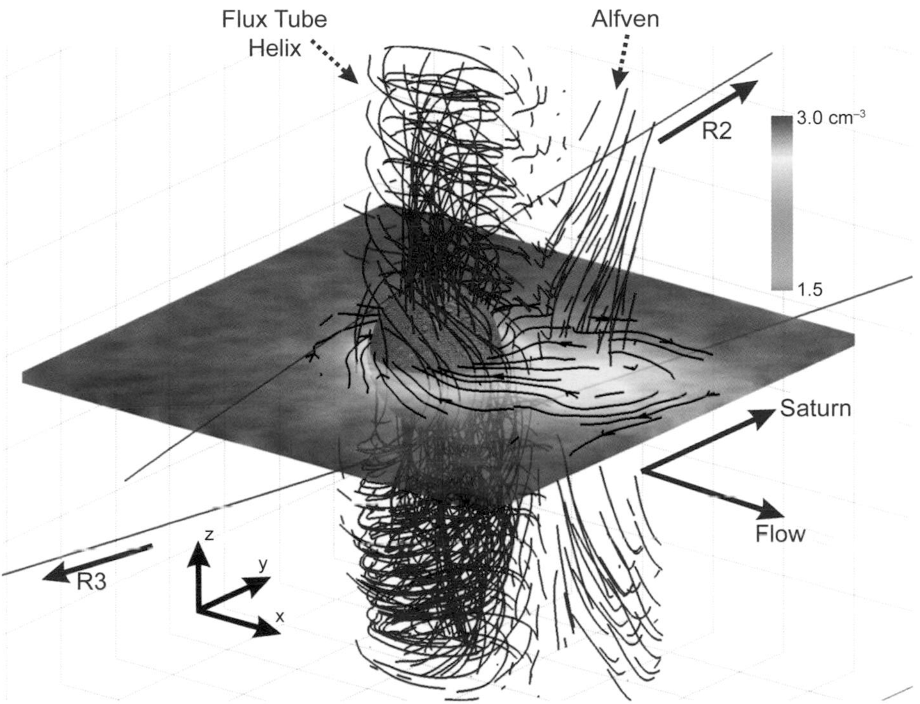

Plate 45. From *Teolis et al.* (2014); their model prediction of the magnetic field perturbation (i.e., total field minus constant $\vec{B}_0$ background) with southward and northward perturbations colored blue and red. Equatorial cross section shows the plasma density (blue-to-red: increasing density) and the diamagnetic current spiraling into Rhea and circling the wake (black streamlines). The helical perturbation is due to the flux tube electron current. Contrary to the Alfvén disturbance from the wake, the flux tube feature has negligible draping angle to the ambient magnetic field, consistent with the high whistler group velocities on the order of 10^3 km s^{-1}.

Accompanies chapter by Teolis et al. (pp. 361–386).

Howett C. J. A., Hendrix A. R., Nordheim T. A., Paranicas C., Spencer J. R., and Verbiscer A. J. (2018) Ring and magnetosphere interactions with satellite surfaces. In *Enceladus and the Icy Moons of Saturn* (P. M. Schenk et al., eds.), pp. 343–360. Univ. of Arizona, Tucson, DOI: 10.2458/azu_uapress_9780816537075-ch017.

Ring and Magnetosphere Interactions with Satellite Surfaces

Carly J. A. Howett
Southwest Research Institute

Amanda R. Hendrix
Planetary Science Institute

Tom A. Nordheim
Jet Propulsion Laboratory

Christopher Paranicas
Johns Hopkins University, Applied Physics Laboratory

John R. Spencer
Southwest Research Institute

Anne J. Verbiscer
University of Virginia

1. INTRODUCTION

The surfaces of Saturn's icy satellites undergo weathering from a large number of external sources, including photons, electrons, protons, ions, interplanetary dust, and ring particles. In this chapter we explore the nature of this alteration on Saturn's major icy satellites: Mimas, Enceladus, Tethys, Dione, Rhea, and Iapetus. Full details on weathering on Saturn's other satellites are given in the chapters by Thomas et al. and Denk et al. in this volume.

All the major icy satellites orbit Saturn within nine saturnian radii (R_S) with the exception of Iapetus (59.09 R_S): Mimas (3.08 R_S), Enceladus (3.95 R_S), Tethys (4.89 R_S), Dione (6.26 R_S), and Rhea (8.75 R_S). Therefore these satellites are imbedded in Saturn's E ring, which extends from 3 to 8 R_S (*Kempf et al.,* 2008; *Juhàsz et al.,* 2007). Here we discuss the effect E-ring grains have on the albedo and composition of Enceladus' neighboring satellites, but not Enceladus itself, as that is the subject of the chapter by Kempf et al. in this volume. We also discuss surface alteration by the Phoebe ring, and that of possible rings around Rhea and Iapetus.

Saturn's inner satellites are also weathered by the plasma and energetic charged particles trapped in Saturn's magnetosphere. The main charged particle species involved are electrons, protons, and water-group ions, and the nature of this weathering depends upon the type and energy of the particles bombarding the surface. All these different weathering mechanisms can competitively occur on the same area of the surface, or affect different regions of the satellites. This leads to a complex pattern of surface alteration across each satellite and between the various satellites. In this chapter we explore the possible ring and magnetospheric sources for surface alteration and outline how they affect the surfaces of Saturn's major satellites.

2. SURFACE-RING PROCESSING

2.1. E-Ring Deposition

It wasn't until Cassini's three close encounters with Enceladus in 2005 that its activity was confirmed (*Dougherty et al.,* 2006; *Hansen et al.,* 2006; *Porco et al.,* 2006; *Spahn et al.,* 2006; *Spencer et al.,* 2006; *Waite et al.,* 2006). The plume, composed of water vapor and small particles (mainly water ice), was discovered to erupt from Enceladus' south polar region (*Hansen et al.,* 2006; *Waite et al.,* 2006; see also the chapters in this volume by Goldstein et al. and Postberg et al.) and sustain Saturn's E ring (*Spahn et al.,* 2006; see also the chapters in this volume by Smith et al. and Postberg et al.). The brightness of the plume, a proxy for the amount of solid material being ejected, is greatest when Enceladus is at apocentre (*Hedman et al.,* 2013). Models predict this increase in density occurs when surface fissures are under tension, and thus may be able to open wider, enabling more material to escape (*Hurford et al.,* 2007, 2009; *Smith-Konter and Pappalardo,* 2008).

In situ measurements of the E ring show it is dominated by water ice (*Hillier et al.,* 2007). This result is supported by the *in situ* measurement of Enceladus' plumes, which maintains the E ring (*Waite et al.,* 2006). Other constituents include carbon dioxide, methane, salts, ammonia, and organics (*Postberg et al.,* 2008, 2009; *Waite et al.,* 2017). The E ring is made up of fine-grained particles ~1 μm in size, with a narrow size distribution between 0.3 and 3 μm (*Nicholson et al.,* 1996). The ring extends from 3 and 8 R_S, with its peak density and narrowest vertical extent near Enceladus' orbit (3.95 R_S) (*Showalter et al.,* 1991; *Hedman et al.,* 2012). More details on the nature of the E ring are given in the chapter in this volume by Kempf et al.

Groundbased observations throughout the 1970s showed that the leading hemispheres of Tethys, Dione, and Rhea are brighter at visible wavelengths than their trailing sides (*McCord et al.,* 1971; *Noland et al.,* 1974; *Cruikshank,* 1979). This result was confirmed by results from the Voyager spacecraft in the early 1980s (*Buratti et al.,* 1990; *Verbiscer and Veverka,* 1989). Voyager observed the opposite hemispheric trends on Mimas and Enceladus: Mimas and Enceladus have brighter trailing and darker leading hemispheres (*Buratti and Veverka,* 1984; *Buratti et al.,* 1990; *Verbiscer and Veverka,* 1992, 1994; *Verbiscer et al.,* 2005). This same trend was also observed by Cassini at other wavelengths and with other techniques, for example, in the far-UV (170 to 190 nm), where Mimas has a brighter trailing hemisphere while Tethys and Dione have brighter leading hemispheres (*Hendrix et al.,* 2012a,b; *Royer and Hendrix,* 2014). Analysis of thermal (visual wavelength) measurements taken by Cassini show Rhea's leading hemisphere has a higher bolometric Bond albedo, 0.63 (0.55 ± 0.08) compared to 0.57 (0.42 ± 0.10) on the trailing hemisphere (*Howett et al.,* 2010; *Pitman et al.,* 2010).

This pattern of satellites interior to and including Enceladus having a brighter trailing hemisphere, while those exterior to Enceladus having a brighter leading one, gave strong support to the hypothesis that Enceladus was somehow brightening its neighbors (*Baum et al.,* 1981; *Pang et al.,* 1984; *Hamilton and Burns,* 1994). The idea was that if material was being ejected from Enceladus then nongravitational forces (predominantly electromagnetism and solar radiation pressure) would excite the eccentricity of the material, causing it to collide with Enceladus' neighboring satellites. This collision would occur near apocenter (pericenter) for those satellites exterior (interior) of Enceladus. It was initially thought that this would lead to it bombarding their leading (trailing) hemispheres (*Horànyi et al.,* 1992). However, recent preliminary modeling work indicates the bombardment pattern of Saturn's icy satellites by E-ring dust may actually be more complicated, with different satellites being bombarded on different hemispheres: Mimas (trailing); Enceladus, Tethys, and Rhea (sub- and anti-saturnian); and Dione (leading) (*Juhàsz and Horànyi,* 2015; see the chapter in this volume by Kempf et al.). In any case, this material is predicted to collide at such high velocities (> 5 km s^{-1}) that it would "sandblast" the surface, uncovering the clean water ice from below the surface and effectively brightening it (*Shkuratov and Helfenstein,* 2001; *Verbsicer et al.,* 2007).

The significant effect of E-ring grain bombardment on Saturn's satellites was highlighted in 2005 when an exact alignment of the Sun, Earth, and Saturn made it possible to observe the Saturn system at "true" opposition, at the minimum phase angle of 0.01°. Mean visual geometric albedos of Epimethus, Janus, Mimas, Enceladus, Tethys, Dione, and Rhea were measured by the Hubble Space Telescope (HST) (Fig. 1) (*Verbiscer et al.,* 2007). The results showed that the satellites have decreasing brightness with increasing distance from Enceladus as well as a strong correlation between the geometric albedo of the satellites and the pole-on radial reflectance profile of the E ring as a function of orbital radius. Thus, the global surfaces of Enceladus' neighboring satellites are being brightened by E-ring grain bombardment, but with one hemisphere being especially so (*Franz and Millis,* 1975; *Koutchmy and Lamy,* 1975; *Domingue et al.,* 1995; *Verbiscer and Veverka,* 1989; *Verbiscer et al.,* 2007; *Buratti et al.,* 1990).

Analyses of the 1.52- and 2.0-μm bands in Cassini Visual and Infrared Mapping Spectrometer (VIMS) measurements show that the particle sizes on Mimas' trailing hemisphere are smaller than its leading one (*Hendrix et al.,* 2012a), which is consistent with it being bombarded by small E-ring grains. The leading hemispheres of Tethys, Dione, and Rhea have deeper H_2O absorption features than their trailing hemispheres (*Stephan et al.,* 2010, 2012, 2016), indicative of fresher ice on their surface possibly due to E-ring grain bombardment. The leading hemisphere of Dione displays a suppressed Fresnel reflection (a characteristic H_2O reflection peak at 3.1 μm observed in crystalline water ice with grain

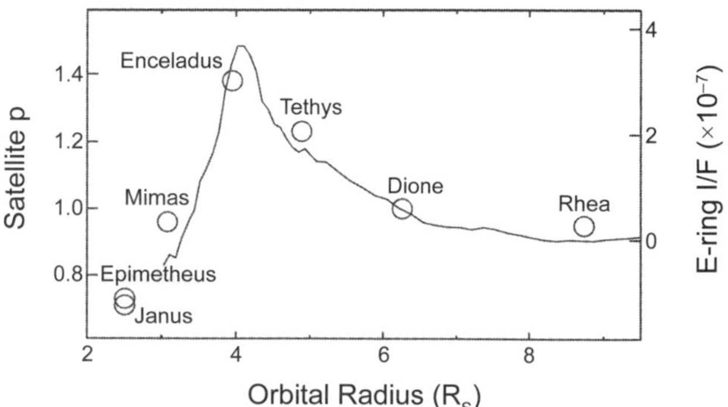

Fig. 1. The mean visual geometric albedo, p (left vertical axis), of satellites embedded in the E ring: Mimas, Enceladus, Tethys, Dione, and Rhea vs. radial distance from Saturn (R_S) mimics the pole-on reflectance (I/F) profile (solid line and right vertical axis) of the E ring. The Hubble Space Telescope's Wide Field and Planetary Camera 2 (WFPC2) observed both the 1995 ring plane crossings and the satellites at true opposition with the same filter (F555W). From *Verbiscer et al.* (2007).

sizes >1 μm), perhaps due to surface recoating by small E-ring grains (*Stephan et al.,* 2010). Finally, it has been proposed that a mismatch between the current models and observed VIMS data in the 3.0- to 3.5-μm region could be due to the layering of E-ring particles on the satellites' surfaces (*Fillachione et al.,* 2012), since the misfit is moderate for Mimas and Tethys and small for Dione, in direct relation to the density of the E ring at those radial locations (see Fig. 1).

While this bombarding or "sandblasting" may increase the purity of the surface ice (and/or alter its grain size), we note that competing surface alteration processes sometimes make the detection of such a signature difficult. Detecting surface alteration by E-ring grains is further complicated by the fact that depending on their impact velocities, E-ring grains may coat rather than sandblast a surface. For example, one analysis of VIMS data was unable to show any evidence for grain-sized or compositional changes on the leading hemisphere of Dione that could be directly related to its bombardment by E-ring particles (*Stephan et al.,* 2010). More recent work has found that the characteristic water ice absorption bands are deeper on the leading hemisphere than on the trailing hemisphere for both Rhea and Dione, indicating that these regions may have been coated with fresh water ice (e.g. from E-ring deposition) (*Scipioni et al.,* 2013, 2014). Furthermore, these authors found that the trailing hemisphere of both moons have the strongest concentration of submicrometer ice grains. However, this same analysis highlighted that Rhea is generally brighter and less contaminated than Dione, which is not what one would expect if E-ring grain bombardment is the only strong source of surface alternation at both moons.

Tethys, Dione, and Rhea have bluer surfaces at visible wavelengths at all latitudes along their 0° and 180° meridians, while their central leading and central trailing hemispheres appear relatively red (*Schenk et al.,* 2011). One possible explanation is that the reddening of one hemisphere is produced by solid- or charged-particle bombardment, and that the other hemisphere was reddened by the same processes before a large impact reoriented the surface (*Chapman and McKinnon,* 1986). However, this explanation requires all three satellites to have undergone a similar reorientation, which seems statistically unlikely. Another proposed explanation is that (on some suitable timescale) Saturn's magnetic dipole reverses, which would switch the Lorentz force on the particles orbiting Saturn, changing the orbital speeds of larger E-ring particles on circular orbits from super-Keplerian to sub-Keplerian — meaning they were swept up by the leading hemisphere instead of bombarding the trailing one (*Schenk et al.,* 2011). Thus, under this scenario, the current reddening on the apex of motion of Tethys, Dione, and Rhea is a relic from an earlier time and will slowly be erased by continued space weathering.

2.2. A Ring Around Rhea?

Two of Cassini's instruments [the Low-Energy Magnetospheric Measurement System (LEMMS) portion of the Magnetospheric Imaging Instrument (MIMI) and the Cassini Plasma Spectrometer (CAPS)] detected an unexpected decrease in the flux of high-energy (>20 keV) magnetospheric electrons around Rhea, continuing out to 8 Rhea radii on either side of the moon (*Jones et al.,* 2008a). It was proposed that a disk of electron-absorbing debris around Rhea was the cause of this depletion and the structure of the depletion could be explained by the presence of discrete rings or arcs within the disk. However, subsequent VIMS and Imaging Science Subsystem (ISS) observations were unable to verify the presence of such a disk in either the visible or near-infrared (*Pitman et al.,* 2008; *Tiscareno et al.,* 2010). Furthermore, additional analysis has not yielded a particle size distribution that could explain both the imaging and charged-particle data (*Tiscareno et al.,* 2010). Thus, the presence of a current ring around Rhea seems doubtful.

However, there is strong evidence on Rhea's surface that it had a ring in its past. There are a series of discrete elongated blue patches that form a great circle across Rhea's leading hemisphere and onto its trailing one, inclined slightly around its equator (*Jones et al.,* 2008b; *Schenk et al.,* 2011). These patches are <10 km wide and up to 50 km long, and are most prominent on local high-topography and steeper (possibly east-facing) slopes. They do not appear to be correlated with any endogenic tectonic process. Infalling debris either from a current or former ring around Rhea may have formed the patches, since the debris would have preferentially impacted high-standing topography as it approached Rhea's surface. If these patches are indeed preferentially located on east-facing slopes, then this implies the ring particles had a retrograde orbit (*Schenk et al.,* 2011).

If all these blue regions were entirely created by primary impacts, then the impacting bodies must have been between 100 m and several kilometers wide (*Schenk et al.,* 2011). Such large objects (and the small particles that would be generated by their collisions) should be visible to Cassini (*Tiscareno et al.,* 2010). Since these populations are not seen, it is perhaps more likely that the features were formed long ago, recently enough so the craters are still visible but long enough ago that the bright rays (which would be likely generated by the impacts) have been erased (*Schenk et al.,* 2011). Alternatively, the ring material could be much smaller (~meter-sized) and simply remobilized regolith in preexisting craters exposing the fresh bluer ice beneath, in which case the formation of these patches could be recent.

It is unknown what cratering event could have formed such a debris ring around Rhea. However, it is unlikely that the impact that formed the Inktomi crater (a young 80-km-wide crater centered at 14°S and 112°W) could have produced the required debris ring (*Schenk et al.,* 2011). This is primarily because Inktomi's ejecta pattern implies a westward impactor (which would not produce the required retrograde-moving ring), and the crater itself is so young that impacts from its debris would still have bright rays (which are not observed) (*Schenk et al.,* 2011).

The inclination of the great circle of blue patches could be explained by polar wander of Rhea (*Schenk et al.,* 2011).

A 100-km-diameter impactor could cause the required reorientation (*Nimmo and Matsuyama,* 2007). However, no obvious craters lie along the required meridians (190°W in the northern hemisphere, or 20°W in the southern one) (*Schenk et al.,* 2011).

2.3. A Ring Around Iapetus?

Voyager 1 and 2 imaging of Iapetus was hampered by poor spatial resolution (8.5 km pixel^{-1} at best) and poor dynamic range (*Smith et al.,* 1981, 1982). However, white dots were observed along Iapetus' equator, and were believed to be mountains. Subsequent limb measurement analysis of these images confirmed mountain-like structures with heights up to 25 km (*Denk et al.,* 2000). Cassini imaging revolutionized our understanding of Iapetus, returning images with spatial resolutions as high as 0.1 km pixel^{-1} (*Porco et al.,* 2005; *CICLOPS,* 2007). These Cassini images revealed a 70-km-wide, 13-km-high equatorial ridge that remains structurally coherent from 23°W to 213°W (*Porco et al.,* 2005; *Giese et al.,* 2008). Thus it was discovered that the equatorial mountains seen by Voyager are in fact a discontinuous extension of this ridge. The ridge lies exactly along Iapetus' equator (and only on the equator), appears heavily cratered, and shows no sign of flexural trough (i.e., the lithosphere was strong enough at the time of the ridge's formation to support its weight without flexing) (*Giese et al.,* 2008; *Dombard and Cheng,* 2008). The cratering record on the ridge indicated that it is comparable in age to other terrains on Iapetus (*Schmedemann et al.,* 2008). No similar ridges are observed elsewhere in our solar system.

Many formation mechanisms have been proposed to explain Iapetus' ridge. Upon its discovery it was suggested that the ridge was formed by compression due to the slowing of Iapetus' rotation, with the caveat that the ridge does not show the characteristic tectonic patterns expected for despinning stresses (*Porco et al.,* 2005; *Castillo-Rogez et al.,* 2007). If the despinning stress was combined with variations in the lithospheric thickness, then formation of the ridge by equatorial extension and diking may have been possible (*Roberts and Nimmo,* 2009; *Melosh and Nimmo,* 2009). Another proposed explanation is that faulting warped the surface up, which would explain why the slopes of the ridge flanks vary and the lower slopes are much less than the angle of repose (*Giese et al.,* 2008). An exogenic origin was also suggested: that the ridge was formed by collisional accretion of a ring remnant (*Ip,* 2006; *Levison et al.,* 2011; *Dombard et al.,* 2012). The exact details of how such a ring was formed, and how that ring was able to make a ridge on Iapetus, are still being debated. However, the general notion is that the ring was formed a long time ago when Iapetus either captured another body into orbit, or was impacted by an object whose debris then coalesced in orbit. Either way, Iapetus for a while had an icy porous subsatellite <100 km in size. At this time Iapetus was still despinning, which had the net effect of pulling Iapetus' subsatellite inside its Roche limit, whereupon it was torn apart creating a thick debris ring. The particle size of the debris was gradually reduced by collisions within the ring, leading to a flatter ring aligned with Iapetus' spin equator. Gradually the ring particles themselves deorbited, hitting Iapetus surface at low speed, thus building up (rather than excavating) the surface. The net result was a large amount of material deposited around Iapetus' equator, building up a ridge.

2.4. Phoebe Ring

A large dust ring of particles orbiting Saturn in the ecliptic plane was discovered using data taken by Spitzer's Space Telescope's Multiband Imaging Photometer (MIPS) (24 μm and 70 μm). The ring extends radially from at least 128 R_S to 207 R_S and 40 R_S in the direction perpendicular to the ring plane, which matches the orbital inclination of Phoebe (one of Saturn's irregular outer satellites) (*Verbiscer et al.,* 2009). Subsequent imaging by the Wide-field Infrared Survey Explorer (WISE) at 22 μm revealed the radial extent of the ring is actually from 100 to 270 R_S (*Hamilton et al.,* 2015). Thus Phoebe, which orbits Saturn at a distance of 215 R_S, is imbedded in the ring. Since the ring is likely composed of dust ejected from Phoebe, and Phoebe is in a retrograde orbit, Phoebe ring particles will also be in retrograde orbits. Therefore, Phoebe ring particles will strike the leading hemisphere of Iapetus, which has tidally locked rotation and is in a prograde orbit at 60 R_S. Phoebe ring particles will also strike Hyperion orbiting at 25 R_S; however, Hyperion's chaotic rotation means that Phoebe ring particles strike the entire surface of Hyperion, not just one hemisphere preferentially, as they do on Iapetus.

The discovery of the Phoebe ring provides strong support to an answer first proposed by *Soter* (1974) to the centuries-old question of why the leading hemisphere of Iapetus is 10 times darker than the trailing. Soter proposed that dark ring particles launched from Phoebe (visible geometric albedo $p_V = 0.09$) preferentially strike and darken the leading hemisphere of Iapetus ($p_V = 0.04$). The albedo contrast may be enhanced by runaway global thermal migration of water ice, which can also explain the presence of bright poles and the extension of the dark material onto the leading side at low latitudes (*Spencer and Denk,* 2010). In their analysis of the spectral signatures of aromatic and aliphatic hydrocarbons on Phoebe, Iapetus, and Hyperion, *Cruikshank et al.* (2014) confirmed the long-held assertion that material originating on Phoebe's surface finds its way to the leading hemisphere of Iapetus and the entire surface of Hyperion (more on this can be found in the chapter by Hendrix et al. in this volume).

Using a photometric surface roughness model to analyze ISS data taken during Cassini's 2005 flyby of Iapetus, *Lee et al.* (2010) determined that the dark, leading hemisphere of Iapetus is much smoother than that of the bright terrains, which is presumably because impacts from Phoebe ring particles have obliterated and subsequently infilled shallow craters and facets. The surface of Iapetus' leading hemisphere (across both the dark and bright terrain) is notably redder than the respective material on its trailing hemisphere

(*Denk et al.,* 2010), suggesting that the reddening has an exogenic origin.

We note that materials from the Phoebe ring are unlikely to be the only source of Iapetus' dark material. Some of Iapetus' spectral features are also seen on objects that Phoebe ring particles are unlikely to reach (because they are swept up by Titan): Dione, Epimetheus, and the Cassini Division in Saturn's rings, and the F ring (*Clark et al.,* 2012). The dark material has spectral properties dominated by ice, nanophase hematite, nanophase and fine grains of metallic iron, and smaller amounts of organic compounds, carbon dioxide, and possible trace ammonia, as well as yet to be identified compounds (*Clark et al.,* 2012). The origin of the material isn't known, but it is postulated that the iron may be from meteoritic dust falling into the Saturn system (e.g., *Chapman,* 2004).

3. SURFACE-MAGNETOSPHERE PROCESSES

3.1. Introduction

In this section we consider how satellite surfaces are altered by interaction with the various types particles in Saturn's magnetosphere. Specifically we discuss how and where different types of charged particles bombard Saturn's major icy satellites (as summarized in Fig. 2) and the effect this bombardment has upon the surface.

3.2. High-Energy Electrons

Voyager 1 observed a dark east-west lens-shaped patch on Tethys' leading hemisphere. The region extends 15° either side of Tethys' equator, and is observed from 0° to 160°W (*Smith et al.,* 1981; and reanalysis by *Stooke,* 1989, 2002). Further analysis showed that it is 5% darker than its surroundings and bluer (*Buratti et al.,* 1990). At that time it was not clear why this surface region is notably different from its surroundings, especially since there is no change to the underlying geology and its margins are gradual.

Color maps of Tethys' surface produced using Cassini ISS images show this bluer surface feature in more detail (*Elder et al.,* 2007; *Schenk et al.,* 2011). The feature is clearly observed in color maps made using ISS' infrared-3 (IR) (0.930 μm), green (0.568 μm), and ultraviolet-3 (UV) (0.338 μm) channels (Fig. 3) (*Schenk et al.,* 2011). It is dark at visible wavelengths, and very prominent in IR/UV color ratio maps, since it is dark in the IR and green but bright in the UV (*Elder et al.,* 2007; *Schenk et al.,* 2011). These

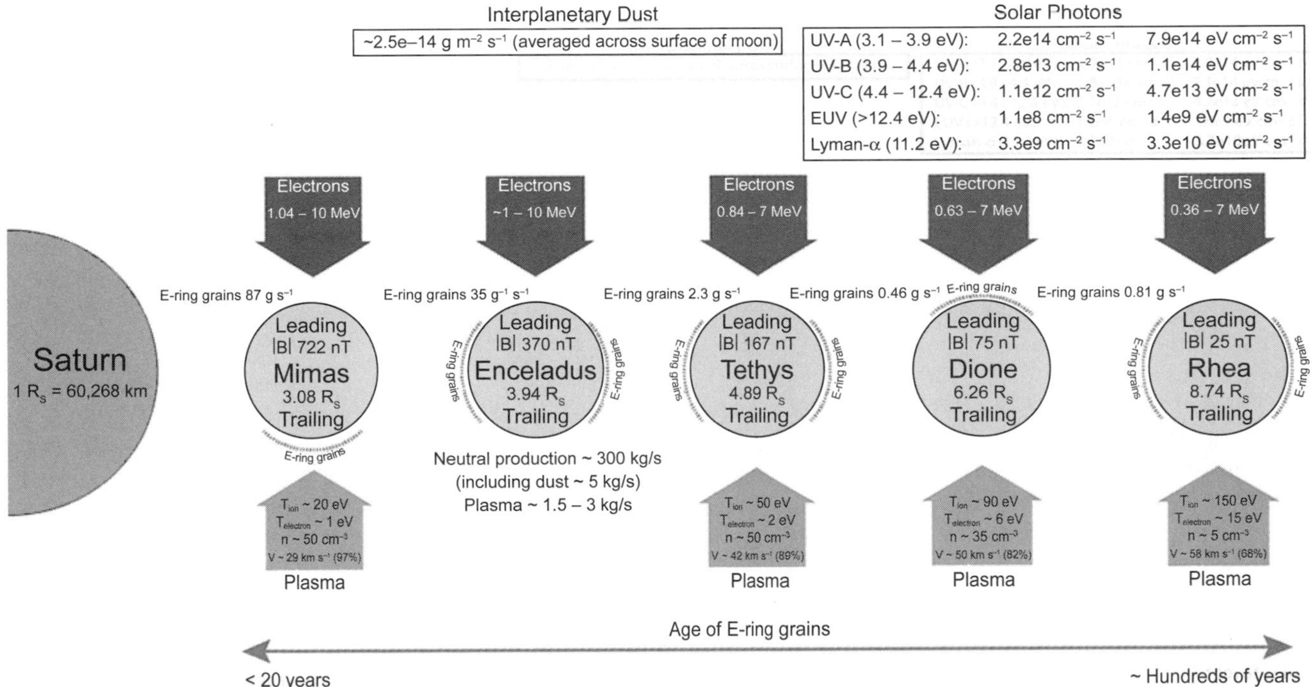

Fig. 2. An overview of the bombardment of Saturn's major icy satellites (excluding Iapetus) by ring and magnetospheric particles. Values are as follows: UV fluxes from *Madey et al.* (2002) but scaled to 9.54 AU; IDP fluxes from *Poppe* (2016); E-ring deposition rates and precipitation patterns from the chapter in this volume by Kempf et al., with grain lifetimes from *Juhàsz and Horànyi* (2015); plasma parameters from *Roussos et al.* (2010); co-rotation fractions from *Mauk et al.* (2005) and *Wilson et al.* (2010) (Rhea); magnetic field from *Khurana et al.* (2008) (all but Mimas) and *Saur and Strobel* (2005) (Mimas); energetic electron resonance energies assume an equatorial pitch angle of 59° and co-rotation fractions from *Mauk et al.* (2005); Enceladus neutral and plasma production rates from *Khurana et al.* (2007) and *Burger et al.* (2007); dust production rates from *Schmidt et al.* (2008).

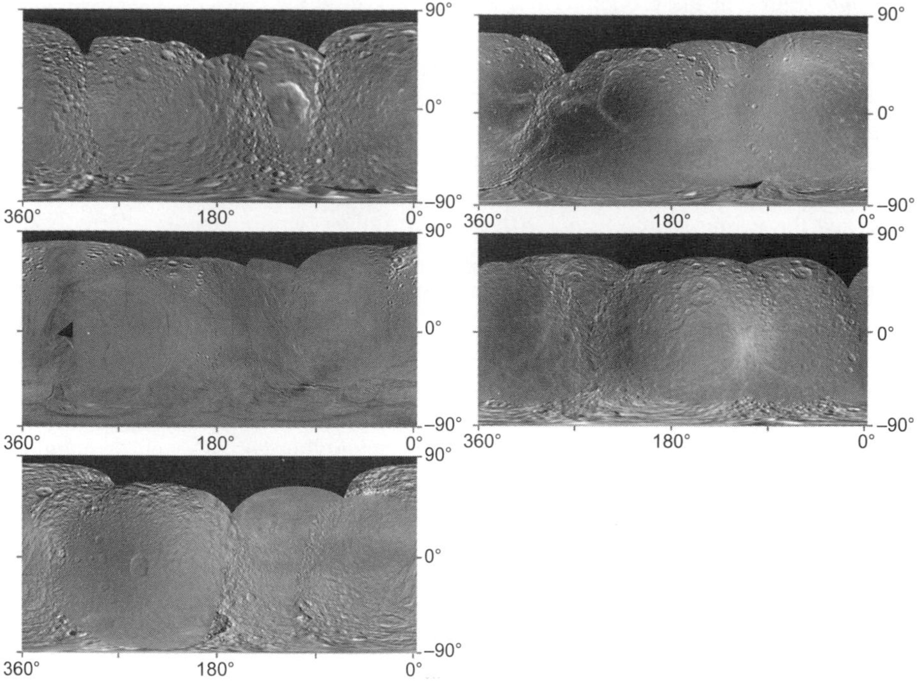

Fig. 3. See Plate 34 for color version. Enhanced Cassini three-color (IR-green-UV filter) global maps of the five inner mid-sized icy satellites of Saturn (from top to bottom: Mimas, Enceladus, Tethys, Dione, Rhea). Maps are in simple cylindrical projection from 90°S to 90°N and from –2°W to 360°W. From *Schenk et al.* (2011).

high-spatial-resolution images revealed the band to actually extend to ~20° north and south of Tethys' equator and across its entire leading hemisphere, being widest at 90°W, and narrowest at the anti-Saturn and Saturn-facing points.

Confirmation of the presence of this feature on Tethys by Cassini was also accompanied by a surprising discovery: A similar feature is present on Mimas (Fig. 3) (*Schenk et al.,* 2011). Like on Tethys, Mimas' feature is lens-shaped and located on its leading hemisphere, but its latitudinal extent is much greater: ~±40° about the equator, and it wraps onto the trailing hemisphere by a few degrees (*Schenk et al.,* 2011). In IR/green/UV color maps the feature appears blue, and is dark in the IR/UV color ratio maps. However, this region is not notably different to its surroundings in green images (unlike on Tethys) (*Schenk et al.,* 2011). The feature does span across Mimas' giant Herschel crater and appears to be less blue inside the crater.

Analysis of Cassini CIRS data showed that color anomalies on Mimas and Tethys were also spatially correlated with a thermal anomaly (Figs. 3 and 4) (*Howett et al.,* 2011, 2012). The bluer lens-shaped regions were cooler in the daytime, and warmer at night, than their surroundings. The reason for this was shown to be a notably higher thermal inertia inside the anomaly region. Thermal inertia describes a surface's ability to store and emit radiation; the higher its value, the slower a surface will be to heat up during the day or cool down at night. Changes in a variety of surface properties can lead to an increase in thermal inertia, since it is defined as being directly proportional to the square root of the thermal conductivity, bulk density, and specific heat capacity of a surface.

On Mimas the thermal inertia inside the anomalous region is 66 ± 23 J m^{-2} K^{-1} $s^{-1/2}$ (units henceforth referred to as MKS), compared to <16 MKS outside it. On Tethys the difference was smaller: 25 ± 3 MKS inside the anomaly, and 5 ± 1 MKS outside it (*Howett et al.,* 2011, 2012). Daytime temperature maps, derived indirectly from the position of the 3.6-µm water ice continuum peak in the Cassini VIMS data, also show the thermal anomaly on Mimas and Tethys (*Filacchione et al.,* 2016). The VIMS data show a band of cooler daytime temperatures at low latitudes on the two satellites. However, VIMS spectra do not show any obvious difference in surface composition (*Hendrix et al.,* 2012a; *Buratti et al.,* 2005; *Clark et al.,* 2008) or grain size (*Hendrix et al.,* 2012a) associated with the anomaly. Rather, grain sizes on Mimas vary with hemisphere, with the leading (trailing) hemisphere having larger (smaller) particles (*Hendrix et al.,* 2012a). There is no evidence in the far-UV (170–190 nm) for a similar anomaly (*Hendrix et al.,* 2012a); rather, the UV data follow a pattern created by E-ring grain accretion and cold plasma bombardment (see sections 3.4 and 3.5 in this chapter and the chapter by Hendrix et al. in this volume).

Another surprise came from the CIRS data: Dione also shows evidence of a thermal anomaly at low latitudes on its leading hemisphere, but Rhea does not (*Howett et al.,* 2014). On Dione the magnitude of the anomaly is lower than both Mimas and Tethys: 11 MKS inside the anomaly, and 8 MKS outside. Furthermore, unlike on Mimas and Tethys,

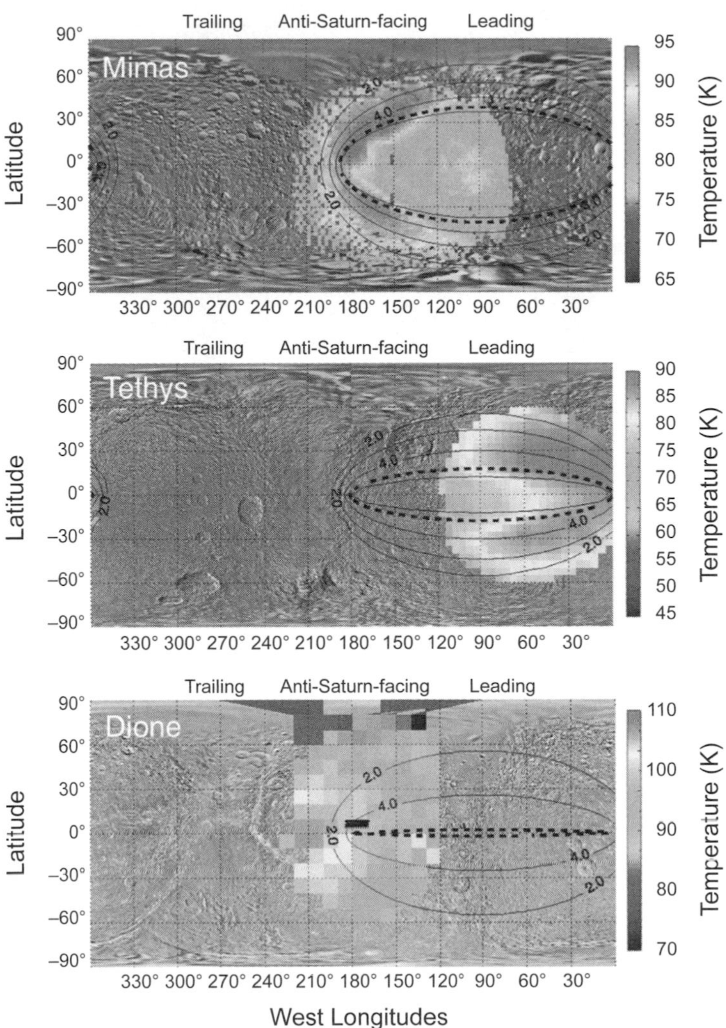

Fig. 4. See Plate 35 for color version. Daytime temperature observations of Mimas, Tethys, and Dione. Overlaid are contours of energetic electron power deposited into the surface per unit area, Q (log10 MeV cm^{-2} s^{-1}), determined using updated results from Cassini's Magnetospheric Imaging Instrument (MIMI). The best-fitting contour to the Mimas (Tethys) color and thermal inertia anomaly boundary (cf. *Howett et al.,* 2011, 2012) is given by the dashed line at 5.6 (1.8) × 10^4 MeV cm^{-2} s^{-1}.

there is no corresponding color change on the surface of Dione (*Schenk et al.,* 2011).

Any formation mechanism proposed to explain these anomalies must explain the shape and location of the anomalies, and also why they decrease in latitudinal extent with increasing distance from Saturn, until eventually at Rhea no anomaly is detected. Finally, it must explain why the anomaly is observed at certain wavelengths (particularly the thermal and UV), but not at others (the far-UV or near-IR).

The shape of the color anomaly provided the first clue as to the cause of the surface alteration. A similarly shaped feature is seen on Europa, in the location of its hydrated species (albeit on the opposite hemisphere). This feature is spatially correlated with the region preferentially bombarded by energetic electrons (*Paranicas et al.,* 2001), which suggests that the hydrates are produced radiolytically by the bombardment of these high-energy electrons (*Carlson et al.,* 1999).

At Saturn, the motion of charged particles in the magnetosphere is dominated by three fundamental processes: longitudinal motion, bounce between magnetic mirror points, and gyration around magnetic field lines. The bouncing motion occurs much faster than the longitudinal one, and within the megaelectronvolt range an electron's gyroradius is much smaller than the radius of any of Saturn's inner mid-sized satellites. Thus, when an electron guiding center field line comes into contact with a moon, the electron is absorbed by that moon (*Paranicas et al.,* 2014). As an example, at Mimas, a 2-MeV electron takes ~1 second to bounce from its north to south mirror point, but 40 seconds to move over a distance equivalent to Mimas' diameter (*Paranicas et al.,* 2014).

The speed of the longitudinal motion is the sum of the drifts due to the co-rotation electric field, and the gradient and curvature of the magnetic field (*Nordheim et al.,* 2017). The typical motion is easterly, causing the electrons to overtake the satellite and bombard their trailing hemisphere (*Roussos et al.,* 2007) (see Fig. 2). However, above a certain energy (known as the Keplerian resonance energy), the oppositely directed gradient-curvature drift becomes more dominant, causing the electrons to bombard a satellite's leading hemisphere (*Roussos et al.,* 2007; *Thomsen and van Allen,* 1980). At the distance of the inner saturnian satellites this change from prograde to retrograde motion occurs between 0.36 and 1.5 MeV (*Roussos et al.,* 2007). Thus, kiloelectronvolt electrons are below the Keplerian resonance energy and bombard the satellite's trailing hemisphere, while megaelectronvolt electrons can be above it and therefore can bombard the satellite's leading hemispheres. These megaelectronvole electrons do not bombard the leading hemisphere of Saturn's satellites uniformly, but rather since the higher-energy megaelectronvole electrons have a greater gradient-curvature drift, they are able to bombard a larger range of latitudes than lower-energy megaelectronvolt ones, which are restricted to equatorial regions (*Paranicas et al.,* 2014; *Nordheim et al.,* 2017). This is the opposite for the still lower-energy kiloelectronvolt electrons that bombard the trailing hemisphere, where the lowest energy electrons bombard a large range of latitudes while the higher ones (close to the resonance energy) are confined near the equator (*Nordheim et al.,* 2017).

These effects combine so that the regions bombarded by electrons on both the leading and trailing hemispheres are lens-shaped, like that first predicted on Europa (*Paranicas et al.,* 2001; *Patterson et al.,* 2012). This shape is largely the result of the very rapid bounce time of electrons along the magnetic field line compared to their slow longitudinal drift around the planet and relative to the satellite's own motion. These lens-shaped patterns have been predicted using several different modeling techniques (*Truscott et al.,* 2011; *Paranicas et al.,* 2001; *Dalton et al.,* 2013). This result is shown in Fig. 5, which depicts the electron power per unit area into the surface of the satellites (Q) (*Paranicas et al.,* 2014; *Schenk et al.,* 2011). A comparison between these contours and the boundary of the color and thermal anomalies reveals that the color anomalies receive $>3.2 \times 10^4$ MeV cm^{-2} s^{-1} (*Schenk et al.,* 2011), while the thermal anomalies receive >5.6 (1.8) $\times 10^4$ MeV cm^{-2} s^{-1} at Mimas (Tethys) (*Howett et al.,* 2011, 2012).

The clear spatial correlation between the locations preferentially bombarded by high-energy electrons on Mimas and Tethys and the locations of their thermal/color anomalies provides strong evidence that high-energy electrons are the cause of the surface alteration. However, by itself this doesn't explain why Dione's thermal anomaly is restricted to along its equator, or why a color anomaly isn't seen there but is seen on Tethys, since the predicted latitudinal extent of Q is similar between Tethys and Dione (Fig. 5).

The explanation may lie in the intensity variation of the different electron energies that combine to form Q. For example, the flux of the highest-energy electrons (greater than a few megaelectronvolts) decreases dramatically from Mimas, to Tethys, to Dione (Fig. 6), while the flux of lower-energy electrons (300 keV to 1 MeV) is greater at Dione than at Tethys. Thus, while Dione and Tethys have similar Q contours there is great variation in the energies of the electrons that

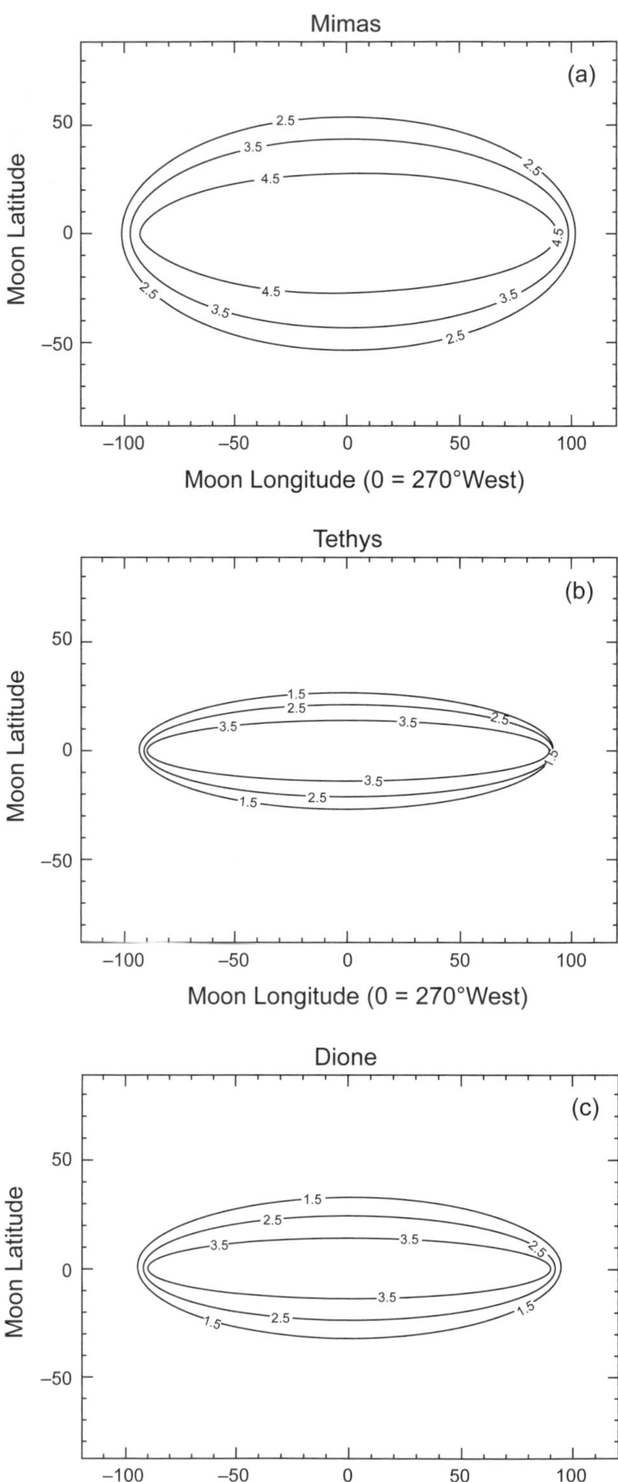

Fig. 5. Contours of Q (energetic electron power deposited into the surface per unit area), with levels in units of log10 (MeV cm^{-2} s^{-1}). The zero longitude in these figures is placed at 270°W, the center of the satellite's trailing hemisphere. From *Paranicas et al.* (2014).

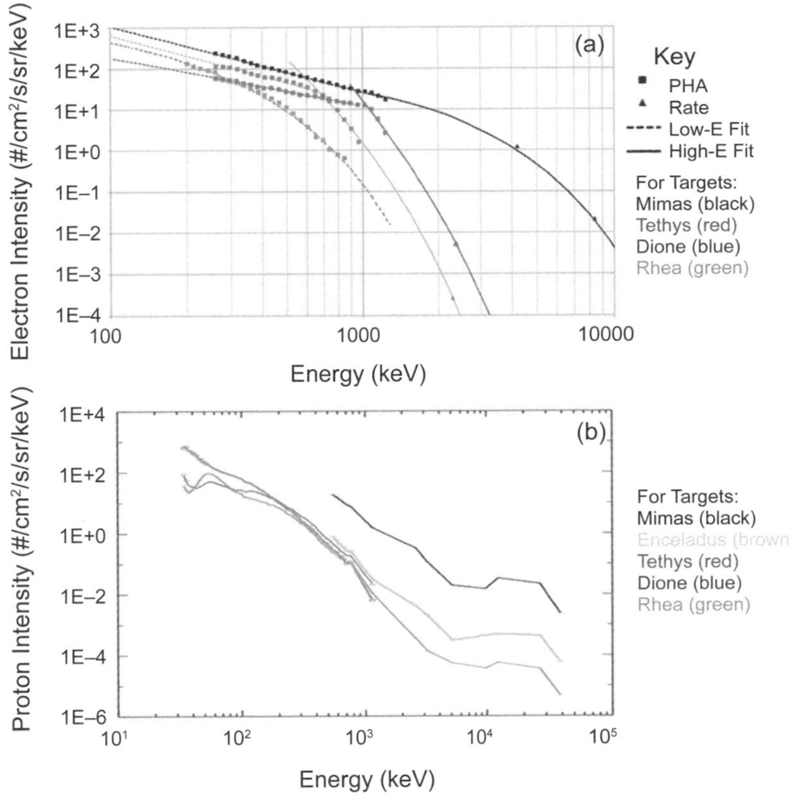

Fig. 6. See Plate 36 for color version. **(a)** Differential energy spectra (electrons per cm^2 s sr keV) at the L shells of four inner satellites of Saturn, adapted from *Paranicas et al.* (2014). Cassini MIMI pulse height analysis (PHA) and (higher-energy) rate data are shown with different plot symbols. Separate fits are created for each moon and energy range (see key for details). **(b)** Differential energy spectra (protons per cm^2 s sr keV) at the L shells of five inner satellites of Saturn.

contribute to Q: On Tethys the electron dose is comprised of more high-energy and fewer low-energy electrons than on Dione. Furthermore, the planetary magnetic field becomes less dipolar with radial distance so the bombardment pattern becomes less specific moving outward from Mimas. Finally, the dust pattern, which competes with the lens formation, is not the same on Tethys, Dione, and Rhea. The dust from Enceladus must contain a radial component in its orbit to reach the outer satellites and this radial component of motion increases for the more distant satellites. This alters the dynamics of the dust collision with the satellite and the bombardment pattern (see the dust deposition pattern shown by the dashed lines in Fig. 2, as well as the chapter by Kempf et al. in this volume).

The size of the lens for each electron energy/satellite may be a critical factor. Electrons near the Keplerian resonance bombard very close to the equator. For energies well below or well above the Keplerian resonance for that satellite, the latitude range of bombardment increases (*Nordheim et al.,* 2017). For example, Fig. 5 shows that highest-energy electrons (>10 MeV) are able to bombard all of Mimas' leading hemisphere, while the lower-energy ones (~1 to 2 MeV) are bound to its equatorial region. However, on the trailing hemisphere it is the lower-energy electrons that can bombard a wider latitudinal range (*Paranicas et al.,* 2014; *Nordheim et al.,* 2017). Thus, Dione's leading-hemisphere surface alteration is more restricted to the equatorial regions compared to Tethys simply because most of the electrons that bombard it have a lower energy, which means they are restricted to its equatorial region. In other words, due to the decreasing intensity of Saturn's radiation belts with radial distance, the more energetic electrons (which would form a lens with a larger latitudinal extent) are present in much lower numbers. If higher-energy electrons are required to form a color anomaly, these differences could explain why a color anomaly is seen on Tethys but not on Dione.

The dose change with depth is also an important consideration. For jovian icy satellites, we have found that in the top 0.01 to 1 m of the surface the dose is dominated by energetic electrons. This is due to the fact that electrons generally have much greater penetration depth than ions of the same energy. Furthermore, the secondary Bremsstrahlung photons that are created as electrons slow down can penetrate to depths of several meters in ice (*Paranicas et al.,* 2002). However, it is important to note that the energetic electron weathering of the saturnian satellites is much weaker than the jovian ones. For example, *Nordheim et al.* (2017) found that secondary Bremsstrahlung photons were much less important at Mimas than at Europa. On Saturn's icy moons the higher-energy megaelectronvolt electrons penetrate approximately centimeter depths into the surface of the leading hemisphere (Fig. 7) (*Paranicas et al.,* 2014; *Nordheim et al.,* 2017; *Schaible et al.,* 2017). This is approximately the same depth sampled by the diurnal thermal wave, thus changes at this depth can effect the thermal emission observed (*Howett et al.,* 2011, 2012, 2014). However, the visible and far-UV reflectances are only affected by changes in the upper few micrometers of the surface (*Schenk et al.,* 2011; *Hendrix et al.,* 2012a; *Paranicas et al.,* 2014). Thus if the megaelectronvolt electrons alter the deeper surface, this would preferentially effect the thermal emission.

Since surface alteration by high-energy electrons could explain so many properties of the color and thermal anomalies (i.e., their shape, location, the depth at which they occur, and their decreasing magnitude with distance from Saturn) it was quickly adopted as the most likely explanation for these anomalies. The precise mechanism by which high-energy electrons are modifying the surface in these anomalous regions is not well understood, however. Such bombardment can lead to physical changes to the ice (amorphization, sintering) as well as chemical changes (creation of new molecules, coloring of ice impurities), but how this translates to increases in the thermal inertia or color changes is not clear (*Baragiola,* 2003; *Paranicas et al.,* 2014; *Ferrari and Lucas,* 2016). However, it is plausible that the high-energy electrons excite and mobilize water molecules

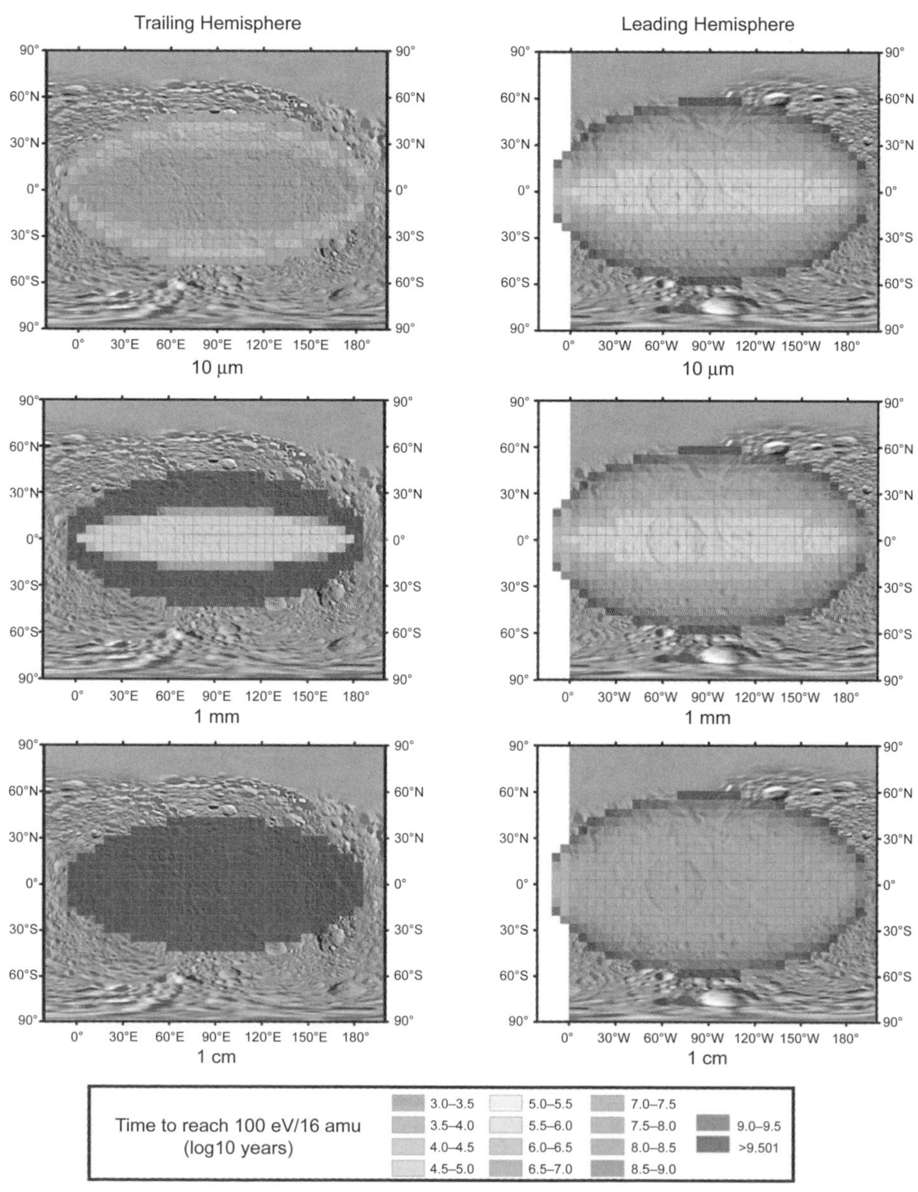

Fig. 7. See Plate 37 for color version. Energy deposition maps at different depths on the trailing hemisphere of Mimas. The energetic electron dose is given in terms of years to reach a significant dose of 100 eV/16 amu, which is equal to a dose of 60.3 G_{rad}. From *Nordheim et al.* (2017).

in their path, and that the mobilized molecules recondense at grain contacts. This process increases the contact area between the grains and thus their thermal conductivity (and hence thermal inertia). In essence, the high-energy electrons glue the grains together (*Howett et al.,* 2011; *Schenk et al.,* 2011). The new defects that form in the ice grains act as scattering centers, increasing the surface's ability to scatter near-UV photons (*Johnson et al.,*1985; *Johnson and Jesser,* 1997; *Schenk et al.,* 2011).

This process of atoms or molecules moving to the contact regions between grains (either due to radiation- or thermal-induction diffusion), and thus increasing grain contacts, is known as sintering (e.g., *Myers,* 1980; *Blackford,* 2007; *Sirono,* 2011). Sintering rates depend strongly upon surface properties (such as temperature, grain size, and the pressure in the void spaces). Recent work has shown that radiation-induced sintering can produce the stable thermal anomalies observed on Mimas, Tethys, and Dione (*Schaible et al.,* 2017). This work shows that surface alteration by radiation-induced sintering occurs faster than competing resurfacing processes (E-ring infall on Tethys and Dione, and interplanetary dust gardening on Mimas).

Recently, *Nordheim et al.* (2017) studied energetic electron and proton bombardment at at Mimas. This work predicted that in addition to the observed leading hemisphere lens (*Howett et al.,* 2011; *Schenk et al.,* 2011; *Paranicas et al.,* 2014), there should be a corresponding, and as of yet unobserved, trailing hemisphere lens (Fig. 8). The trailing lens would be created by electrons with energies below about 1 MeV. However, unlike electrons at the leading hemisphere of Mimas, which reach depths of ~1–2 cm into the ice, trailing hemisphere electrons deposit most of their dose within the top approximately millimeters of the surface (Fig. 7). *Nordheim et al.* (2017) also suggested that the trailing hemisphere lens could be obscured by the competing effect of cold plasma bombardment. However, E-ring grains preferentially bombard the trailing hemisphere of Mimas (*Hamilton and Burns,* 1994; *Juhàsz and Horànyi,* 2015), and therefore it is possible that the effects of electron irradiation are simply obscured by the deposition of E-ring grains there. Electron lens bombardment patterns should also be found at both the leading and trailing hemispheres of Tethys, Dione, and Rhea. It is not presently clear, however, if the effects of the trailing hemisphere lens would be observable at any of the midsized inner satellites.

3.3. High-Energy Protons

Determining the weathering effects of protons at energies above a few hundred kiloelectronvolts on Saturn's icy satellites is somewhat complicated. This is because at the orbits of the inner moons a deep drop-out in proton flux occurs due to the absorption by the satellite (known as a macrosignature) (*Paranicas et al.,* 2008; *Kollmann et al.,* 2013). This means that the rate of surface weathering is regulated by the rate of diffusion into the macrosignature. The ambient flux levels inside the macrosignatures are so low that the weathering is probably not strong (*Nordheim et al.,* 2017). However, the macrosignature structure suggests that there could be more weathering by the associated particles around the sub- and anti-Saturn apex points. Charged particles in the energy range below the radiation belts (<less than ~40 MeV; see Fig. 6) weather all the satellites to some degree, but many singly charged ions in this energy range are lost due to charge exchange with the abundant neutral species present in the inner magnetosphere of Saturn primarily due to plume activity on Enceladus. For this group of energetic particles, e.g., protons between 1 keV and 200 keV, there may be little to no weathering.

3.4. Cold Plasma and Plasma Implantation

Weathering by cold plasma (i.e., ions and electrons of energies less than ~10 keV) occurs principally on the satellite trailing hemispheres because the co-rotating plasma overtakes the satellites in their orbital motion. The cold plasma density is highest close to Saturn (inside 6 R_S), which means it primarily effects Mimas, Enceladus, and Tethys (*Andrè et al.,* 2008). The plasma probably then impacts the satellites with a bullseye-shaped pattern (*Hendrix et al.,* 2012a; *Schenk et al.,* 2011). Irradiation of icy moons can have several effects, including the production of new species (radiolysis, considered further in section 3.6) in the surface and ejecting volatiles from the surface (sputtering or desorption) (e.g., *Johnson et al.,* 2004). Impacting particles exhibit spatial bombardment distributions across a satellite's surface that depend on their energy, mass, and charge (*Johnson et al.,* 2004). For cold plasma, a darkening of the surface is commonly related to radiolysis of the host material. For example, the radiolysis of hydrocarbons can darken them via carbonization (e.g., *Strazulla et al.,* 1995), and *Sack et al.* (1991) demonstrated that irradiation of pure water ice can darken and redden the ice at UV-visible wavelengths

At Mimas, cold plasma bombardment on the trailing hemisphere (potentially a UV darkening effect) is expected to compete with E-ring grain bombardment (a brightening effect). Indeed, such leading-trailing hemisphere brightness differences are measured at Mimas at UV-IR wavelengths (e.g., *Verbiscer and Veverka,* 1992; *Verbiscer et al.,* 2007). UVIS coverage of Mimas' trailing hemisphere (*Hendrix et al.,* 2012a) shows a hint of a darkening in the central trailing hemisphere, which could be consistent with the predicted plasma-related bullseye pattern (*Nordheim et al.,* 2017), although complete observational coverage is lacking.

At Enceladus, effects of plasma bombardment are unclear due to the competing processes of E-ring grain bombardment and plume fallout. Plume fallout is observed, as evidenced by the strong spatial correlation between Enceladus' surface color variation and the regions of predicted high plume fallout (*Kempf et al.,* 2008; *Schenk et al.,* 2011; *Filacchione et al.,* 2016).

Tethys displays a prominent reddish patch on its trailing hemisphere, as shown in ISS IR-UV color ratios (*Schenk et al.,* 2011), and also found to be very low in albedo in

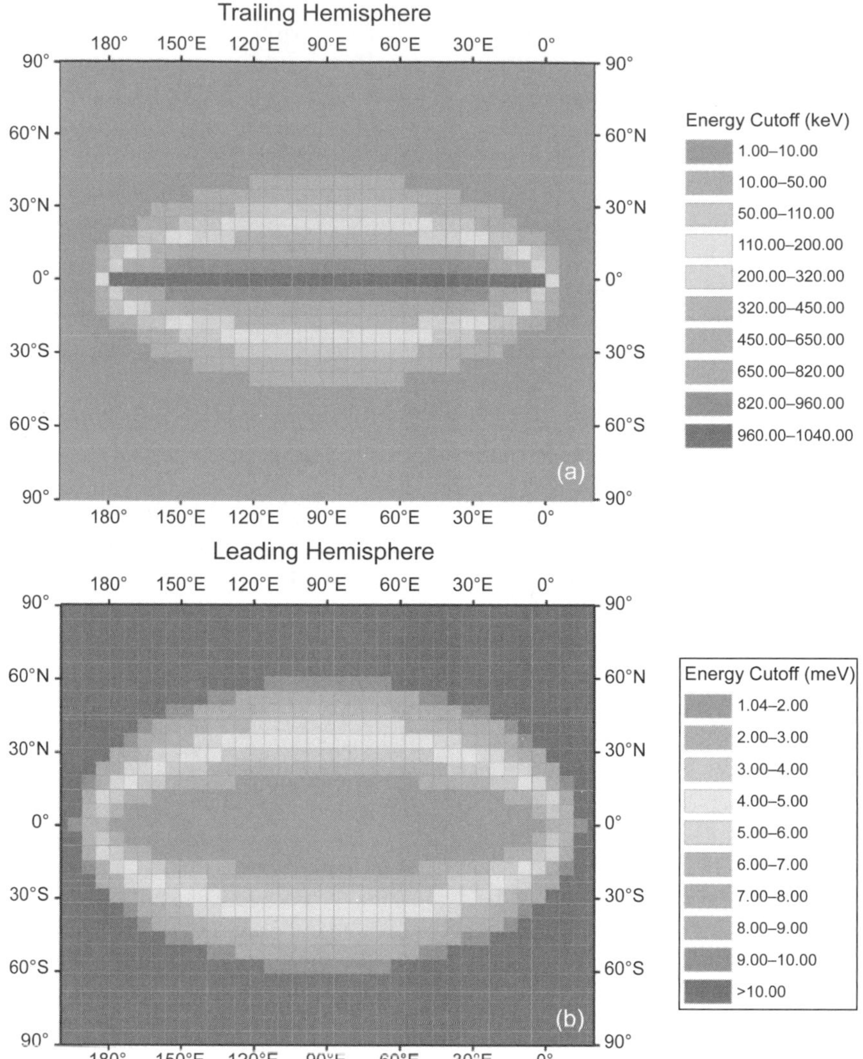

Fig. 8. See Plate 38 for color version. Predicted cutoff energy vs. surface location for Mimas. (a) On the trailing hemisphere, the cutoff energy represents the highest-energy electrons capable of accessing that point on the surface. **(b)** On the leading hemisphere, the cutoff energy represents the lowest-energy electrons capable of accessing that point. From *Nordheim et al.* (2017).

far-UV observations from UVIS (*Hendrix et al.,* 2012b). The shape of the far-UV dark region on the trailing hemisphere is consistent with that expected from cold plasma and nanograin bombardment (*Hendrix et al.,* 2012b). The visible stain has a somewhat different shape, indicating another mechanism must also be at work (*Schenk et al.,* 2011; *Hendrix et al.,* 2012a). One possible explanation is that the far-UV darkening is caused by cold plasma, whose penetration depth is likely similar to that sensed by UVIS, while at the slightly larger depths sensed by ISS are perhaps controlled by a more energetic population of particles (*Hendrix et al.,* 2012a). *Hendrix et al.* (2018) show a relationship between UV absorption depth on the trailing hemispheres of the moons and intensity of approximately tens of kiloelectronvolt electrons. They also show that plasma (colder) electrons and ions do not show such a correlation, suggesting chemical processing of ice and/or non-ice materials by this mid-energy range of electrons, similar to that seen at Europa (*Hand and Carlson,* 2015).

Evidence of surface alteration by plasma implantation is more difficult to find in the saturnian system than the jovian system. This is because in the jovian system plasma is dominated by sulfur ions, which originate from Io's volcanos (*Cassidy et al.,* 2009, and references therein). This plasma bombards and hence can be implanted in the surface of its neighboring icy satellites. A direct result from this surface implantation is that sulfur dioxide (SO_2) has been observed on the surface of Europa, Ganymede, and Callisto (*Showman and Malhotra,* 1999, and references therein). However, plasma in the saturnian system is dominated by H^+, H_2^+, and water group ions (O^+, OH^+, H_2O^+, H_3O^+) with minor constituents of N^+ (~5%) and O_2^+ (~1–2%) (*Young et al.,* 2005; *Sittler et al.,* 2005; *Tokar et al.,* 2005; *Smith et al.,* 2005). Due to the similarity in composition between the plasma and an icy satellite surface, these species are hard to detect on the surface of Saturn's icy satellites. Enceladus is the main mass source for the plasma, and its plume contributes water vapor (96–99%), carbon dioxide (CO_2, 0.3–0.8%), methane (CH_4, 0.1–0.3%), ammonia (NH_3, 0.4–1.3%), and hydrogen (H_2, 0.4–1.4%) (*Waite et al.,* 2017).

3.5. E-Ring Neutrals

Shemansky et al. (1993) discovered a large cloud of neutral OH orbiting Saturn, initially believed and later confirmed to be from activity on Enceladus (*Jurac et al.,* 2002; *Jurac and Richardson,* 2005; *Smith et al.,* 2010). Direct simulation Monte Carlo models of these neutrals show that they primarily bombard the leading (trailing) hemispheres of satellites exterior (interior) to Enceladus (*Cassidy and Johnson,* 2010). These patterns of deposition are similar to that of the E-ring grains (e.g., see the chapter by Kempf et al. in this volume; *Hamilton and Burns,* 1994), which have been shown to brighten the surfaces they bombard (*Verbiscer et al.,* 2007). *Cassidy and Johnson* (2010) conclude that more work is needed to distinguish between the effect of gas and grain bombardment.

Finding evidence of water group contamination on a water ice surface is obviously difficult. However, it is possible to find evidence for these and other trace species on the surface of Saturn's icy satellites, which are possibly due to exogenous delivery (e.g., E-ring bombardment, and other sources internal and external to the Saturn system). For example, carbon dioxide has been directly detected on the surface of Dione (*Clark et al.,* 2008; *Cruikshank et al.,* 2010), Iapetus (*Buratti et al.,* 2005; *Palmer and Brown,* 2011), and Phoebe (*Buratti et al.,* 2008), and carbon-containing material on the surface of Rhea and Dione may explain their UV-redder slopes (*Noll et al.,* 1997). However, carbon alone may not be able to explain the UV reddening (*Clark et al.,* 2008; *Jaumann et al.,* 2009; *Stephan et al.,* 2010). Light organics and CO_2 have been observed near Enceladus' active tiger stripe region, and ammonia (*Emery et al.,* 2005) and ammonia hydrate were possibly detected on Enceladus (*Verbiscer et al.,* 2006). Finally, *Hendrix et al.* (2017) suggest that radiolytic processing of organics within E-ring grains coating the surfaces of the moons can explain their UV absorption and reddish slopes.

3.6. Chemical Alteration by Radiolysis and Photolysis

When a water-ice surface, like those of Saturn's icy satellites, is exposed to charged particles (electrons and ions) and UV photons, a number of processes can occur. Photolysis and radiolysis, due respectively to the action of UV photons and charged particles, may induce chemical changes by ionizing and breaking bonds in surface material. Laboratory measurements show how water ice (H_2O) dissociates to form a number of products including molecular hydrogen and oxygen, hydroxyl (OH), hydrogen peroxide (H_2O_2), hydrogen (H_2), oxygen (O_2), and ozone (O_3) (see *Johnson and Quickenden,* 1997, and references therein). These products can themselves go on to react with other species in the ice, such as other radiolytically produced species, carbonaceous minerals, or organics that maybe present in the surface or deposited by micrometeorid bombardment (*Spencer,* 1998; *Cruikshank et al.,* 2005, 2010; *Clark et al.,* 2008), to form new molecules such as carbon dioxide (CO_2), carbonic acid (H_2CO_3), and sulfur dioxide (SO_2) (e.g., *Hage et al.,* 1998; *Carlson et al.,* 1999; *Hand et al.,* 2007; *Cassidy et al.,* 2009, and references therein).

Lighter species (such as H_2) can diffuse out of the ice to be lost to space, while heavier species (such as O_2) may diffuse out of the very top surface layers but are trapped deeper in the surface (*Johnson and Jesser,* 1997; *Shi et al.,* 2007). However, sputtering (the processes whereby surface molecules are ejected by particle impacts) may cause some of these trapped molecules to escape. The escaped molecules help sustain a satellite's exosphere or atmospheres, and can lead to its surface becoming oxygen-rich.

Radiolysis can significantly alter the composition and hence appearance of a satellite's surface. These changes can occur on relatively short timescales; for example, on the jovian satellites, radiolysis-induced chemical changes can occur at micrometer to meter depths over 10 to 10^9 years respectively (*Cooper et al.,* 2001; *Paranicas et al.,* 2001, 2002). Surface alteration by radiolysis has to compete with other processes, such as regolith erosion of coating due to E-ring grains, interplanetary dust particles, meteoroid gardening, sublimation, and geological resurfacing. It should also be noted that these processes may act to bury material such that chemically altered material can be found at much greater depths than that directly affected by radiation (*Johnson et al.,* 2004).

There are many tracers for radiolysis, several of which have been observed on both the jovian and saturnian icy satellites. The possible signature of ozone at 260 nm on the surfaces of Dione and Rhea (but not on Iapetus) provided an early indication that radiolysis was altering these satellite's surfaces (*Noll et al.,* 1997), although it was curious that the signature appeared on both leading and trailing hemispheres. The amount of absorption was constant across both targets, and corresponded to a column abundance of ozone on both Dione and Rhea of 1–6 × 10^{16} cm^{-2}. This similarity in ozone absorption depths across both targets led to the conclusion that whatever type of particle was causing the radiolysis it must impact both moons isotropically (a 20-keV oxygen ion was proposed).

The detection of peroxides on a satellite's surface provides further evidence of radiolysis. Peroxides absorb very strongly in the UV: Just 0.13% H_2O_2 can cause a four-fold decrease in the 210–330-nm signal (*Carlson et al.,* 1999), and could be responsible for UV surface reddening on Europa, Ganymede, and Callisto (*Hendrix et al.,* 1999; *Noll et al.,* 1997). Photolytically produced hydrogen peroxide (H_2O_2) has been proposed to explain the observed far-UV darkening in the southern summer regions of Mimas (*Hendrix et al.,* 2012a). Hendrix et al. estimated that on Mimas, H_2O_2 had a production timescale of ~1 year, and a destruction timescale of just ~8 years. Thus, while it is likely that H_2O_2 is produced globally, it should be more concentrated on Mimas' summer hemisphere (at the time of observation, this was Mimas' southern hemisphere). Finally (as discussed more deeply in the chapter by Teolis et al. in this volume), exospheres of oxygen and carbon dioxide have been discovered recently on Rhea and Dione (*Teolis and Waite,* 2016; *Teolis et al.,* 2010; *Tokar et al.,* 2012). Their presence is taken as further evidence of surface radiolysis, where sputtering of radiolytically produced products sustain the atmosphere.

4. SUMMARY

Saturn's icy satellites undergo widespread surface alteration by charged particles and ring particles. The major satellites closest to Saturn (Mimas, Enceladus, Tethys, Dione, and Rhea) are all embedded in its E ring, which leads to a net brightening of their surfaces, while in contrast the more distant Phoebe ring bombards Iapetus' leading hemisphere, darkening it. Iapetus and Rhea display surface evidence that they may have had their own rings in the past. Iapetus has a large ridge that lies exactly along its equator that may have been formed by the accumulation of slow-moving deorbiting ring material, while Rhea has a series of blue patches running along a great circle located close to its equator that may have been caused by small-sized ring particles infalling and disrupting the surface, mobilizing the regolith and exposing bluer fresh ice.

The bombardment of Saturn's icy satellites by charged particles leads to changes in the structure of the ice, as well as chemical changes in its composition. For example, the mobilization of ice molecules by high-energy electron bombardment is believed to have caused both thermal and color anomalies on the leading hemisphere of Mimas and Tethys, and a thermal anomaly on Dione. Surface radiolysis may be responsible for the formation of ozone and peroxides on the satellites' surfaces; this includes the surface alteration caused by cold plasma, which primarily bombards the trailing hemispheres of the satellites and leads to a darkening (for some unknown reason) at UV wavelengths. Photolysis is believed to be responsible for the formation of peroxide in the uppermost portion of the satellite surfaces.

Cassini has undoubtedly revolutionized our understanding of the interaction between magnetospheric and ring particles, and the icy satellites in the Saturn system. However, more work still needs to be done to distinguish surface alteration by these different mechanisms, how they vary across and between targets, and how they are changing with time. Many open questions remain; for example, is there any evidence of a thermal anomaly on Mimas' trailing hemisphere due to low-energy electron bombardment? Further analysis of Cassini data will undoubtedly contribute to this effort, particularly those from the remote sensing instruments (notably ISS, CIRS, VIMS, and UVIS). However, with Cassini's demise future observations will have to rely on groundbased and spacebased observatories.

Acknowledgments. C.H. would like to thank the Southwest Research Institute for their support. We appreciate E. Roussos supplying mission-averaged pha data for Fig. 6.

REFERENCES

André N., Blanc M., Maurice S., Schippers P., Pallier E., Gombosi T. I., Hansen K. C., Young D. T., Crary F. J., Bolton S., Sittler E. C., Smith H. T., Johnson R. E., Baragiola R. A., Coates A. J., Rymer A. M., Dougherty M. K., Achilleos N., Arridge C. S., Krimigis S. M., Mitchell D. G., Krupp N., Hamilton D. C., Dandouras I., Gurnett D. A., Kurth W. S., Louarn P., Srama R., Kempf S., Waite H. J., Esposito L. W., and Clarke J. T. (2008) Identification of Saturn's magnetopheric regions and associated plasma processes: Synopsis of Cassini observations during orbit insertion. *Rev. Geophys., 46,* RG4008, DOI: 10.1029/2007RG000238.

Baragiola R. A. (2003) Water ice on outer solar system surfaces: Basic properties and radiation effects. *Planet. Space Sci., 51,* 953–961, DOI: 10.1016/j.pss.2003.05.007.

Baum W. A., Kreidl T., Westphal J. A., Danielson G. E., Seidelmann P. K., Pascu D., and Currie D. G. (1981) Saturn's E ring. *Icarus, 47,* 84–94, DOI: 10.1016/0019-1035(81)90093-2.

Blackford J. R. (2007) Sintering and microstructure of ice: A review. *J. Phys. D. Appl. Phys., 40,* R355–R385, DOI: 10.1088/0022-3727/40/21/R02.

Buratti B. and Veverka J. (1984) Voyager photometry of Rhea, Dione, Tethys, Enceladus and Mimas. *Icarus, 58,* 254–264, DOI: 10.1016/0019-1035(84)90042-3.

Buratti B. J., Mosher J. A., and Johnson T. V. (1990) Albedo and color maps of the saturnian satellites. *Icarus, 87,* 339–357, DOI: 10.1016/0019-1035(90)90138-Y.

Buratti B. J., Hicks M. D., and Davies A. (2005) Spectrophotometry of the small satellites of Saturn and their relationship to Iapetus, Phoebe, and Hyperion. *Icarus, 175,* 490–495, DOI: 10.1016/j.icarus.2004.11.024.

Buratti B. J., Soderlund K., Bauer J., Mosher J. A., Hicks M. D., Simonelli D. P., Jaumann R., Clark R. N., Brown R. H., Cruikshank D. P., and Momary T. (2008) Infrared (0.83–5.1 μm) photometry of Phoebe from the Cassini Visual Infrared Mapping Spectrometer. *Icarus, 193,* 309–322, DOI: 10.1016/j.icarus.2007.09.014.

Burger M. H., Sittler E. C., Johnson R. E., Smith H. T., Tucker O. J., and Shematovich V. I. (2007) Understanding the escape of water from Enceladus. *J. Geophys. Res., 112,* A06219, DOI: 10.1029/2006JA012086.

Carlson R. W., Johnson R. E., and Anderson M. S. (1999) Sulfuric acid on Europa and the radiolytic sulfur cycle. *Science, 286,* 97–99, DOI: 10.1126/science.286.5437.97.

Cassidy T. A. and Johnson R. E. (2010) Collisional spreading of Enceladus' neutral cloud. *Icarus, 209,* 696–703, DOI: 10.1016/j.icarus.2010.04.010.

Cassidy T., Coll P., Raulin F., Carlson R. W., Johnson R. E., Loeffler M. J., Hand K. P., and Baragiloa R. A. (2009) Radiolysis and photolysis of icy satellite surfaces: Experiments and theory. *Space Sci. Rev., 153(1–4),* 299–315, DOI: 10.1007/s11214-009-9625-3.

Castillo-Rogez J. C., Matson D. L., Sotin C., Johnson T. V., Lunine J. I., and Thomas P. C. (2007) Iapetus' geophysics: Rotation rate, shape, and equatorial ridge. *Icarus, 190,* 179–202, DOI:10.1016/j.icarus.2007.02.018.

Chapman C. R. (2004) Space weathering of asteroid surfaces. *Annu. Rev. Earth Planet. Sci., 32,* 539–567, DOI: 10.1146/annurev.earth.32.101802.120453.

Chapman C. R. and McKinnon W. B. (1986) Cratering of planetary satellites. In *Satellites* (J. A. Burns and M. S. Matthews, eds.), pp. 492–580. Univ. of Arizona, Tucson.

CICLOPS (Cassini Imaging Central Laboratory for Operations) (2007) Closest view of Iapetus. PIA 08378, *http://ciclops.org/view/3798/Closest_View_of_Iapetus?js=1.*

Clark R. N., Curchin J. M., Jaumann R., Cruikshank D. P., Brown R. H., Hoefen T. M., Stephan K., Moore J. M., Buratti B. J., Baines K. H., Nicholson P. D., and Nelson R. M. (2008) Compositional mapping of Saturn's satellite Dione with Cassini VIMS and implications of dark material in the Saturn system. *Icarus, 193,* 372–386, DOI: 10.1016/j.icarus.2007.08.035.

Clark R. N., Cruikshank D. P., Jaumann R., Brown R. H., Stephan K., Dalle Ore C. M., Livo K. E., Pearson N., Curchin J. M., Hoefen T. M., Buratti B. J., Filacchione G., Baines K. H., and Nicholson P. D. (2012) The surface composition of Iapetus: Mapping results from Cassini VIMS. *Icarus, 218,* 831–860, DOI:10.1016/j.icarus.2012.01.008.

Cooper J. F., Johnson R. E., Mauk B. H., Garrett H. B., and Gehrels N. (2001) Energetic electrons and ion irradiation of the icy Galilean satellites. *Icarus, 149,* 133–159, DOI: 10.1006/icar.2000.6498.

Cruikshank D. P. (1979) The surfaces and interiors of Saturn's satellites. *Rev. Geophys. Space Phys., 17,* 165–176, DOI: 10.1029/RG017i001p00165.

Cruikshank D. P., Owen T. C., Dalle Ore C., Thomas G. R., Roush T. L., de Bergh C., Sandford S. A., Poulet F., Francois B. K., Gretchen K., and Emery J. P. (2005) A spectroscopic study of the surfaces of Saturn's large satellites: H_2O ice, tholins, and minor constituents. *Icarus, 175,* 268–283, DOI: 10.1016/j.icarus.2004.09.003.

Cruikshank D. P., Meyer A. W., Brown R. H., Clark R. N., Jaumann R., Stephan K., Hibbitts C. A., Sandford S. A., Mastrapa R. M. E., Filacchione G., Dalle Ore C. M., Nicholson P. D., Buratti B. J., McCord T. B., Nelson R. M., Dalton J. B., Baines K. H., and Matson D. L. (2010) Carbon dioxide on the satellites of Saturn: Results from the Cassini VIMS investigation and revisions to the VIMS wavelength scale. *Icarus, 206,* 561–572, DOI: 10.1016/j.icarus.2009.07.012.

Cruikshank D. P., Dalle Ore C. M., Clark R. N., and Pendleton Y. J. (2014) Aromatic and aliphatic organic materials on Iapetus: Analysis of Cassini VIMS data. *Icarus, 233,* 306–315, DOI: 10.1016/j.icarus.2014.02.011.

Dalton J. B., Cassidy T. A., Paranicas C., and Kamp L. W. (2013) Exogenic controls on sulfuric acid hydrate production at the surface of Europa. *Planet. Space Sci., 77,* 45–63, DOI: 10.1016/j.pss.2012.05.013.

Denk T., Neukum G., Roatsch T., Porco C. C., Burns J. A., Galuba G. G., Schmedemann N., Helfenstein P., Thomas P. C., Wagner R. J., and West R. A. (2010) Iapetus: Unique surface properties and a global color dichotomy from Cassini imaging. *Science, 327,* 435–439, DOI: 10.1126/science.1177088.

Dombard A. J. and Cheng A. F. (2008) Constraints on the evolution of Iapetus from simulations of its ridge and bulge. In *Lunar and Planetary Science XXXIX,* Abstract #2262. Lunar and Planetary Institute, Houston.

Dombard A. J., Cheng A. F., McKinnon W. B., and Kay J. P. (2012) Delayed formation of the equatorial ridge on Iapetus from a subsatellite created in a giant impact. *J. Geophys. Res., 117,* E03002, DOI: 10.1029/2011JE004010.

Domingue D. L., Lockwood G. W., and Thompson D. T. (1995) Surface textural properties of icy satellites: A comparison between Europa and Rhea. *Icarus, 115,* 228–249, DOI: 10.1006/icar.1995.1094.

Dougherty M. K., Khurana K. K., Neubauer F. M., Russell C. T., Saur J., Leisner J. S., and Burton M. E. (2006) Identification of a dynamic atmosphere at Enceladus with the Cassini magnetometer. *Science, 311,* 1406–1409, DOI: 10.1126/science.1120985.

Elder C., Helfenstein P., Thomas P., Veverka J., Burns J. A., Denk T., and Porco C. (2007) Tethys' mysterious equatorial band. *Bull. Am. Astron. Soc., 39,* 429.

Emery J. P., Burr D. M., Cruikshank D. P., Brown R. H., and Dalton J. B. (2005) Near infrared (0.8–4.0 μm) spectroscopy of Mimas, Enceladus, Tethys, and Rhea. *Astron. Astrophys., 435,* 353–362, DOI: 10.1051/0004-6361:20042482.

Ferrari C. and Lucas A. (2016) Low thermal inertias of icy planetary surfaces. Evidence for amorphous ice? *Astron. Astrophys., 588,* A133, DOI: 10.1051/0004-6361/201527625.

Filacchione G., Capaccioni F., Ciarniello M., Clark R. N., Cuzzi J. N., Nicholson P. D., Cruikshank D. P., Hedman M. M., Buratti B. J., Lunine J. I., Soderblom L. A., Tosi F., Cerroni P., Brown R. H., McCord T. B., Jaumann R., Stephan K., Baines K. H., and Flamini E. (2012) Saturn's icy satellites and rings investigated by Cassini–VIMS: III — Radial compositional variability. *Icarus, 220,* 1064–1096, DOI: 10.1016/j.icarus.2012.06.040.

Filacchione G., D'Aversa E., Capaccioni F., Clark R. N., Cruikshank D. P., Ciarniello M., Cerroni P., Bellucci G., Brown R. H., Buratti B. J., Nicholson P. D., Jaumann R., McCord T. B., Sotin C., Stephan K., and Dalle Ore C. M. (2016) Saturn's icy satellites investigated by Cassini-VIMS. IV. Daytime temperature maps. *Icarus, 271,* 292–313, DOI: 10.1016/j.icarus.2016.02.019.

Franz O.G. and Millis R. L. (1975) Photometry of Dione, Tethys, and Enceladus on the UBV system. *Icarus, 24,* 433–442, DOI: 10.1016/0019-1035(75)90061-5.

Giese B., Denk T., Neukum G., Roatsch T., Helfenstein P., Thomas P. C., Turtle E. P., McEwen A., and Porco C. C. (2008) The topography of Iapetus' leading side. *Icarus, 193,* 359–371, DOI: 10.1016/j.icarus.2007.06.005.

Hage W., Liedl K. R., Hallbrucker A., and Mayer E. (1998) Carbonic acid in the gas phase and its astrophysical relevance. *Science, 279,* 1332–1335, DOI: 10.1126/science.279.5355.1332.

Hamilton D. P. and Burns J. A. (1994) Origin of Saturn's E ring: Self-sustained, naturally. *Science, 264,* 550–553, DOI: 10.1126/science.264.5158.550.

Hamilton D. P., Skrutskie M. F., Verbiscer A. J., and Masci F. J. (2015) Small particles dominate Saturn's Phoebe ring to surprisingly large distances. *Nature, 522,* 185–187, DOI: 10.1038/nature14476.

Hand K. and Carlson R. W. (2015) Europa's surface color suggests an ocean rich with sodium chloride. *Geophys. Res. Lett., 42,* 3174–3178, DOI: 10.1002/2015GL063559.

Hand K. P., Carlson W. R., and Chyba C. F. (2007) Energy, chemical disequilibrium, and geological constraints on Europa. *Astrobiology, 7,* 1006–1022, DOI: 10.1089/ast.2007.0156.

Hansen C. J., Esposito L., Stewart A. I. F., Colwell J., Hendrix A., Pryor W., Shemansky D., and West R. (2006) Enceladus' water vapor plume. *Science, 311,* 1422–1425, DOI: 10.1126/science.1121254.

Hedman M. M., Burns J. A., Hamilton D. P., and Showalter M. R. (2012) The three-dimensional structure of Saturn's E ring. *Icarus, 217,* 322–338, DOI: 10.1016/j.icarus.2011.11.006.

Hedman M. M, Gosmeye C. M., Nicholson P. D., Sotin C., Brown R. H., Clark R. N., Baines K. H., Buratti B. J., and Showalter M. R. (2013) An observed correlation between plume activity and tidal stresses on Enceladus. *Nature, 500,* 182–184, DOI: 10.1038/nature12371.

Hendrix A. R., Barth C. A., Stewart A. I. F., Hord C. W., and Lane A. L. (1999) Hydrogen peroxide on the icy Galilean satellites. In *Lunar and Planetary Science XXX,* Abstract #2043. Lunar and Planetary Science Institute, Houston.

Hendrix A. R., Cassidy T. A., Buratti B. J., Paranicas C., Hansen C. J., Teolis B., Roussos E., Bradley E. T., Kollmann P., and Johnson R. E. (2012a) Mimas' far-UV albedo: Spatial variations. *Icarus, 220,* 922–931, DOI: 10.1016/j.icarus.2012.06.012.

Hendrix A. R., Cassidy T. A., Paranicas C., Teolis B., and Hansen C. J. (2012b) Seasonal variability on Saturn's moons Mimas and Tethys. *EPSC Abstracts 2012,* 2012-950-1.

Hendrix A. R., Filacchione G., Schenk P., Paranicas C., Clark R., and Scipioni F. (2018) Icy saturnian satellites: Disk-integrated UV-IR characteristics and links to exogenic processes. *Icarus, 300,* 103–114, DOI: 10.1016/j.icarus.2017.08.037, in press.

Hillier J. K., Green S. F., McBride N., Schwanethal J. P., Postberg F., Srama R., Kempf S., Moragas-Klostermeyer G., McDonnell J. A. M., and Grün E. (2007) The composition of Saturn's E ring. *Mon. Not. R. Astron. Soc., 377,* 1588–1596, DOI: 10.1111/j.1365-2966.2007.11710.

Horànyi M., Burns J. A., and Hamilton D. P. (1992) The dynamics of Saturn's E ring particles. *Icarus, 97,* 248–259, DOI: 10.1016/0019-1035(92)90131-P.

Howett C. J. A., Spencer J. R., Pearl J., and Segura M. (2010) Thermal inertia and bolometric Bond albedo values for Mimas, Enceladus, Tethys, Dione, Rhea and Iapetus as derived from Cassini/CIRS measurements. *Icarus, 206,* 573–593, DOI: 10.1016/j.icarus.2009.07.016.

Howett C. J. A., Spencer J. R, Schenk P., Johnson R. E., Paranicas C., Hurford T. A., Verbiscer A., and Segura M.(2011) A high-amplitude thermal inertia anomaly of probable magnetospheric origin on Saturn's moon Mimas. *Icarus, 216,* 221–226, DOI: 10.1016/j.icarus.2011.09.007.

Howett C. J. A., Spencer J. R., Hurford T., Verbiscer A., and Segura M. (2012) PacMan returns: An electron-generated thermal anomaly on Tethys. *Icarus, 221,* 1084–1088, DOI: 10.1016/j.icarus.2012.10.013.

Howett C. J. A., Spencer J. R., Hurford T., Verbiscer A., and Segura M. (2014) Thermophysical property variations across Dione and Rhea. *Icarus, 241,* 239–247, DOI: 10.1016/j.icarus.2014.05.047.

Hurford T. A., Helfenstein P., Hoppa G. V., Greenberg R., and Bills B. G. (2007) Eruptions arising from tidally controlled periodic openings of rifts on Enceladus. *Nature 447,* 292–294, DOI: 10.1038/nature05821.

Hurford T. A, Bills B. G., Helfenstein P., Greenberg R., Hoppa G. V., and Hamilton D. P. (2009) Geological implications of a physical libration on Enceladus. *Icarus, 203,* 541–552, DOI: 10.1016/j.icarus.2009.04.025.

Ip W.-H. (2006) On a ring origin of the equatorial ridge of Iapetus. *Geophys. Res. Lett., 33,* L16203, DOI: 10.1029/2005GL025386.

Johnson R. E. and Jesser W. A. (1997) O_2/O_3 micro-atmospheres in the surface of Ganymede. *Astrophys. J. Lett., 480,* L79–L82, DOI: 10.1086/310614.

Johnson R. E. and Quickenden T. I. (1997) Photolysis and radiolysis of water ice on outer solar system bodies. *J. Geophys. Res., 102,* 10985–10996.

Johnson R. E., Barton L. A., Boring J. W., Jesser J. A., Brown W. L., and Lanzerotti L. J. (1985) Charged particle modification of ices in the saturnian and jovian systems. In *Ices in the Solar System* (J. Klinger, ed.), pp. 301–315. Reidel, Dordrecht.

Johnson R. E., Carlson R. W., Cooper J. F., Paranicas C., Moore M. H., and Wong M. C. (2004) Radiation effects on the surfaces of the Galilean satellites. In *Jupiter — The Planet, Satellites and Magnetosphere* (F. Bagenal et al., eds.), pp. 485–512. Cambridge Univ., Cambridge.

Jones G. H., Roussos E., Krupp N., Beckmann U., Coates A. J., Crary F., Dandouras I., Dikarev V., Dougherty M. K., Garnier P., Hansen C. J., Hendrix A. R., Hospodarsky G. B., Johnson R. E., Kempf S., Khurana K. K., Krimigis S. M., Krüger H., Kurth W. S., Lagg A., McAndrews H. J., Mitchell D. G., Paranicas C., Postberg F., Russell C. T., Saur J., Seiß M., Spahn F., Srama R., Strobel D. F., Tokar R., Wahlund J.-E., Wilson R. J., Woch J., and Young D. (2008a) The dust halo of Saturn's largest icy moon, Rhea. *Science, 319,* 1380–1384, DOI: 10.1126/science.1151524.

Jones G., Roussos E., Krupp N., Krimigis S. M., Young D., Denk T., Tiscareno M. S., Burns J. A., Strobel D. F., and Kempf S. (2008b) A debris disk surrounding Saturn's moon Rhea. *Eos Trans. AGU, 89(53),* Fall Meet. Suppl., Abstract #P32A-04.

Juhász A. and Horányi M. (2015) Dust delivery from Enceladus to the moons of Saturn. Abstract P51B-2060 presented at 2015 Fall Meeting, AGU, San Francisco, California, 14–18 Dec.

Juhász A., Horányi M., and Morfill G. E. (2007) Signatures of Enceladus in Saturn's E ring. *Geophys. Res. Lett., 34,* L09104, DOI: 10.1029/2006GL029120.

Jurac S. and Richardson J. D. (2005) A self-consistent model of plasma and neutrals at Saturn: Neutral cloud morphology. *J. Geophys. Res., 110,* A09220, DOI: 10.1029/2004JA010635.

Jurac S., McGrath M. A., Johnson R. E., Richardson J. D., Vasyliunas V. M., and Eviatar A. (2002) Saturn: Search for a missing water source. *Geophys. Res. Lett., 29,* 2172, DOI: 10.1029/2002GL015855.

Kempf S., Beckmann U., Moragas-Klostermeyer G., Postberg F., Srama R., Economou T., Schmidt J., Spahn F., and Grün E. (2008) The E ring in the vicinity of Enceladus I. Spatial distribution and properties of the ring particles. *Icarus, 193,* 420–437, DOI: 10.1016/j.icarus.2007.06.027.

Khurana K. K, Dougherty M. K., Russell C. T., and Lesiner J. S. (2007) Mass loading of Saturn's magnetosphere near Enceladus. *J. Geophys. Res., 112,* A08203, DOI: 10.1029/2006JA012110.

Khurana K. K, Russell C. T., and Dougherty M. K. (2008) Magnetic portraits of Tethys and Rhea. *Icarus, 193,* 465–474, DOI:10.1016/j.icarus.2007.08.005.

Kollmann P., Roussos E., Paranicas C., Krupp N., and Haggerty D. K. (2013) Processes forming and sustaining Saturn's proton radiation belts. *Icarus, 222,* 323–341, DOI: 1016/j.icarus.2012.10.033.

Koutchmy S. and Lamy P. L. (1975) Study of the inner satellites of Saturn by photographic photometry. *Icarus, 25,* 459–465, DOI: 10.1016/0019-1035(75)90011-1.

Lee J. S., Buratti B. J., Hicks M., and Mosher J. (2010) The roughness of the dark side of Iapetus from the 2004 to 2005 flyby. *Icarus, 206,* 623–630, DOI: 10.1016/j.icarus.2009.11.008.

Levison H. F., Walsh K. J., Barr A. C., and Dones L. (2011) Ridge formation and de-spinning of Iapetus via an impact-generated satellite. *Icarus, 214,* 773–778, DOI: 10.1016/j.icarus.2011.05.031.

Madey T. E., Johnson R. E., and Orlando T. M. (2002) Far-out surface science: Radiation-induced surface processes in the solar system. *Surface Sci., 500,* 838–858, DOI: 10.1016/S0039-6028(01)01556-4.

Mauk B. H., Saur J., Mitchell D. G., Roelof E. C., Brandt P. C., Armstrong T. P., Hamilton D. C., Krimigis S. M., Krupp N., Livi S. A., Manweiler J. W., and Paranicas C. P. (2005) Energetic particle injections in Saturn's magnetosphere. *Geophys. Res. Lett., 32,* L14S05, DOI: 10.1029/2005GL022485.

McCord T. B., Johnson T. V., and Elias J. H. (1971) Saturn and its satellites: Narrow-band spectroscopy (0.3–1.1 μm). *Astrophys. J., 165,* 413–424, DOI: 10.1086/150907.

Melosh H. J. and Nimmo F. (2009) An intrusive dike origin for Iapetus' enigmatic ridge? In *Lunar and Planetary Science XL,* Abstract #2478. Lunar and Planetary Institute, Houston.

Myers S. M. (1980) Ion-beam-induced migration and its effect on concentration profiles. *Nucl. Instr. Meth., 168,* 265–274, DOI: 10.1016/0029-554X(80)91264-1.

Nicholson P. D., Showalter M. R., Dones L., French R. G., Larson S. M., Lissauer J. J., McGhee C. A., Seitzer P., Sicardy B., and Danielson G. E. (1996) Observations of Saturn's ring-plane crossings in August and November 1995. *Science, 272,* 509–515, DOI: 10.1126/science.272.5261.509.

Nimmo F. and Matsuyama I. (2007) Reorientation of icy satellites by impact basins. *Geophys. Res. Lett., 34,* L19203, DOI: 10.1029/2007GL030798.

Noland M., Veverka J., Morrison D., Cruikshank D. P., Lazarewicz A. R., Morrison N. D., Elliot J. L., Goguen J., and Burns J. A. (1974) Six-color photometry of Iapetus, Titan, Rhea, Dione and Tethys. *Icarus, 23,* 334–354, DOI: 10.1016/0019-035(74)90052-9.

Noll K. S., Roush T. L., Cruikshank D. P., Johnson R. E., and Pendleton Y. J. (1997) Detection of ozone on Saturn's satellites Rhea and Dione. *Nature, 388,* 45–47, DOI: 10.1038/40348.

Nordheim T. A., Hand K. P., Paranicas C., Howett C. J. A., Hendrix A. R., Jones G. H., and Coates A. J. (2017) The near surface electron radiation environment of Saturn's moon Mimas. *Icarus, 286,* 56–68, DOI: 10.1016/j.icarus.2017.01.002.

Palmer E. E. and Brown R. H. (2011) Production and detection of carbon dioxide on Iapetus. *Icarus, 212,* 807–818, DOI: 10.1016/j.icarus.2010.12.007.

Pang K. D., Voge C. C., Rhoads J. W., and Ajello J. M. (1984) The E ring of Saturn and satellite Enceladus. *J. Geophys. Res., 89,* 9459–9470, DOI: 10.1029/JB089iB11p09459.

Paranicas C., Carlson R. W., and Johnson R. E. (2001) Electron bombardment of Europa. *Geophys. Res. Lett., 28,* 673–676, DOI: 10.1029/2000GL012320.

Paranicas C., Mauk B. H., Ratliff J. M., Cohen C., and Johnson R. E. (2002) The ion environments near Europa and its role in surface energetics. *Geophys. Res. Lett., 29,* L5, DOI: 10.1029/2001GL014127.

Paranicas C., Mitchell D. G., Krimigis S. M., Hamilton D. C., Roussos E., Krupp N., Jones G. H., Johnson R. E., Cooper J. F., and Armstrong T. P. (2008) Sources and losses of energetic protons in Saturn's magnetosphere. *Icarus, 197,* 519–525, DOI: 10.1016/j.icarus.2008.05.011.

Paranicas C., Roussos E., Decker R. B., Johnson R. E., Hendrix A. R., Schenk P., Cassidy T. A., Dalton J. B. III, Howett C. J. A., Kollmann P., Patterson W., Hand K. P., Nordheim T. A., Krupp N., and Mitchell D. G. (2014) The lens feature on the inner saturnian satellites. *Icarus, 234,* 155–161, DOI: 10.1016/j.icarus.2014.02.026.

Patterson G.W., Paranicas C., and Prockter L. M. (2012) Characterizing electron bombardment of Europa's surface by location and depth. In *Lunar and Planetary Science XLVIII,* Abstract #2447. Lunar and Planetary Institute, Houston.

Pitman K., Buratti B., Mosher J., Bauer J., Momary T., Brown R., Nicholson P., and Hedman M. (2008) First high solar phase angle observations of Rhea using Cassini VIMS: Upper limits on water vapor and geologic activity. *Astrophys. J. Lett., 680,* L65–L68, DOI: 10.1086/589745.

Pitman K. M., Buratti B. J., and Mosher J. M. (2010) Disk-integrated bolometric Bond albedos and rotational light curves of saturnian satellites from Cassini Visual and Infrared Mapping Spectrometer. *Icarus, 206,* 537–560, DOI: 10.1016/j.icarus.2009.12.001.

Poppe A. R. (2016) An improved model for interplanetary dust fluxes in the outer solar system. *Icarus, 264,* 369–386, DOI: 10.1016/j.icarus.2015.10.001.

Porco C. C., Baker E., Barbara J., Beurle K., Brahic A., Burns J. A., Charnoz S., Cooper N., Dawson D. D., Del Genio A. D., Denk

T., Dones L., Dyudina U., Evans M. W., Giese B., Grazier K., Helfenstein P., Ingersoll A. P., Jacobson R. A., Johnson T. V., McEwen A., Murray C. D., Neukum G., Owen W. M., Perry J., Roatsch T., Spitale J., Squyres S., Thomas P. C., Tiscareno M., Turtle E., Vasavada A. R., Veverka J., Wagner R., and West R. (2005) Cassini imaging science: Initial results on Phoebe and Iapetus. *Science, 307,* 1237–1242, DOI: 10.1126/science.1107981.

Porco C. C., Helfenstein P., Thomas P. C., Ingersoll A. P., Wisdom J., West R., Neukum G., Denk T., Wagner R., Roatsch T., Kieffer S., Turtle E., McEwen A., Johnson T. V., Rathbun J., Veverka J., Wilson D., Perry J., Spitale J., Brahic A., Burns J. A., Del Genio A. D., Dones L., Murray C. D., and Squyres S. (2006) Cassini observes the active South Pole of Enceladus. *Science, 311,* 1393–1401, DOI: 10.1126/science.1123013.

Postberg F., Kempf S., Hillier J. K., Srama R., Green S. F., McBride N., and Grün E. (2008) The E-ring in the vicinity of Enceladus. II. Probing the moon's interior — The composition of E-ring particles. *Icarus, 193,* 438–454, DOI: 10.1016/j.icarus.2007.09.001.

Postberg F., Kempf S., Schmidt J., Brilliantov N., Beinsen A., Abel B., Buck U., and Srama R. (2009) Sodium salts in E-ring ice grains from an ocean below the surface of Enceladus. *Nature, 459,* 1098–1101, DOI: 10.1038/nature08046.

Roberts J. H. and Nimmo F. (2009) Tidal dissipation due to despinning and the equatorial ridge on Iapetus. In *Lunar and Planetary Science XL,* Abstract #1927. Lunar and Planetary Institute, Houston.

Roussos E., Jones G. H., Krupp N., Paranicas C., Mitchell D. G., Lagg A., Woch J., Motschmann U., Krimigis S. M., and Dougherty M. K. (2007) Electron microdiffusion in the saturnian radiation belts: Cassini MIMI/LEMMS observations of energetic electron absorption by the icy moons. *J. Geophys. Res., 112,* A06214, DOI: 10.1029/2006JA012027.

Roussos E., Krupp N., Krüger H., and Jones G. H. (2010) Surface charging of Saturn's plasma-absorbing moons. *J. Geophys. Res., 115,* 2156–2202, DOI: 10.1029/2010JA015525.

Royer E. M. and Hendrix A. R. (2014) First far-ultraviolet disk-integrated phase curve analysis of Mimas, Tethys and Dione from the Cassini-UVIS data sets. *Icarus, 242,* 158–171, DOI: 10.1016/j.icarus.2014.07.026.

Sack N. J., Boring J. W., Johnson R. E., Baragiola R. A., and Shi M. (1991) Alteration of the UV-visible reflectance spectra of H_2O ice by ion bombardment. *J. Geophys. Res., 96,* 17535–1753, DOI: 10.1029/91JE01681.

Saur J. and Strobel D. F. (2005) Atmospheres and plasma interactions at Saturn's largest inner icy satellites. *Astrophys. J. Lett., 620,* L115, DOI: 10.1086/428665.

Schaible M. J., Johnson R. E., Zhigilei L. V., and Piqueux S. (2017) High energy electron sintering of icy regoliths: Formation of the PacMan thermal anomalies on the icy saturnian moons. *Icarus, 285,* 211–223, DOI: 10.1016/j.icarus.2016.08.033.

Schenk P. M., Hamilton D. P., Johnson R. E., McKinnon W. B., Paranicas C., Schmidt J., and Showalter M. R. (2011) Plasma, plumes and rings: Saturn system dynamics as recorded in global color patters on its midsize icy satellites. *Icarus, 211,* 740–757, DOI: 10.1016/j.icarus.2010.08.016.

Schmedemann N., Neukum G., Denk T., and Wagner R. J. (2008) Stratigraphy and surface ages on Iapetus and other saturnian satellites. In *Lunar and Planetary Science XXXIX,* Abstract #1391. Lunar and Planetary Institute, Houston.

Schmidt J., Brilliantov N., Spahn F., and Kempf S. (2008) Slow dust in Enceladus' plume from condensation and wall collisions in tiger stripe fractures. *Nature, 451,* 685–688, DOI: 10.1038/nature06491.

Scipioni F., Tosi F., Stephan K., Filacchione G., Ciarniello M., Capaccioni F., and Cerroni P. (2013) Spectroscopic classification of icy satellites of Saturn I: Identification of terrain units on Dione. *Icarus, 226,* 1331–1349, DOI: 10.1016/j.icarus.2013.08.008.

Scipioni F., Tosi F., Stephan K., Filacchione G., Ciarniello M., Capaccioni F., Cerroni P., and the VIMS Team (2014) Spectroscopic classification of icy satellites of Saturn II: Identification of terrain units on Rhea. *Icarus, 234,* 1–16, DOI: 10.1016/j.icarus.2014.02.01.

Shi J., Teolis B. D., and Baragiola R. A. (2007) Irradiation enhanced adsorption and trapping of O_2 on microporous water ice. *AAS/Division for Planetary Sciences Meeting Abstracts, 39,* 38.04.

Shkuratov Y. G. and Helfenstein P. (2001) The opposition effect and the Quasi-fractal structure of regolith: I. Theory. *Icarus, 152,* 96–116, DOI: 10.1006/icar.2001.6630.

Showalter M. R., Cuzzi J. N., and Larson S. M. (1991) Structure and particle properties of Saturn's E ring. *Icarus, 94,* 451–473, DOI: 10.1016/0019-1035(91)90241-K.

Showman A. P. and Malhotra R. (1999) The Galilean satellites. *Science, 286,* 77–84, DOI: 10.1126/science.286.5437.77.

Shemansky D. E., Matheson P., Hall D. T., Hu H.-Y., and Tripp T. M. (1993) Detection of the hydroxyl radical in the Saturn magnetosphere. *Nature, 363,* 329–331, DOI: 10.1038/363329a0.

Sirono S. (2011) The sintering region of icy dust aggregates in a protoplanetary nebula. *Astrophys. J., 735,* 131, DOI: 10.1088/0004-637X/735/2/131.

Sittler E. C. Jr., Thomsen M., Chornay D., Shappirio M. D., Simpson D., Johnson R. E., Smith H. T., Coates A. J., Rymer A. M., Crary F., McComas D. J., Young D. T., Reisenfeld D., Dougherty M., and Andre N. (2005) Preliminary results on Saturn's inner plasmasphere as observed by Cassini: Comparison with Voyager. *Geophys. Res. Lett., 32,* L14S04, DOI: 10.1029/2005GL022653.

Smith B., Soderblom L., Beebe R., Boyce J., Briggs G., Bunker A., Collins S. A., Hansen C. J., Johnson T. V., Mitchell J. L, Terrile R. J., Carr M., Cook A. F. II, Cuzzi J., Pollack J. B., Danielson G. E., Ingersoll A., Davis M. E., Hunt G. E., Masursky H., Shoemaker E., Morrison D., Owen T., Sagan C., Veverka J., Strom R., and Suomi V. E. (1981) Encounter with Saturn: Voyager 1 imaging science results. *Science, 212,* 163–191, DOI: 10.1126/science.212.4491.163.

Smith B., Soderblom L., Batson R., Bridges P., Inge J., Masursky H., Shoemaker E., Beebe R., Boyce J., Briggs G., Bunker A., Collins S. A., Hansen C. J., Johnson T. V., Mitchell J. L., Terrile R. J., Cook A. F. II, Cuzzi J., Pollack J. B., Danielson G. E., Ingersoll A., Davis M. E., Hunt G. E., Morrison D., Owen T., Sagan C., Veverka J., Strom R., and Suomi V. E. (1982) A new look at the Saturn system: The Voyager 2 images. *Science, 215,* 504–537, DOI: 10.1126/science.215.4532.504.

Smith H. T., Shappirio M., Sittler E. C., Reisenfeld D., Johnson R. E., Baragiola R. A., Crary F. J., McComas D. J., and Young D. T. (2005) Discovery of nitrogen in Saturn's inner magnetosphere. *Geophys. Res. Lett., 32,* L14S03, DOI: 10.1029/2005GL022654.

Smith H. T., Johnson R. E., Perry M. E., Mitchell D. G., McNutt R. L., and Young D. T. (2010) Enceladus plume variability and the neutral gas densities in Saturn's magnetosphere. *J. Geophys. Res., 115,* A10252, DOI: 10.1029/2009JA015184.

Smith-Konter B. and Pappalardo R. T. (2008) Tidally driven stress accumulation and shear failure of Enceladus's tiger stripes. *Icarus, 198,* 435–451, DOI: 10.1016/j.icarus.2008.07.005.

Soter S. (1974) Brightness of Iapetus. Presented at Planetary Satellites, August 18–21, Cornell University, Ithaca, New York, IAU Colloquium 28.

Spahn F., Schmidt J., Albers N., Hörning M., Makuch M., Seiß M., Kempf S., Srama R., Dikarev V., Helfert S., Moragas-Klostermeyer G., Krivov A. V., Sremčević M., Tuzzolino A. J., Economou T., and Grün E. (2006) Cassini dust measurements at Enceladus and implications for the origin of the E ring. *Science, 311,* 1416–1418, DOI: 10.1126/science.1121375.

Spencer J. R. (1998) Upper limits for condensed O_2 on Saturn's icy satellites and rings. *Icarus, 136,* 349–352, DOI: 10.1006/icar.1998.6024.

Spencer J. R. and Denk T. (2010) Formation of Iapetus' extreme albedo dichotomy by exogenically triggered thermal ice migration. *Science, 327,* 432–435, DOI: 10.1126/science.1177132.

Spencer J. R., Pearl J. C., Segura M., Flasar F. M., Mamoutkine A., Romani P., Buratti B., Hendrix A., Spilker L. J., and Lopes R. M. C. (2006) Cassini encounters Enceladus: Background and the discovery of a south polar hot spot. *Science, 311,* 1401–1405, DOI: 10.1126/science.1121661.

Stephan K., Jaumann R., Wagner R., Clark R. N., Cruikshank D. P., Hibbitts C. A., Roatsch T., Hoffmann H., Brown R. H., Filiacchione G., Buratti B. J., Hansen G. B., McCord T. B., Nicholson P. D., and Baines K. H. (2010) Dione's spectral and geological properties. *Icarus, 206,* 631–652, DOI: 10.1016/j.icarus.2009.07.036.

Stephan K., Jaumann R., Wagner R., Clark R. N., Cruikshank D. P., Giese B., Hibbitts C. A., Roatsch T., Matz K.-D., Brown R. H., Filiacchione

G., Cappacioni F., Scholten F., Buratti B. J., Hansen G. B., Nicholson P. D., Baines K. H., Nelson R. M., and Matson D. L. (2012) The saturnian satellite Rhea as seen by Cassini VIMS. *Planet. Space Sci., 61,* 142–160, DOI: 10.1016/j.pss.2011.07.019, 2012.

Stephan K., Wagner R., Jaumann R., Clark R. N., Cruikshank D. P., Brown R. H., Giese B., Roatsch T., Filiacchione G., Matson D. L., Dalle Ore C., Cappacioni F., Baines K. H., Rodriguez S., Krupp N., Buratti B. J., and Nicholson P. D. (2016) Cassini's geological and compositional view of Tethys. *Icarus, 274,* 1–22, DOI: 10.1016/j.icarus.2016.03.002.

Stooke P. J. (1989) Tethys: Volcanic and structural geology. In *Lunar and Planetary Science XX,* Abstract #1071. Lunar and Planetary Institute, Houston.

Stooke P. J. (2002) Tethys and Dione: New geological interpretations. In *Lunar and Planetary Science XXXIII,* Abstract #1553. Lunar and Planetary Institute, Houston.

Strazulla G., Casstorina A. C., and Palumbo M. E. (1995) Ion irradiation of astrophysical ices. *Planet. Space Sci., 43,* 1247–1251, DOI: 10.1088/1742-6596/101/1/012002.

Teolis B. D. and Waite J. H. (2016) Dione and Rhea seasonal exospheres revealed by Cassini CAPS and INMS. *Icarus, 272,* 277–289, DOI: 10.1016/j.icarus.2016.02.031.

Teolis B. D., Jones G. H., Miles P. F., Tokar R. L., Magee B. A., Waite J. H., Roussos E., Young D. T., Crary F. J., Coates A. J., Johnson R. E., Tseng W.-L., and Baragiola R. A. (2010) Cassini finds an oxygen-carbon dioxide atmosphere at Saturn's icy moon Rhea. *Science, 330,* 1813–1815, DOI: 10.1126/science.1198366.

Thomsen M. F. and Van Allen J. A. (1980) Motion of trapped electrons and protons in Saturn's inner magnetosphere. *J. Geophys. Res., 85,* 5831–5834, DOI: 10.1029/JA085iA11p05831.

Tiscareno M. S., Burns J. A., Cuzzi J. N., and Hedman M. M. (2010) Cassini imaging search rules out rings around Rhea. *Geophys. Res. Lett., 37,* L14205, DOI: 10.1029/2010GL043663.

Tokar R. L., Johnson R. E., Thomsen M. F., Delapp D. M., Baragiola R. A., Franis M. F., Reisenfeld D. B., Fish B. A., Young D. T., Crary F. J., Coates A. J., Gurnett D. A., and Kurth W. S. (2005) Cassini observations of the thermal plasma in the vicinity of Saturn's main rings and the F and G rings. *Geophys. Res. Lett., 32,* L14S04, DOI: 10.1029/2005GL022690.

Tokar R. L., Johnson R. E., Thomsen M. F., Sittler E. C., Coates A. J., Wilson R. J., Crary F. J., Young D. T., and Jones G. H. (2012) Detection of exospheric O_2^+ at Saturn's moon Dione. *Geophys. Res. Lett., 39,* L03105, DOI: 10.1029/2011GL050452.

Truscott P., Heynderickx D., Sicard-Piet A., and Bourdarie S. (2011) Simulation of the radiation environment near Europa using the GEANT4-based PLANETOCOSMICS-J model. *IEEE Trans. Nucl. Sci., 58,* 6, DOI: 10.1109/TNS.2011.2172818.

Verbiscer A. J. and Veverka J. (1989) Albedo dichotomy of Rhea: Hapke analysis of Voyager photometry. *Icarus, 82,* 336–353, DOI: 10.1016/0019-1035(89)90042-0.

Verbiscer A. J. and Veverka J. (1992) Mimas: Photometric roughness and albedo map. *Icarus, 99,* 63–69, DOI: 10.1016/0019-1035(92)90171-3.

Verbiscer A. and Veverka J. (1994) A photometric study of Enceladus. *Icarus, 110,* 155–164.

Verbiscer A. J., French R. G., and McGhee C. A. (2005) The opposition surge of Enceladus: HST observations 338–1022 nm. *Icarus, 173,* 66–83, DOI: 10.1016/j.icarus.2004.05.001.

Verbiscer A. J., Peterson D. E., Skrutskie M. F., Cushing M., Helfenstein P., Nelson M. J., Smith J. D., and Wilson J. C. (2006) Near-infrared spectra of the leading and trailing hemispheres of Enceladus. *Icarus, 182,* 211–223, DOI: 10.1016/j.icarus.2005.12.008.

Verbiscer A. J., French R., Showalter M., and Helfenstein P. (2007) Enceladus: Cosmic graffiti artist caught in the act. *Science, 315,* 815, DOI: 10.1126/science.1134681.

Verbiscer A. J., Skrutskie M. F., and Hamilton D. (2009) Saturn's largest ring. *Nature 461,* 1098–1100, DOI: 10.1038/nature08515.

Waite J. H., Combi M. R., Ip W.-H., Cravens T. E., McNutt R. L., Kasprzak W., Yelle R., Luhmann J., Niemann H., Gell D., Magee B., Fletcher G., Lunine J., and Tseng W.-L. (2006) Cassini Ion and Neutral Mass Spectrometer: Enceladus plume composition and structure. *Science, 311,* 1419–1422, DOI: 10.1126/science.1121290.

Waite J. H., Glein C. R., Perryman R. S., Teolis B. D., Magee B. A., Miller G., Grimes J., Perry M. E., Miller K. E., Bouquet A., Lunine J. I., Brockwell T., and Bolton S. J. (2017) Cassini finds molecular hydrogen in the Enceladus plume: Evidence for hydrothermal processes. *Science, 356,* 155–159, DOI: 10.1126/science.aai8703.

Wilson R. J., Tokar R. L., Kurth W. S., and Peterson A. M. (2010) Properties of the thermal ion plasma near Rhea as measured by the Cassini plasma spectrometer. *J. Geophys. Res., 115,* A05201, DOI: 10.1029/2009JA014679.

Young D. T., Berthelier J.-J., Blanc M., Burch J. L., Bolton S., Coates A. J., Crary F. J., Goldstein R., Grande M., Hill T. W., Johnson R. E., Baragiola R. A., Kelha V., McComas D. J., Mursula K., Sittler E. C., Svenes K. R., Szegö K., Tanskanen P., Thomsen M. F., Bakshi S., Barraclough B. L., Bebesi Z., Delapp D., Dunlop M. W., Gosling J. T., Furman J. D., Gilbert L. K., Glenn D., Holmlund C., Illiano J.-M., Lewis G. R., Linder D. R., Maurice S., McAndrews H. J., Narheim B. T., Pallier E., Reisenfeld D., Rymer A. M., Smith H. T., Tokar R. L., Vilppola J., and Zinsmeyer C. (2005) Composition and dynamics of plasma in Saturn's magnetosphere. *Science, 307,* 1262–1266, DOI: 10.1126/science.1106151.

Teolis B., Tokar R., Cassidy T., Khurana K., and Nordheim T. (2018) Exospheres and magnetospheric currents at Saturn's icy moons: Dione and Rhea. In *Enceladus and the Icy Moons of Saturn* (P. M. Schenk et al., eds.), pp. 361–386. Univ. of Arizona, Tucson, DOI: 10.2458/azu_uapress_9780816537075-ch018.

Exospheres and Magnetospheric Currents at Saturn's Icy Moons: Dione and Rhea

Ben Teolis
Southwest Research Institute

Robert Tokar
Planetary Science Institute

Tim Cassidy
University of Colorado

Krishan Khurana
University of California, Los Angeles

Tom Nordheim
NASA Jet Propulsion Laboratory/California Institute of Technology

---

The discovery of dynamic sputter-produced O_2 and CO_2 exospheres at Saturn's moons Dione and Rhea, and of unexpectedly intense field-aligned electric currents at Rhea, has transformed understanding of the exospheric physics and magnetospheric interaction of icy, plasma-absorbing objects. Molecular oxygen and carbon dioxide exospheres remotely detected at Jupiter's large icy moons are sustained by radiation chemistry and magnetospheric ion sputtering of their surface ices. Such exospheres have long been posited to exist at Saturn's large moons. During the last decade, the Cassini spacecraft's multiple low-altitude flybys of Dione and Rhea have enabled the first *in situ* reconnaissance of such sputtered-produced exospheres at any solar system icy moon. Two major discoveries by Cassini's Plasma Spectrometer (CAPS) and Ion Neutral Mass Spectrometer (INMS) during these flybys are that (1) the exospheres of Dione and Rhea are seasonal, peaking near Saturn's equinoxes, and (2) exospheric O_2 sputtering from their surfaces is far less efficient and as a result, their exospheres far less dense than expected. Given these results, the detection of field-aligned electric currents by Cassini's magnetometer was surprising, since Rhea was expected to be electromagnetically "inert" at the low measured exospheric densities. Two never-before-observed types of current systems were revealed by modeling of these magnetometer findings: (1) an Alfvénic current system emanating from Rhea's plasma wake, and (2) a flux tube current that maintains charge balance on Rhea's surface. The Rhea magnetic field data are a fantastic demonstration of a general principle that astronomical bodies can generate field-aligned currents both by the well-known induction mechanism (e.g., by motion of a "conductive" atmosphere through the magnetic field), *and* "thermoelectrically" (even with no atmosphere or no relative motion to the field) by the absorption of hot plasma into the "cold" body.

1. INTRODUCTION

As the Enceladus plumes populate Saturn's inner magnetosphere with ionized water vapor, the effects of this process are felt over vast distances across the Saturn system, and especially at Saturn's *other* inner magnetospheric icy moons Mimas, Tethys, Dione, and Rhea. The Enceladian ions are "picked up" and accelerated by stealing rotational energy from Saturn via its magnetic field and are then exchanged (*Bagenal and Delamere,* 2011; *Fleshman et al.,* 2012; *Thomsen,* 2013) between the inner and outer magnetosphere. Some ions impact and deposit this energy into the surfaces of Saturn's large moons, driving surface radiation chemistry and ejecting, or "sputtering," water molecules, radiolytic hydrogen and oxygen, and other species off their surfaces to generate gravitationally bound atmospheres. These "collisionless" atmospheres, or exospheres, are of such low density that the gas molecules rarely collide with each other.

Saturn's moons and their exospheres also interact with the ambient plasma to generate electric currents and magnetic fields. These interactions include (1) ambient ion and electron absorption, and (2) the introduction of new exospheric "pickup" ions into the magnetosphere, which contribute to the momentum loading of the plasma and supply the magnetosphere with additional material. Thus, Saturn provides the energy and Enceladus the material to power, produce, and sustain the magnetospheric generation of exospheres and electric currents at Saturn's other massive icy satellites. Knowledge of the magnetospheric interactions of these moons is therefore essential to understanding the inner workings of the saturnian system as a whole.

In this chapter we describe the Cassini spacecraft's close up measurements of the exospheres and magnetospheric interaction of the two inner magnetospheric moons for which low-altitude flybys have been performed: Dione and Rhea. We discuss the following major findings by Cassini:

1. Exospheres of O_2 and CO_2 at Dione and Rhea that are (a) nonuniform and (b) change seasonally as gas adsorbs onto winter polar latitudes and desorbs near the equinoxes.
2. Low surface sputtering rates that result in lower than expected exospheric gas densities and suggest that the surface material is resistant to sputtering.
3. Two never before observed types of field-aligned electric current systems around Rhea:
    - An induction current system and Alfvén wings generated by the plasma pressure gradient in Rhea's plasma wake. This current is induced by relative motion of Rhea to Saturn's co-rotating magnetosphere.
    - A "thermoelectric" current that flows approximately along field lines connecting to Rhea's surface (the flux tube) and forms to maintain charge balance on the surface.

To help elucidate the significance of these major discoveries by the Cassini mission and their contributions to understanding of the exospheres and magnetospheric interactions of Saturn's icy moons, it is instructive to consider the state of knowledge in the field and the progress made in the decades before Cassini. Therefore, we begin this chapter with a brief history of the field. The remainder of the chapter describes in detail Cassini's observations and their interpretation. We conclude the chapter with a discussion of an outstanding question: What is the source of negative ions observed outflowing from Dione and Rhea?

2. THE PICTURE BEFORE CASSINI

The history of investigation into the presence, properties, and physics of exospheres at solar system icy satellites goes back at least half a century. The ubiquity of water ice on the surfaces of Europa, Ganymede, and other satellites was becoming evident from reflection spectra by the time of the space age (*Pilcher et al.,* 1972) and, even before the first interplanetary spacecraft arrived at Jupiter and Saturn, Earth-based stellar occultation measurements (*Carlson et al.,* 1973) had already suggested the presence of a Ganymede exosphere of at least 10^{-3} mbar surface pressure. Accordingly, *Yung and McElroy* (1977) proposed desorption of surface H_2O as a logical source for Ganymede's exosphere, and they even suggested that the exosphere might consist of O_2 from photochemistry of exospheric H_2O. This explanation, although not quite correct in the details, captures the essential principal that radiation chemistry of H_2O can result in an oxidizing exosphere, as H and H_2 escape the satellite's gravity more rapidly than heavier O-bearing species. However, stellar occultation measurements conducted by the Voyager 1 spacecraft on its arrival at Jupiter did not detect Ganymede's exosphere above an upper limit of ~10^{-10} mbar (*Broadfoot et al.,* 1979), and for the next two decades, until additional measurements could be conducted, the density and composition of the exospheres of outer solar system icy satellites was speculative.

However, around this time new laboratory experiments were leading to a reconsideration of radiation effects on icy satellite surface materials and the exospheric implications. The first such experiments (*Brown et al.,* 1978) revealed that electrical insulators including ice are sputtered by energetic ions orders of magnitude faster than conductive solids. In a major augmentation of the classic picture of sputtering (*Sigmund,* 1969) in which impinging projectile ions initiate collision cascades and sputtering through direct interatomic momentum transfer collisions, *Brown et al.* (1980) suggested that the relative immobility of excited electrons in insulators allows the projectile (or secondary electrons) to produce electronic excitations or ionizations sufficiently long lived to cause interatomic motion and sputtering (*Bringa and Johnson,* 2002). Such efficient "electronic sputtering" was suggested by *Lanzerotti et al.* (1978) to be a major, and possibly dominant, source of material to the exospheres of Ganymede, Europa, and other icy satellites. *Sieveka and Johnson* (1982) and *Spencer* (1987) followed up by modeling the transport and redistribution through adsorption and desorption of such sputtered material across the surfaces of these bodies.

Mirroring the process of exospheric oxygenation by preferential H_2 escape, the laboratory experiments showed an analogous process is possible in the surface ice. These experiments showed that molecular hydrogen emission detected from pure H_2O ice irradiated in the laboratory (*Boring et al.,* 1983) was attributable to solid-state radiolysis and preferential desorption of H_2 over other less-volatile radiolytic oxidants (*Johnson,* 1990). The resulting oxygenation of the ice (*Johnson et al.,* 2005) — largely with H_2O_2 (*Loeffler et al.,* 2006) and trapped O_2 (*Teolis et al.,* 2009) — was seen to lead to increasing radiolytic O_2 ejection as the laboratory samples were irradiated (*Reimann et al.,* 1984), until the required stoichiometric sputtering ratio (*Johnson,* 1990) of ~2:1 H_2-to-O_2 in the ejecta was attained. Accordingly, sputtering of such radiolytically oxygenated ice on the surfaces of icy satellites may eject O_2 into their exospheres (together with H_2O and H_2). *Johnson et al.* (1982) pointed out that unlike H_2O and H_2, which are effectively removed, respectively, by gravitational escape and fallback/readsorption onto the surface, O_2 is both heavy and volatile, which they suggested may lead to the buildup of O_2 exospheres at these satellites.

The 1990s initiated a new era in understanding of icy satellite exospheres, as a new generation of telescopes and spacecraft provided a flood of surface and exospheric observations. In Ganymede's surface ice, *Spencer et al.* (1995) reported the detection of O_2 in reflection spectra acquired by the Lowell Observatory's Perkins 72-inch telescope, based on 577.3- and 627.5-nm visible absorption bands from interacting O_2 molecules as seen in laboratory spectra of solid O_2 (*Landau et al.,* 1962). Additionally, with the Hubble Space Telescope (HST) in orbit and fully operational by this time, *Noll et al.* (1996, 1997) were able to use the HST's Faint Object Spectrograph to identify a broad UV absorption band near 260 nm consistent with the Hartley band of ozone in the reflection spectra of Ganymede and of Saturn's moons Dione and Rhea. *Noll et al.* (1996, 1997), *Spencer et al.* (1995), and *Johnson and Jesser* (1997) suggested that radiolytic O_2 (from which O_3 is synthesized) might be produced and stably trapped in the surface water ice of these bodies in sufficient concentrations to account for the absorption bands. This suggestion, controversial at the time (*Baragiola et al.,* 1999; *Johnson,* 1999; *Johnson and Quickenden,* 1997; *Vidal et al.,* 1997), was later confirmed in the laboratory (*Teolis et al.,* 2005, 2006, 2009). Subsequent telescopic and Galileo spacecraft spectra have shown O_2 (*Spencer and Calvin,* 2002) in the surface ices of Callisto and Europa, in addition to H_2O_2, CO_2, and SO_2 at Europa (*Carlson et al.,* 2009); CO_2 and SO_2 at Callisto (*Hibbits et al.,* 2000); and CO_2 at Ganymede (*Hibbits et al.,* 2003), providing additional evidence that the surface ices of these objects are oxygenated and, as such, possible radiolytic O_2 sources to their exospheres.

The first detection of such an O_2 exosphere was made in 1995 from UV spectra of Europa acquired by the HST's Goddard High Resolution Spectrograph (GHRS). The spectrum contained 135.6- and 130.4-nm UV luminescence features from electronically excited O atoms in a ratio ~2:1, consistent with atomic oxygen formation by electron impact dissociation of gaseous O_2 (*Hall et al.,* 1995) with an estimated 2.4–14 × 10^{18} O_2 m^{-2} vertical column density (*Hall et al.,* 1998). The presence of exospheric O_2 has since also been confirmed with HST data at two other Galilean satellites: at Ganymede (*Hall et al.,* 1998) (1–10 × 10^{18} O_2 m^{-2}), and much more recently, at Callisto by *Cunningham et al.* (2015) (~40 × 10^{18} O_2 m^{-2}). A significant (~8 × 10^{18} m^{-2}) CO_2 abundance in Callisto's exosphere was also found by the Galileo spacecraft's near-infrared mapping spectrometer (*Carlson,* 1999). For multiple follow-up observational studies of the exospheric major species O_2 and H_2O at the Galilean moons, the reader is referred to UV emission studies by *Feldman et al.* (2000), *McGrath et al.* (2013), *Roth et al.* (2014, 2016), and *Shemansky et al.* (2014), as well as radio occultations of ionospheres at Europa (*Kliore et al.,* 1997) and Callisto (*Kliore et al.,* 2002). In the years since, based on these data, extensive exospheric modeling of these moons has been carried out (see, e.g., *Burger and Johnson,* 2004; *Cassidy,* 2008; *Cassidy et al.,* 2007; *Leblanc et al.,* 2005; *Liang et al.,* 2005; *Marconi,* 2007; *Plainaki et al.,* 2010, 2015; *Saur et al.,* 1998; *Shematovich et al.,* 2005; *Smyth and Marconi,* 2006; *Teolis et al.,* 2017b). Multiple other observations by HST, groundbased telescopes, and the Gallileo and Cassini spacecraft have confirmed the presence of a whole suite of other minor europan and ganymedian exospheric constituents. These include atomic H clouds at Ganymede (*Barth et al.,* 1997) and Europa (*Hansen et al.,* 2005; *Lagg et al.,* 2003; *Mauk et al.,* 2003; *Roth et al.,* 2017) from exospheric H_2 or H_2O photodissociation and charge exchange, and sputtered SO_2, Cl (*Volwerk and Khurana,* 2010; *Volwerk et al.,* 2001), Na, and K from surface salts and exogenous ion implantation (*Brown,* 2001; *Brown and Hill,* 1996).

Amid this rapid evolution in understanding of icy moon exospheres from the jovian system observations, the Cassini spacecraft was built and launched in 1997 to Saturn, carrying the most sophisticated and comprehensive assemblage of plasma physics instruments and neutral, plasma, and optical spectrometers ever sent into interplanetary space. The potential was evident at Saturn's large icy satellites for this diverse array of Cassini instruments to probe *in situ* all aspects of their exospheric physics, including exospheric material sources, structure and dynamics, and escape. Exospheric measurements of Saturn's large moons were therefore incorporated into the primary science objectives of multiple Cassini instruments (*Dougherty et al.,* 2004; *Esposito et al.,* 2004; *Waite et al.,* 2004; *Young et al.,* 2005). Highly compelling was the possibility, never before achieved by any spacecraft, to physically capture and analyze exospheric molecules from an icy satellite — including newly formed ions swept downstream of the moons in the magnetospheric flow co-rotating with Saturn (*Mauk et al.,* 2009) and the exospheric neutrals themselves. Except for early hints from Pioneer 11 and Voyager 1 of a Dione exospheric ion source from ion cyclotron wave data (*Barbosa,* 1993; *Smith and Tsurutani,* 1983), little was known about the exospheres of these moons at the time of Cassini's arrival in 2004. In preparation for Cassini's observations, *Sittler et al.* (2004) (at Dione and Enceladus) and *Saur and Strobel* (2005) (at Enceladus, Mimas, Tethys, Dione, and Rhea) modeled the sputtered exospheres of Saturn's large icy satellites. Using the best exospheric ionization rate estimates from Voyager data [photo and ion/electron impact ionization and charge exchange (*Sittler et al.,* 2004)], and the laboratory estimates of ice sputtering available at that time (*Shi et al.,* 1995), *Saur and Strobel* (2005) estimated the exospheric column densities required to balance exospheric sources and losses at each of these moons. They concluded that Rhea's exosphere should be borderline collisional with a column density of 6 × 10^{17} molecules m^{-2} and sufficiently dense to generate detectable Alfvénic magnetic field perturbations. These conclusions, although reasonable, later turned out to be incorrect for some fascinating and unexpected reasons, as we will discuss.

Cassini's initial discoveries at Saturn immediately recast understanding of the magnetosphere, with the earlier concept of a self-sustaining magnetosphere that sputters the satellites to supply itself with more material (*Richardson et al.,* 1986), superseded by the current picture in which the Enceladus plume is the major magnetospheric material source [~200 kg s^{-1} (Hansen et al., 2011) water group ions]. Saturn's rings were also confirmed to be a major material source early in the mission, with O^+ and O_2^+ detections by the CAPS and INMS during Saturn orbit insertion in 2004 (*Tokar et al.,* 2005; *Waite et al.,* 2005) giving evidence for radiolytic O_2 photodesorption from the ring particles (*Johnson et al.,* 2006). The ring O_2 source was later determined to be seasonal, peaking at ~100 kg s^{-1} (comparable in magnitude to the Enceladus plume) at Saturn solstice as the rings are subject to maximal solar illumination (*Elrod et al.,* 2014; *Tseng et al.,* 2010).

However, the role of Saturn's other large icy moons, their exospheres, and their significance as a magnetospheric material source remained as something of a mystery in the early years of the Cassini mission. Some early clues were provided by CAPS. *Burch et al.* (2007) suggested a plasma source from the orbits of Dione and Tethys on the basis of electron pitch angle butterfly distributions measured in that region by the CAPS electron spectrometer (ELS), and *Martens et al.* (2008) reported an O_2^+ source near Rhea's orbit based on CAPS Ion Mass Spectrometer (IMS) data from 23 Cassini orbits of Saturn. However, attempts to remotely detect Rhea's exosphere with Ultraviolet Imaging Spectrograph (UVIS) stellar occultations and Magnetospheric Imaging Instrument (MIMI) energetic neutral atom data were unsuccessful and gave an upper limit on the O_2 column density of 16×10^{17} m^{-2} (*Jones et al.,* 2007b). It was only with the extension of the Cassini mission and the execution of multiple low-altitude flybys (<100 km) of Dione and Rhea beginning in 2010 that their exospheres were finally detected and analyzed by INMS. As we discuss in the remainder of this chapter, data from CAPS and INMS show exospheres of O_2 and CO_2 to be far weaker by 2 orders of magnitude, but also far more dynamic than originally expected. These findings, and those of other Cassini instruments such as the magnetometer, give a glimpse of what we might expect to encounter in future *in situ* exploration of such exospheres at Saturn and other planetary systems.

3. THE PHYSICS OF ICY MOON EXOSPHERES

To supplement the discussion of Cassini's findings at Dione and Rhea (sections 4–7), here we give a brief description of the physical processes that determine the gas abundance, composition, and exospheric structure and evolution at such solar system icy moons. We can start by considering a planetary atmosphere in the simplest terms, i.e., as essentially the result of (1) a competition of gas sources and losses, and (2) dynamical processes that act to transport and exchange the gas between the surface and the atmosphere, and between different geographic regions. Averaged over time (e.g., over an orbit or a season) the total number of molecules in the exosphere is given by the balance of the exospheric sources (e.g., surface radiolysis and sputtering, micrometeoritic impact vaporization, cryovolcanism) with the sinks, or loss processes. In a magnetospherically driven exosphere, such as those of Saturn's airless icy moons, surface sputtering competes with multiple loss processes.

The relative importance of gravitational escape as a loss mechanism depends on the mass of the moon in question and the velocity distribution of the sputtered molecules. Molecules below the escape speed fall back to the surface. Species such as H_2O that are non-volatile (sticky) at the surface temperatures of Dione and Rhea readsorb permanently onto the surface and are effectively "lost" from the exosphere. More volatile species such as H_2, O_2, and CO_2 stick only transiently and eventually desorb from the surface back into the exosphere (section 5). Molecules execute ballistic trajectories on ejection from the surface, falling back to the surface unless they are (1) above the escape speed or (2) dissociated or ionized in flight. Gravitational escape rates from bodies as massive as Dione and Rhea are sufficiently slow that exospheric volatiles have enough time to ballistically "hop" around the body before being lost from the exosphere. Thus, volatile molecules form a globally distributed exosphere, irrespective of where they were initially sputtered or otherwise produced.

Exospheric loss processes other than gravitational escape include dissociation and ionization by ultraviolet solar photons and magnetospheric ions and electrons. As dissociated neutral radicals and atoms tend to have electron volt kinetic energies, those that do not impact the surface will tend to escape. Ionized exospheric molecules, and dissociatively ionized fragments, are subject to acceleration by Saturn's magnetic and co-rotation electric fields (section 4). These ions are "picked up" by Saturn's co-rotating magnetospheric plasma and (if they do not strike the moon's surface first) flow with the ambient plasma "downstream" of the moon. Gravitationally escaping exospheric neutrals are also ionized out in the magnetosphere on timescales of days or weeks, where they contribute ions and electrons to the magnetospheric particle population.

The sputtering of molecular solids such as water ice differs from classical sputtering in that the ejecta often consists mainly of stable volatile molecules resulting from radiolysis. Surface temperatures are often too warm to completely recondense such volatiles out of the exosphere, and therefore sputtering in the outer solar system is able to produce, in principal, relatively dense exospheres compared to the much more tenuous ones of the inner solar system. In the case of the water ice surfaces of the larger jovian and saturnian icy moons, sputter ejecta have high proportions of H_2O, and O_2 and H_2 derived from solid-state radiolysis of ice. O_2 dominates these exospheres because the H_2O ejecta immediately refreezes upon falling back to the surface, while the H_2 is light enough to gravitationally escape. Compared to refractory surfaces, where sputtering ejects particles with typical energies on the order of several electron volts, the particles sputtered from icy surfaces are ejected with modest energies — typically a small fraction of an electron volt. This energy distribution determines the fraction of the initially sputtered molecules that escape gravitationally.

For sputtering of stable, volatile radiation products such as H_2 and O_2 from an ice regolith, a large fraction [~70% (*Cassidy and Johnson,* 2005)] of the sputtered molecules also interact and thermalize with neighboring regolith grains before ever leaving the surface. Approximately 70% of the sputtered material leaves the surface with a thermalized Maxwellian energy distribution at the local surface temperature, and therefore has less tendency to escape gravitationally than the remaining directly sputtered ~30%. Gravitationally bound volatile molecules such as radiolytic O_2 may execute several random "hops" across the surface if the other loss processes, including ionization and pickup, are sufficiently slow and may diffuse across the entire satellite prior to escape. At such slow exospheric loss rates (as for O_2 and CO_2 at Dione and Rhea), the gas spatial distribution is effectively decorrelated from the distribution of surface sputtering (maximal on the moon's trailing hemisphere).

The exospheric source rates for sputtered H_2O and radiolytic H_2 and O_2 can be estimated by convolving incident magnetospheric particle fluxes with estimated yields (molecules produced/sputtered per incident particle) (*Cassidy et al.,* 2013). Nonpenetrating heavy ions (e.g., Saturn's magnetospheric thermal "water group" ions at hundreds of electron volts or low kiloelectron-volt energies) often dominate exospheric O_2 production since the energy deposited in the uppermost angstroms of the surface ice has the largest influence on both the production of O_2 and sputtering. The sputtering yields used in estimating exospheric source rates are adapted from laboratory experiments [see compilations of sputtering yields in *Cassidy et al.* (2013) and *Teolis et al.* (2010a, 2017a)], while ion fluxes impinging onto the surfaces of the satellites are estimated from the observed particle environment [e.g., moments from the CAPS and MIMI instruments on Cassini (*Dialynas et al.,* 2009; *Schippers et al.,* 2008; *Thomsen et al.,* 2010)]. However, as we discuss in this chapter, these widely used calculations appear to overestimate O_2 production at Saturn's moons and do not account for the finding of significant exospheric CO_2 — discrepancies possibly reflecting the impure composition of their surfaces.

4. EXOSPHERIC IONS FROM DIONE AND RHEA

4.1. Cassini's Plasma Spectrometer Detects "Pickup" Ions at Rhea

The first glimpse of a possible exosphere at Rhea came in 2005 and 2007 from the detection by CAPS of outflowing positive ions during the "R1" and "R1.5" Cassini flybys (Table 1). Both flybys took place through Rhea's plasma wake (Fig. 1), formed "downstream" of the moon as Saturn's magnetospheric plasma flows past the moon. Since the entire inner magnetosphere co-rotates with Saturn (*Mauk et al.,* 2009), the plasma flow direction and speed at Rhea are determined approximately by Saturn's rotation rate and Rhea's orbit radius. Positive ions are generated from the exospheric molecules by (1) solar UV photoabsorption, (2) charge exchange with magnetospheric ions, or (3) ion or electron impact. Table 2 gives estimated reaction rates for conversion of exospheric O_2 molecules into ions and ionized and neutral fragments by UV photons, ions, and electrons, taking into consideration the magnetospheric density and energy distributions at the orbits of Dione and Rhea. Depending on their starting location and trajectory, some ionized and neutral products can reimpact the moon (*Sittler et al.,* 2004), inducing secondary surface sputtering. The ions are initially accelerated, or "picked up," by Saturn's co-rotation electric field. Those pickup ions that avoid striking the moon's surface are swept downstream along cycloidal trajectories by the $q(E + v \times B)$ force. Thus, the Cassini R1 and R1.5 trajectories, downstream of Rhea, were well positioned to capture and analyze these exospheric pickup ions (Fig. 1).

Owing to the mass and gravity of Dione and Rhea, the gravitational escape and reaction rates are similar at both moons as shown in Table 2, and pickup ionization therefore constitutes a major exospheric loss process. The dominant ionization channels at Dione and Rhea are electron-impact ionization and charge-exchange (Table 2), which transfer negligible momentum to the newly created exospheric ions. Most pickup ions are therefore born at nearly their original exospheric velocity, which is ≤ 1 km s^{-1} if the exosphere is equilibrated to the local surface temperature. In the pickup process these ions are accelerated to speeds of several tens of kilometers per second on the order of the co-rotation speed, forming a 90° pitch angle cold (slender) ring-shaped distribution in velocity space (*Sittler et al.,* 2004). Only those ion trajectories connecting back to the location of the exosphere can contribute to the ring distribution detected by the spacecraft downstream (Figs. 2 and 3b). For objects like Rhea with diameters similar to the ion gyroradius, this means that the ring distribution is only partly filled ('non-gyrotropic').

During the 2005 (R1) and 2007 (R1.5) flybys, both the CAPS IMS and the ELS were aimed back toward Rhea to measure the energy and angular distribution of the pickup ion flux (Figs. 2 and 3b). To make such measurements, the IMS and ELS each collect particle flux into a "fan" of eight differently-pointed wedge-shaped anodes. During the R1 and R1.5 flybys the fan plane was held at a fixed orientation perpendicular to Saturn's magnetic field (in the plane of the "ring") to sample the pickup ion's ring distribution.

Although IMS has a time-of-flight analyzer to measure ion mass, the Cassini spacecraft's flyby times at Saturn's moons are only a few minutes, which is too short to acquire sufficient signal for a mass spectrum. Instead, to constrain the ion mass at Dione and Rhea, *Tokar et al.* (2012) and *Teolis and Waite* (2016) employed an ion-back-tracing method, plotting (for different mass) the ion trajectories backward from the locations, speed (energy) and anode direction along which they were detected (Figs. 2–4). As shown in Fig. 2 (3) for the R1 (R.1.5) flyby, the trajectories within the anode 3 (7) field of view (FOV) trace back to a region

TABLE 1. Dione and Rhea flyby observations (from *Teolis and Waite,* 2016).

Flyby	Body	Date	UTC	Speed (km s^{-1})	Alt (km)	Detection	Species	References	Description
D1	Dione	11-Oct-05	17:52:00	9.12	500	MAG	—	*Simon et al.* (2011)	Upstream southern flux tube, Saturn inbound
R1	Rhea	26-Nov-05	22:37:38	7.28	500	CAPS	CO_2^+, O^-	*Teolis et al.* (2010a,b); this work	Equatorial wake, toward nightside, Saturn inbound
R1.5	Rhea	30-Aug-07	01:18:55	6.76	5725	CAPS	CO_2^+	This work	Northern wake, toward nightside, Saturn outbound
R2	Rhea	2-Mar-10	17:40:36	8.58	100	INMS	O_2, CO_2	*Teolis et al.* (2010a,b); this work	Low altitude, north polar, toward dayside, Saturn inbound
D2	Dione	7-Apr-10	05:16:11	8.34	500	CAPS	O_2^+	*Tokar et al.* (2012)	Equatorial wake, toward dayside, Saturn inbound
R3	Rhea	11-Jan-11	04:53:25	8.05	72	INMS	O_2	This work	Low altitude, south polar, toward dayside, Saturn outbound
D3	Dione	12-Dec-11	09:39:23	8.73	99	INMS	O_2, CO_2	This work	Low altitude, equatorial wake, toward dayside, Saturn outbound; CAPS offline
R4	Rhea	9-Mar-13	18:17:26	9.29	996	—	none	This work	South-to-north, over anti-Saturn nightside, CAPS offline; altitude too high, and pointing poor, for INMS: non-detection
D4	Dione	16-Jun-15	20:11:52	7.32	516	INMS	O_2	This work	North polar, toward nightside, Saturn outbound; CAPS offline
D5	Dione	17-Aug-15	18:33:25	6.45	476	—	none	This work	North polar, toward nightside, Saturn inbound; CAPS offline

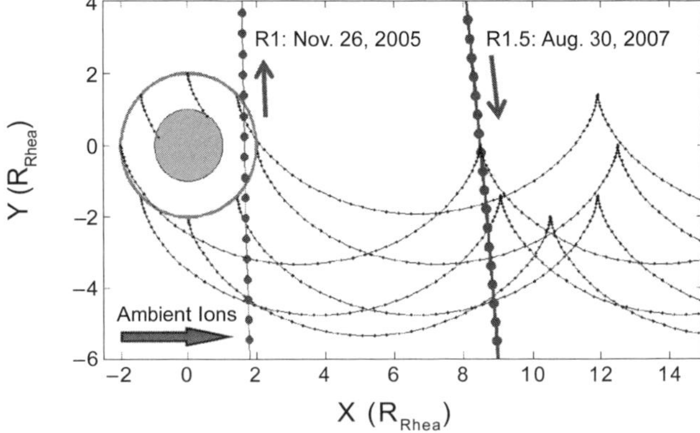

Fig. 1. Diagram showing the Cassini R1 and R1.5 flyby trajectories (north polar view; Y points Saturnward, X points in co-rotation), together with exospheric CO_2^+ pickup ion trajectories from the region (circled) above Rhea's surface.

of space, or "sample zone," the shape and location of which depends on the pickup ion mass. Comparing the timing, the magnitude, and the shape of the detected ion signature along the spacecraft trajectories to the predicted exospheric density and distribution at the times and dates of the flybys, *Teolis and Waite* (2016) constrained the Rhea pickup ion species as CO_2^+. While the energy of the sharp CO_2^+ pickup ion signature lies just above the broad energy spectrum of the thermal co-rotating water group ions, lower-mass pickup ions such as O_2^+, H_2^+, and H_2O^+ entering CAPS at lower energies are swamped by Saturn's co-rotating plasma, rendering them undetectable.

Changes over time in the exospheric molecular abundance are also seen in CAPS-based estimates of the exospheric CO_2 abundance, which are ~50% higher for the 2007 R1.5 than the 2005 R1 flyby. A buildup in the CO_2 exosphere from 2005 to 2007 is consistent with increasing desorption of seasonally cold trapped CO_2 from Rhea's polar region, as discussed in section 5.2. During this time, the solar terminator

TABLE 2. Average exospheric O_2 loss rates (units 10^{-8} s^{-1} mol^{-1}).

Loss Channel	Rhea	Dione	References Cross Sections	Flux Spectra
$O^+ + O_2 \rightarrow O + O_2^+$	0.68	1.5	[5,6]	[1]
$O^+ + O_2 \rightarrow O + O^+ + O$	0.24	0.54	[5,6]	[1]
$O^+ + O_2 \rightarrow O + 2O^+ + e$	0.023	0.041	[5,6]	[1]
$O^+ + O_2 \rightarrow O + O^{++} + O + e$	0.0003	0.00012	[5,6]	[1]
$O^+ + O_2 \rightarrow O^+ + O_2^+ + e$	0.11	0.28	[5,6]	[1]
$O^+ + O_2 \rightarrow O^+ + O^+ + O + e$, $O^+ + O_2 \rightarrow O^+ + 2O+ + 2e$	0.072	0.058	[5,6]	[1]
$O^+ + O_2 \rightarrow O^+ + O^{++} + O + 2e$	0.0081	0.0064	[5,6]	[1]
$H^+ + O_2 \rightarrow H + O_2^+$	0.59	1.1	[5,7]	[1]
$H^+ + O_2 \rightarrow H + O^+ + O$	0.032	0.062	[5,7]	[1]
$H^+ + O_2 \rightarrow H + 2O^+ + e$	0.0084	0.018	[5,7]	[1]
$H^+ + O_2 \rightarrow H + O^{++} + O + e$	0.0004	0.0012	[5,7]	[1]
$H^+ + O_2 \rightarrow H^+ + O_2^+ + e$	0.018	0.044	[5,7]	[1]
$H^+ + O_2 \rightarrow H^+ + O^+ + O + e$, $H^+ + O_2 \rightarrow H^+ + 2O^+ + 2e$	0.018	0.041	[5,7]	[1]
$H^+ + O_2 \rightarrow H^+ + O^{++} + O + 2e$	0.018	0.041	[5,7]	[1]
$e + O_2 \rightarrow O_2^+ + 2e$	2.2	0.55	[4]	[2]
$e + O_2 \rightarrow O^+ + O + 2e$, $e + O_2 \rightarrow 2O^+ + 3e$	0.95	0.19	[4]	[2]
$e + O_2 \rightarrow O^{++} + O + 3e$, $e + O_2 \rightarrow 2O^{++} + 5e$	0.011	0.0013	[4]	[2]
$e + O_2 \rightarrow O + O + e$	0.89	0.42	[4]	[2]
$e + O_2 \rightarrow O + O^-$	0.0022	0.0058	[4]	[2]
$v + O_2 \rightarrow O(3P) + O(3P)$	0.081(0.120)	0.081(0.120)	[3]	
$v + O_2 \rightarrow O(3P) + O(1D)$	2.2(3.5)	2.2(3.5)	[3]	
$v + O_2 \rightarrow O(1S) + O(1S)$	0.022(0.053)	0.022(0.053)	[3]	
$v + O_2 \rightarrow O_2^+ + e$	0.26(0.66)	0.26(0.66)	[3]	
$v + O_2 \rightarrow O + O^+ + e$	0.061(0.192)	0.061(0.192)	[3]	
Gravitational escape	13	38		
Total	22(24)	46(48)		

Solar rates (quiet/active Sun outside/inside parentheses) directly from *Huebner et al.* (1992) scaled to 9.6 A.U. Ion and electron rates estimated by convolving published energy-dependent cross sections and Cassini CAPS/MIMI plasma flux vs. energy spectra. References: [1] *Dialynas et al.* (2009), *Thomsen et al.* (2010), *Wilson et al.* (2008, 2010); [2] *Schippers et al.* (2008); [3] *Huebner et al.* (1992); [4] *Itikawa* (2008); [5] *Luna et al.* (2005); [6] *Sieglaff et al.* (2000); [7] *Cabrera-Trujillo et al.* (2004), *Williams et al.* (1984). Data from *Teolis and Waite* (2016).

was advancing across the north polar terrain, warming the surface as the 2009 spring equinox approached.

The profile of the pickup ion signature vs. time and position along Cassini's trajectory also gives information on the exospheric spatial distribution, as the spacecraft sampled ions from different parts of the exosphere at different points on its trajectory (Figs. 2 and 3). For example, during R1, the pickup ion signal gradually increased as ions from the lower altitudes were sampled (Fig. 2) until the sampling zone intersected Rhea's surface, at which point the signal sharply dropped off. The observed R1 pickup ion profile is consistent (Fig. 2) with a non-uniform "sticky" CO_2 exosphere, which is less dense on the nightside due to CO_2 cryosorption onto the cold night surface (section 5.2).

The double-peaked pickup ion profile along Cassini's R1.5 trajectory (Fig. 3) appears to be sampling the peculiar spatial profile of the exosphere. This distribution, modeled in Fig. 3, results from desorption from a northern CO_2 "frost" cap. The model of a seasonal exosphere discussed in section 5.2 is remarkably successful at capturing not only the exospheric spatial structure as seen in the pickup ion profiles, but also the magnitude of the pickup ion signature, *and* the magnitude and structure of the gas density measured several years later by INMS (section 5.2).

4.2. Oxygen Ions at Dione

The D2 500-km flyby through Dione's wake in 2010 (Table 1) provided the first (and only) opportunity for CAPS to directly detect pickup ions from this saturnian icy satellite. Again, the CAPS instrument was pointed upstream toward Dione to detect outflowing pickup ions, but unlike the Rhea flybys, the anode fan plane was oriented along the magnetic field (rather than perpendicular to it), and only anode 4 was aligned with the gyroplane to capture the pickup ions at a 90° pitch angle.

Despite this limitation, pickup ion detection is actually easier at Dione than it is at Rhea because of the ~20% slower co-rotation flow speed of the magnetosphere (~40 km s^{-1}) at Dione's orbit. The ambient magnetosphere therefore flows more slowly past Dione than Rhea, resulting in colder ion temperatures at Dione. The energy distribution for water group ions drops off sharply above a few hundred electron volts at Dione (Fig. 5), making it possible to discern O_2^+

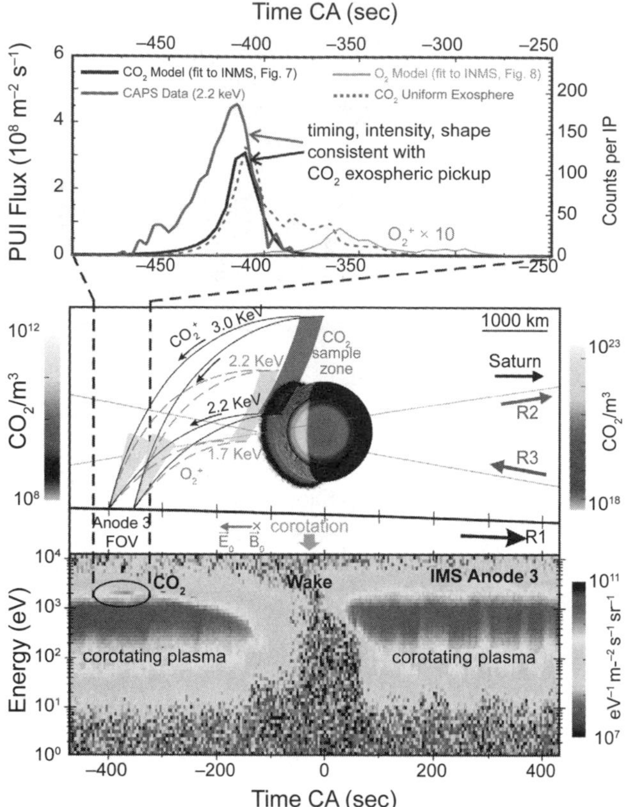

Fig. 2. See Plate 39 for color version. From *Teolis and Waite* (2016). Comparison of the Rhea pickup ion flux and distribution detected by CAPS IMS anode 3 on the R1 flyby, to that anticipated by the model CO_2 and O_2 exospheres, fit to the later INMS R2 and R3 data and projected back to the R1 flyby date and time. (*Bottom*) Anode 3 ion flux energy spectrogram with the CO_2^+ pickup ion signature circled. (*Center*) R1 flyby configuration showing the Rhea exospheric CO_2 density cross section, Rhea surface illumination and adsorbed column density, the region (purple) sampled by anode 3 during the pickup ion detection, the anode 3 FOV at the time and location of the pickup ion detection along the R1 trajectory, and the limiting O_2^+ and CO_2^+ energies (on arrival at Cassini) and trajectories accepted into anode 3. The R1 sampling region is approximated as two-dimensional in the flyby's equatorial plane. (*Top*) Anticipated (dark blue: CO_2^+, orange: O_2^+ scaled × 10) and measured (red) total pickup ion signal vs. time from closest approach, compared to the expected profile (dashed line) for a uniform CO_2 exosphere. As shown the timing, intensity and shape of the signal is not consistent with O_2^+ (nor with fragments such as CO^{n+}, O^{n+}, C^{n+}), but is consistent with CO_2^+. The absence of a shoulder after –350 s is consistent with a dayside exosphere, as anticipated by the CO_2 exosphere model and later observed by INMS.

pickup ions from Dione in the CAPS spectrum. *Tokar et al.* (2012) analyzed the evolution of IMS energy spectra in anode 4 during the flyby and identified the appearance of a shoulder on the edge of the water group ion energy distribution. They showed that the energy (~1 keV) and timing (location) of the shoulder are consistent with O_2^+ pickup ions with trajectories tracing back to the exosphere's location just above Dione's surface (Figs. 4 and 5). The O_2^+ densities were consistent with surface neutral exospheric densities of ~0.6–5 × 10^{10} m^{-3}, in the range estimated on the basis of INMS data (section 5.1, Fig. 6), and consistent with the estimates of *Sittler et al.* (2004).

However, the agreement of the *Sittler et al.* (2004) prediction with the observed density appears to be coincidental. On the one hand, the *Sittler et al.* (2004) study approximated the O_2 source rate from Dione as ~10^{26} s^{-1}, or 10% that estimated for H_2O (*Johnson et al.,* 2008), but this exceeds the ~5 × 10^{22} O_2 s^{-1} source rate later observed by Cassini (Table 3, section 5.3). On the other hand, *Sittler et al.* (2004) applied an analytical exospheric model (*Watson,* 1981) that assumes, implicitly, that the sputtered molecules permanently stick after falling back to the surface. This assumption, appropriate for H_2O, is less applicable to O_2, which is volatile over most of Dione's ~25–95 K surface temperature range (*Howett et al.,* 2014). The two effects — the first increasing and the second decreasing the model O_2 exospheric density — roughly cancel. While *Saur and Strobel* (2005) considered a more realistic "non-sticky" O_2 exosphere, they used a similar O_2 source rate to *Sittler et al.* (2004). They therefore predicted a Dione O_2 column density of 10^{17} m^{-2}, 1–2 orders of magnitude above the subsequent CAPS (*Tokar et al.,* 2012) and INMS measurements (Fig. 6). In the next section we discuss these INMS measurements of the neutral gas densities and composition at both Dione and Rhea.

5. EXOSPHERIC COMPOSITION, STRUCTURE, AND SEASONS

5.1. *In Situ* Exospheric Sampling: Cassini's Ion Neutral Mass Spectrometer at Rhea and Dione

Beginning in 2010 as part of its extended mission, the Cassini spacecraft began a series of close flybys of Rhea and Dione below 100 km altitude, much lower than previous encounters (Table 1). For the first time at any solar system icy satellite, Cassini INMS conducted a series of *in situ* mass spectrometric measurements of the sputtered exospheric density and composition vs. position along the spacecraft trajectory and between flybys at different times and flyby locations. The first two Rhea encounters were over the north and south poles: the R2 and R3 flybys on March 2, 2010, and January 11, 2011, at 100 and 72 km altitude, respectively (Table 1). These were followed by the close (99 km) Dione D3 equatorial flyby on December 12, 2011, over the dusk terminator. Neither sputtered H_2O, nor H_2 from surface radiolysis (section 3), were sufficiently abundant for detection by INMS at either moon. This is expected, as both species are subject to efficient gravitational escape (*Teolis et al.,* 2010a) and H_2O falling back to the surface is permanently readsorbed. However, INMS did detect low-density (collisionless) exospheres of O_2 and CO_2 at both moons that were non-uniformly distributed along Cassini's trajectory and appeared to change between flybys (Fig. 6).

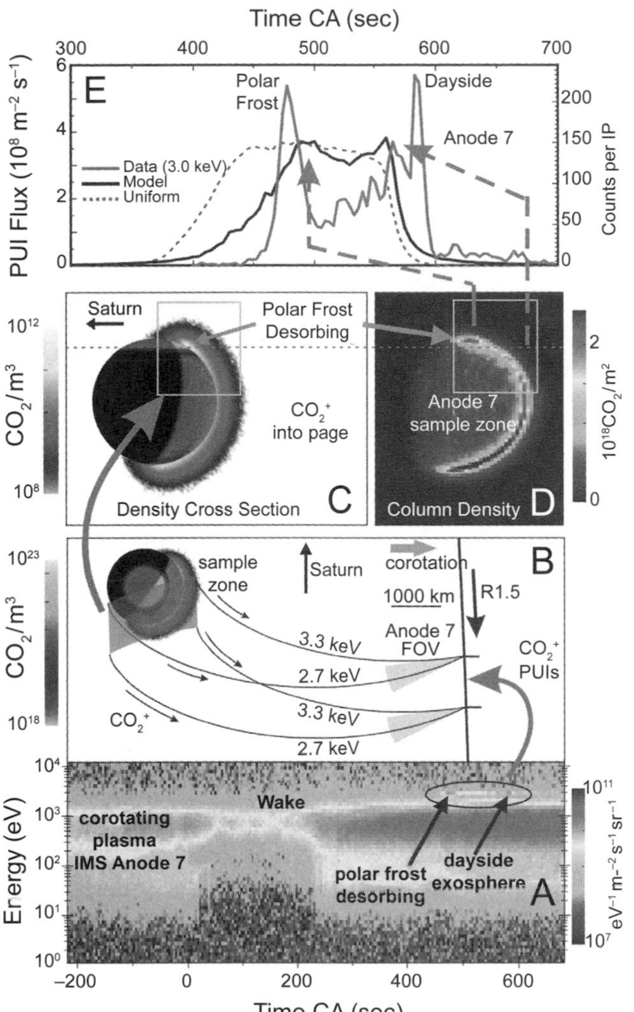

Fig. 3. See Plate 40 for color version. From *Teolis and Waite* (2016). Comparison of the Rhea pickup ion flux and distribution detected by CAPS IMS anode 7 on the R1.5 flyby, to that anticipated by the model CO_2 exosphere (fit to the later INMS R2 and R3 data) projected back to the R1.5 flyby date and time. (A) Anode 7 ion flux energy spectrogram with the CO_2^+ pickup ion signature circled. (B) R1.5 flyby configuration showing Rhea with the model exospheric CO_2 density cross section and adsorbed column density, the surface illumination, the region (purple box) sampled by anode 7 during the pickup ion detection, the anode 7 FOV at the time and location of the pickup ion detection along the R1.5 trajectory, and the limiting CO_2^+ energies (on arrival at Cassini) and trajectories accepted into anode 7. (C) Equatorial view of the model exosphere viewed through the sample box, showing the box's approximate north-south extent if the anode 8.5° FWHM elevation angular acceptance is projected back to Rhea. (D) Exospheric column density from the orientation of (C) as sampled by anode 7, showing the intense feature from polar frost desorption. (E) Anticipated total anode 7 pickup ion signal vs. time from closest approach (red), with arrows showing the correspondence of the double peaked structure to the polar frost desorption, and dayside exospheric signatures seen in (D). As at R1 (Fig. 2), the timing, intensity, and shape of the signal is not consistent with O_2^+ (not shown) nor fragment species, but is consistent with CO_2^+. The observed pickup ion signature (red) shows a similar (even more intense) double peaked structure, which implies that CAPS is observing a CO_2 exosphere with an abundance and spatial structure consistent with INMS based modeling. Also shown for comparison (top, dashed line) is the expected pickup ion profile for a uniform CO_2 exosphere.

More O_2 and CO_2 were detected by INMS over the *dayside* hemispheres of Dione and Rhea as shown in Fig. 6, suggesting that the exospheric scale heights are greater over that hemisphere. Larger dayside O_2 and CO_2 scale heights are consistent with thermal equilibration of these gravitationally bound volatiles species to the ~94/52 max/min day/night surface temperatures (*Howett et al.,* 2014). However, CO_2 is more unevenly distributed between the day and night hemispheres than O_2 (Fig. 6), suggesting that the less-volatile CO_2 molecules are also adsorbing onto the nightside surface.

While exospheric O_2 can be sputtered from an icy satellite as a radiolysis product of the surface water ice, the finding of CO_2 at a similar density to O_2 during the Rhea R2 and Dione D3 flybys was unexpected. Although non-ice surface constituents including CO_2 have been detected at Dione and Rhea (*Cruikshank et al.,* 2010; *Stephan et al.,* 2012), infrared spectra show as expected an overwhelming abundance of H_2O ice in Rhea's surface (*Cruikshank et al.,* 2005). We will return to the question of CO_2 and its possible origin in section 5.3.

Equally mysterious were the changes in Rhea's exosphere between the R2 and R3 flybys as seen in Fig. 6. Following the confirmation of Rhea's exosphere by INMS during the north polar flyby (*Teolis et al.,* 2010b), it was anticipated that the R3 flyby 10 months later (on January 11, 2011) over the south pole would encounter higher gas densities due to the lower flyby altitude (72 vs. 100 km on R3 vs. R3). The detection of 10 times *less* O_2 (~0.5 vs. 5 $\times 10^{10}$ m^{-3} on R3 vs. R2) and no CO_2 above an upper limit of 0.7×10^{10} m^{-3} was therefore a surprise. The unexpected findings suggest either an asymmetrically distributed exosphere between the northern and southern hemispheres and/or drastic changes in the exosphere over time between flybys.

5.2. Exospheric Seasons

The low surface temperature reached by the polar regions of these icy moons during northern/southern winter is a potential major consideration in understanding the apparent exospheric structure and changes over time implied by these Dione and Rhea INMS data. Near the equator, the surfaces of Dione and Rhea cool below 50 K at night, which as mentioned, is already cold enough to cryopump CO_2 out of the exosphere on the nightside. However, the poles get

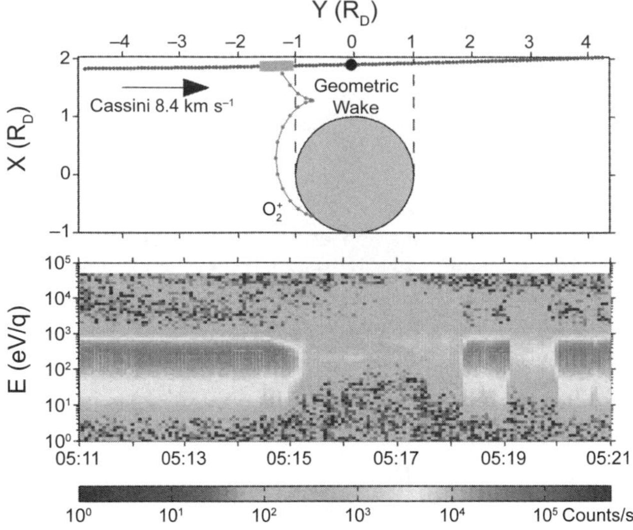

Fig. 4. See Plate 41 for color version. From *Tokar et al.* (2012). (*Bottom*) CAPS ion count rate at 4 s resolution as a function of time and energy per charge. The measurements are for anode 4 of CAPS, sensing flow predominantly in the co-rotation direction. (*Top*) Cassini trajectory during D2 and Dione's geometric wake, with a sample pick-up O_2^+ ion trajectory in the ambient flow plotted at 2 s resolution. The pickup ions are detected by CAPS just before entering the wake, in the green shaded region on the trajectory.

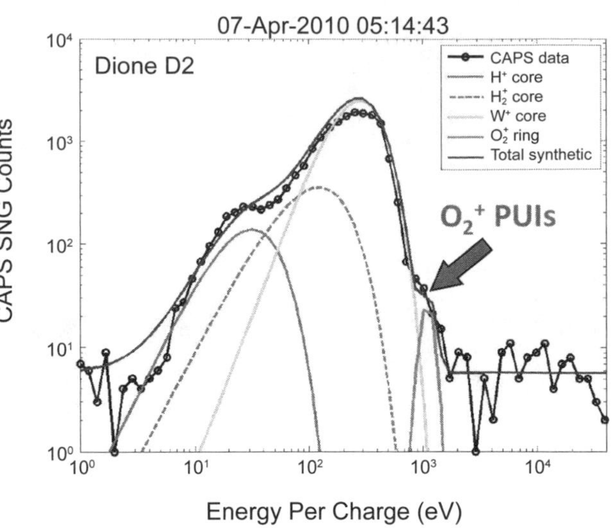

Fig. 5. Adapted from *Tokar et al.* (2012). This figure shows CAPS ion counts vs. energy measured by anode 4 (black circles) close to Dione's wake. In addition, ion counts are simulated for both O_2^+ and H_2O^+ pickup ion rings for the nominal plasma flow, indicating good agreement of the observed pick-up ion energy with the simulated O_2^+ pick-up ion energy.

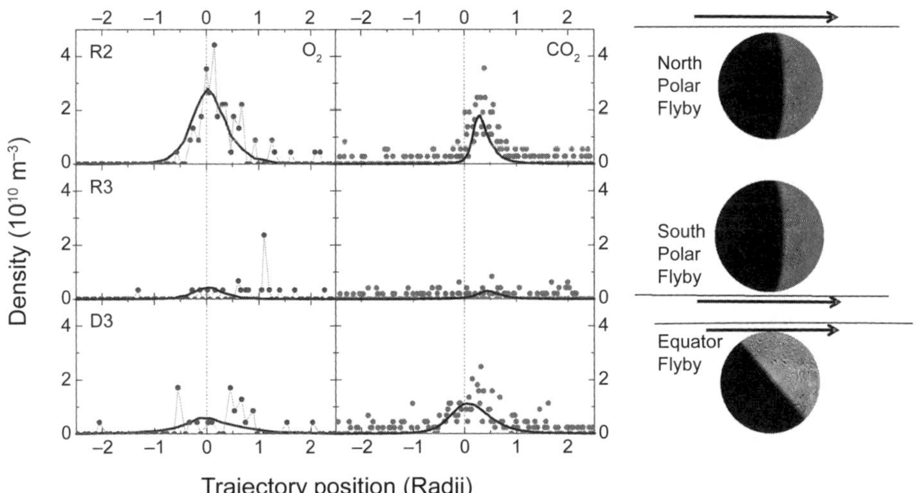

Fig. 6. From *Teolis and Waite* (2016). The Dione and Rhea O_2 and CO_2 exospheric densities measured by Cassini INMS vs. distance (in planetary radii) from closest approach (vertical dashed line) along the night-to-day R2, R3, and D3 trajectories (shown on right). The data are unbinned, such that single counts are visible. Time increases left to right in the plots, with Cassini moving from the nightside to dayside hemisphere on all flybys. As shown, the exospheric profiles are concentrated (especially for CO_2) over the dayside when equivalent inbound-outbound altitudes are compared. Lines are exosphere model results.

much colder. Surface temperatures at the Rhea and Dione winter poles are some of the coldest observed in the solar system, dropping as low as 20 K (*Howett et al.,* 2016). This is cold enough to condense not only CO_2, but also the more volatile O_2 out of the exosphere. However, such polar condensation is *seasonal*, with the northern and southern poles alternately going into winter shadow once per Saturn year (29.5 Earth years). Owing to the Saturn system's 26.7°

TABLE 3. Modeled exospheric sources and sinks.

	Units	Rhea		Dione	
		O_2	CO_2	O_2	CO_2
Avg in-flight abundance*	10^{28} molecules	3.3	3.9	9.7	0.57
Time to escape†	Earth years	2.7	700	2.5	160
Jeans escape rate	10^{-8} s^{-1} mol^{-1}	13	1.0	38	20
Dissociative channels‡		4.7(6.1)		3.8(5.2)	
e-impact ionization		2.2		0.55	
Photo-ionization		0.26(0.66)		0.26(0.66)	
Ion-impact ionization		0.13		0.33	
Charge exchange ioniz		1.3		2.6	
Total ionization¶		3.9(4.3)	3.9§	3.7(4.1)	3.7‡
Total loss rate		22(24)	9.6(11.6)	46(48)	28(30)
Expected source rate**	10^{21} mol s^{-1}	2200		2100	
Observed source rate††		7.2	3.7	45	1.6
Exp sputter yield‡‡		2.7		1.4	
Obs sputter yield‡‡,§§	molecules/ion	0.0088	0.0045	0.029	0.0010
Expected/observed	none	306		47	

* Total number of exospheric molecules, time averaged over a season.
† Includes time stuck to surface.
‡ All dissociation and dissociative ionization channels (Table 2).
§ Taking O_2 rate: Insufficient published cross sections to estimate CO_2 independently to same accuracy as O_2.
¶ Sums e-impact, ion-impact, photo, charge exchange channels, leading directly to O_2^+, CO_2^+.
** For a pure water ice surface in Saturn's magnetosphere, using model of *Teolis et al.* (2010a).
†† Taken to be constant in time. Given by average inflight abundance times total loss rate.
‡‡ Global average over all ions and energies, using ambient plasma surface ion bombardment fluxes from *Teolis et al.* (2010a).
§§ Treats sputtering by ions as the dominant source process as *Teolis et al.* (2010b) found for pure water ice. Even lower (ion) sputtering yields result if photo, electron, or micrometeoritic impact-driven production/desorption contributions from the altered layer are appreciable.

O_2 loss rates summed from Table 2. Row groupings: same units. Rates for quiet/active Sun are outside/inside parentheses. Data from *Teolis and Waite* (2016).

tilt to the ecliptic, large swaths of terrain in polar night are subject to years of radiative cooling.

To elucidate the implications of seasonal O_2 and CO_2 polar condensation on the exospheric spatial distribution and dynamics measured by INMS, *Teolis and Waite* (2016) implemented a Monte Carlo exospheric simulation in which the diurnal and seasonal variability of both surface temperature and exospheric adsorption and desorption were included. The model took into consideration (1) the exospheric loss rates by both gravitational escape and pickup ionization (Tables 2 and 3), (2) the moon's rotation that transports gas adsorbed onto the nightside across the dawn terminator where it desorbs, (3) the ±26.7° seasonal change in solar declination over 29.5 (Earth)-year cycles, and (4) gas diffusion into the porous surface regolith. The model is largely successful at fitting the data from both CAPS (Figs. 2 and 3) and INMS (Figs. 7 and 8), and provides plausible explanations for several key observations, as follows.

1. *Observation*: The finding (Fig. 6) that CO_2 is more concentrated on the daysides of Dione and Rhea than O_2, which is captured by the model [Figs. 7, 8, and 11 (top)].

- *Interpretation*: Unlike O_2, the less-volatile CO_2 sticks transiently to the nightsides of Dione and Rhea and desorbs after sunrise from their dawn terminators. Therefore, the CO_2 exosphere exhibits a far more pronounced day-night asymmetry (Fig. 7) than O_2 (Fig. 8).

2. *Observation*: The lack of CO_2 detection by INMS over Rhea's southern hemisphere (Fig. 6) during the R3 flyby, implying CO_2 densities no more than a tenth of those detected at R2 over the north.

- *Interpretation*: The south polar region was cooling at the times of the R2 and R3 flybys, which took place shortly past the August 2009 southern autumn (northern spring) equinox. While frozen CO_2 was evaporating into the exosphere from the north polar terrain, the opposite was occurring in the south, with CO_2 being "cryopumped" out of the southern exosphere onto the cooling surface. Hence the exosphere was, according to the model, less dense in the south at the times of the flybys (Fig. 7).

3. *Observation*: Like CO_2, lower O_2 densities were measured by INMS in the south than in the north (Fig. 6).

- *Interpretation*: Unlike CO_2, the loss of O_2 between the flyby dates, 10 months apart, appears indicative of a rapid collapse of the global O_2 exosphere between R2 and R3. The collapse occurred as the supply of O_2

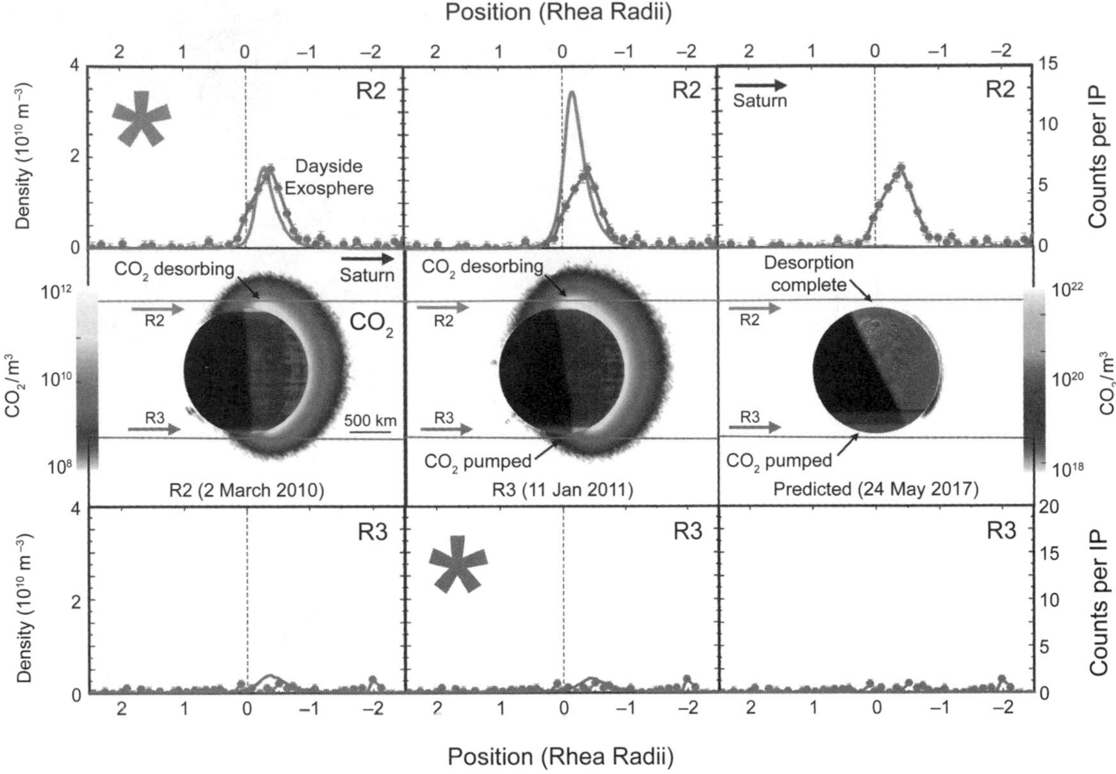

Fig. 7. See Plate 42 for color version. From *Teolis and Waite* (2016). Rhea CO_2 exospheric model results, showing (center row) the surface solar illumination (dawn terminator as viewed over the equator), and predicted gas density cross sections and surface frost column density at the R2 (left) and R3 (center column) flyby dates and times, and prediction (right) for the 2017 Saturn solstice. (*Top*) Comparison of the observed (red points, data binned) and predicted (green line) CO_2 densities vs. position from closest approach along the R2 trajectory (x-axis, in Rhea radii, dashed line: closest approach). (*Bottom*) Same for the R3 flyby trajectory. Asterisks denote model-data comparisons with the same flyby date; all other plots show model projections to different dates. The model assumed a 3.7 × 10^{21} CO_2 s^{-1} source rate as required to match the observed peak density on both flybys. The observation of CO_2 concentrated on the dayside is consistent with the modeled exospheric structure due to nightside and polar cryopumping. CO_2 was desorbing from the north, and cryopumping onto the southern polar surface during the flybys. Accordingly, as anticipated by the model (bottom center), CO_2 was undetectable in the south on R3. As shown (right), exhaustion of the northern frost cap should have resulted in the collapse of the CO_2 exosphere by the time of the 2017 solstice.

evaporating into the exosphere from the north polar terrain was exhausted, according to the model (Fig. 8). In contrast to the less-volatile CO_2, for which evaporation from one pole overlaps in time (for ~2–3 Earth years) with cryopumping by the other, most O_2 has already desorbed from the spring pole prior to significant cryopumping at the autumn pole. The flybys, it appears, were fortuitously timed to catch the O_2 exosphere's collapse.

Drastic exospheric seasons at Dione and Rhea are a distinctive feature of these exospheric models, with both the O_2 and CO_2 exospheric densities peaking at the saturnian equinoxes. At the low ~20 K winter polar surface temperatures, both CO_2 and O_2 are predicted by the model to accumulate seasonally at the winter poles, and then desorb the following spring, resulting in a transient enhancement of the O_2 and CO_2 exospheres near every equinox. In this way, frozen O_2 and CO_2 are transferred between the northern and southern terrains on a seasonal basis (Fig. 9), while the modeled total exospheric gas abundances (Fig. 10) vary by 2–3 orders of magnitude between solstice and equinox.

The CAPS pickup ion profiles detected in Rhea's plasma wake, discussed in section 4.1, are also successfully fit by the model of a seasonal exosphere as shown in Figs. 2 and 3. Here, the modeled pickup ion profiles along Cassini's R1 and R1.5 trajectories are estimated (*Teolis and Waite,* 2016) by (1) taking into consideration the expected pickup ion production rates (Tables 2 and 3) and (2) "rewinding" the exospheric model (fit to INMS) back to the dates and times (Table 1) of the CAPS pickup ion measurements. One can see in Fig. 10 that the global exospheric O_2 and CO_2 abundances — extrapolated on the basis of the INMS and CAPS data by scaling the model exosphere to best match the data on each observation — show the anticipated seasonal evolution vs. time between flybys. INMS data from

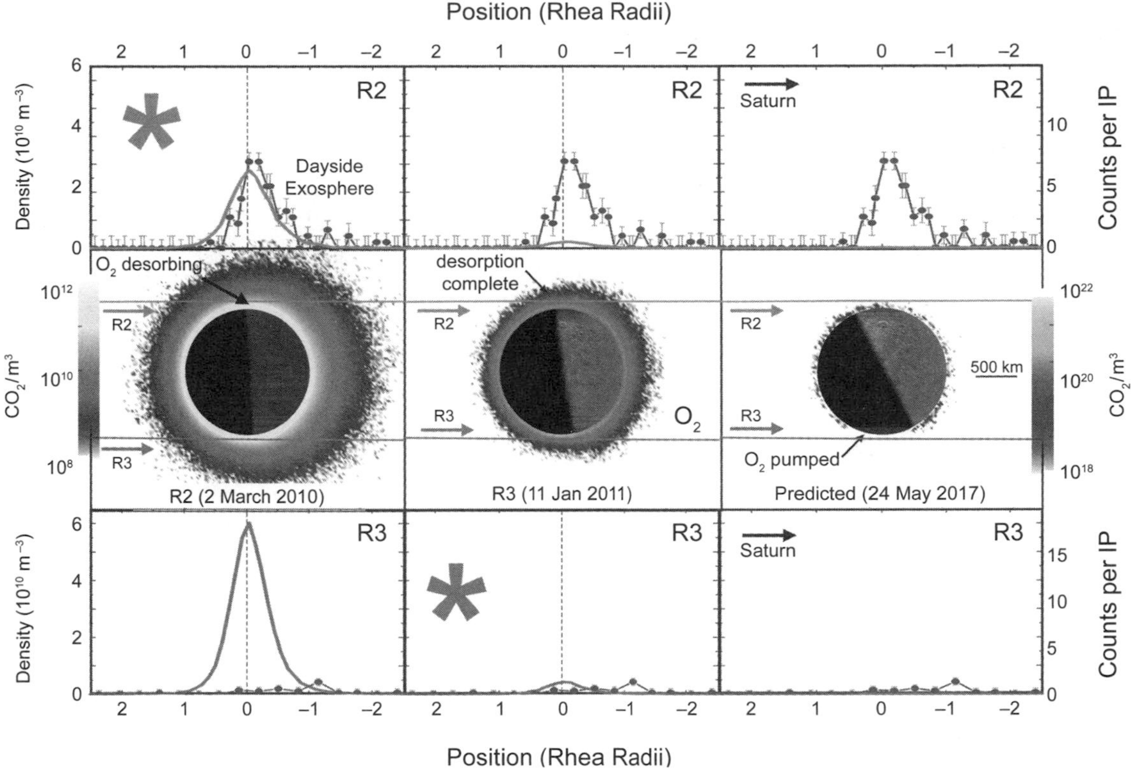

Fig. 8. See Plate 43 for color version. From *Teolis and Waite* (2016). Rhea O_2 exospheric model results, showing (center row) the surface solar illumination (dawn terminator as viewed over the equator), and predicted gas density cross sections and surface frost column density at the R2 (left) and R3 (center) flyby dates and times, and prediction (right) for the 2017 Saturn solstice. (*Top*) Comparison of the observed (data binned) and predicted O_2 densities vs. position from closest approach along the R2 trajectory (x-axis, in Rhea radii, dashed line: closest approach). (*Bottom*) Same for the R3 flyby trajectory. Asterisks denote model-data comparisons with the same flyby date; all other plots show model projections to different dates. The model assumed a 7.2×10^{21} O_2 s^{-1} source rate as required to match the observed peak density on both flybys. An O_2 tail is detected on the dayside, consistent with the anticipated greater dayside scale height. Unlike CO_2, the southern latitudes are not yet cold enough to cryopump O_2 from the southern exosphere on the R2 and R3 flybys dates. However, the modeled O_2 frost cap is exhausted between the R2 and R3 flybys, resulting in an anticipated collapse of the global exosphere, consistent with the reduced O_2 detection on R3.

the 2015 D4 and D5 flybys (see *Teolis and Waite,* 2016) are also consistent with the predictions of the seasonal model. Contrary to the 2011 D3 flyby, during which both O_2 and CO_2 were observed (Fig. 11), only O_2 was detected at D4. The loss of CO_2 between flybys is consistent with the predicted collapse of the CO_2 exosphere between D3 and D4.

The gas permeability through the porous surface regoliths of Dione and Rhea was also found by *Teolis and Waite* (2016) to be a consideration in the seasonal variability of the exospheric models. For the limiting case of impermeable ice on Dione and Rhea (i.e., with no surface regolith porosity) they concluded that the rate of surface sputtering was too great for polar O_2 and CO_2 to stay condensed for an entire winter (~14 Earth years). In this limiting case sputtering prevents condensate from buildup up near the poles, thereby limiting the seasonal effect on the exosphere. Rather, to reproduce the observed seasonal variability of the exospheric O_2 and CO_2 at Dione and Rhea, the models *require* that gas be allowed to diffuse a short distance (no more than a few meters) into the polar regolith to "protect" it against resputtering. Such regolith diffusion may occur, e.g., by diffusion of adsorbed gas on the surface of the regolith grains, together with desorption from one regolith grain to an adjacent grain (*Cassidy and Johnson,* 2005).

The responsiveness of the Dione and Rhea exospheres to transient variations of the magnetospheric sputtering rate is also, according to the model, effectively "damped out" by exospheric adsorption and desorption to/from the surface. The reason is that the dominant gas source to the exosphere is (for most of the seasonal cycle) evaporation of gas previously condensed onto the nightside or polar terrain, with only a small fraction consisting of newly sputtered material. Consequently the average "lifetimes" of O_2 and CO_2, from their original ejection from the surface to their loss to space, are long — years for O_2 and hundreds of years for CO_2 (Table 3) — as molecules spend substantially more time stuck to the surface than they do as a gas. Thus the exospheric density and molecular abundance tends to

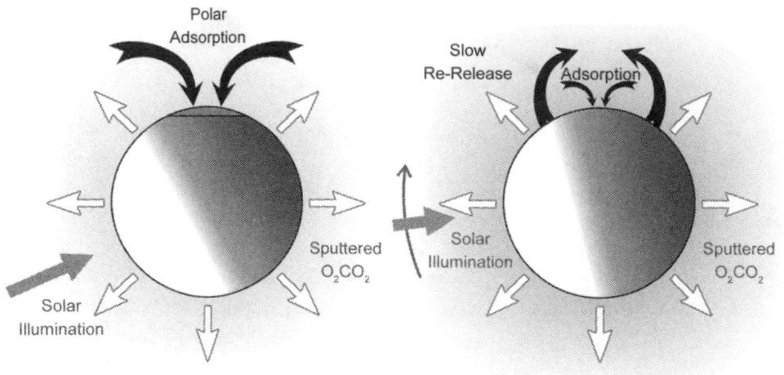

Fig. 9. From *Teolis and Waite* (2016). Schematic showing the mechanism responsible for the exospheric seasonal variability. The exosphere is cryopumped at the solstices onto the cold winter polar surface. As equinox approaches, the Sun rises over the winter latitudes, warming the surface and desorbing the frost cap to produce a transient augmentation of the exosphere.

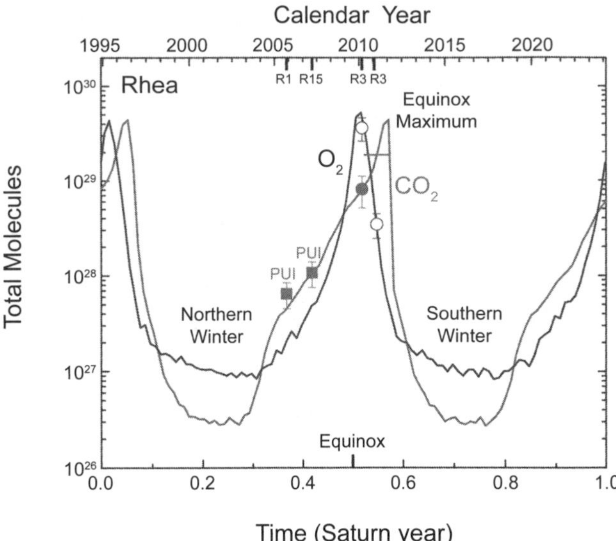

Fig. 10. From *Teolis and Waite* (2016). Rhea modeled total O_2 and CO_2 exospheric molecular abundance vs. time over a Saturn season, showing the periodic exospheric maxima and minima at the equinoxes and solstices, respectively. The INMS (circles) and CAPS IMS (squares) abundances are extrapolations obtained by scaling the model exosphere (with predicted spatial structure at the flyby dates and times) to the measurements. Dash: CO_2 upper limit from R3. The CO_2 frost cap has greater longevity than that of O_2, and therefore the CO_2 exospheric abundance peaks later in time.

reflect the *mean* surface sputtering rate time averaged over many years. An exception to this is near the solstices when desorption from the polar terrain and the exospheric density are minimal. During these time periods, the exospheres *are* expected to respond to magnetospheric variability because a significant fraction of the exospheric gas does consist of newly sputtered material.

5.3. Inefficient Sputtering and Possible Origin of Carbon Dioxide

The Dione and Rhea exospheric densities are sufficiently small that feedbacks from nonlinear exospheric drivers [i.e., (1) intermolecular collisions, (2) secondary sputtering by surface impingent pickup ions, and (3) disruption by the exosphere of the ambient magnetosphere (which sputters the surface)] are not significant effects. For these "linear" exospheres the model exospheric density, spatial distribution, and total molecular abundance simply scale in proportion to the (time-averaged) magnetospheric sputtering rate from the surface. The average exospheric O_2 and CO_2 source rates at Dione and Rhea can therefore be estimated by scaling the exospheric model to obtain agreement with the gas densities and pickup ion fluxes measured by INMS and CAPS on all flybys (Figs. 2, 3, 7, 8, and 10).

However, the yields of O_2 radiolysis and sputtering from irradiated ice vs. projectile species, energy, and ice temperature are already well known from laboratory experiments (section 3). These yields can be used, together with knowledge of the particle flux vs. energy impacting the surface, to estimate the expected O_2 sputtering rates from an irradiated icy moon. For the first time at any icy moon, Cassini's data have enabled these laboratory-based estimates of O_2 from ice to be directly compared to the O_2 source rate obtained by scaling exospheric models to *in situ* spacecraft data.

Remarkably, the *observed* Dione and Rhea O_2 source rates are drastically lower than expected (Table 3) — by factors of ~50 and ~300 — for a pure water ice surface. *Teolis and Waite* (2016) suggested that these low O_2 source rates, together with the production of comparatively substantial CO_2 (Table 3), may be evidence of significant carbonaceous surface refractory sputter-resistant surface impurities. The source of such materials may be carbon-bearing species endogenic to the surface material, or implanted into the surface by chondritic micrometeoritic particles. Carbonaceous impurities may become ultra-concentrated in the topmost "altered" molecular layer (perhaps angstroms thick) of the exposed surface regolith grains as volatile constituents (i.e., H_2O molecules) are preferentially sputtered. The sputtered composition (i.e., O_2 and CO_2, and presumably H_2O and H_2) will reflect the endogenous stoichiometric composition of the surface material below the altered layer (*Johnson,* 1990). However, the O_2 and CO_2 could be radiolyzed and sputtered (in the required stoichiometric proportion) from a surface coated with an atomic scale "lag" of highly oxygenated graphitic material [e.g., graphitic oxide (*Hou et al.,* 2015; *Matsumoto et al.,* 2011)]. Such material is perhaps too thin for spectroscopic detection by Cassini VIMS. For a sufficiently sputter-resistant surface, micrometeorite impacts may in fact control the source rate of sputtered O_2 and H_2O to the exosphere by reexposing fresh water ice at the impact sites (*Killen et al.,* 2004).

Whether such low surface radiolytic O_2 source rates should be expected at other solar system icy satellites is

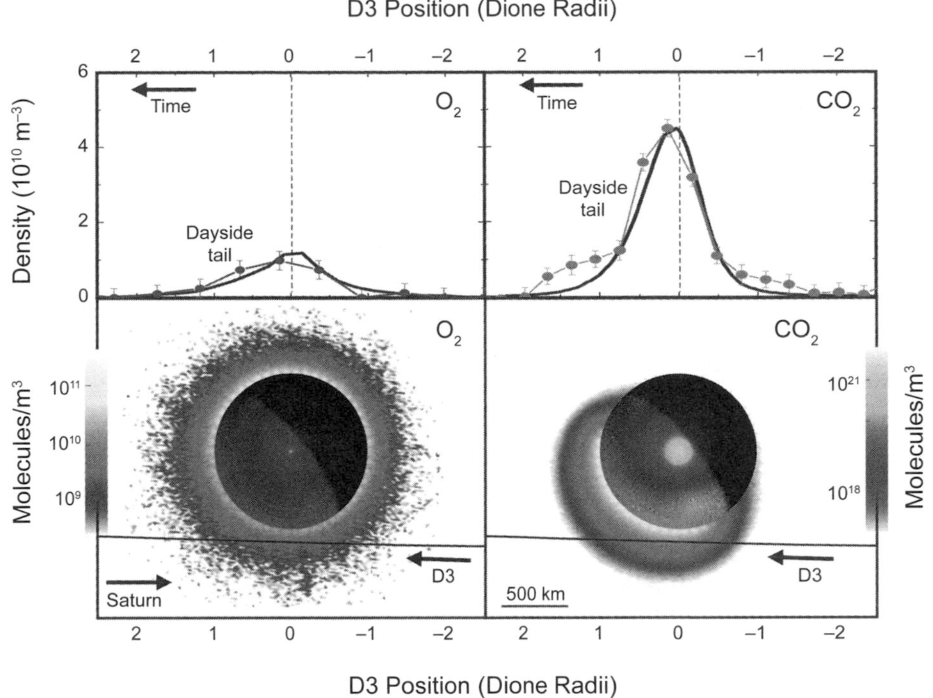

Fig. 11. See Plate 44 for color version. From *Teolis and Waite* (2016). Dione O_2 (left) and CO_2 (right) exospheric model results, showing (bottom) the surface solar illumination (north polar view), and predicted equatorial gas density cross sections and surface frost column density at the December 12, 2011, D3 flyby date and time. (*Top*) Comparison of the observed (blue points: O_2, red points: CO_2, data binned) and predicted (black lines) O_2 and CO_2 densities vs. position from closest approach (dotted lines) along the D3 trajectory (x-axis, in Dione radii). The model assumed 45 and 1.6 × 10^{21} s^{-1} O_2 and CO_2 source rates, as required to match the observed peak density. CO_2 and, to a lesser degree, O_2 are concentrated over the dayside as predicted by the model. As observed at Rhea, the Dione exospheric model predicts seasonal exospheric variability.

an interesting question. At Enceladus or Europa (*Roth et al.,* 2014), continuous fallout of cryovolcanic plume ice grains onto the surface may maintain a "refreshed" surface composition with relatively pure H_2O ice, resulting in O_2 source rates closer to laboratory-based predictions for ice. However, for a small object like Enceladus, the low gravity prevents such O_2 from accumulating in a bound exosphere detectable by INMS.

Conversely, the large masses of the Galilean satellites, 1–2 orders of magnitude greater than Dione and Rhea, results in drastically lower gravitational escape rates for both exospheric O_2 *and* sputtered H_2O. At Ganymede (Europa) only ~20 (30)% of sputtered H_2O molecules escape gravitationally (*Johnson et al.,* 1983) [compared to 90 (83)% at Dione (Rhea) (*Teolis and Waite,* 2016)], and therefore the fallback of these H_2O molecules may maintain fresh H_2O concentrations on their surfaces. For example, at Europa, O_2 source rates (*Milillo et al.,* 2016; *Plainaki et al.,* 2012; *Teolis et al.,* 2017b) inferred from modeling on the basis of HST observations agree within an order of magnitude with the predicted ~10^{26} s^{-1} O_2 source rate from radiolytic sputtering of pure water ice on the surface (*Cassidy et al.,* 2013) [although the presence of sulfates on the trailing hemisphere (*Carlson et al.,* 2005) and possible chlorates on the leading hemisphere (*Brown and Hand,* 2013) complicates the comparison].

5.4. Dione and Rhea as Magnetospheric Material Sources

Measuring the amount and composition of material supplied to Saturn's magnetosphere by sputtering of its icy satellites (*Johnson et al.,* 2008; *Jurac et al.,* 2001) was a major objective of the Cassini mission. From large moons such as Dione and Rhea, this material consists of pickup ions produced from either (1) the gravitationally bound exosphere itself, or (2) directly sputtered neutrals or thermalized exospheric volatiles that escape the moons gravity *before* being ionized. On the basis of the CAPS and INMS data and exospheric modeling examined above, these rates can now be accurately estimated and compared to other magnetospheric material sources (i.e., ring photo-sputtering, Enceladus plumes). Figure 12 shows the estimated time evolution of exospheric loss at Dione and Rhea, and the resulting magnitude and seasonal variability of the pickup sources to the magnetosphere.

Approximately 83 (59)% of the O_2, and 71 (10)% of the CO_2 lost from Dione (Rhea) escapes gravitationally to

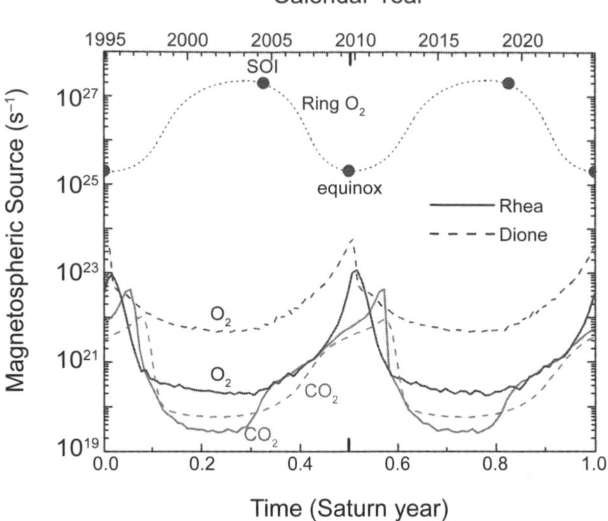

Fig. 12. From *Teolis and Waite* (2016). Dione (dashed lines) and Rhea (solid lines) O_2 and CO_2 total exospheric loss rates, estimated by scaling the model exospheric abundances (Fig. 10 for Rhea) to the total O_2 and CO_2 loss rates (Table 3) compared to (filled circles) the ring O_2 source rate from *Tseng et al.* (2013) (dotted line to guide the eye). As shown in Table 3, ~83 (59)% of O_2 and 71 (10)% of CO_2 lost from Dione (Rhea) escapes gravitationally to supply Saturn's magnetosphere with neutral O_2 and CO_2, while ~8 (18)% and ~13 (41)% respectively is converted to O_2^+ and CO_2^+ pickup ions. From Table 2, the conversion factors of O_2 to O^+, O^{++}, and O are –2 (5), 0.1 (0.2), and 17 (37) %, where the negative 2% Dione O^+ value represents net removal of magnetospheric O^+ by charge exchange with O_2 (Table 2, loss channel 1). Note that Dione and Rhea should also be nonseasonal sources of sputtered gravitationally escaping H_2O and radiolytic H_2 (*Teolis et al.*, 2010a). The ring O_2 source has opposite seasons to the icy satellites, peaking at the solstices rather than the equinoxes.

supply Saturn's magnetosphere with neutral O_2 and CO_2. These molecules are eventually ionized and/or dissociated far from the moons out in the magnetosphere, but the rates are slow (see Tables 2 and 3), on the order of weeks to months per molecule. *Teolis and Waite* (2016) estimated (using the rates in Table 2) that ~8 (18)% and ~13 (41)% of the O_2 and CO_2 lost from Dione (Rhea), respectively, is eventually converted to O_2^+ and CO_2^+, with the rest dissociated into fragment ions and neutrals. The CO_2^+ pickup ion source to the magnetosphere is also expected to exhibit a weak ±15% diurnal variability as the pickup ions from the dayside CO_2 exospheres of Dione and Rhea are obstructed by the moons when their daysides are Saturn-facing. Dione and Rhea are also expected to be nonseasonal sources of H_2O^+ and H_2^+ pickup ions (and H_2O and H_2 fragment ions) from exospheric and gravitationally escaping sputtered H_2O and H_2.

It is interesting that the Dione and Rhea exospheres, as well as their pickup ion sources to the magnetosphere, are 180° out of phase (Fig. 12) with the photodesorbed O_2 source from Saturn's rings, which peaks at solstice (rather than equinox) when the solar illumination flux to the rings is maximal (*Christon et al.*, 2013; *Elrod et al.*, 2014; *Tseng et al.*, 2013). However, as shown in Fig. 12, the O_2 source rates to the magnetosphere from Dione and Rhea are 1–3 orders of magnitude below the ring O_2 source, and therefore the rings should dominate the O_2 supply to the magnetosphere [with the possible exception, near equinox, of the vicinity of Rhea's L shell (*Martens et al.*, 2008)].

6. ELECTRIC CURRENT SYSTEMS AT DIONE AND RHEA

6.1. Cassini's Magnetometer: Could It See Exospheres at Dione and Rhea?

The Cassini magnetometer (MAG) is a demonstrated tool for the detection of exospheres in the Saturn system, having been in part responsible for the initial discovery of the Enceladus plume. As exospheric neutral gas produces pickup ions and disrupts the plasma flow through momentum transfer collisions and charge exchange, electrical conductivity and field-aligned currents are produced. These currents generate standing Alfvén waves, or "wings," that are characterized by a "draped" magnetic field perturbation pointed in the plasma (co-rotation) flow direction. This perturbation's magnitude can be used to infer the presence of an atmosphere and to estimate its density, location and extent, and molecular abundance (*Dougherty et al.*, 2006). A major effort was therefore made to detect the exospheres of Saturn's other large icy satellites with the magnetometer, especially since the pre-Cassini exospheric modeling of *Saur and Strobel* (2005) had predicted Dione and Rhea exospheres with sufficient magnitude to enable detection.

However, the exospheric abundances detected by INMS and inferred by CAPS at Dione and Rhea turned out to be lower by 2 orders of magnitude than these early predictions, due to the unexpectedly low rates of O_2 sputtering from their surfaces discussed in section 5.3. At the low exospheric densities on the order of 10^{10} m^{-3} found by CAPS and INMS, the Dione and Rhea exospheres have conductivities too low, by 2 orders of magnitude, for detection by MAG (*Simon et al.*, 2012). At these low exospheric gas densities, Dione and Rhea are expected to be "inert" plasma absorbers merely acting to draw in magnetic field lines to compensate the absorption of plasma pressure. Hence, the detection (discussed below) of significant magnetic field components along the plasma flow direction over high Dione and Rhea latitudes, indicative of substantial *field-aligned* currents, was a surprise, as CAPS and INMS did not find exospheres sufficiently dense to produce such currents.

6.2. What Did Cassini's Magnetometer Detect at Dione?

The first such detection of flow-directed magnetic field perturbations came at Dione in 2005 during the Dione D1 flyby past the edge of the southern flux tube (Fig. 13). These MAG data suggested a Dione exosphere with a molecular abundance of ~6 × 10^{29} molecules (*Simon et al.,* 2011), which is greater than the exospheric abundances suggested by CAPS and INMS (Table 3). Although INMS was favorably pointed during the D1 flyby to allow detection of the exosphere inferred from MAG, no exosphere was detected by INMS

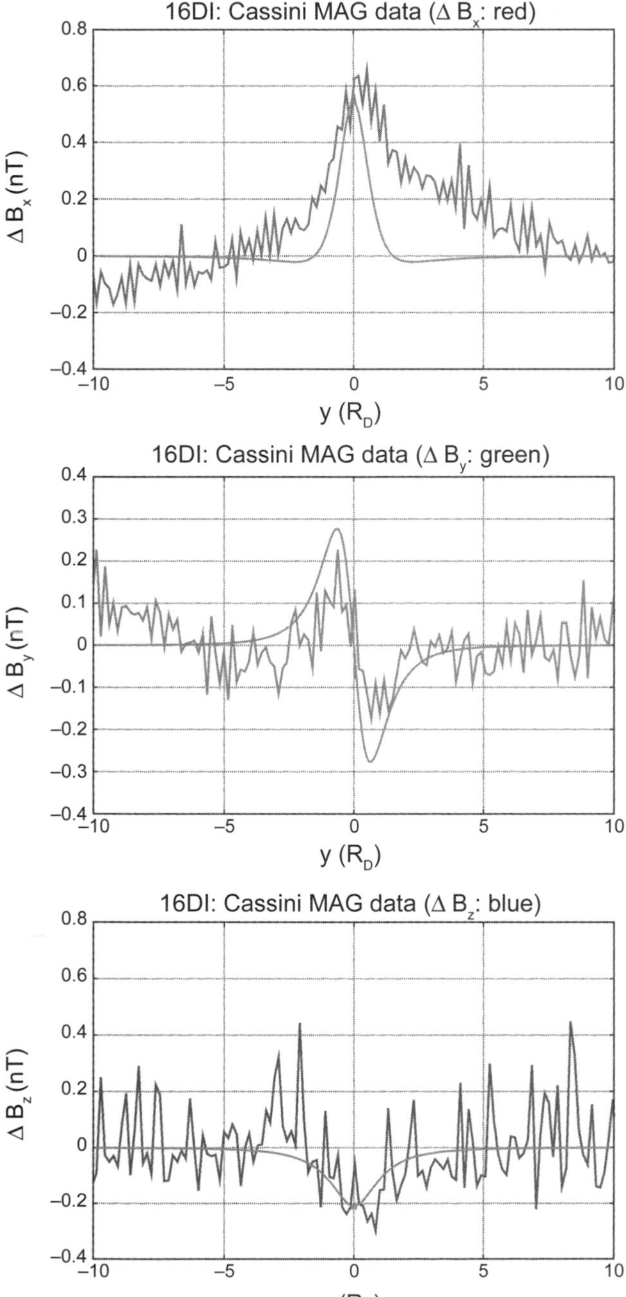

Fig. 13. From *Simon et al.* (2011). Observed magnetic field perturbation and modeling results for the D1 flyby (Table 1). The figure displays the magnetic field components measured during D1 with the background field subtracted (ΔBx: red, ΔBy: green, ΔBz: blue). The curves show the magnetic field perturbations obtained from the *Simon et al.* (2011) analytical exospheric interaction model.

during D1. This implied an upper limit of only 1 (0.8) × 10^{29} O_2 (CO_2) on Dione's exospheric molecular abundance (*Teolis and Waite,* 2016), significantly below the ~6 × 10^{29} molecules suggested by MAG. A high abundance would also be contrary to expectations if the exosphere is seasonal, given that the exosphere would have been near its minimum on the D1 flyby date close to Saturn solstice (Table 1).

Among possible explanations of this apparent contradiction between instruments is an Enceladus-like water vapor plume in Dione's northern hemisphere, i.e., the hemisphere opposite to the D1 flyby. Plume water vapor ejected below the 0.5 km s^{-1} Dione escape speed [as at Europa (*Roth et al.,* 2014)] would fall back to the surface and recondense, producing a detectable magnetic signature, while avoiding detection by INMS. However, such a plume (*Buratti et al.,* 2011) would have to be intermittent, as neither INMS nor CAPS detected evidence of plumes during other Dione flybys, including D4 and D5 (Table 1), which were in the north. Alternatively, Dione's flow-directed perturbations may be unrelated to the exosphere. Although investigation is still ongoing, Dione's magnetic signature may turn out to have a similar explanation, as we now discuss, as those found at Rhea.

6.3. Rhea's Wake Alfvén Wings

At Rhea, a new understanding of fundamental aspects of the interaction of plasma absorbing bodies with space plasma has come from MAG data from the R1 flyby through the plasma wake, from the R2 and R3 polar flyby's through Rhea's flux tube, and from two distant downstream flybys below and above Rhea's orbital plane. The results of these flybys and their interpretation has been the subject of a series of recent papers (*Khurana et al.,* 2017; *Roussos et al.,* 2008; *Simon et al.,* 2012; *Teolis et al.,* 2014), which we discuss here and in section 6.4.

We start with Fig. 14 showing the Rhea MAG data from the R2 and R3 flux tube traversals. On top of the expected enhancement of the southward (–Z directed) saturnian magnetic field, the readings show unusual perturbations in the Y (Saturn) direction and, especially, the X (plasma flow) direction. These data were first analyzed by *Simon et al.* (2012), who pointed out that the exospheric densities on the order of 10^{10} m^{-3} detected by INMS on the same flybys (Fig. 6) are 2 orders of magnitude too low to produce flow-directed field perturbations of the observed magnitude.

However, two distinctive aspects of Rhea's plasma interaction (and, to a lesser degree, Dione's) are (1) the large ion plasma beta (~1.1) of the magnetospheric water group ions and (2) the large gyroradius of these ions (~400 km average) relative to the diameter (1528 km) of the moon. Under these conditions ion kinetic and gyroradius effects are important. One such effect is rapid infilling within a few Rhea radii (*Roussos et al.,* 2008) of Rhea's downstream plasma wake, due to thermal ion motion along the magnetic field lines. *Simon et al.* (2012) applied hybrid magnetohydrodynamic simulations [particle ions, fluid electrons (*Müller et al.,*

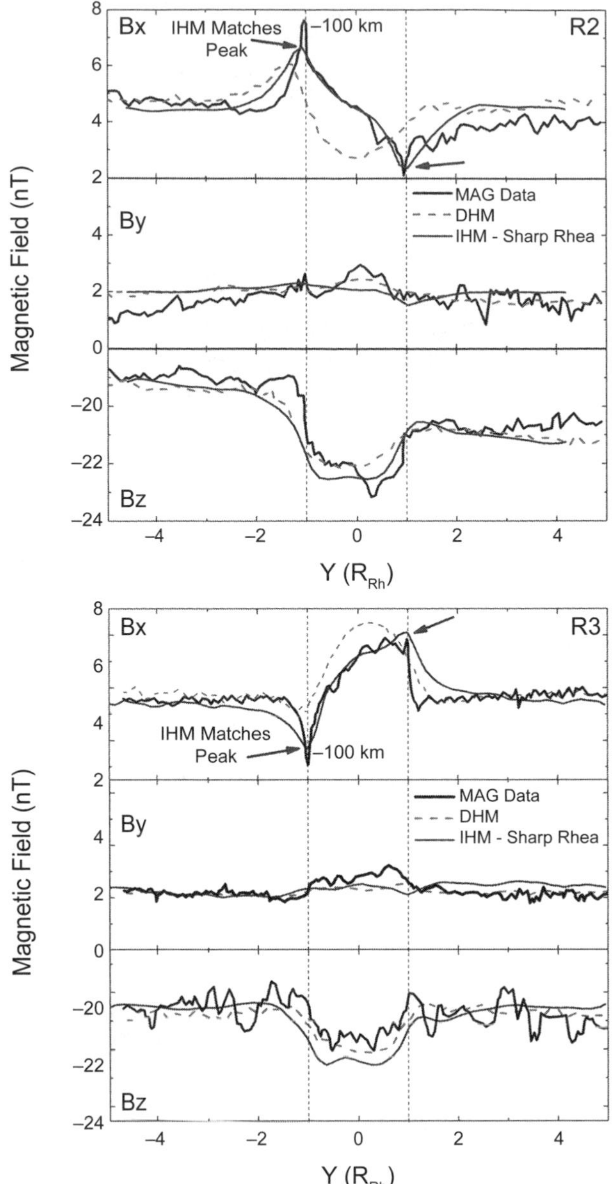

Fig. 14. From *Teolis et al.* (2014). Rhea MAG $|\vec{B}_x|$, $|\vec{B}_y|$, and $|\vec{B}_z|$ measurements vs. spacecraft Y-coordinate for the R2 (top) and R3 (bottom) flybys through the northern and southern flux tubes, with the differential [dashed line, from *Simon et al.* (2012)] and integral [lighter solid line, from *Teolis et al.* (2014)] hybrid models (IHM) also shown. The R2 $|\vec{B}_x|$ prediction resembles the model of a "wire" current in the flux tube. The IHM successfully predicts the magnitude and alignment of the $|\vec{B}_x|$ peaks with the edge of the flux tube at Y ~ ± 1.

2011)] to capture these kinetic aspects and obtained partial agreement with the MAG data as shown in Fig. 14.

Their simulations revealed a new type of current system, resulting from the combination of (1) ion kinetic effects in a "hot" plasma and (2) relative motion between the plasma and an "inert" absorbing body. They discovered that the infilling of the wake and the resulting plasma pressure gradient directed back toward Rhea causes diamagnetic current closure across the wake and perpendicular to the co-rotation flow. The plasma pressure gradient exerts a force directed toward Rhea and "mimics" a real exosphere by generating a field-aligned Alfvénic current system that extracts momentum from the co-rotating plasma outside the wake and transfers it to the wake. The Alfvén wings from the wake (*Khurana et al.,* 2017) produce flow-directed field perturbations north and south of Rhea's equatorial plane, detectable at the locations of R2 and R3 and also during the two distant [102 R_H and 54 R_H altitude on June 3 and October 17, 2010 (*Khurana et al.,* 2012)] downstream flybys (Fig. 15). The observation of this never-before-observed current system nicely illustrates the utility of Cassini's multiple instruments (i.e., INMS, CAPS) to provide exospheric constraints on the interpretation of magnetic field data.

6.4. Rhea's "Thermoelectric" Flux Tube Current

However, it turns out that even this picture is not complete. *Another* field-aligned current system is also generated in Rhea's flux tube that, remarkably, does not in principal require motion relative to the plasma.

This current system was investigated by *Teolis et al.* (2014), who were motivated by what seemed at the time to be a minor mystery: the sharp peaks detected (Fig. 14) in the X-component of the magnetic field at the edges of Rhea's flux tube. As Rhea lacks an intrinsic magnetic field, the flux tube is a roughly north-to-south cylindrical region of saturnian magnetic field lines that intersect Rhea's surface. Magnetic field peaks were detected on both the northern and southern R2 and R3 flybys as shown in Fig. 14. The reversal of sign in these peaks on opposite sides of the flux tube indicates a field twisting about the flux tube, consistent with a "wire" of flux tube current flowing away, north and south, from Rhea (*Santolik et al.,* 2011; *Simon et al.,* 2012). However, their small ~100-km thickness, much less than the ~500-km ion inertial length, makes these features too small to be Alfvénic. As shown in Fig. 14, the *Simon et al.* (2012) hybrid model did not fit these features.

A solution to this problem is to solve the magnetic diffusion equation numerically with a set of integral equations, which *Teolis et al.* (2014) did at Rhea by developing an alternative implementation of the hybrid model's algorithm. The more standard approach of solving a differential equation runs into difficulty at Rhea's surface because the surface is a discontinuity, meaning that the electrical resistivity exhibits a "sharp" jump from negligible outside Rhea, to a high value inside. Unlike planetary bodies surrounded by a significant "smooth" atmosphere, Rhea lacks an exosphere sufficiently dense to smooth its resistivity profile. However, it is trivial to integrate across the surface step. The integration method is therefore more successful at implementing the surface current balance condition at the "sharp" surface of a plasma-absorbing object, such as Rhea, with a "negligible" exosphere. This condition requires equality of time-averaged ion and electron flows into the nonconducting surface, which

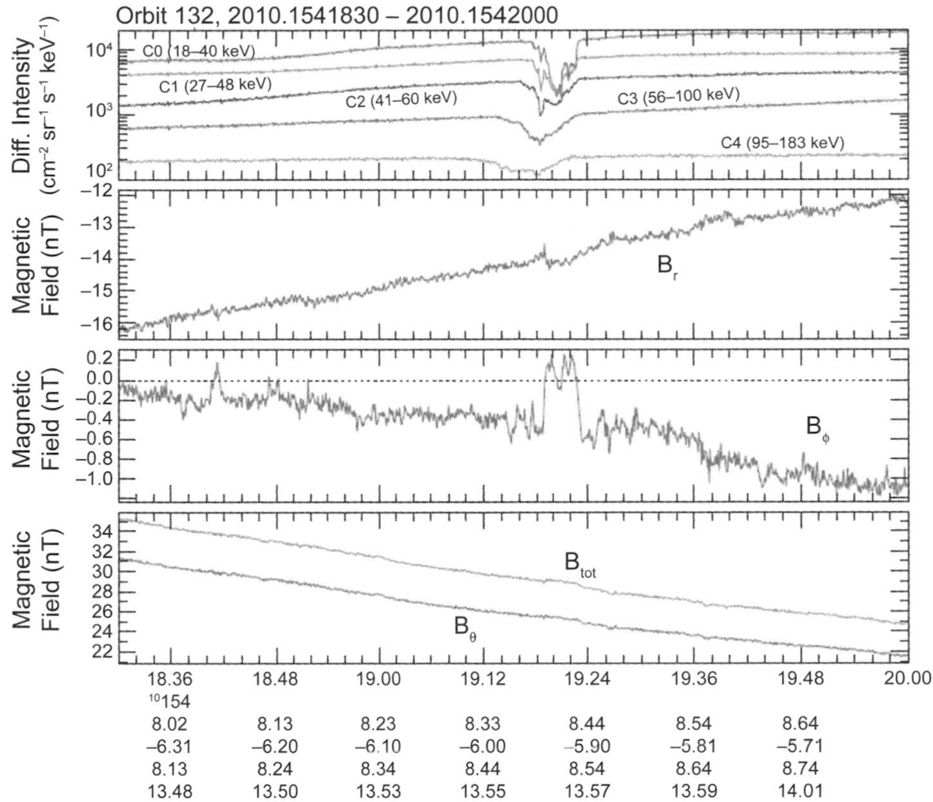

Fig. 15. From *Khurana et al.* (2012). MAG and MIMI-LEMMS data from Cassini's June 3, 2010, distant Rhea flyby, through Rhea's southern Alfvén wing, with closest approach 102 R_H from Rhea and 66 R_H, south of the equatorial plane. Top panel: Electron fluxes in the MIMI-LEMMS energy channels, showing a flux dropout as energetic electrons are scattered by the Alfvén wing. Bottom three panels: Magnetic field magnitude (bottom), and directional components in Saturn-centric spherical coordinates. The abrupt increase of B_ϕ between ~19:15 and ~19:18 shows the rotation of the field into the corotation flow direction as Cassini traversed the Alfvén wing.

is a necessary requirement for surface charge balance (*Nordheim et al.,* 2014; *Roussos et al.,* 2010). As shown in Fig. 14, this "integral hybrid model" is largely successful at capturing the sharp perturbations at the edge of the flux tube, thereby revealing the physics responsible for the flux tube current.

The mechanism, shown schematically in Fig. 16 (and modeled in Fig. 17), results from (1) the requirement to balance ion and electron currents at Rhea's nonconductive sharp surface, and (2) the difference of gyroradius between the positive (ion) and negative (electron) charge carriers. The ions that (owing to their larger gyroradius) discharge into the plasma absorber's surface from all directions are balanced by the electrons that are constrained to flow along the magnetic field lines. More magnetic field lines connect to the planetary equator due to the oblique angle (Fig. 16), and therefore the electron current is most intense at the edge of the flux tube. The resulting magnetic field perturbation circles about the flux tube and is maximal at the edges, as observed by the Cassini spacecraft.

Teolis et al. (2014) gave a detailed discussion of the plasma physics responsible for the generation, maintenance, and transmission of the currents up the field lines and propose that the field perturbations may consist of oblique semi-standing whistler waves, perhaps similar to those observed leading planetary bow shocks (*Gary and Mellott,* 1985; *Orlowski et al.,* 1995), or to so-called "whistler wings" as seen in laboratory experiments (*Stenzel,* 1999) and magnetized asteroids (*Omidi et al.,* 2002; *Simon et al.,* 2006). A fundamental property of this flux tube current system is its proportionality only to the ion flux into Rhea, i.e., the plasma density times the ion thermal speed — and its independence of the motion of the moon's conductive atmosphere through the magnetosphere.

Such a motion-independent current system is a stark departure from the classic "induction" picture, wherein motion of a conductor (e.g., the atmosphere) through the planetary magnetic field drives field-aligned currents. In driving induction currents, a planetary atmosphere acts something like an electrical generator: a device for converting the energy of mechanical *motion* to electric power. The mechanism illustrated in Fig. 16 is quite different; it involves no atmosphere nor relative motion to the magnetosphere. *Only*

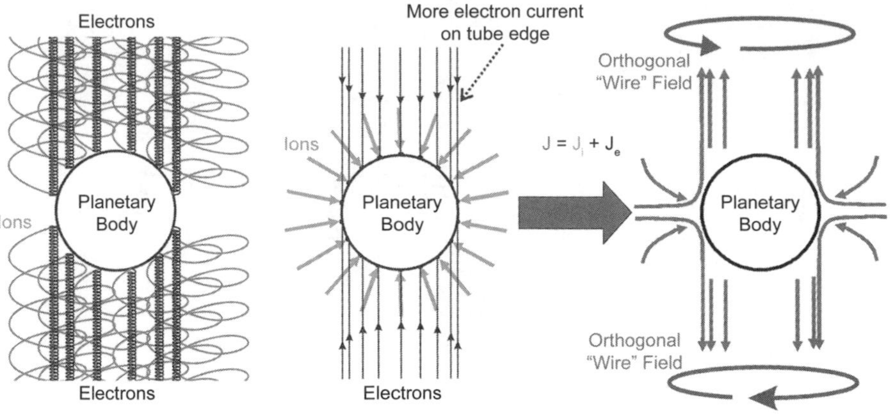

Fig. 16. From *Teolis et al.* (2014). Schematic of the Rhea flux tube current system. Ions discharge into the body approximately uniformly (the figure neglects for simplicity the leading/trailing hemispherical difference due to co-rotation flow), while electrons are confined to flow along the field lines due to the smaller gyroradius. The requirement to balance ion and electron currents on the sharp surface yields a flux tube current flowing away from Rhea. The current is maximum on the flux tube edge due to the oblique angle of the magnetic field to the planetary equator.

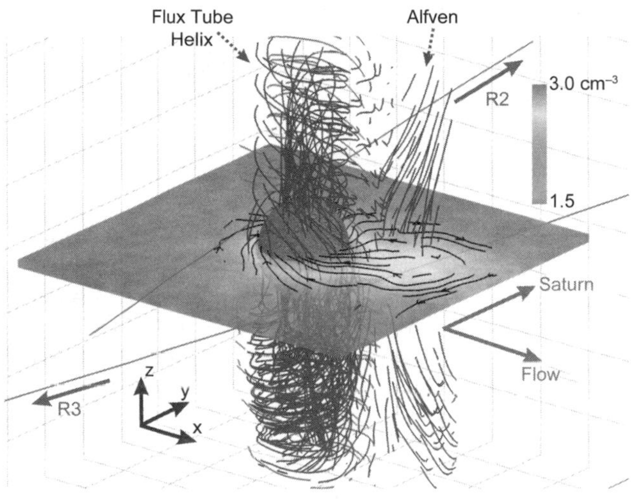

Fig. 17. See Plate 45 for color version. From *Teolis et al.* (2014); their model prediction of the magnetic field perturbation (i.e., total field minus constant $\vec{B}_0$ background) with southward and northward perturbations colored blue and red. Equatorial cross section shows the plasma density (blue-to-red: increasing density) and the diamagnetic current spiraling into Rhea and circling the wake (black streamlines). The helical perturbation is due to the flux tube electron current. Contrary to the Alfvén disturbance from the wake, the flux tube feature has negligible draping angle to the ambient magnetic field, consistent with the high whistler group velocities on the order of 10^3 km s^{-1}.

the flow of hot plasma into the "cold" absorbing object is required to generate the flux tube current. Rhea's flux tube current therefore has much more in common with a heat engine — a device for converting *heat* energy to useful work. A thermoelectric device operates on a similar principal by converting the heat flow between two materials at different temperatures to an electric current.

Hence, a spectacular, albeit unexpected outcome of Cassini's exospheric measurements at Rhea has been the discovery from the magnetometer that, in addition to the classic inductively driven current systems, plasma-absorbing objects may also generate field-aligned currents "thermoelectrically." Currents of this type require only absorption of hot plasma by the body and do not, in principal, require relative motion between the object and the plasma. *Teolis et al.* (2014) analyzed the conditions — i.e., specifically the particle gyroradii and inertial lengths and the size of the plasma absorber — under which such flux tube currents may form. They concluded that this previously unknown thermoelectric type flux tube current system should occur in a wide range of space plasma environments, and thus represents a fundamental aspect of the interaction of airless astrophysical bodies with space plasmas.

7. THE QUESTION OF NEGATIVE PICKUP IONS

A currently unresolved question at both Rhea and Dione is the source of negative pickup ions, which were observed by CAPS ELS during the Rhea R1 and Dione D2 flybys as shown in Figs. 18 and 19. Applying the same ion backtracing approach used to identify the positive pickup from Rhea as CO_2^+ (section 4.1), *Teolis et al.* (2010b) found Rhea's negative pickup ions to be consistent with O^- ions. The source of the ions is not completely understood. One obvious process, dissociative electron attachment to exospheric O_2, does not have sufficient cross sections (*Itikawa,* 2008) at the density and energy of the magnetospheric plasma electrons to account for the O^- flux inferred from ELS. As shown in Fig. 18, the pickup ion energy increases slightly vs. time during the flyby, causing the backtracked ion trajectories to focus onto Rhea's surface. This could imply a surface mediated negative ion generation process, such as dissociative

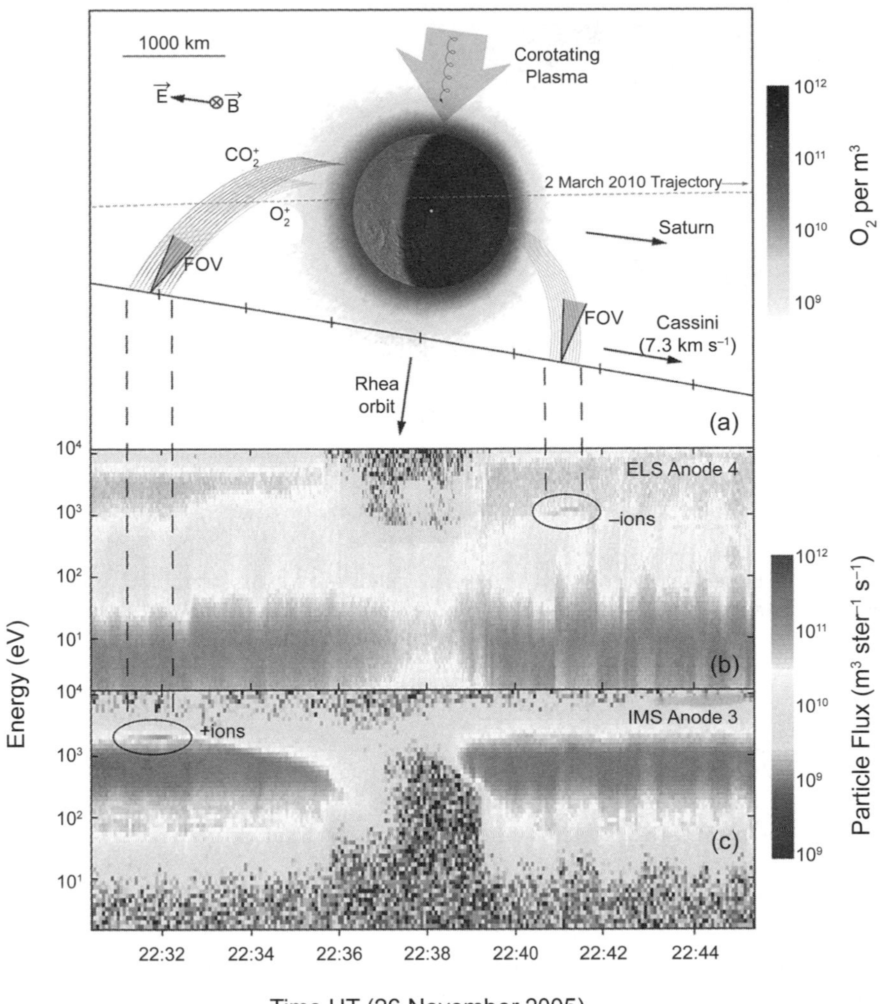

Fig. 18. From *Teolis et al.* (2010b). **(a)** Diagrammatic Rhea north polar view with the R1 flyby trajectory. The timescale is matched to **(b)** and **(c)**. The day/night hemispheres are shown during closest approach. A model prediction of the O_2 density (226 km south cross section) is also shown. The O_2^+ and O^- (orange) and CO_2^+ trajectories (blue) are those required to enter anodes 4 and 3 (32) of the CAPS electron spectrometer (ELS) and ion mass spectrometer (IMS) at the time and energy of the ion signatures. **(b)** ELS negative particle flux spectrogram from anode 4 (20° FOV), which had optimal pointing. Negative pickup ions are indicated by the sharp feature near 22:41 UT (±0.35 min) and 1.14 (±0.15) keV over the electron background. **(c)** Positive ions from IMS anode 3: Pickup ions produce the sharp 22:32 UT (±0.5 min), 2.06 (±0.2) keV signature over the background of (mostly) co-rotating H^+/W^+.

attachment reactions involving O_2 or CO_2 adsorbed onto the night hemisphere (*Teolis and Waite,* 2016), the hemisphere from which the negative ions are observed to originate (Fig. 18). Another hypothesis (*Desai et al,* 2018) is that ions somewhat heavier than O^-, such as C_2^- or C_2H^- ejected from carbonaceous surface material, might also contribute to Rhea's negative pickup ion signature.

Dione's negative ion signature, as observed by CAPS ELS during the D2 flyby through the co-rotation plasma wake, is shown in Fig. 19. The moon's plasma wake is clearly visible in the ELS data as a depletion of thermal and suprathermal electrons due to geometric shielding from Dione. About halfway through the wake passage and near the closest approach to Dione, an additional negative particle feature was observed. This feature was observed at ~600 eV in CAPS-ELS anodes 3 and 4, which were pointed toward the disk of the moon and sensitive to pitch angles near 90° (Nordheim et al., in preparation; *Tokar et al.,* 2012). This was suggestive of newly born pickup ions, which would be expected to have a ring velocity distribution centered around a 90° pitch angle as discussed by *Tokar et al.* (2012) (section 4). However, as discussed by Nordheim et al. (in preparation) and *Tokar et al.* (2012), the D2 flyby also coincided with a number of intense enhancements in the background hot plasma population, characteristic of magnetospheric injection events (*Hill et al.,* 2005; *Rymer et al.,* 2009). The presence of the injection events complicated the analysis of the D2 observations as it was not immediately

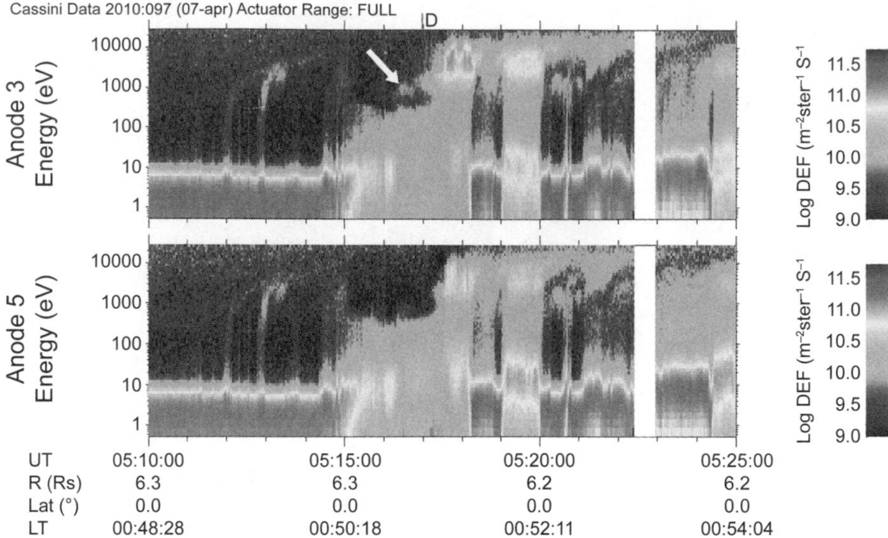

Fig. 19. CAPS ELS anodes 3 (top) and 5 (bottom) flux energy spectrograms vs. time during the 2010 500-km Dione D2 flyby (Table 1). Flux drops between ~5:15:00 and ~5:18:00 UT as Cassini traverses Dione's plasma wake. Abrupt changes on the outbound portion of the flyby past ~5:18:00 UT are magnetospheric injection events. White arrow shows a feature consistent with O^- pickup ions (Nordheim et al., in preparation).

clear whether the observed features were related to Dione or simply due to background magnetospheric processes. To investigate this feature further, Nordheim et al. (in preparation) carried out back-tracing calculations to investigate the possible origin of these negative particles. These calculations were found to be consistent with a source of negative pick-up ions near Dione's surface with a mass in the range 15–28 Da. Based on these findings, it was concluded that the observed negative particle signature was likely due to O^- pickup ions, possibly originating from Dione's tenuous exosphere or surface, similar to the observation of O^- pickup ions reported by *Teolis et al.* (2010b) at Rhea.

8. CONCLUSIONS

A major milestone of Cassini's extended tour of Saturn's icy moons has been the discovery of exospheres around Rhea and Dione (*Simon et al.,* 2011; *Teolis et al.,* 2010b; *Tokar et al.,* 2012), produced by magnetospheric radiolysis and sputtering of the surfaces, and held in place by gravity. The exospheres of Dione and Rhea have a similar density and chemical composition, consisting of carbon dioxide and molecular oxygen, which suggests similar exospheric mechanisms at both moons. Over multiple close flybys of these moons at different dates and over different geographic locations, Cassini CAPS and INMS measurements have led to a remarkably detailed new understanding of these icy satellite exospheres, revealing that (1) the exospheres stick to the cold nightside and polar surfaces, (2) the gas diffuses into the regolith, and (3) the exospheres are seasonal.

Exospheres at Saturn's icy moons had long been predicted (*Jurac et al.,* 2001; *Saur and Strobel,* 2005; *Sittler et al.,* 2004), first on the basis of laboratory studies confirming radiolytic O_2 to be a major component of the material sputtered from ion and electron irradiated water ice (*Brown et al.,* 1982) and then from the discovery by the HST of O_2 exospheres at the irradiated Galilean icy moons (*Cunningham et al.,* 2015; *Hall et al.,* 1995, 1998). However, exospheres at the saturnian satellites proved more difficult to detect than anticipated, with several unsuccessful attempts at exospheric detection at Rhea (*Jones et al.,* 2007a) by way of UVIS occultation and MIMI energetic neutral atom data. The explanation was later provided by Cassini CAPS and INMS on several close Rhea and Dione flybys. These data showed that the amount of molecular oxygen sputtered from the surfaces of Dione and Rhea into their exospheres is lower than anticipated on the basis of known sputtering and radiolysis yields from water ice in the laboratory.

Another surprise came from MAG, which measured magnetic field perturbations at Rhea (*Simon et al.,* 2012) and Dione (*Simon et al.,* 2011) too intense to be explained by exospheres of the low density found by CAPS and INMS (*Teolis and Waite,* 2016). At Rhea the apparent inconsistency was resolved by two new fundamental discoveries in the plasma physics of moon-magnetosphere interactions: (1) the formation of Alfvén wings from the plasma wake (*Khurana et al.,* 2014; *Simon et al.,* 2012), and (2) a field-aligned thermoelectric current system formed due to plasma absorption by the moon (*Teolis et al.,* 2014). Analysis of the Dione results is ongoing but, in addition to possible similar current systems as seen at Rhea, another possibility is that MAG detected episodic plume activity, although this remains unconfirmed.

The presence of exospheric CO_2 in almost equal amounts to O_2 at both Rhea and Dione was not expected, and suggests significant carbon-bearing molecular constituents in the surface material, a finding at odds with VIMS infrared spectra indicating a surface of almost pure H_2O. Together with the observation of surface sputtering rates far below expectations, such a high carbon content suggests a possible optically thin refractory altered surface layer built up by preferential sputtering of H_2O (*Teolis and Waite,* 2016), possibly only nanometers thick. This raises the possibility that micrometeorite impacts (which may themselves be a source of carbon to the surface material) may expose fresh ice from below this surface layer, and possibly control the surface sputtering rate, and the O_2 source, to the exosphere.

These fundamental discoveries, resulting from only a handful of Cassini spacecraft flybys, may have major ramifications for the understanding of exospheric physics and magnetospheric interactions at other solar system icy moons. One major lesson learned is the significant potential of surface condensation, regolith diffusion, and evaporation to drastically affect the exospheric structure and dynamics at these objects. The Jupiter system, unlike that of Saturn, is almost aligned with the ecliptic, and thus the exospheres of the Galilean satellites do not exhibit significant seasons. However, the nightsides of Ganymede and Europa cool to 90 and 86 K (*Orton et al.,* 1996; *Spencer et al.,* 1999), which is cold enough to condense CO_2 out of their nightside exospheres. Plume vapor fallout at Europa might contain a number of "semi-volatile" species including CO_2, which may condense near the plumes at night and "burst" off the surface at dawn (*Teolis et al.,* 2017b). A similar effect might also occur seasonally at Enceladus, since the plume vapor contains an ~0.5% molecular abundance of CO_2 (*Bouquet et al.,* 2015), which may condense near Enceladus' surface vents over the 14-year winter night and then rapidly desorb at southern spring (an event that, unfortunately, Cassini will not be present to witness). Cassini's finding of inefficient sputtering at Dione and Rhea, together with the substantial relative abundance of CO_2 in their exospheres, suggests there is much to be learned at other icy moons about the relationship of sputtered flux and exospheric composition to the surface composition. Understanding this relationship (e.g., through laboratory studies) will be an important issue over the next decade, as future spacecraft arrive at Ganymede and Europa and begin making *in situ* measurements of their exospheric composition. Finally, the physics responsible for Rhea's wake Alfvén wings and its thermoelectric flux tube current will certainly occur at other solar system planetary bodies. How this manifests, e.g., in the magnetic field topology near Jupiter's moons, has hardly been explored and much work remains to be done. However, clearly, the spectacular results from Cassini's encounters of Dione and Rhea augurs well for the science potential for future spacecraft missions to other solar system icy moons.

REFERENCES

Bagenal F. and Delamere P. A. (2011) Flow of mass and energy in the magnetospheres of Jupiter and Saturn. *J. Geophys. Res., 116,* A05209.

Baragiola R. A., Atteberry C. L., Bahr D. A., and Peters M. (1999) Reply. *J. Geophys. Res., 104,* 14183–14187.

Barbosa D. D. (1993) Theory and observations of electromagnetic ion cyclotron waves in Saturn's inner magnetosphere. *J. Geophys. Res., 98(A6),* 9345–9350, DOI: 10.1029/93JA00476.

Barth C. A., Hord C. W., Stewart A. I. F., Pryor W. R., Simmons K. E., McClintock W. E., Ajello J. M., Naviaux K. L., and Aiello J. J. (1997) Galileo ultraviolet spectrometer observations of atomic hydrogen in the atmosphere at Ganymede. *Geophys. Res. Lett., 24,* 2147.

Boring J. W., Johnson R. E., Reimann C. T., Garret J. W., Brown W. L., and Marcantonio K. J. (1983) Ion-induced chemistry in condensed gas solids. *Nucl. Instr. Meth., 218(1–3),* 707–711.

Bouquet A., Mousis O., Waite J. H., and Picaud S. (2015) Possible evidence for a methane source in Enceladus' ocean. *Geophys. Res. Lett., 42,* 1334.

Bringa E. M. and Johnson R. E. (2002) Coulomb explosion and thermal spikes. *Phys. Rev. Lett., 88,* 165501.

Broadfoot A. L. et al. (1979) Extreme ultraviolet observations from Voyager 1 encounter with Jupiter. *Science, 204,* 979–982.

Brown M. E. (2001) Potassium in Europa's atmosphere. *Icarus, 151,*190.

Brown M. E. and Hand K. P. (2013) Salts and radiation products on the surface of Europa. *Astron. J., 145,* 110.

Brown M. E. and Hill R. E. (1996) Discovery of an extended sodium atmosphere around Europa. *Nature, 380,* 229.

Brown W. L., Lanzerotti L. J., Poate J. M., and Augustyniak W. M. (1978) 'Sputtering' of ice by MeV light ions. *Phys. Rev. Lett., 40(15),* 1027–1030.

Brown W. L., Augustyniak W. M., Lanzerotti L. J., Johnson R. E., and Evatt R. (1980) Linear and nonlinear processes in the erosion of H_2O ice by fast light ions. *Phys. Rev. Lett., 45(20),* 1632–1635.

Brown W. L., Augustyniak W. M., Simmons E., Marcantonio K. J., Lanzerotti L. J., Johnson R. E., Boring J. W., Reimann C. T., Foti G., and Pirronello V. (1982) Erosion and molecule formation in condensed gas films by electronic energy loss of fast ions. *Nucl. Instr. Meth., 198(1),* 1–8.

Buratti B. J., Faulk S. P., Mosher J., Baines K. H., Brown R. H., Clark R. N., and Nicholson P. D. (2011) Search for and limits on plume activity on Mimas, Tethys, and Dione with the Cassini Visual Infrared Mapping Spectrometer (VIMS). *Icarus, 214,* 534.

Burch J. L., Goldstein J., Lewis W. S., Young D. T., Coates A. J., Dougherty M. K., and Andre N. (2007) Tethys and Dione as sources of outward-flowing plasma in Saturn's magnetosphere. *Nature, 447,* 833.

Burger M. H. and Johnson R. E. (2004) Europa's neutral cloud: Morphology and comparisons to Io. *Icarus, 171(2),* 557–560.

Cabrera-Trujillo R., Ohrn Y., Deumens E., and Sabin J. R. (2004) Absolute differential and total cross sections for direct and charge-transfer scattering of keV protons by O_2. *Phys. Rev. A, 70,* 042702.

Carlson R. W. (1999) A tenuous carbon dioxide atmosphere on Jupiter's Moon Callisto. *Science, 283,* 820.

Carlson R. W., Bhattacharyya J. C., Smith B. A., Johnson T. V., Hidayat B., Smith S. A., Taylor G. E., O'Leary B., and Brinkmann R. T. (1973) An atmosphere on Ganymede from its occultation of SAO 186800 on 7 June 1972. *Science, 182,* 53–55.

Carlson R. W., Anderson M. S., Mehlman R., and Johnson R. E. (2005) Distribution of hydrate on Europa: Further evidence for sulfuric acid hydrate. *Icarus, 177,* 461.

Carlson R. W., Calvin W. M., Dalton J. B., Hansen G. B., Hudson R. L., Johnson R. E., McCord T. B., and Moore M. H. (2009) Europa's surface composition. In *Europa* (R. T. Pappalardo et al., eds.), pp. 283–327. Univ. of Arizona, Tucson.

Cassidy T. A. (2008) Europa's tenuous atmosphere. Ph.D. thesis, University of Virginia, Charlottesville.

Cassidy T. A. and Johnson R. E. (2005) Monte Carlo model of sputtering and other ejection processes within a regolith. *Icarus, 176(2),* 499–507.

Cassidy T. A., Johnson R. E., Mcgrath M. A., Wong M. C., and Cooper J. F. (2007) The spatial morphology of Europa's near-surface O_2 atmosphere. *Icarus, 191,* 755.

Cassidy T. A., Paranicas C., Shirley J. H., Dalton J. B., Teolis B. D., Johnson R. E., Kamp L., and Hendrix A. R. (2013) Magnetospheric ion sputtering and water ice grain size at Europa. *Planet. Space Sci., 77,* 64.

Christon S. P., Hamilton D. C., DiFabio R. D., Mitchell D. G., Krimigis S. M., and Jontof-Hutter D. S. (2013) Saturn suprathermal O_2^+ and mass-28^+ molecular ions: Long-term seasonal and solar variation. *J. Geophys. Res.–Space Phys., 118(6),* 3446–3463.

Cruikshank D. P., Owen T. C., Ore C. D., Geballe T. R., Roush T. L., de Bergh C., Sandford S. A., Poulet F., Benedix G. K., and Emery J. P. (2005) A spectroscopic study of the surfaces of Saturn's large satellites: H_2O ice, tholins, and minor constituents. *Icarus, 175,* 268.

Cruikshank D. P. et al. (2010) Carbon dioxide on the satellites of Saturn: Results from the Cassini VIMS investigation and revisions to the VIMS wavelength scale. *Icarus, 206,* 561.

Cunningham N. J., Spencer J. R., Feldman P. D., Strobel D. F., France K., and Osterman S. N. (2015) Detection of Callisto's oxygen atmosphere with the Hubble Space Telescope. *Icarus, 254,* 178.

Desai R. T., Taylor S. A., Regoli L. H., Coates A. J., Nordheim T. A., Cordiner M. A., Teolis B. D., Thomsen M. F., Johnson R. E., Jones G. H., Cowee M. M., and Waite J. H. (2018) Cassini CAPS identification of pickup ion compositions at Rhea. *Geophys. Res. Lett., 45,* DOI: 10.1002/2017GL076588.

Dialynas K., Krimigis S. M., Mitchell D. G., Hamilton D. C., Krupp N., and Brandt P. C. (2009) Energetic ion spectral characteristics in the saturnian magnetosphere using Cassini/MIMI measurements. *J. Geophys. Res., 114,* A01212.

Dougherty M. K. et al. (2004) The Cassini Magnetic Field Investigation. *Space Sci. Rev., 114,* 331.

Dougherty M. K., Khurana K. K., Neubauer F. M., Russel C. T., Saur J., Leisner J. S., and Burton M. E. (2006) Identification of a dynamic atmosphere at Enceladus with the Cassini Magnetometer. *Science, 311,* 1406.

Elrod M. K., Tseng W.-L., Woodson A. K., and Johnson R. E. (2014) Seasonal and radial trends in Saturn's thermal plasma between the main rings and Enceladus. *Icarus, 242,* 130.

Esposito L. W. et al. (2004) The Cassini Ultraviolet Imaging Spectrograph Investigation. *Space Sci. Rev., 115(1),* 299–361, DOI: 10.1007/s11214-004-1455-8.

Feldman P. D., McGrath M. A., Strobel D. F., Moos H. W., Retherford K. D., and Wolven B. C. (2000) HST/STIS ultraviolet imaging of Polar Aurora on Ganymede. *Astrophys. J., 535,* 1085–1090.

Fleshman B. L., Delamere P. A., Bagenal F., and Cassidy T. (2012) The roles of charge exchange and dissociation in spreading Saturn's neutral clouds. *J. Geophys. Res., 117,* E05007.

Gary S. P. and Mellott M. M. (1985) Whistler damping at oblique propagation: Laminar shock precursors. *J. Geophys. Res., 90,* 99.

Hall D. T., Strobel D. F., Feldman P. D., McGrath M. A., and Weaver H. A. (1995) Detection of an oxygen atmosphere on Jupiter's moon Europa. *Nature, 373(6516),* 677–679.

Hall D. T., Feldman P. D., McGrath M. A., and Strobel D. F. (1998) The far-ultraviolet oxygen airglow of Europa and Ganymede. *Astrophys. J., 499(1),* 475–481.

Hansen C. J., Shemansky D. E., and Hendrix A. R. (2005) Cassini UVIS observations of Europa's oxygen atmosphere and torus. *Icarus, 176,* 305.

Hansen C. J. et al. (2011) The composition and structure of the Enceladus plume. *Geophys. Res. Lett., 38,* L11202.

Hibbits C. A., McCord T. B., and Hansen G. B. (2000) Distributions of CO_2 and SO_2 on the surface of Callisto. *J. Geophys. Res., 105,* 22541.

Hibbits C. A., Pappalardo R. T., Hansen G. B., and McCord T. B. (2003) Carbon dionxide on Ganymede. *J. Geophys. Res., 108,* 5036.

Hill T. W., Rymer A. M., Burch J. L., Crary F. J., Young D. T., Thomsen M. F., Delapp D., André N., Coates A. J., and Lewis G. R. (2005) Evidence for rotationally driven plasma transport in Saturn's magnetosphere. *Geophys. Res. Lett., 32(14),* L14S10, DOI: 10.1029/2005GL022620.

Hou W.-C., Chowdhury I., Goodwin D. G., Henderson W. M., Fairbrother D. H., Bouchard D., and Zepp R. G. (2015) Photochemical transformation of graphene oxide in sunlight. *Environ. Sci. Tech., 49,* 3435.

Howett C. J. A., Spencer J. R., Hurford T., Verbiscer A., and Segura M. (2014) Thermophysical property variations across Dione and Rhea. *Icarus, 241,* 239.

Howett C. J. A., Spencer J. R., Hurford T., Verbiscer A., and Segura M. (2016) Thermal properties of Rhea's poles: Evidence for a meter-deep unconsolidated subsurface layer. *Icarus, 272,* 140.

Huebner W. F., Keady J. J., and Lyon S. P. (1992) Solar photo rates for planetary atmospheres and atmospheric pollutants. *Astrophys. Space Sci., 195,* 1.

Itikawa Y. (2008) Cross sections for electron collisions with oxygen molecules. *J. Phys. Chem. Ref. Data 38(1),* DOI: 10.1063/1.3025886.

Johnson R. E. (1990) *Energetic Charged-Particle Interactions with Atmospheres and Surfaces*. Springer-Verlag, New York. 232 pp.

Johnson R. E. (1999) Comment on "Laboratory studies of the optical properties and stability of oxygen on Ganymede" by Raul A. Baragiola and David A. Bahr. *J. Geophys. Res., 104,* 14179–14182.

Johnson R. E. and Jesser W. A. (1997) O_2/O_3 microatmospheres in the surface of Ganymede. *Astrophys. J. Lett., 480,* L79.

Johnson R. E. and Quickenden T. I. (1997) Photolysis and radiolysis of water ice on outer solar system bodies. *J. Geophys. Res., 102,* 10985–10996.

Johnson R. E., Lanzerotti L. J., and Brown W. L. (1982) Planetary applications of ion induced erosion of condensed-gas frosts. *Nucl. Instr. Meth., 198(1),* 147–157.

Johnson R. E., Boring J. W., Reimann C. T., Barton L. A., Sieveka E. M., Garrett J. W., and Farmer K. R. (1983) Plasma ion-induced molecular ejection on the Galilean satellites: Energies of ejected molecules. *Geophys. Res. Lett., 10,* 892.

Johnson R. E., Cooper P. D., Quickenden T. I., Grieves G. A., and Orlando T. M. (2005) Production of oxygen by electronically induced dissociations in ice. *J. Chem. Phys., 123,* 184715.

Johnson R. E. et al. (2006) Production, ionization and redistribution of O_2 in Saturn's ring atmosphere. *Icarus, 180,* 393.

Johnson R. E., Fama M. A., Liu M., Baragiola R. A., Sittler E. C., and Smith H. T. (2008) Sputtering of ice grains and icy satellites in Saturn's inner magnetosphere. *Planet. Space Sci., 56,* 1238.

Jones G. H. et al. (2007a) The dust halo of Saturn's largest icy moon, Rhea. *Science, 319,* 1380.

Jones G. H. et al. (2007b) The dust halo of Saturn's largest icy moon, Rhea. *Science (Supporting Online Material), 319,* 1380.

Jurac S., Johnson R. E., Richardson J. D., and Paranicas C. (2001) Satellite sputtering in Saturn's magnetosphere. *Planet. Space Sci., 49(3–4),* 319–326.

Khurana K. K., Krupp N., Kivelson M. G., Roussos E., and Dougherty M. K. (2012) Cassini's flyby through Rhea's distant Alfvén wing. *European Planet. Sci. Congr. 2012,* EPSC2012-2309.

Khurana K. K., Roussos E., Krupp N., Holmstrom M., Lindkvist J., Dougherty M. K., and Russel C. T. (2014) How are Rhea's Alfvén wings generated. *Eos Trans. AGU, 95,* Fall Meeting 2014, Abstract #P43B-3986.

Khurana K. K., Fatemi S., Kindkvist J., Roussos E., Krupp N., Holmstrom M., Russell C. T., and Dougherty M. K. (2017) The role of plasma slowdown in the generation of Rhea's Alfvén wings. *J. Geophys. Res., 122,* 1778–1788, DOI: 10.1002/2016JA023595.

Killen R. M., Sarantos M., Potter A. E., and Reiff P. (2004) Source rates and ion recycling rates for Na and K in Mercury's atmosphere. *Icarus, 171,* 1.

Kliore A. J., Hinson D. P., Flasar F. M., and Cravens T. E. (1997) The ionosphere of Europa from Galileo radio occultations. *Science, 277,* 355.

Kliore A. J., Anabtawi A., Herrera R. G., Asmar S. W., Nagy A. F., Hinson D. P., and Flasar F. M. (2002) Ionosphere of Callisto from Galileo radio occultation observations. *J. Geophys. Res., 107(A11),* SIA 19-11 to SIA 19-17, DOI: 10.1029/2002JA009365.

Lagg A., Krupp N., Woch J., and Williams J. D. (2003) In-situ observations of a neutral gas torus at Europa. *Geophys. Res. Lett., 30(11),* 1556.

Landau A., Allin E. J., and Welsh H. L. (1962) The absorption spectrum of solid oxygen in the wavelength region from 12,000 A to 3300 A. *Spectrochim. Acta, 18(1),* 1–19.

Lanzerotti L. J., Brown W. L., Poate J. M., and Augustyniak W. M. (1978) On the contribution of water products from Galilean satellites to the jovian magnetosphere. *Geophys. Res. Lett., 5(2),* 155, DOI: 10.1029/GL005i002p00155.

Leblanc F., Potter A. E., Killen R. M., and Johnson R. E. (2005) Origins of Europa Na cloud and torus. *Icarus, 178,* 367.

Liang M.-C., Lane B. F., Pappalardo R. T., Allen M., and Yung Y. L. (2005) Atmosphere of Callisto. *J. Geophys. Res., 110,* E02003, DOI: 10.1029/2004JE002322.

Loeffler M. J., Raut U., Vidal R. A., Baragiola R. A., and Carlson R. W. (2006) Synthesis of hydrogen peroxide in water ice by ion irradiation. *Icarus, 180,* 265.

Luna H., McGrath C., Shah M. B., Johnson R. E., Liu M., Latimer C. J., and Montenegro E. C. (2005) Dissociative charge exchange and ionization of O_2 by fast H^+ and O^+ ions: Energetic ion interactions in Europa's oxygen atmosphere and neutral torus. *Astrophys. J., 628,* 1086.

Marconi M. L. (2007) A kinetic model of Ganymede's atmosphere. *Icarus, 190,* 155.

Martens H. R., Reisenfeld D. B., Williams J. D., Johnson R. E., and Smith H. T. (2008) Observations of molecular oxygen ions in Saturn's inner magnetosphere. *Geophys. Res. Lett., 35,* L20103.

Matsumoto Y., Koinuma M., Ida S., Hayami S., Taniguchi T., Hatakeyama K., Tateishi H., Watanabe Y., and Amano S. (2011) Photoreaction of graphene oxide nanosheets in water. *J. Phys. Chem. C, 115,* 19280.

Mauk B. H. et al. (2009) Fundamental plasma processes in Saturn's magnetosphere. In *Saturn from Cassini-Huygens* (M. K. Dougherty et al., eds.), p. 281. Springer, New York.

Mauk B. H., Mitchell D. G., Krimigis S. M., Roelof E. C., and Paranicas C. P. (2003) Energetic neutral atoms from a trans-Europa gas torus at Jupiter. *Nature, 421(6926),* 920–922.

McGrath M. A., Jia X., Retherford K., Feldman P. D., Strobel D. F., and Saur J. (2013) Aurora on Ganymede. *J. Geophys. Res., 118,* 2043.

Milillo A., Plainaki C., De Angelis E., Mangano V., Massetti A., Mura A., Orsini S., and Rispoli R. (2016) Analytical model of Europa's O_2 exosphere. *Planet. Space Sci., 130,* 3–13.

Müller J., Simon S., Motschmann U., Schule J., Glassmeier K., and Pringle G. (2011) A.I.K.E.F.: Adaptive hybrid model for space plasma simulations. *Comp. Phys. Commun., 182,* 946.

Noll K. S., Johnson R. E., Lane A. L., Domingue D. L., and Weaver H. A. (1996) Detection of ozone on Ganymede. *Science, 273(5273),* 341–343.

Noll K. S., Roush T. L., Cruikshank D. P., Johnson R. E., and Pendleton Y. J. (1997) Detection of ozone on Saturn's satellites Rhea and Dione. *Nature, 388(6637),* 45–47.

Nordheim T. A., Jones G. H., Rousos E., Leisner J. S., Coates A. J., Kurth W. S., Khurana K. K., Krupp N., Dougherty M. K., and Waite J. H. (2014) Detection of a strongly negative surface potential at Saturn's moon Hyperion. *Geophys. Res. Lett., 41,* 7011.

Omidi N., Blanco-Cano X., Russell C. T., Karimabadi H., and Acuna M. (2002) Hybrid simulations of solar wind interaction with magnetized asteroids: General characteristics. *J. Geophys. Res., 107(A12),* 1487.

Orlowski D. S., Russell C. T., Krauss-Varban D., Omidi N., and Thomsen M. F. (1995) Damping and spectral formation of upstream whistlers. *J. Geophys. Res., 100,* 17117.

Orton G. S., Spencer J. R., Travis L. D., Martin T. Z., and Tamppari L. K. (1996) Galileo photopolarimeter-radiometer observations of Jupiter and the Galilean satellites. *Science, 274,* 389–391.

Pilcher C. B., Ridgway S. T., and McCord T. B. (1972) Galilean satellites: Identification of water frost. *Science, 178(4065),* 1087, DOI: 10.1126/science.178.4065.1087.

Plainaki C., Milillo A., Mura A., Orsini S., and Cassidy T. (2010) Neutral particle release from Europa's surface. *Icarus, 210,* 385.

Plainaki C., Milillo A., Mura A., Orsini S., Massetti S., and Cassidy T. (2012) The role of sputtering and radiolysis in the generation of Europa exosphere. *Icarus, 218,* 956.

Plainaki C., Milillo A., Massetti S., Mura A., Jia X., Orsini S., Mangano V., De Angelis E., and Rosanna R. (2015) The H_2O and O_2 exospheres of Ganymede: The result of a complex interaction between the jovian magnetospheric ions and the icy moon. *Icarus, 245,* 306.

Reimann C. T., Boring J. W., Johnson R. E., Garrett J. W., Farmer K. R., Brown W. L., Marcantonio K. J., and Augustyniak W. M. (1984) Ion-induced molecular ejection from D_2O ice. *Surf. Sci., 147,* 227.

Richardson J. D., Eviatar A., and Siscoe G. L. (1986) Satellite tori at Saturn. *J. Geophys. Res., 91(A8),* 8749–8755, DOI: 10.1029/JA091iA08p08749.

Roth L., Saur J., Retherford K. D., Strobel D. F., Feldman P. D., McGrath M. A., and Nimmo F. (2014) Transient water vapor at Europa's south pole. *Science, 343,* 171.

Roth L., Saur J., Retherford K. D., Strobel D. F., Feldman P. D., McGrath M. A., Spencer J. R., Blöcker A., and Ivchenko N. (2016) Europa's far ultraviolet oxygen aurora from a comprehensive set of HST observations. *J. Geophys. Res., 121(3),* 2143–2170, DOI: 10.1002/2015JA022073.

Roth R., Retherford K. D., Ivchenko N., Schlatter N., Strobel D. F., Becker T. M., and Grava C. (2017) Detection of a hydrogen corona in HST Lyα images of Europa in transit of Jupiter. *Astron. J., 153(2),* 67, DOI: 10.3847/1538-3881/153/2/67.

Roussos E., Muller J., Simon S., Boswetter A., Motschmann U., Krupp N., Franz M., Woch J., Khurana K. K., and Dougherty M. K. (2008) Plasma and fields in the wake of Rhea: 3-D hybrid simulation and comparison with Cassini data. *Ann. Geophys., 26,* 619.

Roussos E., Krupp N., Kruger H., and Jones G. H. (2010) Surface charging of Saturn's plasma-absorbing moons. *J. Geophys. Res., 115,* A08225.

Rymer A. M. et al. (2009) Cassini evidence for rapid interchange transport at Saturn. *Planet. Space Sci., 57(14–15),* 1779, DOI: 10.1016/j.pss.2009.04.010.

Santolik O., Gurnett D. A., Jones G. H., Schippers P., Crary F. J., Leisner J. S., Hospodarsky G. B., Kurth W. S., Russell C. T., and Dougherty M. K. (2011) Intense plasma wave emissions associated with Saturn's moon Rhea. *Geophys. Res. Lett., 38,* L19204.

Saur J. and Strobel D. F. (2005) Atmospheres and plasma interactions at Saturn's largest inner icy satellites. *Astrophys. J. Lett., 620,* L115.

Saur J., Strobel D. F., and Neubauer F. M. (1998) Interaction of the jovian magnetosphere with Europa: Constraints of the neutral atmosphere. *J. Geophys. Res., 103,* 19947.

Schippers P. et al. (2008) Multi-instrument analysis of electron populations in Saturn's magnetosphere. *J. Geophys. Res., 113,* A07208.

Shemansky D. E., Yung Y. L., Liu X., Yoshi J., Hansen C. J., Hendrix A. R., and Esposito L. W. (2014) A new understanding of the Europa atmosphere and limits on geophysical activity. *Astrophys. J., 797(2),* 84.

Shematovich V. I., Johnson R. E., Cooper J. F., and Wong M. C. (2005) Surface-bounded atmosphere of Europa. *Icarus, 173,* 480.

Shi M., Baragiola R. A., Grosjean D. E., Johnson R. E., Jurac S., and Schou J. (1995) Sputtering of water ice surfaces and the production of extended neutral atmospheres. *J. Geophys. Res., 100(E12),* 26387–26395.

Sieglaff D. R., Lindsay B. G., Merrill R. L., Smith K. A., and Stebbings R. F. (2000) Absolute differential and integral cross sections for charge transfer of keV O^+ ions with O_2. *J. Geophys. Res., 105,* 10631.

Sieveka E. M. and Johnson R. E. (1982) Thermal- and plasma-induced molecular redistribution on the icy satellites. *Icarus, 51(3),* 528–548.

Sigmund P. (1969) Theory of sputtering. I. Sputtering yield of amorphous and polycrystalline targets. *Phys. Rev., 184,* 383.

Simon S., Bagdonat T. B., Motschmann U., and Glassmeier K.-H. (2006) Plasma environment of magnetized asteroids: A 3-D hybrid simulation study. *Ann. Geophys., 24,* 407.

Simon S., Saur J., Neubauer F. M., Wennmacher A., and Dougherty M. K. (2011) Magnetic signatures of a tenuous atmosphere at Dione. *J. Geophys. Res., 38,* L15102.

Simon S., Kriegel H., Saur J., Wennmacher A., Neubauer F. M., Roussos E., Motschmann U., and Dougherty M. K. (2012) Analysis of Cassini magnetic field observations over the poles of Rhea. *J. Geophys. Res., 117,* A07211.

Sittler E. C., Johnson R. E., Jurac S., Richardson J. D., McGrath M. A., Crary F., Young D. T., and Nordholt J. E. (2004) Pickup ions at Dione and Enceladus: Cassini plasma spectrometer simulations. *J. Geophys. Res., 109,* A01214.

Smith E. J. and Tsurutani B. T. (1983) Saturn's magnetosphere: Observations of ion cyclotron waves near the Dione L shell. *J. Geophys. Res., 88(A10),* 7831–7836, DOI: 10.1029/JA088iA10p07831.

Smyth W. H. and Marconi M. L. (2006) Europa's atmosphere, gas tori, and magnetospheric implications. *Icarus, 181,* 510.

Spencer J. R. (1987) Thermal segregation of water ice on the Galilean satellites. *Icarus, 69(2),* 297–313.

Spencer J. R. and Calvin W. M. (2002) Condensed O_2 on Europa and Callisto. *Astron. J., 124,* 3400–3403.

Spencer J. R., Calvin W. M., and Person M. J. (1995) CCD spectra of the Galilean satellites: Molecular oxygen on Ganymede. *J. Geophys. Res., 100,* 19049–19056.

Spencer J. R., Tamppari L. K., Martin T. Z., and Travis L. D. (1999) Temperatures on Europa from Galileo PPR: Nighttime thermal anomalies. *Science, 284,* 1514–1516.

Stenzel R. L. (1999) Whistler waves in laboratory plasmas. *J. Geophys. Res., 104,* 14379.

Stephan K. et al. (2012) The saturnian satellite Rhea as seen by Cassini VIMS. *Planet. Space Sci., 61,* 142.

Teolis B. D. and Waite J. H. (2016) Dione and Rhea seasonal exospheres revealed by Cassini CAPS and INMS. *Icarus, 272,* 277.

Teolis B. D., Vidal R. A., Shi J., and Baragiola R. A. (2005) Mechanisms of O_2 sputtering from water ice by keV ions. *Phys. Rev. B, 72,* 245422.

Teolis B. D., Loeffler M. J., Raut U., Fama M., and Baragiola R. A. (2006) Ozone synthesis on the icy satellites. *Astrophys. J. Lett., 644,* L141.

Teolis B. D., Shi J., and Baragiola R. A. (2009) Formation, trapping, and ejection of radiolytic O_2 from ion-irradiated water ice studied by sputter depth profiling. *J. Chem. Phys., 130,* 134704.

Teolis B. D. et al. (2010a) Cassini finds an oxygen-carbon dioxide atmosphere at Saturn's icy moon Rhea. *Science (Supporting Online Material), 330,* 1813.

Teolis B. D. et al. (2010b) Cassini finds an oxygen-carbon dioxide atmosphere at Saturn's icy moon Rhea. *Science, 330,* 1813.

Teolis B. D., Sillanpää I., Waite J. H., and Khurana K. K. (2014) Surface current balance and thermoelectric whistler wings at airless astrophysical bodies: Cassini at Rhea. *J. Geophys. Res., 119,* 8881.

Teolis B. D., Plainaki C., Cassidy T., and Raut U. (2017a) Water ice O_2, H_2 and H_2O_2 radiolysis yields for any projectile species, energy or temperature: A model for icy astrophysical bodies. *J. Geophys. Res., 122,* DOI: 10.1002/2017JE005285.

Teolis B. D., Wyrick D. Y., Bouquet A., Magee B. A., and Wate J. H. (2017b) Plume and surface feature structure and compositional effects on Europa's global exosphere: Preliminary Europa mission predictions. *Icarus, 284,* 18.

Thomsen M. F. (2013) Saturn's magnetospheric dynamics. *Geophys. Res. Lett., 40,* 5337.

Thomsen M. F., Reisenfeld D. B., Delapp D., Tokar R. L., Young D. T., Crary F. J., Sittler E. C., McGraw M. A., and Williams J. D. (2010) Survey of ion plasma parameters in Saturn's magnetosphere. *J. Geophys. Res., 115,* A10220.

Tokar R. L. et al. (2005) Cassini observations of the thermal plasma in the vicinity of Saturn's main rings and the F and G rings. *Geophys. Res. Lett., 32,* L14S04.

Tokar R. L., Johnson R. E., Thomsen M. F., Sittler E. C., Coates A. J., Wilson R. J., Crary F. J., Young D. T., and Jones G. H. (2012) Detection of exospheric O_2^+ at Saturn's moon Dione. *J. Geophys. Res., 39,* L03105.

Tseng W.-L., Ip W.-H., Johnson R. E., Cassidy T. A., and Elrod M. K. (2010) The structure and time variability of the ring atmosphere and ionosphere. *Icarus, 206,* 382.

Tseng W.-L., Johnson R. E., and Elrod M. K. (2013) Modeling the seasonal variability of the plasma environment in Saturn's magnetosphere between main rings and Mimas. *Planet. Space Sci., 77,* 126.

Vidal R. A., Bahr D., Baragiola R. A., and Peters M. (1997) Oxygen on Ganymede: Laboratory studies. *Science, 276(5320),* 1839–1842.

Volwerk M. and Khurana K. K. (2010) Ion pick-up near the icy Galilean satellites. In *Pickup Ions Throughout the Heliosphere and Beyond: Proceedings of the 9th Annual International Astrophysics Conference,* pp. 263–269. AIP Conf. Proc. 1302, DOI: 10.1063/1.3529982.

Volwerk, M., Kivelson M. G., and Khurana K. K. (2001) Wave activity in Europa's wake: Implications for ion pickup. *J. Geophys. Res., 106,* 26033.

Waite J. H. et al. (2004) The Cassini Ion and Neutral Mass Spectrometer (INMS) investigation. *Space Sci. Rev., 114,* 113.

Waite J. H. et al. (2005) Oxygen ions observed near Saturn's A ring. *Science, 307,* 1260–1262.

Watson C. C. (1981) The sputter-generation of planetary coronae: Galilean satellites of Jupiter. *Proc. Lunar Planet. Sci. 12B,* pp. 1569–1583.

Williams I. D., Geddes J., and Gilbody H. B. (1984) Electron capture, loss and axcitation in collisions of H^+, H(1s), H(2s) and H^- in atomic oxygen. *J. Phys. B: Atom. Mol. Phys., 17,* 1547.

Wilson R. J., Tokar R. L., Henderson M. G., Hill T. W., Thomsen M. F., and Pontius D. H. (2008) Cassini plasma spectrometer thermal ion measurements in Saturn's inner magnetosphere. *J. Geophys. Res., 113,* A12218.

Wilson R. J., Tokar R. L., Kurth W. S., and Persoon A. M. (2010) Properties of the thermal ion plasma near Rhea as measured by the Cassini plasma spectrometer. *J. Geophys. Res., 115,* A05201.

Young D. T. et al. (2005) Composition and dynamics of plasma in Saturn's magnetosphere. *Science, 307(5713),* 1262–1266.

Yung Y. L. and McElroy M. B. (1977) Stability of an oxygen atmosphere on Ganymede. *Icarus, 20,* 97.

Thomas P. C., Tiscareno M. S., and Helfenstein P. (2018) The inner small satellites of Saturn and Hyperion. In *Enceladus and the Icy Moons of Saturn* (P. M. Schenk et al., eds.), pp. 387–408. Univ. of Arizona, Tucson, DOI: 10.2458/azu_uapress_9780816537075-ch019.

The Inner Small Satellites of Saturn, and Hyperion

P. C. Thomas
Cornell University

M. S. Tiscareno
SETI Institute

P. Helfenstein
Cornell University

The small (<150 km mean radius) satellites that orbit Saturn interior to Iapetus occupy distinct dynamical niches and have physical properties peculiar to each niche. Morphologies range from nearly lunar-like cratered surfaces to smooth surfaces indicative of fluid-like behavior. Results of interaction with external materials range from surficial color and albedo effects to deep regolith coverings. Mean densities can be as low as ~300 kg m^{-3}. Although morphologies correlate with dynamical positions, in large measure it remains unclear how these correlations arise.

1. INTRODUCTION

Saturn's collection of satellites, at least 62 in number, can be classified in different ways and in varying detail. The broadest classification comprises two groups: the regular moons, which orbit near Saturn's equatorial plane from within the ring system at about 2 Saturn radii (Rs) to Iapetus at 61 Rs, and the irregular satellites, which have orbital inclinations throughout the unit sphere and orbital distances ranging from ~190 Rs to over 400 Rs. The outer irregular satellites are treated in the chapter by Denk *et al.* in this volume. Over 95% of the mass of the regular satellites is contained in Titan; most of the remaining mass is in the mid-sized (200–700-km radii) moons, which are sufficiently large to have relaxed to nearly equilibrium ellipsoidal forms. Scattered at positions from within the main rings to beyond Titan, a retinue of smaller (radii <150 km) regular moons can be placed in separate dynamical categories; these moons are the subject of this chapter.

In this chapter we review the data available and the basic physical information on the small satellites, arranged by dynamical position. We then summarize information relating to interior structures. This is followed by an overview of the impact cratering record. Then in section 6 we present data on albedos, colors, and their possible implications. We then discuss the orbital characteristics and rotational properties. Finally, in section 8, we list some of the remaining open questions.

Hyperion and several of the other small satellites were discovered before spacecraft exploration, but essentially all information on the small satellites derives from spacecraft observations, and the greatest part of that information comes from the Cassini mission, especially from the Imaging Science Subsystem (ISS) (*Porco et al.,* 2004), the Visual and Infrared Mapping Spectrometer (VIMS) (*Brown et al.,* 2004), and the Cassini Ultraviolet Imaging Spectrograph (UVIS) (*Esposito et al.,* 2004). The quality and volume of data are constrained by the geometry of the occasional passes of the Cassini spacecraft in the vicinity of these satellites. Cassini's orbital period and inclination varied greatly during the mission, and data-taking resources were easily oversubscribed in the periapse periods, the times relevant for observing these objects. Thus the coverage of longitude, phase angle, and resolution is neither uniform nor continuous over the satellite surfaces. Figure 1 shows views of all the small inner satellites at differing, globally useful, scales.

Table 1 gives basic orbital information and the best ISS pixel scales (as of late 2016) for the satellites as a crude indicator of the kinds of data returned. The Narrow Angle Camera (NAC) has a pixel scale of 6×10^{-6}, or 600 m at 100,000-km range (*Owen,* 2003). Many combinations of filters, from UV to infrared, are available (*Porco et al.,* 2004), and provide broadband color information. VIMS spectral mapping is at a spatial scale 50× coarser than that of the NAC, but provides information on compositions and grain size not possible from the ISS filter combinations.

Details of the methods of shape modeling, mapping, photometry, and rotational states can be found in *Thomas et al.* (2013); spectral methods are discussed in *Brown et al.* (2004) and *Buratti et al.* (2005, 2010).

Fig. 1. Survey of the inner small satellites and Hyperion. Images (Cassini Narrow Angle Camera, NAC) are sized individually considering the detail of interest and available resolution. Note individual scale bars in kilometers. First row, left to right: Pan: N1867604669, Daphnis: N1863267232, Atlas: N1870698933, Prometheus: N1643263159, Pandora: N18670790629. Second row, left to right: Janus: N1627319647, Epimetheus: N1575363079, Methone (top): N1716192103, Pallene middle, two views: with Saturn shine: N1694657087, and in transit: N1665947247, Aegaeon (bottom right of second row): N1643264379. Third row left to right: Telesto (top): N1507754947, Calypso (bottom): N1644755335, Helene: N1687120624, Hyperion: N1506376183.

TABLE 1. Orbital parameters and best imaging.

Object	a (km)	a (Rs)	e	I (°)	Period (d)	Best Res (km)
Pan	133585	2.21			0.575	1.269
Daphnis	136504	2.26			0.594	0.438
Atlas	137666	2.28	0.001	0.003	0.602	0.190
Prometheus	139378	2.31	0.002	0.007	0.613	0.201
Pandora	141713	2.35	0.004	0.05	0.629	0.312
Epimetheus	151452*	2.51	0.010	0.35	0.695	0.154
Janus	151452*	2.51	0.007	0.16	0.695	0.178
Aegaeon	167491	2.78	0.0003	0.002	0.808	0.091
Methone	194230	3.22	0.001	0.01	1.010	0.027
Anthe	197650	3.28	0.001	0.01	1.036	5.3
Pallene	212283	3.52	0.004	0.18	1.154	0.220
Telesto	294675	4.88	0.0002	1.2	1.888	0.063
Calypso	294675	4.88	0.0005	1.5	1.888	0.136
Polydeuces	377400	6.26	0.019	0.17	2.737	0.411
Helene	377400	6.26	0.007	0.21	2.737	0.042
Hyperion	1480000	24.5	0.1	1.1	21.28	0.012

* Both Janus and Epimetheus trade places around this central semimajor axis every 4.00 yr, Janus with an amplitude of ±11 km, Epimetheus with an amplitude of ±38 km.

2. SIZES, SHAPES, AND MEAN DENSITIES OF THE INNER SMALL SATELLITES AND HYPERION

Summary results for the sizes and mean densities are given in Table 2. Some derived quantities are given in Table 3. We include Phoebe in Table 2 for completeness; other discussion of Phoebe is in the chapter by Denk et al. in this volume. Of the inner satellites, Janus and Epimetheus have the highest mean densities, ~640 kg m^{-3}. The lowest mean density of these satellites may be at or below 310 kg m^{-3} (Methone and Pallene); surface gravity on most of these objects is below 10 mm s^{-2}. This small acceleration means that on the Moon, equivalent regolith pressures are reached at roughly 2 orders of magnitude smaller depths. The net accelerations can vary greatly across the surfaces of those low-density, rapidly spinning, tidally affected objects. Assuming that these objects are homogeneous and using the current nominal mean densities, we find that all have inwardly directed gravity vectors, but accelerations (and escape velocities) can be very low on the sub- and anti-Saturn regions.

Mean density for the inner small satellites with determined masses or well-constrained geophysically inferred mean densities are shown in Fig. 2. The co-orbitals' densities are distinct from the shepherds', and are likely distinct from those of Methone and Pallene. Hyperion is less dense than Janus and Epimetheus and lies well within the collective mean densities of the small satellites. The weak radial trend in density from ring satellites to the co-orbitals is almost entirely dependent upon the co-orbital density values. Probably the most significant result from the data in Fig. 2 is the near identity of the densities of members of the shepherd and co-orbital pairs, and the suggestion that the members of the ring-arc objects might also be more similar to each other than to any other group. The similarity of the densities of the shepherds is interesting, especially if a significant portion of Prometheus has been stripped off a core that in some way is distinct from the outer parts (section 3).

3. SURFACE FEATURES BY DYNAMICAL GROUP

3.1. Main-Ring Satellites: Pan, Daphnis, Atlas

Pan and Daphnis orbit within gaps (respectively 320 km and 35 km wide) in the outer part of the main rings, while Atlas orbits ~900 km off the outer edge of the main rings. These moons have dynamical interactions with nearby gap edges in the rings that help define both the masses of the satellites and properties of the rings (*Porco et al.,* 2005; *Weiss et al.,* 2009). Pan and Atlas have unusual equatorial bulges (Fig. 1) that likely result from enhanced accretion of ring particles at those latitudes [both moons are much larger in mean radius than the vertical thickness of the rings (*Charnoz et al.,* 2007)]. These bulges constitute more than 10% of each satellite's volume; in contrast, the equatorial ridge on Iapetus (*Ip,* 2006) is on the order of 0.1% the volume of that satellite. Daphnis has a less prominent equatorial bulge, and suggestions of elongated ridges at intermediate latitudes. The best available data are for Atlas and show three main characteristics: a smooth equatorial band; a rougher, lumpy, "core" at latitudes above ~30°; and five subdued impact

TABLE 2. Sizes and mean densities.

Object	Group	a	b	c	Rm	Mass (g)	ρ (kg m^{-3})
Pan	Ring	17.2 ± 1.9	15.5 ± 1.3	10.4 ± 0.8	14.1 ± 1.3	0.495E + 19 ± 0.70E + 18	424 ± 137
Daphnis	Ring	4.7 ± 0.8	4.4 ± 0.4	2.6 ± 0.7	3.8 ± 0.6	0.770E + 17 ± 0.15E + 17	338 ± 205
Atlas	Ring	20.3 ± 1.2	17.6 ± 0.7	9.3 ± 0.5	14.9 ± 0.8	0.659E + 19 ± 0.45E + 18	473 ± 83
Prometheus	Shep.	68.2 ± 0.5	41.2 ± 1.9	28.1 ± 0.4	42.9 ± 0.7	0.159E + 21 ± 0.15E + 19	481 ± 25
Pandora	Shep.	52.3 ± 1.8	40.7 ± 2.0	31.6 ± 0.9	40.6 ± 1.5	0.137E + 21 ± 0.19E + 19	487 ± 56
Epimetheus	Co-O	64.8 ± 0.7	57.0 ± 1.6	53.4 ± 0.2	58.2 ± 0.7	0.526E + 21 ± 0.30E + 18	639 ± 22
Janus	Co-O	101.7 ± 1.1	93.0 ± 0.2	74.4 ± 0.2	89.0 ± 0.4	0.190E + 22 ± 0.12E + 19	643 ± 9
Aegaeon	DR	0.7 ± 0.0	0.3 ± 0.1	0.2 ± 0.0	0.3 ± 0.0	0.724E + 14 ± 0.30E + 14	540 ± 140
Methone	DR	1.9 ± 0.0	1.3 ± 0.0	1.2 ± 0.0	1.4 ± 0.0	0.392E + 16 ± 0.10E + 16	310 ± 30
Anthe	DR				<1 km?		
Pallene	DR	2.9 ± 0.4	2.1 ± 0.3	1.8 ± 0.3	2.2 ± 0.3	0.115E + 17 ± 0.40E + 16	250 ± 70
Telesto	Troj.	15.9 ± 0.3	11.7 ± 0.3	9.8 ± 0.2	12.2 ± 0.3		
Calypso	Troj.	14.6 ± 0.3	9.3 ± 2.2	6.4 ± 0.2	9.5 ± 0.6		
Polydeuces	Troj.	1.5 ± 0.3	1.3 ± 0.3	1.0 ± 0.2	1.3 ± 0.3		
Helene	Troj.	22.6 ± 0.1	19.5 ± 0.1	13.3 ± 0.0	18.1 ± 0.1		
Hyperion		164.1 ± 4.0	130.1 ± 4.0	107.1 ± 4.1	135.1 ± 4.0	0.562E + 22 ± 0.50E + 20	544 ± 50
Phoebe		108.4 ± 0.4	109.3 ± 0.9	101.8 ± 0.2	106.4 ± 0.4	0.829E + 22 ± 0.90E + 19	1642 ± 18

Semiaxes are of ellipsoids fit to shape models and rescaled to volume of the model. Rm (mean radius) is the radius of a sphere of equivalent volume. Masses for Janus, Epimetheus, Atlas, Prometheus, and Pandora are from *Jacobson et al.* (2008). Mass of Pan is from *Porco et al.* (2005); mass of Daphnis from *Porco et al.* (2007). Mass of Hyperion from *Thomas et al.* (2007). Mass of Phoebe is from *Jacobson et al.* (2006). Masses of Aegaeon, Pallene, and Methone are estimates from equilibrium shape interpretations (*Thomas et al.,* 2013). Masses of Telesto, Calypso, Polydeuces, and Helene are not known; possible ranges are discussed in the text. Data received after late 2016 have not been included in these results.

TABLE 3. Derived gravitational quantities.

Object	ρ (kg m^{-3})	g (mm s^{-2})	Δdh	Δdh/Rm	Slope	Best Res
Pan	424 ± 137	0.1–1.8	7.95	0.56	12.46	1.269
Daphnis	330 ± 247	0.1–0.4	2.24	0.59	14.23	0.438
Atlas	473 ± 83	0.2–1.9	10.87	0.73	14.55	0.190
Prometheus	481 ± 25	1.1–5.8	19.86	0.46	12.22	0.201
Pandora	487 ± 56	2.3–5.9	11.30	0.28	8.34	0.312
Epimetheus	639 ± 22	6.3–11.0	17.01	0.29	11.23	0.154
Janus	643 ± 9	10.9–16.9	19.47	0.22	9.93	0.178
Aegaeon	540 ± 140	0.0–0.0	0.01	0.03	3.50	0.091
Methone	310 ± 30	0.1–0.1	0.01	0.01	1.01	0.027
Anthe						5.290
Pallene	250 ± 70	0.1–0.2	0.06	0.03	1.18	0.220
Telesto	500	0.7–1.8	2.64	0.22	12.09	0.063
Calypso	701	0.8–2.0	3.40	0.36	10.86	0.136
Polydeuces						0.411
Helene	550	2.4–2.8	5.86	0.32	12.96	0.012
Hyperion	544 ± 50	17.1–20.7	41.67	0.31	11.68	0.064

Derived values for Helene, Calypso, Telesto use assumed densities from *Thomas et al.* (2013), based on topographic features. Calculated slopes for Helene are not highly dependent on assumed density; rotation effects are more for Telesto and Calypso, and thus slopes for these objects are highly uncertain.
g varies due to shape and tidal and rotational effects.
Δdh is range of dynamic heights in kilometers.
Slope is area averaged slope in degrees.
Best res is best ISS NAC pixel scale.

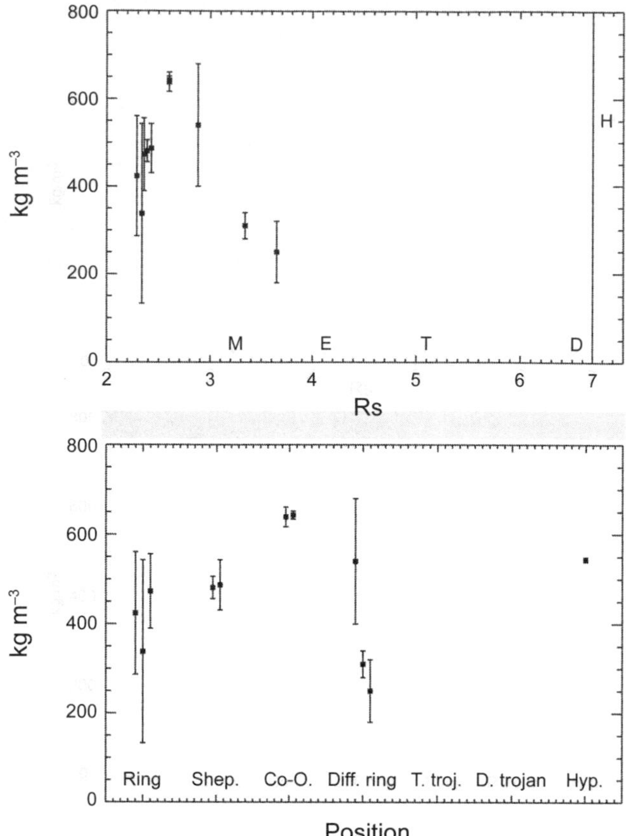

Fig. 2. Mean densities of inner small satellites. These are plotted as a function of distance from Saturn in Saturn radii (using 58232 km) on the top, and by dynamical position on the bottom. Positions of Tethys and Dione Trojans are indicated, but there are no mass determinations for these objects. For the top plot Hyperion's value is included in a separate box at its density because its distance is 25.4 Rs, far off the nominal edge of the plot. Densities of the mid-sized satellites would plot off the top of this plot (985 ± 3 for Tethys to 1478 ± 3 kg m^{-3} for Dione), so only their radial positions are noted. Data received after late 2016 have not been included in analysis.

craters more than 1 km in diameter on the core (*Thomas et al.,* 2013). The topography of both Pan and Atlas supports substantial slopes [20° in places (*Thomas et al.,* 2013)]. The geometric and gravitational shapes of these satellites are shown in Fig. 3. The gravitational shape is calculated by adding relative gravitational heights (*Vanicek and Krakiwsky,* 1986; *Thomas,* 1993) to a sphere of the object's mean radius.

Pan's equatorial band constitutes ~10% of its volume; Atlas' band is about 25% of its volume. These estimates depend upon extrapolation of the surface of the cores (in this chapter we use "core" geometrically with no implication of thermal or chemical differentiation), and thus are crude values at best. The substantial topography along the equatorial bands (see the lower row of cross sections in Fig. 3) indicates that gravitationally driven mass movement is not the primary control on formation of the bands. *Charnoz et al.* (2007) found that accretion from ring particles, dependent in part upon inclination of the satellites' orbits relative to the rings, might explain the geometry of the bands. This scenario has specific requirements of timing of ring and satellite formation. It should be noted that these smooth, rounded bands are quite unlike some sharply terminated equatorial bands present on some rapidly rotating asteroids (*Ostro et al.,* 2006; *Harris et al.,* 2009) that result from material sliding to the gravitationally low equator. Parts of the bands on Pan and Atlas are gravitationally higher than the rest of the satellite, and have rounded, flat, and sharply terminated equatorial topography. Thus, at least a two-stage development of these objects seems highly likely: formation of the "cores" and subsequent accretion of ring materials.

The small density of craters on Atlas, and their subdued morphology, suggest relatively young surface ages compared to those of Mimas and the other mid-sized icy satellites. Currently there is no consensus on the ages of the mid-sized objects as well as the ring system (*Ćuk et al.,* 2016; see also the chapter by Kirchoff et al. in this volume), thus great uncertainty also attends the ages of all the small satellites occupying positions within or between the rings or mid-sized satellites.

3.2. Shepherds: Prometheus, Pandora

The F-ring shepherds, Prometheus (inner) and Pandora (outer), were discovered in Voyager images in 1980. Cassini

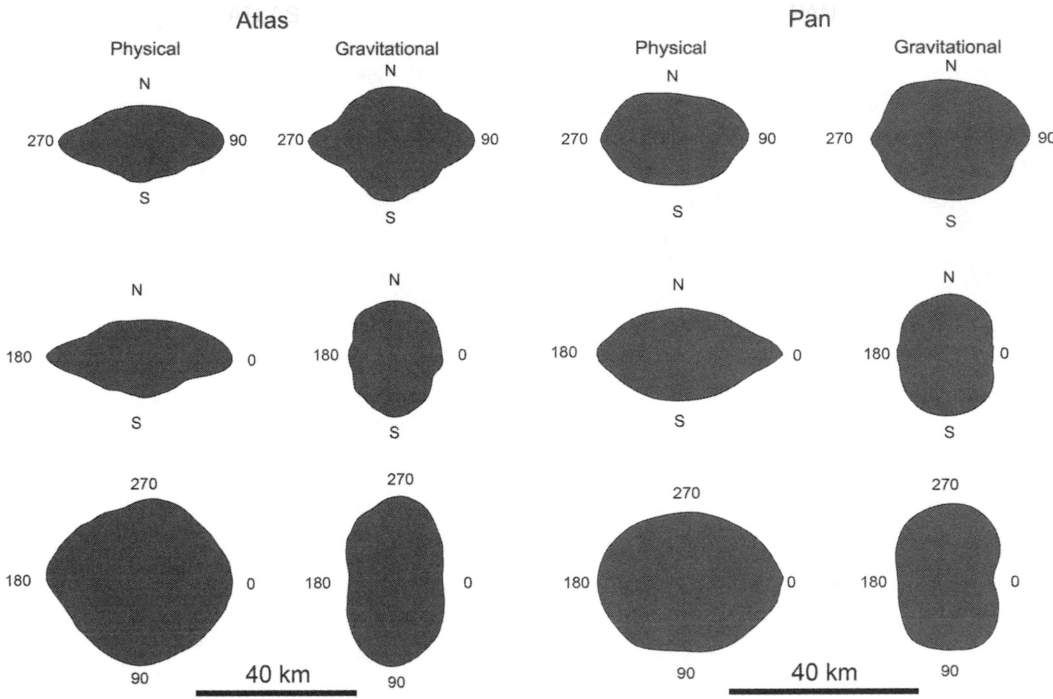

Fig. 3. Physical and gravitational shapes of Atlas and Pan. Top row: view from anti-Saturn direction. Second row: view from leading point. Bottom row: view from north pole. Gravitational shapes are the dynamic topography [potential energy at surface divided by average surface acceleration (*Thomas,* 1993)] added to a sphere of the object's mean radius.

imaging is sufficiently well distributed to map most of the surfaces and closely constrain the shapes of the shepherding satellites (Table 1). These satellites have similar volumes, and similar mean densities, but Prometheus has a more elongated shape. Prometheus' terrain is suggestive of a core partly exposed by delamination of a few kilometers of material (Fig. 1) (see also Fig. 3 in *Thomas et al.,* 2013). The possible core is exposed over two ranges of longitude, ~20°–100°W and 170°–260°W. (We use west longitudes for these saturnian satellites (*Archinal et al.,* 2011).]. The volume of the core is estimated (*Thomas et al.,* 2013) to be about two-thirds that of the entire (current) object, being approximated by an ellipsoid of 60-, 35-, and 25-km semi-axes. The margins of the putative remnant of an outer layer are defined by two bounding scarps totaling ~300 km in length. The only vaguely comparable scarp on other small objects is the smooth/rough transition on Atlas, although it would appear to be very different in origin. Prometheus' core is heavily cratered (see section 5), which indicates that if this is indeed a surface exposed by spallation, it happened early in the retained cratering record of these moons.

If Prometheus has indeed lost a noticeable amount of mass, is that mass now residing in the F ring, or does it perhaps constitute Pandora? The lost material can be crudely estimated by extending the shape model over areas visually interpreted as core surface; this increases the model volume by~ 60,000 km^3. Alternatively, simply adding an estimated 6 km to stripped areas (~7300 km^2) gives an estimate of lost material of 44,000 km^3. Can 40,000 to 60,000 km^3 (<20% of Prometheus) supply the mass of Pandora or the F ring? Pandora's volume is ~280,000 km^3, so it seems Pandora is unlikely to be a reassembly of the outer parts of Prometheus as they are currently interpreted (*Thomas et al.,* 2013). The F-ring mass is estimated to be less than 25% of the mass of Prometheus (*Murray et al.,* 1996), thus it could conceivably be made up of debris from Prometheus. Explanation of Prometheus and the F ring really requires also deciphering the association with Pandora. Could there have been a co-orbital arrangement disrupted by a near-catastrophic collision on Prometheus?

Prometheus also has several elongated depressions merging into pit chains, usually termed grooves when observed on asteroids and other small satellites. The largest is over 3 km wide and 40 km in length. These are discussed below in section 4.

Pandora shows no sign of a stripped core. It is heavily cratered, including some shallow, elongate forms similar to some craters on Hyperion (*Thomas et al.,* 2007, 2013), and has an extensive set of grooves with patterns similar to that hypothesized to reflect tidal stresses (*Morrison et al.,* 2009). (See section 5 for a summary of these forms.)

The mean densities of Prometheus and Pandora are effectively identical, being constrained to ~10% uncertainty (481 ± 25 and 487 ± 56 kg m^{-3}).

3.3. Co-Orbitals: Janus, Epimetheus

The co-orbitals Janus and Epimetheus swap orbits every 4.00 Earth years (*Yoder et al.,* 1989). They contain nearly 90% of the inner small satellite mass apart from Hyperion (which has twice the total mass of Janus and Epimetheus combined). The co-orbitals' mean densities are essentially identical (639 ± 22 and 643 ± 9 kg m^{-3}) and are the greatest of the satellites considered in this chapter. Both objects are heavily cratered with a range of crater morphologies indicative of long-term interaction between crater formation and erosive processes. A distinguishing characteristic of this pair of objects is the segregation of debris (regolith) in local topographic lows (*Morrison et al.,* 2009), as opposed to the style of debris accumulation on the other small satellites: global or hemispheric accumulations on the Trojans (section 3.5), latitudinal accumulations on ring satellites, global smoothing on ring-arc objects, gradational (ill-defined) covering on the shepherds, or the mass movement accumulations on Hyperion (Fig. 4). The debris in craters on Janus and Epimetheus is relatively far less voluminous than that on the Trojan moons (see section 3.5 below). On both Janus and Epimetheus the higher areas (crater rims and ridges that may be degraded rims of large craters) are brighter and have discrete, commonly serrated, boundaries with material that has moved downslope into craters and other low areas. In the higher-resolution image data for Epimetheus, at least two stages of accumulation in craters can be discerned, also with discrete boundaries and slight color differences (*Morrison et al.,* 2009). These discrete boundaries of materials suggest that any diffusive processes of degredation are strongly affected by differences in material properties, such as cohesion, between components of the regolith or between the regolith and underlying consolidated material. These characteristics suggest the co-orbitals may have different regolith thickness and/or stratigraphy from those on the other small satellites.

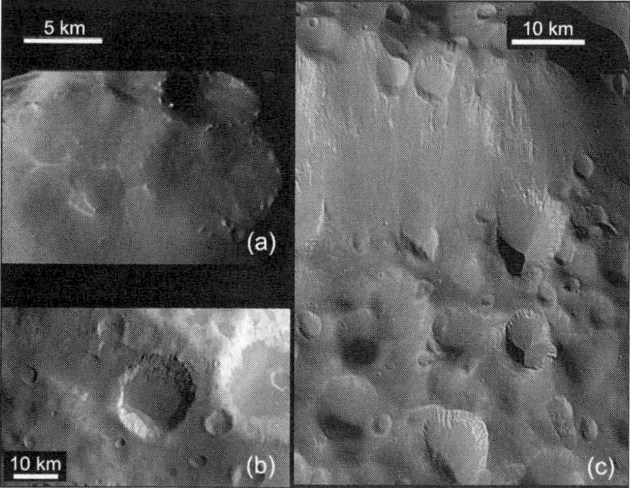

Fig. 4. Downslope motion of regolith on small saturnian satellites. Upper left, Telesto: N1507754947. Here debris nearly fills large, degraded craters with slopes converging in branching troughs. Lower left, Epimetheus: N1575363079. Relatively smooth fill has ponded in craters, leaving crater walls relatively undisturbed or coated. Right, Hyperion: N1506391566. The top of this view is the wall of a large crater, and debris has moved downslope partly filling some smaller craters at the base. Other craters also show more local filling, giving topography somewhat between that of Telesto's deep fill, and the crater floor ponded deposits on Epimetheus.

Epimetheus has an extensive collection of grooves, many of which were visible only with higher-resolution data than were available for the maps presented in *Morrison et al.* (2009). These grooves are mapped with the full dataset and discussed in section 4.

3.4. Satellites in Ring-Arcs and Diffuse Rings

Aegaeon orbits interior to Mimas and within the G ring (*Hedman et al.,* 2010). Methone, Anthe, and Pallene orbit between Mimas and Enceladus and have associated rings or ring arcs (*Hedman et al.,* 2009).

Methone, Pallene, and Aegaeon share characteristics otherwise absent in all other small solar system objects so far observed: smooth ellipsoidal shapes. The best image data are for Methone at 27 m/pixel. This object has albedo patterns probably indicative of interaction with charged particles (*Thomas et al.,* 2013), but probably has no topography greater than ~20 m based on the limb coordinate fits and small variations in brightness near the terminator and the shape of the terminator even with extreme image contrast stretching. Pallene data are lower resolution than those for Methone (220 m/pixel vs. 27 m/pixel), but several views including silhouettes and ones with Saturn shine allow good characterization of its size and shape (Fig. 1) and confirm that it is unusually smooth for a small object. Anthe is not spatially resolved and is probably <1 km in radius.

Ellipsoidal shapes invite geophysical interpretation, and can be precisely determined from ellipses fit to subpixel-measured limb coordinates (*Dermott and Thomas,* 1988; *Thomas,* 1998). Details of the geophysical interpretion of the ellipsoidal shapes of the ring-arc and diffuse ring satellites are given in *Thomas et al.* (2013). The shape solution for Methone is particularly robust and interesting. Its shape is consistent with an equilibrium ellipsoid form, i.e., one with the surface at a uniform potential energy (*Chandrasekhar,* 1969; *Dermott,* 1979). The measured shape of Methone implies a mean density of 310^{+20}_{-30} kg m^{-3} (~1σ). A particular equilibrium ellipsoid is consistent with a lower mean density if the object is centrally condensed than if it is homogeneous (*Dermott,* 1979). Thus, if the interpretation of an equilibrium form is correct, Methone would have some material with a density less than 310 kg m^{-3} if it has a denser core. While no model can uniquely satisfy a shape, the extremely smooth limbs and excellent conformity to equilibrium shapes [specific relation of the three axes (*Chandrasekhar,* 1969; *Thomas et al.,* 2013)] is strong evidence that these objects have very-low-density materials that do not support shear stresses on geologic timescales.

The low density and low gravity on the ring-arc moons produce surface environments much different from those on planets or on most satellites or asteroids. The central pressure of a 1.4-km sphere with a density of 310 kg m^{-3} would be about 260 Pa, or about that at 1 cm depth in a lunar regolith of density 1500 kg m^{-3} (*Carrier et al.,* 1991; *Hapke and Sato,* 2016). Pressures a few hundred meters at depth in Methone would match those a few meters depth on Janus. A bulk density of 310 kg m^{-3} implies, for ice particles, a porosity of ~67%. While within the porosity range of familiar regoliths (*Hapke and Sato,* 2016), this porosity is low for new terrestrial snow and other accumulations of highly irregular particles. Some fresh snow can have densities less than one-tenth that of Methone's inferred value (*Judson and Doesken,* 1999). Thus the possible porosity of ice particle accumulations in outer parts of these objects with over 70% porosity is entirely plausible. Such high porosity is consistent with the implied very low shear strengths on these objects. High porosities imply minimal grain contacts, such that processes such as charged particle bombardment, thermal or electrostatics effects, or micrometeoritc impacts might be efficient at mobilizing surface particles, thus allowing low effective shear strength.

One might consider comets, as low-density, low-gravity, active objects, for possible insights into the surface processes of the ring-arc moons. However, the comets are erosionally very active (e.g., *A'Hearn et al.,* 2011; *Sierks et al.,* 2015), with very rough topography and low albedos indicative of the roles of both volatile and nonvolatile materials. The ring-arc moons (and the other small satellites of Saturn) may have relatively more ice than most comets, but with temperatures below 100 K, volatile release is not a major factor in their morphology. The surface of Methone is certainly not a product of erosion. Such an equilibrium shape indicates some minimum depth of material that cannot support stress, and the likely bombardment history suggests that mobility of this material rapidly erases craters.

3.5. Trojans: Telesto, Calypso, Polydeuces, Helene

Telesto and Calypso orbit ahead and behind Tethys; Helene and Polydeuces share Dione's orbit. Polydeuces is only marginally resolved in Cassini images; the other Trojans are imaged at sufficient resolution to show surface features. Helene has by far the best data with pixel scales as small as 40 m. Their masses are unknown because no other object is known to be measurably affected by their tiny gravity.

The distinguishing surface characteristic of all the well-resolved Trojan moons is a near-global debris covering subject to downslope motion over lengths of tens of degrees of arc. Substantial slopes remain: The area-averaged slopes are slightly over 10° on all three objects (Table 3). These slopes are similar to those on most small, irregularly shaped satellites and asteroids (*Richardson and Thomas,* 2010; *Richardson and Bowling,* 2014). On Calypso, the debris covering, shown by relatively smooth surfaces and by elongate albedo and topographic markings, is lower than the bounding topography, such as crater rims (lower left, Fig. 1) (see also Fig. 12 in *Thomas et al.,* 2013). Downslope motion is indicated by the convergence of the lineation patterns in gravitationally low regions. The optical contrast on Calypso is especially large in the UV, and suggests that this downslope transport exposes materials with different amounts of space weathering effects (see section 6). On Telesto, the debris moves downslope within degraded craters and in regions of overlapping degraded craters, resulting in branching networks of valleys (Figs. 1 and 4). On Helene, the best imaged of these satellites, a darker (especially in the UV) unit 10–20 m thick has covered nearly all the well-imaged surface and subsequently has been eroded/is being eroded at scarps (Fig. 5). The material eroded from the scarps appears very mobile as smooth surfaces of transport abut sharply against the scarps. The albedo and topographic patterns indicative of debris covering on Helene appear to extend at least from 350°W westward to 220°W. Image coverage of the other regions is lower resolution, although there is a suggestion of similar albedo patterns between 310° and 350°W. This erosion has resulted in elongate mesas and smooth, lower surfaces that extend along slopes of ~6°–9° to regional low points. Possible properties of the downslope motion have been examined by *Umurhan et al.* (2015), in particular the likelihood that non-steady-state processes are needed to explain the liberation of material into prominent flow lanes. The images (Figs. 1 and 5) and the crater data (section 5) indicate a non-steady-state history with rapid covering followed by downslope removalof loose material.

The amount of debris on these objects can only be crudely estimated from crater morphologies and populations. *Thomas et al.* (2013) estimated 40–180 m depths on Helene, and 100–400 m possible deposition on Telesto. Calypso imaging resolution and the character of its morphology does not

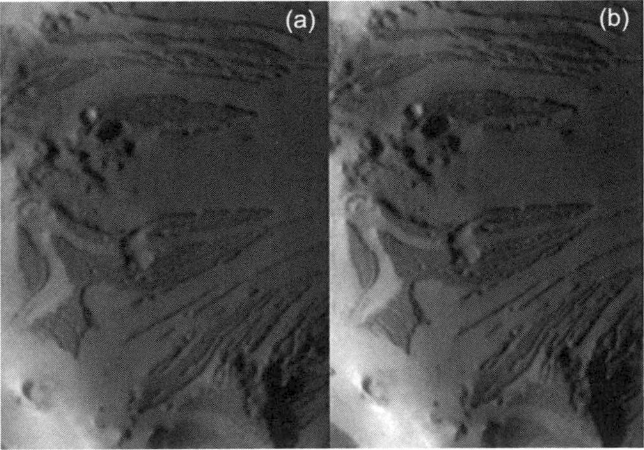

Fig. 5. Stereo view of mesas and lower smooth material on Helene. Frame width 6.8 km. N1687119973; N1687119756.

facilitate reliable estimates of debris depths. *Hirata et al.* (2014) estimated 10–300 m cover on Helene, and more on Calypso and Telesto. *Thomas et al.* (2013) made comparisons to visible crater populations and possible locally derived ejecta, and concluded there is an excess of material on all the Trojans. Thus the question of external sources for the debris on the Trojans arises. Ejecta exchange in the Saturn system has been examined by *Alvarellos et al.* (2005) and *Dobrovolskis et al.* (2009), who find that ejecta from large craters on Tethys can reach the libration points. However, only ~0.1% of debris from crater Penelope on Tethys is predicted to intersect Telesto or Calypso; depending on estimates of the excavated volume, these particular results suggest less than 100 m covering is possible on Tethys' Trojans from the largest impacts on Tethys. There appears to be considerable possible modeling remaining of oblique impacts and combined debris from several satellites that might provide a source for extra debris at the libration points. *Hirata et al.* (2014) suggested E-ring particle deposition could account for the Trojan coverings. The discrete nature of the Helene covering, and its distance from Enceladus, lead us to conclude that such a source is unlikely to be responsible for a substantial fraction of the debris cover on Helene.

3.6. Hyperion

Hyperion orbits beyond Titan in a 4:3 resonance. It has been chiefly noted for its chaotic rotation (*Klavetter,* 1989a,b; *Black et al.,* 1995) (see also section 7) and its sponge-like appearance (*Thomas et al.,* 2007; *Howard et al.,* 2012). The chaotic rotation changes the spin vector within the body of the satellite as well as its orientation in the sky, and it also changes the instantaneous spin rate. Consequently there is no predictable seasonality for insolation, and there is no way to predict body-centered coordinates for spacecraft flyby conditions. Despite the many flybys, a few regions of this satellite have been imaged only at high emission and/or high incidence angles.

Hyperion has a crudely ellipsoidal shape, fit by semi-axes of 164, 130, and 107 km. Its shape does not appear to be related to any formerly relaxed form. Phoebe, which is smaller with a mean radius of 107 km, has a much higher mean density and possibly retains a vestige of rotational equilibrium shape (*Castillo-Rogez et al.,* 2012). Although larger than Janus, Hyperion's mean density (535 ± 39 kg m^{-3}) is less than that of the larger co-orbital (643 ± 39 kg m^{-3}). Such a density indicates a mean porosity of ~40% if composed solely of water ice. A presence of denser rocky components would increase the required mean porosity. The surface composition of Hyperion inferred from Cassini VIMS data is primarily crystalline H_2O ice, CO_2 complexed with other materials, and organic matter (*Cruikshank et al.,* 2007, 2012; *Filacchione et al.,* 2010; *Dalton and Cruikshank,* 2007). The albedo varies by a factor of ~3, with reduced H_2O abundance in the darker areas, largely concentrated in topographically lower areas (Fig. 1). The body of Hyperion is likely largely H_2O and CO_2 ice; the surface of Hyperion seems consistent with mixing of H_2O ice and the dark red material on the leading side of Iapetus (*Cruikshank et al.,* 2007).

Typical Hyperion morphology is shown in Figs. 1 and 4 compared to surfaces of other small satellites. The sponge-like appearance is the geometric result of slightly crenulated depression rims merging into a network of crests. Low areas commonly have substantially darker material (*Cruikshank et al.,* 2007; *Howard et al.,* 2012). Some craters are elongate and shallow, but commonly these retain distinct rims and thus likely were formed by impacts different from typical hypervelocity ones rather than being the result of erosion or filling. The formation of the unusual spongy-appearing topography has been attributed to possible effects of a high-porosity target that reduces the amount of ejecta produced (*Thomas et al.,* 2007; based in part on *Housen and Holsapple,* 2003). More detailed analysis by *Howard et al.* (2012) focuses on a combination of diffusive slope mass wasting and on sublimation-induced retreat of crater walls. The low solar flux and ambient temperatures of the regolith (*Howard et al.,* 2012) suggest that sublimation of H_2O ice would be ~10 m/Ga for the dark floors, and 0.15 m/Ga for the brighter, colder surfaces. These rates are far too low to cause the several hundred meters of modification of craters needed to match the current appearance. A CO_2 component would sublimate many orders of magnitude more rapidly than would water ice, but its likely small mass fraction of Hyperion's materials (*Cruikshank et al.,* 2007) would mean that its sublimation role would be chiefly as a disaggregation agent, rather than as a significant mass removal from the landscape (*Howard et al.,* 2012). Thermal feedback from the darker material in the low parts of depressions may enhance the albedo contrasts by increased volatile loss, but the gradational boundaries of the dark regions suggest that other, more diffusive processes affect Hyperion's surface than are in play in the sharper light-dark boundaries on Iapetus (*Spencer and Denk,* 2010).

Hyperion's mean radius is 135 ± 4 km. It supports (Fig. 1) a crater with diameters of ~260 × 190 km that includes a broad central mound ~50 km across and ~6 km high. This crater is as large compared to the host object as the likely impact scar on Deimos (a ~10-km crater on a 6.2-km mean radius object). The density of craters within this large basin suggest it is ancient, although some rim areas have relatively crisp topography and appear to be undergoing active mass movement (Fig. 4). The active scarp erosion is occurring along the longest diameter of the basin, a condition that suggests a maximum crater diameter at formation somewhat less than the 260 km currently measured. Even allowing for a few kilometers of scarp retreat, the mean diameter of this basin is well over 200 km on an object of mean radius 135 km, making it one of the largest relative sized craters known, but still within bounds of previously observed bodies, and thus apparently not grossly different from others in terms of global responses to cratering despite the unusual spongy-appearing surface morphology at smaller scales.

4. INTERIORS AND STRUCTURES

Detecting internal structures of regolith-covered small satellites and asteroids has proven challenging but possible (e.g., *Buczkowski et al.,* 2008). Atlas and Pan exhibit obvious global structure in their cores and equatorial bands. As noted above, there is a likely core partially exposed on Prometheus. Other inferences about interior density structure of small satellites might be made by measurements of libration (section 7), or by other measurement of moments of inertia. The latter is available only for a few larger satellites for which Cassini has made multiple, close flybys while collecting radio science data for gravity field mapping. Surface grooves might help in deducing some internal structure or other body physical properties (Fig. 6). Grooves are long, approximately linear depressions in the regoliths of small bodies that have been attributed to a variety of causes, chiefly body fracturing expressed in an overlying regolith or secondary impacts in a variety of scenarios. Their occurrence on small saturnian moons in the context of the presence of grooves on other satellites and asteroids is summarized in *Morrison et al.* (2009) using primary Cassini mission data. More recent coverage of Epimetheus and Prometheus (prior to late 2016) has expanded the geography of known grooves on these objects. Maps and stereographic views of the small satellite grooves are given in Fig. 7.

A few of the grooves on Epimetheus have the morphology of graben (*Morrison et al.,* 2009), and trend approximately perpendicular to the long axis. Such a pattern may be expected from tidal elongation and is seen on Phobos (*Soter and Harris,* 1977; *Weidenschilling,* 1979; *Morrison et al.,* 2009, *Hurford et al.,* 2016; and others). However, Phobos is within the synchronous distance and tidal stresses are increasing as it evolves inward. Thus Phobos might be expected to more easily display recent or surviving effects of tidal fractures than would Epimetheus, which orbits beyond the synchronous point. But contemporary librations forced by the eccentricity of its orbit (*Tiscareno et al.,* 2009) might impose sufficient tensional stresses to form fractures. Fractures such as those suggested by the graben forms on Epimetheus are not slices through the whole body; rather, they show failure of an upper layer brought on by extension of the whole body. For graben, the maximum stress is normal to the surface from self-gravity, thus any occurrence of such forms on these objects with very small gravity implies fracturing of very weak materials. Extension of the whole body under the influence of tides may fracture a relatively more brittle, but still very weak, near-surface layer (*Hurford et al.,* 2016). The typical rounded cross profiles of grooves may be attributed to the fracture failure occurring in material that is capped by some depth of less-consolidated regolith. Of particular interest on Epimetheus are areas that have linear markings in the brighter, relatively higher areas that appear to lack the loose regolith cover seen in low areas. These markings are aligned with the major pattern of grooves elsewhere on Epimetheus that appear to be formed in loose regolith, the usual appearance of small-body grooves. This association, and the occurrence of some graben morphology, is strong evidence that at least some of the Epimetheus grooves are indeed related to fractures. It is strong evidence that similar patterns on other small objects are also related to fracturing (*Morrison et al.,* 2009).

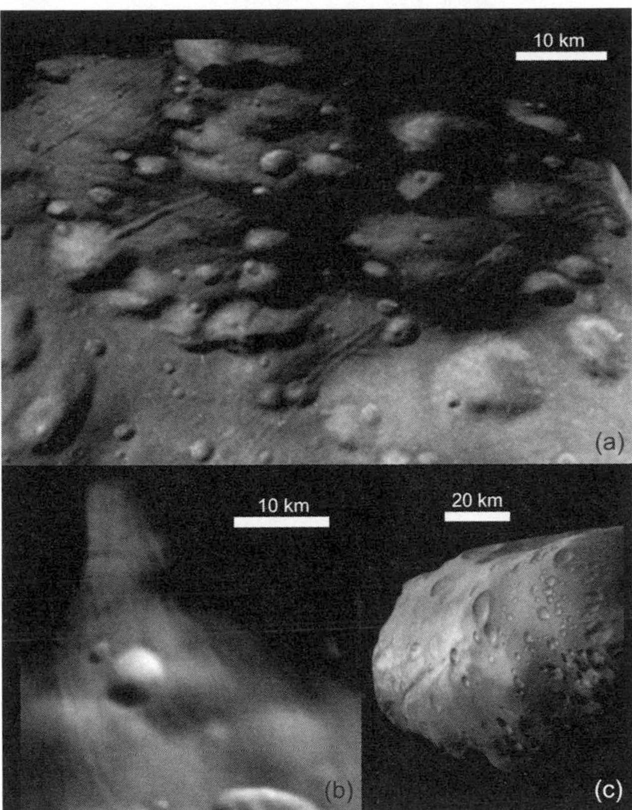

Fig. 6. Grooves on small satellites. Upper, Epimetheus, N1828124444; lower left, Pandora, N1504613517; lower right, Prometheus, N1828134697.

The well-organized grooves on Pandora are aligned ~30° to the Saturn-pointing axis, and thus may require more complicated origins than elongation from simple tidal stress. The grooves on Prometheus almost certainly include some that were formed from either oblique impacts or from ejecta (such as those prominent in Fig. 6: wide, not part of sets, and with gradual tapering of width). Pending final mapping of the groove morphologies on Prometheus and Pandora, the role of secondary or primary cratering in these features is not yet defined. Some morphologically distinct grooves on Phobos have been modeled as "sesquinary" catenae formed by material ejected into Mars orbit for short periods that re-impacts in patterns that are not easily tied to source craters as are classical secondaries (*Nayak and Asphaug,* 2015). Such a mechanism may explain some of the small saturnian satellite grooves, but specific application of sesquinary models to the small satellites in the saturnian system still awaits.

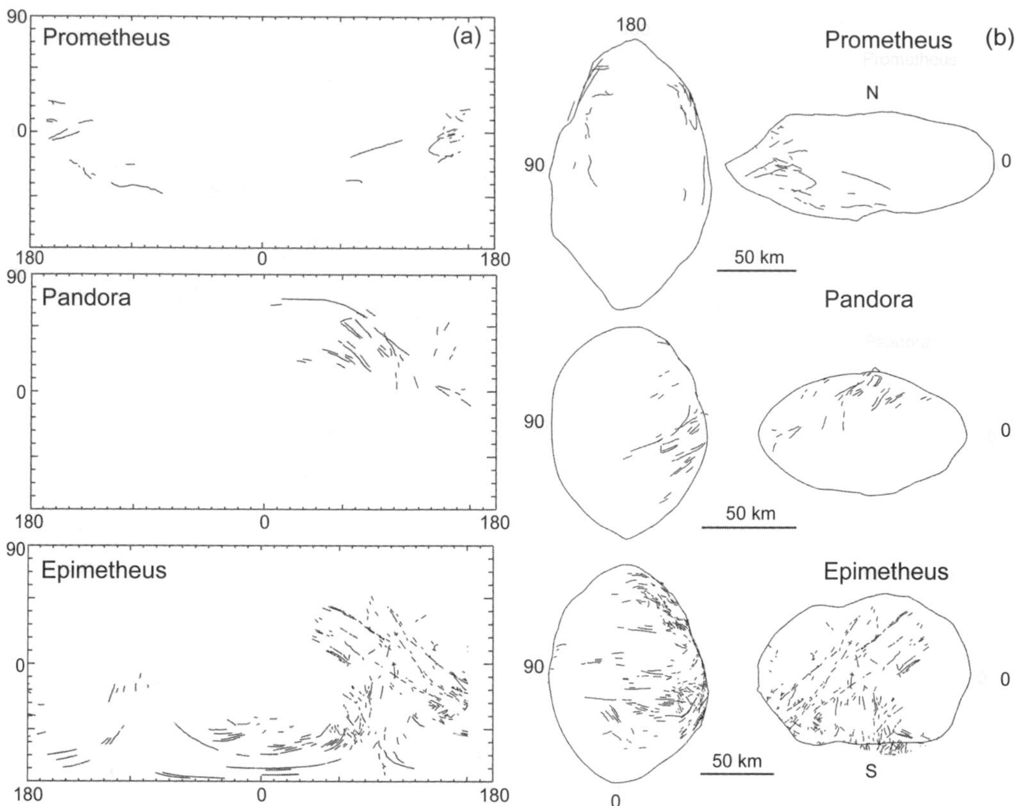

Fig. 7. (a) Maps of grooves on Prometheus, Pandora, and Epimetheus. Simple cylindrical projections; 0 longitude is sub-Saturn meridian. Lines are the centers of mapped grooves; no width information is included. These maps do not include data from December 2016 and later. **(b)** Orthographic views of groove traces on Prometheus, Pandora, and Epimetheus. Only groove center traces are recorded, no widths or other morphology is represented, and projections include far side grooves. Bounding shape model surfaces are the equator (left column) and 0°–180° longitudes right. Some grooves project outside these reference lines. Views on the left are from north with Saturn toward bottom; views on right are from the leading side, north toward top; selected longitudes and directions are marked.

The presence of grooves in patterns that are suggestive of tidal fracturing encourages two inferences: These objects are made of very weak materials, and there are differences in rigidity between near-surface and deeper parts (*Morrison et al.,* 2009; *Hurford et al.,* 2016). Although distinct two-layer modeling has understandably been employed for computational reasons (e.g., *Horstman and Melosh,* 1989; *Hurford et al.,* 2016), there would seem to be little likelihood that these objects have distinct, uniform outer layers given the almost certain history of ejecta coverings of varying thickness and the observations of regolith differences across Epimetheus and Janus (section 3.3). The scarcity of good graben morphology among grooves may reflect erosional smoothing and the effects of formation in loose regolith of substantial depth.

5. CRATERS

5.1. Crater Size Distributions

Crater density data are fundamental tools for the study of planetary surfaces, primarily as age markers and as probes of material properties. A specific survey of the Saturn system may be found in the chapter by Kirchoff et al. in this volume. Surface age estimates derived from crater density data are most potent where there are radiometric dates that can calibrate ages, or where there is a firm understanding of the impacting population and of the relation of crater size to the impactor energies. There are of course no samples of the saturnian satellites, and at present there is no consensus on the impactor populations that have been effective in the Saturn system. After the Voyager flybys, common interpretations invoked two different populations of impactors with a variety of possible sources (*Shoemaker and Wolfe,* 1982; *Smith et al.,* 1981, 1982; *Strom and Woronow,* 1982; *Horedt and Neukum,* 1984; *Plescia and Boyce,* 1985; *Chapman and McKinnon,* 1986; *Strom,* 1987; *Kargel and Pozio,* 1996; *Dones et al.,* 2009). Recently as many as four impactor populations have been considered: planetocentric bodies from satellite collisions and primary crater ejecta, secondary impactors, comets, and asteroids (*Kirchoff and Schenk,* 2010). The large range of distances from Saturn among the small satellites means that variations in fluxes and velocities of any group of impactors will be considerable, adding to the complexities of deriving relative ages among different objects. Additionally, there is uncertainty in the response of different surfaces at low gravity to different energies of impactors. Furthermore, the irregular shapes of these objects mean that most views do not present uniform lighting or viewing, in effect further reducing the areas suitable for accumulating high-quality crater density statistics.

Despite the limitations in the data, crater size frequency distributions on the small satellites are determined well enough to show at least two instances of similarity to other crater records in the Saturn system: a fall-off in relative crater densities for diameters less than 10 km, and in even more limited occurrences, a rise in the relative densities at diameters less than ~1 km. Additionally, crater densities on Helene document a two-stage surface history (*Thomas et al.,* 2013; also in this section below), and imaging data for Helene and Janus are good enough over sufficient longitude ranges to allow testing for leading/trailing asymmetry (although uniform global coverage is not achieved).

Data for the size frequency distribution of craters are summarized as R plots ["relative" plots: the differential size distribution is plotted relative to a power law, usually of –3 value (*Crater Analysis Techniques Working Group,* 1979)] in Fig. 8. Statistics of craters >10 km diameter for small objects are, of course, minimal. Between diameters of 10 and 2 km these crater density curves show roughly the same groupings of R values that decrease at smaller diameters (Fig. 8a), as do R plots for Tethys, Dione, Rhea, Iapetus, and Phoebe (*Kirchoff and Schenk,* 2010) (Mimas data go only to 4 km, but are in the same band of R values below 10 km size). Given the different objects, the different data-takers, and the spread of sizes and orbital positions, the replication of the fall in R values on all these objects from 10 to ~2 km is somewhat surprising, and might suggest crater-size-dependent equilibrium processes instead of highly similar impact histories.

Most surprising among the size frequency distributions is the minimum in the Phoebe data at ~2 km, which is in agreement with data in *Kirchoff and Schenk* (2010), therein termed a "dip," and the suggestion of a minimum at only slightly smaller diameters on Helene and Hyperion (Fig. 8b). The origin of this small group of minima is not known (*Kirchoff and Schenk,* 2010), but secondaries are not a significant factor given the very low escape velocities (*Bierhaus et al.,* 2012). Sesquinaries, those resulting from debris temporarily ejected into Saturn orbit (*Zahnle et al.,* 2008; *Nayak and Asphaug,* 2016), might add to the record, and could be present on some of the small satellites, but we cannot yet extract those components.

The multipart history of the surface of Helene is shown in Fig. 5 and expressed in the data shown in Fig. 8c. The debris covering and the smooth, erosional product of its removal are substantially less cratered than is the remainder of the object. The effects of the discrete debris covering on Helene explain at least part of the Helene crater density data, as is examined further in section 5.2.

Figure 9 illustrates the surfaces of four satellites at 42 m/pixel, effectively the best available for Helene. With only these data one might infer that the material moving

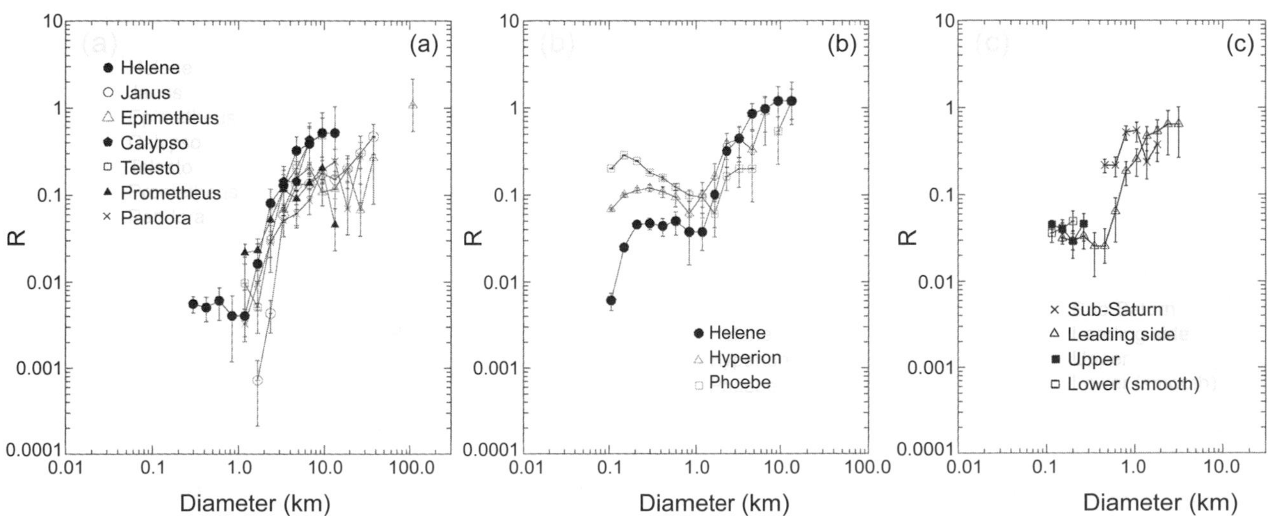

Fig. 8. Crater densities on small saturnian satellites. **(a)** R plots for several satellites. Best data available, lower sizes cut off at ~7 pixels, the practical measurement size limit (*Thomas et al.,* 2013; see also the chapter by Kirchoff et al. in this volume). **(b)** High-resolution R plots for Phoebe, Hyperion, and Helene illustrating the minimum in Phoebe data near 1 km diameter, and somewhat comparable shaped curves for Hyperion and Helene. Data are shown to 100 m size to illustrate where resolution effects start. **(c)** Portions of Helene. Lower sizes are cut off at ~7 pixels. "Upper" is the dark higher areas (see Fig. 5). "Smooth" is the lower, brighter areas in Fig. 5. Leading side is for lower-resolution data that also include the upper and smooth areas.

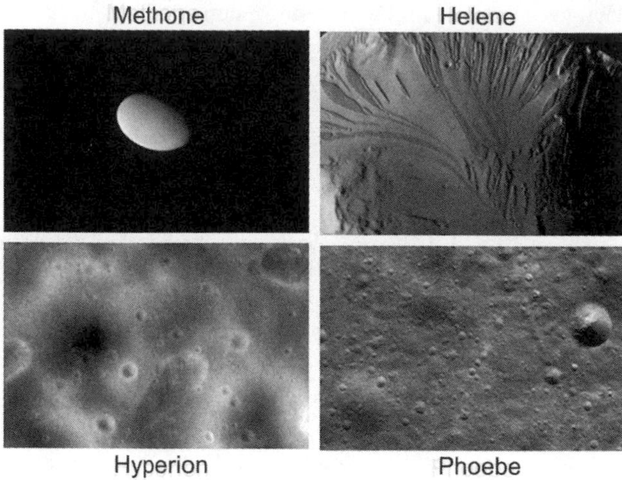

Fig. 9. Small satellites at 42 m/pixel. Methone: N1716192103; Helene: N1567129584; Hyperion: N1506393440. All panels are 13 km wide.

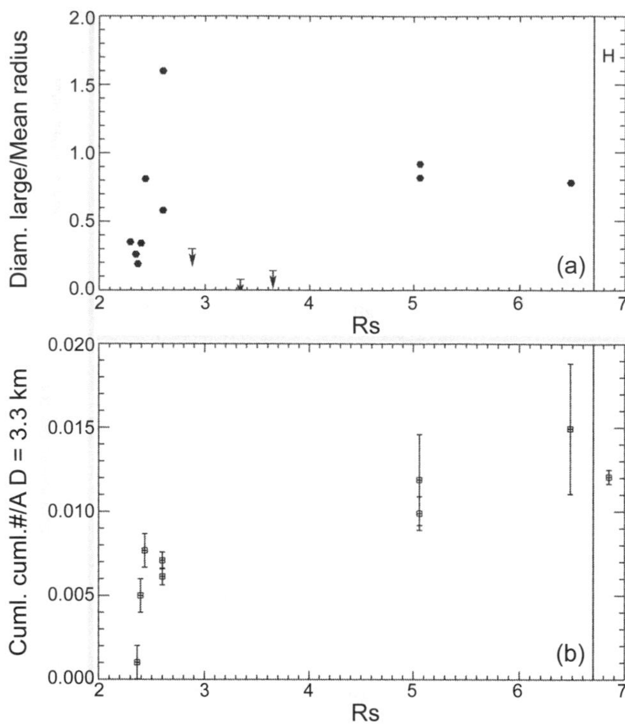

Fig. 10. Characteristics of large craters on small saturnian satellites as a function of distance from Saturn. **(a)** Ratio of largest crater to object mean radius. Maximum values for Pallene, Methone, and Aegaeon, shown by downward arrows, are based on estimates of the largest crater that could have escaped detection. Largest craters should be a function of age, impactor distributions, and mechanics of the target, "age" possibly being time since a major disruption or resurfacing, with variation due to statistics of small numbers. **(b)** Cumulative crater densities at 3.3 km diameter, the smallest well-detected sizes. Rs is Saturn radii (using 58232 km).

downslope on Helene is nearly as smooth as the surface of Methone. However, the Methone materials support essentially no slopes, in a lower gravity environment, while the Helene materials have not drained from the underlying topography. The Helene smooth materials also have accumulated a few craters, further reducing likely commonality with the Methone surface processes.

Given the different surfaces and suggestions of some variety in surface processes we are limited to very simple attempts at detecting differences in global ages that focus on the few largest craters on each object. We use two measures of relatively large craters: cumulative number of craters per square kilometer at the largest common well-measured sizes, here using 3.3 km diameter, and the relative size of the single largest crater on each object. The latter may be a measure of the age of the object, or of the mechanical properties, or a combination of those factors. These measures are plotted in Fig. 10 as a function of distance from Saturn. There is no clear radial pattern of relative size of the largest crater. The southern hemisphere depression on Epimetheus is the largest among these. The satellites closest to Saturn have much smaller relative sizes for their largest craters. None of the values in Fig. 10 is out of line of those for other small objects (*Thomas,* 1999).

There is a modest upward trend in cumulative density of craters 3.3 km in diameter and larger outward from Saturn (Fig. 10b). This trend would seem more consistent with older surfaces farther from Saturn than with differences in the cratering population, inasmuch as any heliocentric component would increase toward Saturn by focusing and by velocity increase. Older objects farther from Saturn might give some support to models of formation of the small satellites from ring materials followed by outward migration (*Charnoz et al.,* 2010); however, most of the trend in Fig. 10b is determined by the values for the Trojan moons, which may not be part of the scenario proposed by *Charnoz et al.* (2010). If the Trojans are indeed older, or at least have older surfaces, the possibly younger surfaces of the inner moons might still be related to their formation times. However, the present uncertainty in converting crater populations to cratering ages in this system means the significance of the crater populations will emerge only with convergence of models of orbital histories, cratering mechanics, and impactor dynamics.

5.2. Leading/Trailing Crater Density Data

Crater densities on the leading and trailing sides of synchronously rotating satellites have been used to test for an influence of possible different impactor populations (*Zahnle et al.,* 2001; *Hirata,* 2016), in particular, planetocentric impactors. Figure 11 shows cumulative plots of leading and trailing side data for the best imaged of the small inner satellites, using areas appropriate for the actual image coverage.

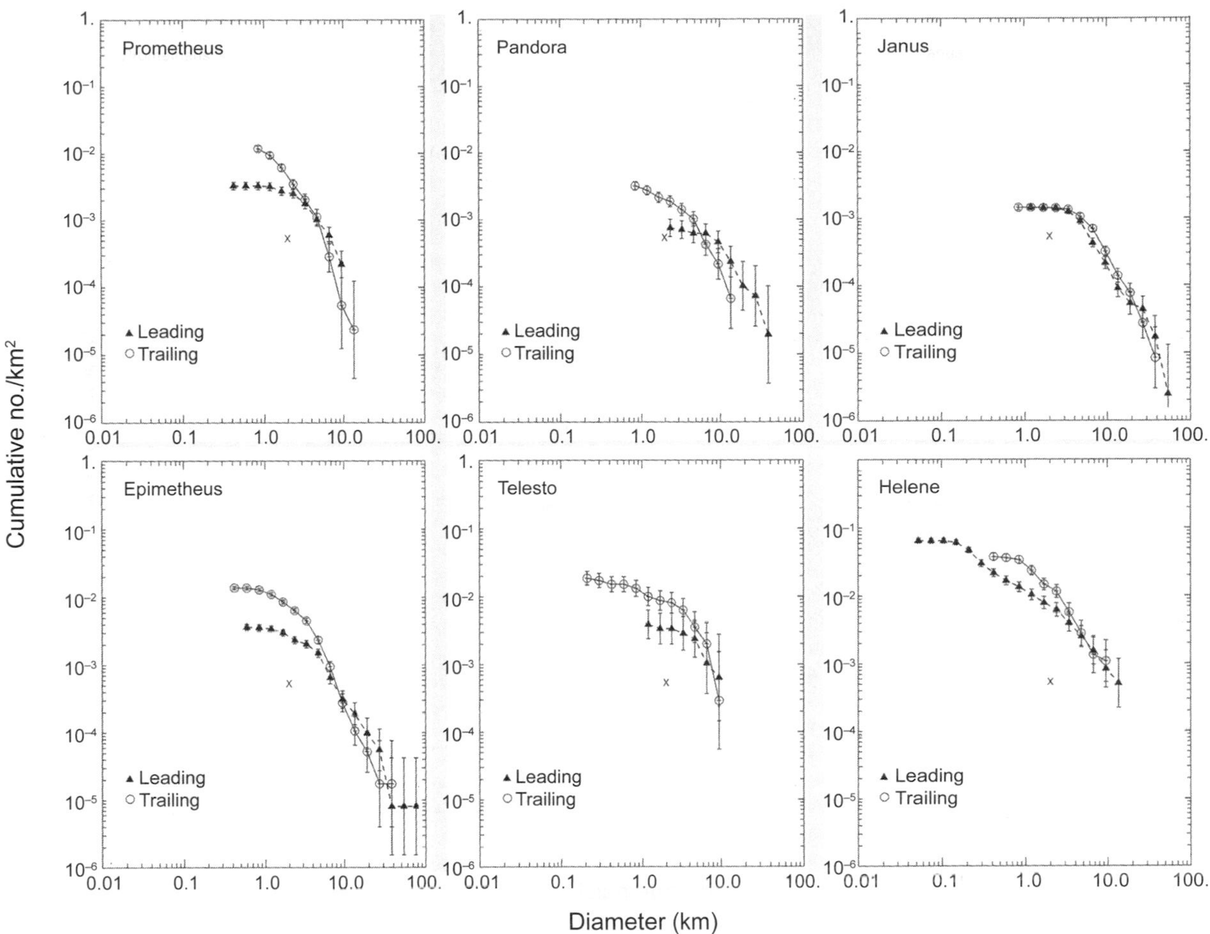

Fig. 11. Comparison of leading-trailing cumulative crater densities on several small satellites (from data in *Thomas et al.,* 2013). Helene is the only example where differences occur well outside formal uncertainties and in size ranges not affected by resolution. "x" is an arbitrary common visual reference value for all plots.

For Prometheus, Pandora, Epimetheus and Telesto, no firm conclusion on similarity or differences between leading and trailing crater densities can be made. For Janus and Helene at sizes larger than 5 km, densities on the two sides are indistinguishable. However, on Helene between diameters of 1 and 5 km, the trailing side clearly has more craters, and because of the lower resolution of the trailing side data, this difference could extend to diameters as small as 0.2 km. As noted in section 3 above, differences in crater density on Helene can be confidently ascribed, within the area of high-resolution image coverage, to a debris covering that hides many of the older, smaller craters. Figure 11 suggests that this process is concentrated on the leading side of Helene, although images suggest that such debris covering extends somewhat beyond 0° and 180° longitude (*Thomas et al.,* 2013). *Hirata et al.* (2014) found a similar, but not identical, relative distribution of craters on Helene.

The mid-sized icy satellites have no clear leading-trailing crater density differences (see the chapter by Kirchoff et al. in this volume). The far better statistics on these objects, and their lack of clear leading-trailing differences, are consistent with the apparent association on Helene of a modest hemispherical difference in crater density being due to hemispherical variations in the excess ejecta cover rather than to differences in the external cratering flux.

6. ALBEDO, COLOR, AND SPECTRAL PROPERTIES

The albedo, color, and spectral properties of airless planetary bodies can reflect an object's gross composition, materials deposited from other objects, weathering phenomena, high-energy particle bombardment, or any combination of these properties and processes. The exchange of matter between the saturnian rings and the moons is a topic of special interest and complexity, as it involves multiple ring systems at different distances from Saturn and satellites of different sizes and geological properties.

In this section we examine the implications of the albedo, color, and spectral characteristics of the small inner satellites for their makeup and interactions with the environment, especially that of the ring systems. We have updated the

TABLE 4. Whole-disk albedo and color properties.

Body	Symbol	CL1/CL2 Whole-Disk Albedo		IR3/UV3		
		α = 0°	α = 10°	Global	Leading	Trailing
Pan	PN		0.52 ± 0.04	2.50±0.17		
Daphnis	DA		0.56 ± 0.04	2.30±0.25		
Atlas	AT		0.49 ± 0.08	2.42±0.06		
Prometheus	PM		0.67 ± 0.07	2.14 ± 0.18‡		
Pandora	PA		0.62 ± 0.08	1.85 ± 0.13‡		
Janus	JA	0.71 ± 0.02*	0.37 ± 0.03	1.68 ± 0.08‡		
Epimetheus	EP	0.73 ± 0.03*	0.41 ± 0.04	1.76 ± 0.07‡		
Aegaeon	AG		0.25 ± 0.23			
Mimas	MI	0.96 ± 0.01*	0.62 ± 0.06	1.18 ± 0.02	1.17	1.20
Methone	ME		0.42 ± 0.05	1.26 ± 0.02‡		
Pallene	PL		0.29 ± 0.11	1.17 ± 0.02		
Enceladus	EN	1.38 ± 0.01*	1.00 ± 0.05	1.03 ± 0.02	1.06	1.03
Telesto	TL		0.61 ± 0.04	1.06 ± 0.04‡		
Calypso	CP	1.34 ± 0.10*	0.69 ± 0.14	1.23 ± 0.07‡		
Tethys	TE	1.23 ± 0.01*	0.84 ± 0.19	1.25 ± 0.05	1.20	1.30
Helene	HE	[1.67 ± 0.20]*	0.74 ± 0.15	1.09 ± 0.09‡		
Polydeuces	PO		0.86 ± 0.06	1.17 ± 0.07		
Dione	DI	1.00 ± 0.01*	0.66 ± 0.05	1.29 ± 0.04	1.33	1.25
Rhea	RH	0.95 ± 0.01*	0.63 ± 0.07	1.58 ± 0.03	1.55	1.61
Hyperion	HY		0.24 ± 0.07	2.10 ± 0.24		
Iapetus	IA		0.17 ± 0.05	1.36 ± 0.11	1.47	1.24
Phoebe	PH	0.085 ± 0.002§	0.034 ± 0.002†	0.85 ± 0.07		

* *Verbiscer et al.* (2007).
† *Simonelli et al.* (1999).
‡ *Thomas et al.* (2013).
§ *Miller et al.* (2011).

Formal errors are given for telescopic albedos at α = 10°; Cassini values at α = 10° include a 5% uncertainty in absolute brightness scale (*West et al.,* 2010).

color and albedo information from *Thomas et al.* (2013) on the small inner satellites using techniques in *Helfenstein et al.* (1994) and *Thomas et al.* (2013). In Table 4 we report published opposition albedos and our new measurements obtained for a larger range of satellites. We adopt the ratio of the IR3/UV3 (930 nm, 341 nm) whole-disk brightness as a measure of the intensity of color as a simple survey device to detect differences among the satellites.

6.1. Albedo

Verbiscer et al. (2007) first demonstrated a strong spatial correlation between the geometric albedos of the satellites and the brightness of the E ring from the orbits of Janus and Epimetheus to the orbit of Rhea. They concluded that Enceladus's geysers were feeding the E ring, which subsequently coats satellites with ice particles. The effect is strongest near Enceladus and diminishes with increasing range to either side of it, vanishing at the orbits of Janus and Epimetheus within Enceladus's orbit and at Rhea on the outside. Geometric albedos obtained range from 0.085 ± 0.003 for Phoebe to 1.375 ± 0.008 for Enceladus. Verbiscer et al. determined albedo exactly at opposition (phase angle α = 0°), which includes the "opposition effect," a pronounced surge in brightness near α = 0°. Due to coherent backscatter (cf. Chapter 9 in *Hapke,* 2012), geometric albedo can exceed unity (see the chapter by Verbiscer et al. in this volume). This geometry is generally not available from Cassini data, so our expanded suite of data is at, or extrapolated to, 10° phase. As a check that our 10° whole-disk values are reasonable relative approximations to corresponding true geometric albedos, Table 4 values indicate that of the small satellites Janus and Epimetheus fall along a trend defined by the larger icy satellites and Phoebe, with Helene and Calypso somewhat off that trend, both of which are not as well constrained in *Verbiscer et al.*'s (2007) observations.

Our expanded whole-disk albedo survey at 10° phase, shown in Fig. 12, is broadly consistent with that shown in Verbiscer et al., but it is more detailed. Enceladus defines a maximum albedo. Moving out from Enceladus, the values decrease to Phoebe. Moving inward from Enceladus, Mimas and the co-orbitals show a decrease in albedo similar to the pattern in Verbiscer et al. However, Aegaeon, Methone, and Pallene do not present a single radial pattern and are all notably darker than Mimas, whose orbit they straddle. Progressing closer to Saturn in the main ring system, the

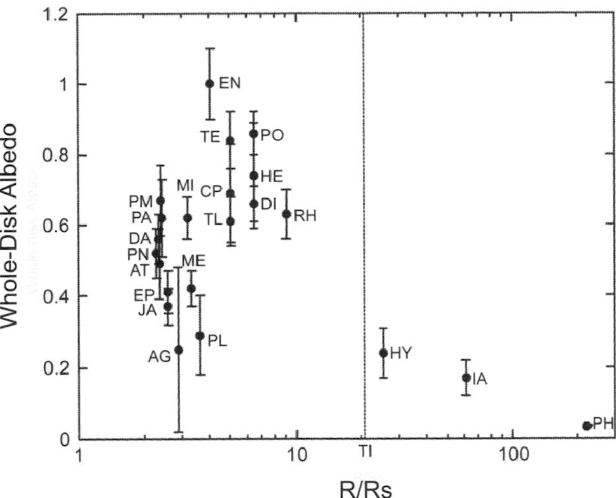

Fig. 12. Whole-disk albedo measured at 10° phase angle vs. distance from Saturn (R/Rs) in Saturn radii. Orbital distance of Titan is marked. Data points from Table 4. PA: Pan, AT: Atlas, DA: Daphnis, PM: Prometheus, PA: Pandora, EP: Epimetheus, JA: Janus, AG: Aegaeon, MI: Mimas, ME: Methons, PL: Pallene, EP: Epimetheus, TE: Tethys, CP: Calypso, TL: Telesto, DI: Dione, PO: Polydueces, HE: Helene, TI: Titan, HY: Hyperion, IA: Iapetus, PH: Phoebe.

albedos generally increase reaching a secondary maximum at the innermost satellite, Pan. New ISS observations of Pan, Atlas, Pandora, Daphnis, and Epimetheus obtained between December 2016 and May 2017 as well as a detailed investigation of the photometric phase curves of the small satellites will provide a more accurate representation of the preliminary trends shown here.

Three aspects of Fig. 12 that show departures from the simple pattern in Verbiscer et al. merit discussion. First, the lower albedos of Aegaeon, Methone, and Pallene suggest a different response to accumulated ice particles than on Mimas. This may be consistent with the apparent young age and very mobile surface materials (section 3.4) that are likely necessary to explain the almost fluid-like geophysics of these bodies. Additionally, these objects are embedded in ring arcs, and sharing or exchanging those particles may negate or modify any E-ring interaction. Although Methone has a darker oval region on its leading side, as does Mimas (cf. *Verbiscer and Veverka,* 1992), its geometry appears somewhat more complex.

Second, the sharp rise in albedo from the co-orbitals through the ring moons suggests entirely different influences. These are discussed further with the color data below.

Third, the Trojan moons simply do not fit the pattern. On a geological scale these objects have significant depths of what may be externally derived ejecta (section 3.5) that may interact differently with infalling E-ring materials than do the surfaces of other satellites.

Farther outward from Rhea, the E ring disappears and the presence of Titan may produce a gap between the outer edge of the E ring and the inner edge of the Phoebe ring (*Verbiscer et al.,* 2009). However, from Hyperion outward to Phoebe, the average whole-disk albedos continue to decrease and reach a minimum at the orbit of Phoebe, the presumed source of dark particles that populate the Phoebe ring (cf. *Verbiscer et al.,* 2009; *Hamilton et al.,* 2015).

6.2. Color

The ISS NAC uses two broadband filter wheels with filters that are used in pairs. Because of the rapidity of close satellite flybys, the number of color filter pairs that could practically be used was often limited. Most often, the CL1:UV3 filter (341 nm) and CL1:IR3 filter (930 nm) were used to bracket observations over the spectral range of the camera. We use the term IR3/UV3 to represent the ratio of observed brightnesses of pixels in each of the broadband filters. Figure 13 shows that within the orbit of Titan, the IR3/UV3 ratios of satellite surfaces are grossly similar from Mimas to Dione and are substantially higher farther from Saturn. Within the group of mid-sized objects, Enceladus, the presumed source of ice particles that mute colors on other satellites, has the lowest IR3/UV3 ratio (1.03 ± 0.02); Rhea, the most distant from Enceladus, has a ratio of 1.58 ± 0.03. Among the small objects from Methone to the Trojans, there is no overall relation of color to distance from Enceladus. The Trojan objects may provide insights into effects other than those deriving from radial position. The trailing Trojans (Calypso and Polydeuces) have larger IR3/UV3 ratios than the leading Trojans (Telesto and Helene; Fig. 13). The trailing Trojans are also most similar in color to their parent bodies, while the leading Trojans have the most muted colors. This relation suggests that the leading objects are more strongly affected by E-ring particle bombardment, but it is not clear why this should be so. Both Tethys and Dione have leading/trailing hemisphere asymmetries in albedo and color. For Tethys, the IR3/UV3 ratio is about 8% higher on the trailing side than the leading side, whereas for Dione the IR3/UV3 ratio is about 6% higher on the leading side.

Inside Mimas' orbit (Fig. 13) the color trend is a rapid increase in the IR3/UV3 ratio toward Saturn. Geologically there is an association of the inner three satellites with the rings and thus possibly with effects of material deposited from the ring. While Atlas lies outside the A ring, Pan and Daphnis are embedded in gaps within the ring. Closest to Saturn, Pan's average IR3/UV3 ratio of 2.5 ± 0.2 is significantly smaller than the value of 3.3 ± 0.2 of the A ring on either side of the Encke Gap that bounds it. Further out, the A ring IR3/UV3 ratio decreases from 2.7 ± 0.2 on the inside of the Keeler Gap to 2.2 ± 0.3 on the outside. The mean value is not statistically different from the value of 2.3 ± 0.3 of Daphnis itself. It is not surprising that these inner satellites follow the same qualitative trend as the rings, but they do not have exactly the same colors. The equatorial bands on the ring satellites may be very old (*Charnoz et al.,* 2007) and the colors most likely reflect geologically recent and ongoing processes. Farther outward, Janus and Epimetheus, along with nearby Prometheus and Pandora, are

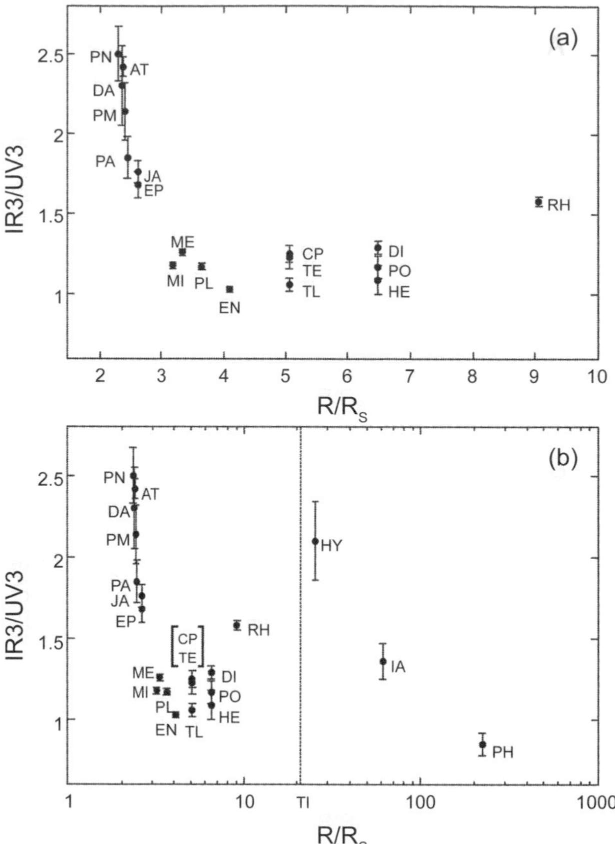

Fig. 13. Average IR3/UV3 (930 nm and 338 nm effective wavelengths) ratios for the small and mid-sized satellites from Table 4. Values for Prometheus, Pandora, Janus, Epimetheus, Methone, Telesto, Calypso, and Helene are from *Thomas et al.* (2013); values for Pan, Atlas, Daphnis, Mimas, Pallene, Enceladus, Polydeuces, Dione, Rhea, Hyperion, Iapetus, and Phoebe are new. Plotted vs. distance (R) relative to Saturn radius (R_S). PA: Pan, AT: Atals, DA: Daphnis, PM: Prometheus, PA: Pandora, EP: Epimetheus, JA: Janus, AG: Aegaeon, MI: Mimas, ME: Methons, PL: Pallene, EP: Epimetheus, TE: Tethys, CP: Calypso, TL: Telesto, DI: Dione, PO: Polydueces, HE: Helene, TI: Titan, HY: Hyperion, IA: Iapetus, PH: Phoebe.

heavily cratered and show no morphologic burial by main ring materials. Thus, the colors of these objects likely reflect active processes related to the whole ring system; the spectra (section 6.3) specifically link these objects to ring materials.

The orbital location of Titan (Fig. 13b) appears to be a pivotal point at which the trend of coloration with distance from Enceladus reverses progressively through Hyperion, Iapetus, and Phoebe. Phoebe exhibits the smallest IR3/UV3 ratio of any of these satellites, even lower than that of Enceladus.

Neither albedo nor IR3/UV3 color alone uniquely relates the satellite properties to their orbital taxonomic group. Albedos drop off to either side of Enceladus in Fig. 12 in overlapping ranges. At the same time, by coincidence the lowest-albedo and highest-albedo satellites have broadly similar IR3/UV3 ratios, despite their obvious differences in surface composition.

Albedo and IR3/UV3 ratios together can be used to map a relationship between albedo and color to satellite taxonomy. Figure 14 shows that nearly all the moons map into distinct regions of the plot that correlate with their ring family group. Moons that are strongly affected by the E-ring map into a region of low IR3/UV3 values and moderate to high albedos. Moons that are embedded in the dark Phoebe ring map into a region in which moons that have progressively higher albedos also have systematically higher IR3/UV3 ratios. The ring shepherds map into a region of moderate albedo and high IR3/UV3 ratios distinct from other objects interior to Titan and not occupying any possible mixture of the other surfaces interior to Titan. This color distinction accompanies their morphologic distinction and reinforces the idea their surface processes, and possibly their materials, are unique in at least this system. The co-orbitals map into a domain that is in between all others — they have moderately low albedos and intermediate values of UV3/IR3 ratios. Rhea is the only moon that falls outside the defined domains in the plot. It appears to be the satellite that is least affected by any interaction with any ring.

6.3. Spectral Results: Compositional Variations

Cassini's VIMS and ISS obtained spectra of the small moons and mid-sized satellites that span 250–5100 nm spectral range. The measurements (*Filacchione et al.,* 2010, 2012; *Buratti et al.,* 2010) provide details about moon-to-moon variations in composition that range from the water-ice-rich, nearly uncontaminated surfaces of Enceladus and Calypso to the metal/organic composition of dark materials on Iapetus' leading hemisphere and Phoebe. The data generally show that ring spectra have more intense 1500–2000-nm band depths than the embedded satellites, but they also appear redder than the icy moons in the visible range.

6.3.1. Ring shepherds and co-orbitals. Spectra of these satellites clearly establish that they are compositionally linked to the A ring. They are mostly composed of water ice with <5% of dark, likely hydrocarbon contaminants. CO_2 is absent on the small inner satellites such that their low-albedo contaminants are distinct from those on the Phoebe-ring moons (see below). Prometheus and Pandora both orbit close to the F ring, but they differ in surface composition in that Prometheus is water ice-rich and very red at VIS wavelengths, similar to A- and B-ring materials. However, Pandora is distinctly bluer. The co-orbitals are spectrally similar to each other; however, Epimetheus appears to have a thicker coating of ring material than Janus.

6.3.2. E-ring moons. Coating of these satellites by E-ring frost causes them to be bluer than the inner satellites but with variations related to their distance from Enceladus and attendant density of the E ring itself. The Trojan satellites of Tethys, like Enceladus, are primarily composed of water ice. However, Calypso is even more water-ice-rich and bluer than Telesto.

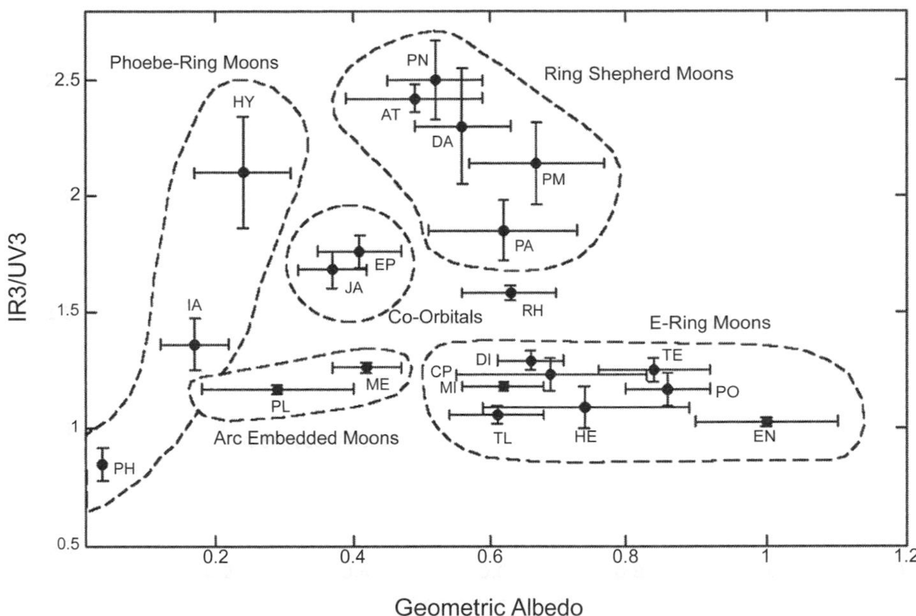

Fig. 14. Satellite whole-disk values of Cassini NAC ISS IR3/UV3 color ratio vs. corresponding 10° phase albedo from Table 4. Dashed curves identify different ring families in which the moons are embedded. See text for discussion.

6.3.3. Phoebe-ring moons. Iapetus and Hyperion spectra show signs of a diagnostic absorption line at 680 nm that *Vilas et al.* (1996) attribute to hydrated iron minerals. Among Phoebe-ring satellites there is a progressive linear trend in both decreasing water ice content and increasing reddening. Hyperion is the reddest object in the group.

7. DYNAMICS: ORBITAL CHARACTERISTICS, ROTATIONAL STATES, CONSTRAINTS ON AGES AND PROCESSES

7.1. Orbital Characteristics

The groupings introduced in section 3 have their origins in orbital characteristics.

The ring moons are intimately entwined with Saturn's main rings. Pan orbits within the Encke Gap, a 320-km sharp-edged gap in the outer part of the A ring. Daphnis orbits within the Keeler Gap, a much narrower 35-km sharp-edged gap even closer to the A ring outer edge. Atlas orbits ~900 km outward of the A-ring outer edge. Pan and Daphnis both have very low eccentricities, in keeping with the very low eccentricities of continuum ring particles, although vertical structure in the ring near the Keeler Gap, which was seen to cast shadows during the 2009 equinox (*Weiss et al.,* 2009), indicate that Daphnis likely has a nonzero inclination. Atlas exhibits changes in its semimajor axis of magnitude ±2 km that are only quasi-periodic, due to perturbations from Prometheus and Pandora (*Spitale et al.,* 2006). Daphnis appeared to have an orderly circular orbit until about 2011, after which its semimajor axis appears to have changed by a few hundred meters (*Jacobson,* 2014).

The so-called "shepherd" moons, Prometheus and Pandora, are situated on either side of the F ring, although it may be that they stir up the F ring as much as they confine it. Prometheus and Pandora have a semichaotic interaction that occurs when their periapses become anti-aligned, which occurs every 6.2 yr, although chaotic changes may occur at other times also (*Cooper et al.,* 2015, and references therein).

The "co-orbital" moons, Janus and Epimetheus, are unique in that they share an orbit but are comparable in mass to each other. Every 4.00 yr, the inner moon begins to catch up to the outer moon. Heuristically, the inner moon's gravity pulls on the outer moon, slowing it down so that its semimajor axis decreases, and vice versa. When the interaction is complete, the new inner moon pulls ahead of the new outer moon, continuing until it "laps" it another 4.00 yr later. Because Janus is ~3× more massive than Epimetheus, its semimajor axis changes by a smaller amount during their conjunctions.

Aegaeon, Methone, and Anthe are embedded within dusty ring arcs and are also in mean-motion resonances with Mimas (*Hedman et al.,* 2009). Pallene is not in a first-order resonance, and correspondingly its dusty ring appears to be azimuthally uniform rather than concentrated in an arc.

The mid-sized moons Tethys and Dione each have a pair of Trojan moons orbiting 60° behind and ahead of them. Telesto and Calypso orbit behind and ahead of Tethys, respectively. Polydeuces and Helene orbit behind and ahead of Dione, respectively. Unlike the case of Janus and Epimetheus, the Trojans are so small that they exert no discernible effects upon Tethys or Dione.

Hyperion is in a 4:3 mean-motion resonance with Titan, which pumps up Hyperion's orbital eccentricity to a large value.

7.2. Rotational States

Hyperion is the only object in the solar system that is known to have a chaotic tumbling rotation state. This can be understood as being due to the combined effect of its nonspherical shape and its noncircular orbit (i.e., high orbital eccentricity). Other objects are known to be more nonspherical than Hyperion but to not tumble because their orbital eccentricities are low enough to allow for orderly rotation. Other objects have higher orbital eccentricities than Hyperion, but do not tumble because their shapes are spherical enough (or, as we shall see, elongated enough so as to go past the unstable region in parameter space) to allow for orderly rotation.

Following the common progression of discovery, Hyperion was visible to observers as only a point of light for decades following its discovery. A standard procedure for objects in this situation is to repeatedly measure the brightness of that point of light as a function of time (the "lightcurve") and to look for periodicities in the lightcurve from which the rotation period can be inferred. This method can be effective because of albedo features, which cause the total brightness to vary depending on whether they face toward or away from the observer; it can also be effective because of a significantly nonspherical shape, which causes the face-on area of the object as seen by the observer to vary. Sometimes albedo features and/or nonsphericity are very subtle, such that periodicities in the lightcurve are difficult to detect. Hyperion, however, was remarkable in that its lightcurve showed strong variations, but those variations did not exhibit a stable periodicity. This was shown by *Wisdom et al.* (1984, 1987) to be due to a chaotic rotation state, under which Hyperion's rotation may appear "quasi-regular" for relatively short periods of time but becomes unstable with a Lyapunov time of a few weeks (see also *Klavetter,* 1989a,b; *Black et al.,* 1995). Imaging of Hyperion by Cassini during close flybys led to some constraints on the moments of inertia, but did not yield a unique rotation model (*Harbison et al.,* 2011).

A moon on an eccentric orbit moves faster along its orbit at periapse and slower at apoapse (this is Kepler's Second Law). If the moon rotates at a rate that is synchronized with the mean orbital period, then, to first order in the eccentricity, its prime meridian lags behind the direction to the planet from periapse to apoapse, and leads it from apoapse to periapse (the prime meridian faces the planet directly only at periapse and apoapse). This variation is the *optical libration,* and is shown as ϕ in Fig. 15. Consequentially, if the moon is not spherical but rather oriented with its long axis pointed toward the mean location of the planet, the planet will have a lever-arm with which to torque the moon's rotation. The standard physics of a driven oscillator (for details, see *Murray and Dermott,* 1999) then yields the picture shown in Fig. 15, in which the moon oscillates ("librates") about the mean orientation with period that matches the mean orbital rate. The driven portion of the libration is called the *physical libration,* and is shown as γ in Fig. 15; the actual

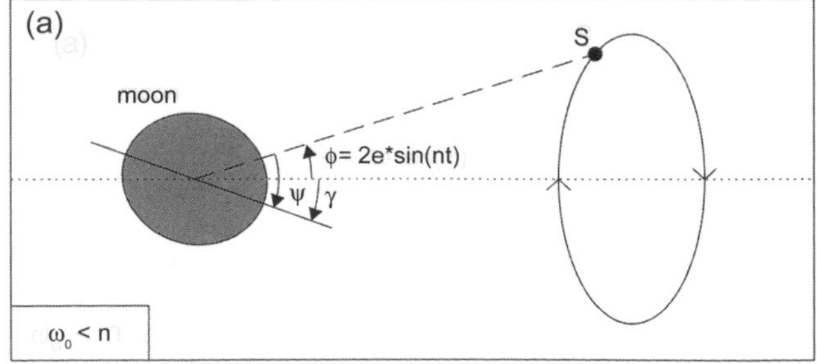

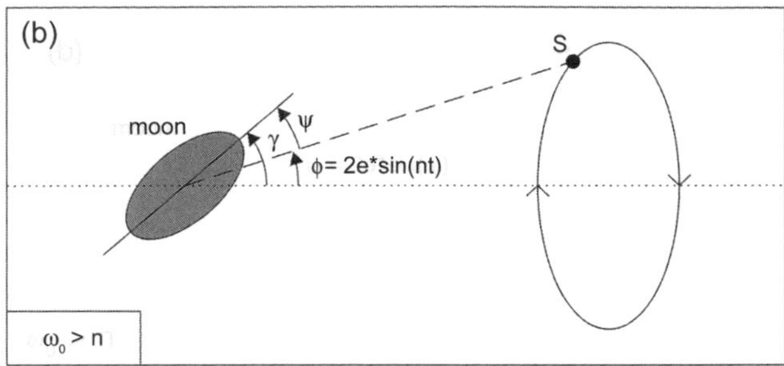

Fig. 15. The orientation of a moon's long axis under forced libration for a **(a)** near spherical or **(b)** elongated moon, measured from the direction toward Saturn (ψ) and from the direction towards the empty focus of the moon's orbit (γ). The latter is the deviation from synchronous rotation for $e \ll 1$. All angles are positive in the counterclockwise direction, so both ψ and γ are negative in **(a)**. The optical libration angle ϕ is determined by the eccentricity and orbital phase. The natural frequency ω_0, which is determined by the shape parameter $\Gamma \equiv (B-A)/C$ (equation (1)), determines whether the moon's long axis points away from or past the planet, as seen from the empty focus of the moon's orbit (towards which a synchronously rotating moon would face). The transition at $\omega_0 \sim n$ occurs when $\Gamma \sim 1/3$. Figure adapted from *Tiscareno et al.* (2009).

angle between the planet and the moon's long axis might be called the *tidal libration,* $\psi = \phi + \gamma$.

The moon's natural libration frequency, at which it would oscillate if the forcing were to vanish, turns out to be

$$\omega_0 = n\sqrt{3\left(\frac{B-A}{C}\right)} \qquad (1)$$

where n is the mean orbital rate and $A < B < C$ are the principal moments of inertia. Because $\omega_0 \propto n$, the orbital rate drops out of the equation and the libration amplitude depends only on the shape parameter $\Gamma \equiv (B-A)/C$. Specifically (e.g., *Tiscareno et al.,* 2009)

$$\gamma_{max} = \frac{2e}{1-1/3\Gamma} \qquad (2)$$

A resonance arises when the moment of inertia ratio Γ approaches 1/3, at which the denominator of equation (2) approaches zero. Both Epimetheus and Hyperion are close to this resonance; in fact, Epimetheus is even closer than Hyperion (Fig. 16), and measurements of its libration amplitude of 6° enabled a much finer determination of its moments of inertia than could be accomplished from measuring its shape (*Tiscareno et al.,* 2009). Hyperion, on the other hand, is close enough to the resonance that its 10× higher eccentricity (due to its 4:3 mean-motion resonance with Titan; see section 7.1) causes its amplitude as predicted by our linear theory to grow to the point that our small-amplitude approximation no longer applies. This heuristic discussion might provide an illuminating means of understanding the origin of Hyperion's chaotic rotation.

An overview of the parameters relevant to rotation is given for Saturn's inner moons in Fig. 17. Most significantly, the physical librations of both Mimas and Enceladus have been measured to be much larger than predicted by linear theory, which has been interpreted in both cases as evidence of a subsurface global ocean that causes the dead weight of the moon's core to have no connection to the surface shell (*Tajeddine et al.,* 2014; *Thomas et al.,* 2016). Enceladus, in particular, has a low orbital eccentricity in addition to being quite close to spherical in shape; only the very large number of images taken over the decade-plus extent of the Cassini mission made it possible to measure its libration amplitude of 0.1° with convincing precision.

Among the moons with predicted libration amplitude γ large enough that it might be measured, Telesto stands out as being even closer to the $\Gamma \sim 1/3$ resonance than Epimetheus — i.e., Telesto's (B–A)/C (triangular point in Fig. 17) is closer to the horizontal dashed line than is Epimetheus' — although its low e means its predicted γ is modest. Pandora, Pallene, and Polydeuces are better candidates, with relatively higher eccentricities that bring their predicted γ into the vicinity of 1°. Prometheus, Aegaeon, Methone, and Calypso have $\Gamma > 1/3$, which means they should be librating in the opposite sense (Fig. 15b), a state that has never been detected in nature. However, Aegaeon appears to be so elongated that $\Gamma \gg 1/3$, which means its physical libration will likely be as difficult to detect as those of near-spherical moons.

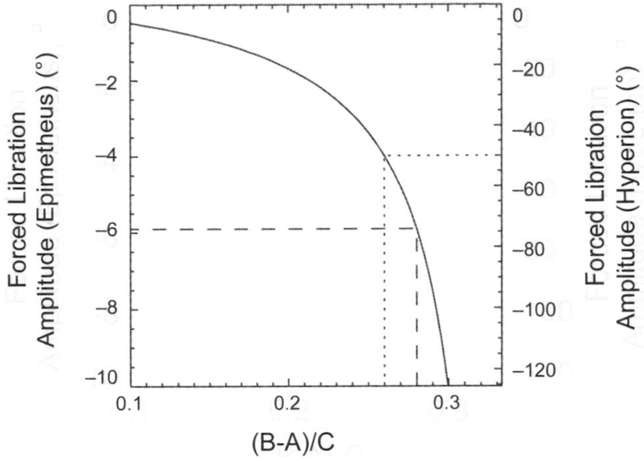

Fig. 16. Predicted libration amplitude γ for Epimetheus (left axis, dashed lines) and Hyperion (right axis, dotted lines), according to linear theory. The different scalings are due to different values of e in equation (2). Epimetheus is even closer than Hyperion to the resonance at (B–A)/C ~ 1/3, but Hyperion's higher e causes it to be the one whose predicted amplitude is so high that it violates the assumptions of the small-amplitude model, such that its actual rotation state is chaotic tumbling.

7.3. Constraints on Ages and Processes

Consensus on the dynamical history of the small and mid-sized moons of Saturn is currently in flux. Although origin scenarios are easier to construct when set near the beginning of solar system history, longtime suspicions that Saturn's rings are only ~100 m.y. old have recently intensified due to realizations that the rings are less massive than previously thought (*Hedman and Nicholson,* 2016) and that the pollution of the rings' pristine water ice by interplanetary dust is faster than previously thought (*Kempf et al.,* 2015). The simultaneous realization that orbital evolution of moon orbits due to Saturn's internal friction is faster than previously thought (*Lainey et al.,* 2015) has led some to conclude that even the mid-sized moons (e.g., Enceladus) cannot be more than 100 m.y. old (*Ćuk et al.,* 2016). Genesis from a young massive ring, rather than from the protosatellite disk, was previously discussed for the small (*Charnoz et al.,* 2010) and mid-sized (*Crida and Charnoz,* 2012) moons and should be revisited in light of these new results.

8. MAJOR OPEN QUESTIONS

After 12 years of Cassini observations, and after other missions that have investigated small objects, our appreciation of their surface processes and the fundamentals of their history remains tenuous. Much of our expectations for these irregularly shaped, low-gravity, often low-density bodies are guided by the larger and much more detailed study of the Moon and other "large" objects (including Earth). Yet even some fundamental processes and rates on the Moon remain poorly understood, for example, the rate and mechanism of surface regolith turnover (*Speyerer et al.,* 2016). Additionally, detailed study of comet nuclei has yet to provide convincing baselines for processes on small rocky and icy but nonsublimating bodies. The association of different surface morphologies with different dynamical positions suggests histories that are dependent upon the system as a whole, in particular, the dynamics of material exchange between objects. Thus, the most important question is how dynamics and the surface processes interact. Overlapping this question is the longstanding effort to derive the impactor history of the system, and even more generally, the age of the system.

We may translate the above into some more specific queries, such as:

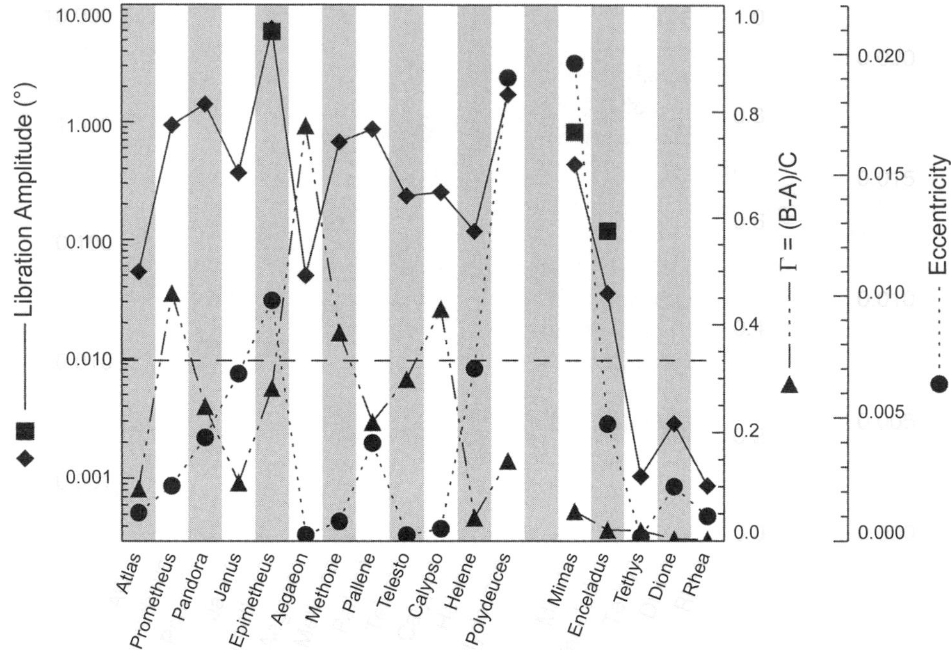

Fig. 17. Orbital eccentricity (circles connected by dotted lines), shape parameter Γ (triangles connected by dot-dot-dash lines), and libration amplitude γ from linear theory (circles connected by solid lines) and from observation (squares) for selected moons of Saturn. The horizontal dashed line indicates the resonance at Γ ~ 1/3.

- What are the ages of the satellites?
- Why are the surfaces different in each dynamical group?
- What processes shape the ring-arc moons, and why are they apparently unique?
- How many, and what, source or exchange processes control the variations in color radially from Saturn?
- Most generally, how are these objects integrated into, and in turn illuminate, an overall model of the development of the Saturn satellite and ring system?

We have the basic information that will stand for many years; interpretation of all the data may require that much time as well.

Acknowledgments. P.T. and P.H. are supported by the Cassini Project. M.S.T. acknowledges funding from NASA grant NNX-15AL21G and the Cassini project. We thank the myriad people who designed, built, and flew the Cassini spacecraft. Helpful reviews were provided by H. J. Melosh and an anonymous reviewer, and R. Dotson provided excellent manuscript support.

REFERENCES

A'Hearn M. F. et al. (2011) EPOXI at Comet Hartley 2. *Science, 332,* 1396.

Alvarellos J. L., Zahnle K. J., Dobrovolskis A. R., and Hamill P. (2005) Fates of satellite ejecta in the Saturn system. *Icarus, 178,* 104–123.

Archinal B. A. et al. (2011) Report of the IAU Working Group on cartographic coordinates and rotational elements: 2009. *Cel. Mech. Dyn. Astron., 109,* 101–135.

Bierhaus E. B., Dones L., Alvarellos J. L., and Zahnle K. (2012) The role of ejecta in the small crater populations on the mid-sized saturnian satellites. *Icarus, 218,* 602–621.

Black G. J., Nicholson P. D., and Thomas P. C. (1995) Hyperion: Rotational dynamics. *Icarus, 117,* 149–161.

Brown R. H. et al. (2004) The Cassini Visual and Infrared Mapping Spectrometer (VIMS) Investigation. *Space Sci. Rev., 115,* 111–168.

Buczkowski D. L., Barnouin-Jha O. S., and Prockter L. M. (2008) 433 Eros lineaments: Global mapping and analysis. *Icarus, 193,* 39–52.

Buratti B. J., Hicks M. D., and Davies A. (2005) Spectrophotometry of the small satellites of Saturn and their relationship to Iapetus, Phoebe, and Hyperion. *Icarus, 175,* 490–495.

Buratti B. J. et al. (2010) Cassini spectra and photometry 0.25–5.1 μm of the small inner satellites of Saturn. *Icarus, 206,* 524–536.

Carrier W. III, Ohlcroft G., and Mendell W. (1991) Physical properties of the lunar surface. In *Lunar Sourcebook: A Users Guide to the Moon* (G. Heiken et al., eds.), pp. 475–595. Cambridge Univ., Cambridge.

Castillo-Rogez J. C., Johnson T. V., Thomas P. C., Choukroun M., Matson D. L., and Lunine J. I. (2012) Geophysical evolution of Saturn's satellite Phoebe, a large planetesimal in the outer solar system. *Icarus, 219,* 86–109.

Chandrasekhar S. (1969) *Ellipsoidal Figures of Equilibrium.* Yale Univ., New Haven. 264 pp.

Chapman C. R. and McKinnon W. B. (1986) Cratering of planetary satellites. In *Satellites* (J. A. Burns and M. S. Matthews, eds.), pp. 492–580. Univ. of Arizona, Tucson.

Charnoz S., Brahic A., Thomas P. C., and Porco C. C. (2007) The equatorial ridges of Pan and Atlas: Terminal accretionary ornaments? *Science, 318,* 1622–1624.

Charnoz S., Salmon J., and Crida A. (2010) The recent formation of Saturn's moonlets from viscous spreading of the main rings. *Nature, 465,* 752–754.

Cooper N. J., Renner S., Murray C. D., and Evans M. W. (2015) Saturn's inner satellites: Orbits, masses, and the chaotic motion of Atlas from new Cassini imaging observations. *Astron. J., 149,* 27.

Crater Analysis Techniques Working Group (1979) Standard techniques for presentation and analysis of crater size-frequency data. *Icarus, 37,* 467–474.

Crida A. and Charnoz S. (2012) Formation of regular satellites from ancient massive rings in the solar system. *Science, 338,* 1196.

Cruikshank D. P. et al. (2007) Surface composition of Hyperion. *Nature, 448,* 54–56.

Cruikshank D. P., Dalle Ore C. M., Pendleton Y. J., and Clark R. N. (2012) Hydrocarbons on Phoebe, Iapetus, and Hyperion: Quantitative analysis. *AAS/DPS Meeting Abstracts #44,* 104.03.

Ćuk M., Dones L., and Nesvorny D. (2016) Dynamical evidence for a late formation of Saturn's moons. *Astrophy. J., 820,* 97.

Dalton J. B. and Cruikshank D. P. (2007) Compositional mapping of Hyperion with Cassini VIMS. *Bull. Am. Astron. Soc., 39,* 11.10.

Dermott S. F. (1979) Shapes and gravitational moments of satellites and asteroids. *Icarus, 37,* 575–586.

Dermott S. F. and Thomas P. C. (1988) The shape and internal structure of Mimas. *Icarus, 73,* 25–65.

Dobrovolskis A. R., Alvarellos J. L., Zahnle K. J., and Lissauer J. J. (2009) Fates of ejecta from co-orbital satellites: Tethys, Telesto, and Calypso. Abstract P51C-1146 presented at the 2009 Fall Meeting, AGU, San Francisco, Calif., 14–18 Dec.

Dones L., Chapman C. R., and McKinnon W. B. (2009) Icy satellites of Saturn: Impact cratering and age determination. In *Saturn from Cassini-Hyugens* (M. K. Dougherty et al., eds.), pp. 613–635. Springer, Dordrecht.

Esposito L. W. et al. (2004) The Cassini Ultraviolet Imaging Spectrograph investigation. *Space Sci. Rev., 115,* 299–361.

Filacchione G. et al. (2010) Saturn's icy satellites investigated by Cassini-VIMS. II. Results at the end of nominal mission. *Icarus, 206,* 507–523.

Filacchione G. et al. (2012) Saturn's icy satellites and rings investigated by Cassini-VIMS: III — Radial compositional variability. *Icarus, 220,* 1064–1096.

Hamilton D. P. et al. (2015) Small particles dominate Saturn's Phoebe ring to surprisingly large distances. *Nature, 522,* 185–187.

Hapke B. (2012) *Theory of Reflectance and Emittance Spectroscopy.* Cambridge Univ., New York. 513 pp.

Hapke B. and Sato H. (2016) The porosity of the upper lunar regolith. *Icarus, 273,* 75–83.

Harbison R. A., Thomas P. C., and Nicholson P. D. (2011) Rotational modeling of Hyperion. *Cel. Mech. Dyn. Astron., 110,* 1–16.

Harris A. W., Fahnestock E. G., and Pravec P. (2009) On the shapes and spins of "rubble pile" asteroids. *Icarus, 199,* 310–318.

Hedman M. M. and Nicholson P. D. (2016) The B ring's surface mass density from hidden density waves: Less than meets the eye? *Icarus, 279,* 109–124.

Hedman M. M., Murray C. D., Cooper N. J., Tiscareno M. S., Beurle K., Evans M. W., and Burns J. A. (2009) Three tenuous rings/arcs for three tiny moons. *Icarus, 199,* 378–386.

Hedman M. M., Cooper L. J., Murray C. D., Beurle K., Evans M. W., Tiscareno M. S., and Burns J. A. (2010) Aegaeon (Saturn LIII), a G-ring object. *Icarus, 207,* 433–447.

Helfenstein P. et al. (1994) Galileo photometry of asteroid 951 Gaspra. *Icarus, 107,* 37–60.

Hirata N. (2016) Differential impact cratering of Saturn's satellites by heliocentric impactors. *J. Geophys. Res., 121,* 111–117.

Hirata N., Miyamoto H., and Showman A. P. (2014) Particle deposition on the saturnian satellites from ephemeral cryovolcanism on Enceladus. *Geophys. Res. Lett., 41,* 4135–4141.

Horedt G. P. and Neukum G. (1984) Planetocentric versus heliocentric impacts in the jovian and saturnian satellite system. *J. Geophys. Res., 89,* 10405–10410.

Horstman K. C. and Melosh H. J. (1989) Drainage pits in cohesionless materials — Implications for the surface of Phobos. *J. Geophys. Res., 94,* 12433–12441.

Housen K. R. and Holsapple K. A. (2003). Impact cratering on porous asteroids. *Icarus, 163,* 102–119.

Howard A. D., Moore J. M., Schenk P. M., White O. L., and Spencer J. (2012) Sublimation-driven erosion on Hyperion: Topographic analysis and landform simulation model tests. *Icarus, 220,* 268–276.

Hurford T. A., Asphaug E., Spitale J. N., Hemingway D., Rhoden A. R., Henning W. G., Bills B. G., Kattenhorn S. A., and Walker M. (2016) Tidal disruption of Phobos as the cause of surface fractures. *J. Geophys.Res., 121,* 1054–1065.

Ip W. H. (2006) On a ring origin of the equatorial ridge of Iapetus. *Geophys. Res. Lett., 33,* L16203.

Jacobson R. A. (2014) The small saturnian satellites: Chaos and conundrum. *AAS/DDA Meeting Abstracts #45,* 304.05.

Jacobson R. A. et al. (2006) The gravitational field of Saturn and the masses of its major satellites. *Astron. J., 132,* 2520–2526.

Jacobson R. A., Spitale J., Porco C. C., Beurle K., Cooper N. J., Evans M. W., and Murray C. D. (2008) Revised orbits of Saturn's small inner satellites. *Astron. J., 135,* 261–263.

Judson A. and Doesken N. (1999) *Density of Freshly Fallen Snow in the Central Rocky Mountains.* Climatology Report 99-2, Colorado State University, Fort Collins. 36 pp.

Kargel J. S. and Pozio S. (1996) The volcanic and tectonic history of Enceladus. *Icarus, 119,* 385–404.

Kempf S., Horanyi M., Srama R., and Altobelli N. (2015) Exogenous dust delivery into the Saturnian system and the age of Saturn's rings. *EPSC Abstracts, Vol. 10,* 411.

Kirchoff M. R. and Schenk P. (2010) Impact cratering records of the mid-sized, icy saturnian satellites. *Icarus, 206,* 485–497.

Klavetter J. J. (1989a) Rotation of Hyperion. I — Observations. *Astron. J., 97,* 570–579.

Klavetter J .J. (1989b) Rotation of Hyperion. II — Dynamics. *Astron. J., 98,* 1855–1874.

Lainey V. et al. (2015) New constraints on Saturn's interior from Cassini astrometric data. *Icarus, 281,* 286.

Miller C. et al. (2011) Comparing Phoebe's 2005 opposition surge in four visible light filters. *Icarus, 212,* 819–834.

Morrison S. J., Thomas P. C., Tiscareno M. S., Burns J. A., and Veverka J. (2009) Grooves on small saturnian satellites and other objects: Characteristics and significance. *Icarus, 204,* 262–270.

Murray C. D. and Dermott S. F. (1999) *Solar System Dynamics.* Cambridge Univ., Cambridge. 575 pp.

Murray C. D. and Giuliatti Winter S. M. (1996) Periodic collisions between the moon Prometheus and Saturn's F ring. *Nature, 380,* 139–141.

Nayak M. and Asphaug E. (2015) Sesquinary catanae on Phobos from reaccretion of ejected material. *AAS/DPS Meeting Abstracts #47,* 201.08.

Nayak M. and Asphaug E. (2016) Sesquinary catenae on the martian satellite Phobos from reaccretion of escaping ejecta. *Nature Commun., 7,* 12591.

Ostro S. J. et al. (2006) Radar imaging of binary near-Earth asteroid (66391) 1999 KW4. *Science, 314,* 1276–1280.

Owen W. M. Jr. (2003) *Cassini ISS Geometric Calibration.* JPL Interoffice Memorandum 312, E-2003-001.

Plescia J. B. and Boyce J. M. (1985) Impact cratering history of the saturnian satellites. *J. Geophys. Res., 90,* 2029–2037.

Porco C. C. et al. (2004) Cassini imaging science: Instrument characteristics and anticipated scientific investigations at Saturn. *Space Sci. Rev., 115,* 363–497.

Porco C. C. et al. (2005) Cassini imaging science: Initial results on Saturn's rings and small satellites. *Science, 307,* 1226–1236.

Porco C. C., Thomas P. C., Weiss J. W., and Richardson D. C. (2007) Saturn's small inner satellites: Clues to their origins. *Science, 318,* 1602.

Richardson J. E. and Bowling T. J. (2014) Investigating the combined effects of shape, density, and rotation on small body surface slopes and erosion rates. *Icarus, 234,* 53–65.

Richardson J. E. and Thomas P. C. (2010) Uncovering the saturnian impactor population via small satellite cratering records. In *Lunar and Planetary Science XLI,* Abstract #1523. Lunar and Planetary Institute, Houston.

Shoemaker E. M. and Wolfe R. F. (1982) Cratering time scales for the Galilean satellites. In *Satellites of Jupiter* (J. A. Burns and M. S. Matthews, eds.), pp. 277–339. Univ. of Arizona, Tucson.

Sierks H. et al. (2015) On the nucleus structure and activity of omet 67P/Churyumov-Gerasimenko. *Science, 347,* aaa1044.

Simonelli D.P. et al. (1999) Phoebe: Albedo map and photometric properties. *Icarus, 138,* 249–258.

Smith B. A. et al. (1981) Encounter with Saturn — Voyager 1 imaging science results. *Science, 212,* 163–191.

Smith B. A. et al. (1982) A new look at the Saturn system — The Voyager 2 images. *Science, 215,* 504–537.

Soter S. and Harris A. (1977) Are striations on Phobos evidence for tidal stress? *Nature, 268,* 421.

Speyerer E. J., Povilaitis R. Z., Robinson M. S., Thomas P. C., and Wagner R. V. (2016) Quantifying crater production and regolith overturn on the Moon with temporal imaging. *Nature, 538,* 215–218.

Spencer J. R. and Denk T. (2010) Formation of Iapetus's extreme albedo dichotomy by exogenically triggered thermal ice migration. *Science, 327,* 432.

Spitale J. N., Jacobson R. A., Porco C. C., and Owen W. M. Jr. (2006) The orbits of Saturn's small satellites derived from combined historic and Cassini imaging observations. *Astron. J., 132,* 692–710.

Strom R. G. (1987) The solar system cratering record — Voyager 2 results at Uranus and implications for the origin of impacting objects. *Icarus, 70,* 517–535.

Strom R. G. and Woronow A. (1982) Solar system cratering populations. In *Lunar and Planetary Science XIII,* pp. 782–783. Lunar and Planetary Institute, Houston.

Tajeddine R., Rambaux N., Lainey V., Charnoz S., Richard A., Rivoldini A., and Noyelles B. (2014) Constraints on Mimas' interior from Cassini ISS libration measurements. *Science, 346,* 322–324.

Thomas P. C. (1993) Gravity, tides, and topography on small satellites and asteroids — Application to surface features of the martian satellites. *Icarus, 105,* 326–334.

Thomas P. C. (1998) Ejecta emplacement on the martian satellites. *Icarus, 131,* 78–106.

Thomas P. C. (1999) Large craters on small objects: Occurrence, morphology, and effects. *Icarus, 142,* 89–96.

Thomas P. C. et al. (2007) Hyperion's sponge-like appearance. *Nature, 448,* 50–56.

Thomas P. C., Burns J. A., Hedman M., Helfenstein P., Morrison S., Tiscareno M. S., and Veverka J. (2013) The inner small satellites of Saturn: A variety of worlds. *Icarus, 226,* 999–1019.

Thomas P. C., Tajeddine R., Tiscareno M. S., Burns J. A., Joseph J., Loredo T. J., Helfenstein P., and Porco C. (2016) Enceladus's measured physical libration requires a global subsurface ocean. *Icarus, 264,* 37.

Tiscareno M. S., Thomas P. C., and Burns J. A. (2009) The rotation of Janus and Epimetheus. *Icarus, 204,* 254–261.

Umurhan O. M., Howard A. D., Moore J. M., Schenk P. M., and White O. L. (2015) Reconstructing Helene's surface history, plastics and snow. In *Lunar and Planetary Science XLVI,* Abstract #2400. Lunar and Planetary Institute, Houston.

Vanicek P. and Krakiwsky E. J. (1986) *Geodesy: The Concepts,* 2nd edition. Elsevier, New York. 714 pp.

Verbiscer A. and Veverka J. (1992) Mimas: Photometric roughness and albedo map. *Icarus, 99,* 63–69.

Verbiscer A., French R., Showalter M., and Helfenstein P. (2007) Enceladus: Cosmic graffiti artist caught in the act. *Science, 315,* 815.

Verbiscer A., Skrutskie M., and Hamilton D. P. (2009) Saturn's largest ring. *Nature, 461,* 1098–1100.

Vilas F., Larsen S. M., Stockstill K. R., and Gaffley M. J. (1996) Unraveling the zebra: Clues to the Iapetus dark material composition. *Icarus, 124,* 262–267.

Weidenschilling S. J. (1979) A possible origin for the grooves of Phobos. *Nature, 282,* 697.

Weiss J. W., Porco C. C., and Tiscareno M. S. (2009) Ring edge wakes and the masses of nearby satellites. *Astron. J., 138,* 272–286.

West R. A. et al. (2010). In-flight calibration of the Cassini imaging science sub-system cameras. *Planet. Space Sci., 58,* 1475–1488.

Wisdom J., Peale S. J., and Mignard F. (1984) The chaotic rotation of Hyperion. *Icarus, 58,* 137–152.

Wisdom J. (1987) Rotational dynamics of irregularly shaped natural satellites. *Astron. J., 94,* 1350–1360.

Yoder C. F., Synnott S. P., and Salo H. (1989) Orbits and masses of Saturn's co-orbiting satellites, Janus and Epimetheus. *Astron. J., 98,* 1875–1889.

Zahnle K., Schenk P., Sobieszczyk S., Dones L., and Levison H. F. (2001) Differential cratering of synchronously rotating satellites by ecliptic comets. *Icarus, 153,* 111–129.

Zahnle K., Alvarellos J. L., Dobrovolskis A., and Hamill P. (2008) Secondary and sesquinary craters on Europa. *Icarus, 194,* 660–674.

Denk T., Mottola S., Tosi F., Bottke W. F., and Hamilton D. P. (2018) The irregular satellites of Saturn. In *Enceladus and the Icy Moons of Saturn* (P. M. Schenk et al., eds.), pp. 409–434. Univ. of Arizona, Tucson, DOI: 10.2458/azu_uapress_9780816537075-ch020.

The Irregular Satellites of Saturn

Tilmann Denk
Freie Universität Berlin

Stefano Mottola
Deutsches Zentrum für Luft- und Raumfahrt

Federico Tosi
Istituto di Astrofisica e Planetologia Spaziali

William F. Bottke
Southwest Research Institute

Douglas P. Hamilton
University of Maryland, College Park

With 38 known members, the outer or irregular moons constitute the largest group of satellites in the saturnian system. All but exceptionally big Phoebe were discovered between the years 2000 and 2007. Observations from the ground and from near-Earth space constrained the orbits and revealed their approximate sizes (~4 to ~40 km), low visible albedos (likely below ~0.1), and large variety of colors (slightly bluish to medium-reddish). These findings suggest the existence of satellite dynamical families, indicative of collisional evolution and common progenitors. Observations with the Cassini spacecraft allowed lightcurves to be obtained that helped determine rotational periods, coarse shape models, pole-axis orientations, possible global color variations over their surfaces, and other basic properties of the irregulars. Among the 25 measured moons, the fastest period is 5.45 h. This is much slower than the disruption rotation barrier of asteroids, indicating that the outer moons may have rather low densities, possibly as low as comets. Likely non-random correlations were found between the ranges to Saturn, orbit directions, object sizes, and rotation periods. While the orbit stability is higher for retrograde objects than for progrades very far away from Saturn, a compelling physical cause for size and spin relations to orbital elements is not yet known. The large moon Phoebe was resolved by Cassini during a close flyby in June 2004, showing numerous craters of all sizes on a surface composed of water ice and amorphous carbon. While the origin of the irregulars is still debated, capture of comets via three-body interactions during giant planet encounters do the best job thus far at reproducing the observed orbits. This chapter gives a summary of our knowledge of Saturn's irregular moons as of the end of 2017.

1. INTRODUCTION

The outer or irregular moons of Saturn are a class of objects that is very distinct from the other satellites treated in this book. It not only has more objects (38 are presently known) than the class of the inner moons (24), but also occupies a much larger volume within the Hill sphere of Saturn. On the other hand, almost all "irregulars" are quite small: Besides the large moon Phoebe (213 km diameter), 37 objects of sizes on the order of ~40 km down to ~4 km are known (the uncertainties of the diameter values are still substantial). Additional ones smaller than 4 km certainly exist as well. Therefore, they significantly contribute to the total number, but not to the overall mass of Saturn's satellite system (less than 0.01%).

The discrimination between regular (or inner) and irregular (or outer) satellites is through distance to the center planet, orbit eccentricities, and inclinations. The irregulars reach ranges to Saturn between 7.6 × 10^6 km (at the periapsis of Kiviuq; ~12% of Saturn's Hill sphere radius of ~65 × 10^6 km) and 33 × 10^6 km (~50% of the Hill radius at the apoapsis of Surtur). They require between 1.3 and 4.1 years (Ijiraq and Fornjot, respectively) for one revolution around Saturn. As a comparison, Iapetus, the regular satellite farthest from Saturn, has an apoapsis distance of 3.6×10^6 km and requires just 0.22 years for one orbit. The separation distance is the so-called critical semimajor axis (*Goldreich*, 1966; *Burns*, 1986), which depends on the mass and radius of Saturn, the quadrupole gravitational harmonic J_2' of the saturnian system (e.g., *Tremaine et al.,* 2009), as well as the distance Saturn-to-Sun and the mass of the Sun. This range marks the location where the precession of the satellite's orbital plane is dominated by the Sun rather than by the planet's oblateness (e.g., *Shen and Tremaine*, 2008). For Saturn, $a_{crit} \sim 3.4 \times 10^6$ km or ~5% of its Hill sphere; this is close to the orbit of Iapetus (which is not considered as an irregular moon although many mysteries about its origin still exist). The orbit eccentricities of Saturn's irregulars vary between 0.11 and 0.54, while the orbits of the regular moons, except for Hyperion, are almost circular ($e < 0.03$). The inclinations of the regular moons against the local Laplace plane are very close to zero (Iapetus with $i \sim 8°$ deviates most), while those of the irregulars may by principle vary between 0° and 180°, indicating that they might reside on retrograde paths (planetocentric coordinate system). At Saturn, 9 prograde and 29 retrograde irregular moons have been discovered so far. All were found in the stable dynamical region that surrounds Saturn (*Carruba et al.*, 2002; *Nesvorný et al.*, 2003; *Shen and Tremaine*, 2008).

As with all satellites of Saturn, the irregular moons cannot be seen by the naked eye. Therefore, the irregulars were unknown to the ancients, and Phoebe was discovered only 119 years before the publication of this book (*Pickering*, 1899a,b). Nevertheless, its discovery contained several "firsts": Phoebe was not just the first saturnian irregular moon to be discovered, but also the first outer (far distant) moon of any planet. It was the first-ever discovery of a moon through photography and the first moon in the solar system of which the direction of motion is opposite to the other moons of the common planet (*Pickering*, 1905).

The other irregulars of Saturn are even much fainter than Phoebe. Siarnaq and Albiorix barely scratch the 20-mag mark (V-band magnitude), and the others do not exceed 21 to 25 mag even under ideal observation conditions, making them difficult targets for Earth-based observers. An additional issue for observing these objects from Earth is the proximity of the very bright planet. For example, an orbital radius of 13×10^6 km translates into a maximum elongation of 30′ from Saturn, which becomes a challenge — especially for small objects — due to the planet's straylight. All the other 37 irregular moons of Saturn were thus discovered rather recently (Table 1). Their discovery became possible with the introduction of highly sensitive large CCDs in combination with very large telescopes and the ability to process large volumes of data (*Gladman et al.*, 1998, 2001; *Nicholson et al.*, 2008). Furthermore, the impending arrival of the Cassini spacecraft was a major driver for the initiation of this search.

Saturn is not the only home of irregular satellites. All four large planets of our solar system host a large number of outer moons that revolve around their planet at large distances of many million kilometers on eccentric and inclined orbits. Similarly to the Jupiter Trojan asteroids, comets, Centaurs, Plutinos, classical Kuiper belt objects, etc., they constitute a distinct group of numerous objects residing in the outer solar system. As of the end of 2017, 114 outer moons were known in the solar system. Sixty-one of these moons orbit Jupiter, 38 orbit Saturn, 9 orbit Uranus, and 6 orbit Neptune. Most of them (96) were discovered between 10 and 20 years before the publication of this book, all through direct imaging from Earth. Before 1998, just 12 outer moons were known: Phoebe, Neptune's Nereid (discovered in 1949), the two uranian moons Sycorax and Caliban (discovered in 1997), and the eight "classical" irregulars of Jupiter (discovered between 1904 and 1974). A comprehensive summary of the different systems of irregular satellites is provided in *Nicholson et al.* (2008). Other nice summaries were written by *Sheppard* (2006), *Jewitt et al.* (2006), and *Jewitt and Haghighipour* (2007).

After the decade of extensive discoveries, almost no additional objects have been reported. The reason is presumably that few are left undiscovered at the accessible brightness ranges (e.g., *Hamilton,* 2001), but also that the required large telescopes are highly contested. Besides providing our summary of the state of the knowledge of the saturnian irregular moons, one task of this chapter is thus to motivate and encourage a new generation of solar system astronomers to initiate new search programs to discover the still fainter irregular moons and to help to further complete their inventory.

Ironically, the satellite discovery boom of the first decade of this century allowed the irregular moons to outnumber the regular planetary moons. While the ratio between inner and outer satellites of the giant planets was 48:11 in 1996, it is now 58:114. Thus, "irregular" is the rule, and "regular" the minority, at least by number of objects. However, outrivaling both groups are the known or suspected moons of the minor bodies. As of April 2018, 348 potential companions of 331 asteroids and transneptunian objects (TNOs) were listed (*Johnston,* 2018).

What do we know about the irregular moons of Saturn, and how did we learn it? What we know quite well for a large majority of them are astronomical properties like the orbital elements that were determined from Earth and for which a brief description is given in section 2.1. Roughly known or estimated are physical properties like absolute magnitudes, albedos (and combined the approximate sizes), or colors for the brightest objects. This knowledge comes from photometric measurements of groundbased observation data and from the Spitzer and Near-Earth Object Wide-field Infrared Survey Explorer (NEOWISE) missions and is presented in section 2.2. An important tool to obtain physical information, especially many rotational periods, was Cassini's Imaging Science Subsystem (ISS) (*Porco et al.,* 2004). Although operating at Saturn, Cassini was still too far away from the irregulars to resolve their surfaces (except for Phoebe). Section 3, which is mainly based on the work of Denk and Mottola (in preparation, 2018, hereafter *DM18*), summarizes some of the Cassini-based ongoing research of

Saturn's irregulars. Phoebe, discussed in section 4, was the sole irregular moon of Saturn where disk-resolved images of the surface were obtained by Cassini, mainly during the close flyby. The origin of the irregular moons is still debated. Several mechanisms were proposed, and the status quo is briefly described in section 5. The chapter ends with a summary of the most important missing information as well as prospects for future exploration (section 6).

TABLE 1. Discovery circumstances of Saturn's 38 irregular moons.

Moon Name	IAU Number	Provisional Designation	SPICE ID*	Observation Date†	Discoverer Group‡	IAU Circ. No.	IAU Circ. Issued§	Moon Abbrev.¶
Phoebe	IX	—	609	16 Aug 1898	Pickering	—	17 Mar 1899	Pho
Ymir	XIX	S/2000 S 1	619	07 Aug 2000	Gladman	7512	25 Oct 2000	Ymi
Paaliaq	XX	S/2000 S 2	620	07 Aug 2000	Gladman	7512	25 Oct 2000	Paa
Siarnaq	XXIX	S/2000 S 3	629	23 Sep 2000	Gladman	7513	25 Oct 2000	Sia
Tarvos	XXI	S/2000 S 4	621	23 Sep 2000	Gladman	7513	25 Oct 2000	Tar
Kiviuq	XXIV	S/2000 S 5	624	07 Aug 2000	Gladman	7521	18 Nov 2000	Kiv
Ijiraq	XXII	S/2000 S 6	622	23 Sep 2000	Gladman	7521	18 Nov 2000	Iji
Thrymr	XXX	S/2000 S 7	630	23 Sep 2000	Gladman	7538	07 Dec 2000	Thr
Skathi	XXVII	S/2000 S 8	627	23 Sep 2000	Gladman	7538	07 Dec 2000	Ska
Mundilfari	XXV	S/2000 S 9	625	23 Sep 2000	Gladman	7538	07 Dec 2000	Mun
Erriapus	XXVIII	S/2000 S 10	628	23 Sep 2000	Gladman	7539	07 Dec 2000	Err
Albiorix	XXVI	S/2000 S 11	626	09 Nov 2000	Holman	7545	19 Dec 2000	Alb
Suttungr	XXIII	S/2000 S 12	623	23 Sep 2000	Gladman	7548	23 Dec 2000	Sut
Narvi	XXXI	S/2003 S 1	631	08 Apr 2003	Sheppard	8116	11 Apr 2003	Nar
S/2004 S 7		S/2004 S 7	65035	12 Dec 2004	Jewitt	8523	04 May 2005	4S7
Fornjot	XLII	S/2004 S 8	642	12 Dec 2004	Jewitt	8523	04 May 2005	For
Farbauti	XL	S/2004 S 9	640	12 Dec 2004	Jewitt	8523	04 May 2005	Far
Aegir	XXXVI	S/2004 S 10	636	12 Dec 2004	Jewitt	8523	04 May 2005	Aeg
Bebhionn	XXXVII	S/2004 S 11	637	12 Dec 2004	Jewitt	8523	04 May 2005	Beb
S/2004 S 12		S/2004 S 12	65040	12 Dec 2004	Jewitt	8523	04 May 2005	4S12
S/2004 S 13		S/2004 S 13	65041	12 Dec 2004	Jewitt	8523	04 May 2005	4S13
Hati	XLIII	S/2004 S 14	643	12 Dec 2004	Jewitt	8523	04 May 2005	Hat
Bergelmir	XXXVIII	S/2004 S 15	638	12 Dec 2004	Jewitt	8523	04 May 2005	Ber
Fenrir	XLI	S/2004 S 16	641	13 Dec 2004	Jewitt	8523	04 May 2005	Fen
S/2004 S 17		S/2004 S 17	65045	13 Dec 2004	Jewitt	8523	04 May 2005	4S17
Bestla	XXXIX	S/2004 S 18	639	13 Dec 2004	Jewitt	8523	04 May 2005	Bes
Hyrrokkin	XLIV	S/2004 S 19	644	12 Dec 2004	Sheppard	8727	30 Jun 2006	Hyr
S/2006 S 1		S/2006 S 1	65048	04 Jan 2006	Sheppard	8727	30 Jun 2006	6S1
Kari	XLV	S/2006 S 2	645	04 Jan 2006	Sheppard	8727	30 Jun 2006	Kar
S/2006 S 3		S/2006 S 3	65050	05 Jan 2006	Sheppard	8727	30 Jun 2006	6S3
Greip	LI	S/2006 S 4	651	05 Jan 2006	Sheppard	8727	30 Jun 2006	Gre
Loge	XLVI	S/2006 S 5	646	05 Jan 2006	Sheppard	8727	30 Jun 2006	Log
Jarnsaxa	L	S/2006 S 6	650	05 Jan 2006	Sheppard	8727	30 Jun 2006	Jar
Surtur	XLVIII	S/2006 S 7	648	05 Jan 2006	Sheppard	8727	30 Jun 2006	Sur
Skoll	XLVII	S/2006 S 8	647	05 Jan 2006	Sheppard	8727	30 Jun 2006	Sko
Tarqeq	LII	S/2007 S 1	652	16 Jan 2007	Sheppard	8836	11 May 2007	Taq
S/2007 S 2		S/2007 S 2	65055	18 Jan 2007	Sheppard	8836	11 May 2007	7S2
S/2007 S 3		S/2007 S 3	65056	18 Jan 2007	Sheppard	8836	11 May 2007	7S3

* For SPICE, see JPL's NAIF web page (*https://naif.jpl.nasa.gov/*).

† Object was first spotted in an image taken on that date. Some of the 2006 objects were later found in 2004 data as well.

‡ The discoverer groups included the following: 1899: W. H. Pickering, Stewart; 2000: Gladman, Kavelaars, Petit, Scholl, Holman, Marsden, Nicholson, Burns; 2003: Sheppard; 2005, 2006, 2007: Sheppard, Jewitt, Kleyna.

§ Day of official announcement of the discovery. Phoebe was announced in a handwritten Bulletin of the Harvard College Observatory (*Pickering,* 1899a; the first IAU circular was published no earlier than October 1922).

¶ Abbreviations of moon names used in the figures.

2. RESULTS FROM (NEAR) EARTH OBSERVATIONS

2.1. Orbital Properties

Figure 1, based on the numbers of Table 2, shows the "orbital architecture" of the irregular-moon system of Saturn. The mean orbital elements a and i are displayed in a polar-coordinate plot, with the apoapsis-periapsis excursion being shown as thin bars for each object as a proxy for the eccentricity e. The inclinations are measured against the local Laplace plane, which is very close to the orbit plane of Saturn about the Sun for all irregulars. From Fig. 1, it becomes obvious that a fundamental classification of the irregular moons is the discrimination into objects with prograde and retrograde motions about Saturn in planetocentric coordinates. Furthermore, many of the moons cluster around similar a-e-i values and are thus likely parts of a "family" or the partners of a "pair." Possible relations are marked in Fig. 1 and Table 2.

Members of an object family share similar orbital elements and are genetically related, but not gravitationally bound anymore to each other. Among asteroids, families are believed to have formed through catastrophic collisions (e.g., *Margot et al.,* 2015). On the other hand, a pair contains just two objects that originally co-orbited, but for whatever reason separated in the past. Among Saturn's irregular moons, there are several objects that share their orbital elements with just one other moon. In this chapter, we will use the term "pair" for them although their origin is not known. They might be true orbital pairs, but also of collisional origin from a single object where smaller family members were simply not yet discovered. The existence of families among Saturn's irregulars was suspected soon after the first discoveries were made (*Gladman et al.,* 2001; *Grav et al.,* 2003). Their presence indicates that the individual irregulars we observe today might not have been captured independently, but could be remnants of originally larger moons.

Among the prograde objects, there exist two distinct inclination groups. The Gallic group (Albiorix, Tarvos, Erriapus, and Bebhionn, named after Gallic mythology characters) [for satellite naming, see, e.g., *Blunck* (2010)] is well clustered in the a-e-i space (Fig. 1, Table 2) and thus likely represents a collisional family; *Turrini et al.* (2008) modeled a dispersion velocity of ~130 m s^{-1}. For the other five prograde moons (Siarnaq, Paaliaq, Kiviuq, Ijiraq, Tarqeq), dubbed the Inuit group (named after characters from the Inuit folklore and mythology), a common origin is rather questionable. They share an inclination value of i ~ 46°, but their semimajor axes (a ~ 11.4 to 18.2 × 10^6 km) and eccentricities (e ~ 0.17 to 0.33) are quite different. However, within the Inuit group, *Turrini et al.* (2008) found a dispersion velocity of only ~100 m s^{-1} for the Ijiraq/Kiviuq satellite pair, making a common progenitor very plausible for these two objects. Siarnaq and Tarqeq might also form a pair, while no partner is known for Paaliaq.

The mean orbital elements of the 29 known retrograde objects, sometimes called the Norse group, widely range from a = 12.9 to 25.2 × 10^6 km, e = 0.11 to 0.52, and i = 145° to almost 180°. Herein, the orbit of large irregular moon Phoebe is very different from that of all other objects, and a clustering as for the progrades is not immediately obvious. In the work of *Turrini et al.* (2008), only one cluster of seven objects, but otherwise only small groups of two or three moons, show rather moderate dispersion velocities below 170 m s^{-1}. These potential families include only half of the

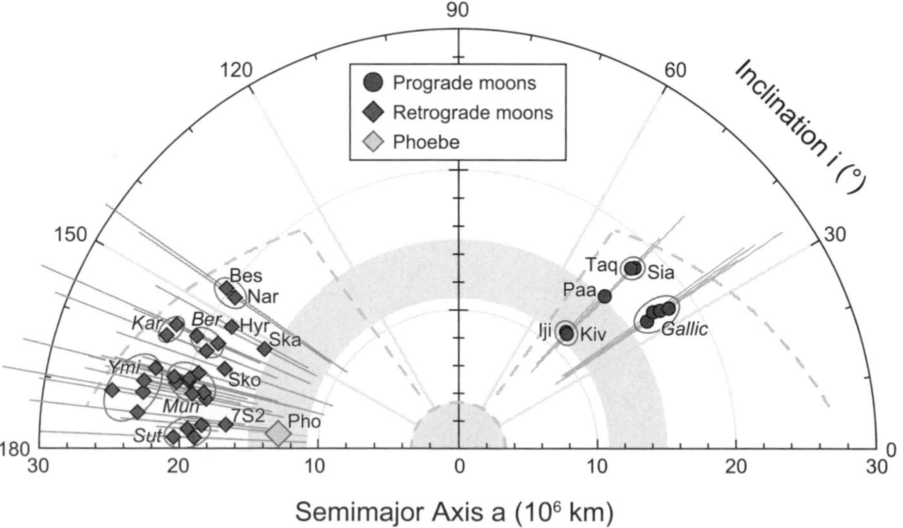

Fig. 1. Polar plot of the a-i space for the 38 irregular moons of Saturn. The thin bars are proxies for the eccentricities, by showing periapsis and apoapsis distances of each object. The light-gray band indicates the sphere of influence of Phoebe. The dashed lines show a_{crit} and apparent outer boundaries of the semimajor axes and inclinations for the irregulars. Most of the potential families or pairs are encircled. Individual moons are labeled except for members of families where only the name of the family is given (according to Table 2).

known retrogrades. In this context, it must be cautioned that some orbital elements used at the time of Turrini's work differed considerably from current, updated values, and that the orbits of six retrogrades are still so poorly determined that they are considered lost [objects S/2006 S 1, S/2004 S 7, S/2004 S 17, S/2007 S 3, S/2007 S 2, and S/2004 S 13 (*Jacobson et al.*, 2012)]. We thus suggest a slightly modified grouping with about half a dozen pairs and families where just 3–6 retrogrades (plus potentially Phoebe) were left as stand-alone moons (Fig. 1, Table 2).

A noticeable difference between the prograde and the retrograde moons are their average distances to Saturn. The

TABLE 2. Astronomical properties and sizes of Saturn's 38 irregular moons.

Moon Name	Group Member	*Family/* Pair*	×†	a (10^6 km)‡	e‡	i (°)‡	i' (°)‡	P (a)‡	H (mag)§	App. Mag.¶	Size (km)**	Moon Abbrev.
Phoebe	retro	?	—	12.95	0.16	175.2	4.8	1.50	6.6	16	213	Pho
Kiviuq	Inuit	(Iji)	×	11.38	0.33	46.8	46.8	1.23	12.6	22.0	17	Kiv
Ijiraq	Inuit	(Kiv)	×	11.41	0.27	47.5	47.5	1.24	13.2	22.6	13	Iji
Paaliaq	Inuit	–	×	15.20	0.33	46.2	46.2	1.88	11.7	21.3	25	Paa
Tarqeq	Inuit	(Sia)	×	17.96	0.17	46.3	46.3	2.43	14.8	23.9	6	Taq
Siarnaq	Inuit	(Taq)	×	18.18	0.28	45.8	45.8	2.45	10.6	20.1	42	Sia
Albiorix	Gallic	*Gallic*	×	16.39	0.48	34.1	34.1	2.15	11.1	20.5	33	Alb
Bebhionn	Gallic	*Gallic*	×	17.12	0.47	35.1	35.1	2.29	15.0	24.1	6	Beb
Erriapus	Gallic	*Gallic*	×	17.60	0.47	34.5	34.5	2.38	13.7	23.0	10	Err
Tarvos	Gallic	*Gallic*	×	18.24	0.54	33.7	33.7	2.53	12.9	22.1	15	Tar
Narvi	retro	(Bes)	×	19.35	0.43	145.7	34.3	2.75	14.4	23.8	7	Nar
Bestla	retro	(Nar)	×	20.21	0.51	145.1	34.9	2.98	14.6	23.8	7	Bes
Skathi	retro	–	×	15.64	0.27	152.6	27.4	1.99	14.3	23.6	8	Ska
S/2007 S 2	retro	–	×	16.7	0.18	174	6	2.21	15.3	24.4	5	7S2
Skoll	retro	–	×	17.67	0.46	161.0	19.0	2.40	15.4	24.5	5	Sko
Hyrrokkin	retro	(Gre?)	×	18.44	0.34	151.5	28.5	2.55	14.3	23.5	8	Hyr
Greip	retro	*Sut?* (Hyr?)	×	18.46	0.32	174.8	5.2	2.56	15.4	24.4	5	Gre
S/2007 S 3	retro	*Sut*		18.9	0.19	178	2	2.68	15.8	24.9	4	7S3
Suttungr	retro	*Sut*		19.47	0.11	175.8	4.2	2.78	14.5	23.9	7	Sut
Thrymr	retro	*Sut?* (4S7?)	×	20.42	0.47	177.7	2.3	3.00	14.3	23.9	8	Thr
S/2004 S 13	retro	*Mun*	×	18.4	0.26	169	11	2.56	15.6	24.5	4	4S13
Mundilfari	retro	*Mun*	×	18.65	0.21	167.4	12.6	2.61	14.5	23.8	7	Mun
Jarnsaxa	retro	*Mun*		19.35	0.22	163.6	16.4	2.76	15.6	24.7	4	Jar
S/2004 S 17	retro	*Mun*		19.4	0.18	168	12	2.78	16.0	25.2	4	4S17
Hati	retro	*Mun*	×	19.87	0.37	165.8	14.2	2.99	15.3	24.4	5	Hat
S/2004 S 12	retro	*Mun*	×	19.89	0.33	165.3	14.7	2.86	15.7	24.8	4	4S12
Aegir	retro	*Mun*		20.75	0.25	166.7	13.3	3.06	15.5	24.4	4	Aeg
S/2004 S 7	retro	*Mun?* (Thr?)	×	21.0	0.53	166	14	3.12	15.2	24.5	5	4S7
S/2006 S 1	retro	*Ber*		18.8	0.14	156	24	2.64	15.5	24.6	4	6S1
Bergelmir	retro	*Ber*		19.34	0.14	158.6	21.4	2.75	15.2	24.2	5	Ber
Farbauti	retro	*Ber (?)*		20.39	0.24	156.5	23.5	2.98	15.7	24.7	4	Far
Kari	retro	*Kar*	×	22.09	0.48	156.1	23.9	3.37	14.8	23.9	6	Kar
S/2006 S 3	retro	*Kar*	×	22.43	0.38	158.6	21.4	3.36	15.6	24.6	4	6S3
Fenrir	retro	*Ymi*		22.45	0.13	165.0	15.0	3.45	15.9	25.0	4	Fen
Surtur	retro	*Ymi*	×	22.94	0.45	169.7	10.3	3.55	15.8	24.8	4	Sur
Loge	retro	*Ymi*		23.06	0.19	167.7	12.3	3.59	15.3	24.6	5	Log
Ymir	retro	*Ymi*		23.13	0.33	173.5	6.5	3.60	12.3	21.7	19	Ymi
Fornjot	retro	*Ymi*		25.15	0.21	170.4	9.6	4.09	14.9	24.6	6	For

* Suggestions according to Fig. 1 (see also text). For just two objects, the partner is given in parantheses.
† Checked if periapsis range of moon is smaller than apoapsis range of Phoebe.
‡ Orbital semimajor axis a, eccentricity e, inclination i, inclination supplemental angle i' = 90°– |90°–i|, orbit period P. Planetocentric coordinates; from JPL's solar-system dynamics website (*https://ssd.jpl.nasa.gov/*).
§ Absolute magnitude H; the numbers may be uncertain by several tenths of magnitude. From MPC ephemeris service (*http://www.minorplanetcenter.net/iau/NatSats/NaturalSatellites.html*).
¶ Apparent optical magnitude (R-band) from Earth; from S. Sheppard's satellite and moon web page (*https://home.dtm.ciw.edu/users/sheppard/satellites/*).
** Calculated from H and assumed albedo A = 0.06 through $D = 2 \times 1 \text{ au} \times A^{-0.5} \times 10^{-0.2 \cdot (H - M_{\odot})}$; with $M_{\odot} = -26.71 \pm 0.02$ mag (*Pecaut and Mamajek*, 2013). Note that the errors may be large. Phoebe's value is from *Castillo-Rogez et al.* (2012).

average semimajor axis of the progrades is ~16 × 10^6 km and thus clearly smaller than the ~20 × 10^6 km for the retrogrades. That more "space" is used by retrograde objects appears to be a common phenomenon among the irregular-moon systems of all four giant planets (e.g., *Carruba et al.,* 2002; *Nicholson et al.,* 2008; work on this issue goes back to *Hénon*, 1969). Orbital stability curves in *Nesvorný et al.* (2003) (see also *Shen and Tremaine,* 2008) show a distinct asymmetry between prograde and retrograde, which they interpret as being due to the asymmetric location of a phenomenon called the evection resonance. In addition, the curved dashed line in Fig. 1 indicates that objects with inclinations near 180°, i.e., close to Saturn's orbit plane, may have larger semimajor axes than objects on highly tilted orbits. Approximate inclination limits for long-term stable orbits (*Shen and Tremaine,* 2008) are also shown in Fig. 1 by straight dashed lines.

The lack of orbits with inclinations ~55° < i < ~125° is likely a consequence of the Lidov-Kozai effect (*Carruba et al.,* 2002; *Nesvorný et al.,* 2003) where solar perturbations cause oscillations of the inclination and eccentricity. Originally extremely inclined orbits become unstable because the Lidov-Kozai oscillation makes the eccentricity so high that these irregulars reach the inner moon systems or even Saturn itself with the consequence of removal from the system through scattering or collision, or, near apoapsis, reach the edge of the Hill sphere and may escape. Such an inclination gap is found in all irregular-moon systems of the giant planets (*Nicholson et al.,* 2008) (see also section 5). Interestingly, there is also a gap at 0° < i < ~25°, for which the reason is not yet known. This gap is very obvious in the a,e,i-plot in Plate 8 of *Nicholson et al.* (2008); the sole exception is Neptune's Nereid. However, Nereid is very unusual since it contains approximately twice the mass of all other outer moons of the giant planets combined. It might even not be a captured object, but a former regular moon. *Goldreich et al.* (1989) and *Ćuk and Gladman* (2005) hypothesized in this direction, while *Nogueira et al.* (2011) provided arguments why Nereid could never have been a regular satellite.

A major player in at least the inner parts of Saturn's irregular-moon system appears to be the dominating moon Phoebe. In a sense behaving like a major planet, Phoebe with its periapsis-to-apoapsis range from 10.8 to 15.1 × 10^6 km (Fig. 1) partially cleared its surroundings. The semimajor axes for all but two irregulars are actually farther away at 15.2 to 25.2 × 10^6 km (Table 2), indicating that they are outside the realm of Phoebe for most of the time. Since gravitational scattering among Saturn's irregulars should be negligible even for close encounters with Phoebe, this "sweeping effect" is likely due to the much larger size of Phoebe and thus its much larger collisional cross-section. *Nesvorný et al.* (2003) calculated mutual collision rates between the 13 individual irregulars known at that time. While for all combinations not including Phoebe, the collision rate per 4.5 Ga is 0.02 or less, for all objects with orbits potentially crossing Phoebe's orbit it is ≥0.2. For Kiviuq, Ijiraq, and Thrymr, this collision rate is even >1, indicating that these moons will likely not survive another 4.5 b.y. From Fig. 1, it is obvious why these moons are in particular danger of collision with Phoebe: Ijiraq and Kiviuq reside within the ranges of Phoebe for most of the time, thus it is just a question of time as to when one of these moons will pass within ~110 km of Phoebe. For Thrymr, the low tilt compared to Phoebe's orbit lengthens the "corridor" where a collision might take place during the epochs where nodes of the two moons are very close to each other. Among the objects that were not yet known to *Nesvorný et al.* (2003), relatively-low-tilt objects Greip and S/2007 S 2 should be in high danger of eventually colliding with Phoebe as well. Table 2 marks all objects with current periapses lower than Phoebe's apoapsis as a first-order criterion for "being in danger of collision with Phoebe." Actually, two-thirds of the known irregulars of Saturn qualify as future Phoebe impactors.

2.2. Physical Properties

Besides orbital properties, groundbased observations and observations from Infrared telescopes close to Earth also revealed fundamental physical properties like approximate sizes, albedos, and colors. The first photometric survey from *Grav et al.* (2003) obtained BVRI (blue, visual, red, infrared) photometry of three Gallic moons (Albiorix, Tarvos, Erriapus), three Inuits (Siarnaq, Paaliaq, Kiviuq), and two retrograde moons (Phoebe, Ymir). The colors were found to vary between "neutral/gray" and "light red"; the asteroidal analogs are C-type and P-/D-type. One rationale behind this research was the hypothesis that objects from similar dynamical families should exhibit the same color if the progenitor object was not differentiated. *Grav et al.* (2003) found the observed Gallic moons in good "color agreement" to each other, and the same for the Inuits. Only Phoebe and Ymir were found to be significantly differently colored, from which they concluded that Ymir should not have been a part of Phoebe in the past.

Through JHK (near-infrared) photometry, *Grav and Holman* (2004) extended the measurements of Phoebe, Siarnaq, Albiorix, and Paaliaq into the near-infrared. Their seven-color spectra were again consistent with C-, P-, or D-type objects. In this context, it must be cautioned that the use of the terms "C, P, or D type" refers to a color classification originally introduced in the context of asteroids, but that its usage for the irregulars does not necessarily imply that these moons are asteroids that originated in the asteroid belt, nor that they have the same surface composition as asteroids. The terminology for transneptunian objects (TNOs) or Centaurs includes "neutral/gray," "red," or "ultra-red" and also simply describes the spectral slopes of the irregulars, but again not the origin region and object type (*Grav et al.,* 2015).

In January 2005, the Saturn opposition led to unusually low phase angles for the irregular moons, with the lowest value of 0.01° reached for Ymir, and 0.03° to 0.11° for six other objects (*Bauer et al.,* 2006). During the same apparition, *Miller et al.* (2011) observed Phoebe's opposition surge in four color filters. Tarvos and Albiorix, the two Gallic moons, again showed a common behavior, but their phase curves near 0° were much shallower than for Paaliaq, Ijiraq, and Phoebe. *Bauer et al.* (2006) proposed that the cause of the subdued opposition surge observed for Tarvos and Albiorix may be a higher compaction state of the surface.

The most extensive study of the saturnian irregulars before Cassini was the "deeper look" published by *Grav and Bauer* (2007) (hereafter refered to as *GB07*). They presented broadband four-color photometry at wavelengths between ~420 and ~820 nm of the 13 brightest objects. These included 3 of the 4 known Gallic moons (Albiorix, Tarvos, Erriapus), 4 of the 5 Inuits (Siarnaq, Paaliaq, Kiviuq, Ijiraq), and 6 of the 29 retrogrades (Skathi, Mundilfari, Thrymr, Suttungr, Ymir; plus earlier data from Phoebe). The four-point spectra were compared through mean spectral slopes. The goal of the work was again to detect correlations between dynamical families and spectral properties. While the colors and spectral slopes were mostly found to be consistent with C-, P-, and D-type objects, the measured diversity in surface colors was surprising. The spectral slope range was found to vary between ~−5 and ~+20$^{\%}/_{100\ nm}$ (Table 3).

Among the Gallic moons, homogeneous colors were found (slopes ~+5$^{\%}/_{100\ nm}$; P-type) except for two of the three Albiorix measurements. *GB07* suggest that the color of this moon varies over the surface. The Inuit moons were also found to be quite homogeneous in color (~+12$^{\%}/_{100\ nm}$; D-type), except for Ijiraq (~+20$^{\%}/_{100\ nm}$; "red"). The spectral slope of Ijiraq is redder than what is known from Jupiter Trojans, Hildas, or main-belt asteroids, possibly suggesting that Ijiraq originated in the realm of the Kuiper belt objects. Puzzling in this context is that the colors of Ijiraq and Kiviuq appear to be very different, while these two moons are the prime example for a dynamical relation (Fig. 1; Table 2). Among the retrogrades, the results for Ymir showed strong variations at short wavelengths between different apparitions, and a large peak-to-peak amplitude of ~0.3 mag at very low phase angles. *GB07* attribute these properties to significant surface variegations and an irregular shape. While the latter is well confirmed by Cassini observations, the former is not (see section 3.2). The measured spectral slopes vary between ~+6 and ~+8$^{\%}/_{100\ nm}$, putting Ymir at the boundary between P- and D-type. Other retrograde moons investigated by *GB07* are Suttungr and Thrymr, which are possibly members of the same dynamical family. Their colors appear neutral/gray (~−3$^{\%}/_{100\ nm}$; C-type), as does Mundilfari, the object with the "least-reddish" color (−5$^{\%}/_{100\ nm}$; C-type). Since Phoebe (−2.5$^{\%}/_{100\ nm}$; C-type) is also gray, *GB07* speculate that Mundilfari might be a piece from Phoebe from a collision with an impact velocity of ~5 km s^{-1}. If true, this scenario may work for an interplanetary impactor, but is unlikely for a planetocentric one, and must have occurred very likely in the early history of the solar system.

The lack of so-called ultra-red matter (spectral slope >+25$^{\%}/_{100\ nm}$) among the irregulars (while common among TNOs) is evident and challenges the hypotheses that assume the transneptunian region is the origin area of the irregular moons. *GB07* mention two possible solutions to this apparent contradiction. One is that space weathering closer to the Sun might fade them, the other that an increased cratering rate in the saturnian environment might make the surfaces less red. Support for the second idea comes from observational and collisional-modeling work of the Trojan asteroids by *Wong and Brown* (2016), who argue that the red and less-red colors are byproducts of the presence or absence of H_2S ice, which is lost from the surfaces during later collisions. Since collisional evolution was so predominant for the irregulars of Saturn (*Bottke et al.,* 2010), few objects would still be expected to be red. A third possibility might be that the irregulars did not form that far out. *GB07* conclude from the high variegation among the irregular moons of Saturn that they might have two distinct origin regions. Some (the grayish ones) might be former main-belt objects, while the others (the reddish ones) might come from the outer solar system. However, they note a caveat for this scenario: Phoebe, a grayish or even slightly bluish object, was proposed by *Johnson and Lunine* (2005) to originate from the transneptunian region. Consequently, a consensus among the scientists on the origin question still lies ahead of us. Additional aspects of this problem concerning orbital dynamics are given in section 5.

The brightest irregular moons of Saturn were also observed with the Spitzer Space Telescope (*Mueller et al.,* 2008) and with NEOWISE (*Grav et al.,* 2015). Spitzer data at 24 μm of the Gallic moons Albiorix, Tarvos, and Erriapus; of the Inuits Siarnaq, Paaliaq, Kiviuq, and Ijiraq; and of retrograde moons Phoebe and Ymir showed that the albedos should be generally low, probably less than 0.1, similar to cometary nuclei, Jupiter Trojans, and TNOs. Thermal data available for three of the moons indicate rather low thermal inertias, suggestive of regolith-covered surfaces (*Mueller et al.,* 2008). NEOWISE data could be extracted for the three largest moons Phoebe (at 3.4, 12, and 24 μm), Siarnaq, and Albiorix (both at 24 μm through the technique of data stacking). Recording of Paaliaq and Tarvos was also attempted, but these two objects were too faint for a signal to be detected. Siarnaq's size was determined to 39.3 ± 5.9 km for an albedo of 0.050 ± 0.017 at a signal-to-noise ratio (SNR) of ~7. The results given for Albiorix are 28.6 ± 5.4 km, 0.062 ± 0.028, and SNR ~ 3 (*Grav et al.,* 2015). For Phoebe, the size and albedo determinations were accurate to ~5% and ~20%, respectively, to values determined from the Cassini and Voyager spacecraft.

Approximate sizes of all of Saturn's irregular moons were also determined from groundbased photometry. The first estimates were given in the International Astronomical Union Circulars (IAUCs) and Minor Planet Electronic Circulars (MPECs) issued by the Central Bureau for Astronomical Telegrams of the International Astronomical Union (IAU) and by the IAU Minor Planet Center, respectively (see also Table 1). Since the irregular moons are unresolved in the data, their sizes cannot be measured directly, but can be estimated from their brightness and by assuming their visible albedos. Values calculated for Saturn's irregular moons are summarized in Table 2. From their survey, *GB07* determined approximate radii for 12 irregulars. Their values for

TABLE 3. Physical properties and Cassini observations of 25 saturnian irregular moons.[*]

Moon Name	†	Rotational period (h)‡	$(a/b)_{min}$[§]	LC¶	Spectral Slope ($\%/_{100\ nm}$)[**]	No. of Obs.[††]	Cassini Imaging Observations: First–Last (mm/yy)[††]	Best Mag.[‡‡]	Phase (°)[§§]	Moon Abbrev.
Phoebe	(a)	9.2735 ± 0.0006	1.01	1	−2.5	8	08/04–01/15	5.1	3–162	Pho
Kiviuq	(b)	21.97 ± 0.16	2.32	2	+11.8	24	06/09–08/17	12.0	4–136	Kiv
Ijiraq	(c)	13.03 ± 0.14	1.08	2	+19.5	11	01/11–04/16	11.8	40–104	Iji
Paaliaq	(d)	18.79 ± 0.09	1.05	4	+10.0	12	11/07–01/17	11.2	21–112	Paa
Tarqeq	(e)	76.13 ± 0.04	1.32	2		10	08/11–01/17	15.0	15–49	Taq
Siarnaq	(f)	10.18785 ± 0.00005	1.17	3	+13.0	8	03/09–02/15	10.8	4–143	Sia
Albiorix	(g)	13.33 ± 0.03	1.34	2,3	+12.5[**]	13	07/10–01/17	9.5	5–121	Alb
Bebhionn	(h)	16.33 ± 0.03	1.41	2		9	03/10–07/17	14.6	19–79	Beb
Erriapus	(i)	28.15 ± 0.25	1.51	2	+5.1	15	02/10–12/16	13.6	26–116	Err
Tarvos	(j)	10.691 ± 0.001	1.08	2,3	+5.4	9	07/11–10/16	12.8	1–109	Tar
Narvi	(k)	10.21 ± 0.02		3		4	03/13–01/16	15.6	54–80	Nar
Bestla	(l)	14.6238 ± 0.0001	1.47	2		14	10/09–11/15	13.5	30–96	Bes
Skathi	(m)	11.10 ± 0.02	1.27	2	+5.2	8	03/11–08/16	15.1	15–77	Ska
Skoll	(n)	7.26 ± 0.09 (?)	1.14	3		2	11/13–02/16	15.5	42–47	Sko
Hyrrokkin	(o)	12.76 ± 0.03	1.27	3		7	03/13–03/17	14.4	20–82	Hyr
Greip	(p)	12.75 ± 0.35 (?)	1.18	2 (?)		1	09/15	15.4	27	Gre
Suttungr	(q)	7.67 ± 0.02	1.18	2,3	−3.2	5	05/11–11/16	15.4	12–72	Sut
Thrymr	(r)	38.79 ± 0.25 (?)	1.21	2 (?)	−3.0	9	11/11–09/17	14.9	13–105	Thr
Mundilfari	(s)	6.74 ± 0.08	1.43	2	−5.0	1	03/12	15.3	36	Mun
Hati	(t)	5.45 ± 0.04	1.42	2		6	02/13–12/15	15.3	14–73	Hat
Bergelmir	(u)	8.13 ± 0.09	1.13	2		2	10/10–09/15	15.9	16–26	Ber
Kari	(v)	7.70 ± 0.14		3		1	10/10	14.8	56	Kar
Loge	(w)	6.9 ± 0.1 ?	1.04	2 ?		2	10/11–02/15	16.2	12	Log
Ymir	(x)	11.92220 ± 0.00002	1.37	3	+8.1	9	04/08–07/15	13.2	2–102	Ymi
Fornjot	(y)	9.5 ?	1.11	3 ?		2	03/14–04/14	16.4	17–30	For

[*] Cassini high-level observation descriptions and processed data of irregular moons are provided on T. Denk's "Outer Moons of Saturn" web page (*https://tilmanndenk.de/outersaturnianmoons/*).
[†] Corresponding character in the itemization in section 3.2.
[‡] Rotational periods from *DM18* and unpublished data. A question mark indicates that the period is not completely unambiguous. The Phoebe value is from *Bauer et al.* (2004).
[§] Minimum ratio of the equatorial axes of a reference ellipsoid of uniform albedo with dimensions a and b (derived from the lightcurve amplitudes). The Phoebe value is from *Castillo-Rogez et al.* (2012).
[¶] Amount of maxima and minima in the lightcurves; see text.
[**] Spectral slope S_2' from Table 3 of *GB07*. Positive values indicate reddish, negative bluish spectra. *GB07* give errors between ±0.3 and ±2.8$\%/_{100\ nm}$. Individual measurements for Albiorix (+5.3, +12.9, +14.9$\%/_{100\ nm}$) vary much more.
[††] Number of Cassini imaging observation "requests" ("visits") where data of the object were achieved. The targeted flyby of Phoebe (June 2004) and data from optical navigation are not included in the counts.
[‡‡] Best magnitude of the object as seen from Cassini at a time where data were acquired.
[§§] Lowest and highest observation phase angles during Cassini observations. Phase angles from Earth are always <7°.

the absolute magnitude H are systematically higher by ~0.2 to ~0.6 mag compared to the values from the Minor Planet Center. However, for consistency throughout all 38 objects, we use the values of the Minor Planet Center in Table 2, but with the albedos set to 0.06 for the diameter calculations to stay close to the diameter values by *GB07*, who assumed albedos of 0.08. An assumed albedo of 0.06 is also consistent with the NEOWISE measurements of Albiorix and Siarnaq described above, for which the determined diameters are within the error bars of the NEOWISE results and of Table 2. These numbers imply that 7 of the 9 known progrades should be between ~10 and ~40 (maybe up to ~50) km in size, while only 2 of 29 retrograde moons are larger than 10 km (Phoebe and Ymir).

With respect to the rotational periods, very little work from the ground has been published. For the relatively bright object Phoebe, the spin rate has been known for a long time. It was first determined through data from Voyager 2 from *Thomas et al.* (1983) to 9.4 ± 0.2 h, a few years later from Earth by *Kruse et al.* (1986) to 9.282 ± 0.015 h, and eventually more accurately by *Bauer et al.* (2004) to 9.2735 ± 0.0006 h. For the other saturnian irregulars, only fragmentary lightcurve observations of Siarnaq (*Buratti et al.,* 2005; *Bauer et al.,* 2006) and Ymir (*GB07*) indicated perceptible amplitudes, but no reliable periods were given.

3. CASSINI RESEARCH

3.1. Observing Irregular Moons with Cassini

With the Cassini spacecraft en route to Saturn and then orbiting the planet during and after the time of the discovery of the irregular-moon system, a set of small telescopes and spectrometers was well placed close to the irregular moons for many years. Among these instruments, the Narrow Angle Camera (NAC) of the ISS experiment (*Porco et al.*, 2004) was best suited to perform the task of recording these objects. Except during the first orbit, Cassini was rarely more than ~4 × 10^6 km away from Saturn, and most of the time even closer than ~2 × 10^6 km and thus always inside the orbits of the irregulars. One consequence of this inside location of Cassini was that a dedicated irregular-moon search was not promising because the area to consider filled half the sky.

Observations were performed between ~43 and ~275 times closer to the irregular saturnian moons than Earth. The closest range to an irregular moon of Saturn that has been used for data recording was 4.8 × 10^6 km (Ijiraq in September 2014), and the average observation distance was on the order of 14 × 10^6 km. From this distance, the spatial resolution of the NAC is below 80 km pxl^{-1}, too low to resolve the surfaces of the outer moons. The prime observation goal with the NAC was thus to obtain lightcurves to determine rotational periods and other physical parameters, and to improve the orbit ephemeris. During the second half of the mission when most observations took place, the apparent brightness of the irregulars varied between ~10 and ~37 mag (Phoebe reached up to ~5 mag). While the brighter values (compared to Earth-based observers) were a consequence of the smaller distances, the large variation in brightness was a result of the changing phase angles of the irregulars as seen from Cassini, which may in principle have reached any value between 0° and 180°. The large phase-angle range achievable from the Cassini location represented a great advantage for constraining an object's shape from lightcurve inversion, and for the characterization of its photometric properties. Other advantages of the spacecraft-based observations were the absence of Earth-related effects like a day/night cycle (which greatly reduced the period aliasing problem), of the annual opposition cycle, or of weather effects except for storms or rain at the Deep Space Network stations.

From the 38 known irregular moons of Saturn, all 9 progrades and 16 of the 29 retrogrades were successfully observed with Cassini (*DM18*). Thirteen objects were not observed because of inaccurate ephemeris or because they were too faint. Many moons were targeted repeatedly. As seen from Cassini, almost all the irregular saturnian moons occasionally became brighter than ~16.5 mag, a practical limiting magnitude for useful lightcurve studies with the NAC. The apparent brightness at a particular epoch was mainly a function of object size, range, and phase angle. Except for the largest objects, the visibility windows were thus rather limited, and many objects were only observable for a few weeks or months at the frequency of their orbit periods. The competition between the individual irregular moons was indeed so big that not every object that might have been observed during its visibility window has actually been observed.

Measured and derived numbers are compiled in Table 3 and include the following quantities: (1) Rotational periods — these can be determined through proper phasing of repeating lightcurve patterns (e.g., *Harris and Young*, 1989; *Mottola et al.,* 1995); (2) minimum values $(a/b)_{min}$ for the ratios between the two equatorial diameters of a reference ellipsoid; and (3) the amount of maxima and minima of the measured lightcurves during one rotational cycle. Two maxima and 2 minima ("2-max/2-min") or 3 maxima and 3 minima ("3-max/3-min") was found in almost all lightcurves; some objects "switch" when being observed at different phase angles.

The whole irregular-moon planning and observation process was initiated and performed by one of the authors (T.D.). The use of the Cassini camera was the first use of an interplanetary spacecraft for a systematic photometric survey of a relatively large group of solar system objects.

3.2. Individual Objects

In this section, selected results for each observed moon from the work of *DM18* is presented. A subset of their lightcurves is reproduced in Fig. 2 (again, see Table 3 for a listing of the measured quantities).

(a) Phoebe is covered in section 4, but briefly mentioned here because two Cassini observations were designed to obtain lightcurves at low and high phase angles. These are probably the only lightcurves from Cassini that are not shape-driven, but rather exclusively albedo-driven. Figure 2 shows the Phoebe curve taken at 109° phase angle. The prominent maximum is attributed to bright ice excavated at the rims of large craters Erginus and Jason.

(b) Kiviuq lightcurves taken at low and high phase angle are shown in Fig. 2. The rotational period was determined to 21.97 h ± 0.16 h. The extreme amplitude of 1.7 mag at 31° phase is unique among the observed saturnian irregulars. It indicates that Kiviuq is a very elongated object with an $(a/b)_{min}$ of at least 2.3, possibly higher. The clear and quite symmetric 2-max/2-min pattern even at high phase is also unique. The lightcurve obtained at 108° phase shows an amplitude of 2.5 mag (or about a factor of 10 in brightness); this is a record among all lightcurves from Cassini measured so far. Kiviuq might even be a contact-binary or

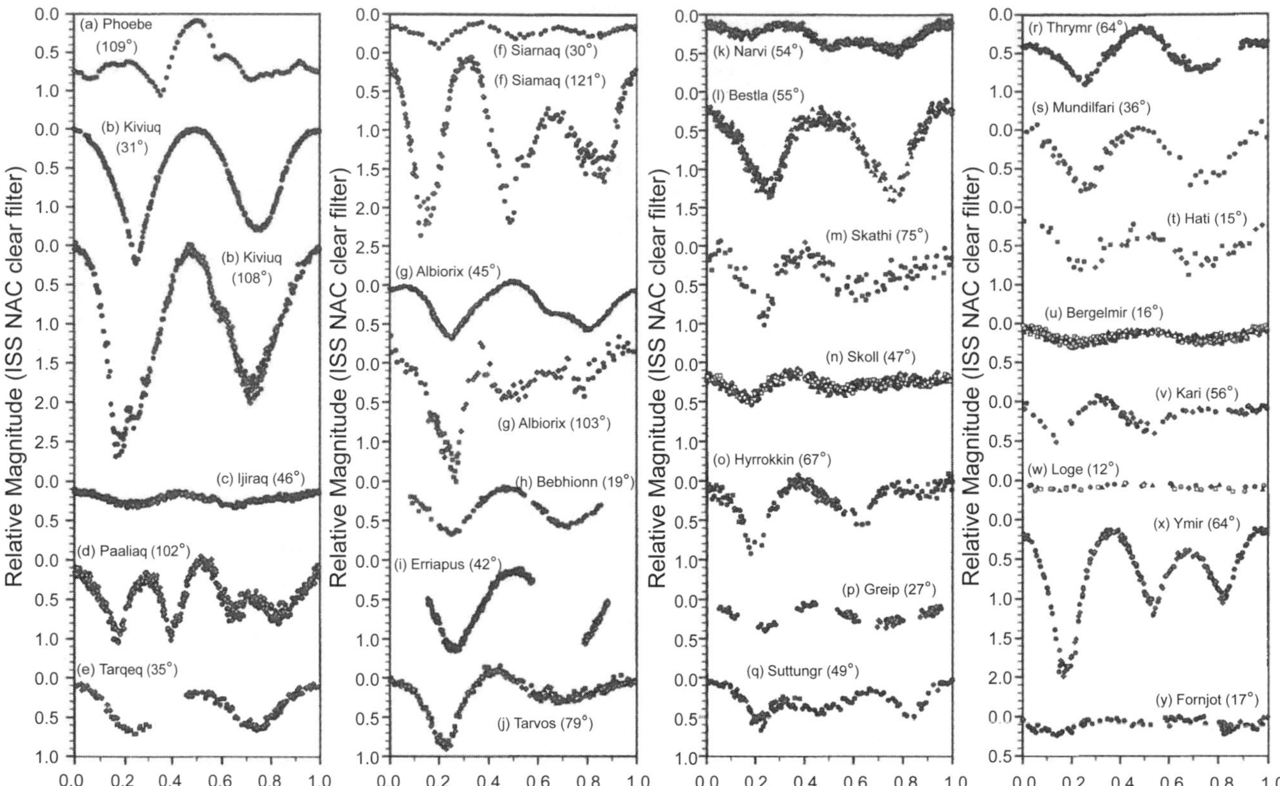

Fig. 2. Lightcurves of 25 irregular moons of Saturn, taken with Cassini ISS NAC at the annotated phase angles. Different symbols indicate different rotational cycles from the same observation or from another observation close in time. Details [moon; observation start MJD (start calendar date, UTC); Cassini orbit (rev = revolution); subspacecraft RA/Dec]:

- *Phoebe:* 53707.48882 (03 Dec 2005); rev 18; 233°/−18°
- *Kiviuq (31°):* 55438.32268 (30 Aug 2010); rev 137; 173°/+27°
- *Kiviuq (108°):* 57202.62406 (29 Jun 2015); rev 218; 143°/+40°
- *Ijiraq:* 56472.45195 (29 Jun 2013); rev 193; 235°/+30°
- *Paaliaq:* 56504.47605 (31 Jul 2013); rev 195; 330°/−24°
- *Tarqeq:* 57550.74095 (11 Jun 2016); rev 236; 232°/+10°
- *Siarnaq (30°):* 56656.16083 (30 Dec 2013); rev 200; 239°/−42°
- *Siarnaq (121°):* 56398.54837 (16 Apr 2013); rev 186; 83°/+60°
- *Albiorix (45°):* 55420.18990 (12 Aug 2010); rev 136; 180°/+45°
- *Albiorix (103°):* 56455.05408 (12 Jun 2013); rev 192; 138°/−40°
- *Bebhionn:* 55322.42646 (06 May 2010); rev 130; 201°/−4°
- *Erriapus:* 55229.41932 (02 Feb 2010); rev 125; 142°/+22°
- *Tarvos:* 56407.48038 (25 Apr 2013); rev 187; 304°/−47°
- *Narvi:* 57396.12153 (09 Jan 2016); rev 230; 284°/−70°
- *Bestla:* 56177.41095 (07 Sep 2012); rev 171; 263°/+6°
- *Skathi:* 55636.50419 (16 Mar 2011); rev 146; 124°/+31°
- *Skoll:* 57440.74498 (22 Feb 2016); rev 232; 210°/+9°
- *Hyrrokkin:* 56359.61829 (08 Mar 2013); rev 183; 146°/−11°
- *Greip:* 57274.82415 (09 Sep 2015); rev 221; 217°/−10°
- *Suttungr:* 57720.50040 (28 Nov 2016); rev 250; 208°/−7°
- *Thrymr:* 57013.87652 (22 Dec 2014); rev 210; 170°/−4°
- *Mundilfari:* 55995.26586 (09 Mar 2012); rev 162; 167°/−6°
- *Hati:* 56351.92441 (28 Feb 2013); rev 182; 222°/+2°
- *Bergelmir:* 57278.15338 (13 Sep 2015); rev 221; 227°/−26°
- *Kari:* 55498.75138 (29 Oct 2010); rev 140; 144°/+36°
- *Loge:* 57074.80992 (21 Feb 2015); rev 212; 228°/−9°
- *Ymir:* 56048.42537 (01 May 2012); rev 165; 147°/+22°
- *Fornjot:* 56762.10922 (14 Apr 2014); rev 203; 245°/−18°

binary moon with similarly sized components in a doubly synchronous state; see section 3.3.2 for discussion.

(c) Ijiraq rotates once every 13.03 ± 0.14 h, and all measured data even at high phase and various subspacecraft locations show relatively shallow lightcurves, implying that Ijiraq has a relatively circular equatorial cross-section.

(d) Paaliaq was the first discovery of a prograde irregular saturnian moon. Its name is not taken from the Inuit mythology, but from a fictional Inuit character of a modern novel (*Kusugak*, 2006). Paaliaq's rotational period is 18.79 ± 0.09 h. Its high-phase lightcurves show a distinct 4-max/4-min pattern, which was not seen in any other lightcurve of the irregulars. Eight relatively clear extrema during one rotation is very rare throughout the solar system and likely indicative of an unusual shape.

(e) Tarqeq is the smallest known prograde outer moon and may well be a shard from Siarnaq. From Cassini observations between 2014 and 2017, the rotational period was determined to 76.13 ± 0.04 h. This is by far the longest period of all measured irregular moons, and only a few of the tidally influenced regular moons rotate more slowly. Intriguingly, this period is also very close to the 1:5 resonance of the orbit of Titan (382.690 h), raising the question of tidal alteration of Tarqeq's rotation. The difference between the Tarqeq period and one-fifth of the Titan period is only ~0.4 h or 0.5% of Tarqeq's period.

(f) Siarnaq is probably the largest prograde irregular moon of Saturn (*Grav et al.,* 2015) (see also Table 2). The lightcurves show a rather unusual 3-max/3-min pattern with equally spaced but uneven extrema. The amplitudes in high-phase-angle observations are very large (exceeding 2 mag in a 121° phase observation), but shallow at low phases (~0.3 mag in a 30° phase observation; see Fig. 2). Images taken in four different broadband color filters of the NAC at wavelengths between 440 and 860 nm revealed no color variations on the surface. A convex-shape model from seven observations is shown in Fig. 3; Siarnaq resembles a triangular prism in the model. Note that the technique to determine convex-shape models cannot reproduce concavities like craters or other constrictions. The rotational period, which is a sidereal period here, was determined to 10.18785 h ± 0.2 s. The pole axis points toward λ,β = 98°/−23° ± 15°; this rather low latitude indicates that Siarnaq experiences extreme seasons, somewhat reminiscent of the regular satellites of Uranus. During summer solstice — the most recent on the northern hemisphere approximately occurred in mid-2018 — the Sun might reach the zenith at a surface latitude of 60° to 70° on one hemisphere, while the other remains in a darkness lasting for many years.

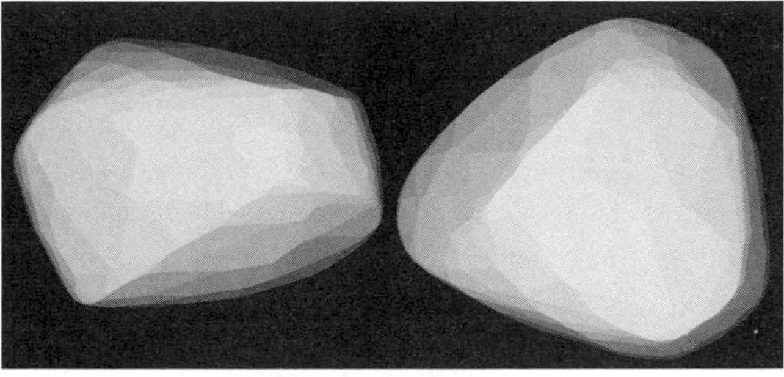

Fig. 3. Convex-shape model of Siarnaq. *Left:* Equatorial view (north up); *right:* north-pole view (rotated around the horizontal axis).

(g) Albiorix is the only moon besides Phoebe that became brighter than 10 mag for Cassini. Its lightcurve amplitude at 45° phase is relatively large, indicative of an elongated object. The period was determined to 13.33 ± 0.03 h. Another observation obtained at a high phase angle of 103° revealed a very different lightcurve shape. While one minimum with an amplitude of 1.5 mag remained prominent, the other disappeared and instead separated into two shallow ones (Fig. 2). The best resolved images have a resolution of 34 km pxl^{-1}, which is very close to the diameter of this moon. No secondary object was detected in these data.

(h) Bebhionn shows a "nice" 2-max/2-min lightcurve with a quite large amplitude at low phase (Fig. 2), indicative of an elongated object with an $(a/b)_{min} > 1.4$. From data obtained in 2017, the period was determined to 16.33 ± 0.03 h.

(i) Erriapus revealed lightcurves with 2-max/2-min patterns and relatively large amplitudes (Fig. 2), indicative of a prolate or possibly contact-binary or binary moon. The rotational period was determined to 28.15 h ± 0.25 h.

(j) Tarvos has a high orbit eccentricity of 0.54, which led to extreme differences in its range to the Cassini spacecraft (6 to 32 × 10^6 km) at the orbit-period frequency of 2.5 years. Since Tarvos' periapsis is locked toward the direction of the Sun, the range and phase angle effects canceled each other out for most of the time, resulting in a remarkably constant apparent magnitude seen from Cassini, between ~13 and ~14.5 during the mission. The rotation period of Tarvos is 10.691 h ± 3s and thus quite close to the rotation period of Saturn itself (10.53 to 10.79 h) (*Helled et al.*, 2015). Figure 2 shows a lightcurve from an observation where Tarvos was tracked over more than three rotation cycles at 79° phase. One of the two minima is very pronounced, while the other one is broad and shallow (~0.9 and ~0.4 mag amplitude, respectively).

(k) Narvi was observed with Cassini during two observation campaigns, and a rotation period of 10.21 ± 0.02 h was found. The lightcurves show a very clear 3-max/3-min pattern including the one-sixth rotational-phase spacing of the extrema seen at Siarnaq and several other moons. A unique feature is that the brightest minimum is brighter than the lowest maximum (Fig. 2).

(l) Bestla is one of the most unusual irregular moons of Saturn because of its extreme orbit elements. The eccentricity of 0.52 is among the highest in the saturnian system, the periapsis distance of 9.7 × 10^6 km is among the lowest for retrograde moons, and the orbit inclination of ~145° is the lowest of all retrograde moons. Bestla is also the retrograde moon that came closest to outermost regular moon Iapetus during the Cassini mission (2.3 × 10^6 km in November 2003). This is more than three times closer than the closest distance between Iapetus and Phoebe, which was 7.4 × 10^6 km. Bestla was favorable for Cassini observations every three years. Figure 2 shows a 2-max/2-min lightcurve with a quite large amplitude of 1.1 mag. The sidereal rotation period was determined to 14.6238 h ± 0.4 s. The pole orientation with an ecliptic latitude $\beta = -85° \pm 15°$ is pointing anti-parallel to the normal of the ecliptic, indicating that the rotation is also retrograde.

(m) Skathi has the smallest semimajor axis of any retrograde moon of Saturn except Phoebe, and no other moon is known that shares its orbital elements. From several Cassini observations, its period was determined to 11.10 ± 0.02 h.

(n) Skoll is another "lonely" moon; it has been observed twice with Cassini. With the second observation from 2016, the period was determined to 7.26 ± 0.09 h (Fig. 2). The pattern of the extrema is less pronounced than for Siarnaq, but also contains the "one-sixth spacing" found for the 3-max/3-min lightcurves. For unknown reasons, the 2013 observation could not be fitted well with this period. A secondary frequency from wobble might be the cause, but this is very speculative and cannot be addressed with the available data. The Skoll period thus remains tentative.

(o) Hyrrokkin was observed several times, and its period was determined to 12.76 ± 0.03 h. Hyrrokkin shows clear

3-max/3-min lightcurves at mid and high phase similar to Ymir and Siarnaq, but with a lower amplitude, only around 1 mag.

(p) Greip was successfully observed in 2015, and the period was determined to 12.75 ± 0.35 h. Since the observation time base of ~17.5 h was too short to unambiguously confirm the 2-max/2-min pattern, an ~19-h period is considered less likely, but cannot be ruled out for Greip. The 12.75-h solution is within the errors identical to the Hyrrokkin period. Interestingly, Greip and Hyrrokkin are not just "twins" with respect to the rotational period, but also to their semimajor axes and orbit eccentricities (Table 2), and thus orbit periods, periapsis distances, and apoapsis ranges. Only the orbit inclinations are quite different. The latter makes a joint progenitor less plausible at first glance, but very similar a and e values allow for a scenario where the separation of a progenitor object or binary occurred very close to one of the orbital nodes. Future color observations might indicate if Greip is related to Hyrrokkin or rather to the Suttungr group.

(q) Suttungr was observed in 2011 and 2016, and its rotational period was determined to 7.67 ± 0.02 h. At low phase, the lightcurve shows a 2-max/2-min pattern, while 3-max/3-min were observed at mid phase (Fig. 2).

(r) Thrymr was observed in 2011/2012 and 2014 (Fig. 2), and again in 2017 during the very last Cassini observation of an irregular moon. Because of the long spin period, complete coverage of the rotation could not be achieved during a single Cassini observing request. Assuming the most probable case of a 2-max/2-min lightcurve, the rotation period is 38.79 ± 0.25 h, with the mid-phase lightcurves being reminiscent of Skathi's or Tarvos'. Thrymr is by far the slowest rotator among the retrograde moons. Somewhat similar to the situation of Hyrrokkin and Greip, the orbital elements a and e, but not i, of Thrymr are close to those of S/2004 S 7, potentially qualifying these two objects as a pair as well. Especially the eccentricities of Thrymr and S/2004 S 7 are much higher than those of the other objects in the Suttungr and Mundilfari families, respectively, of which these two moons might be members as well (Table 2).

(s) Mundilfari was observed by Cassini once over a time span of ~9 h at a phase angle of 36°. This was sufficient to determine the rotational period to 6.74 ± 0.08 h. The lightcurve shows a clear 2-max/2-min pattern. Its amplitude of 0.78 mag is substantial, indicating a quite elongated object.

(t) Hati, a small moon with a size of ~5 km, might be a member of the Mundilfari cluster. The period of 5.45 ± 0.04 h is the fastest among all 25 measured moons. In the first observation in 2013, Hati was detected near the very edge of a 1×2 mosaic, and the Cassini data helped to improve the orbital elements significantly. The low-phase lightcurve has a relatively high amplitude of 0.55 mag, resulting in a high $(a/b)_{min}$ of 1.4 to 1.7. Hati's rotational period is also the fastest reliably known period of all moons in the solar system, including moons of asteroids.

(u) Bergelmir was observed twice with Cassini, and the period was determined to 8.13 ± 0.09 h for a 2-max/2-min repetitive lightcurve pattern. The lightcurve amplitude of ~0.2 mag implies a rather circular equatorial cross-section.

(v) Kari was targeted once with Cassini. With this 16-h observation, the Kari rotational period was determined to 7.70 ± 0.14 h. The lightcurve was taken at 56° phase angle and has an amplitude of ~0.5. The 3-max/3-min pattern with two rather deep minima and one shallow minimum resembles Hyrrokkin's lightcurve.

(w) Loge is orbiting near the edge of the Saturn system and might be part of a family including Ymir and Fornjot (Fig. 1). Its small size and large distance made this moon very faint for Cassini. An observation in 2015 at 12° phase potentially covered ~4.7 rotation cycles, but the SNR of the data was quite low. The amplitude of only ~0.07 mag was the shallowest lightcurve measured with Cassini (Fig. 2). The proposed period of 6.9 ± 0.1 h from a 2-max/2-min lightcurve must be considered uncertain.

(x) Ymir has a diameter of ~19 km and is, besides Phoebe, the dominant retrograde object in the saturnian system. A relatively large peak-to-peak lightcurve amplitude noted by *GB07* from groundbased observations was confirmed with Cassini low-phase data. Ymir is another object that exhibits the 3-max/3-min lightcurve pattern with homogeneously spaced extrema (Fig. 2). The convex-shape model (*Denk and Mottola*, 2013) reveals a triangular equatorial cross-section very similar to Siarnaq. The pole-axis of Ymir points close to the south-ecliptic pole, indicating a retrograde spin. Ymir's sidereal period is 11.92220 h ± 0.1 s. Observations through four color filters of the NAC showed no deviation of any color lightcurve from the shape of the clear-filter lightcurve; Ymir is thus uniformly colored on a global scale.

(y) Fornjot is the object with the largest semimajor axis of all known moons of Saturn. It was difficult to observe by Cassini because of its large distance and small size and because of its proximity to the galactic plane, which meant a substantial increase in the number of background stars, during the used opportunities. A period of ~6.9 h might work for a 2-max/2-min lightcurve, but the fit is not good. About 9.5 h for 3-max/3-min looks good (Fig. 2), but the lightcurve overlap is very small. This leaves Fornjot as the object with the most uncertain period result among the observed satellites. From the modest lightcurve amplitude, a rather circular equatorial cross-section is expected.

3.3. Patterns and Correlations

Starting from lightcurves and rotational periods obtained for Saturn's irregular satellites, we searched for statistical patterns; examined the potential of binary and contact-binary objects; looked for potential correlations between the object spins, sizes, and orbit parameters; made bulk-density considerations; and compared the irregular-moon sizes and periods to those of other minor bodies. Orbit dynamical and physical properties are also visualized in an a,i'-diagram that includes information on Saturn distances, orbit tilts, object movement directions, object sizes, spin rates, lightcurve shapes, and potential binarity of the irregulars (Fig. 4).

Herein, the inclination supplemental angle i' is equal to the orbit inclination i for prograde moons, and to 180°−i for retrograde satellites.

3.3.1. Lightcurve "end-members." The number of lightcurve extrema gives a hint on basic shapes of the moons. Satellites Ymir, Siarnaq, Hyrrokkin, Skoll, Kari, Suttungr, Narvi, and possibly Fornjot share the 3-max/3-min lightcurve pattern with *six equally spaced* extrema at medium or even low phase angles. For this group of objects, a near-triangular equatorial cross-section is suggested as a proxy for a "convex-shape end-member," very similar to the model shapes of Siarnaq (Fig. 3) and Ymir. Opposite to this finding, many moons show strict 2-max/2-min lightcurves (Kiviuq, Bestla, Erriapus, Bebhionn, Hati, Mundilfari, Tarqeq) with *four* extrema quite symmetrically spaced. If this symmetry remains even at high phase angles, a regular ellipsoid is a good first-order shape approximation, and it is suggested that these two model-shape ensembles — ellipsoids and near-triangular prisms — represent some sort of convex-shape end-members among the saturnian moons. Of course, it must be cautioned that the potential variety of shapes is so high that this interpretation has to be considered as first-order only as long as no additional shape information is available, and it also does not include significant concavities or variations between the northern and southern hemispheres. Nevertheless, it gives a rough idea on some object shapes where no shape model is available. The moons with the 3-max/3-min lightcurves at mid or low phase angles are also tagged in Fig. 4, where they appear to show a random distribution. The only obvious common property is that all have rotation periods faster than 13 h.

3.3.2. Binary candidates. It has repeatedly been proposed and is plausible that irregular moons with very similar orbit parameters (a,e,i) were once a single object that has been separated into two or more pieces by a collision (Fig. 1 and Table 2; see also section 5 for details). In this context and also per se, an interesting question is if there also exist "moons with a moon" that were formed in the course of a collision where the separation conditions of the collisional remnants were such that they remained gravitationally bound to each other, without complete reaccretion into one body. Such "double moons" might exist as binaries with different or equal sizes, or as contact binaries. The outer satellites of the giant planets might be the best places to search for binary configurations among moons because the sizes of their Hill spheres (on the order of ~100 to ~300 satellite radii; this depends on the moon's density) are much larger than for the regular satellites (almost all below ~20 moon radii).

The review by *Margot et al.* (2015) about binary asteroids indicates that three distinct types of double objects exist among this group of minor bodies: (1) Large asteroids (D > 90 km) with small satellites; (2) small, doubly synchronous binaries with similarly sized components and rather long rotation periods ≥13 h; and (3) small (primary-component diameter D_1 < 11 km), very rapidly spinning primaries with much smaller secondary components. For types 1 and 3,

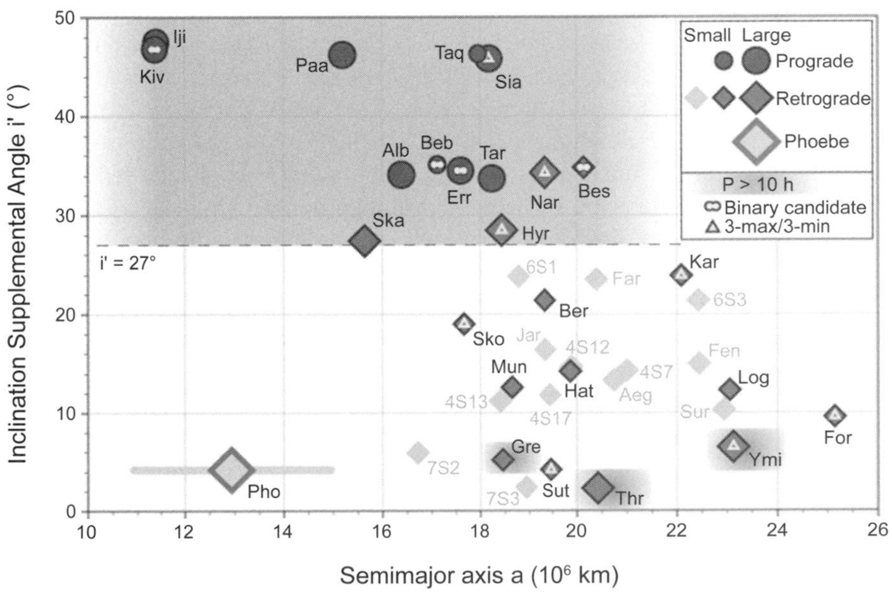

Fig. 4. a-i' plot for 25 irregular moons of Saturn, showing the distributions of object ranges to Saturn (through a), orbit tilts (through inclination supplemental angle i'), orbital senses of motion (pro-/retrograde; through symbols), object sizes (as two equally-sized bins separating into "large" and "small" at absolute magnitude H ~ 14.4 mag), and object spin rates (as two bins separated at 2.4 d^{-1}; the slow rotators are highlighted by a gray background). The light-gray bar at Phoebe indicates its periapsis and apoapsis distance to Saturn. Objects with 3-max/3-min lightcurves at mid or low phase angles as well as binary candidates are marked through insets. The 13 objects not observed by Cassini are included for reference (pale diamonds). They all fall into the categories "small," "retrograde," and "low i'," and all but one are part of the "farther satellite group." Their rotational periods are unknown.

only the smaller component rotates synchronously. Binaries of type 1 are usually detected through direct imaging, if their separation allows the components to be resolved. Under particular illumination and viewing geometries, all binaries experience mutual events (occultations and eclipses) that produce signatures in their lightcurves that can be detected from disk-integrated photometry. Such signatures are more easily recognizable in type 3 binaries since the shape-induced brightness variations and those due to mutual events are characterized by different periodicities (rotational and orbital periods, respectively). For this reason, most of the asteroid binaries discovered so far belong to type 3. In type 2 systems, given their full synchronicity, a binary nature is more difficult to demonstrate as the lightcurve signatures due to mutual events are more difficult to distinguish from those due to shape or albedo variations. However, there are lightcurve features that, although not a proof of binarity, can hint at a double nature of a type 2 object (*Lacerda and Jewitt*, 2007; *Margot et al.*, 2015; *Sonnett et al.*, 2015). Such features include high amplitudes, long rotation periods, flat ("plateau-shaped") extrema, peaked minima, minima of different depth with maxima being equal, highly structured minima, or lightcurve slopes with kinks (e.g., *Lacerda and Jewitt*, 2007). For asteroid binaries, mutual events take place at least every few years. For potential binaries in the saturnian system, "solar-eclipse seasons" occur every ~14–15 years when the orbital nodes of the two bodies line up with the Sun direction. Their durations depend on the rotation-axis orientation as well as the range, the sizes, and the oblatenesses of the components.

Based on these indications, the best candidates for contact-binary or doubly synchronous binary moons from Cassini ISS data are Bestla, Kiviuq, Erriapus, and very possibly Bebhionn. Bestla, with subtle kinks on the lightcurve flanks and a broad plateau-like maximum, and Kiviuq, with its smooth and symmetric 2-max/2-min lightcurves of extreme amplitudes which possess equal maxima but different minima (Fig. 2), are especially promising candidates. Figure 4 shows that these moons all have orbits with a high inclination supplemental angle and are parts of an undoubted satellite pair or family. Final proof of binarity, however, requires accurate modeling of the binary system that is capable of exactly reproducing the observed lightcurve features. This effort is undergoing at the time of publication of this chapter.

Figure 5 shows the rotational periods of the primaries P_1 over the diameters of the primaries D_1 for asteroids larger than 1 km. Interestingly, there are large gaps between the three binary types described by *Margot et al.* (2015) that are not necessarily only due to observational bias. Almost all objects of type 3 have a component-diameter ratio $D_2/D_1 \leq 0.5$ (mass ratio <0.2), and none is known showing doubly synchronous rotation. Contrary to this, almost all type 2 asteroids have a component-diameter ratio $D_2/D_1 \geq 0.8$. Adding the saturnian irregular moons to Fig. 5 strengthens the suspicion that Kiviuq, Erriapus, Bestla, or Bebhionn may be doubly synchronous binaries, because they fall in the region of the type 2 objects in the P_1-D_1 diagram. Contrary to this, most of the other moons occupy regions not typical for binaries. Thrymr and Tarqeq do fit the slow-period and the 2-max/2-min extrema criteria, but only show moderate lightcurve amplitudes and are thus still possible, but less obvious binary candidates. Paaliaq might also rotate slowly enough, but the shapes of its lightcurves do not suggest a binary system with two separated components, although a contact binary might be a viable option for this moon. Hyrrokkin, Albiorix, and Ijiraq presumably rotate too fast for being binaries, and their lightcurve shapes or amplitudes do not point in this direction as well. Greip's rotation may also be too fast.

Among the fast rotators of Saturn's outer moons, no one really falls well within the cluster of the small, fast rotating asteroids with small-mass secondaries (type 3). From Fig. 5, it cannot be ruled out that Hati or another fast rotator possesses a small satellite that was possibly formed through mass shedding, but this cannot be revealed with Cassini data because the time coverage and/or SNR of the available data is insufficient. Therefore, such a search remains a task for a future investigation, possibly with a camera on a future spacecraft mission to Saturn, or through observations of stellar occultations.

3.3.3. Orbit parameters, object sizes, and rotational periods. Among Saturn's irregular moons, orbital semimajor axes a, orbital senses of motion (through inclinations i), orbit tilts (inclination supplemental angles i'), and object sizes (through absolute magnitudes H) appear to be correlated to some degree. The prograde moons are on average closer to Saturn, on more inclined orbits, and larger than the retrograde objects (Fig. 4). As discussed in section 2.1, the differences for the mean distances between prograde and retrograde moons are partly explained by the modeling result that at large distances, retrograde orbits are more stable than prograde orbits. Furthermore, the ability of Phoebe to collisionally destroy relatively large objects likely explains the lack of moons with smaller semimajor axes and lower tilts, with only some high-i' objects closer to Saturn having been able to escape a collision so far. The orbit-tilt differences between the progrades and the retrogrades might be an origin effect — that the net direction of motion was retrograde for the progenitors of the low-i' objects, but prograde for the high-i' objects, and that the present configuration just reflects the one of the few original progenitors. In case the hypothesis is correct that several dynamical families originate from Phoebe, this might explain the predominance of objects on retrograde orbits. The mainly larger sizes of the progrades are harder to understand; possibly Phoebe cleared a higher amount of low-i' objects' mass and thus of retrograde objects. Alternatively, the progenitors of the prograde moons might simply have been larger by chance.

The Cassini measurements show that correlations also exist between rotational periods P and the semimajor axes, orbital senses of motion, orbital tilts, and sizes. We divided the group of objects for which rotation periods were available in two samples, comprising objects having semimajor axes smaller and larger than 18.45×10^6 km, respectively. This specific semimajor axis threshold was chosen because

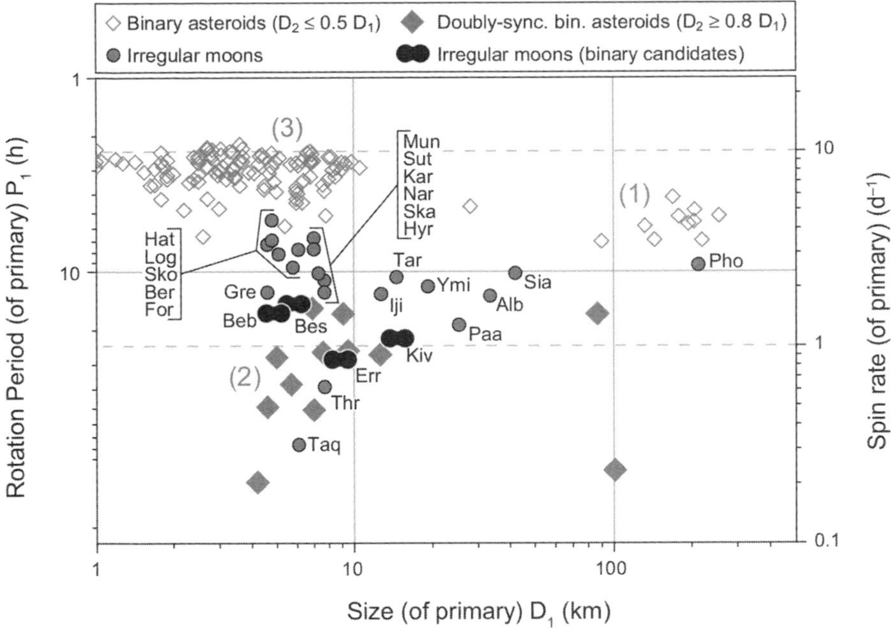

Fig. 5. Primary-component rotation period P_1 over primary-component diameter D_1 for known binary asteroids and for Saturn's irregular moons. For the binary candidates among Saturn's irregulars, the left and right edges of the elongated symbols indicate two diameters as follows: The right edges correspond to the diameters D_1 of the moons in case the moon is no binary ($D_2 = 0$); the left edges show D_1 in the case the objects consists of equally-sized binary components ($D_1 = D_2$). The data for the asteroids are from the LCDB (asteroid lightcurve data base) (*Warner et al.,* 2009, accessed on September 5, 2016); the saturnian-moon data are from *DM18*.

it splits the total sample in two equal-sized groups of 12 members each. (Phoebe was excluded because of its special position with respect to mass and collisional history.) It was found that the median spin rates were 1.8 d^{-1} and 2.7 d^{-1} for the closer and farther satellite group, respectively. A formal Mann-Whitney non-parametric test for median comparison confirmed that the difference is significant at the 95% confidence level. A similar subdivision may also be made for i' and H. For the orbital senses of motion, this may be done as well; only the number of objects in the two samples is now different (9 progrades, 15 retrogrades). In all cases, the rotation periods of the two samples are clearly different from each other. On average, the moons on prograde orbits, at high i', closer to Saturn, and of larger size show longer rotational periods, while most of the smaller, retrograde, low-i', and more-distant objects are the fast rotators. As shown in Fig. 4, there is not even a single object at high i' known in the saturnian system to have a fast spin rate >2.4 d^{-1}; this includes all prograde moons. Note that all known high-i' moons were measured by Cassini. Opposite to this, rotational periods >10 h are rare among the measured lower-i' retrogrades.

No good explanation has been found for these correlations with the periods so far. Since a, i', the direction of motion, and possibly the size are likely not completely independent, it is plausible that just one or two physical mechanisms may explain all correlations with the spin rate. Tidal dissipation may be considered for the relation between a and P since the despinning timescale is directly proportional to the sixth power of the moon's distance to the center planet (*Peale,* 1977). However, the timescale is also inversely proportional to the fifth power of the object size, making tidal despinning as a controlling process for the spin rates of the irregulars unlikely. In fact, despite its large size and closer distance to Saturn, even the despinning of Iapetus is not easy to explain (see, e.g., *Castillo-Rogez et al.,* 2007; *Levison et al.,* 2011).

3.3.4. Bulk densities of fast rotators and potential binaries. The shortest rotation period P_{min} of an object is limited because it would otherwise break into pieces or at least lose material. P_{min} depends on the bulk density ρ, the ratio of the equatorial dimensions a/b, and the internal tensile strength; indepth discussions of this problem can be found, e.g., in *Davidsson* (2001) or *Thomas* (2009). The fastest known rotators in the saturnian system, Hati and Mundifari, are both rather elongated objects with an a/b of at least 1.4 and possibly up to 1.7 (Hati) or even up to 2.0 (Mundilfari). These a/b values correspond to minimum densities of ~500 to ~700 kg m^{-3} (Hati) or ~300 to ~500 kg m^{-3} (Mundilfari). For all other irregulars, the minimum bulk densities from rotations are well below 300 kg m^{-3}.

A low density opens up the possibility that the object may be of cometary structure. For example, the density of 67P/Churyumov-Gerasimenko was determined to 533 ± 6 kg m^{-3} (*Pätzold et al.,* 2016), and that of 19P/Borrelly to a really lightweight 180–300 kg m^{-3} (*Davidsson and Gutiérrez,* 2004). For an object with ρ ~ 200 kg m^{-3}, the rotational period cannot be faster than 7.4 h if its shape is an oblate spheroid with negligible tensile strength. For elongated objects, the critical period increases noticeably. For contact-binaries where the objects are connected near the tips of the single components, a/b (of the equivalent single ellipsoid) mostly exceeds 2 and the minimum rotational period is well above ~10 h. For hypothetical binary moons with almost equally sized components (type 2) and with a surface-to-suface minimum distance of one-half the larger component's diameter (for a discussion of this limit, see *Margot et al.,* 2015), the minimum density may be assessed if the orbit period is known and if the component sizes can be reasonably guessed. Depending on the overall configuration, the binary candidates discussed in section 3.3.2 might all be low-ρ objects because they rotate so slowly.

3.3.5. Size and spin compared to other minor bodies. Comparing average spin rates among different groups of solar system objects is interesting because it might reveal fundamental differences in their compositions or evolutions. The average rotation period for 22 saturnian irregulars of the size range 4 km < D < 45 km (all presented in section 3.2 except Phoebe, Loge, and Fornjot) is 11.4 ± 0.1 h (spin rate 2.10 ± 0.02 d^{-1}). This is quite slow compared to asteroids at this size range in the inner solar system, but maybe not much different from the Jupiter Trojans or objects beyond Saturn. Figure 6 shows the spin rate over diameter for these objects at the chosen size range. The distributions and average values differ noticeably between the groups.

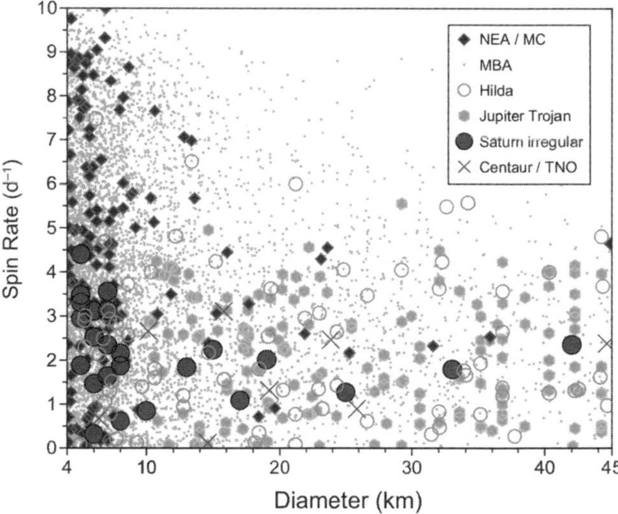

Fig. 6. Object diameters vs. spin rates for various groups of solar-system objects in comparison to the saturnian irregular moons. Shown are objects with sizes between 4 km and 45 km, and with spin rates below 10 d^{-1}. The plotted data, except for the saturnian irregular moons, are from the LCDB (*Warner et al.,* 2009, accessed on September 5, 2016). The average rotational periods for this size range are as follows (in brackets for sizes 10 to 45 km): near-Earth asteroids and Mars crossers: 5.4 h [6.1 h]; main-belt asteroids: 6.4 h [7.0 h]; Hilda asteroids: 10.2 h [10.2 h]; Jupiter Trojans: N/A [11.3 h]; saturnian irregulars: 11.4 h [14.2 h]; Centaurs and TNOs: ~13 h.

However, there are different observational biases within these numbers. While the value for the main-belt asteroids should be reliable, the observational cutoff at the small end is ~5 km for the Hildas, and ~10 km for the Trojans. Small Hildas are thus missing in a statistically significant way, and small Trojans are completely absent in the sample. For the Centaurs/TNOs, only nine objects are in the plot. Even the record of the saturnian moons is incomplete; for the sizes below ~6 km (H ≥ 15 mag), more than half the objects are missing, not counting so-far-undetected moons larger than 4 km.

Nevertheless, the deficit or absence of fast rotators among the outer solar system objects is evident. The figure indicates an increasing upper limit for the spin rate with increasing distance to the Sun. For the Hildas, this boundary appears near ~4 to 4.5 h, for the Jupiter Trojans around 5 h. For the saturnian irregulars, as discussed, the fastest rotator (Hati) has a period of ~5.5 h, and the larger ones even require >10 h for one rotation. About one-third of the asteroids of the size range 4 km ≤ D ≤ 45 km rotate faster than Hati, and more than 60% of the asteroids of the size range ~10 km ≤ D ≤ 45 km rotate faster than 10 h. Two reasons are proposed for the differences of average spin rates between inner and outer solar system objects. One is the potential efficiency (or lack) of the YORP effect [anisotropic emission of thermal radiation causes a tiny torque to the asteroid (see, e.g., *Bottke et al.,* 2006)], which slowly but steadily changes spin rates of smaller objects in the inner solar system. The other is the supposed lower bulk density of the outer objects, which does not allow them to sustain fast spins as long as they cannot withstand strong tensile stress. Since YORP is not efficient at pushing objects at large heliocentric distances toward the boundary of physical disruption, and because their bulk densities as well as equatorial-diameter ratios should not be homogeneous throughout the populations, and possibly because the number of measured objects might be too low, their spin barriers do not appear as sharp as for the asteroids.

3.3.6. Merging the observations: Thoughts on the physics. The determination of rotational periods for almost two-thirds of Saturn's known irregular moons offered new insights into fundamental properties and poses the question about the physical reasons behind the findings. From the comparison with other small solar system objects (Fig. 6), it appears quite likely that the irregular moons of Saturn, except Phoebe, are rather low-density objects. It is plausible that the real disruption spin barrier for the moons is somewhere between ~5 h and ~6 h and not near 2.2 h as for the asteroids. The equivalent bulk density of this period range is ~300 to ~450 kg m^{-3} and thus cometary in nature.

Answers to the question about the physical reasons for the found correlations between rotational periods on the one hand and object sizes, semimajor axes, and orbit tilts on the other remain speculative. Due to the lack of straightforward hypotheses, we just offer some thoughts. The first one suggests that the pattern observed and illustrated in Fig. 4 might be due to different physical characteristics of progenitor objects. The destruction and reaccretion processes of the prograde progenitors might have formed rubble-pile type objects consisting of considerable amounts of water ice with high porosity and, consequently, of a particularly low bulk density. For some reasons, most (~92%) of the current irregular-moon mass outside Phoebe resides in the highly inclined prograde orbits. For the minority of mass farther away from Saturn on lower-tilted retrograde orbits, the different rotational-period properties might be attributed to different physical properties of the progenitors of the different orbit-dynamical families. In particular, the family where Hati is a member might have had a progenitor with a somewhat higher bulk density. This family also contains the second-fastest rotator, Mundilfari, and it would be interesting to see if most of the other six members are also as fast rotators as these two moons are, and if their colors match the neutral color of Mundilfari. A different bulk density and color compared to members of other families may be indicative for a different region of origin in the solar system, somewhat in line with a similar speculation by *GB07* based on the object colors.

The second thought is a corollary to the first one, speculating that many of the retrograde irregular moons might be ejecta from violent impacts on Phoebe. The high-dispersion velocities on the order of 0.5 km s^{-1} do not argue for Phoebe being the origin of some other irregulars, but since Phoebe has survived violent impacts forming craters with sizes up to 100 km, which would easily have pulverized the other moons, debris escaping at much higher speed than from impacts on the smaller moons might in principle be possible. Since Phoebe's density is ~1.6 g cm^{-3}, irregular moons originating from Phoebe should have a significantly higher bulk density than the "cometary" irregulars, even if they are entirely rubble piles. Consequently, the spin barrier for these potential "Phoebe-debris moons" would be at a shorter rotational period than for the "cometary moons." In this scenario, the Mundilfari and the Suttungr families, plus possibly S/2007 S 2 or even Skoll (see Figs. 1 and 4) might result from just a few large impacts on Phoebe, and about half the known retrograde moons might thus have formed this way. The color measurements by *GB07* are consistent and even supportive for such a scenario (Table 3). Although nice, it is incomplete because the fast rotators in the Bergelmir, Kari, and Ymir families are not covered by this explanation unless these were also remnants from Phoebe. It is also generally questionable if ejecta debris from violent impacts that form up to 100-km-sized craters may actually lead to objects with diameters of many kilometers.

The third thought is whether the capture events for the progenitors of the prograde objects might have been different from that of the retrograde moons with respect to the formation regions of the different progenitors. While the objects captured into prograde orbits, by whatever reason, were of low, cometary density, those captured into retrograde orbits had a somewhat higher density. Interesting in this scenario is that only Thrymr and Bestla are the really slow rotators among the numerous retrogrades — the next-slowest retrograde moons, Hyrrokkin and Greip, already have quite fast periods of 12.75 h. Thrymr and Bestla might simply be some kind of outliers. The situation is exactly opposite for the prograde moons, where all but two objects have rotational periods slower than 13 h.

Finally, the observed distribution of rotational periods vs. sizes, Saturn distances, and inclinations might simply be random. Although we consider this unlikely because the measured differences were found to be significant at a high confidence level, the number of objects available for good statistical considerations is still too small to rule this out. In any case, the Cassini measurements yielded a lot of new information that helps to further characterize these objects and to set further constraints on their formation and evolution.

4. PHOEBE

Phoebe (Fig. 7, Table 4) is the only large irregular satellite of Saturn and the best investigated object of all outer moons in the solar system. It was discovered by William Pickering in 1899 from photographic plates and announced by his brother Edward Pickering (*Pickering*, 1899a,b). Phoebe orbits Saturn at an average distance of ~13 million kilometers

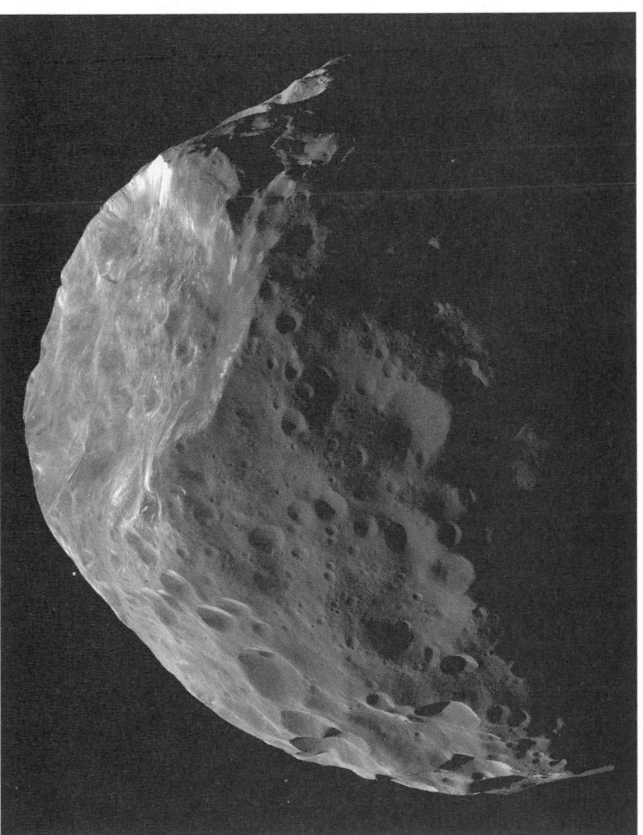

Fig. 7. Global eight-panel mosaic of Phoebe, taken with the Cassini NAC on June 11, 2004, about 0.5 h after closest approach from a distance of ~10,800 km at a phase angle of ~83°. North is up. The large ~100-km-sized crater to the left is Jason. The large wall of exposed water ice at the top left of the image belongs to the ~38-km-sized crater Erginus, which is located within Jason. See also Fig. 9 in the chapter by Thomas et al. in this volume for a high-resolution view of Phoebe's surface.

TABLE 4. Phoebe by numbers.

Orbit *

Semimajor axis...12.948 × 10^6 km — Eccentricity...0.163 — Orbit direction...retrograde

Period...548.02 d — Inclination...175.24° — Mean orbit velocity...1.71 km s^{-1}

Body †

Mean radius...106.4 ± 0.4 km — Ellipsoidal radii (a × b × c)...109.3 ± 0.9 km × 108.4 ± 0.4 km × 101.8 ± 0.2 km

Volume...5.06 ± 0.20 × 10^6 km^3 — Mass..829.2 ± 1.0 × 10^{16} kg — Mean density...1642 ± 18 kg m^{-3}

Surface gravity...0.038 to 0.050 m s^{-2} (~1/200 to ~1/260 g) — Escape velocity... 89 to 108 m s^{-1}

Rotation ‡

Period...9.2735 ± 0.0006 h — Spin direction...prograde — J2000 spin-axis...RA/Dec = 356.6° ± 0.3°/+78.0° ± 0.1°

Surface §

Geometric albedos:

B-band...0.0855 ± 0.0031 — V-band...0.0856 ± 0.0023 — R-band...0.0843 ± 0.0020 — I-band...0.0839 ± 0.0023

Bolometric Bond albedo (disk-averaged)...0.023 ± 0.007 — Range of normal reflectances (Cassini ISS)...0.07 to 0.3

Opposition surge (brightness increase from 2° to 0° phase)...0.33 ± 0.02 mag

Spectral slope...−2.5 ± 0.4%/$_{100\ nm}$ — Phase integral... 0.29 ± 0.03 — Macroscopic roughness...33° ± 3°

Thermal inertia...~25 $Jm^{-2}K^{-1}$ $s^{-½}$ — Temperatures (low-latitude)...82 K before dawn to 112 K near subsolar point

Global mosaic and standard map sheet ¶

Scale...1:1,000,000 — Resolution...8 pxl deg^{-1} or 0.233 km pxl^{-1} — Grid system... Planetocentric lat., west lon.

Interplanetary spacecraft observations

Voyager 2**...date: 03/04 Sep 1981 — range: 2 × 10^6 km — spatial resolution: 20 km pxl^{-1}

Cassini targeted...date: 10–12 Jun 2004 — minimum altitude: 2183 km (to Phoebe center) — no. of images: 552

best spatial resolution (ISS): 13 m pxl^{-1} — solar phase angles: 24°–92°

Cassini non-targeted††...dates: 06 Aug, 07 Oct 2004; 05 Oct, 06 Nov, 03, 14 Dec 2005; 04 Jan 2006; 07 Jan 2015

no. of observations: 8 — spatial resol. (ISS): 37–150 km pxl^{-1} — solar phase angles: 3°–162°

* From JPL solar-system dynamics website (*http://ssd.jpl.nasa.gov*); orbit direction from *Pickering* (1905).

† Chapter by Thomas et al. (this volume), *Castillo-Rogez et al.* (2012), *Jacobson et al.* (2006), *Porco et al.* (2005).

‡ *Bauer et al.* (2004), *Colvin et al.* (1989), *Giese et al.* (2006).

§ *Miller et al.* (2011), *Buratti et al.* (2008), *Porco et al.* (2005), *Thomas* (2010), *GB07*, *Flasar et al.* (2005).

¶ *Roatsch et al.* (2006); see also IAU planetary nomenclature (*https://planetarynames.wr.usgs.gov/Page/PHOEBE/target*).

** *Thomas et al.* (1983).

†† Does not include optical navigation observations.

every 1.5 years in a retrograde orbit, which suggests that it was captured some time in the past (*Pollack et al.,* 1979; *Burns*, 1986). In 1981, Phoebe was the very first irregular moon from which disk-resolved images were obtained by Voyager 2. The images had shown the satellite to be dark, with relatively small regions that were somewhat brighter (*Simonelli et al.,* 1999), and to have an essentially flat spectrum at visible wavelengths or even a slightly negative spectral slope, i.e., exhibiting a higher reflectance toward the blue and ultraviolet (UV) wavelengths (*Tedesco et al.,* 1989; *Simonelli et al.,* 1999).

On June 11, 2004, during the first and so far only close flyby of an irregular moon, the Cassini-Huygens spacecraft on its way to Saturn orbit insertion came as close as 2070 km to Phoebe's surface to enable high-resolution remote sensing observations and the determination of the satellite's mass and field-and-particle environment. The ISS cameras obtained images at better than 2 km pxl^{-1} over slightly more than three Phoebe rotations. The highest-resolution ISS images have a pixel scale of 13 m. The mean diameter of Phoebe was determined to 213 km (*Thomas*, 2010; *Castillo-Rogez et al.,* 2012). Phoebe's global shape is close to an oblate spheroid, with a = b to within the uncertainties of the data. The calculated volume, combined with the mass determined from radio tracking of the spacecraft, yields a mean density of about 1.64 g cm^{-3} (*Thomas,* 2010). Phoebe's impact craters range in diameter from the lower limit imposed by the ISS image resolution up to ~100 km (crater Jason), which is almost the maximum limit imposed by the size of Phoebe itself. There are more than 130 craters with diameters >10 km, and about 20 craters are between 25 and 50 km across (*Porco et al.,* 2005).

Bright material on Phoebe appears to be exposed on flat areas and gentle slopes by cratering, and by mass wasting of steep scarps. Bright spots are associated with craters ranging from below the image resolution to ~1 km in size. Material excavated by impacts typically comes from depths <0.1 crater diameters (*Gladman et al.,* 2001); thus, the bright crater deposits represent material from a few meters to ~100 m in depth. Bright exposures also occur in landslide debris, which represents a mixture of materials from a variety of depths. Therefore, the brighter, ice-rich material occurs at both shallow and deeper depths in widespread geographic and geologic settings. However, only a small fraction of craters (<10%) in a limited size range (diameters ≤1 km) presently display bright materials. This observation suggests that bright materials either darken or are covered as they age by processes such as (1) infall of dark material from impacts among other small, outer satellites (*Melosh,* 1989), (2) deposition (regional or global) of debris excavated from elsewhere on Phoebe; (3) sublimation of ice from the bright component, or, possibly, (4) photochemical darkening of impurities in the brighter material.

Cassini ISS found a range of normal albedo from 0.07 to 0.3 (*Porco et al.,* 2005). Local albedo variations on the surface of Phoebe having contrast factors of 2 to 3 as measured by ISS are manifested chiefly as brighter downslope streamers and bright annuli, rays, or irregular bright areas around small craters (*Porco et al.,* 2005). These contrast ratios suggest normal reflectances of ~30% or less, values incompatible with clean ice. Thus, although most of the brighter outcrops are volatile-rich, they are "dirty" (contaminant fraction could still be small) and could evolve to darker lag deposits that mantle Phoebe's surface through sublimation and thermal degradation processes related to insolation, sputtering, and impact cratering (*Porco et al.,* 2005). Maps of normal reflectance show the existence of two major albedo regimes in the infrared, with gradations between the two regimes and much terrain with substantially higher albedos (*Buratti et al.,* 2008).

Rotational lightcurves derived from the pre-flyby and post-flyby data displayed no substantial variations of whole-disk color with longitude (*Porco et al.,* 2005). The phase curve suggests that Phoebe is overall covered by a mantle of fine particles, resulting from its collisional history or from outgassing (*Buratti et al.,* 2008; *Mueller et al.,* 2008, *Miller et al.*, 2011). The surficial macroscopic roughness of Phoebe was found to be above 30° (*Buratti et al.,* 2008), which is significantly higher than that estimated for many other small bodies (~20°) and is consistent with a violent collisional history (*Nesvorný et al.,* 2003; *Turrini et al.,* 2009).

Thermal infrared data of Phoebe acquired by the Composite Infrared Spectrometer (CIRS) onboard the Cassini orbiter (*Flasar et al.,* 2004) during the closest approach phase at a spatial resolution >8 km showed that temperatures on Phoebe, measured at low latitudes, vary between 82 K and 112 K (*Flasar et al.,* 2005). This diurnal variation yields a thermal inertia of about 25 J m^{-2} K^{-1} $s^{-1/2}$ (*Flasar et al.,* 2005), which is comparable to that of many small bodies (asteroids and comets) and is overall indicative of a fine-grained surface regolith.

Based on visible and near-infrared spectra out to 2.4 μm, *Owen et al.* (2001) identified the bulk composition of Phoebe to be water ice and amorphous carbon. The first identification of non-water-ice volatiles and minerals on Phoebe came from spatially resolved hyperspectral images acquired by the Cassini Visual and Infrared Mapping Spectrometer (VIMS), covering the spectral range 0.35–5.1 μm (*Brown et al.,* 2004). These data allowed the identification and mapping of previously detected water ice (*Owen et al.,* 1999), ferrous iron-bearing minerals, bound water, trapped CO_2, trapped H_2, and organics (*Clark et al.*, 2005, 2012). Phoebe's organic-rich composition is unlike any surface yet observed in the inner solar system, strengthening the possibility that Phoebe is coated by material of cometary or outer solar system origin (*Clark et al.,* 2005).

VIMS mapping indicates that water ice is distributed over most of Phoebe's observed surface, but generally shows stronger spectral signatures toward the southern polar region (*Clark et al.,* 2005). Spectral parameters measured by VIMS suggest that Phoebe's CO_2 is native to the body as part of the initial inventory of condensates and now exposed on the surface, strongly mixed with water ice and hydrocarbons (*Cruikshank et al.*, 2010; *Filacchione et al.*, 2010). In medium-resolution VIMS data (4–8 km pxl^{-1}), water ice exposed in the wall of the 38-km crater Erginus (Fig. 7) showed increased abundance of CO_2 and aromatic hydrocarbons, suggesting that these compounds are strongly associated with water ice (*Coradini et al.*, 2008). On the other hand, pixels where CO_2 is depleted showed higher concentrations of non-ice compounds, and vice versa. However, this anti-correlation is not sharp, but rather smooth, with some regions showing an intermediate situation (*Coradini et al.*, 2008). In the highest-resolution data, crater interiors display less exposed ice than the surrounding terrain. The ice-rich layer exposed in crater walls just below Phoebe's surface and the blue peak seen in the highest-resolution Phoebe spectra imply that the abundance of dark material is low (less than about 2%), and is probably only a surface coating (*Clark et al.,* 2005, 2008).

While Phoebe's topography, relative to an equipotential surface, is within the range of other small objects and is much higher than that for clearly relaxed objects, the nearly oblate spheroid shape of Phoebe may retain characteristics of an early, relaxed object (*Thomas*, 2010). Phoebe's global shape is actually close to that for a hydrostatic-equilibrium spheroid rotating with Phoebe's spin period (*Castillo-Rogez et al.*, 2012). In particular, the *(a-c)* value for Phoebe is most compatible with some degree of mass concentration toward the center (*Johnson et al.,* 2009). Phoebe's low bulk porosity and near-spherical shape suggest that the satellite was not disrupted and subsequently reaccreted. If it were, then the disruption and reaccretion conditions would have had to be very exceptional for Phoebe not to end up as a rubble pile (*Castillo-Rogez et al.,* 2012).

The volatile-rich composition of Phoebe should also reflect, to some extent, the composition of the region where it accreted. One reasonable possibility is that it could have accreted in the volatile-rich outer solar nebula where the Kuiper belt objects (KBOs) originated. Physical properties that Phoebe shares with KBOs include the presence of water ice and an overall low surface albedo (e.g., *Brown et al.,* 1999; *Jewitt and Luu*, 2004; *Pinilla-Alonso et al.*, 2007). The ratio of water ice to other materials in its interior, inferred from the average density of the object, might help determine whether Phoebe originated in the solar nebula or in the circum-saturnian disk (*Johnson and Lunine*, 2005). Mean density is a function of both the sample density and the body's overall porosity. The mean density of Phoebe is higher than the average density of the regular (inner) satellite system of Saturn (excluding Titan), which is ~1300 kg m^{-3}. Based on the average surface composition as derived from VIMS, the plausible range of porosity of Phoebe is from near zero to about 50%. If Phoebe were derived from the same compositional reservoir as Pluto and Triton (whose uncompressed density is ~1900 kg m^{-3}), its present porosity would have to be ~0.15 to attain the same material density (*Johnson and Lunine*, 2005). Even if the porosity of Phoebe were zero, its Cassini-derived density would be 1σ above that of the regular icy saturnian satellites. Therefore, Phoebe appears to be compositionally different from the mid-sized regular satellites of Saturn, ultimately supporting the evidence that it is a captured body (*Johnson and Lunine*, 2005). *Miller et al.* (2011) also conclude that Phoebe should originate from the outer solar system. Their photometric study from the ground at very low phase angles indicates that Phoebe's geometric albedo and opposition-surge magnitude are a good match to outer solar system bodies. Regardless of its origin, Phoebe's diverse mix of surface materials probably samples primitive materials in the outer solar system.

Since 1974, it has been postulated that Phoebe may be the source of the low-albedo material coating the leading side of Saturn's moon Iapetus (e.g., *Soter*, 1974; *Mendis and Axford*, 1974; *Cruikshank et al.*, 1983; *Bell et al.*, 1985; *Burns et al.*, 1996; *Jarvis et al.*, 2000; *Buratti et al.*, 2005; *Verbiscer et al.*, 2009; *Tosi et al.*, 2010; *Tamayo et al.*, 2011; *Cruikshank et al.*, 2014). Even a single impact event on Phoebe, like the one that originated the medium-sized crater Hylas [diameter 28 km, depth 4.7 km (*Giese et al.*, 2006)], could in principle supply nearly the amount of material needed to darken the leading side of Iapetus (*Verbiscer et al.*, 2009; *Tosi et al.*, 2010). Statistically, a significant fraction of the retrograde dust particles of grain sizes greater than 1 μm impact Iapetus while migrating inward (*Tosi et al.*, 2010).

However, the data indicate that dark material from Phoebe cannot be the only cause of the *albedo dichotomy* of Iapetus as observed today, for four reasons. First, Phoebe's surface is essentially gray and spectrally flat in the visible range and thus substantially different from the reddish color exhibited by the dark material on Iapetus (*Owen et al.*, 2001; *Buratti et al.*, 2002). However, on the basis of Cassini/VIMS data, a higher degree of similarity is found in the near-infrared range between Iapetus and Phoebe, as suggested by the detection of spectral signatures at 2.42 μm, 2.97 μm, and 3.29 μm on both of these bodies (*Cruikshank et al.*, 2008; *Clark et al.*, 2012). Second, there is spectral evidence from Cassini ISS and VIMS suggesting that the dark material on Iapetus has two distinct compositional classes, whose spatial distribution reveals that at least two separate events or mechanisms were responsible for the darkening (*Denk et al.*, 2010; *Dalle Ore et al.*, 2012). Third, infalling dust would form fuzzy instead of the observed sharp albedo boundaries on Iapetus (*Denk et al.*, 2010). And finally, since dark material from Phoebe almost exclusively hits the leading side of Iapetus, there should be no prominent local dark areas on the trailing side of Iapetus if no other dark material were present. Thus, "other" dark material on Iapetus, presumably intimately mixed in the water ice, should have been a constituent of the surface before the formation of the albedo dichotomy. An (ongoing) deposition of (Phoebe) dust on the leading side of Iapetus was then sufficient to trigger a thermal segregation process, globally redistributing water ice through a self-sustaining process away from low- and mid-latitudes of the leading side toward high latitudes and the trailing side, but no such globally acting process removes water ice from the trailing side or the poles (*Spencer and Denk*, 2010). The two color classes mentioned above show an unsharp transition near the boundary between the leading and the trailing side of Iapetus in the ISS data (aka near ~0°W and ~180°W longitudes); *Denk et al.* (2010) dubbed it the *color dichotomy* of Iapetus. This feature is consistent with an origin from Phoebe or the other irregular moons.

An interesting feature related to Phoebe and possibly to the other irregular moons is the so-called *Phoebe ring*, a very faint but huge dust torus around Saturn discovered in Spitzer data at wavelengths of 24 and 70 μm as the result of a dedicated search (*Verbiscer et al.*, 2009). Subsequent observations with the Wide-field Infrared Survey Explorer (WISE) spacecraft at 22 μm showed a radial extension up to 16.3 × 10^6 km, which is beyond Phoebe's apoapsis distance (15.1 × 10^6 km) (*Hamilton et al.*, 2015). The Phoebe ring was also observed through dedicated Cassini ISS observations with the Wide Angle Camera (*Tamayo et al.*, 2014). *Kennedy et al.* (2011) examined Spitzer data for dust from the non-Phoebe irregulars. From their non-detection, they derive limits for the size distribution of the micrometer-sized dust and conclude that the strength properties of the irregulars should be more porous than expected for asteroids and therefore more akin to comets.

5. ORIGIN OF THE IRREGULAR SATELLITES

The irregular satellites of Saturn have defied conventional origin models for some time. As a population, their orbital properties, size-frequency distributions, and colors are roughly similar to the irregular satellites surrounding Jupiter, Uranus, and Neptune. This means that any scenario that hopes to describe how Saturn's irregular satellites reached their current state might be broadly applicable to those worlds as well. Here we briefly summarize the properties of these populations and refer to the reviews by *Sheppard* (2006), *Jewitt and Haghighipour* (2007), and *Nicholson et al.* (2008) for more details.

Orbits. The irregular satellites fill the stable orbital zones that exist around each giant planet (*Carruba et al.*, 2002; *Nesvorný et al.*, 2003; *Shen and Tremaine*, 2008). They are located on prograde and retrograde orbits with semimajor axes values between 0.1 and 0.5 times the distance to their respective planets' Hill sphere. They also have a wide range of eccentricity e and inclination i values ($0.1 < e < 0.7$; $25° < i < 60°$ or $130° < i < 180°$; Nereid: $e = 0.75$, $i = 7°$). Families of fragments with similar orbital properties can also be found, as discussed in previous sections.

Size-frequency distributions (SFDs). The shapes of their cumulative SFDs are unlike any small body population seen anywhere in the solar system (*Bottke et al.,* 2010). The combined prograde and retrograde SFDs at each planet have a cumulative power-law index of $q \sim -1$ for diameters $20 < D < 200$ km. This means that outside the largest objects (e.g., Phoebe at Saturn), they tend to have very little mass. The power-law slopes then change, in some cases dramatically, for D < 10 to 30 km. The steepest slopes are seen at Saturn, where $q \sim -3.3$ between $6 < D < 8$ km.

Colors. The irregular satellites have colors that are consistent with dark C-, P-, and D-type asteroids, which are common among the Hilda and Trojan asteroid populations (e.g., *Grav et al.,* 2003, 2004; *GB07*; *Grav and Holman,* 2004) as well as in the outer asteroid belt (e.g., *Vokrouhlický et al.,* 2016). These objects are also a good match to the observed dormant comets.

There have been many capture scenarios described in the literature. The various methods of formation could be grouped into three broad categories: drag capture, collision-induced capture, and three-body capture. For drag capture, we have several flavors: capture due to the sudden growth of the gas giant planets, which is often referred to as the "pull-down" capture method (*Heppenheimer and Porco,* 1977); capture of planetesimals due to the dissipation of their orbital energy via gas drag (*Pollack et al.*, 1979; *Astakhov et al.*, 2003; *Ćuk and Burns*, 2004; *Kortenkamp*, 2005); and capture during resonance-crossing events between primary planets at a time when gas drag was still active (*Ćuk and Gladman*, 2006). Capture by collisions between icy planetesimals is characterized by *Colombo and Franklin* (1971) and *Estrada and Mosqueira* (2006). Three-body capture has been considered using three-body exchange reactions between a binary planetesimal and the primary planet (*Agnor and Hamilton*, 2006; *Vokrouhlický et al.*, 2008; *Philpott et al.*, 2010), and capture in three-body interactions during close encounters between the gas giant planets within the framework of the so-called Nice model (*Nesvorný et al.*, 2007, 2014; and see below).

Reviews of these different scenarios can be found in *Jewitt and Haghighipour* (2007), *Nesvorný et al.* (2007), *Nicholson et al.* (2008), and *Vokrouhlický et al.* (2008). The arguments they present suggest that nearly all the models above are problematic at some level because they suffer from one or more of the following problems: They are schematic and underdeveloped, they do not really match what is known about planet formation processes and/or planetary physics, their capture efficiency is too low to be viable, they require fine and probably unrealistic timing in terms of gas accretion processes or the turning on/off of gas drag, they cannot reproduce the observed orbits of the irregular satellites, or they can produce satellites around some but not all gas giants. Moreover, if the outer planets migrate after the capture of the irregular satellites, the satellites themselves will be efficiently removed by the passage of larger planetesimals or planets through the satellite system. This suggests that while different generations of irregular satellites may have existed at different times, the irregular satellites observed today were probably captured in a gas-free environment.

The most successful model at reproducing constraints thus far is from *Nesvorný et al.* (2007, 2014), with giant planetary migration taking place after the gas disk is gone acting as the conduit for satellite capture. This model takes advantage of the Nice model (*Tsiganis et al.*, 2005; *Gomes et al.*, 2005), a family of solutions where the giant planets started in a different configuration and then experienced migration in response to a dynamical instability. The most successful Nice model simulations are discussed in *Nesvorný* (2011), *Batygin et al.* (2012), and *Nesvorný and Morbidelli* (2012). They assume that the giant planets were once surrounded by a primordial disk of comet-like bodies comprising at least ~20 Earth masses. This system was stable for a few tens to a few hundreds of millions of years (e.g., see *Bottke and Norman,* 2017; *Kaib and Chambers*, 2016), but eventually became dynamically unstable. This drove Uranus-Neptune into the primordial disk, where their migration through it not only created the observed Kuiper belt and related populations (e.g., Jupiter/Neptune Trojans) (*Morbidelli et al.*, 2005, *Nesvorný and Vokrouhlický*, 2009, *Nesvorný et al.*, 2013; *Gomes and Nesvorný*, 2016), but also scattered most of the disk's mass into the giant planet region. A consequence of this migration is that giant planets were encountering one another while surrounded by comet-like objects. This allowed many comets to be captured onto stable giant planet orbits via three-body gravitational interactions (*Nesvorný et al.,* 2007, 2014). Intriguingly, model results show the captured population surviving on stable orbits can reproduce the observed prograde and retrograde irregular satellites remarkably well.

A potential problem with the *Nesvorný et al.* (2007) model, however, is that it predicts that the irregular satellites should potentially show SFDs similar to those of the Trojan asteroids. Trojan asteroids, having roughly a factor of 5 fewer diameter >100-km bodies than the main asteroid belt, have most likely experienced less collisional evolution than the main belt since being captured ~4–4.5 G.y. ago. As such, the shape of the Trojan size distribution should be fairly close to the shape it had immediately after capture took place (*Wong et al.,* 2014; *Wong and Brown,* 2015). An important aspect of irregular satellites, however, is that they were captured into a relatively tiny region of space with short orbital periods around Saturn. This makes high-velocity collisions between irregular satellites unavoidable. Moreover, collision probabilities between typical irregular satellites are 4 orders of magnitude higher than those found among main-belt asteroids (*Bottke et al.,* 2010). An analogy used by *Bottke et al.* (2010) would be to consider the rate of car crashes occurring along the empty back roads of the

American West (asteroids) to rush-hour traffic in Los Angeles (irregular satellites). Accordingly, their SFDs should undergo rapid and extensive collisional evolution.

Collisional simulations of *Bottke et al.* (2010) show that the prograde and retrograde SFDs quickly grind themselves down from a high-mass into a low-mass state. When low masses are reached, the SFDs may stay in a quasi-collisional steady state for hundreds of millions of years until a stochastic disruption event produces a swarm of new smaller fragments with diameters D < 1–7 km. This fragment trail will then itself slowly grind away on timescales of tens of millions of years until the next stochastic disruption or cratering event on one of the larger remaining satellites produces new ejecta. In this manner, the D < 7 km SFDs for both Saturn's prograde and retrograde populations wave up and down again and again over billions of years. The implication is that smaller irregular satellites are mostly fragments of larger bodies, and all have been subjected to numerous impacts since they formed. The orbit-dynamical and compositional families as discussed in sections 2 and 3 (Fig. 1) are an observational support of this one-dimensional model. The prediction is that these populations have lost ~99% of their mass via impacts, making them the most collisionally evolved populations in the solar system. Overall, the model does a good job of explaining the SFD of Saturn's irregular satellites, as well as of those found around the other giant planets. It is also in good agreement with the interpretation of the measured rotational-period distribution vs. object sizes from section 3.3.

A curious yet intriguing prediction for these collisional models concerns the ~99% of the mass lost to comminution ovcr 4 G.y. at cach giant planet. The question is what could have happened to all this material. Since the small particles are affected by solar radiation pressure forces and Poynting-Robertson (P-R) drag, they preferentially move inward (*Burns et al.,* 1979). The implication is that the outermost regular satellites of the gas giants could have potentially been blanketed by large amounts of dust produced by a collisional cascade among the irregular satellites, a scenario which in particular the jovian moon Callisto seems to fit (*Bottke et al.,* 2013). For the Saturn system, vast amounts of dust have traveled from Phoebe and the irregular satellites to Iapetus, much visible today on Iapetus' leading hemisphere and possibly hidden beneath upper surface ices. Dust from Phoebe and the irregular satellites also reached Titan. It is speculated that the equatorial longitudinal dune fields that cover a significant fraction of Titan's surface are made of this exogenic material.

6. MISSING INFORMATION AND OUTLOOK

In this chapter, we have briefly summarized the state of knowledge of Saturn's irregular satellites such as orbits, sizes, colors, rotational periods, rough shapes, or bulk-density constraints. In addition, conclusions were discussed such as grouping into dynamical families and possibly pairs similar to the main-belt asteroids; correlations between orbital elements, object spin rates, and sizes; then hints of low bulk densities, hints of possible binarity, or the corroboration of the hypothesis that these objects were once formed in other regions of the solar system, later captured and eventually further shredded through collisions over aeons. Research results for Phoebe, in particular some outcome from the unprecedented flyby of Cassini, were presented.

However, despite the advancement in the exploration of the irregular moons of Saturn by Cassini from its vantage point inside the system, many questions remain unanswered and are likely to remain so even when the Cassini dataset will be fully analyzed. For example, we see patterns in the distribution of the lightcurve features and correlations in rotational and dynamical properties (discussed in sections 3.3.3 and 3.3.5; see also Fig. 4) that are not yet explained and that likely relate to the formation process. Furthermore, the capture process for the progenitors of the irregulars is not well understood (section 5). Also, hints for low bulk densities were derived (section 3.3.4), but conclusive measurements with useful error bars do not exist. Another example is the question about the potential existence of binary or contact-binary moons where the Cassini data reach their limits (section 3.3.2).

Although apparently not glamorous science at first glance, probably most important to further increase our understanding of the irregulars is a steady buildup of statistics for both physical and dynamical properties. Discovering more objects and getting more periods and colors (or even spectra) would reveal the described (and probably additional, so far unnoticed) patterns and their (high or low) significance more clearly. As a consequence, boundary conditions for modeling the capture process and the subsequent evolution would become available, which might allow revealing of the history of these objects and thus of a crucial part of the history of the solar system. Ultimately, it is not just the saturnian system that is of interest, but also the irregular-moon systems of Jupiter, Uranus, and Neptune, which are practically unexplored so far.

Due to their faintness and proximity to the bright host planet, most of the irregular moons are challenging targets for physical characterization via Earth-bound observations. As already done for the brightest objects, large telescopes may again be used for targeted observations that do not require vast amounts of telescope time. The situation could further improve in the near future when really giant telescopes, like the Extremely Large Telescope (E-ELT), Thirty Meter Telescope (TMT), or Giant Magellan Telescope (GMT), will be brought into service. These new optical tools might demonstrate their power by barely resolving the largest irregulars and the binary candidates. With these discovery machines on the ground, and in particular with the James Webb Space Telescope (JWST), which is planned to launch to Earth's L_2 Lagrangian point in 2021, spectrophotometry and spectroscopy from the visible to the near-infrared will become possible for numerous irregular satellites. Such studies will enable characterization of the surface composition of these bodies, and will put them in the context of other populations inhabiting the outer solar system, thereby contributing to addressing the question of their origin.

However, a comprehensive groundbased survey of the rotational properties and convex shapes of the faint irregulars remains difficult due to the large amounts of observing time needed on large telescopes. Amateur astronomers may start to play a role if they can get regular access to telescopes of the 3-m class; adaptive optics, coronographs, and outstanding observation sites would definitely help. Automated surveys like those planned for the Large Synoptic Survey Telescope (LSST) may slowly but steadily sample data, which could eventually provide sparse-sampled lightcurves. Despite the new astrometric catalogs from Gaia, observing stellar occultations to explore possible binary moons is still a big challenge because of the small sizes of the objects and thus very short occultation times — on the order of 1 s or less — and because of the small ground tracks. However, mid-2017 occultation observations of (486958) 2014 MU69 ("Ultima Thule"), the target of the New Horizons spacecraft after the Pluto-Charon flyby, were a comparable challenge, yet they were successful (*Buie et al.*, 2017).

All in all, space missions coming close to the giant planets likely remain key. Even though Cassini was not specifically designed to study irregulars (remember that all but Phoebe were undiscovered when Cassini was launched), and although the spacecraft was more than busy with tasks related to the regular moons (see all other chapters in this book) and other components of the saturnian environment, it made terrific strides in this direction. This achievement was mainly due to two factors: (1) the excellence of the onboard payload, which was not designed to simply "minimum requirements," but represented the best achievable within the available resources; and (2) the ingenuity and *spirit* of the management, the science, the navigation, and the operations teams, who excelled in designing innovative trajectories and acquisition sequences to seize every possible opportunity to enable new science. The lesson from Cassini should be taken up by every future spacecraft mission: Implement flexibility both in the instrument and in the mission design, in order to be able to readily respond to discoveries that will undoubtedly come. Considering the sum of all advantages and benefits of spacecraft-based research of irregular satellites, we strongly recommend implementing a program to observe these objects in every spacecraft mission to the giant planets.

This final paragraph dares a short look into the far future when humankind might experience crewed spacecraft to the outer solar system. The year is 2118. Curious scientists are eager to explore, impatient tourists are waiting in the wings. Your cargo ship, carrying the future "Kiviuq Base One," gently approaches its final destination. With the foundation of a station and hotel on this natural stable platform millions of kilometers above the ringed planet lying in your hands, your task to pave the way for the rising era of human expeditions into the wilderness areas of the inner saturnian moons is close to completion. So close to Kiviuq, you head to the main observation deck *just to get the view* (Fig. 8). Sparkling like a brilliant jewel against the blackness of space, Saturn and its main rings, fully four times larger than the full Moon as seen from Earth, just emerge from behind the small irregular shard and remind you how dreamful, excited, weak, and lonesome a human can feel...

Fig. 8. Saturn comes into view behind Kiviuq while hovering at 400 km altitude during the first-ever landing approach on January 17, 2118, to establish the "Kiviuq Base One" as the gate to the saturnian system for coming scientific and tourist exploration.

Acknowledgments. We thank the reviewers JJ Kavelaars and P. Thomas, as well as editor A. Verbiscer, for their very useful comments and suggestions, which improved the chapter significantly. T.D. thanks the Cassini Imaging team and the Cassini Project at JPL for their impressive support while planning and performing observations with the Cassini-ISS cameras. T.D. gratefully acknowledges the support by the Deutsches Zentrum für Luft- und Raumfahrt (DLR) in Bonn (Germany) (grant no. 50 OH 1503). F.T. gratefully acknowledges the support by the Agenzia Spaziale Italiana (ASI) in Rome (Italy), ASI-INAF grant no. I/2017-10-H.O.

The participation of W.F.B. was supported by NASA's SSERVI program "Institute for the Science of Exploration Targets (ISET)" through institute grant number NNA14AB03A. D.P.H. gratefully acknowledges support from NASA's Cassini Data Analysis Program. T.D. extends a special thanks to G. Neukum (1944–2014), member of the Cassini Imaging team, who allowed him to have access to the Cassini mission and provided the opportunity to help design many observations, especially of the moons of Saturn.

REFERENCES

Agnor C. B. and Hamilton D. P. (2006) Neptune's capture of its moon Triton in a binary-planet gravitational encounter. *Nature, 441*, 192–194.

Astakhov S. A., Burbanks A. D., Wiggins S., and Farrelly D. (2003) Chaos-assisted capture of irregular moons. *Nature, 423*, 264–267.

Batygin K., Brown M. E., and Betts H. (2012) Instability-driven dynamical evolution model of a primordially five-planet outer solar system. *Astrophys. J. Lett., 744(1),* L3, DOI: 10.1088/2041-8205/744/1/L3.

Bauer J. M., Buratti B. J., Simonelli D. P., and Owen W. M. (2004) Recovering the rotational light curve of Phoebe. *Astrophys. J., 610*, L57–L60, DOI: 10.1086/423131.

Bauer J. M., Grav T., Buratti B. J., and Hicks M. D. (2006) The phase curve survey of the irregular saturnian satellites: A possible method of physical classification. *Icarus, 184*, 181–197.

Bell J. F., Cruikshank D. P., and Gaffey M. J. (1985) The composition and origin of the Iapetus dark material. *Icarus, 61*, 192–207.

Blunck J. (2010) *Solar System Moons — Discovery and Mythology.* Springer-Verlag, Berlin. 142 pp. DOI 10.1007/978-3-540-68853-2.

Bottke W. F. and Norman M. (2017) The late heavy bombardment. *Annu. Rev. Earth Planet. Sci., 45(1),* 619–647.

Bottke W. F., Vokrouhlický D., Rubincam D., and Nesvorný D. (2006) The Yarkovsky and YORP effects: Implications for asteroid dynamics. *Annu. Rev. Earth Planet. Sci., 34*, 157–191, DOI: 10.1146/annurev.earth.34.031405.125154.

Bottke W. F., Nesvorný D., Vokrouhlický D., and Morbidelli A. (2010) The irregular satellites. The most collisionally evolved population in the solar system. *Astron. J., 139*, 994–1014.

Bottke W. F., Vokrouhlický D., Nesvorný D., and Moore J. M. (2013) Black rain: The burial of the Galilean satellites in irregular satellite debris. *Icarus, 223*, 775–795.

Brown R. H, Cruikshank D. P., and Pendleton Y. (1999) Water ice on Kuiper belt object 1996 TO_{66}. *Astrophys. J. Lett., 519(1),* L101–L104.

Brown R. H., Baines K. H., Bellucci G., Bibring J.-P., Buratti B. J., Capaccioni F., Cerroni P., Clark R. N., Coradini A., Cruikshank D. P., Drossart P., Formisano V., Jaumann R., Langevin Y., Matson D. L., McCord T. B., Mennella V., Miller E., Nelson R. M., Nicholson P. D., Sicardy B., and Sotin C. (2004) The Cassini Visual and Infrared Mapping Spectrometer (VIMS) investigation. *Space Sci. Rev., 115(1–4),* 111–168.

Buie M. W., Porter S. B., Terrell D., Tamblyn P., Verbiscer A. J., Soto A., Wasserman L. H., Zangari A. M., Skrutskie M. F., Parker A., Young E. F., Benecchi S., Stern S. A., and the New Horizons MU69 Occultation Team (2017) Overview of the strategies and results of the 2017 occultation campaigns involving (486958) 2014 MU69. *AAS/Division for Planetary Sciences Meeting Abstracts, 49,* #504.01, *http://adsabs.harvard.edu/abs/2017DPS....4950401B.*

Buratti B. J., Hicks M. D., Tryka K. A., Sittig M. S., and Newburn R. L. (2002) High-resolution 0.33–0.92 μm spectra of Iapetus, Hyperion, Phoebe, Rhea, Dione, and D-type asteroids: How are they related? *Icarus, 155*, 375–381.

Buratti B. J., Hicks M. D., and Davies A. (2005) Spectrophotometry of the small satellites of Saturn and their relationship to Iapetus, Phoebe, and Hyperion. *Icarus, 175*, 490–495.

Buratti B. J., Soderlund K., Bauer J., Mosher J. A., Hicks M. D., Simonelli D. P., Jaumann R., Clark R. N., Brown R. H., Cruikshank D. P., and Momary T. (2008) Infrared (0.83–5.1 μm) photometry of Phoebe from the Cassini Visual Infrared Mapping Spectrometer. *Icarus, 193*, 309–322.

Burns J. A. (1986) The evolution of satellite orbits. In *Satellites* (J. A. Burns and M. S. Matthews, eds.), pp. 117–158. Univ. of Arizona, Tucson.

Burns J. A., Lamy P. L., and Soter S. (1979) Radiation forces on small particles in the solar system. *Icarus, 40*, 1–48.

Burns J. A., Hamilton D. P., Mignard F., and Soter S. (1996) The contamination of Iapetus by Phoebe dust. In *Physics, Chemistry, and Dynamics of Interplanetary Dust* (B. Å. S. Gustafson and M. S. Hanner, eds.), pp. 179–182. ASP Conf. Ser. 104, Astronomical Society of the Pacific, San Francisco.

Carruba V., Burns J. A., Nicholson P. D., and Gladman B. J. (2002) On the inclination distribution of the jovian irregular satellites. *Icarus, 158*, 434–449.

Castillo-Rogez J. C., Matson D. L., Sotin C., Johnson T. V., Lunine J. I., and Thomas P. C. (2007) Iapetus' geophysics: Rotation rate, shape, and equatorial ridge. *Icarus, 190*, 179–202.

Castillo-Rogez J. C., Johnson T. V., Thomas P. C., Choukroun M., Matson D. L., and Lunine J. I. (2012) Geophysical evolution of Saturn's satellite Phoebe, a large planetesimal in the outer solar system. *Icarus, 219*, 86–109, DOI: 10.1016/j.icarus.2012.02.002.

Clark R. N. and 25 colleagues (2005) Compositional maps of Saturn's moon Phoebe from imaging spectroscopy. *Nature, 435*, 66–69.

Clark R. N., Curchin J. M., Jaumann R., Cruikshank D. P., Brown R. H., Hoefen T. M., Stephan K., Moore J. M., Buratti B. J., Baines K. H., Nicholson P. D., and Nelson R. M. (2008) Compositional mapping of Saturn's satellite Dione with Cassini VIMS and implications of dark material in the Saturn system. *Icarus, 193(2),* 372–386.

Clark R. N., Cruikshank D. P., Jaumann R., Brown R. H., Stephan K., Dalle Ore C. M., Livo K. E., Pearson N., Curchin J. M., Hoefen T. M., Buratti B. J., Filacchione G., Baines K. H., and Nicholson P. D. (2012) The surface composition of Iapetus: Mapping results from Cassini VIMS. *Icarus, 218(2),* 831–860.

Colombo G. and Franklin F. A. (1971) On the formation of the outer satellite groups of Jupiter. *Icarus, 15*, 186–189.

Colvin T. R., Davies M. E., Rogers P. G., and Heller J. (1989) *Phoebe: A Preliminary Control Network and Rotational Elements.* NASA STI/Recon Technical Report N-2934-NASA.

Coradini A., Tosi F., Gavrishi A. I., Capaccioni F., Cerroni P., Filacchione G., Adriani A., Brown R. H., Bellucci G., Formisano V., D'Aversa E., Lunine J. I., Baines K. H., Bibring J.-P., Buratti B. J., Clark R. N., Cruikshank D. P., Combes M., Drossart P., Jaumann R., Langevin Y., Matson D. L., McCord T. B., Mennella V., Nelson R. M., Nicholson P. D., Sicardy B., Sotin C., Hedman M. M., Hansen G. B., Hibbitts C. A., Showalter M., Griffith C., and Strazzulla G. (2008) Identification of spectral units on Phoebe. *Icarus, 193(1),* 233–251.

Cruikshank D. P., Bell J. F., Gaffey M. J., Brown R. H., Howell R., Beerman C., and Rognstad M. (1983) The dark side of Iapetus. *Icarus, 53*, 90–104.

Cruikshank D. P., Wegryn E., Dalle Ore C. M., Brown R. H., Baines K. H., Bibring J.-P., Buratti B. J., Clark R. N., McCord T. B., Nicholson P. D., Pendleton Y. J., Owen T. C., Filacchione G., and the VIMS Team (2008) Hydrocarbons on Saturn's satellites Iapetus and Phoebe. *Icarus, 193*, 334–343.

Cruikshank D. P., Meyer A. W., Brown R. H., Clark R. N., Jaumann R., Stephan K., Hibbitts C. A., Sandford S. A., Mastrapa R. M. E., Filacchione G., Dalle Ore C. M., Nicholson P. D., Buratti B. J., McCord T. B., Nelson R. M., Dalton J. B., Baines K., and Matson D. L. (2010) Carbon dioxide on the satellites of Saturn: Results from the Cassini VIMS investigation and revisions to the VIMS wavelength scale. *Icarus, 206(2),* 561–572.

Cruikshank D. P., Dalle Ore C. M., Clark R. N., and Pendleton Y. J. (2014) Aromatic and aliphatic organic materials on Iapetus: Analysis of Cassini VIMS data. *Icarus, 233*, 306–315.

Ćuk M. and Burns J. A. (2004) On the secular behavior of irregular satellites. *Astron. J., 128*, 2518–2541.

Ćuk M. and Gladman B. J. (2005) Constraints on the orbital evolution of Triton. *Astrophys. J. Lett., 626*, L113–L116.

Ćuk M. and Gladman B. J. (2006) Irregular satellite capture during planetary resonance passage. *Icarus, 183*, 362–372.

Dalle Ore C. M., Cruikshank D. P., and Clark R. N. (2012) Infrared spectroscopic characterization of the low-albedo materials on Iapetus. *Icarus, 221(2),* 735–743.

Davidsson B. J. R. (2001) Tidal splitting and rotational breakup of solid biaxial ellipsoids. *Icarus, 149*, 375–383.

Davidsson B. J. R. and Gutiérrez P. J. (2004) Estimating the nucleus density of Comet 19P/Borrelly. *Icarus, 168*, 392–408.

Denk T. and Mottola S. (2013) Irregular saturnian moon lightcurves from Cassini-ISS observations: Update. *AAS/Division for Planetary Sciences Meeting Abstracts, 45,* #406.08. *http://adsabs.harvard.edu/abs/2013DPS....4540608D.*

Denk T., Neukum G., Roatsch Th., Porco C. C., Burns J. A., Galuba G. G., Schmedemann N., Helfenstein P., Thomas P. C., Wagner R. J., and West R. A. (2010) Iapetus: Unique surface properties and a global color dichotomy from Cassini imaging. *Science, 327(5964),* 435–439, DOI: 10.1126/science.1177088.

Estrada P. R. and Mosqueira I. (2006) A gas-poor planetesimal capture model for the formation of giant planet satellite systems. *Icarus, 181,* 486–509.

Filacchione G., Capaccioni F., Clark R. N., Cuzzi J. N., Cruikshank D. P., Coradini A., Cerroni P., Nicholson P. D., McCord T. B., Brown R. H., Buratti B. J., Tosi F., Nelson R. M., Jaumann R., and Stephan K. (2010) Saturn's icy satellites investigated by Cassini–VIMS: II. Results at the end of nominal mission. *Icarus, 206,* 507–523.

Flasar F. M. and 44 colleagues (2004) Exploring the Saturn system in the thermal infrared: The Composite Infrared Spectrometer. *Space Sci. Rev., 115(1–4),* 169–297.

Flasar F. M. and 45 colleagues (2005) Temperatures, winds, and composition in the saturnian system. *Science, 307,* 1247–1251.

Giese B., Neukum G., Roatsch Th., Denk T., and Porco C. C. (2006) Topographic modeling of Phoebe using Cassini images. *Planet. Space Sci., 54,* 1156–1166.

Gladman B., Nicholson P. D., Burns J. A., Kavelaars J. J., Marsden B. G., Williams G. V., and Offutt W. B. (1998) Discovery of two distant irregular moons of Uranus. *Nature, 392,* 897–899.

Gladman B., Kavelaars J. J., Holman M., Nicholson P. D., Burns J. A., Hergenrother C. W., Petit J.-M., Marsden B. G., Jacobson R., Gray W., and Grav T. (2001) Discovery of 12 satellites of Saturn exhibiting orbital clustering. *Nature, 412,* 163–166.

Goldreich P. (1966) History of the lunar orbit. *Rev. Geophys. Space Phys., 4,* 411–439.

Goldreich P., Murray N., Longaretti P. Y., and Banfield D. (1989) Neptune's story. *Science, 245,* 500–504.

Gomes R. and Nesvorny D. (2016) Neptune Trojan formation during planetary instability and migration. *Astron. Astrophys., 592,* A146.

Gomes R., Levison H. F., Tsiganis K., and Morbidelli A. (2005) Origin of the cataclysmic late heavy bombardment period of the terrestrial planets. *Nature, 435,* 466–469.

Grav T. and Bauer J. (2007) [GB07] A deeper look at the colors of the saturnian irregular satellites. *Icarus, 191,* 267–285.

Grav T. and Holman M. J. (2004) Near-infrared photometry of the irregular satellites of Jupiter and Saturn. *Astrophys. J. Lett., 605,* L141–L144.

Grav T., Holman M. J., Gladman B. J., and Aksnes K. (2003) Photometric survey of the irregular satellites. *Icarus, 166,* 33–45.

Grav T., Bauer J. M., Mainzer A. K., Masiero J. R., Nugent C. R., Cutri R. M., Sonnett S., and Kramer E. (2015) NEOWISE: Observations of the irregular satellites of Jupiter and Saturn. *Astrophys. J., 809(3),* 9 pp.

Hamilton D. P. (2001) Saturn: Saturated with satellites. *Nature, 412,* 132–133.

Hamilton D. P., Skrutskie M. F., Verbiscer A. J., and Masci F. J. (2015) Small particles dominate Saturn's Phoebe ring to surprisingly large distances. *Nature, 522,* 185–187.

Harris A. W. and Young J. W. (1989) Asteroid lightcurve observations from 1979–1981. *Icarus, 81,* 314–364.

Helled R., Galanti E., and Kaspi Y. (2015) Saturn's fast spin determined from its gravitational field and oblateness. *Nature, 520,* 202–204, DOI: 10.1038/nature14278.

Hénon M. (1969) Numerical exploration of the restricted problem. V. Hill's Case: Periodic orbits and their stability. *Astron. Astrophys., 1,* 223–238.

Heppenheimer T. A. and Porco C. C. (1977) New contributions to the problem of capture. *Icarus, 30,* 385–401.

Jacobson R. A., Antreasian P. G., Bordi J. J., Criddle K. E., Ionasescu R., Jones J. B., Mackenzie R. A., Pelletier F. J., Owen W. M. Jr., Roth D. C., and Stauch J. R. (2006) The gravitational field of Saturn and the masses of its major satellites. *Astron. J., 132,* 2520–2526.

Jacobson R., Brozović M., Gladman B., Alexandersen M., Nicholson P. D., and Veillet C. (2012) Irregular satellites of the outer planets: Orbital uncertainties and astrometric recoveries in 2009–2011. *Astron. J., 144(132),* 8 pp., DOI: 10.1088/0004-6256/144/5/132.

Jarvis K. S., Vilas F., Larson S. M., and Gaffey M. J. (2000) Are Hyperion and Phoebe linked to Iapetus? *Icarus, 146,* 125–132.

Jewitt D. and Haghighipour N. (2007) Irregular satellites of the planets: Products of capture in the early solar system. *Annu. Rev. Astro. Astrophys., 45,* 261–295.

Jewitt D. and Luu J. X. (2004) Crystalline water ice on the Kuiper belt object (50000) Quaoar. *Nature, 432,* 731–733.

Jewitt D., Sheppard S. S., and Kleyna J. (2006) The strangest satellites in the solar system. *Sci. Am., 295,* 40–47.

Johnson T. V. and Lunine J. I. (2005) Saturn's moon Phoebe as a captured body from the outer solar system. *Nature, 435,* 69–71.

Johnson T. V., Castillo-Rogez J. C., Matson D. L., and Thomas P. C. (2009) Phoebe's shape: Possible constraints on internal structure and origin. In *Lunar and Planetary Science XL,* Abstract #2334. Lunar and Planetary Institute, Houston.

Johnston W. R. (2018) Asteroids with Satellites. *http://www.johnstonsarchive.net/astro/asteroidmoons.html* [accessed April 29, 2018].

Kaib N. A. and Chambers J. E. (2016) The fragility of the terrestrial planets during a giant-planet instability. *Mon. Not. R. Astron. Soc., 455,* 3561–3569.

Kennedy G. M., Wyatt M. C., Su K. Y. L., and Stansberry J. A. (2011) Searching for Saturn's dust swarm: Limits on the size distribution of irregular satellites from km to micron sizes. *Mon. Not. R. Astron. Soc., 417,* 2281–2287.

Kortenkamp S. J. (2005) An efficient, low-velocity, resonant mechanism for capture of satellites by a protoplanet. *Icarus, 175,* 409–418.

Kruse S., Klavetter J. J., and Dunham E. W. (1986) Photometry of Phoebe. *Icarus, 68,* 167–175.

Kusugak M. (2006) *The Curse of the Shaman.* Harper Trophy Canada, Toronto. 158 pp.

Lacerda P. and Jewitt D. C. (2007) Densities of solar system objects from their rotational lightcurves. *Astron. J., 133,* 1393–1408.

Levison H. F., Walsh K. J., Barr A. C., and Dones L. (2011) Ridge formation and de-spinning of Iapetus via an impact-generated satellite. *Icarus, 214,* 773–778.

Margot J.-L., Pravec P., Taylor P., Carry B., and Jacobson S. A. (2015) Asteroid systems: Binaries, triples, and pairs. In *Asteroids IV* (P. Michel et al., eds.), pp. 355–374. Univ. of Arizona, Tucson, DOI: 10.2458/azu_uapress_9780816532131-ch019.

Melosh H. J. (1989) *Impact Cratering: A Geologic Process.* Oxford, New York. 253 pp.

Mendis D. A. and Axford W. I. (1974) Satellites and magnetospheres of the outer planets. *Annu. Rev. Earth Planet. Sci., 2,* 419–474.

Miller C. A., Verbiscer A. J., Chanover N. J., Holtzman J. A., and Helfenstein P. (2011) Comparing Phoebe's 2005 opposition surge in four visible light filters. *Icarus, 212,* 819–834.

Morbidelli A., Levison H. F., Tsiganis K., and Gomes R. (2005) Chaotic capture of Jupiter's Trojan asteroids in the early solar system. *Nature, 435,* 462–465.

Mottola S., De Angelis G., Di Martino M., Erikson A., Hahn G., and Neukum G. (1995) The Near-Earth Objects Follow-Up Program: First results. *Icarus, 117,* 62–70.

Mueller M., Grav T., Trilling D., Stansberry J., and Sykes M. (2008) Size and albedo of irregular saturnian satellites from Spitzer observations. *AAS/Division for Planetary Sciences Meeting Abstracts, 40,* #61.08, *http://adsabs.harvard.edu/abs/2008DPS....40.6108M.*

Nesvorný D. (2011) Young solar system's fifth giant planet? *Astrophys. J. Lett., 742,* L22.

Nesvorný D. and Morbidelli A. (2012) Statistical study of the early solar system's instability with four, five, and six giant planets. *Astron. J., 144,* 117.

Nesvorný D. and Vokrouhlický D. (2009) Chaotic capture of Neptune trojans. *Astron. J., 137(6),* 5003–5011.

Nesvorný D., Alvarellos J. L. A., Dones L., and Levison H. F. (2003) Orbital and collisional evolution of the irregular satellites. *Astron. J., 126,* 398–429.

Nesvorný D., Vokrouhlický D., and Morbidelli A. (2007) Capture of irregular satellites during planetary encounters. *Astron. J., 133,* 1962–1976.

Nesvorný D., Vokrouhlický D., and Morbidelli A. (2013) Capture of trojans by jumping Jupiter. *Astrophys. J., 768(1),* DOI: 10.1088/0004-637X/768/1/45.

Nesvorný D., Vokrouhlický D., and Deienno R. (2014) Capture of irregular satellites at Jupiter. *Astrophys. J., 784,* 22.

Nicholson P. D., Ćuk M., Sheppard S. S., Nesvorný D., and Johnson

T. V. (2008) Irregular satellites of the giant planets. In The Solar System Beyond Neptune (M. A. Barucci et al., eds.), pp. 411–424. Univ. of Arizona, Tucson.

Nogueira E., Brasser R., and Gomes R. (2011) Reassessing the origin of Triton. *Icarus, 214,* 113–130.

Owen T. C., Cruikshank D. P., Dalle Ore C. M., Geballe T. R., Roush T. L., and de Bergh C. (1999) Detection of water ice on Saturn's satellite Phoebe. *Icarus, 139(2),* 379–382.

Owen T. C., Cruikshank D. P., Dalle Ore C. M., Geballe T. R., Roush T. L., de Bergh C., Meier R., Pendleton Y. J., and Khare B. N. (2001) Decoding the domino: The dark side of Iapetus. *Icarus, 149,* 160–172.

Pätzold M., Andert T., Hahn M., Asmar S. W., Barriot J.-P., Bird M. K., Häusler B., Peter K., Tellmann S., Grün E., Weissman P. R., Sierks H., Jorda L., Gaskell R., Preusker F., and Scholten F. (2016) A homogeneous nucleus for Comet 67P/Churyumov-Gerasimenko from its gravity field. *Nature, 530,* 63–65, DOI: 10.1038/nature16535.

Peale S. J. (1977) Rotation histories of the natural satellites. In *Planetary Satellites* (J. A. Burns, ed.), pp. 87–112. Univ. of Arizona, Tucson.

Pecaut M. J. and Mamajek E. E. (2013) Intrinsic colors, temperatures, and bolometric corrections of pre-main-sequence stars. *Astrophys. J. Suppl., 208(*9), 22 pp.

Philpott C., Hamilton D. P., and Agnor C. B. (2010) Three-body capture of irregular satellites: Application to Jupiter. *Icarus, 208,* 824–836.

Pickering E. C. (1899a) A new satellite of Saturn. *Harvard College Observatory Bulletin, No. 49,* March 17, 1899.

Pickering E. C. (1899b) A new satellite of Saturn. *Harvard College Observatory Circular, No. 43,* 3 pp. (Also reprinted in *Astronomische Nachrichten, 3562,* 189–192; *Astrophys. J., 9(4),* 274–276; *Nature, 60(1540),* 21–22; *Pop. Astron., 7,* 233–235.)

Pickering W. H. (1905) The ninth satellite of Saturn. *Ann. Harvard College Observatory, 53(3),* 45–73.

Pinilla-Alonso N., Licandro J., Gil-Hutton R., and Brunetto R. (2007) The water ice rich surface of (145453) 2005 RR_{43}: A case for a carbon-depleted population of TNOs?. *Astron. Astrophys., 468,* L25–L28.

Pollack J. B., Burns J. A., and Tauber M. E. (1979) Gas drag in primordial circumplanetary envelopes: A mechanism for satellite capture. *Icarus, 37,* 587–611.

Porco C. C. West R. A., Squyres S. W., McEwen A. S., Thomas P. C., Murray C. D., Del Genio A., Ingersoll A. P., Johnson T. V., Neukum G., Veverka J., Dones L., Brahic A., Burns J. A., Haemmerle V., Knowles B., Dawson D., Roatsch Th., Beurle K., and Owen W. (2004) Cassini imaging science: Instrument characteristics and capabilities and anticipated scientific investigations at Saturn. *Space Sci. Rev., 115,* 363–497, DOI: 10.1007/s11214-004-1456-7.

Porco C. C., Baker E., Barbara J., Beurle K., Brahic A., Burns J. A., Charnoz S., Cooper N., Dawson D. D., Del Genio A. D., Denk T., Dones L., Dyudina U., Evans M. W., Giese B., Grazier K., Helfenstein P., Ingersoll A. P., Jacobson R. A., Johnson T. V., McEwen A. S., Murray C. D., Neukum G., Owen W. M., Perry J., Roatsch T., Spitale J., Squyres S. W., Thomas P. C., Tiscareno M., Turtle E. P., Vasavada A. R., Veverka J., Wagner R., and West R. (2005) Cassini imaging science: Initial results on Phoebe and Iapetus. *Science, 307,* 1237–1242.

Roatsch Th., Wählisch M., Scholten F., Hoffmeister A., Matz K.-D., Denk T., Neukum G., Thomas P. C., Helfenstein P., and Porco C. C. (2006) Mapping of the icy saturnian satellites: First results from Cassini-ISS. *Planet. Space Sci., 54,* 1137–1145.

Shen Y. and Tremaine S. (2008) Stability of the distant satellites of the giant planets in the solar system. *Astron. J., 136,* 2453–2467.

Sheppard S. S. (2006) Outer irregular satellites of the planets and their relationship with asteroids, comets and Kuiper belt objects. In *Asteroids, Comets, Meteors (ACM)* (D. Lazzaro et al., eds.), pp. 319–334. IAU Symp. No. 229, Cambridge Univ., Cambridge.

Simonelli D. P., Kay J., Adinolfi D., Veverka J., Thomas P. C., and Helfenstein P. (1999) Phoebe: Albedo map and photometric properties. *Icarus, 138,* 249–258.

Sonnett S., Mainzer A., Grav T., Masiero J., and Bauer J. (2015) Binary candidates in the Trojan and Hilda populations from NEOWISE light curves. *Astrophys. J., 799,* 191, 20 pp.

Soter S. (1974) Brightness of Iapetus. Paper presented at IAU Colloquium 28, Cornell University, Ithaca.

Spencer J. R. and Denk T. (2010) Formation of Iapetus's extreme albedo dichotomy by exogenically-triggered thermal migration of water ice. *Science, 327(5964),* 432–435, DOI: 10.1126/science.1177132.

Tamayo D., Burns J. A., Hamilton D. P., and Hedman M. M. (2011) Finding the trigger to Iapetus' odd global albedo pattern: Dynamics of dust from Saturn's irregular satellites. *Icarus, 215,* 260–278.

Tamayo D., Hedman M. M., and Burns J. A. (2014) First observations of the Phoebe ring in optical light. *Icarus, 233,* 1–8.

Tedesco E. F., Williams J. G., Matson D. L., Weeder G. J., Gradie J. C., and Lebofsky L. A. (1989) A three-parameter asteroid taxonomy. *Astron. J., 97,* 580–606.

Thomas N. (2009) The nuclei of Jupiter family comets: A critical review of our present knowledge. *Planet. Space Sci., 57,* 1106–1117, DOI: 10.1016/j.pss.2009.03.006.

Thomas P. C. (2010) Sizes, shapes, and derived properties of the saturnian satellites after the Cassini nominal mission. *Icarus, 208,* 395–401, DOI: 10.1016/j.icarus.2010.01.025.

Thomas P. C., Veverka J., Davies M. E., Morrison D., and Johnson T. V. (1983) Phoebe: Voyager 2 observations. *J. Geophys. Res., 88,* 8736–8742.

Tosi F., Turrini D., Coradini A., Filacchione G., and the VIMS Team (2010) Probing the origin of the dark material on Iapetus. *Mon. Not. R. Astron. Soc., 403(3),* 1113–1130.

Tremaine S., Touma J., and Namouni F. (2009) Satellite dynamics on the Laplace surface. *Astron. J., 137,* 3706–3717.

Tsiganis K., Gomes R., Morbidelli A., and Levison H. F. (2005) Origin of the orbital architecture of the giant planets of the solar system. *Nature, 435,* 459–461.

Turrini D., Mazari F., and Beust H. (2008) A new perspective on the irregular satellites of Saturn — I. Dynamical and collisional history. *Mon. Not. R. Astron. Soc., 391,* 1029–1051.

Turrini D., Mazari F., and Tosi F. (2009) A new perspective on the irregular satellites of Saturn — II. Dynamical and physical origin. *Mon. Not. R. Astron. Soc., 392,* 455–474.

Verbiscer A. J., Skrutskie M. F., and Hamilton D. P. (2009) Saturn's largest ring. *Nature, 461,* 1098–1100, DOI: 10.1038/nature08515.

Vokrouhlický D., Nesvorný D., and Levison H. F. (2008) Irregular satellite capture by exchange reactions. *Astron. J., 136,* 1463–1476.

Vokrouhlický D., Bottke W. F., and Nesvorný D. (2016) Capture of trans-neptunian planetesimals in the main asteroid belt. *Astron. J., 152,* 39.

Warner B. D., Harris A. W., and Pravec P. (2009) The asteroid lightcurve database. *Icarus, 202,* 134–146, DOI: 10.1016/j.icarus.2009.02.003.

Wong I. and Brown M. E. (2015) The color-magnitude distribution of small Jupiter Trojans. *Astron. J., 150,* 174.

Wong I. and Brown M. E. (2016) A hypothesis for the color bimodality of Jupiter Trojans. *Astron. J., 152,* 90.

Wong I., Brown M. E., and Emery J. P. (2014) The differing magnitude distributions of the two Jupiter Trojan color populations. *Astron. J., 148,* 112.

Part 4:

Astrobiology and Exploration of Enceladus

McKay C. P., Davila A., Glein C. R., Hand K. P., and Stockton A. (2018) Enceladus astrobiology, habitability, and the origin of life. In *Enceladus and the Icy Moons of Saturn* (P. M. Schenk et al., eds.), pp. 437–452. Univ. of Arizona, Tucson, DOI: 10.2458/azu_uapress_9780816537075-ch021.

Enceladus Astrobiology, Habitability, and the Origin of Life

C. P. McKay and A. Davila
NASA Ames Research Center

C. R. Glein
Southwest Research Institute

K. P. Hand
Jet Propulsion Laboratory

A. Stockton
Georgia Institute of Technology

Analyses of the plume of icy material emanating from the south polar terrain of Enceladus by the Cassini spacecraft point to a habitable subsurface ocean. Icy particles in the plume, originating from an ocean of liquid water in contact with a rocky core, contain organic molecules, biologically available nitrogen, sodium, potassium, hydrogen, and carbon dioxide. The hydrogen is inferred to originate from alkaline hydrothermal activity and is present at a level adequate to support methanogenic microorganisms. The biogenic elements (C, H, N, O, P, S) have been detected or are expected to be present due to the rock–water interactions. The strong indication of habitable conditions contrasts with the question of the origin of life. If deep-ocean alkaline hydrothermal vents are suitable sites for the origin of life, then the most basic conditions are met for a second genesis (*McKay,* 2001) of life on Enceladus, and evidence of life could be found in the ocean. However, if life requires surface environments for its origin, and if panspermia has not been effective in the case of Enceladus, then this world could be habitable but uninhabited. Thus, near-term missions to Enceladus will try to shed light on these possible scenarios by searching for biomolecules in plume materials. Evidence for life might appear as a molecular pattern indicative of biological production; e.g., a set of several amino acids with common chirality that could represent a basis for protein assembly. Lacking analog environments on Earth that have all the features of the Enceladus ocean, a better understanding of possible ecosystems on Enceladus can still be constructed piecewise. Relevant features of analog environments on Earth include dark, anoxic water bodies virtually sealed by ice, anaerobic chemoautotrophic microbial ecosystems, and low-temperature alkaline hydrothermal vents.

1. INTRODUCTION

Enceladus is likely to be a world habitable for a range of Earth microorganisms. The plume emanating from the south polar region was discovered by Cassini in 2006 (*Hansen et al.,* 2006; *Porco et al.,* 2006; *Spencer et al.,* 2006; *Waite et al.,* 2006) and has been investigated extensively as the spacecraft flew through it repeatedly. The data from these flybys indicate that the plume originates from a global ocean below a surface layer of ice (*Iess et al.,* 2014; *Thomas et al.,* 2016). The Ion and Neutral Mass Spectrometer (INMS) on Cassini detected hydrogen, carbon dioxide, and organic compounds up to C_6 (*Waite et al.,* 2009, 2017). The hydrogen and carbon dioxide form a redox couple suitable for supporting methanogens (*McKay et al.,* 2008). Nitrogen is present in the form of ammonia, and sulfur has been tentatively detected in the form of hydrogen sulfide (*Waite et al.,* 2009). If the preliminary analysis of *Khawaja et al.* (2015) is correct, then the Cosmic Dust Analyzer (CDA) instrument detected biologically available nitrogen in the form of amines. The CDA also detected organic compounds of high molecular weight in E-ring particles derived from the plume (*Khawaja et al.,* 2015; *Postberg et al.,* 2017). Sodium and other salts detected in the E ring and lower parts of the plume indicate that the ocean salinity is about 0.5% to 2% dominated by NaCl (*Postberg et al.,* 2009, 2011) — a water activity near unity that is suitable for life as we know it. The detection of nanometer-sized silica particles in the E ring, and inferred to be derived from the plume, suggest that the ocean hosts hydrothermal vents in which water is in contact with the rocky core at temperatures of at least ~100°C and that the pH of the ocean is between 8.5 and 10.5 (*Hsu et al.,* 2015; *Sekine et al.,* 2015). All the major elements needed for life (C, H, N, O, P, S) have been detected or are expected to be present due to the rock–water interaction. Thus there is good indication that the ocean on Enceladus is habitable and that the icy particles in the plume are samples of that habitable water.

However, *habitable* does not necessarily mean *inhabited* (*Hand et al.,* 2009; *Cockell,* 2011). A search for signatures of life in the habitable ocean of Enceladus directly confronts different hypotheses for the origin of life (*Davis and McKay,* 1996). The astrobiology, habitability, and origin of life questions pertaining to Enceladus motivate future missions to sample icy grains from the plume and search for signatures of life, using comparisons with life on Earth. In this chapter we expand on these topics in more detail, reviewing our current knowledge of the astrobiology of Enceladus and emphasizing the big questions that lie ahead.

2. THE ORIGIN OF LIFE

A significant, if often overlooked, challenge in considering the astrobiological potential of Enceladus is the possibility of origin of life events. The difficulty arises from the lack of a consensus regarding how life originates. In the case of life on Earth, the only fact we have about the origin of life is that it happened more than 3.5 G.y. (e.g., *Djokic et al.,* 2017). We do not know where, when, or how life originated, or how long it took. The hypothesis that the one example of life we see on Earth originated on Earth and/or took a long time is not supported, nor excluded, by any geological evidence.

The scientific inquiry into the origin of life is further hampered by the lack of any necessary connection between environments that can support life and environments in which life can originate. As discussed above, there is compelling evidence that the ocean on Enceladus is an environment that can support life (it is likely habitable for many Earth microorganisms), but we have no criteria with which to judge if it is an environment that can promote the origin of life.

Similar to the case of life on Earth, there are two broad classes of hypotheses for an origin of life on Enceladus: (1) those that postulate a local origin of life event; and (2) those that postulate the transfer of life — panspermia — from other sources in the solar system, or from sources outside the solar system (i.e., presolar grains), even prior to its formation. The transfer of life to Enceladus from sources in the inner solar system (i.e., Earth or Mars) is unlikely due to its relative isolation from impact debris created in the inner solar system (*Worth et al.,* 2013). The transfer of life to Enceladus from sources in the outer solar system (i.e., other icy moons) might have been more likely, but has not yet been thoroughly addressed in the literature. The arrival of presolar grains carrying life forms that became seeds of life on Enceladus (and Earth) cannot be dismissed *a priori,* and would have resulted in a common biochemistry across the solar system — a hypothesis that can only be tested when life is discovered outside the Earth.

While different forms of panspermia are worth considering, more attention is directed toward hypotheses that life originated on Earth itself. Here the debate of interest to Enceladus astrobiology is the role of submarine hydrothermal vents in the origin of life. *Russell* (2003) and *Russell et al.* (2014) have suggested that life originated on Earth in alkaline hydrothermal vents similar to the Lost City system; the evidence for hydrothermal activity on Enceladus is consistent with the origin of life there as well (see also *Barge and White,* 2017). In direct contrast with this hypothesis, *Deamer et al.* (2006), *Deamer and Georgiou* (2015), and *Hud et al.* (2013) have argued that the origin of life requires periodic drying to concentrate biologically relevant molecules in order to form biopolymers and enclosed vesicles. They suggest that the origin of life occurred on Earth in a land environment that experienced repeated cycles of wetting and drying. *Deamer and Damer* (2017) explicitly state that their work suggests that Enceladus may be habitable but would be uninhabited.

Even if the geochemical environment that could promote the origin of life was understood, the question of duration would still be relevant. The record of life on Earth provides very little information on how long it took for life to start on this planet; the first appearance of life occurred within 100 to 500 m.y. after the formation of Earth. Clearly this is an upper limit set by the resolution of the historical record, and the origin of life may have been much faster. Considering this question, *Lazcano and Miller* (1994) suggested that "in spite of the many uncertainties involved in the estimates of time for life to arise and evolve to cyanobacteria, we see no compelling reason to assume that this process, from the beginning of the primitive soup to cyanobacteria, took more than 10 million years." However, *Orgel* (1998) criticized this argument on the grounds that we do not understand the steps that lead to life and consequently we cannot predict the time required: "Attempts to circumvent this essential difficulty are based on misunderstandings of the nature of the problem." Thus the problem of the time required for the emergence of life, like the question of its location, remains unsolved with the current data.

The age of Enceladus is unknown, as is the persistence of the ocean over that age. Conventionally, it is assumed that Saturn's moons are nearly as old as the planet, and they formed near the end of the planet's formation (*Canup and Ward,* 2006). However, recent dynamical modeling may imply that the inner moons, including Enceladus, are only ~100 m.y. old (*Ćuk et al.,* 2016). If the origin of life does occur at low-temperature hydrothermal vents, would it have occurred in the ocean of Enceladus in 100 m.y.? We have no basis with which to answer this question, emphasizing again the importance of an empirical approach to the origin of life on other worlds.

The lack of scientific understanding of the origin of life even on Earth implies that we cannot state with confidence that a search for life in the habitable ocean of Enceladus will be fruitful. But we can state that results of a search for life on Enceladus, whether positive or negative, will provide important constraints on our models for origins of life anywhere. The science that is currently limited to the single data point of the "habitable and inhabited Earth" would finally have a second data point, providing an enormous increase in available information and perhaps enabling the long-sought-after comprehensive theory for the emergence of life on Earth (*Lazcano and Hand,* 2012).

3. HABITABILITY: SUSTAINABILITY AND LIMITS FOR LIFE

In a comprehensive review of the requirements and limits for life in the context of exoplanets, *McKay* (2014a) provided three tables: one on the ecological requirements for life, a second on the elemental abundances by mass used in life, and a third on the ecological limits to life. These have been variously used and defined from previous studies (*McKay,* 2014a, and references therein). It is useful therefore to reiterate the entries in these tables and add a column for Enceladus. This is shown in Table 1.

At the basic level of the requirements for life — liquid water, energy, carbon, and other key elements — Enceladus meets all the requirements. The second set of requirements specifies the elements needed for life in detail. Here we find that Enceladus is *likely* to meet all the requirements, but the presence of P and Ca are inferred, not direct. Phosphorus is of particular importance and is expected based on the composition of meteoritic materials and especially its detection in comets (*Altwegg et al.,* 2016). However, the low solubility of phosphate minerals in alkaline water could limit the availability of P in the Enceladus ocean (*Zolotov,* 2012). The third set of requirements is that used to characterize the limits of life in extreme environments. Here, the ocean of Enceladus easily meets all requirements except for the level of sunlight needed for photosynthesis. Thus, the habitability of Enceladus is contingent on the presence of chemical redox couples or alternative sources of energy.

Although the habitability of Enceladus is formally established — pending the expected confirmation of the elements P and Ca, metals such as Fe, and a firmer detection of S — there are still gaps in our understanding of the physical and chemical characteristics of the Enceladus ocean that need to be addressed. In terms of the bulk ocean properties, the key values of pH and salinity are only crudely constrained. The pH in particular is uncertain by 4 orders of magnitude (in terms of H^+ activity) (*Glein et al.,* 2015). Direct measurements of pH to ±1 and salinity to ±10% are needed to refine the assessment of the habitability of Enceladus in terms of possible types of microbial ecosystems that could be present, and availability of nutrients and reaction pathways. The only potential biological energy source clearly indicated by the Cassini INMS results is the $CO_2 + H_2$ redox reaction (*Waite et al.,* 2017). It is interesting to note that this reaction is only used by Archaean methanogens, and if this is the only form of biological energy available in the ocean of Enceladus, it points to a particular type of microbial ecosystem as the basis for all primary biological production. An interesting note is the high requirement for the element Ni observed in methanogens on Earth. This requirement has been suggested to have global ecological implications over Earth's history (*Konhauser et al.,* 2009). The concentration of Ni in the ocean of Enceladus is unknown, but may be restricted by the low solubility of Ni-bearing sulfide minerals in a reduced ocean (*Zolotov,* 2012).

To obtain a more quantitative understanding of the habitability of Enceladus' ocean, an initial step is to assess the amount of chemical (free) energy that could be harnessed if metabolic reactions were performed by organisms in the ocean. The Cassini INMS instrument detected CO_2, H_2, and CH_4 in the plume gas (*Waite et al.,* 2017), which enables an evaluation of the energy yield from the methanogenesis reaction

$$CO_2\left(aq\right) + 4H_2\left(aq\right) \rightarrow CH_4\left(aq\right) + 2H_2O\left(l\right) \qquad (1)$$

The amount of Gibbs free energy that would be released from the reaction is equivalent to the chemical affinity of the reaction (A), which can be calculated as

$$A = 2.3026RT\left(\log K - \log Q\right) \qquad (2)$$

where R denotes the gas constant (8.3145×10^{-3} kJ mol^{-1} K^{-1}), T ≈ 273 K for Enceladus' ocean, K designates the equilibrium constant, and Q the reaction quotient. The equilibrium constant can be computed from widely available standard-state thermodynamic data (i.e., Gibbs energies of formation). Here, we use log K = 37.44 (*Waite et al.,* 2017). The reaction quotient for the methanogenesis reaction can be approximated as

$$\log Q \approx \log\left(\frac{CH_4}{CO_2}\right) - 4\log\left[H_2\right] \qquad (3)$$

for dilute (ideal) aqueous solutions, where the CH_4/CO_2 ratio corresponds to the ocean and $[H_2]$ represents the molal concentration of H_2 in the ocean. *Waite et al.* (2017) assumed that the CH_4/CO_2 ratio in the ocean is similar to that in the plume (~0.4) for a scenario of rapid degassing of these volatiles from droplets of ocean water. They developed an approach for estimating the dissolved concentration of H_2 by assuming that the ratio of H_2/CO_2 is also conserved between the ocean and plume. This ratio can be expressed as H_2/CO_2 = $(H_2/H_2O)/(CO_2/H_2O)$, in which $CO_2/H_2O \approx 0.005$ (*Waite et al.,* 2017). Multiplying the H_2/CO_2 ratio times the concentration of CO_2 in the ocean gives the concentration of H_2. The concentration of CO_2 was constrained using a speciation model (*Glein et al.,* 2015) with input parameters of pH and total dissolved carbonate [nominally 0.03 molal HCO_3^- + CO_3^{-2} (*Postberg et al.,* 2009)]. This approach allows the affinity for methanogenesis to be calculated as a function of pH and the relative abundance of H_2 in the plume (Fig. 1).

TABLE 1. Habitability of Enceladus.

Requirement	Limit (from *McKay,* 2014a)	Enceladus Ocean
Ecological Requirements for Life		
Energy		
Predominately light	Photosynthesis at 100 AU light levels	No
Chemical energy	e.g., $H_2 + CO_2 \rightarrow CH_4 + H_2O$	Only H_2, CO_2 confirmed
Carbon	Common as CO_2 and CH_4	CH_4, organics to C_6 + CO_2
Liquid water	Rare, certain on Earth	Firm detection of water of suitable salinity
N		Present as NH_3 and possibly amines
P, S, Na, and other elements	Likely to be common	P possible from water-rock interaction S tentatively as H_2S Na, K, Cl detected
Elemental Abundances by Number in E. coli		
O	68%	As H_2O
C	15	CO_2, organics to C_6 +
H	10.2	As H_2O, H_2
N	4.2	Present as NH_3 and possibly amines
P	0.83	Inferred from Na
K	0.45	Detected
Na	0.40	Detected
S	0.30	Tentatively detected as H_2S
Ca	0.25	Inferred from Na
Cl	0.12	Detected
Ecological Limits for Life		
Lower temperature	~−15°C	Above limit (~0°C)
Upper temperature	122°C	Below limit in bulk ocean
Low light	$\sim 10^{-4}$ S_o	Zero light
pH	0–11	5–10, within limits
Salinity	Saturated NaCl	3% salt, below limit
Water activity	0.6 (yeasts) 0.8 (bacteria)	0.97 above limit
UV	≥ 1000 J m^{-2}	Zero UV, below limit
Radiation	50 G.y./hr	Crustal radiation only, from U, Th, K on the order of ~0.2 rad/yr, below limit

For the present best constraints of pH ~9–11 and H_2 mixing ratio [~0.4–1.4% (see *Waite et al.,* 2017)], the affinity based on the preceding model lies between ~50 and ~120 kJ per mole of CH_4. This is larger than the free energy required for methanogens to grow, as determined experimentally by *Kral et al.* (1998) to be about 40 kJ mole^{-1}. It is much larger than the theoretical maintenance energy of ~10 kJ mole^{-1} (*Hoehler,* 2004; *Hoehler et al.,* 2007), which may be the minimum energy that is required to permit the persistence of organisms (at least on Earth). The excess of chemical energy at Enceladus provides a strong case that its ocean is energetically habitable.

An alternative energy source motivated by the possible presence of complex organics detected in the plume is fermentation. Fermentation is a general term that refers to the acquisition of biological energy that does not require an external oxidizing or reducing agent, but instead the organic compound undergoes self-disproportionation (fragmentation to more oxidized and more reduced species). The familiar example of this is the process that converts glucose to smaller organic molecules. Typically fermentation does not proceed to completion and, for example, one glucose molecule can be converted into two ethanol molecules and two carbon dioxide molecules

$$C_6H_{12}O_6 \rightarrow 2C_2H_5OH + 2CO_2 \qquad (4)$$

in a reaction, well known and often banned. As another example to note, bacteria such as *Clostridium pasteurianum* ferment glucose, producing carbon dioxide and hydrogen gas plus other organics via reactions such as (see review by *Thauer et al.,* 1977)

$$\begin{aligned} &C_6H_{12}O_6 + 4H_2O \rightarrow \\ &2CH_3COO^- + 2HCO_3^- + 4H^+ + 4H_2 \end{aligned} \qquad (5)$$

There is no evidence of glucose in the plume of Enceladus, but even if present, it would not have been detectable by the instrumentation on Cassini because of its relatively high mass and its many isomers.

Schink (1985) showed that C_2H_2 can be the basis of a fermentation metabolism by the organism *Pelobacter acetylenicus*. This is perhaps not surprising given the chemical energy stored in the carbon-carbon triple bond. The reaction generates acetaldehyde by the reaction

$$C_2H_2 + H_2O \rightarrow CH_3CHO \quad (6)$$

Oremland and Voytek (2008) have suggested C_2H_2 fermentation as a possible metabolism for life in Enceladus and CH_3CHO as a potential biomarker. However, *Waite et al.* (2009) reported an upper limit of 7×10^{-4} on the mixing ratio of C_2H_4O in the plume gas. Detection of C_2H_2 in the plume of Enceladus remains ambiguous partly because C_2H_2 (and other "high-energy" organics) could be produced by impact chemistry of complex organic materials in the titanium antechamber of the Cassini INMS (*Waite et al.,* 2009). *McKay et al.* (2012) suggested that if C_2H_2 is present in Enceladus, it would reflect an accreted source of this compound because thermal, hydrothermal, or biological processes would not be expected to produce it. C_2H_2 and other unsaturated hydrocarbons such as C_2H_6 can release energy by reaction with H_2, as has been suggested for Titan (*McKay and Smith,* 2005).

A further energy source could be the direct use of geophysically released electrons. Recent studies have shown that some microorganisms can directly use electrons for their metabolism and biomass production (*Nielsen et al.,* 2010; *Rosenbaum et al.,* 2011; *Bose et al.,* 2014). Electrons can be made available within seafloor sediments and hydrothermal fields (*Nielsen et al.,* 2010; *Yamamoto et al.,* 2017).

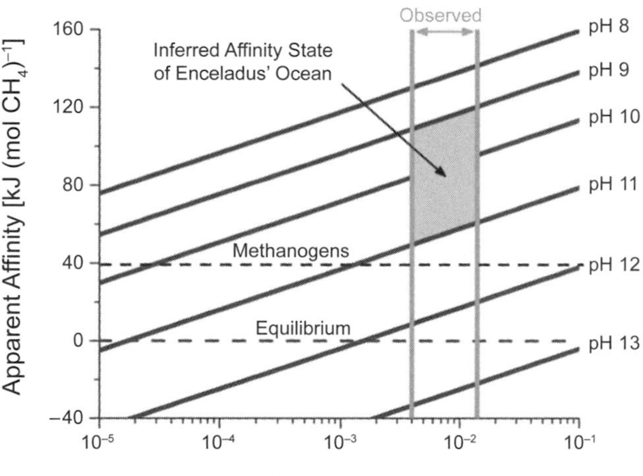

Fig. 1. Reaction energy for the reaction of hydrogen and carbon dioxide in the ocean of Enceladus as a function of the H_2/H_2O mixing ratio in the plume for various values of pH. The dotted lines show energy of 0 and energy of 40 kJ/mole — the minimum for methanogens based on experiments by *Kral et al.* (1998). The inferred state of Enceladus' ocean is shown by the shaded regions and indicate sufficient energy to support methanogens. Modified from *Waite et al.* (2017).

The persistence of H_2 in the ocean is key to available energy for life. As discussed above, the H_2 and CO_2 present in the ocean provide a basis for methanogens that form CH_4 and biomass, and are the primary producers in the microbial community that potentially form biofilms at the core-ocean interface (or in fractures inside the core). Clearly, with no way to recycle this CH_4 and biomass, the system would eventually die down.

In addition to being produced by rock-water interactions, H_2 on Enceladus can be regenerated from CH_4 at high temperature — effectively cycling biological carbon. Figure 2 shows a schematic of how H_2 could be recycled in the hot core of Enceladus, providing a continuing fuel for a methanogenic ecosystem in the ocean. Water flowing through the subsurface could reform H_2 and CO_2 from CH_4 and H_2O. For simple cometary ratios of C, H, and O (e.g., *McKay and Borucki,* 1997; *Kress and McKay,* 2004), the shift in thermodynamic stability with temperature favors CO_2 above about 500°C. It is not clear that the core of Enceladus could be this hot at present. Thermodynamic calculations (*Glein et al.,* 2008) imply lower temperatures in the presence of mineral redox buffers that set the equilibrium oxidation state. For magnetite and hematite (i.e., the MH buffer) the minimum temperature for recycling CH_4 to CO_2 is as low as 25°C if the abiotic reaction kinetics are suitable (*Glein et al.,* 2008, Fig. 1). However, the formation of hematite usually requires highly oxidizing environments (e.g., for Earth, an O_2-rich atmosphere). In contrast, pyrrhotite-pyrite-magnetite (PPM) or fayalite-magnetite-quartz (FMQ) buffers are more representative of submarine hydrothermal systems on Earth. For these buffers, the required temperatures for a significant amount of CO_2 reformation exceed ~250°C (*Glein et al.,* 2008, Fig. 1). A key question then is the maximum temperature of ocean water as it circulates through the core of Enceladus and how this relates to the observed (high) levels of H_2 in the plume (*Waite et al.,* 2017) and the inferred conditions of the hydrothermal system (*Hsu et al.,* 2015; *Sekine et al.,* 2015).

McCollom (1999) developed a quantitative model of a methanogen-based biota for Europa and a similar analysis has been applied to Enceladus (*Steel et al.,* 2017). Steel

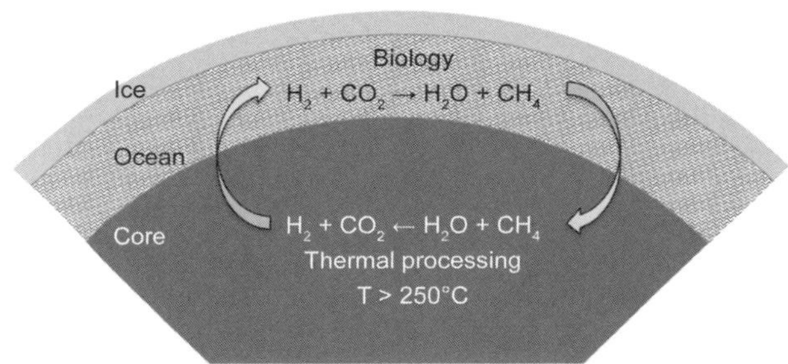

Fig. 2. Possible carbon cycle on Enceladus based on methane recycling by thermal reactions in the core.

et al. estimated, based on the energy flux observed at the south pole and the inferred internal hydrothermal activity, that H_2 production is 0.6–34 mol s^{-1} from serpentinization, sufficient to sustain abiotic and biotic amino acid synthesis of 0.005–0.25 g s^{-1} and 1–52 g s^{-1}, respectively. Assuming methanogens consume virtually all the H_2 implies up to 90 μM concentrations of amino acids and cell concentrations of 80–4250 cells cm^{-3} in the plume and the ocean (*Steel et al.,* 2017). Abiotic processes alone imply glycine, alanine, α-aminoisobutyric acid (AIB), and glutamic acid in the plume and in the ambient ocean would all be above 0.01 μM. No enantiomeric excess is expected in the ocean in either case, because racemization timescales are short compared to production timescales. Clearly the H_2 concentrations reported by *Waite et al.* (2017) of 0.4% to 1.4% in the plume are not consistent with the assumption that biology consumes all the H_2 produced, suggesting either the lack of biology or greatly reduced efficiency of H_2 consumption, due perhaps to a limited spatial extent of the methanogen biofilms. For example, *Kral et al.* (1998) showed a variety of methanogens will consume H_2 to low levels, to partial pressures as low as ~4 Pa. Given the utility of H_2 in all energy schemes discussed above, its abundance in the ocean at levels well above 4 Pa [~20–~10^4 Pa (*Waite et al.,* 2017)] would seem inconsistent with biological consumption — high levels of H_2 may be an anti-biomarker.

The high levels of H_2 in the plume (*Waite et al.,* 2017) have been attributed to production by hydrothermal reactions (e.g., through serpentinization) within the core or, as discussed above, by thermal processing in the core. However, models of these processes do not predict ongoing production (e.g., *Malamud and Prialnik,* 2013). Furthermore, the mineralogy of the core and, in particular, indications that it has not undergone igneous differentiation (*Sekine et al.,* 2015) may further suggest a short timescale for H_2 production by consumption of unaltered, primary minerals by hydration within Enceladus. These explanatory issues notwithstanding, the direct measurement of H_2 in the plume motivates the study of this microbial energy source.

The movement of ocean water through the core of Enceladus could also recycle nitrogen. Organic matter from carbonaceous chondrites releases NH_3 when heated to 300°C (*Pizzarello and Williams,* 2012). For this moderate level of heating, NH_3 dissociation into N_2 and H_2 is kinetically inhibited (*Sekine et al.,* 2015). For much higher temperatures (e.g., ~500°C), where CH_4 and H_2O are effectively converted into CO_2 and H_2, NH_3 also could dissociate into N_2 and H_2 (*Matson et al.,* 2007). However, such high-temperature processing is not expected and would be inconsistent with the observations of low N_2 in the plume (*Hansen et al.,* 2011; *Waite et al.,* 2017).

4. ANALOGS ON EARTH

An astrobiology analog on Earth of the ocean on Enceladus would be (1) a dark, anoxic water body virtually sealed by ice; (2) an environment containing an anaerobic chemoautotrophic microbial ecosystem; and (3) a recirculating, low-temperature, alkaline, hydrothermal vent. No known environment on Earth has all three of these features, but there are systems that have one of the three characteristics, and these systems can help inform our understanding of the astrobiological potential of Enceladus.

Physical analogs for Enceladus are ecosystems sealed under ice, but such systems are not common. Most of the lakes in the Dry Valleys (Lakes Vanda, Hoare, Fryxell, Bonney, and Joyce) are ice-covered, not ice-sealed. The ice floats, and each summer a moat forms. Indeed, the energy balance that sets the thickness of the ice cover and the area of these lakes depends on summer melt water flowing into the lake (e.g., *McKay et al.,* 1985; *Chinn,* 1993). The lakes found beneath kilometers of ice on the polar plateau (e.g., Lake Whillans and probably Lake Vostok) are also not sealed — they are part of an extensive network of flow underneath the plateau (see, e.g., *Fricker and Scambos,* 2009; *Vincent and Laybourn-Parry,* 2008).

The best known example of an ice-sealed ecosystem is Lake Vida in the Dry Valleys of Antarctica (*Murray et al.,* 2012). This system receives no material flow and no sunlight. Early observers and models of Lake Vida assumed it was frozen to its base (*Calkin and Bull,* 1967; *McKay et al.,* 1985). However, radar revealed a constant highly reflecting interface about 19 m below the surface of the ice in the interior of the lake. Drilling through the ice revealed the presence of a brine layer (~245‰ NaCl) in the lake extending from about 16 m downward. Radiocarbon dating of organic matter sampled at 12 m in the lake ice cover suggests that the brine has been isolated for more than 2800 years (*Doran et al.,* 2003). Analysis of microbial assemblages within the perennial ice cover of the lake revealed a diverse array of bacteria in this sealed brine ecosystem at –13°C (*Murray et al.,* 2012).

A surprising result of the analysis of the material in the brine was the mix of oxidized and reduced species. For example, nitrate, nitrite, and ammonium are present at concentrations of about 900, 23, and 4000 μM/L, respectively. Iron (Fe) is present at high levels (>300 μM L^{-1}) and there are also high levels of dissolved organic carbon (580 mg L^{-1}). Most surprisingly, perchlorate is present at 50 μg L^{-1}, the highest concentration of any Dry Valley lake. Due to this large perchlorate level, the ratio of nitrate to perchlorate (~10^3) is lower than anywhere else on Earth, other than the Atacama Desert — typical values of nitrate to perchlorate ratio on Earth range from 10^4 to 10^5 (*Jackson et al.,* 2015). Even allowing for the slowdown of metabolism with temperature, the expectation from other closed systems is that this system would have long ago run down, depleting energy reserves and going down the redox couple ladder toward sulfate reduction and methanogenesis (see, e.g., *Nealson,* 1997). Perchlorate and nitrate would have long ago been used to oxidize organic material, making their presence alongside relatively high levels of organic carbon interesting, but challenging to explain. *Murray et al.* (2012) suggested that the high levels of dissolved and gaseous nitrogen compounds,

Fe and H_2, suggest that rock-water reactions may be occurring at the base of the brine, producing H_2 and maintaining redox disequilibrium. How this works is not understood, and the rate of such reactions at the low temperatures involved is also unknown, but if such a source of H_2 is confirmed and is playing a role in maintaining the redox mix in the Lake Vida brine, this may have interesting implications for the chemistry, and habitability, of the ocean on Enceladus.

The Lake Vida brine is rich with evidence of microbial life, dominated by bacteria (~10^7 cells ml^{-1}) (*Murray et al.,* 2012). Eight bacterial phyla were identified from a 16S rRNA gene clone library from brine collected by filtration on 0.2-μm pore-sized filters: *Proteobacteria* (classes γ, δ, and ε), *Lentisphaerae*, *Firmicutes*, *Spirochaeta*, *Bacteroidetes*, *Actinobacteria*, *Verrucomicrobia*, and Candidate Division TM7. This is a rather large diversity for a low-temperature brine environment, but may not be a good biological analog for Enceladus. Methanogenic archaea, expected to be a useful analog organism for the base of any microbial ecosystem present on Enceladus (as discussed above), comprise a negligible fraction of the organisms detected in the Lake Vida brine and methane is only found at trace levels.

Lake Untersee, also in Antarctica, is an ultra-oligotrophic lake located at –71.342°, 13.473° in Dronning Maud Land, in the region due south of Africa. Lake Untersee occupies a basin dammed by the terminus of the Anuchin Glacier. The lake is 563 m above sea level, with an area of 11.4 km^2, and is among the largest surface lakes in East Antarctica. This lake is also effectively sealed by ice cover from the atmosphere, which transmits ~5% of the visible wavelengths of sunlight (*Andersen et al.,* 2011). The lake has two subbasins; the largest, 169 m deep, lies adjacent to the glacier face and is separated by a sill at 50 m depth from a smaller, 100-m-deep basin in the southwest corner. The deep basin is well mixed and oxygenated to the bottom. In contrast, the shallow basin is density-stratified below the sill depth and is anoxic at its base (*Wand et al.,* 2006). The deep basin and upper part of the anoxic basin have similar water chemistry and are well-mixed. Two unusual features of the lake are a high concentration of methane (>20 mmol l^{-1}) in the deep part of the anoxic basin, and a pH of ≥10.4 in the mixed layer of the lake.

The primary source of methane in the anoxic basin of Lake Untersee is microbial and provides a plausible ecological model of a methanogen-based ecosystem on Enceladus. The microbial reactions occur at the bottom, ~100 m depth, in the anoxic zone in Lake Undersee, resulting in the production of CH_4 from H_2 and CO_2 (*Wand et al.,* 2006). The CH_4, and other biogenic products such as NH_3 produced at the bottom, diffuse upward through the stagnant water column in the anoxic trough. There are no sources or sinks for these biogenic gases until they reach the oxygen-rich layer and are consumed by microbial oxidation. This oxidation begins at about 80 m depth and is complete by ~75 m depth (*Wand et al.,* 2006). With the exception of the oxidation layer, this is a model in miniature of the situation hypothesized for Enceladus based on the detection of H_2 and CO_2 (as well as CH_4) in the plume. It is presumed that a putative methanogenic ecosystem on Enceladus could be operating at the interface between a core and the ocean. Hydrogen released from reactions in the core is carried outward and would be consumed by the methanogens using CO_2 from the water column — analogous to the bottom of the anoxic zone in Lake Untersee. The CH_4 produced, as well as other biogenic products, would be carried upward from the source region into the plume without much further loss other than by dilution (*Steel et al.,* 2017) because unlike Lake Untersee, Enceladus does not, to the best of our knowledge, have an upper O_2-rich layer. Thus, the ocean and plume of Enceladus corresponds to the zone between 80 m and 100 m depth in Lake Untersee. It is interesting to note that, as on Enceladus, NH_3 as well as CH_4 are present. Direct sampling of the water in the CH_4 flow region (80–100 m depth) of Lake Untersee for biogenic gases and microbially produced biomarkers could provide a detailed basis for defining target biosignatures, and expected concentrations, in the search for life in the plume of Enceladus.

The high pH in the oxic waters of Lake Untersee is not fully explained, but an important clue to its cause is the fact that carbon entering the lake does not accumulate as carbonate, but rather as biologically produced organic carbon on the lake bottom (*Andersen et al.,* 2011). The photosynthetic microbial mats are carbon-starved and draw down CO_2. It has been shown that consumption of CO_2 by phototrophs can cause the pH to be as high as 10 when there is limited gas exchange with the atmosphere (see, e.g., *Talling,* 1976). Thus, the high pH in Lake Untersee appears to be a biological effect. While photosynthesis is not likely a viable niche on Enceladus, it is important to consider how various biological processes can mediate environmental parameters, such as pH.

Considering the available evidence, relevant analog systems for microbiology on Enceladus are, for the most part, anaerobic chemoautotrophic closed systems. As discussed above, H_2 is a likely "fuel" molecule on Enceladus and methanogens could be the primary producers. Most of the surface and subsurface biosphere on Earth is based, ultimately, on photosynthesis directly, or on heterotrophic decomposition of organic material produced at the surface, usually reacting with oxygen produced from the surface. This is not relevant to the subsurface of worlds that have no surface biosphere or other source of oxidants. There are three isolated ecosystems reported in the literature that are anaerobic chemoautotrophic, two of which are based on methanogens that use H_2 derived from rock-water reactions (*Stevens and McKinley,* 1995; *Chapelle et al.,* 2002), and a third based on sulfur-reducing bacteria that use redox couples produced ultimately by radioactive decay (*Lin et al.,* 2006).

The first example of a microbial community completely independent of surface photosynthesis and O_2 was reported by *Stevens and McKinley* (1995). The system is deep below the surface of the Columbia River basalts and H_2 is produced by the serpentinization of olivine in the rock. *Chapelle et al.*

(2002) found a similar system in the massive basalts in the Twin Falls area of Idaho. These two systems are useful examples of strongly metabolically constrained but biologically active ecosystems within rock-hosted environments on Earth.

There is one clear analog on Earth for hydrothermal vents in Enceladus' ocean: the Lost City field along the Atlantic Massif to the west of the mid-Atlantic ridge (Fig. 3) (*Kelley et al.,* 2001). The measurements made by the Cassini spacecraft support the conclusion that the subsurface liquid water of Enceladus is cycling through warm, alkaline hydrothermal systems at the bottom of the ocean. On Earth the discovery of such systems within our own ocean occurred only relatively recently, although their existence had been predicted based on geological observations (*Kelley et al.,* 2001). Even before their discovery it had been argued that alkaline hydrothermal systems could have been important locales for the origin of life (*Russell et al.,* 1988, 1989; *Russell,* 2003; *Barge and White,* 2017).

Russell (2003) and *Hanczyc et al.* (2003) showed that alkaline environments such as Lost City could provide a geochemical interface that supports lipid vesicle formation. Importantly, however, a moderate to high pH is problematic for the stability of RNA and to a lesser extent for DNA, but the functionality of RNA may have been of much greater importance to the origin of life as we know it (e.g., *Joyce,* 1989). High salinity, and in particular the prevalence of divalent cations such as magnesium, also poses a problem for the formation and stability of RNA and other polymers (*Monnard et al.,* 2002; *Hand and Chyba,* 2007). *Kelley et al.* (2001) measured magnesium at Lost City to be in the range of 9–19 mmol kg^{-1}, which is considerably lower than seawater (54 mmol kg^{-1}) but significantly higher than high-temperature, low-pH, axial hydrothermal systems (~0 mmol kg^{-1}).

The Lost City site has become the canonical example of an active, off-axis, serpentinizing hydrothermal system, generating low-temperature (~70°C) alkaline fluids. Several additional sites have been discovered over a range of seafloor depths and host-rock conditions (see, e.g., *German et al.,* 2010; *Schrenk et al.,* 2013). *Brazelton et al.* (2006) measured the hydrogen and methane concentrations at active venting sites within the Lost City field to be 1 to 15 mmol kg^{-1} and 1.28 to 1.98 mmol kg^{-1}, respectively. The corresponding pH range was measured to be between 9 and 11, with temperature variations between approximately 40°C and 90°C. Carbon dioxide has been found to be largely absent (*Kelley et al.,* 2005), which restricts the availability of inorganic carbon in these systems.

The exothermic (heat-releasing) serpentinization reactions produce Ca-OH fluids that mix with seawater, leading to the formation of carbonate chimneys, several of which have grown to ~50 m in height over a lifetime of approximately 30,000 years (*Fruh-Green et al.,* 2003). As measured by *Brazelton et al.* (2006, 2011), within the carbonates are microbial populations feeding off the vent fluid, generating a biomass of 3.4×10^6 to 1.4×10^9 cells g^{-1} of carbonate. Considering fluids sampled from active vents, these numbers

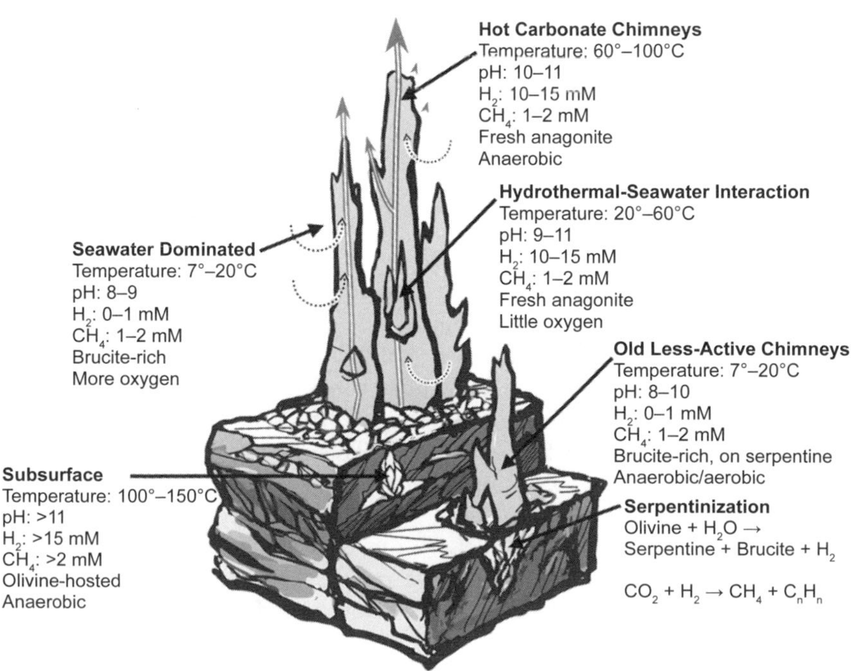

Fig. 3. Cross-section diagram of geological, geochemical, and biological zonation within the Lost City hydrothermal vents. The towers are several tens of meters in height and the system is powered by the exothermic serpentinization reaction that occurs between the ultramafic peridotite and gabbro of the Atlantis Massif and the surrounding seawater. The combination of reductants such as H_2 and CH_4 with oxidants in seawater helps to support a diverse array of microbial populations within Lost City. LCMS refers to the methanotrophic archaea Methanosarcinales. Adapted from *Brazelton et al.* (2006).

drop to 4.7 × 10^4 to 3.9 × 10^5 cells cm^{-3} of vent fluid. Within and beneath the chimneys, H_2 and CH_4 directly derived from serpentinization and fluid circulation drive microbial activity, predominantly in the form of methane-oxidizing archaea. Given the observation that some methane-oxidizing microbes also carry genes for methanogenesis (*Hallam et al.,* 2004), there is some debate as to the magnitude of biological processes as a source and sink for methane at Lost City. Further up the chimneys, and closer to the exterior, seawater cycling brings oxidants such as oxygen and sulfate into the fluids, permitting additional metabolic pathways for microbial populations.

The putative hydrothermal systems of Enceladus are clearly not as well characterized as the Lost City vents, but there appears to be a basis for the Lost City analogy. Alkaline systems on Enceladus may have lower concentrations of Mg than Lost City fluids, because submarine fluids on Earth are mixed with Mg-rich seawater. Based on the 0.4% to 1.4% H_2 in the plumes and the assumption that the H_2/CO_2 ratio is conserved between the ocean and plume, *Waite et al.* (2017) estimated that the dissolved H_2 in the ocean on Enceladus ranges from 1 × 10^{-4} to 2 × 10^{-7} mol kg^{-1}, for pH from 9 to 11, respectively. This is considerably lower than the Lost City values of 1 to 1.5 × 10^{-2} mol kg^{-1}, but the Lost City values were made from fluids collected at the vent field and do not reflect significant dilution, as could be expected for the case of Enceladus' ocean.

A fundamental difference between the Lost City vents and the possible vents on Enceladus is the energy available for life through various metabolic pathways. The Enceladus plume chemistry indicates that only H_2 and CO_2 are definitely available as a redox energy source. Sulfate and O_2 are not observed, and are perhaps unlikely to be present. Thus, given our current understanding of Enceladus' chemistry, oxidation reactions using these species, while important on Earth, may not be relevant for Enceladus.

If the hydrothermal outflow on Enceladus has a concentration of cells similar to that seen in the Lost City fluids, ~10^5 cells cm^{-3}, biological material could be detectable in a sample collected by a spacecraft flying through the plume, even if the hydrothermal outflow is diluted by 10 to 1 with ambient ocean water, as determined by two-dimensional fluid flow calculations of *Steel et al.* (2017). In a related study, *Porco et al.* (2017) estimated that microbial concentrations at hydrothermal vents on Enceladus could be comparable to those on Earth, ~10^5 cells cm^{-3}, by scaling the average geothermal flux into the sea beneath Enceladus' south polar terrain to that of the average Atlantic ocean, and assuming energy and metabolic partitioning is the same on both worlds.

The analogs discussed above — two physical analogs based on ice-sealed Antarctic lakes, two biogeochemical analogs based on methanogens, and the Lost City hydrothermal system — each have features that may resemble the ocean on Enceladus. These analogs have not yet been adequately studied in relation to how they might inform ecological models of Enceladus, but such work would be useful as missions to Enceladus are developed.

5. LIFE DETECTION INSTRUMENTATION APPROACHES

The Viking Missions to Mars in 1976 were the first, and to date only, direct search for life on another world. Each of the two landers carried three biology experiments designed to detect metabolic activity in samples from the top few centimeters of the martian soil. The pyrolytic release experiment (*Horowitz and Hobby,* 1977) detected the capability to incorporate radioactively labeled carbon dioxide in the presence of sunlight (i.e., photosynthesis). The labeled release (LR) experiment (*Levin and Straat,* 1977) attempted to detect life by the release of radioactively labeled carbon initially incorporated into organic compounds in a nutrient solution. The gas exchange experiment (GEx) (*Oyama and Berdahl,* 1977) was designed to determine if martian life could metabolize and exchange gaseous products in the presence of water vapor and in a nutrient solution. In all three cases the Viking Biology Experiments required that living organisms were present in the sample and that these organisms would respond to the conditions and nutrients provided. The results were considered negative (*Klein,* 1978, 1979, 1999) but controversy remains over the LR results (*Levin and Straat,* 2016).

Another key instrument on the Viking landers was the gas chromatograph/mass spectrometer (GCMS), which searched for organics in the soil. The most surprising result of the Viking mission was thc apparent inability to detect organics in surface samples, and from samples below the surface (*Biemann et al.,* 1977). The explanation has emerged that the lack of detection was due to perchlorate in the soil (*Hecht et al.,* 2009; *Quinn et al.,* 2013) and the reactivity of perchlorates when heated (*Navarro-González et al.,* 2010; *Glavin et al.,* 2013).

Life detection on future missions will not follow the approach of the Viking Biology Experiments and expose samples to nutrient solutions (see, e.g., *Hand et al.,* 2017). Since Viking, the microbiology community has discovered that the ability of soil microorganisms to metabolize or grow in an experimental nutrient solution (e.g., culturing) is severely limited (*Lok,* 2015). On Earth the methods for detecting and characterizing microorganisms is based on non-culture methods. Such methods include the detection of key biomolecules such as phospholipids, Adenosine triphosphate (ATP), and DNA, and upon the direct sequencing of DNA and RNA to infer phylogeny and gene expression.

Analysis of specific biological molecules (e.g., ATP, chlorophyll) and genetic and protein sequencing are powerful methods but they are quite specific to Earth biology. Indeed, they would be essentially blind to an alternative type of life here on Earth — the proposed shadow biosphere (*Davies et al.,* 2009). They do not provide general tools for the search for life that may differ even slightly from life on Earth.

Instruments have been designed to detect signs of life on Enceladus and other solar system targets based on multiple biosignatures (Fig. 4), including cellular morphology and motility (*Lindensmith et al.,* 2016), and large biopolymers

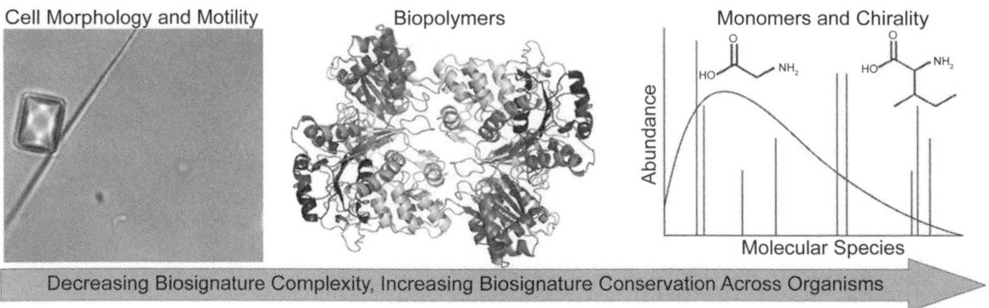

Fig. 4. Biosignatures, from highly complex but not well-conserved cell morphology and motility, to simple and well-conserved patterns of small organic molecules.

like DNA and proteins (*Parro et al.,* 2011; *Sims et al.,* 2005; *Benner,* 2017). However, searching for cells or large biopolymers carries important Earth-centric assumptions and suffers from a low probability of conservation across independent origins. An alternative approach to search for life on other worlds is to search for the building blocks of these large molecules. The key observation is that biology uses a small set of monomers for the construction of the large polymers that are used for structure, function, and information content. Thus, polymers such as DNA, RNA, polysaccharides, lipids, and proteins are constructed of specific monomers without the use of chemically similar compounds. *McKay* (2004) referred to this as the "Lego Principle" and suggested that the search for life on other worlds in the solar system could be based on the identification of this biological selectivity (Fig. 5). Perhaps the best example of this is the proteinogenic amino acids. Life on Earth uses 20 amino acids to make proteins, with small variations. Ten of the proteinogenic amino acids (the structurally simpler ones) are commonly found in abiotic chemistry and were likely present in the prebiotic world. The other 10 amino acids (structurally more complex) are only known as byproducts of biochemical synthesis. Life elsewhere arising in liquid water environments such as the Enceladus ocean would likely use some of the same 10 simple amino acids, but the chances of alien life making the exact same choices for the large amino acids are small (*Davila and McKay,* 2014). Indeed, while selective pressures clearly shape the set of amino acids life will use, the set of 20 in Earth life is not unique. *Philip and Freeland* (2011), and more recently *Ilardo et al.* (2015), show through computational analysis that the set of 20 amino acids found within the standard genetic code is the result of considerable natural selection, but that alternative choices for the large amino acids are possible. In addition, to optimally form proteins, all the amino acids used must have the same handedness (chirality). Life on Earth uses only L-amino acids to form proteins; life on Enceladus could have made the same choice, or could have chosen the mirror image of the entire suite of compounds. This selectivity of biology with respect to the possible amino acids is shown in Fig. 6.

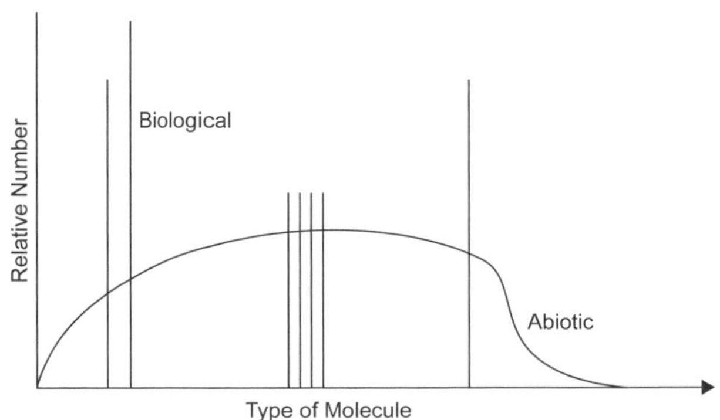

Fig. 5. Comparison of biogenic with abiogenic distributions of organic material. Nonbiological processes produce smooth distributions of organic compounds, illustrated here by the curve. Biology, in contrast, selects and uses only a few distinct molecules, shown here as spikes (e.g., the 20 L-amino acids on Earth) and builds up complex biomolecules from the combination of these few building blocks. From *McKay* (2004).

When considering a search for signs of life on other worlds there is often an implicit assumption that the choices will be binary: life or no life. However, it may be possible that we discover chemistry on its way to life that is arguably neither abiotic chemistry nor life undergoing Darwinian evolution, such as a protometabolic chemical system. The existence of such a "missing link" between chemical evolution and Darwinian evolution is logically required, but the nature of this link is unclear. Fundamental aspects of living systems, such as homochirality, may have been present in this intermediate stage even before Darwinian evolution, and hence life, became operative (e.g., *Benner et al.,* 2017). Future investigations of Enceladus or other habitable worlds should be prepared to investigate this question — a question in many ways just as interesting as the discovery of a second genesis of life.

6. EXPLORATION TECHNOLOGIES

To further investigate Enceladus' habitability and potential inhabitants, robotic exploration will someday lead to additional sampling of the plume material, landing on

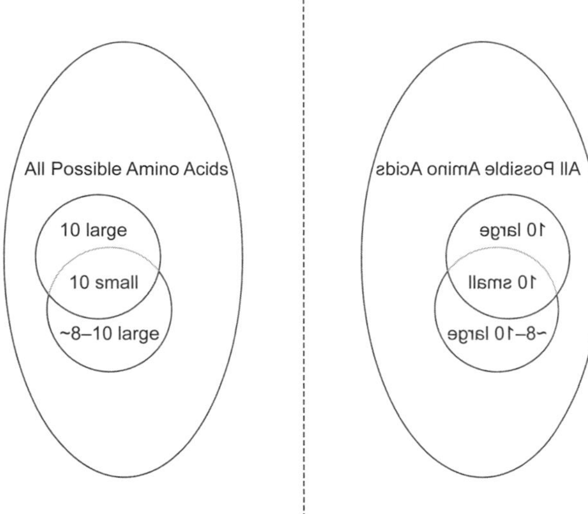

Fig. 6. The "Lego Principle" applied as the basis for a search for alien biochemistry using amino acids. Life on Earth appears to use 10 simple amino acids inherited from prebiotic processes and 10 large amino acids that were selected by evolution and biosynthesized from precursors not incorporated from prebiotic production directly (lefthand panel). Life elsewhere arising in liquid water environments would likely use the same 10 simple amino acids, but the chances of alien life making the same choices for the large amino acids are small. In addition to this choice, life can choose the mirror image of the entire site — as represented by the righthand panel. Amino acids are found naturally in meteorites and comets, but the distribution and chirality are distinct from a proteinogenic set. From *McKay* (2014b).

Enceladus' active south polar terrain, and perhaps eventually navigating into plume fractures, enabling exploration through the ice shell and into the ocean (Fig. 7).

Enceladus has a plume that launches samples into space, which makes flyby *in situ* missions possible in the near term. Flyby systems are relatively low cost and benefit from the heritage of particulate sample collection enabled by sample return missions like Stardust and Hayabusa. Instrumentation to conduct quantitative compositional and chiral analyses of trace (parts-per-million or lower) organic molecules in the plume requires a highly sensitive detection system and a separation method to resolve different species and enantiomers. Highly sensitive detection systems include mass spectrometry (MS), which is currently the predominant organic detection system for space flight, and laser-induced fluorescence (LIF), which has been in development for space flight for a decade and is capable of very low limits of detection [sub-parts-per-trillion (*Chiesl et al.,* 2009; *Creamer et al.,* 2016; *Mathies et al.,* 2017)]. However, for compositional and chiral resolution, a front-end separation method is typically needed. While gas chromatography has been used with success on Mars missions, capillary electrophoresis (CE) is showing new promise. The Enceladus Organic Analyzer is an instrument concept from the University of California (UC) Berkeley and the Berkeley Space Science Laboratory (*Butterworth et al.,* 2016; *Mathies et al.,* 2017) that uses microcapillary CE (μCE-LIF) for quantitative compositional and chiral analysis of amino acids and would be suitable for a flyby mission. The Mathies group at UC Berkeley has already demonstrated μCE-LIF for quantitative compositional analysis of amines, amino acids, dipeptides, carboxylic acids, aldehydes, ketones, and polycyclic aromatic hydrocarbons in astrobiologically relevant samples including the Murchison meteorite (*Chiesl et al.,* 2009), the Rio Tinto (*Stockton et al.,* 2009a), the Atacama Desert (*Chiesl et al.,* 2009), hydrothermal surface pools (*Stockton et al.,* 2010), submarine hydrothermal vents (*Stockton et al.,* 2009b), etc., and have proposed this instrument for Enceladus (*Mathies et al.,* 2017). They additionally successfully dated the age of the Atacama Desert using chiral analysis of amino acids with a field-portable μCE-LIF instrument (*Skelley et al.,* 2007). *Fujishima et al.* (2016) demonstrated CE with MS to resolve biopolymers and chiral amino acids, and *Creamer et al.* (2016) reported on the use of microcapillary CE to detect biological amino acids in the brine of Mono Lake.

Benner (2017) has proposed a concept instrument that could detect any linear biopolymer (in water) that would be able to encode information and thus support Darwinism. The detection method is based on the conclusion that any such linear molecule would be a polyelectrolyte. The essential feature of this approach is the view that linear biomolecules

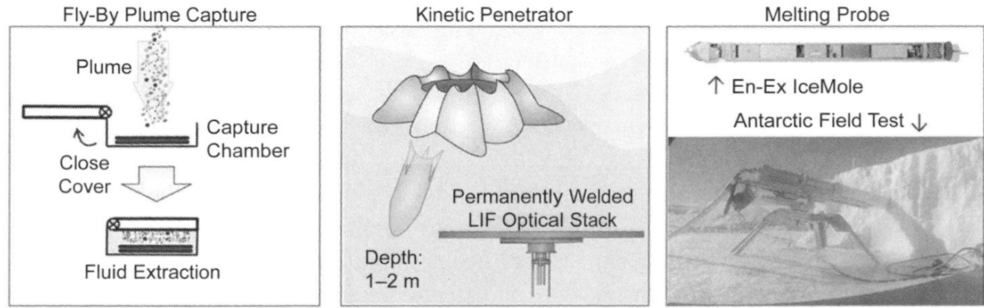

Fig. 7. Instrument concepts for accessing subsurface samples at Enceladus. A flyby mission can access samples from the interior via the plume, while kinetic penetrators and melt probes are landed missions to the icy surface itself. Melt probes may be able to access the subsurface liquid water ocean.

would necessarily precede the development of chirality or other aspects of molecular selectivity (amino acids and lipids) — the hypothetical RNA world is an example (e.g., *Joyce,* 1989).

In the longer timeframe, surface missions might be possible on Enceladus. Relatively fresh biomarkers could be accessible in the "snow" falling out of the plume onto the surface in regions near the base of the jets, in the south polar terrain. Drilling or melting into the subsurface is an obvious route to access near-surface samples, and recently kinetic impactors have provided an innovative, and potentially lower-cost, avenue to these sought-after samples. Kinetic impactors must survive up to a 5 km s^{-1} impact with cryo-ice (roughly the hardness of granite) or about 50,000 g. *Stockton et al.* (2016) proposed a microfluidic processor with a LIF detector for this application. Currently the results of models indicate that the systems can penetrate to 1–2 m depth, and testing indicates that the monolithic optical stack for LIF will survive these extreme impact conditions. Planned and ongoing testing at high impact forces will help inform and constrain designs for this ambitious scheme for low-cost subsurface access.

However, direct investigations of the ocean would require deep-subsurface access. Considering only the pressure conditions on Enceladus, numerous existing technologies and submersible systems could be used, were they somehow transported there and delivered through at least a few kilometers of ice. The pressure beneath the ice of Enceladus is only a few megapascals (for the global average crust thickness) or less (for the crust thickness at the south pole regions) and at the seafloor (assuming a total ice + ocean depth of ~100 km), the pressure is only ~10 MPa. Within Earth's ocean this range corresponds to a depth of 1–2 km, or roughly twice the depth of Lost City [~700–900 m depth (*Kelley et al.,* 2005)]. The German Enceladus Explorer (EnEx) initiative uses a miniaturized probe to access subsurface liquid pockets near the surface via melting (*Kowalski et al.,* 2016). The probe uses an ice screw at the head and differential heating along its surfaces to enable steerable subsurface operation, and has been demonstrated under laboratory tests. Two prototypes were successfully tested on glaciers in Switzerland and Iceland, demonstrating directional melting in all degrees of freedom, as well as curve driving and dirt-layer penetration. The EnEx-IceMole probe enables obstacle avoidance and was tested on Canada Glacier and Taylor Glacier in Antarctica. During the Taylor Glacier deployment, the probe successfully returned samples from the source of Blood Falls, demonstrating the capability for clean access, sampling, and return from highly protected areas. The redesign and modeling of a lighter-weight probe is underway.

As a first site for investigation, the ice-water interface may supply sufficient chemical disequilibrium so as to sustain an ecosystem within, or just beneath, the ice shell. Robotic vehicles that provide mobility and sampling at this interface would be desirable. One such solution is a small, low-mass, low-power system that operates as a roving vehicle buoyantly supported against the ice-water interface (*Leichty et al.,* 2013). Getting through the ice and autonomously navigating to scientifically compelling sites within the subsurface ocean will require a host of additional technologies. *Stone et al.* (2010, 2014) and *Richmond et al.* (2011) describe the development and successful deployment of robotic vehicles in Antarctic subice environments with some of the capabilities that would be needed for an eventual Enceladus vehicle. Additional work by those teams continue to advance both the power and communication relays' need for penetrating through kilometers of ice and communicating data back to the surface.

Spectral analysis to detect organics, and even biomarkers, in the plume may be an important contribution by spacecraft flying through the plume or telescopes at Earth. On missions that ultimately sample the plume, a pre-encounter phase may utilize plume characterization by spectroscopy. Low-cost missions are also under consideration that consist only of high-speed flybys of the plume. Large telescopes at Earth deployed for exoplanetary research, and the associated methods of transit analysis, could also detect organics in the plume.

Using data from the Visual and Infrared Mapping Spectrometer (VIMS) on the Cassini spacecraft, *Clark et al.* (2005) mapped organic signatures attributed to aromatic or cycloalkane hydrocarbons on Saturn's satellite Phoebe. *Cruikshank et al.* (2008) demonstrated the detection of organics on the surface of Iapetus and refined the *Clark et al.* (2005) interpretation of organics on Phoebe, reporting aromatic and aliphatic units in complex macromolecular carbonaceous material with a kerogen- or coal-like structure, similar to that in carbonaceous meteorites on both satellites. *Dhingra et al.* (2017) reported on VIMS observations of the plume of Enceladus. Their results suggest that organic features in the 3-μm range could be detected in the plume if the signal to noise could be increased by about a factor of 10.

Drabek-Maunder et al. (2017) report a methanol (CH_3OH) detection in the vicinity of the plume. The abundance (>0.5% with regard to H_2O) exceeds the observed abundance in the direct vicinity of the vents (~0.01%), suggesting CH_3OH is produced chemically in the plume, probably from the photolysis products of CH_4 and H_2O. Radiolysis of more complex organics mixed with ice might provide an additional source of CH_3OH.

Judge (2017) presents a detailed analysis of the possibilities of organic and biomarker detection by next generation groundbased and orbital telescopes. An important opportunity in this regard is that Enceladus will transit the face of Saturn in spring 2022. The sophisticated methods of transit analysis developed for exoplanet research may be fruitful when applied to this transit of Enceladus.

Beyond organic detection it is possible that spectral methods could detect biomarkers. For example, identification of specific amino acids may be possible with reflectance spectroscopy and possibly even detection of chirality [see the U.S. Geological Survey dataset (*Kokaly et al.,* 2017)]. In another example, current studies of flyby missions suggest that UV fluorescence can be used to detect aromatic amino acids and to distinguish them from polycyclic aromatic hydrocarbons. The spacecraft would fly by the plume and shine a pulsed UV laser on the plume at close range. Potential target molecules include tyrosine, tryptophan, and phenylalanine (*Johnson et al.,* 2011; *Smith et al.,* 2014; *Abbey et al.,* 2017). A challenge for remote sensing is the unambiguous identification of biomarker targets if they are present as complex mixtures with other organic spectral signatures.

7. CONCLUSIONS

Measurements of the plume of Enceladus by the Cassini spacecraft have clearly indicated that there is an ocean of liquid water below the ice and that this ocean is likely to be habitable for a range of terrestrial microorganisms. The samples of this habitable ocean coming out in the plume provide a unique opportunity for astrobiology. Future missions can refine our understanding of the habitability of the ocean perhaps to the level that we can specify, in detail, the type of ecosystem that can be expected. If the ocean is rich in biomarkers, and these biomarkers are similar to those from Earth biology, then *in situ* studies of the plume may reveal the presence and biochemical nature of life on Enceladus. If the signs of life are at a low level, swamped by non-biological sources of the same molecules, or if the molecules produced by life on Enceladus are profoundly different from those associated with life on Earth, then *in situ* investigations are likely to be inadequate and a sample return will be required. Whatever the outcome, the investigation of the habitable ocean on Enceladus will inform us not just about life on that world, but also about the nature of the origin of life in general and the prevalence of habitable environments in the universe.

REFERENCES

Abbey W. J., Bhartia R., Beegle L. W., DeFlores L., Paez V., Sijapati K., Sijapati S., Williford K., Tuite M., Hug W., and Reid R. (2017) Deep UV Raman spectroscopy for planetary exploration: The search for *in situ* organics. *Icarus, 290,* 201–214.

Altwegg K., Balsiger H., Bar-Nun A., Berthelier J. J., Bieler A., Bochsler P., Briois C., Calmonte U., Combi M. R., Cottin H., and De Keyser J. (2016) Prebiotic chemicals — amino acid and phosphorus — in the coma of Comet 67P/Churyumov-Gerasimenko. *Sci. Adv., 2(5),* e1600285.

Andersen D. T., Sumner D. Y., Hawes I., Webster-Brown J., and McKay C. P. (2011) Discovery of large conical stromatolites in Lake Untersee, Antarctica. *Geobiology, 9(3),* 280–293.

Barge L. M. and White L. M. (2017) Experimentally testing hydrothermal vent origin of life on Enceladus and other icy/ocean worlds. *Astrobiology, 17,* 820–833.

Benner S. A. (2017) Detecting Darwinism from molecules in the Enceladus plumes, Jupiter's moons, and other planetary water lagoons. *Astrobiology, 17,* 840–851.

Biemann K., Oro J., Toulmin P., Orgel L. E., Nier A. O., Anderson D. M., Simmonds P. G., Flory D., Diaz A. V., Rushneck D. R., and Biller J. E. (1977) The search for organic substances and inorganic volatile compounds in the surface of Mars. *J. Geophys. Res., 82(28),* 4641–4658.

Bose A., Gardel E. J., Vidoudez C., Parra E. A., and Girguis P. R. (2014) Electron uptake by iron-oxidizing phototrophic bacteria. *Nature Commun., 5,* 3391.

Brazelton W. J., Schrenk M. O., Kelley D. S., and Baross J. A. (2006) Methane- and sulfur-metabolizing microbial communities dominate the Lost City hydrothermal field ecosystem. *Appl. Environ. Microbiol., 72(9),* 6257–6270.

Brazelton W. J., Mehta M. P., Kelley D. S., and Baross J. A. (2011) Physiological differentiation within a single-species biofilm fueled by serpentinization. *MBio, 2(4),* e00127-11.

Butterworth A. L., Kim J., Stockton A. M., Turin P., Ludlam M., and Mathies R. A. (2016) Instrument for capturing and analyzing trace organic molecules from plumes for ocean worlds missions. In *Third International Workshop on Instrumentation for Planetary Missions,* Abstract #4100. Lunar and Planetary Institute, Houston.

Calkin P. E. and Bull C. (1967) Lake Vida, Victoria Valley, Antarctica. *J. Glaciol., 6(48),* 833–836.

Canup R. M. and Ward W. R. (2006) A common mass scaling for satellite systems of gaseous planets. *Nature, 441(7095),* 834–839.

Chapelle F. H., O'Neill K., Bradley P. M., Methe B. A., Ciufo S. A., Knobel L. L., and Lovley D. R. (2002) A hydrogen-based subsurface microbial community dominated by methanogens. *Nature, 415,* 312–315.

Chiesl T. N., Chu W. K., Stockton A. M., Amashukeli X., Grunthaner F., and Mathies R. A. (2009) Enhanced amine and amino acid analysis using Pacific Blue and the Mars Organic Analyzer Microchip Capillary Electrophoresis System. *Anal. Chem., 81(7),* 2537–2544.

Chinn T. J. (1993) Physical hydrology of the dry valley lakes. *Antarct. Res., 59,* 1–51.

Clark R. N., Brown R. H., Jaumann R., Cruikshank D. P., Nelson R. M., Buratti B. J., McCord T. B., Lunine J., Baines K. H., Bellucci G., Bibring J.-P., Capaccioni F., Cerroni P., Coradini A., Formisano V., Langevin Y., Matson D. L., Mennella V., Nicholson P. D., Sicardy B., Sotin C., Hoefen T. M., Curchin J. M., Hansen G., Hibbits K., and Matz K.-D. (2005) Compositional maps of Saturn's moon Phoebe from imaging spectroscopy. *Nature, 435,* 66–69, DOI: 10.1038/nature03558.

Cockell C. S. (2011) Vacant habitats in the universe. *Trends Ecol. Evol., 26(2),* 73-80.

Creamer J. S., Mora M. F., and Willis P. A. (2016) Enhanced resolution of chiral amino acids with capillary electrophoresis for biosignature detection in extraterrestrial samples. *Anal. Chem., 89(2),* 1329–1337.

Cruikshank D. P., Wegryn E., Dalle Ore C. M., Brown R. H., Bibring J. P., Buratti B. J., Clark R. N., McCord T. B., Nicholson P. D., Pendleton Y. J., and Owen T. C. (2008) Hydrocarbons on Saturn's satellites Iapetus and Phoebe. *Icarus, 193(2),* 334–343.

Ćuk M., Dones L., and Nesvorný D. (2016) Dynamical evidence for a late formation of Saturn's moons. *Astrophys. J., 820(2),* 97.

Davies P. C., Benner S. A., Cleland C. E., Lineweaver C. H., McKay C. P., and Wolfe-Simon F. (2009) Signatures of a shadow biosphere. *Astrobiology, 9(2),* 241–249.

Davis W. L. and McKay C. P. (1996) Origins of life: A comparison of theories and application to Mars. *Origins Life Evol. Biosph., 26(1),* 61–73.

Davila A. F. and McKay C. P. (2014) Chance and necessity in biochemistry: Implications for the search for extraterrestrial biomarkers in Earth-like environments. *Astrobiology, 14(6),* 534–540.

Deamer D. and Damer B. (2017) Can life begin on Enceladus? A perspective from hydrothermal chemistry. *Astrobiology, 17,* 834–839.

Deamer D. W. and Georgiou C. D. (2015) Hydrothermal conditions and the origin of cellular life. *Astrobiology, 15(12),* 1091–1095.

Deamer D., Singaram S., Rajamani S., Kompanichenko V., and Guggenheim S. (2006) Self-assembly processes in the prebiotic environment. *Philos. Trans. R. Soc. London, Ser. B, Biol. Sci., 361(1474),* 1809–1818.

Dhingra D., Hedman M. M., Clark R. N., and Nicholson P. D. (2017) Spatially resolved near infrared observations of Enceladus' tiger stripe eruptions from Cassini VIMS. *Icarus, 292,* 1–12.

Djokic T., Van Kranendonk M. J., Campbell K. A., Walter M. R., and Ward C. R. (2017) Earliest signs of life on land preserved in ca. 3.5 Ga hot spring deposits. *Nature Commun., 8,* DOI: 10.1038/ncomms15263.

Doran P. T., Fritsen C. H., McKay C. P., Priscu J. C., and Adams E. E. (2003) Formation and character of an ancient 19-m ice cover and underlying trapped brine in an "ice-sealed" east Antarctic lake. *Proc. Natl. Acad. Sci., 100(1),* 26–31.

Drabek-Maunder E., Greaves J., Fraser H. J., Clements D. L., and Alconcel L. N. (2017) Ground-based detection of a cloud of

methanol from Enceladus: When is a biomarker not a biomarker? *Intl. J. Astrobiology,* 1–8, DOI: 10.1017/S1473550417000428.
Fricker H. A. and Scambos T. (2009) Connected subglacial lake activity on lower Mercer and Whillans ice streams, West Antarctica, 2003–2008. *J. Glaciol., 55(190),* 303–315.
Fruh-Green G. L., Kelley D. S., Bernasconi S. M., Karson J. A., Ludwig K. A., Butterfield D. A., Boschi C., and Proskurowski G. (2003) 30,000 years of hydrothermal activity at the Lost City vent field. *Science, 301,* 495–498.
Fujishima K., Dziomba S., Takahagi W., Shibuya T., Takano Y., Guerrouache M., Carbonnier B., Takai K., Rothschild L., and Yano H. (2016) A fly-through mission strategy targeting peptide as a signature of chemical evolution and possible life in Enceladus plumes (abstract). In *Enceladus and the Icy Moons of Saturn,* Abstract #3085. Lunar and Planetary Institute, Houston.
German C. R., Bowen A., Coleman M. L., Honig D. L., Huber J. A., Jakuba M. V., Kinsey J. C., Kurz M. D., Leroy S., McDermott J. J., de Lépinay B. M., Nakamura K., Seewald J. S., Smith J. L., Sylva S. P., Van Dover C. L., Whitcomb L. L., and Yoerger D. R. (2010) Diverse styles of submarine venting on the ultraslow spreading Mid-Cayman Rise. *Proc. Natl. Acad. Sci., 107(32),* 14020–14025.
Glavin D. P., Freissinet C., Miller K. E., Eigenbrode J. L., Brunner A. E., Buch A., Sutter B., Archer P. D., Atreya S. K., Brinckerhoff W. B., and Cabane M. (2013) Evidence for perchlorates and the origin of chlorinated hydrocarbons detected by SAM at the Rocknest aeolian deposit in Gale Crater. *J. Geophys. Res.–Planets, 118(10),* 1955–1973.
Glein C. R., Zolotov M. Y., and Shock E. L. (2008) The oxidation state of hydrothermal systems on early Enceladus. *Icarus, 197(1),* 157–163.
Glein C. R., Baross J. A., and Waite J. H. (2015) The pH of Enceladus' ocean. *Geochim. Cosmochim. Acta, 162,* 202–219.
Hallam S. J., Putnam N., Preston C. M., Detter J. C., Rokhsar D., Richardson P. M., and DeLong E. F. (2004) Reverse methanogenesis: Testing the hypothesis with environmental genomics. *Science, 305,* 1457–1462.
Hanczyc M. M., Fujikawa S. M., and Szosta J. W. (2003) Experimental models of primitive cellular compartments: Encapsulation, growth, and division. *Science, 302(5645),* 618–622.
Hand K. P. and Chyba C. F. (2007) Empirical constraints on the salinity of the Europan ocean and implications for a thin ice shell. *Icarus, 189(2),* 424–438, DOI: 10.1016/j.icarus.2007.02.002.
Hand K. P., Chyba C. F., Priscu J. C., Carlson R. W., and Nealson K. H. (2009) Astrobiology and the potential for life on Europa. In *Europa* (R. T. Pappalardo et al., eds.), pp. 589–629. Univ. of Arizona, Tucson.
Hand K. P., Murray A. E., Garvin J. B., Brinckerhoff W. B., Christner B. C, Edgett K. S, Ehlmann B. L., German C. R., Hayes A. G., Hoehler T. M., Horst S. M., Lunine J. I., Nealson K. H., Paranicas C., Schmidt B. E., Smith D. E., Rhoden A. R., Russell M. J., Templeton A. S., Willis P. A., Yingst R. A., Phillips C. B, Cable M. L., Craft K. L., Hofmann A. E., Nordheim T. A., Pappalardo R. T., and the Project Engineering Team (2017) *Report of the Europa Lander Science Definition Team,* available online at *https://solarsystem.nasa.gov/docs/Europa_Lander_SDT_Report_2016.pdf.*
Hansen C. J., Esposito L., Stewart A. I. F., Colwell J., Hendrix A., Pryor W., Shemansky D., and West R. (2006) Enceladus' water vapor plume. *Science, 311(5766),* 1422–1425.
Hansen C. J., Shemansky D. E., Esposito L. W., Stewart A. I. F., Lewis B. R., Colwell J. E., Hendrix A. R., West R. A., Waite J. H., Teolis B., and Magee B. A. (2011) The composition and structure of the Enceladus plume. *Geophys. Res. Lett., 38(11),* DOI: 10.1029/2011GL047415.
Hecht M. H., Kounaves S. P., Quinn R. C., West S. J., Young S. M. M., Ming D. W., Catling D. C., Clark B. C., Boynton W. V., Hoffman J., and DeFlores L. P. (2009) Detection of perchlorate and the soluble chemistry of martian soil at the Phoenix lander site. *Science, 325(5936),* 64–67.
Hoehler T. M. (2004) Biological energy requirements as quantitative boundary conditions for life in the subsurface. *Geobiol., 2(4),* 205–215.
Hoehler T. M., Amend J. P., and Shock E. L. (2007) A "follow the energy" approach for astrobiology. *Astrobiology, 7,* 819–823.
Houtkooper J. M. and Schulze-Makuch D. (2007) A possible biogenic origin for hydrogen peroxide on Mars: The Viking results reinterpreted. *Intl. J. Astrobiol., 6,* 147–152.
Horowitz N. H. and Hobby G. L. (1977) Viking on Mars: The carbon assimilation experiments. *J. Geophys. Res., 82,* 4659–4662.
Hud N. V., Cafferty B. J., Krishnamurthy R., and Williams L. D. (2013) The origin of RNA and "my grandfather's axe." *Chem. Biol., 20(4),* 466–474.
Hsu H. W., Postberg F., Sekine Y., Shibuya T., Kempf S., Horányi M., Juhász A., Altobelli N., Suzuki K., Masaki Y., and Kuwatani T. (2015) Ongoing hydrothermal activities within Enceladus. *Nature, 519(7542),* 207–210.
Iess L., Stevenson D. J., Parisi M., Hemingway D., Jacobson R. A., Lunine J. I., Nimmo F., Armstrong J. W., Asmar S. W., Ducci M., and Tortora P. (2014) The gravity field and interior structure of Enceladus. *Science, 344(6179),* 78–80.
Ilardo M., Meringer M., Freeland S., Rasulev B., and Cleaves H. J. II (2015) Extraordinarily adaptive properties of the genetically encoded amino acids. *Sci. Rept., 5,* 9414.
Jackson W. A., Böhlke J. K., Andraski B. J., Fahlquist L., Bexfield L., Eckardt F. D., Gates J. B., Davila A. F., McKay C. P., Rao B., and Sevanthi R. (2015) Global patterns and environmental controls of perchlorate and nitrate co-occurrence in arid and semi-arid environments. *Geochim. Cosmochim. Acta, 164,* 502–522.
Johnson P. V., Hodyss R., Bolser D. K., Bhartia R., Lane A. L., and Kanik I. (2011) Ultraviolet-stimulated fluorescence and phosphorescence of aromatic hydrocarbons in water ice. *Astrobiology, 11(2),* 151–156.
Joyce G. F. (1989) RNA evolution and the origins of life. *Nature, 338(6212),* 217–224.
Judge P. (2017) A novel strategy to seek biosignatures at Enceladus and Europa. *Astrobiology, 17(9),* 852–861.
Kelley D. S., Karson J. A., Blackman D. K., Früh-Green G. L., Butterfield D. A., Lilley M. D., Olson E. J., Schrenk M. O., Roe K. K., Lebon G. T., and Rivizzigno P. (2001) An off-axis hydrothermal vent field near the Mid-Atlantic Ridge at 30 N. *Nature, 412(6843),* 145–149.
Kelley D. S., Karson J. A., Früh-Green G. L., Yoerger D. R., Shank T. M., Butterfield D. A., Hayes J. M., Schrenk M. O., Olson E. J., Proskurowski G., and Jakuba M. (2005) A serpentinite-hosted ecosystem: The Lost City hydrothermal field. *Science, 307(5714),* 1428–1434.
Khawaja N., Postberg F., Reviol R., Klenner F., Nölle L., and Srama R. (2015) Organic compounds from Enceladus' sub-surface ocean as seen by CDA. *EPSC Abstracts, Vol 10,* EPSC2015-652.
Klein H. P. (1978) The Viking biological experiments on Mars. *Icarus, 34,* 666–674.
Klein H. P. (1979) The Viking mission and the search for life on Mars. *Rev. Geophys. Space Phys., 17,* 1655–1662.
Klein H. P. (1999) Did Viking discover life on Mars? *Orig. Life. Evol. Biosph., 29,* 62531.
Kokaly R. F., Clark R. N., Swayze G. A., Livo K. E., Hoefen T. M., Pearson N. C., Wise R. A., Benzel W. M., Lowers H. A., Driscoll R. L., and Klein A. J. (2017) *USGS Spectral Library Version 7: U.S. Geological Survey Data Series 1035.* 61 pp., available online at *https://doi.org/10.3133/ds1035.*
Konhauser K. O., Pecoits E., Lalonde S. V., Papineau D., Nisbet E. G., Barley M. E., Arndt N. T., Zahnle K., and Kamber B. S. (2009) Oceanic nickel depletion and a methanogen famine before the Great Oxidation Event. *Nature, 458(7239),* 750–753.
Kowalski J., Linder P., Zierke S., von Wulfen B., Clemens J., Konstantinidis K., Ameres G., Hoffmann R., Mikucki J., Tulaczyk S., and Funke O. (2016) Navigation technology for exploration of glacier ice with maneuverable melting probes. *Cold Regions Sci. Tech., 123,* 53–70.
Kral T. A., Brink K. M., Miller S. L., and McKay C. P. (1998) Hydrogen consumption by methanogens on the early Earth. *Orig. Life Evol. Biosph., 28(3),* 311–319.
Kress M.. and McKay C. P. (2004) Formation of methane in comet impacts: Implications for Earth, Mars, and Titan. *Icarus, 168(2),* 475–483.
Lazcano A. and Hand K. P. (2012) Astrobiology: Frontier or fiction. *Nature Forum, 488,* 7410, 160–161.

Lazcano A. and Miller S. L. (1994) How long did it take for life to begin and evolve to cyanobacteria? *J. Mol. Evol., 39,* 546–554.

Leichty J. M., Klesh A. T., Berisford D. F., Matthews J. B., and Hand K. P. (2013) Positive-Buoyancy Rover for under ice mobility. In *NASA Tech Briefs, December 2013, 13,* 20130014512, available online at *https://www.techbriefs.com/tb/archive/2009/3-ntb/tech-briefs/mechanics-and-machinery/18752-npo-48863.*

Levin G. V. and Straat P. A. (1977) Recent results from the Viking Labeled Release Experiment on Mars. *J. Geophys. Res., 82,* 4663–4667.

Levin G. V. and Straat P. A. (2016) The case for extant life on Mars and its possible detection by the Viking Labeled Release Experiment. *Astrobiology, 16(10),* 798–810.

Lin L.-H., Wang P.-L., Rumble D., Lippmann-Pipke J., Boice E., Pratt L. M., Sherwood Lollar B., Brodie E. L., Hazen T. C., Andersen G. L., DeSantis T. Z., Moser D. P., Kershaw D., and Onstott T. C. (2006) Long-term sustainability of a high-energy, low-diversity crustal biome. *Science, 314,* 479–482.

Lindensmith C. A., Rider S., Bedrossian M., Wallace J. K., Serabyn E., Showalter G. M., Deming J. W., and Nadeau J. L. (2016) A submersible, off-axis holographic microscope for detection of microbial motility and morphology in aqueous and icy environments. *Plos One, 11(1),* DOI: 10.1371/journal.pone.0147700.

Lok C. (2015) Mining the microbial dark matter. *Nature, 522(7556),* 270–273.

Malamud U. and Prialnik D. (2013) Modeling serpentinization: Applied to the early evolution of Enceladus and Mimas. *Icarus, 225(1),* 763–774.

Mathies R. A., Razu M. E., Kim J., Stockton A., Turin P., and Butterworth A. (2017) Feasibility of detecting bioorganic compounds in Enceladus plumes with the Enceladus organic analyzer. *Astrobiology, 17,* 902–912.

Matson D. L., Castillo J. C., Lunine J., and Johnson T. V. (2007) Enceladus' plume: Compositional evidence for a hot interior. *Icarus, 187(2),* 569–573.

McCollom T. M. (1999) Methanogenesis as a potential source of chemical energy for primary biomass production by autotrophic organisms in hydrothermal systems on Europa. *J. Geophys. Res.–Planets, 104(E12),* 30729–30742.

McKay C. P. (2001) The search for a second genesis of life in our solar system. In *First Steps in the Origin of Life in the Universe* (J. Chela-Flores et al., eds.), pp. 269–277, Kluwer, Dordrecht.

McKay C. P. (2004) What is life — and how do we search for it in other worlds? *PLOS Biology, 2(9),* e302.

McKay C. P. (2014a) Requirements and limits for life in the context of exoplanets. *Proc. Natl. Acad. Sci., 111(35),* 12628–12633.

McKay C. P. (2014b) The search for life on other worlds: Second genesis. *The Biochemist, 36(6),* 16–19.

McKay C. P. and Borucki W. J. (1997) Organic synthesis in experimental impact shocks. *Science, 276(5311),* 390–392.

McKay C. P. and Smith H. D. (2005) Possibilities for methanogenic life in liquid methane on the surface of Titan. *Icarus, 178,* 274–276.

McKay C. P., Clow G. D., Wharton R. A., and Squyres S. W. (1985) Thickness of ice on perennially frozen lakes. *Nature, 313,* 561–562.

McKay C. P., Porco C. C., Altheide T., Davis W. L., and Kral T. A. (2008) The possible origin and persistence of life on Enceladus and detection of biomarkers in the plume. *Astrobiology, 8,* 909–919.

McKay C. P., Khare B. N., Amin R., Klasson M., and Kral T. A. (2012) Possible sources for methane and C2–C5 organics in the plume of Enceladus. *Planet. Space Sci., 71(1),* 73–79.

Monnard P.-A., Apel C. L., Kanavarioti A., and Deamer D. W. (2002) Influence of ionic solutes on self-assembly and polymerization processes related to early forms of life: Implications for a prebiotic aqueous medium. *Astrobiology, 2,* 213–219.

Murray A. E., Kenig F., Fritsen C. H., McKay C. P., Cawley K. M., Edwards R., Kuhn E., McKnight D. M., Ostrom N. E., Peng V., and Ponce A. (2012) Microbial life at –13°C in the brine of an i ce-sealed Antarctic lake. *Proc. Natl. Acad. Sci., 109(50),* 20626–20631.

Navarro-González R., Vargas E., de La Rosa J., Raga A. C., and McKay C. P. (2010) Reanalysis of the Viking results suggests perchlorate and organics at midlatitudes on Mars. *J. Geophys. Res.–Planets, 115,* E12010, DOI: 10.1029/2010JE003599.

Nealson K. H. (1997) The limits of life on Earth and searching for life on Mars. *J. Geophys. Res., 102,* 23675–23686.

Nielsen L. P., Risgaard-Petersen N., Fossing H., Christensen P. B., and Sayama M. (2010) Electric currents couple spatially separated biogeochemical processes in marine sediment. *Nature, 463,* 1071–1074.

Oremland R. S. and Voytek M. A. (2008) Acetylene as fast food: Implications for development of life on anoxic primordial Earth and in the outer solar system. *Astrobiology, 8,* 45–58.

Orgel L. E. (1998) The origin of life — how long did it take? *Orig. Life. Evol. Biosph., 28,* 91–96.

Oyama V. I. and Berdahl B. J. (1977) The Viking gas exchange experiment results from Chryse and Utopia surface samples. *J. Geophys. Res., 82,* 4669–4676.

Parro V., de Diego-Castilla G., Rodriguez-Manfredi J. A., Rivas L. A., Blanco-Lopez Y., Sebastian E., Romeral J., Compostizo C., Herrero P. L., Garcia-Marin A., Moreno-Paz M., Garcia-Villadangos M., Cruz-Gil P., Peinado V., Martin-Soler J., Perez-Mercader J., and Gomez-Elvira J. (2011) SOLID3: A multiplex antibody microarray-based optical sensor instrument for *in situ* life detection in planetary exploration. *Astrobiology, 11(1),* 15–28.

Philip G. K. and Freeland S. J. (2011) Did evolution select a nonrandom "alphabet" of amino acids? *Astrobiology, 11(3),* 235–240.

Pizzarello S. and Williams L. B. (2012) Ammonia in the early solar system: An account from carbonaceous meteorites. *Astrophys. J., 749(2),* 161.

Porco C. C., Helfenstein P., Thomas P. C., Ingersoll A. P., Wisdom J., West R., Neukum G., Denk T., Wagner R., Roatsch T., Kieffer S., Turtle E., McEwen A., Johnson T. V., Rathbun J., Veverka J., Wilson D., Perry J., Spitale J., Brahic A., Burns J. A., Delgenio A. D., Dones L., Murray C. D., and Squyres S. (2006) Cassini observes the active south pole of Enceladus. *Science, 311,* 1393–1401.

Porco C. C., Dones L., and Mitchell C. (2017) Could it be snowing microbes on Enceladus? Assessing conditions in its plume and implications for future missions. *Astrobiology, 17(9),* 876-901.

Postberg F., Kempf S., Schmidt J., Brilliantov N., Beinsen A., Abel B., Buck U., and Srama R. (2009) Sodium salts in E-ring ice grains from an ocean below the surface of Enceladus. *Nature, 459(7250),* 1098–1101.

Postberg F., Schmidt J., Hillier J., Kempf S., and Srama R. (2011) A salt-water reservoir as the source of a compositionally stratified plume on Enceladus. *Nature, 474,* 620–622.

Postberg F., Khawaja N. A., Kempf S., Waite J. H., Glein C., Hsu H. W., and Srama R. (2017) Complex organic macromolecular compounds in ice grains from Enceladus. In *Lunar and Planetary Science XLVIII,* Abstract #1401. Lunar and Planetary Institute, Houston.

Quinn R. C., Martucci H. F. H., Miller S. R., Bryson C. E., Grunthaner F. J., and Grunthaner P. J. (2013) Perchlorate radiolysis on Mars and the origin of martian soil reactivity. *Astrobiology, 13,* 515–20.

Richmond K., Febretti A., Gulati S., Flesher C., Hogan B. P., Murarka A., Kuhlman G., Mohan S., Johnson A., Stone W. C., Priscu J., and Doran P. (2011) Sub-ice exploration of an Antarctic lake: Results from the ENDURANCE project. Available online at *https://pdfs.semanticscholar.org/d453/536f64f20b781c1b0738bb046055ad84e9dc.pdf.*

Rosenbaum M., Aulenta F., Villano M., and Angenent L. T. (2011) Cathodes as electron donors for microbial metabolism: Which extracellular electron transfer mechanisms are involved? *Bioresource Tech., 102(1),* 324–333.

Russell M. J. (2003) Geochemistry: The importance of being alkaline. *Science, 302(5645),* 580–581.

Russell M. J., Hall A. J., Cairns-Smith A. G., and Braterman P. S. (1988) Submarine hot springs and the origin of life. *Nature, 336(6195),* 117.

Russell M. J., Hall A. J., and Turner D. (1989) In vitro growth of iron sulphide chimneys: Possible culture chambers for origin-of-life experiments. *Terra Nova, 1,* 238–241.

Russell M. J., Barge L. M., Bhartia R., Bocanegra D., Bracher P. J., Branscomb E., Kidd R., McGlynn S., Meier D. H., Nitschke W., and Shibuya T. (2014) The drive to life on wet and icy worlds. *Astrobiology, 14(4),* 308–343.

Schink B. (1985) Fermentation of acetylene by an obligate anaerobe, *Pelobacter acetylenicus* sp. nov. *Arch. Microbiol., 142,* 295–301.

Sekine Y., Shibuya T., Postberg F., Hsu H. W., Suzuki K., Masaki Y., Kuwatani T., Mori M., Hong P. K., Yoshizaki M., and Tachibana S. (2015) High-temperature water-rock interactions and hydrothermal environments in the chondrite-like core of Enceladus. *Nature Commun., 6,* 8604.

Schrenk M. O., Brazelton W. J., and Lang S. Q. (2013) Serpentinization, carbon, and deep life. *Rev. Mineral. Geochem., 75(1),* 575–606.

Sims M. R., Cullen D. C., Bannister N. P., Grant W. D., Henry O., Jones R., McKnight D., Thompson D. P., and Wilson P. K. (2005) The specific molecular identification of life experiment (SMILE). *Planet. Space Sci., 53(8),* 781–791.

Skelley A. M., Aubrey A. D., Willis P. A., Amashukeli X., Ehrenfreund P., Bada J. L., Grunthaner F. J., and Mathies R. A. (2007) Organic amine biomarker detection in the Yungay region of the Atacama Desert with the Urey instrument. *J. Geophys. Res.–Biogeosci., 112(G4),* 04S11.

Smith H. D., McKay C. P., Duncan A. G., Sims R. C., Anderson A. J., and Grossl P. R. (2014) An instrument design for non-contact detection of biomolecules and minerals on Mars using fluorescence. *J. Biol. Eng., 8(1),* 16.

Spencer J. R., Pearl J. C., Segura M., Flasar F. M., Mamoutkine A., Romani P., Buratti B. J., Hendrix A. R., Spilker L. J., and Lopes R. M. C. (2006) Cassini encounters Enceladus: Background and the discovery of a south polar hot spot. *Science, 311(5766),* 1401–1405.

Stevens T. O. and McKinley J. P. (1995) Lithoautotrophic microbial ecosystems in deep basalt aquifers. *Science, 270,* 450–454.

Steel E. L., Davila A., and McKay C. P. (2017) Formation of prebiotic and biotic organic building blocks in Enceladus' ocean. *Astrobiology, 17,* 862–875.

Stockton A. M., Chiesl T. N., Lowenstein T. K., Amashukeli X., Grunthaner F., and Mathies R. (2009a) Capillary electrophoresis analysis of organic amines and amino acids in saline and acidic samples using the Mars Organic Analyzer. *Astrobiology, 9(9),* 823.

Stockton A. M., Chiesl T. N., Scherer J. R., and Mathies R. A. (2009b) Polycyclic aromatic hydrocarbon analysis with the Mars Organic Analyzer Microchip Capillary Electrophoresis System. *Anal. Chem., 81(2),* 790–796.

Stockton A. M., Chandra Tjin C., Huang G. L., Benhabib M., Chiesl T. N., and Mathies R. A. (2010) Analysis of carbonaceous biomarkers with the Mars Organic Analyzer Microchip Capillary Electrophoresis System: Aldehydes and ketones. *Electrophoresis, 31(22),* 3642–3649.

Stockton A., Duca Z., Tan G., Cantrell T., Van Enige M., Dorn M., Cato M., Putman P., Kim J., and Mathies R. A. (2016) Development of an Extraterrestrial Organic Analyzer (EOA) for highly sensitive organic detection on a kinetic penetrator. In *Enceladus and the Icy Moons of Saturn,* Abstract #3087. Lunar and Planetary Institute, Houston.

Stone W., Hogan B., Flesher C., Gulati S., Richmond K., Murarka A., Kuhlman G., Sridharan M., Siegel V., Price R. M., Doran P. T., and Priscu J. (2010) Design and deployment of a four-degrees-of-freedom hovering autonomous underwater vehicle for sub-ice exploration and mapping. *Proc. Inst. Mech. Eng., Part M: J. Eng. Maritime Environ., 224(4),* 341–361.

Stone W. C., Hogan B., Siegel V., Lelievre S., and Flesher C. (2014) Progress towards an optically powered cryobot. *Ann. Glaciol., 55(65),* 1–13.

Talling J. F. (1976) The depletion of carbon dioxide from lake water by phytoplankton. *J. Ecol., 64(1),* 79–121.

Thauer R. K., Jungermann K., and Decker K. (1977) Energy conservation in chemotrophic anaerobic bacteria. *Bacteriol. Rev., 41(1),* 100–180.

Thomas P. C., Tajeddine R., Tiscareno M. S., Burns J. A., Joseph J., Loredo T. J., Helfenstein P., and Porco C. (2016) Enceladus's measured physical libration requires a global subsurface ocean. *Icarus, 264,* 37–47.

Vincent W. F. and Laybourn-Parry J., eds. (2008) *Polar Lakes and Rivers: Limnology of Arctic and Antarctic Aquatic Ecosystems.* Oxford Univ., New York.

Waite J. H., Combi M. R., Ip W. H., Cravens T. E., McNutt R. L., Kasprzak W., Yelle R., Luhmann J., Niemann H., Gell D., and Magee B. (2006) Cassini ion and neutral mass spectrometer: Enceladus plume composition and structure. *Science, 311(5766),* 1419–1422.

Waite J. H., Lewis W. S., Magee B. A., Lunine J. I., McKinnon W. B., Glein C. R., Mousis O., Young D. T., Brockwell T., Westlake J., Nguyen M. J., Teolis B. D., Niemann H. B., McNutt R. L., Perry M., and Ip W. H. (2009) Liquid water on Enceladus from observations of ammonia and 40A in the plume. *Nature, 460,* 487–490.

Waite J. H., Glein C. R., Perryman R. S., Teolis B. D., Magee B. A., Miller G., Grimes J., Perry M. E., Miller K. E., Bouquet A., and Lunine J. I. (2017) Cassini finds molecular hydrogen in the Enceladus plume: Evidence for hydrothermal processes. *Science, 356(6334),* 155–159.

Wand U., Samarkin V. A., Nitzsche H. M., and Hubberten H. W. (2006) Biogeochemistry of methane in the permanently ice-covered Lake Untersee, central Dronning Maud Land, East Antarctica. *Limnology and Oceanography, 51(2),* 1180–1194.

Worth R. J., Sigurdsson S., and House C. H. (2013) Seeding life on the moons of the outer planets via lithopanspermia. *Astrobiology, 13,* 1155–1165.

Yamamoto M., Nakamura R., Kasaya T., Kumagai H., Suzuki K., and Takai K. (2017) Spontaneous and widespread electricity generation in natural deep-sea hydrothermal fields. *Angewandte Chem. Intl. Edition, 56(21),* 5725–5728.

Zolotov M. Y. (2012) Aqueous fluid composition in CI chondritic materials: Chemical equilibrium assessments in closed systems. *Icarus, 220(2),* 713–729.

Lunine J. I., Coustenis A., Mitri G., Tobie G., and Tosi F. (2018) Future exploration of Enceladus and other saturnian moons. In *Enceladus and the Icy Moons of Saturn* (P. M. Schenk et al., eds.), pp. 453–468. Univ. of Arizona, Tucson, DOI: 10.2458/azu_uapress_9780816537075-ch022.

Future Exploration of Enceladus and Other Saturnian Moons

Jonathan I. Lunine
Cornell University

Athena Coustenis
Paris Observatory

Giuseppe Mitri
International Research School of Planetary Sciences, Università G. d'Annunzio

Gabriel Tobie
Université de Nantes

Federico Tosi
Istituto Nazionale di Astrofisica

The final flythrough of the plume of Enceladus by the Cassini Orbiter on October 21, 2015, marked the end of Cassini's *in situ* investigations of this remarkable small moon of Saturn. Over a decade of flybys and seven flythroughs of the plume, Cassini determined that Enceladus has a global ocean of salt water, organics, and a possible hydrothermal system or systems at the ocean's base. This makes Enceladus a key target for future exploration, as one of the best bodies to search for extant life. The first step of flying again through the plume with more advanced instruments, to do either *in situ* measurements or collect a sample for return to Earth, may be accomplished near-term with a medium-class principal investigator-led mission. More ambitious plans to land, penetrate the ocean, or explore multiple moons will require larger-scale missions, possibly embedded within a thematic "Ocean Worlds" program.

1. RATIONALE FOR FUTURE EXPLORATION DEVOTED TO HABITABILITY AND LIFE

This book is a celebration of a remarkable moon, one that is small and yet geologically active. Beneath its ice surface lies a global water ocean from which material is ejected into space, and which was sampled *in situ* seven times by the Cassini spacecraft. No other planetary mission has made so much progress in going from discovery of geologic activity on an object to the determination that the basic requirements of habitability are satisfied by an environment within that body. And in no other case does future direct sampling for signs of life seem so straightforward.

Enceladus must therefore be considered a primary target in the search for life elsewhere in the solar system. Objections to putting Enceladus first have included (1) the possibility that the tidal heating source supporting the ocean, and therefore the ocean itself, might be too short-lived or episodic to allow life to begin and be sustainable; (2) the notion that Enceladus' ocean is too small in volume to either have had life begin within it or for life to be present today; and (3) the claim that the origin of life requiring a region where water is absent — dry land, in effect — rules out Enceladus as a likely location for life and hence as an attractive target in the search for life.

The lifetime of Enceladus' ocean at the core of objection (1) is an active area of research and the answer may depend upon the details of the rheology of the water ice, orbital evolution of the moons, and possibly other factors — but some choices of parameters do allow for a long-lived ocean (*Shoji et al.,* 2013; *Choblet et al.,* 2017), and even the removal of tidal heating would not result in immediate freezing (*Iess et al.,* 2014); Enceladus may remain active during at least 20–30 m.y. (*Choblet et al.,* 2017). However, more fundamental is that the "requirement" for geologic longevity of a habitable environment to allow life to originate is far from established, and often-quoted required timescales of tens or hundreds of millions of years are speculations that lack an empirical or even theoretical foundation. Objection (2) is based solely on direct analogy with the terrestrial oceanic environment, and to some extent on a comparison with Europa, where the subsurface ocean holds much more liquid water than does Earth's ocean — by contrast, due to the much smaller size of Enceladus, its ocean is only equivalent to one-tenth of the Indian Ocean (*Čadek et al.,* 2016). However, the water/rock volume fraction is much larger than on Europa and Earth, reaching about 0.4 of the total satellite volume, with potentially up to 20% of water content in the porous core (*Choblet et al.,* 2017). While one may expect there to be a minimum ecosystem volume associated with the origin of life and another with habitability, we have no evidence that either of these is close to or larger than the size of Enceladus' ocean. Likewise, objection (3) forms part of a lively, ongoing, and unresolved debate between proponents of the origin of life in hydrothermal systems, at the base of Earth's primitive ocean, and those who favor formation in environments at the interface between liquid water and dry land — or where the liquids dry up and are replenished episodically.

Regardless of where one stands on the above discussion, the more general point is that no environment in the solar system tagged as presently "habitable" closely resembles Earth's inhabited air, water, surface, and subsurface. "Habitable" for the solar system *de facto* means that, were one to introduce terrestrial microorganisms into the environment, they might be viable. Beyond that, little else can be said. Perhaps the only quantitative concern regarding the habitability of the Enceladean ocean is the high pH inferred from Cassini data (*Glein et al.,* 2015), although even here there is disagreement over both the numbers (*Postberg et al.,* 2009) and whether this is a global oceanic value or instead reflects the presence of a hydrothermal complex under the south polar fracture zone. Reflecting instead on the aspects of Enceladus' ocean that make it habitable — salt water, organic molecules, and evidence for water circulating through hot rock — one must conclude that this is an extremely attractive environment to explore for life. And given the accessibility of the ocean via the plume, Enceladus is likely the cheapest and easiest option for a life detection mission among any of the targets in the solar system.

Although much remains to be done to understand the origin, evolution, geology, and oceanography of Enceladus, both the importance of searching for life, and the properties of Enceladus itself, argue for the primacy of a biological goal in a future space mission: to find life. As this chapter describes, either ambitious missions that probe all aspects of the body, or less-expensive missions focused on the subset of objectives related to life, are possible ways to accomplish this. The potential implications of the discovery of biological activity for our understanding of life's place in the cosmos, as well as the impact on public awareness and support of exploration, are so substantial that it would be difficult to imagine an exploration program bereft of the search for life itself.

2. FROM VOYAGER THROUGH CASSINI

Before the Voyager missions in the 1980s, our view of Enceladus improved little from pioneering groundbased observations. Only its orbital characteristics were known, with estimations of its mass, density, and albedo. The Voyager 1 and 2 missions observed Enceladus in November 1980 and August 1981 at distances of about 200,000 km and 87,000 km from the surface respectively (e.g., *Smith et al.,* 1982). The images thus obtained, despite their coarse spatial resolution, were surprising for scientists, since models predicted that such small icy worlds could not possibly sustain activity in their interior. Indeed, parts of Enceladus' surface close to the equator and toward the southern regions appeared quite young, due to the absence of impact craters and the presence of tectonic fractures and ridges. The anomalously high albedo of Enceladus was also another puzzle, which was interpreted as the possible result of fresh frost deposits, suggesting recent eruption activities (*Squyres et al.,* 1983; see also the chapter by Dougherty et al. in this volume).

Thus already the Voyager flybys put to the test the then-established theories on the formation and evolution of icy moons and even allowed scientists to predict the interactions among the satellites and with the rings (*Squyres et al.,* 1983). The demonstration was made that *in situ* measurements were needed to improve our understanding of essential problems related to the outer solar system.

It should then not have come as a surprise that the Cassini mission arriving in the saturnian system in 2004 would further revolutionize our perception of the satellite system — and yet surprised we were. The Voyager flybys had provided a very nice description of the surface morphology, but had never hinted at the south polar activity or the plumes, neither of which were visible during the Voyager flybys (e.g., *Kargel and Pozio,* 1996). The first hints of abnormal activity were provided by the Cassini orbiter's magnetometer, which detected distortion of the magnetic field lines during the first flybys in February and March 2005 (*Dougherty et al.,* 2006). In parallel, the Imaging Science Subsystem (ISS) camera detected an abnormal brightness over the south pole, which was understood only after the first close flyby over the south pole on July 14, 2005, once higher-resolution images were acquired (*Porco et al.,* 2006). Several of the other instruments onboard the orbiter contributed to the discovery of the Enceladus plume and the south polar terrain (SPT) activity (*Brown et al.,* 2006; *Hansen et al.,* 2006; *Spahn et al.,* 2006; *Spencer et al.,* 2006; *Waite et al.,* 2006). During the subsequent flybys, these instruments have monitored the activity and provided important insights on the plume composition, the internal structure, and the composition of the subsurface ocean, and provided key constraints on the origin of the jets.

Rapidly, it was established that the ejecta from the south pole were connected to the fractures and ultimately became the E ring. Whether this jet activity was connected to a subsurface water reservoir had been debated for several years. The debate has been closed by the detection of a global subsurface ocean, first suggested by the gravity measurements (*Iess et al.,* 2014; *McKinnon,* 2015), and then confirmed by imaging observations of Enceladus' unexpectedly large amplitude of physical libration (*Thomas et al.,* 2016).

Plunging through the plumes seven times constitutes a groundbreaking technological achievement for the Cassini Saturn orbiter, giving access to the first direct samples of an extraterrrestrial ocean. The Ion and Neutral Mass Spectrometer (INMS) sampled the gas phase and provided its composition: mostly water vapor with minor amounts of methane, carbon dioxide, and simple organics, possibly resulting from fragmentation of complex organic compounds (*Waite et al.,* 2009). By analyzing grains originating from the E ring and grains directly in the plumes, the Cosmic Dust Analyzer (CDA) revealed that the icy grains emitted by Enceladus contain salts (*Postberg et al.,* 2009, 2011) as well as nanometric silica particles (*Hsu et al.,* 2015), suggesting active hydrothermal processes (*Sekine et al.,* 2015). The occurrence of hydrothermal processes at present was confirmed by the detection of native H_2 (*Waite et al.,* 2017) and appears consistent with the reduced ice shell thickness over the south pole (*Čadek et al.,* 2016; *Choblet et al.,* 2017).

Cassini's *in situ* investigations of Enceladus ended with the discovery of molecular hydrogen (*Waite et al.,* 2017) during a seventh and final flythrough of the plumes on October 21, 2015, after a decade of flybys revealing Enceladus as a key target for future exploration. However, the nature of that future exploration is uncertain. Improvements in both remote sensing and *in situ* instrumentation since the development of Cassini mean that any future mission will provide qualitatively new information on this moon. As a Flagship mission, Cassini was able to bring a broad array of remote sensing and *in situ* instruments that worked together to provide a comprehensive picture of the geology, chemistry, and habitability of Enceladus. Future Flagship missions could greatly extend this comprehensive picture, but given that the next outer solar system large missions [Flagship for NASA; L-class for the European Space Agency (ESA)] are being developed to go to the Jupiter system, and NASA's advanced mission studies at the time of this writing are oriented toward Uranus or Neptune as the next Flagship target based on decadal survey recommendations (*National Research Council,* 2011), the next step in exploring Enceladus will either have to be a less ambitious mission (New Frontiers or Discovery for NASA, medium-class for ESA), or be postponed to the middle of the century. We argue here that medium-class missions, carefully constructed, can provide fundamental new information on Enceladus, particularly regarding the question of whether life exists there.

3. POST-CASSINI: PLUME FLYTHROUGH PROBES

The plume structure and dynamics have been studied by a suite of instruments onboard the Cassini spacecraft during successive close flybys (see the chapter by Postberg et al. in this volume). Cassini investigations provided first insights on the plume structure, the composition of the vapor and icy grain components (also called the dust in the following), their mass ratio, the speed and size distributions of the constituents, and the interaction with the saturnian corotational plasma, as well as on the replenishment of the magnetosphere and E-ring region with fresh plasma and dust particles. Future *in situ* plume investigations by a spacecraft performing multiple flybys or by an Enceladus orbiter will allow for a complete characterization of the plume structure, temporal evolution, and plume material composition, which will be crucial to understand the connections with the subsurface water reservoir and to assess its astrobiological potential.

Although geyser-like plumes and transient water vapor activity have been reported on Triton (*Soderblom et al.,* 1990) and on Europa (*Roth et al.,* 2014; *Sparks et al.,* 2016) respectively, Enceladus is the only icy world in the solar system proven to have continuous endogenic activity, very likely associated with ongoing hydrothermal systems (*Hsu et al.,* 2015; *Sekine et al.,* 2015). Sampling this material freshly erupting from its subsurface offers a unique possibility to analyze an extraterrestrial aqueous environment, possibly hosting life. Several mission concepts have been proposed and studied in recent years to perform such *in situ* analysis. Below we detail the main science investigations and the technical challenges of such mission scenarios.

3.1. Science Investigations by a Plume Flythrough Probe

The Titan Saturn System Mission (TSSM) to Titan and Enceladus was studied by ESA and NASA jointly as a large mission in 2008–2009, prior to some of the recent discoveries that motivate *in situ* exploration of Enceladus and its plume, but would have carried suitable instrumentation for a search for life on both Titan and Enceladus. This concept was the result of the merging of two space mission proposals. In 2007, the Titan and Enceladus Mission (TandEM) was proposed as an ESA Large-Class (L1) mission in response to ESA's Cosmic Vision 2015–2025 Call and was selected for further study (*Coustenis et al.,* 2009). TandEM was eventually merged with NASA's Titan Explorer study (*Lorenz et al.,* 2008) to create the joint ESA-NASA large Flagship Titan Saturn System Mission (TSSM) international concept. TSSM would have had a minimum of seven close encounters with Enceladus (*Reh et al.,* 2009). The space agencies decided to first proceed with a Jupiter system mission as a priority, and this led to more constrained proposals under other programs to perform specific portions of TSSM's investigations.

A proposal in 2011 from the Jet Propulsion Laboratory (JPL) with international collaborations to fly a Discovery-class Saturn orbiter repeatedly past Enceladus and Titan, called Journey to Enceladus and Titan (JET), would have carried a Rosetta-spare mass spectrometer and a near-infrared mapping spectrometer, and would have conducted experiments in gravity mapping of the two bodies (*Sotin et al.,* 2011). The intent was to provide an astrobiological assessment of both Saturn moons, together with Titan mapping and geophysical objectives. The spacecraft power supply was the Advanced Stirling Radioisotopic Generators (ASRGs) then under development but with an uncertain future. The proposal was not selected.

In 2015 a second JPL Discovery proposal, again with international collaboration, refocused the objectives on those that could be accomplished entirely with mass spectrometry. This concept, called Enceladus Life Finder (ELF), was to fly a solar-powered spacecraft into Saturn orbit and to make multiple deep flythroughs of the Enceladus plume (*Lunine et al.,* 2015). The payload consisted of one mass spectrometer to study the gaseous species in the plume, and a second that would have analyzed ice grain spectra, both with an exceptional mass range, sensitivity, and substantial resolution improvement over Cassini. In addition to assessing the habitability of the ocean by determining key parameters such as redox state, pH, and temperature of the ocean, ELF would conduct three tests looking for biological activity in the ocean. The proposal was not selected.

A sample return concept involving a flythrough of the plume and capture of material for return to Earth was also developed at JPL. Called Life Investigation for Enceladus (LIFE), the mission would also sample and analyze the plume *in situ* (*Tsou et al.,* 2012). The concept was too ambitious for Discovery but conceivable as a New Frontiers-class mission. Key challenges surrounding this mission are the ability to capture and preserve in a documentable fashion the delicate ices and volatiles to be collected from the plume, both during the collection process and during return to Earth, and the acceptability of robotically returning a potentially biologically viable sample to Earth against the possibility of a landing anomaly and capsule breach.

There have been parallel efforts in Europe to continue the exploration of Enceladus. In 2013, a white paper proposing a large-class mission concept involving a Saturn Titan orbiter, a Titan balloon, and multiple Enceladus flybys for the further exploration of Titan and Enceladus was submitted in response to ESA's Call for White Papers for the definition of the L2 and L3 missions in the ESA Science Program (*Tobie et al.,* 2014). In 2016, the Explorer of Enceladus and Titan (E^2T) mission was proposed in response to the ESA medium-class (M5) call to assess the evolution and habitability of these saturnian icy moons using a solar-electric-powered spacecraft in orbit around Saturn, performing multiple flythroughs of Enceladus' plume (*Mitri et al.,* 2017). The E^2T mission payload consists of two time-of-flight mass spectrometers and a high-resolution infrared camera.

All these continuing studies and proposals to the agencies demonstrate on the one hand the unrelenting interest of the science community for further exploration of the saturnian moons and in particular Enceladus, and on the other the ingenuity of the engineers in inventing technologically spectacular solutions to achieve this challenging investigation.

Indeed, detailed investigations of Enceladus' plume require multiple passes at low altitudes (<150 km) in order to collect plume material in sufficient quantity. These can be achieved either by a Saturn orbiter performing multiple flybys of Enceladus over its south pole, or by an Enceladus orbiter on polar orbits. For cost reasons, the mission concepts proposed to NASA Discovery calls (JET in 2010, ELF in 2015), New Frontiers (ELF and a similar mission called ELSAH in 2017), and the ESA M-class call (E^2T in 2016) were focused on multiple flyby concepts using a Saturn orbiter. In these multi-flyby mission concepts, the spacecraft's orbit crosses the orbit of Enceladus (and Titan), resulting in 5–10 encounters with the moon. Other concepts intended, but not formally proposed, for Discovery have been published (*Sotin et al.,* 2011).

3.2. Life Detection Approaches on Plume Flythrough Probes

With nanograms of material collectable on each plume flythrough, it is molecules and not cells that are the relevant currency, because few if any of the latter will be present in the plume sample. Because there is no single, "universal" test for life on the molecular level, key to the viability of a plume flythrough probe as a means of determining whether life is present in the ocean of Enceladus is a carefully structured strategy that includes multiple tests for life. Furthermore, this strategy should be relatively agnostic with respect to the biomolecular foundation of enceladean life, although the presence of salty water and organic molecules provides some constraint on the basic building blocks. It is probably a reasonable assumption that amino acids are involved in the structure of any water- and carbon-based life, given the ubiquity of these molecules in a variety of abiotic environments (see the chapter in this volume by McKay et al.). However, one cannot be certain that the particular set of amino acids used by terrestrial life will also be the set used by a putative life form within Enceladus, although some arguments have been advanced that about half the terrestrial amino acids might well be universal for life (*Davila and McKay,* 2014). Similar reasoning holds for hydrocarbons that form the tails of membrane-forming molecules; i.e., one might imagine lipid-precursors occurring as key components of cells in Enceladus as on Earth (*Georgiou and Deamer,* 2014).

Beyond a few possibly essential biomolecules, little more can be said on specific recipes for life, except that there will be certain properties distinct from those of abiotically produced chemical systems. Taking again the example of amino acids, the abundance distribution in abiotic systems is dictated by energies-of-formation and kinetics (*Higgs and Pudritz,* 2009), while that in biological systems must be driven by adaptive utility. Thus, glycine and alanine are the easiest amino acids to make and the most abundant in abiotic systems, in contrast to biological systems. A similar concept holds for the distribution of carbon number in the hydrocarbon tails of lipid or lipid-like cellular membrane molecules. In abiotic systems, hydrocarbon abundance vs. carbon number is distributed in a random fashion, whereas, because biology synthesizes such membrane-forming molecules from a given starting molecule (the "Lego" principle, a term coined by C. McKay), one expects and sees a pattern strikingly different from a Poisson distribution in carbon number: carbon numbers predominantly even, or divisible by five, in terrestrial life forms. The pattern may be different for Enceladus life if a different basis molecule is used, but it will still be distinctly non-random.

Some tests are useful but require supporting measurements to ensure that supposedly biotic signatures really are biotic; one example is isotopic ratios in carbon (*Lunine et al.,* 2015). Others are "classic" life tests but in fact are not by themselves diagnostic: the ratio of left- vs. right-handed chiral molecules, in particular amino acids, can be confounded both by modest enantiomeric excesses in abiotically produced amino acids (*Elsila et al.,* 2016) and by racemization processes in aqueous media (*Monroe et al.,* 2017). However, it is the multiplicity of tests, enabling consistency checks among the outcomes of various tests, that makes it possible to design life detection protocols when one is sampling not cells but the molecular signatures left behind by biology in the Enceladus ocean.

A key issue for the detection of life by plume flythrough is the appropriate speed for plume sampling. At Enceladus, the encounters occur at speeds (>4 km s^{-1}) favorable for icy grain analysis by dust detector, because the grains are broken apart by a plate at the inlet to the particle mass spectrometer. The fragmentation of the molecules within the gas mass spectrometer is potentially advantageous because the resulting fragmentation pattern allows discrimination among species with the same integer mass value (number of protons and neutrons), provided that chemical reactions associated with the collisions do not occur. This in turn depends on the choice of material for the chamber. Cassini INMS measurements during flythroughs ranging from 7.5 to 17 km s^{-1} provide strong empirical documentation on the acceptability of flyby speeds at and below 7.5 km s^{-1} (see supplemental information in *Waite et al.,* 2009). Even with a limited number of flybys, most of the science investigations listed above can be achieved with the exception of a detailed determination of the three-dimensional plume structure and its temporal evolution. An advantage of the Saturn orbiter is that it allows for the sampling of E-ring particles and thus provides an additional way to constrain the composition of icy grains emitted from Enceladus' south pole.

4. ENCELADUS ORBITER MISSIONS

An Enceladus orbiter is a further level up in complexity and hence cost. These missions would use the leveraging tour to lower the spacecraft's Saturn orbit to allow orbital insertion at Enceladus. Because Enceladus has a very small gravity and is very close to Saturn, polar orbits would be unstable, making study of the south polar plumes difficult. The plumes would either have to be studied in the low-speed flybys prior to insertion or from brief excursions from stable orbits around Enceladus. An orbiter would allow detailed studies of the surface and interior of Enceladus up to latitudes of ~65°, providing the opportunity to study the extent of any interior oceans and the surface history of this moon, but would not provide useful constraints on the plume composition or structure.

Further complicating the orbiter story is the need to avoid contaminating the surface of Enceladus, since the fractures in the south polar region likely connect directly or indirectly to the ocean beneath (*Kite and Rubin,* 2016). Unlike a Saturn orbiter, which can be disposed of straightforwardly in Saturn's atmosphere or by collision with a geologically inactive moon, an Enceladus orbiter must be brought back into Saturn orbit, which is energetically difficult, not to mention potentially risky should a failure occur prior to or during deorbit. One alternative, which is to sterilize the spacecraft to the level at which entry of components into the Enceladus ocean might be acceptable, is very expensive. Another, which merits consideration, is to impact the orbiter in the ancient terrains in the equatorial region around the subsaturnian point where the crust is thick (30–40 km) (*Čadek et al.,* 2016; *Beuthe et al.,* 2016; see also the chapter by Hemingway et al. in this volume). In either case, an Enceladus orbiter could be classified as a Category IV mission from the planetary protection point of view: a mission operating in, or eventually ending up in, an environment capable of hosting life.

5. SAMPLE RETURN MISSIONS

To ascertain the habitability conditions and potential for life on Enceladus is a complex undertaking. *McKay et al.* (2014) points out that it is "unlikely that we can plan in advance the analytical steps in an organic and life search in the plume of Enceladus because the subsequent steps in the search will depend on the results of the previous steps." It is argued that a sample return mission to Enceladus with the goal of collecting samples of the volatile-rich plume and returning them to Earth for detailed analysis would be the best way to look for life given the uncertainties in our knowledge of what such life might be like. Samples returned to Earth can be analyzed and repeatedly studied by multiple laboratories with different technologies and state-of-the-art instrumentation not available at the time of mission instrument design. For example, a sample return mission that collected cometary coma particles was shared with over 175 scientists all over the world (*Sandford et al.,* 2011). Recently, this collaborative study of cometary coma particles confirmed the presence of cometary glycine after an international three-year effort (*Elsila et al.,* 2009), demonstrating that at least some amino acids from an extraterrestrial body can be successfully returned to Earth in a flyby mission without special preservation techniques (*Tsou et al.,* 2012). In addition, samples can be maintained for future developments in instrumentation and laboratory techniques. Furthermore, important results can be replicated by independent techniques and anomalous results can be checked without the need of a new mission. On the other hand, the utility of such a returned sample will depend on the ability to preserve its isotopic, molecular, and stereochemical composition, which places extraordinary requirements on the conditions during retrieval, return cruise, entry, and subsequent handling and curation. These requirements are different from those levied on the handling of lunar rocks, and even the Stardust comet samples, which were not expected to contain delicate or highly volatile material.

Sample return missions have a relatively long history in space exploration. The first sample returns of lunar soil and rock samples to Earth were achieved by the Apollo missions

(11, 12, 14, 15, 16, and 17). The Soviet Union also conducted sample return missions to the Moon: the Luna 16, 20, and 24 missions. Following a 20-year absence of this type of mission, multiple sample return missions to study a number of solar system bodies were launched utilizing new technologies, including the NASA Stardust mission (1999; returning cometary coma dust and particles from Jupiter-family comet Wild 2 and interstellar dust stream) and the NASA Genesis mission [2001; returning solar wind dust from the Earth-Sun Lagrange 1 (L1) point] (*Brownlee et al.,* 2004; *Burnett et al.,* 2003, 2013). Stardust was the first sample return mission to travel beyond the Moon. Figures 1 and 2 show the sample return capsules recovered from the Stardust mission and from the Genesis mission. Unfortunately, the Genesis capsule crashed upon landing in 2004 due to technical difficulties and was severely damaged, liberating some of its contents into the surrounding terrestrial environment and contaminating the samples themselves. Some samples, although contaminated by impact debris, were still retrieved and studied; however, had the capsule contained volatiles, the information would have been lost. Had it contained biologically active agents, these might have been released into the terrestrial environment.

The Japan Aerospace Exploration Agency (JAXA) launched Hayabusa 1 in 2003 to sample the surface of asteroid 25143 Itokawa and returned to Earth in 2010, making it the first mission to perform a sample return mission on a solar system body other than the Moon. Unfortunately, Hayabusa 1 encountered many technical difficulties, but even though its sampling mechanism did not work, at least 1534 particles in the 1–200-μm range were recovered by accident when the spacecraft landed briefly on the surface of the asteroid in 2005 (*Yano et al.,* 2006; *Tsuchiyama et al.,* 2011). In 2014, JAXA launched the improved Hayabusa 2 to obtain samples from the carbon-rich asteroid 162173 Ryugu with the intention of returning asteroid samples by 2020 (*Tsuda et al.,* 2013). In 2016, NASA launched its first asteroid sample return mission, the Origins, Spectral Interpretation, Resource Identification, and Security Regolith Explorer (OSIRIS-REx), a New Frontiers mission to obtain samples from the near-Earth asteroid 101955 Bennu; samples are expected to be returned to Earth by 2023 (*Lauretta et al.,* 2012).

Since the lunar sample return missions of Apollo and Luna, there have been great advances in sample-collection processes. Perhaps the most compelling technology for an Enceladus plume sample and return mission, which would require high-speed passes through the plume, is an advanced material used in the Stardust and OSIRIS-REx missions to decelerate and capture material, a silicon-based product that is 99.8% air called aerogel. The aerogel prevented the silicate grains and volatile organic particles collected by Stardust from being destroyed as the particle's kinetic energy was converted to heat, at impact speeds of ~6 km s^{-1}. An array configuration of aerogel allows for the collection of samples to be compartmentalized for a specific flyby or orbit. Depending on projected sample capture velocities, aerogel densities can be customized for the flyby or orbit (*Carr et al.,* 2013). Indeed, the LIFE sample return concept studied (but not proposed) for NASA's thirteenth Discovery mission in 2015 suggested reducing the aerogel density by a factor of 5 compared to the Stardust aerogel density to reduce initial shock energy and enable a less-destructive sample capture (*Tsou et al.,* 2012). Figure 3 shows the aerogel collector array used in the Stardust mission. When a particle hits this substance, it penetrates the material, creating a track that scientists use to find the minuscule particles as shown in Fig. 4. Tracks lengths depend on the size and structure of the impacting particle (*Sandford et al.,* 2011).

Given the success of previous sample return missions, the accessibility of subsurface ocean materials offered by Enceladus' eruption activities, and improved sample acquisition technologies, a sample return mission to Enceladus seems opportune. In 2007, the NASA-funded Titan and Enceladus Mission Feasibility Study (*Reh et al.,* 2007) and

Fig. 1. Genesis sample return capsule recovered at Utah Test Training Range, USA, in 2004. Image courtesy of NASA/JPL.

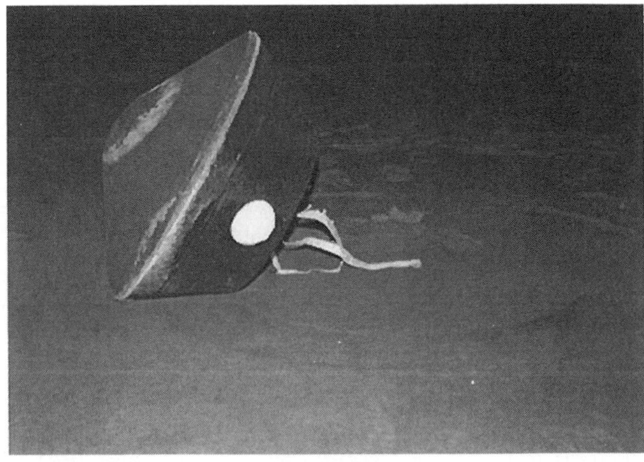

Fig. 2. Stardust sample return capsule recovered at Utah Test Training Range, USA, in 2006. Image courtesy of NASA/JPL.

the Enceladus Flagship Mission Concept Study (*Razzaghi et al.,* 2007), as well as the 2010 Enceladus mission study funded by the Satellites Panel of the National Research Council (NRC) Planetary Science Decadal Survey (*Adler et al.,* 2011; *Moeller et al.,* 2011), studied the option of a sample return mission to Enceladus. All concluded that while the science return would be high, the risks were also high for several reasons: (1) the sample capture velocities of the spacecraft estimated to be between 7 and 10 km s^{-1} for a free-return single flyby mission make preservation of the volatiles difficult; (2) the long mission duration assumes a roundtrip time of >18 years; and (3) returned sample handling issues include the need for curation facilities capable of safely handling alien life; in fact, sample return missions to Mars, Europa, and Enceladus are considered to be a Restricted Category V mission under the Committee on Space Research (COSPAR); and (4) the high cost of the mission. In addition, as *Reh et al.* (2007) pointed out, there is a nonzero possibility that a sample return mission to Enceladus could return without any useful samples.

There were three types of trajectories considered for the Enceladus sample return mission: (1) free-return with a single plume passage, including a Saturn orbiter with free-return sample return and an Enceladus orbiter with free-return; (2) a Saturn orbiter that would permit multiple plume passages; and (3) an Enceladus orbiter that would include multiple plume passages. While a Saturn or Enceladus orbiter would allow multiple plume passages at lower sample capture speeds, the mission durations were longer than the estimated 18–26 years for a free-return mission, and the mission complexity and cost increased.

Despite the risks, *Tsou et al.* (2012, 2014) have shown that a sample return mission to Enceladus could be viable and might overcome many of the technical hurdles that were deemed high risk by previous feasibility studies. The LIFE mission studied in 2015 would use a Saturn orbit trajectory with gravity assists from Titan, which would allow for multiple plume passages in order to achieve sample capture velocities as low as 3–4.5 km s^{-1} (*Tsou et al.,* 2012). The proposed mission would also sample particles in Saturn's E ring and Titan's upper atmosphere. In order to minimize sample damage, LIFE would employ several strategies: (1) reduce aerogel density to reduce initial shock energy; (2) maintain samples at low temperatures (<150 K) and pressure, and (3) use a volatiles trapping and sealing deposition collector with Hayabusa heritage. In order to deal with planetary protection issues, LIFE could take advantage of a proposed construction by the Japan Agency for Marine-Earth Science and Technology (JAMSTEC) of a Bio Safety Level 4 facility in international waters for future sample return missions to bring samples from possible habitable bodies in the solar system onboard the research ship *Chikyu* (*Takano et al.,* 2014). While previous mission durations were predicted to be long, LIFE has a 13.5-year roundtrip trajectory, including an approximately 8-year journey to Saturn with another 5-year trip back to Earth following sample collection. In order to reduce the size of the launch vehicle and the mission duration, LIFE would need to take advantage of a Jupiter-gravity-assist opportunity, which would have required it to be launched by 2019; unfortunately, the next opportunity for a Jupiter gravity assist is in 2058. The astrobiological community continues to be interested in an

Fig. 3. Silica aerogel array used in Stardust mission to investigate cometary coma of Wild 2, a Jupiter-family comet. Image courtesy of NASA/JPL.

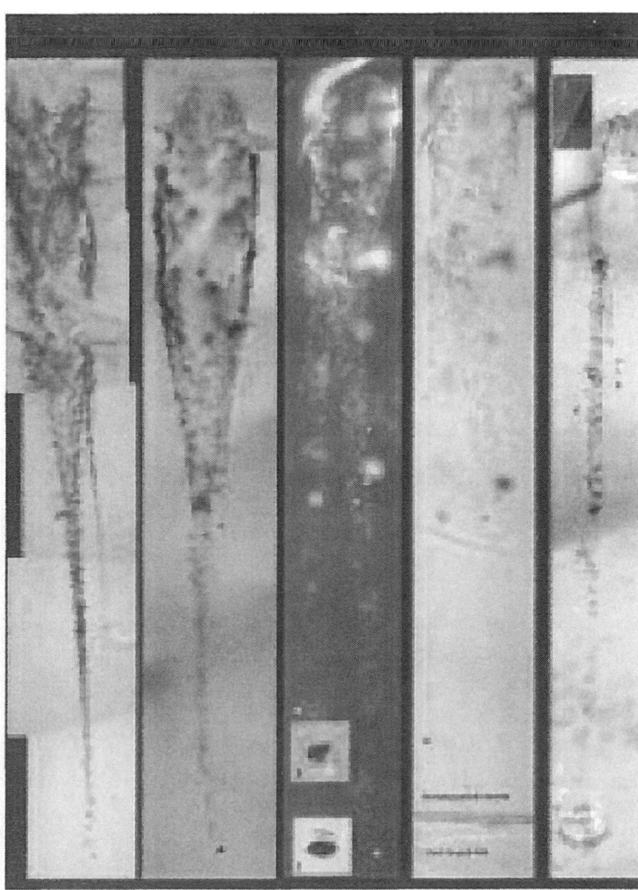

Fig. 4. Tracks of comet particles trapped in silica aerogel from the Stardust mission. Image courtesy of NASA/JPL.

Enceladus sample return mission given its high science potential and the success of previous sample return missions.

6. SURFACE LANDERS AND SUBMARINE MISSIONS

A lander mission to Enceladus' SPT would be an ideal way to investigate some of the processes at work on this geologically active body (see the chapter by Patterson et al. in this volume). A lander would be able to chemically analyze the surface, including the fallout of ejected plume material; directly observe cryovolcanic processes and tectonism; as well as investigate the processes occurring in the subsurface, with instrumentation such as seismometers, a magnetometer, and/or a ground-penetrating radar (GPR). Furthermore, a lander would be able to provide "ground truth" for remote sensing measurements. In addition to geological activity, Enceladus is considered one of the best candidates in the solar system for extraterrestrial habitability due to its warm organic-rich subsurface ocean. A lander mission would be uniquely qualified to investigate habitability thanks to its ability to sample and conduct *in situ* astrobiology analysis near or close to the jets' source.

The eruption activity of Enceladus offers a unique opportunity to easily sample fresh material emerging from a salty subsurface ocean in a flyby or in an orbit around Enceladus, as has happened with the Cassini mission, which was limited by its low mass range and resolution and inability to identify complex molecules. A new flyby or orbiter mission to Enceladus would be able to analyze plume constituents at higher resolutions and mass ranges and identify important biomarkers. However, the best way to assess the biological potential of Enceladus would be to sample the plume subsurface liquids before potential biosignatures are degraded or destroyed by exposure to the colder temperatures and UV radiation, gamma and cosmic rays present on or near the surface and in outer space.

In order to sample the subsurface ocean, the lander would need a means of accessing either (1) the subsurface ocean directly, or (somewhat easier) (2) the near-surface water-bearing fractures surrounding the active plume vents (see the chapter in this volume by Spencer et al. and Hemingway et al.). *Porco et al.* (2014) suggest that the top of the liquid water table within the fractures or cracks supplying the jets at the SPT may lie within a few kilometers of the surface, while liquid water in small quantities may in fact reach all the way to the surface (Fig. 5). *Čadek et al.* (2016), analyzing the data of topography and gravity (*Iess et al.,* 2014) returned by the Cassini mission, have indicated that the depth of the liquid reservoirs in the SPT is less than 5 km. Either way, an autonomous water-ice penetrator, also termed a "cryobot" (*Zimmerman et al.,* 2001; *French et al.,* 2001), capable of penetrating large distances through the ice via heat-induced melting and gravity, must be used in concert with the lander in order to sample the subsurface. A lander could be delivered to the SPT by either a flyby or orbiter mission or even as a combined orbiter-lander.

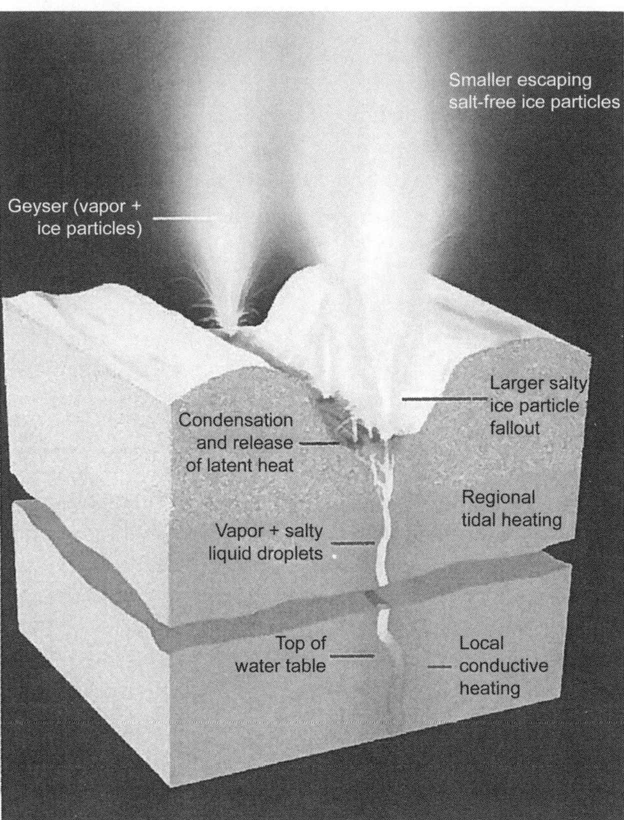

Fig. 5. Conceptual model of fractures surrounding the active plume vents on Enceladus (*Porco et al.,* 2014).

A lander mission would be classified as a Category IV mission from the planetary protection point of view, but unlike an orbiter (section 5), could not be removed from Enceladus and would therefore need to be sterilized to extremely high levels to avoid contamination of the ocean by terrestrial organisms.

6.1. Challenges for a Lander Mission

A lander mission to Enceladus' SPT would be challenged by the rugged topography and surface properties present on SPT as well as by Enceladus' low gravity. This geologically active region, surrounded by a continuous chain of folded ridges and troughs, is dominated by four linear depressions known as the Tiger Stripes hosting ~100 distinct narrow jets or geysers (*Porco et al.,* 2014). Figure 6 shows rugged terrain typical of the SPT with close-up views of one of linear depressions making up the Tiger Stripes, Damascus Sulcus. Damascus Sulcus features two large parallel ridges 100–150 m high with a medial trough 200–250 m deep between them. This trough is the primary source of numerous jets in Damascus Sulcus. As seen in Fig. 6, the plains between the depressions making up the Tiger Stripes present a rugged terrain resembling crumpled or folded topography. The landing spot would likely be one of the valley floors of the Tiger Stripes and need to be a safe distance away from active vents, but not too far if the intention is to sample subglacial

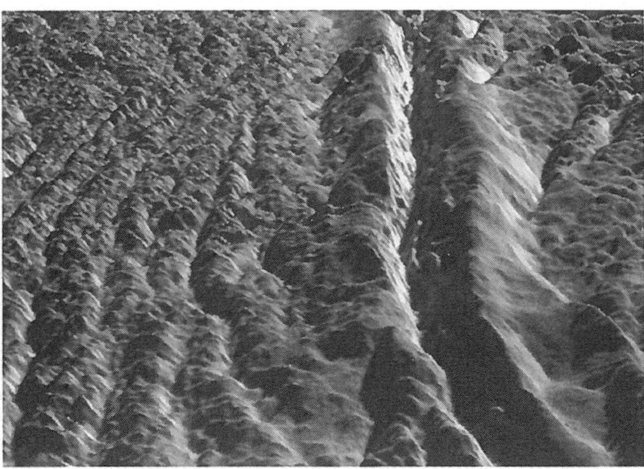

Fig. 6. Damascus Sulcus on Enceladus. This perspective view was generated using high-resolution images of Enceladus acquired in August 2008 at 12 to 30 m resolution, together with a new topographic map of the region produced by P. Schenk. For scale, the width of Damascus Sulcus, which cuts through the image right of center, is roughly 5 km across from ridgetop to ridgetop. The ridges are each 100 to 150 m high, while the medial trough between the ridges is 200 to 250 m deep. Relief has been exaggerated by a factor of 10 for clarity. PIA image 12207, image courtesy of NASA/JPL/Space Science Institute/Universities Space Research Association/Lunar and Planetary Institute.

water-bearing fractures depending on the capabilities of the cryobot probe. Thermal emissions and subglacial morphology obtained from either (1) an orbiter or (2) previous mission data could be used to determine the best landing site in terms of identifying the most accessible water-bearing fractures. In addition, images showing evidence of fresh material from plume fallout can be used. Prior to landing, the lander would need to locate an area relatively free of steep slopes, boulders, and other hazardous terrain. This would require high landing accuracy and the ability of the lander to autonomously detect and avoid hazards, given that communication with an orbiter is compromised due to signal delays resulting from communication distances and landscape obstacles. Thus, remote control of the landing is not a viable option.

The surface near the plume accumulates the greatest amount of ejected plume fallout, approximating super-fine snow, and the larger grains tend to accumulate nearer to the plumes (see the chapter by Goldstein et al. in this volume). The particle deposition rates can reach up to 0.5 mm yr^{-1}, which implies a fallout thickness layer of possibly tens of meters, assuming the plumes have been active in the past million years (*Kempf et al.,* 2010). A thick snow-like, or even "cotton-candy," surface cover could require the lander feet to have a very large surface area while the possibility of touching down on solid ice would indicate the need for ice screws in the lander's feet similar to that of Rosetta's Philae. Given that surface properties can range from fine snow to solid crustal ice, the uncertainty regarding surface properties in the SPT must be considered in mission planning. Cassini Composite InfraRed Spectrometer (CIRS) measurements have shown that most of Enceladus' heat emission emanates from the Tiger Stripe fractures in the SPT (*Spencer et al.,* 2006; *Spitale and Porco,* 2007; *Howett et al.,* 2011); the greatest thermal emission along these fractures correlates with regions undergoing the most active jetting and undergoing the greatest tidal stresses (*Porco et al.,* 2014). These hot spot regions have temperatures approaching 200 K (*Howett et al.,* 2011). Thus the lander and its instrument payload will need to be able to accommodate the widely varying temperature changes as it travels from the inner solar system to a valley within one of the Tiger Stripes. Radiation levels on Enceladus' surface ice are much less than on the Galilean moons, particularly Europa, which has the most extreme radiation environment of any icy body in the solar system.

Low gravity levels present on Enceladus dictate that multiple bouncing of the lander should be minimized. Low gravity levels in combination with geological activity around the landing site would require that anchoring of the lander may need to be considered. The trace atmosphere on Enceladus means that an aeroshell is unnecessary. Finally, strict planetary protection measures must be incorporated into any lander mission design to ensure that lander/probe measurements reflect conditions on Enceladus rather than terrestrial contamination, and that terrestrial microbes are not present to contaminate Enceladus' habitable environments.

6.2. Lander Types

There are several types of landers that could be considered for a mission to Enceladus: (1) soft lander, (2) hard lander, and (3) mobile lander. However, before considering what type of lander should be used, the overall system architecture of the lander mission needs to be considered. The first consideration is whether the lander should be part of a flyby mission or an orbiter. An orbiter mission could either consist of a separate orbiter and lander or else a single-element orbiter/lander similar to the Hayabusa mission in 2003, which used an orbiter/lander in a sample return mission to the Itokawa asteroid. In the first instance, an orbiter around Saturn or Enceladus would act as communication relay to Earth. In the second instance, the combined orbiter/lander would need to communicate directly with Earth. A drawback of a flyby mission and a combined orbiter/lander mission is that the lander would have to communicate directly with Earth. Direct communication can be complicated by the fact that Earth may be visible only at low elevation angles; in addition, local topography near the landing spot could comprise the signal. A separate orbit/lander mission ensures more favorable communication with Earth and allows the orbiter to gather information necessary to determine the optimum landing site of the lander.

Soft landers use some type of deceleration mechanism such as parachutes, thrusters, or airbags to slow the descent velocity to ensure a soft landing. A hard lander, also known as an impactor or penetrator, achieves a hard landing with the surface via impact and places itself at some depth in the subsurface. The majority of missions have used soft landers. The classical lander, similar to Rosetta's Philae, is a soft lander with landing legs or platform (Fig. 7). Hard landers such as penetrators were used in the Russian Mars 96 mission and NASA's 1999 Mars Polar Lander without success. Another option for soft landers is mobile landers. The science return of any lander mission can be increased with mobility. A mobile lander not only is able to access multiple sites of interest, but may also have the option to change its initial landing site should it prove not to be optimum. Due to the low-gravity environment of Enceladus, a lander with wheels may not be able to generate enough friction at contact with the ground to cause motion; however, for low-gravity environments a propulsion system or mechanically triggered hopping may provide alternate means of mobility (*Ulamec and Beale,* 2006). PROF-F of the Russian Phobos-2 mission in 1988 was the first, although ultimately unsuccessful, attempt at a mobile lander on Mars, and it relied on a hopping mechanism for mobility. The 2003 Hayabusa probe, Minerva, although also not successful, relied on a hopping method of mobility. Learning from previous lessons of the first Hayabusa, Hayabusa 2 in 2014 included a mobile lander called the Mobile Asteroid Surface Scout (MASCOT), and built by the German Aerospace Center (DLR) in cooperation with the French space agency Centre National d'études Spatiales (CNES); Hayabusa 2 is expected to reach the asteroid 162173 Ryugu in 2020. In 2011, NASA chose the Comet Hopper (CHopper) as one of three Discovery program semi-finalists, although ultimately the Interior Exploration using Seismic Investigations, Geodesy and Heat Transport (InSight) mission to Mars was selected. Had it been selected, CHopper would have orbited and landed multiple times on Comet 46P/Wirtanen (*B. C. Clark et al.,* 2008). Another proposed mission to the NASA Discovery program that included a mobile lander was the Mars Geyser Hopper (MGH). MGH would have performed two surface hops after landing in the south pole of Mars to investigate the springtime geyser phenomena (*Landis et al.,* 2012).

Fig. 7. Rosetta Philae Lander artistic concept of nominal touchdown. Image courtesy of the European Space Agency.

6.3. Lander Power

The power technology both MGH and CHopper had intended to use was an ASRG. MGH was a design reference mission to study this technology to clarify the issues involved in the use of an ASRG power source in a planetary surface mission. The ASRG would be able to provide an adequate power supply despite the low Sun angle of the polar environment and the period of low sunlight during the martian winter (*Landis et al.,* 2012; *B. C. Clark et al.,* 2008). However, in 2013 NASA canceled further work on ASRG due to technical and budgetary issues. Whether a standard radioisotope thermoelectric generator (RTG), with several times the heat output per watt of power generated compared to the ASRG, could be used instead is an open question.

While solar power has been used for many landers, a drawback of a lander mission on Enceladus, particularly one that would require extra power necessary for mobility is the low insolation of Enceladus, as solar panels large enough to produce necessary power may be too large and bulky for such a mission. Battery power for a lander would likely limit the lifetime of such a mission to days or weeks given current technology.

6.4. Lander Probes and Instrumentation

Investigation of cryobot-type probes, small steerable robotic probes capable of penetrating the ice by melting it and bringing scientific instrumentation in contact with water-bearing fractures or a subsurface ocean, have been investigated by both ESA and NASA and are currently being tested in extreme environments on Earth such as subglacial liquid bodies, deep-sea hydrothermal vents, and underwater cave and karst systems. Building upon a tethered ice-melting Philbert probe invented in the 1960s, engineers at JPL investigated a Cryo-Hydro Integrated Robotic Penetrator System (CHIRPS) for subsurface ice and water investigations with a focus on Mars' polar caps and Europa. This melt probe was the basis of a 2007 Mars north polar cap penetration mission proposed, unsuccessfully, under the Mars Scout Program (*Zimmermann et al.,* 2001, 2002). In 2012, the German Aerospace Center's (DLR) Enceladus Explorer (EnEx) program began investigation of the IceMole subglacial probe to be used in a potential Enceladus lander mission (*Konstantinidis et al.,* 2015). The IceMole was successfully able to extract an uncontaminated, subglacial water sample and bring it back to the surface in Blood Falls glacier in southern Antarctica in 2014, breaking through the glacier crust to penetrate into the liquid water underneath (Fig. 8). Both the CHIRP and IceMole designs would use nuclear power in an interplanetary mission; however, an alternate method of power has been proposed for these cryobots. Another ice probe design being investigated is the Very-deep Autonomous Laser-powered Kilowatt-class Yo-yoing Robotic Ice Explorer (VALKYRIE), which is powered by a 5-kW laser carried through a fiber optic wire and is being developed with support from NASA. VALKYRIE has

Fig. 8. IceMole Probe during testing in Antarctica. Image courtesy of the German Aerospace Center (DLR).

undergone testing in Matanuska Glacier, Alaska, in 2014 and 2015. The next step in this process is to build on this design to test in Antarctica (*Stone et al.,* 2014).

7. TOURING THE OTHER SATELLITES OF SATURN

Exploration of Saturn's satellites was among the main objectives of the Cassini-Huygens mission, and while Enceladus proved to be perhaps the most fascinating in terms of its activity, the other moons proved interesting in their own right (see the chapter by Schenk et al. in this volume). Cassini's tour of the saturnian moons began in July 2004, and the satellites were investigated in many flybys during the nominal mission, which ended in 2008, then during the first extended mission ("equinox mission") between 2008 and 2010, and finally during the second extended mission ("solstice mission") between 2010 and 2017. This overall campaign enabled a thorough coverage of those surfaces, including the northern parts, that were not illuminated during the nominal mission.

Because of their importance with respect to the overall scientific objectives of the Cassini mission, a total of 126 and 22 close targeted flybys occurred at Titan and Enceladus, achieving minimum altitudes of 880 km and 25 km, respectively. However, most of the mid-sized moons were investigated by the Cassini Orbiter at least once at relatively close range (<5000 km): Tethys (one close flyby at minimum altitude of 1500 km), Dione (five close flybys, minimum altitude 99 km), Rhea (four close flybys, minimum altitude 76 km), Hyperion (one close flyby, minimum altitude 500 km), and Iapetus (one close flyby, minimum altitude 1227 km). As a result, atlases of the medium-sized icy moons Mimas, Enceladus, Tethys, Dione, and Rhea were produced at a resolution of 1:1,000,000, 1:1,000,000, 1:1,000,000, 1:500,000, and 1:1,000,000, respectively (*Roatsch et al.,* 2009, 2013). A mosaic map of Iapetus was also generated in a scale of 1:3,000,000 (*Roatsch et al.,* 2009).

Among the smaller icy moons, the Cassini flyby of Phoebe was a unique and remarkable opportunity because (1) it was the very first close look at a giant planet's outer irregular satellite; (2) Phoebe is believed to have accreted in the volatile-rich outer solar nebula where the Kuiper belt objects (KBOs) originated (*Johnson and Lunine,* 2005), thus representing the very first close observation of a former KBO; and (3) it occurred even prior to Cassini-Huygens's orbit insertion at Saturn. Cassini made numerous non-targeted observations, providing new or improved coverage with respect to previous Voyager exploration, including observations of Mimas, Polydeuces, Methone, Pallene, Janus, Calypso, Helene, Telesto, Epimetheus, Prometheus, Aegeon, Pandora, Daphnis, Pan, and Atlas.

As a key target unto itself for the Cassini-Huygens mission, Titan, the largest moon, was extensively explored. Its dense and thick nitrogen-rich atmosphere hides an incredible world made up of seas and lakes of liquid methane and ethane, hydrocarbon-rich sand dunes, and possible cryovolcanic features. While this is a huge leap compared to the Voyager era, results obtained by Cassini-Huygens on Titan have already shaped future mission concepts with a set of investigations meant to cover specific remaining gaps (e.g., *Tobie et al.,* 2014; *Mitri et al.,* 2014, 2017). The key goals were to unveil and map the surface composition (inventory of organic constituents and presence/absence of ammonia), which provide clues for the origin of life; shed light on the methane cycle and the methane reservoirs; assess whether cryovolcanism and tectonism are actively ongoing or are relics of a more active past; clarify the existence of a magnetic field and of a subsurface ocean; and shed light on the chemistry that occurs in the upper atmosphere between 400 and 900 km, which remains poorly explored after Cassini. In addition, seasonal changes of the atmosphere at all levels, and the long-term escape of constituents to space, are so far only marginally understood. Finally, the presence of a deep internal liquid-water ocean and surface seas and lakes of liquid methane and other hydrocarbons renders Titan an important target in the search for life. While communication between the deep ocean (50–100 km below the surface) and the surface is questionable, sampling of the surface sea or lake environments to test the limits of life (in this case, whether life can evolve and survive in a non-aqueous environment) is an important future goal (*National Research Council,* 2007).

Future scientific investigations on the mid-sized icy moons should include, but not be limited to, (1) improving high-resolution coverage in those regions where this is still relatively limited, (2) confirming previous Cassini measurements where necessary, and (3) shedding light on processes that remain unclear. We close this section with a brief discussion of some of the outstanding issues for these moons.

Data returned by the Visual and Infrared Mapping Spectrometer (VIMS) (*Brown et al.,* 2004) indicate similar spectral properties throughout the Saturn system on icy surfaces with observed gradients in abundances. The surface composition of the icy moons is overall dominated by ice, nanophase and fine grains of metallic iron, nanophase hematite, and smaller amounts of carbon dioxide, organic compounds, possible trace ammonia, and yet to be identified compounds (*R. N. Clark et al.,* 2008; *Clark et al.,* 2012). The outer satellites have more dark material and CO_2 on their surfaces, while inner satellites and the rings contain little to no detectable CO_2 and less dark material (*Clark et al.,* 2012; *Filacchione et al.,* 2013). Fortuitous measurements of the saturnian satellites other than Titan, carried out by Cassini's RADAR (*Elachi et al.,* 2004), confirmed this trend down to depths of at least one to several decimeters (*Ostro et al.,* 2006). While CO_2 may be synthesized by radiolysis of water-derived O atoms with endogenic and/or implanted organics (*R. N. Clark et al.,* 2008; *Cruikshank et al.,* 2010), the process responsible for a greater abundance of dark, organic-rich contaminants in the outer saturnian system is not known.

Iapetus is particularly puzzling. Since its discovery, Iapetus is known to show a dichotomy in brightness between its leading hemisphere (very low albedo, compatible with carbonaceous chondrite material) and its trailing hemisphere (albedo much higher and consistent with relatively pure water ice). Cassini data, aided by the telescopic detection of a dust ring located on Phoebe's orbit (*Verbiscer et al.,* 2009), were able to substantially confirm that Phoebe is one source of the dark material coating the leading side of Iapetus (e.g., *Tosi et al.,* 2010), and that this coating is almost everywhere no thicker than a meter (*Denk et al.,* 2010). Moreover, exogenically triggered global thermal segregation of bright and dark material on Iapetus is a likely explanation for both the extreme amplitude and the shape of Iapetus' albedo dichotomy (*Spencer and Denk,* 2010). However, the composition of the dark material of Iapetus as inferred from Cassini's VIMS instrument (*Clark et al.,* 2012) is not fully addressed. For example, there is evidence that at least two separate events were responsible for the darkening (*Dalle Ore et al.,* 2010). Detection of ammonia still awaits confirmation from future measurements carried out at higher spatial and spectral resolution. Multiple flybys of Iapetus on its leading hemisphere at distances closer than the one achieved by Cassini in its single flyby, using modern *in situ* instruments, would be key to quantifying the magnitude of the exogenous contamination not only on Iapetus but also on inner satellites such as Titan and Hyperion.

Cassini data suggested that Dione, the second-largest inner moon of Saturn, might have been as active as Enceladus in the past and could possibly be active now, but to a smaller extent than Enceladus. Dione's trailing hemisphere is marked by an extensive network of troughs and bright lineaments that stand out with respect to a relatively dark background. These features, once known as "wispy terrains," are in fact bright ice cliffs, some of which are several hundred meters high, created by tectonic fractures (chasmata), therefore indicating past global tectonic activity. Using stereo-generated topography, it was possible to reveal ancient, inactive fractures at Dione similar to those seen at Enceladus that currently spray water ice and organic particles (*Hammond et al.,* 2013). Folds in the crust in one mountainous region suggested that the ice once had to be "hot," suggestive of an ocean under the crust at the time when the reliefs are formed. In parallel, results obtained by the Cassini Plasma Spectrometer (CAPS) (*Young et al.,* 2004) indicated that Dione is ejecting streams of particles into space, therefore establishing it as an important source of outward-flowing plasma in Saturn's magnetosphere (*Burch et al.,* 2007). Finally, data returned by CIRS (*Flasar et al.,* 2004) show hemispheric bolometric Bond albedo asymmetries on Dione, which deserve further investigation (*Howett et al.,* 2010).

8. PUTTING THE MISSIONS TOGETHER: AN ENCELADUS EXPLORATION TIMELINE/DECISION TREE

While we have considered the various mission types in isolation from each other, in fact it is possible to put together an exploration program for Enceladus that involves a sequential use of missions of increasing complexity that build on the results of previous ones. An example of such a program from *Sherwood* (2016) is shown in Fig. 9. The flythrough orbiter represents both Cassini and a possible follow-on plume mission designed to make *in situ* measurements to provide more detail on habitability and to look for life. Positive results from such a mission would then stimulate a sample return mission, or possibly a lander or *in situ* submarine. The *in situ* submarine might precede the sample return if the aim is to know the level of sophistication of enceladean biology (in order, for example, to protect Earth, should returned samples get dispersed), or might follow the sample return mission based on the much greater technological challenge of the submarine mission. Whether a lander or submarine would be used might also be a technological decision, supplemented by an assessment of the feasibility of collecting samples from a lander given the variable locations of individual jets. Alternatively, should a plume flythrough mission provide a compelling case against life and even against habitability (that, for example, the whole ocean has very high pH), one might truncate the roadmap for Enceladus and focus on other solar system bodies. For example, the results of a plume flythrough mission could be gotten at roughly the same time the Europa Clipper mission determines the habitability of Europa.

While the logical flow of such a roadmap is inarguable, the timescale to implement it is sobering. Without a large launch vehicle like the Space Launch System (SLS), a coupled plume flythrough probe followed by sample return would stretch into the middle of the century. However, absent a compelling case for a Flagship mission to be developed now, so that it could be launched in the mid-2020s, it seems that Enceladus exploration will have to be a multi-decade affair.

9. ENCELADUS EXPLORATION IN THE CONTEXT OF AN OCEAN WORLDS EXPLORATION PROGRAM

We are at the beginning of a new era in the exploration of the outer solar system. Spacecraft have visited each of the giant planets and made detailed observations of the satellites

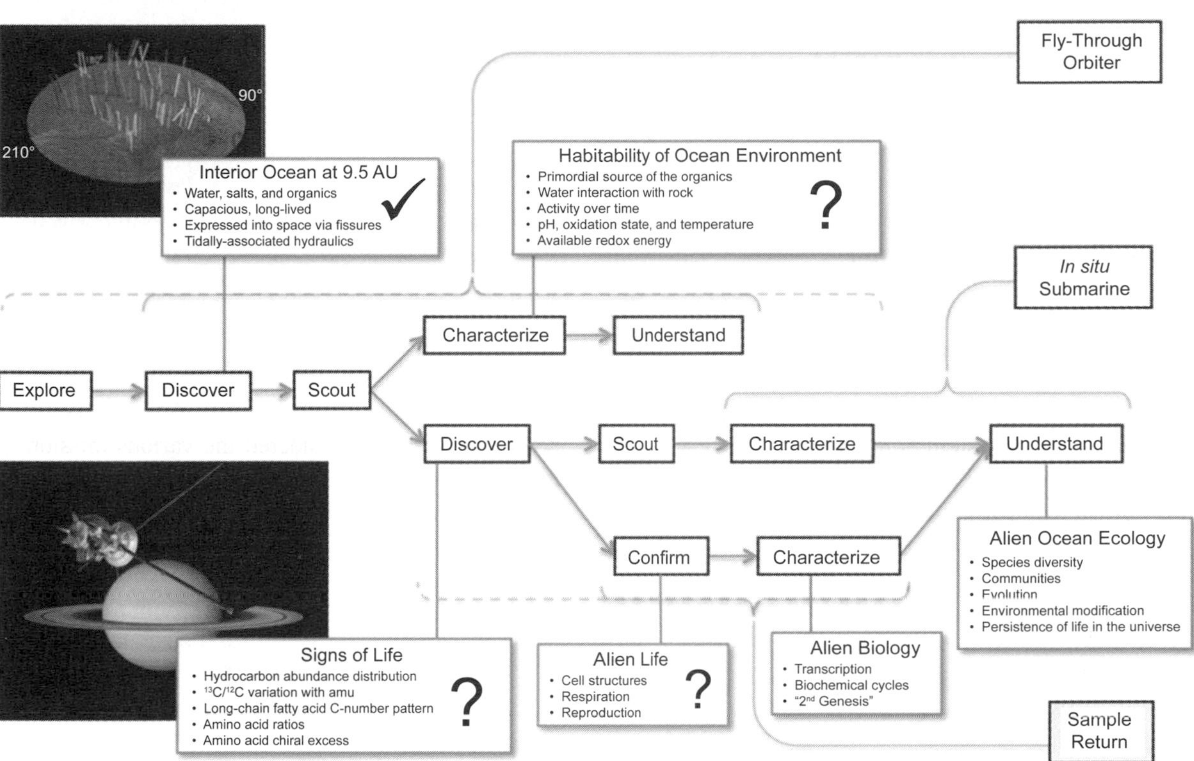

Fig. 9. Example roadmap for exploring Enceladus, in which successive positive results lead to ever more ambitious follow-up missions. From *Sherwood* (2016).

of both Jupiter and Saturn. On September 15, 2017, the Cassini spacecraft ended its highly successful 13-year mission in the saturnian system, Juno reached Jupiter in 2016, the New Horizons spacecraft has revealed the Pluto/Charon system in 2015 and will be exploring next the Kuiper belt, and future major outer solar system exploration activities are now under detailed study by the major space agencies, in particular NASA and ESA. The Jupiter ICy moons Explorer (JUICE) is ESA's first selected Cosmic Vision Large mission currently under development, with the purpose of exploring the jovian system — in particular its satellites Ganymede, Europa, and Callisto — to study the emergence of habitable worlds around gas giants. It is planned to be launched in 2022 for arrival in 2030. NASA is developing the Europa Clipper for launch in the early 2020s to comprehensively investigate through multiple flybys the characteristics and potential habitability of the subsurface liquid water ocean and the co-evolution of Europa's surface, crust, and ocean. NASA is also studying a possible lander to search for biosignatures at a site to be selected with the help of Europa Clipper data. Several other proposals with exciting concepts have been or will be submitted to the space agencies in response to calls issued for medium-sized missions (e.g., Discovery and New Frontiers for NASA, M5 calls for ESA).

These investigations all tie into the astrobiological significance of the outer solar system. In looking for habitable worlds, the criteria are liquid water; the biogenic elements carbon, hydrogen, nitrogen, oxygen, phosphorus, and sulfur ("CHNOPS"); energy sources; and a stable environment. First, organic materials are present in considerable abundance in many locations in the outer solar system. While this is particularly true of Titan and Enceladus around Saturn, the spectral characteristics of the tenuous atmospheres around many other icy and rocky bodies, in particular around Jupiter's moons, suggest the presence of organic compounds. Second, many of the giant planets' satellites are geologically active today. This is apparent on several large satellites, specifically Io (silicate volcanism), Europa (youthful and complexly modified surface), Ganymede (active magnetic field implying a hot core), Titan (active hydrological cycle based on methane and aeolian processes), and Triton (active geysers); moreover, activity occurs even in tiny Enceladus (as we have seen). Third, and very important, it is now recognized that a large number of the icy satellites have the potential to possess (or have possessed in either the recent or ancient past) subsurface liquid water (*Lazcano and Hand,* 2012). As a consequence, it is possible that the surfaces and interiors of many outer solar system bodies have undergone significant chemical processing over the past 4.5 b.y. Thus, based on our current knowledge, liquid water may exist not just inside Europa, Enceladus, Ganymede, Callisto, and Titan, but also inside Mimas, and possibly Dione, Triton, and Pluto (*Lunine,* 2016).

The combination of water, organic materials, and internal geological activity suggest that significant results pertinent to astrobiology and to our understanding of the habitable conditions in the solar system, including clues for the emergence and survival of life on our planet, are likely to emerge from further exploration of these objects. Understanding the astrobiological potential of other planetary systems, or the range of astrobiological circumstances within the solar system, requires investigation of the origin and evolution of these objects, the extent and type of chemical processing that has occurred (*Hand and Carlson,* 2015), and the potential habitability of diverse solar system environments. In addition, the architecture of the outer solar system itself has important implications for our understanding of astrobiology: the nature and composition of the giant planets, their locations in the solar system, their relationship to KBOs, and the properties of planetary rings all help us to understand the processes responsible for evolution of the solar system as a whole.

Astrobiological studies of the outer solar system have a significance that extends outside the confines of our planetary system. There may exist innumerable icy worlds around giant planets, harboring habitable environments. In turn, many worlds that were once like Titan or Europa must surround parent stars that have now become red giants, raising temperatures to the point where formerly icy worlds might support liquid water at their surfaces.

In order to make progress in our investigations of these subjects, future missions to the so-called "ocean worlds" are mandatory. Both ESA and NASA have inserted the possibility for such missions in their space programs. The 2016 appropriations bill from the U.S. House Appropriations Committee, for instance, called for the creation of an "Ocean Worlds Exploration Program" that would fund new missions to Europa, Enceladus, and Titan. From the bill (*http://appropriations.house.gov/uploadedfiles/hrpt-114-hr-fy2016-cjs.pdf*): "Many of NASA's most exciting discoveries in recent years have been made during the robotic exploration of the outer planets. The Cassini mission has discovered vast oceans of liquid hydrocarbons on Saturn's moon Titan and a submerged saltwater sea on Saturn's moon Enceladus. The Committee directs NASA to create an Ocean World Exploration Program whose primary goal is to discover extant life on another world using a mix of Discovery, New Frontiers, and Flagship-class missions consistent with the recommendations of current and future Planetary Decadal surveys." This new program renders possible the investigation of these worlds within the "Vision and Voyages for Planetary Science in the Decade 2013–2022" Decadal Survey and has therefore been positively received by the scientific community.

Many ideas exist to investigate worlds that have the potential to harbor undersurface liquid water oceans and the implications as described above. The future missions would need to explore various aspects such as:

- Understanding the geochemical, geological, and geophysical histories of the satellites and other outer solar system bodies; the history and location of liquid water (or other liquids); and the potential for habitability.
- Determining the sources and reservoirs of organic materials and processes (particularly those involving liquid water) by which carbon compounds are formed, processed, and destroyed on the surfaces or within the interiors of outer solar system bodies.
- Bringing together knowledge of the composition and dynamics of the gas giants and related minor bodies in the outer solar system to help us understand the observed architecture of our solar system, and the prospects for habitats to have formed in our outer solar system and other planetary systems.
- Establishing whether life exists or has ever existed on any of these objects.

Future missions require state-of-the-art payloads in order to address the questions Cassini has left us about Enceladus. Therefore sustained investments are required to bring relevant technological developments in instrumentation as well as spacecraft to required levels of maturity. Future missions will also have to take into account stringent planetary protection policies as they apply to outer solar system bodies, and may necessitate preparatory work to identify areas where revisions of these policies will be needed in the light of recent discoveries and improved understanding.

Future space missions would further benefit from opportunities for Earth-based laboratory measurements, telescopic studies, or theoretical analyses that would enhance their scientific return and maximize the chances of success in the ambitious arena of the search for habitable environments and — for Enceladus at least — life.

Acknowledgments. F.T. gratefully acknowledges the support of the Italian Space Agency (ASI) in Rome (Italy). ASI-INAF grant No. I/2017-10-H.O.

REFERENCES

Adler M. and 13 colleagues (2011) Rapid mission architecture trade study of Enceladus mission concepts. *2011 Aerospace Conference,* pp. 1–13. IEEE, DOI: 10.1109/AERO.2011.5747289.

Beuthe M., Rivoldini A., and Trinh A. (2016) Enceladus's and Dione's floating ice shells supported by minimum stress isostasy. *Geophys. Res. Lett., 43,* 10088–10096.

Brown R.H., et al. (2004) The Cassini Visual and Infrared Mapping Spectrometer (VIMS) investigation. *Space Sci. Rev., 115,* 111–168.

Brown R. H., Clark R. N., et al. (2006) Composition and physical properties of Enceladus' surface. *Science, 311,* 1425–1428.

Brownlee D. E., Horz F., Newburn R. L., Zolensky M., Duxbury T. C., Sandford S., Sekanina Z., Tsou P., Hanner M. S., Clark B. C., Green S. F., and Kissel J. (2004) Surface of young Jupiter family Comet 81P/Wild 2: View from the Stardust spacecraft. *Science, 304,* 1764–1769.

Burch J. L. et al. (2007) Tethys and Dione as sources of outward-flowing plasma in Saturn's magnetosphere. *Nature, 447,* 833–835.

Burnett D.S. (2013) The Genesis solar wind sample return mission: Past, present, and future. *Meteoritics & Planet. Sci., 48,* 2351–2370.

Burnett D. S. and 10 colleagues (2003) The Genesis Discovery mission: Return of solar matter to Earth. *Space Sci. Rev., 105,* 509–534.

Čadek O., Tobie G., Van Hoolst T., Masse M., Choblet G., Lefevre A., Mitri G., Baland R.-M., Behounkova M., Bourgeois O., and Trinh A. (2016) Enceladus's internal ocean and ice shell constrained from Cassini gravity, shape and libration data, *Geophys. Res. Lett., 43,* 5653–5660, DOI: 10.1002/2016GL068634.

Carr C. E., Zuber M. T., and Ruvkun G. (2013) Life detection with the Enceladus Orbiting Sequencer. *2013 Aerospace Conference,* pp. 1–12. IEEE, DOI: 10.1109/AERO.2013.6497129.

Choblet G., Tobie G., Sotin C., Behounkova M., Cadek O., Postberg F., and Soucek O. (2017) Prolonged hydrothermal activity inside

Enceladus. *Nature Astron.,* 1–7, DOI: 10.1038/s41550-017-0289-8.
Clark B. C., Sunshine J. M., A'Hearn M. F., Cochran A. L., Farnham T. L., Harris W. M., McCoy T. J., and Veverka J. (2008) Comet Hopper: A mission concept for exploring the heterogeneity of comets. In *Asteroids, Comets, Meteors 2008,* Abstract #8131. Lunar and Planetary Institute, Houston.
Clark R. N. et al. (2008) Compositional mapping of Saturn's satellite Dione with Cassini VIMS and implications of dark material in the Saturn system. *Icarus, 193,* 372–386.
Clark R. N. et al. (2012) The surface composition of Iapetus: Mapping results from Cassini VIMS. *Icarus, 218,* 831–860.
Coustenis A., Atreya S. K., Balint T., Brown R. H., Dougherty M. K., Ferri F., Fulchignoni M., Gautier D., Gowen R. A., Griffith C. A., Gurvits L. I., Jaumann R., Langevin Y., Leese M. R., Lunine J. I., McKay C. P., Moussas X., Mueller-Wodarg I., Neubauer F., Owen T. C., Raulin F., Sittler E. C., Sohl F., Sotin C., Tobie G., Tokano T., Turtle E. P., Wahlund J., Waite J. H., Baines K. H., Blamont J., Coates A. J., Dandouras I., Krimigis T., Lellouch E., Lorenz R. D., Morse A., Porco C. C., Hirtzig M., Saur J., Spilker T., Zarnecki J. C., Choi E., Achilleos N., Amils R., Annan P., Atkinson D. H., Bénilan Y., Bertucci C., Bézard B., Bjoraker G. L., Blanc M., Boireau L., Bouman J., Cabane M., Capria M. T., Chassefière E., Coll P., Combes M., Cooper J. F., Coradini A., Crary F., Cravens T., Daglis I. A., de Angelis E., de Bergh C., de Pater I., Dunford C., Durry G., Dutuit O., Fairbrother D., Flasar F. M., Fortes A. D., Frampton R., Fujimoto M., Galand M., Grasset O., Grott M., Haltigin T., Herique A., Hersant F., Hussmann H., Ip W., Johnson R., Kallio E., Kempf S., Knapmeyer M., Kofman W., Koop R., Kostiuk T., Krupp N., Küppers M., Lammer H., Lara L. M., Lavvas P., Le Mouélic S., Lebonnois S., Ledvina S., Li J., Livengood T. A., Lopes R. M., Lopez-Moreno J. J., Luz D., Mahaffy P. R., Mall U., Martinez- Frias J., Marty B., McCord T., Menor Salvan C., Milillo A., Mitchell D. G., Modolo R., Mousis O., Nakamura M., Neish C. D., Nixon C. A., Nna Mvondo D., Orton G., Paetzold M., Pitman J., Pogrebenko S., Pollard W., Prieto-Ballesteros O., Rannou P., Reh K., Richter L., Robb F. T., Rodrigo R., Rodriguez R., Romani P., Ruiz Bermejo M., Sarris E. T., Schenk P., Schmitt B., Schmitz N., Schulze-Makuch D., Schwingenschuh K., Selig A., Sicardy B., Soderblom L., Spilker L. J., Stam D., Steele A., Stephan K., Strobel D. F., Szego K., Szopa C., Thissen R., Tomasko M. G., Toublanc D., Vali H., Vardavas I., Vuitton V., West R. A., Yelle R., and Young E. F. (2009) TandEM: Titan and Enceladus mission. *Exp. Astron., 23,* 893–946.
Cruikshank D. P. et al. (2010) Carbon dioxide on the satellites of Saturn: Results from the Cassini VIMS investigation and revisions to the VIMS wavelength scale. *Icarus, 206,* 561–572.
Dalle Ore C. M. et al. (2010) Infrared spectroscopic characterization of the low-albedo materials on Iapetus. *Icarus, 221,* 735–743.
Davila A. F. and McKay C. P. (2014) Chance and necessity in biochemistry: Implications for the search for extraterrestrial biomarkers in Earth-like environments. *Astrobiology, 14,* 534–540.
Denk T. et al. (2010) Iapetus: Unique surface properties and a global color dichotomy from Cassini imaging. *Science, 327,* 435–439.
Dougherty M. K., Khurana K. K., et al. (2006) Identification of a dynamic atmosphere at Enceladus with the Cassini Magnetometer. *Science, 311,* 1406–1409.
Elachi C. et al. (2004) Radar: The Cassini Titan Radar Mapper. *Space Sci. Rev., 115,* 71–110.
Elsila J., Glavin D. P., and Dworkin J. P. (2009) Cometary glycine detected in samples returned by Stardust. *Meteoritics & Planet. Sci., 44,* 1323–1330.
Elsila J., Aponte J. C., Blackmond D. G., Burton A. S., Dworkin J. P., and Glavin D. P. (2016) Meteoritic amino acids: Diversity in compositions reflects parent body histories. *ACS Cent. Sci., 2,* 370–379.
Filacchione G. et al. (2013) The radial distribution of water ice and chromophores across Saturn's system. *Astrophys. J., 766,* Article ID 76.
Flasar F.M., et al. (2004) Exploring the Saturn system in the thermal infrared: The Composite Infrared Spectrometer. *Space Sci. Rev., 115,* 169–297.
French L., Anderson F. S., Carsey F., French G., Lane A. L., Shakkottai P., and Zimmerman W. (2001) Cryobots: An answer to subsurface mobility in planetary icy environments. In *Proceedings of the 6th International Symposium on Artificial Intelligence and Robotics & Automation in Space: i-SAIRAS 2001,* Canadian Space Agency, St-Hubert, Quebec, Canada.
Georgiou C. D. and Deamer D. W. (2014) Lipids as universal biomarkers of extraterrestrial life. *Astrobiology, 14,* 541–549.
Glein C., Baross J. A., and Waite J. H. Jr. (2015) The pH of Enceladus' ocean. *Geochim. Cosmochim. Acta, 162,* 202–219.
Hammond N. P. et al. (2013) Flexure on Dione: Investigating subsurface structure and thermal history. *Icarus, 223,* 418–422.
Hand K. P. and Carlson R. W. (2015) Europa's surface color suggests an ocean rich with sodium chloride. *Geophys. Res. Lett., 42,* 3174–3178.
Higgs R. E. and Pudritz R. E. (2009) A thermodynamic basis for prebiotic amino acid synthesis and the nature of the first genetic code. *Astrobiology, 9,* 483–490.
Howett C. J. A. et al. (2010) Thermal inertia and bolometric Bond albedo values for Mimas, Enceladus, Tethys, Dione, Rhea and Iapetus as derived from Cassini/CIRS measurements. *Icarus, 206,* 573–593.
Howett C. J. A., Spencer J. R., Pearl J., and Segura M. (2011) High heat flow from Enceladus' south polar region measured using 10–600 cm^{-1} Cassini/CIRS data. *J. Geophys. Res., 116,* E03003.
Hsu H.-W., Postberg F., et al. (2015) Ongoing hydrothermal activities within Enceladus. *Nature, 519,* 207–210.
Iess L. and 10 colleagues (2014) The gravity field and interior structure of Enceladus. *Science, 344,* 78–80.
Johnson T. V. and Lunine J. I. (2005) Saturn's moon Phoebe as a captured body from the outer solar system. *Nature, 435,* 69–71.
Kargel J. and Pozio S. (1996). The volcanic and tectonic history of Enceladus. *Icarus, 119,* 385–404.
Kempf S., Beckmann U., and Schmidt J. (2010) How the Enceladus dust plume feeds Saturn's E ring. *Icarus, 206,* 446–457.
Kite E. S. and Rubin A. M. (2016) Sustained eruptions on Enceladus explained by turbulent dissipation in tiger stripes. *Proc. Natl. Acad. Sci., 113,* 3972–3975.
Konstantinidis K., and 10 colleagues (2015) A lander mission to probe subglacial water on Saturn's moon Enceladus for life. *Acta Astron., 106,* 63–89.
Landis G. A., Oleson S. J., and McGuire M. (2012) ASRG Mars Geyser Hopper. In *Concepts and Approaches for Mars Exploration,* Abstract #4219. Lunar and Planetary Institute, Houston.
Lauretta D. S. , and the OSIRIS-REx Team (2012) An overview of the OSIRIS-REx asteroid sample return mission. In *Lunar and Planetary Science XLIII,* Abstract #2491. Lunar and Planetary Institute, Houston.
Lazcano A. and Hand K. P. (2012) Astrobiology: Frontier or fiction. *Nature, 488,* 160–161.
Lorenz R. D., Leary J. C., Lockwood M. K., and Waite J. H. (2008) Titan explorer: A NASA Flagship mission concept. In *Space Technology and Applications International Forum — STAIF* 2008 (M. S. El-Genk, ed.), pp. 380–387. AIP Conf. Proc. 969, Albuquerque, New Mexico.
Lunine J. I. (2016) Ocean worlds exploration. *Acta Astron., 131,* 123–130.
Lunine J. I. et al. (2015) Enceladus life finder: The search for life in a habitable moon. In *Lunar and Planetary Science XLVI,* Abstract #1525. Lunar and Planetary Institute, Houston.
McKay C. P., Anbar A. D., Porco C., and Tsou P. (2014). Follow the plume: The habitability of Enceladus. *Astrobiology, 14,* 352–355.
McKinnon W.B. (2015) Effect of Enceladus's rapid synchronous spin on interpretation of Cassini gravity. *Geophys. Res. Lett., 42,* 2137–2143.
Mitri G., Coustenis A., Fanchini G., Hayes A. G., Iess L., Khurana K., Lebreton J.-P., Lopes R. M., Lorenz R. D., Meriggiola R., Moriconi M. L., Orosei R., Sotin C., Stofan E., Tobie G., Tokano T., and Tosi F. (2014) The exploration of Titan with an orbiter and a lake probe. *Planet. Space Sci., 104,* 78–92, DOI: 10.1016/j.pss.2014.07.009.
Mitri G., Postberg F., Soderblom J. M., Wurz P., Tortora P., Abel B., Barnes J. W., Berga M., Carrasco N., Coustenis A., de Vera J. P., D'Ottavio A., Ferri F., Hayes A. G., Hayne P. O., Hillier J. K., Kemp S., Lebreton J.-P., Lorenz R. D., Martelli A., Orosei R., Petropoulos A. E., Reh K., Schmidt J., Sotin C., Srama R., Tobie G., Vorburger A., Vuitton V., Wong A., and Zannoni M. (2017) Explorer of Enceladus and Titan (E^2T): Investigating ocean worlds' evolution and habitability in the solar system. *Planet. Space Sci.,* DOI: 10.1016/j.pss.2017.11.001, in press.
Moeller R. C., Borden C., Spilker T., Smythe W., and Lock R. (2011) Space missions trade space generation and assessment using the JPL Rapid Mission Architecture (RMA) Team approach. *2011 Aerospace Conf. 2011,* Abstract # 1599, Version 3, Updated January 4, 2011.

Monroe A., Glein C., Anbar A. D., Shock E. L., and Lunine J. I. (2017) Amino acid destruction considerations for *in situ* measurements of Enceladus and other ocean worlds. *Astrobiology Science Conference,* Abstract #3319, Lunar and Planetary Institute, Houston.

National Research Council (2007) *The Limits of Organic Life in Planetary Systems*. National Academies, Washington, DC, DOI: 10.17226/11919.

National Research Council (2011) *Vision and Voyages for Planetary Science in the Decade 2013–2022.* National Academies, Washington, DC, DOI: 10.17226/13117.

Ostro S. et al. (2006) Cassini RADAR observations of Enceladus, Tethys, Dione, Rhea, Iapetus, Hyperion, and Phoebe. *Icarus, 183,* 479–490.

Porco C. et al. (2006) Cassini observes the active south pole of Enceladus. *Science, 311,* 1393–1401.

Porco C., DiNino D., and Nimmo F. (2014) How the geysers, tidal stresses, and thermal emission across the south polar terrain of Enceladus are related. *Astron. J., 148,* 45.

Postberg F., Kempf S., Schmidt J., Brilliantov N., Beinsen A., Abel B., Buck U., and Srama R. (2009) Sodium salts in E-ring ice grains from an ocean below the surface of Enceladus. *Nature, 459,* 1098–1101.

Postberg F., Schmidt J., et al. (2011) A salt-water reservoir as the source of a compositionally stratified plume on Enceladus. *Nature, 474,* 620–622.

Razzaghi A. (2007) *Enceladus Flagship Mission Concept Study.* White paper submitted to NASA Visions and Voyages Decadal Survey, National Research Council, Washington, DC.

Reh K et al. (2007) *Titan and Enceladus $1B Mission Feasibility Study Report.* JPL D-37401 B, Jet Propulsion Laboratory, Pasadena, California.

Reh K., Magner T., Matson D., Coustenis A., Lunine J., Lebreton J-P., Jones C., and Sommerer J. (2009) *Titan Saturn System, Mission Study 2008: Final Report.* NASA/ESA, available online at *https://solarsystem.nasa.gov/docs/08_TSSM_Final_Report_Public_Version.pdf.*

Roatsch T., Wahlisch M., Hoffmeister A., Kersten E., Matz K-D., Giese B., Scholten F., Wagner R., Denk T., Neukum G., Helfenstein P., and Porco C. (2009) High resolution atlases of Mimas, Tethys and Iapetus derived from Cassini-ISS images. *Planet. Space Sci., 57,* 83–92.

Roatsch T., Kersten E., Hoffmeister A., Wahlisch M., Matz K-D., and Porco C. C. (2013) Recent improvements of the saturnian satellites atlases: Mimas, Enceladus, and Dione. *Planet. Space Sci., 77,* 118–125.

Roth L., Saur J., Retherford K. D., Strobel D., Feldman P., and McGrath M. (2014) Transient water vapor at Europa's south pole. *Science, 343,* 171–174.

Sandford S.A. (2011) The power of sample return missions — Stardust and Hayabusa. In *The Molecular Universe* (J. Cernicharo and R. Bachiller, eds.), pp. 275–287. IAU Symp. 280, Cambridge Univ., Cambridge.

Sekine Y., Shibuya T., Postberg F., Hsu H-W., Suzuki K., Masaki Y., Kuwatani T., Mori M., Hong P. K., Yoshisazi M., Tachibana S., and Sirono S. (2015) High temperature water-rock interactions and hydrothermal environments in the chondrite-like core of Enceladus. *Nature Commun., 6,* 8604.

Sherwood B. (2016) Strategic map for exploring the ocean world Enceladus. *Acta Astron., 126,* 52–58.

Shoji D., Hussman H., Kurita K., and Sohl F. (2013) Ice rheology and tidal heating of Enceladus. *Icarus, 226*, 10–19.

Smith B. A., Soderblom L., et al. (1982) A new look at the Saturn system: The Voyager 2 images. *Science, 215,* 504–537.

Soderblom L. A., Kieffer S. W., Becker T. L., Brown R. H., Cook A. F. II, Hansen C. J., Johnson T. V., Kirk R. L., and Shoemaker E. M. (1990) Triton's geyser-like plumes: Discovery and basic characterization. *Science, 250,* 410–415.

Sotin C., Altwegg K., Brown R. H., Hand K., Lunine J. I., Soderblom J., Spencer J., Tortora P., and the JET Team (2011) JET: Journey to Enceladus and Titan. In *Lunar and Planetary Science XLII, Abstract* #1326. Lunar and Planetary Institute, Houston.

Spahn F., Schmidt J., Albers N., Hörning M., Makuch M., Seiss M., Kempf S., Srama R., Dikarev V., Helfert S., Moragas-Klostermeyer G., Krivov A. V., Sremcevic M., Tuzzolino A. J., Economou T., and Grün E. (2006) Cassini dust measurements at Enceladus and implications for the origin of the E ring. *Science, 311,* 1416–1418.

Sparks W. B., Hand K. P., McGrath M. A., Bergeron E., Cracraft M., and Deustua S. E. (2016) Probing for evidence of plumes on Europa with HST/STIS. *Astrophys. J., 829,* 121.

Spencer J. R. and Denk T. (2010) Formation of Iapetus' extreme albedo dichotomy by exogenically triggered thermal ice migration. *Science, 327,* 432–435.

Spencer J. R., Pearl J. C., Segura M., Flasar F. M., Mamoutkine A., Romani P., Buratti B. J., Hendrix A. R., Spilker L. J., and Lopes R. M. C. (2006) Cassini encounters Enceladus: Background and the discovery of a south polar hot spot. *Science, 311,* 1401–1405.

Spitale J. N. and Porco C. C. (2007) Association of the jets of Enceladus with the warmest regions on its south-polar fractures. *Nature, 449,* 695–697.

Squyres S. W. (1983) Planetary science: 1979–1982. *Rev. Geophys. Space Phys., 21,* 139–142.

Stone W. C., Hogan B., Siegel V., Lelievre S., and Flesher C. (2014) Progress towards an optically powered cryobot. *Ann. Glaciol., 55,* 2–13, DOI: 10.3189/2014AoG65A200.

Takano Y. et al. (2014) Planetary protection on international waters: An onboard protocol for capsule retrieval and biosafety control in sample return mission. *Adv. Space Res., 53,* 1135–1142.

Thomas P., Tajeddine R., Tiscareno M. S., Burns J. A., Joseph J., Loredo T. J., Helfenstein P., and Porco C. (2016) Enceladus's measured physical libration requires a global subsurface ocean. *Icarus, 264,* 37–47.

Tobie G., Teanby N., Coustenis A., Jaumann R., Raulin F., Schmidt J., Carrasco N., Coates A., Cordier D., De Kok R., Geppert W., Lebreton J-P., Lefevre A., Livengood T., Mandt K., Mitri G., Nimmo F., Nixon C., Norman L., Pappalardo R., Postberg F., Rodriguez S., Schulze-Makuchm D., Soderblom J., Solomonido A., Stephan K., Stofan E., Turtle E., Wagner R., West R., and Westlake J. (2014) Science goals and mission concept for the future exploration of Titan and Enceladus. *Planet. Space Sci., 104,* 59–77.

Tosi F. et al. (2010) Probing the origin of the dark material on Iapetus. *Mon. Not. R. Astron. Soc., 403,* 1113–1130.

Tsou P. and 15 colleagues (2014) LIFE: Enceladus plume sample return via Discovery. In *Lunar and Planetary Science L,* Abstract #2192. Lunar and Planetary Institute, Houston.

Tsou P., Brownlee D. E., McKay C. P., Anbar A. D., Yano H., Altwegg K., Beegle L. W., Dissly R., Strange N. J., and Kanik I. (2012) LIFE: Life Investigation for Enceladus — A sample return mission concept in search for evidence of life. *Astrobiology, 12,* 730–742.

Tsuchiyama A., Uesugi M., Matsushima T., Michikama T., Kadono T., et al. (2011) Three dimensional structure of Hayabushi samples: Origin and evolution of Itokawa regolith. *Science, 333,* 1125–1128.

Tsuda Y., Yoshikawa M., Abe M., Minamino H., and Nakazawa S. (2013) System design of the Hayabusa 2: Asteroid sample return mission to 1999 JU3. *Acta Astron., 91,* 356–362.

Ulamec S. and Biele J. (2006) From the Rosetta Lander Philae to an asteroid hopper: Lander concepts for small bodies missions. *International Planetary Probe Workshop 7, January 2006.*

Verbiscer A. et al. (2009) Saturn's largest ring. *Nature, 461,* 1098–1100.

Waite J. H., Combi M. R., et al. (2006) Cassini Ion and Neutral Mass Spectrometer: Enceladus plume composition and structure. *Science, 311,* 1419–1422.

Waite J. H. Jr., Lewis W. S., Magee B. A., Lunine J. I., McKinnon W. B., Glein C. R., Mousis O., Young D. T., Brockwell T., Westlake J., Nguyen M.-J., Teolis B. D., Niemann H. B., McNutt R. L., Perry M., and Ip W.-H. (2009) Liquid water on Enceladus from observations of ammonia and ^{40}Ar in the plume. *Nature, 460,* 487–490.

Waite J. H., Glein C. R., Perryman R. S., Teolis B. D., Magee B. A., Miller G., Grimes J., Perry M. E., Miler K. E., Bouquet A., Lunine J. I., Brockwell T., and Bolton S. J. (2017) Cassini finds molecular hydrogen in the Enceladus plume: Evidence for hydrothermal processes. *Science, 356,* 155–159.

Yano H. and 19 colleagues (2006) Touchdown of the Hayabusa spacecraft at the Muses Sea on Itokawa. *Science, 312,* 1350–1353.

Young D. T. et al. (2004) Cassini Plasma Spectrometer investigation. *Space Sci. Rev., 114,* 1–112.

Zimmerman W., Bryant S., Zitzelberger J., and Nesmith B. (2001) A radioisotope powered cryobot for penetrating the Europan ice shell. In *Space Technology and Applications International Forum — 2001* (M. S. El-Genk and M. J. Bragg, eds.), pp. 707–715. AIP Conf. Ser. 552, American Institute of Physics, Melville, New York.

Zimmermann and 7 colleagues (2002) The Mars '07 North Polar Cap deep penetration cryo-scout mission. *2009 Aerospace Conf.,* DOI: 10.1109/AERO.2002.1036850.

Index

Page numbers refer to specific pages on which an index term or concept is discussed. “ff” indicates that the term is also discussed on the following pages.